Advances in Design Optimization

T0199647

Other titles from E & FN Spon and Chapman & Hall

The Behaviour and Design of Steel Structures
N.S. Trahair and M.A. Bradford

Boundary Element Methods in Elastodynamics
G.D. Manolis and D.E. Beskos

Boundary Element Methods in Solid Mechanics
S.L. Crouch and A.M. Starfield

Computer-based Design Process
A.J. Medland

Computer Methods in Structural Analysis
J.L. Meek

Developments in Structural Engineering
Edited by B.H.V. Topping

Evaluating Supercomputers
Edited by A.J. van der Steen

Expert Systems in Construction and Structural Engineering
Edited by H. Adeli

Flexural-Torsional Buckling of Structures
N.S. Trahair

Neural Computing Architectures
Edited by I. Aleksander

Parallel Processing and Data Management
Edited by P. Valduriez

Probabilistic Methods in Structural Engineering
G. Augusti, A. Baretta and F. Casciati

Structural Dynamics for the Practising Engineer
H.M. Irvine

For details of these and other titles, contact The Promotion Department, E & FN Spon, 2–6 Boundary Row, London SE1 8HN, Telephone 071 865 0066

Advances in Design Optimizaton

Edited by

PROFESSOR HOJJAT ADELI

The Ohio State University, USA

CRC Press
Taylor & Francis Group
Boca Raton London New York

CRC Press is an imprint of the
Taylor & Francis Group, an **informa** business

A CHAPMAN & HALL BOOK

CRC Press
Taylor & Francis Group
6000 Broken Sound Parkway NW, Suite 300
Boca Raton, FL 33487-2742

First issued in paperback 2019

© 1994 by Taylor & Francis Group, LLC
CRC Press is an imprint of Taylor & Francis Group, an Informa business

No claim to original U.S. Government works

Typeset in $9\frac{1}{2}/11$ pt Times by Thomson Press (India) Ltd., New Delhi

ISBN-13: 978-0-412-53730-1 (hbk)
ISBN-13: 978-0-367-86402-6 (pbk)

This book contains information obtained from authentic and highly regarded sources. Reasonable efforts have been made to publish reliable data and information, but the author and publisher cannot assume responsibility for the validity of all materials or the consequences of their use. The authors and publishers have attempted to trace the copyright holders of all material reproduced in this publication and apologize to copyright holders if permission to publish in this form has not been obtained. If any copyright material has not been acknowledged please write and let us know so we may rectify in any future reprint.

Except as permitted under U.S. Copyright Law, no part of this book may be reprinted, reproduced, transmitted, or utilized in any form by any electronic, mechanical, or other means, now known or hereafter invented, including photocopying, microfilming, and recording, or in any information storage or retrieval system, without written permission from the publishers.

For permission to photocopy or use material electronically from this work, please access www.copyright.com (http://www.copyright.com/) or contact the Copyright Clearance Center, Inc. (CCC), 222 Rosewood Drive, Danvers, MA 01923, 978-750-8400. CCC is a not-for-profit organization that provides licenses and registration for a variety of users. For organizations that have been granted a photocopy license by the CCC, a separate system of payment has been arranged.

Trademark Notice: Product or corporate names may be trademarks or registered trademarks, and are used only for identification and explanation without intent to infringe.

A catalogue record for this book is available from the British Library

Library of Congress Cataloging-in-Publication data

Advances in design optimization / edited by Hojjat Adeli.—1st ed.
    p. cm.
  Includes bibliographical references and index.
  ISBN 0-412-53730-3 (alk. paper)
    1. Engineering design—Mathematical models. I. Adeli, Hojjat, 1950–
TA174.A415   1994
620'.0042'011—dc20   93-32183
                                                                        CIP

Visit the Taylor & Francis Web site at
http://www.taylorandfrancis.com

and the CRC Press Web site at
http://www.crcpress.com

Contents

Preface

Design being an open-ended problem, the selection of the 'best' or 'optimum' design has always been a major concern of designers. In recent years, we have seen a growing interest in the application of mathematical optimization techniques in engineering design. Optimization algorithms are no longer considered esoteric tools used for solution of theoretical problems only. They can be in fact an effective tool for design of large and complicated engineering and industrial systems. Commercial computer-aided design packages with optimization capabilities have recently become widely available.

A number of leading optimization researchers and 'practitioners' have contributed to this volume. It attempts to summarize advances in a number of fundamental areas of optimization with application in engineering design. Introductory materials are also included in each chapter in an attempt to make the book as self-contained as possible. Thus, the book can be used as a textbook in a graduate level course on design optimization.

We hope this book will generate further interest in the subject and help engineers in their quest to create more efficient designs.

Hojjat Adeli
Columbus, Ohio

About the editor

Hojjat Adeli is Professor of Civil Engineering and a member of Center of Cognitive Science at The Ohio State University. A contributor to 38 research journals he has authored nearly 240 research and scientific publications and edited 10 books in various fields of computer science and engineering since 1976 when he received his PhD from Stanford University. He is the author of *Interactive Microcomputer-aided Structural Steel Design* and co-author of *Expert Systems for Structural Design* both published by Prentice-Hall in 1988, and *Parallel Processing in Structural Engineering* published by Chapman & Hall in 1993. He is the Editor-in-Chief of the international journal *Microcomputers in Civil Engineering* which he founded in 1986. He has been an organizer or member of advisory boards of over 30 national and international conferences and a contributor to over 80 conferences held in 23 different countries. He was a Keynote and Plenary Lecturer at international knowledge engineering and computing conferences held in Italy (1989), Mexico (1989), Japan (1991), China (1992), Canada (1992), Portugal (1992), USA (1993) and Germany (1993). He has received numerous academic, research, and leadership awards, honors, and recognitions. His recent awards include The Ohio State University College of Engineering 1990 Research Award in Recognition of Outstanding Research Accomplishments and Lichtenstein Memorial Award for Faculty Excellence. In 1990 he was selected as Man of the Year by the American Biographical Institute.

Biographies of contributors

JASBIR ARORA received his PhD in the area of Structural Mechanics from the University of Iowa in 1971. He is currently Professor of Civil Engineering and Mechanical Engineering at the University of Iowa. He is also Director of the Optimal Design Laboratory in the College of Engineering. Dr Arora has published over 120 research papers. He is the author of the textbook *Introduction to Optimum Design*. McGraw-Hill, 1989, and co-author of a graduate level textbook, *Applied Optimal Design*, Wiley–Interscience, 1979. His current research interests include software development for design optimization, numerical optimization algorithms for large-scale problems, practical applications of optimization, design sensitivity analysis and optimization of nonlinear systems, optimal design under dynamic loads, and optimal control.

SCOTT A. BURNS received his PhD from the University of Illinois at Urbana-Champaign in 1985. He is currently an Associate Professor of General Engineering at the University of Illinois, Urbana. Dr Burns is a 1989 recipient of the National Science Foundation Presidential Young Investigator award.

DAN M. FRANGOPOL received his doctorate from the University of Liege, Belgium, in 1976. From 1969 to 1973 and from 1977 to 1979, he taught at the Institute of Civil Engineering, Bucharest, Romania. He was a project engineer in the Consulting Group A. Lipski (Brussels, Belgium) from 1979 to 1983. Since March 1983 he has been a professor of Civil Engineering at the University of Colorado, Boulder. He has performed research and consulting sponsored by the National Science Foundation, Federal Highway Administration, Pennsylvania Department of Transportation, Transportation Research Board, US Department of the Interior and US Army Corps of Engineers. His research interests include structural optimization, structural reliability, bridge engineering, and earthquake engineering.

D.E. GRIERSON received his PhD from the University of Waterloo, Canada, in 1968. He is currently Professor of Civil Engineering at the University of Waterloo. He is the co-author of structural optimization software for the design of steel structures that is used throughout North America in professional design offices and academic institutions.

SANTIAGO HERNÁNDEZ received his doctorate from University of Cantabria, Spain, in 1982. He is currently Professor of Structural Analysis at the University of La Coruna, Spain. He has written a book, co-authored two other books, and edited four books on structural optimization. He has organized international conferences on optimization and European training programs on intelligent design in engineering and architecture.

KAZUTO HORIMATSU received his Master of Engineering degree from Keio University, Tokyo, Japan, in 1988, and worked at Sony Corporation as a software engineer for design and production system division. He was a research scientist at the Computational Mechanics Laboratory at the University of Michigan during 1990–1992. He is currently with the Sony Corporation of America.

MANOHAR KAMAT received his PhD from Georgia Institute of Technology, Atlanta, Georgia, in 1972. From 1972 to 1985 he was on the faculty of the Department of Engineering Science and Mechanics at the Virginia Polytechnic and State University, Blacksburg, Virgina. He is currently Professor of Aerospace Engineering at Georgia Institute of Technology. He is the co-author of the text *Elements of Structural Optimization* and Editor of ASCE Journal of Aerospace Engineering. He is also the Editor of AIAA's Progress on Aeronautics and Astronautics Series volume *Structural Optimization – Status & Promise*.

NARBEY KHACHATURIAN received his PhD from the University of Illinois at Urbana-Champaign in 1952. He is now an Emeritus Professor of Civil Engineering at the University of Illinois, Urbana-Champaign. He was Associate Head of Department of Civil Engineering there during 1983–1989. He has performed research and published extensively on structural optimization during the last three decades. In 1992 he was elected an Honorary Member of the American Society of Civil Engineers.

NARENDA S. KHOT received his PhD from the University of Cincinnati in 1964. He is currently Senior Aerospace Engineer at the Flight Dynamics Directorate of Wright Patterson Air Force Base, Ohio. He has performed research and published extensively in the area of structural optimization, in particular the optimality criteria approach and structural control. He is the past associate editor of AIAA Journal. At present, he is on the editorial board of the journals of *Computational Mechanics* and *Structural Optimization*.

NOBORU KIKUCHI received his PhD from the Department of Aerospace Engineering and Engineering Mechanics at the University of Texas at Austin in 1977. He joined the Department of Mechanical Engineering and Applied Mechanics at the University of Michigan in 1979 where he is currently a professor. He has authored two books in the areas of contact mechanics and finite element method and over 80 journal papers on computational mechanics and structural optimization.

JUHANI KOSKI received his doctorate in Mechanical Engineering from Tampere University of Technology, Finland, in 1984. He is currently an Associate Professor of Structural Mechanics at Tampere University of Technology.

LUIS MESQUITA received his PhD from the Virginia Polytechnic Institute and State University, Blacksburg, Virginia, in 1985. He taught at the University of Nebraska at Lincoln from 1985 to 1990. Since 1991 he has been working as Project Engineer at Stevens & Associates in Woburn, Massachusetts.

FRED MOSES received his PhD in Civil Engineering from Cornell University in 1963 and served on the faculty of Case Western Reserve University until 1992. Since then, he has been Professor and Chairman of the Department of Civil Engineering at the University of Pittsburgh. He has served as a visiting faculty member at the Norwegian Institute of Technology, Israel Institute of Technology, Imperial College, and the Federal University of Lausanne, Switzerland. His current research interests include structural system reliability, optimization, load models for bridges, safety codes for offshore structures, design and evaluation of highway bridges, fatigue life prediction, and testing.

PANOS Y. PAPALAMBROS received his PhD from the Design Division of Mechanical Engineering Department at Stanford University in 1979. He joined the University of Michigan faculty in 1979 and has been teaching design there since then. He is currently Chairman of Department of Mechanical Engineering and Applied Mechanics at the University of Michigan. He has published over 50 book and journal articles primarily in the field of design automation, and has co-authored the text *Principles of Optimal Design* with D.J. Wilde. He is a past associate editor for Design Automation for the ASME Mechanical Design Transactions Journal.

GEORGE I.N. ROZVANY received his PhD from Monash University, Melbourne, Australia, in 1967. Since 1985 he has been Professor of Structural Design at the University of Essen, Germany. In the 1970s he was the closest research associate of the late Professor William Prager of Brown University. His over 190 publications on structural optimization include two books, three conference proceedings, and numerous principal lectures at other conferences. He has organized three international meetings on structural optimization and held visiting positions at Oxford (England), Waterloo (Canada), Stuttgart (Germany), Bangalore (India), and Warsaw (Poland). He is the Editor of journal of Structural Optimization which he founded in 1988, and Chairman of the Executive Committee of the International Society of Structural and Multidisciplinary Optimization.

ERIC SANDGREN received his PhD from Purdue University in 1977. He is currently Director of Design Engineering at TRW Steering and Suspension Systems in Sterling Heights, Michigan. Previously, he was an Associate Professor in the School of Mechanical Engineering at Purdue University in West Lafayette, Indiana. He has industrial experience with IBM and General Motors Corporation as well as additional academic experience at the University of Missouri in Columbia, Missouri. He was the recipient of the US National Science Foundation Presidential Young Investigator Award in 1984.

OLE SIGMUND received his MSc degree from the Technical University of Denmark in 1991. During 1991–1992 he performed research at the University of Essen under the sponsorship of German Research Foundation (DFG). He has published in the areas of structural optimization, dynamics, and active control. He is currently employed at the Department of Solid Mechanics at the Technical University of Denmark.

MING ZHOU received his Dr-Ing degree from Essen University, Germany, in 1992. He has been a research associate at the Essen University since 1988. He has published over thirty papers on different topics of structural optimization.

Metric/imperial equivalents for common units

Basic conversion factors

The following equivalents of SI units are given in imperial and, where applicable, metric technical units.

1 mm = 0.039 37 in	1 in = 25.4 mm	1 m² = 1.196 yd²	1 yd² = 0.8361 m²
1 m = 3.281 ft	1 ft = 0.3048 m	1 hectare = 2.471 arces	1 acre = 0.4047 hectares
= 1.094 yd	1 yd = 0.9144 m	1 mm³ = 0.000 061 02 in³	1 in³ = 16 390 mm³
1 km = 0.6214 mile	1 mile = 1.609 km	1 m³ = 35.31 ft³	1 ft³ = 0.028 32 m³
1 mm² = 0.001 55 in²	1 in² = 645.2 mm²	= 1.308 yd³	1 yd³ = 0.7646 m³
1 m² = 10.76 ft²	1 ft² = 0.0929 m²	1 mm⁴ (M of I) = 0.000 002 403 in⁴	1 in⁴ = 416 200 mm⁴

Force

1 N = 0.2248 lbf = 0.1020 kgf	1 kN = 0.1004 tonf = 102.0 kgf = 0.1020 tonne f
4.448 N = 1 lbf = 0.4536 kgf	9.964 kN = 1 tonf = 1016 kgf = 1.016 tonne f
9.807 N = 2.205 lbf = 1 kgf	9.807 kN = 0.9842 tonf = 1000 kgf = 1 tonne f

Force per unit length

1 N/m = 0.068 52 lbf/ft = 0.1020 kgf/m	1 kN/m = 0.0306 tonf/ft = 0.1020 tonne f/m
14.59 N/m = 1 lbf/ft = 1.488 kgf/m	32.69 kN/m = 1 tonf/ft = 3.333 tonne f/m
9.807 N/m = 0.672 lbf/ft = 1 kgf/m	9.807 kN/m = 0.3000 tonf/ft = 1 tonne f/m

Force per unit area

1 N/mm² = 145.0 lbf/in² = 10.20 kgf/cm²	1 N/mm² = 0.064 75 tonf/in² = 10.20 kgf/cm²
0.006 895 N/mm² = 1 lbf/in² = 0.0703 kgf/cm²	15.44 N/mm² = 1 tonf/in² = 157.5 kgf/cm²
0.098 07 N/mm² = 14.22 lbf/in² = 1 kgf/cm²	0.098 07 N/mm² = 0.006 350 tonf/in² = 1 kgf/cm²

1 N/m² = 0.020 89 lbf/ft² = 0.102 kgf/m²	1 N/mm² = 9.324 tonf/ft² = 10.20 kgf/cm²
47.88 N/m² = 1 lbf/ft² = 4.882 kgf/m²	0.1073 N/mm² = 1 tonf/ft² = 1.094 kgf/cm²
9.807 N/m² = 0.2048 lbf/ft² = 1 kgf/m²	0.098 07 N/mm² = 0.9144 tonf/ft² = 1 kgf/cm²

Force per unit volume

1 N/m³ = 0.006 366 lbf/ft³ = 0.102 kgf/m³	1 kN/m³ = 0.002 842 tonf/ft³ = 0.1020 tonne f/m³
157.1 N/m³ = 1 lbf/ft³ = 16.02 kgf/m³	351.9 kN/m³ = 1 tonf/ft³ = 35.88 tonne f/m³
9.807 N/m³ = 0.0624 lbf/ft³ = 1 kgf/m³	9.807 kN/m³ = 0.027 87 tonf/ft³ = 1 tonne f/m³

1 kN/m³ = 0.003 684 lbf/in³ = 0.1020 tonne f/m³
271.4 kN m³ = 1 lbf/in³ = 27.68 tonne f/m³
9.807 kN/m³ = 0.036 13 lbf/in³ = 1 tonne f/m³

Moment

1 Nm = 8.851 lbf in = 0.7376 lbf ft = 0.1020 kgf m
0.1130 Nm = 1 lbf in = 0.083 33 lbf ft = 0.011 52 kgf m
1.356 Nm = 12 lbf in = 1 lbf ft = 0.1383 kgf m
9.807 Nm = 86.80 lbf in = 7.233 lbf ft = 1 kgf m

1

Mathematical theory of optimum engineering design

S. HERNÁNDEZ

1.1 INTRODUCTION

Mathematical formulation of optimum design in engineering is usually written as

$$\min F(X) \tag{1.1}$$

subject to

$$g_j(X) \leqslant 0 \quad i = 1, \ldots, m \tag{1.2}$$

In this expression

X is the vector of problem variables
$F(X)$ is the objective function
$g_j(X)$ are the set of problem constraints

In structural optimization (Schmit, 1960) the most usual design variables are mechanical parameters of structure elements, size of cross section internal components or coordinates of nodes which define the structure. The objective function $F(X)$ is generally the structural weight, but more complicated functions are chosen in some problems. The constraints $g_j(X)$ represent state variables of structural response for each loading case: stresses, internal forces, displacements, natural frequencies or buckling loads.

All points which satisfy the constraints $g_j(X)$ are called feasible design and form the feasible region. Figure 1.1 shows several shapes of feasible regions for a two-variables design problem and the problem solution in each case.

- In this case several local minima exist inside the feasible region. All restraints are passive and the problem can be considered an unconstrained optimization (Fig. 1.1(a)).
- There are local minima due to the shape of the objective function, and some constraints $g_j(X)$ are activated (Fig. 1.1(b)).
- Local minima appear because of the geometry of constraints. The optimum is on the boundary with some $g_j(X) = 0$ (Fig.1.1(c)).
- This is a nonlinear problem having a unique constrained minimum as solution (Fig. 1.1(d)).
- This case corresponds to a linear problem, which always has only one minimum in situations related to engineering design (Fig. 1.1(e)).

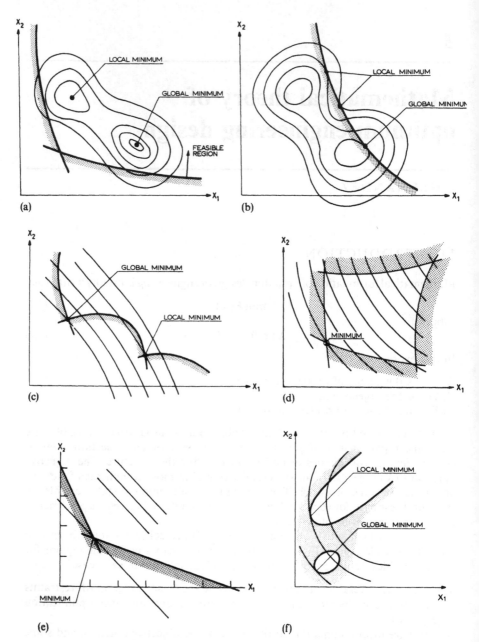

Fig. 1.1 Shape of feasible region in minimization problems: (a) nonlinear problem with two minima in the domain (nonconvex objective function); (b) nonlinear problem with two minima on the boundary (nonconvex objective function); (c) nonlinear problem with two minima on the boundary (nonconvex feasible region); (d) nonlinear problem with one minimum; (e) linear problem with one minimum; (f) nonlinear problem with two minima (disjoint feasible regions). (Parts (a)–(c) from U. Kirsch, *Optimum Structural Design*, © 1981 McGraw-Hill, Inc. Used with permission of the publisher.)

● This is a problem with disjoint feasible regions. For each of them any of the aforementioned cases can occur (Fig. 1.1(f)).

Among situations represented in Fig. 1.1 only the case in (a) does not occur for properly posed real problems, but cases (b)–(f) are associated with many structural optimization problems. There is an important difference among them: while in cases (d) and (e) only one minimum exists, several local minima are found in cases (b) and (c) and the minimum providing the lowest value of the objective function $F(X)$ turns out to be the solution of the problem. In case (f) the feasible region is formed by various subregions, and the same considerations should be taken into account.

1.2 PATTERN OF SHAPE OF THE FEASIBLE REGION

In this section it is intended to classify the most important structural optimization problems according to the shape of the feasible region originated.

1.2.1 Feasible region bounded by straight lines

For this kind of problem the geometry of the objective function is very important. If $F(X)$ is a straight line the solution is unique and placed at a vertex of the feasible region (Fig. 1.2(a)); the optimum can be found out by using any method of linear programming such as the very well-known simplex method (Dantzig, 1963) or a more recent technique given by Karmarker (1984). By contrast, in cases when the objective function is nonlinear several local minima may exist as indicated in Fig. 1.2(b)).

A number of problems exist in structural optimization with a linear objective function and also a linear set of constraints, thus leading to a linear programming problem. A few examples follow.

(a) Optimum design of trusses under plastic behaviour

Design variables are the cross-sectional areas of elements and the objective function chosen is the structural weight (Dorn, Gomory and Greenberg, 1964; Goble and LaPay, 1971; Russell and Reinschmidt, 1964).

An example is the symmetric four-bar truss of Fig. 1.3, with two design variables x_1, x_2 and loaded with two isolated forces P.

$$\min F(X) = \frac{4l}{3^{1/2}} x_1 + \frac{4l}{2^{1/2}} x_2 \tag{1.3}$$

subject to

$$2^{1/2}\sigma x_2 \geqslant P \tag{1.4}$$

$$3^{1/2}\sigma x_1 \geqslant P \tag{1.5}$$

$$\sigma x_1 \geqslant P \tag{1.6}$$

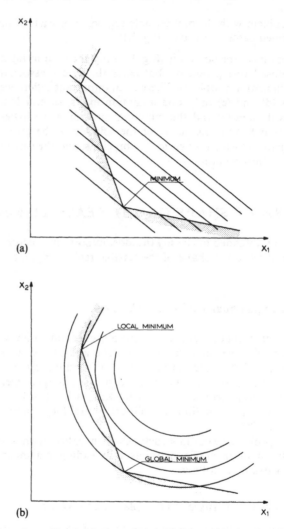

Fig. 1.2 Feasible region with straight boundaries: (a) linear objective function; (b) nonlinear objective function.

This example is represented in Fig. 1.4 and the solution is reached at values

$$F = \left(\frac{4}{3^{1/2}} + 2\right)\frac{Pl}{\sigma} \quad x_1 = \frac{P}{\sigma} \quad x_2 = \frac{P}{2^{1/2}\sigma} \tag{1.7}$$

(b) Optimum design of planar frames formed by prismatic elements, considering plastic collapse related only to bending moment

Design variables are linked to the inertia of bars and the weight is the objective function (Cohn, Ghosh and Parini, 1972; Horne and Morris, 1973). An example

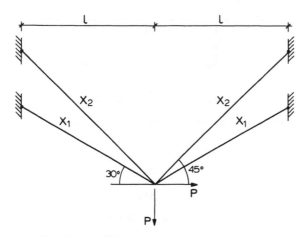

Fig. 1.3 Four-bar truss with two design variables.

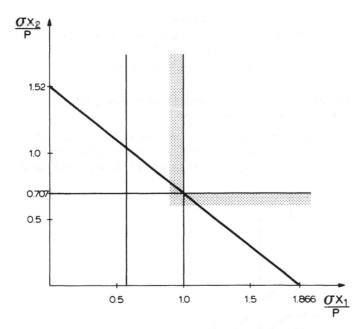

Fig. 1.4 Graphical representation of four-bar truss problem.

is the frame shown in Fig. 1.5. A set of constraints is established by using the kinematic analysis that compares the work produced by external loads and the internal energy of plastic hinges in each collapse mode. Also, size constraints are set up in equation (1.11) to avoid the disappearance of bars.

$$F(X) = 2lx_1 + 3lx_2 \tag{1.8}$$

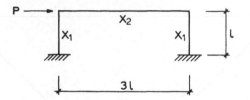

Fig. 1.5 A simple portal frame with two design variables.

subject to

$$2x_1 + 2x_2 \geqslant 1 \qquad (1.9)$$

$$4x_1 \geqslant 1 \qquad (1.10)$$

$$x_1, x_2 \geqslant 0.1 \qquad (1.11)$$

The problem is drawn in Fig. 1.6 and the optimal solution is

$$F = 1.1l \quad x_1 = 0.4 \quad x_2 = 0.1 \qquad (1.12)$$

If optimum design of frames is carried out considering not plastic behavior but linear static analysis, the problem turns out to be nonlinear. In addition to that, many other sets of constraints need to be considered in real structures, including local buckling, lateral torsion buckling, and code-allowable stresses.

(c) Optimization of prestressing forces and tendon configuration in continuous beams, grillages or frames

Design variables are the prestressing forces and tendon positions throughout the structure. Constraints can be related to stresses, displacements or tendon curvature (Goble and LaPay, 1971; Kirsch, 1973; Maquoi and Rondal, 1977).

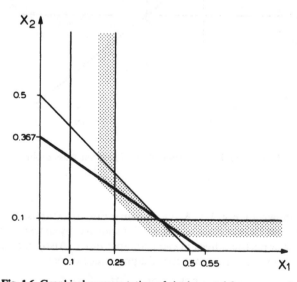

Fig. 1.6 Graphical representation of single portal frame example.

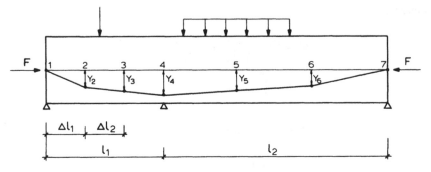

Fig. 1.7 Prestressed beam example.

Consider a continuous beam of several spans with $i = 1, \ldots, C$ applied loading conditions (Fig. 1.7). Tendon configuration can be approximated to a polygonal line defined by the eccentricities y_j at each section studied.

The prestressing force F shown in Fig. 1.7 produces a bending moment M_k and a displacement u_k at a coordinate x_h that can be written as

$$M_k = F \sum_{j=1}^{K} m_{jk} y_j \quad u_k = F \sum_{j=1}^{K} u_{jk} y_j \qquad (1.13)$$

where m_{jk} and u_{jk} are influence coefficients of eccentricity y_k at coordinate x_k.

The aim in this problem is to minimize the amount of prestressing force F. The usual formulation includes constraints for stresses σ_k and τ_k, displacement u_k, tendon configuration y_k and curvature ρ_k.

Assuming Navier theory the stresses σ_k and τ_k at a coordinate x_k may be evaluated as

$$\sigma_k = F \left(\sum_{j=1}^{K} \frac{m_{jk}}{W_k} y_j - \frac{1}{A_k} \right) + \sigma_{Pk} \qquad (1.14)$$

$$\tau_k = F \sum_{j=1}^{K} \tau_{jk} y_j + \tau_{Pk} \qquad (1.15)$$

Similarly the displacement u_k is

$$u_k = F \sum_{j=1}^{K} u_{jk} y_j + u_{Pk} \qquad (1.16)$$

where W_k and A_k are the section modulus and cross-sectional area: τ_{jk} and u_{jk} are influence coefficients of shear stresses and displacements; σ_{Pk}, τ_{Pk} and u_{Pk} are normal stress, shear stress and displacement produced by the applied external loads.

Considering constraints for tendon eccentricity y_j and curvature ρ_j, and setting up lower and upper limits for all constraints, the problem formulation turns out to be

$$\min F \qquad (1.17)$$

subject to

$$\sigma^L \leqslant F\left(\sum_{j=1}^{K} \frac{m_{jk}}{W_k} y_j - \frac{1}{A_k}\right) + \sigma_{Pk} \leqslant \sigma^U \tag{1.18}$$

$$\tau^L \leqslant F \sum_{j=1}^{K} \tau_{jk} y_j + \tau_{Pk} \leqslant \tau^U \tag{1.19}$$

$$u^L \leqslant F \sum_{j=1}^{K} u_{jk} y_j + u_{Pk} \leqslant u^U \tag{1.20}$$

$$y^L \leqslant y_j \leqslant y^U \quad j = 1, \ldots, K \tag{1.21}$$

$$\frac{1}{\rho^U} \leqslant \frac{1}{\rho_j} \leqslant \frac{1}{\rho^L} \tag{1.22}$$

Prestressing losses through the tendon can also be included in the formulation given by equations (1.17)–(1.22). An expression which approximates quite adequately to the actual variation of prestressing force F is (Hernández, 1990)

$$F = F_0 e^{-\mu x_k} \qquad 0 \leqslant x_k \leqslant L/2 \tag{1.23}$$

$$F = F_0 e^{-\mu(L - x_k)} \quad L/2 \leqslant x_k \leqslant L \tag{1.24}$$

where F_0 is the prestressing force at both ends of the beam, L is the beam length and μ is a parameter of value $0.1 \leqslant \mu \leqslant 0.4$, depending on the prestressing procedure.

By carrying out a change of variables

$$F_0 = Y_0 \tag{1.25}$$

$$F y_j = Y_j \tag{1.26}$$

the problem formulated by equations (1.17)–(1.22) turns out to be linear, so it can be solved easily with techniques of linear programming.

1.2.2 Problems with a nonlinear feasible region

Most problems on structural optimization belong to this kind of situation. Problems such as optimum design of structures under elastic behavior or optimization of cross sections to obtain least weight lead to a feasible region bounded by constraints defined by curve lines. The following examples describe such situations.

The bar shown in Fig. 1.8 has the volume as objective function, and dimensions b, e of its cross-section as design variables (Hernández, 1990a). Collapse modes considered are

1. yield stress σ_e,
2. buckling stress σ_e/ω,
3. maximum value of slenderness ratio $\lambda = l/i$,
4. geometrical ratio $b/e \leqslant 30$, and
5. minimum value of $e \geqslant 0.3$.

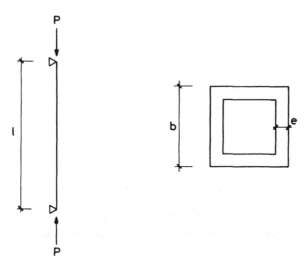

Fig. 1.8 Single bar under compression load. $P = 800 \, \text{kN}$; $l = 3 \, \text{m}$; $E = 2.1 \times 10^4 \, \text{kN cm}^{-2}$; $\sigma_4 = 2.6 \, \text{kN cm}^{-2}$.

Buckling is incorporated via the parameter ω (Spanish code NBE-MV 103 (1978)), producing equation (1.29) where σ_E is Euler's buckling stress

$$\sigma_E = \frac{\pi^2 EI}{l^2}$$

The slenderness ratio in equation (1.30) is limited to an upper value of 200 by the Spanish code mentioned earlier. Constraints 4 and 5 refer to design considerations in real structures.

Formulation of this problem can be written as

$$\min F = 4ebl \tag{1.27}$$

subject to

$$\frac{P}{4be} \leqslant \sigma_e \tag{1.28}$$

$$\frac{2P}{4be} \frac{\sigma_e}{1.3\sigma_e + \sigma_E - [(1.3\sigma_e + \sigma_E)^2 - 4\sigma_e\sigma_E]^{1/2}} \leqslant \sigma_e \tag{1.29}$$

$$2.45\frac{l}{b} \leqslant 200 \tag{1.30}$$

$$\frac{b}{e} \leqslant 30 \tag{1.31}$$

$$-e \leqslant -0.3 \tag{1.32}$$

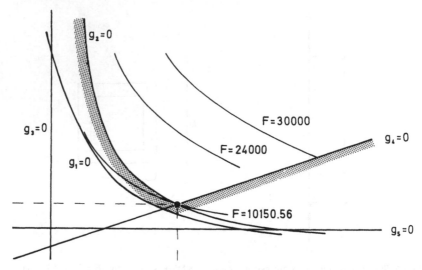

Fig. 1.9 Graphical representation of single-bar problem.

The objective function and the set of constraints which define a feasible region with curved geometry are shown in Fig. 1.9. The solution is at the following values:

$$e = 0.53\,\text{cm} \quad b = 15.96\,\text{cm} \quad F = 10\,150.56\,\text{cm}^3$$

Another example of optimum design of structures is the grillage of Fig. 1.10. It is loaded by two distributed loads q_1, q_2, and an isolated load P (Moses and Onoda, 1969). The objective function to be minimized is the volume of beams, considering constraints of normal stress at points of maximum bending moments. Design variables are x_1, x_2 and other mechanical parameters, such as the strength modulus W and inertia modulus I, are usually linked as in equation (1.32) in

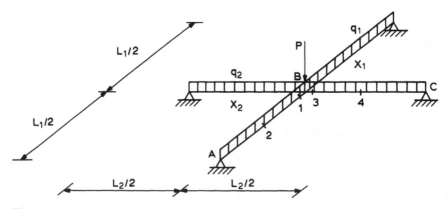

Fig. 1.10 Four-bar grillage. (From Moses and Onoda, *International Journal of Numerical Methods in Engineering*, 1969, Vol. 1, pp. 311–331. © 1969 John Wiley & Sons Ltd. Reprinted by permission of the publisher.)

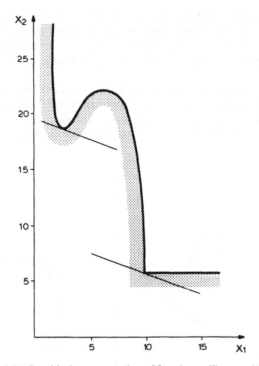

Fig. 1.11 Graphical representation of four-bar grillage problem.

order to reduce the number of variables of the problem.

$$W_i = \left(\frac{x_i}{1.48}\right)^{1.82} \qquad I_i = 1.007\left(\frac{x_i}{1.48}\right)^{2.65}. \tag{1.33}$$

When the values are

$$-\sigma^L = \sigma^U = 20\,\text{kN cm}^{-2} \quad L_1 = 100\,\text{cm} \quad L_2 = 300\,\text{cm} \tag{1.34}$$

$$q_1 = q_2 = 0.1\,\text{kN m}^{-1} \qquad P = 1\,\text{kN} \tag{1.35}$$

the feasible region has the shape indicated in Fig. 1.11. Two local minima exist and correspond to the following values:

$$x_1 = 2.2\,\text{cm} \quad x_2 = 18.8\,\text{cm} \quad V = 5860\,\text{cm}^3 \tag{1.36}$$

$$x_1 = 9.7\,\text{cm} \quad x_2 = 6.1\,\text{cm} \quad V = 2800\,\text{cm}^3 \tag{1.37}$$

The major difference between the last two examples is the number of local minima which appear in each one. This difference is due to the shape of the feasible region.

A domain D is called convex if, given two points X_1 and X_2 contained in it, the line segment joining them is also contained completely in the domain. In other words, the domain D is convex if for $0 < \alpha < 1$ any point $X = \alpha X_1 + (1 - \alpha)X_2$ is also inside the domain.

Examples of convex and nonconvex domains are shown in Figs 1.12(a) and
1.12(b) respectively. The intersection of convex domains is also convex, as in
Fig. 1.13(a); domains bounded by straight lines, as for the example in Fig. 1.13(b),
are always convex. Convex or nonconvex domains may be bounded or
unbounded.

A minimization problem is called convex if the objective function $F(X)$ is convex
and the set of constraints $g_j(X)$ bounds a convex domain; otherwise the problem
is nonconvex. In accordance with that, linear problems are always convex. If a
minimization problem is convex there is only one minimum which is the solution.

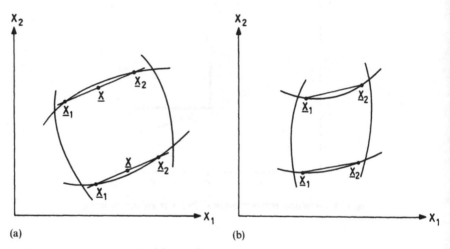

Fig. 1.12 Examples of optimization domains: (a) convex domain; (b) nonconvex domain.

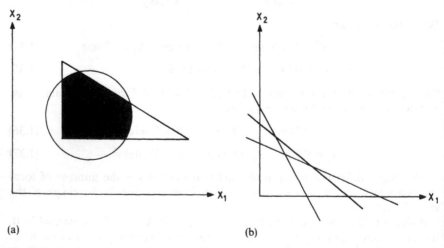

Fig. 1.13 Examples of convex domains; (a) intersection of two convex domains; (b) unbounded
convex domain.

On the contrary, in nonconvex problems several local minima may exist and the least value among them corresponds to the global minimum and the desired solution.

Some constraints considered in structural optimization problems lead to convex problems. Svanberg (1984, 1991) has identified the following cases.

- *Symmetric displacement constraint.* A constraint is said to be a symmetric displacement when the displacement vector u_k of a node has the same direction as the external load P_k acting at this node. Usually an upper bound is included for u_k.

$$u_k = \mu P_k \leqslant u_k^U \tag{1.38}$$

- *Global displacement constraint.* This consists of imposing a maximum value limit u^U to any component of the displacement vector u of the structure.

$$\max u_j \leqslant u^U \quad j = 1, \ldots, N \tag{1.39}$$

- *Lower limit of the smallest eigenvalue.* The constraint is intended to set up a lower bound for the smallest eigenvalue of the stiffness matrix K.

$$\lambda_m \leqslant \lambda_j \quad j = 1, \ldots, N \tag{1.40}$$

Many other constraints, such as stress- or displacement-related constraints, produce nonconvex feasible regions. Usually, in any structural optimization problem many kinds of constraints are considered and the feasible region can be expected not to be convex, and some local minima may appear.

1.2.3 Problems with disjoint feasible regions

This kind of situation arises in structures undergoing dynamic loading in the absence of damping. It was first described by Cassis (1974) in planar frames subjected to horizontal vibrations at the foundation. In the case shown in Fig. 1.14 the frame is subjected to a distributed load and a harmonic vibration of ground acting during half a period. Design variables are the inertia moduli of columns I_1 and I_2 and the weight is the objective function. The feasible design domain is represented in Fig. 1.14(a) and it appears to be split into two disjoint regions, a greater, unbounded one and another that is smaller, placed as a hole in the infeasible solutions domain.

Other authors (Johnson, 1976; Johnson et al., 1976; Mills-Curran and Schmit, 1985) have pointed out that in a dynamic system with N degrees of freedom undergoing C harmonic loads, the maximum number of disjoint feasible regions is $n(C + 1)$.

Another well-known example is a thin-walled cantilever tubular beam shown in Fig. 1.15 subjected to a torsional moment of amplitude M_0 and frequency ω_e, acting along its length. The objective is to minimize the beam weight; the sets of constraints are the yield stress of material and the lower limits of the design values. Again, the feasible region is constituted by two regions as shown in Fig. 1.16 for two different loading values. In both cases each region has a local minimum. The solution at point B, that corresponds to a larger value of e_1, gives

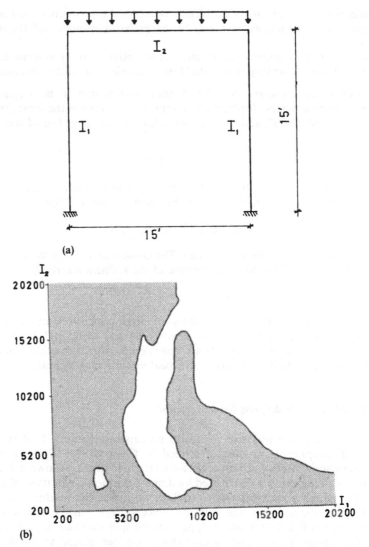

Fig. 1.14 Single portal frame example: (a) single frame under distributed load and vibrating foundation; (b) two disjoint feasible regions arising in single portal frame example.

a structure is which the first eigenvalue ω_1 satisfies $\omega_e < \omega_1$; the minimum at point A with smaller values of e_1 has the opposite characteristic.

1.3 NUMERICAL METHODS TO OBTAIN LOCAL MINIMA

There are several techniques to solve the minimization problem expressed by equations (1.1) and (1.2) but they can only guarantee that local minima are reached. A suitable division of these methods is

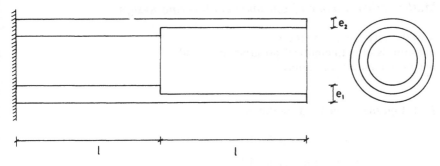

Fig. 1.15 Cantilever tubular example.

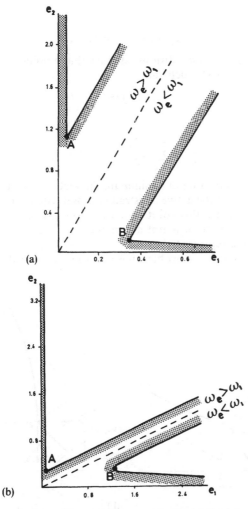

Fig. 1.16 Feasible domains in cantilever tubular example: (a) global minimum with $\omega_e < \omega_l$; (b) global minimum with $\omega_e > \omega_l$. (From E.H. Johnson, *AIAA Journal*, 1976, Vol. 14, pp. 259–261. © 1976 American Institute of Aeronautics and Astronautics. Reprinted by permission of the publisher.)

- optimality criteria methods,
- methods of mathematical programming, and
- approximation techniques.

1.3.1 Optimality criteria methods

(a) Kuhn–Tucker-based methods

Let us consider a problem defined by

$$\min F(X) \tag{1.41}$$

subject to

$$g_j(X) \leqslant 0 \quad i = 1, \ldots, m \tag{1.42}$$

Kuhn–Tucker necessary conditions establish that a point X is a local minimum if a set of scalars λ_j exists such that

$$\lambda_j \geqslant 0 \quad j = 1, \ldots, J \tag{1.43}$$

and

$$\frac{\partial F}{\partial x_i} + \sum_{j=1}^{J} \lambda_j \frac{\partial g_j}{\partial x_i} = 0 \quad i = 1, \ldots, n \tag{1.44}$$

Equation (1.44) explains that at a minimum the vector $-\nabla F$ is a linear combination of the gradients of the active constraints subset with positive components for all gradients ∇g_j active at this point. In the opposite case, when some components λ_j are nonpositive the point is not a minimum. In Fig. 1.17 two different cases of this situation are represented.

To solve the minimization problem it is necessary to find values of the Lagrange

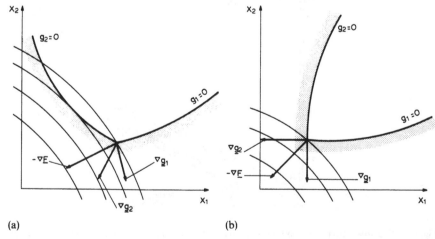

(a)   (b)

Fig. 1.17 Graphical explanation of Kuhn–Tucker condition: (a) point accomplishing KT condition; (b) point not accomplishing KT condition. (From U. Kirsch, *Optimum Structural Design*, © 1981 McGraw-Hill, Inc. Used with permission of the publisher.)

multipliers λ and design variables X. An iterative process is usually carried out which requires

- a recurrence relation between design variables X, and
- identifying the subset of active constraints at the optimum.

The recurrence relation is set up taking into account that at the solution

$$\frac{\partial F}{\partial x_i} + \sum_{j=1}^{m} \lambda_j \frac{\partial g_j}{\partial x_i} = 0 \quad i = 1,\ldots,n \tag{1.45}$$

assuming that

$$\frac{\partial F}{\partial x_i} \neq 0$$

$$-\sum_{j=1}^{m} \lambda_j \frac{\partial g_j}{\partial x_i} \bigg/ \frac{\partial F}{\partial x_i} = 1 \tag{1.46}$$

Throughout the process the left-hand side of equation (1.46) has the value

$$-\sum_{j=1}^{m} \lambda_j \frac{\partial g_j}{\partial x_i} \bigg/ \frac{\partial F}{\partial x_i} = T_i \quad i = 1,\ldots,n \tag{1.47}$$

and at the minimum will take the value $T_i = 1 \ (i = 1,\ldots,n)$. At the kth iteration equation (1.46) will be

$$-\sum_{j=1}^{m} \lambda_{jk} \frac{\partial g_{jk}}{\partial x_i} \bigg/ \frac{\partial F_k}{\partial x_i} = T_{ik} \tag{1.48}$$

From equation (1.46) a recurrence relation may be established between values of any element x_i of design variables X_k and X_{k+1} at two consecutive iterations. A commonly used relation is (Khot, Berke and Venkayya, 1979)

$$x_{ik+1} = [\alpha + (1-\alpha) T_{ik}] x_{ik} \quad i = 1,\ldots,n \tag{1.49}$$

where α is a relaxation factor controlling the step size. It usually takes values $\alpha = 0.5$ or $\alpha = 0.75$.

To obtain the Lagrange multipliers λ the following procedure is carried out. The set of active constraints $g_j(X)$ is linearized (Kinsalaas, 1972; Rizzi, 1976):

$$g_{jk+1} \approx g_{jk} + \sum_{i=1}^{n} \frac{\partial g_{jk}}{\partial x_i} (x_{ik+1} - x_{ik}) = 0 \quad j = 1,\ldots,J \tag{1.50}$$

introducing the expression for x_{ik+1} given by equation (1.49)

$$g_{jk} + (1-\alpha) \sum_{i=1}^{n} \frac{\partial g_{jk}}{\partial x_i} (T_{ik} - 1) x_{ik} = 0 \quad j = 1,\ldots,J \tag{1.51}$$

If T_{ik} is substituted by its value in equation (1.47) it turns out that for $j = 1,\ldots,J$

$$\sum_{i=1}^{n} \frac{\partial g_{jk}}{\partial x_i} x_{ik} \sum_{j=1}^{m} \lambda_j \frac{\partial g_{jk}}{\partial x_i} \bigg/ \frac{\partial F_k}{\partial x_i} = \frac{g_{jk}}{1-\alpha} - \sum_{i=1}^{n} \frac{\partial g_{jk}}{\partial x_i} x_{ik} \tag{1.52}$$

Equation (1.52) is a system of linear equations with $\lambda_j \ (j = 1,\ldots,J)$ as unknowns. Each of them that has a nonnegative value corresponds to active constraints, according to the Kuhn–Tucker conditions.

Briefly, the whole process is carried out by repeating the following steps.

1. Define an initial design X.
2. Solve the system of equations to obtain the vector λ.
3. Calculate values of T_i considering only the subset of constraints associated with nonnegative λ_j.
4. Go to recurrence relation (1.21) and go back to step 2.

Steps 2 through 4 should be repeated until absolute or relative convergence is achieved. If the subset of active constraints $g_j(X)$ $(j = 1, \ldots, J)$ changes frequently from one iteration to the next, convergence may become difficult. The same happens if the initial design is very dissimilar to the solution.

(b) Method based on information theory

In this method (Templeman and Li, 1985, 1987; Templeman, 1989) the optimization problem defined by equation (1.1) is substituted by an equivalent problem:

$$\min F(X) \tag{1.53}$$

subject to

$$\sum_{j=1}^{m} \lambda_j g_j(X) = 0 \tag{1.54}$$

$$\sum_{j=1}^{m} \lambda_j = 1 \tag{1.55}$$

$$\lambda_j \geqslant 0 \tag{1.56}$$

Each multiplier λ_j may be considered as the probability for a constraint $g_j(X)$ to be active at the optimum. Hence, for each constraint the following condition stands:

$$g_j \lambda_j = 0 \tag{1.57}$$

Consequently, the condition (1.53) is satisfied at the solution. It is well known (Everett, 1963; Brooks and Geoffrion, 1966; Gould, 1969; Greenberg and Pierskalla, 1970) that the problem defined by equations (1.53)–(1.56) has the same solution as the problem defined by equations (1.1) and (1.2).

As in the methods mentioned earlier the vectors X and λ constitute the set of unknowns. In this technique the probabilities λ_j are obtained by using techniques not biasing the iterative process. The innovative approach taken by Templeman and coworkers is that they do this by using techniques based on information theory such as the maximum entropy formalism. This formalism, enunciated by Jaynes (1957), indicates that the least biased values for probabilities λ_j at each iteration are those which maximize the entropy information of the process, which is measured according to the function defined by Shannon (1948):

$$\max \frac{S(\lambda)}{K} = -\sum_{j=1}^{m} \lambda_j \ln \lambda_j \tag{1.58}$$

$$\sum_{j=1}^{m} \lambda_j = 1 \tag{1.59}$$

The optimization procedure is carried out in the following way.

1. An initial design is defined and the probabilities vector λ is obtained through the maximization problem of equations (1.58) and (1.59). The solution of the problem turns out to be

$$\lambda = \left[\frac{1}{m}, \ldots, \frac{1}{m}\right] \quad S = K \ln m \tag{1.60}$$

2. By solving the minimization problems of equations (1.53)–(1.56) a new design X_2 is obtained.
3. Beginning at iteration $k = 2$ and thereafter the elements of the vector λ_k at the kth iteration are updated to λ_{k+1} by taking in account that

$$\sum_{i=1}^{m} \lambda_{jk} g_j(X_k) = 0 \quad \text{and} \quad \sum_{i=1}^{m} \lambda_{jk+1} g_j(X_k) = \varepsilon_{k+1} \tag{1.61}$$

where $\varepsilon_{k+1} \neq 0$ because the vectors λ_{k+1} and X_k belong to different iterations in the procedure. As the procedure approaches convergence the value of ε_{k+1} should decrease until it cancels out, in theory, at the optimum. Because of that, in order not to bias the undating of λ_k, and to maintain its normality, the probability vector λ_{k+1} is found by solving the following maximization problem:

$$\max \frac{S(\lambda)}{K} = -\sum_{j=1}^{m} \lambda_{jk+1} \ln \lambda_{jk+1} \tag{1.62}$$

subject to

$$\sum_{j=1}^{m} \lambda_{jk+1} = 1 \tag{1.63}$$

$$\sum_{j=1}^{m} \lambda_{jk+1} g_j(X_k) = \varepsilon_{k+1} \tag{1.64}$$

where ε_{k+1} is a user-selected parameter the expression of which must tend to zero.
4. Then, the process goes back to steps 2 and 3 until convergence is achieved. The solution usually arrives from outside the feasible region, so if the process is stopped at an intermediate design some constraints are violated, and thus it is necessary to continue the method until a good ratio of convergence is obtained.

1.3.2 Methods based on mathematical programming

(a) Methods of linear programming

The most popular method for linear problems is the simplex method (Dantzig, 1963) and it is explained in every optimization book. A different approach started by Karmarkar (1984) has been claimed to be more efficient than the simplex method for large size problems. The main reason is that the simplex method requires an effort to obtain the solution that increases exponentially for the worst case, and in Karmarkar's methods the effort increases polynomially. There is a large number of real problems of many classes solved very efficiently by the simplex method while experience with Karmarkar's method is still limited. However, it is indeed a very promising procedure and it is explained next. Let us remember that a very usual formulation for linear problems in design optimization is

$$\min F = \sum_{j=1}^{n} c_j x_j \tag{1.65}$$

subject to

$$\sum_{j=1}^{n} a_{ij} x_j \geqslant b_i \quad i = 1, \ldots, m \tag{1.66}$$

$$x_j \geqslant 0 \quad j = 1, \ldots, n \tag{1.67}$$

This formulation is called a primal problem. There exists a dual problem, in the space of dual variables $\lambda_i (i = 1, \ldots, m)$, that can be formulated as

$$\max G = \sum_{i=1}^{m} b_i \lambda_i \tag{1.68}$$

subject to

$$\sum_{i=1}^{m} a_{ji} \lambda_i \leqslant c_j \tag{1.69}$$

$$\lambda_i \geqslant 0 \tag{1.70}$$

It is very well known that

$$\min F = \max G \tag{1.71}$$

The method of Karmarkar is intended to solve problems formulated as

$$\min K = \sum_{j=1}^{N+1} e_j' y_j' \tag{1.72}$$

subject to

$$\sum_{j=1}^{N+1} k_{ij}' y_j' = 0 \quad i = 1, \ldots, M \tag{1.73}$$

$$\sum_{j=1}^{N+1} y_j' = 1 \tag{1.74}$$

where the objective function K is zero at the optimum.

As the problem defined by equations (1.65)–(1.67) is different from the problem solved by Karmarkar, the first required step is to convert it to a formulation suitable for the method of Karmarkar. The way to do this is to combine the primal and dual formulations of equations (1.64)–(1.69) to set up a new linear problem. Using matrix notation it can be written

$$\min(F - G) = \min K = C^T X - B^T \lambda \tag{1.75}$$

subject to

$$AX \geqslant B \tag{1.76}$$

$$A^T \lambda \leqslant C \tag{1.77}$$

$$X, \lambda \geqslant 0 \tag{1.78}$$

if slack variables are introduced

$$\min(C^T X - B^T \lambda) \tag{1.79}$$

subject to

$$AX - X' = B \tag{1.80}$$

$$A^T \lambda + \lambda' = C \tag{1.81}$$

$$X, X', \lambda\lambda' \geqslant 0 \tag{1.82}$$

and also

$$\min E^T Y \tag{1.83}$$

subject to

$$KY = 0 \tag{1.84}$$

$$Y \geqslant 0 \tag{1.85}$$

where

$$\underset{(1 \times N)}{Y^T} = \left[\underset{(1 \times n)}{Y^T} \; \underset{(1 \times m)}{X'^T} \; \underset{(1 \times m)}{\lambda^T} \; \underset{(1 \times n)}{\lambda'^T} \right] \tag{1.86}$$

$$\underset{(1 \times N)}{E^T} = \left[\underset{(1 \times n)}{C^T} \; \underset{(1 \times m)}{0} \; \underset{(1 \times m)}{-B^T} \; \underset{(1 \times n)}{0} \right] \tag{1.87}$$

$$\underset{(1 \times M)}{D^T} = \left[\underset{(1 \times m)}{B^T} \; \underset{(1 \times n)}{C^T} \right] \tag{1.88}$$

$$\underset{(M \times N)}{K} = \left[\begin{array}{cccc} \underset{(m \times n)}{A} & \underset{(m \times m)}{-I} & \underset{(m \times m)}{0} & \underset{(m \times n)}{0} \\ \underset{(n \times n)}{0} & \underset{(n \times m)}{0} & \underset{(n \times m)}{A^T} & \underset{(n \times n)}{I} \end{array} \right] \tag{1.89}$$

Then, starting from an initial design Y_1, a projective transformation is carried out

$$y'_j = \frac{y_j}{y_{1j}} \bigg/ \left(1 + \sum_{i=1}^{N} \frac{y_i}{y_{1i}} \right) \quad j = 1, \dots, N \tag{1.90}$$

and a variable y'_{N+1} is defined

$$y'_{N+1} = 1 - \sum_{j=1}^{N} y'_j \tag{1.91}$$

Equation (1.90) may be written as

$$y'_j = \frac{y'_{N+1} y_j}{y_{1j}} \tag{1.92}$$

or inversely

$$y_j = \frac{y_{1j} y'_j}{y'_{N+1}} \quad j = 1, \dots, N \tag{1.93}$$

The transformation defined by equation (1.90) transforms points of an N-coordinate open space into points of an $(N + 1)$-coordinate closed space. It can be easily proved that the point Y_1 is transformed into the point

$$Y'_1 = \left(\frac{1}{N+1}, \dots, \frac{1}{N+1} \right)$$

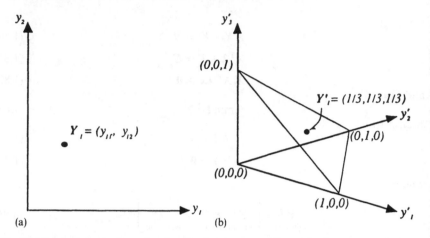

Fig. 1.18 Projective transformation for $N = 2$: (a) space with coordinate system y_j; (b) space with coordinate system y'_j.

the center of the new coordinate system. This transformation is shown in Fig. 1.18 for an example with $N = 2$.

If we put equation (1.93) into equations (1.83)–(1.85) we obtain

$$\min \frac{e_j y_{1j} y'_j}{y'_{N+1}} \tag{1.94}$$

subject to

$$\frac{k_{ij} y_{1j} y'_j}{y'_{N+1}} = d_i \quad i = 1, \ldots, M \tag{1.95}$$

$$\sum_{j=1}^{N+1} y'_j = 1 \tag{1.96}$$

$$y'_j \geq 0 \quad j = 1, \ldots, N+1 \tag{1.97}$$

defining

$$e'_j = e_j y_{1j} \tag{1.98}$$

$$k'_{ij} = k_{ij} y_{1j} \quad j = 1, \ldots, N \tag{1.99}$$

$$k'_{iN+1} = -d_i \tag{1.100}$$

Examining equations (1.90) and (1.91) it can be concluded that the variable y'_{N+1} always satisfies $y'_{N+1} > 0$. As the objective function K is zero at the optimum, the variable y'_{N+1} can be eliminated from equation (1.94). After all these changes the formulation of the problem is

$$\min K = E'^{\mathrm{T}} Y' \tag{1.101}$$

subject to

$$K' Y' = 0 \tag{1.102}$$

$$\sum_{j=1}^{N+1} y'_j = 1 \tag{1.103}$$

$$y'_j \geq 0 \quad j = 1, \ldots, N+1 \tag{1.104}$$

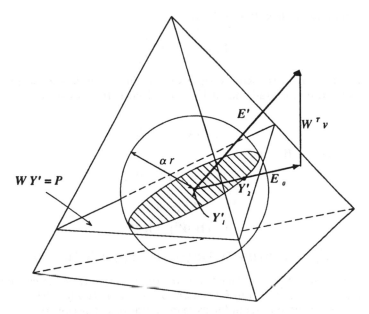

Fig. 1.19 Karmarkar minimization phase.

that coincides with the class of problem solved by the method of Karmarkar. Constraints of equations (1.102) and (1.103) may be grouped in the form

$$WY' = \begin{bmatrix} K' \\ 1 \end{bmatrix} Y' = \begin{bmatrix} 0 \\ 1 \end{bmatrix} = P \tag{1.105}$$

The approach of Karmarkar for obtaining the minimum is to carry out an iterative procedure that will be explained with the example that appears in Fig. 1.19. In this example $N + 1 = 4$, and thus the coordinate system is represented by a tetrahedron and the set of constraints by a plane. The projective transformation takes the initial design Y_1 into Y_1', the center of the tetrahedron. From geometrical considerations it can be concluded that the tetrahedron inscribes a ball of radius

$$r = \frac{1}{[N(N+1)]^{1/2}}$$

Instead of moving directly to the minimum, Karmarkar tries to improve the current design Y_1' by obtaining another point Y_2' such that

- it is placed on the boundary of a ball centered at Y_1', of radius αr, with $0 < \alpha < 1$, and
- it decreases the objective function K.

As the point Y_2' minimizes the objective function of equation (1.101) it will lie along E_P' which is the projection of the gradient vector $-E'$ onto the null space of the constraint surface $WY' = P$, so

$$Y'_2 = Y'_1 - \alpha r \frac{E'_P}{\| E'_P \|} \tag{1.106}$$

To identify $-E'_P$ it is useful to observe that the vector $E' - E_P$ will be in the space spanned by the gradients of $WY' = P$. Hence, there exists a vector $\boldsymbol{v}$ such that

$$W\boldsymbol{v} = E' - E'_P \tag{1.107}$$

Premultiplying by W^T gives

$$WW^T\boldsymbol{v} = WE' - WE'_P \tag{1.108}$$

since

$$WE'_P = 0 \tag{1.109}$$

$$\boldsymbol{v} = (WW^T)^{-1} WE' \tag{1.110}$$

and

$$E'_P = E' - W(WW^T)^{-1}WE' = \{I - W(WW^T)^{-1}W\}^{-1}E \tag{1.111}$$

After the point Y'_2 is obtained, the vector of variables Y_2 is calculated by reversing the projective transformation of equation (1.90). Then, the whole procedure is carried out again for as many iterations as necessary until convergence is achieved.

The four-bar structure in Fig. 1.3 may be used to explain numerically Karmarkar's procedure. Assuming numerical values

$$P = 2 \quad \sigma = 2 \quad l = 1 \tag{1.112}$$

Equations (1.3)–(1.6) become

$$\min F(X) = 2.31x_1 + 2.82x_2 \tag{1.113}$$

subject to

$$1.41x_2 \geqslant 1 \tag{1.114}$$

$$1.73x_1 \geqslant 1 \tag{1.115}$$

$$x_1 \geqslant 1 \tag{1.116}$$

Combining the primal and dual formulation gives

$$\min 2.31x_1 + 2.82x_2 - \lambda_1 - \lambda_2 - \lambda_3 \tag{1.117}$$

subject to

$$
\begin{array}{lll}
1.41x_2 - x_3 & = 1 & \text{(1.118)} \\
1.73x_1 \qquad\quad - x_4 & = 1 & \text{(1.119)} \\
x_1 \qquad\qquad\quad - x_5 & = 1 & \text{(1.120)} \\
1.73\lambda_2 + \lambda_3 + \lambda_4 & = 2.31 & \text{(1.121)} \\
1.41\lambda \qquad\qquad\quad + \lambda_5 & = 2.82 & \text{(1.122)}
\end{array}
$$

Choosing Y_1 as initial design

$$Y_1 = (2, 1.41, 2, 4.92, 2, 0.5, 0.233, 0.646, 0.5, 1) \tag{1.123}$$

Table 1.1 Numerical values for iterations in the Karmarkar procedure

N_{ite}	x_1	x_2	K	F
1	1.69	1.15	0.44	7.15
7	1.01	0.75	0.098	4.58
15	1.00	0.71	0.008	4.33

and carrying out the operations indicated by equations (1.98)–(1.100) the following formulation of Karmarkar is obtained:

$$\min 4.62y'_1 + 3.976y'_2 - 0.5y'_6 - 0.233y'_7 - 0.646y'_8 \qquad (1.124)$$

subject to

$$2y'_2 - 2y'_3 \qquad\qquad\qquad = 0 \quad (1.125)$$
$$3.46y'_1 \qquad\quad - 4.92y'_4 \qquad\qquad = 0 \quad (1.126)$$
$$2y'_1 \qquad\qquad - 2y'_5 \qquad\qquad = 0 \quad (1.127)$$
$$0.40y'_7 + 0.646y'_8 + 0.5y'_9 \qquad = 0 \quad (1.128)$$
$$0.70y'_6 \qquad\qquad\qquad + y'_{10} = 0 \quad (1.129)$$

Numerical values for several iterations are shown in Table 1.1.

(b) Methods based on mathematical programming

All these methods try to obtain a minimum of the optimization problem defined by equations (1.1) and (1.2). Starting from an initial design X_1 the process is carried out in an iterative scheme producing intermediate designs X_k which improve the objective function and bring the problem to the solution.

Design modification is done at each iteration along a search direction S_k as indicated in

$$X_{k+1} = X_k + \alpha S_k \qquad (1.130)$$

Almost every method has a different approach to obtaining the direction S_k, and the most important techniques are explained next.

In the method of **feasible directions** (Zoutendijk, 1960; Vanderplaats, 1984a,b) the choice of S_k depends on the position of point X_k.

1. If X_k is inside the feasible region the gradient of the objective function is chosen as indicated below

$$S_k = -\nabla F_k \qquad (1.131)$$

2. If X_k is on the boundary, at one or more constraints, the aim is to improve the current design X_k keeping the new point X_{k+1} inside the feasible region as much as possible. To do that the direction S_k is calculated by solving a linear programming problem stated by equations (1.132)–(1.135):

$$\max \beta \qquad (1.132)$$

subject to

$$S_k^T \nabla g_j + \theta_j \beta \leqslant 0 \quad j = 1, \ldots, J \tag{1.133}$$

$$S_k^T \nabla F + \beta \leqslant 0 \tag{1.134}$$

$$-1 \leqslant S_k \geqslant 1 \tag{1.135}$$

There are two possibilities for placing the point X_{k+1}: it can be in the domain, or at a new constraint. Both possibilities are shown in Fig. 1.20. If X_{k+1} is located in the domain the method proceeds according to step 1. If X_{k+1} is on the boundary then step 2 is carried out.

The constraints $g_j(X)$ in equation (1.135) are those which contain the current design and thus $g_j(X) = 0$. The scalars θ_j are usually given by $\theta_j = 1$.

Other methods change the problem of equations (1.1) and (1.2) into an unconstrained optimization problem. This is how the **sequential unconstrained minimization technique** (SUMT) acts. This approach, also called the **penalty function method** (Fiacco and McCormick, 1963, 1968; Zangwill, 1967; Kavlie, 1971; Haftka and Starnes, 1976) creates a penalty function in the following way:

$$\phi(X, r) = F(X) + r \sum_{j=1}^{m} p(g_j) \tag{1.136}$$

The penalty function $P(g_j)$ is formed by using the set of constraints and the parameter r is a control parameter of convergence. There are two variations of this technique.

- *Exterior penalty function.* In this formulation, equation (1.136) is written as

$$\phi(X, r) = F(X) + r \sum_{j=1}^{m} \langle g_j \rangle^{\gamma} \tag{1.137}$$

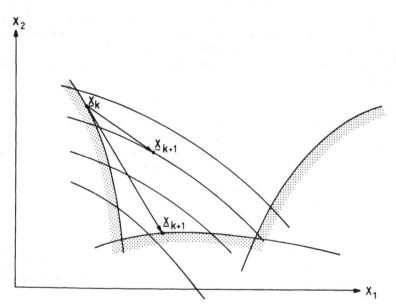

Fig. 1.20 Possible situations for the improved design after linear programming phase.

In equation (1.137) the expression $\langle g_j \rangle$ symbolizes

$$\langle g_j \rangle = g_j \quad \text{if} \quad g_j(X) > 0 \tag{1.138}$$

$$\langle g_j \rangle = 0 \quad \text{if} \quad g_j(X) \leqslant 0 \tag{1.139}$$

and usually $\gamma = 2$.

The starting point must be an infeasible design. The function $\phi(X, r)$ is minimized repeatedly for increasing values of the control parameter r. This produces intermediate designs X_k that reduce the values of the violated constraints at each iteration and thus leads to the solution. The starting point must be an infeasible design. Minimization of the function $\phi(X, r)$ forces each violated constraint to become smaller. The minimum $\phi(X, r)$ is also the minimum of $f(X)$ and it is reached from outside the feasible region.

- *Interior penalty function.* In this case equation (1.136) becomes

$$\phi(X, r) = F(X) - r \sum_{j=1}^{m} \frac{1}{g_j(X)} \tag{1.140}$$

In this technique the starting point must be a feasible one, so any constraint is negative. The function defined by equation (1.140) is minimized iteratively for decreasing values of r. This causes intermediate designs to be obtained that produce lower bounds of every constraint in order to reduce $\phi(X, r)$. As in the previous technique the minimum of $\phi(X, r)$ is the same as the minimum of the problem of equations (1.1) and (1.2). The solution is obtained from inside the feasible region, so each intermediate design is a valid one.

In both categories of penalty function methods it is necessary to solve several times an unconstrained optimization problem, which is accomplished according to equation (1.130). There are three groups of techniques to obtain S_k.

- *Order zero techniques.* One of the most useful algorithms is the **conjugate directions** (Powell, 1964) approach. This method requires a set of conjugate directions to be identified. A set of directions are called conjugate when they satisfy the condition

$$S_i^T A S_j = 0 \quad \text{for} \quad i \neq j \tag{1.141}$$

A is a positive definite quadratic form. This technique only needs to evaluate the objective function and, when the objective function of the unconstrained problem is quadratic, the minimum is obtained by optimizing once along each direction of the conjugate directions set. In any other case it could be necessary to repeat this technique several times before reaching the optimum.

- *Order one techniques.* These are usually called gradient methods (Fletcher and Reeves, 1964) and they require the gradient of the objective function to be calculated. The direction S_k is defined as

$$S_k = -\nabla F_k + \beta_k S_{k-1} \tag{1.142}$$

The value of β_k depends on each specific formulation. A pair of more well-known cases are **steepest descent**, that is

$$\beta_k = 0 \tag{1.143}$$

and **conjugate gradient**, that is

$$\beta_k = \frac{\nabla F^T \nabla F_k}{\nabla F_{k-1}^T \nabla F_{k-1}} \qquad (1.144)$$

- *Order two methods.* These are denominated Newton methods, and, in addition to the objective function $F(X)$ and its gradient, they require the hessian matrix $H(X)$ to be evaluated. Each direction S_k is defined as

$$S_k = -H_k^{-1} \nabla F_k \qquad (1.145)$$

Some variants of this method (Davidon, 1959; Fletcher and Powell, 1963) obtain the matrix $H(X)$ by using approximation schemes.

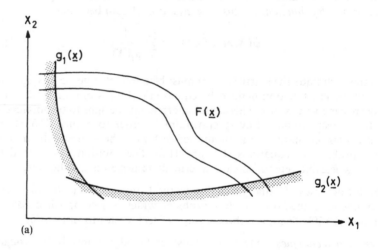

(a)

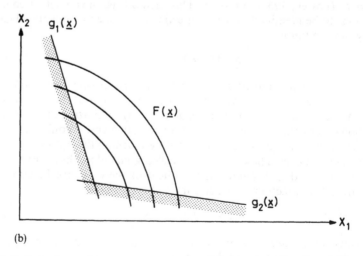

(b)

Fig. 1.21 Optimization problem and quadratic approximation: (a) actual optimization problem; (b) quadratic approximation.

1.3.3 Approximation techniques

Finally, another type of method solves the optimization problem by writing equations (1.1) and (1.2) using a Taylor series expansion, which is truncated after two or three terms, thus creating an approximate problem. The most efficient techniques are as follows.

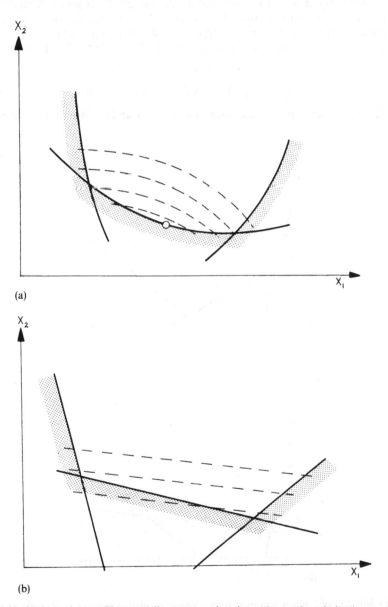

(a)

(b)

Fig. 1.22 Optimization problem and linear approximation: (a) actual optimization problem; (b) linear approximation. (From E. Atrek, R.H. Gallagher, K.M. Ragsdell and O.C. Zienkiewicz (eds), 1984, *New Directions in Optimum Structural Design*, John Wiley & Sons. Reprinted by permission of John Wiley & Sons, Ltd.)

- *Sequence of quadratic problems.* Once an initial design X_1 is chosen, the problem of equations (1.1) and (1.2) is approximated to

$$F(X) \approx F(X_1) + \nabla F(X_1)^T(X - X_1) + \tfrac{1}{2}(X - X_1)^T H(X_1)(X - X_1) \quad (1.146)$$

$$g_j(X) \approx g_j(X_1) + \nabla g_j^T(X_1)(X - X_1) \leqslant 0 \quad j = 1,\ldots,m \quad (1.147)$$

By solving equations (1.146)–(1.147) a new design X_2 is given. Then, these expressions are again calculated at the new point X_2 and the optimization is carried out repeatedly until convergence is obtained (Fig. 1.21).

- *Sequence of linear problems.* In this approach (Fig. 1.22) the formulation is

$$\min F(X) \approx F(X_1) + \nabla F_1^T(X - X_1) \quad (1.148)$$

$$g_j(X) \approx g_j(X_1) + \nabla g_j^T(X_1)(X - X_1) \leqslant 0 \quad (1.149)$$

Several variants of this technique exist (Cheney and Goldstein, 1959; Kelley,

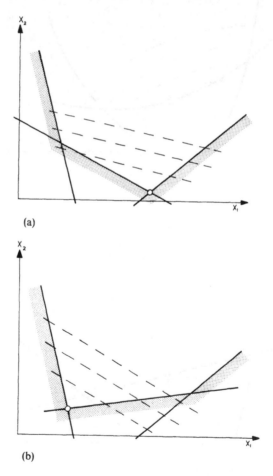

(a)

(b)

Fig. 1.23 Solution of a linear approximation problem oscillating between two points. (From E. Atrek, R.H. Gallagher, K.M. Ragsdell and O.C. Zienkiewicz (eds), 1984, *New Directions in Optimum Structural Design*, John Wiley & Sons. Reprinted by permission of John Wiley & Sons, Ltd.)

1960), and sometimes a few inconveniences arise in this method. If the solution of the real problem formulated by equations (1.1) and (1.2) is placed on a unique constraint, the linear approximation may oscillate indefinitely between the two vertices nearer the real optimum, as shown in Fig. 1.23. There are improvements on this method which can overcome this drawback (Baldur, 1972; Hernández and Ceron, 1987) by stating some **move limits** to set bounds to the variation of the design variables at each iteration.

These move limits are intended to avoid the oscillating behavior of linear approximations by preventing any design variable from changing too much between two consecutive iterations. Let us suppose that x_i^L, x_i^U are lower and upper limits of variable x_i; if this variable has the value x_{ik} at the kth iteration and a move limit Δx_{ik} it set up, the value of the design variable x_{ik+1} would be limited by

$$\max(x_i^L, x_{ik} - \Delta x_{ik}) \leqslant x_{ik+1} \leqslant \min(x_i^U, x_{ik} + \Delta x_{ik}) \qquad (1.150)$$

Move limits can be set up independently for each variable with different values for lower and upper limits. Additionally, they can also vary between iterations.

1.4 THE TUNNELING METHOD IN GLOBAL OPTIMIZATION

Given a function $F(X)$ each optimization technique mentioned previously starts by defining an initial design X_1 and tries to obtain another point X_1^* which satisfies

$$F(X_1^*) \leqslant F(X_1) \qquad (1.151)$$

$$F(X_1^*) \leqslant F(X_1^* + \Delta X) \qquad (1.152)$$

where ΔX is an increment of X_1^*.

If the problem considered is a convex one the value of $F(X_1^*)$ satisfying equations (1.151) and (1.152) is the optimal solution and the global minimum of $F(X)$, but in the opposite case several points X_i^* may be found under conditions (1.151) and (1.152), so there is a set of local minima containing the global one. If points corresponding to local minima are ordered according to their objective function value the following series can be written:

$$F(X_1^*) \geqslant F(x_2^*) \geqslant \cdots \geqslant F(X_n^*) \qquad (1.153)$$

In this section a method for obtaining the global minimum in nonconvex problems is presented, and numerical results corresponding to practical examples are included to show the performance of the method.

1.4.1 The tunneling method of unconstrained optimization

This method has been developed by Gómez and Levy (1982) and Levy and Gómez (1984) and is intended to identify, in a systematic way, each of the local minima of a nonconvex function. Along the procedure each new minimum obtained produces an equal or lower objective function value $F(X)$ than the previous one.

Given a function $F(X)$, the aim of the method is obtain a point X^* that for any vector Y satisfies

$$F(X^*) \leqslant F(X) \tag{1.154}$$

$$\nabla F(X^*) = 0 \quad Y^T H(X^*) Y > 0 \tag{1.155}$$

The method has two different phases.

1. *Minimization phase.* The purpose of this phase is to obtain a local minimum satisfying equation (1.154). Starting from an initial point X_1 the first local minimum X_1^* is identified as shown in Fig. 1.24. In order to do it, any numerical optimization technique already mentioned may be used.
2. *Tunneling phase.* Departing from X^*, this phase aims to idenify a point X_2, indicated in Fig. 1.24, having $F(X_1^*) = F(X_2)$ and not being a local minimum.

After the tunneling phase the process returns to phase 1. By carrying out these two phases repeatedly a set of local minima producing decreasing objective function values is obtained, and eventually the global minimum X_G is reached.

The method is conceptually similar to a miner excavating a horizontal tunnel until finding a sloping cavern to descend into, and having to proceed several

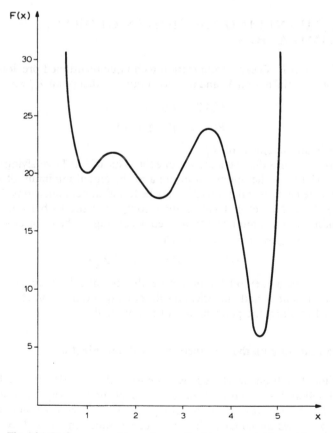

Fig. 1.24 Performance of tunneling method in one variable problem.

times repeating the same task until arriving at the lowest point. Because of that this technique is called the **tunneling method**.

Phase 1 is not worth specific comment as it is solved by techniques discussed before. However, we need to explain how the method works in the tunneling phase. Given a local minimum X_i^* the following tunneling function is defined:

$$T(X, X_i^*, s_i) = \frac{F(X) - F(X_i^*)}{[(X - X_i^*)^{\mathrm{T}}(X - X_i^*)]^{s_i}} \qquad (1.156)$$

In equation (1.156) the parameter s_i is the order of the root of the function $F(X) - F(X_i^*) = 0$ at $X = X_i^*$. In consequence, the tunneling function is $T(X, X_i^*, s_i) \neq 0$ at this point and thus the function $T(X, X_i^*, s_i)$ can be used to obtain the new root at $X = X_2$. The parameter s_i is not known and must be identified. In order to do this the function T is evaluated several times at $X = X_i^*$ for increasing values of s_i; it is useful to begin with $s_i = 0$ and to add increments Δ_s such as $0.01 \leqslant \Delta_s \leqslant 0.05$. When for a value s_i it is found that

$$T(X_i^*, X_i^*, s_i) \neq 0 \text{ and } T(X_i^* + \Delta X, X_i^*, s_i) \leqslant T(X_i^*, X_i^*, s_i) \qquad (1.157)$$

the current value of the parameter s_i is the root order. The procedure proceeds by trying to calculate another root of the function T which will provide the

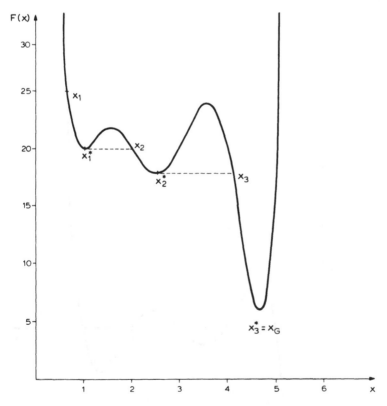

Fig. 1.25 Example of one-variable function.

design X_{i+1} desired, because according equation (1.156) if

$$T(X_{i+1}, X_i^*, s_i) = 0 \tag{1.158}$$

$$F(X_{i+1}) = F(X_i^*) \tag{1.159}$$

An example of a function $F(x)$ with one variable may be used to explain the method (Fig. 1.25):

$$F(x) = x^6 - 16x^5 + 100x^4 - 310x^3 + 499x^2 - 394x + 140$$

The function $F(x)$ has a local minimum at $x = 1$ where $F(1) = 20$. The function $T(x, 1, s_1)$ is represented in Figs. 1.26 and 1.27 for the cases $s_1 = 0.5$ and $s_1 = 1.0$.

$$T(x, 1, 0.5) = \frac{F(x) - F(1)}{[(x - 1)(x - 1)]^{0.5}} = x^5 - 15x^4 + 85x^3 - 225x^2 + 274x - 120$$

$$T(x, 1, 1) = \frac{F(x) - F(1)}{(x - 1)(x - 1)} = x^4 - 14x^3 + 71x^2 - 154x + 120$$

It turns out that the parameter s_1 has the value 1.0. The root of the function $T(x, 1, 1)$ is obtained at $x_2 = 2.0$ as shown in Fig. 1.27. By carrying out from this point another minimization phase, a local minimum at $x_2^* = 2.517$ will be reached (Fig. 1.28).

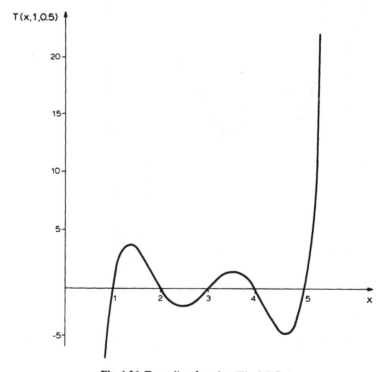

Fig. 1.26 Tunneling function $T(x, 1, 0.5)$.

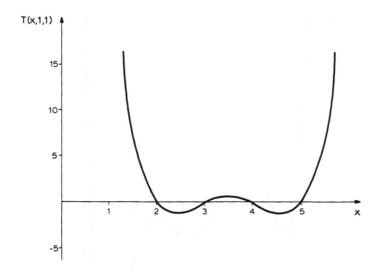

Fig. 1.27 Tunneling function $T(x, 1, 1)$.

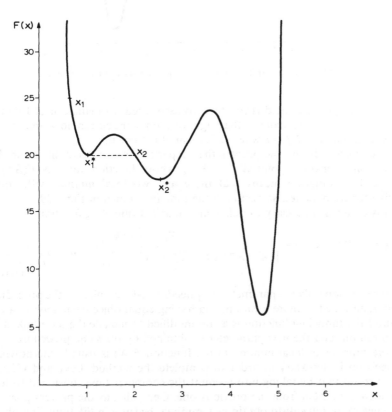

Fig. 1.28 Tunneling phase in example of one-variable function.

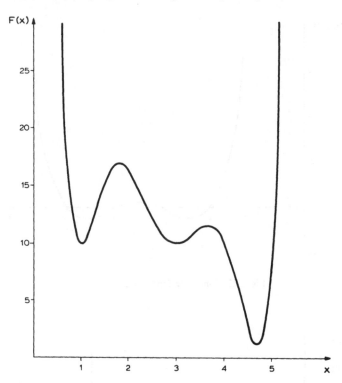

Fig. 1.29 Example with two minima at equal objective function values.

To avoid in practice the division by zero associated with equation (1.156) the tunneling function is evaluated in the neighborhood of the current local minimum, namely at point $X_i = X_i^* + \varepsilon$ where ε is a small increment.

It may happen during the process that departing from a local minimum X_i^* the next phase gives a point $X_{i+1} = X_{i+1}^*$ whose function value is $F(X_i^*) = F(X_{i+1}^*)$. This situation means that there are two local minima with equal objective function values, as occurs in the function shown in Fig. 1.29.

In this case it is necessary to define an enhanced tunneling function

$$T(X, X_i^*, s_i, X_{i+1}^*, s_{i+1}) = \frac{F(X) - F(X_i^*)}{[(X - X_i^*)^{\mathrm{T}}(X - X_i^*)]^{s_i}[(X - X_{i+1}^*)^{\mathrm{T}}(X - X_{i+1}^*)]^{s_{i+1}}}$$

(1.160)

In this function only the parameter s_{i+1} needs to be identified by the procedure already mentioned. If more local minima having equal objective function values are found the tunneling function is again modified to include the term linked to each minimum and the new parameter is obtained by the same procedure.

As the number of local minima of the function $F(X)$ is usually unknown, a user decision is necessary in order to complete the method. Levy and Gómez propose to do that by taking into account the computer time consumed to find a new local minimum from the previous one. Generally, as the process goes on it becomes more difficult to obtain new minima, because in the tunneling phase it is more difficult to identify points with the objective function value of the current minimum. Hence, it is suggested that the procedure be ended when no local minimum is found in computer time $10t$, t being the time needed to obtain the current one.

1.4.2 The tunneling method in structural optimization

The tunneling method may also be a useful tool to solve constrained optimization problems by linking it with penalty function techniques. This approach may be implemented as follows.

1. From the constrained problem defined in equations (1.1) and (1.2),

$$\min F(X) \tag{1.161}$$

subject to

$$g_j(X) \leqslant 0 \quad i = 1, \ldots, m \tag{1.162}$$

an unconstrained problem via a penalty function is created:

$$\phi(X, r) = F(X) + r \sum_{j=1}^{m} P(g_j) \tag{1.163}$$

In the problem defined in equation (1.163) a local minimum X_1^* may be obtained by using any mentioned method.
2. The tunneling function T is defined at point X_1^* and the tunneling phase is worked out to find a point X_2 having

$$F(X_1^*) = F(X_2) \tag{1.164}$$

3. Steps 1 and 2 are repeated until it is not possible to find a new local minimum as indicated in the previous section.

The optimum design of the grillage shown in Fig. 1.10 has been solved as indicated, by using the quadratic extended interior penalty function. This technique extends the penalty function defintion to the infeasible region (Haftka and Starnes, 1976) and improves performance of the interior penalty function method near the optimum.

$$\phi(X, r) = F(X) - r \sum_{j=1}^{m} P(g_j) \tag{1.165}$$

$$P(g_j) = \begin{cases} 1/g_j & \text{if } g_j \leqslant \varepsilon & (1.166) \\ [(g_j/\varepsilon)^2 - 3g_j/\varepsilon + 3]/\varepsilon & \text{if } g_j > \varepsilon & (1.167) \end{cases}$$

The results obtained are the following:

initial design	$X_1 = (5, 25)$	$F_1 = 8000$
local minimum	$X_1^* = (2.25, 19.52)$	$F_1^* = 6082$
tunneling phase	$X_2 = (9.16, 13.98)$	$F_2 = 6076$
global minimum	$X_2^* = (9.72, 6.07)$	$F_2^* = 2795$

The point X_2^* is the global minimum of the problem. Other examples of structural optimization may be found in Hernández (1990b, 1991).

REFERENCES

Baldur, R. (1972) Structural optimization by inscribed hyperspheres. *Proceedings ASCE*, **98**, (EM3), 503–18.

Brooks, R. and Geoffrion, A. (1966) Finding Everett's Lagrange multipliers by lineal programming. *Operations Research*, **14**(61), 1149–53.

Cassis, J.H. (1974) Optimum Design of Structures Subjected to Dynamic Loads. *UCLA-ENG-7451*, UCLA School of Engineering and Applied Science, CA.

Cheney, E.W. and Goldstein, A.A. (1959) Newton's method for convex programming on Tchebycheff approximation. *Numerische Mathematik*, **1**, 253–68.

Cohn, N.Z., Ghosh, S.K. and Parini, S.R. (1972) Unified approach to the theory of plastic structure. *Journal of Engineering Mechanical Division, ASCE*, **98** (EM5), 1133–58.

Dantzig, G. (1963) *Linear Programming and Extensions*, Princeton University Press, Princeton, N.J.

Davidon, W.C. (1959) Variable Metric Method for Minimization. *ANL-5990 Rev.*, Argonne National Laboratory.

Dorn, W.S., Gomory, R.E. and Greenberg, H.J. (1964) Automatic design of optimal structure. *Journal de Mecanique*, **3**(1), 25–52.

Everett, H. (1963) Generalized Lagrange multiplier method for solving problems of optimum allocation of resources. *Operations Research*, **11**(2), 397–417.

Fiacco, A.V. and McCormick, G.P. (1963) Programming under Nonlinear Constraints by Unconstrained Minimization: A Primal-Dual Method. *Tech. Paper RAC-TP-96*, Research Analysis Corporation, Bethesda, MD.

Fiacco, A.V. and McCormick, G.P. (1968) *Nonlinear Programming Sequential Unconstrained Minimization Techniques*. Wiley, New York.

Fletcher, R. and Powell, M.J.D. (1963) A rapidly convergent method for minimization. *Computer Journal*, **6**(2), 163–8.

Fletcher, R. and Reeves, C.M. (1964) Function minimization by conjugate gradients. *Computer Journal*, **7**(2), 149–54.

Goble, G. and LaPay, W.S. (1971) Optimum design of prestressed beams. *Journal ACI*, **68**, 712–18.

Gómez, S. and Levy, A.V. (1982) The tunneling method for solving the constrained global optimization problems with several nonconnected feasible regions, in *Lecture Notes in Mathematics no. 909* (ed. J.P. Hennart), Springer, Berlin, Heidelberg.

Gould, F.J. (1969) Extensions of Lagrange multipliers in nonlinear programming. *SIAM Journal of Applied Mathematics*, **13**(6), 1280–97.

Greenberg, H.J. and Pierskalla, W.P. (1970) Surrogate mathematical programming. *Operations Research*, **18**(5), 924–39.

Haftka, R.T. and Starnes, J.H. (1976) Application of a quadratic extended interior penalty function for structural optimization. *AIAA Journal*, **14**, 718–24.

Hernández, S. (1990a) *Metodos de diseño óptimo de estructuras*, Colegio de Ingenieros de Caminos, Canales y Puertos (in Spanish).

Hernández, S. (1990b) El método de tunelización. Una técnica de optimización global en problemas no convexos en la ingeniería. IX Congreso de Ingeniería Mecánica, Zaragoza.

Hernández, S. (1991) A new mathematical procedure for global optimization of nonconvex problems. NATO/DFG ASI Seminar on Optimization of Large Structural Systems, Berchtesgaden, Germany.

Hernández, S. and Ceron, C. (1987) A sequence of linear problems with move limits for shape optimization of truss structures. SIAM Conference on Optimization, Houston, TX.

Horne, M.R. and Morris, L.I. (1973) Optimum design of multistorey rigid frames, in *Optimum Structural Design* (eds. R.H. Gallagher and O.C. Zienkiewicz), Wiley, New York.

Jaynes, E.T. (1957) Information theory and statistical mechanics. *The Physical Review*, **106**, 620–30, **108**, 171–90.

Johnson, E.H. (1976) Disjoint design spaces in the optimization of harmonically excited structures. *AIAA Journal*, **14**(2), 259–61.

Johnson, E.H., Rizzi, P., Ashley, H., *et al.* (1976) Optimization of continuous one-dimensional structures under steady harmonic excitation. *AIAA Journal*, **14**(2), 1690–8.

Karmarkar, N. (1984) A new polynomial-time algorithm for linear programming. *Combinatórica*, **4**, 373–95.

Kavlie, D. (1971) Optimum design of statically indeterminate structures. Ph.D. Thesis, University of California, Berkeley, CA.

Kelley, H.J. (1960) The cutting plane method for solving complex programs. *SIAM Journal*, **8**, 703–12.

Khot, N.S., Berke, L. and Venkayya, V.B. (1979) Comparison of optimality criteria. Algorithms for minimum wieght design of structures. *AIAA Journal*, **17**, 182–90.

Kinsalaas, J. (1972) Minimum Weight Design of Structures via Optimality Criteria. *NASA TND-7115*.

Kirsch, U. (1973) Optimized prestressing by linear programming. *International Journal of Numerical Methods in Engineering*, **7**, 125–6.

Kuhn, H.W. and Tucker, A.W. (1951) Nonlinear programming, in *Proceedings of 2nd Derkeley Symposium on Mathematical Statistics and Probability*, University of California Press, pp. 481–90.

LaPay, W.S. and Goble, G.G. (1971) Optimum design of trusses for ultimate loads. *Journal of Structural Division ASCE*, **97**(ST1), 157–74.

Levy, A.V. and Gómez, S. (1984) The tunneling method applied to global optimization, in *Numerical Optimization 1984*, (eds. P.T. Boggs, R.H. Byrd and R.B. Schnabel), SIAM Conference.

Maquoi, R. and Rondal, J. (1977) Approache realiste du dimensionnement optimal des ponts précontraints hyperstatiques. *Annales des Travaux Publics de Belgique*, (3), 197–215.

Mills-Curran, W.C. and Schmit, L.A. (1985) Structural optimization with dynamic behaviour constraints. *AIAA Journal*, **23**(1), 136–8.

Moses, F. and Onoda, S. (1969) Minimum weight design of structures with application to elastic grillages. *International Journal of Numerical Methods in Engineering*, **1**, 311–31.

NBE-MV103 (1978) *Cálculo de las estructuras de acero laminado en la edificación* (in Spanish).

Powell, M.J.D. (1964) An efficient method of finding the minimum of a function of several variables without calculating derivatives. *Computer Journal*, **7**(4), 303–7.

Rizzi, P. (1976) Optimization of multiconstrained structures based on optimality criteria. *17th AIAA/ASME/SAE Conference*, King of Prussia, PA.

Russell, A.D. and Reinschmidt, K.F. (1964) Discussion of optimum design of trusses for ultimate loads. *Journal of Structural Division ASCE*, **3**(1), 25–52.

Schmit, L.A. (1960) Structural design by systematic synthesis. Second Conference on Electronic Computation, ASCE, Pittsburgh, PA, pp. 105–32.

Shannon, C.E. (1948) A mathematical theory of communication. *Bell System Techanical Journal*, **27**(3), 379–428.

Svanberg, K. (1984) On local and global minima in structural optimization, in *New Directions in Optimum Design* (eds. E. Atrek *et al.*), Wiley, New York.

Svanberg, K. (1991) On local and global optima. NATO/DFG ASI Seminar on Optimization of Large Structural Systems, Berchtesgaden, Germany.

Templeman, A.B. (1989) Entropy-based minimax applications in shape-optimal design, in *Lecture Notes on Engineering, No. 42: Discretization Methods and Structural Optimization – Procedures and Applications* (eds. H.A. Eschenauer and G. Thierauf), Springer, Berlin, Heidelberg, pp. 335–42.

Templeman, A.B. and Li, X. (1985) Entropy duals. *Engineering Optimization*, **9**, 107–20.

Templeman, A.B. and Li, X. (1987) A maximum entropy approach to constrained nonlinear programming. *Engineering Optimization*, **12**, 191–205.

Vanderplaats, G.N. (1984a) An efficient feasible directions algorithm for design synthesis, *AIAA Journal*, **22**(11), 1633–40.

Vanderplaats, G.N. (1984b) *Numerical Optimization Techniques for Engineering Design: with Applications*, McGraw-Hill, New York.

Zangwill, W.I. (1967) Nonlinear programming via penalty functions. *Management Sciences*, **13**(5), 344–58.

Zoutendijk, G. (1960) *Methods of Feasible Directions*, Elsevier, Amsterdam.

2

Optimality criteria methods for large discretized systems

G.I.N. ROZVANY and M. ZHOU

2.1 INTRODUCTION: WHY OPTIMALITY CRITERIA METHODS?

Research into new optimality crieria methods was motivated by the currently existing discrepancy between analysis capability and optimization capability in structural design, which was pointed out repeatedly by Berke and Khot (e.g. Berke and Khot, 1987). Reasons for this discrepancy can be understood better if we consider the quantities that determine the optimization capability of various methods used in structural design, which are summarized in Table 2.1 in the context of static problems.

Whereas in primal mathematical programming (MP) methods the critical quantity is the number of variables (N), in dual and traditional discretized optimality criteria (DOC) methods it is the number of active behavioral constraints which, for static problems, includes the number (m_d) of active deflection constraints and number (m_s) of active stress constraints. Finally, in optimality criteria methods combined with fully stressed design for stress constraints (DOC–FSD) and in the new optimality criteria methods (DCOC) introduced recently (Zhou and Rozvany, 1992; Zhou, 1992), the critical quantity is the number (m_d) of active deflection (or other global) constraints. It is important to note, however, that the DOC–FSD method is known to lead in general to a nonoptimal solution, although the error is often relatively small. For this reason, DOC–FSD should be disqualified in a comparison of correct methods of structural optimization. In the case of very large systems, many thousand stress constraints may be active, whereas the number of active global constraints (such as deflection, natural frequency or system stability constraints, involving the entire system) can be rather small. For this reason, the proposed DCOC method achieves an improvement in our optimization capability by several orders of magnitude, if typical large structural systems are being optimized.

2.2 HISTORICAL BACKGROUND AND GENERAL FEATURES OF THE DCOC ALGORITHM

Historically, the DCOC algorithm can be regarded as a unification of discretized optimality criteria (DOC) techniques (e.g. Berke, 1970; Venkayya, Khot and

Table 2.1 Quantities determining the optimization capability of various methods

	Primal MP	DOC, Dual	DOC–FSD	COC DCOC
Critical quantity	N	$m_d + m_s$	m_d	m_d
Solution	Optimal	Optimal	Non-optimal	Optimal

Berke, 1973; for reviews see for example Berke and Khot, 1974, 1987, 1988) and of continuum-based optimality criteria (COC) methods derived by the analytical school of structural optimization (e.g. Prager and Shield, 1967; Prager and Taylor, 1968; for a comprehensive review see a book by Rozvany, 1989). The latter were applied indirectly to large finite element systems under the term 'iterative discretized COC methods' (Rozvany *et al.*, 1989; Rozvany, Zhou and Gollub, 1990; Rozvany and Zhou, 1991), in which the optimality criteria were derived for continua in terms of stress resultants for each cross section and then discretization was carried out for computations by FE methods. Moreover, iterative COC methods used expressions which were exact only for a continuous variation of the design variable along the members or elements involved. For these reasons, it represented a considerable improvement of the continuum-type approach when Zhou (1992) reformulated the COC method directly for discretized systems in the matrix notation of finite element analysis. This new discretized continuum-type optimality criteria technique has been termed the DCOC method for historical reasons and it is hoped that it will help in bridging over the communication gap between the numerical and analytical schools of structural optimization.

Before discussing DCOC in detail, we shall summarize briefly the main differences between the formulation of this technique and that of earlier discretized optimality criteria (DOC) methods (Table 2.2). In DCOC the variables in the formulation include not only design variables but also the real forces and the virtual forces used in work equations for the deflection constraints. Whereas in DOC the stress constraints are converted into global (displacement) constraints involving the entire system, in DCOC they are expressed in terms of real forces

Table 2.2 Main differences between the DOC and COC–DCOC methods

	DOC	COC–DCOC
Variables	Design variables	Design variables, real and virtual forces
Formulation of stress constraints	Through equivalent displacements	Directly in terms of real forces
Equilibrium	Implicit	Explicit equality constraints
Compatibility	Implicit	Not included, implied by optimality
Calculation of Lagrangians for stress constraints	Iteratively, at system level	Explicitly, at element level

involving only the element concerned. Whilst in DOC equilibrium and compatibility are included in the formulation implicitly through the structural analysis, in DCOC the equilibrium constraint is included explicitly, but compatibility (kinetic admissibility) is relaxed initially and turns out to be an optimality condition, if any one displacement condition is active. This feature of DCOC simplifies the formulation considerably. The most important advantage of DCOC lies in the calculation of the Lagrange multipliers for stress constraints. In DOC the latter have been converted into relative displacement constraints and hence the corresponding Lagrange multipliers (i) require an analysis for a virtual load for each stress constraint and (ii) are coupled with each other at the system level. The evaluation of the Lagrangians in DOC therefore involves a complicated and expensive iterative procedure. In DCOC, these Lagrange multipliers can be calculated explicitly at the element level for certain simpler types of structures (e.g. trusses, frames and membranes). Another important advantage of DCOC is the fact that some of the optimality criteria are interpreted in terms of the so-called adjoint structure, which has the same analysis equations as the real structure. Analysis of the adjoint structure replaces the usual sensitivity analysis of other methods and involves only a simple substitution of the decomposed stiffness matrix, already computed in the analysis of the real structure. This means that the adjoint analysis in DCOC requires a relatively insignificant amount of computer time.

The difference between DCOC and DOC–FSD is that the adjoint system (virtual force, i.e. dummy load system) in DCOC is subject to prestrains (initial relative displacements) due to active stress constraints. This ensures correctness of the optimal solution, whereas DOC–FSD usually leads to an incorrect solution for statically indeterminate structures. The above statements will be demonstrated on test examples later (section 2.5).

Because DCOC is more useful than COC for practical applications, the formulation and derivation of this method is given in detail in section 2.3, and its computational implementation discussed in section 2.4. Illustrative examples are presented in section 2.5. The COC method, which is important for understanding the theoretical background and necessary for the derivation of exact analytical optimal layouts, will be discussed in section 2.6 and illustrated with examples in section 2.7.

Although the computational implementation of DCOC uses the stiffness method, derivation of the optimality criteria is based on the flexibility method, because this simplifies considerably the mathematical procedure involved.

2.3 DERIVATION OF OPTIMALITY CRITERIA FOR THE DCOC ALGORITHM

For didactic reasons, the optimality criteria for the DCOC method will be first derived in section 2.3.1 for a very simple class of problems, namely for the case of stress constraints and one displacement constraint. This type of derivation was originally proposed by Zhou (1992). Various extensions of DCOC are derived in sections 2.3.2 to 2.3.5.

2.3.1 Stress constraints and one displacement constraint

(a) Problem formulation

The considered problem can be stated as follows:

$$\min \Phi = \sum_{e=1}^{E} w^e(\{\mathbf{x}^e\}) \tag{2.1}$$

subject to the following constraints. The displacement constraint is

$$\{\hat{\mathbb{F}}\}^{\mathrm{T}}[\mathbf{D}]\{\mathbb{F}\} + \{\hat{\mathbb{F}}\}^{\mathrm{T}}\{\mathbf{\Delta^*}\} - t \leqslant 0 \tag{2.2}$$

The stress constraints are

$$\theta_{hr}^e(\{\mathbf{n}_h^e\}) - \sigma_{hr}^e = \theta_{hr}^e([\mathbf{\omega}_h^e]\{\mathbf{ff}^e\} + \{\mathbf{n}_{oh}^e\}) - \sigma_{hr}^e \leqslant 0$$
$$(h = 1, \ldots, H^e; r = 1, \ldots, R_h^e; e = 1, \ldots, E) \tag{2.3}$$

The side constraints are

$$-x_i^e + x_i^e{\downarrow} \leqslant 0 \quad (i = 1, \ldots, I^e; e = 1, \ldots, E) \tag{2.4}$$

$$x_i^e - x_i^e{\uparrow} \leqslant 0 \quad (i = 1, \ldots, I^e; e = 1, \ldots, E) \tag{2.5}$$

The equilibrium condition for the real forces is

$$\{\mathbf{P}\} - [\mathbf{B}]\{\mathbb{F}\} = \mathbf{0} \tag{2.6}$$

The equilibrium condition for the virtual forces is

$$\{\hat{\mathbf{P}}\} - [\mathbf{B}]\{\hat{\mathbb{F}}\} = \mathbf{0} \tag{2.7}$$

In relations (2.1)–(2.7), Φ is the objective function, which in this case can be the total weight of the structure; $e = 1, \ldots, E$ identify the elements; w^e is the objective function (weight) for the element e; $\{\mathbf{X}\} = \lfloor \mathbf{x}^1, \ldots, \mathbf{x}^e, \ldots, \mathbf{x}^E \rfloor^{\mathrm{T}}$ with $\mathbf{x}^e = \lfloor x_1^e, \ldots, x_i^e, \ldots, x_{I^e}^e \rfloor^{\mathrm{T}}$ the design variables (cross-sectional dimensions); $\hat{\mathbb{F}}$ denotes the virtual nodal forces in flexibility formulation, equilibrating the virtual loads $\{\hat{\mathbf{P}}\}$ (unit dummy load for a deflection constraint at one point or loads corresponding to the weighting factors if a weighted combination of deflections is prescribed), $[\mathbf{D}]$ is the flexibility matrix, $\{\mathbb{F}\}$ denotes the real nodal forces in flexibility formulation, $\{\mathbf{\Delta^*}\} = \lfloor \delta^{1*}, \ldots, \delta^{e*}, \ldots, \delta^{E*} \rfloor^{\mathrm{T}}$ are the real initial relative displacements caused by (i) loads within an element, (ii) temperature strains or (iii) prestrains; t is the permissible value of the displacement; θ_{hr}^e is a stress condition for the type of stress or stress location r at the cross section h of the element e; $\{\mathbf{n}_h^e\} = \lfloor n_{h1}^e, \ldots, n_{hv}^e, \ldots, n_{hV^e}^e \rfloor^{\mathrm{T}}$ are the stress resultants at the cross section h of the element e; $[\mathbf{\omega}_h^e]$ is a matrix converting the nodal forces $\{\mathbf{ff}^e\}$ to stress resultants at the cross section h; $\{\mathbf{n}_{oh}^e\}$ are the stress resultants caused by the loads acting on the interior of the element e; σ_{hr} is the permissible value of the r type stress at the cross section h; $x_i^e{\downarrow}$ and $x_i^e{\uparrow}$ are the lower and upper limits on the value of the design variable x_i^e; $\{\mathbf{P}\}$ are the real nodal loads at the free degrees of freedom; $[\mathbf{B}]$ is the statics matrix. In general, upper-case letters denote quantities at the system level and lower-case letters quantities at the element level.

The relaxed problem formulation in equations (2.1)–(2.7) requires statical admissibility of the real and virtual forces by relations (2.6) and (2.7) but includes no provision for compatibility conditions. It will be seen later that compatibility for the real forces is ensured by the Kuhn–Tucker conditions. In representing a displacement via a work equation in relation (2.2), the virtual forces need to be statically admissible only. However, the Kuhn–Tucker conditions will also introduce compatibility conditions for the strain caused by the virtual forces, making them thereby also kinematically admissible. The virtual force system, together with compatibility conditions and initial relative displacements due to active stress constraints, will represent a fictitious structure which will be termed the 'adjoint structure'.

The stress constraint in relation (2.3) is often linear with respect to the stress resultants, in which case it can be represented as

$$\{s_{hr}^e\}^T\{n_h^e\} - \sigma_{hr}^e = \{s_{hr}^e\}([\omega_h^e]\{ff^e\} + \{n_{oh}^e\}) - \sigma_{hr}^e \leqslant 0 \qquad (2.8)$$

where the vector $\{s_{hr}^e\}$ involves design variables and converts the stress resultants at the cross section h of element e into a stress of type and/or location r at the same cross section.

(b) Example illustrating the problem formulation: plane frame element

Since the above formulation may be unfamiliar to some of the readers, we shall use a simple example to illustrate the concepts involved.

Figure 2.1(a) shows a plane frame element and its rectangular cross section of constant width b and variable depth x_1^e is indicated in Fig. 2.1(b). The element objective function shall be in this case the element weight

$$w^e(x_1^e) = bL\rho x_1^e \qquad (2.9)$$

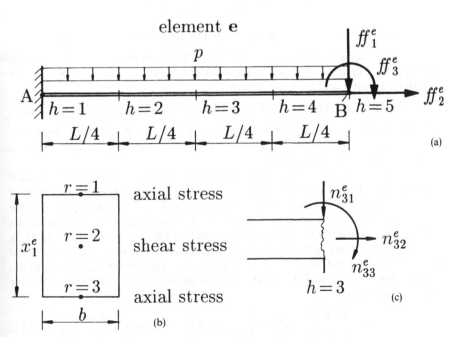

Fig. 2.1 Illustrative example; plane frame element.

where L is the length of the element and ρ is the specific weight of the frame material. In this example, there is only one design variable per element.

Considering the displacement constraint in equation (2.2), the following should be clarified. In the flexibility formulation, each element must be supported in a stable, statically determinate manner (Fig. 2.1(a)). For a plane frame element, it is most convenient to clamp the element at the left end (A) and leave the right end (B) free. Then the three nodal forces $\{\mathbf{ff}^e\} = \lfloor ff^e_1, ff^e_2, ff^e_3 \rfloor^T$ represent the vertical and horizontal forces and the moment at the right end (B) of the beam element (Fig. 2.1(a)). The virtual forces $\{\hat{\mathbf{ff}}^e\} = \lfloor \hat{ff}^e_1, \hat{ff}^e_2, \hat{ff}^e_3 \rfloor^T$ are similar but they equilibrate the virtual loads instead of the real loads.

The element flexibility matrix $[\mathbf{d}^e]$ maps the nodal forces $\{\mathbf{ff}^e\}$ of the element e into relative displacements $\{\boldsymbol{\delta}^e\}$ which have the same location and direction as the former:

$$\{\boldsymbol{\delta}^e\} = [\mathbf{d}^e]\{\mathbf{ff}^e\} + \{\boldsymbol{\delta}^{*e}\} \tag{2.10}$$

where $\{\boldsymbol{\delta}^{*e}\}$ are the initial relative nodal displacements.

It is easy to check by elementary statical considerations that the frame element flexibility matrix is

$$[\mathbf{d}^e] = \frac{1}{E} \begin{bmatrix} \dfrac{L^3}{3I} & 0 & \dfrac{L^2}{2I} \\ 0 & \dfrac{L}{A} & 0 \\ \dfrac{L^2}{2I} & 0 & \dfrac{L}{I} \end{bmatrix} = \frac{1}{Eb} \begin{bmatrix} \dfrac{4L^3}{(x^e_1)^3} & 0 & \dfrac{6L^2}{(x^e_1)^3} \\ 0 \cdot & \dfrac{L}{x^e_1} & 0 \\ \dfrac{6L^2}{(x^e_1)^3} & 0 & \dfrac{12L}{(x^e_1)^3} \end{bmatrix} \tag{2.11}$$

where I is the moment of inertia and A is the area of the cross section.

The initial relative displacement $\{\boldsymbol{\delta}^{e*}\}$ for the element in Fig. 2.1(a) are the three displacement components at the right end B of the element caused by the distributed load p and have the values

$$\{\boldsymbol{\delta}^{e*}\} = \lfloor pL^4/8EI, 0, pL^3/6EI \rfloor^T = \lfloor 3pL^4/2Eb(x^e_1)^3, 0, 2pL^3/Eb(x^e_1)^3 \rfloor^T \tag{2.12}$$

If the entire element were subject to a temperature change of τ, then the second entry of $\{\boldsymbol{\delta}^{e*}\}$ in relation (2.12) would change from zero to $\tau\alpha L$, where α is the coefficient of thermal expansion.

Considering the stress constraints in equation (2.3), the cross sections for which stress values are constrained are $h = 1, 2, 3, 4, 5$ in Fig. 2.1(a) and the type of stress and location within a cross section are $r = 1, 2, 3$ in Fig. 2.1(b). For the considered element, the stress constraints are linear in terms of the stress resultants and hence equation (2.8) can be used instead of the general formulation (2.3). The stress resultants $\{\mathbf{n}^e_h\} = \lfloor n^e_{h1}, n^e_{h2}, n^e_{h3} \rfloor^T$ represent the shear force, axial force and bending moment, respectively (Fig. 2.1(c)). For calculating the stresses at the three points $r = 1, 2, 3$ in Fig. 2.1(b) from the stress resultant, we have the following transformation vectors:

$$\{\mathbf{s}^e_{h1}\} = \left\lfloor 0, \frac{1}{A}, \frac{1}{Q} \right\rfloor^T, \quad \{\mathbf{s}^e_{h2}\} = \left\lfloor \frac{1.5}{bx^e_1}, 0, 0 \right\rfloor^T, \quad \{\mathbf{s}^e_{h3}\} = \left\lfloor 0, \frac{1}{A}, -\frac{1}{Q} \right\rfloor^T \tag{2.13}$$

where A is the cross-sectional area $(A = bx_1^e)$ and Q is the section modulus for calculating flexural stresses in the extreme fiber $(Q = b(x_1^e)^2/6)$. The factor $1.5/bx_1^e$ converts a shear force into a shear stress at the neutral axis $(r = 2)$.

The transformation matrix $[\omega_h^e]$ for calculating the stress resultants $\{n_h^e\}$ from the nodal forces $\{ff^e\}$ takes the following form for $h = 3$, for example:

$$[\omega_3^e] = \begin{bmatrix} 1 & 0 & 0 \\ 0 & 1 & 0 \\ \dfrac{L}{2} & 0 & 1 \end{bmatrix} \tag{2.14}$$

since the moment n_3^e at the cross section $h = 3$ is the sum of the vertical nodal force ff_1^e multiplied by the lever arm $L/2$ and the nodal moment ff_3^e.

Finally, the stress resultants $\{n_{oh}^e\} = \lfloor n_{oh1}^e, n_{oh2}^e, n_{oh3}^e \rfloor^T$ caused by the distributed transverse load p are for $h = 3$

$$\{n_{o3}^e\} = \lfloor pL/2, 0, pL^2/8 \rfloor^T \tag{2.15}$$

In the above example, the stress constraint was linear with respect to the stress resultants. An example of a nonlinear stress constraint is a membrane element (plate in plane stress) for which the Mises stress is constrained:

$$\sigma_1^2 + \sigma_2^2 - \sigma_1\sigma_2 - \sigma_{hr}^2 = \sigma_x^2 + \sigma_y^2 - \sigma_x\sigma_y + 3\tau_{xy}^2 - \sigma_{hr}^2 \leqslant 0 \tag{2.16}$$

where σ_{hr} is the Mises stress for the considered point. Denoting the axial and shear forces (usually N_x, N_y, N_{xy}) at a cross section (point) h of the membrane element by $\{n_h^e\} = \lfloor n_{h1}^e, n_{h2}^e, n_{h3}^e \rfloor^T$, we have the stress constraint

$$[(n_{h1}^e)^2 + (n_{h2}^e)^2 - n_{h1}^e n_{h2}^e + 3(n_{h3}^e)^2]/x_1^e - \sigma_{hr}^2 \leqslant 0 \tag{2.17}$$

where x_1^e is the thickness of the membrane element.

(c) Necessary conditions for optimality

The Lagrange multipliers for the constraints in relations (2.2)–(2.7) will be denoted by

$$v, \lambda_{hr}^e, \beta_i^e\downarrow, \beta_i^e\uparrow, \{\bar{L}\}, \{L\} \tag{2.18}$$

For the problem in relations (2.1)–(2.8), we have the following Kuhn–Tucker conditions with respect to the variables stated below.

Design variables

$$\frac{\partial w^e}{\partial x_i^e} + v\{\hat{ff}^e\}^T \left[\frac{\partial d^e}{\partial x_i^e}\right]\{ff^e\} + v\{\hat{ff}^e\}^T \left\{\frac{\partial \delta^{*e}}{\partial x_i^e}\right\} + \sum_{h=1}^{H^e} \sum_{r=1}^{R_h^e} \lambda_{hr}^e \frac{\partial \theta_{hr}^e(\{n_h^e\})}{\partial x_i^e} - \beta_i^e\downarrow + \beta_i^e\uparrow = 0$$

$$(i = 1, \ldots, I^e, e = 1, \ldots, E) \tag{2.19}$$

Real nodal forces $\mathbb{F}$

$$v[D]\{\hat{\mathbb{F}}\} + \{\bar{\Delta}^*\} - [B]^T\{\bar{L}\} = 0 \tag{2.20}$$

with

$$\{\bar{\Delta}^*\} = \lfloor \bar{\delta}^{*1}, \ldots, \bar{\delta}^{*e}, \ldots, \bar{\delta}^{*E} \rfloor^T \tag{2.21}$$

and

$$\{\bar{\delta}^{*e}\} = \sum_{h=1}^{H^e} \sum_{r=1}^{R_h^e} \lambda_{hr}^e \{\omega_h^e\}^T \left\{ \frac{\partial \theta_{hr}^e}{\partial \mathbf{n}_h^e} \right\} \tag{2.22}$$

where the partial derivatives in curly brackets can also be represented as $\text{grad}_{\mathbf{n}_h^e} \theta_{hr}^e$.

For linear stress constraints of the type given in equation (2.8), we have

$$\{\bar{\delta}^{*e}\} = \sum_{h=1}^{H^e} \sum_{r=1}^{R_h^e} \lambda_{hr}^e [\omega_h^e]^T \{\mathbf{s}_{hr}^e\} \tag{2.23}$$

Virtual (adjoint) nodal forces $\widehat{\mathbb{F}}$

$$v([\mathbf{D}]\{\mathbb{F}\} + \{\mathbf{\Delta}^*\}) - [\mathbf{B}]^T\{\mathbf{L}\} = 0 \tag{2.24}$$

Moreover, the Lagrangians, λ_{hr}^e, v, $\beta_i^e\downarrow$ and $\beta_i^e\uparrow$ must be nonnegative, and they are nonzero only if the corresponding constraint is satisfied as an equality. Since the expression in parentheses after v in equation (2.24) represents the real relative displacements $\{\mathbf{\Delta}\}$ and $[\mathbf{B}]^T$ is the kinematics matrix mapping the nodal displacements $\{\mathbf{U}\}$ into relative displacements $\{\mathbf{\Delta}\}$, we can see that $\{\mathbf{L}\}/v$ represents the real nodal displacements:

$$\frac{1}{v}\{\mathbf{L}\} = \{\mathbf{U}\} \tag{2.25}$$

Then it follows from equation (2.24) that kinematic admissibility, which was not included in the relaxed formulation (2.1)–(2.7), is automatically enforced as an optimality condition, provided that the displacement condition (2.2) is active. The fact that optimality with respect to an active displacement constraint ensures kinematic admissibility for any statically admissible real and virtual stress fields was also shown much earlier by Shield and Prager (1970) and Huang (1971).

Moreover, by equation (2.20) we can interpret the Lagrange multipliers $\{\bar{\mathbf{L}}\}$ as the nodal displacements of a fictitious structure termed the adjoint structure

$$\{\bar{\mathbf{L}}\} = \{\bar{\mathbf{U}}\} \tag{2.26}$$

and the corresponding relative displacements in equation (2.20), i.e.

$$\{\bar{\mathbf{\Delta}}\} = v([\mathbf{D}]\{\widehat{\bar{\mathbf{F}}}\} + \{\bar{\mathbf{\Delta}}^*\}) \tag{2.27}$$

as the relative displacements of the adjoint structure. We can see from the foregoing that for stress constraints and a simple displacement constraint the adjoint structure has the following properties:

• the internal forces $\{\bar{\mathbb{F}}\}$ equilibrate the virtual loads multiplied by v, that is $\{\bar{\mathbf{P}}\} = v\{\widehat{\mathbf{P}}\}$;

• the elements are subject to initial relative displacements given by relation (2.22) or (2.23) for elements with active stress constraints.

Necessary conditions for optimality of a discretized elastic system with stress constraints, one displacement constraint and one loading are summarized in Fig. 2.2. This representation is called a "conceptual scheme" because for the

computational implementation the necessary conditions are modified somewhat
with a view to using a stiffness method.

(d) Solution algorithm

Since the above equations cannot be solved simultaneously for most problems of
a practical nature, it is convenient to adopt an iterative procedure consisting of
the following two major steps in each iteration:

- analysis of the real and adjoint structures;
- updating the Lagrange multipliers and the design variables $\{x\}$ on the basis
  of the relations at the top of Fig. 2.2.

Further details of the computational implementation of DCOC will be explained
in section 2.4.

2.3.2 Stress constraints, several displacement constraints and one load condition

If we consider several displacement constraints, identified by the subscripts
$k = 1, \ldots, K$, then equation (2.2) is replaced by

$$\{\hat{\mathbb{F}}_k\}^{\mathrm{T}}[\mathbf{D}]\{\mathbb{F}\} + \{\hat{\mathbb{F}}_k\}^{\mathrm{T}}\{\Delta^*\} - t_k \leqslant 0 \quad (k = 1, \ldots, K) \tag{2.28}$$

where $\{\hat{\mathbb{F}}_k\}$ equilibrates the virtual load associated with the kth displacement
constraints and t_k is the limiting value of the kth displacement. In addition, the
equilibrium condition for the virtual forces becomes

$$\{\hat{\mathbf{P}}_k\} - [\mathbf{B}]\{\hat{\mathbb{F}}_k\} = 0 \tag{2.29}$$

Moreover, instead of the Lagrange multiplier v and $\{\mathbf{L}\}$ in equation (2.18) we
have

$$v_k, \{\mathbf{L}_k\} \quad (k = 1, \ldots, K) \tag{2.30}$$

and then the second and third terms in equation (2.19) become

$$\cdots + \sum_{k=1}^{K} v_k \{\hat{\mathbf{f}}^e\}^{\mathrm{T}} \left[\frac{\partial \mathbf{d}^e}{\partial x_i^e}\right]\{\mathbf{f}^e\} + \sum_{k=1}^{K} v_k \{\hat{\mathbf{f}}_k^e\}^{\mathrm{T}} \left\{\frac{\partial \delta^{*e}}{\partial x_i^e}\right\} + \cdots \tag{2.31}$$

Finally, equations (2.20) and (2.24) are replaced by

$$\sum_{k=1}^{K} v_k [\mathbf{D}]\{\hat{\mathbb{F}}_k\} + \{\bar{\Delta}^*\} - [\mathbf{B}]^{\mathrm{T}}\{\bar{\mathbf{L}}\} = 0 \tag{2.32}$$

$$v_k([\mathbf{D}]\{\mathbb{F}\} + \{\Delta^*\}) - [\mathbf{B}]^{\mathrm{T}}\{\mathbf{L}_k\} = 0 \quad (k = 1, \ldots, K) \tag{2.33}$$

Introducing

$$\frac{1}{v_k}\{\mathbf{L}_k\} = \{\mathbf{U}\} \tag{2.34}$$

equation (2.33) with any one k value implies kinematical admissibility of the real
nodal displacements and relative displacements. By equation (2.32) the adjoint

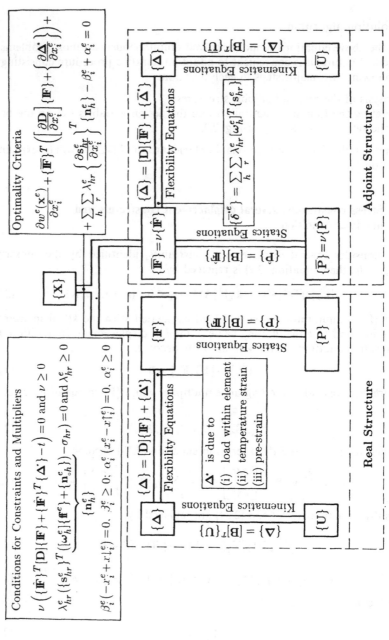

Fig. 2.2 DCOC algorithm for linearly elastic structures with stress constraints, one displacement constraint and one load condition: conceptual scheme.

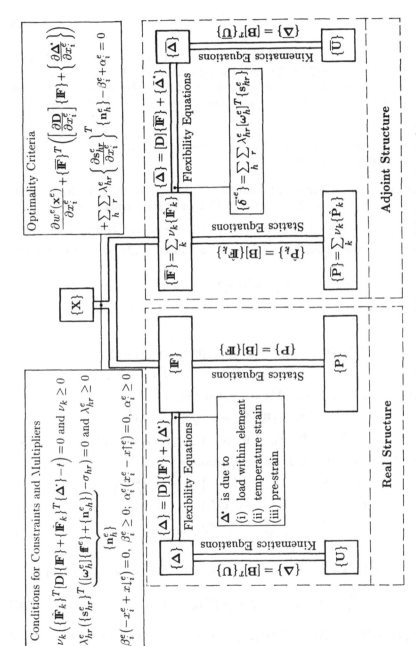

Fig. 2.3 DCOC algorithm: extension of the conceptual scheme to several displacement constraints.

structure is subject to the loading

$$\{\bar{\mathbf{P}}\} = \sum_{k=1}^{K} v_k \{\hat{\mathbf{P}}_k\} \tag{2.35}$$

and fulfills compatibility, subject to the initial relative displacements $\{\bar{\mathbf{\Delta}}^*\}$ given by equation (2.22) or (2.23) for active stress constraints. For the case of linear stress constraints of the type shown in equation (2.8), the necessary conditions of optimality for the considered class of problems are given in Fig. 2.3.

2.3.3 Stress constraints, several displacement constraints and several load conditions

We assume that for each load condition $l = 1, \ldots, L$ we have K_l displacement conditions: $k = 1, \ldots, K_l$. Then equation (2.2) or (2.28) is replaced by

$$\{\hat{\mathbf{F}}_{kl}\}^T [\mathbf{D}] \{\mathbf{F}_l\} + \{\hat{\mathbf{F}}_{kl}\}^T \{\mathbf{\Delta}_l^*\} - t_{kl} \leqslant 0 \quad (k = 1, \ldots, K_l; l = 1, \ldots, L) \tag{2.36}$$

Moreover, the stress constraints in equations (2.3) and (2.8) become, respectively,

$$\theta_{hr}^e(\{\mathbf{n}_{hl}^e\}) - \sigma_{hr}^e = \theta_{hr}^e([\mathbf{\omega}_h^e]\{\mathbf{ff}_l^e\} + \{\mathbf{n}_{ohl}^e\}) - \sigma_{hr}^e \leqslant 0$$

$$(h = 1, \ldots, H; r = 1, \ldots, R_h^e; l = 1, \ldots, L; e = 1, \ldots, E) \tag{2.37}$$

and

$$\{\mathbf{s}_{hr}^e\}^T \{\mathbf{n}_{hl}^e\} - \sigma_{hr}^e = \{\mathbf{s}_{hr}^e\}^T ([\mathbf{\omega}_h^e]\{\mathbf{ff}_l^e\} + \{\mathbf{n}_{ohl}^e\}) - \sigma_{hr}^e \leqslant 0$$

$$(h = 1, \ldots, H; r = 1, \ldots, R_h^e; l = 1, \ldots, L; e = 1, \ldots, E) \tag{2.38}$$

Finally, equations (2.6) and (2.7) are replaced by

$$\{\mathbf{P}_l\} - [\mathbf{B}]\{\mathbf{F}_l\} = \mathbf{0} \quad (l = 1, \ldots, L) \tag{2.39}$$

$$\{\hat{\mathbf{P}}_{kl}\} - [\mathbf{B}]\{\hat{\mathbf{F}}_{kl}\} = \mathbf{0} \quad (k = 1, \ldots, K_l; l = 1, \ldots, L) \tag{2.40}$$

Then the optimality condition (2.19) becomes

$$\frac{\partial w^e}{\partial x_i^e} + \sum_{k=1}^{K_l} \sum_{l=1}^{L} v_{kl} \{\hat{\mathbf{ff}}_{kl}^e\}^T \left(\left[\frac{\partial \mathbf{d}^e}{\partial x_i^e} \right] \{\mathbf{ff}_l^e\} + \left\{ \frac{\partial \mathbf{\delta}_l^{*e}}{\partial x_i^e} \right\} \right)$$

$$+ \sum_{h=1}^{H} \sum_{r=1}^{R} \sum_{l=1}^{L} \lambda_{hrl}^e \{\mathbf{\omega}_h^e\}^T \frac{\partial \theta_{hr}^e(\{\mathbf{n}_{hl}^e\})}{\partial x_i^e} - \beta_i^e \downarrow + \beta_i^e \uparrow = 0 \tag{2.41}$$

The generalized form of equations (2.20) and (2.24) can be represented as

$$\sum_{k=1}^{K_l} v_{kl} [\mathbf{D}] \{\hat{\mathbf{F}}_{kl}\} + \bar{\mathbf{\Delta}}_l^* - [\mathbf{B}]^T \{\bar{\mathbf{L}}_k\} = \mathbf{0} \quad (l = 1, \ldots, L) \tag{2.42}$$

$$v_{kl} ([\mathbf{D}] \{\mathbf{F}_l\} + \{\mathbf{\Delta}_l^*\}) - [\mathbf{B}]^T \{\mathbf{L}_{kl}\} = \mathbf{0} \quad (k = 1, \ldots, K_l; l = 1, \ldots, L) \tag{2.43}$$

Introducing

$$\frac{1}{v_{kl}} \{\mathbf{L}_{kl}\} = \{\mathbf{U}_l\} \quad (l = 1, \ldots, L) \tag{2.44}$$

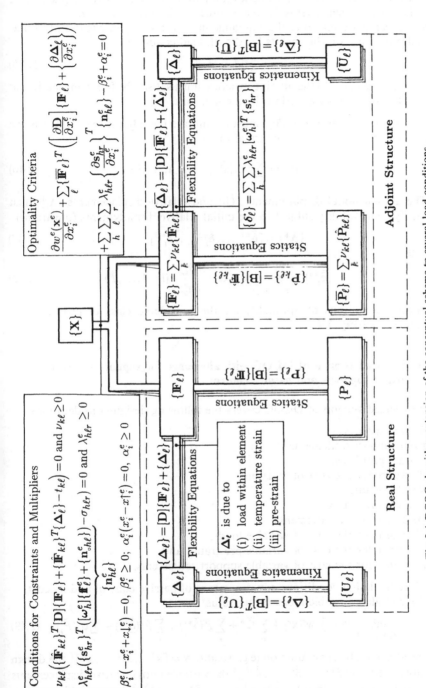

Fig. 2.4 DCOC algorithm: extension of the conceptual scheme to several load conditions.

equation (2.43) with any k value implies compatibility of the relative displacements and the nodal displacements for the loading conditions $l = 1, \ldots, L$.

Moreover, if we interpret the Lagrange multipliers $\{\bar{\mathbf{L}}_l\}$ in equation (2.42) as the adjoint nodal displacements for the lth load condition

$$\{\bar{\mathbf{L}}_l\} = \{\bar{\mathbf{U}}_l\} \tag{2.45}$$

then by equation (2.42) the adjoint structure must satisfy the following static and kinematic conditions for various load conditions.

- The adjoint nodal forces $\{\bar{\mathbb{F}}_l\}$ must equilibrate the adjoint load which is the factored combination of the virtual loads

$$\{\bar{\mathbf{P}}_l\} = \sum_{k=1}^{K_l} v_{kl} \{\hat{\mathbf{P}}_{kl}\} \quad (l = 1, \ldots, L) \tag{2.46}$$

- The adjoint nodal displacements $\{\bar{\mathbf{U}}_l\}$ and relative displacements $\{\bar{\mathbf{\Delta}}_l\}$ must satisfy compatibility subject to the initial relative displacements (prestrains)

$$\{\mathbf{\Delta}_l^*\} = [\bar{\boldsymbol{\delta}}_l^{*1}, \ldots, \bar{\boldsymbol{\delta}}_l^{*e}, \ldots, \bar{\boldsymbol{\delta}}_l^{*E}]^{\mathrm{T}} \tag{2.47}$$

with

$$\{\bar{\boldsymbol{\delta}}_l^{*e}\} = \sum_{h=1}^{H^e} \sum_{r=1}^{R_h^e} v_{hrl} [\boldsymbol{\omega}_h^e]^{\mathrm{T}} \frac{\partial \theta_{hr}^e}{\partial \mathbf{n}_{hl}^e} \quad (l = 1, \ldots, L) \tag{2.48}$$

Necessary conditions of optimality for the considered class of problems are summarized in Fig. 2.4.

2.3.4 Extended version of DCOC with allowance for support settlements, cost of supports and passive control

This extended version of DCOC covers the following design considerations:

- elastic supports,
- given support settlements,
- contact problems,
- allowance for the cost of reactions,
- variable loads,
- variable prestrains,
- variable support settlements,
- allowance for the cost of variable loads,
- allowance for the cost of variable prestrains, and
- allowance for the cost of variable support settlements.

The last six design conditions concern means of passive control and its cost.

The considered problem can be stated as

$$\min_{\mathbf{x}, \tilde{\mathbf{\Delta}}^*, \tilde{\mathbf{\Delta}}^*, \tilde{\mathbf{P}}} \Phi = \sum_e w^e(\mathbf{x}^e) + \sum_n \breve{\Omega}^n + \sum_e \chi^e(\tilde{\boldsymbol{\delta}}^{*e}) + \sum_n \vartheta^n(\tilde{\underline{\boldsymbol{\delta}}}^{*n}) + \sum_j \psi_j(\tilde{P}_j) \tag{2.49}$$

where Φ is the objective function (e.g. weight), $\mathbf{x} = [\mathbf{x}^1, \ldots, \mathbf{x}^e, \ldots, \mathbf{x}^E]^{\mathrm{T}}$ the design variables, $\{\tilde{\mathbf{\Delta}}^*\} = [\tilde{\boldsymbol{\delta}}^{*1}, \ldots, \tilde{\boldsymbol{\delta}}^{*e}, \ldots, \tilde{\boldsymbol{\delta}}^{*E}]^{\mathrm{T}}$ the variable initial relative displacements (due to variable prestrain), $\{\tilde{\underline{\mathbf{\Delta}}}^*\} = [\tilde{\boldsymbol{\delta}}^{*1}, \ldots, \tilde{\boldsymbol{\delta}}^{*n}, \ldots, \tilde{\boldsymbol{\delta}}^{*N}]^{\mathrm{T}}$ the variable support settlements, $\{\tilde{\mathbf{P}}\} = [\tilde{P}_1, \ldots, \tilde{P}_j, \ldots, \tilde{P}_J]^{\mathrm{T}}$ the variable loads, w^e the objective function (e.g. weight) for the element e, $\check{\Omega}^n = \max_l \Omega^n(\mathbf{r}_l^n)$ is the highest cost requirement (out of all load conditions) for the reaction element (support) n, $\Omega^n(\ )$ is the support cost function for the reaction element n, $\{\mathbf{r}^n\} = \lfloor r_1^n, \ldots, r_q^n, \ldots, r_{Q^n}^n \rfloor^T$ the reactions in the support n, χ^e the cost of variable initial relative displacements (variable pre-strain) in element e, $\{\tilde{\boldsymbol{\delta}}^{*e}\} = \lfloor \tilde{\delta}_1^{*e}, \ldots, \tilde{\delta}_m^{*e}, \ldots, \tilde{\delta}_{M^e}^{*e} \rfloor^T$ the adjustable initial relative displacements in element e, ϑ^n the cost of variable support settlements in support n, $\{\tilde{\boldsymbol{\delta}}^{*n}\} = \lfloor \tilde{\delta}_1^{*n}, \ldots, \tilde{\delta}_q^{*n}, \ldots, \tilde{\delta}_{Q^n}^{*n} \rfloor^T$ the variable settlements at support n and ψ_j the cost of variable load at the degree of freedom j. The above minimization problem is subject to the following constraints.

The displacement constraints are

$$\begin{Bmatrix} \hat{\mathbb{F}}_{kl} \\ \hat{\mathbb{R}}_{kl} \end{Bmatrix}^T \begin{bmatrix} \mathbf{D} & \\ & \underline{\mathbf{D}} \end{bmatrix} \begin{Bmatrix} \mathbb{F}_l \\ \mathbf{R}_l \end{Bmatrix} + \begin{Bmatrix} \hat{\mathbb{F}}_{kl} \\ \hat{\mathbb{R}}_{kl} \end{Bmatrix}^T \left(\begin{Bmatrix} \Delta_l^* \\ \underline{\Delta}_l^* \end{Bmatrix} + \begin{Bmatrix} \tilde{\Delta}^* \\ \underline{\tilde{\Delta}}^* \end{Bmatrix} \right) - t_{kl} \leqslant 0$$

$$(k = 1, \ldots, K_l; l = 1, \ldots, L) \tag{2.50}$$

where the subscripts k and l, respectively, identify a displacement constraint and a load condition, $\{\hat{\mathbb{F}}_{kl}\}$ and $\{\hat{\mathbb{R}}_{kl}\}$, respectively, are the virtual nodal forces (flexibility notation) and virtual reactions for the displacement constraint k and load condition l, $\{\mathbb{F}_l\}$ and $\{\mathbf{R}_l\}$ are, respectively, real nodal forces and real reactions for the load condition l, $[\mathbf{D}]$ is the flexibility matrix and $[\underline{\mathbf{D}}]$ the reaction element (support) flexibility matrix, $\{\Delta_l^*\}$ and $\{\underline{\Delta}_l^*\}$ are, respectively, given initial relative displacements and support settlements for the load condition l, $\{\tilde{\Delta}^*\}$ and $\{\underline{\tilde{\Delta}}^*\}$ are variable initial relative displacements and support settlements and t_{kl} is the prescribed limit in the displacement constraint k for the load condition l. Since here we are dealing with passive control, the variable initial relative displacements and variable support settlements are the same for all load conditions.

The stress conditions are as in equation (2.37) or (2.38).

The side constraints are as in equations (2.4) and (2.5).

The equilibrium condition for the real forces is

$$\{\mathbf{P}_l\} + \begin{Bmatrix} \mathbf{0} \\ \tilde{\mathbf{P}} \end{Bmatrix} - [\mathbf{B}] \begin{Bmatrix} \mathbb{F}_l \\ \mathbf{R}_l \end{Bmatrix} = 0 \quad (l = 1, \ldots, L) \tag{2.51}$$

where $\{\mathbf{P}_l\}$ are the real loads for the load condition l, $\{\tilde{\mathbf{P}}\}$ are the variable loads (for some of the degrees of freedom, but the same for all load conditions), $\{\mathbf{0}\}$ denotes zero variable load for some of the degrees of freedom, $[\mathbf{B}]$ is the statics matrix, $\{\mathbb{F}_l\}$ are the real nodal forces (flexibility formulation) for the load condition l and $\{\mathbf{R}_l\}$ are the real reactions for the load condition l.

The equilibrium condition for the virtual forces is

$$\{\hat{\mathbf{P}}_{kl}\} - [\mathbf{B}] \begin{Bmatrix} \hat{\mathbb{F}}_{kl} \\ \hat{\mathbf{R}}_{kl} \end{Bmatrix} = 0 \quad (k = 1, \ldots, K_l; l = 1, \ldots, L) \tag{2.52}$$

where $\{\hat{\mathbf{P}}_{kl}\}$ are the virtual loads (e.g. a unit dummy load for a displacement constraint), $\{\hat{\mathbb{F}}_{kl}\}$ the virtual nodal forces and $\{\hat{\mathbf{R}}_{kl}\}$ the virtual reactions, all three for the displacement constraint k under the load condition l.

The support cost inequality is

$$\Omega^n(\mathbf{r}_l^n) - \check{\Omega}^n \leqslant 0 \quad (l = 1, \ldots, L; n = 1, \ldots, N) \tag{2.53}$$

where $\breve{\Omega}^n$ is the maximum cost requirement for the support n. It is assumed that each support element is governed by one design parameter. Alternatively, a reaction element may have several design variables and stress constraints. In that case, it can be treated the same way in the DCOC procedure as an ordinary element.

The Lagrange multipliers for the constraints (2.50), (2.37), (2.4), (2.5), (2.51), (2.52) and (2.53) will be denoted, respectively, by

$$v_{kl}, \lambda_{hrl}^e, \beta_i^e\downarrow, \beta_i^e\uparrow, \{\bar{\mathbf{L}}_l\}, \{\mathbf{L}_{kl}\}, \rho_l^n \tag{2.54}$$

For the above problem, we have the following Kuhn–Tucker conditions for the variation of the variables stated below. For the design variables x_i^e,

$$\frac{\partial w^e}{\partial x_i^e} + \sum_k \sum_l v_{kl} \{\hat{\mathbf{ff}}_{kl}\}^{\mathrm{T}} \left(\left[\frac{\partial \mathbf{d}^e}{\partial x_i} \right] \{\mathbf{ff}_l^e\} + \left\{ \frac{\partial \boldsymbol{\delta}_l^{*e}}{\partial x_i^e} \right\} \right)$$

$$+ \sum_h \sum_r \sum_l \lambda_{hrl}^e \frac{\partial \theta_{hr}^e \{\mathbf{n}_{hl}^e\}}{\partial x_i^e} - \beta_i^e\downarrow + \beta_i^e\uparrow = 0 \quad (i = 1,\dots,I^e; e = 1,\dots,E) \tag{2.55}$$

For the real nodal forces $\{\mathbf{F}_l\}$ and real reactions $\{\mathbf{R}_l\}$,

$$\sum_k v_{kl} \begin{bmatrix} \mathbf{D} & \\ & \underline{\mathbf{D}} \end{bmatrix} \begin{Bmatrix} \hat{\mathbf{F}}_{kl} \\ \hat{\mathbf{R}}_{kl} \end{Bmatrix} - \begin{Bmatrix} \bar{\mathbf{\Delta}}_l^* \\ \underline{\bar{\mathbf{\Delta}}}_l^* \end{Bmatrix} + [\mathbf{B}]^{\mathrm{T}}\{\bar{\mathbf{L}}_l\} = 0 \quad (l = 1,\dots,L) \tag{2.56}$$

where

$$\{\bar{\mathbf{\Delta}}_l^*\} = \lfloor \bar{\boldsymbol{\delta}}_l^{*1},\dots,\bar{\boldsymbol{\delta}}_l^{*e},\dots,\bar{\boldsymbol{\delta}}_l^{*E} \rfloor^{\mathrm{T}} \tag{2.57}$$

$$\{\underline{\bar{\mathbf{\Delta}}}_l^*\} = \lfloor \underline{\bar{\boldsymbol{\delta}}}_l^{*1},\dots,\underline{\bar{\boldsymbol{\delta}}}_l^{*n},\dots,\underline{\bar{\boldsymbol{\delta}}}_l^{*N} \rfloor^{\mathrm{T}} \tag{2.58}$$

$$\bar{\boldsymbol{\delta}}_l^{*e} = \sum_h \sum_r \lambda_{hrl}^e \left\{ \frac{\partial \theta_{hr}^e}{\partial \mathbf{n}_{hl}^e} \right\} \tag{2.59}$$

$$\underline{\bar{\delta}}_{ql}^{*q} = \rho_l^n \frac{\partial \Omega^n}{\partial r_{ql}^n}, \quad \sum_l \rho_l^n = 1 \tag{2.60}$$

For the virtual nodal forces $\{\hat{\mathbf{F}}_{kl}\}$ and virtual reactions $\{\hat{\mathbf{R}}_{kl}\}$,

$$v_{kl} \left(\begin{bmatrix} \mathbf{D} & \\ & \underline{\mathbf{D}} \end{bmatrix} \begin{Bmatrix} \mathbf{F}_l \\ \mathbf{R}_l \end{Bmatrix} + \begin{Bmatrix} \mathbf{\Delta}_l^* \\ \underline{\mathbf{\Delta}}_l^* \end{Bmatrix} + \begin{Bmatrix} \tilde{\mathbf{\Delta}}^* \\ \underline{\tilde{\mathbf{\Delta}}}^* \end{Bmatrix} \right) + [\mathbf{B}]^{\mathrm{T}}\{\mathbf{L}_{kl}\} = 0$$

$$(k = 1,\dots,K; l = 1,\dots,L) \tag{2.61}$$

Since the expression in parentheses after v_{kl} in equation (2.61) represents the real relative nodal displacements, and $[\mathbf{B}]^{\mathrm{T}}$ is the kinematics matrix, we can see that

$$\frac{1}{v_{kl}}\{\mathbf{L}_{kl}\} = \{\mathbf{U}_l\} \tag{2.62}$$

where $\{\mathbf{U}_l\}$ denotes the real nodal displacements. Then it follows from equation (2.61) that kinematic admissibility, which was not included in the original formulation above, is automatically satisfied as an optimality condition.

Moreover, we interpret the Lagrange multiplier $\{\bar{\mathbf{L}}_l\}$ in (2.56) as the adjoint nodal displacement,

$$\{\bar{\mathbf{L}}_l\} = \{\bar{\mathbf{U}}_l\} \tag{2.63}$$

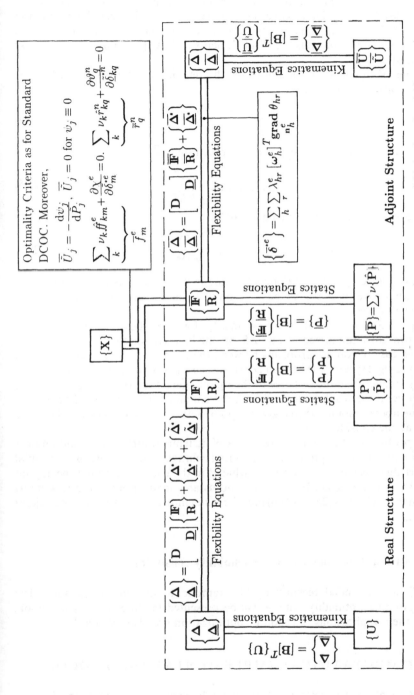

Fig. 2.5 DCOC algorithm: extension of the conceptual scheme to elastic supports, allowance for support costs, variable loads and variable prestrains – one loading condition.

and the corresponding relative displacements ($\{\bar{\Delta}\}$ at system level and $\{\bar{\delta}^e\}$ at element level) as adjoint relative displacements. It can be seen from equation (2.56) that the adjoint system has the same flexibility matrix as the real structure, but is subject to initial relative displacements $\lfloor\Delta_l^*, \Delta_i^*\rfloor^T$ owing to active stress constraints.

Further Kuhn–Tucker conditions for the remaining variables are as follows. For the variable loads $\tilde{P}_j$,

$$\sum_l \bar{U}_{jl} = -\frac{\mathrm{d}\psi_j}{\mathrm{d}\tilde{P}_j} \tag{2.64}$$

For the variable relative displacements (prestrains) $\{\tilde{\delta}^{*e}\}$,

$$-\sum_k \sum_l v_{kl}\hat{f}^e_{klm} = \frac{\partial \chi^e}{\partial \tilde{\delta}^{*e}_m} \tag{2.65}$$

For the variable support settlements $\{\underline{\tilde{\delta}}^{*n}\}$,

$$-\sum_k \sum_l v_{kl} f^n_{klq} = \frac{\partial \vartheta^n}{\partial \underline{\tilde{\delta}}^{*n}_q} \tag{2.66}$$

Moreover, all Lagrange multipliers for inequality constraints in equation (2.54) are nonnegative and can be positive only if the corresponding constraint is satisfied as an equality.

The necessary conditions of optimality for the above extended DCOC algorithm are summarized in Fig. 2.5 for one load condition.

In the case of contact problems between the elastic structure to be optimized and some rigid system, the contact constraint consists of two parts. If the displacement is smaller than the distance to the rigid system, then the contact constraint is inactive. If the displacement without the contact constraint would be greater than the above distance, then a given support settlement type constraint becomes active at the considered node.

For problems with given prestrain, lack of fit, temperature changes and support settlements, the appropriate value of initial relative displacements δ^{*e} or initial support displacements $\underline{\delta}^{*e}$ must be included in the real structure but the adjoint structure is not affected. In the case of adjustable loads, prestrains and support settlements, relations (2.64)–(2.66) must be taken into consideration for the adjoint structure.

2.3.5 Natural frequency and system stability constraints

DCOC was extended recently by O. Sigmund to dynamic constraints. The corresponding optimality criteria will be presented in chapter 10 of this book, where their applications in topology optimization are also discussed.

2.4 COMPUTATIONAL IMPLEMENTATION OF DCOC

As was mentioned in section 2.3, the flexibility method is suitable for deriving optimality criteria for DCOC, but in actual computations it is more convenient and more economical to use the stiffness method.

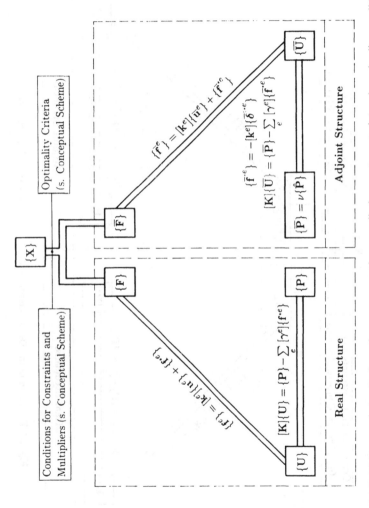

Fig. 2.6 DCOC computational scheme for stress constraints, one displacement constraint and one loading condition.

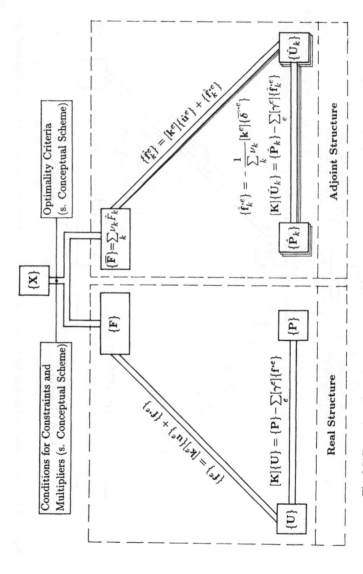

Fig. 2.7 Extension of the DCOC computational scheme to several displacement constraints.

For the simple case of stress constraints, one displacement constraint and one loading condition, the computational scheme based on the stiffness method is shown in Fig. 2.6. For the adjoint system the initial relative displacements in the conceptual scheme (Fig. 2.2) are first extended to the stiffness formulation $\{\bar{\delta}^{o*e}\}$ and then converted via the element stiffness matrices $[\mathbf{k}^e]$ into fixed end forces $\{\bar{\mathbf{f}}^{*e}\}$. As mentioned before, the decomposed stiffness matrix for the real structure can also be used for the adjoint structure. For one displacement constraint and one design variable per element, the Lagrange multiplier v can be updated explicitly from the displacement constraint (section 2.4.1(c)). The update of the design parameters is based on the optimality criteria given in Fig. 2.2.

Figure 2.7 shows the extension of the computational scheme to several displacement constraints. Although conceptually the adjoint structure is subject to the loads $\{\bar{\mathbf{P}}\} = \sum_k v_k \hat{\mathbf{P}}_k$, it is more convenient for the calculation of the Lagrange multipliers v_k to analyze the structure for each virtual load $\hat{\mathbf{P}}_k$. The relative initial displacements $\{\hat{\delta}^{*e}\}$ due to active stress constraints, or the corresponding fixed end forces $\{\hat{\mathbf{f}}_k^{*e}\}$, can be distributed arbitrarily to force and displacement systems associated with each virtual load $\hat{\mathbf{P}}_k$. One such distribution is given in Fig. 2.7. For further details see section 2.4.4.

2.4.1 Computational procedure for problems with one load condition, one displacement constraint, one design variable per element and flexibility matrices/stress constraints that are linear in the reciprocal variable $1/x^e$ and in the stress resultants

We consider a subclass of problems with

- one load condition and one displacement constraint,
- one design variable x^e per element ($I^e = 1$ for $e = 1, \ldots, E$),
- one stress constraint $(\theta^e(\mathbf{n})^e - \sigma^e \leqslant 0)$ per element ($H^e = 1$ and $R_h^e = 1$ for $e = 1, \ldots, E$),
- a linear element objective function

$$w^e(x^e) = c^e x^e \tag{2.67}$$

where c^e is a given constant,

- element flexibility matrices of the type

$$[\mathbf{d}^e] = [\mathbf{d}_0^e]/x^e, \tag{2.68}$$

where $[\mathbf{d}_0^e]$ is a given matrix,

- stress constraints that in equations (2.3) and (2.8), respectively, can be expressed as

$$\theta^e(\{\mathbf{n}^e\}) - \sigma^e = \frac{\theta_0^e(\{\mathbf{n}^e\})}{x^e} - \sigma^e \leqslant 0 \tag{2.69}$$

and

$$\{\mathbf{s}^e\}^T\{\mathbf{n}^e\} - \sigma^e = \frac{\{\mathbf{s}_0^e\}^T}{x^e}\{\mathbf{n}^e\} - \sigma^e \leqslant 0 \tag{2.70}$$

where θ_0^e and $\{\mathbf{s}_0^e\}$ are independent of the design variable x^e, and
- only lower side constraints.

For simplicity, the subscripts representing $i = 1$, $h = 1$ and $r = 1$ are omitted in equations (2.67)–(2.70) ($x_1^e \rightarrow x^e$, $\theta_{11}^e \rightarrow \theta^e$, $\mathbf{n}_1^e \rightarrow \mathbf{n}^e$).

For the considered class of problems, each element belongs to one of the following sets termed 'regions':

R_σ the stress constraint is active
R_s the lower side constraint is active
$R_{\sigma s}$ both the stress constraint and the lower side constraint are active
R_d none of the local (stress or side) constraints is active; the element is controlled by the displacement constraint

A flowchart for the main steps of the DCOC algorithm is shown in Fig. 2.8. Some explanatory comments on various steps are given below.

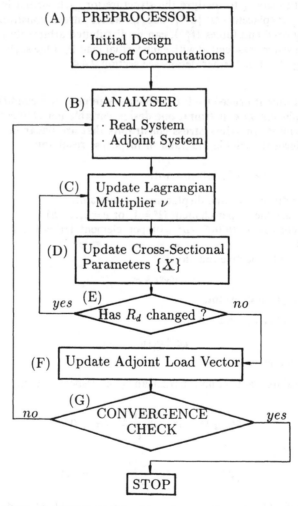

Fig. 2.8 Flowchart for the DCOC algorithm: one design variable and one stress constraint per element, and one displacement constraint.

(a) Pre-processor

The input includes the definition of the problem, including types of elements, geometrical and material properties, loading, permissible stresses and displacements as well as initial values of the design variables. One-off computations include calculations of certain geometrical and mechanical properties of the elements, bandwidth of the stiffness matrix, etc.

(b) Analysis of the real and adjoint systems

As can be seen from Fig. 2.6, the stiffness equation for the real structure is of the form

$$[\mathbf{K}]\{\mathbf{U}\} = \{\mathbf{P}\} - \sum_e [\gamma^e]\{\mathbf{f}^{*e}\} \tag{2.71}$$

where $[\gamma^e]$ is a transformation matrix from element level to system level and $\{\mathbf{f}^{*e}\}$ are 'fixed end forces' due to loads on the interior of an element or prestrains or temperature strains. Using an assembly technique known as the direct stiffness method, the stiffness matrix at the system level can be generated by using the relation

$$[\mathbf{K}] = \sum_{e=1}^{E} [\gamma^e]^T [\mathbf{k}^e][\gamma^e] \tag{2.72}$$

It can also be seen from Fig. 2.6 that the nodal forces can be calculated from the relations

$$\{\mathbf{f}^e\} = [\mathbf{k}^e]\{\mathbf{u}^e\} + \{\mathbf{f}^{*e}\} \tag{2.73}$$

at the element level.

The calculation of the fixed-end forces $\{\mathbf{f}^{*e}\}$ for the real structure is a well-known operation. However, the fixed-end forces $\{\bar{\mathbf{f}}^{*e}\}$ for the adjoint system must be calculated from the fictitious relative displacement $\{\bar{\delta}^{*e}\}$ given by equation (2.22) or (2.33):

$$\{\bar{\mathbf{f}}^{*e}\} = -[\mathbf{k}^e]\{\bar{\delta}^{o*e}\} \tag{2.74}$$

where $\{\bar{\delta}^{o*e}\}$ is the extended version of $\{\bar{\delta}^{*e}\}$ for the stiffness formulation which has a greater number of components than the relative displacement vector for the flexibility formulation (zero components are added for supported degrees of freedom in the flexibility formulation).

The analysis of the adjoint system involves only the forward and backward substitution of the adjoint load vector (Fig. 2.6)

$$\{\bar{\mathbf{P}}\} = \sum_e [\gamma^e]\{\bar{\mathbf{f}}^{*e}\} \tag{2.75}$$

using the decomposed stiffness matrix, which is already available from the analysis of the real system. Since in the DCOC algorithm the analysis of the adjoint system, in effect, replaces sensitivity analysis of other methods, this is a computationally very efficient feature of DCOC.

In the analysis of the adjoint system during the first iteration, the adjoint load vector consists of the virtual loads corresponding to the displacement constraints. During subsequent iterations, the adjoint load vector is the sum of the virtual load vector and of the equivalent nodal loads (reversed fixed end forces) caused by the adjoint initial relative displacements due to active stress constraints (equation (2.74)).

(c) Updating the Lagrange multiplier v

For elements within the set R_d the Lagrange multipliers λ_{hr}^e, $\beta_i^e \downarrow$ and $\beta_i^e \uparrow$ take on a zero value and hence equation (2.19) with equations (2.67) and (2.68) reduces to

$$\frac{\partial w^e}{\partial x^e} + v\{\hat{\mathbf{ff}}^e\}^T \left[\frac{\partial \mathbf{d}^e}{\partial x^e}\right]\{\mathbf{ff}^e\} = c^e - v\{\hat{\mathbf{ff}}^e\}^T[\mathbf{d}_o^e]\{\mathbf{ff}^e\}/(x^e)^2 = 0 \qquad (2.76)$$

or

$$x^e = v(\{\hat{\mathbf{ff}}^e\}^T[\mathbf{d}_o^e]\{\mathbf{ff}^e\}/c^e)^{1/2} \qquad (2.77)$$

Then the displacement constraint can be expressed as

$$t = \sum_{e=1}^{E}(\{\hat{\mathbf{ff}}^e\}^T[\mathbf{d}^e]\{\mathbf{ff}^e\}) = \sum_{e \in R_d} \frac{\{\hat{\mathbf{ff}}^e\}[\mathbf{d}_o^e]\{\mathbf{ff}^e\}}{(v\{\hat{\mathbf{ff}}^e\}^T[\mathbf{d}_o^e]\{\mathbf{ff}^e\}/c^e)^{1/2}} + \sum_{e \notin R_d}(\{\hat{\mathbf{ff}}^e\}^T[\mathbf{d}^e]\{\mathbf{ff}^e\}) \qquad (2.78)$$

implying

$$v^{1/2} = \frac{\displaystyle\sum_{e \in R_d}(c^e\{\hat{\mathbf{ff}}^e\}[\mathbf{d}_o^e]\{\mathbf{ff}^e\})^{1/2}}{t - \displaystyle\sum_{e \notin R_d}(\{\hat{\mathbf{ff}}^e\}^T[\mathbf{d}^e]\{\mathbf{ff}^e\})} \qquad (2.79)$$

The values of c^e and $[\mathbf{d}_o^e]$ are given for some simple elements in Table 2.3, where ρ is the specific weight of the structural material, E is Young's modulus, μ is Poisson's ratio and other symbols are defined in Fig. 2.9. The superscript e is omitted for simplicity after the symbols h, L, ρ, b, E, d and μ.

(d) Updating the cross-sectional variables $\{X\}$

This is done on the basis of equation (2.77) for elements of the R_d region. For elements contained in R_σ, R_s or $R_{\sigma s}$ regions, the stress or side constraint determines the value of x^e.

(e) Recycling if R_d has changed

The reason for this step is as follows. If the displacement-controlled region has changed, then equation (2.79) would give an incorrect estimate of v and hence steps (c) and (d) must be repeated.

(f) Updating the adjoint load vector

The equivalent element nodal loads for $e \in R_\sigma$ are the reversed fixed end forces calculated from equations (2.74) and (2.22) or (2.23). The Lagrange multipliers $\lambda_{11}^e \rightarrow \lambda^e$ for elements of R_σ are determined from equation (2.19) with $H_e = 1$,

Table 2.3 Design variable, weight constant and constant part of the flexibility matrix for some simple elements

Type of element	x^e	c^e	$[\mathbf{d}_o^e]$
Truss elements (Fig. 2.9(a))	Cross-sectional area	$L\rho$	$\dfrac{L}{E}$
Beam element, rectangular, cross section, variable width (Fig. 2.9(b))	Width	$hL\rho$	$\begin{bmatrix} \dfrac{4L^3}{Eh^3} & \dfrac{6L^2}{Eh^3} \\ \dfrac{6L^2}{Eh^3} & \dfrac{12L}{Eh^3} \end{bmatrix}$
Frame element, rectangular cross section, variable width (Fig. 2.9(c))	Width	$hL\rho$	$\begin{bmatrix} \dfrac{4L^3}{Eh^3} & 0 & \dfrac{6L^2}{Eh^3} \\ 0 & \dfrac{L}{Eh} & 0 \\ \dfrac{6L^?}{Eh^3} & 0 & \dfrac{12L}{Eh^3} \end{bmatrix}$
Triangular, constant-strain membrane element (Fig. 2.9(d))	Thickness	$\dfrac{bh\rho}{2}$	$\dfrac{2}{Ebh}\begin{bmatrix} b^2 & bd & -\mu hd \\ bd & 2(1+\mu)h^2 + d^2 & -\mu hd \\ -\mu hd & -\mu hd & h^2 \end{bmatrix}$

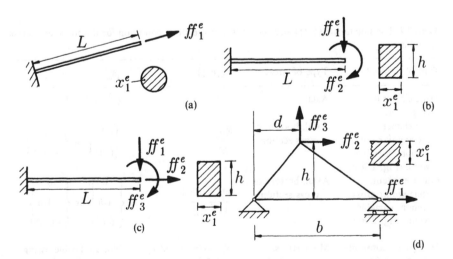

Fig. 2.9 Examples of the simplest class of elements for the DCOC method (see also Tables 2.3 and 2.4).

$R_h^e = 1$, $\beta_i^e \downarrow = 0$ and $\beta_i^e \uparrow = 0$ (see also equations (2.67)–(2.70)):

$$\lambda^e = \frac{(x^e)^2 c^e - v\{\bar{\mathbf{ff}}^e\}^T[\mathbf{d}_o^e]\{\mathbf{ff}^e\}}{\theta_o^e(\{\mathbf{n}^e\})} \tag{2.80}$$

or

$$\lambda^e = \frac{(x^e)^2 c^e - v\{\bar{\mathbf{ff}}^e\}[\mathbf{d}_o^e]\{\mathbf{ff}^e\}}{\{\mathbf{s}_o\}^T\{\mathbf{n}^e\}} \tag{2.81}$$

For the element types considered in Table 2.3 and Fig. 2.9, the function $\theta_o^e(\{\mathbf{n}^e\})$ and the vector $\{\mathbf{s}_o\}$ are given in Table 2.4. It can be seen from equation (2.8) that, in equations (2.80) and (2.81), $\{\mathbf{n}^e\}$ is given by $[\mathbf{\omega}^e]\{\mathbf{ff}^e\}$.

In Table 2.4, the stress resultants n_1^e, n_2^e and n_3^e correspond to the nodal forces ff_1^e, ff_2^e and ff_3^e in Fig. 2.9, but at some given cross section of the element e. For the first three cases, we have two transformation vectors in the last column, which correspond to tensile and compressive permissible stresses (σ_T^e, σ_C^e), respectively. This represents two stress constraints per element, out of which only one at a time can be active.

(g) The convergence criterion

The simplest criterion is

$$\frac{|\Phi_{new} - \Phi_{old}|}{\Phi_{new}} \leqslant T \tag{2.82}$$

where T is a given tolerance value. One may also use the following additional convergence tests:

$$\frac{|x_{i,new}^e - x_{i,old}^e|}{x_{i,new}^e} \leqslant T_1 \quad (i = 1, \ldots, I^e; \quad e = 1, \ldots, E) \tag{2.82a}$$

Table 2.4 The function $\theta_o^e(\{\mathbf{n}^e\})$ and vector $\{\mathbf{s}_o^e\}$ in the stress condition for the considered class of problems

Type of element	Type of stress	$\theta_o^e(\{\mathbf{n}^e\})$	$\{\mathbf{s}_o^e\}$
Truss elements (Fig. 2.9a)	Axial	$\|n_1^e\|$	$\{1\}$ or $\{-1\}$
Beam element, rectangular cross section, variable width (Fig. 2.9(b))	Axial, bottom of cross section	$\dfrac{6\|n_2^e\|}{h^2}$	$\begin{Bmatrix} 0 \\ 6/h^2 \end{Bmatrix}$ or $\begin{Bmatrix} 0 \\ -6/h^2 \end{Bmatrix}$
Frame element, rectangular cross section, variable width (Fig. 2.9(c))	Axial, bottom of cross section	$\dfrac{\|n_2^e\|}{h} + \dfrac{6\|n_3^e\|}{h^2}$	$\begin{Bmatrix} 0 \\ 1/h \\ 6/h^2 \end{Bmatrix}$ or $\begin{Bmatrix} 0 \\ -1/h \\ -6/h^2 \end{Bmatrix}$
Triangular, constant-strain membrane element (Fig. 2.9(d))	Mises stress, equation (2.17)	$(n_1^e)^2 + (n_2^e)^2 - n_1^e n_2^e + 3(n_3^e)^2$	Not applicable owing to nonlinearity

In most test examples in section 2.5, very stringent tolerance values (e.g. $T = 10^{-8}$) will be used with a view to demonstrating the accuracy of this method. In practical applications a tolerance value of $T = 10^{-3}$ or $T = 10^{-4}$ is usually sufficient.

2.4.2 A simple application illustrating the computational procedure for the class of problems in section 2.4.1: trusses

The design variable, element objective function, element flexibility matrix and stress constraints for a truss element were already given in equations (2.67)–(2.70), together with Tables 2.3 and 2.4. Denoting the two ends of the truss element by A and B, respectively, we have the element force vector and displacement vector in the stiffness formulation as

$$\{f^e\} = \lfloor f_A^e, f_B^e \rfloor^T, \quad \{u^e\} = \lfloor u_A^e, u_B^e \rfloor^T \tag{2.83}$$

and the force–displacement relation

$$\{f^e\} = [k^e]\{u^e\} \tag{2.84}$$

with the element stiffness matrix

$$[k^e] = \frac{E^e x^e}{L^e} \begin{bmatrix} 1 & -1 \\ -1 & 1 \end{bmatrix} \tag{2.85}$$

in which x^e is the cross-sectional area (Table 2.3) and other symbols were defined earlier. The element–system coordinate transformation matrix for a space truss can be represented as

$$[\gamma^e] = \begin{bmatrix} \cos \alpha^e & \cos \beta^e & \cos \zeta^e & 0 & 0 & 0 \\ 0 & 0 & 0 & \cos \alpha^e & \cos \beta^e & \cos \zeta^e \end{bmatrix} \tag{2.86}$$

where α^e, β^e and ζ^e are the angles between the longitudinal axis of the element e and the global (system) coordinate axes. In the flexibility formulation, the end A of the truss element has a rigid support and end B can move freely. Then the displacement–force relation contains only scalar quantities:

$$\delta^e = d^e f\!f^e \tag{2.87}$$

where δ^e and $f\!f^e$ refer to the relative displacement (element elongation = displacement at end B) and axial force at the end B, respectively. Moreover, the flexibility matrix reduces to

$$d^e = \frac{L^e}{E^e x^e} \tag{2.88}$$

as in Table 2.3 with equation (2.68). The element objective function w^e is also given by Table 2.3 with equation (2.67) as

$$w^e(x^e) = c^e x^e = L\rho^e x^e \tag{2.89}$$

If (i) we express the stress constraint using the function $\theta^e(\{n^e\})$ and (ii) the permissible stresses in tension and compression have the same value (σ^e), then we have (Table 2.4)

$$\theta^e(\{n^e\}) - \sigma^e = |n^e|/x^e - \sigma^e \leqslant 0 \tag{2.90}$$

If (i) we express the stress constraint using a transformation vector $\{s^e\}$ and (ii) we have different permissible stresses (σ_T^e, σ_C^e) in tension and compression, then we have (Table 2.4)

$$\{s^e\}^T\{n^e\} - \sigma_T^e = n^e/x^e - \sigma_T^e \leqslant 0 \tag{2.91}$$

$$-\{s^e\}^T\{n^e\} - \sigma_C^e = -n^e/x^e - \sigma_C^e \leqslant 0 \tag{2.92}$$

Since for trusses $ff^e = n^e$ or $[\omega] = 1$, by equation (2.81) we have the following value for the Lagrange multiplier of elements with an active stress constraint $(e \in R_\sigma)$ and equal permissible stresses (σ) in tension and compression:

$$\lambda^e = \frac{\rho^e(x^e)^2 - v\,ff^e\,\hat{ff}^e/E^e}{|ff^e|}\,L^e \tag{2.93}$$

or

$$\lambda^e = \frac{L^e}{\sigma^e}\left(\rho^e x^e - \frac{v\,\hat{ff}^e ff^e}{x^e E^e}\right) \tag{2.94}$$

The adjoint initial relative displacement by equations (2.22) and (2.90) with $n^e = ff^e$ becomes

$$\bar{\delta}^{*e} = \lambda^e\,\mathrm{sgn}\,ff^e/x^e \tag{2.95}$$

The adjoint initial relative displacements, which are extended for the stiffness method, are given by (Fig. 2.6)

$$\{\bar{\delta}^{o*e}\} = \begin{Bmatrix} \delta^{*e} \\ 0 \end{Bmatrix} \tag{2.96}$$

and the corresponding fixed end forces by

$$\{\bar{f}^{*e}\} = -[k^e]\{\delta^{o*e}\} \tag{2.97}$$

where $[k^e]$ is given by equation (2.85). For the updating of the Lagrange multiplier v, equation (2.79) implies with Table 2.3

$$v^{1/2} = \frac{\sum\limits_{e \in R_d} L^e(\rho^e\{\hat{ff}^e ff^e/E^e\}^{1/2}}{t - \sum\limits_{e \notin R_d}(L^e\,\hat{ff}^e ff^e/E^e x^e)} \tag{2.98}$$

In equation (2.98), the value of x^e for $e \in R_\sigma$ is based on the latest information using the real forces from the analysis step and equation (2.101): for $e \in R_\sigma$

$$L^e\,\hat{ff}^e ff^e/(E^e x^e) = L^e\,\hat{ff}^e ff^e/(E^e|ff^e|/\sigma^e) = L^e\,\hat{ff}^e\sigma^e\,\mathrm{sgn}\,ff^e/E^e \tag{2.98a}$$

For $e \in R_s$ we have $x^e = x^e\!\downarrow$.

In step (D) of the flow-chart in Fig. 2.8, we have the following re-sizing formula:

$$(x^e)^2 = \max\{(x_d^e)^2, (x_\sigma)^2, (x^e\!\downarrow)^2\} \tag{2.99}$$

where by equation (2.77) and Table 2.3

$$(x_d^e)^2 = v\,\frac{\hat{ff}^e ff^e}{E^e\rho} \tag{2.100}$$

and by equation (2.90)

$$x_\sigma^e = \frac{|f\!f^e|}{\sigma^e} \tag{2.101}$$

2.4.3 Certain computational difficulties and methods for removing them

(a) Truss element with active stress and side constraints $(e \in R_{\sigma s})$

If both the stress and the side constraints are active for a truss element, then by equations (2.19), (2.67)–(2.70) and Tables 2.3 and 2.4 we have the relation for the Lagrange multiplier λ^e (assuming that v is given):

$$c^e - v\hat{f\!f}^e f\!f^e L^e / E^e (x^e)^2 - \lambda^e |f\!f^e| / (x^e)^2 - \beta^e \!\downarrow \, = 0 \tag{2.102}$$

which cannot be solved uniquely because there are two unknowns $(\lambda^e$ and $\beta^e\!\downarrow)$ in one equation.

(b) No displacement constraint is active for a truss

We recall that, in the derivation of the DCOC algorithm in section 2.3.1, compatibility of the displacements was not included in the formulation and it turned out to be an optimality criterion, provided that a displacement condition was active. If only stress constraints are active, then the optimality criteria derived above are not valid and the solution algorithm must be modified.

(c) One method for avoiding the above difficulties: upgrading stress
    constraints to global constraints

If either of the difficulties outlined under (a) or (b) is encountered then, respectively,

- the stress constraint for all elements with $e \in R_{\sigma s}$, or
- any one stress constraint

can be upgraded into a global (displacement) constraint. This removes the difficulties outlined above, but the introduction of an extra global constraint also requires additional computer time.

Considering a truss element, for example, a stress constraint can be replaced by a constraint on the relative displacement between the two ends of the considered truss element (i.e. on the elongation of the element). Using the corresponding virtual loads $\{\hat{\mathbf{P}}_k\} \to \{\hat{\mathbf{P}}\}$ shown in Fig. 2.10, this additional displacement constraint becomes

$$\sum_{e=1}^{E} \frac{\hat{f\!f}_k^e f\!f^e L^e}{x^e E^e} - t_k \leqslant 0 \tag{2.103}$$

with

$$t_k = \frac{L^e \sigma^e}{E^e} \tag{2.104}$$

where e refers to the element for which the stress constraint is upgraded into a displacement constraint.

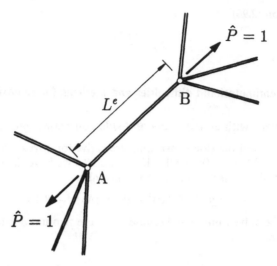

Fig. 2.10 Method for upgrading a stress constraint for trusses into a displacement constraint.

(d) Another possibility: forcing uniqueness of the adjoint strains by corner-rounding at constraint intersections using KS functions

The nonuniqueness of the multiplier λ^e in equation (2.102) can easily be understood by considering the cross-sectional area–member force relation $x^e(ff^e)$ for this problem (thick line in Fig. 2.11(a)). The above relation is based on the inequalities

$$x^e \geqslant x^e{\downarrow}, \quad x^e \geqslant ff^e/\sigma^e, \quad x^e \geqslant -ff^e/\sigma^e \tag{2.105}$$

with at least one of these three constraints satisfied as an equality.

If both the stress and the side constraints are active, for example at the point R in Fig. 2.11(a), then the subgradient of $x^e(ff^e)$ is clearly nonunique, ranging from zero to $1/\sigma_e$. This causes, in turn, the nonuniqueness of the corresponding Lagrange multiplier λ_i.

The above difficulty can be avoided by replacing the exact $x^e(ff^e)$ relation by a smooth envelope function (SEF).

Applications of Kreisselmeier–Steinhauser functions in optimization problems were discussed recently (Chang, 1992; Sobieszczanski-Sobieski, 1992). On the basis of equation (2) in the latter reference, we have the following KS functions for the three linear segments in Fig. 2.11(a): for $|ff^e| \geqslant \sigma^e x^e{\downarrow}$,

$$\mathrm{KS} = |ff^e|/\sigma^e + \frac{1}{\eta}\ln\left[1 + \mathrm{e}^{-2\eta|ff^e|/\sigma^e} + \mathrm{e}^{\eta(x^e{\downarrow} - |ff^e|/\sigma^e)}\right] \tag{2.106}$$

for $|ff^e| \leqslant \sigma^e x^e{\downarrow}$,

$$\mathrm{KS} = x^e{\downarrow} + \frac{1}{\eta}\ln\left[1 + \mathrm{e}^{\eta(ff^e/\sigma^e - x^e{\downarrow})} + \mathrm{e}^{\eta(-ff^e/\sigma^e - x^e{\downarrow})}\right] \tag{2.107}$$

where η is a given constant.

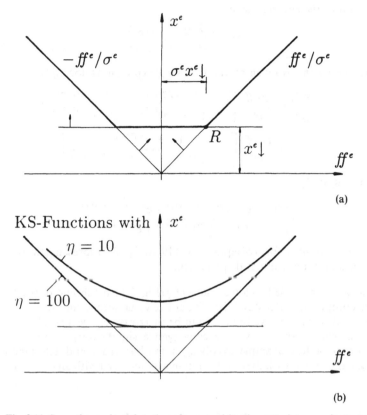

(a)

(b)

Fig. 2.11 Smooth envelope functions for several local constraints per element.

The above KS functions for $\eta = 10$ and $\eta = 100$ are shown in Fig. 2.11(b). For $\eta = 1000$, the graph of the KS function would be on this scale indistinguishable from the three straight line segments in Fig. 2.11(a). The expressions in both equations (2.106) and (2.107) represent the same function for all values of ff^e, but these separate expressions are necessary to avoid an overflow in the computer due to too large values of the exponentials.

Using the SEF in equations (2.106) and (2.107) or in Fig. 2.11(b), we can reformulate the Lagrangian function for the problem in equations (2.1)–(2.7) in the following form (see also the restrictions in equations (2.67)–(2.70)):

$$\min \Phi = \sum_{e=1}^{E} c^e x^e + v \sum_{e=1}^{E} \frac{\hat{ff}^e \, ff^e L^e}{E^e x^e} + \sum_{e=1}^{E} \lambda^e [\mathrm{KS}(ff^e) - x^e]$$

$$+ \{\bar{\mathbf{L}}\}^{\mathrm{T}}(\{\mathbf{P}\} - [\mathbf{B}]\{\mathbf{F}\}) + \{\mathbf{L}\}(\{\hat{\mathbf{P}}\})^{\mathrm{T}} - [\mathbf{B}]\{\hat{\mathbf{F}}\}) \qquad (2.108)$$

Then we have the following Kuhn–Tucker conditions.

- *Variation of the design variables:* x^e

$$c^e - \frac{v \hat{ff}^e \, ff^e L^e}{E^e (x^e)^2} = \lambda^e \qquad (2.109)$$

- *Variation of the real forces:* ff^e

$$\bar{\delta}^{*e} = \frac{v\hat{ff}^e L^e}{E^e x^e} + \lambda^e \frac{\partial(KS)}{\partial ff^e} \tag{2.110}$$

where $\partial(KS)/\partial ff^e$ is given by the following expressions: for $|ff^e| \geqslant \sigma^e x^e \downarrow$,

$$\frac{\partial(KS)}{\partial ff^e} = \frac{\operatorname{sgn} ff^e}{\sigma^e}\left[1 - \frac{2e^{-2\eta|ff^e|/\sigma^e} + e^{\eta(x^e\downarrow - |ff^e|/\sigma^e)}}{1 + e^{-2\eta|ff^e|/\sigma^e} + e^{\eta(x^e\downarrow - |ff^e|/\sigma^e)}}\right]$$

$$= \frac{\operatorname{sgn} ff^e}{\sigma^e}\left[\frac{1 - e^{-2\eta|ff^e|/\sigma}}{1 + e^{-2\eta|ff^e|/\sigma^e} + e^{\eta(x^e\downarrow - |ff^e|/\sigma^e)}}\right] \tag{2.111}$$

for $|ff^e| \leqslant \sigma^e x^e \downarrow$,

$$\frac{\partial(KS)}{\partial ff^e} = \frac{1}{\sigma^e}\frac{e^{\eta(ff^e/\sigma^e - x^e\downarrow)} - e^{\eta(-ff^e/\sigma^e - x^e\downarrow)}}{1 + e^{\eta(ff^e/\sigma^e - x^e\downarrow)} + e^{\eta(-ff^e/\sigma^e - x^e\downarrow)}} \tag{2.112}$$

and then a modified form of equations (2.19) and (2.22) represents the optimality criteria and adjoint initial displacement.

In using the above SEF, a lower value of η (Fig. 2.11(b)) should be used initially to avoid computational instabilities. The above value should then be progressively increased. Moreover, move limits are to be progressively decreased during later iterations, which slows down the convergence significantly. Using the above procedure in a test example involving a ten-bar truss and KS functions, a four-digit agreement with the exact solution was achieved with an η value of 1000.

2.4.4 Computational procedure for problems with several displacement constraints

As mentioned at the beginning of section 2.4, the adjoint load for several displacement constraints (Fig. 2.3) is $\{\bar{P}\} = \sum_{k=1}^{K} v_k\{\hat{P}_k\}$ together with relative initial displacements $\{\bar{\delta}^{*e}\}$ given by equations (2.21)–(2.23) for elements with active stress constraints. Since for inactive displacement constraints $v_k = 0$, we can replace the above relation by

$$\{\bar{P}\} = \sum_{k=1}^{K_A} v_k\{\hat{P}_k\} \tag{2.113}$$

where K_A is the number of active displacement constraints.

Zhou (1992) has proposed the following algorithm for calculating the Lagrange multipliers v_k, which is presented here in a somewhat modified form. First a structural analysis is carried out for each virtual load system $\{\hat{P}_k\}$ with a fixed fraction

$$\{\hat{\delta}_k^{*e}\} = \{\bar{\delta}^{*e}\}\bigg/\left(\sum_{k=1}^{K_A} v_k\right) \tag{2.114}$$

of the initial relative displacement $\{\bar{\delta}^{*e}\}$ allocated to each such analysis. The corresponding fixed end forces are given by (Fig. 2.7)

$$\{\overline{\mathbf{f}}_k^{*e}\} = -[\mathbf{k}^e]\{\hat{\boldsymbol{\delta}}_k^{*e}\} = -\frac{1}{\sum\limits_{k=1}^{K_A} v_k}[\mathbf{k}^e]\{\overline{\boldsymbol{\delta}}^{\circ *e}\} \qquad (2.115)$$

where the superscript 0 simply means that these initial displacements have been extended from flexibility formulation to stiffness formulation by adding some zeros to these vectors. The analysis is carried out by means of the stiffness method (Fig. 2.7):

$$[\mathbf{K}]\{\hat{\mathbf{U}}_k\} = \{\hat{\mathbf{P}}_k\} - \sum_{e=1}^{E} [\boldsymbol{\gamma}^e]\{\hat{\mathbf{f}}_k^{*e}\} \qquad (2.116)$$

where $[\boldsymbol{\gamma}^e]$ is the transformation matrix from local to global coordinates. The internal nodal forces $\{\hat{\mathbf{f}}_k^e\}$ for the above virtual systems can then be calculated at the element level (Fig. 2.7):

$$\{\hat{\mathbf{f}}_k^e\} = [\mathbf{k}^e]\{\hat{\mathbf{u}}_k^e\} + \{\hat{\mathbf{f}}_k^{*e}\} \qquad (2.117)$$

On the basis of this analysis, the correct adjoint internal forces

$$\{\overline{\mathbf{f}}^e\} = \sum_{k=1}^{K_A} v_k\{\hat{\mathbf{f}}_k^e\} \qquad (2.118)$$

are calculated. These can be used in the re-design formula based on the optimality criterion (2.19) with the extension (2.31), also shown at the top of Fig. 2.3. It can be easily seen that the sum of the initial displacements in equation (2.114), after multiplication by v_k in equation (2.117) adds up to the correct value

$$\sum_{k=1}^{K_A} v_k\{\hat{\boldsymbol{\delta}}_k^{*e}\} = \sum_{k=1}^{K_A} v_k\left(\frac{1}{\sum\limits_{k=1}^{K_A} v_k}\{\overline{\boldsymbol{\delta}}^{*e}\}\right) = \{\overline{\boldsymbol{\delta}}^{*e}\} \qquad (2.119)$$

In Zhou's algorithm, the K_A Lagrange multipliers v_k ($k = 1, \ldots, K_A$) for the displacement constraints are calculated from work equations representing active displacement constraints (equation (2.28) with an equality sign). In the above equations, the flexibility matrix $[\mathbf{D}]$ and the initial relative displacements (if any) $\{\boldsymbol{\Delta}^*\}$ depend on the design variables $\{\mathbf{X}\}$ and the latter depend, via the optimality criteria (relation (2.19) with the extension in relation (2.31)) on v_k. Thus the K_A work equations (2.28) depend implicitly on the K_A Lagrange multipliers $v_k (k = 1, \ldots, K_A)$. Zhou's algorithm uses the approximation that the real and virtual forces ($\{\mathbf{F}\}, \{\hat{\mathbf{F}}_k\}$) are based on the prior analysis and are not supposed to change as a result of variations of $\{\mathbf{v}\} = \lfloor v_1, \ldots, v_k \rfloor^T$ and $\{\mathbf{X}\}$. The Lagrange multipliers can be determined within the above approximation by Newton's method, using the recurrence relation

$$\{\mathbf{v}\}_{t+1} = \{\mathbf{v}\}_t - [\mathbf{J}]_t^{-1}\{\boldsymbol{\varphi}\}_t \qquad (2.120)$$

where $\{\boldsymbol{\varphi}\} = \lfloor \varphi_1, \ldots, \varphi_{K_A} \rfloor^T$ are the LHSs of the active displacement constraints in equation (2.28), $[\mathbf{J}]$ is the Jacobian matrix containing derivatives of $\{\boldsymbol{\varphi}\}$ with respect to $\{\mathbf{v}\}$ and t is the Newton iteration number. The apparent disadvantage of this algorithm is K_A analyses for the virtual loads $\{\hat{\mathbf{P}}_k\}$ instead of one analysis for the adjoint load $\{\bar{\mathbf{P}}\} = \sum_{k=1}^{K_A} v_k\{\hat{\mathbf{P}}\}$, but this is more than compensated by the efficient calculation of the Lagrange multipliers v_k, which does not involve any further structural analyses.

2.4.5 Several design variables and several stress constraints per element

In the general formulation of DCOC, we have I^e design variables per element ($\{\mathbf{x}^e\} = \lfloor x_1,\ldots,x_i,\ldots,x_{I^e} \rfloor^T$) and R_h^e stress conditions at each of the cross sections $h = 1,\ldots,H^e$. Assuming that the Lagrange multipliers for the displacement constraints (v_k) have already been determined (see step (C) in Fig. 2.8) for the update at the element level, the unknowns for a given element e are the I^e design variables and the Lagrange multipliers, say λ_r^e $(r = 1,\ldots,R_A^e)$, for the active local (i.e. stress or side) constraints where R_A^e is the number of such active constraints. The number of equations available is also $I^e + R_A^e$, consisting of the I^e optimality criteria (2.19) and of the R_A^e stress constraint satisfied as an equality.

In the above general case one should be able to determine all unknowns required for the update operation. A difficulty arises when the number of active local (stress and side) constraints exceeds the number of design variables ($R_A^e > I^e$) because this means that some of the R_A^e active constraints with I^e unknowns are not independent and they introduce R_A^e unknown Lagrange multipliers, whilst determining only I^e-design variables. This means that the number of independent equations is $2I_e^e$ and the number of variables is $R_A^e + I^e$, with $R_A^e > I^e$, resulting in nonuniqueness of the solution. A special case of this problem was discussed in sections 2.4.3(a), 2.4.3(c) and 2.4.3(d), in which a truss element had active stress and side constraints ($I^e = 1$, $R_A^e = 2$). The above difficulty can be removed by upgrading the stress constraints involved into a displacement constraint (section 2.4.3(c)) or by introducing smooth envelope functions (SEFs, section 2.4.3(d)).

2.5 TEST EXAMPLES SOLVED BY THE DCOC METHOD

All computational results reported in this section were obtained on an HP-9000 workstation with double precision (FORTRAN 77). The dual method used for verification of the results is based on quadratic approximation of the objective function and linear approximation of the behavioral constraints in terms of reciprocal variables (Fleury, 1979; Zhou, 1989).

2.5.1 An elementary example: three-bar truss

The aim of this example is mainly educational: owing to the extreme simplicity of the problem, the principles involved are not obscured by computational complexities and the solution can be obtained even by hand calculations. By working his/her way laboriously over the steps of the DCOC procedure, the reader can achieve a better intuitive insight into this method.

We consider the truss shown in Fig. 2.12(a), in which the three bars may take on different cross-sectional areas ($\{\mathbf{X}\} = \lfloor x^{(1)}, x^{(2)}, x^{(3)} \rfloor^T$). The truss is subject to a horizontal load of $P_1 = P$ and a vertical load of $P_2 = kP$ at its free joint. The

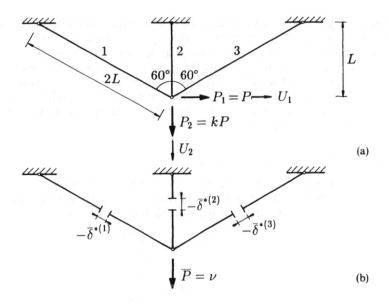

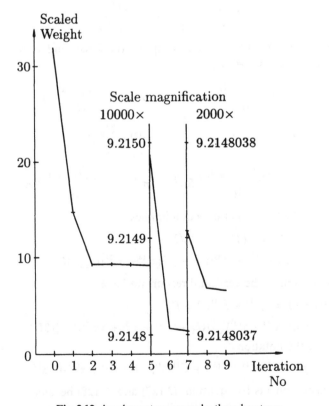

Fig. 2.12 An elementary example: three-bar truss.

structural weight is to be minimized, subject to a single displacement constraint $U_2 \leqslant t$ and constraints on all stresses, with $\sigma_C = c\sigma_T$ where σ_C and σ_T are the permissible stresses in compression and tension, respectively, for all members ($e = 1, 2, 3$). Young's modulus (E) and the specific weight ρ are the same for all three bars. The adjoint load $\bar{P} = v$ for the displacement constraint is shown in Fig. 2.12(b). Superscripts representing element numbers are given in parentheses, in order to indicate that they are not exponents. For simplicity, we shall use the stiffness notation for bar forces (f^e).

(a) Analysis equations for the real truss

The nodal displacement–bar elongation–bar force relations can be easily determined by projecting the nodal displacements onto the member directions (e.g. Haftka *et al.* (1990), who used the variable linking $x^{(1)} = x^{(3)}$):

$$\frac{\sqrt{3}U_1}{2} + \frac{U_2}{2} = \delta^{(1)} = \frac{2f^{(1)}L}{x^{(1)}E} \tag{2.121}$$

$$U_2 = \delta^{(2)} = \frac{f^{(2)}L}{x^{(2)}E} \tag{2.122}$$

$$-\frac{\sqrt{3}U_1}{2} + \frac{U_2}{2} = \delta^{(3)} = \frac{2f^{(3)}L}{x^{(3)}E} \tag{2.123}$$

From equations (2.121)–(2.123) we can express the forces and stresses in terms of the displacements $\{U\} = \lfloor U_1, U_2 \rfloor^T$:

$$f^{(1)} = \frac{\sqrt{3}U_1 + U_2}{4L} Ex^{(1)}, \qquad \sigma^{(1)} = \frac{\sqrt{3}U_1 + U_2}{4L} E \tag{2.124}$$

$$f^{(2)} = \frac{U_2}{L} Ex^{(2)}, \qquad \sigma^{(2)} = \frac{U_2}{L} E \tag{2.125}$$

$$f^{(3)} = \frac{-\sqrt{3}U_1 + U_2}{4L} Ex^{(3)}, \qquad \sigma^{(3)} = \frac{-\sqrt{3}U_1 + U_2}{4L} E \tag{2.126}$$

Equilibrium in the horizontal direction implies

$$P_1 = P = \sqrt{3}(f^{(1)} - f^{(3)})/2$$
$$= E[3(x^{(1)} + x^{(3)})U_1 + \sqrt{3}(x^{(1)} - x^{(3)})U_2]/8L \tag{2.127}$$

and by equilibrium in the vertical direction we have

$$P_2 = kP = (f^{(1)} + f^{(3)})/2 + f^{(2)}$$
$$= E[(x^{(1)} - x^{(3)})\sqrt{3}U_1 + (x^{(1)} + x^{(3)} + 8x^{(2)})U_2]/8L \tag{2.128}$$

Introducing the notation

$$q = 2x^{(2)}(x^{(1)} + x^{(3)}) + x^{(1)}x^{(3)}, \quad t = x^{(1)} + x^{(3)} + 8x^{(2)} \tag{2.129}$$

the nodal displacements by equations (2.127) and (2.128) become

$$U_2 = (2PL/E)[k(x^{(1)} + x^{(3)}) - (x^{(1)} - x^{(3)})/\sqrt{3}]/q \tag{2.130}$$

$$U_1 = (2PL/3E)[t - \sqrt{3}k(x^{(1)} - x^{(3)})]/q \tag{2.131}$$

Then by equations (2.124)–(2.126) the bar forces are given by

$$f^{(1)} = Px^{(1)}[2kx^{(3)} + (t + x^{(3)} - x^{(1)})/\sqrt{3}]/2q \tag{2.132}$$

$$f^{(2)} = 2Px^{(2)}[k(x^{(1)} + x^{(3)}) - (x^{(1)} - x^{(3)})/\sqrt{3}]/q \tag{2.133}$$

$$f^{(3)} = Px^{(3)}[2kx^{(1)} + (x^{(3)} - x^{(1)} - t)/\sqrt{3}]/2q \tag{2.134}$$

It can be easily checked that equations (2.130) and (2.131) reduce to known relations (Haftka, Gürdal and Kamat, 1990, p. 184) for $x^{(1)} = x^{(3)}$ and $k = 8$.

(b) Analysis equations for the adjoint truss

We consider now the adjoint truss in Fig. 2.12(b), with the load $\bar{P} = v$ (i.e. v times the virtual load $\hat{P} = 1$) and initial relative displacements caused by the active stress constraints (equations (2.93)–(2.95)). The adjoint bar elongation $\bar{\delta}^e$–nodal displacement U_j–bar force $\bar{f}^e$ relations can be derived easily from compatibility:

$$\bar{\delta}^{(1)} = \frac{\bar{U}_1\sqrt{3}}{2} + \frac{\bar{U}_2}{2} = \bar{\delta}*^{(1)} + \frac{\bar{f}^{(1)}2L}{x^{(1)}E} \tag{2.135}$$

$$\bar{\delta}^{(2)} = \bar{U}_2 = \bar{\delta}*^{(2)} + \frac{\bar{f}^{(2)}L}{x^{(2)}E} \tag{2.136}$$

$$\bar{\delta}^{(3)} = -\frac{\bar{U}_1\sqrt{3}}{2} + \frac{\bar{U}_2}{2} = \bar{\delta}*^{(3)} + \frac{\bar{f}^{(3)}2L}{x^{(3)}E} \tag{2.137}$$

Expressing the adjoint bar forces $\bar{f}^e$ from equations (2.135)–(2.137), we have

$$\bar{f}^{(1)} = Ex^{(1)}\frac{\sqrt{3}\bar{U}_1 + \bar{U}_2 - 2\bar{\delta}*^{(1)}}{4L} \tag{2.138}$$

$$\bar{f}^{(2)} = Ex^{(2)}\frac{\bar{U}_2 - \bar{\delta}*^{(2)}}{L} \tag{2.139}$$

$$\bar{f}^{(3)} = Ex^{(3)}\frac{-\sqrt{3}\bar{U}_1 + \bar{U}_2 - 2\bar{\delta}*^{(3)}}{4L} \tag{2.140}$$

Equilibrium in the horizontal direction implies

$$0 = \bar{f}^{(1)} - \bar{f}^{(3)} \tag{2.141}$$

or

$$0 = x^{(1)}(\sqrt{3}\bar{U}_1 + \bar{U}_2 - \bar{\delta}*^{(1)}) - x^{(3)}(-\sqrt{3}\bar{U}_1 + \dot{U}_2 - 2\bar{\delta}*^{(3)}) \tag{2.142}$$

Moreover, by equilibrium in the vertical direction and equation (2.141), we have

$$v = \frac{\bar{f}^{(1)} + \bar{f}^{(3)}}{2} + \bar{f}^{(2)} = \bar{f}^{(1)} + \bar{f}^{(2)} \tag{2.143}$$

or

$$4Lv/E = x^{(1)}(\sqrt{3}\bar{U}_1 + \bar{U}_2 - 2\bar{\delta}*^{(1)}) + 4x^{(2)}(\bar{U}_2 - \bar{\delta}*^{(2)}) \tag{2.144}$$

Introducing the notation

$$q = x^{(1)}x^{(3)} + 2x^{(2)}(x^{(1)} + x^{(3)}), \quad r = \bar{\delta}^{*(1)}x^{(1)} - \bar{\delta}^{*(3)}x^{(3)}$$

$$s = 2Lv/E + \bar{\delta}^{*(1)}x^{(1)} + 2\bar{\delta}^{*(2)}x^{(2)} \tag{2.145}$$

the nodal displacements become

$$\bar{U}_2 = [(x^{(1)} + x^{(3)})s - x^{(1)}r]/q \tag{2.146}$$

$$\bar{U}_1 = -[(x^{(1)} - x^{(3)})s - (x^{(1)} + 4x^{(2)})r]/\sqrt{3}q \tag{2.147}$$

and the bar forces are given by

$$\bar{f}^{(1)} = (Ex^{(1)}/2L)[(x^{(3)}s + 2x^{(2)}r)/q - \bar{\delta}^{*(1)}] \tag{2.148}$$

$$\bar{f}^{(2)} = (Ex^{(2)}/L)\{[(x^{(1)} + x^{(3)})s - x^{(1)}r]/q - \bar{\delta}^{*(2)}\} \tag{2.149}$$

$$\bar{f}^{(3)} = (Ex^{(3)}/2L)\{[(x^{(1)}s - (x^{(1)} + 2x^{(2)})r]/q - \bar{\delta}^{*(3)}\} \tag{2.150}$$

(c) Initial design and first iteration

We consider the problem in Fig. 2.12(a) in a normalized form with $P = t = L = E = \rho = 1, P_2 = 8, P_1 = 1$ (i.e. $k = 8$) and $\sigma_T = \sigma_C = 1.3 = \sigma$ (i.e. $c = 1$). The initial design

$$x^{(1)} = x^{(2)} = x^{(3)} = 6.4 \tag{2.151}$$

has the following features:

- all members have the same cross-sectional area, and
- they have been scaled to give exactly the prescribed displacement.

Then by equations (2.130)–(2.134) we have

$$U_1 = 0.208\,333\cdots, \quad U_2 = 1.0, \quad f^{(1)} = 2.177\,350, \quad f^{(2)} = 6.4, \quad f^{(3)} = 1.022\,650 \tag{2.152}$$

As mentioned in section 2.4.1(b), the analysis of the adjoint truss in the first iteration may be based on the virtual load without initial relative displacements for active stress constraints because at the start of the first iteration the value of the Langrange multipliers is not yet known. By equations (2.148)–(2.150), we have for an initial adjoint load of $v = 1$ and $\bar{\delta}^{*1} = \bar{\delta}^{*2} = \bar{\delta}^{*3} = 0$

$$\bar{f}^{(1)} = 0.2, \quad \bar{f}^{(2)} = 0.8, \quad \bar{f}^{(3)} = 0.2 \tag{2.153}$$

At this stage it cannot be decided on the basis of equations (2.99)–(2.101) which members belong to the regions R_d, R_σ and R_s, but it is clear that a truss element with $\bar{f}^e f^e \leqslant 0$ cannot be part of the displacement controlled set R_d. For this reason, we can assume for the first iteration that

$$e \in R_d \quad (\text{if } \bar{f}^e f^e > 0), \quad e \notin R_d \quad (\text{if } \bar{f}^e f^e \leqslant 0) \tag{2.154}$$

giving for this case

$$1 \in R_d, \quad 2 \in R_d, \quad 3 \in R_d \tag{2.155}$$

The values of c^e and $[d_o]$ by Table 2.3 take on the values $c^e = L\rho = 1$ and

$[\mathbf{d}_o] = L/E = 1$. Hence equations (2.98) and (2.98a) with (2.152), (2.153) and (2.155) yield

$$\sqrt{v} = 2\sqrt{2.177\,350 \times 0.2} + \sqrt{6.4 \times 0.8} + 2\sqrt{1.022\,650 \times 0.2} = 4.487\,045,$$
$$v = 20.133\,569 \qquad (2.156)$$

The comparison values in equations (2.100) and (2.101) then become

$$(x_d^{(1)})^2 = v\bar{f}^{(1)}f^{(1)} = 8.767\,565, \quad (x_d^{(2)})^2 = 103.083\,873, \quad (x_d^{(3)})^2 = 4.117\,919$$

$$(x_\sigma^{(1)})^2 = \left(\frac{f^{(1)}}{1.3}\right)^2 = 2.805\,238, \quad (x_d^{(2)})^2 = 24.236\,686, \quad (x_d^{(3)})^2 = 0.618\,824 \quad (2.157)$$

Adopting a prescribed minimum cross-sectional area of $x\!\downarrow = 0.01$, equations (2.156) and (2.157) confirm the regions assumed in equation (2.155). It follows then from equations (2.156) and (2.157) that the design variables at the end of the first iteration are

$$x^{(1)} = 2.961\,007, \quad x^{(2)} = 10.153\,023, \quad x^{(3)} = 2.029\,266 \qquad (2.158)$$

The initial design had the weight

$$\Psi_{\text{initial}} = 6.4(1 + 2 \times 2) = 32 \qquad (2.159)$$

and the unscaled truss weight after the first iteration has been reduced to

$$\Phi(1) = x^{(2)} + 2(x^{(1)} + x^{(3)}) = 20.133\,569 \qquad (2.160)$$

The initial relative displacements $\bar{\delta}^{*e}$ for the adjoint truss in the second iteration take on a zero value because no stress constraints are active at the end of the first iteration.

(d) Second iteration

Using the x^e values in equations (2.158), (2.130)–(2.134) yields the following displacements, forces and stresses for the real truss:

$$U_1 = 0.455\,269, \quad U_2 = 0.733\,813, \quad f^{(1)} = 1.126\,932,$$
$$f^{(2)} = 7.450\,419, \quad f^{(3)} = -0.027\,769 \quad \sigma^{(1)} = 0.380\,591,$$
$$\sigma^{(2)} = 0.733\,813, \quad \sigma^{(3)} = -0.013\,684 \qquad (2.161)$$

It can be seen from relations (2.161) that for a feasible solution with at least one active constraint, we can multiply the x^e values in equation (2.158) by 0.733 813, thereby obtaining the following scaled designs:

$$x^{(1)} = 2.172\,825, \quad x^{(2)} = 7.450\,420, \quad x^{(3)} = 1.489\,102 \qquad (2.162)$$

The corresponding scaled weight value becomes

$$\Phi(1)_{\text{scaled}} = 0.733\,813\,\Phi(1) = 14.774\,274 \qquad (2.163)$$

Substituting the design variables f^e from equation (2.162) into equations (2.148)–(2.150) with a virtual load of $v = 1$ and $\bar{\delta}^{*e} = 0$ (for all e), we obtain the virtual forces $\hat{f}^e = \bar{f}^e/v$:

$$\hat{f}^{(1)} = 0.055\,977, \quad \hat{f}^{(2)} = 0.944\,023, \quad \hat{f}^{(3)} = 0.055\,977 \qquad (2.164)$$

Using the assumption in equation (2.154), we have the following regions for the calculation of v:

$$1 \in R_d, \quad 2 \in R_d, \quad 3 \in R_\sigma \qquad (2.165)$$

Then by equations (2.98), (2.98a), (2.161) and (2.164) we have

$$\sqrt{v} = \frac{2\sqrt{1.126\,932 \times 0.055\,977} + \sqrt{7.450\,419 \times 0.944\,023}}{1 + 0.055\,977 \times 1.3 \times 2} = 2.753\,612,$$

$$v = 7.582\,384 \qquad (2.166)$$

The comparison values under equations (2.100) and (2.101) then become

$$(x_d^{(1)})^2 = 0.478\,316, \quad (x_d^{(2)})^2 = 53.329\,675, \quad (x_d^{(3)})^2 < 0$$
$$(x_\sigma^{(1)})^2 = 0.751\,465, \quad (x_\sigma^{(2)})^2 = 32.845\,410, \quad (x_\sigma^{(3)})^2 = 0.000\,456, \qquad (2.167)$$

which indicates the following optimal regions:

$$1 \in R_\sigma, \quad 2 \in R_d, \quad 3 \in R_\sigma \qquad (2.168)$$

The corresponding Lagrange multiplier v as given by equations (2.98) and (2.98a) becomes

$$\sqrt{v} = \frac{\sqrt{7.450\,419 \times 0.944\,023}}{1 - (0.055\,977 - 0.055\,977)1.3 \times 2} = \sqrt{7.450\,419 \times 0.944\,023} = 2.652\,050,$$

$$v = 7.033\,367 \qquad (2.169)$$

The following comparison values are given by this new Lagrange multiplier:

$$(x_d^{(1)})^2 = 0.443\,681, \quad (x_d^{(2)}) = 49.468\,247, \quad x_d^3 < 0 \qquad (2.170)$$

Then equation (2.99) with equations (2.167) and (2.170) implies

$$x^{(1)} = 0.866\,871, \quad x^{(2)} = 7.033\,367, \quad x^{(3)} = 0.021\,361 \qquad (2.171)$$

The corresponding unscaled weight becomes

$$\Phi(2) = x^{(2)} + 2(x^{(1)} + x^{(3)}) = 8.809\,831 \qquad (2.172)$$

By equations (2.94) and (2.95) the initial relative displacements corresponding to active stress constraints are given by

$$\bar{\delta}^{*e} = \frac{L^e}{1.3} \left[1 - \frac{v \hat{f}^e f^e}{(x^e)^2} \right] \operatorname{sgn} f^e \qquad (2.173)$$

implying

$$\bar{\delta}^{*(1)} = \frac{2}{1.3} \left(1 - \frac{7.033\,367 \times 0.055\,977 \times 1.126\,932}{0.866\,871^2} \right) = 0.630\,122$$

$$\bar{\delta}^{*(2)} = \frac{1}{1.3} \left(1 - \frac{7.033\,367 \times 0.944\,023 \times 7.450\,419}{7.033\,367^2} \right) = 0$$

$$\bar{\delta}^{*(3)} = -\frac{2}{1.3} \left(1 + \frac{7.033\,367 \times 0.055\,977 \times 0.027\,769}{0.021\,361^2} \right) = -38.400\,269 \qquad (2.174)$$

Since $2 \in R_d$, a zero relative initial displacement was to be expected for $e = 2$.

(e) Third iteration

Using the x^e values in equation (2.171), by equations (2.130)–(2.134) we have for the real truss

$$U_1 = 2.420\,910, \quad U_2 = 1.057\,729, \quad f^{(1)} = 1.137\,957,$$
$$f^{(2)} = 7.439\,393, \quad f^{(3)} = -0.016\,744 \quad \sigma^{(1)} = 1.312\,718,$$
$$\sigma^{(2)} = 1.057\,729, \quad \sigma^{(3)} = -0.783\,867 \tag{2.175}$$

Since the constraint violation for the displacement constraint is 1.057 729, the scaled design variables and scaled weight become

$$x^{(1)} = 0.916\,915, \quad x^{(2)} = 7.439\,396, \quad x^{(3)} = 0.022\,594,$$
$$\Phi(2)_{\text{scaled}} = 1.057\,729\,\Phi(2) = 9.318\,413 \tag{2.176}$$

The analysis of the adjoint truss is based on equations (2.148)–(2.150), with the load v from equation (2.169), the initial displacements from equation (2.174) and the scaled design variables from equation (2.176):

$$\bar{f}^{(1)} = 0.426\,220, \quad \bar{f}^{(2)} = 6.607\,147, \quad \bar{f}^{(3)} = 0.426\,220 \tag{2.177}$$

The next step is the calculation of the Lagrange multiplier from equation (2.98), in which the virtual forces $\hat{f}^e$ can be obtained by dividing the adjoint forces $\bar{f}^e$ in equation (2.177) by the old Lagrange multiplier $v = 7.033\,367$. The regions in equation (2.168) are assumed to be still valid. As in equation (2.169), the denominator in equation (2.98) becomes unity:

$$\sqrt{v} = \sqrt{7.439\,393 \times 6.607\,147/7.033\,367} = 2.643\,590, \quad v = 6.988\,568 \tag{2.178}$$

The test on the basis of equations (2.99)–(2.101) confirms the regions in equation (2.168).

We shall not continue the above procedure by hand calculations beyond this point, but show in Table 2.5 the output from a standard DCOC program, which carries out the same operations but uses a general-purpose stiffness method for the analysis of the real and adjoint trusses. We can see that there is an almost complete agreement with the iterations calculated by hand. A small difference in the last digit in some values is due to the fact that the hand calculations were

Table 2.5 Iteration history of the three-bar truss example

Iteration number	Unscaled weight	Scaled weight	Scaled design variables		
			$x^{(1)}$	$x^{(2)}$	$x^{(3)}$
0	32	32	6.4	6.4	6.4
1	20.133 568	14.774 273	2.172 827	7.450 418	1.489 102
2	8.809 826	9.318 408	0.916 915	7.439 393	0.022 593
3	8.765 021	9.321 990	0.930 976	7.432 643	0.013 698
4	9.012 294	9.260 719	0.904 816	7.430 536	0.010 276
5	9.214 932	9.214 988	0.882 170	7.430 648	0.010 000
$\vdots$	$\vdots$	$\vdots$	$\vdots$	$\vdots$	$\vdots$
9	9.214 804	9.214 804	0.882 077	7.430 649	0.010 000

rounded up for six decimal digits whereas the computer program used double precision with a much higher accuracy.

The iteration history of the computer solution is also shown in Fig. 2.12(c). We can see that the convergence is monotonic, apart from the third iteration which shows a slight increase in the scaled weight. This is because initial relative displacements due to active stress constraints appear in that iteration for the first time in the iteration history. However, in subsequent iterations the convergence is incredibly fast, which can be seen from iteration 4 onwards in Fig. 2.12(c), in spite of scale magnifications by 10 000 and 2000 at iterations 5 and 7, respectively.

In iteration 9, the structural weight changed only from 9.214 803 724 2 to 9.214 803 723 9, thereby satisfying the convergence criterion in equation (2.82) with an extremely stringent tolerance value of $T = 10^{-10}$. Naturally, for practical applications such a high accuracy is not necessary. It can be seen from Table 2.5 that already after two iterations the weight differs only by 1% from the optimal weight.

2.5.2 Ten-bar truss with stress constraints and one displacement constraint

The ten-bar truss shown in Fig. 2.13 is a well-known example (Haftka, Gürdal and Kamat, 1990). A modified version of the ten-bar truss is considered herein, insofar as the displacement constraint is $|U_{V6}| \leqslant 5.0$ in, where U_{V6} is the vertical displacement at node 6. The purpose of this modification is to set a problem, for which the optimal solution contains an R_σ region but no $R_{\sigma s}$ region. The material properties are as follows:

$$E^e = 10^7 \text{ lbf in}^{-2}, \quad \rho^e = 0.1 \text{ lb in}^{-3}, \quad \sigma_a^e = 25\,000 \text{ lbf in}^{-2},$$
$$x^e \downarrow = 0.1 \text{ in}^2 \quad \text{(for } e = 1,...,10) \tag{2.179}$$

Since the purpose of this example is to verify the validity and accuracy of the DCOC method, a convergence tolerance value of $T = 10^{-12}$ was employed for

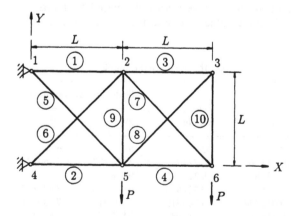

$$L = 360 \text{ in}, P = 10^5 \text{ } lb$$

Fig. 2.13 Ten-bar truss. $L = 360$ in; $P = 10^5$ lbf.

Table 2.6 Results for the ten-bar truss in section 2.5.2

	Cross-sectional area (in^2)		
e	DCOC	Dual	DOC–FSD
1	12.161 173 957	12.161 173 956	12.126 576 172
2	8.707 029 023	8.707 029 026	8.827 450 732
3	0.100 000 000	0.100 000 000	0.100 000 000
4	6.040 579 884	6.040 579 884	6.046 585 281
5	5.560 164 853	5.560 164 853	5.564 322 434
6	8.573 640 198	8.573 640 196	8.497 882 192
7	8.542 669 996	8.542 669 996	8.551 162 911
8	0.100 000 000	0.100 000 000	0.100 000 000
9	0.100 000 000	0.100 000 000	0.100 000 000
10	0.100 000 000	0.100 000 000	0.100 000 000
Number of analyses	24	18	25
Weight (lb)	2139.104 979 978 1	2139.104 979 977 9	2139.197 925 706 7

the DCOC method as well as for the DOC–FSD and dual methods used for comparisons. The results are given in Table 2.6, which shows that the optimal weight obtained by the DCOC method has 13 significant digits agreement with that of the dual method, and the design variables obtained by these two methods show an at least 9 significant digits agreement. As expected, the DOC–FSD method does not yield the same solution as the other two methods, although it is very close to them.

2.5.3 Clamped beam with rectangular cross-section of variable width

The beam shown in Fig. 2.14(a) has a rectangular cross-section with a constant depth and a variable width. Since the beam and loads are symmetric, the half beam shown in Fig. 2.14(b) can be considered. Normalized parameters are as follows:

$$\rho = 1, \quad E = 12, \quad d = 1, \quad q = 1, \quad t = 1, \quad a = 1 \qquad (2.180)$$

where ρ is the specific weight, E Young's modulus, d the constant beam depth, q the uniformly distributed load, t the permissible displacement prescribed at midspan and a the half beam span. The flexural and shear stress constraints are represented by $x^e \leqslant k_1 |M^e_{max}|$ and $x^e \leqslant k_2 |F^e_{Y\,max}|$, where M^e_{max} and $F^e_{Y\,max}$ are, respectively, the maximum moment and shear force in the eth element, and $k_1 = 0.23$ and $k_2 = 0.03$ are given constants.

The half beam is discretized into prismatic beam elements. Two models consisting of 100 and 1000 elements are considered. For the 100 element model, a convergence tolerance of $T = 10^{-8}$ was used (equation (2.82)) for all computations concerned. The optimal weight and CPU times for the DCOC, DOC–FSD and dual methods are given in Table 2.7. The CPU times given for 'analysis' include those for the analysis of the adjoint system (DCOC), analysis for the virtual load

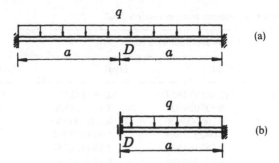

Fig. 2.14 Clamped beam example.

Table 2.7 Results for the clamped beam example

	100 element model		
	DCOC	DOC–FSD	Dual
Optimal weight	0.064 202 389	0.064 203 018	0.064 202 389
Number of analyses	29	28	9
CPU times (s)			
Optimization	7.84	3.63	535.28
Analysis	116.95	112.70	117.41
Total	124.79	116.33	706.69
	1000 element model		
	DCOC	DOC–FSD	
Optimal weight	0.063 996 543	0.063 999 993	
Number of Analyses	18	20	
CPU times (s)			
Optimization	57.35	34.98	
Analysis	928.59	1126.41	
Total	985.84	1161.39	

system (DOC–FSD) and sensitivity analysis (dual method). The optimum weight obtained by the DCOC method has an eight significant digits agreement with that of the dual method, and the optimal design variables for these two methods have shown an at least five significant digits agreement (Zhou, 1992). Again, the solution of the DOC–FSD method results in a higher weight than the other two methods. The CPU time needed for the optimization phase of the dual method is much more than that for the DCOC method, since 76 stress constraints and 1 deflection constraint are active at the optimum. It can be seen that a much larger number of analyses were needed for the DCOC and DOC–FSD methods. This is because, in those two methods, the resizing rule for the members with active stress constraints is based on forces given by the prior computation step

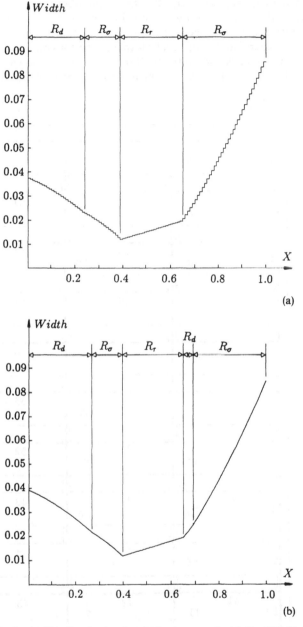

Fig. 2.15 Optimal width distribution for the beam example: (a) $E = 100$; (b) $E = 1000$.

and the dependence of the forces on design variables is not considered. For this reason, the treatment of stress constraints in the resizing procedure of the DCOC and DOC–FSD methods can be regarded as a zero-order approximation. On the other hand, the dual method is based on first-order approximation which achieves a faster convergence. Table 2.7 shows that about the same analysis time was needed for the dual method with only 9 iterations as for the DCOC method (with 29 iterations). This is because the sensitivity analysis needed by the dual method is very expensive. The optimal width distributions obtained by the DCOC method for both models with 100 and 1000 elements are shown in Fig. 2.15.

For the 1000 element model, for which a convergence tolerance of $T = 10^{-6}$

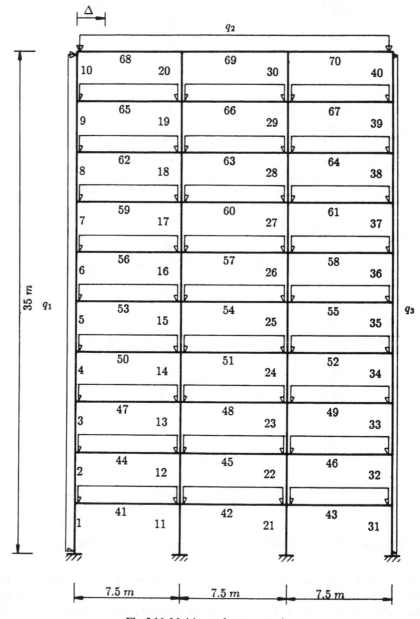

Fig. 2.16 Multistory frame example.

was used, the results of the DCOC and DOC–FSD methods are also given in Table 2.7. The displacement constraint was active in the solutions by both methods and 730 stress constraints were active in the solution by the DCOC method. Table 2.7 shows that for both DCOC and DOC–FSD methods the CPU time needed for the optimization phase is a fraction of that needed for the analysis phase. The 1000 element model was not solved by the dual method because, as a result of over 700 active constraints, the computer time would have been prohibitively high.

2.5.4 Ten-storey, three-bay frame

Figure 2.16 shows the geometry and loading for the considered frame, which has 70 elements with rectangular cross sections of variable width. The material properties are as follows:

$$E = 2.1 \times 10^7 \, \text{kN m}^{-2}, \quad \sigma_a = 3.0 \times 10^5 \, \text{kN m}^{-2}, \quad \tau_a = 5.0 \times 10^4 \, \text{kN m}^{-2} \quad (2.181)$$

where E is Young's modulus, σ_a the permissible flexural stress and τ_a the permissible shear stress. A horizontal displacement at the left top corner of the frame is constrained to a value of $U_H \leq 0.3 \, \text{m}$. The given constant depth is $d = 0.3 \, \text{m}$ for all elements. The values of the distributed loads in Fig. 2.16 are:

$$q_1 = 10.0 \, \text{kN m}^{-1}, \quad q_2 = 20.0 \, \text{kN m}^{-1}, \quad q_3 = 5.0 \, \text{kN m}^{-1} \quad (2.182)$$

With the initial design variables of $x^e = 0.6 \, \text{m}$ for all elements, a stress constraint violation of 47.6% occurs. A convergence tolerance value of $T = 10^{-4}$ was used. The results of the DCOC method are compared with those of the DOC–FSD and dual methods in Table 2.8. The reason for small constraint violations in the results of the DCOC and DOC–FSD methods is the zero-order treatment of stress constraints discussed in section 2.5.3. The optimal width distribution is shown in Fig. 2.17. For detailed results the reader is referred to Zhou's dissertation (Zhou, 1992).

2.5.5 Ten-bar truss with stress constraints and two displacement constraints

The ten-bar truss considered in section 2.5.2 is used here again as a test example for problems with multiple displacement constraints. The displacement con-

Table 2.8 Results for ten-story and three-bay frame

	DCOC	DOC–FSD	Dual
Optimal volume (m^3)	53.51	53.57	53.50
Constraint violation	0.19%	0.74%	0.00%
Number of analyses	13	8	12
CPU times (s)			
Optimization	2.63	1.20	108.55
Analysis	49.61	34.67	197.20
Total	52.24	35.87	305.75

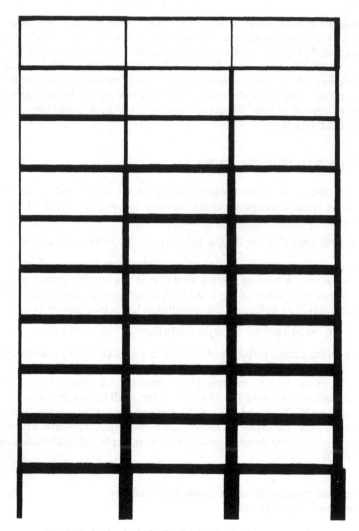

Fig. 2.17 Optimal width distribution in the frame example.

straints are $|U_{H6}| \leqslant 1.0$ in and $|U_{V6}| \leqslant 5.0$ in, where U_{H6} and U_{V6} are, respectively, horizontal and vertical displacement of node 6. The other conditions are the same as described in section 2.5.2. It was shown by the dual method that both displacement constraints and the stress constraint at the 5th bar are active in the optimal design. The results from the DCOC, dual and DOC–FSD methods are indicated in Table 2.9, for which a convergence tolerance value of $T = 10^{-12}$ was used.

2.5.6 Ten-bar truss with active stress and side constraints for one member

The technique of upgrading stress constraints into global constraints is illustrated in this subsection by another version of the ten-bar truss considered in section

Table 2.9 Results for the ten-bar truss in section 2.5.5

e	Cross-sectional area (in²)		
	DCOC	Dual	DOC–FSD
1	10.827 889 1	10.827 889 1	10.836 227 6
2	12.295 024 3	12.295 024 3	12.331 033 0
3	0.100 000 0	0.100 000 0	0.100 000 0
4	8.602 843 0	8.602 843 0	8.569 172 0
5	5.641 706 0	5.641 706 0	5.643 366 7
6	7.619 254 7	7.619 254 7	7.567 562 9
7	7.605 251 3	7.605 251 3	7.648 485 4
8	0.100 000 0	0.100 000 0	0.100 000 0
9	0.100 000 0	0.100 000 0	0.100 000 0
10	0.100 000 0	0.100 000 0	0.100 000 0
Number of analyses	23	19	29
Weight (lb)	2220.352 475 375	2220.352 475 375	2220.390 773 364

2.5.2. The single displacement constraint is changed to $|U_{V6}| \leqslant 4.0$ in where U_{V6} represents the vertical displacement of node 6. Other design conditions are exactly the same as those given in section 2.5.2. As a result of this modification, the optimal solution contains both R_σ and $R_{\sigma s}$ type regions. It was found by using the dual method that the displacement constraint as well as the stress constraints for the 5th and 9th elements are active and the cross-sectional area of the 9th bar takes the lower side value. Therefore, the DCOC method including the technique of upgrading stress constraints in the $R_{\sigma s}$ region must be applied.

A convergence tolerance value of $T = 10^{-10}$ is used. The optimum results of the DCOC, DOC–FSD and dual methods are given in Table 2.10, which shows

Table 2.10 Results for the ten-bar truss in section 2.5.6

e	Cross-sectional area (in²)		
	DCOC	Dual	DOC–FSD
1	14.973 877 3	14.973 877 3	15.550 021 6
2	11.717 729 4	11.717 729 4	11.222 496 6
3	0.100 000 0	0.100 000 0	0.100 000 0
4	7.700 225 6	7.700 225 6	7.717 639 3
5	5.531 657 0	5.531 657 0	5.083 493 4
6	10.188 680 2	10.188 680 2	10.966 889 5
7	10.889 763 5	10.889 763 5	10.914 390 2
8	0.100 000 0	0.100 000 0	0.100 000 0
9	0.100 000 0	0.100 000 0	0.416 827 5
10	0.100 000 0	0.100 000 0	0.100 000 0
Number of analyses	18	16	191
Weight (lb)	2608.762 283 67	2608.762 283 67	2641.764 762 99

that the optimum design obtained by the DCOC method shows a 12 digit agreement with that by the dual method. The displacement constraint, stress constraints for the 5th and 9th bars and the side constraint of the 9th bar are active in the solutions of the DCOC and dual methods. The DOC–FSD method gives a significantly different solution in which the side constraint of the 9th element is not active. A much larger number of analyses is needed by the DOC–FSD method for the considered example.

Note
The number of iterations used in most test examples of this chapter is relatively high owing to the unusually stringent convergence criteria used, resulting in an eight to twelve digits agreement between weight values of the DCOC and dual methods. For an accuracy required in practical problems, a much smaller number of iterations is sufficient (e.g. the multistorey frame example in section 2.5.4).

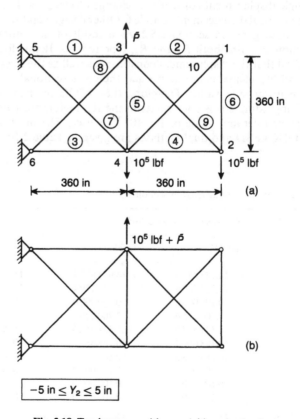

Fig. 2.18 Ten-bar truss with a variable point load.

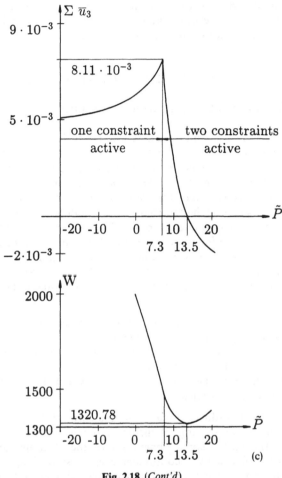

Fig. 2.18 (*Cont'd*)

2.5.7 Ten-bar truss with a variable point load

In order to verify some recent extensions of DCOC introduced in section 2.3.4, we consider a ten-bar truss with the two load conditions given in Figs. 2.18(a) and 2.18(b), including a variable vertical point load $\tilde{P}$ at the joint 3. The material properties are $E^e = 10^7$ lbf in^{-2}, $\rho^e = 0.1$ lb in^{-3} and the minimum cross-sectional area $x^e \downarrow = 0.1$ in^2 for all members. The vertical displacement at joint 6 is restricted in both directions to ± 5 in. It is assumed that the variable load has a zero cost, $\psi \equiv 0$. Then it follows from equation (2.64) that the sum of the vertical adjoint displacements at joint 3 must take on a zero value for optimality:

$$\sum_{l=1}^{2} \bar{U}_{3vl} = 0$$

Figures 2.18(c) and 2.18(d) show, respectively, the variation of $\sum \bar{U}_3$ and the structural weight W (which is optimal within a given P value) in dependence of the variable load $\tilde{P}$. It can be seen that, indeed, the minimum weight corresponds to the fulfillment of the above optimality condition.

2.6 A BRIEF REVIEW OF THE COC ALGORITHM

2.6.1 Continuum formulation in structural mechanics

In sections 2.1–2.5 we used the so-called **discretized formulation**, in which the structure is divided into finite elements and then at all points of the structure stresses, strains and displacements can be derived from the displacements at certain points of the elements termed 'nodes'. In the so-called **continuum formulation**, we represent the translational and rotational displacements $\mathbf{u} = [u_1, \ldots, u_h, \ldots, u_H]$ along the centroidal axis or middle surface (the brackets { } are omitted for vectors in sections 2.6 and 2.7) of a one- or two-dimensional structural component as a function of the spatial coordinates ξ, i.e $\mathbf{u} = \mathbf{u}(\xi)$. Then stresses and strains at any arbitrary point of a cross section can be derived from given displacement functions $\mathbf{u}(\xi)$. It follows from the above fundamental simplification that, in the continuum approach to structural mechanics, structures may be treated as

- one-dimensional continua (e.g. bars, beams, arches, rings, frames, trusses, beam-grids or grillages, shell-grids, cable nets),
- two-dimensional continua (plates, disks, structures subject to plane stress or plane strain, shells, folded plates, etc.) or
- three-dimensional continua (stress systems for which the above simplifying idealizations are not possible).

Special cases of two-dimensional continua are grid-type or perforated structures in which the spacing of members or the microstructure is small compared with the macroscopic dimensions of the structure and hence the structure can be replaced by a continuum using a smoothing-out process or homogenization. In the case of grid-type structures, Prager and the first author (e.g. Prager and Rozvany, 1977) used terms such as 'truss-like continua' and 'grillage-like continua' for such homogenized structures.

In sections 2.6 and 2.7, we use basically Prager's notation and terminology, but some of the symbols shall be made compatible with the notation of the numerical school. Stress resultants (or, in Prager's terminology, generalized stresses) at a given cross section shall be denoted by $\boldsymbol{\sigma} = [\sigma_1, \ldots, \sigma_j, \ldots, \sigma_J]^T$. Examples of such stress resultants are bending moments, shear forces, etc. For three-dimensional continua, $\boldsymbol{\sigma}$ denotes the local stress components (σ_x, τ_{xy}, etc.). In the considered approach to structural mechanics we can use cross-sectional strains or generalized strains $\boldsymbol{\varepsilon} = [\varepsilon_1, \ldots, \varepsilon_j, \ldots, \varepsilon_J]^T$ which refer to spatial derivatives of displacements $\mathbf{u}$ of the centroidal axis or middle surface. For example, if the deflection of a beam is given by $u_1(\xi)$, where ξ is the distance along the beam, then the curvature $\varepsilon_1 = \kappa$ of the same beam is

$$\kappa = -\frac{d^2 u_1}{d\xi^2} \tag{2.183}$$

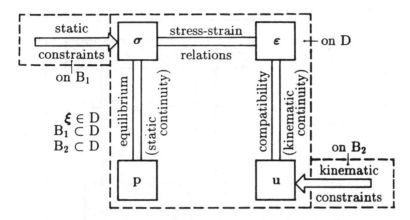

Fig. 2.19 Fundamental relations of structural mechanics.

The term 'cross-sectional strain' is justified because κ denotes the rate of spatial change per unit length of the angular rotation of the cross-sections (which are assumed to remain plane). The loads at a given cross-section ξ may have three force and three moment components and are denoted by $\mathbf{p} = [p_1, \ldots, p_h, \ldots, p_H]^T$. The union of all admissible values of the spatial coordinates is termed the structural domain D, with $\xi \in D$.

The fundamental relations of structural mechanics are summarized in Fig. 2.19. On the structural domain D, it is necessary to satisfy static continuity (equilibrium) conditions $(\mathbf{p}, \boldsymbol{\sigma})$, kinematic continuity (compatibility) conditions $(\mathbf{u}, \boldsymbol{\varepsilon})$ and generalized stress–strain relations $(\boldsymbol{\sigma}, \boldsymbol{\varepsilon})$. On certain subsets of the structural domain, $B_1 \subset D$, $B_2 \subset D$, respectively, static and kinematic constraints (or 'boundary conditions') must be fulfilled. As a simple illustrative example, we consider a Bernoulli cantilever beam. In that case, the generalized stress is the bending moment $\sigma_1 \to M$, the generalized strain is the curvature $\varepsilon_1 \to \kappa$, the load is the transverse load $p_1 \to p$ and the displacement the transverse displacement $u_1 \to u$. The fundamental relations listed above are

$$d^2 M / d\xi^2 = -p, \quad d^2 u / d\xi^2 = -\kappa, \quad EI\kappa = M \tag{2.184}$$

where E is Young's modulus and I is the moment of inertia. The boundary conditions at the fixed and free ends (B_2, B_1) are, respectively,

$$u = du/d\xi = 0, \quad M = dM/d\xi = 0 \tag{2.185}$$

if the free end has no concentrated load or moment acting on it.

In the remaining sections it is assumed that the cross-sectional geometry is partially defined so that a finite number of design variables $\mathbf{x} = [x_1 \cdots x_i \cdots x_I]^T$ fully determine the cross section. Examples of such design variables, which are functions of the spatial coordinates ξ, are given in Fig. 2.20.

The specific cost (cost per unit length, area or volume) ψ at a cross section ξ depends on the design variables

$$\psi = \psi[\mathbf{x}(\xi)] \tag{2.186}$$

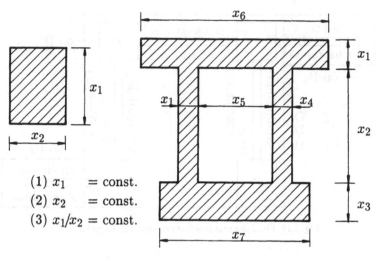

Fig. 2.20 Examples of design variables representing cross-sectional dimensions.

The total cost or objective functional is then given by

$$\Phi = \int_D \psi[\mathbf{x}(\xi)]\,d\xi \tag{2.187}$$

Considering a beam of rectangular cross section having width x_1 and depth x_2 (Fig. 2.20(a)), the objective functional being the total beam weight, the specific cost function becomes

$$\psi = \rho x_1 x_2 \tag{2.188}$$

where ρ is the specific weight of the beam material.

The generalized stress–strain relations for linearly elastic structures can be represented as

$$\mathbf{\varepsilon} = [\mathbf{d}]\mathbf{\sigma}, \tag{2.189}$$

where $[\mathbf{d}]$ denotes the generalized flexibility matrix.

Considering, for example, a plate in plane stress, the strain–stress relation in equation (2.189) takes the form

$$\begin{Bmatrix} \varepsilon_x \\ \varepsilon_y \\ \gamma_{xy} \end{Bmatrix} = \frac{1}{x_1 E} \begin{bmatrix} 1 & -\mu & 0 \\ -\mu & 1 & 0 \\ 0 & 0 & 2(1+\mu) \end{bmatrix} \begin{Bmatrix} N_x \\ N_y \\ N_{xy} \end{Bmatrix} \tag{2.190}$$

where N_x, N_y and N_{xy} are the stress resultants for the x and y directions, ε_x, ε_y and γ_{xy} are the usual strain components (which can also refer to an entire cross section), x_1 is the design variable representing the plate thickness and μ is Poisson's ratio.

2.6.2 Historical background

As mentioned already in section 2.2, discretized optimality criteria (DOC) methods were developed by a group of aerospace scientists (Berke, 1970; Gellatly and Berke, 1971; Venkayya, Khot and Berke, 1973; Berke and Khot, 1974; for reviews see Berke and Khot, 1987, 1988; Khot and Berke, 1984; Haftka, Gürdal and Kamat, 1990), whereas continuum-type optimality criteria (COC) methods were introduced mostly by the anaytical school of structural optimization around W. Prager. The fundamental ideas behind the COC approach first appeared in the literature, in the context of least-weight trusses, around the turn of the century in a paper by the versatile Australian inventor A.G.M. Michell (Michell, 1904), whose interests ranged from machine design (crankless engine, thrust bearing) to hydraulics and lubrication as well as structural mechanics. Michell's optimality condition for least-weight trusses with a single load condition and a constraint on the stresses consisted of the following requirement on strains ε

$$\varepsilon^K = c \operatorname{sgn} N^S \ (\text{for } N > 0), \quad |\varepsilon^K| \leqslant c \ (\text{for } N = 0) \tag{2.191}$$

where the superscripts K and S, respectively, denote kinematic and static admissibility, c is a given constant and N is the axial force along a line element of the available space. An interesting feature of the above optimality criteria, which will have a particular significance in layout optimization (Chapter 10), is that the strains are also restricted along vanishing members (of zero cross-sectional area).

Another important early development was the uniform energy dissipation principle by Drucker and Shield (1956), which states that for a simple class of cost functions expressed in terms of the generalized stresses σ and having the property $\psi(\lambda\sigma) = \lambda\psi(\sigma)$, a plastically designed structure takes on a minimum cost ψ if

$$\text{on } D, \varepsilon\sigma/\psi = \text{constant} \tag{2.192}$$

Considering the optimal plastic design of beams and frames with segment-wise constant cross sections, Foulkes (1954) obtained the following optimality condition:

$$\left(\sum_{D_i} |\theta|\right) \bigg/ L_i = \text{constant} \tag{2.193}$$

where $D_i = (i = 1, \ldots, n)$ are segments having the length L_i and θ denotes plastic hinge rotations.

For beams and frames of freely varying cross section, with a specific cost function $\psi = c|M|$, where c is a given constant and M is the bending moment, Heyman (1959) introduced the optimality condition

$$\kappa^K = c \operatorname{sgn} M^S \ (\text{for } M > 0), \quad |\kappa^K| \leqslant c \ (\text{for } M = 0) \tag{2.194}$$

where c is a given constant, the superscripts S and K denote static and kinematic admissibility and κ are curvature rates. A comparison of equations (2.191) and (2.194) shows that Heyman, in effect, extended Michell's optimality criteria from trusses to frames.

A general optimality condition for the plastic design of structures was introduced by Prager and Shield (1967), who extended a more restricted optimality criterion by Marcal and Prager (1964). The same optimality condition can be derived from earlier criteria by Mroz (1963). In the Prager–Shield condition, the specific cost is expressed in terms of the generalized stresses $\psi(\sigma)$,

$$\bar{\varepsilon}^K = \text{grad } \psi(\sigma^S) \tag{2.195}$$

where $\bar{\varepsilon}$ denotes a fictitious strain field now termed as adjoint strain field, but is also one possible velocity field at the collapse of the plastically designed structure. It was shown by the first author (e.g. Rozvany, 1989, p. 42) that for slope discontinuities of $\psi(\sigma)$, the gradient in equation (2.195) is replaced by the subgradient, representing a convex combination of the gradient values for the adjacent stress regimes.

The first continuum-type optimality criterion for a given elastic deflection was derived for beams of varying width by Barnett (1961) in the form

$$\psi(\xi) = c(M\bar{M})^{1/2} \tag{2.196}$$

where ψ is the cross-sectional area, c is a constant, M is the bending moment for the external load and $\bar{M}$ is the bending moment caused by a virtual load (e.g. unit dummy load). The same optimality condition was expressed in terms of constant mutual complementary energy by Shield and Prager (1970) and Huang (1971). For sandwich beams with a given compliance (total external work) and for sandwich columns of given Euler buckling load, the following optimality condition was derived by Prager and Taylor (1968):

$$|\kappa^K| = \text{constant} \tag{2.197}$$

Optimality criteria for eigenvalue problems have been used extensively by Olhoff (1981).

A comprehensive treatment of optimality criteria for a variety of design conditions in both elastic and plastic design is given in a recent book by the first author (Rozvany, 1989). The same text includes a large number of analytically solved illustrative examples and a bibliography with some 1200 references. An original contribution in the above book is a general treatment of linearly elastic structures with stess and displacement constraints, for which continuum-type optimality criteria (COC) are reproduced here in Fig. 2.21. It will be seen that the conceptual schemes of DCOC in Figs. 2.2–2.5 represent discretized equivalents of the COC approach. The optimality criteria in Fig. 2.21 are based on stress constraints of the type

$$S_r[\mathbf{x}(\xi), \sigma(\xi)] \leqslant 0 \text{ (for all } \xi \in D) \tag{2.198}$$

and displacement constraints

$$\int_D \bar{\sigma}_k^T[\mathbf{d}]\sigma \, d\xi \leqslant t_k \tag{2.199}$$

which are expressed in terms of mutual work involving the real stresses $\sigma(\xi)$ and the virtual generalized stresses $\bar{\sigma}_k(\xi)$ equilibrating the virtual loads $\bar{\mathbf{P}}_k(\xi)$ associated with the kth displacement condition. The virtual load consists of a unit (dummy) load for a constrained deflection at one point and of the weighting factors if the constraint limits a weighted combination of several displacements.

As exemplified by Fig. 2.21, an intrinsic feature of the COC formulation is that the optimality criteria are expressed in terms of a fictitious structure termed

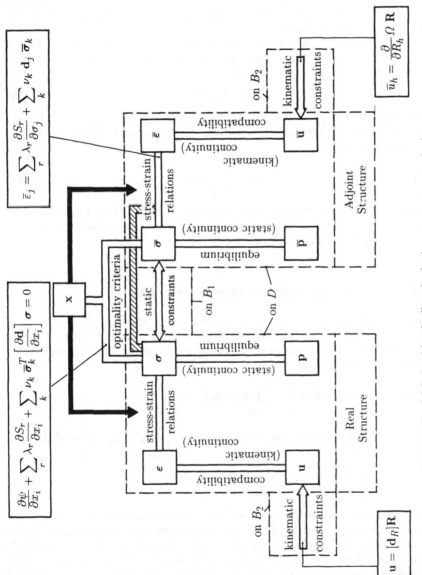

Fig. 2.21 The COC algorithm for linearly elastic systems: conceptual scheme.

the adjoint structure (Rozvany and Zhou, 1993). The equilibrium and compatibility conditions for the real and adjoint structures are usually identical, as is one part $(\bar{\varepsilon}_j = \sum_k v_k \mathbf{d}_j \bar{\sigma}_k)$ of the generalized strain–stress relations, where $\mathbf{d}_j$ denotes the jth row of the generalized flexibility matrix $[\mathbf{d}]$. However, if stress constraints are active for the considered cross section v, then an additional strain component, which depends on the real generalized stresses $[\sum_r \lambda_r(\partial S_r/\partial \sigma_j)]$ also influences the adjoint strains.

The kinematic boundary conditions are the same for both real and adjoint structures if the supports are rigid and costless. In the case of elastic supports, however, the real support displacements are

$$\mathbf{u} = [\mathbf{d}_R]\mathbf{R} \qquad (2.200)$$

where $[\mathbf{d}_R]$ is the support flexibility matrix and $\mathbf{R} = (R_1, \ldots, R_h, \ldots, R_H)$ are the reactions. Moreover, in the case of allowance for support costs the adjoint support displacements become

$$u_h = \frac{\partial}{\partial R_h}\Omega(\mathbf{R}) \qquad (2.201)$$

where $\Omega(\mathbf{R})$ is the support cost function. Finally, the design variables (cross-sectional dimensions) $\mathbf{x}$ are related to the real and adjoint generalized stresses through the optimality criteria.

2.6.3 Derivation of the optimality criteria in Fig. 2.21

Using the Lagrangians $\bar{\theta}, \theta_k(k = 1, \ldots, K), v^k(k = 1, \ldots, K)$ and $\lambda_r(\xi)$ $(r = 1, \ldots, R)$, and the nonnegative slack functions $\alpha_r(\xi)$ $(r = 1, \ldots, R)$ and slack variables β_k $(k = 1, \ldots, K)$, the augmented (Lagrangian) functional for the considered problem can be stated as

$$\Phi^L = \int_D \psi[\mathbf{x}(\xi)]\,\mathrm{d}\xi + \sum_k v_k \left(\int_D \bar{\sigma}_k^T[\mathbf{d}]\sigma\,\mathrm{d}\xi - t_k + \beta_k \right)$$
$$+ \int_D \lambda_r(\xi)\{S_r[\mathbf{x}(\xi), \sigma(\xi)] + \alpha_r(\xi)\}\,\mathrm{d}\xi + \bar{\theta}\left[\int_D \mathbf{p}^T\bar{\mathbf{u}}^K\,\mathrm{d}\xi - \int_D \sigma^T\bar{\varepsilon}^K\,\mathrm{d}\xi \right]$$
$$+ \sum_k \theta_k \left[\int_D \bar{\mathbf{p}}_k^T\mathbf{u}_k^K\,\mathrm{d}\xi - \int_D \bar{\sigma}_k^T\varepsilon_k^K\,\mathrm{d}\xi \right] \qquad (2.202)$$

In equation (2.202), statical admissibility of the real loads and generalized stresses $(\mathbf{p}, \sigma)$ and of the adjoint loads and stresses $(\bar{\mathbf{p}}, \bar{\sigma})$ is expressed via the principle of virtual displacements, in which at this stage $(\bar{\mathbf{u}}^K, \bar{\varepsilon}^K)$ and $(\mathbf{u}_k^K, \varepsilon_k^K)$ represent any kinematically admissible small displacement system which is not necessarily related to the real and adjoint loads $(\mathbf{p}, \bar{\mathbf{p}})$. This is a relaxed form of the considered optimization problem, because in the original problem the real strains and displacements $(\varepsilon, \mathbf{u})$ must also fulfill kinematic admissibility.

The problem in equation (2.202) is a so-called variational problem with unknown functions which are to be optimized. Necessary conditions of optimality are provided by the Euler–Lagrange equations, which for variation of the design variables $x_i(i = 1, \ldots, I)$ and real generalized stresses $\sigma_j(j = 1, \ldots, J)$ yield, respectively, the optimality criteria and adjoint stress–strain relations at the top of Fig. 2.21. It can be seen from the latter that, although in displacement calculations the virtual forces and stresses need to satisfy only statical admissibility, in the COC algorithm they must also fulfill kinematic admissibility via the above adjoint stress–strain relations. Finally, Euler–Lagrange equations for variation of $\bar{\sigma}_{kj}$ take the form

$$\theta_k \varepsilon_{kj}^K = v_k \mathbf{d}_j \boldsymbol{\sigma} \tag{2.203}$$

where $\mathbf{d}_j$ is the jth row of the generalized flexibility matrix $[\mathbf{d}]$. This means that ε_{kj}^K, which until now represented any kinematically admissible strain field, by means of this optimality condition has become equal to elastic strains (factored by v_k/θ_k) caused by the statically admissible real stress field $\boldsymbol{\sigma}$.

This shows that kinematic admissibility, which was relaxed in the original formulation in equation (2.202), is now satisfied through the optimality criterion in equation (2.203).

In most practical problems, it is not possible to find a closed-form analytical solution satisfying simultaneously all the relations shown in Fig. 2.21. It is therefore necessary to adopt an iterative procedure consisting of the same main steps as DCOC:

- analysis of the real and adjoint structures;
- updating the design variables using optimality criteria.

2.7 APPLICATIONS OF THE ITERATIVE COC ALGORITHM

2.7.1 Classes of problems solved by COC

The COC algorithm was tested on a number of test examples (e.g. Rozvany *et al.*, 1989; Rozvany, Zhou and Gollub, 1990; Zhou and Rozvany, 1993) which included the following types of problems:

- prescribed lower and upper bounds on the cross-sectional dimensions,
- segment-wise constant cross sections,
- allowance for the cost of supports,
- allowance for selfweight,
- nonlinear and nonseparable specific cost and flexibility functions,
- two-dimensional structural systems (plates).

For example in the case of allowance for selfweight, it is sufficient to multiply the RHS of the adjoint stress–strain relation (top right corner of Fig. 2.21) by $1 + \bar{u}$, a surprisingly simple extension.

In this section we shall illustrate the iterative COC algorithm with only one test example.

2.7.2 Beam of independently variable width x_1 and depth x_2 with bending in both horizontal and vertical directions

(a) Optimality criteria

For the above class of problems we have

$$\sigma = (M_1, M_2), \quad \bar{\sigma} = (\bar{M}_1, \bar{M}_2), \quad [\mathbf{d}] = \begin{bmatrix} \dfrac{1}{rx_1^3 x_2} & 0 \\ 0 & \dfrac{1}{rx_1 x_2^3} \end{bmatrix}, \quad \psi = cx_1 x_2, \quad r = E/12 \tag{2.204}$$

where M_1 and M_2 are the real bending moments in the horizontal and vertical directions, $\bar{M}_1$ and $\bar{M}_2$ are the corresponding adjoint moments, $[\mathbf{d}]$ is the specific generalized matrix, $c = \rho$ is the specific weight and E is Young's modulus for the beam material. We consider a single deflection constraint which limits the sum of the horizontal and vertical displacements

$$t \geqslant \int_D \left(\frac{M_1 \bar{M}_1}{rx_1^3 x_2} + \frac{M_2 \bar{M}_2}{rx_1 x_2^3} \right) d\xi \tag{2.205}$$

and the cross-sectional dimensions are constrained from below

$$x_1 \geqslant x_1 \downarrow, \quad x_2 \geqslant x_2 \downarrow \tag{2.206}$$

For the above problem, the following optimality conditions have been derived from both the general formulas in Fig. 2.21 (with $\lambda_r = 0$ for all r and $v_1 \to v$) and by a variational derivation (Rozvany, Zhou and Gollub, 1990).

● Case A: $x_1 > x_1 \downarrow$, $x_2 > x_2 \downarrow$.

$$x_1^6 = 4v \frac{(M_1 \bar{M}_1)^2}{M_2 \bar{M}_2}, \quad x_2^6 = 4v \frac{(M_2 \bar{M}_2)^2}{M_1 \bar{M}_1} \tag{2.207}$$

The conditions for the validity of case A and equation (2.207) are

$$M_2 \bar{M}_2 > 0, \quad M_1 \bar{M}_1 > 0, \quad x_1^6 \downarrow < 4v \frac{(M_1 \bar{M}_1)^2}{M_2 \bar{M}_2}, \quad x_2^6 \downarrow < \frac{(M_2 \bar{M}_2)^2}{M_1 \bar{M}_1} \tag{2.208}$$

● Cases B and C: $x_i = x_i \downarrow$, $x_k > x_k \downarrow$ ($i = 1$, $k = 2$ and $i = 2$, $k = 1$, respectively).

$$x_k^2 = \frac{v M_i \bar{M}_i}{2x_i^4 \downarrow} + \left[\left(\frac{v M_i \bar{M}_i}{2x_i^4 \downarrow} \right)^2 + \frac{3v M_k \bar{M}_k}{x_i^2 \downarrow} \right]^{1/2} \tag{2.209}$$

The conditions for the validity of case B or C and equation (2.209) are

$$x_i^4 \downarrow x_k^4 > v(3M_i \bar{M}_i x_k^2 + M_k \bar{M}_k x_{ia}^2), \quad M_k \bar{M}_k > 0 \tag{2.210}$$

● Case D: $x_1 = x_1 \downarrow$, $x_2 = x_2 \downarrow$. We have the following conditions for this case:

$$x_1^4 \downarrow x_2^4 \geqslant v(3M_2 \bar{M}_2 x_1^2 \downarrow + M_1 \bar{M}_1 x_2^2 \downarrow), \quad x_1^4 \downarrow x_2^4 \downarrow \geqslant v(M_2 \bar{M}_2 x_1^2 \downarrow + 3M_1 \bar{M}_1 x_2^2 \downarrow) \tag{2.211}$$

The calculation of the Lagrangian from the deflection condition in equation (2.205) requires a Newton–Raphson procedure.

(b) Solutions by the iterative COC and SQP methods

Considering a clamped beam having a span of 2.0 and a uniformly distributed load of unit itensity in the vertical direction and a central unit point load in the horizontal direction (Fig. 2.22), a deflection value of $\Delta = 3$ in equation (2.205) and $c = r = 1$ were adopted.

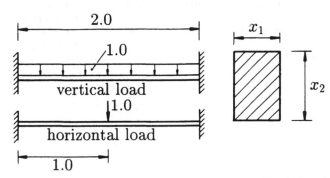

Fig. 2.22 Test example for the COC algorithm: beam of variable width and depth.

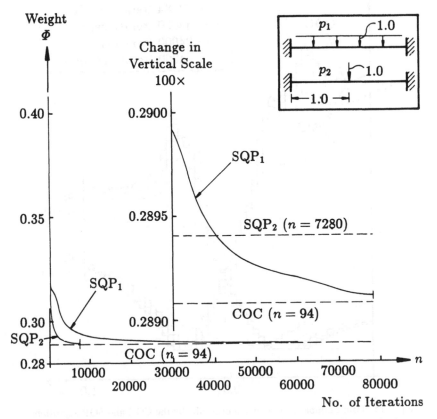

Fig. 2.23 Beam example: a comparison of iteration histories by various methods.

Table 2.11 A comparison of weight values obtained by the SQP and COC methods after convergence and percentage differences

	Weight Φ	$\Delta\%$	Number of iterations	CPU time (s)
COC	0.289 081	0	94	57
SQP_1	0.289 114	0.011	77 564	1 507 451
				(17.45 days)
SQP_2	0.289 408	0.113	7280	176 746
				(2.05 days)

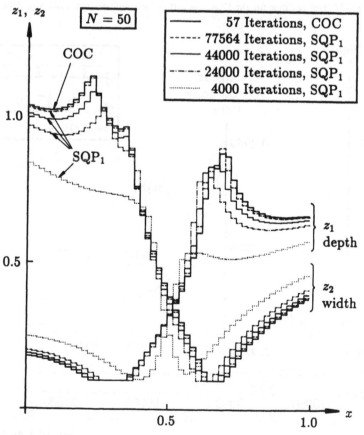

Fig. 2.24 Beam example: a comparison of results by the COC and SQP algorithms.

Using the COC method with only 50 elements and 100 variables (for a comparison with the SQP method), and a tolerance criterion of $(\Phi_{new} - \Phi_{old})/\Phi_{new} \leqslant 10^{-8}$, a weight value of $\Phi = 0.289\,081$ was obtained after 94 iterations, with a total CPU time (analysis plus optimization) of 57 s. On repeating the COC procedure with 10 000 elements and 20 000 variables, the above convergence criterion was reached after 103 iterations with a CPU time of 12 096 s and a weight of $\Phi = 0.288\,782$.

The same problem with 50 elements and 100 variables was also computed using the SQP method with reciprocal variables. The above convergence criterion was satisfied after 77 564 iterations, requiring a CPU time of 1 507 451 s (17.45 days). In Fig. 2.23, the iteration history of this calculation is compared with the result given by the COC method after convergence.

Since the convergence was extremely slow, a modified procedure was used in which the coupling effect was relaxed through multiplying the mixed second derivatives by various factors. After testing several such factors, a value of 0.2 was adopted, which resulted in a much faster convergence (SQP$_2$ in Fig. 2.23).

The SQP$_2$ procedure reached the convergence criterion after 7280 iterations, requiring a CPU time of 'only' 176 746 s (2.05 days). However, because of small instabilities in the convergence, satisfaction of the convergence criterion was somewhat accidental giving a fairly large error in comparison to the COC and SQP$_1$ results (enlarged right-hand part of Fig. 2.22). The results of the COC, as well as the unmodified and modified SQP methods, are compared in Table 2.11. The width and depth distributions given by the two SQP methods after various numbers of iterations are compared with those obtained by the COC method in Fig. 2.24. It can be concluded from these diagrams and Table 2.11 that the SQP method fully confirms the results by the COC algorithm. However, the more accurate SQP method (SQP$_1$) requires almost 30 000 times as much time as the COC procedure.

2.8 CONCLUDING REMARKS

A new discretized optimality criteria method (DCOC) was discussed in detail in this chapter, together with a brief review of the corresponding continuum-type optimality criteria (COC) method. It was shown that DCOC increases our optimization capability by several orders of magnitude if a large number of stress constraints are active, which is usual in structural design, and hence this method should become an indispensable tool in the optimization of very large structural systems. Applications of the DCOC and COC algorithms in topology optimization will be discussed in Chapter 10.

ACKNOWLEDGEMENTS

The authors are greatly indebted to the Deutsche Forschungsgemeinschaft for financial support (Projects Ro 744/1-1 and Ro 744/1-2), to Sabine Liebermann for processing the text, preparing some diagrams and checking algebraically part of the equations, to Peter Moche for the remaining diagrams, and to Susann Rozvany for editing and correcting the text.

APPENDIX 2A   NOTATION

2A.1   Notation – discretized systems

A	cross-sectional area
b	given width of cross section
$[\mathbf{B}], [\mathbf{B}]^{\mathrm{T}}$	statics and kinematics matrices
c^e	cost constant for element e
$[\mathbf{D}], [\mathbf{d}^e]$	flexibility matrix
$[\underline{\mathbf{D}}], [\underline{\mathbf{d}}^n]$	reaction element (support) flexibility matrix
$e = 1, \ldots, E$	superscript identifying elements
$[\mathbf{d}_o^e]$	constant part of element flexibility matrix
$\{\mathbf{F}\}, \{\mathbf{f}^e\}$	nodal forces (stiffness method)
$\{\mathbb{F}\}, \{\mathbb{f}^e\}$	nodal forces (flexibility method)
$\{\hat{\mathbb{F}}\}, \{\hat{\mathbb{f}}^e\}$	virtual nodal forces (flexibility method)
$\{\bar{\mathbb{F}}\}, \{\bar{\mathbb{f}}^e\}$	adjoint nodal forces (flexibility method)
$\{\mathbf{F}^*\}, (\mathbf{f}^{*e}\}$	fixed end forces
$\{\bar{\mathbf{F}}^*\}, \{\bar{\mathbf{f}}^{*e}\}$	adjoint fixed end forces
h	given depth of cross section
$h = 1, \ldots, H^e$	subscript identifying cross sections along element e
$i = 1, \ldots, I^e$	subscript identifying design variables for element e
I	moment of inertia
$j = 1, \ldots, J$	subscript identifying degrees of freedom (joint displacement and loads)
$k = 1, \ldots, K$	subscript identifying displacement constraints
$k = 1, \ldots, K_l$	subscript identifying displacement constraints (for the lth load condition)
$[\mathbf{K}], [\mathbf{k}^e]$	stiffness matrix
L	length of member
$\{\bar{\mathbf{L}}_l\}, \{\mathbf{L}_{kl}\}$	Lagrange multipliers
$l = 1, \ldots, L$	subscript identifying load conditions
$m = 1, \ldots, M^e$	subscript identifying nodal forces and relative displacements at nodes of element e
$n = 1, \ldots, N$	superscript identifying reaction elements (supports)
$\{\mathbf{n}\}$	stress resultants at an arbitrary cross section
$\{\mathbf{n}_h^e\}$	stress resultants at cross section h of element e
$\{\mathbf{n}_{oh}^e\}$	stress resultants at cross section h of element e caused by the load acting on the interior of element e
$\{\mathbf{P}\}$	real loads
$\{\hat{\mathbf{P}}\}$	virtual loads
$\{\bar{\mathbf{P}}\}$	adjoint loads
$\{\tilde{\mathbf{P}}\}$	variable loads
$q = 1, \ldots, Q^n$	subscript identifying reaction components and displacements at support n
Q	section modulus

$r = 1, \ldots, R_h^e$	subscript identifying a type and/or location of a stress or stress condition at a cross section h of element e
R_σ	set of elements for which a stressed constraint is active
R_s	set of elements for which a side constraint is active
$R_{\sigma s}$	set of elements for which both a stress constraint and a side constraint are active
R_d	set of elements controlled by a deflection constraint
$\{\mathbf{R}\}, \{\mathbf{r}^n\}$	real reactions
$\{\bar{\mathbf{R}}\}, \{\bar{\mathbf{r}}^n\}$	adjoint reactions
$\{\mathbf{s}_{hr}^e\}$	transformation vector relating stress resultants $\{\mathbf{n}_h^e\}$ to stresses σ_{hr} of type and/or location r at cross section h
$\{\mathbf{s}_0^e\}$	constant part of stress transformation vector
t, t_k, t_{kl}	specified value of deflections
T, T_1	tolerance values in convergence criteria
$\{\mathbf{U}\}$	real nodal displacements
$\{\bar{\mathbf{U}}\}$	adjoint nodal displacements
$\{\tilde{\mathbf{U}}\}$	adjoint nodal displacements corresponding to variable loads
$v = 1, \ldots, V_h^e$	subscript identifying stress resultants at the cross section h
$W(\mathbf{X}), w^e(\mathbf{x})$	objective functions associated with elements
$\{\mathbf{X}\}, \{\mathbf{x}^e\}$	design variables
$\{\mathbf{X}\downarrow\}, \{\mathbf{x}\downarrow^e\}$	lower limit on design variables
$\{\mathbf{X}\uparrow\}, \{\mathbf{x}\uparrow^e\}$	upper limit on design variables
α	coefficient of thermal expansion
$\beta_i^e\downarrow, \beta_i^e\uparrow, v_{kl}, \lambda_{hrl}^e, \rho_e^n$	Lagrange multipliers
$[\boldsymbol{\Gamma}], \{\boldsymbol{\gamma}^e\}$	transformation matrix from element level to system level
$\{\boldsymbol{\Delta}\}, \{\boldsymbol{\delta}^e\}$	real relative displacements
$\{\bar{\boldsymbol{\Delta}}\}, \{\bar{\boldsymbol{\delta}}^e\}$	adjoint relative displacements
$\{\boldsymbol{\Delta}^*\}, \{\boldsymbol{\delta}^{*e}\}$	real initial relative displacements
$\{\bar{\boldsymbol{\Delta}}^*\}, \{\bar{\boldsymbol{\delta}}^{*e}\}$	adjoint initial relative displacements
$\{\boldsymbol{\Delta}^{o*}\}, \{\boldsymbol{\delta}^{o*e}\}$	real initial relative displacements extended for stiffness method
$\{\bar{\boldsymbol{\Delta}}^{o*}\}, \{\bar{\boldsymbol{\delta}}^{o*e}\}$	adjoint initial relative displacements extended for stiffness method
$\{\boldsymbol{\Delta}\}, \{\boldsymbol{\delta}^n\}$	displacements of a reaction element (support) n
$\{\boldsymbol{\Delta}^*\}, \{\boldsymbol{\delta}^{*n}\}$	prescribed initial displacements of a reaction element n (support settlements)
$\{\tilde{\boldsymbol{\Delta}}^*\}, \{\tilde{\boldsymbol{\delta}}^{*e}\}$	variable initial relative displacements (due to prestrain)
$\{\tilde{\boldsymbol{\Delta}}^*\}, \{\tilde{\boldsymbol{\delta}}^{*n}\}$	variable settlements at support n
$\{\underline{\bar{\boldsymbol{\Delta}}}\}, \{\underline{\bar{\boldsymbol{\delta}}}^n\}$	displacements at supports of adjoint structure
$\{\underline{\bar{\boldsymbol{\Delta}}}^*\}, \{\underline{\bar{\boldsymbol{\delta}}}^{*n}\}$	initial displacements at supports of adjoint structure
η	constant in Kreisselmeier–Steinhauser functions
$\theta_{hr}^e(\mathbf{n}_h^e)$	stress conditions of type r (function of stress resultants $\{\mathbf{n}_h^e\}$) at a cross section h of element e
θ_o^e	constant part of stress condition
$\vartheta^n(\tilde{\underline{\delta}}^{*n})$	cost of variable support settlement
μ	Poisson's ratio

ρ^e	specific weight of structural material of element e
σ_{hr}^e	local stress of type and/or location r at cross-section h of element e
τ	magnitude of temperature change
Φ	objective function
$\chi^e(\tilde{\delta}^{*e})$	cost of variable initial relative displacement
$\psi_j(\tilde{P}_j)$	cost of variable load $\tilde{P}_j$
$[\omega_h^e]$	matrix relating nodal forces $\{\mathbf{ff}^e\}$ and stress resultants $\{\mathbf{n}_h^e\}$ at cross section h of element e
$\Omega^n(\mathbf{r}^n)$	reaction cost function for the reaction element (support) n

Brackets used for matrices

$\{\ \}$	column matrix (vector)
$\lfloor\ \rfloor$	row matrix
$[\ ]$	rectangular or square matrix

Explanatory notes on the notation for discretized systems

Bold upper-case letters represent vectors and matrices at system level and bold lower-case letters vectors and matrices at element level. Symbols in italics denote scalar quantities. $\mathbb{F}$ and $\mathbf{ff}$ refer to nodal forces in flexibility formulation whilst $\mathbf{F}$ and $\mathbf{f}$ denote nodal forces in stiffness formulation. An asterisk * signifies initial relative displacements or fixed end forces. A tilde ~ indicates quantities that are variable by passive control (variable prestrains or loads). Symbols with an underline _ refer to reaction elements (supports). An overline ‾ indicates an adjoint quantity and a hat ˆ a quantity associated with a virtual load.

2A.2   Notation – structural continua

B_1	subset of the structural domain D, on which static boundary conditions are given
B_2	subset of the structural domain, on which kinematic boundary conditions are given
D	structural domain ($\xi \in D$)
$[\mathbf{d}]$	generalized flexibility matrix
$h = 1, \ldots, H$	subscript identifying load components p_h and displacement components u_h at a given cross section
$i = 1, \ldots, I$	subscript identifying design variables x_i for a given cross section
$j = 1, \ldots, J$	subscript identifying stress resultants σ_j and generalized strains ε_j at a given cross section
$k = 1, \ldots, K$	subscript identifying displacement constraints
$\mathbf{p}(\xi)$	loads at a given cross section
$\bar{\mathbf{p}}(\xi)$	adjoint loads
$r = 1, \ldots, R$	subscripts identifying stress constraints
$S_r(\mathbf{x}, \sigma) \leqslant 0$	stress constraints
$\mathbf{u}(\xi)$	displacements at a given cross section

$\bar{u}(\xi)$	adjoint displacements
$x(\xi)$	design variables at a given cross section
$\varepsilon(\xi)$	cross-sectional strains (generalized strains, e.g. beam curvature)
$\bar{\varepsilon}(\xi)$	adjoint generalized strains
$\lambda_r(\xi)$	Lagrange multiplier for stress constraints
μ	Poisson's ratio
v_k	Lagrange multiplier for displacement constraints
$\xi \in D$	spatial coordinates
σ	stress resultant (generalized stress)
$\bar{\sigma}$	adjoint generalized stress
$\Phi = \int_D \psi[x(\xi)]\, d\xi$	total cost, objective functional
$\psi[x(\xi)]$	specific cost at a given cross section

REFERENCES

Barnett, R.L. (1961) Minimum-weight design of beam for deflection. *Journal of Engineering Mechanics ASCE* **87**, 75–109.

Berke, L. (1970) An efficient approach in the minimum weight design of deflection limited structures. *AFFDL-TM-70-FDTR*.

Berke, L. and Khot, N.S. (1974) Use of optimality criteria methods for large scale systems. *AGARD LS-70*.

Berke, L. and Khot, N.S. (1987) Structural optimization using optimality criteria, in *Computer Aided Optimal Design: Structural and Mechanical Systems* (ed. C.A. Mota Soares), Springer, Berlin, pp. 271–312.

Berke, L. and Khot, N.S. (1988) Performance characteristics of optimality criteria methods, in *Structural Optimization* (eds. G.I.N. Rozvany and B.L. Karihaloo). Proceedings of the IUTAM Symposium, Melbourne, 1988, Kluwer, Dordrecht, pp. 39–46.

Chang, K.J. (1992) Optimality criteria methods using K–S functions. *Structural Optimization*, **4**, 213–17.

Drucker, D.C. and Shield, R.T. (1956) Design for minimum weight. Proceedings of the 9th International Congress on Applied Mechanics, Brussels, 1956, Vol. 5, pp. 212–22.

Fleury, C. (1979) Structural weight optimization by dual methods of convex programming. *International Journal of Numerical Methods in Engineering*, **14**, 1761–83.

Foulkes, J. (1954) The minimum-weight design of structural frames. *Proceedings of the Royal Society*, **223**, 482–94.

Gellatly, R.A. and Berke, L. (1971) Optimal structural design. *AFFDL-TR-70-165*.

Haftka, R., Gürdal, Z. and Kamat, M.P. (1990) *Elements of Structural Optimization*, Kluwer, Dordrecht.

Heyman, J. (1959) On the absolute minimum weight design of framed structures. *Quarterly Journal of Mechanics and Applied Mathematics*, **12**, 314–24.

Huang, N.-C. (1971) On the principle of stationary mutual complementary energy and its application to structural design. *Zeitschrift für angewandte Mathematik und Physik*, **22**, 608–20.

Khot, N.S. and Berke, L. (1984) Structural optimization using optimality criteria methods, in *New Directions in Optimum Structural Design* (eds. E. Atrek, R.H. Gallagher, K.M. Ragsdell and O.C. Zienkiewicz). Wiley, Chichester, pp. 47–74.

Marcal, P.V. and Prager, W. (1964) A method of optimal plastic design. *Journal de Mécanique*, **3**, 509–30.

Michell, A.G.M. (1904) The limits of economy of material in frame-structures. *Philosophical Magazine*, **8**, 589–97.

Mroz, Z. (1963) Limit analysis of plastic structures subject to boundary variations. *Archiwum Mechaniki Stosowanej*, **15**, 63–76.

Olhoff, N. (1981) Optimization of columns against buckling. Optimization of transversely vibrating beams and rotating shafts, in *Optimization of Distributed Parameter Structures* (eds. E.J. Haug and J. Cea). Proceedings of NATO ASI, Iowa, 1980, Sijthoff and Noordhoff, Alphen aan der Rijn, pp. 152–76, 177–99.

Prager, W. and Rozvany, G.I.N. (1977) Optimization of structural geometry, in *Dynamical Systems* (eds. A.R. Bednarek and L. Cesari). Academic Press, New York, pp. 265–93.

Prager, W. and Shield, R.T. (1967) A general theory of optimal plastic design. *Journal of Applied Mechanics*, **34**, 184–6.

Prager, W. and Taylor, J. (1968) Problems of optimal structural design. *Journal of Applied Mechanics*, **5**, 102–6.

Rozvany, G.I.N. (1989) *Structural Design via Optimality Criteria*, Kluwer, Dordrecht.

Rozvany, G.I.N. and Zhou, M. (1991) The COC algorithm, part I: cross-section optimization or sizing (presented 2nd World Congr. on Computational Mechanics, Stuttgart, 1990). *Computer Methods in Applied Mechanics and Engineering*, **89**, 281–308.

Rozvany, G.I.N. and Zhou, M. (1993) Continuum-based optimality criteria (COC) methods: an introduction, in *Optimization of Large Structural Systems*, (ed. G.I.N. Rozvany). Proceedings of NATO/DFG ASI, Berchtesgaden, 1991, Kluwer, Dordrecht, pp. 1–26.

Rozvany, G.I.N., Zhou, M. and Gollub, W. (1990) Continuum-type optimality criteria methods for large finite element systems with a displacement constraint – part II. *Structural Optimization*, **2**, 77–104.

Rozvany, G.I.N., Zhou, M., Rotthaus, M., Gollub, W. and Spengemann, F. (1989) Continuum-type optimality criteria methods for large systems with a displacement constraint – part I. *Structural Optimization*, **1**, 47–72.

Shield, R.T. and Prager, W. (1970) Optimal structural design for given deflection. *Zeitschrift für angewandte Mathematik und Physik*, **21**, 513–23.

Sobieszczanski-Sobieski, J. (1992) A technique for locating function roots and for satisfying equality constraints in optimization. *Structural Optimization*, **4**, 241–3.

Venkayya, V.B., Khot, N.S. and Berke, L. (1973) Application of optimality criteria approaches to automated design of large practical structures. *AGARD-CP-123*, pp. 3.1–3.9.

Zhou, M. (1989) Geometrical optimization of trusses by a two-level approximation concept. *Structural Optimization*, **1**, 235–40.

Zhou, M. (1992) A new discretized optimality criteria method in structural optimization. Doctoral Dissertation, DVI-Verlag, Düsseldorf.

Zhou, M. and Rozvany, G.I.N. (1992) A new discretized optimality criteria method in structural optimization. Proceedings of 33rd AIAA/ASME/ASCE/AHS/ASC Structures, Structural Dynamics and Materials Conference, Dallas, TX, 1992, AIAA, Washington, DC, pp. 3106–20.

Zhou, M. and Rozvany, G.I.N. (1993) Iterative COC methods – parts I and II, in *Optimization of Large Structural Systems* (ed. G.I.N. Rozvany). Proceedings of NATO/DFG ASI, Berchtesgaden, 1991, Kluwer, Dordrecht, pp. 27–75.

3

Model reduction and verification

PANOS Y. PAPALAMBROS

3.1 INTRODUCTION

A successful optimization study depends strongly on the appropriateness of the underlying mathematical model used for performing optimization. Indeed in most cases the model will determine whether a solution may or may not be achieved. Numerical algorithms are very sensitive to the precise mathematical form in which the model is cast. Moreover, preliminary models resulting from early modeling efforts may be poorly bounded so that either no finite optimum exists or an optimum is located by artificially forcing activity of constraints that have little or no engineering meaning. Even when a model is properly bounded the solution may be obtained, some times trivially, by forcing activity of as many constraints as the number of degrees of freedom in the model, thus making further optimization unnecessary. This tends to happen in preliminary model formulations with objective and constraint functions depending monotonically on most of the design variables.

Proper modeling requires experience and good understanding of the numerical algorithms that will be used for computing solutions. There are also some simple and rigorous principles that can be applied to exploit monotonicity properties of the model. This chapter reviews some of the basic ideas for proper model formulation and illustrates by example how they may be applied during an optimization study. Modeling considerations prior to embarking on numerical computations are examined first. Next, an extended example illustrates how a model may be reduced and verified systematically. Finally some automation efforts for such modeling considerations are discussed.

3.2 MODELING CONSIDERATIONS PRIOR TO COMPUTATION

At some point during the design process, it becomes possible to represent design decisions by a precise mathematical optimization statement. Usually a nonlinear programming (NLP) model formulation, such as

$$
\begin{aligned}
&\text{minimize} \quad f(x) && x \in X \\
&\text{subject to} \quad h_j(x) = 0 && j = 1, 2, \ldots, m_1 \\
&\phantom{\text{subject to}} \quad g_j(x) \leqslant 0 && j = m_1 + 1, m_1 + 2, \ldots, m_1 + m_2 \quad (3.1)
\end{aligned}
$$

is adequate. In equation (3.1), f, h and g are the scalar objective, equality and inequality constraints respectively, x is the vector of design variables $x_1, x_2, \ldots, x_n$, and the set X is a subset of the n-dimensional real space R^n, included in the statement to indicate that there may exist other restrictions on the variables that are not explicit in the constraints, for example, integer values. (Note that boldface characters indicate vector quantities.) Here the objective is shown as a single one, since multiobjective formulations are usually transformed to a single objective. The above mathematical statement will be referred to as the **design optimization model** cast into the so-called **negative null form**, a standard form where the objective is to minimize, while equality and inequality constraints are written with zero and less than or equal to zero, respectively, placed at the right-hand side of the equation.

3.2.1 Explicit vs implicit models

The design model implicitly assumes the existence of an **analysis model** used to evaluate the functions f, h, g. For example, the analysis models may be finite element analyses of structural behavior, system simulation, or curve-fitted experimental data. Analysis models may be explicit or implicit. In an **explicit** model, the model function forms are explicitly represented by mathematical, usually algebraic, functions; in an **implicit** model, the model functions are usually represented by a procedure or subroutine, for example, an iterative numerical solution of a set of differential equations.

It is generally preferable to deal with explicit models since then it is easier to make assertions about optimality, for example, by detecting mathematical properties such as monotonicity or convexity (a property that guarantees a unique minimum); also, it is often possible to manipulate the model in order to extract additional information or properties, or to recast it in a more convenient form, as we will see below. From a practical viewpoint, the models in many important classes of design optimization problems are implicit, which poses a substantially increased burden on ascertaining reliability of the optimization results. In practice one seldom seeks a mathematical optimum. Rather, one aims only at so-called **satisfycing** designs (Wilde, 1978), namely, design solutions with a value of the objective function within a known acceptable distance from the (unknown) true optimal value, and frequently only at designs improved by a certain percentage over current ones. Thus the disadvantage of implicit models is significantly tempered.

From an operational viewpoint, it is useful to distinguish between two phases in the optimization process. At the point where the optimal design model has been precisely formulated, the **modeling phase** of the optimization procedure is concluded. What follows constitutes the **solution phase** of the procedure. For most realistic models, the solution phase will include an iterative procedure that should converge to the optimum. An iteration begins by selecting an initial design corresponding to a starting value for the design variables and executing the calculation represented by the analysis model(s), so the values for objective and constraint functions are obtained. The cost of subsequent calculations for evaluating the next iterant design increases with the size of the problem, as measured by the number of variables and constraints.

When an implicit analysis model is computationally expensive, the iterative nature of numerical solution strategies makes the use of such a model unattractive. In a preliminary design effort it may be possible to use a simplified explicit model instead of the more accurate implicit one. The process could be similar to the design of physical experiments, such as Taguchi methods. These methods use a carefully planned efficient set of experiments to discover how design variables relate to each other. In the present context the experiments are performed computationally using the 'full' model and the results are curve fitted. After an initial optimization study has been successfully completed with the simplified model, a few iterations using the full model and the identified optimum as a starting point may complete the study.

3.2.2 Design parameters vs variables

The optimization model for any given design problem is more accurately described as

$$\begin{array}{lll} \text{minimize} & f(x,p) & x \in X, p \in P \\ \text{subject to} & h_j(x,p) = 0 & j = 1, 2, \ldots, m_1 \\ & g_j(x,p) \leqslant 0 & j = m_1 + 1, m_1 + 2, \ldots, m_1 + m_2 \end{array} \qquad (3.2)$$

where p is a vector of parameters, $p = (p_1, p_2, \ldots, p_q)$, and P is a subset of R^q. The quantities $p_1, p_2, \ldots, p_q$ are the **design parameters** and they are considered fixed to specific values during the optimization process. Variables and parameters together give a complete description of the design, and the constraint functions expressed as equalities and inequalities in terms of variables and parameters give a complete description of the **design space**, i.e. the space that contains all acceptable solutions. The selection of variables and parameters is generally a modeling decision that depends on the position of the designer in the hierarchy of decision-making and on expediency for actually solving eventually the underlying mathematical problem.

When posing the design problem one must consider the position one occupies in the decision-making hierarchy. The higher is the position the fewer are the quantities that one considers as fixed; hence more quantities will be considered as variables than as parameters. A more subtle consideration in classifying parameters and variables comes from the relative ease of solution that results from this classification. For example, material selection is a typical key decision in design. In an optimization model, one material will correspond to several quantities in the model, e.g. density, yield strength, modulus of elasticity. Treating the material as a variable will require inclusion of additional relations in the model that link these quantities, such relations usually given at best in some tabular form. In subsequent numerical processing table look-up or curve-fitting will be then required. Table look-up should not be used as it may introduce discontinuities and nondifferentiabilities that could defy many NLP algorithms. Even with curve-fitting the problem will still contain discrete variables corresponding to the different materials, so continuous NLP algorithms cannot be used and the problem becomes considerably more difficult. Thus the best choice is to treat material as a parameter and to find the optimal solutions for different materials separately. The strategy will be appropriate as long as the number of available materials is not very large.

Even when there is no other good reason to treat a variable as a parameter, one may wish to do so just to make the problem simpler to solve. Generally, the smaller the number of design variables is, the easier is the problem to solve. A variable may appear in complicated nonlinear expressions and fixing it as a parameter may make the problem substantially easier, perhaps linear. Finally, there may be engineering evidence that some variables have little influence on the problem, so fixing them as parameters may be a good idea, at least in early studies.

3.2.3 Constraint activity

The concept of constraint activity is very important in design optimization. Loosely speaking, an active constraint corresponds to a 'critical' design requirement, i.e. one whose presence determines where the optimum will be. In traditional structural design, 'fully stressed' designs would correspond to finding minimum weight designs with all stress constraints active. Formally, a constraint (equality or inequality) is called (globally) **active** if and only if removing it from the model changes the set of (globally) optimal solutions; it is **inactive** otherwise. An optimal solution is called **constraint bound** if and only if there are exactly as many active constraints as there are design variables. A constraint is called **tight** if and only if it is satisfied as a strict equality at the optimum. For continuous variables, active constraints will also be tight. However, not all tight constraints are active. A feasible point is called **constraint bound** if and only if at this point there are as many tight constraints as there are design variables. As mentioned earlier, in a constraint-bound optimum one needs only to find the active constraints. Thus, constraint-bound points are defined by solving a system of n equations in n unknowns, these equations constituting a **working (active) set**. Of course, not all such solutions are constraint-bound optima. For some sufficiency criteria see Papalambros (1988a). If the point is optimal, then the working set is the (true) active set. The term 'active' is sometimes used for local minima, but then all relevant properties must be considered valid within an appropriate neighborhood or region. A more comprehensive discussion of activity is given in Pomrehn and Papalambros (1992). Identified active inequalities are indicated with the symbol $\leqq$ instead of $\leqslant$.

An active constraint is a direct contributor to optimality and corresponds either to an equality constraint (such as equilibrium) or to a critical design requirement (such as setting the maximum stress equal to the strength of the material used). When a subset of the original constraints is selected for further processing, these constraints together with the objective function constitute a **submodel** of the original problem. The submodel contains only those constraints that are known to be active or that have been judiciously selected as having a high likelihood to become active. An **active set strategy** is a set of rules which are used to decide which constraints may be active at the optimum. An **active set** (at a given point) is a set that contains only those constraints that are considered as being definitely active at the current iteration, which usually means that they are satisfied as equalities (at that point). An **extended active set** is a set that contains both the currently active constraints and additional ones that are considered candidates for activity in one or more subsequent iterations. Use of such sets is often necessary in order to control computational costs associated with having a small number of variables but a very large number of inequality constraints (only few of which can possibly be active).

Clearly, at a given point the active set is a subset of the extended one. The rules in an active set strategy will in general apply to both an active set and an extended one. In the mathematical programming literature the term 'active set strategy' applies only to the active set, while in the structural optimization literature the term usually applies to the extended active set.

The main idea here is to emphasize that knowing the active constraints as early as possible may be very advantageous both in locating and in verifying an optimal solution.

3.2.4 Monotonicity principles

In many preliminary design models, usually because of simplification of trade-offs, the objective and constraint functions are **monotonic** with respect to the design variables. A continuous differentiable function $f(x)$ is increasing (decreasing) with respect to (wrt) a design variable x_i, if $\partial f/\partial x_i > 0$ ($\partial f/\partial x_i < 0$). We say that f is coordinate-wise monotonic wrt x_i, or that x_i is a monotonic variable (in f). This concept can be extended to other functions but this is not necessary here. We will also assume that all design variables are strictly positive. This is usually the case in design problems; if not, a simple shift in the datum can make the variables positive, e.g. using an absolute temperature scale rather than Farenheit or Celsius. Monotonicity analysis is based on two simple principles (see Papalambros and Wilde (1988) for further details).

> *First monotonicity principle (MP1)*   In a well-constrained objective function every increasing (decreasing) variable is bounded below (above) by at least one active constraint.
>
> *Second monotonicity principle (MP2)*   Every monotonic variable not occurring in a well-constrained objective function is either irrelevant and can be deleted from the problem together with all constraints in which it occurs, or is relevant and bounded by two active constraints, one from above and one from below.

We will now illustrate how these principles can be used to identify a model that is not well constrained or even to obtain an optimal solution in simple cases. The modeling example of a linear actuator will later demonstrate the approach in a more complicated situation.

Functional relations in a model usually must be rearranged in order to cast them into a standard form, such as equation (3.1). In general, a given relation may be 'equivalent' to several standard forms, some of which may appear monotonic and some not. Therefore, monotonicity should be examined in the original formulation of the problem and, if algebraic manipulations are employed, care should be taken not to disguise monotonic properties. However, once a relation has identified monotonicities, these are invariant in the sense that any standard form resulting in identified monotonicities will have exactly the same types of monotonicities. The following implicit function theorem specifically stated for monotone functions is useful in studying how monotonicity properties may be inherited when equalities are eliminated through variable substitution.

Implicit function theorem Let $X_i \subseteq R$, $i = 1, \ldots, n$, be n subsets (finite or infinite) of R and let $X = \{x \,|\, x = (x_1, \ldots, x_n), \; x_i \in X_i, \; i = 1, \ldots, n\}$. Let $F : X \to R$ be a function coordinate-wise monotonic on X. Then, for each s in the range of F and for each $i = 1, \ldots, n$, there exists a function $\phi(i, s; x_i')$ of the variable vector $x_i' = (x_1, \ldots, x_{i-1}, x_{i+1}, \ldots, x_n)$ such that $\phi(i, s; x_i')$ is (coordinate-wise) monotonic wrt x_i. Furthermore, for $1 \leqslant j \leqslant n$ and $i \neq j$, if F is monotonic in the same (opposite) sense wrt x_i and x_j, then $\phi(i, s; x_i')$ is decreasing (increasing) wrt x_j.

Monotonicity principles can be applied directly when the model contains only inequalities, i.e.

$$\begin{array}{ll} \text{minimize} & f(x) \qquad\qquad x \in X \\ \text{subject to} & g_j(x) \leqslant 0 \qquad j = 1, 2, \ldots, m \end{array} \tag{3.3}$$

When active constraints (equalities or inequalities) have been identified, a new round of monotonicity analysis can be performed if the active constraints are eliminated through variable substitution. In the new reduced model the exact functions may not be known but the relevant monotonicities may be implicitly defined using the above theorem.

Before proceeding with some examples, it is interesting to recall that in the presence of differentiability and the usual constraint qualifications the Karush–Kuhn–Tucker (KKT) necessary optimality conditions for the model in equation (3.3) are

$$\nabla f + \mu^T \nabla g = 0, g \leqslant 0$$
$$\mu^T g = 0, \mu \geqslant 0 \tag{3.4}$$

where ∇g is the Jacobian matrix of the vector function $g(x)$. For this case, the two monotonicity principles can be derived trivially from equation (3.4). The principles are more general necessary conditions than the KKT conditions, as they apply globally and do not assume continuity or differentiability. On the other hand, they assume (global or regional) monotonicity, a property that could be considered more restrictive. Their utility is based on the argument that many early design optimization models do have extensive monotonicity properties (see for example, Papalambros and Wilde (1988)). The principles frequently allow a drastic reduction of the number of cases needed to be examined under the complementary slackness conditions of equation (3.4). This can be seen in some simple examples (Papalambros and Li, 1983). Note that when an implicit function is used, a positive (negative) superscript sign for a variable indicates increasing (decreasing) function wrt that variable.

EXAMPLE 3.1

Consider the problem

$$\begin{array}{ll} \text{minimize} & f = (x_1 - 3)^2 + (x_2 - 3)^2 \\ \text{subject to} & g_1 = -x_1 \leqslant 0 \\ & g_2 = -x_2 \leqslant 0 \\ & g_3 = x_1 + x_2 - 4 \leqslant 0 \end{array} \tag{3.5}$$

There are eight possible cases dictated by the complementary slackness conditions. The objective is nonmonotonic but we observe that if $x_i \geqslant 3, i = 1, 2$, there is no solution since g_3 is violated. Thus $x_i < 3$ for at least one $i, i = 1, 2$, and the objective is decreasing wrt at least one i, which then makes g_3 always active by the first monotonicity principle (MP1). The problem has now one degree of freedom and the solution can be obtained quickly using constrained derivatives (Wilde and Beightler, 1967). If f_1 is the reduced objective resulting from elimination of x_1 and g_3 we have

$$\mathrm{d}f_1/\mathrm{d}x_2 = (\partial f/\partial x_1)(\partial \phi/\partial x_2) + \partial f/\partial x_2 = 0 \tag{3.6}$$

where $\phi(x_2) \equiv x_1 = 4 - x_2$. Evaluation of equation (3.6) gives $x_1^* = 2, x_2^* = 2, f_o^* = 2$.

EXAMPLE 3.2

Consider the problem

$$\begin{aligned}
\text{minimize} \quad & f = (1/2)(x_1^2 + x_2^2 + x_3^2) - (x_1 + x_2 + x_3) \\
\text{subject to} \quad & g_1 = x_1 + x_2 + x_3 - 1 \leqslant 0 \\
& g_2 = 4x_1 + 2x_2 - 7/3 \leqslant 0 \\
& x_i \geqslant 0, i = 1, 2, 3
\end{aligned} \tag{3.7}$$

From the KKT conditions there are 32 cases to be examined. The objective appears nonmonotonic, but in fact $\partial f/\partial x_i = x_i - 1 \leqslant 0$ because g_1 implies $x_i \leqslant 1, i = 1, 2, 3$. Hence, by MP1 wrt x_3 we have two cases: if $x_3 = 1$ then $x_2 = x_1 = 0$ (g_1 is tight) and $f = -1/2$; if $x_3 < 1$, then g_1 is active. Defining $x_3 = 1 - x_1 - x_2 = \phi_1(x_1^-, x_2^-)$ we can calculate the constrained derivative of the reduced objective f_1 wrt one of the remaining variables, say x_1.

$$\partial f_1/\partial x_1 = (\partial f/\partial x_3)(\partial \phi_1/\partial x_1) + \partial f/\partial x_1 = (1/2)(4x_1 + 2x_2 - 2) \tag{3.8}$$

Now we observe the following: (a) if $4x_1 + 2x_2 < 2$, then g_2 is inactive with $f_1(x_1^-, x_2^-)$ and no solution exists; (b) if $4x_1 + 2x_2 = 2$, then g_2 is inactive and the remaining one-degree-of-freedom problem has the solution $x_1 = x_2 = x_3 = \frac{1}{3}$, $f = -\frac{5}{6}$; (c) if $4x_1 + 2x_2 > 2$, then the problem becomes $\{\min f_1(x_1^+, x_2)$, subject to $g_2(x_1^+, x_2^+) \leqslant 0$ with $x_1 \geqslant 0, x_2 \geqslant 0, x_1 + x_2 \leqslant 1\}$ and the solution is $x_1 = 0$, $x_2 > 1$ which is not permitted. The conclusion is a global optimum at $x_i = \frac{1}{3}, i = 1, 2, 3$.

EXAMPLE 3.3

Consider the problem

$$\begin{aligned}
\text{maximize} \quad & f = 0.0201 \, d^4 \, wn^2 \\
\text{subject to} \quad & g_1 = d^2 w - 675 \leqslant 0 \\
& g_2 = d - 36 \leqslant 0 \\
& g_3 = n - 125 \leqslant 0 \\
& g_4 = n^2 d^2 - (0.419)(10^7) \leqslant 0 \\
& d, n, w \geqslant 0
\end{aligned} \tag{3.9}$$

This problem involves maximizing the stored energy in a flywheel and is reduced to the above-mentioned form after some manipulations (Siddall, 1972). Constraint g_1 is active by MP1 wrt w. Note that the standard negative null form is obtained by using the objective {minimize $-f$}. Elimination of w from the objective gives $f = (0.0201)(675)d^2n^2$. Clearly g_4 will be active except if g_2, g_3 combined impose a stricter bound on d^2n^2. For the given numbers this is not the case, so there are infinite solutions:

$$\max f^* = (5.68)(10^7), \, w_* = 675d_*^{-2}, \, n_* = 2047d_*^{-1}$$

$$16.4 \leqslant d_* \leqslant 36 \tag{3.10}$$

EXAMPLE 3.4

Consider now the problem

$$\begin{aligned}
\text{maximize} \quad & f = x_1^{-2} + x_2^{-2} + x_3^{-2} \\
\text{subject to} \quad & g_1 = 1 - x_1 - x_2 - x_3 \leqslant 0 \\
& g_2 = x_1^2 + x_2^2 - 2 \leqslant 0 \\
& g_3 = 2 - x_1 x_2 x_3 \leqslant 0 \\
& x_i \geqslant 0, i = 1, 2, 3
\end{aligned} \tag{3.11}$$

By MP1 constraint g_2 must be active providing upper bounds on x_1 and x_2. However, x_3 is unbounded from above since f, g_1, and g_3 are all decreasing wrt x_3. The problem has no solution unless the objective and/or constraint functions are appropriately modified by remodeling. Any nonredundant equality constraint would also serve. The usual practice in such cases is to add a simple inequality constraint, e.g. $x_3 \leqslant a, a > 0$. However, this constraint will be always active by construction, which means that the optimum is in fact determined by this artificial bound. In real problems, such information must be consciously considered because the optimum is essentially arbitrarily fixed by the modeler. We will explore this further in the linear actuator example.

3.2.5 Modeling decisions

It is good practice as one embarks on an optimization study to make a checklist of all the decisions that are required for developing a model. As an example, Table 3.1 shows what this checklist might look like for a structural optimization problem using finite elements for the analysis model (Papalambros, 1988b). This list is certainly not complete and it may change depending on the problem. With experience, the need for such lists may diminish. Yet even experienced users will benefit by going through a detailed itemization of all the decisions they make as they go. It is not uncommon for some overlooked decision to have a strong bearing on the optimization results, particularly when something goes wrong. Furthermore, such decisions can be increasingly automated with the use of artificial intelligence techniques; see, for example, Thomas (1990). This is a particularly attractive approach for developing quickly and accurately optimization

Table 3.1 Some decisions related to modeling (finite element based models)

General
    model cost level (what is an 'expensive' model)
        size of model
            number of elements, constraints, and variables
        complexity of geometry
        preparation time for FEM model
    identification of problem type
        shape or size variables
    problem decomposition (reduction of dimensionality)
        substructuring
        separability of objective
        multilevel model
        variable linking
Objective
    single
    multiple
        transformation to single objective
            weight factors
Constraints
    selection of 'appropriate' constraints
    selection of parametric bounds
        hard and soft bounds
    constraint interaction, compatibility rules
    consistency (non-empty feasible domain)
Design variables
    selection
    transformation
    continuous or discrete values
        maximum number of discrete variable values
        maximum number of coordinate variables in an element
    starting point values (initial design)
Design parameters
    selection
    values for current application
    range of values
Submodel
    active set strategy (many decisions, part of them model related)
Program parameters related to model
    maximum number of nets
    maximum number of nodes
    maximum number of active elements
    maximum number of element variables
    maximum number of composite stacks
    maximum number of master variables
    maximum number of layers in stack
Scaling
    functions
    variables
Finite differencing
    step size for material angle variables
    step size for coordinate variables
    step size for thickness variables

models within a specific domain, for example, optimal trusses or plate girders in bridge construction (Adeli and Balasubramanyam, 1988a; Adeli and Mak, 1988). For further discussion on the use of expert systems in structural design see Adeli and Balasubramanyam (1988b).

3.3 MODELING EXAMPLE: A LINEAR ACTUATOR

In this extensive example we will go through various modeling steps using several of the ideas mentioned earlier and some new ones. The procedure followed is typical for models with relatively small number of variables and explicit functions, but can be used in other situations to advantage.

3.3.1 Model setup

The example presented here is part of a larger classroom project dealing with the optimal design of a drive screw linear actuator (Alexander and Rycenga, 1990). The drive screw is part of a power module assembly that converts rotary to linear motion, so that a given load is linearly oscillated at a specified rate. Such a device is used is some household appliances. The assembly consists of an electric motor, drive gear, pinion (driven) gear, drive screw, load-carrying nut, and a chassis providing bearing surfaces and support. The present example addresses only the design of the drive screw, schematically shown in Fig. 3.1.

The objective function in the design model is to minimize product cost consisting of material and manufacturing costs. Machining costs for a metal drive screw or injection molding costs for a plastic one are considered fixed for relatively small changes in the design, hence only material cost is taken as the objective to minimize, namely

$$f_o = (C_m \pi/r)(d_2^2 L_1 + d_2^2 L_2 + d_3^2 L_3) \tag{3.12}$$

Here C_m is the material cost ($\$ \text{lb}^{-1}$), d_1, d_2 and d_3 are the diameters of gear–drive screw interface, threaded and bearing surface segments, respectively, L_1, L_2 and L_3 being the respective segment lengths.

There are operational, assembly, and packaging constraints. For strength against bending during assembly we set

$$M c_1/I \leqslant \sigma_{\text{all}} \tag{3.13}$$

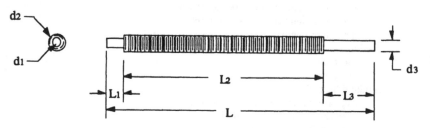

Fig. 3.1 Schematic of drive screw design.

where the bending moment $M = F_a L/2$, F_a being the force required to snap the drive screw into the chassis during assembly, and L being the total length of the component:

$$L = L_1 + L_2 + L_3 \tag{3.14}$$

Furthermore, $c_1 = d_2/2, I = \pi d_2^4/64$ is the moment of inertia, and σ_{all} is the maximum allowable bending stress for a given material.

During operation, a constraint against fatigue failure in shear must be imposed:

$$KTc_3/J \leqslant \tau_{\text{all}} \tag{3.15}$$

Here, K is a stress concentration factor, T is the applied torque, $C_3 = d_1/2$, $J = \pi d_1^4/32$ is the polar moment of inertia, and τ_{all} is the maximum allowable shear stress. The torque is computed from the equation

$$T = T_m C_2 N_S/N_m \tag{3.16}$$

where T_m is the motor torque, $c_2 = 1/16 \, (\text{lb oz}^{-1})$ is a conversion factor, and N_s and N_m are the number of teeth on the screw (driven) and motor (drive) gear, respectively.

To meet the specified linear cycle rate of oscillation, a speed constraint is imposed:

$$c_4 N_m S_m/N_s N_T \leqslant S \tag{3.17}$$

where $c_4 = 60^{-1}$ (number of threads rev^{-1})(min s^{-1}) is a conversion factor, S_m is the motor speed (rev min^{-1}), N_T is the number of threads per inch, and S is the specified linear cycle rate (in s^{-1}).

In order for the screw to operate in a drive mode the following constraint must be satisfied (Juvinall, 1983).

$$\frac{Wd_2}{2} \frac{\pi f d_2 + N_T^{-1} \cos \alpha_n}{\pi d_2 \cos \alpha_n - N_T^{-1} f} \leqslant T \tag{3.18}$$

Here W is the drive screw load, f is the friction coefficient, N_T^{-1} is the lead of screw threads, and α_n is the thread angle measured in the normal plane. There is also an upper bound on the number of threads per inch imposed by mass production considerations,

$$N_T \leqslant 24 \tag{3.19}$$

From gear design considerations, particularly avoidance of interference, limits on the numbers of gear teeth are imposed:

$$N_m \geqslant 8 \qquad N_s \leqslant 52 \tag{3.20}$$

Finally, there are some packaging and geometric considerations that impose constraints:

$$8.75 \leqslant L_1 + L_2 + L_3 \leqslant 10.0 \tag{3.21}$$

$$7.023 \leqslant L_2 \leqslant 7.523 \tag{3.22}$$

$$1.1525 \leqslant L_3 \leqslant 1.6525 \tag{3.23}$$

$$d_1 \leqslant d_2, d_3 \leqslant d_2, d_2 \leqslant 0.625 \tag{3.24}$$

Note that several assumptions were invoked in the model above: manufacturing costs remain fixed; a high volume production is planned; Standard Unified Threads are used; the assembly force for the drive screw is concentrated at the midpoint; frictional forces are considered only between threads and load nut, and all others are assumed negligible.

3.3.2 Model validity constraints

During the early stages of developing a mathematical optimization model many assumptions are made in order to obtain reasonably simple expressions for the objective and constraint functions. One must always check whether subsequent results from optimization conform to these assumptions, lest they are violated. This would indicate that the model used is inappropriate for the optimal design obtained and the optimization results are at least suspect and possibly erroneous. The remedy is usually a more accurate, probably also more complicated, mathematical model of the phenomenon under question. For example, equation (3.13) is valid only if the length/diameter ratio is more than 10.

3.3.3 Material choice as a parameter

A final observation on the initial drive screw model is that a significant trade-off exists on the choice of material: stainless steel vs plastic. A steel screw will have higher strength and be smaller in size but will require secondary processing, such as rolling of threads and finishing of bearing surfaces. A plastic screw would be made by injection molding in a one-step process that is cheaper but more material would be used because of lower strength, plastic having a higher cost per pound than steel. Specialty plastics with high strength would be even more expensive and less moldable. Thus the choice of material must be based on a model that contains more information than the current one. The constant term representing manufacturing costs should be included in the objective. Indeed a more accurate cost objective should include capital investment costs for manufacturing.

Nevertheless, substantial insight can be gained from the present model if we include material as a parameter; in fact each material is represented by four parameters $C_m, \sigma_{all}, \tau_{all}, f$. In the model analysis that follows we keep these parameters in the model with their symbols, rather than giving numerical values insofar as possible. The goal is to derive as many additional results as possible independently of the material used. This will substantially facilitate a post-optimal parametric study on the material. It would be much more difficult to treat material as a variable, because then we would have four additional variables with discrete values and implicitly linked, perhaps through a table of material properties. As mentioned earlier, this would destroy the continuity assumed in nonlinear programming formulations.

3.3.4 Standard null form

The model is now summarized in the negative null form, all parameters represented by their numerical values (Table 3.2), except for material parameters.

Table 3.2 List of parameters (material values for stainless steel)

C_m	material cost ($/lb)
f	friction coefficient (0.35 for steel on plastic)
F_a	force required to snap drive screw into chassis during assembly (6 lbf)
K	stress concentration fator (3)
L_1	length of gear/drive screw interface segment (0.405 in)
S	linear cycle rate (0.0583 in s^{-1})
S_m	motor speed (300 rpm)
T_m	motor torque (2 in oz)
W	drive screw load (3 lb)
a_n	thread angle in normal plane (60°)
σ_{all}	maximum allowable bending stress (20 000 lbf in^{-2})
τ_{all}	maximum allowable shear stress (22 000 lbf in^{-2})

All 'intermediate' variables defined through equalities are eliminated together with the associated equality constraints by direct substitution. This should be always done when possible, in order to arrive at a model with only inequality constraints, thus facilitating subsequent monotonicity analysis. This is a model reduction step, since the number of design variables is reduced. Note, however, that this may not be always a model simplification step, as the resulting expressions may become more complex with undetermined monotonicities. Some judgement must be exercised here. Occasionally, 'directing' an equality may be useful (see further below) in avoiding direct elimination. In the model below the intermediate variables M, L, T, I, c_1, c_3 and J together with the corresponding defining equalities have been eliminated. The variables are $d_1, d_2, d_3, L_2, L_3, N_m,$ N_S, N_T.

Model 1

minimize $f_0 = (C_m \pi/4)(0.405 d_1^2 + L_2 d_2^2 + L_3 d_3^2)$
subject to

$g_1 = 38.88 + 96 L_2 + 96 L_3 - \pi \sigma_{all} d_2^3 \leqslant 0$

$g_2 = 6 N_s/N_m - \pi \tau_{all} d_1^3 \leqslant 0$

$g_3 = 8.345 - L_2 - L_3 \leqslant 0 \qquad g_4 = -9.595 + L_2 + L_3 \leqslant 0$

$g_5 = L_2 - 7.523 \leqslant 0 \qquad\qquad g_6 = 7.023 - L_2 \leqslant 0$

$g_7 = L_3 - 1.6525 \leqslant 0 \qquad\quad g_8 = 1.1525 - L_3 \leqslant 0$

$g_9 = d_2 - 0.625 \leqslant 0 \qquad\qquad g_{10} = d_3 - d_2 \leqslant 0$

$g_{11} = d_1 - d_2 \leqslant 0 \qquad\qquad g_{12} = 5 N_m/N_S - 0.0583 N_T \leqslant 0$

$g_{13} = 1.5 d_2 \dfrac{f\pi d_2 + 0.5 N_T^{-1}}{0.5 \pi d_2 - f N_T^{-1}} - 0.125 N_S/N_m \leqslant 0$

$g_{14} = N_T - 24 \leqslant 0 \qquad\qquad g_{15} = 8 - N_m \leqslant 0$

$g_{16} = N_S - 52 \leqslant 0 \tag{3.25}$

As there are no equality constraints, there are eight degrees of freedom corresponding to the eight design variables. Three of these variables, N_m, N_T and N_S, must take integer values, so this problem is in fact a mixed continuous–integer variable nonlinear programming problem and standard numerical NLP methods will not work. We will see later how this is dealt with in the particular example.

3.3.5 Feasibility checking

Before embarking on analyzing the model, it is a good idea to check that the feasible domain is not empty, i.e. there exists at least one proven feasible point. From a mathematical viewpoint this may be a hard problem (possibly as hard as the optimization itself), but from an engineering viewpoint past experience can be a guide. In the drive screw example, an existing design using stainless steel has the following values: $d_1 = 0.1875$ in, $d_2 = 0.3125$ in, $d_3 = 0.2443$ in, $L_2 = 7.273$ in, $L_3 = 1.4025$ in, $N_m = 8, N_S = 48, N_T = 18$. The design is feasible with an objective function value of $f_0 = 0.635 C_m$. Now we can proceed knowing that an optimization attempt is possible.

3.3.6 Monotonicity

Looking at model 1 one notes that most constraints were written in a form that requires no divisions. This is always advisable, since in subsequent numerical processing a denominator may become zero and cause an abrupt termination by overflow error. This can happen even if the imposed constraints exclude the relevant variable values, because many numerical algorithms will temporarily operate in the infeasible domain. In model 1, constraint g_{13} has not been rewritten yet because of concern that this might obscure its monotonicity wrt d_2 and N_T. Let us examine this more carefully. Assuming a strictly positive denominator of the first term in g_{13} we multiply both sides by it and collect terms:

$$g_{13}: \; 1.5 f\pi d_2^2 + 0.75 N_T^{-1} d_2 - 0.0625\pi(N_S/N_m)d_2 + 0.125 f(N_S/N_m)N_T^{-1} \leqslant 0$$
(3.26)

Clearly g_{13} decreases wrt N_T but is nonmonotonic wrt d_2. In fact,

$$\partial g_3 / \partial d_2 = 3\pi f d_2 + 0.75 N_T^{-1} - 0.0625\pi N_S/N_m$$
(3.27)

which can be positive or negative depending on the variable values. All remaining functions in the constraints have obvious monotonicities.

3.3.7 Variable transformation

The model can be further simplified by a variable transformation. We observe that the two variables N_S, N_m appear together as a ratio everywhere except in the simple bounds g_{15}, g_{16}. We can define a new variable R,

$$R = N_m/N_S$$
(3.28)

which indeed is the reduction ratio of the gear drive, and eliminate variable N_m

using $N_m = RN_S$. The new model, including the reformulated constraint g_{13}, is as follows.

Model 2

$$\min f_0 = (C_m \pi/4)(0.405 d_1^2 + L_2 d_2^2 + L_3 d_3^2)$$

subject to

$$g_1 = 38.88 + 96 L_2 + 96 L_3 - \pi \sigma_{all} d_2^3 \leqslant 0$$

$$g_2 = 6 - \pi \tau_{all} R d_1^3 \leqslant 0$$

$$g_3 = 8.345 - L_2 - L_3 \leqslant 0 \qquad g_4 = -9.595 + L_2 + L_3 \leqslant 0$$

$$g_5 = L_2 - 7.523 \leqslant 0 \qquad g_6 = 7.023 - L_2 \leqslant 0$$

$$g_7 = L_3 - 1.6525 \leqslant 0 \qquad g_8 = 1.1525 - L_3 \leqslant 0$$

$$g_9 = d_2 - 0.625 \leqslant 0 \qquad g_{10} = d_3 - d_2 \leqslant 0$$

$$g_{11} = d_1 - d_2 \leqslant 0 \qquad g_{12} = 5 R - 0.0583 N_T \leqslant 0$$

$$g_{13} = 1.5 \pi f R d_2^2 + 0.75 R N_T^{-1} d_2 - 0.0625 \pi d_2 + 0.125 f N_T^{-1} \leqslant 0$$

$$g_{14} = N_T - 24 \leqslant 0 \qquad g_{15} = 8 - R N_S \leqslant 0$$

$$g_{16} = N_S - 52 \leqslant 0 \qquad\qquad\qquad\qquad\qquad (3.29)$$

Note that the requirement of integer values for N_m is now converted to one of rational values for R.

3.3.8 Repairing a model

Model 2 is now used to perform the first cycle of monotonicity analysis. The monotonicity table is a convenient tool to do this: Table 3.3. The columns are the design variables and the rows are the objective and constraint functions, the entries in the table being the monotonicities of each function with respect to each variable. Positive (negative) sign indicates increasing (decreasing) function, U indicates undetermined or unknown monotonicity. An empty entry indicates that the function does not depend on the respective variable, so the table acts also as an incidence table. (Items in parentheses will be explained in the next subsection.)

Monotonicity principles can be quickly applied by inspection using the monotonicity table. Looking at Table 3.3, by MP1 wrt d_3 we see that model 2 is not well constrained because no lower bound exists for d_3. Note that $d_3 > 0$ is not an appropriate bound because of the strict inequality. If the model were treated numerically as is, no convergence would occur if the algorithm was successful or an erroneous result would be found if the algorithm was led astray. Examining the engineering meaning of this model deficiency we see that an adequate thrust surface must be provided to keep the shaft from wearing into the bearing support, so we accept the simple remedy of adding a new constraint

$$g_{17} = 0.1875 - d_3 \leqslant 0 \qquad\qquad\qquad\qquad (3.30)$$

Poor boundedness is cause for concern, so the above deficiency triggers also

Table 3.3 Monotonicity table for model 2 (with model repairs in parentheses)

Functions	Variables							
	d_1	d_2	d_3	L_2	L_3	R	N_S	N_T
f_0	+	+	+	+	+			
g_1		−		+	+			
g_2	−						−	
g_3				−	−			
g_4				+	+			
g_5				+				
g_6				−				
g_7					+			
g_8					−			
g_9		+						
(g_{10})		−	+					
g_{11}	+	−						
g_{12}						+		−
g_{13}		U				+		−
g_{14}								+
g_{15}						−	−	
g_{16}							+	
(g_{17})		−						
(g_{18})		−	+					−

a closer examination of constraint g_{10}, the only other one containing g_3: $d_3 \leqslant d_2$. On reflection, constraint activity for g_{10} should not be allowed, i.e. $d_2 > d_3$. Examining the geometry of the screw more closely a new constraint is discovered:

$$g_{18}: d_2 \geqslant d_3 + 1.2990/N_T \tag{3.31}$$

3.3.9 Optimality rules

At this point we decide to add constraints $g_{17}(d_3^-)$ and $g_{18}(d_2^-, d_3^+, N_T^-)$ to the model and delete g_{10} as redundant. These changes are shown in parentheses in Table 3.2 and applying monotonicity analysis to this model we now obtain the following results, which represent necessary rules for optimality.

(R1) By MP1 wrt d_1, g_2 is active.
(R2) By MP1 wrt d_2, at least one constraint from the set $\{g_1, g_{11}, g_{13}\}$ is active.
(R3) By MP1 wrt L_2, at least one constraint from the set $\{g_3, g_6\}$ is active.
(R4) By MP1 wrt L_3, at least one constraint from the set $\{g_3, g_8\}$ is active.
(R5) By MP2 wrt R, either all constraints g_2, g_{12}, g_{13} and g_{15} are inactive or at least one from each of the sets $\{g_2, g_{15}\}, \{g_{12}, g_{13}\}$ is active.
(R6) By MP2 wrt N_S, either g_{15} and g_{16} are both active or they are both inactive.
(R7) By MP2 wrt R, if g_2 is active then at least one of $\{g_{12}, g_{13}\}$ is active. Then, by MP2 wrt N_T and (R1), g_{14} is active.

The original eight degrees of freedom have now been reduced to five, because of the identified active constraints:

$$g_2 : \pi \tau_{all} R d_1^3 = 6$$

$$g_{14} : N_T = 24$$

$$g_{17} : d_3 = 0.1875 \tag{3.32}$$

The remaining rules above give only conditional activity and in order to identify a single constraint as active dominance arguments are required. We will proceed with these later below. Thus, because of (R1),(R5) must be modified as

(R5′) By MP2 wrt R, at least one constraint from the set $\{g_{12}, g_{13}\}$ is active.

This interaction was also used in deriving (R7).

3.3.10 Active constraint elimination

The three active constraints, equations (3.32), are used to eliminate three variables from model 2, namely d_1, d_3 and N_T. There are two reasons for this. One is that monotonicity analysis on the new reduced model may reveal additional activity requirements. Another is that dominance arguments will be simpler in the reduced model. Which variables to eliminate is a judicious choice, based on what may be algebraically simpler and what may be a desirable form of the reduced model. The new model is as follows.

Model 3

minimize $f_0 = 0.25 \pi C_m [0.405(6/\pi \tau_{all} R)^{2/3} + L_2 d_2^2 + L_3 (0.1875)^2]$

subject to

$$g_1 = 38.88 + 96 L_2 + 96 L_3 - \pi \sigma_{all} d_2^3 \leqslant 0$$

$$g_3 = 8.345 - L_2 - L_3 \leqslant 0 \qquad g_4 = -9.595 + L_2 + L_3 \leqslant 0$$

$$g_5 = L_2 - 7.523 \leqslant 0 \qquad g_6 = 7.023 - L_2 \leqslant 0$$

$$g_7 = L_3 - 1.6525 \leqslant 0 \qquad g_8 = 1.1525 - L_3 \leqslant 0$$

$$g_9 = d_2 - 0.625 \leqslant 0 \qquad g_{11} = (6/\pi \tau_{all} R)^{1/3} - d_2 \leqslant 0$$

$$g_{12} = R - 0.2798 \leqslant 0$$

$$g_{13} = 1.5 \pi f R d_2^2 + 0.0313 R d_2 - 0.0625 \pi d_2 + 0.0052 f \leqslant 0$$

$$g_{15} = 8 - R N_S \leqslant 0 \qquad g_{16} = N_S - 52 \leqslant 0$$

$$g_{18} = 0.2416 - d_2 \leqslant 0 \tag{3.33}$$

The monotonicity table for model 3 is shown in Table 3.4. No new results are obtained from this table. So dominance arguments must be sought in order to clarify the previously stated conditional activities and to obtain further model reduction.

Table 3.4 Monotonicity table for model 3

Functions	d_2	L_2	L_3	R	N_s
f_0	+	+	+	−	
g_1	−	+	+		
g_3		−	−		
g_4		+	+		
g_5		+			
g_6		−			
g_7			+		
g_8			−		
g_9	+				
g_{11}	−			−	
g_{12}				+	
g_{13}	U			+	
g_{15}				−	−
g_{16}					+
g_{18}	−				

3.3.11 Activity map and dominance

Consider constraints g_3, g_6 and g_8 and the derived rules (R3) and (R4). An **activity map**, as shown in Fig. 3.2, can assist in dominance analysis. In this map all possible activity combinations for the three constraints are examined. Only constraint numbers are shown for simplicity, the overbar on a number indicating an inactive constraint while a plain number indicates an active one. The crosshatched areas indicate combinations that are not possible. In Fig. 3.2, three combinations are excluded since they violate rules (R3) and/or (R4), as indicated. The combination 368 is excluded by the maximal activity principle (Papalambros and Wilde, 1988), which basically says that the number of active constraints cannot exceed the number of variables in them. Finally, if g_3 is inactive then both g_6 and g_8 must be active giving $L_2 = 7.023$, $L_3 = 1.1525$, and $L_2 + L_3 = 8.1755$, which violates g_3. Hence this case is infeasible and excluded. From Fig. 3.2, it is now obvious that g_3 must be active irrespective of the activity of g_6 and g_8. This also makes g_4 and at least three constraints from the set $\{g_5, g_6, g_7, g_8\}$ be inactive. The conditional inactivity can be resolved easily as L_2, L_3 appear only in a small part of the model.

Consider eliminating $L_2 (= 8.345 - L_3)$ from the model. The submodel containing L_2 and L_3 becomes now

$$\text{minimize } f_0 = 0.25\pi C_m \{0.405(6/\pi\tau_{all} R)^{2/3} + 8.345 d_2^2 + L_3[(0.1875)^2 - d_2^2]\}$$

$$g_1 = 38.88 + 96(8.345) - \pi\sigma_{all} d_2^3 \leqslant 0$$

$$g_5 = 0.822 - L_3 \leqslant 0 \qquad g_6 = L_3 - 1.322 \leqslant 0$$

$$g_7 = L_3 - 1.6525 \leqslant 0 \qquad g_8 = 1.1525 - L_3 \leqslant 0 \qquad (3.34)$$

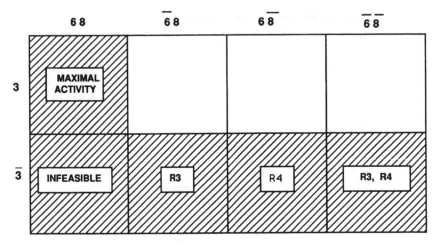

Fig. 3.2 Activity map for constraints g_3, g_6 and g_8.

In the objective, the monotonicity wrt L_3 is determined by the sign of the quantity

$$(0.1875)^2 - d_2^2 \qquad (3.35)$$

which is negative since $d_2 \geqslant 0.2416$ from g_{18}. Hence, the objective is decreasing wrt L_3 and an upper bound is required by MP1. Constraint g_6 is obviously the dominant and active one, so

$$L_{3*} = 1.322, L_{2*} = 7.023 \qquad (3.36)$$

and g_5, g_7, g_8 are inactive.

The above results lead to yet another further reduced model with only three degrees of freedom.

Model 4

minimize $f_0 = 0.25\pi C_m[0.405(6/\pi\tau_{all}R)^{2/3} + 7.023d_2^2 + 1.322(0.1875)^2]$

subject to

$$g_1 = 840 - \pi\sigma_{all}d_2^3 \leqslant 0 \qquad\qquad g_9 = d_2 - 0.625 \leqslant 0$$

$$g_{11} = (6/\pi\tau_{all}R)^{1/3} - d_2 \leqslant 0 \qquad g_{12} = R - 0.2798 \leqslant 0$$

$$g_{13} = 1.5\pi f R d_2^2 + 0.0313 R d_2 - 0.0625\pi d_2 + 0.0052 f \leqslant 0$$

$$g_{15} = 8 - R N_s \leqslant 0 \qquad\qquad g_{16} = N_s - 52 \leqslant 0$$

$$g_{18} = 0.2416 - d_2 \leqslant 0 \qquad\qquad\qquad (3.37)$$

The monotonicity table for this model is shown in Table 3.5. Rules (R2), (R5') and (R6) are still the only results derived from monotonicity principles.

At this point, the unknown monotonicity of g_{13} wrt d_2 prevents us from continuing the reduction process. Indeed,

$$\partial g_{13}/\partial d_2 = 3\pi f R d_2 + 0.0313 R - 0.0625\pi \qquad (3.38)$$

Table 3.5 Monotonicity table for model 4

Functions	Variables		
	d_2	R	N_S
f_0	+	−	
g_1	−		
g_9	+		
g_{11}	−	−	
g_{12}		+	
g_{13}	U	+	
g_{15}		−	−
g_{16}			+
g_{18}	−		

and for g_{13} to be increasing wrt d_2 we would need to have (for $f = 0.34$)

$$d_2 \geqslant 0.0595 R^{-1} - 0.0033 \triangleq D_2(R^-) \tag{3.39}$$

for the intire feasible range of d_2, R. We note that

$$\max D_2(R^-) = 0.0595 R_{\min}^{-1} - 0.0033 = 0.0595 N_{\text{Smax}}/8 - 0.0033$$
$$= (0.0595)(52)/8 - 0.0033 = 0.3983 \tag{3.40}$$

This last number falls within the known feasible range of $d_2, 0.2416 \leqslant d_2 \leqslant 0.625$, so g_{13} appears really nonmonotonic in the feasible domain.

3.3.12 Parametric models and case decomposition

The idea of **regional monotonicity** could be used here, i.e. try to identify in what interval of values of d_2 is g_{13} monotonic and examine each case separately, comparing the results at the end. This would be unnecessarily complicated here. Instead, a simple problem decomposition can be applied: case A with g_{13} inactive, and case B with g_{13} active. We can examine these two cases separately and compare the results.

Before we proceed with the cases, it is instructive to recast model 4 in a simplified parametric form, by introducing the parameters $K_0', K_i, i = 0, \ldots, 4$, and rearranging (Table 3.6). The revised model is as follows. Note that all parameters K_i relate to material properties.

Model 5

minimize $f_0 = K_0 R^{2/3} + K_0'[7.023 d_2^2 + 1.322(0.1875)^2]$

subject to

$$g_1 = K_1 - d_2 \leqslant 0 \qquad\qquad g_9 = d_2 - 0.625 \leqslant 0$$

$$g_{11} = K_2 R^{1/3} - d_2 \leqslant 0 \qquad g_{12} = R - 0.2798 \leqslant 0$$

$$g_{13} = K_3 R d_2^2 + 0.0313 R d_2 - 0.1963 d_2 + K_4 \leqslant 0$$

Table 3.6 Parameter definitions for model 5

$$K'_0 = 0.25\pi C_m$$
$$K_0 = K'_0 (0.405)(6/\pi\tau_{all})^{2/3}$$
$$K_1 = (840/\pi\sigma_{all})^{1/3}$$
$$K_2 = (6/\pi\tau_{all})^{1/3}$$
$$K_3 = 1.5\pi f$$
$$K_4 = 0.0052f$$

$$g_{15} = 8 - RN_S \leqslant 0 \qquad g_{16} = N_S - 52 \leqslant 0$$

$$g_{18} = 0.2416 - d_2 \leqslant 0 \tag{3.41}$$

Consider now case A with g_{13} inactive. Then g_{12} is active from rule (R5') and $R_* = 0.2798$. From rule (R2) we now have

$$d_{2*} = \max(K_1, K_2 R_*^{1/3}, 0.2416) \tag{3.42}$$

and from rule (R5) we have

$$8/R_* \leqslant N_{S*} \leqslant 52 \tag{3.43}$$

Note that one degree of freedom remains at the optimum, as N_m, N_S can be selected to give the largest rational number not exceeding 0.2798 and satisfying the range in equation (3.43). Any solution thus obtained must be checked that it satisfies the remaining inactive constraints. This **feasibility** check is frequently overlooked in the application of monotonicity analysis, leading to erroneous conclusions.

Next, consider case B with g_{13} active. Rules (R2) and (R5') are satisfied and no new activity results can be obtained. Note that now, locally, g_{13} must be decreasing wrt d_2, i.e. $g_{13}(d_2^-, R^+) \leqq 0$, while $f_0(d_2^+, R^-)$. An implicit solution of g_{13} gives $d_2 = \varphi_{13}(R^+)$ and substitution in the objective gives $f_0(d_2^+, R^-) = f_0[\varphi_{13}^+(R^+), R^-] = f_0(R)$ with R having unclear monotonicity. There are two degrees of freedom left, but a one-dimensional search in d_2 would suffice if g_{13} is solved explicitly for R and the objective is expressed as a function of d_2 only. Constraints g_{15} and g_{16} could be replaced by

$$R \geqslant 8N_S^{-1} \geqslant 8/52 = 0.1538 \tag{3.44}$$

There is a lingering concern, though, regarding the physical meaning of g_{13} being active. Essentially, friction forces and applied forces on an equivalent inclined plane would be equal and motion would be impending. This would not represent a stable design for the lead screw, albeit possibly an optimal one. The designer must then examine the implications on the appropriateness of the model and/or the parameter values selected. Also, satisfaction of model validity for the beam stress formula (3.13) should be checked. The results obtained numerically using SQP code NLPQL (Schittkowski, 1984) with $C_m = 10.0$ and continuous values for the integer variables indicate constraints $\{g_2, g_3, g_6, g_{14}, g_{17}\}$ and $\{g_{12}, g_{18}\}$ are active. The activity of the first five was the one discovered *a priori* by analysis. Constraints $\{g_1, g_4, g_5, g_7, g_8, g_9, g_{10}, g_{11}, g_{13}, g_{15}, g_{16}\}$ are inactive.

3.3.13 Directing an equality

It is interesting to note that g_{18} is active in the final solution. Indeed, according to Spotts (1985) the required relation is an equality:

$$h_{10}: d_3 = d_2 - 1.2990/N_T \qquad (3.45)$$

rather than the inequality (3.31). Instead of using g_{18} in the model we could have used h_{10}, which would prevent direct application of MP1.

However, an equality constraint can be viewed as an active inequality that has been 'properly directed' (Papalambros and Wilde, 1988). One way to determine such a direction is to use the monotonicity principles. Ignoring equation (3.30) for the moment, and with equation (3.45) replacing g_{10}, we see that by MP1 wrt d_3 a lower bound is required. Then h_{10} can provide this bound if directed as $d_3 \geq d_2 - 1.2990 N^{-1}$ or $d_2 - 1.2990 N_T^{-1} - d_3 \leq 0$.

Not much inference is possible when the directed h_{10} is used in the model instead of g_{18}. One could proceed by assuming h_{10} inactive and include g_{17} which would then be active. This would take us essentially through the same steps as before, checking h_{10} for violation in the final solution. Interestingly, numerical results obtained for such a scenario using NLPQL indicate h_{10} is satisfied in the final solution with a zero multiplier value.

3.4 MODELING AUTOMATION

Traditional numerical solutions rely almost exclusively on iterative use of local information, such as values of functions, gradients or composition of locally active sets. Knowledge accumulated during such an iterative process may be used heuristically to speed up the solution process, for example, as discussed in Arora and Baenzinger (1986) and Adeli and Balasubramanyam (1988a, b). Analytical techniques aiming at active constraint identification, such as monotonicity analysis, use primarily **global** information, which is true for all points in the design space. A production system using rules for global information processing was first introduced in Li and Papalambros (1985a). Combined qualitative and quantitative reasoning for optimization in an AI environment has also been discussed in Agogino and Almgren (1987), Hammond and Johnson (1987, 1988), Hansen, Jaumard and Lu (1988a, b), and Watton and Rinderle (1991). The need for developing new types of algorithms using both global and local knowledge was explored in Li (1985) and Li and Papalambros (1988). Artificial intelligence (AI) techniques, including symbolic manipulations, have been proposed as modeling tools; see, for example, Li and Papalambros (1985b) and Choy and Agogino (1986).

One possibility offered by the use of AI techniques in modeling is to generate automatically global information about constraint activity and model boundedness, which can then be given directly to a global or local–global strategy, as described in the above references. This will remove, at least in part, the difficulty that relatively unskilled users may encounter in discovering rigorous global knowledge. Moreover, it eliminates tedious procedures prone to errors, when manually performed. The system **PRIMA** (production system for implicit elimination in monotonicity analysis) was implemented in the OPS5/OPS87 production system development tool (Forgy, 1981).

In PRIMA (Rao and Papalambros, 1987) the monotonicity table forms the very basic abstract data structure in the problem domain. Rules based on the monotonicity principles are applied to obtain **facts** about the design mode. These facts, called **state predicates**, being derived from necessary conditions, must be satisfied by every optimum solution, called a **state**. As these facts are derived, the system automatically generates LISP-like macros in the form of predicate functions, each of which should evaluate to true at every optimal design or state. In the presence of monotonicity, additional facts can be derived after one or more active constraints have been implicitly eliminated from the model (together with one or more corresponding variables). The result is again a new derived monotonicity table that may generate new state predicates. This process continues until all active constraints are eliminated, or no useful monotonicities remain.

The primary task, then, is to search an 'implicit elimination tree' (Fig. 3.3), where (i) visiting a node, i.e. monotonicity table, consists of applying mathematically rigorous necessary conditions implemented in the form of **rules**, (ii) the tree arcs (connecting one node to another) correspond to the process of implicit elimination, and (iii) the choice of **pivots** used in elimination (specific variable and constraint that will be eliminated) corresponds to selective expansion of nodes and is based on heuristic rules. This classification of rigorous rules (at

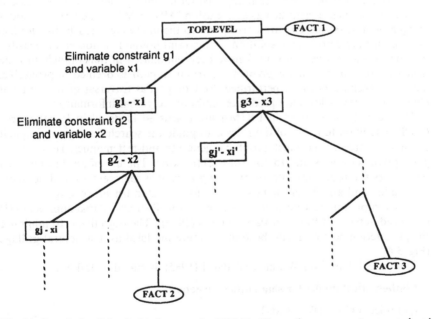

Fig. 3.3 Generic implicit elimination tree in PRIMA. The nodes $g_1 - x_1$ etc. are reduced monotonicity tables obtained by implicit elimination. Facts 1, 2, 3 etc. could typically be obtained at any node in this tree.

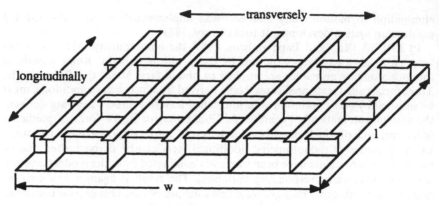

Fig. 3.4 Ship bottom structure.

the nodes) and heuristic rules (at the arcs) can be considered a characteristic of this problem domain.

The model of the desired automation procedure is as follows. A tree data structure, called the **implicit elimination tree**, Fig. 3.3, is built in the space of feasible designs. Each node is a monotonicity table containing all the boundedness and constraint activity information that characterizes it. The root of the tree is the given top-level monotonicity table. The arcs of the tree structure correspond to a pivot used during the implicit elimination process. The root node has a degree of freedom (dof) greater than or equal to unity, which might decrease if more constraints are identified as active using MP1 or MP2. The terminal nodes along a depth-first search path correspond to the following cases: (i) the dof has been reduced to zero (i.e. the solution is constraint bound in a non-empty feasible domain); (ii) all active constraints have been eliminated in that branch; (iii) the heuristics do not provide a pivot, the particular path is deemed unprofitable, and backtracking must be performed. Full or partial traversal of this implicit elimination tree yields state predicates embodying global information.

The heuristic rules used in PRIMA are discussed in detail in Papalambros (1988). The main function of the rules is to guide the search through an implicit elimination tree in the most profitable generate-and-test manner, i.e. to make good pivot selections and to control backtracking. The use of an AI approach here offers the usual advantages of having a flexible representation of heuristic knowledge and an efficient data structure for repeated list processing.

An example application is repeated here from Rao and Papalambros (1991) (a revised version of Rao and Papalambros (1987)). The objective is to minimize the production cost of a ship bottom structure modeled as a structural grillage (Fig. 3.4).

The model, based on Winkle and Baird (1985), is stated as follows.

Mathematical model for ship bottom structure

minimize $VC = MW + k\,MH$
subject to

h_1: SM_{freqd} $= (4.74 c_{freqd} h_{freqd} s_{floor} l_{floor}^2)/100^3$
h_2: c_{freqd} $= 0.9$

h_3:	s_{floor}	$= s_{frame}$
h_4:	A_{kr}	$= w_{kr}t_{kr}$
h_5:	A_{tp}	$= w_{tp}t_{tp}$
h_6:	l_{floor}	$= B - 0.750$
h_7:	MH	$= f_1(l, w, \text{etc.})$
h_8:	MW	$= w_{ms}f_2(l, w, \text{etc.})$
h_9:	n_{floor}	$= l/s_{floor}$
h_{10}:	$n_{keelson}$	$= w/s_{keelson} - 1$
h_{11}:	SM_{floor}	$= f_3(l, w, \text{etc.})$
g_1:	t_{keel}	$\geqslant t_{bmin} + 0.0015$
g_2:	t_{cg}	$\geqslant (0.063L + 5)/1000$
g_3:	A_{tp}	$\geqslant (0.168L^{3/2} - 8)/10000$
g_4:	$s_{keelson}$	$\leqslant 2.13$
g_5:	h_{cg}	$\geqslant h_{keelson}$
g_6:	$h_{keelson}$	$\geqslant h_{floor}$
g_7:	$t_{keelson}$	$\geqslant (0.063L + 4)/1000$
g_8:	A_{kr}	$\geqslant (0.038L^{3/2} + 17)/10000$
g_9:	h_{freqd}	$\geqslant d$
g_{10}:	h_{freqd}	$\geqslant 0.66D$
g_{11}:	$w_{fr}(h_{floor} + t_{fr})$	$\geqslant w_{fr}0.0625\, l_{floor}$
g_{12}:	h_{floor}	$\geqslant 0.0625l_{floor}$
g_{13}:	$w_{fr}t_{floor}$	$\geqslant w_{fr}[(h_{floor} + t_{fr})/100 + 0.003]$
g_{14}:	t_{floor}	$\geqslant h_{floor}/100 + 0.003$
g_{15}:	t_{floor}	$\leqslant 0.0115$
g_{16}:	SM_{floor}	$\geqslant SM_{freqd}$
g_{17}:	s_{frame}	$\leqslant (2.08L + 438)/1000$
g_{18}:	D_s	$\geqslant D$
g_{19}:	t_{bottom}	$\geqslant t_{bmin}$
g_{20}:	t_{bmin}	$\geqslant (s_{frame}/519)[(L - 19.8)\max(d/D_s, 0.65)]^{1/2} + 0.0025$
g_{21}:	t_{bmin}	$\geqslant (s_{frame})(L + 45.73)/(25L + 6082)$
g_{22}:	w_{keel}	$\geqslant 0.750$

$$(3.46)$$

Functions f_1, f_2, and f_3 are defined using man–hour estimates, total volume of structural elements, and plate–stiffener combination section moduli respectively. The problem has 11 equality constraints and 20 degrees of freedom. The design variables are shown in Table 3.7.

Equality constraints are explicitly eliminated and a top-level monotonicity table is obtained where PRIMA deduces six different facts. The first monotonicity principle is repeatedly applied eight times and constraints $g_1, g_3, g_4, g_5, g_8, g_{22}$ are identified as being active. These six active constraints and the variables appearing in each of them provide several possibilities for selecting implicit elimination pivots, and the heuristic rules are now used. Starting with the top-level monotonicity table, the sequence of pivots $\{g_1 - x_8, g_3 - x_{20}, g_8 - x_{19}, g_5 - x_2, g_{22} - x_{18}, g_4 - x_7\}$ leads to the table shown in Fig. 3.5, where the following additional facts or rules are obtained.

Fact 1: At least one of g_{16}, g_{13} and g_{11} is active.
Fact 2: At least one of g_{21} and g_{20} is active.
Fact 3: At least one of g_{16}, g_{14} and g_{13} is active.
Fact 4: At least one of g_{16} and g_{19} is active.
Fact 5: At least one of g_{21}, g_{20}, g_{17} and g_{16} is active.

Table 3.7 Variables in the ship structure design problem

x_1	A_{kr}	Cross-sectional area of keelson rider plate
x_2	A_{tp}	Horizontal top plate area at midship
x_3	c_{freqd}	A constant, dependent upon ship type
x_4	D_s	Scantling depth
x_5	h_{cg}	Height of center girder
x_6	h_{floor}	Height of vertical floor plate
x_7	h_{freqd}	Height for calculating floor section modulus
x_8	$h_{keelson}$	Height of vertical keelson plate
x_9	l_{floor}	Length of floor
x_{10}	MH	Manhours of labor
x_{11}	MW	Material weight
x_{12}	n_{floor}	Number of floors
x_{13}	$n_{keelson}$	Number of keelsons
x_{14}	s_{floor}	Floor spacing
x_{15}	s_{frame}	Frame spacing
x_{16}	$s_{keelson}$	Keelson spacing
x_{17}	SM_{floor}	Floor section modulus
x_{18}	SM_{freqd}	Floor required section modulus
x_{19}	t_{bmin}	Minimum bottom shell thickness
x_{20}	t_{bottom}	Bottom shell thickness
x_{21}	t_{cg}	Center girder thickness at midships
x_{22}	t_{floor}	Floor thickness
x_{23}	t_{fr}	Floor rider plate thickness
x_{24}	t_{keel}	Keel plate thickness
x_{25}	$t_{keelson}$	Keelson thickness
x_{26}	t_{kr}	Keelson rider plate thickness
x_{27}	t_{tp}	Top plate thickness
x_{28}	w_{fr}	Width of floor rider plate
x_{29}	w_{keel}	Width of keel plate
x_{30}	w_{kr}	Keelson rider plate width
x_{31}	w_{tp}	Width of top plate

These not very obvious facts are not sufficient for identifying an optimum, but they can disqualify any existing solution that does not satisfy them.

Another use of AI technology in optimal design modeling is the selection and assembly of appropriate optimization models from a database of available configurations within a design class. For example, in the design of truss structures an expert system can be used to determine the appropriate truss configuration, the associated parameter values, such as various bounds, and to invoke the mathematical model for subsequent numerical optimization (Adeli and Bala-subramanyam, 1988a, b).

A generic model for developing a knowledge-based methodology to aid design optimization formulations was proposed by Balachandran and Gero (1987) and further articulated by Balachandran (1991). Many other authors have commented on the general complementary nature of mathematical optimization and knowledge-based systems. The common idea is the obvious one: if a precise analytical model can be generated with resonable effort, then a decision model based on mathematical optimization is preferable. In the absence of analytical

MONOTONICITY TABLE toplevel--> g1--x8-->g3--x20-->g8--x19-->g5--x2-->g22--x18-->g4--x7

	x1	x3	x4	x5	x6	x9	x10	x11	x12	x13	x14	x15	x16	x17
obj	U	U	U	U	-	+	-	+	+	+	U	U	U	+
g2														
g6		+	-											
g7											-			
g9			-											
g10				-										
g11		-		-					-					-
g12		-						-						
g13		+			+			-	+					-
g14		+						+						
g15								+						U
g16	U	U			U	U		U	U					
g17				+	+	U								
g18	-													
g19					+	-				+				
g20	-	-			+					-				
g21	-	-			+					-				

THE FOLLOWING FACT IS DERIVED W.R.T. VAR x17
AT LEAST ONE OF THE FOLLOWING CONSTRAINTS IS ACTIVE- g16--g13--g11

THE FOLLOWING FACT IS DERIVED W.R.T. VAR x13
AT LEAST ONE OF THE FOLLOWING CONSTRAINTS IS ACTIVE- g21--g20

THE FOLLOWING FACT IS DERIVED W.R.T. VAR x11
AT LEAST ONE OF THE FOLLOWING CONSTRAINTS IS ACTIVE- g16--g14--g13

THE FOLLOWING FACT IS DERIVED W.R.T. VAR x9
AT LEAST ONE OF THE FOLLOWING CONSTRAINTS IS ACTIVE- g19--g16

THE FOLLOWING FACT IS DERIVED W.R.T. VAR x6
AT LEAST ONE OF THE FOLLOWING CONSTRAINTS IS ACTIVE- g21--g20--g17--g16

Fig. 3.5 Final PRIMA results for ship bottom structure problem.

models codification of domain knowledge may be possible in the form of rules or other AI data structures, which can be then manipulated to make useful deductions.

Of special interest here is also work by Schittkowski on an intelligent modeling and mathematical programming package (Schittkowski, 1985), and the available commercial package GAMS (Brooke, Kendrick and Meeraus, 1988). In Schittkowski's system a set of rules assists the user in selecting algorithms and parameters within algorithms, based on an evaluation of the type of model that needs to be solved. The rules are a result of numerical analysis knowledge and experimentation with actual problems. Each rule has a level of certainty or veracity probability associated with it. These are propagated and indicated in the deductions produced by the rules. The system keeps a record of all major decisions made and attendant results every time a new problem is being solved. This information is used to modify or sharpen the rules and their associated veracity probabilities. Thus a learning function is supported. GAMS (General Algebraic Modeling System) is a high level language that integrates features of a relational database in formulating explicit (algebraic) mathematical programming models. Originally it was conceived to facilitate easy data entry and modification of very large linear programming models. Current extensions can handle nonlinear and binary (zero–one) linear programming models. The major advantage of a GAMS environment in the present context is the ability to generate quickly modifications of models by adding, deleting or changing the expressions of constraints without worrying about FORTRAN reprogramming. Such functions for optimization computations are now beginning to appear also in general-purpose packages such as MATHEMATICA (Wolfram, 1988) and MATLAB (Grace, 1991).

3.5 CONCLUDING REMARKS

When an optimization study is undertaken, a careful model analysis will save a lot of subsequent effort in trying to obtain good and reliable results from numerical computations. An initial model must be properly cast into the standard form for numerical processing and possible unboundedness should be checked. A clear understanding of the physical meaning of each constraint is necessary, as well as an exploration of how constraints may interact with each other. Model validity limitations must be always checked to ascertain that they are indeed satisfied by the optimal values of the design variables. Although we have demonstrated these ideas on relatively small explicit models, such ideas are important also when implicit models are used. Proper modeling is somewhat of an art but the tactics involved are grounded frequently on a rigorous understanding of how optimization methods work. Many of these ideas will continue to become increasingly automated.

ACKNOWLEDGEMENTS

Thanks are due to the University of Michigan graduate students B. Alexander and B. Rycenga for posing the linear actuator problem and R.P. Krishnamachari for helping with the final form of the manuscript.

REFERENCES

Adeli, H. and Balasubramanyam, K.V. (1988a) A knowledge-based system for design of bridge trusses. *Journal of Computing in Civil Engineering*, **2** (1), 1–20.

Adeli, H. and Balasubramanyam, K.V. (1988b) *Expert Systems for Structural Design – A New Generation*, Prentice-Hall, Englewood Cliffs, NJ.

Adeli, H. and Mak, K.Y. (1988) Architecture of a coupled expert system for optimum design of plate girder bridges. *Engineering Applications of AI*, **1**, 277–85.

Agogino, A.M. and Almgren, A.S. (1987) Techniques for integrating qualitative reasoning and symbolic computation in engineering optimization. *Engineering Optimization*, **12** (2), 117–35.

Alexander, B. and Rycenga, B. (1990) Optimization of a linear actuator. *Rep. UM-MEAM-555-91*, Department of Mechanical Engineering and Applied Mechanics, The University of Michigan, Ann Arbor, MI.

Arora, J.S. and Baenziger, G. (1986) Uses of artificial intelligence in design optimization. *Computer Methods in Applied Mechanics and Engineering*, **54**, 303–23.

Balachandran, M. (1991) *Knowledge Based Optimum Design*, Computational Mechanics Publications, Southampton.

Balachandran, M. and Gero, J.S. (1987) A knowledge-based approach to mathematical design, modeling, and optimization. *Engineering Optimization*, **12**, 91–116.

Brooke, A., Kendrick, D. and Meeraus, A. (1988) *GAMS – A User's Guide*, The Scientific Press, Redwood City, CA.

Choy, J.K. and Agogino, A.M. (1986) SYMON: automated symbolic monotonicity analysis for qualitative design optimization, in *Proc. ASME 1986 Int. Comp. in Eng. Conf.*, Chicago, IL, pp. 207–12.

Forgy, C.L. (1981) *OPS5 User's Manual*, Department of Computer Science, Carnegie Mellon University, Pittsburgh, PA.

Grace, A. (1991) *Optimization Toolbox for Use with MATLAB*, The Math Works Inc.

Hammond, C.R. and Johnson, G.E. (1987) A general approach to constrained optimal design based on symbolic mathematics, in *Advances in Design Automation – 1987* (ed. S.S. Rao), DE. Vol. 10-1, ASME, New York, pp. 31–40.

Hammond, C.R. and Johnson, G.E. (1988) The method of alternate formulations, an automated strategy for optimal design. *Journal of Mechanisms, Transmissions, and Automation in Design*, **110**, 459–63.

Hansen, P., Jaumard, B. and Lu, S.H. (1988a) A framework for algorithms in globally optimal design. *Journal of Mechanisms, Transmissions, and Automation in Design*, **111** (3), 353–60.

Hansen, P., Jaumard, B. and Lu, S.H. (1988b) An automated procedure for globally optimal design. *Journal of Mechanisms, Transmissions, and Automation in Design*, **111** (3), 361–7.

Juvinall, R.C. (1983) *Fundamentals of Machine Components*, Wiley, New York.

Li, H.L. (1985) Design optimization strategies with global and local knowledge. PhD Dissertation, Department of Mechanical Engineering and Applied Mechanics, University of Michigan, Ann Arbor, MI.

Li, H.L. and Papalambros, P. (1985a) A production system for use of global optimization knowledge. *Journal of Mechanisms, Transmissions, and Automation in Design*, **107** (2), 277–84.

Li, H.L. and Papalambros, P. (1985b) REDUCE applications in design optimization, in CAD/CAM Robotics and Auto, Int. Conf. Proc., Tucson, A.

Li, H.L. and Papalambros, P. (1988) A combined local–global active set strategy for nonlinear design optimization. *Journal of Mechanisms, Transmissions, and Automation in Design*, **110** (4), 464–471.

Papalambros, P. (1988a) Remarks on sufficiency of constraint-bound solutions in optimal design, in *Advances in Design Automation – 1988* (ed. S.S. Rao), ASME, New York. Also (1993) *Journal of Mechanical Design*, **115** (3), 374–9.

Papalambros, P. (1988b) Enhancements in design optimization problem solving: a knowledge-based approach. *Rep. TKHS-88.79*, SAAB-Scania aircraft Division, Linköping.

Papalambros, P. and Li, H.L. (1983) Notes on the operational utility of monotonicity in optimization. *Journal of Mechanisms, Transmissions, and Automation in Design,* **105** (2), 174–81.

Papalambros, P.Y. and Wilde, D.J. (1988) *Principles of Optimal Design,* Cambridge University Press, New York.

Pomrehn, L.P. and Papalambros, P.Y. (1992) Constraint activity revisited: application to global and discrete design optimization, in *Advances in Design Automation* (ed. D.A. Hoeltzel), DE-Vol.44-1, ASME, New York, pp. 223–9.

Rao, J.R. and Papalambros, P. (1987) Implementation of semi-heuristic reasoning for boundedness analysis of design optimization models, in *Advances in Design Automation – 1987* (ed. S.S. Rao), ASME, New York.

Rao, J.R. and Papalambros, P. (1991) PRIMA: a production-based implicit elimination system for monotonicity analysis of optimal design models. *Journal of Mechanical Design,* **113** (4), 408–15.

Schittkowski, K. (1984) NLPQL: a FORTRAN subroutine solving constrained nonlinear programming problems. *Tech. Rep.,* Institute für Informatik, University of Stuttgart.

Schittkowski, K. (1985) EMP: an expert system for mathematical programming. *Mathematical Institute Rep.,* University of Bayreuth.

Siddall, J.N. (1972) *Analytical Decision-Making in Engineering Design,* Prentice-Hall, Englewood Cliffs, NJ.

Spotts, M.F. (1985) *Design of Machine Elements,* Prentice-Hall, Englewood Cliffs, NJ.

Thomas, R. (1990) Development of an intelligent front-end for design optimization programs. Diploma Thesis, *Rep. LiTH-IKP-R-604,* Department of Mechanical Engineering, University of Linköping.

Watton, J.D. and Rinderle, J.R. (1991) Symbolic design optimization: a computer aided method to increase monotonicity through variable reformulation. *Rep. 24-56-91,* Engineering Design Research Center, Carnegie Mellon University, Pittsburgh, PA.

Wilde, D.J. (1978) *Globally Optimal Design,* Wiley–Interscience, New York.

Wilde, D.J. and Beightler, C.S. (1967) *Foundations of Optimization,* Prentice-Hall, Englewood Cliffs, NJ.

Winkle, I.E. and Baird, D. (1985) Towards more effective structural design through synthesis and optimization of relative fabrication costs. Proceedings of the Joint Meetings of the Royal Institute of Naval Architecture (RINA) and the Institute of Engineers and Shipbuilders, Glasgow.

Wolfram, S. (1988) *Mathematica: A System for Doing Matematics by Computer,* Addition-Wesley, Redwood City, CA.

4

Generalized geometric programming and structural optimization

SCOTT A. BURNS and NARBEY KHACHATURIAN

4.1 INTRODUCTION

This chapter contains a brief presentation of some of the new developments in optimal structural design which have proven to be effective tools in design of structures, and which have the potential of introducing major changes in our approach to optimization of structures. Specifically, our attention will be directed to the generalized geometric programming (GGP) method and its application to problems with both inequality and equality constraints. This represents typical problems in optimization of structures.

The GGP method represents a significant development in mathematical programming. It is an extension of classical geometric programming, which originally concerned problems with positive coefficients on all terms only. In its generalized form, it reduces a nonlinear problem to a stable sequence of linear problems, and permits the coefficients to be either positive or negative. The method provides a general and convenient approach for nonlinear optimization problems (Avriel, Dembo and Passy, 1975). The method has been applied successfully to many areas of engineering; several dozen engineering applications of geometric programming are reported in a 1978 survey paper (Rijckaert and Martens, 1978), and more recently, geometric programming algorithms have been developed for optimization of space structures and nonprismatic plate girders (Adeli and Kamal, 1986; Abuyounes and Adeli, 1986; Adeli and Chompooming, 1989). With recent extensions to include equality constraints in the problem formulation (Burns, 1987b), GGP becomes particularly well suited to the optimization of structures (Burns, 1985).

A second development is the integrated approach to structural optimization, in which a mathematical programming problem is posed in the mixed space of design variables (such as dimension) and behavior variables (such as forces and displacements). This integrated formulation contains both equality and inequality constraints. The equality constraints interrelate the structural behavior variables and perform the dual function of a structural analysis and an implicit sensitivity analysis; the inequality constraints establish the limits of safety and serviceability of the structure.

In the conventional approach of structural optimization by mathematical programming, the solution process is carried out in several iterations of analysis and design. The analysis phase is concerned with the equality constraints assuring that the equilibrium, compatibility, and constitutive equations are satisfied. This phase is also concerned with determining sensitivity information, such as changes in stress with respect to changes in geometric proportion. This information is passed to the design phase, which is concerned with adjusting the design variables so that an objective function is optimized while inequality constraints on behavior remain satisfied. The nature of the two-phase approach is suoh that a restricted amount of information is passed between the analysis and design phases. The integrated formulation eliminates this bottleneck, yielding significant improvements in performance, as discussed in subsequent sections.

The integrated formulation has appeared in the literature periodically. In early work, Fox and Schmit applied the penalty function and conjugate gradient methods to solve the integrated formulation of several trusses, giving rise to nonlinear programs with 30–78 design variables (Schmit and Fox, 1965; Fox and Schmit, 1966). Later, Fuchs (1982) developed an integrated formulation for optimization of trusses using three different sets of analysis constraints within the integrated formulation. Haftka (1985) applied a preconditioned conjugate gradient method to solve the integrated formulation of a 72-bar space truss (120 design variables), and Newton's method to optimize a damage tolerant wing-box structure (45 design variables). Soeiro and Hoit (1987) have applied the integrated formulation to optimize a piecewise prismatic cantilever beam and a one-story one-bay frame. Chibani, Burns and Khachaturian (1989) apply decomposition techniques to the problem of member sizing of truss structures under alternative loading conditions using the integrated formulation.

So far, the research effort in structural optimization has been predominantly in the solution of problems in which the geometry is given and the optimal proportioning of the members is sought. Shape optimization has many applications although they are generally of a higher level of complexity than the problem of member sizing. This chapter extends the integrated formulation to the problem of shape optimization of continuum structures modeled by finite elements.

In the discussions that follow, the GGP method will be reviewed briefly for application to problems with both equality and inequality constraints. Two examples will be presented to demonstrate the application of the GGP method to different problems in structural optimization. Then an integrated formulation will be developed for shape optimization of continuum structures and an example will be presented.

4.2 GENERALIZED GEOMETRIC PROGRAMMING

This section presents a mathematical programming technique to solve large-scale, nonlinear, mixed equality–inequality constrained problems. It is based on a method called generalized geometric programming (GGP) (Avriel, Dembo and Passy, 1975), with a modification to include equality constraints (Burns, 1987b).

Consider the following three algebraic constructions: monomial,

$$u(x) = c \prod_{j=1}^{n} x_j^{a_j}$$

for example

$$u = 3.2x_1^2x_2^{1.5}x_3^{-0.5}$$

posynomial,

$$p(x) = \sum_{i=1}^{t} u_i(x)$$

for example

$$p = 3.2x_1^2x_2^{1.5}x_3^{-0.5} + 5x_2^{-1}$$

polynomial,

$$g(x) = p^+(x) - p^-(x)$$

for example

$$g = 4x_1^2x_2^2 - 3.2x_1^2x_2^{1.5}x_3^{-0.5} - 5x_2^{-1}$$

All monomial coefficients, c, are required to be positive, but the exponents, a, are unrestricted in sign. The variables, x, are assumed to be strictly positive in value. Note that a posynomial is simply a polynomial with all coefficients positive.

GGP solves mathematical programs that are posed in polynomial form, as defined above. It operates on the posynomial parts of the polynomial objective and constraint functions using a process called 'condensation', as described in the next section.

4.2.1 Condensation

Condensation is the process of approximating a posynomial function with a monomial function (Duffin, 1970; Avriel and Williams, 1970; Pascual and Ben-Israel, 1970; Passy, 1971). It is based on the weighted arithmetic–geometric (A–G) mean inequality (Hardy, Littlewood and Polya, 1959; Beckenbach and Bellman, 1961),

$$\sum_i u_i \geqslant \prod_i (u_i/\delta_i)^{\delta_i}$$

where u_i are positive values, the δ_i are positive weights, and $\sum \delta_i = 1$. For example, if $u = \{1, 2, 3\}$ and $\delta = \{1/3, 1/3, 1/3\}$, then the sum of these three numbers, 6, is greater than the weighted product of the numbers, 5.451. If the u_i in this inequality are chosen to represent monomial functions instead of numbers, then the left-hand side of this inequality becomes a posynomial. The right-hand side is a product of weighted monomials, which is itself a monomial. It is easily shown that when all of the u_i/δ_i are equal, the inequality holds as an equality.

When applied to a posynomial, the A–G inequality converts the posynomial into an approximating monomial. The monomial produced is dependent on the selection of weights, which can be any set of positive values that sum to unity. One very useful choice is to set the weights equal to the fraction that each monomial term of the posynomial contributes to the total value of the posynomial, when evaluated at some operating point, $\bar{x}$:

$$\delta_i = \frac{u_i(\bar{x})}{p(\bar{x})} \tag{4.1}$$

It can be seen that all u_i/δ_i are equal when u is evaluated at the operating point.

Thus the approximating monomial is equal in value to the posynomial when evluated at $\bar{x}$.

The process of condensing a posynomial to a monomial may be represented symbolically as

$$C[p(x), \bar{x}] = \prod_{i=1}^{t} [u_i(x)/\delta_i]^{\delta_i}$$

where the δ_i are defined by equation (4.1). Consider the following example. Condense $2x_1^2 x_2 + 4x_1^{-1}$ at $\bar{x} = (2, 1)$.

$$p(\bar{x}) = 2(2)^2(1) + 4(2)^{-1} = 10$$

$$\delta_1 = 2(2)^2(1)/10 = 0.8$$

$$\delta_2 = 4(2)^{-1}/10 = 0.2$$

$$\prod_{i=1}^{t} [u_i(x)/\delta_i]^{\delta_i} = \left(\frac{2x_1^2 x_2}{0.8}\right)^{0.8} \left(\frac{4x_1^{-1}}{0.2}\right)^{0.2} = 3.79 x_1^{1.4} x_2^{0.8}$$

Thus

$$C[2x_1^2 x_2 + 4x_1^{-1}, \bar{x} = (2, 1)] = 3.79 x_1^{1.4} x_2^{0.8}$$

4.2.2 Problem statement

GGP is concerned with minimizing a polynomial objective function subject to polynomial inequality constraints:

$$\begin{aligned}
\text{minimize} \quad & g_0(x) \\
\text{subject to} \quad & g_k(x) \leqslant 0 && k = 1, 2, \ldots, m \\
& 0 < x_j^{\text{LB}} \leqslant x_j && j = 1, 2, \ldots, n
\end{aligned}$$

where x_j^{LB} are positive lower bounds on the variables, and the polynomial functions, g, are differences of posynomial functions. The GGP method first introduces a new variable so that the objective function becomes linear. The old nonlinear objective function is absorbed as an additional constraint:

$$\begin{aligned}
\text{minimize} \quad & x_0 \\
\text{subject to} \quad & g_0(x) \leqslant x_0 \\
& g_k(x) \leqslant 0 && k = 1, 2, \ldots, m \\
& 0 < x_j^{\text{LB}} \leqslant x_j && j = 1, 2, \ldots, n
\end{aligned}$$

Next, each polynomial is separated into its positive and negative parts, giving differences of pairs of posynomials:

$$\begin{aligned}
\text{minimize} \quad & x_0 \\
\text{subject to} \quad & p_0^+(x) - p_0^-(x) \leqslant x_0 \\
& p_k^+(x) - p_k^-(x) \leqslant 0 && k = 1, 2, \ldots, m \\
& 0 < x_j^{\text{LB}} \leqslant x_j && j = 1, 2, \ldots, n
\end{aligned}$$

Finally, all of the negative terms are brought to the right-hand side of the inequalities and then divided through to yield a quotient form:

Quotient form

minimize x_0

subject to $\dfrac{p_0^+(x)}{p_0^-(x) + x_0} \leqslant 1$

$\dfrac{p_k^+(x)}{p_k^-(x)} \leqslant 1$ $k = 1, 2, \ldots, m$

$0 < x_j^{LB} \leqslant x_j$ $j = 1, 2, \ldots, n$ (4.2)

Note that because the variables are positive, there is no danger of dividing by zero when the negative terms are divided into the positive terms. The problem in quotient form has the same feasible region and solutions as the original GGP problem.

From this point on, the following problem will be used to demonstrate the operation of the GGP method:

$$\text{minimize}\quad x_2$$
$$\text{subject to}\quad x_1^2 - 10x_1 - 2x_2 + 29 \leqslant 0$$
$$-x_1^2 + 8x_1 + 4x_2 - 40 \leqslant 0$$
$$-x_1^2 + 12x_1 - x_2 - 32 \leqslant 0$$
$$x_1 \geqslant 1, x_2 \geqslant 1 \qquad (4.3)$$

The feasible region of this problem is the shaded area in Fig. 4.1. Here, x_1 and x_2 correspond to the horizontal and vertical axes, respectively. The objective function is chosen to be linear from the start to facilitate a two-dimensional

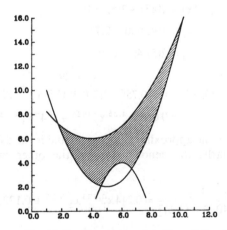

Fig. 4.1 Feasible region of problem (4.3).

graphical representation. This problem in quotient form is

$$\text{minimize} \quad x_2$$

$$\text{subject to} \quad \frac{x_1^2 + 29}{10x_1 + 2x_2} \leqslant 1$$

$$\frac{8x_1 + 4x_2}{x_1^2 + 40} \leqslant 1$$

$$\frac{12x_1}{x_1^2 + x_2 + 32} \leqslant 1$$

$$x_1 \geqslant 1, x_2 \geqslant 1$$

4.2.3 Condensation of the denominators

Solution of the GGP problem is a two-level iterative process. The first level consists of condensing the denominators of each of the constraints at the operating point. This results in a posynomial divided by an approximating monomial. The monomial can be divided into the posynomial in the numerator, yielding an approximating posynomial that is always greater than or equal to the parent quotient form:

$$\frac{p^+(x)}{C[p^-(x), \bar{x}]} \geqslant \frac{p_k^+(x)}{p_k^-(x)}$$

Since this inequality relationship holds for all positive values of x, the feasible side of the approximating posynomial constraint is a subset of the feasible side of the parent quotient constraint.

Now consider the example problem. If the operating point is selected to be $\bar{x} = (3, 5)$, the approximating posynomial constraint corresponding to the first quotient constraint is

$$p^-(\bar{x}) = 10(3) + 2(5) = 10$$

$$\delta_1 = 10(3)/40 = 0.75$$

$$\delta_2 = 2(5)/40 = 0.25$$

$$\frac{p^+(x)}{C[p^-(x), \bar{x} = (3, 5)]} = \frac{x_1^2 + 29}{(10/0.75)^{0.75}(2/0.25)^{0.25} x_1^{0.75} x_2^{0.25}} \leqslant 1$$

$$= 0.0852 x_1^{1.25} x_2^{-0.25} + 2.47 x_1^{-0.75} x_2^{-0.25} \leqslant 1$$

The feasible region of the approximating posynomial is the darker shaded region in Fig. 4.2(a). Similarly, the denominators of the other two constraints are condensed to give

$$\frac{8x_1 + 4x_2}{C[x_1^2 + 40, \bar{x} = (3, 5)]} = 0.244 x_1^{0.633} x_2^{-0.25} + 0.122 x_1^{-0.367} x_2 \leqslant 1$$

$$\frac{12x_1}{C[x_1^2 + x_2 + 32, \bar{x} = (3, 5)]} = 0.478 x_1^{0.609} x_2^{-0.109} \leqslant 1$$

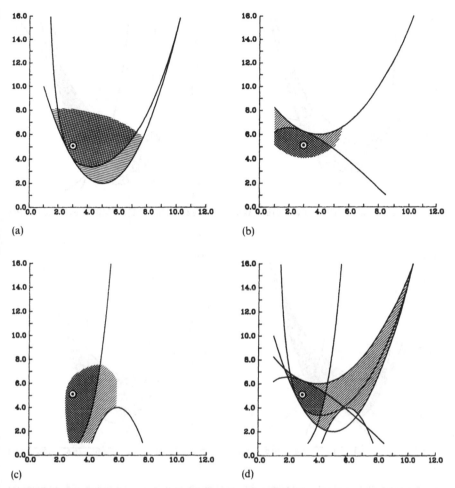

Fig. 4.2 (a) Feasible region of constraint 1 of problem (4.3), original (ruled) and with denominator condensed at $\bar{x} = (3, 5)$ (cross-hatched). (b) Feasible region of constraint 2 of problem (4.3), original (ruled) and with denominator condensed at $\bar{x} = (3, 5)$ (cross-hatched). (c) Feasible region of constraint 3 of problem (4.3), original (ruled) and with denominator condensed at $\bar{x} = (3, 5)$ (cross-hatched). (d) Feasible region of all constraints of problem (4.3), original (ruled) and with denominators condensed at $\bar{x} = (3, 5)$ (cross-hatched).

The corresponding feasible regions are shown in Figs. 4.2(b) and 4.2(c), respectively. The feasible region resulting from the intersection of these three approximations is shown in Fig. 4.2(d). Notice that it is a subset of the original feasible region.

It is interesting to observe how the posynomial subregion changes with respect to different operating points. Figure 4.3 shows four different posynomial sub-regions corresponding to four different operating points. Notice that when the operating point falls within the feasible region of the parent problem, it also falls within the feasible region of the posynomial subregion. Also note that when the operating point falls outside of the parent feasible region, a well-defined posynomial

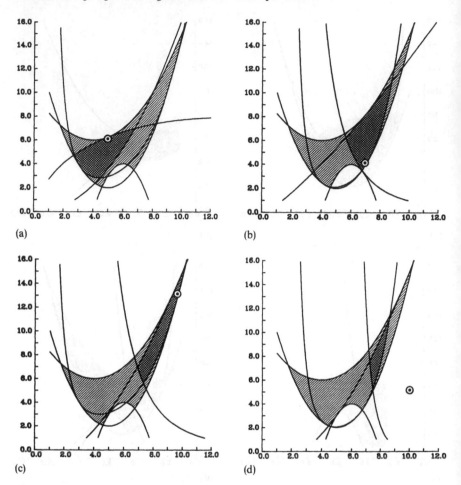

Fig. 4.3 (a) Posynomial subregion (cross-hatched) production by condensation of denominators at $\bar{x} = (5, 6)$ for problem (4.3). (b) Posynomial subregion (cross-hatched) produced by condensation of denominators at $\bar{x} = (7, 4)$ for problem (4.3). (c) Posynomial subregion (cross-hatched) produced by condensation of denominators at $\bar{x} = (9.7, 13)$ for problem (4.3). (d) Posynomial subregion (cross-hatched) produced by condensation of denominators at $\bar{x} = (10, 5)$ for problem (4.3).

subregion still exists within the parent feasible region. Although this latter case will not always be true (certain operating points outside of the parent feasible region will produce a set of posynomial approximations with the null set as their intersection), it indicates that in some cases an infeasible operating point can be used as a starting point in the GGP process. Regardless, an operating point that satisfies the parent constraints will always produce a posynomial subregion that contains the operating point and that is completely contained within the parent feasible region.

The condensation of denominators gives rise to the following posynomial subproblem:

Posynomial subproblem

minimize x_0

subject to $\dfrac{p_0^+(x)}{C[p_0^-(x) + x_0, \bar{x}]} \geqslant 1$

$\dfrac{p_k^+(x)}{C[p_k^-(x), \bar{x}]} \geqslant 1$ $k = 1, 2, \ldots, m$

$0 < x_j^{\text{LB}} \leqslant x_j$ $j = 1, 2, \ldots, n$ (4.4)

This subproblem is contained completely within the parent problem (4.2). The solution of this problem is, therefore, feasible to the parent problem.

4.2.4 The outer loop

In the previous section, it was demonstrated how a problem in quotient form is converted to a posynomial-constrained problem by condensing the denominators at some operating point. The outer loop of the two-level GGP process consists of the formation of a series of such posynomial subproblems, each condensed at the solution of the previous posynomial problem.

The outer loop is initiated by choosing a starting point for the variables. It is assumed that the initial operating point is feasible to the GGP problem. This starting point is used to generate a posynomial subproblem of the form (4.4). This subproblem is solved by the inner loop process described in the next section.

Once the solution to the posynomial subproblem has been obtained, it is used as the new operating point and a new posynomial subproblem is generated from it. This subproblem is solved and the process is repeated until convergence to a solution has been achieved. This process is shown graphically in Fig. 4.4. Aviel and Williams (1970) have shown that this process will converge to a point satisfying the Kuhn–Tucker necessary conditions for optimality if mild regularity conditions are satisfied.

All that remains now is the solution of the sequence of posynomial subproblems generated by the outer loop. This takes place in the inner loop, and it is where the bulk of the computation occurs. One may question the merit of solving a series of posynomial problems instead of a single polynomial problem. The advantage of doing so lies in the structure of the posynomial subproblem. The parent problem is generally nonconvex and may possess several local minima. The posynomial subproblem has only one solution because it is convex when transformed by a logarithmic change of variables (Duffin, Peterson and Zener, 1967). This convexity gives rise to very efficient solution strategies for the posynomial subproblem.

4.2.5 The inner loop

This section presents a linear-programming-based solution technique for the posynomial subproblem generated by the outer loop of the GGP solution process. It is an adaptation of the Kelly cutting plane method for convex problems (Kelly, 1960).

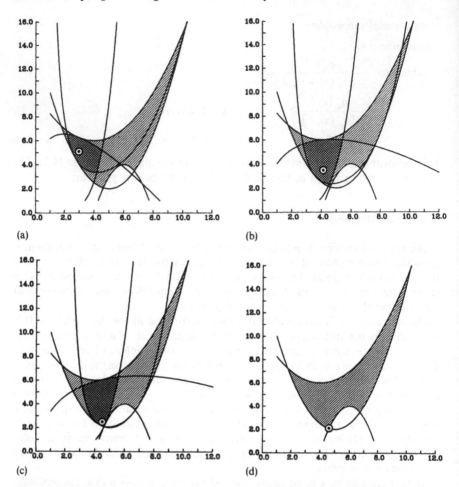

Fig. 4.4 (a) First iteration of outer loop sequence for problem (4.3). (b) Second iteration of outer loop sequence for problem (4.3). (c) Third iteration of outer loop sequence for problem (4.3). (d) Fourth iteration of outer loop sequence and final solution for problem (4.3).

The inner loop uses a second operating point, $\hat{x}$, that is initially set equal to $\bar{x}$. The first step of the inner loop is to condense each polynomial constraint of the subproblem at $\hat{x}$. Recall that the denominators of the parent quotient constraints have already been condensed at $\bar{x}$ to arrive at the posynomial constraints. This second round of condensations results in a set of monomial approximations to the posynomial constraints. Observing the direction of the inequality sign in the A–G inequality, it is apparent that the feasible side of the monomial approximation completely contains the feasible region of the posynomial constraint. Thus, the intersection of this set of monomial approximations results in a feasible region that completely contains the posynomial subproblem. Since the logarithm of a monomial function is a linear function, the logarithm of the monomial problem gives rise to the following linear program (LP), which is an exterior approximation of the posynomial subproblem:

Linear (log monomial) program

minimize x_0

subject to $\ln\left(C\left[\dfrac{p_0^+(x)}{C[p_0^-(x)+x_0,\bar{x}]}, \hat{x}\right]\right) \leqslant 0$

$\ln\left(C\left[\dfrac{p_k^+(x)}{C[p_k^-(x),\bar{x}]}, \hat{x}\right]\right) \leqslant 0 \qquad k = 1, 2, \ldots, m$

$0 < x_j^{LB} \leqslant x_j \qquad\qquad\qquad j = 1, 2, \ldots, n \qquad (4.5)$

This linear program is put into standard form once the variables are transformed into log space according to

$$z_j = \ln\left(\frac{x_j}{x_j^{LB}} \right) \quad j = 1, 2, \ldots, n \qquad (4.6)$$

Now the lower bounds on all z variables are zero.

This LP problem is solved, and the solution is assigned to be the new $\hat{x}$. Since the solution of an LP always lies on the boundary of the LP feasible region, $\hat{x}$ will generally not be feasible to the posynomial subproblem. Therefore, the most violated of the posynomial constraints is condensed to a monomial, this time at the new $\hat{x}$. It is then transformed into log space and appended to the previous LP. The solution of this new LP becomes the new $\hat{x}$ and another check is made to determine whether all posynomial constraints are satisfied (to within some prespecified tolerance). If not, the most violated posynomial is recondensed at the new $\hat{x}$ and appended to the growing LP. This process is repeated until an acceptable solution to the posynomial subproblem is reached.

To clarify this inner loop process, consider again the example problem (4.3). As a result of the first round of denominator condensations, the posynomial subproblem, which was shown in Fig. 4.4(a), is produced:

minimize x_2

subject to $0.0852x_1^{1.25}x_2^{-0.25} + 2.47x_1^{-0.75}x_2^{-0.25} \leqslant 1$

$0.244x_1^{0.633}x_2^{-0.25} + 0.122x_1^{-0.367}x_2 \leqslant 1$

$0.478x_1^{0.609}x_2^{-0.109} \leqslant 1$

$x_1 \geqslant 1, x_2 \geqslant 1$

Condensation of each of these posynomial constraints at the operating point $\hat{x} = (3, 5)$ yields

minimize x_2

subject to $1.92x_1^{-0.276}x_2^{-0.25} \leqslant 1$

$0.355x_1^{0.178}x_2^{0.455} \leqslant 1$

$0.478x_1^{0.609}x_2^{-0.109} \leqslant 1$

$x_1 \geqslant 1, x_2 \geqslant 1$

The feasible region produced by the intersection of these three monomials is shown in Fig. 4.5. The feasible region of the posynomial problem is also shown in this figure. Note that it is a subset of the monomial feasible region. Also note that the third posynomial is already in monomial form, and is unchanged by the condensation process.

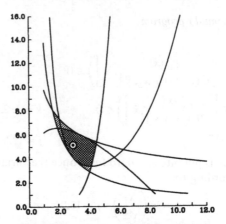

Fig. 4.5 Inner loop monomial problem.

After making the logarithmic change of variables in equation (4.6), the following LP is produced:

$$\begin{aligned}
\text{minimize} \quad & z_2 \\
\text{subject to} \quad & 0.276z_1 + 0.250z_2 \geqslant 0.652 \\
& 0.178z_1 + 0.455z_2 \leqslant 1.04 \\
& 0.609z_1 - 0.109z_2 \leqslant 0.738 \\
& z_1 \geqslant 0, z_2 \geqslant 0
\end{aligned}$$

The solution of this LP, after transforming out of log space, is $(x_1, x_2) = (4.07, 2.91)$. The posynomial constraints are each evaluated at this point, and it is determined that the first constraint is the most violated of the three:

$$0.0852(4.07)^{1.25}(2.91)^{-0.25} + 2.47(4.07)^{-0.75}(2.91)^{-0.25} = 1.04$$

$$0.244(4.07)^{0.633}(2.91)^{-0.25} + 0.122(4.07)^{-0.367}(2.91) = 0.805$$

$$0.478(4.07)^{0.609}(2.91)^{-0.109} = 1.00$$

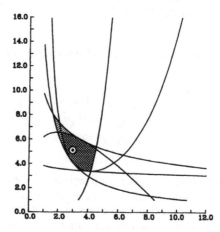

Fig. 4.6 Monomial problem with first cutting plane.

Recondensing the first posynomial constraint at $\hat{x} = (4.07, 2.91)$, and transforming to log space, gives the cutting plane $0.0221z_1 + 0.250z_2 \leqslant 0.336$. The monomial version of this cutting plane is shown in Fig. 4.6. Note that it 'cuts off' a portion of the feasible region that is infeasible to the posynomial problem. Solving the appended LP gives the new solution $(x_1, x_2) = (4.18, 3.36)$. This satisfies the posynomial constraints sufficiently well and therefore, the inner loop procedure has been completed. To continue with this problem, one would return to the outer loop and generate a new posynomial problem (that shown in Fig. 4.4(b)) by condensing the denominators of the parent quotient constraints at the point $\bar{x} = (4.18, 3.36)$.

The inner loop would be very inefficient if the LP were to be solved from scratch for each additional cutting plane. Fortunately, this is not necessary because the dual formulation of the LP has special properties that result in significant computational savings.

The LP (4.5) may be expressed in a more standard form as

$$
\begin{aligned}
\text{minimize} \quad & z_0 \\
\text{subject to} \quad & -\mathbf{A}\mathbf{z} \geqslant \mathbf{b} \\
& \mathbf{I}\mathbf{z} \geqslant \mathbf{0} \\
& -\mathbf{I}\mathbf{z} \geqslant -\mathbf{z}^{\text{UB}} \\
& -\mathbf{A}^{\text{CP}}\mathbf{z} \geqslant \mathbf{b}^{\text{CP}}
\end{aligned}
\tag{4.7}
$$

where $\mathbf{A}$ is the exponent matrix of the monomial problem, $\mathbf{b}$ is the vector of logarithms of the coefficients of the monomials, $\mathbf{I}$ is the identity matrix, $\mathbf{z}^{\text{UB}}$ is a vector of upper bounds on the variables transformed into log space, and $\mathbf{A}^{\text{CP}}$ and $\mathbf{b}^{\text{CP}}$ are similar to $\mathbf{A}$ and $\mathbf{b}$, but relate to the cutting planes. For every cutting plane added to the problem, an additional row is added to the LP. Since the new cutting plane excludes the previous LP solution from the new feasible region, feasibility must be regained before the new LP can be solved.

Recall from the duality theory of linear programming that associated with an LP of the form

$$
\begin{aligned}
\text{minimize} \quad & \mathbf{c}\mathbf{z} \\
\text{subject to} \quad & \mathbf{A}\mathbf{z} \geqslant \mathbf{b} \\
& \mathbf{z} \geqslant \mathbf{0}
\end{aligned}
$$

there is a dual formulation of the form

$$
\begin{aligned}
\text{maximize} \quad & \mathbf{b}\mathbf{y} \\
\text{subject to} \quad & \mathbf{A}^{\text{T}}\mathbf{y} \leqslant \mathbf{c} \\
& \mathbf{y} \geqslant \mathbf{0}
\end{aligned}
$$

Thus, the dual formulation of the LP (4.7) is

$$
\text{maximize} \quad [b \quad 0 \quad -z_j^{\text{UB}} \quad b_i^{\text{CP}}]\{y\}
$$

$$
\text{subject to} \quad [-\mathbf{A}^{\text{T}} \quad \mathbf{I} \quad -\mathbf{I} \quad -(\mathbf{A}^{\text{CP}})^{\text{T}}]\{y\} \leqslant \begin{Bmatrix} 1 \\ 0 \\ \vdots \\ 0 \end{Bmatrix}
$$

The dual variables, y, are associated with each constraint and cutting plane. Note that a basic feasible solution is always available for this problem, namely the zero vector. No 'phase I' procedure is necessary to initiate the LP solution process. Also, since the dual problem solves from the infeasible side inward, each cutting plane makes the previous LP solution feasible to the dual problem, and the new LP solution can be obtained after only a few additional pivot operations. Therefore, the inner loop solution process consists of the solution of an initial LP and a few additional pivot operations as cutting planes are added, until the posynomial constraints are satisfied by the LP solution.

4.2.6 Equality constraints

The GGP procedure outlined in the previous sections has concerned only inequality-constrained problems. The method can be extended to include equality constraints with minor changes to the algorithm. The equality constraints are first put in quotient form and the denominators are condensed to make posynomial equality constraints. Then, in the inner loop, the equality constraints are again condensed to form monomial constraints. These monomial equality constraints are transformed into log space and added to the linear program. Up to this point, the only difference between the treatment of the equality and inequality constraints is that, in the dual LP, the equality constraints appear as free variables instead of as variables bounded below by zero.

Once the LP solution has been found, only the posynomial inequalities are checked to determine which one is most violated. In other words, the equality constraints appear only once in the LP and do not produce cutting planes. It has been shown that this modification to the GGP process is equivalent to a Newton–Raphson solution of the equality constraints in log space conducted simultaneously with the cutting plane treatment of the inequality constraints (Burns, 1987b). More recent developments have shown that this treatment of the equality constraints has special invariance properties not shared by the Newton–Raphson method that give rise to enhanced performance. Further details are available in Burns and Locascio (1991). Figure 4.7 presents a flowchart of the modified equality–inequality constrained GGP algorithm.

To demonstrate the GGP method discussed in this section, the following two examples are presented.

EXAMPLE 4.1 SYMMETRICAL ONE-STORY, ONE-BAY FRAME

Figure 4.8 shows a symmetrical one-story, one-bay frame subjected to a uniformly distributed load of 0.6 klbf ft^{-1}. Each member is prismatic and the columns have the same section properties. The bases of the columns are assumed to be free to rotate. For convenience, the shape of all members will be assumed to be square.

The allowable stress in the beam is given as 1.8 klbf in^{-2} and the allowable stress in the columns is given as 1.6 klbf in^{-2}.

It is intended to determine the dimensions of the beam and the column for each local minimum. The objective function is the volume of the structure.

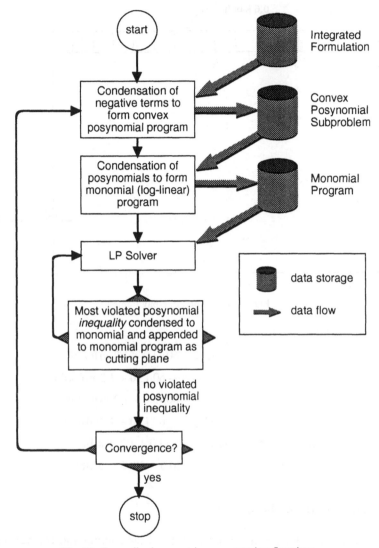

Fig. 4.7 Generalized geometric programming flowchart.

The assumption that the cross section of each member is square is made for convenience. With this simplification, it becomes unnecessary to assume a depth for each section, making comparisons more direct. However, the method can be used for any other shape in the nondiscrete formulation of the problem. In particular, for I-sections of given depth, the section of the smallest acceptable web thickness with the highest flexural efficiency can be selected.

In this simple example the analysis is carried out in a general form in order to obtain relationships between the forces, dimensions and section properties of the structure.

Assuming $I_b/I_c = m$, the analysis of the frame by the slope–deflection method

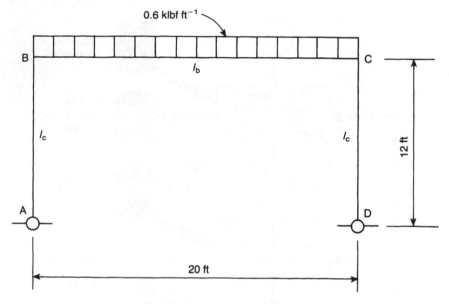

Fig. 4.8 One-story, one-bay frame.

will yield the following forces:

$$M_{BA} = 60/(3.0 + 1.2m)\,\text{ft klbf}$$
$$M_{BC} = -\,60/(3.0 + 1.2m)\,\text{ft klbf}$$
$$M_{CB} = 60/(3.0 + 1.2m)\,\text{ft klbf}$$
$$M_{CD} = -\,60/(3.0 + 1.2m)\,\text{ft klbf}$$
$$R_H\,(\text{horizontal reaction}) = 5/(3.0 + 1.2m)\,\text{klbf}$$
$$R_V\,(\text{vertical reaction}) = (0.6 \times 20)/2 = 6\,\text{klbf}$$

Formulation of the problem

Assume the dimensions of the cross sections of the column and the beam to be x_1 and x_2 respectively:

$$m = I_b/I_c = x_2^4/x_1^4$$

The optimization problem can be described as follows:

$$\text{minimize} \quad V = 24x_1^2 + 20x_2^2$$

subject to

1. stress at the top of the column should be less than $1.60\,\text{klbf in}^{-2}$:

$$6/x_1^2 + 4320x_1^4/(3x_1^4 + 1.2x_2^4)x_1^3 \leqslant 1.6\,\text{klbf in}^{-2}$$

2. stress at the end of the beam should be less than $1.80\,\text{klbf in}^{-2}$:

$$5x_1^4/(3x_1^4 + 1.2x_2^4)x_2^2 + 4320x_1^4/(3x_1^4 + 1.2x_2^4)x_2^3 \leqslant 1.8\,\text{klbf in}^{-2}$$

3. stress at the midspan of the beam should be less than $1.80\,\mathrm{klbf\,in^{-2}}$:

$$5x_1^4/(3x_1^4 + 1.2x_2^4)x_2^2 + 2160/x_2^3 - 4320x_1^4/(3x_1^4 + 1.2x_2^4)x_2^3 \leqslant 1.8\,\mathrm{klbf\,in^{-2}}$$

For this simple problem there are only inequality constraints, since the equations relating the forces to the loads and relative stiffnesses have been conveniently included in the constraint expressions.

Solution of the above optimization problem by geometric programming results in two solutions corresponding to the following two local minima:

local minimum 1

$$x_1 = 2.654\,\mathrm{in}$$
$$x_2 = 10.604\,\mathrm{in}$$
$$V = 2417.95\,\mathrm{in^2\,ft}$$

local minimum 2

$$x_1 = 8.933\,\mathrm{in}$$
$$x_2 = 8.480\,\mathrm{in}$$
$$V = 3353.37\,\mathrm{in^2\,ft}$$

Figure 4.9 shows the two local minima.

Local minimum 1 is also the global minimum. However, it is clear that the columns are subjected mainly to axial forces. The moment at the top of the column, or at the end of the beam, is small and as a result the beam acts almost as a simply supported beam resting on piers subjected to axial compressive force.

Specifically, the following are the forces and stresses for local minimum 1:

$$H\,(\text{horizontal reaction}) = 0.0162\,\mathrm{klbf}$$

$$M_{BA}\,(\text{moment, top of column}) = 0.1944\,\mathrm{ft\,klbf}$$

$$M_M\,(\text{moment, midspan}) = 29.81\,\mathrm{ft\,klbf}$$

$$f_{BA}\,(\text{stress, top of column}) = 1.600\,\mathrm{klbf\,in^{-2}}$$

$$f_M\,(\text{stress, midspan}) = 1.800\,\mathrm{klbf\,in^{-2}}$$

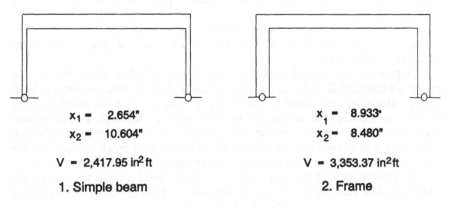

$x_1 =$ 2.654"	$x_1 =$ 8.933"
$x_2 =$ 10.604"	$x_2 =$ 8.480"
V = 2,417.95 in²ft	V = 3,353.37 in²ft
1. Simple beam	2. Frame

Fig. 4.9 The two local minima.

The solution would suggest that if the frame were subjected only to a vertical uniformly distributed load, a better design would be a simply-supported beam resting on two piers. Local minimum 1 is fully stressed.

Local minimum 2 corresponds to a structure that is heavier than that in local mimimum 1 by about 39%. However, in this case the structure behaves as a frame and the column is subjected to a large moment. Since the column is relatively long, to make the structure behave as a frame requires a large increase in the size of the column which results in an increased weight.

Specifically, the following are the forces and stresses for local minimum 2:

$$H \text{ (horizontal reaction)} = 1.258 \text{ klbf}$$

$$M_{BA} \text{ (moment, top of column)} = 15.10 \text{ ft klbf}$$

$$M_M \text{ (moment, midspan)} = 14.90 \text{ ft klbf}$$

$$f_{BA} \text{ (stress, top of column)} = 1.600 \text{ klbf in}^{-2}$$

$$f_M \text{ (stress, midspan)} = 1.777 \text{ klbf in}^{-2}$$

In this example since the column is long the local minimum 1 corresponds to a lighter structure. However, if the column were very short in relation to the girder, local minimum 2 would become the global minimum. In this case local minimum 2 is not fully stressed, since the computed stress in the girder is somewhat below the allowable values.

EXAMPLE 4.2 TWO-SPAN NONPRISMATIC CONTINUOUS BEAM

Figure 4.10 shows a two-span nonprismatic continuous beam subjected to a uniformly distributed fixed load of 1.8 klbf ft^{-1}. The beam is fabricated in three sections that are to be spliced in the field at the points of inflection of the uniformly distributed load. Section B–B shows the cross section of the beam in the negative moment region and Section A–A shows the cross section in the positive moment region. The width of flange, which is given as 8 in, and the 50 in $\times \frac{9}{16}$ in web are the same for the entire structure.

Assume that the uniformly distributed load includes the weight of the beam, and that the increased weight in the negative moment region need not be considered.

The flexural stress in the structure cannot exceed 20 klbf in^2.

Calculate α, and the flange thicknesses t_1 and t_2 such that the volume of the structure is minimum.

This problem is constrained so that the points of splice where the section properties change fall at the points of inflection of the uniformly distributed load. This constraint is introduced to reduce the level of stress at the point of splice in order to increase the fatigue life of the structure. Although this constraint results in reversal of stress at the point splice, the stress level is so low that allowable range stress is not exceeded. It should be pointed out, however, that this constraint does not contribute to the optimal design of a continuous beam. It results in an increase in the weight of the structure since it tends to increase the value of α. The moment at the interior support can be expressed in the following

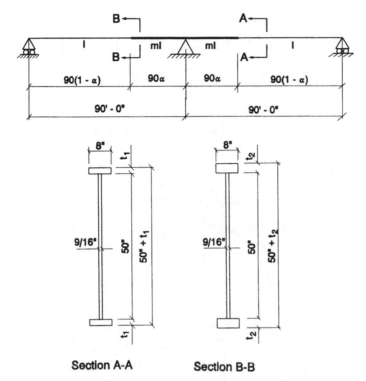

Section A-A   Section B-B

Fig. 4.10 Two-span continuous beam.

form:

$$M = CwL^2/8$$

where

$$C = [m - (3\alpha^4 - 8\alpha^3 + 6\alpha^2)(m - 1)]/[1 + (1 - \alpha)^3(m - 1)]$$

and m is the ratio of the moment of inertia of the section in the negative moment region to that of the positive moment region. Figure 4.11 shows the variation of C with m and α.

For the splice to be at the point of inflection, the following expressions must hold:

$$m = (\alpha^4 - 4\alpha^3 + 6\alpha^2)/(1 - \alpha)^4$$

and

$$M_2 \text{ (bending moment at interior support)} = 4\alpha wL^2/8$$

The following are the section properties and forces: cross-sectional area of the beam in the positive moment region (in^2),

$$A_1 = 8(50 + 2t_1) - 50(8 - 9/16) = 16t_1 + 225/8$$

moment of inertia of the beam in the positive moment region (in^4),

$$I_1 = 8(50 + 2t_1)^3/12 - (50)^3(8 - 9/16)/12$$
$$= 16(25 + t_1)^3/3 - 77\,474$$

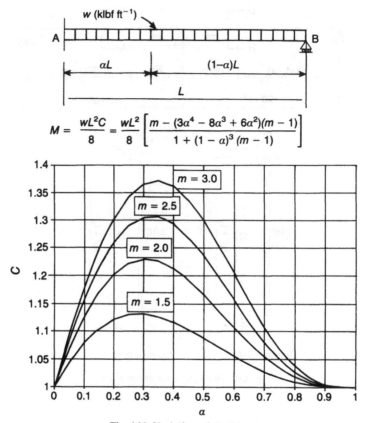

$$M = \frac{wL^2 C}{8} = \frac{wL^2}{8}\left[\frac{m - (3\alpha^4 - 8\alpha^3 + 6\alpha^2)(m-1)}{1 + (1-\alpha)^3(m-1)}\right]$$

Fig. 4.11 Variation of C with α.

distance from the neutral axis to the extreme fiber in the positive moment region (in),

$$c_1 = 25 + t_1$$

maximum positive moment in each span (in klbf),

$$M_1 = w[90(1-\alpha)]^2/8 = (12 \times 1.8 \times 8100)(1-\alpha)^2/8 = 21\,870(1-\alpha)^2$$

cross-sectional area of the beam in the negative moment region (in²),

$$A_2 = 8(50 + 2t_2) - 50(8 - 9/16) = 16t_2 + 225/8$$

moment of inertia of the beam in the negative moment region (in⁴),

$$I_2 = 8(50 + 2t_2)^3/12 - (50)^3(8 - 9/16)/12$$
$$= 16(25 + t_2)^3/3 - 77\,474$$

distance from the neutral axis to the extreme fiber in the negative moment region (in),

$$c_2 = 25 + t_2$$

maximum negative moment at the interior support (in klbf)

$$M_2 = wL^2(4\alpha)/8 = (12 \times 1.8)(90)^2(4\alpha)/8 = 21\,870(4\alpha)$$

The objective function is taken as U which is a measure of the volume of the structure, as follows:

$$\text{minimize } U = A_1(1 - \alpha) + A_2(\alpha)$$

$$= (16t_1 + 225/8)(1 - \alpha) + (16t_2 + 225/8)\alpha$$

$$= 16t_1 + 16\alpha(t_2 - t_1) + 28.125$$

The inequality constraints are

1. the stress due to the maximum positive moment should be less than $20 \, \text{klbf in}^{-2}$:

$$M_1 c_1/I_1 \leqslant 20 \, \text{klbf in}^{-2}$$

or

$$M_1 c_1 - 20 I_1 \leqslant 0$$

$$21\,870(1 - \alpha)^2(25 + t_1) - 20[16(25 + t_1)^3/3 - 77\,474] \leqslant 0$$

and

$$(1 - \alpha)^2(25 + t_1) - (25 + t_1)^3/205 + 70.85 \leqslant 0$$

2. and stress due to the maximum negative moment should be less than $20 \, \text{klbf in}^{-2}$:

$$M_2 c_2/I_2 \leqslant 20 \, \text{klbf in}^{-2}$$

$$M_2 c_2 - 20 I_2 \leqslant 0$$

$$21\,870(4\alpha)(25 + t_2) - 20[16(25 + t_2)^3/3 - 77\,474] \leqslant 0$$

and

$$(4\alpha)^2(25 + t_2) - (25 + t_2)^3/205 + 70.85 \leqslant 0$$

The equality constraint is, since the point of splice is at the point of inflection, the following equation must hold:

$$m = (\alpha^4 - 4\alpha^3 + 6\alpha^2)/(1 - \alpha)^4$$

or

$$[(25 + t_2)^3 - 14\,526](1 - \alpha)^4 - [(25 + t_1)^3 - 14\,526][(1 - \alpha)^4 - 4(1 - \alpha) + 3] = 0$$

The optimization problem can now be stated as

$$\text{minimize} \quad 16t_1 + 16\alpha(t_2 - t_1) + 28.125$$

subject to

$$(1 - \alpha)^2(25 + t_1) - (25 + t_1)^3/205 + 70.85 \leqslant 0$$

$$(4\alpha)(25 + t_2) - (25 + t_2)^3/205 + 70.85 \leqslant 0$$

$$[(25 + t_2)^3 - 14\,526](1 - \alpha)^4 - [(25 + t_1)^3 - 14\,526]$$

$$\times [(1 - \alpha)^4 - 4(1 - \alpha) + 3] = 0$$

It should be noted that the first two constraints are inequalities and the third constraint is an equation.

Solution of the above problem by geometric programming will yield the following answers:

$$\alpha = 0.351$$

$$t_1 = 0.718 \, \text{in}$$

$$t_2 = 3.303 \, \text{in}$$

$$U = 54.13 \, \text{in}^2$$

and

$$V = 54.13 \times 90 \times 12 = 58\,460\,\text{in}^3$$

$$M_1 = 9211.7\,\text{in klbf}$$

$$I_1 = 13\,248\,\text{in}^4$$

$$c_1 = 25.718\,\text{in}$$

s_1 (maximum stress in the positive moment region)

$$= (9211.7 \times 25.718)/(13\,248)\,\text{klbf in}^{-2}$$

$$= 17.88\,\text{klbf in}^{-2} < 20.00\,\text{klbf in}^{-2}$$

$$M_2 = 30\,705\,\text{in klbf}$$

$$I_2 = 43\,445\,\text{in}^4$$

$$c_2 = 28.303\,\text{in}$$

s_2 (stress at the interior support)

$$= (30\,705.5 \times 28.303)/43\,445 = 20.00\,\text{klbf in}^{-2}$$

It should be noted that in this case the maximum stress in the region of positive moment is less than the allowable stress. However, the maximum stress at the first interior support is equal to the allowable stress. The optimum design is not fully stressed.

4.3 APPLICATION TO SHAPE OPTIMIZATION

In this section, the integrated formulation is applied to the shape optimization of continuum structures modeled by finite elements. The stiffness matrix relating nodal forces to nodal displacements is expressed explicitly as a set of nonlinear equality constraints. The stiffness equations are nonlinear because they explicitly include a representation of the shape of the structure, as defined by the element nodal coordinates. The variables in this formulation are chosen to be the nodal coordinates of each element, the displacements of each node, and the stresses in each element. This selection of variables permits the stiffness matrix equations to be expressed in a convenient explicit form.

One advantage of this integrated formulation is that an explicit sensitivity analysis is not required, which eliminates the two-phase nature of typical shape optimization methods. The problem is related as one large mathematical program in the mixed space of shape variables and structural behavior variables, related to one another through a large set of equality and inequality constraints, such as the nonlinear stiffness equations. The sensitivity information, that is the changes in structural behavior with respect to changes in shape, are implicit in the constraint functions and need not be explicitly calculated.

The integrated formulation requires that very large mathematical programs be solved with each iteration of the method. Because of this, extreme care must be taken in implementing a solution technique to minimize the computational effort required for a solution. The GGP method is extremely efficient for solving large-scale nonlinear problems because of the special properties of condensation (Burns and Locascio, 1991), and because of its reliance on linear programming for the bulk of the computational effort.

In the following sections, the integrated shape optimization formulation is developed and examples are presented. Then, issues relating to practical computer-based implementation are addressed.

4.3.1 Integrated formulation for shape optimization

This formulation considers the class of continuum structures modeled by two-dimensional finite elements. The objective is to determine the shape of the boundary of the structure which minimizes some quantity, such as total volume, while satisfying side constraints on stress, displacement, shape, etc. throughout the structure.

The specific finite element considered in this formulation is the six-noded linear strain triangle (LST) element. This element has a quadratically varying displacement field and linearly varying stress and strain fields over the face of the element. Isoparametric elements are not considered because they require numerical integration and an explicit analytical representation of the inverse of the Jacobian matrix is cumbersome (Babu and Pinder, 1984). The formulation is easily extendable, however, to all nonisoparametric rectangular and triangular families of elements, including three-dimensional finite elements. The particular choice of the LST element is based on its ability to model flexure well, without an excessive number of elements, and because its triangular shape provides a good fit to a wide variety of shapes. This element has been found to work well for shape optimization (Pedersen, 1981).

The objective function is chosen to be volume, although any algebraic function of the design variables may be used instead. The shape variables are taken as the x, y coordinates of all element corner nodes in the finite element mesh. This allows the generality of a piecewise linear boundary shape. The coordinates of the interior corner nodes are tied to the coordinates of the boundary corner nodes by equality constraints. This avoids the problems of insensitivity of the objective function to interior node movement. These equality constraints are also chosen to maintain a well-proportioned mesh as the shape of the boundary changes.

Behavioral design variables consist of the displacement components at each node for each load case and the stress components at each corner of each element for each load case. It will be shown that this selection of design variables leads to a convenient formulation.

Constraints may be placed on stress, displacement, shape, or any algebraic functions of the design variables: nodal coordinates, nodal displacements, and element stresses. Alternate load cases are considered in the formulation. All load cases are independent; combinations of loads must be specified as additional independent load cases.

(a) Development of the stiffness equality constraints

In this section, the explicit stiffness equality constraints are derived. They relate nodal displacements to externally applied nodal forces, utilizing an intermediate

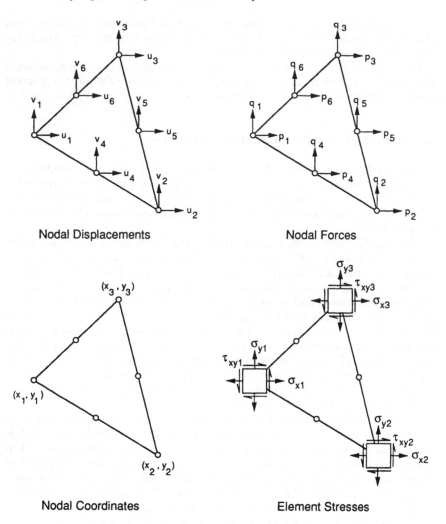

Fig. 4.12 Linear strain triangle finite element.

set of variables to simplify the formulation. These constraints may be viewed as being equivalent to individual rows of the stiffness matrix, but with an explicit representation of the nonlinearities that are introduced by the nodal coordinates of the finite elements being treated as variables.

The assumed quadratic displacement component fields within the LST element are uniquely determined by the two nodal displacement components of the three corner nodes and three midside nodes. These twelve degrees of freedom are shown in Fig. 4.12. The displacement component fields and the twelve nodal displacements are related by shape functions, **N**:

$$\begin{Bmatrix} u \\ v \end{Bmatrix} = [\mathbf{N}]\{u_1\,u_2\,u_3\,u_4\,u_5\,u_6\,v_1\,v_2\,v_3\,v_4\,v_5\,v_6\}^{\mathrm{T}} \tag{4.8}$$

where

$$[\mathbf{N}] = \begin{bmatrix} f_1 f_2 f_3 f_4 f_5 f_6 \, 0 \, 0 \, 0 \, 0 \, 0 \, 0 \\ 0 \, 0 \, 0 \, 0 \, 0 \, 0 \, f_1 f_2 f_3 f_4 f_5 f_6 \end{bmatrix}$$

and

$$f_1 = L_1(2L_1 - 1), f_2 = L_2(2L_2 - 1), f_3 = L_3(2L_3 - 1),$$

$$f_4 = 4L_1L_2, f_5 = 4L_2L_3, f_6 = 4L_3L_1$$

$$\begin{Bmatrix} L_1 \\ L_2 \\ L_3 \end{Bmatrix} = \frac{1}{2A} \begin{bmatrix} x_2y_3 - x_3y_2 & y_2 - y_3 & x_3 - x_2 \\ x_3y_1 - x_1y_3 & y_3 - y_1 & x_1 - x_3 \\ x_1y_2 - x_2y_1 & y_1 - y_2 & x_2 - x_1 \end{bmatrix} \begin{Bmatrix} 1 \\ x \\ y \end{Bmatrix}$$

In these equations, u and v are the displacement components at an arbitrary location (x, y) within the element. The nodal displacement components in the x and y directions are u_i and v_i respectively, for $i = 1, 2, \ldots, 6$. The x coordinates and y coordinates of the corner nodes are x_i and y_i for $i = 1, 2$, and 3. The symbol A represents the area of the element. Note that matrix $\mathbf{N}$ has elements that are nonlinear functions of the nodal coordinates.

The strains within the element are determined using the usual strain–displacements relationships, $\varepsilon_x = \partial u / \partial x, \varepsilon_y = \partial v / \partial y$, and $\gamma_{xy} = \partial u / \partial y + \partial v / \partial x$, which results in the strain–displacement matrix, $\mathbf{B}$:

$$\begin{Bmatrix} \varepsilon_x \\ \varepsilon_y \\ \gamma_{xy} \end{Bmatrix} = [\mathbf{B}]\{u_1 \, u_2 \, u_3 \, u_4 \, u_5 \, u_6 \, v_1 \, v_2 \, v_3 \, v_4 \, v_5 \, v_6\}^{\mathrm{T}}$$

where

$$[\mathbf{B}] = \begin{bmatrix} \partial/\partial x & 0 \\ 0 & \partial/\partial y \\ \partial/\partial y & \partial/\partial x \end{bmatrix} [\mathbf{N}]$$

Here, ε and γ are the components of strain within the element.

Stresses within the element are related to strains by the stress–strain matrix, $\mathbf{D}$:

$$\begin{Bmatrix} \sigma_x \\ \sigma_y \\ \tau_{xy} \end{Bmatrix} = [\mathbf{D}] \begin{Bmatrix} \varepsilon_x \\ \varepsilon_y \\ \gamma_{xy} \end{Bmatrix}$$

where

$$[\mathbf{D}] = \frac{E}{1 - v^2} \begin{bmatrix} 1 & v & 0 \\ v & 1 & 0 \\ 0 & 0 & (1-v)/2 \end{bmatrix}$$

Here, σ and τ are the components of stress throughout the element, E is the modulus of elasticity, and v is Poisson's ratio. Because the stresses vary linearly throughout the LST finite element, nine stress component quantities are sufficient to uniquely define the state of stress throughout the element. These nine quantities are shown in Fig. 4.12.

The element stiffness matrix is produced by a volume integral over a product of matrices $\mathbf{B}$ and $\mathbf{D}$ (Zienkiewicz, 1977). The element stiffness matrix relates nodal displacements to internal nodal forces, p_i and q_i:

$$\{p_1 \cdots p_6 q_1 \cdots q_6\}^T = \left[\iiint_V [\mathbf{B}]^T [\mathbf{D}][\mathbf{B}]\, dV \right] \{u_1 \cdots u_6 v_1 \cdots v_6\}^T$$

If the shape of the structure were not permitted to vary, then the nodal coordinates would be constant, yielding linear equilibrium equations as rows of the stiffness matrix. In this case, however, the nodal coordinates are permitted to vary and **B** becomes a matrix of nonlinear functions of the nodal coordinates. Much of the algebraic complexity of expressing the rows of the stiffness matrix in explicit form may be circumvented by introducing an additional set of variables, the stress components at each corner of each element. This allows the element stiffness equations to be decomposed into two sets of equations:

$$\{p_1 p_2 p_3 p_4 p_5 p_6 q_1 q_2 q_3 q_4 q_5 q_6\}^T = \iiint_V [\mathbf{B}]^T \begin{Bmatrix} \sigma_x \\ \sigma_y \\ \tau_{xy} \end{Bmatrix} dV$$

and

$$\begin{Bmatrix} \sigma_x \\ \sigma_y \\ \tau_{xy} \end{Bmatrix} = [\mathbf{D}][\mathbf{B}]\{u_1\, u_2\, u_3\, u_4\, u_5\, u_6\, v_1\, v_2\, v_3\, v_4\, v_5\, v_6\}^T$$

For example, the explicit form of the x component of nodal force at node $1, p_1$, is

$$p_1 = \frac{t}{6}(-y_2\sigma_{x1} + y_3\sigma_{x1} - x_3\tau_{xy1} + x_2\tau_{xy1})$$

and the explicit form of the x component of normal stress at node $1, \sigma_{x1}$, is

$$\sigma_{x1} = \frac{E}{2A(1-v^2)}(3y_2u_1 - 3y_3u_1 - y_3u_2 + y_1u_2 - y_1u_3 + y_2u_3 + 4y_3u_4 - 4y_1u_4$$
$$+ 4y_1u_6 - 4y_2u_6 + 3vx_3v_1 - 3vx_2v_1 - vx_1v_2 + vx_3v_2 - vx_2v_3$$
$$+ vx_1v_3 + 4vx_1v_4 - 4vx_3v_4 + 4vx_2v_6 - 4vx_1v_6)$$

The entire set of explicit stiffness expressions for the LST element has been compiled in Burns (1987a).

Equilibrium is maintained by summing all internal forces acting on each node and setting the sum equal to the externally applied, work-equivalent nodal loads, F_x and F_y:

$$F_x = \sum_{inc} p \quad \text{and} \quad F_y = \sum_{inc} q$$

Here, 'inc' represents a summation over all elements incident to the node. The summation is equivalent to the process of assembly of the global stiffness matrix.

(b) Stress limits

The introduction of stress components as intermediate variables not only simplifies the stiffness equations, but also provides a convenient means of constraining the stress magnitudes throughout the structure. Principal stresses are easily computed from the stress components,

$$\sigma_{1,2} = \frac{\sigma_x + \sigma_y}{2} \pm \left[\left(\frac{\sigma_x - \sigma_y}{2} \right)^2 + \tau_{xy}^2 \right]^{1/2}$$

and inequality constraints which limit the principal stresses throughout the structure can be imposed:

$$\left| \frac{\sigma_x + \sigma_y}{2} \pm \left[\left(\frac{\sigma_x - \sigma_y}{2} \right)^2 + \tau_{xy}^2 \right]^{1/2} \right| \leqslant \sigma_{\text{allowable}}$$

This is equivalent to the maximum normal stress failure theory. Other failure theories, such as the maximum distortion energy failure theory, could be implemented instead, when appropriate.

(c) Problem formulation

The final step in formulating the integrated shape optimization problem is to relate nodal coordinates to element areas. This is done for two reasons. First, the objective function formulation is greatly simplified, and, second, the shape function expressions (4.8) are also greatly simplified when element area is treated as a single quantity. Combining all of the previous equality and inequality constraints, the final integrated formulation is

minimize

$$\sum_i A_i t_i \qquad\qquad \text{total volume (t is element thickness)}$$

subject to

$$2A = x_3 y_1 - x_2 y_1 + x_1 y_2 - x_3 y_2 + x_2 y_3 - x_1 y_3 \qquad\qquad \text{element area}$$

$$\begin{Bmatrix} \sigma_x \\ \sigma_y \\ \tau_{xy} \end{Bmatrix} = [\mathbf{D}][\mathbf{B}] \{ u_1 u_2 u_3 u_4 u_5 u_6 v_1 v_2 v_3 v_4 v_5 v_6 \}^{\mathrm{T}} \qquad\qquad \text{stress–displacement}$$

$$\{ p_1 p_2 p_3 p_4 p_5 p_6 q_1 q_2 q_3 q_4 q_5 q_6 \}^{\mathrm{T}} = \iiint_V [\mathbf{B}]^{\mathrm{T}} \begin{Bmatrix} \sigma_x \\ \sigma_y \\ \tau_{xy} \end{Bmatrix} dV \qquad\qquad \text{force–stress}$$

$$F_x = \sum_{\text{inc}} p \quad \text{and} \quad F_y = \sum_{\text{inc}} q \qquad\qquad \text{equilibrium}$$

$$\left| \frac{\sigma_x + \sigma_y}{2} \pm \left[\left(\frac{\sigma_x - \sigma_y}{2} \right)^2 + \tau_{xy}^2 \right]^{1/2} \right| \leqslant \sigma_{\text{allowable}} \qquad\qquad \text{stress limits}$$

$$f(x, y) = 0 \qquad\qquad \text{coordinate linking}$$

$$u_{\min} \leqslant u \leqslant u_{\max}$$
$$v_{\min} \leqslant v \leqslant v_{\max} \qquad\qquad \text{support conditions and displacement limits}$$

$$x_{\min} \leqslant x \leqslant x_{\max}$$
$$y_{\min} \leqslant y \leqslant y_{\max} \qquad\qquad \text{envelopes on geometry and move limits}$$

The 'coordinate linking' constraints in the above formulation serve several purposes. The most important is the control of the location of interior nodes. A piecewise linear boundary shape is determined by the locations of the boundary nodes only. The interior nodes must be related to these boundary nodes so that a well-proportioned finite element mesh is maintained as the shape of the structure evolves during the optimization process. This is accomplished through a set of constraints that specify the locations of the interior nodes relative to other nodes, or that specify an upper bound on the aspect ratio of each finite element.

Another use of the coordinate linking constraints is to regulate the shape of the boundary. For example, it may be desirable to restrict a portion of the boundary to be flat, or to limit the curvature of a portion of the boundary. Algebraic constraint functions of the nodal coordinates may be imposed to do this. Envelopes on the allowable geometry may exist, leading to inequality coordinate linking constraints.

Of course, other constraints may be added to the formulation, and they need not be restricted to the nodal coordinate variables. Any algebraic function of the stress, displacement, coordinate, and element area variables can be specified. For example, displacement-dependent stress limits could be imposed.

The support conditions on the structure are handled through the bounds on the displacement variables. A fixed node would have its displacement degrees of freedom fixed between identical upper and lower bounds. More exotic support conditions are implemented with constraint functions relating the displacements to the other variables. Constraints on maximum displacement are enforced as bounds on the corresponding displacement variables. The bounds on the nodal coordinate variables can serve as move limits if the change in shape of the structure is too rapid from one iteration to the next.

EXAMPLE 4.3 VARIABLE DEPTH FIXED-ENDED BEAM

This section presents an example which illustrates the application of the integrated formulation to the shape optimization of a fixed-ended beam of variable depth under the action of a unformly distributed load. The beam has a 60 in span length, a selfweight of $150 \, \mathrm{lb \, ft^{-3}}$, and a distributed load of $1 \, \mathrm{klbf \, in^{-1}}$ applied along the top surface. The beam has a uniform thickness of 12 in. The mesh consists of 48 elements but symmetry is used to reduce it to 24. The x coordinates of all corner nodes are fixed. The y coordinates of the interior corner nodes are constrained to lie at the centroid of the corner nodes above and below them. The y coordinates of the corner nodes on the top and bottom surface of the beam are unrestricted. Principal stresses are constrained between $-4 \, \mathrm{klbf \, in^{-2}}$ and $4 \, \mathrm{klbf \, in^{-2}}$ everywhere throughout the beam; no constraints are placed on maximum displacement. The modulus of elasticity and Poisson's ratio of this hypothetical material are specified as $4000 \, \mathrm{klbf \, in^{-2}}$ and 0.15, respectively.

Figure 4.13 illustrates the iterative solution sequence using GGP to solve the integrated formulation. The solution is obtained in six iterations, with each iteration requiring an average of 1.7 min of computation on a Apple Macintosh IIfx computer.

A second solution of this problem was performed, this time with the allowable tensile stress reduced from $4 \, \mathrm{klbf \, in^{-2}}$ to $3 \, \mathrm{klbf \, in^{-2}}$. The allowable compressive stress was kept at $-4 \, \mathrm{klbf \, in^{-2}}$. The sequence of iterations is presented in

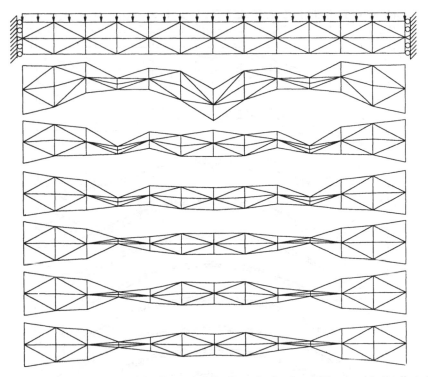

Fig. 4.13 First six iterations of the shape optimization of a fixed-ended beam with distributed load.

Fig. 4.14. This time, the structure evolves into an arch which experiences uniform axial compression everywhere. The solution process has completely avoided the more restrictive allowable tensile stress constraints, which are not active in this solution at all.

4.3.2 Discussion of example problem results

It is interesting to note that the second solution above has far less volume than the first solution. Since the second problem is more highly constrained than the first, this indicates that the first solution was not a global minimum. Tightening the allowable tensile stress constraints forced the solution out of the symmetric solution and into the arch-shaped global minimum. Because the allowable tensile stress constraints are not binding in the arch solution, this indicates that the arch solution is an alternative solution to the first problem with equal tensile and compressive allowable stresses. Had the first problem been initiated with a starting configuration more closely resembling the arch-shaped solution, the solution process could have converged to the arch solution instead. A third optimal solution to the first problem is a downward sweeping arch in uniform

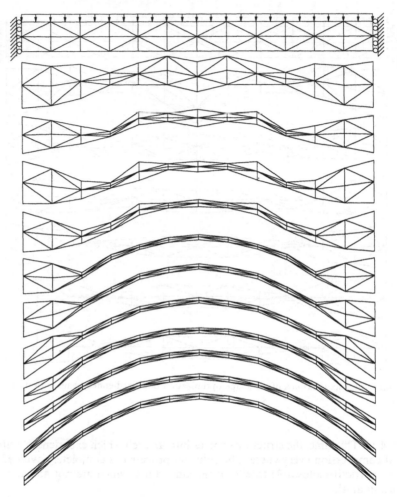

Fig. 4.14 First eleven iterations of the shape optimization of a fixed-ended beam with distributed load and tightened allowable tensile stress constraints.

axial tension. This problem suggests that local minima are certainly possible, and probably quite prevalent, in shape optimization.

This example also raises another important issue. When the two-phase approach of shape optimization is used, it becomes necessary to define 'shape functions' which limit the range of possible shapes the structure may assume during the course of optimization. Shape functions provide a means of specifying the shape of the entire structure through the use of a relatively small number of shape parameters. This simplifies the sensitivity analysis because derivatives of behavior are calculated only with respect to the shape parameters. Unfortunately, the final shape becomes highly dependent on the particular choice of shape functions. Extreme care must be taken to ensure that the shape functions will not overly constrain the solution, possibly excluding useful shapes that are not expressible by the shape functions. With the integrated formulation, each nodal coordinate can be taken as a shape variable, if desired. This results in a general piecewise linear boundary shape. In areas of the structure where it is desirable to have less variability in shape, for example, on a surface that is desired to be flat, shape functions can easily be imposed as additional constraints in the problem formulation.

This suggests a potentially useful interactive strategy. Initially, all boundary nodes can be allowed to vary, with the exception of perhaps the location of the supports. Then as the shape optimization progresses, those areas where the shape is evolving in an undesirable way can be constrained interactively. This strategy improves the likelihood of discovering useful, yet unexpected, optimal shapes without having to build expectations into a set of imposed shape functions.

4.4 IMPLEMENTATION OF GGP FOR SHAPE OPTIMIZATION

Solution of the large mathematical program produced by the integrated formulation requires a great deal of computational effort and also substantial computer storage. Each finite element requires nine stress variables per load case, and each node requires two displacement variables per load case. In addition, each corner node requires two coordinate variables. Each stress variable produces an equality constraint relating it to displacement and nodal coordinate variables. There is an additional pair of equality constraints for each node that maintains equilibrium in the x and y directions at that node for each load case. Finally, there are two inequality constraints for each corner node, for each load case, constraining the principal stress magnitudes. The example presented in the previous section had 412 variables and 517 constraints.

Figure 4.7 illustrates that three separate problem formulations coexist during the GGP solution process – the original integrated formulation, the posynomial subproblem, and the LP. These three formulations must be represented internally in the computer, which may require a great deal of storage capacity. The bulk of the computational effort lies with the solution of the LP. Although LPs are routinely solved with thousands of variables, the computational effort is not small. Having to solve a new LP for each outer loop iteration may increase the required effort by an order of magnitude. This section addresses the issues of storage and computational effort and suggests implementation strategies for both.

4.4.1 Constraint emitters

This section concerns the issue of efficient use of computer memory. Up to this point, it has been assumed that the integrated formulation is supplied in explicit form. This, of course, would be impractical to do by hand for even a small shape optimization problem. It would be possible to construct, in software, a system to generate the explicit formulation from a standard structural database describing the structure to be optimized. While this is certainly a step in the right direction, it still requires that all three problem formulations shown in Fig. 4.7 be created.

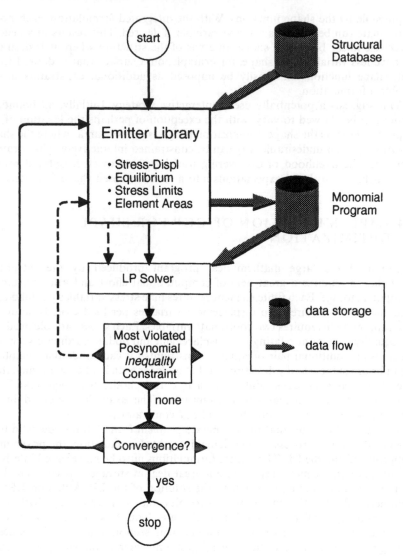

Fig. 4.15 Constraint emitters in the solution process.

A more useful strategy makes use of what are termed 'constraint emitters'. Since the format of the constraints in the integrated formulation is known in advance, a general-purpose condensation routine is not needed. This activity is more efficiently performed by special-purpose software routines called 'constraint emitters', which generate the log–monomial linearizations directly from the structural database, given the two operating points belonging to the outer and inner loops, $\bar{x}$ and $\hat{x}$. Consequently, the only problem formulation that ever exists in memory is the LP problem. The integrated formulation and the posynomial subproblems never exist explicitly in memory. They are replaced by a much smaller structural database describing the problem being solved (nodal coordinates, element connectivity, loads, supports, etc.). Figure 4.15 shows the modified program flowchart after the emitters have been incorporated. One emitter exists for each class of constraint in the integrated formulation and together they form a library of emitter subroutines.

The constraint emitters may be viewed as producing a 'convex linearization' directly from the general structural description. This can be viewed as a form of approximate structural modeling, a strategy that has appeared frequently in shape optimization literature (Schmit and Farshi, 1974; Bennett, 1981; Kirsch, 1981; Kirsch and Hofman, 1981; Schmit, 1981; Prasad, 1983).

4.4.2 Preferred pivots

The second strategy concerns computational effort rather than efficient use of computer memory. Recall that for each outer loop iteration, a complete cutting plane problem is solved. The only information passed from .one outer loop iteration to the next is the updated set of design variable values. A considerable saving in computational effort is possible if additional information is passed between outer loop iterations.

At the end of any given cutting plane procedure, the solution of a posynomial subproblem exists. However, in addition, an extended LP tableau that gave rise to this solution exists, including a set of columns of the LP tableau that form a square basis matrix. Recall that the dual formulation is being solved. Each column of the tableau is associated with one constraint of the original problem formulation, and each row is associated with one of the design variables. Through careful bookkeeping, it is possible to keep track of which constraints of the original problem formulation are associated with the specific columns that constitute the basis matrix.

The 'preferred pivot' strategy makes use of this bookkeeping. The set of constraints associated with the basis matrix, or the 'active set', tends to be similar from one upper-level iteration to the next, particularly toward later iterations. If, at the start of an inner loop iteration, the LP is initialized with a basis consisting of the columns derived from the active set of the previous cutting plane problem, then far less pivoting is required to solve the new LP than would otherwise be required by starting with a standard basis of slack variables. It has been demonstrated that a 50% saving in total computation can result from the preferred pivot strategy (Burns, 1985).

Some risk is associated with this strategy, however. It is possible, particularly during the early iterations when the active sets tend to be less similar, that a basis matrix formed from a previous active set may violate initial feasibility of the dual problem. In this case, the method must revert to the safe approach of starting with a basis of slack variables. A useful hybrid strategy is to give preference to the active-set constraints during pivoting operations in the early upper-level iterations instead of forcing them into the initial basis. In other words, the hybrid method would start with a basis of slack variables, but when searching for a column to enter the basis, would be restricted to the active-set constraints until no further improvement can be made, at which time the other columns would be included. During these early upper-level iterations, the active set would be monitored. When successive active sets start to become similar, then the forced-basis strategy would take over.

4.5 CONCLUSIONS

The generalized geometric programming method has been presented as an effective tool for optimizing nonlinear problems with equality and inequality constraints. It is demonstrated how the method reduces the optimization problem to a stable sequence of linear programs through a convex linearization scheme. Other linearization-based methods tend to experience difficulties when treating nonlinear problems. By establishing the convex subproblem prior to linearization, the GGP method avoids these difficulties.

In addition, an integrated formulation for shape optimization has been presented as an alternative to the standard two-phase procedure. The integrated approach has several notable attributes: (1) an explicit sensitivity analysis is not needed; (2) the bulk of the computational effort is in the solution of linear programs, for which highly developed and efficient computer codes are readily available; (3) shape functions are no longer necessary to make the sensitivity analysis computationally economical – useful optimal shapes are more likely to be found that might otherwise be disallowed by poorly chosen or overly restrictive shape functions.

APPENDIX 4A CONVERSION FACTORS BETWEEN US CUSTOMARY AND SI UNITS

To convert	To	Multiply by
in	m	0.02540
ft	m	0.3048
lbf	N	4.448
klbf	kN	4.448
$klbf\,ft^{-1}$	$kN\,m^{-1}$	14.59
$klbf\,in^{-2}$	MPa	6.894

REFERENCES

Abuyounes, S. and Adeli, H. (1986) Optimization of steel plate girders via general geometric programming. *Journal of Structural Mechanics*, **14**(4), 501–24.

Adeli, H. and Chompooming, K. (1989) Interactive optimization of nonprismatic girders. *Computers and Structures*, **31**(4), 505–22.

Adeli, H. and Kamal, O. (1986) Efficient optimization of space trusses. *Computers and Structures*, **24**(3), 501–11.

Avriel, M. and Williams, A. (1970) Complementary geometric programming. *SIAM Journal of Applied Methematics*, **19**(1), 125–41.

Avriel, M., Dembo, R. and Passy, U. (1975) Solution of generalized geometric programs. *International Journal of Numerical Methods in Engineering*, **9**, 149–68.

Babu, D.K. and Pinder, G.F. (1984) Analytical integration formulae for linear isoparametric finite elements. *International Journal of Numerical Methods in Engineering*, **20**, 1153–66.

Beckenbach, E. and Bellman R. (1961) *An Introduction to Inequalities*, Singer.

Bennett, J.A. (1981) Application of linear constraint approximations to frame structures.

Proceedings of the 11th ONR Naval Structural Mechanics International Symposium on Optimum Structural Design, Tucson, AZ, pp. 7-9-7-15.

Burns, S.A. (1985) Structural optimization using geometric programming and the integrated formulation. Ph.D. Thesis, University of Illinois at Urbana-Champaign.

Burns, S.A. (1987a) Simultaneous design and analysis using geometric programming and the integrated formulation. Proceedings of the Swanson Analysis Systems, Inc. ANSYS 1987 Conference and Exhibition, Newport Beach, CA.

Burns, S.A. (1987b) Generalized geometric programming with many equality constraints. *International Journal of Numerical Methods in Engineering*, **24**(4), 725–41.

Burns, S.A. and Locascio, A. (1991) A monomial-based method for solving systems of nonlinear algebraic equations. *International Journal of Numerical Methods in Engineering*, **31**(7), 1295–318.

Chibani, L., Burns, S.A. and Khachaturian, N. (1989) A move coordination method for alternative loads in structural optimization. *International Journal of Numerical Methods in Engineering*, **28**, 1041–60.

Duffin, R.J. (1970) Linearizing geometric programs. *SIAM Review*, **12**(2), 211–27.

Duffin, R.J., Peterson, E. and Zener, C. (1967) *Geometric programming*, Wiley, New York.

Fox, R.L. and Schmit, L.A. Jr (1966) Advances in the integrated approach to structural synthesis. *Journal of Spacecraft*, **3**(6), 858–66.

Fuchs, M.B. (1982) Explicit optimum design. *International Journal of Solids and Structures*, **18**(1), 13–22.

Haftka, R.T. (1985) Simultaneous analysis and design. *AIAA Journal*, **23**, 1099–103.

Hardy, G., Littlewood, J. and Polya, G. (1959) *Inequalities*, Cambridge University Press, Cambridge.

Kelly, J.E. (1960) The cutting plane method for solving convex problems. *Journal of the Society for Industrial and Appled Mathematics*, **8**(4), 703–12.

Kirsch, U. (1981) *Optimum Structural Design*, McGraw-Hill, New York, p. 18.

Kirsch, U. and Hofman, B. (1981) Approximate behavior models for optimum structural design. Proceeding of the 11th ONR Naval Structural Mechanics International Symposion on Optimum Structural Design, Tucson, AZ, pp. 7-17-7-26.

Pascual, L. and Ben-Israel, A. (1970) Constrained maximization of posynomials by geometric programming. *Journal of Optimization Theory and Applications*, **5**, 73–86.

Passy, U. (1971) Generalized weighted mean programming. *SIAM Journal of Applied Mathematics*, **20**, 763–78.

Pedersen, P. (1981) The integrated approach of FEM–SLP for solving problems of optimal design, in *Optimization of Distributed Parameter Structures*, Vol. 14 (eds E. Haug and J. Cea), Sijthoff and Noordhoff, NATO-ASI, Series E, No. 49, pp. 757–80.

Prasad, B. (1983) Explicit constraint approximation forms in structural optimization. *Computer Methods in Applied Mechanical Engineering*, **40**, 1–26.

Rijckaert, M.J. and Martens, X.M. (1978) Bibliographical note on geometric programming. *Journal of optimization Theory and Applications*, **26**(2), 325–37.

Schmit, L.A. (1981) Structural optimization, some key ideas and insights. Proceedings of the 11th ONR Naval Structural Mechanics International Symposium on Optimum Structural Design, Tucson, AZ, pp. 1–7.

Schmit, L.A. and Farshi, B. (1974) Some approximation concepts for structural synthesis. *AIAA Journal*, **12**(5), 692–9.

Schmit, L.A., Jr, and Fox, R.L. (1965) An integrated approach to structural synthesis and analysis. *AIAA Journal*, **3**, 1104–12.

Soeiro, A. and Hoit, M. (1987) Sizing optimization, in *Computer Applications in Structural Engineering* (ed. D.R. Jenkins), American Society of Civil Engineers, New York, pp. 342–56.

Zienkiewicz, O.C. (1977) *The Finite Element Method*, 3rd edn, McGraw-Hill (UK) Ltd., London, p. 104.

5

Nonlinear mixed integer programming

MANOHAR P. KAMAT and LUIS MESQUITA

5.1 INTRODUCTION

Many variables entering into engineering design problems are restricted to assume only discrete values. There are two important reasons for this. The first one could be due to manufacturing constraints or commercial viability. Examples would be pipes, I-beams, screws etc. that are commercially available only in certain discrete sizes. The second reason is the nature of the design variables. Examples would be number of items, such as the number of stiffeners on a plate, the number of transverse bulkheads on a ship, the number of bars in a truss, the number of blades in a gas turbine engine etc. Practical engineering problems could be such that some of the design variables can take any finite continuous values while others may take certain integer values. Furthermore, if the objective function for such problems is nonlinear then the problem falls into the class of nonlinear mixed integer programming. Most optimization packages assume that the design variables can take any continuous values subject to certain design constraints while the objective function is minimized. In problems where the design variables can only take integer values, the practice is to round off these variables in the final continuous solution thereby obtaining only an approximate solution to the truly mixed integer programming problem.

Integer programming involves the use of an algorithm that forces the solution obtained to integer values. Techniques used therein involve the solution of a series of optimization subproblems. The subproblems are obtained by imposing additional constraints on the original problem. The conventional way of obtaining the integer solution by rounding off the continuous solution involves making a decision as to which design variables should be rounded off to the next lower or higher integer. Usually there is no rational way to make this decision. Secondly, the rounded-off solution may be in the infeasible region and moving from the infeasible region to the feasible region could be difficult. More importantly the true integer solution to the nonlinear mixed integer programming problem has a better chance of being globally optimal when the solution space is concave. This follows from the fact that the subproblems solved have disjoint feasible spaces.

The computational effort required to solve a continuous optimization problem followed by rounding off the integer variables is almost always far less than the effort to solve the same problem using integer programming. If only a few of the design variables are required to be integers, then it may be possible to obtain a solution by simple enumeration. However, if the number of integer variables is large (say greater than five) then the number of possible combinations is prohibitively large. For example, if we have six integer design variables each of which can take on integer values in the range from 1 to 10, the possible number of combinations is one million. In instances such as these, integer programming is the only viable option.

5.2 NONLINEAR MIXED INTEGER PROGRAMMING

A general optimization problem could be stated as follows:

$$\text{minimize} \quad f(\mathbf{X}), \mathbf{X} = (x_1 \, x_2 \cdots x_n)^t \tag{5.1}$$

$$\text{subject to} \quad h_i(\mathbf{X}) = 0 \qquad i = 1, 2, \dots, n_e \tag{5.2}$$

$$g_j(\mathbf{X}) > 0 \qquad j = 1, 2, \dots, n_g \tag{5.3}$$

The vector $\mathbf{X}$ will be referred to as the design variable vector. For a structural optimization problem, it normally represents some characteristics of structural elements such as cross-sectional diameters, areas, etc. If f, h_i, and g_j are all linear functions of the design variable vector $\mathbf{X}$ the problem is a linear programming problem (LP). If any of the f, h_i or g_j are nonlinear functions of $\mathbf{X}$, a nonlinear programming problem (NLP) results. If all the design variables are such that they can assume any real values within specified bounds then one has the classical continuous variable optimization problem. In several structural design problems, however, some or all of the design variables may assume only integer or discrete values. Such problems can be formulated as nonlinear mixed integer programming problems (NMIP).

Integer programming techniques fall into two broad types: (1) search methods and (2) cutting plane methods. The first type is motivated by the fact that the integer solution can be regarded as consisting of a finite number of points. In its simplest form, search methods seek the enumeration of all such points. This would be equivalent to a simple exhaustive enumeration. However, what makes search methods more promising than simple exhaustive enumeration is that they have been developed to enumerate only a portion of all candidate solutions while automatically discarding the remaining points because they are either infeasible or inferior to points that are retained. The efficiency of the resulting search algorithm depends on the power of such a technique to discard nonpromising solution points. Thus, search methods primarily include implicit enumeration techniques and branch-and-bound techniques. The first type is mostly suited for the 'zero–one' problem, wherein the integer design variables may take on values of either zero or one. Implicit enumeration techniques are not suited for discrete or integer problems which are not of the zero–one type as posed.

Cutting plane methods were the first systematic techniques available for the solution of linear integer optimization problems. The early work of Dantzig, Fulkerson and Johnson (1954) directed the attention of researchers to the importance of solving linear integer programming problems. Markovitz and

Manne (1957) considered the more general case of discrete variables. Dantzig (1959) was the first to propose the cutting plane method for linear integer solution in a finite number of steps. The first cutting plane method that guaranteed an integer solution in a finite number of steps is due to Gomory (1958). Though this algorithm converged in a finite number of steps, it ran into difficulties caused by machine round-off. A new algorithm was developed by Gomory (1960a) to overcome the roundoff problem. In his subsequent work, Gomory (1960b) extended the above methods to cover mixed integer problems. Glover (1965) introduced a new type of cutting plane method known as the bound escalation method. Young (1971), Balas (1971) and Glover (1971) developed what is known as the 'convexity' cut method.

The above cutting plane methods have the drawback that the solution is not available until the algorithm converges. A primal algorithm was first introduced by Ben-Israel and Charnes (1962); however, the first finite primal algorithm (a problem that converges in a finite number of steps) was developed by Young (1965). Improvements on these algorithms were given by Young (1968) and Glover (1968).

Cutting plane methods cannot be used for nonlinear problems as they are based on linearity assumptions. Branch and bound algorithms do not rely on linearity and can be used to solve nonlinear integer programming problems. The branch and bound procedure does not deal directly with the integer problem. Rather, it considers a continuous problem obtained by relaxing the integer restrictions on the variables. Thus the solution space of the integer problem is only a subset of the continuous space. The prime reason for dealing with the continuous problem is that it is simpler to manipulate such a problem. Furthermore, one can draw upon the tremendous advances in the theory of continuous optimization methods. If the optimal continuous solution is integer, then it is also optimum for the integer problem. The branch and bound technique consists of two basic operations:

- *Branching.* This operation partitions the continuous solution spaces into subspaces (subproblems), which are also continuous. The purpose of partitioning is to eliminate part of the continuous space that is not feasible for the integer problem. This elimination is achieved by imposing (mutually exclusive) constraints that are necessary conditions for producing integer solutions, but in a way that no feasible integer point is eliminated. In other words, the resulting collection of subproblems defines every feasible integer point of the original problem. Because of the nature of the partitioning operations it is called branching.
- *Bounding.* Assuming the original problem is of the minimization type the optimal objective values for each subproblem created by branching set a lower bound on the objective value associated with any of its integer feasible values. This bound is essential for 'ranking' the optimum solutions of the subsets, and hence in locating the optimum integer solution.

The first known branch and bound algorithm was developed by Land and Doig (1960) as an application to the mixed and pure integer problems. Dakin (1965) modified the Land and Doig algorithm to facilitate computer implementation, and also to make it applicable to nonlinear mixed integer programming problems. Gupta (1980) and Gupta and Ravindran (1983) use numerical results to show that Dakin's algorithm is well suited to solve relatively large, nonlinear mixed integer programming problems.

In addition to the above two basic classes of methods, several researchers have devised other variations of what may be called primal and dual methods primarily with a view to solving structural optimization problems involving integer or discrete variables along with continuous variables.

Schmit and Fleury (1980) use approximate concepts and dual methods to solve structural problems involving a mix of discrete and continuous sizing type of design variables. The optimization problem is converted into a sequence of explicit approximate primal problems of separable form. These problems are solved by constructing continuous explicit dual functions, which are maximized subject to simple nonnegativity constraints. Dual methods are not guaranteed to yield an optimum solution to the discrete problem, because of the nonconvexity involved. However, encouraging results are presented by Schmit and Fleury (1980).

Imai (1979) uses the above concepts for the problem of material selection. Hua (1983) uses an implicit enumeration scheme for finding the minimum weight design of a structure with discrete member sizes. A simplifying assumption is made, namely that, if reducing the size of a member causes an originally feasible solution to become infeasible, further reducing the size will not make it feasible. Johnson (1981) uses a branch and bound algorithm to find the minimum weight design of a rigid plastic structure with discrete sizes. Linking of elements is done to reduce the number of design variables. The objective function and the constraints are linear functions of the design variables, making it possible to use linear programming techniques to obtain the solution to the continuous problem.

Gisvold and Moe (1971) use a penalty method approach to minimize the weight of a structure with two discrete variables. The objective function that is to be minimized is augmented by a term that is positive everywhere and vanishes at the discrete points.

Templeman and Yates (1983) use a segmental method for the discrete design of a minimum weight structure. The problem of finding the discrete sizes of the cross-sectional areas is converted to one of finding the continuous lengths of element segments each having one of the two possible discrete cross-sectional areas. The resulting continuous problem is solved using linear programming. Linearity considerations are then used to force the discrete variables to assume discrete values.

Olsen and Vanderplaats (1989) present a numerical method for the solution of nonlinear discrete optimization problems. The method uses approximation techniques to create subproblems suitable for linear mixed integer programming methods.

Dong, Gurdal and Griffin (1990) have used a penalty formulation to solve a nonlinear integer optimization problem. The penalty approach is used for converting a constrained optimization problem into a sequence of unconstrained optimization problems. To force the solution toward discrete values sine function penalty terms are introduced. The penalty terms add zero penalty at discrete points and nonzero penalty at nondiscrete values.

In this chapter we discuss and illustrate the application of Dakin's modification of Land and Doig's algorithm for the solution of nonlinear mixed integer programming problems. The Land and Doig algorithm is unsuitable for general nonlinear mixed integer programming problems on two counts. Firstly, it employs a tree structure that requires excessive computer storage (Little *et al.*, 1960) and secondly the branching rules are tied to the assumption of linearity (Handy, 1975).

Dakin's modification of the Land and Doig algorithm makes the branching rule independent of the linearity condition. The algorithm starts by finding a solution to the continuous problem wherein the integrality requirements are relaxed. If this solution is integer, then it is the optimal solution to the given discrete problem. If the solution is noninteger then at least one integer variable, say x_k, can be divided into integral and fractional parts $[b]$ and c respectively as

$$b_k = [b] + c \tag{5.4}$$

where $[b]$ is integral and $0 < c < 1$. If x_k has to take an integer value then either one of the following two conditions must be satisfied.

$$x_k \leqslant [b] \tag{5.5}$$

or

$$x_k \geqslant [b] + 1 \tag{5.6}$$

Constraint conditions expressed by equations (5.5) and (5.6) have nothing to do with linearity and as such can be used to solve nonlinear mixed integer programming problems. Equations (5.5) and (5.6) lead to two subproblems each satisfying one of the above conditions.

Subproblem (1)

$$\text{minimize} \quad f(\mathbf{X}), \mathbf{X} = (x_1\, x_2 \cdots x_n)^t \tag{5.7}$$

$$\text{subject to} \quad h_i(\mathbf{X}) = 0 \quad i = 1, 2, \ldots, n_e \tag{5.8}$$

$$g_i(\mathbf{X}) > 0 \quad j = 1, 2, \ldots, n_g \tag{5.9}$$

$$x_k \leqslant [b] \tag{5.10}$$

Subproblem (2)

$$\text{minimize} \quad f(\mathbf{X}), \mathbf{X} = (x_1\, x_2 \cdots x_n)^t \tag{5.11}$$

$$\text{subject to} \quad h_i(\mathbf{X}) = 0 \quad i = 1, 2, \ldots, n_e \tag{5.12}$$

$$g_i(\mathbf{X}) > 0 \quad j = 1, 2, \ldots, n_g \tag{5.13}$$

$$x_k \geqslant [b] + 1 \tag{5.14}$$

In the subproblems, the integrality requirement on the design variables has been removed. Furthermore, the two subproblems have removed the space $[b] < x_k < [b] + 1$ from the feasible region as this space is not allowable for an integral solution. It should also be noted that none of the integer feasible solutions has been eliminated. Each of these subproblems is then solved again as a continuous problem and the information regarding the optimal solution is stored at a node with corresponding value of the objective function, and also the lower and upper bounds on the design variables.

The foregoing procedure of branching and solving a sequence of continuous problems is continued until a feasible integer solution to one of the continuous problems is found. This integer feasible solution becomes an upper bound on the objective function. At this point, all nodes that have a value of the objective function higher than the upper bound are eliminated from further consideration, and the corresponding nodes are said to be fathomed.

The procedure of branching and fathoming is repeated for each of the unfathomed nodes. When a feasible integer solution is found, and when the value of the objective function is less than the upper bound to date, it becomes the new upper bound on the objective function. A node is fathomed (i.e. no further consideration is required) in any of the following cases.

- The continuous solution is a feasible integer solution.
- The continuous solution is infeasible.
- The optimal value (of the continuous problem) is higher than the current upper bound.

This search for the optimal solution terminates when all nodes are fathomed. The current best integer solution gives the optimal solution to the given discrete optimization problems.

This process is illustrated by the following simple example involving two variables:

$$\text{minimize} \quad f(n_1, n_2) = 4n_1^2 - 5n_1 n_2 + n_2^2 + 20 \tag{5.15}$$

subject to the constraints

$$n_1 + n_2 \leqslant 10 \tag{5.16}$$

$$n_1 > 0 \tag{5.17}$$

$$n_2 > 0 \tag{5.18}$$

where n_1, n_2 are integers.

Solving the above problem yields a continuous solution:

$$n_1 = 3.8461, \quad n_2 = 6.1538, \quad f(n_1, n_2) = -1.3014$$

Choosing the first variable n_1 for branching, we obtain two subproblems.

Subproblem (1)

$$\text{minimize} \quad f(n_1, n_2) = 4n_1^2 - 5n_1 n_2 + n_2^2 + 20$$
$$\text{subject to} \quad n_1 + n_2 \leqslant 10$$
$$n_1 > 0$$
$$n_2 > 0$$
$$n_1 \leqslant 3 \tag{5.19}$$

Subproblem (2)

$$\text{minimize} \quad f(n_1, n_2) = 4n_1^2 - 5n_1 n_2 + n_2^2 + 20$$
$$\text{subject to} \quad n_1 + n_2 < 10$$
$$n_1 > 0$$
$$n_2 > 0$$
$$n_1 \geqslant 4 \tag{5.20}$$

Conditions (5.19) and (5.20) are the additional constraints imposed to branch the solution. Solution of subproblem (1) leads to $n_1 = 3$, $n_2 = 7$, with the objective function $f(n_1, n_2) = 0$. Solution of subproblem (2) leads to $n_1 = 4$, $n_2 = 6$ with the objective function $f(n_1, n_2) = 0$. Solutions to subproblems (1) and (2) are both integer feasible solutions. In this case both integer solutions have the same value for the objective function, namely 0. The above solutions are obtained by branching the first design variable n_1. By branching the second design variable n_2, leads to two subproblems that are denoted as subproblems (3) and (4).

Subproblem (3)

$$\text{minimize} \quad f(n_1, n_2) = 4n_1^2 - 5n_1 n_2 + n_2^2 + 20$$

$$\text{subject to} \quad n_1 + n_2 \leqslant 10$$

$$n_1 > 0$$

$$n_2 > 0$$

$$n_2 \leqslant 6 \qquad\qquad (5.21)$$

Subproblem (4)

$$\text{minimize} \quad f(n_1, n_2) = 4n_1^2 - 5n_1 n_2 + n_2^2 + 20$$

$$\text{subject to} \quad n_1 + n_2 \leqslant 10$$

$$n_1 > 0$$

$$n_2 > 0$$

$$n_2 \geqslant 7 \qquad\qquad (5.22)$$

Relations (5.21) and (5.22) are additional constraints due to the branching. Solution of subproblem (3) leads to $n_1 = 3.75$, $n_2 = 6$, with the objective function $f(n_1, n_2) = -0.25$. Solution of subproblem (4) leads to $n_1 = 3$, $n_2 = 7$ with the objective function $f(n_1, n_2) = 0$. Solution of subproblem (4) is an integer feasible solution, whereas the solution to subproblem (3) is not. We need to investigate subproblem (3) further as the value of the objective function (-0.25) is lower than the least value amongst the integer feasible solutions (the least value of the integer feasible solutions is zero).

Once again branching the variable n_1, we obtain two subproblems (5) and (6):

Subproblem (5)

$$\text{minimize} \quad f(n_1, n_2) = 4n_1^2 - 5n_1 n_2 + n_2^2 + 20$$

$$\text{subject to} \quad n_1 + n_2 \leqslant 10$$

$$n_1 > 0$$

$$n_2 > 0$$

$$n_2 \leqslant 6$$

$$n_1 \leqslant 3 \qquad\qquad (5.23)$$

Subproblem (6)

$$\text{minimize} \quad f(n_1, n_2) = 4n_1^2 - 5n_1 n_2 + n_2^2 + 20$$

$$\text{subject to} \quad n_1 + n_2 \leqslant 10$$

$$n_1 > 0$$

$$n_2 > 0$$

$$n_2 \leqslant 6$$

$$n_1 \geqslant 4 \qquad\qquad (5.24)$$

Solution of subproblem (5) leads to $n_1 = 3$, $n_2 = 6$, with the objective function $f(n_1, n_2) = 2$. Solution of subproblem (6) leads to $n_1 = 4$, $n_2 = 6$, with the objective function $f(n_1, n_2) = 0$. These are no further nodes to fathom. Integer feasible solutions are $n_1 = 4$, $n_2 = 6$, $f(n_1, n_2) = 0$ and $n_1 = 3$, $n_2 = 7$, $f(n_1, n_2) = 0$. This is very characteristic of real problems wherein multiple optimum solutions often exist. The strength of the Land–Doig algorithm lies in its ability to find all possible optimum solutions.

5.3 APPLICATION TO A VIBRATING STIFFENED COMPOSITE PLATE

A simply supported laminated composite rectangular plate of dimensions L_x and L_y is considered. The laminate is assumed to be of a symmetric construction of n_1 plies at 90°, n_2 plies at 60°, n_3 plies at 45°, n_4 plies at 30° and n_5 at 0°. Each ply has a uniform thickness, t_0, and $n_1, n_2, ..., n_5$ form the five integer variables of the problem. For the purpose of analysis, the plate is discretized using a mesh of 4 by 4, 8-noded, shear deformable, isoparametric, penalty plate bending elements (Reddy, 1979). The plate is reinforced by two stiffeners placed symmetrically with respect to the laminate midplane. Each of the two stiffeners is discretized using 8 frame elements. The cross-sectional area of the 16 elements are reduced to 8 independent unknowns through linking. The linking reduces the total number of continuous design variables from 16 to 8. For each frame element the cross-sectional area, A_k, is assumed to be related to its cross-sectional moment of inertia, I_k, by

$$I_k = \alpha(A_k)^n \tag{5.25}$$

where

$$\alpha = \frac{I_0}{A_0^n}; \quad n = 3 \tag{5.26}$$

To obtain plates of different aspect ratios, the dimension L_y was set to 50 units while the dimension L_x was varied from 40 to 80 units. A nonstructural mass equal to 2% of the mass of the plate was assumed as its center.

Finally, the nondimensional material constants of the plate were assumed as $E_1/E_2 = 40$, $G_{12}/E_2 = 0.6$, $G_{23}/E_2 = 0.5$, $G_{31}/E_2 = 0.6$, $v_{13} = v_{12} = 0.25$ while the shear correction factors $k_{12} = k_{22}$ were assumed to be $\frac{5}{6}$. With these preliminaries the optimization problem for the simply supported plate of Fig. 5.1 can be stated as

$$\text{maximize} \quad f(n_i, A_j) = \frac{\lambda_1}{\lambda_{01}} l_1 + \frac{\lambda_2}{\lambda_{02}} l_2 + \frac{\lambda_3}{\lambda_{03}} l_3 \tag{5.27}$$

$$i = 1, 2, ..., 5, \quad j = 1, 2, ..., 8$$

$$\text{subject to} \quad \sum n_i \leqslant 10 \tag{5.28}$$

$$\lambda_k > \lambda_{0k_1}, \ k = 1, 2, 3 \tag{5.29}$$

$$0.1 < \frac{A_j}{A_0} < 0.6 \tag{5.30}$$

$$\sum_{r=1}^{2} m_{sr} < 0.006 m_{\text{plate}} \tag{5.31}$$

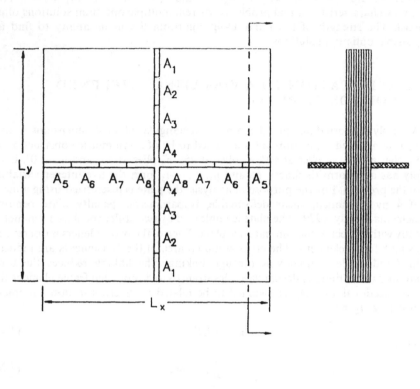

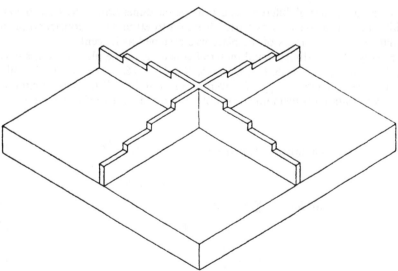

Fig. 5.1 Different views of the stiffened laminated composite plate with a definition of the design variables linking.

where λ_k is the kth eigenvalue of the plate obtained by solving the eigenvalue problem

$$\mathbf{Kq} = \mathbf{Mq} \tag{5.32}$$

λ_{0k} is the kth eigenvalue of the original unoptimized plate, l_1, l_2, l_3 are decision variables that take on the values of either zero or unity. Equation (5.28) restricts the total number of plies to ten or less. Equations (5.29) are the frequency separation constraints which are intended for obtaining coalescence-free designs, but as posed they do not guarantee them. Equation (5.30) specifies upper and lower bounds on the continuous design variables with A_0 being a reference area. Equation (5.31) ensures that the total mass of the two stiffeners is less than 0.6% of the mass of the plate. For the initial design all the elements of the two stiffeners are assumed to have the same area

$$A_s = 0.006 \frac{L_x L_y h}{L_x + L_y} \tag{5.33}$$

h being the total thickness of the plate.

In the process of applying Dakin's algorithm to the solution of the optimization problem as posed, an algorithm for the solution of the continuous optimization subproblem is necessary. Each of the continuous optimization subproblems is solved using the projected Lagrangian or variable metric method for constrained optimization (VMCON) proposed by Powell (1978). This method has been found to be ideal for small problems of the type being considered herein. It is based on a theorem which states that the optimum is a minimum of the Lagrangian function in the subspace of vectors orthogonal to the active constraint gradients. It employs a quadratic approximation to the Lagrangian in the subspace with the Hessian of the Lagrangian function being based on a BFGS (Broyden–Fletcher–Goldfarb–Shanno) update. Finally, the direction-seeking algorithm requires the solution of a quadratic programming problem (quadratic objective function and linearized constraints).

VMCON requires derivatives of the objective function and the constraints which involve the plate frequencies. It can be easily verified that (Haftka and Kamat, 1984) the derivatives of the square of the kth frequency, i.e. of λ_k, with respect to the jth design variable are given by

$$\frac{\partial \lambda_k}{\partial x_j} = \mathbf{q}_k^t \left(\frac{\partial \mathbf{K}}{\partial x_j} - \lambda_k \frac{\partial \mathbf{M}}{\partial x_j} \right) \mathbf{q}_k \tag{5.34}$$

Calculation of the frequency derivatives using analytical formulae is computationally cheap and does not require the solution of any additional eigenvalue problems other than the one for computing the frequencies and modeshapes. In equation (5.34) $\mathbf{q}_k$ is the kth eigenvector, $\mathbf{K}$ is the stiffness matrix and $\mathbf{M}$ is the mass matrix. The global $\mathbf{K}$ and $\mathbf{M}$ matrices are assembled from the element matrices. To differentiate $\mathbf{K}$ and $\mathbf{M}$, it is sufficient to assemble the matrices obtained by differentiating the corresponding element matrices. The number of plies of a given fiber orientation and the stiffener areas form the two sets of design variables. The derivatives of the frame element stiffness and mass matrices with respect to the cross-sectional areas or moments of inertia are straightforward and hence are not presented here. Of interest here are the expressions for sensitivity derivatives of the stiffness and mass matrices of an individual laminated composite plate element.

Calculation of $\partial \mathbf{K}^e / \partial x_j$ for the eth element requires the differentiation of matrices $\mathbf{A}$ and $\mathbf{D}$ defined by

$$(A_{ij}, D_{ij}) = \int_{-h/2}^{h/2} \sum_m \bar{Q}_{ij}^m (1, Z^2) \, dZ, \quad i, j = 1, 2, 6 \tag{5.35}$$

$$A_{ij} = k_\alpha k_\beta \int_{-h/2}^{h/2} \bar{Q}_{ij} \, dZ, \quad i, j = 4, 5 \tag{5.36}$$

$$\alpha = 6 - i,$$
$$\beta = 6 - j$$

where $\bar{Q}_{ij}^m$ are the stiffness coefficients for the mth ply in a standard laminate analysis. The differentiation of matrices $\mathbf{A}$ and $\mathbf{D}$ defined above will now be illustrated for a six-ply symmetric laminate with

$$Z_1 = -(n_1 + n_2 + n_3)t_0 \tag{5.37}$$
$$Z_2 = -(n_2 + n_3)t_0$$
$$Z_3 = -n_3 t_0$$
$$Z_4 = 0$$

with t_0 being the individual ply thickness and h being the total thickness of the laminate. Thus

$$A_{ij} = 2[\bar{Q}_{ij}^1 (Z_2 - Z_1) + \bar{Q}_{ij}^2 (Z_3 - Z_2) + \bar{Q}_{ij}^3 (Z_4 - Z_3)] \tag{5.38}$$

$$D_{ij} = \tfrac{2}{3}[\bar{Q}_{ij}^2 (Z_2^3 - Z_1^3) + \bar{Q}_{ij}^2 (Z_3^3 - Z_2^3) + \bar{Q}_{ij}^3 (Z_4^3 - Z_3^3)] \tag{5.39}$$

From which the relations

$$\frac{\partial A_{ij}}{\partial n_k} = 2\left[\bar{Q}_{ij}^1 \left(\frac{\partial Z_2}{\partial n_k} - \frac{\partial Z_1}{\partial n_k} \right) + \bar{Q}_{ij}^2 \left(\frac{\partial Z_3}{\partial n_k} - \frac{\partial Z_2}{\partial n_k} \right) + \bar{Q}_{ij}^3 \left(\frac{\partial Z_4}{\partial n_k} - \frac{\partial Z_2}{\partial n_k} \right) \right] \tag{5.40}$$

and

$$\frac{\partial D_{ij}}{\partial n_k} = 2\left[\bar{Q}_{ij}^1 \left(Z_2^2 \frac{\partial Z_2}{\partial n_k} - Z_1^2 \frac{\partial Z_1}{\partial n_k} \right) + \bar{Q}_{ij}^2 \left(Z_3^2 \frac{\partial Z_3}{\partial n_k} - Z_2^2 \frac{\partial Z_2}{\partial n_k} \right) \right.$$
$$\left. + \bar{Q}_{ij}^3 \left(Z_4^2 \frac{\partial Z_4}{\partial n_k} - Z_3^2 \frac{\partial Z_3}{\partial n_k} \right) \right] \tag{5.41}$$

are obtained wherein the derivatives for $\partial Z_1 / \partial n_k$ are known from equation (5.37).

To calculate $\partial M^e / \partial x_j$ for the eth element the expression for p and I defined by

$$(p, I) = \int_{-h/2}^{h/2} \sum_m (1, z^2) \rho^m \, dz \tag{5.42}$$

must be differentiated. ρ^m in equation (5.42) is the material density of the mth layer. For a six-ply symmetric laminate

$$p = 2[(Z_2 - Z_1)\rho^1 + (Z_3 - Z_2)\rho^2 + (Z_4 - Z_3)\rho^3] \tag{5.43}$$

and
$$I = \tfrac{2}{3}[(Z_2^3 - Z_1^3)\rho^1 + (Z_3^3 - Z_2^3)\rho^2 + (Z_4^3 - Z_3^3)\rho^3]$$

Hence the derivatives of p and I are again given by equations (5.40) and (5.41) respectively with $\bar{Q}_{ij}^m$ being replaced by ρ^m.

5.3.1 Measures to accelerate the subproblem optimization

Depending on the number of integer design variables and their bounds, it is necessary to solve a large number of continuous subproblems. A number of steps are taken to make the solution of the subproblems computationally efficient. These steps are discussed below.

- Each of the subproblems inherits the constraints of the 'parent' problem (relations (5.7)–(5.9) or (5.11)–(5.13)) plus the additional constraints (5.10) and (5.14). While equation (5.10) lowers the upper bound constraint, equation (5.14) increases the lower bound on x_k leading ultimately to coalescence of the upper and lower bounds for a given integer design variable. A tightening of the bounds on a given integer variable leads to slight reduction of the solution times for the two subproblems. The corresponding variable is then removed from the design space. This reduces the number of design variables that are removed. Reduction in the problem size results in drastic reduction in solution times for the subproblem.
- As the geometry of the structural model does not change, certain integration results computed during the initial assembly of the structure stiffness and mass matrices can be stored for later use. This reduces the time for assembly of the two matrices by a factor of about six. It should be noted that the assembly of the two matrices needs to be performed once each time the frequencies are computed. Also, during the computation of sensitivity derivatives one needs to perform the assembly for each design variable with respect to which the derivative is required.
- It is possible to make the computation of frequencies extremely efficient. For the very first time the frequencies are computed using the method of subspace iteration in conjunction with Jacobi diagonalization. For subsequent calculations of frequencies, the sensitivity derivative information can be used to predict approximate values of the new frequencies on the basis of a Taylor series expansion. The approximate frequencies are then used to shift the stiffness matrix. Inverse iteration on the shifted stiffness matrix yields the required frequencies. This method preserves the skyline structure of the stiffness and mass matrices.
- The VMCON algorithm uses the identity matrix as an initial approximation estimate to the Hessian matrix of the Lagrangian function. The solution time for the continuous subproblem depends on how close the initial estimate is to the actual Hessian matrix at the optimal point. The Hessian matrix at the end of the previous optimization subproblem serves as an excellent initial approximation to the Hessian matrix of the current subproblem. Indeed, the number of design variables may change from one subproblem to the next. Only those elements of the Hessian matrix corresponding to design variables of the current subproblem that were also present in the previous subproblem

are retained. For those new design variables which enter the design space all the elements of the columns and rows corresponding to these design variables except the diagonal elements are set to zero. The new diagonal elements are set to unity.

5.3.2 Discussion of results

To obtain plates of differing aspect ratios, L_y is set to 50 units and L_x is varied over the three values 40, 50, and 80 units. Six different designs are considered in Tables 5.1–5.5. Three of these are labeled 1(a), 1(b) and 1(c) for $l_1 = l_2 = l_3 = 0$ and the remaining three are labeled 2(a), 2(b) and 2(c) for $l_1, l_2, l_3 = 1$.

Table 5.1 compares the first three eigenvalues for the initial and optimum designs together with those for the first continuous solution for the six cases. An examination of Table 5.3 indicates that the rounded-off design is clearly inferior to the optimal design in all the cases except for case 2(c) where the optimal design is infeasible. In case 2(b) the rounded-off design not only has a smaller value for the objective function compared with the optimal design but is also infeasible. A further examination of Table 5.3 indicates that every integer feasible optimal design has an objective function that is larger than that for the continuous design for all the cases except case 2(c). This behavior is due to the fact that the design space is nonconvex and has multiple optimal solutions. The integer programming algorithm therefore greatly increases the probability of locating the global

Table 5.1 Eigenvalues for the different designs

Case number	λ_1^*	λ_2^*	λ_3^*	Type of design
1(a)	0.404 36	1.6609	2.5336	*
	0.616 95	1.7947	3.0845	**
	0.666 97	1.8460	3.9932	***
1(b)	0.300 77	0.981 90	2.2262	*
	0.407 75	1.1855	2.4987	**
	0.417 65	1.5373	2.2236	***
1(c)	0.200 70	0.393 10	0.842 50	*
	0.252 18	0.419 94	0.926 07	**
	0.253 91	0.396 39	0.851 92	***
2(a)	0.404 36	1.6609	2.5336	*
	0.628 42	1.977 91	4.1896	**
	0.631 05	1.9910	4.1398	***
2(b)	0.300 77	0.981 90	2.2262	*
	0.359 82	1.3395	2.222 62	**
	0.347 91	1.4392	2.224 65	***
2(c)	0.200 70	0.393 10	0.842 50	*
	0.202 91	0.536 48	1.4154	**
	0.201 45	0.553 69	1.3654	***

*, initial unoptimized structure; **, first continuous solution; ***, optimal design; λ_i^*, $i = 1, 2, 3$, normalized eigenvalues.

Table 5.2 Summary of structural optimization results

Converged design variable vector	Initial design variable vector	Case 1(a) $A_s = 0.2667$	Case 1(b) $A_s = 0.3000$	Case 1(c) $A_s = 0.3692$	Case 2(a) $A_s = 0.2667$	Case 2(b) $A_s = 0.3000$	Case 2(c) $A_s = 0.3692$
n_1 (90°)	2	1	1	2	1	3	1
n_2 (60°)	2	0	0	4	0	0	0
n_3 (45°)	2	1	2	2	1	2	3
n_4 (30°)	2	2	3	1	2	0	0
n_5 (0°)	2	6	4	1	6	5	6
A_1	A_s	0.100	0.600	0.600	0.100	0.100	0.600
A_2	A_s	0.100	0.100	0.100	0.100	0.100	0.100
A_3	A_s	0.100	0.100	0.100	0.100	0.100	0.100
A_4	A_s	0.100	0.100	0.100	0.600	0.100	0.100
A_5	A_s	0.600	0.600	0.600	0.600	0.600	0.600
A_6	A_s	0.600	0.600	0.600	0.100	0.200	0.100
A_7	A_s	0.600	0.200	0.538	0.600	0.600	0.600
A_8	A_s	0.100	0.100	0.100	0.600	0.600	0.538

Table 5.3 Objective function comparison for the different designs

Case number	Initial design	First continuous design	Rounded-off design	Optimal design	
1(a)	1.0	1.525	1.548	1.649	*
	0	0	0	0	**
1(b)	1.0	1.355	1.350	1.388	*
	0	0	0	0	**
1(c)	1.0	1.256	1.245	1.265	*
	0	0	0	0	**
2(a)	3.0	4.399	4.393	4.393	*
	0	0	0	0	**
2(b)	3.0	3.560	3.574	3.631	*
	0	0	1	0	**
2(c)	3.0	4.055	4.591	4.033	*
	0	0	1	0	**

*, value of objective function to be maximized; **, number of constraints violated.

Table 5.4 Iteration time history for problem 1(c)

Subproblem number	Number of design variables	Time (s)	$\dfrac{T}{T_0} \times 100$
1	13	$T_0 = 792$	100
2	13	237	30
3	12	60	7
4	12	150	19
5	11	39	5
6	11	40	5
7	10	34	4
8	12	45	6
9	12	81	10
10	12	95	12
11	12	82	10
12	12	45	6
13	12	81	10

optimum because it searches for an optimal solution in a number of disjoint design spaces (each of the two subproblems generated is disjoint).

Table 5.4 provides the CPU time history for the subproblems of case 1(c). It is evident that the total CPU time is less than proportional to the total number of subproblems solved. For example in Table 5.5 for case 1(c), the CPU time to solve 15 subproblems is only 4.7 times the CPU time for the first continuous

Table 5.5 Solution statistics

Case number	Number of continuous solutions	CPU time for first continuous solution T_0(s)	CPU time for optimal solution T(s)	$\dfrac{T}{T_0}$
1(a)	11	1148	7332	6.4
1(b)	12	420	2049	4.9
1(c)	15	792	3727	4.7
2(a)	6	1504	5717	3.8
2(b)	17	5243	27243	5.2
2(c)	7	7152	9647	1.4

solution. The additional subproblems are therefore obtained very cheaply because of the measures taken to accelerate the optimization process.

5.4 SOME RECENT METHODS FOR NONLINEAR MIXED INTEGER STRUCTURAL OPTIMIZATION PROBLEMS

5.4.1 Sequential linear discrete programming (SLDP)

This method proposed by Olson and Vanderplaats (1989) uses SLDP to solve the discrete optimization problem. The optimization problem tackled can be expressed as

minimize $f(\mathbf{X})$
subject to

$$g_j(\mathbf{X}) \leqslant 0 \qquad j = 1, \ldots, m$$
$$h_k(\mathbf{X}) = 0 \qquad k = 1, \ldots, l \tag{5.44}$$
$$\mathbf{X} = (x_1 \cdots x_p \, x_{p+1} \cdots x_n)^{\mathrm{t}}$$
$$x_i \in (d_{i1}, d_{i2}, \ldots, d_{iq}) \qquad i = 1, \ldots, p$$

where

$$x_k^l \leqslant x_k \leqslant x_k^u \qquad k = p + 1, \ldots, n$$

where p is the number of discrete design variables, n the total number of design variables, d_{ij} is the jth discrete value for design variable i, x_k^l the lower bound for design variable k and x_k^u the upper bound for design variable k. The corresponding linear programming problem at $\mathbf{X}^0$ becomes

$$f(\mathbf{X}) = f(\mathbf{X}^0) + \nabla F(\mathbf{X}^0)\, \delta \mathbf{X}$$

Subject to

$$g_j(\mathbf{X}) = h_k(\mathbf{X}^0) + \nabla h_k(\mathbf{X}^0)\, \delta \mathbf{X} \leqslant 0 \qquad j = 1, \ldots, m$$
$$h_k(\mathbf{X}) = h_k(\mathbf{X}^0) + \nabla h_k(\mathbf{X}^0)\, \delta \mathbf{X} = 0 \qquad k = 1, \ldots, l \tag{5.45}$$

where

$$\delta \mathbf{X} = \mathbf{X} - \mathbf{X}^0$$
$$\mathbf{X} = (x_1 \cdots x_p x_{p+1} \cdots x_n)^t$$
$$x_i^0 + \delta x_i \in (d_{i1}, d_{i2}, \ldots, d_{iq}) \qquad i = 1, \ldots, p$$
$$x_k^l \leqslant x_k^0 + \delta x_k \leqslant x_k^u \qquad k = p + 1, \ldots, n$$

The following construct is made to incorporate the discrete design:

$$x_i = z_{i1} d_{i1} + z_{i2} d_{i2} + \cdots + z_{iq} d_{iq} \tag{5.46}$$

with

$$z_{i1} d_{i1} + z_{i2} d_{i2} + \cdots + z_{iq} d_{iq} = 1 \tag{5.47a}$$

and

$$z_{ij} = 0 \quad \text{or} \quad 1 \qquad j = 1, \ldots, q \tag{5.47b}$$

Equations (5.46) and (5.47) ensure that the design variable will have one of the discrete values d_{iq}. The above equations for x_i can be inserted into the linear approximate problem, yielding

minimize

$$f(\mathbf{X}) \cong f(\mathbf{X}^0) + \sum_{i=1}^{p} \left[\left(\sum_{j=1}^{q} z_{ij} d_{ij} \right) - x_k^0 \right] + \sum_{k=p+1}^{n} \frac{\partial F}{\partial x_k} (x_k - x_k^0) \tag{5.48}$$

subject to

$$g_j(\mathbf{X}^0) \cong g_j(\mathbf{X}^0) + \sum_{i=1}^{p} \frac{\partial g_j}{\partial x_i} \left[\left(\sum_{m=1}^{q} z_{im} d_{im} \right) - x_i^0 \right] + \sum_{k=p+1}^{n} \frac{\partial g_j}{\partial x_k} (x_k - x_k^0) \leqslant 0$$

$$\sum_{l=1}^{q} z_{il} = 1 \qquad i = 1, \ldots, p$$

$$z_{ij} = 0 \quad \text{or} \quad 1 \qquad \text{all } i \text{ and } j$$

$$x_k^l < x_k^0 + \sum x_k x_k^u \qquad k = p + 1, \ldots, n$$

To choose the initial guess X^0, problem (5.48) can be solved using a continuous optimization package and rounding off the solution to the discrete values. While rounding off, the gradient information can be used to decide whether to round to the higher or lower discrete value.

The continuous problem is solved using the ADS program (Olson and Vanderplaats, 1989). The LINDO program uses a branch and bound approach to solve a discrete linear optimization problem. The values of z_{iq} returned from the LINDO program are converted to the design variables, using equation (5.46), and used as an initial guess for a subsequent ADS–LINDO iteration.

5.4.2 Penalty approach for nonlinear optimization with discrete design variables

This method proposed by Dong, Gurdal and Griffin (1990) uses a sequential unconstrained minimization technique (SUMT) whereby the constrained optimization is transformed into a sequence of unconstrained minimization problems.

To ensure that the design variables take integral values a penalty term is defined to take on a zero value at the discrete points. At nondiscrete points the penalty term assumes a nonzero value. This forces the optimal solution to be at the selected discrete points.

The SUMT technique transforms the original constrained optimization problem into a sequence of unconstrained optimization problems. The sequence of the unconstrained problems tends to the solution of the original constrained problem. Suppose the original constrained optimization problem is defined by the following relations.

$$\text{minimize} \quad f(\mathbf{X})$$
$$\text{such that} \quad g_j(\mathbf{X}) \geqslant 0, \quad j = 1, 2, \ldots, n_g \tag{5.49}$$

where $\mathbf{X} = (x_1 \, x_2 \cdots x_n)^t$ and n is the total number of design variables and n_g is the total number of constraints.

The constrained optimization problem is replaced by the following unconstrained minimization problem:

$$\text{minimize} \quad \Phi(\mathbf{X}, r) = f(\mathbf{X}) + r \sum_{j=1}^{n_g} y(g_j) \tag{5.50}$$

where $y(g_j)$ is a constraint violation penalty function that can be defined in a number of ways to force the solution of the unconstrained problem (5.50) towards the solution of the constrained optimization problem (5.49). The positive multiplier, r, in equation (5.50) controls the contribution of the constraint penalty terms. If $r = 0$, the penalty terms do not contribute to the solution of equation (5.50) at all. If the solution obtained (with $r = 0$) does not violate any constraints, then it is the solution to equation (5.49) and no further computations are necessary. In general, this will not be the case, as solution of equation (5.50) with $r = 0$ will violate a number of constraints. As the value of r is increased, solution of equation (5.50) will tend toward the solution of equation (5.49).

SUMT with discrete design variables

$$\text{minimize} \quad f(\mathbf{X})$$
$$\text{such that} \quad g_j(\mathbf{X}) \geqslant 0, \quad j = 1, 2, \ldots, n_g$$
$$\text{where} \quad \mathbf{X} = (x_1 \, x_2 \cdots x_n)^t \tag{5.51}$$
$$x_i = (d_{i1} \, d_{i2} \cdots d_{iq})^t, \quad i = 1, 2, \ldots, n_d$$

and n_d is the number of discrete design variables, d_{ik} the kth discrete value of the ith design variable and q the number of discrete values for each design variable.

To account for the discrete design variables, the objective function in equation (5.50) is modified to

$$\psi(\mathbf{X}, r, s) = f(\mathbf{X}) + r \sum_{j=1}^{n_g} p(g_j) + s \sum_{i=1}^{n_d} \Phi_d^i(\mathbf{X}) \tag{5.52}$$

Different forms for $\Phi_d^i(\mathbf{X})$ are possible, but Dong, Gurdal and Griffin (1990) recommend the following sine function:

$$\Phi_d^i(\mathbf{X}) = \frac{1}{2} \left\{ \sin \frac{2\pi [x_i - (1/4)(d_{ij+1} + 3d_{ij})]}{d_{ij+1} - d_{ij}} + 1 \right\} \tag{5.53}$$

where $d_{ij} < x_i < d_{ij+1}$. The proposed function $\Phi_d^i(\mathbf{X})$ penalizes only nondiscrete design variables and assures the continuity of the first derivatives of the modified pseudo-function at the discrete values of the design variables.

This method has the advantage that a discrete optimization problem can be solved using a computer program for continuous structural optimization, such as NEWSUMT (Dong, Gurdal and Griffin, 1990). Only the additional penalty term defined by the above equation needs to be specified as an addition to the objective function. The numerical examples described by Dong, Gurdal and Griffin also seem to indicate that the method is computationally very efficient in terms of the number of iterations required to converge. The drawback of the approach is that it makes the design space highly nonconvex and this increases the chances that the solution will become stuck in a local optimum. Great care would have to be exercised to ensure that the solution does not converge to a local optimum that is far from the global optimum. This can be done by carefully controlling the penalty multiplier during the optimization process.

5.5 CONCLUDING REMARKS

In this chapter, the authors have reviewed some of the techniques for nonlinear mixed integer programming and tried to demonstrate how they can be a viable structural optimization alternative to the conventional practice of rounding off a continuous optimum solution. The latter practice leads not only to suboptimal but often to infeasible solutions. With the recent advent of parallel distributed processing, it would appear that the branch and bound algorithm may be nicely parallelizable in which case computation times could be drastically reduced making the use of nonlinear mixed integer programming algorithms even more popular and routine.

REFERENCES

Balas, E. (1971) Intersection cuts – a new type of cutting planes method for integer programming. *Operations Research*, **19**, 19–39.

Ben-Israel, A. and Charnes, A. (1962) On some problems in Diophantine programming. *Cahiers du Centre d'Etudes de Recherche Operationelle*, **4**, 215–80.

Dakin, R.J. (1965) A tree search algorithm for mixed integer programming problems. *Computer Journal*, **8**(3), 250–5.

Dantzig, G.B. (1959) Notes on solving linear problems in integers. *Nadal Research Logistics Quarterly*, 75–6.

Dantzig, G.B., Fulkerson, D.R. and Johnson, S.M. (1954) Solution of large scale travelling salesman problem. *Operations Research*, **2**, 393–410.

Dong, K.S., Gurdal, Z. and Griffin, O.H., Jr (1990) A penalty approach for nonlinear optimization with discrete design variables. *Engineering Optimization*, **16**, 29–42.

Gisvold, K.M. and Moe, M. (1971) A method for nonlinear mixed-integer programming and its application to design problems. *Journal of Engineering for Industry*, **94**, 353–64.

Glover, F. (1965) A bound escalation method for the solution of integer linear programs. *Cahiers du Centre d'Etudes de Recherche Operationelle*, **6** 131–68.

Glover, F. (1968) A new foundation for a simplified primal integer programming algorithm. *Operations Research*, **16**, 727–40.

Glover, F. (1971) Convexity cut and search. *Operations Research*, **21**, 123–4.

Gomory, R.E. (1958) Outline of an algorithm for integer solutions to linear problems. *Bulletin of the American Mathematical Society*, **64**, 275–8.

Gormory, R.E. (1960a) All integer programming algorithm. *RC-189*, IBM, Yorktown Heights, NY.

Gormory, R.E. (1960b) An algorithm for the mixed integer problem. *RM-2597*, Rand Corp., Santa Monica, CA.

Gupta, O.K. (1980) Branch and bound experiments in nonlinear integer programming. PhD Thesis, Purdue University, West Lafayette, IN.

Gupta, O. and Ravindran, A. (1983) Nonlinear integer programming and discrete optimization. *Journal of Mechanisms, Transmission, and Automation in Design*, **105**, 160–4.

Haftka, R.T. and Kamat, M.P. (1984) *Elements of Structural Optimization*, Martinus Nijhoff, Netherlands.

Handy, A.T. (1975) *Interger Programming*, Academic Press, New York.

Hua, H.M. (1983) Optimization of structures of discrete size elements. *Computers and Structures*, **17**(3), 327–33.

Imai, K. (1979) Structural optimization by material selection. Information Processing Center, Kajima Corp.

Johnson, R.C. (1981) Rigid plastic minimum weight plane frame design using hot rolled shapes. Proceedings of the International Symposium on Optimal Structural Design, Tucson, AZ, pp. 9-1–9-6.

Land, A.H. and Doig, A.G. (1960) An automatic method for solving discrete programming problems. *Econometrica*, **28**, 497–520.

Little, J.D.C., Murthy, K.G., Sweeney, D.W. and Kavel, C. (1960) An algorithm for the travelling salesman problem. *Operations Research*, **11**, 979–89.

Markovitz, H.M. and Manne, A.S. (1975) On the solution of discrete programming problems. *Econometrica*, **25**, 84–110.

Olson, G.R. and Vanderplaats, G.N. (1989) Methods for nonlinear optimization with discrete design variables. *AIAA Journal*, **27**(11), 1584–9.

Powell, M.J.D. (1978) A fast algorithm for nonlinearly constrained optimization calculations. Proceedings of the Biennial, 1977 Dundee Conference on Numerical Analysis, *Lecture Notes in Mathematics*, Vol. 630, Springer, Berlin, pp. 144–57.

Reddy, J.N. (1979) Simple finite elements with related continuity for nonlinear analysis of plates. Proceedings of the Third International Conference in Australia on Finite Element Methods, University of New South Wales, Sydney.

Schmit, L.A. and Fleury, C. (1980) Discrete–continuous variable structural synthesis using dual method. *AIAA Journal*, **18**(12), 1515–24.

Templeman, A.B. and Yates, D.F. (1983) A segmented method for the discrete optimum design of structures. *Engineering Optimization*, **6**, 145–55.

Young, R.D. (1965) A primal (all integer) integer programming algorithm. *Journal of Research of the National Bureau of Standards, Section B*, **69**, 213–50.

Young, R.D. (1968) A simplified primal (all integer) integer programming algorithm. *Operations Research*, **16**, 750–82.

Young, R.D. (1971) Hypercylindrically deduced cuts in 0–1 integer programming. *Operations Research*, **19**, 1393–405.

6

Multicriterion structural optimization

JUHANI KOSKI

6.1 INTRODUCTION

Ideally structural mechanics can be viewed as a discipline which comprises the whole interactive process of analysis and design resulting in an optimal structure. After the advent of the modern computer the finite element method and optimization have gradually developed into essential parts of this process. In optimization the research has mainly been focused on the numerical solution of problems with a constantly increasing number of design variables and constraints. Usually a scalar-valued objective function, which in most cases is the weight of the structure, is optimized in the feasible set defined by the equality and inequality constraints. In practical applications, however, the weight rarely represents the only measure of the performance of a structure. In fact, several conflicting and noncommensurable criteria usually exist in real-life design problems. This situation forces the designer to look for a good compromise design by performing trade-off studies between the conflicting requirements. Consequently, he must take a decision-maker's role in an interactive design process where generally several optimization problems must be solved. Multicriterion optimization offers one flexible approach for the designer to treat this overall decision-making problem in a systematic way.

Multicriterion (multicriteria, multiobjective, Pareto, vector) optimization has recently achieved an established position also in structural design. One reason for the introduction of this approach is its natural property of allowing participation in the design process after the formulation of the optimization problem. It is generally considered that multicriterion optimization in its present sense originated towards the end of the last century when Pareto (1848–1923) presented a qualitative definition for the optimality concept in economic problems with several competing criteria (Pareto, 1896). Some other even earlier contributors have been discussed for example in Stadler (1987). A wider interest in this subject concerning the fields of optimization theory, operations research and control theory was aroused at the end of the 1960s and since then the research has been very intensive also in engineering design (Osyczka, 1984; Stadler, 1988; Eschenauer, Koski and Osyczka, 1990). Especially in structural optimization, the first applications in the English-language literature appeared in the late 1970s (Stadler,

1977; Leitmann, 1977; Stadler, 1978; Gerasimov and Repko, 1978; Koski, 1979), giving an impetus to emphasizing the decision-maker's viewpoint also in the design of load-supporting structures.

The purpose of this chapter is to introduce the basic concepts and methods used in multicriterion structural optimization. Also a brief literature review is given to describe the present research in the field. Special attention has been paid to those fundamental matters which are common to most of the published applications. These general ideas have been illustrated by some example problems where the emphasis is rather on the multicriterion view than on the numerical solution techniques. Specifically, the multicriterion problem formulation and the generation of Pareto optima are considered in these examples. The decision-making process for finding the best Pareto optimal design is also briefly discussed. On the whole, this chapter should both clarify the logical structure of the multicriterion approach and describe its diverse possibilities as well as recent advances in the field.

6.2 PARETO OPTIMAL DESIGN

6.2.1 Criteria and conflict

Any designer who applies optimization is faced by the question of which criteria are suitable for measuring the economy and performance of a structure. Such a quantity that has a tendency to improve or deteriorate is actually a criterion in nature. On the other hand, those quantities which must only satisfy some imposed requirements are not criteria but they can be treated as constraints. Most of the commonly used design quantities have a criterion nature rather than a constraint nature because in the designer's mind they usually have better or worse values. As an example of a strict constraint the structural analysis equations or any physical laws governing the system can be mentioned. They represent equality constraints whereas different official regulations and norms generally impose inequality constraints. For example such matters as space limitations, strength and manufacturing requirements are often treated as inequality constraints. One difficulty appears in choosing the allowable constraint limits which may be rather fuzzy in real-life problems. If these allowable values cannot be determined it seems reasonable to treat the quantity in question as a criterion. In general, the separation of the criteria and constraints presupposes a balance between two goals; the computational feasibility and the flexibility in the design process.

One important basic property in the multicriterion problem statement is a conflict between the criteria. Only those quantities which are competing should be treated as criteria whereas the others can be combined as a single criterion or one of them may represent the whole group. In the literature the concept of the conflict has deserved only a little attention while on the contrary the solution procedures have been studied to a great extent. In the problem formulation, however, it is useful to consider the conflict properties because it helps to create a good optimization model. For example in Cohon (1978) this topic is discussed in general terms and in Koski (1984b), where truss design is studied, the concepts of a local and global conflict have been proposed. According to the latter presentation the local conflict between two criteria can be defined as follows.

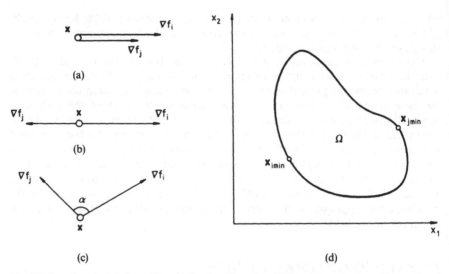

Fig. 6.1 Local and global conflict: (a) local collinearity; (b) local complete conflict; (c) local conflict; (d) global conflict in the case of two design variables. Points x_{imin} and x_{jmin} represent the individual minima of the criteria f_i and f_j in Ω, respectively.

Definition 6.1 Functions f_i and f_j are called **collinear** with no conflict at point x if there exists $c > 0$ such that $\nabla f_i(x) = c\nabla f_j(x)$. Otherwise, the functions are called **locally conflicting** at x.

This definition states that any two criteria are locally conflicting at a point in the design space if their maximum improvements are achieved in different directions. The angle between the gradients can be used as a natural measure of the local conflict. Even if two criteria are locally conflicting almost everywhere in the design space they still can achieve their optimum value at the same point. Thus it seems necessary to consider separately the concept of the global conflict where also the feasible set is involved.

Definition 6.2 Functions f_i and f_j are called **globally conflicting** in Ω if the optimization problems $\min_{x \in \Omega} f_i(x)$ and $\min_{x \in \Omega} f_j(x)$ have different solutions.

Both of these concepts have been illustrated in Fig. 6.1 where the relevant situations in the design space are shown. In structural optimization, usually the weight and any chosen displacement are both strongly locally and globally conflicting quantities. Displacements often achieve their minima at the same point but still they may be locally conflicting in that part of Ω where the best design locates.

6.2.2 Multicriterion problem

In formulating an optimization problem the choice of the design variables, criteria and constraints certainly represents the most important decision. The designs which will be available in the continuation are fixed at this very early stage. For example in scalar optimization the minimization of a single criterion in the feasible set usually results in one optimal solution only. Certainly numerical computations are needed to get that optimum design but, as a matter of fact, all the decisions have been made already in the problem formulation.

The multicriterion problem inherently offers a possibility to perform a systematic sensitivity analysis for the chosen criteria. As was pointed out earlier, the difference between criteria and constraints is that the designer wants to improve the value of a criterion whereas this kind of desire is not associated with the constraints. As a natural consequence of the separation of the criteria $f_i, i = 1, 2, \ldots, m$, and the constraints the following multicriterion problem is obtained:

$$\min_{\mathbf{x} \in \Omega} [f_1(\mathbf{x}) f_2(\mathbf{x}) \cdots f_m(\mathbf{x})]^T \tag{6.1}$$

Here $\mathbf{x} = [x_1 \, x_2 \cdots x_n]^T$ represents a design variable vector and Ω is the feasible set in design space R^n. It is defined by inequality and equality constraints in the form

$$\Omega = \{\mathbf{x} \in R^n | \mathbf{g}(\mathbf{x}) \leqslant \mathbf{0}, \mathbf{h}(\mathbf{x}) = \mathbf{0}\} \tag{6.2}$$

By using the notation $\mathbf{f}(\mathbf{x}) = [f_1(\mathbf{x}) f_2(\mathbf{x}) \cdots f_m(\mathbf{x})]^T$ for the vector objective function, which contains the m conflicting and possibly noncommensurable criteria as the components, the image of the feasible set in criterion space R^m is expressed as

$$\Lambda = \{\mathbf{z} \in R^m | \mathbf{z} = \mathbf{f}(\mathbf{x}), \mathbf{x} \in \Omega\}. \tag{6.3}$$

This is called the attainable (criteria) set and apparently it is more interesting for the decision maker than the feasible set. Usually there exists no unique point which would give an optimum for all m criteria simultaneously. Thus the common optimality concept used in scalar optimization must be replaced by a new one, especially adapted to the multicriterion problem.

Only a partial order exists in criterion space R^m and thus the concept of Pareto optimality offers the most natural solution in this context.

Definition 6.3 A vector $\mathbf{x}^* \in \Omega$ is **Pareto optimal** for problem (6.1) if and only if there exists no $\mathbf{x} \in \Omega$ such that $f_i(\mathbf{x}) \leqslant f_i(\mathbf{x}^*)$ for $i = 1, 2, \ldots, m$ with $f_j(\mathbf{x}) < f_j(\mathbf{x}^*)$ for at least one j.

In words, this definition states that $\mathbf{x}^*$ is Pareto optimal if there exists no feasible vector $\mathbf{x}$ which would decrease some criterion without causing a simultaneous increase in at least one other criterion. In the literature also some other terms have been used instead of the Pareto optimality. For example words such as nondominated, noninferior, efficient, functional-efficient and EP-optimal solution have the same meaning. Here only the mathematical programming problem has been shown and applied but the corresponding control theory formulation can be found for example in Stadler (1988) and Eschenauer, Koski and Osyczka (1990).

Two different spaces R^n and R^m, called the design and the criterion space, appear in a multicriterion problem. In order to avoid any confusion it is necessary to distinguish the optimal solutions in these separate spaces. Consequently, the vector $\mathbf{z}^* = \mathbf{f}(\mathbf{x}^*)$, which represents the image of the Pareto optimum $\mathbf{x}^*$ in the criterion space, is called the **minimal solution**. Optimality concepts in both spaces have been illustrated in Fig. 6.2 where the bicriterion case has been considered. So-called weak solutions, which are also shown in the figure, and their existence

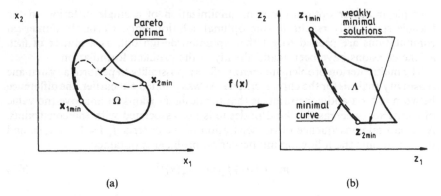

Fig. 6.2 Optimality in multicriterion problem: (a) feasible set and Pareto optimal curve; (b) attainable set and minimal curve. In this bicriterion case both the Pareto optimal and minimal set are represented by curves. Points $x_{1\,min}$ and $x_{2\,min}$ in the design space correspond to points $z_{1\,min}$ and $z_{2\,min}$ in the criterion space.

in structural optimization have been discussed in Koski and Silvennoinen (1987). In scalar optimization, one optimal solution is usually characteristic of the problem, whereas there generally exists a set of Pareto optima as a solution to the multicriterion problem. Mathematically, problem (6.1) can be regarded as solved immediately after the Pareto optimal set has been determined. In practical applications, however, it is necessary to order this set further because only one final solution is wanted by the designer. Thus he must take a decision-maker's role and introduce his own preferences to find the best compromise solution among Pareto optima.

6.2.3 Plate bending problem

As an introductory example a bicriterion plate bending problem shown in Fig. 6.3 is considered. The weight of this simply supported uniformly loaded plate and the vertical displacement of the central point A should be simultaneously minimized. These two criteria have been denoted by W and Δ, respectively. The plate thicknesses t_i, $i = 1, 2, \dots, 6$, are used as the design variables and they have been organized as zones offering a realistic design also from the manufacturing point of view. The Mindlin–Reissner thick plate element is applied in the analysis of the structure and the finite element mesh is shown in Fig. 6.3(b). The design variable zones, which correspond to certain groups of the elements, are presented in Fig. 6.3(c). By introducing constraints for the stresses and plate thicknesses the following bicriterion optimization problem is obtained:

$$\min \left[W(\mathbf{x})\, \Delta(\mathbf{x}) \right]^{\mathsf{T}} \tag{6.4}$$

subject to

$$\sigma_i^{M}(\mathbf{x}) \leqslant \bar{\sigma}, \quad i = 1, 2, \dots, 36$$

$$\underline{t} \leqslant t_i \leqslant \bar{t}, \quad i = 1, 2, \dots, 6$$

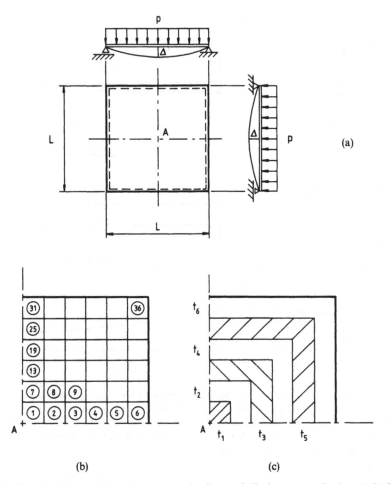

Fig. 6.3 Plate bending problem: (a) structure, loading and displacement criterion Δ; (b) finite element model; (c) design variables $t_1 \cdots t_6$ which represent the plate thicknesses. Design data for the problem: $p = 0.4\,\text{N mm}^{-2}$, $E = 206 \times 10^3\,\text{N mm}^{-2}$, $v = 0.3$, $\rho = 7800\,\text{kg m}^{-3}$, $L = 600\,\text{mm}$, $\bar{\sigma} = 140\,\text{MPa}$, $\underline{t} = 2\,\text{mm}$, $\bar{t} = 40\,\text{mm}$.

Here $\mathbf{x} = [t_1\, t_2 \cdots t_6]^T$ is the design variable vector and σ_i^M represents the von Mises stress computed at the lower surface middle point of each element. Notations $\bar{\sigma}$, $\underline{t}$ and $\bar{t}$ are used for the allowable stress, the lower and upper limit of the design variables, respectively. The numerical design data for this problem are given in the figure legend.

Several methods discussed in the next section are available for the generation of Pareto optima for this plate problem. Ten Pareto optimal plates have been computed by using the constraint method and the sequential quadratic programming optimization algorithm. The corresponding minimal solutions and three Pareto optima, which represent the minimum weight plate, the design at point 5 and the minimum displacement structure, are shown in Fig. 6.4. Also the curve

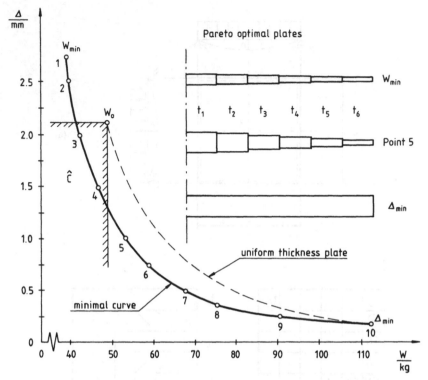

Fig. 6.4 Solution to bicriterion plate bending problem. Three Pareto optimal plates, corresponding to minimal points 1,5 and 10, are shown here. The curve of the uniform thickness plate has been depicted for comparison. For example, minimal points 3 and 4 located in the negative cone $\hat{C}$ are better designs than point W_0. The minimal curve represents the set of the minimal solutions in the bicriterion case.

Table 6.1 Pareto optimal plate thicknesses and criteria values at the minimal points shown in Fig. 6.4 (here millimeter and kilogram units are used)

Point	t_1	t_2	t_3	t_4	t_5	t_6	W	Δ
1	20.6	19.7	18.4	16.4	13.8	8.6	39.4	2.73
2	26.1	20.8	18.4	16.4	13.8	8.6	40.0	2.50
3	30.2	26.1	20.6	16.4	13.8	8.6	42.4	2.00
4	31.0	28.9	24.7	19.4	14.1	8.6	46.8	1.50
5	37.3	34.3	26.8	22.1	16.3	9.8	53.3	1.00
6	40.0	37.1	30.2	24.0	18.3	10.8	58.8	0.75
7	40.0	40.0	36.4	27.8	21.0	12.8	67.6	0.50
8	40.0	40.0	40.0	32.6	24.6	14.4	75.6	0.375
9	40.0	40.0	40.0	40.0	33.5	20.5	90.8	0.25
10	40.0	40.0	40.0	40.0	40.0	40.0	112.3	0.1746

representing the uniform thickness designs has been depicted in the criterion space for comparison. This curve is located inside the attainable set everywhere except at the minimum displacement point. Accordingly, the minimal curve gives better designs because every point except Δ_{min} on the uniform thickness curve is dominated by a minimal curve segment, as is shown by the negative cone $\hat{C}$ located at the lightest uniform design point W_0. The stiffest structure and the plate which is cheapest to fabricate seem to unite in this case, but in general it is reasonable to treat the manufacturing cost and displacement Δ as separate criteria because they may be locally conflicting elsewhere. This leads to a three-criterion problem which is not considered here. All the ten Pareto optima and the corresponding minimal points are presented numerically in Table 6.1.

6.3 GENERATION OF PARETO OPTIMA

6.3.1 Linear weighting method

Different methods for generating Pareto optimal solutions to a multicriterion optimization problem have been developed. Usually their application leads to the solution of several scalar problems which include certain parameters. Typically, each parameter combination corresponds to one Pareto optimum and by varying their values it is possible to generate the Pareto optimal set or its part. In the sequel those fundamental methods, which have been applied repeatedly in the structural optimization literature, are briefly described. They are also illustrated graphically in the criterion space in order to show the reasons for their different potential to cover the Pareto optimal set.

The linear weighting method combines all the criteria into one scalar objective function by using the weighted sum of the criteria. If the weighting coefficients are denoted by $w_i, i = 1, 2, \ldots, m$, this scalar optimization problem takes the form

$$\min_{x \in \Omega} \sum_{i=1}^{m} w_i f_i(x) \tag{6.5}$$

where the normalization

$$\sum_{i=1}^{m} w_i = 1 \tag{6.6}$$

can be used without losing generality. By varying these weights it is now possible to generate Pareto optima for problem (6.1). The main disadvantage of this method is the fact that only in convex problems it can be guaranteed to generate the whole Pareto optimal set. According to the author's experience, such nonconvex cases where the weighting method fails to generate all Pareto optima are not typical of structural optimization. Some simple truss examples, which demonstrate that this phenomenon really exists in applications, have been reported in the literature (Koski, 1985). The geometrical interpretation of the weighting method in a bicriterion problem is shown in Fig. 6.5(a) where it corresponds to the case $p = 1$. It is interesting to notice that problem (6.5) expressed in the criterion space has the form

$$\min_{z \in \Lambda} \sum_{i=1}^{m} w_i z_i \tag{6.7}$$

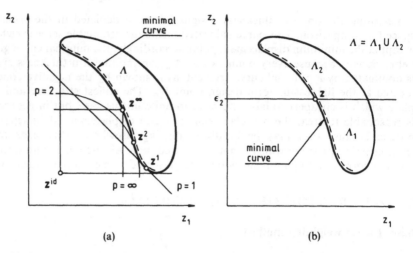

Fig. 6.5 Geometrical interpretation of distance and constraint methods in bicriterion case: (a) linear weighting method ($p = 1$), weighted quadratic ($p = 2$) and minimax ($p = \infty$) methods illustrated in the criterion space; (b) constraint method where $f_1(\mathbf{x})$ is chosen as the scalar objective function and $f_2(\mathbf{x})$ is removed into the constraints.

where $z_i = f_i(\mathbf{x})$ for $i = 1, 2, \ldots, m$. Thus a linear objective function is minimized in the attainable set.

6.3.2 Minimax approach and use of distance function

The distance methods are based on the minimization of the distance between the attainable set and some chosen reference point in the criterion space. In the literature they have also been called metric, norm, and global criterion methods. The resulting scalar problem is

$$\min_{\mathbf{x} \in \Omega} d_p(\mathbf{x}) \tag{6.8}$$

where the distance function

$$d_p(\mathbf{x}) = \left\{ \sum_{i=1}^{m} w_i [f_i(\mathbf{x}) - \hat{z}_i]^p \right\}^{1/p} \tag{6.9}$$

has been widely used in structural optimization. The reference point $\hat{z} \in R^m$ may be chosen by the designer and often the so-called ideal or utopia point

$$\mathbf{z}^{id} = [f_{1\,min}\, f_{2\,min} \cdots f_{m\,min}]^T \tag{6.10}$$

can be found in the applications. This ideal vector contains all the individual minima of the criteria in Ω as components. Thus it is necessary to solve m scalar optimization problems

$$\min_{\mathbf{x} \in \Omega} f_i(\mathbf{x}), \quad i = 1, 2, \ldots, m \tag{6.11}$$

if z^{id} is used as a reference point $\hat{z}$. The normalization given in equation (6.6) is also applicable for the weights w_i here. Usually $\hat{z}$ and p are fixed and w_i are the only parameters but also other possibilities exist (Koski and Silvennoinen, 1987). The extreme case $p = \infty$ in equation (6.9) corresponds to the weighted minimax problem

$$\min_{x \in \Omega} \max_i [w_i f_i(x)], \quad i = 1, 2, \ldots, m \tag{6.12}$$

which is capable of generating all Pareto optima also in nonconvex problems. The other extreme case $p = 1$ can be interpreted as the linear weighting method if the origin is used as a reference point, i.e. $\hat{z} = 0$. Correspondingly, the case $p = 2$ might be called as a weighted quadratic method. All these three cases have been illustrated together in Fig. 6.5(a). In practical applications, where the numerical values of the noncommensurable criteria may have huge variations with respect to each other, it is useful to normalize all the criteria before computations. One possibility is to use the formula

$$\tilde{f}_i(x) = \frac{f_i(x) - f_{imin}}{f_{imax} - f_{imin}} \tag{6.13}$$

where all the nondimensional criteria are limited to an equal range, i.e. $\tilde{f}_i(x) \in [0, 1]$, $i = 1, 2, \ldots, m$. The quadratic case $p = 2$ seems to be the most popular choice in the literature but also both of the extreme cases have been used frequently in structural design applications.

6.3.3 Constraint method

One possibility is to replace the original multicriterion problem by a scalar problem where one criterion f_k is chosen as the objective function and all the other criteria are removed into the constraints. By introducing parameters ε_i into these new constraints an additional feasible set

$$\Omega_k(\varepsilon_i) = \{x \in R^n | f_i(x) \leqslant \varepsilon_i, i = 1, 2, \ldots, m, i \neq k\} \tag{6.14}$$

is obtained. If the resulting feasible set is denoted by $\bar{\Omega}_k = \Omega \cap \Omega_k$, the parametrized scalar problem can be expressed as

$$\min_{x \in \bar{\Omega}_k} f_k(x) \tag{6.15}$$

Here each parameter combination yields a separate problem usually corresponding to one Pareto optimum. This technique, called the **constraint method**, can generate the whole Pareto optimal set also in nonconvex cases and it has been applied to some extent in structural optimization. If so-called weak solutions (Koski and Silvennoinen, 1987) shown in Fig. 6.2 exist, the constraint method can be modified to cover that case as well but then equality constraints appear and several new scalar problems must be solved for just one Pareto optimum. The geometric interpretation of the method is given in Fig. 6.5(b) where z_1 is minimized in the set Λ_1 which is the image of the set $\bar{\Omega}_1$.

6.4 ANALYTIC SOLUTION FOR STATICALLY DETERMINATE TRUSSES

The necessary conditions of Pareto optimality offer one more possibility to solve the multicriterion optimization problem (6.1). In structural optimization this approach has been used mainly in problems where the formulation is based on the optimal control theory and generally analytic solutions to certain bicriterion problems have been obtained (Stadler, 1988, Leitmann, 1977). Naturally, different numerical solution techniques, corresponding to the successful optimality criteria methods in scalar optimization of structures, could be developed for multicriterion problems. Thus far, however, only a few structural design problems (Bendsøe, Olhoff and Taylor, 1983–1984) have been solved by using directly the necessary conditions of the nonlinear programming problem (6.1). Next an application, where an analytic solution to a class of truss problems has been obtained in this way, is described.

Statically determinate and indeterminate structures are often considered separately in structural mechanics. In analyzing determinate trusses the member forces can be solved directly from the static equilibrium equations of the nodes only whereas for indeterminate structures the compatibility equations are also needed to couple the elongations of the members at each node. This property of obtaining the member forces directly is worth utilizing in optimization as well. Next a relatively general multicriterion problem is formulated for statically determinate trusses and the analytic solution presented in Koski and Silvennoinen (1982) is considered in broad outline.

The material volume and several arbitrary nodal displacements of a structure are chosen as the criteria to be minimized. Member areas A_i are used as the design variables and upper limits $\bar{A}_i$ as well as lower limits $\underline{A}_i$ for them can be imposed. If the number of members is denoted by k, the design variable vector will be

$$\mathbf{x} = [A_1\, A_2 \cdots A_k]^{\mathrm{T}} \tag{6.16}$$

Both material and geometrical linearity are assumed in analyzing a truss. Because all member forces N_i can be solved directly from the nodal equilibrium equations, new lower limits for the design variables are obtained immediately from the stress constraints as follows:

$$\underline{A}_i = \max_j (N_i/\sigma_i^{\mathrm{a}})_j, \quad i = 1, 2, \ldots, k, \quad j = 1, 2, \ldots, q$$

$$\sigma_i^{\mathrm{a}} = \bar{\sigma}_i \quad \text{for tension members} \tag{6.17}$$

$$\sigma_i^{\mathrm{a}} = \underline{\sigma}_i \quad \text{for compression members}$$

where the allowable stresses σ_i^{a} are chosen for every member i and q is the number of loading conditions. The local instability of the compression members in the elastic range may be prevented by applying the Euler buckling constraints in the expression

$$\sigma_i \leqslant \frac{\pi^2 E I_i}{n A_i L_i^2}, \quad i = 1, 2, \ldots, k \tag{6.18}$$

which states that the compressive stress σ_i in member i must be less than or equal to the Euler buckling stress divided by the safety factor n. On the right-hand side of this inequality EI_i and L_i are the bending rigidity and the length of member i, respectively. In order to convert the constraint into the form which has the member areas A_i as the only design variables the well-known relation $I_i = cA_i^p$ is introduced. By this substitution the buckling constraints (6.18) yield another lower limit for each compression member in addition to the imposed ones and those resulting from the stress constraints (6.17). When the most severe of these lower limits is chosen for every design variable subject to compression, the feasible set

$$\Omega = \{x \in R^k \mid \underline{A}_i \leqslant A_i \leqslant \bar{A}_i, i = 1, 2, \ldots, k\} \tag{6.19}$$

is obtained. This region consists of a rectangular prism in the design space generated by the member areas. It is further assumed that all the lower limits of the member areas are strictly positive, i.e. $\underline{A}_i > 0$ for $i = 1, 2, \ldots, k$, as is natural in real statically determinate applications. The vector objective function consists of the material volume V and $m - 1$ arbitrary nodal displacements Δ_i which all can be written in an explicit form in this determinate case. The multicriterion optimization problem for statically determinate trusses can thus be stated as

$$\min_{x \in \Omega} [V(x)\, \Delta_1(x)\, \Delta_2(x) \cdots \Delta_{m-1}(x)]^T \tag{6.20}$$

where

$$V = \sum_{i=1}^{k} a_i A_i, \quad a_i > 0$$

$$\Delta_j = \sum_{i=1}^{k} \frac{\alpha_i^j}{A_i}, \quad \alpha_i^j \gtreqless 0, \quad j = 1, 2, \ldots, m - 1$$

This explicit problem is not convex, because displacement criteria are not convex functions of the member areas, but it can be transformed into a convex one by replacing the design variables by their inverses. This property can be utilized to prove that the necessary conditions of Pareto optimality are also sufficient for this problem. It turns out that these conditions can be solved exactly in the case $m = 2$, i.e. if the bicriterion problem

$$\min_{x \in \Omega} [V(x)\, \Delta(x)]^T \tag{6.21}$$

where

$$V = \sum_{i=1}^{k} a_i A_i, \quad a_i > 0, \quad \Delta = \sum_{i=1}^{k} \frac{\alpha_i}{A_i}, \quad \alpha_i \in R$$

is considered. The necessary conditions and the following mathematical proof, which is rather lengthy, is skipped and only the results are presented here. First a theorem (Koski and Silvennoinen, 1982) that gives a complete solution to the above bicriterion problem is given.

Theorem 6.1 The set of Pareto optima for problem (6.21) consists of a connected polygonal line $l_1 \cup l_2 \cup \cdots \cup l_N$. The consecutive line segments l_n,

$n = 1, 2, \ldots, N$, have the parametric equations

$$A_i = \bar{A}_i, \qquad i \in I_n$$
$$A_j = \underline{A}_j, \qquad j \in J_n$$
$$A_s = c_s^{-1} t, \qquad s \in K \backslash (I_n \cup J_n) \tag{6.22}$$
$$t_{n-1} \leqslant t \leqslant t_n$$

where $K = \{1, 2, \ldots, k\}$, $Q = \{i \in K \,|\, \alpha_i \leqslant 0\}$, $Q \neq K$, $N = \min\{n \in \mathbb{N} \,|\, I_{n+1} = K \backslash Q\}$ and $c_s = a_s^{1/2} \alpha_s^{-1/2}$ for $s \in K \backslash Q$. Further, $I_0 = \varnothing$, $J_0 = K$, $t_0 = \min\{c_j \underline{A}_j \,|\, j \in K \backslash Q\}$ and for $n = 1, 2, \ldots, N$

$$I_n = I_{n-1} \cup \{s \in K \backslash (I_{n-1} \cup J_{n-1}) \,|\, c_s \bar{A}_s = t_{n-1}\}$$
$$J_n = J_{n-1} \backslash \{j \in J_{n-1} \backslash Q \,|\, c_j \underline{A}_j = t_{n-1}\} \tag{6.23}$$
$$t_n = \min\{c_s \bar{A}_s, c_j \underline{A}_j \,|\, s \in K \backslash (I_n \cup J_n), j \in J_n \backslash Q\}$$

The notation $\mathbb{N} = \{1\,2\,3\ldots\}$ is used for the set of all positive integers and the set difference is denoted by $K \backslash Q = \{i \in K \,|\, i \notin Q\}$ in this theorem. It shows that the set of all Pareto optima will be a polygonal line in the design space. Index sets I and J change from one Pareto optimal line segment to another, whereas index set Q remains constant. After solving the Pareto optimal set of problem (6.21), the corresponding minimal solutions can be found easily by substituting relations (6.22) into the expressions of V and Δ. It is also possible to eliminate parameter t, resulting in an analytic presentation of the function $V(\Delta)$. Even if the expressions (6.22) and (6.23) look complicated, their application is very easy as is illustrated by the following small example.

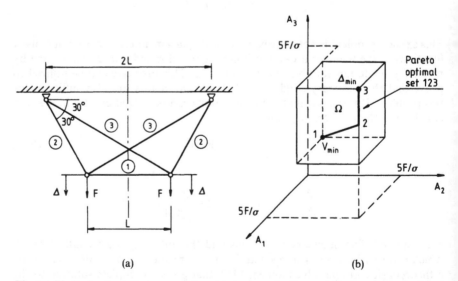

Fig. 6.6 Bicriterion isostatic truss problem: (a) structure, loading and displacement criterion Δ; (b) feasible set Ω in the design space and the Pareto optimal polygonal line 1–2–3. Allowable stresses are σ in tension and $-\sigma$ in compression. Only upper limits $\bar{A} = 5F/\sigma$ have been imposed for all member areas whereas the lower limits are obtained from the stress constraints.

A five-bar truss shown in Fig. 6.6 should be optimized by using only three design variables A_1, A_2 and A_3, owing to the symmetry. The structure is subjected to one loading condition and the equal vertical displacement of both the loaded nodes is chosen as criterion Δ in problem (6.21). Stress constraints $-\sigma \leqslant \sigma_i \leqslant \sigma$ for all members and upper limits $\bar{A} = 5F/\sigma$ for the member areas are imposed in this example whereas buckling constraints are excluded. The lower limits for the member areas can be computed directly from the stress constraints because no other lower limits have been imposed. After determining the member forces the following explicit bicriterion problem is obtained:

$$\min \begin{bmatrix} (A_1 + 2A_2 + 2\sqrt{3}A_3)L \\ \left(\dfrac{3/2}{A_1} + \dfrac{3}{A_2} + \dfrac{\sqrt{3}}{A_3} \right) \dfrac{FL}{E} \end{bmatrix} \tag{6.24}$$

subject to

$$\sqrt{3}F/\sigma \leqslant A_1 \leqslant 5F/\sigma$$
$$\sqrt{3}F/\sigma \leqslant A_2 \leqslant 5F/\sigma$$
$$F/\sigma \leqslant A_3 \leqslant 5F/\sigma$$

Next the Pareto optimal polygonal line is generated for this problem by using the scheme given in relations (6.22) and (6.23). In this case set $Q = \varnothing$ because all $\alpha_i > 0$ in the displacement criterion. Two Pareto optimal line segments, which are given as functions of parameter t, are obtained for this problem.

First it is practical to remove constants L and FL/E from the expressions of the criteria for a while. From the dimensionless coefficients a_i and α_i, $i = 1, 2, 3$, appearing in the criteria the numbers $c_1 = \sqrt{6}/3$, $c_2 = \sqrt{6}/3$ and $c_3 = \sqrt{2}$ are directly obtained. At the beginning the index sets are $I_0 = \varnothing$, $J_0 = K = \{1, 2, 3\}$ and parameter $t_0 = \sqrt{2}F/\sigma$. This Pareto optimum, where all the member areas are at their lower limits, corresponds to the minimum volume solution.

Line segment 12 lies in the interior of Ω because $I_1 = J_1 = \varnothing$ and $K\backslash(I_1 \cup J_1) = \{1, 2, 3\}$. It is given by

$$A_1 = \frac{\sqrt{6}}{2}t, \quad A_2 = \frac{\sqrt{6}}{2}t, \quad A_3 = \frac{\sqrt{2}}{2}t, \quad \sqrt{2}\frac{F}{\sigma} \leqslant t \leqslant \frac{5\sqrt{6}}{3}\frac{F}{\sigma} \tag{6.25}$$

The second line segment 23 lies on that edge of Ω where $I_2 = \{1, 2\}$, $J_2 = \varnothing$ and $K\backslash(I_2 \cup J_2) = \{3\}$. Its expression is

$$A_1 = 5F/\sigma, \quad A_2 = 5F/\sigma, \quad A_3 = \frac{\sqrt{2}}{2}t, \quad \frac{5\sqrt{6}}{3}\frac{F}{\sigma} \leqslant t \leqslant 5\sqrt{2}\frac{F}{\sigma} \tag{6.26}$$

These parametric equations (6.25) and (6.26) represent the Pareto optimal set of problem (6.24). This polygonal line, starting from the minimum volume solution and ending at the point where Δ achieves its minimum value, has been depicted in Fig. 6.6(b). In this specific case no Pareto optima are located on the faces of the feasible set. In large-scale problems more line segments appear but the solution procedure is quite insentive to the size of the problem.

The original multicriterion problem (6.20) is convex in the reciprocal variables and thus all Pareto optima for it can be generated by the linear weighting method.

This leads to one scalar optimization problem for each weight combination. In the present case, however, it is preferable to convert problem (6.20) into a parametric bicriterion problem. Considerable advantage is obtained in this way compared with the weighting method or any scalarization technique. First, the number of parameters is reduced by one and, secondly, each parameter combination gives a large set of Pareto optima instead of only one point. In addition, the Pareto optimal set of each bicriterion problem is known exactly and no approximate optimization procedure, possibly involving high computation costs and difficulties in convergence, is needed. The parametric bicriterion problem

$$\min_{\mathbf{x} \in \Omega} \left[V(\mathbf{x}) \sum_{j=1}^{m-1} \lambda_j \Delta_j(\mathbf{x}) \right]^{\mathrm{T}} \tag{6.27}$$

includes a linear combination of the original displacement criteria as the second component of the vector objective function. This combined displacement criterion again has the general form

$$\sum_{j=1}^{m-1} \lambda_j \Delta_j(\mathbf{x}) = \sum_{i=1}^{k} \frac{\beta_i}{A_i}, \quad \beta_i \in R \tag{6.28}$$

which is similar to the expression of the displacement criterion $\Delta(\mathbf{x})$ in problem (6.21). Here the coefficients β_i can be computed from the corresponding expressions of the original displacement criteria given in problem (6.20) as follows:

$$\beta_i = \lambda_1 \alpha_i^1 + \lambda_2 \alpha_i^2 + \cdots + \lambda_{m-1} \alpha_i^{m-1}, \quad i = 1, 2, \ldots, k \tag{6.29}$$

For every fixed combination of parameters $\lambda_j, j = 1, 2, \ldots, m - 1$, it is possible to formulate problem (6.27) which can be solved completely by the given theorem. Apparently, if these parameters have the following properties

$$\sum_{i=1}^{m-1} \lambda_j = 1, \quad \lambda_j \geq 0, \quad j = 1, 2, \ldots, m - 1 \tag{6.30}$$

the solution set of each bicriterion problem with fixed λ_j values represents also a Pareto optimal polygonal line of the original problem (6.20). It can be further proved that the whole Pareto optimal set of problem (6.20) is obtained in this way by going through all the possible parameter combinations.

By combining the preceding results a numerical method for generating Pareto optima for the m-criterion problem (6.20) may be constructed. First the original problem is converted into a bicriterion problem by choosing parameters λ_j, $i = 1, 2, \ldots, m - 1$, and combining the original displacement criteria into one. The given theorem can then be applied to compute the corresponding Pareto optimal polygonal line in the design space. The solution scheme uses no approximate optimization technique and thus any accuracy wanted for the results may be achieved. The free parameter t is used for convenience in order to attain a clear representation form for the results and to enable an easy movement along the polygonal line in the decision-making stage. The whole Pareto optimal set of problem (6.20) can be generated as the union of the polygonal lines, each corresponding to one bicriterion problem (6.27) with fixed parameters λ_j.

The present method has the capacity of generating Pareto optima at a relatively high speed because each parameter combination gives an entire polygonal line. Moreover, the procedure can easily be coded as a computer program capable of obtaining any Pareto optimal polygonal line by a finite number of calculation steps without computational difficulties even in large problems. In order to solve a traditional scalar problem, where the material volume of an isostatic truss should be minimized subject to stress, elastic buckling and displacement constraints, an approximate optimization procedure is usually needed. This may be the optimality criteria approach or some numerical nonlinear programming algorithm. Instead of applying them, it is advantageous to remove the constrained displacements into the vector objective function in addition to the volume and to use the described method. Plenty of Pareto optimal solutions, including the wanted one, are obtained very quickly with any accuracy.

6.5 DECISION-MAKING PROCESS

Pareto optima represent solutions which are mathematically better than the other feasible points. Thus it is natural to concentrate merely on Pareto optima in choosing the final design called usually the best compromise solution. Comparisons between different Pareto optima can be made by introducing the designer's personal preferences based on his experience and some other factors not included in the multicriterion problem formulation (6.1). The mathematical optimization model hardly ever contains all the detailed requirements existing in practical design applications. In the multicriterion approach the decision making is concentrated on the choice of the relative importances of the competing criteria. Accordingly, this entirety consisting of the search of the best compromise solution among Pareto optima is called the multicriterion decision-making process.

Optimum structural design differs considerably from synthesis problems in other fields, such as social welfare theory, resource planning and management, where the multicriterion approach has been widely applied. Structural analysis equations are very reliable and the optimization problem is generally well defined, offering an excellent starting point for the decision maker. According to the classification especially adapted to structural optimization and proposed in Koski (1992) the multicriterion design process can be divided into three main phases. They are the problem formulation, generation of Pareto optima and decision making. The last phase can be further separated into four classes: interactive methods, choice by comparisons, procedures that use *a priori* fixed parameters and other methods. Different interactive approaches, where the designer proceeds from one Pareto optimum to a better one until a satisfactory solution is achieved, have been published. Another possibility is to generate a large collection of Pareto optima and to choose the best one by direct comparisons. The third class includes methods where some parameters have been fixed in advance. A typical example of this class is the so-called utility function methods, where some chosen scalar combination of the criteria is optimized. Also some other methods, such as goal programming and game theory approach for example, have been applied in structural optimization.

Plenty of relevant information about the system can be obtained by optimization calculations. In order to make good decisions Pareto optimal solutions and some trade-off information associated with them are usually necessary. The term trade-off has been used quite freely in the literature and it seems to take different forms depending on the context. A common feature in most trade-off analyses is that they are concerned with the advantage gained for one criterion by making concessions in another criterion. At any Pareto optimum $\mathbf{x}^*$ a trade-off number, defined symbolically by $\partial f_k / \partial f_i$, can be defined as a change of f_k per unit change in f_j on the minimal surface when all the other criteria remain unchanged. Naturally, these trade-off numbers are closely related to the Kuhn–Tucker multipliers of any corresponding scalar problem (6.5), (6.8), (6.12) and (6.15) at $\mathbf{x}^*$. For example in the constraint method

$$\frac{\partial f_k}{\partial f_i}(\varepsilon_j) = - u_{ki}^* \qquad (6.31)$$

where u_{ki}^* is the Kuhn–Tucker multiplier associated with the constraint $f_i(\mathbf{x}) \leqslant \varepsilon_i$ and $\varepsilon_j, j = 1, 2, \ldots, m, j \neq k$, represent the constraint limits corresponding to Pareto optimum $\mathbf{x}^*$. Geometrically speaking, the computation of a trade-off number at any Pareto optimum provides the designer with useful information about the tangent plane of the minimal surface in the criterion space at $\mathbf{z}^*$. If the whole minimal set is known, then complete trade-off information is directly available. In large-scale applications, however, the decision maker needs trade-off numbers at those few Pareto optima which have been computed.

6.6 LITERATURE REVIEW

Most of the citations used in the preceding text are the author's own contributions because they form a unified basis for this presentation. Altogether, however, they represent only a small part of the versatile production found in the field today.

According to a state-of-the-art study (Koski, 1993), about 70 articles dealing with multicriterion structural optimization have been published in English since the late 1970s. Stadler (1977, 1978) applied the control theory approach to different bicriterion problems with weight and stored energy as criteria and obtained analytical solutions for some structures, calling the results 'natural structural shapes'. Also, Leitmann (1977) used control theory in solving some bicriterion problems in viscoelasticity. Gerasimov and Repko (1978) optimized a truss and a framed plate by applying the mathematical programming approach. Koski (1979) formulated a vector optimization problem, where the material volume and several chosen nodal displacements of a truss were chosen as the criteria, and generated Pareto optima for several truss examples.

In the 1980s the number of publications increased considerably including many different academic and industrial applications. For example, such authors as Adali (1983a, b), Baier (1983), Bühlmeier (1990), Bendsøe, Olhoff and Taylor (1983), Carmichael (1980), Dhingra, Rao and Miura (1990), Diaz (1987a, b), Eschenauer and his coworkers (Eschenauer, 1983; Eschenauer, Kneppe and Stenvers, 1986; Eschenauer, Schäfer and Bernau, 1988; Eschenauer and Vietor, 1990; Eschenauer, Koski and Osyczka, 1990), Fu and Frangopol (1990), Fuchs *et al.* (1988), Hajela and Shih (1990), Jendo, Marks and Thierauf (1985), Koski (1981, 1984a), Lógó and Vásárhelyi (1988), Nafday, Corotis and Cohon (1988), Olhoff (1989), Osyczka (1981, 1984), Post (1990), Rao and his coworkers (Rao, 1984, 1987; Rao and Hati, 1986; Rao, Venkayya and Khot, 1988), Rozvany (1989),

Stadler (1977, 1978, 1987, 1988, 1989), Tseng and Lu (1990) and Yoshimura *et al.* (1984) have contributed in this field. The choice of the criteria varies from one article to another but usually physical quantities rather than economic are used. The most commonly used criterion is the weight or the material volume of a structure and the second in number is the flexibility which is usually measured by displacements or some energy expression. These two criteria are usually strongly conflicting and they offer an excellent possibility to study the multi-criterion problem with the lowest dimension ($m = 2$), which also allows the graphic presentation of the minimal curve in the criterion space. For certain specific problems the Pareto optimal set has been determined analytically but commonly some numerical optimization procedure has been applied. Interactive methods especially adapted to structural optimization are rare for the present. In addition, stochastic and fuzzy optimization problems have been treated to some extent. Industrial applications (Eschenauer, Koski and Osyczka, 1990) have increased considerably during the last decade and a multicriterion module has been added to some structural optimization software packages. In addition to different design problems the multicriterion approach has also been utilized in the plastic analysis of frames (Nafday, Corotis and Cohon, 1988a, b).

6.7 INDUSTRIAL APPLICATION AND ILLUSTRATIVE EXAMPLES

6.7.1 Shape optimization of ceramic piston crown

Different new materials and their composites have recently appeared to replace such traditional alternatives as construction steels and cast iron for example. Among these, ceramics undoubtedly deserve special attention in applications where high temperature capability, wear and corrosion resistance, and certain electrical or optical properties are needed. In combustion engines, bearings, cutting parts of tools and nozzles, just to mention some industrial products, ceramic materials have been used successfully. Most evidently the application of ceramics offers totally new solution possibilities for many traditional construc-tions. The economic potential associated with ceramics seems large, but severe restrictions appear if a high reliability for load-supporting components is required. Especially, shape optimization becomes an important part in the design process because the reliability is closely associated with the geometric shape of the structure. The shape optimization of ceramic components differs considerably from the design of most conventional construction materials. This is due to the fact that no clear allowable stresses exist but the strength of the ceramic component must be evaluated by the statistical approach. Furthermore, it seems impractical to replace the stress constraints by the reliability constraints because the choice of the constraint limit is extremely difficult in practical circumstances. Especially, in structural design, where both economic and human losses are involved, it is reasonable to consider the reliability as a criterion instead of a constraint function. In the following an industrial application (Koski and Silvennoinen, 1990), where the multicriterion approach is used instead of the traditional scalar problem with stress constraints, is considered.

A two-parametric Weibull distribution is used to describe the results of the strength tests for ceramics. The two material parameters, which are the mean fracture stress σ and the Weibull modulus m, can both be obtained from the material tests. Also, the volume effect, which appears such that the strength decreases when the size of the specimen and the loading are proportionally increased, should be considered in the statistical approach. The mean fracture stress σ is associated with a certain reference volume V_0 which usually comprises that part of the specimen where tensile stresses occur. The other material parameter, the Weibull modulus m, describes the narrowness of the strength distribution. The larger the value of m, the smaller is the variation of the fracture stress. From the stress field determined usually by the finite element method it is possible to compute the probability of failure from the expression

$$P_f = 1 - \exp\left[-\left(\frac{1}{m}!\right)^m \left(\frac{1}{\sigma}\right)^m \left(\frac{1}{V_0}\right) \iiint_V (\sigma_1^m + \sigma_2^m + \sigma_3^m)\, dV \right] \qquad (6.32)$$

where the integration is extended over the component volume. Here σ_1, σ_2 and σ_3 are the principal stresses which are included in the volume integral at each material point only if they are tensile, i.e. $\sigma_i > 0$. $\bar{\sigma}$, V_0 and m are the material parameters explained earlier and

$$\left(\frac{1}{m}!\right) = \Gamma\left(\frac{1}{m} + 1\right) = \int_0^\infty t^{1/m} e^{-t}\, dt \qquad (6.33)$$

The numerical computation of P_f is integrated directly to the finite element analysis. From the volume integral in equation (6.32) it may be observed that the crucial parts of the component are those where high tensile stresses occur.

The efficiency of a diesel engine is one chief factor in the competition of the market shares in such applications as the main or auxiliary engines of ships, locomotives and land-based power plants. A standard way to improve the efficiency is the raising of the working temperature in the cylinders. Good thermal insulation and strength properties are needed for the materials in those parts which are subjected to the elevated temperatures. Ceramics are a very potential alternative for this purpose. For example, in the Wärtsilä Vasa 22 HF medium speed diesel engine a considerable benefit is expected by using a ceramic piston crown instead of the traditional cast iron construction. It is uneconomic to use a monolithic ceramic piston because at the moment the material is expensive and extreme difficulties appear in trying to achieve an accurate shape during the sintering process. Thus, a ceramic material is used only in the crown where good thermal insulation and strength properties at these high temperatures are needed. The prototype piston construction including the ceramic crown, which is the object of the shape optimization here, is shown in Fig. 6.7. From the two alternative materials, zirconia (ZrO_2) and silicon nitride (Si_3N_4), the latter was chosen mainly because the density and thermal expansion coefficient values of zirconia are too large. Silicon nitride has a considerable strength at high temperatures and in spite of its small thermal insulation capability compared with zirconia it allows the elevation of the temperature in the cylinder. In addition to the possible improvement in the efficiency, another advantage is obtained simultaneously. This is the possibility of dropping the quality of the heavy fuel used in this engine because of the better ignition properties due to the higher surface temperature.

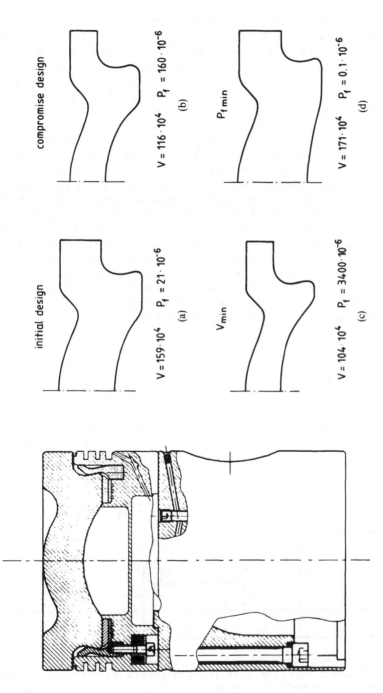

Fig. 6.7 Prototype piston of the Wärtsilä Vasa 22 diesel engine and some crown shapes: (a) initial non-optimal design; (b) Pareto optimal crown where the two criteria V and P_f are compromised; (c) minimum material volume design; (d) maximum reliability design. The numerical values of the two criteria for every shape are presented for a comparison. The volume criterion is given in mm^3; P_f has no units.

In addition, the cooling system of the piston becomes unnecessary and it can be left out.

Shape optimization has been performed for the ceramic piston crown only, leaving the other parts of the piston out. It is important to minimize the material volume of the crown for two reasons. The first is the high price of the silicon nitride itself. The second reason is associated with the inertia forces which should be at the same relatively low level as they are in the traditional cast iron pistons. The other criterion to be minimized is the probability of failure of the crown which should be as small as possible in order to avoid expensive damage during operation. The combustion conditions in the cylinder determine the shape of the upper surface of the crown whereas the shape of the lower surface and the location of the axial support, which are shown in Fig. 6.8, can be varied during optimization. The curve which defines the shape of the lower surface of the crown is represented by two B splines of order 2 with the control nodes P_0, P_1, P_2 and Q_0, Q_1, Q_2. Instead of the nodal coordinates of the finite elements on the boundary it is advantageous to use the B splines to describe the two parts of the lower surface and to apply the coordinates of the control nodes as the design variables. The initial shape and all the eight components of the design variable vector $\mathbf{x} = [x_1 \, x_2 \cdots x_8]^{\mathrm{T}}$ have been depicted in Fig. 6.8. It is natural to use side constraints to determine the allowable space for the component. Thus the feasible set Ω is defined by the inequality constraints imposed for the design variables, i.e. $\Omega = \{\mathbf{x} \in R^8 \, | \, \underline{x}_i \leqslant x_i \leqslant \bar{x}_i, \, i = 1, \ldots, 8\}$. From the practical viewpoint, it seems reasonable to apply the side constraints because the space limits can be reliably

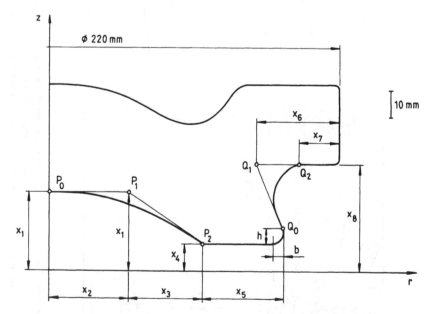

Fig. 6.8 Design variables of the bicriterion pistion crown optimization problem. The locations of the control points P_0, P_1, P_2 and Q_0, Q_1, Q_2, which define the shape of the crown, are the design variables x_i. The fixed curve segment near Q_0 determines the values $h = 5\,\text{mm}$ and $b = 3\,\text{mm}$. The support surface must be a horizontal plane.

assessed and the designer may concentrate on the trade-off between the two conflicting physical criteria in making his final choice. The resulting bicriterion shape optimization problem for the ceramic piston crown is

$$\min \, [V(\mathbf{x}) \, P_f(\mathbf{x})]^T \qquad\qquad (6.34)$$

subject to

$$10 \leqslant x_1 \leqslant 35 \qquad 10 \leqslant x_5 \leqslant 30$$
$$25 \leqslant x_2 \leqslant 45 \qquad 25 \leqslant x_6 \leqslant 35$$
$$10 \leqslant x_3 \leqslant 40 \qquad 10 \leqslant x_7 \leqslant 20$$
$$0 \leqslant x_4 \leqslant 10 \qquad 30 \leqslant x_8 \leqslant 40$$

where $V(\mathbf{x})$ is the material volume of the crown and $P_f(\mathbf{x})$ is its probability of failure given in equation (6.32). Upper and lower limits for the design variables are given in millimeters. The purpose is to compute some Pareto optimal shapes for the designer who can use his own experience in the decision-making phase where the manufacturing cost, the compatibility with the rest of the piston and many other details are considered. The choice of the strictly physical criteria in the vector objective function in problem (6.34) offers a sound basis for the optimum design because they offer Pareto optimal solutions which may be utilized even if the manufacturing techniques or the surrounding construction changes during the design process.

The piston crown is subjected to several loadings during every working cycle. A large amount of heat is transferred from the combustion gas through the crown to the cylinder liner and the cooling water. This causes thermal stresses which can be added to the stresses caused by other loadings. Inertia forces are proportional to the accelerations and their value as well as the gas pressure depend on the position of the piston. A mounting ring is used to fix the crown to the intermediate steel part and it causes a static compressive force along the connection circumference. In the optimization it seems sufficient to consider only the loading case where the thermal charge and the connection force are acting simultaneously.

As a result, an axisymmetric linear static analysis problem is obtained. All the numerical analyses during the optimization, each associated with some modified shape, have been performed by the general-purpose FEM package ANSYS. Only the computation of the probability of failure must be performed outside the program because it was not included in the package. Also, the stationary heat transfer problem for the new shape at every iteration step was calculated by ANSYS in order to determine the corresponding temperature distribution. Thus for each new shape several analyses, including thermal and structural problems, must be performed. An adaptive mesh of isoparametric quadrilateral and triangular ring elements was generated for the crown. By applying a certain parametrization the mesh was adapted for the modified shapes which were obtained during the optimization.

The constraint method was used for the generation of Pareto optimal solutions to problem (6.34). The sequential quadratic programming (SQP) method, which uses the gradients of the objective and constraint functions in the iterations, was chosen and combined with the FEM program ANSYS. A few hundred finite element analyses were needed to compute one Pareto optimum by using an arbitrary starting shape. It was possible to reduce the number of iterations and analyses by choosing the former Pareto optimum as a starting point. The minimum volume and the minimum probability of failure crowns, which represent the extreme Pareto optima for problem (6.34), are shown in Fig. 6.7. A more practical Pareto optimal crown shape and the initial design have also been depicted in the same figure. These three Pareto optimal shapes only illustrate the proposed approach whereas for real decision making many more Pareto optimal crowns were generated.

6.7.2 Composite beam

In traditional lightweight structures, such as aircraft and spacecraft for example, composites have become very popular because of their high strength–weight ratio. Recently these advanced materials have emerged into some terrestrial applications as well mainly because light components in the design offer the possibility to make the construction cheaper elsewhere. Remarkable variations in both the mechanical and the thermal behavior of the composite can be achieved by varying layer thicknesses and fiber orientations but simultaneously making an intuitive guess of a good design becomes extremely difficult. Also, the manufacturing cost may become high and the common idea that the weight represents the material cost is not valid any more. This is due to the large variations in the prices of the different materials used in the layers. Also, in the composite design the multicriterion approach seems suitable because several conflicting criteria usually appear in practical applications.

In the present example it is demonstrated that weight and material cost are competing objectives in the composite design. A simply supported beam shown in Fig. 6.9(a) is optimized subject to failure and side constraints. Carbon and glass fiber layers are located in turn forming a symmetric eight-layer cross section according to Fig. 6.9(b). Because of the symmetry only four layer thicknesses $t_1, \ldots, t_4$ are used as the design variables whereas the fiber orientations have been fixed to $0°$ for the four inner layers (1 and 2) and $90°$ for the remaining outer layers (3 and 4). The finite element analysis model is based on the eight-node rectangular plane-stress element and the Tsai–Hill failure criterion has been applied in the stress constraints. The notations C_c and C_g have been used for the prices of the carbon and glass layers, respectively. In Table 6.2 these prices, which have been presented in relative rather than absolute values, as well as the material properties are given.

The total weight of the beam, denoted here by $W = m_c + m_g$, is chosen as the first criterion to be minimized. The symbols m_c and m_g represent the masses of all the carbon and glass layers, respectively. The total material cost, defined correspondingly as $C = C_c m_c + C_g m_g$, is used as the second criterion which should also be minimized. The notations $\mathbf{x} = [t_1\, t_2\, t_3\, t_4]^T$ and

$$f_i(\mathbf{x}) = \left(\frac{\sigma_1}{X}\right)^2 - \frac{\sigma_1 \sigma_2}{X^2} + \left(\frac{\sigma_2}{Y}\right)^2 + \left(\frac{\tau_{12}}{S}\right)^2 \tag{6.35}$$

are used for the design variable vector and the stress function appearing in the Tsai–Hill failure constraint of layer i. In the latter expression the allowed stresses in tension (X_t and Y_t) and in compression (X_c and Y_c) usually have different

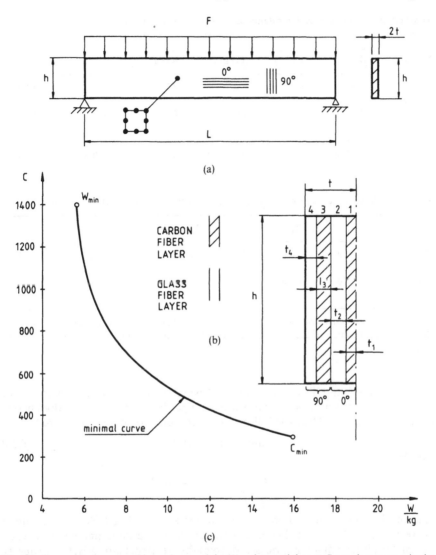

Fig. 6.9 Composite beam problem where weight W and material cost C are chosen as criteria: (a) structure, loading, fiber directions ($0°$ for layers 1 and 2, $90°$ for layers 3 and 4) and plane-stress element used in the analysis; (b) half of the symmetric cross section and the four design variables t_1, t_2, t_3, t_4; (c) minimal curve in the criterion space. The load resultant $F = 180\,\text{kN}$, $L = 1200\,\text{mm}$ and $h = 200\,\text{mm}$.

values. In Table 6.3 the numerical values for them as well as for the limit shear stress S are given. Subscripts 1 and 2 as well as strength notations X and Y in equation (6.35) refer to the longitudinal and transverse directions of the layer. Also, the allowable values $\underline{t}$ and $\bar{t}$, which are the same for all the design variables, are shown in the table.

The optimization problem consists of minimizing the two chosen criteria

Table 6.2 Prices and material constants of carbon and glass layers

Material	C (kg^{-1})	ρ (kg m^{-3})	E_1 (GPa)	E_2 (GPa)	G_{12} (GPa)	ν_{12}
Glass layer	18	1600	22	8.7	5.7	0.278
Carbon layer	250	1600	125	6.2	4.8	0.270

Subscripts 1 and 2 refer to the fiber and the transverse directions of the layer, respectively.

Table 6.3 Allowable values used in the constraints of the composite beam problem

Material	X_t	X_c	Y_t	Y_c	S	$\underline{t}$	$\bar{t}$
Glass layer	500	480	37	176	51	5×10^{-3}	50
Carbon layer	1750	1100	35	140	62	5×10^{-3}	50

Stresses are given in megapascals and thicknesses in millimeters. X refers to the fiber and Y to the transverse direction of the layer. Subscripts t and c are used for tension and compression, respectively.

subject to the Tsai–Hill failure constraint which is applied separately for every layer in each element. Also, the lower and upper limits for the plate thicknesses have been imposed. Thus the following bicriterion problem is considered:

$$\min \left[W(\mathbf{x})\, C(\mathbf{x}) \right]^{\mathrm{T}} \tag{6.36}$$

subject to

$$f_i(\mathbf{x}) \leqslant 1 \qquad i = 1, 2, \ldots, 4$$
$$\underline{t} \leqslant t_i \leqslant \bar{t} \qquad i = 1, 2, \ldots, 4$$

The results have been depicted in the criterion space in Fig. 6.9(c). This minimal curve confirms the assumed conflict between the two criteria immediately when two materials with different prices and strengths are used. Naturally, this property can be expected also in more complicated problems where fiber orientations are additional design variables. A trade-off between these two criteria C and W really seems useful in the decision-making phase because the variations of the criteria values are considerable along the minimal curve.

6.7.3 Nonconvex frame problem

In the preceding design examples the minimal curve has been smooth in the criterion space and the corresponding region of the attainable set has been convex. Thus the Pareto optimal set of every problems could be generated by using any of the basic techniques. Sometimes the attainable set may, however, possess peculiar properties. Especially, the case where that part of the set Λ, which is connected to the minimal curve, has concave parts is interesting. Then some of the basic methods, the linear weighting method for example, can fail in generating the Pareto optimal set. These cases have been reported also in structural optimization (Koski, 1985) but, generally speaking, they seem to be relatively rare. Next a simplified plane frame problem, where all the inessential complexity is reduced as much as possible, is treated to illustrate the nonconvexity phenomenon.

A three-member frame under one loading condition and the kinematic analysis model, which includes also the axial deformation of the members, are shown in Fig. 6.10. The three member areas are the design variables, i.e. $\mathbf{x} = [A_1\,A_2\,A_3]^T$,

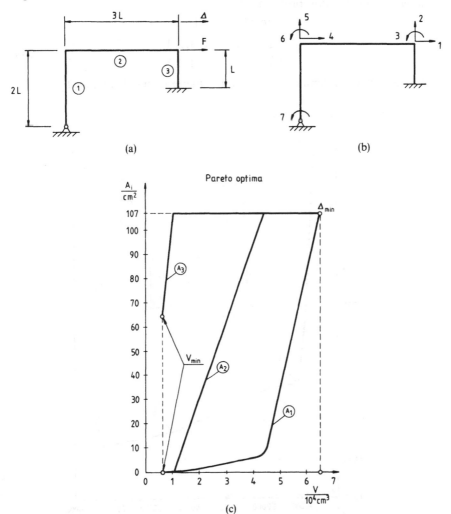

Fig. 6.10 Nonconvex plane frame problem: (a) structure, loading and displacement criterion Δ; (b) nodal degrees of freedom; (c) Pareto optimal member areas. The numerical design data for the problem are as follows: $F = 80\,\text{kN}$, $L = 100\,\text{cm}$, $E = 200\,\text{GPa}$, $\bar{\sigma} = -\underline{\sigma} = 140\,\text{MPa}$, $\bar{A} = 107\,\text{cm}^2$, $\underline{A} = 0\,\text{cm}^2$.

and the stress as well as the side constraints have been imposed for the problem. In the latter constraints the lower limits of the member areas are zero, which allows the removal of the members. In order to reduce the number of design variables to three the well-known relations

$$I_i = 2.1A_i^2, \quad i = 1, 2, 3$$
$$Z_i = 1.1A_i^{3/2}, \quad i = 1, 2, 3 \tag{6.37}$$

are used for the section moment of inertia and for the section modulus, respectively. Both the material volume of the frame and the horizontal displacement of the loaded node, denoted by V and Δ, should be minimized simultaneously. As a result, optimization problem

$$\min [V(\mathbf{x}) \Delta(\mathbf{x})]^T$$

$$\underline{\sigma} \leqslant \sigma_i \leqslant \bar{\sigma}, \quad i = 1, 2, 3$$
$$0 \leqslant A_i \leqslant \bar{A}, \quad i = 1, 2, 3 \tag{6.38}$$

where the constraint limits are given in the figure legend, is obtained.

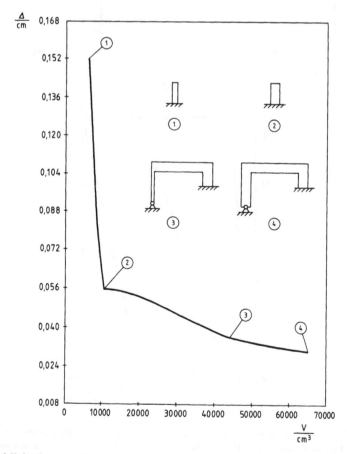

Fig. 6.11 Minimal curve of plane frame problem. Also four Pareto optimal structures corresponding to minimal points 1...4 are shown. A different scale is used for the lengths and the widths of the members.

The Pareto optimal member areas are shown in Fig. 6.10(c) and the corresponding minimal curve is presented in Fig. 6.11. Also, some Pareto optimal structures associated with the most interesting minimal points have been depicted in the latter figure. Only the end points of the curve segment 2–3 can be achieved by the weighting method whereas the intermediate part is lost. On the other hand, the curve segments 1–2 and 3–4 may be generated by this method. Another interesting feature found here is point 2 on the minimal curve, which corresponds to the change of the topology of the frame. According to the numerical computations this seems to be a vertex where the derivative $\partial\Delta/\partial V$ has a discontinuity. This in turn may be troublesome for the decision maker who has not generated the whole curve and is not expecting a jump in the trade-off number. If for example only one minimal solution located near point 2 has been determined, the trade-off study at this Pareto optimum is misleading.

6.8 CONCLUSION

The history of multicriterion structural optimization is relatively short starting from the few articles published in the late 1970s. Now that an early stage of this discipline is still at hand, it seems reasonable to include all the existing publications to represent the recent advances in the field. From this viewpoint it has been natural to concentrate on those fundamental matters which seem especially important in the future and which have been used throughout the published applications. Instead of repeating different cases treated in the articles, the whole decision-making process, starting from the problem formulation and ending at the choice of the best compromise solution, has been discussed in this chapter. The computation of Pareto optima and some characteristic features of a minimal set have been illustrated by several structural design examples.

The vector optimization problem and the Pareto optimality concept offer a clear and sound basis for any further development in the multicriterion optimum design theory. The proper control of the decision-making process becomes possible in this way because then the choice can be restricted to designs which are optimal in an undisputed mathematical sense. Also, the comparison of different methods is facilitated by these concepts. For example the question of which technique should be applied in a certain problem to generate good designs, may be often answered by comparing how many Pareto optima different methods can achieve. This matter was explained in section 6.3 and in Fig. 6.5 the functioning of the most popular methods in different problems was geometrically described. Whatever method the designer may choose, he should check whether it computes Pareto optima or something else. In the latter case the method is useless, provided that the problem formulation corresponds well to reality, and in the former case it should be able to generate as many Pareto optima as possible. If the designer wants to generate good points by just varying the constraint limits in a scalar problem, he has several possibilities. Both equality and inequality constraints can be applied in many different ways. The choice among these alternative methods can be made reliably only by checking which technique gives Pareto optima. The choice of the method by intuition may lead to undesired designs. Sometimes it has been suspected that the solution of a multicriterion problem depends on the chosen decision-making procedure. This is true in the sense that usually experience is being acquired by the decision maker during the design process. If the decision-maker's preferences remain unchanged during the process and the Pareto optimality concept is used, then any available procedure should give the same result provided that the designer makes consistent decisions. The natural requirement is that the chosen method must be able to generate also the best compromise solution. Thus in the case of unchanged preferences the result can be regarded as a function of the person making decisions rather than a function of the design procedure.

The future of the multicriterion approach looks promising also in structural optimization where it has recently reached industrial applications. A multicriterion module, which generates Pareto optimal solutions, can be expected to be an essential part in any new optimization-oriented finite element package. The computation of a covering collection of Pareto optima may be expensive for large-scale problems because several structural and sensitivity analyses must be performed for each solution. This matter together with the graphic or numerical representation of the results, which should be modified into a form that is suitable for the decision making, will be one challenge in the field. Most existing applications deal with physical criteria only but also some economic measures, such as the manufacturing cost for example, might be expected in the vector objective function. New tempting designs become available along with the multicriterion optimization which, on the other hand, requires consistent and competent input information, thus emphasizing the role of the decision maker.

REFERENCES

Adali, S. (1983a) Pareto optimal design of beams subjected to support motions. *Computers and Structures*, **16**, 297–303.

Adali, S. (1983b) Multiobjective design of an antisymmetric angle-ply laminate by nonlinear programming. *Journal of Mechanisms, Transmissions, and Automation in Design*, **105**, 214–19.

Baier, H. (1983) Structural optimization in industrial environment, in *Optimization Methods in Structural Design* (eds H. Eschenauer and N. Olhoff), Proceedings of the Euromech-Colloquium 164, University of Siegen, Bibliographisches Institut A6, Zürich, pp. 140–5.

Bendsøe, M.P., Olhoff, N. and Taylor, J.E. (1983–1984) A variational formulation for multicriteria structural optimization. *Journal of Structural Mechanics*, **11**, 523–44.

Bühlmeier, J. (1990) Design of laminated composites under time dependent loads and material behaviour, in *Engineering Optimization in Design Processes* (eds H.A. Eschenauer, C. Mattheck and N. Olhoff), Proceedings of the International Conference, Karlsruhe Nuclear Research Centre, Lecture Notes in Engineering 63, Springer, Berlin, pp. 107–25.

Carmichael, D.G. (1980) Computation of Pareto optima in structural design. *International Journal for Numerical Methods in Engineering*, **15**, 925–9.

Cohon, J.L. (1978) *Multiobjective Programming and Planning*, Academic Press, New York.

Dhingra, A.K., Rao, S.S. and Miura, H. (1990) Multiobjective decision making in a fuzzy environment with applications to helicopter design. *AIAA Journal*, **28**, 703–10.

Diaz, A. (1987a) Sensitivity information in multiobjective optimization. *Engineering Optimization*, **12**, 281–97.

Diaz, A. (1987b) Interactive solution to multiobjective optimization problems. *International Journal for Numerical Methods in Engineering*, **24**, 1865–77.

Eschenauer, H. (1983) Vector optimization in structural design and its application on antenna structures, in *Optimization Methods in Structural Design* (eds H. Eschenauer and N. Olhoff), Proceedings of the Euromech-Colloquium 164, University of Siegen, Bibliographisches Institut A6, Zürich, pp. 146–55.

Eschenauer, H., Kneppe, G. and Stenvers, K.H. (1986) Deterministic and stochastic multiobjective optimization of beam and shell structures. *Journal of Mechanisms, Transmissions, and Automation in Design*, **108**, 31–7.

Eschenauer, H.A., Schäfer, E. and Bernau, H. (1988) Application of interactive vector optimization methods with regard to problems in structural mechanics, in *Discretization Methods and Structural Optimization – Procedures and Applications* (eds H.A. Eschenauer and G. Thierauf), Proceedings of a GAMM-Seminar, Siegen, Lecture Notes in Engineering 42, Springer, Berlin, pp. 110–17.

Eschenauer, H.A. and Vietor, T. (1990) Some aspects on structural optimization of ceramic structures, in *Engineering Optimization in Design Processes* (eds H.A. Eschenauer, C. Mattheck and N. Olhoff), Proceedings of the International Conference, Karlsruhe Nuclear Research Centre, Lecture Notes in Engineering 63, Springer, Berlin, pp. 145–54.

Eschenauer, H., Koski, J. and Osyczka, A. (eds) (1990) *Multicriteria Design Optimization – Procedures and Applications*, Springer, Berlin.

Fu, G. and Frangopol, D.M. (1990) Reliability-based vector optimization of structural systems. *ASCE Journal of Structural Engineering*, **116**, 2143–61.

Fuchs, W.J., Karandikar, H.M., Mistree, F. and Eschenauer, H.A. (1988) Compromise: an effective approach for designing composite conical shell structures, in *Advances in Design Automation* (ed. S.S. Rao), ASME Design Technology Conference, **14**, 279–86.

Gerasimov, E.N. and Repko, V.N. (1978) Multicriterial optimization. *Soviet Applied Mechanics*, **14**, 1179–84.

Hajela, P. and Shih, C.J. (1990) Multiobjective optimum design in mixed integer and discrete design variable problems. *AIAA Journal*, **28**, 670–5.

Jendo, S., Marks, W. and Thierauf, G. (1985) Multicriteria optimization in optimum structural design. *Large Scale Systems*, **9**, 141–50.

Koski, J. (1979) *Truss Optimization with Vector Criterion*, Tampere University of Technology, Publication No. 6.

Koski, J. (1981) Multicriterion optimization in structural design, in *Proceedings of the 11th ONR Naval Structural Mechanics Symposium on Optimum Structural Design*, University of Arizona, Tucson, AZ.

Koski, J. (1984a) Multicriterion optimization in structural design, in *New Directions in Optimum Structural Design* (eds E. Atrek, R.H. Gallagher, K.M. Ragsdell and O.C. Zienkiewicz), Wiley, New York, pp. 483–503.

Koski, J. (1984b) Bicriterion optimum design method for elastic trusses. Acta Polytechnica Scandinavica, Mechanical Engineering Series No. 86, Dissertation, Helsinki.

Koski, J. (1985) Defectiveness of weighting method in multicriterion optimization of structures. *Communications of Applied Numerical Methods*, **1**, 333–6.

Koski, J. (1993) Multicriterion structural optimization – state of the art, in *Proceedings of the NATO Advanced Study Institute: Optimization of Large Structural Systems*, September 23–October 4, 1991, Berchtesgaden, Kluwer, Dordrecht.

Koski, J. and Silvennoinen, R. (1982) Pareto optima of isostatic trusses. *Computer Methods in Applied Mechanics and Engineering*, **31**, 265–79.

Koski, J. and Silvennoinen, R. (1987) Norm methods and partial weighting in multicriterion optimization of structures. *International Journal for Numerical Methods in Engineering*, **24**, 1101–21.

Koski, J. and Silvennoinen, R. (1990) Multicriteria design of ceramic piston crown. *Engineering Costs and Production Economics*, **20**, 175–89.

Leitmann, G. (1977) Some problems of scalar and vector-valued optimization in linear viscoelasticity. *Journal of Optimization Theory and Applications*, **23**, 93–9.

Lógó, J. and Vásárhelyi, A. (1988) Pareto optima of reinforced concrete frames, *Periodica Polytechnica, Civil Engineering*, **32**(1–2).

Nafday, A.M., Corotis, R.B. and Cohon, J.L. (1988a) Multiparametric limit analysis of frames: part I – model. *ASCE Journal of Engineering Mechanics*, **114**, 377–86.

Nafday, A.M., Corotis, R.B. and Cohon, J.L. (1988b) Multiparametric limit analysis of frames: part II – computations. *ASCE Journal of Engineering Mechanics* **114**, 387–403.

Olhoff, N. (1989) Multicriterion structural optimization via bound formulation and mathematical programming. *Structural Optimization*, **1**, 11–17.

Osyczka, A. (1981) An approach to multi-criterion optimization for structural design, in Proceedings of the 11th ONR Naval Structural Mechanics Symposium on Optimum Structural Design, Tucson, AZ.

Osyczka, A. (1984) *Multicriterion Optimization in Engineering*, Ellis Horwood, Chichester.

Pareto, V. (1896–1897) *Cours d'Economic Politique*, Volumes 1 and 2, Rouge, Lausanne.

Post, P.U. (1990) Optimization of the long-term behaviour of composite structures under hygrothermal loads, in *Engineering Optimization in Design Processes* (eds H.A. Eschenauer, C. Mattheck and N. Olhoff), Proceedings of the International Conference, Karlsruhe Nuclear Research Centre, Lecture Notes in Engineering 63, Springer, Berlin, pp. 99–106.

Rao, S.S. (1984) Multiobjective optimization in structural design with uncertain parameters and stochastic processes. *AIAA Journal*, **22**, 1670–8.

Rao, S.S. (1987) Game theory approach for multiobjective structural optimization. *Computers and Structures*, **25**, 119–27.

Rao, S.S. and Hati, S.K. (1986) Pareto optimal solutions in two criteria beam design problems. *Engineering Optimization*, **10**, 41–50.

Rao, S.S., Venkayya, V.B. and Khot, N.S. (1988) Optimization of actively controlled structures using goal programming techniques. *International Journal for Numerical Methods in Engineering*, **26**, 183–97.

Rozvany, G.I.N. (1989) *Structural Design via Optimality Criteria*, Kluwer, Dordrecht.

Stadler, W. (1977) Natural structural shapes of shallow arches. *Journal of Applied Mechanics*, **44**, 291–8.

Stadler, W. (1978) Natural structural shapes (the static case). *Quarterly Journal of Mechanics and Applied Mathematics*, **31**, 169–217.

Stadler, W. (1987) Initiators of multicriteria optimization – recent advances and historical development of vector optimization, in Proceedings of an International Conference on Vector Optimization (eds J. Jahn and W. Krabs), Darmstadt, August 1986, Lecture Notes in Economics and Mathematical Systems 294, Springer, Berlin.

Stadler, W. (ed.) (1988) *Multicriteria Optimization in Engineering and in the Sciences*, Mathematical Concepts and Methods in Science and Engineering 37, Plenum, New York.

Stadler, W. and Krishnan, V. (1989) Natural structural shapes for shells for revolution in the membrane theory of shells. *Structural Optimization*, **1**, 19–27.

Tseng, C.H. and Lu, T.W. (1990) Minimax multiobjective optimization in structural design, *International Journal for Numerical methods in Engineering*, **30**, 1213–28.

Yoshimura, M., Hamada, T., Yura, K. and Hitomi, K. (1984) Multiobjective design optimization of machine-tool spindles. *Journal of Mechanisms, Transmissions, and Automation in Design*, **106**, 46–53.

7

Multicriteria design optimization by goal programming

ERIC SANDGREN

7.1 INTRODUCTION

A good design is certainly optimal in some sense, but it may not always be possible to express the attributes of such a design within the context of a single objective, nonlinear programming problem. The traditional formulation of a nonlinear programming problem requires the specification of a scalar objective function with all other factors being included as constraints. Nonlinear programming methods also force the computer to operate on a mathematical abstraction of the 'real' design problem with no knowledge of the trade-offs which are being made during the optimization process. The human mind can operate easily in an abstract mode and as such can consider a wide range of design alternatives. In contrast, the use of conventional computer-aided design and analysis tools forces the user to deal with design issues at a very specific level. If one considers design as a process which proceeds from the general to the specific, then the computer cannot have the desired level of impact early in the process where the benefits are potentially the greatest. The result is generally a solution which is optimal with respect to the mathematical formulation applied but far from optimal in a practical design sense. This deficiency has been a major factor in limiting the number of practical applications of design optimization which is unfortunate as the potential of the concept is significant.

The difficulty associated with a conventional, single objective formulation may be observed by considering a hypothetical example. If a structure is optimized for minimum weight subject to constraints imposed on stress, maximum deflection, buckling and natural frequency, the result will be a design in which several of the constraints are active. This means that the final design will border on several failure modes at the solution, subject to the safety factors imposed. In addition, the solution will very likely be extremely sensitive to changes in loading or restraints. When imposed design loads and restraint specifications are only estimates of actual conditions, one must question the validity of a design produced by this process. The value added by computer-aided design lies in the ability to investigate a number of alternate designs which would not be possible by hand in the time available. This means that potential trade-offs which might be extremely valuable to the final design must not be removed from consideration by either the formulation of the problem or the solution process. For example a design which has a maximum stress 1% higher than that allowed by the formulation may produce an additional 10% weight reduction with virtually the same safety factors but this point would never be discovered. Most real design optimization problems are multimodal. There are generally a number of different designs that can adequately perform the intended task which are scattered throughout the design space. Any solution process which simply refines the starting design estimate in a local region will not consider other, perhaps far more attractive design alternatives.

A more natural approach to the formulation of design optimization problems is required which allows potential trade-offs to be evaluated in the widest possible context. Multiobjective optimization provides additional freedom compared with a conventional nonlinear programming approach and therefore is a logical candidate for use as a realistic design tool. A valuable computational design tool should allow the user to function in a manner which closely parallels the actual design process. When a nonlinear programming algorithm is coupled with a finite element code, a specific design is input and a design which is similar in nature to this starting design is output. A more natural process would be to define the desired aspects of performance and then to select a design with both specific and general characteristics which best meet these performance levels. In addition to a multiobjective capability, the design environment should address topological issues as well as deal effectively with the difference between hard and soft constraint specifications.

Topological issues are those design factors which extend beyond a single classification. A topological choice is more than a simple geometric change in a design variable. They generally represent discrete choices rather than continuous ones. In a structural design involving trusses, for instance, the alteration of cross-sectional areas or even the movement of nodal locations does not constitute a topological change. Topological issues involve fundamental aspects such as the number of truss elements used in the structure, the use of beam instead of truss elements or a change in material choice. These changes can have a dramatic impact on design performance and often move the optimization search to different areas of the design space. For example when a material change from steel to aluminum is made, the dimensions of the design structure will increase when subject to the same load conditions. The end result is that topological choices tend to introduce local minima throughout the design space. Of course, many design problems have local minima, but topological choice generally magnifies the problem. This makes the optimization problem more difficult to solve in the global sense, but the resulting design improvement is most often worth the effort.

The difference between hard and soft constraints is subtle but very important. A hard constraint specification involves a condition that absolutely must be satisfied. On the other hand, a soft constraint specification represents a condition which is desirable but may be violated by a small amount without drastically altering the value of the design. Traditional linear and nonlinear programming methods treat all constraints as hard constraints. This is an inherent limitation in the formulation which is required for the mathematically derived solution procedures but one which often delegates the optimum located to be different from what is desired. The issue is easily seen by considering a simple example.

A failure criterion such as a yield stress in one of the members would be correctly considered as a hard constraint. The problem arises when a safety factor is imposed on the yield stress. Say that a safety factor of 2 is prescribed so that the stress constraint allows the maximum stress found in any member to be only one-half of the yield stress of the material. The question to be asked is whether or not the constraint is still a hard constraint. The answer is that it is not. When a safety factor is included, the stress constraint becomes a soft constraint. Whether the safety factor is 2 or only 1.95 is actually of little consequence, particularly if a small sacrifice in the safety factor is very beneficial to the final design. A traditional, single objective formulation with hard constraints has difficulty making this trade-off. In reality, the majority of constraints used in design optimization problem formulations are soft rather than hard constraints. Treating them as hard constraints will generally lead to suboptimal results.

Many different approaches to multiobjective optimization are possible. The intent herein is to present the topic in the widest possible context and then to deal with specific ways of implementing the concept. Before the multiobjective problem is dealt with, however, a discussion of the various classifications of optimization problems is necessary and how these problem classes are important in the multiobjective formulation and solution process. The inclusion of noise in the formulation is also addressed in order to deal with uncertainty in the model. Several multiobjective problem solution strategies are presented, but concentration is focused on a multilevel, nonlinear goal programming approach. The first level deals with the topological aspects of the design and locates promising solution regions. The next level of the process deals with the trade-offs which are present in each of these design regions in order to achieve the best final design. The procedure is demonstrated on several simple examples. Two of the examples considered involve the design of structures with truss elements, and one deals with the design of a planar mechanism.

7.2 PROBLEM CLASSIFICATION

Design is basically a decision process and as such it can be represented in a tree structure. A possible generic design structure is shown in Fig. 7.1. The tree represents the design of a system which may be composed of various subsystems. The branching process is used to decompose the design so that each subsystem is reduced to its lowest functional level. The system and subsystem level nodes allow information to be exchanged which in turn allows system as well as component objectives to be optimized. Design specifications are imposed as required at any level of the tree. Any node which joins subsystem level nodes is an assembly node which contains information on the various ways the subsystems may be joined.

The level directly below the subsystems in the tree structure represents candidate designs which are the various topological options selected for meeting the subsystem design specifications. Each candidate design has a set of design parameters associated with it which may either be fixed or allowed to become design variables in the subsequent optimization. Candidate designs may have subclasses associated with them and these subclasses may themselves span several levels in the tree. Each subclass has material and manufacturing process nodes

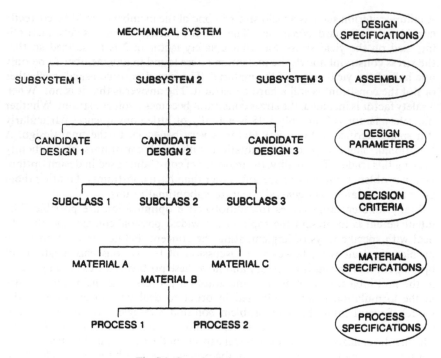

Fig. 7.1 Generic design tree structure.

associated with it. These nodes contain the specifications, including tolerances, for each material and manufacturing process.

Finally, a set of decision criteria is imposed. The decision criteria are those factors which will be used to evaluate the candidate designs. These criteria, or goals as they become in the solution process, may also be associated with any level of the tree. Together with the design parameters for each candidate design, the decision criteria form the basis of the optimization problem formulation which in turn allows the bottom level nodes to be compared on a common basis. The relationship between the parent and child nodes may be specified as any functional expression involving design variables and parameters. The design and parameter values may be selected at any level of the design tree and lower-level nodes may or may not inherit these values depending on the relationship imposed.

When viewed in this manner it becomes possible to develop classifications for the variety of problems commonly encountered. The design, material, assembly and process specifications in conjunction with the decision criteria form a set of design objectives or goals and possibly a constraint set as well. The design parameters form the set of design variables which define how a design can be modified in order better to achieve the goals. The key here is that these variables must represent discrete or topological choices as well as continuous choices. In order to reduce the scope of the discussion, the emphasis will be concentrated on the aspects of shape. The shape optimal design problem may be defined as the determination of the form which best satisfies the design objectives subject to the imposed design constraints. Now the concept of problem classification may be reduced to that of defining the different levels of shape or form. These levels are best described within an application area, thus the initial discussion will be directed at structural design. This will be followed by a more general view of problem classification.

When the term shape optimization is used, the general inclination is immediately to think in terms of structural optimization. Certainly it is true that a large body of experience has been developed in structural design over the last 30–40 years and the definition of shape plays a dominant role in such design applications. The term structural optimization is itself a very general term and is often used in various contexts to refer to different classes of optimization problems. These classes are best defined by the types of design variables encompassed by the formulation. The most elementary level, and that most widely considered, is the modification of the geometric parameters of the elements themselves. This requires the prespecification of the number and type of each element used as well as their placement in space. This is quite a limited framework within which to operate but it is often dictated by the computational effort required to produce a solution. In order to provide maximum flexibility, however, all aspects of structural design must be considered. These aspects include topological as well as geometric factors. In order to understand the problem better in its totality, a brief description of each of the aspects of shape optimization is in order.

Geometric optimization refers to either the direct modification of specific dimensions which define a design or an indirect modification of shape by altering the boundary form. In structural design, the lowest level of geometric optimization is cross-sectional optimization. Cross-sectional optimization refers to the specific geometric dimensions for a preselected design class such as the thickness of a plate or the diameter of a circular truss element. This class of problems has been under investigation since the pioneering work of Schmidt in the early 1960s (Schmidt, 1960). Applications include the design of automotive structural components (Bennett and Botkin, 1981) and truss structures such as electrical transmission towers (Sheppard and Palmer, 1972). For problems involving truss structures, the cross-sectional areas of the elements form the entire set of design variables. For structures involving beam elements, specific cross-sectional dimensions which are sufficient to describe the principal moments of inertia are required in addition. With plate or shell structures, a thickness dimension is needed, either at the element level or for groups of elements (Abuyounes and Adeli, 1986; Adeli and Chompooming, 1989). Generally, a preselected member class is chosen such as an I-beam or channel section in order to relate design variables representing specific dimensions to cross-sectional areas and principal moments of inertia. Variable linking is often employed in order to limit the number of design variables as solution time is a strong function of the number of variables. Design sensitivities which may be used to generate gradient information are now commonly available in many commercial finite element codes.

Other forms of geometric optimization introduce design variables which allow for movement of the boundary in some fashion. The boundary may be altered either by the movement of actual node locations or by moving control points on a spline curve or surface representing the boundary. Because of the increase in complexity over cross-sectional optimization, the geometric change is generally limited to a small region of the design such as around a fillet or a hole. The shape change is also closely linked to the analysis technique employed. Both the finite element method (Pramila and Virtanen, 1986) and the boundary element method (Sandgren and Wu, 1989) have been successfully applied to the geometric shape optimization problem. A geometric change will often require continual remeshing for the analysis routine and this greatly complicates the task of generating valid gradient based search directions. While geometric optimization has been applied with some success, much work remains in the areas of geometric modeling, analysis and design optimization methods before geometric shape optimization can become an integral part of the design process.

The third class of shape optimization problems involves topological issues as well as geometric and cross-sectional ones. Relatively little work has been directed toward topological optimization despite the importance of the concept. Optimizing element cross sections or nodal geometry does little more than produce a local optimum based on the preselected topology. That is to say the optimal ten-bar truss for a defined set of loading conditions may be a good design compared with other ten-bar truss designs but it may be a relatively poor design when compared with other topologies such as a five- or six-bar truss. This issue becomes particularly significant when dealing with new materials or manufacturing processes where no design history is available to help select an appropriate topology. When consideration is being given to a material substitution, it is quite likely that the basic topology of the part must be significantly altered in order to reap potential weight and cost benefits. Traditional nonlinear programming approaches have met with limited success in dealing with topological issues. This fact seems to indicate that a nontraditional optimization approach may be required in order to deal with the structural design problem in the broadest sense (Adeli and Mak, 1988). A method which is capable of taking advantage of new computing hardware architecture such as distributed or parallel processors is also preferred over one relegated to conventional sequential processing (Adeli and Kamal, 1992). This can have an enormous impact on solution time and directly affect the size and complexity of the problem being addressed.

The geometric and topological design of other mechanical systems is possible as well. For example in the design of a mechanism, geometric aspects include items such as link lengths and pivot locations while topological aspects involve the selection of the class of mechanism (i.e. four- or six-link mechanism). For the design of a cam to drive a dynamic system, geometric aspects include the size of the cam (base circle radius), the radius of the follower and even the profile of the cam itself while topological choices include the type of follower (flat faced, roller, etc.). The same identification process may be applied to virtually any shape related problem. Similarly a set of design objectives and constraints may be defined for each application or application class. What remains to be developed is a robust solution algorithm which can deal with the wide variety of practical shape optimization problems.

7.3 PROBLEM FORMULATION

The conventional nonlinear programming problem formulation may be stated as follows:

$$\text{minimize } f(\mathbf{x}); \mathbf{x} = [x_1\, x_2\, x_3, \ldots, x_N]^\mathrm{T} \tag{7.1}$$

subject to

$$g_j(\mathbf{x}) \geqslant 0; \, j = 1, 2, \ldots, J \qquad (7.2)$$

$$h_k(\mathbf{x}) = 0; \, k = 1, 2, \ldots, K \qquad (7.3)$$

and

$$x_i^{(l)} \leqslant x_i \leqslant x_i^{(u)}; \, i = 1, 2, \ldots, N \qquad (7.4)$$

Here, $f(\mathbf{x})$ represents a scalar objective function which defines the merit of a particular selection of design variables. The design variables contained in the vector $\mathbf{x}$ are those parameters which may be adjusted in order to improve the design. The functions $g(\mathbf{x})$ and $h(\mathbf{x})$ form a set of hard constraints which delimit the feasible design space. Upper and lower bounds are also imposed if required.

The problem formulation represented by equations (7.1)–(7.4) is depicted graphically in Fig. 7.2. In this situation, design variables are inputs to the system model which results in a value for the objective function and each of the constraints. This is a very simplistic view of the design optimization process. A more realistic view of the process is presented in Fig. 7.3. Note that multiple design objectives are now allowed. Also, in addition to the design variables, there are both inner and outer noises present. Outer noises represent uncontrollable variations in design parameters while inner noises are unavoidable but in part controllable variations. In a structural optimization context, inner noises might include factors such as tolerances or material properties while outer noises might include the loads applied to the structure. The difference is that some control over tolerances and material properties is possible but the loads applied in the

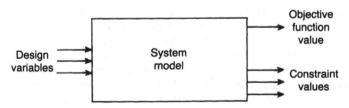

Fig. 7.2 Standard nonlinear programming model.

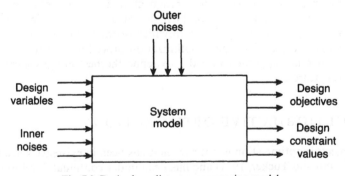

Fig. 7.3 Desired nonlinear programming model.

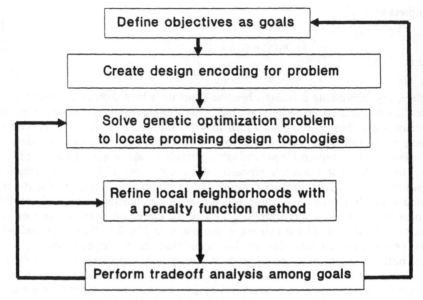

Fig. 7.4 Flowchart for design optimization procedure.

use of the structure are not generally controllable. Both inner and outer noises can play an important role in the success or failure of a design. The generalization which can be made is that a design which is insensitive to inner and outer noises is generally a better or at least a more robust design.

The multiobjective formulation, in conjunction with the presence of inner and outer noises, more closely parallels the actual design process in that it deals with the trade-offs associated with each design evaluated. By formulating design issues as objectives rather than constraints, soft constraints may be included in a natural fashion. Since it is difficult to encode the exact character of the 'real' problem in the computer, it is appropriate that the user should remain an integral part of the optimal design process. This interaction is difficult to implement in many of the multiobjective formulations derived to date, but it is essential that the user be able to overcome the precise mathematical nature of the formulation to ensure the process converges to a practical design. The solution approach constructed for solution of the problem formulation, shown in Fig. 7.4, is a nonlinear goal programming algorithm which operates initially at a topological level through the use of a genetic optimization procedure and then switches to a penalty function approach in order to refine the best designs located in the topological search.

7.4 MULTIOBJECTIVE OPTIMIZATION

Multiobjective optimization is not new as it has been in existence since the early 1950s (Kuhn and Tucker, 1951) and has been under continual development ever since. Various approaches are available with which one can address a multiobjective formulation. These approaches may be divided into two fundamental categories where the primary differentiation is made as to whether the problem is treated as the optimization of a vector of design objectives or a single scalarized equation which contains contributions from all of the objectives. Reviews of multiobjective techniques are available in a number of sources (Stadler, 1979; Evans, 1984; Hwang and Masud, 1979). Most of the techniques which deal with the problem in vector form search for a Pareto optimal point for which no decrease in any objective is possible without a subsequent increase in one or more of the other objectives. In the scalarized form, a commonly implemented approach is to form a weighted sum of the individual objectives. Several of the widely used techniques for multiobjective optimization will be reviewed but the primary concentration will be placed on the method of goal programming which has its original roots in the area of linear programming. This is not to say that goal programming is superior to the outer approaches. It is, however, a straightforward and easily understood approach and makes the examples presented easier to follow.

The first class of methods to be reviewed consists of the vector optimization techniques. The nonlinear programming formulation for this problem class becomes

$$\text{minimize } f(\mathbf{x}) = \{f_1(\mathbf{x}), f_2(\mathbf{x}), \dots, f_m(\mathbf{x})\}^T \tag{7.5}$$

subject to the same constraint sets of the traditional nonlinear programming problem described by equations (7.2)–(7.4). Possible approaches for solving the problem include the min–max method and the e constraint method. The min–max method simply minimizes the maximum deviation of the design from all design objectives subject to the additional constraint that the solution is a Pareto optimal point. The e constraint method seeks to minimize one of the selected objectives with all other objectives constrained to stay within a distance e from their desired goal values.

The techniques which form the problem as a scalar objective include game theory, compromise programming and goal programming. Game theory treats each design objective as a player seeking to optimize his own payoff. The players may be noncooperative where each cares only about his own standing or cooperative where players are willing to compromise their objectives if it improves the state of other players. Compromise programming is based on locating a set of solutions which lie as close as possible to an ideal point, where the ideal point is defined as having each objective at a value equal to that which it would have if it were minimized as a single objective function. Finally, goal programming seeks to minimize the under- or overachievement of a set of objectives or goals which are minimized as a weighted sum in the objective function.

7.5 GOAL PROGRAMMING

In a goal programming formulation, the design objectives or goals are defined and a priority is assigned to each one. The algorithm then attempts to satisfy as many of the goals as possible, starting with the highest priority goal. The inclusion of priority takes the algorithm a step beyond a simple weighted sum of objectives. This approach removes much of the difficulty of problem formulation and more closely follows the normal design pocess. The technique is also capable of operating when there is no feasible design space. It allows for an evaluation of trade-offs in order to determine a more realistic set of goals. The application of goal programming to multiobjective structural design has been demonstrated by several investigators (Shupe and Mistree, 1987; El Sayed, Ridgley and Sandgren, 1989).

The formulation of a goal programming problem is different from that of a nonlinear programming problem. In a goal programming formulation, there is no exact counterpart to the objective function in a nonlinear programming formulation. Design constraints may be included as goal constraints or as hard constraints. The goal constraints are best thought of as multiple objectives, while hard constraints are directly equivalent to the constraints in a nonlinear programming formulation. The solution process revolves around the minimization of the positive and/or negative deviations from the specified goals.

Goal constraints may take the form of inequality constraints (either ≥ 0 or ≤ 0) or equality constraints. The common forms of goal constraints are listed below.

$$G(x_j) + d_i^- = b_i \qquad (7.6)$$

$$G(x_j) - d_i^+ = b_i \qquad (7.7)$$

$$G(x_j) + d_i^- - d_i^+ = b_i \qquad (7.8)$$

The x_j are the design variables and $G(x_j)$ are nonlinear functions relating the design variables to the desired goal values, b_i. The d_i^- and d_i^+ are the under- and overachievement deviational variables which may be thought of as a form of slack variables. With this analogy, the first two goal constraint forms are seen to represent less than or equal constraints respectively. If all deviational variables are only allowed to assume positive values, the specific form of each goal constraint is easy to verify. In the first constraint, the underachievement deviational variable represents the amount the current design falls below the right-hand side value b_i. Since only values below or equal to b_i are possible, the constraint is equivalent to the $\leq$ form. The same logic may be applied to the second constraint form. Since the positive deviation is subtracted in the equation, it allows the constraint to attain only values greater than or equal to the right-hand side, b_i, and is therefore equivalent to a $\geq$ form.

The final goal constraint form represents an equality constraint. This form may be likened to a fuzzy equality constraint. It allows for both positive and negative deviations from the right hand side value as it contains both a d_i^- and a d_i^+ variable. How much of a deviation is allowed is governed by the objective function form and the scaling present among the goal constraints. It should be noted that only one of the deviational variables will be allowed to be simultaneously nonzero in any goal constraint equation. Using these three forms of goal constraints, virtually any problem can be formulated. Traditional hard inequality and equality constraints, may be added as well to help define the feasible region when appropriate.

In a goal programming formulation, only deviational variables are allowed to appear in the objective function. The actual design variables (x_j) are indirectly included by the fact that the values of the design variables determine the over- or underachievement of each goal constraint. The objective function, like the goal constraints, may take on a number of forms depending on how priority levels and weighting factors within priority levels are assigned.

The most basic form of the objective function is given as

$$\text{minimize } f(\mathbf{x}) = \sum_{i=1}^{I} (d_i^- + d_i^+) \tag{7.9}$$

Here, the subscript i represents a particular goal constraint and I is the total number of goal constraints. This objective function would be used when the deviational variables are not distinguished by priority or weighting. In effect, this formulation has the unitary goal of minimizing the total deviation from the specified goal levels. It should be noted that both the d_i^- and d_i^+ deviational variables need not be included in the objective function for each constraint. If only a positive or only a negative deviation results in a suboptimal design, then only the corresponding deviational variable is included for that goal.

The next level of complexity would be to include priorities among the goal constraints. This results in an objective function of the form

$$\text{minimize } f(\mathbf{x}) = \sum_{i=1}^{I} P_k\{d_i^- + d_i^+\} \text{ for } k = 1, 2, \dots, K \tag{7.10}$$

This form allows K different priority levels within the I goal constraints. Here, P_k represents the rank ordering of the goal priorities. Allowances are now possible for a ranking or ordering of goals (i.e. some goals are more important than others). The deviational variables within each priority level have the same weighting and any number of goals may be grouped at any priority level.

The most general form of objective function allows weighting factors to be assigned within individual priority levels. This form is represented by

$$\text{minimize } f(\mathbf{x}) = \sum_{i=1}^{I} W_{kj} P_k\{d_i^- + d_i^+\} \quad \text{for } k = 1, 2, \dots, K; j = 1, 2, \dots, J \tag{7.11}$$

Here for priority level k, a total of j individual weights (W_{kj}) may be assigned within that priority. This form provides the most flexibility by allowing goals at the same priority to have different weights such as positive deviational variables vs negative deviational variables. The solution algorithm still operates on a mathematical abstraction of the real problem, but the flexibility allowed in the goal formulation allows this abstraction to represent better the true design situation. The formulation of a goal programming problem is considered in each of the three example problems presented following the development of the genetic optimization algorithm.

7.6 GENETIC OPTIMIZATION

The solution procedure employed to solve a multiobjective or goal programming problem formulation is generally built around a conventional linear or nonlinear programming algorithm. These methods have undergone continual development over the past 30 years but the algorithms available to date are not particularly robust and no one algorithm performs uniformly well on broad classes of problems. In addition, as a general design tool, nonlinear programming methods do not possess the characteristics of intelligent methods in that they fail to make use of or to learn from information generated at previous stages in the search (Adeli and Balasubramanyam, 1988). This is very different from the iterative, manual design process performed by an engineer who is constantly gathering and updating information throughout the process. Virtually all successful algorithms make use of gradient information to guide the search and as such they tend to be trapped by the first local minimum encountered.

Procedures designed to locate global optima such as simulated annealing have been developed (Vecchi and Kirkpatrick, 1983), but an efficiency trade-off is required and the approach is quite problem specific. The methods which hold some promise as an intelligent design optimization tool are termed genetic optimization methods. These methods emulate the natural selection process of nature and operate on a principle of survival of the fittest. Genetic algorithms have been applied to a wide variety of problems over the past ten years (Goldberg, 1987; Davis and Ritter, 1987; Fourman, 1985). Their utility in solving mechanical and structural problems has recently been demonstrated by a number of investigators (Sandgren and Venkataraman, 1989; Sandgren and Jensen, 1992). The goal of a genetic algorithm is to discover the fundamental building blocks of a good design and to combine these building blocks in such a way as to create the best possible design. To accomplish this task, a design must be encoded in such a way so that the representation of the design is closely linked to its physical structure or function. This turns out to be a relatively straightforward task for structural design applications. The actual topology of the design is encoded and operated on in order to produce new designs based on features of existing designs. Not only is the process stable, but it requires no gradient information and produces multiple optima rather than a single, local optimum. It relies on the randomness present in natural selection, but quickly exploits information gathered in order to produce an efficient design procedure. Additional speed may be achieved as a result of the parallel nature of the genetic process.

A design to a genetic algorithm is an abstract representation termed a chromosome which is directly analogous to a chromosome in a living organism. The chromosome is composed of a number of genes, each of which may assume one of a number of possible values or states. While in an organism, a gene might represent sex or hair color, the gene in the design encoding represents a feature of the design itself. The genetic algorithm operates by manipulating the coding of the set of gene values. This is a fundamental distinction from traditional methods as the coding of the parameter set is altered rather than the parameter values themselves. Other differences between genetic and nonlinear programming methods are that the genetic algorithm operates on a set or population of designs rather than a single design point and the rules which govern the transition from one set of designs to the next are probabilistic rather than deterministic. These differences may not seem that great but they produce an algorithm which exhibits certain aspects of intelligence which might be demonstrated by a human designer. The randomness in the transition from generation to generation also allows for the consideration of factors such as tolerances and uncertainty.

The overall suitability of a chromosome, that is the performance of a specific design, is termed its fitness. This fitness property, which may be related to any function or functions of the design, determines the probability of the chromosome being a parent for the next generation of designs. The chromosomes possessing the greatest fitness have the highest chance of being selected as parents. Parents are combined using various genetic operators to produce new designs for the next generation. The process is continually repeated with the expectation that the fitness of the individual designs as well as the average fitness of the population will gradually improve until the optimum design or designs are produced. In order to initiate the algorithm, a coded representation of the design must be generated from which the fitness may be evaluated. Additionally, an initial population of designs must be generated and how parents are selected and combined in order to form offspring must be defined. Special operators such as mutation are also introduced in order to guard against the loss of important design information which is particularly important when using a relatively small population. The collection of all of these processes form an algorithm and one such algorithm is described in the following section.

A flowchart depicting the major operations taking place within a genetic algorithm is pictured in Fig. 7.5. A multitude of possible implementations are possible and often a small modification can significantly enhance or degrade the performance on a particular problem class. The algorithm described here will be a very basic one but one that is capable of solving a large number of very real problems. The first issue to be considered is the encoding of the design. This

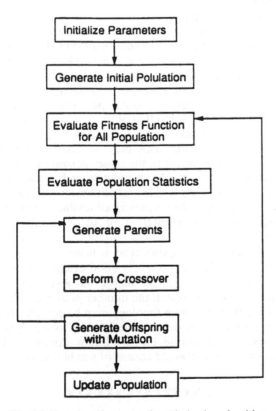

Fig. 7.5 Flowchart for a genetic optimization algorithm.

section will be followed by a discussion of the fitness function, the generation of the initial population and the principle genetic operators.

7.6.1 Design encoding

The representation of a design or the structure of the chromosome which embodies the design in a genetic algorithm may assume a number of different forms. A chromosome is an ordered matrix of values which in some fashion represents the fundamental nature of the design. Traditionally, the chromosome is a one-dimensional vector consisting solely of the binary digits zero or one. This convention allows for a statistical evaluation of the probabilities of various length strings of values within a chromosome to be propagated into future generations of designs. The structure considered here will contain additional flexibility in both the dimensionality and in the number of values which may be associated with a chromosomal position. These extensions allow for a more natural design representation for most structural design problems with a reasonably small chromosome. On the other hand, the ability to perform a statistical analysis on the structures passed on to future generations is more limited. This trade-off is justified on the basis of the results achieved in experimentation on a wide variety of problem formulations.

A one-dimensional example of a chromosome structure would be for a parameter which can assume any integer value between 0 and 31 (a number of choices equal to 2 raised to a power is ideal). This parameter could be represented by a binary string of five digits. Each positional value has only two states and as such, the encoding exactly parallels the manner in which a number is represented by a computer. For instance, a parameter value of 21 would be represented by the string 10101. If another parameter is included, its binary representation could be appended to that of the first parameter. A more effective approach, however, is to couple the encoding of the design as closely as possible with the structure of the design. This may be accomplished by linking the dimensionality of the chromosome to the dimensionality of the problem.

Consider the tapered cantilever beam pictured in Fig. 7.6. The beam is fixed on the left end and has its length divided into five elements. Each of the elements must be assigned a thickness between prescribed minimum and maximum values. A possible chromosomal representation of the beam may be formed by discretizing the thickness of each element into a fixed number of uniform levels. It should be noted that genetic optimization is being used to locate the regions in the design space which hold potentially good designs and not to produce a final design with exact values of all design variables. In this respect, the discretization need not be too fine. If the number of discrete thickness values is selected to be eight, then the beam topology may be completely described by a string of integers ranging from zero to seven with zero signifying the minimum thickness and seven representing the maximum thickness. The traditional one-dimensional chromosome would consist of five blocks of three binary digits each:

$$\text{XXX XXX XXX XXX XXX}$$

A two-dimensional chromosome for the same design representation would be a

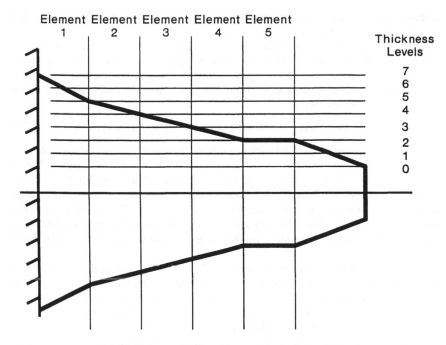

Fig. 7.6 Tapered cantilever beam example for encoding.

matrix of the form

$$
\begin{array}{c}
XXX \\
XXX \\
XXX \\
XXX \\
XXX
\end{array}
$$

Alternatively, the design could be encoded with multiple value chromosomal positions as

$$X\,X\,X\,X\,X$$

where each position assumes a value from the set of integers ranging from zero to seven. In all three forms, there is a direct link between the coding of the chromosome and the physical shape of the beam. The difference between the various encodings will be in the way the algorithm generates new designs or offspring from two parent designs.

7.6.2 The fitness function

The fitness function in a genetic algorithm closely parallels the objective function in a nonlinear programming formulation. It is used to characterize chromosomes in the population so that the best designs have the highest probability of being selected as parents for the next generation. This means that the traits of the fittest designs will be combined in order to form better designs. The fitness function may be a single scalar evaluator or it may be any weighted combination of factors. Within this framework it is a fairly straightforward process to implement a number of multiobjective techniques. Additionally, constraints may be factored in through the use of a penalty function. The most basic form of a fitness function would then be given as

$$\text{base value} - \text{evaluator value} - P[R, g(\mathbf{x}), h(\mathbf{x})] \tag{7.12}$$

Here, the base value is used in order to keep the fitness level positive, the evaluator value is the design objective and $P[R, g(\mathbf{x}), h(\mathbf{x})]$ is a penalty term to handle inequality and equality constraints. The form presented in equation (7.12) will minimize the evaluator value as the most common form of a genetic optimization algorithm performs the maximization of a function. For structural design problems the fitness function will handle objectives or goals on weight, stress, displacement, natural frequency and buckling, when coupled with a finite element code.

As with any penalty function method, the scaling of the constraints relative to the evaluator function is important. Constraint scaling for structural problems is a fairly straightforward and well-understood issue. In a genetic optimization algorithm, however, there is the additional issue of the range of fitness function values within a population. If there is a significant range in the best to worst value, the probability of being selected as a parent for the next generation tends to be skewed to very likely candidates over very unlikely candidates. This will lead to domination by a few design encodings which is not necessarily a good strategy, particularly in the early portion of the optimization process. In order to allow that each chromosome in the current population has at least a reasonable chance of being selected as a parent, the range in the fitness function is scaled by some appropriate procedure. One possibility would be to compute the scaled fitness as

$$\text{fitness} = [0.25 + (\text{evaluator value} - \text{worst value})/(\text{best value} - \text{worst value})]^2 \tag{7.13}$$

The best value and worst value are the lowest and highest evaluator values in the current population. This means that the range of fitness values will lie in the interval of

$$(0.25)^2 \leqslant \text{fitness} \leqslant (1.25)^2 \quad \text{or} \quad 0.0625 \leqslant \text{fitness} \leqslant 1.5625 \tag{7.14}$$

Additionally, it is not uncommon to set a limit on the allowable worst value to prevent too wide a discrepancy among the practical designs. This is particularly true when dealing with an evaluator value which is formed from a penalty function where a design which violates a constraint may have a value which is several orders of magnitude different from that of other feasible design points.

7.6.3 The initial population

For well-posed problems, virtually any combination of values in a chromosome corresponds to a valid design. In structural applications there are exceptions for topological design optimization when one must guard against unconnected elements, but this may be accounted for by a preprocessing operation before any analysis is performed. In order to span the design space completely, every possible combination of chromosomes must be able to be formed by combining traits from members of the initial population. This is generally not possible, as the population size required to accomplish this task would be enormous, leading to unacceptable solution times. The best approach, then, is to generate the initial population randomly. That is, each position of each chromosome for each design is assigned a value at random from its permissible range. This provides a broad range of initial design possibilities, although the likelihood of many good designs being present in the initial population derived in this manner is low. Since locating a good design topology is the primary goal of the optimization, this is of little consequence. In any case, the larger the population, the more representative the members become of the complete design space. The difficulty, particularly with a computationally expensive analysis required for each design, is that only a small population is realistic. Fortunately, a mutation operator can be applied in order to maintain a finite probability for the creation of any arbitrary combination of variables in the design space. The mutation operator is discussed in the following section.

7.6.4 Genetic operators

The specific types of operators used in the implementation of a genetic optimization algorithm depend on both the encoding scheme of the chromosome and on the particular type of problem being solved. Regardless of these factors, several basic genetic operators which are always present include the selection of parents, crossover and mutation. The selection of parents has been alluded to previously and is probabilistically based on the fitness function value. The term crossover refers to how the parent chromosomes are combined in order to generate offspring (new designs). Finally, mutation is a means of allowing purely random alterations in the chromosome values with a small but finite probability in order to introduce design characteristics into the current population which either were lost or were never present in any previous population. The entire genetic optimization process is heuristic in nature which allows a wide range of choices in a particular implementation. The operators described here are fairly elementary in nature and better results may be obtained with more complicated approaches. The basic flow of the algorithm is, however, much easier to comprehend when these basic operator forms are considered.

The parent selection process may be likened to the spinning of a weighted roulette wheel. The weighting is utilized so that the fittest members of the population occupy the largest area on the wheel and the total population fitness is scaled so that the sum of all slot areas fills the wheel. This gives every member of the population a finite chance of being selected but the probability of being selected is directly proportional to the fitness of the design. The wheel is spun twice and two population members are identified as parents. This parent set may be used to generate any number of offspring, but the generally accepted procedure is to produce two offspring from two parents. This means that the parent selection operation is performed one half as many times as the number of chromosomes in the population for each stage of the algorithm. It may be beneficial to limit the number of times a single chromosome may be a parent in any one generation in order to avoid dominance by a particular design. This may also be accomplished by other selection schemes such as by tournament selection (Goldberg, 1990). The way in which the children are formed from the parent chromosome values is the crossover operation.

The crossover operation is a function of the dimensionality of the chromosome. For a one-dimensional chromosome, a position within the length of the string is selected randomly and the parent chromosomes are switched from this point on. For example, if the parent chromosomes are strings of five values as

$$12345 \text{ and } 54321$$

and the crossover point is selected as three, then the offspring generated have the same first three chromosome values as the parent designs, but the remaining two positions are switched. This results in the offspring

$$12321 \text{ and } 54345$$

Whether or not these designs are better than the parent designs will determine their chance of being selected as parents for future generations. Many implementations have a means of propagating the best designs from one generation unchanged into the next generation and this is generally a good idea. Additionally, a single chromosome is not allowed to be selected as both parents.

For two- and three-dimensional chromosomes, both a matrix position and a crossover direction are selected randomly in order to generate offspring. There are four crossover directions for the two-dimensional case (NW, NE, SE, SW) and eight (octant directions) for the three-dimensional case. The reason for using higher-dimensional chromosomes in a shape optimal design problem is to couple the dimensionality of the chromosome with that of the encoding. This means that, for a three-dimensional problem, the crossover operation is actually combining the parent designs in three-dimensional blocks. While no extensive testing has been conducted to date on the effect of the dimension of the chromosome encoding, the rule of thumb to follow is always to link the encoding to the physical structure of the design topology as closely as possible.

The final genetic operator is that of mutation. Mutation randomly alters a position in a chromosome with a fairly low probability. This can at times destroy a good design but it can also introduce design characteristics which could not be achieved through crossover with the current population. A high mutation probability will slow convergence but too low a probability will truncate the design space in a potentially harmful way. The selection of all algorithm parameters is problem dependent and requires some experimentation. This is no different than with a traditional nonlinear programming algorithm. Additional detail on genetic algorithms is presented by Goldberg (1989).

EXAMPLE 7.1 A THREE-BAR TRUSS

The best way to demonstrate the difference between a traditional nonlinear programming problem approach and a multiobjective programming problem approach is to consider an example. This problem involves the design of a

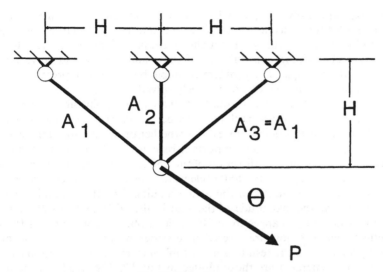

Fig. 7.7 Three-bar truss.

three-bar truss and is used to demonstrate how a multiobjective formulation can drive the solution from one topology to another which might not be located by a single objective formulation. The issue of uncertainty in the magnitude and direction of the applied loading is also considered. The three-bar truss as shown in Fig. 7.7 is to be designed to withstand a specified load of 20 units with the minimum amount of material. The allowable stress in each bar is constrained to be less than 20. The nodal locations are fixed, so the design variables are the cross-sectional area of each truss element. In order to maintain symmetry, both bars 1 and 3 are required to have the same cross-sectional area. This reduces the problem to two design variables which is convenient from the point of view that the results can be displayed graphically. The problem in this form will be addressed as a nonlinear programming problem, and then an uncertainty in the exact load magnitude and direction will be introduced and a multiobjective formulation and solution will be considered.

The nonlinear programming problem formulation for the original problem is expressed as follows:

$$\text{minimize volume} = 2\sqrt{2}x_1 + x_2; \mathbf{x} = \begin{Bmatrix} A_1, & A_3 \\ & A_2 \end{Bmatrix}$$

subject to

$$g_1(\mathbf{x}) = 20 - |\text{stress}_1| \geqslant 0$$
$$g_2(\mathbf{x}) = 20 - |\text{stress}_2| \geqslant 0$$
$$g_2(\mathbf{x}) = 20 - |\text{stress}_3| \geqslant 0$$

Here, A_i represents the cross-sectional area of the ith truss element and stress$_i$ represents the stress in the ith element. The solution to this problem formulation, generated by a standard penalty function algorithm, is $\mathbf{x}^* = \{0.788 \ 0.408\}^T$ with a resulting volume of 2.638 and stresses in the bars of 20, 14.65 and -5.35 units.

The stresses are easily computed by assembling the element stiffness matrices and solving for the deflection of the free node and then calculating the element forces and stresses. It would be a straighforward procedure to include deflection, frequency and buckling constraints, but these additions would serve only to complicate the example. The problem is simple but yet it is representative of a wide class of problems which are routinely solved.

The normal solution process would terminate at this point since the optimum design for this particular formulation has been obtained. The question which should be asked at this point, however, is whether or not the solution generated is indeed a good solution. One important issue is the effect on the design of a change in the magnitude or direction of the applied load. This is often a critical factor as loads applied to finite element models are rarely more than approximations or estimates of the actual loading. For this example, consider what happens as the magnitude of the load is altered from 15 to 25 units and the loading direction changes from 30° to 60° from the horizontal. If the load magnitude and direction are altered independently from the baseline of the original problem and the resulting new problem formulation is reoptimized, the new solutions generated are those plotted in Fig. 7.8. The weight of the design as well as the cross-sectional areas of the elements of the resulting design change on the order of $\pm 25\%$. Even though the change in the direction of the load had a minor impact on the weight for the optimal design, the resulting cross-sectional areas underwent considerable change.

The next question to ask is which, if any, of these designs is optimal? The answer could easily be either all of them or none of them. Each is an optimal design with respect to a specific problem formulation. On the other hand, none is a true optimal design for the real issues in designing the truss. At best, the results point out the region of potential design solutions as represented by the rectangle in Fig. 7.8. The designer can use the information derived from the optimization runs, along with his insight into the problem, and make the final design decisions.

For this two-dimensional problem which is not computationally expensive, it is easy to investigate the solution space around each potential optimum. For larger problems, this is not the case and the procedure becomes difficult if not impossible to apply. In addition, in this particular example, the uncertainty in the loading was well defined. If the uncertainty is known then this runs contrary to the concept of uncertainty. What would be far more beneficial would be a more natural approach to optimization which can deal with uncertainty as well as other important design issues.

In the absence of any specific information concerning the possible range of design uncertainty, there are several courses of action one might take. Two such courses are either to include a margin of safety in the design or to try and make the design as insensitive to changes in uncertain parameters as possible. The inclusion of a safety factor is the traditional approach, but, as was mentioned earlier, this converts the hard stress constraints to soft constraints which are not dealt with effectively by a nonlinear programming problem formulation and solution. The second approach is attractive, but may not in itself lead to a realistic design and is again not easily incorporated into a traditional problem formulation and solution procedure. Nonlinear goal programming may be applied to

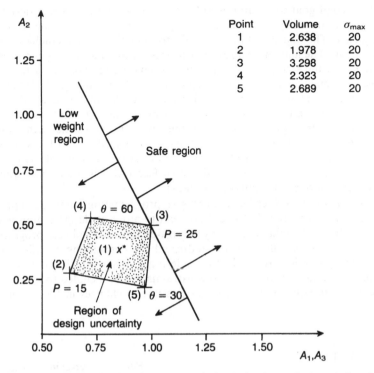

Point	Volume	σ_{max}
1	2.638	20
2	1.978	20
3	3.298	20
4	2.323	20
5	2.689	20

Fig. 7.8 Optimal design for three-bar truss (variation in load magnitude and direction).

formulate a design optimization problem which deals with either or both of these approaches to modifying the problem.

From the margin of safety point of view, one must consider what decreases the chance of a failure of the design in service. For this example, the maximum stress present in the design provides a good indicator. A design which has a very low peak stress is unlikely to fail even if the load magnitude is increased. Of course, this will produce a design which is counter to the original objective of a light weight design. The trade-off between a safe design and a lightweight structure may be evaluated easily through a multiobjective formulation as a nonlinear goal programming problem. Consider the formulation given as

$$\text{minimize } f(\mathbf{x}) = \{ W_1 d_1^+ + W_2 d_2^+ \}$$

with goal constraint of the form

$$\text{volume} + d_1^- - d_1^+ = 0$$

and

$$\max\{\text{stress}_1, \text{stress}_2, \text{stress}_3\} + d_2^- - d_2^+ = 0$$

This formulation will try and strike a compromise between the minimization of the volume and the minimization of the peak stress. The first goal constraint sets forth the goal of a zero weight design, the ultimate lightweight structure, and the second goal constraint establishes the goal of a zero stress design. It is obvious that no design can satisfy the goal constraints, but this is of little consequence. The objective function seeks to minimize the underachievement of the two goals. The maximum stress constraints are maintained to guarantee their satisfaction at the solution. The path the optimal design takes as a function of the relative weights placed on the design goals is shown in Fig. 7.9. When a low weight is assigned to the stress goal relative to the weight goal, the design generated will be equivalent to the nonlinear programming result. As the relative weight on the stress goal is increased, the optimum traces the path shown in Fig. 7.9. As the weight increases, the design moves farther into the safe region. This path provides a wealth of trade-off information. For problems of higher dimension, this path becomes a trade-off surface which is somewhat equivalent to a response surface in multivariate statistics.

The information produced by the first goal programming formulation provided useful design information but it did not lead to any new design region which is potentially better than that located by the straightforward nonlinear programming solution. This issue will now be addressed by considering a different goal programming solution. Instead of considering a design which has low stresses in all of its structural members, goal constraints will be included to try and make the peak design stress as insensitive as possible to changes in the applied load

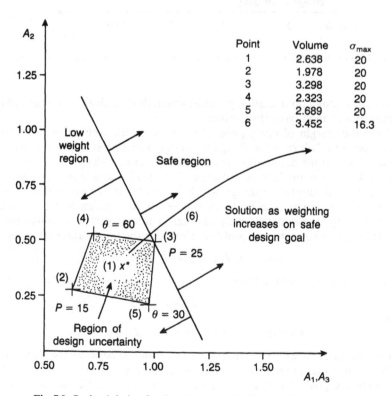

Point	Volume	σ_{max}
1	2.638	20
2	1.978	20
3	3.298	20
4	2.323	20
5	2.689	20
6	3.452	16.3

Fig. 7.9 Optimal design for three-bar truss for first goal formulation.

(both magnitude and direction). A possible goal programming formulation for this problem is given as

$$\text{minimize } f(\mathbf{x}) = \{ W_1 d_1^+ + W_2(d_2^+ + d_3^+ + d_3^-) \}$$

with goal constraints

$$\text{volume} + d_1^- - d_1^+ = 0$$

$$\max\{\text{stress}_1, \text{stress}_2, \text{stress}_3\}_{P=25} - \max\{\text{stress}_1, \text{stress}_2, \text{stress}_3\}_{P=15}$$
$$+ d_2^- - d_2^+ = 0$$

$$\max\{\text{stress}_1, \text{stress}_2, \text{stress}_3\}_{\theta=60°} - \max\{\text{stress}_1, \text{stress}_2, \text{stress}_3\}_{\theta=30°}$$
$$+ d_3^- - d_3^+ = 0$$

The objective function now has one weight for the volume or minimum weight goal and a common goal for the design sensitivity of the peak stress relative to a change in load magnitude or direction. Both d_3^+ and d_3^- are included in the objective function as it is not known beforehand what sign the result will take. The first two goal constraints may only take on possitive values, so only the corresponding positive deviational variable is included in the objective function. It would be a minor matter to include different weights for the stress sensitivity

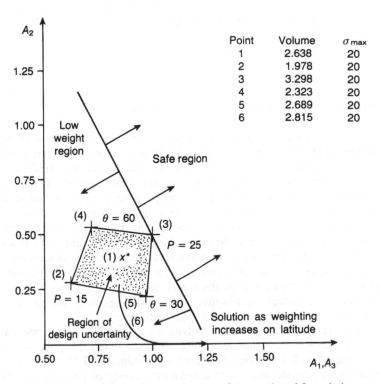

Point	Volume	σ_{max}
1	2.638	20
2	1.978	20
3	3.298	20
4	2.323	20
5	2.689	20
6	2.815	20

Fig. 7.10 Optimal design for three-bar truss for second goal formulation.

with respect to load magnitude and direction, but this would increase the number of weighting parameters.

The results from this formulation are presented in Fig. 7.10. As with the previous results, the path which the solution follows as the relative weights between the weight goal and the sensitivity goals change is plotted. It is interesting to note that a topological change in the design is made by reducing the area of the middle element to zero. This is due to the fact that by using the material from the second element to increase the cross-sectional areas of the outer elements, a design is generated which is less sensitive to load variation. This result, although perhaps intuitive, may not have been achieved without the investigation of this alternative formulation. The most interesting aspect of the result is that the two-bar truss design moves toward the safe design region with less of a weight penalty than did the three-bar truss. This is simply a reminder that topological changes may result in alternative optimal design regions.

Which of the presented formulations produces the best design is not as important as the fact that much useful design information was derived by considering the goal programming formulations. A combination of both the goal formulations would perhaps be even more productive. On the other hand, this problem is quite simple and the approach taken in the solution may not be attractive for more difficult problems. The evaluation of multidimensional trade-offs between competing design criteria is not a simple task. The flexibility of the multiobjective problem formulation, however, is apparent as is the fact that it is a mistake to concentrate on one small portion of the feasible design space. Better design solutions involving topology changes may well be available and should at least be considered as possibilities.

EXAMPLE 7.2 MECHANISM DESIGN

The topological change considered in the first example involved an alteration in the number of structural elements present. Other classes of design problems benefit from a multiobjective formulation as well as from consideration of topological modifications. The problem considered here belongs to the class of planar mechanism design. The basic challenge is to produce a design which possesses a number of specific characteristics including position, velocity, cost and small sensitivities to inner and outer noise. The optimization of mechanisms is difficult owing to a number of factors including a high degree of nonlinearity, the possibility of divergence in the iterative solution process for determining link positions and velocities and the presence of a multitude of local minima. The issue of hard vs soft design specifications also comes into play. Hard constraints such as the requirement of assembly are always present but most other considerations are best treated as soft specifications.

The particular problem at hand involves the design of a planar mechanism which can produce motion that accurately traces a straight line path from point A with coordinates (5, 4) to point B located at coordinates (2, 2). In addition, the mechanism is to provide as close to a constant velocity of 100 units s^{-1} during the motion from point A to B. The pivot point of the driving link for the

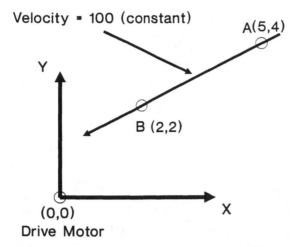

Fig. 7.11 Design specifications for linkage design problem.

mechanism is required to be located at the origin as depicted in Fig. 7.11. The linkage is to be driven at a constant angular velocity. Other design considerations include the cost of the mechanism as well as its sensitivity to small changes in either the angular velocity of the input link or the various link geometries (i.e. tolerances on link lengths and position). The mechanism should also remain far from a position in which it will jam or not assemble. At this point in the process it is not at all obvious which design criteria will be most influential in defining the solution, and as such, it provides an excellent multiobjective optimization example.

There are an extremely large and varied assortment of mechanisms which could be investigated for their potential in meeting the design criteria. This is typical of most design endeavors in the early stage. For the purpose of this example, three mechanism classes will be considered, a four-bar mechanism as illustrated in Fig. 7.12, an inverted slider crank mechanism as shown in Fig. 7.13 and a six-bar mechanism as pictured in Fig. 7.14. There actually are many other configurations of six-bar mechanisms, but the one selected, which is commonly referred to as a Stephenson type III linkage, is assumed to represent the general performance of the class. Each of the mechanisms selected possesses a single degree of freedom in that, for a specified input link motion, the position, velocity and acceleration of each link in the mechanism are completely specified. This allows the design objectives to be evaluated in a direct fashion from the mechanism geometry.

The design variables are the link lengths, ground pivot locations and coupler point location. The coupler point is the position on one of the links which is to provide the displacement and velocity specified in the problem statement. A ground pivot is simply a point where the linkage is fixed in translation but not in rotation. The design variables are defined for each of the three selected design topologies in Figs. 7.12–7.14. This example points out the fact that the number of design variables for each design topology need not be the same as for another. In this case, the inverted slider crank has nine design variables, the four-bar has

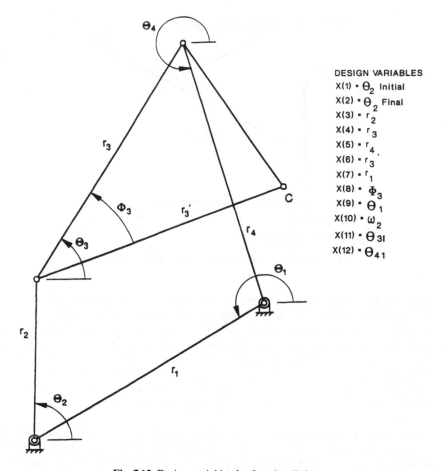

Fig. 7.12 Design variables for four-bar linkage.

ten design variables and the six-bar has sixteen design variables. Cross-sectional link dimensions are not included as no force or stress analysis will be considered in this example. Additional variables were added which estimated the initial link positions for the displacement and velocity analysis which resulted in a total of eleven, thirteen and twenty design variables for the three cases.

The design objectives are formed into the goal constraints based on the criteria listed below.

1. The distance between the actual motion generated by the mechanism and the desired linear path should be as small as possible.
2. The difference between the desired velocity and the actual coupler point velocity should be as small as possible over the range of motion.
3. The coupler point should start as close to point A as possible at the beginning of the required motion range.
4. The coupler point should be located as close to point B as possible at the end of the required motion range.

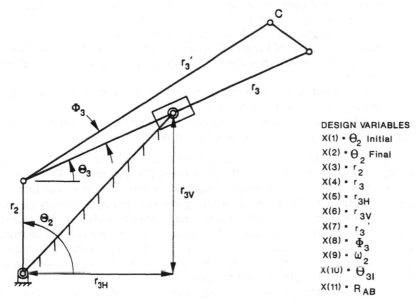

DESIGN VARIABLES

X(1) = Θ_2 Initial
X(2) = Θ_2 Final
X(3) = r_2
X(4) = r_3
X(5) = r_{3H}
X(6) = r_{3V}
X(7) = r_3'
X(8) = Φ_3
X(9) = ω_2
X(10) = Θ_{3I}
X(11) = R_{AB}

Fig. 7.13 Design variables for inverted slider–crank mechanism.

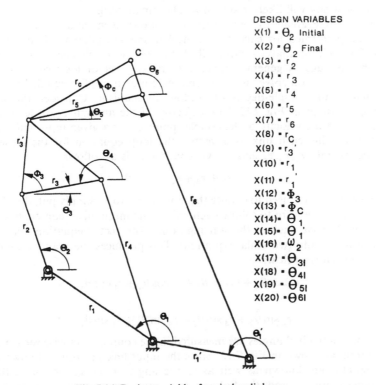

DESIGN VARIABLES

X(1) = Θ_2 Initial
X(2) = Θ_2 Final
X(3) = r_2
X(4) = r_3
X(5) = r_4
X(6) = r_5
X(7) = r_6
X(8) = r_C
X(9) = r_3'
X(10) = r_1
X(11) = r_1'
X(12) = Φ_3
X(13) = Φ_C
X(14) = Θ_1
X(15) = Θ_1'
X(16) = ω_2
X(17) = Θ_{3I}
X(18) = Θ_{4I}
X(19) = Θ_{5I}
X(20) = Θ_{6I}

Fig. 7.14 Design variables for six-bar linkage.

5. The mechanism should remain as far as possible from a position where the mechanism will not assemble over the required motion range.
6. The size and weight of the linkage should be as low as possible.
7. The change in velocity of the coupler point over the required range of motion should be as insensitive as possible to small fluctuations in the angular velocity of the input link.
8. The coupler point path traced should be as insensitive as possible to a small change in the length of the input link.
9. The coupler point path traced should be as insensitive as possible to a small change in the location of the coupler point position.

The goals, when defined in this manner, are easily understood by the designer and apply equally well to any of the mechanism classes selected. This is a key point as placing goals on the design defines the problem but does not limit the range of solutions to the problem. These goals are applied uniformly to each design class. Of course there may be other design goals to include or the existing goals could be defined in an alternative fashion, but the goals listed certainly capture the majority of important design criteria for this example. The goals defined for this example may appear to be rather severe, but, at this point in the process, no concern is placed over the possibility that no solution exists which can even begin to satisfy all of the design goals. Trade-off evaluations and goal redefinition should occur at a later point in the process. Keep in mind that design should proceed from the general to the specific and that any attempt to alter this philosophy will likely result in a suboptimal design.

In order to evaluate the goal constraints, a displacement and velocity analysis must be performed for each of the mechanisms considered. Many techniques exist for such an analysis, especially for a specific mechanism class such as a four-bar linkage. A general approach employing vector loop equations and a nonlinear equation solver is considered herein as it applies equally well to any selected planar, single-degree-of-freedom mechanism. Consider the four-bar linkage pictured in Fig. 7.12. The objective of the analysis is to determine the position and the velocity of the coupler point C, for a given input link position, θ_2. Considering each link as a vector, the loop equation for the closed loop formed by links r_2, r_3, r_4 and the ground link r_1 is as follows:

$$r_2 + r_3 + r_4 + r_1 = 0$$

The bold print is used to denote the terms as vectors. Simply put, this equation states that if one were to follow each link in the mechanism consecutively, then one would arrive back at the starting point. The vector equation may now be broken down into two scalar equations. This produces the x and y component equations shown below:

$$r_2 \cos \theta_2 + r_3 \cos \theta_3 + r_4 \cos \theta_4 + r_1 \cos \theta_1 = 0$$

and

$$r_2 \sin \theta_2 + r_3 \sin \theta_3 + r_4 \sin \theta_4 + r_1 \sin \theta_1 = 0$$

where the individual angles are measured in the counterclockwise sense from the horizontal as shown in Fig. 7.12. Since the input link position is known and all link lengths are known as well as as the angle of the ground link, the only unknowns are θ_3 and θ_4. Solving two equations for two unknowns is a straightforward process, with the only difficulty being the fact that the equations are nonlinear. A Newton–Raphson technique is employed to solve the equations. A starting estimate of the unknown angles is required and, if these estimates are not good enough, the solution process can diverge. Program checks must be inserted in order to flag this situation.

Once angles θ_3 and θ_4 are determined, it becomes a simple matter to locate the position of the coupler point C. A vector loop equation can be written beginning from the input link and ending at the coupler point. This equation takes the following form:

$$C = r_2 + r'_3$$

where C refers to the location of the coupler point relative to the origin. Again this vector equation may be broken down into its x and y component equations which produces

$$C_x = r_2 \cos \theta_2 + r'_3 \cos (\theta_3 + \phi_3)$$

and

$$C_y = r_2 \sin \theta_2 + r'_3 \sin (\theta_3 + \phi_3)$$

Since there are no unknowns on the right-hand sides of the equations, the x and y coordinates of the coupler point are completely determined.

The determination of the velocity of the coupler point is achieved by differentiating the x and y component equations resulting from the vector loop equation for the mechanism. This produces

$$- r_2 \omega_2 \sin \theta_2 - r_3 \omega_3 \sin \theta_3 - r_4 \omega_4 \sin \theta_4 = 0$$

and

$$r_2 \omega_2 \cos \theta_2 + r_3 \omega_3 \cos \theta_3 + r_4 \omega_4 \cos \theta_4 = 0$$

The only two unknowns in these equations are the angular velocities ω_3 and ω_4. The input velocity, ω_2, is constant and is known. These equations are linear in the unknowns, so the solution is trivial. Once the angular velocity of all links is known, the velocity of the coupler point may be determined by differentiation of the component equations which define the position of the coupler point. This results in

$$V_{x\text{point } C} = - r_2 \omega_2 \sin \theta_2 - r'_3 \omega_3 \sin (\theta_3 + \phi_3)$$

and

$$V_{y\text{point } C} = r_2 \omega_2 \cos \theta_2 + r'_3 \omega_3 \cos (\theta_3 + \phi_3)$$

Again, all right-side terms are known which allows for the direct solution of the velocity components.

The solution processes for the displacement and velocity for the inverted slider crank mechanism and the six-bar mechanism follow a similar vector loop process. It is always possible to write a vector loop displacement equation which will result in only two unknowns. For the six-bar linkage, two such vector loops are required for a displacement solution, but this adds little to the difficulty. Once the displacement equations are solved, direct differentiation allows for a velocity solution. The process is repeated for a number of input link positions ranging from the initial position to the final position of interest. A total of ten intermediate positions was arbitrarily selected for this example. This provides sufficient velocity and displacement information for the evaluation of any selected geometric link dimensions for any of the mechanism classes considered.

The first goal constraint concerning the specific·linear motion requirement was formed by summing the distance of the coupler point from the line segment connecting points A and B for all ten intermediate positions. The second goal constraint which specifies the constant velocity requirement is formed by summing the square of the difference in the coupler point velocity and the desired velocity at each intermediate position. The next two goals which state that the coupler point should start at point A and end at point B are formed as the distance of the coupler point from point A at the initial motion position and from point B at the end of the specified motion. The desired goal value for each of these constraints is zero which would produce a mechanism with the exact positional and velocity characteristics.

The fifth goal constraint states that at each intermediate position the linkage should be reasonably far from a position in which it would not assemble. This keeps the mechanism away from a position where it either might lock up or would require a large force to maintain the desired operation. For ease of computation, this constraint is specified as a distance from a nonassembly position. In the case of the four-bar linkage, this distance is simply the difference in the sum of the length of links 3 and 4 and the distance from the ground point of link 4 and the current position of the end of link 2. As long as this difference is positive, the mechanism will assemble. The desired goal value for this constraint was specified as a minimum of one distance unit for the worst case over the range of motion. The satisfaction of this goal directly effects the transmission angles in the mechanism as well. The assembly goal for the inverted slider crank mechanism is that link 3 will always extend beyond the slider pivot point by at least one distance unit. For the six-bar mechanism, two separate assembly conditions are imposed, one for each vector loop used in the displacement solution.

The sixth goal constraint is imposed as a complexity or, in a sense, a cost measure. For the same level of satisfaction of the other goal constraints, a smaller mechansim or one with fewer links is preferred. This type of linkage will require less material and as such should have a cost advantage over a larger and more complex mechanism. The goal constraint is computed as the sum of the lengths of all links, including coupler links for each design. A more precise cost goal could be assigned, but this formulation will serve to point out potential cost differences among the design choices. A desired goal value of zero is assigned to the goal signifying a smaller is better condition.

The final three goal constraints are intended to encourage a design which has performance which is as insensitive as possible to small changes in dimensions or the input velocity of the driving motor. Any inexpensive motor will have some speed fluctuation which is dependent on the load being driven. The dimensional tolerances of the links can be controlled accurately but there is a cost associated with the accuracy required. If a design is found which is insensitive to these changes it is a robust design and this characteristic is highly desirable. The goals are computed as the difference in the velocity accuracy as specified by goal 2 caused by a 5% increase in the angular velocity of the input link, the length of the input link and in the position of the coupler point. The desired goal level for the sensitivity constraints is set to zero which specifies the ideal design to be completely insensitive to these changes. The sensitivity with respect to other link dimensions and ground attachment points could be considered as well but the nine goal constraints specified are adequate to define a rather challenging design problem.

The fact that no design will be able to meet all of the design criteria is of no consequence to the approach. This is a fundamental difference in comparison with traditional nonlinear programming methods. These methods would have the null space defined for the search and would fail to find a feasible point. This would not allow for an easy evaluation of trade-offs involved in meeting the design goals. All that remains for the goal formulation is the selection of a set of goal priorities and weights. The goals are separated into three levels of priorities. The initial and final position requirements for the coupler point have the highest goal priority. The linear motion and constant velocity requirements have the second and third highest priorities. Finally, all remaining goals are set at the lowest priority. The selection of individual goal weights is a scaling issue and the idea is to keep the contribution of each goal constraint approximately

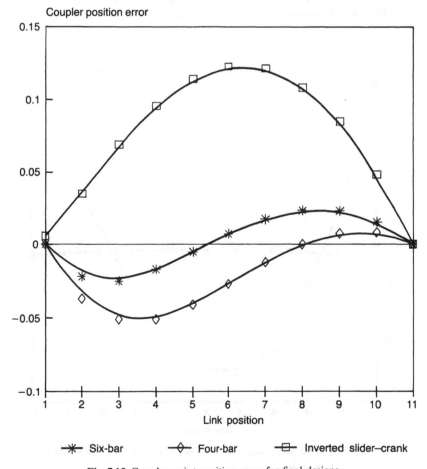

Fig. 7.15 Coupler point position error for final designs.

equal at the starting point selected. The weights may be fine tuned at a later point to modify the final design configuration as deemed appropriate.

The assignment of priorities was not performed in an arbitrary fashion but was in fact meant to mimic the path a human designer might follow. The design search would most likely be for a coupler point on the mechanism which had good displacement characteristics, followed by a modification to achieve the desired velocity requirement. At this point, the designer would consider factors such as weight, cost and design sensitivity. In truth, it is unlikely that a human designer would get past the second or third goal levels as very few design topologies would be investigated because of the time required and the difficulty involved in the process. This is not true, however, of the multiobjective computational process where thousands of designs may be investigated.

For this particular problem both a genetic optimization algorithm and a gradient based nonlinear programming algorithm were employed. Since the link dimensions are continuous, the nonlinear programming algorithm is more efficient in locating a minimum than the genetic algorithm. The genetic method, however, is not trapped by local minima and as such it is used to perform the topological search which defines the mechanism class as well as the starting

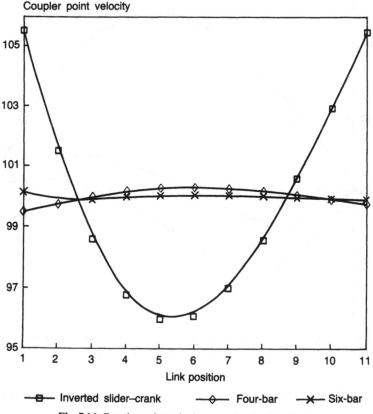

Fig. 7.16 Coupler point velocity error for final designs.

Table 7.1 Unweighted goal values for linkage design problem

Goal	Inverted slider crank	Four bar	Six bar
1	0.804	0.235	0.156
2	126.733	9.144	2.931
3	5.621×10^{-3}	3.000×10^{-4}	2.861×10^{-4}
4	5.376×10^{-5}	5.229×10^{-4}	8.692×10^{-4}
5	2.138	3.206	0.512
6	40.531	23.615	26.480
7	220.942	182.322	182.605
8	0.306	0.118	0.196
9	1.032	0.785	0.512

dimensions for the link lengths which are passed to the nonlinear programming code. Even the initial angle estimates for the iterative solution of the vector loop equations were determined by the genetic algorithm which gave the various alternate assembly positions an equal chance of being investigated. The inclusion of the initial position estimates as variables accounts for an increase in two design variables for the four-bar and inverted slider crank mechanisms and four variables for the six-bar linkage. All goal constraints were treated as soft constraints to allow the maximum amount of freedom in the solution.

Table 7.2 Optimal design variable values for linkage design problem

Design variables	Inverted slider crank	Four bar	Six bar
$X(1)$	97.430	78.745	50.910
$X(2)$	131.762	113.643	87.602
$X(3)$	5.229	4.266	2.833
$X(4)$	15.000	4.098	5.298
$X(5)$	4.919	5.114	3.065
$X(6)$	13.640	4.171	1.274
$X(7)$	5.803	5.966	5.440
$X(8)$	291.664	273.392	4.022
$X(9)$	21.248	224.015	1.967
$X(10)$	56.506	−16.856	0.662
$X(11)$	10.138	84.081	1.919
$X(12)$		306.416	204.147
$X(13)$			323.663
$X(14)$			155.269
$X(15)$			168.177
$X(16)$			17.725
$X(17)$			270.118
$X(18)$			112.954
$X(19)$			36.451
$X(20)$			275.024

In short, the combined genetic–nonlinear goal solution provided several excellent solutions to what would be considered a most difficult design problem by any other approach. One solution from each of the selected mechanism classes is presented. Fig. 7.15 plots the position error produced by the final design for each class while Fig. 7.16 displays the velocity error over the range of motion. From these results it is clearly evident that there are indeed designs which satisfy the positional and velocity goals very precisely. The six-bar mechanism achieves the best results, followed closely by the four-bar mechanism and finally by the inverted slider crank mechanism.

When the remaining goals are factored into the evaluation, it is still fairly easy to distinguish among the various design alternatives. Table 7.1 presents the final values of the unweighted goal values for each of the three final designs. The six-bar mechanism performed well in all meeting all goals with the only caution arising in the assembly condition which is met but not by as much as for the other designs. The performance of the four-bar is certainly adequate unless the mechanism is required to have a very high level of precision. The design variables for the final designs are presented in Table 7.2. It should be noted that other designs were located in addition to the three documented in the final results. Two other six-bar mechanisms and one other four-bar mechanism which had excellent performance characteristics were obtained in the final population as well. Additional trade-offs could be made for each of the final designs to fine tune them as well.

EXAMPLE 7.3 A TEN-BAR TRUSS

The literature concerning structural optimization contains a multitude of problem solutions, particularly in the area of truss structures. Several examples have appeared time and time again to the point that the solutions presented are fairly well accepted as being correct. The ten-bar truss presented in Fig. 7.17 is one such example. The design objective is to minimize the weight of the structure which is subjected to two equal vertical loads applied to nodes 2 and 4 of 10 000 units. The maximum allowable stress in a member (tension or compression) is 25 000. A maximum displacement constraint of 2 units is imposed for nodes 2 and 4 as well. The design variables are simply the cross-sectional area of each of the ten truss members. Lower variable bound limits are placed on all areas to be greater or equal to 0.1. The solution to this problem formulation, generated by a gradient based penalty function method, is presented in the first column of Tables 7.3 and 7.4. The question to be addressed, however, is whether this is indeed the solution or even represents a good design.

If the minimum weight structure is desired and the topology of the structure is held constant (i.e. ten members with fixed nodal location), then this is indeed the optimal point. The solution may be verified by regenerating the solution from a number of starting points. While the formulation presented fits very nicely into the standard structural optimization format, the solution raises several important design issues. First of all, why was a ten-bar truss selected? From the loading specified it seems plausible that several members could be removed and

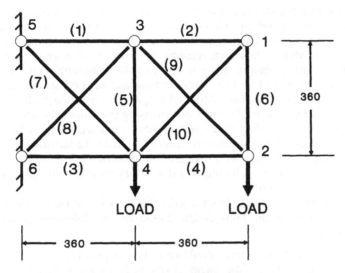

Fig. 7.17 Ten-bar truss design.

Table 7.3 Unweighted goal values for structural design problem

Goal	Design 1	Design 2	Design 3	Design 4	Design 5
1	5062	6689	2317	3067	3082
2	25000	8616	24940	12490	12510
3	2.000	2.000	2.002	1.502	1.499
4	3303	742	2973	2082	2080
5	0.255	0.257	0.407	0.338	0.336
6	7409	2319	5174	1648	1708
7	10	10	6	6	3

Table 7.4 Optimal design variable values for structural design problem

Variable	Design 1	Design 2	Design 3	Design 4	Design 5
Area (1)	30.154	23.720	13.550	17.700	17.960
Area (2)	0.100	12.440	–	–	–
Area (3)	23.440	24.620	10.590	13.730	13.640
Area (4)	15.397	16.100	9.890	12.800	13.640
Area (5)	0.100	9.430	–	–	–
Area (6)	0.493	8.240	–	–	–
Area (7)	7.402	17.450	2.010	2.920	2.920
Area (8)	20.878	17.950	3.840	3.470	2.920
Area (9)	21.764	16.340	13.300	17.700	17.960
Area (10)	0.100	12.780	–	–	–
Node 5(Y)	360.00	360.00	510.24	560.00	560.00
Node 6(Y)	0.00	0.00	−151.72	−200.00	−200.00
Node 3(X)	360.00	360.00	367.38	393.00	389.85
Node 3(Y)	360.00	360.00	317.34	311.00	314.47

not detract from the ability of the structure to meet all design requirements. This perception is warranted by the solution which has three of the member areas set at the lower limit. Whether or not the removal of members would produce a lower weight design, however, remains unknown. It also seems possible that allowing the node points to move in space could be beneficial to the solution. Added to this is the fact that the stress and displacement constraints are treated as hard constraints. What if an additional 10% weight reduction could be achieved by increasing the limit to 2.1 units or increasing the allowable stress to 26 000 units? Alternately, if safety is a consideration, what weight penalty would be imposed if the maximum stress were reduced? Finally the sensitivity of member stresses to small changes in load magnitude or direction is unknown. All of these design issues may be critically important, but they remain uniformly unresolved with the current solution.

Approaching the problem from a strict design point of view with no predisposition as to what the final design should be, the following goals deserve consideration.

1. The weight of the structure should be as low as possible.
2. The maximum stress of any member should be as low as possible.
3. The maximum displacement of any node should be as small as possible.
4. The maximum stress should be as insensitive as possible to a change in loading magnitude or direction.
5. The maximum displacement should be as insensitive as possible to a change in loading magnitude or direction.
6. The level of stress in each member should be as uniform as possible.
7. The number of different beam sections should be as few as possible.

These seven goals encompass the majority of objectives for structural design. The first goal is equivalent to the minimum weight objective in a conventional formulation. It is formed by summing the volume of material used in each member and multiplying the sum by the specific weight of the material. The desired goal value of this constraint was set at zero which is a lighter is better formulation. The specific weight was set at 0.100 and the elastic modulus, E, was set at 10^6.

The second and third goal constraints are equivalent to the inequality constraints imposed in the traditional formulation. The advantage of formulating them as goal constraints is that they now become soft constraints and can be violated slightly if other goals can benefit significantly from the violation. These two goal constraints are normalized by dividing by the desired goal values (25 000 for stress and 2 units for displacement) and subtracting the ratio from 1. In this form, if the stress or displacement goal value is exceeded, the goal constraint becomes negative.

The fourth and fifth goal constraints are to ensure a robust design in that both the maximum stress and displacement should be as insensitive as possible to reasonable changes in the loading. This is an important requirement as exact loading conditions are seldom known precisely and the load seen in actual usage may be far from the load case imposed on the problem. These constraints are computed as the change in peak stress and displacement for a 10% increase in the load. The directional uncertainty of the load was also included by adding a 10 000 unit load in the positive x direction as well. The desired goal level for both these constraints is zero which says the perfect design is completely insensitive to a load change.

The sixth goal constraint seeks to locate a design in which all members are subject to as equal a stress as possible. From a safety point of view this means that no one member is more likely to fail than any other. The goal is calculated by computing the standard deviation of the stresses in all members. The desired goal value is zero which can be achieved if all members have identical stress levels.

The final goal constraint in one which is rarely considered in structural optimization but is of great practical importance. In constructing a truss structure, common sized components are beneficial from both a cost and an assembly point of view. For example consider assembling a 200-bar electrical transmission tower with 200 different sizes of beams. This constraint seeks to produce the fewest groups of elements with common cross-sectional areas. No length consideration is given but it could be imposed if desired (assuming the nodal locations are allowed to float). The constraint is formed by adding the number of elements having cross-sectional areas differing by more than 2%. The desired goal value is one which states that the best design has all elements with the same area.

Other goals are possible including one which keeps the number of truss elements as low as possible or one which maintains the peak stress at a reasonable level even if one element fails. Actually these goals are somewhat accounted for in the current formulation. The minimum number of elements is somewhat equivalent to the minimum weight constraint and the peak stress under the failure of a member would require a larger number of analysis runs for each design although the uniform stress goal has some impact on this criteria. The advantage of this formulation is that all goals may be explained in plain English and they are all computationally inexpensive to evaluate.

Each analysis run was performed using a finite element formulation for truss elements. The solution time required for each analysis of all goal constraints for a design was approximately 10 s on a SUN 4-470 workstation. This allowed many different designs to be investigated. As in the second example, a hybrid solution method was employed with the genetic code used to generate the number of truss elements and the number of different cross-sectional areas to consider and a gradient based penalty function method to perform the local search.

The goal priorities were set with the first goal having the highest priority, the second and third goals having the second priority, the forth and fifth goals having the third priority and the last two goals having the lowest priority. A number of different solutions were generated by making design trade-offs during the solution process. For some cases, the nodal locations were fixed and for others were allowed to float. When the nodal locations were allowed to float, a limit of ± 200 units was imposed as a hard constraint and the ground nodes were only allowed to move in a vertical direction. This allowed for structural optimization with consideration of cross-sectional, geometric and topological design parameters.

Five separate solutions are presented in Tables 7.3 and 7.4. Compared with the results for the traditional ten bar truss design, several of the designs generated by the multiobjective approach are better in every aspect. The majority of the designs turned out to have six truss members in the configuration pictured in Fig. 7.18. This seems reasonable, given the loading applied. The stresses and displacements are lower and the weight is significantly reduced (over 50% in

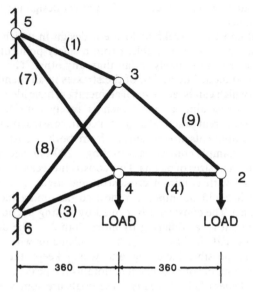

Fig. 7.18 Six-bar truss design.

some cases). Even when only three different cross-sectional areas were used, the total weight was still quite attractive. Precise tuning of the design solution is possible, but would require a more detailed understanding of the intended application. The important point to note is that the multiobjective approach did indeed generate solutions which were far superior to the previously published results. The goal formulation allows the designer to concentrate on real design issues rather than a rigid prescribed mathematical formulation.

7.7 SUMMARY AND CONCLUSIONS

The need for computer-aided design tools which can have a significant impact on a design at an early stage of the design process is evident. Traditional design optimization via a single objective nonlinear programming formulation does not always service this need in the best fashion. A goal programming, multiobjective problem formulation and solution methodology have been presented and demonstrated on three design examples. The key attributes of the procedure are the ability to handle multiple, competing design objectives and that the process is applicable at an early stage of the design process where it can have the greatest impact on shaping the design solution. The hybrid approach of a genetic algorithm for handing topological choices and a traditional nonlinear programming algorithm for refining locally optimum solutions is shown to be an effective design optimization tool. The inclusion of topological design issues is important as multiobjective formulations often have many locally optimal solutions spread over the design space.

In the nonlinear goal programming formulation, the user defines a set of design objectives or goals and decides which goals have the highest priority. Goals at the same priority can be differentiated by weighting factors. Design constraints may be imposed as goals where they can be treated as hard or soft design specifications or as inequality or equality constraints where they are treated as hard constraints. Design variables involving topological decisions are separated from the rest and a genetic algorithm is applied to handle these issues. The remaining variables are manipulated by a nonlinear goal programming algorithm which is built on a gradient based penalty function method. Whether or not a feasible design space is defined has no effect on the ability to generate a solution. Starting with the highest priority goals, the algorithm satisfies them to the best extent possible. From this point, the lower priority goals are factored in and satisfied to the extent that they do not significantly effect the higher priority goals.

The user plays a key role in the optimization process, not only in defining goals, priorities and weighting factors, but in providing insight to the solution process which would be difficult or impossible to codify. Various graphical techniques may be employed to help evaluate and modify solutions. Mathematically precise multiobjective optimization methods which seek a Pareto optimal solution can make poor design trade-offs. This is due to the fact that, in most applications, design objectives or goals have priorities and increasing the goal satisfaction of a high priority goal at the expense of one or more lower priority goals may be beneficial even though the result is not a Pareto optimal point. Also in the formulation of design constraints it is important to recognize the difference between hard and soft specifications. Traditional methods deal almost exclusively with hard constraints which may again mask important design trade-offs which could be made.

The three-bar truss example was used to demonstrate that a multiple objective formulation could include criteria such as design sensitivities which in turn could lead to topological design changes with the appropriate weighting. The second example involving the design of a planar mechanism points out how topological issues can be combined with a gradient based goal programming algorithm to generate valid design solutions to an extremely difficult problem. The final example of a ten-bar truss design is used to highlight the fact that formulating and solving a conventional design optimization problem does not deal effectively with important design issues which when considered can generate a far superior design. Again the inclusion of topological issues in the goal formulation is the key to success.

The hybrid method presented does have some drawbacks. The solution time required is generally well beyond that of a traditional single or multiobjective optimization solution. Generally, many designs must be evaluated and the topological solution is nongradient based to achieve a truly global search and cannot benefit from design sensitivity information supplied by the analysis code. On the positive side, the algorithm is parallel in nature and can achieve virtually a linear time reduction with the number of processors available. Parallel machines can be configured with computing power equal to a supercomputer at a fraction of the cost, but the power of such a machine can only be accessed by software which can run efficiently in parallel. Few conventional optimization techniques can take full advantage of such a machine.

In the final analysis, it is not so much a question of how much computer time was used in generating a solution, but how good a solution was generated. The inclusion of topological issues allows a diverse set of designs to be investigated. This is critical as a poor design optimized will generally result in a poor design. The methodology presented allows for a formulation which is natural and parallels the manual design process better than conventional methods do. The approach is continually undergoing development and is still in the infancy. The potential of the method from the problems solved to date however, is very high.

REFERENCES

Abuyounes, S. and Adeli, H. (1986) Optimization of steel plate girders via general geometric programming. *Journal of Structural Mechanics*, **14**(4), 501–24.

Adeli, H. and Balasubramanyam, K.V. (1988) A synergic man–machine approach to shape optimization of structures. *Computers and Structures*, **30**(3), 553–61.

Adeli, H. and Chompooming, K. (1989) Interactive optimization of nonprismatic girders. *Computers and Structures*, **31**(4), 505–22.

Adeli, H. and Kamal, O. (1992) Concurrent optimization of large structures: part I – algorithms. *ASCE Journal of Aerospace Engineering* (in press).

Adeli, H. and Mak, K.Y. (1989) Application of a coupled system for optimum design of plate girder bridges. *Engineering Applications of AI*, **2**, 72–6.

Bennett, J.A. and Botkin, M.E. (1981) Automated design for automotive structures. Progress on Engineering Optimization, ASME Technical Conference, Hartford, CT.

Davis, L. and Ritter, F. (1987) Schedule optimization with probabilistic search. Proceedings of the 3rd IEEE Conference on Artifical Intelligence Applications, 1987, pp. 231–6.

El Sayed, M., Ridgley, B. and Sandgren, E. (1989) Structural design by nonlinear goal programming. *Computers and Structures*, **32**, 69–73.

Evans, G.W. (1984) An overview of techniques for solving multiobjective mathematical programs. *Management Science*, **30**(11), 1268–82.

Fourman, M.P. (1985) Compaction of symbolic layout using genetic algorithms. Proceedings of International Conference on Genetic Algorithms and their Applications, 1985, pp. 141–53.

Goldberg. D.E. (1987) Computer-aided gas pipeline operation using genetic algorithms and machine learning parts I and II. *Engineering with Computers*, 35–45.

Goldberg, D.E. (1989) *Genetic Algorithms in Search, Optimization and Machine Learning*, Addison-Wesley, Reading, MA.

Goldberg, D.E. (1990) A note on Boltzman tournament selection for Genetic algorithms and population-oriented simulated annealing. *Complex Systems*, (4), 445–60.

Hwang, C.L. and Masud, A.S.M. (1979) *Multiple Objective Decision-making Methods and Applications*, Springer, Berlin.

Kuhn, H.W. and Tucker, A.W. (1951) Nonlinear programming. Proceedings of the Second Berkeley Symposium on Mathematical Statistics and Probability (ed. J. Neyman), University of California, Berkeley, CA, 1951, pp. 481–91.

Pramila, A. and Virtanen, S. (1986) Surface of minimum area by FEM. *International Journal of Numerical Methods in Engineering*, **23**, 1669–77.

Sandgren, E. and Jensen, E. (1992) Automotive structural design employing a genetic optimization algorithm. *SAE Tech. Pap. Ser. 920772*.

Sandgren, E. and Venkataraman, S. (1989) Straight line path generation for a planar robotic manipulator. Proceedings of the First National Applied Mechanisms and Robotics Conference, Cincinnati, OH, November 1989 Vol. 1, pp. 3A-2-1–3A-2-7.

Sandgren, E. and Wu, S.J. (1989) Shape optimization using the boundary element method with substructuring. *International Journal of Numerical Methods in Engineering*, **26**(9), 1913–24.

Schmidt, L.A. (1960) Structural design by symmetric synthesis. Proceedings of the Second ASCE Conference on Electronic Computation, Pittsburgh, PA, 1960, pp. 105–22.

Sheppard, K.H. and Palmer, A.C. (1972) Optimal design of a transmission tower by dynamic programming. *Computers and Structures*, **2**(4), 445–68.

Shupe, J.A. and Mistree, F. (1987) Compromise: an effective approach for the design of damage tolerant structural systems. *Computers and Structures*, **27**(3), 407–15.

Stadler, W. (1979) A survey of multicriteria optimization or the vector maximum problem. *Journal of Optimization Theory and Applications*, **29**, 1–52.

Vecchi, M.P. and Kirkpatrick, S. (1983) Global wiring by simulated annealing. *IEEE Transactions on Computer Aided Design*, **2**(4).

8

Optimization of controlled structures

N.S. KHOT

8.1 INTRODUCTION

The topic of this chapter is to introduce the subject of designing structural and control systems simultaneously by using an integrated approach. The fields of structural optimization and optimum control have reached a stage of maturity. The methods developed under these fields have been extensively used in different engineering fields. However, the development of a design procedure to combine structural and control design has been quite recent and it is emerging as a new field in the integrated approaches of the design process. The literature shows that there are various formulations proposed by different investigators. The problem is highly nonlinear because of the nature of the objective functions and the constraints and therefore it can be solved only by using iterative numerical techniques even for a small problem with few design variables. For fairly large problems, obtaining a solution becomes computationally intensive. The optimum solutions obtained by an integrated process may be nonunique because of the nonlinearity of the problem.

The integration of the structural and control design processes is necessary for flexible structures where the effects of control forces react with the dynamic behavior of the structure and it is required to improve overall performance of the system. For rigid structures, integration of the design process might not yield any advantage. From the references cited in this chapter it can be observed that the integrated design approaches are being developed for civil engineering, mechanical engineering, and aerospace engineering applications even though the maximum work may be found related to flexible space structures. These structures must have minimum weight and need an active control system to suppress vibration during maneuvering. In an increasingly competitive world, it is not enough to design a system which can perform well but it is essential to have the best operating system, and this can be achieved only by using integrated methods, making maximum use of the available numerical design tools and the use of computers.

The conventional approach in designing a controlled structure is to optimize the structure alone and then to design a control for the optimum structural design. This process can then be reiterated until the acceptable design is obtained. If the structure is rigid, then it may not be necessary to iterate the design process because the control system would not significantly interact with the structural dynamic behavior. This can be achieved by limiting the upper frequency of the controller bandwidth to be less than the fundamental frequency of the structural vibrations. However, if the structure is flexible then the control system and structure would interact and the control system would affect the vibration characteristics of the structure and also introduce loads which would create stresses and displacements beyond the allowable limits.

The objective in structural optimization generally is to design a structure with the objective of minimizing the cost or weight with constraints on stresses, displacements, buckling load, frequency distribution, minimum and maximum size requirements etc. subjected to static and dynamic loads. The nature of the loading depends on the functionality of the structure. The design variables would be geometry, cross-sectional areas of the members and materials. In the case of the control design, the constraints would be on the closed-loop frequency distribution, active damping, total available energy, maximum output response, robustness characteristics, and other quantities affecting the performance of the control. The control design variables would be gains, number and location of actuators and sensors, and weighting parameters. A proper selection of the control approach is one of the important design considerations. In the integrated design formulation, the objective functions, constraints and design variables would be selected from both disciplines and treated at the same time.

Most of the structures at present are designed by using finite element methods leading to the large degrees of freedom. The main input to the control design from the structural response is the vibration frequencies and the modes. The control is normally designed for a small number of degrees of freedom. This makes it essential to take into consideration the effect of low order control on the higher structural vibration modes. This is known as the spillover effect.

In the case of a flexible structure, the principal source of concern is the interaction between structure and control due to the control bandwidth overlapping the structural modes. One of the solutions would be to design a stiffer structure, but this adds weight to the system. The best or optimal balance between the structural and control design is essential. To find this balance, the conventional approach of compartmental design is not efficient. A design method which simultaneously takes into consideration the objectives of the structural and control design and satisfies requirements for both disciplines is necessary.

The process of obtaining an efficient design can be considered as a mathematical optimization problem, where single or multiple objective functions are to be minimized and constraints satisfied. The number of design variables and constraints may be large depending on the complexity of the problem.

The standard constrained optimization problem can be stated as follows. Find a vector $\{a\} = (a_1 \, a_2 \cdots a_{n_1})$ of n_1 design variables in order to minimize a cost function

$$f(\boldsymbol{a}) = f(a_1, a_2, \ldots, a_{n_1}) \tag{8.1}$$

subject to p_1 equality constraints

$$h_j(\boldsymbol{a}) = h_j(a_1, a_2, \ldots, a_{n_1}) = 0 \quad j = 1, \ldots, p_1 \tag{8.2}$$

and m_1 inequality constraints

$$g_i(\boldsymbol{a}) = g_i(a_1, a_2, \ldots, a_{n_1}) \leqslant 0 \quad i = 1, \ldots, m_1 \tag{8.3}$$

where p_1 is the total number of equality constraints and m_1 is the total number of inequality constraints. Any design problem can be transcribed into a mathematical model. The method of obtaining the solution depends on the linearity and nonlinearity of the objective function and the constraints. The sensitivities of these quantities with respect to the design variables are required. If they cannot be evaluated analytically, it may be necessary to use a finite difference scheme. Except for a few cases, the constraints can be expressed as an algebraic expression in terms of the design variables. Most of the design problems need elaborate numerical calculations before the objective function, constraints or sensitivities can be calculated. In the definition of the problem in equations (8.1)–(8.3), it was assumed that there is only one objective function to be minimized. One of the purposes of optimization is to enhance one's understanding of the real problem to be modeled.

The first step in formulating a real problem is the selection of the proper objective functions and equality and inequality constraint functions that must be used to define the design problem. In practical design problems, it may be even necessary to consider more than one objective function. In this case, methods such as Pareto optimization or goal programming might have to be used. A detailed discussion on optimization methods may be found in books by Morris (1982), Arora (1989), and Haftka, Gurdal and Kamat (1990).

The following section contains a summary of the literature published in this field. The section on analysis contains basic dynamic equations of a structure, the linear quadratic regulator (LQR) control theory and the concept of robust control. Books by D'Azzo and Houpis (1988) and Meirovitch (1989) contain in-depth discussion on the control analysis. The sensitivities of the closed-loop eigenvalues and the spectral radius are then derived. The application of the integrated approach is illustrated by solving two problems. This is followed by a short discussion on the optimization programs NEWSUMT-A (Thareja and Haftka, 1985), ADS (Vanderplaats, Sugimoto and Sprague, 1983), IDESIGN (Arora, Thanedar and Tseng, 1985), and VMCON (Crane, Hillstrom and Minkoff, 1980) available in the public domain and are frequently used for simultaneous design of the structure and control systems.

8.2 LITERATURE REVIEW

A literature search on the simultaneous design of the structure and control system indicates that most of the papers in this field have been published within the last fifteen years. Research work reported in these references differs in the selection of the objective functions, nature of constraints, control theory and the approximations made in developing the algorithms. Most of the research work is oriented towards understanding of the problem and solving it within a limited scope of the objective functions, number of constraints and design variables, and the type of control.

Kirsch and Moses (1977) have presented the idea of structural optimization with control. For optimization, four different types of control systems were studied: (a) passive system without control forces, (b) passive system with pre-stressing forces, (c) passively controlled system, and (d) actively controlled system. Each method is applied to a 'continuous' beam with intermediate springs and supports. The beam cross-sectional area is minimized subject to compatibility, equilibrium and stress constraints. The paper is intended for civil engineering beam structures.

Haftka, Martinovic and Hallauer (1984) minimized the single rate feedback viscous damping coefficient subject to limits on the real part of the closed-loop system eigenvalues and constraints on the magnitude of design variable changes. A finite element model was developed for a laboratory structure consisting of a beam and cables. Sensitivities of eigenvalues and damping rates were determined analytically with respect to changes in concentrated mass and element thicknesses. The experimental and theoretical damping and frequency values were compared qualitatively for changes in the mass.

A two-step method of determining the optimal control cost for a given set of design variables and minimizing the linear control cost by iteratively modifying the design variables was presented by Messac and Turner (1984). A modal formulation with reduced order modeling was implemented. The constant mass was the constraint requirement on the design. The method was applied to 20 extensional (only) finite elements of equal length to study the effect of reduced order modeling and to allow physical interpretation of the results. The dynamic response of the optimal design was found to be smoother than the initial design with all design variables of the same size.

Khot, Venkayya and Eastep (1984) discussed the effect of optimum structural modifications on the dynamic behavior of the ACOSS-FOUR structure with active controls. The nominal design of the ACOSS-FOUR was modified by using two optimization approaches. In the first case, the static displacements associated with the line-of-sight error were minimized by changing the distribution of the cross-sectional areas of the members for the same weight of the structure as the nominal design. In the second case, a constraint was imposed on the fundamental frequency and the weight of the structure was minimized. The dynamic responses of different optimum designs were then compared. The weighting matrix for state variables used for the LQR control is a function of the structural frequencies. The design, with a constraint on the displacements, showed improvement in the response of the structure to the initial disturbance.

Hale, Lisowski and Dahl (1985) have considered an idealized four-boom structure attached at the center to a rigid hub. The structure was modeled by finite elements to obtain an optimal design for a rest-to-rest (two-point boundary value problem) single-axis rotational maneuver (slew maneuver) due to reduced order optimal control. For the optimization formulation, a weighted nonnegative structural cost (independent of control parameters) and the state vector cost were augmented by penalties due to control derivatives. The optimum structural parameters are obtained by overall cost minimization with control forces and their derivatives treated to be independent of the parameters. The combined structural and control optimization was solved iteratively for the finite time maneuver.

A set-theoretic analysis procedure was described by Hale (1985) for the integrated structural–control synthesis. The procedure approximates box (absolute magnitude) constraints on the state vector, output, and control forces by ellipsoidal sets. Matrices in the ellipsoidal sets are governed by a matrix differential equation; the steady state solution is correspondingly solved by a Lyapunov equation. The input disturbance is assumed to be bounded with an amplitude bound. The set-theoretic analysis procedure providing steady state bounds directly replaces the computations required in a transient analysis procedure. The procedure was applied to a single finite element with structural and nonstructural end masses undergoing only axial motion. The design variable was chosen as the cross-sectional area which is analytically related to the first flexible natural frequency.

Bodden and Junkins (1985) systematically presented the pole placement type eigenvalue optimization formulation for a linear output feedback including the position, velocity and acceleration measurements. The sensitivities – first and second derivatives – of the closed-loop eigenvalues were obtained. The method was applied to the Draper–RPL configuration with four flexible beams attached to a hub at the center. A constrained optimization strategy was developed. Three cases of eigenvalue placement were considered. The objective of all the cases was to minimize the control gains subjected to constraints.

Structural optimization of an optimally controlled structure was performed by Khot, Venkayya and Eastep (1986) using an LQR control law and the VMCON optimization subroutine based on Powell's algorithm. The objective was to minimize the structural weights with equality constraints on the closed-loop damping ratio and lower bounds on cross-sectional areas. The procedure was applied to a two-bar and the ACOSS-FOUR trusses. The gradients of the weight were determined analytically, whereas those of the constraints were evaluated by using a finite difference scheme. The line-of-sight (LOS) error and performance index (PI) time histories show considerable improvements as a result of optimally controlling an optimum structure. Lower modes are more heavily damped.

Eastep, Khot and Grandhi (1987) studied the effects of different percentages of passive damping on an actively controlled structure with minimum weight objective with specific application to the ACOSS-FOUR model. Imaginary parts of the closed-loop eigenvalues and closed-loop modal damping factors for LQR modal formulation were constraints. Analytical sensitivities were incorporated in NEWSUMT-A optimization software. The results include the LOS time histories for various values of passive damping. The effect of percentage damping on the minimum performance index shows an inverse relationship.

A study was performed for weight and/or Frobenius norm minimization on a two-bar and the ACOSS-FOUR trusses by Khot, Grandhi and Venkayya (1987). Constraints were imposed on the damping factors and imaginary part of the closed-loop eigenvalues. Analytical sensitivity expressions were used in the VMCON, NEWSUMT and IDESIGN optimization routines. Comparative evaluation of these routines was tabulated. The main objective of this paper was to indicate the different performance of various optimization packages when used as black boxes. NEWSUMT-A was found to give a better optimum.

Lim and Junkins, (1987) presented a design algorithm using robustness measures. The three cost functions, total mass, stability robustness and eigenvalue sensitivities were minimized with design variables which included structural as well as control parameters. The locations of actuators and sensors were also considered as design variables. The robustness measure proposed by Patel and Toda was used. The optimization algorithm based on the homotopy approach with sequential linear programming algorithms was utilized. A hypothetical structure consisting of a free–free flexible beam with rigid body attached to the center of the beam by a pinpoint and torsional spring was designed with different objective-functions and constraints.

Miller and Shim (1987) performed minimization of combined structural mass and controlled system energy. The objective functions were weighted by positive scalar weights to balance the dimensions. The major emphasis was on the gradient projection and penalty function technique. A ten-bar cantilevered truss was designed for two initial conditions and different constraints on the lowest open loop frequency.

The actual cost of the structure and control system was considered as a non-dimensional objective function by Onoda and Haftka (1987) with constants representing a normalizing mass and weightings. A beam-like spacecraft experiencing a steady state white noise disturbance force was optimized for stiffness distribution, location of controller and control gains. The NEWSUMT-1 program was employed for the unconstrained minimization. The sensitivity derivatives were calculated by finite differences. Both noninertial and inertial disturbances were chosen. The results showed improvements as a result of the simultaneous optimization.

Bendsoe, Olhoff and Taylor (1987) formulated a system design problem based on the criterion to minimize the spillover effect with respect to the stiffness and distribution and variations in feedback gains and actuator positions. Examples of simple mass–spring–damper systems were analytically solved to illustrate the basic approach of structural and control design. It was shown that in the distributed parameter structure the technique of modal control gave rise to design problems which are similar to eigenfrequency optimization. The sensitivities were calculated by using analytical expressions.

Junkins and Rew (1988) discussed the sources of uncertainties or imperfectness in the modeling process. These are (1) ignorance of actual properties, geometric parameters and boundary conditions, (2) discretization and truncation error, (3) ignoring the nonlinearities, (4) ignorance of disturbances – random or deterministic, and (5) sensors and actuator modeling. With this background, the robust eigenstructure assignment methods were discussed. A multicriterion approach was presented to study the trade-offs between robustness and competing performance measures.

Mesquita and Kamat (1988) considered the effect of structural optimization on the improvement of the control design. A laminated composite plate structure was optimized with constraints on the frequency distribution and the structural weight. Two optimization objectives were considered. In one case the fundamental frequency was maximized with separation constraints on other frequencies. In the second case the sum of the square of the differences between the structural frequencies and a certain target frequency was minimized. The design variables were the number of plies in a specified orientation and the areas of the stiffeners. The problem was solved with VMCON.

Belvin and Park (1988) showed that the LQR control with stiffeners and mass dependent weighting matrices to the state and control variable and the full rank actuator influence matrix have a closed-form solution to the Riccati equation. The (undamped) approximate cost function was explicitly stated in terms of structural parameters, two scalar gain variables and the assumed model force coefficients. This objective function was found to be inversely proportional to the frequency cubed suggesting that the low frequency modes contribute most to the cost index. Minimization of the function with constraints on mass was considered for a clamped–pinned beam and a cantilever beam model The constrained optimization was performed using the ADS program.

Khot (1988a) presented two weight minimization problems differing in the constraints and the structure used for the analysis. One, with constraints on the first two closed-loop eigenvalues and four constraints on the closed-loop damping, was applied to a tetrahedral truss, and another with constraints on closed-loop eigenvalues as well as the control gain norm, specifically the Frobenius norm, was applied to a two-bar truss.

Khot (1988b) considered minimization of weight of the structure subjected to constraints on closed-loop modal damping parameter and imaginary part of the closed-loop systems is achieved. A modal formulation with LQR control law is used. Structural optimization is performed by using VMCON routine. Sensitivity derivatives required for the procedure are evaluated analytically. Comparison of nominal and optimum integrated design is made by observing LOS transient response and performance index time history. In addition, frequencies, closed-loop eigenvalues, and the modal damping parameters are also compared.

Khot *et al.* (1988) developed a simultaneous structural–control optimization procedure in which weight as well as the control effort measure, Frobenius norm, were minimized subject to constraints on the closed-loop eigenvalues, with lower bound on cross-sectional areas and upper bound on the norm. A linear full state feedback was used with the state space formulation. Cross-sectional areas and gains were considered as design variables to be evaluated by the simultaneous minimization of the weight and Frobenius norm. The gradients of the objective functions and the constraints were evaluated analytically. The procedure was applied to a two-bar truss.

Khot and Grandhi (1988) discussed the existence of a nonunique solution to this highly nonlinear problem in this paper. The minimization of weight or Frobenius norm was conducted with the LQR control law and modal state space form. Four different optimum designs of the ACOSS-FOUR model were obtained differing in the constraints and objective function. It was indicated that different designs could be obtained satisfying all the constraints with different distributions of design variables. NEWSUMT-A was used as a black box for optimization.

A combined optimization of the structure and control system was presented by Salama *et al.* (1988) with the sum of the weight and the control system performance index as an objective function. An LQR optimal control was scaled to match the dimensions of the objective function for combined optimization. With this operation, the problem is locally converted to the minimization of a function with only the structural design variables present. The constrained optimization was converted to the unconstrained optimization by the use of Lagrange multipliers. The gradients are all evaluated analytically. The procedure was applied to a cantilever beam with a tip control force.

Cheng and Pantelides (1988) discussed structural optimization with (i) non-optimal control, (ii) optimal control, (iii) critical mode control, and (iv) optimal location of controllers applied to the civil engineering/building models. The control was achieved through the active tendons and active mass damper, and the structure was subjected to a random earthquake acceleration. For the non-optimal control, a transfer matrix approach was used in the frequency domain. The optimal control–structural optimization implementation scheme was classified into (i) instantaneously open loop in which the controller nullifies the earthquake excitation in the time domain, (ii) instantaneous closed loop in which the control forces are regulated by the full state feedback at the instant, and (iii) instantaneous open–closed loop control in which the effect of excitation is nullified and the instantaneous control-response performance index is minimized. The problem of weight minimization with the displacement and control forces was solved for an eight-story building using the three different instantaneous controls.

Rao (1988) considered the problem of combined structural and control optimization with different objective functions. The objective functions selected for these parametric studies were structural weight, quadratic performance index, Frobenius norm and the effective damping response time. Designs for a two-bar truss and ACOSS-FOUR were obtained by minimizing one objective function and constraining other functions. The sensitivities of the objective functions to the design variables were investigated. The optimization was carried out by using the VMCON subroutine.

A cooperative game theory approach was described by Rao, Venkayya and Khot (1988) for a combined structure–control optimization problem. The multiobjective functions were structural weight, the quadratic performance index for LQR, the Frobenius norm of the control effort, and the effective damping response time. The objective was to obtain a Pareto-optimal set combining these functions for specified constraints on the closed-loop damping ratios and bounds on the cross-section areas of truss members. The method was applied to a two-bar truss and a twelve-bar ACOSS-FOUR truss. The required sensitivities were determined numerically. The optimization was carried out by using VMCON.

Structural and controller parameter optimization was considered by Lim and Junkins (1989) in the presence of system uncertainty with three independent cost functions. These were expressed as (i) eigenvalue sensitivity minimization, (ii) stability robustness maximization, and (iii) minimization of mass. Direct output feedback controllers were exercised on a rigid body with an attached flexible beam. The eigenvalue sensitivity objective function was expressed as the weighted quadratic sum of the individual eigenvalue sensitivities. The upper bound of the robustness measure was defined as the minimum of the negative real part of the closed-loop state matrix eigenvalues. The design variables were the thickness of the flexible beam and the mass density of the rigid body. The constraints were imposed on the actuator location plant parameters, eigenvalues and local step size allowables for the structural parameters.

Manning and Schmit (1990) considered the composite objective function involving the structural weight and control effort wth behavior constraints and constraints on weight and control effort separately for the two cases. A second-order modal formulation including excitation forces and control forces was used and the equations were integrated by the Wilson-O explicit time integration method. The first two cases of the planar grillage involved weight minimization with control effort constraint, whereas the last two were for control effort minimization with constraint on weight. The constraints were imposed on the displacements and accelerations at some locations. The limitations on dynamic response and actuator force levels were satisfied by treating them as direct behavior constraints.

Onoda and Watanabe (1989) proposed a method to design an LQG type optimal controller with regulator and observer the reduce the spillover effect. A method was proposed to design an LQG type controller with a regulator and observer by minimizing the performance index involving controlled modes and residual modes. The approach consisted of two steps; first, the maximum value of the real parts of the eigenvalues of the closed-loop matrix is minimized, then the minimization of the performance index is carried out. The integrated approach to design the control and structural system was presented in the end where a total cost objective function, which represents the sum of the control cost and structural cost, was minimized with constraints on the response amplitude. The design variables were the observer and regulator gains, and the cross-sectional areas.

Thomas and Schmit (1989) studied the effect of noncollocated sensor–actuator pairs by applications to a cantilever beam and the Draper–RPL (four beams with central hub) structure. A direct output feedback type of controller was used which essentially modifies the effective damping and stiffness properties of the structure leading to a free vibration type of problem. The minimization of either mass or the control effort was performed along with the constraints on the control force, open-loop frequency, closed-loop eigenvalues, dynamic displacements and the damping ratio, appropriately chosen for different cases. Structural parameters and control gains were treated as the design variables.

Grandhi (1989) studied the performance of optimization programs VMCON, NEWSUMT and IDESIGN to design the ACOSS-FOUR structure with constraints on closed-loop eigenvalues and damping. The objective function was the weight of the structure. The objective of this paper was to study the performance of various optimization programs.

Rao, Pan and Venkayya (1990) discussed a multiobjective optimization for improving the robust characteristics of an actively controlled structure. This was achieved by structural modifications with a modal space formulation and LQR control law. Both the performance and stability robustness were considered and appropriate measures defined. The objective functions chosen were the stability robustness, the performance robustness, and the total structural weight which were simultaneously optimized.

Haftka (1990) presented an overview of some of the relevant issues of the integrated structure–control optimization of space structures, namely the accuracy of the sensitivity derivatives, effects due to reduced ordered models, different objective functions and integrated design formulations.

Thomas and Schmit (1990) proposed the integrated design problem for the objective function equal to the weighted sum of the masses, a sum of the amplitudes of the dynamic responses and the sum of the amplitude of the control forces. A noncollocated direct output feedback control was applied to a cantilever beam and a structure having four arms and a central rigid hub. The approximate optimization problem was solved using CONMIN. The design variables used were the thicknesses of different elements, concentrated masses and the controller gains and positions.

Padula *et al.* (1990) applied controls–structures optimization methodology to a general space structure synthesizing the analysis, optimization and control software. The problem was decomposed into a structural optimization, control optimization and system optimization in which both the optimal structural and control steps were included leading to a minimization of mass and power consumption. The constraints included limits on the frequency spacing, fundamental bending frequency, buckling loads, damping factors etc. The design variables were treated at different levels.

Zeiler and Gilbert (1990) proposed a method based on Sobieski's multilevel decomposition approach. This decomposed the problem into subsystem optimization and integrated the effects for the structure–control optimization. The formulation incorporated a structural nodal form and LQR control. The ADS optimization code was used for the top level and structural subsystem, and control law subsystem optimization was achieved using the ORACLS software. The method was applied to the ACOSS-FOUR structure to investigate the feasibility of the proposed method.

Grandhi, Haq and Khot (1990) considered the optimization of structural parameters and control gains (LQR) by maximizing the stability robustness measure in the presence of parametric uncertainty. The measure was proposed by Qiu and Davidson and is based on the closed-loop stability under structured perturbations. The inverse of the robustness was chosen to be proportional to the maximum spectral radius of a matrix related to the closed-loop nominal matrix. The NEWSUMT-A optimization software was used with the finite difference approach to calculate the sensitivities.

Sepulveda and Schmit (1990) formulated a standard form of the modal state space equations with an optimal regulator and observer, and time-dependent external forces. A composite objective function was decomposed into a control subproblem along with a standard structural optimization on the outer loop. The multiobjective functions involving target values of mass control efforts and number of actuators was considered. The optimal placement of actuators and sensors was determined by the (0, 1) design variables in the control input and measurement matrices. The concepts were applied to a cantilever beam problem and an antenna structure.

McLaren and Slater (1990) solved the problem of minimizing the mass of the structure subject to a set of prescribed stochastic disturbances with limitations on the available control energy and a set of allowable output responses. The three controller types considered were (1) direct output feedback control, (2) output filter feedback control, and (3) positive real output feedback control. The mathematical optimization method based on continuation methods to impose nonlinear constraints coupled with sequential linear programming was used to solve the illustrative problems.

Khot and Veley (1990) considered integrated structure–control optimization under the presence of structured uncertainties in the plant matrix. A full state LQR control law was implemented with the formulation given in modal form. The sensitivity of various quantities with respect to the design variables was expressed analytically. Weight minimization, with constraints on the imaginary part of closed-loop eigenvalues, damping factors, spectral radii bound, and cross-sectional areas, was performed by using the NEWSUMT-A software. The method was applied to the tetrahedral truss.

Suzaki and Matsuda (1990) applied a goal programming approach to an active flutter suppression system to perform combined structure–control optimization. A typical wing section with the control surface undergoing a plunging pitching motion was modeled in an unsteady aerodynamic field and a random gust velocity disturbance. A sensitivity analysis for all the dependent variables such as eigenvalues, gains and covariance matrix with respect to the design variables was performed analytically. The design variables were the design airspeed, elastic axis to mass center distance, hinge line to control surface mass center, elastic axis location and the hinge line location. The constraints were imposed on the open-loop characteristics, closed-loop characteristics, control surface deflection angle and structural design changes.

Grandhi (1990) performed weight minimization of the ACOSS-FOUR subject to constraints on the imaginary part of closed-loop eigenvalues and closed-loop modal damping parameters. The objective was to study the quadratic performance index, optimum weight, and active control effort as a function of the passive damping. The range of passive damping chosen was 0–10%.

A model error sensitivity suppression method was implemented for a structure–control optimization problem by Khot (1990) with weight as the primary objective function. The reduced order LQR optimal control problem was then solved for weight minimization with constraints on the closed-loop damping factors, frequencies and structural parameters. The NEWSUMT-A computer package was used to solve the constrained minimization problem. The sensitivities of the weight, damping factor, closed-loop eigenvalues, closed-loop matrix, Riccati matrix, structural frequencies and eigenvectors were evaluated analytically.

Khot and Veley (1991) studied the effect of structured uncertainties on the performance of an optimally controlled closed-loop structure. Weight minimization with constraint on the imaginary part of the closed-loop eigenvalues and on the spectral radius was achieved by using the NEWSUMT-A program. The cross-sectional areas were considered as structural design variables and parameters multiplying the weighting matrices in the quadratic performance index were treated as control design variables. Sensitivity expressions with respect to these design variables were calculated analytically.

Dracopoulos and Öz (1992) proposed an integrated approach for the aeroelastic control of laminated composite lifting surfaces. The lifting surface was modeled as a laminated composite cantilevered plate in an airstream. The geometry of the plate and the ply thickness were assumed constant with ply orientation angles as the structural design variables, and the control design speed was used as a control variable. The optimization problem was solved by using optimization routines, IDESIGN and NEWSUMT-A. The problem formulation incorporated the Rayleigh–Ritz energy method and two-dimensional incompressible unsteady aerodynamic theory.

8.3 ANALYSIS

The basic equations needed for the dynamic analysis of a structure, control design, and sensitivities of the response functions are derived in this section.

8.3.1 Dynamic equations of motion

The dynamic equations of motion for an elastic system with finite degrees of freedom with no external disturbance can be written as

$$\mathbf{M}\ddot{u} + \mathbf{E}\dot{u} + \mathbf{K}u = \mathbf{D}f \tag{8.4}$$

where **M** is the mass matrix, **E** is the proportional damping matrix, and **K** is the total stiffness matrix. These matrices are $n = n$ where n is the number of degrees of freedom of the structure. In equation (8.4), u and f are the displacement and control force vectors of dimensions n and p respectively, and are functions of time. The matrix **D** is $n \times p$ and its elements consists of direction cosines and zeros. This matrix relates the control force vector f in its local coordinates to the global coordinate system in which the equilibrium equations are written.

Introducing the coordinate transformation

$$u = [\phi]\eta \qquad (8.5)$$

where η is the modal coordinates and $[\phi]$ is the $n \times n$ modal matrix, equation (8.4) can be transformed into n uncoupled equations. The vector η is a function of time. The columns of the square matrix $[\phi]$ are the normalized eigenvectors of the homogeneous set of equations obtained by setting the right-hand side equal to zero in equation (8.4) and assuming the dampling matrix **E** is also zero. The uncoupled equations can be written as

$$\bar{\mathbf{M}}\ddot{\eta} + \bar{\mathbf{E}}\dot{\eta} + \bar{\mathbf{K}}\eta = [\phi]^{\mathrm{T}}\mathbf{D}f \qquad (8.6)$$

where

$$\bar{\mathbf{M}} = \mathbf{I} = [\phi]^{\mathrm{T}}\mathbf{M}[\phi] \qquad (8.7)$$

$$\bar{\mathbf{E}} = [2\zeta\omega] = [\phi]^{\mathrm{T}}\mathbf{E}[\phi] \qquad (8.8)$$

$$\bar{\mathbf{K}} = [\omega^2] = [\phi]^{\mathrm{T}}\mathbf{K}[\phi] \qquad (8.9)$$

The matrices $\bar{\mathbf{M}}$, $\bar{\mathbf{E}}$ and $\bar{\mathbf{K}}$ are diagonal square matrices. ω is the vector of structural frequencies which is equal to the eigenvalues of the homogeneous equations of motion. ζ is the vector of modal damping. Equations (8.6) are second-order uncoupled equations.

8.3.2 Control analysis

A system in control theory is defined as an assemblage of components that act together to perform a certain function. The dynamics of the system is the response of the system to the specified disturbance. The system is also known as a plant and the disturbance is termed as input and the response of the system as an output. A typical uncontrolled system which generally occurs in nature is shown in Fig. 8.1, where we have no control of the output. A good example of the uncontrolled system is the behavior of a building during an earthquake, where the building represents the plant and the motion of the foundation is the input. In a controlled system, shown in Fig. 8.2, the desired output acts as an input to the controller and the output from the controller acts as an input to the plant.

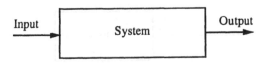

Fig. 8.1 Uncontrolled system.

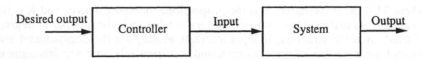

Fig. 8.2 Controlled system.

A good example of this open-loop control system is a heating system in a building which is set to start or shut down at specific times without consideration of the temperature inside the building. A feedback or closed-loop heating system is one in which starting and shutting down of the system is controlled by temperature sensors. The closed-loop system is shown in Fig. 8.3.

In control theory, the equations describing the behavior of the system are written in terms of state variables. The state of a system is a mathematical statement containing state variables, such that the initial values of these state variables and the system inputs at a specified time are sufficient to uniquely describe the behavior of the system in the future. The state equations of the system are a set of first-order differential equations equal to the number of independent states.

The second-order uncoupled equations (8.6) can be reduced to first-order equations by using the transformation

$$x_{2n} = \begin{bmatrix} \eta \\ \dot{\eta} \end{bmatrix}_{2n} \tag{8.10}$$

where x is the state variable vector of size $2n$. This gives

$$\dot{x} = Ax + Bf \tag{8.11}$$

where A is a $2n \times 2n$ plant matrix and B is a $2n \times p$ input matrix and f is $p \times 1$ control vector. The plant and input matrices are given by

$$A = \left[\begin{array}{c|c} 0 & I \\ \hline -\omega^2 & -2\zeta\omega \end{array} \right] \tag{8.12}$$

$$B = \left[\begin{array}{c} 0 \\ \hline \phi^T D \end{array} \right] \tag{8.13}$$

In equation (8.12) 0 is a null matrix and I is a diagonal matrix with elements equal to unity. The bottom left quadrant consists of diagonal elements equal to

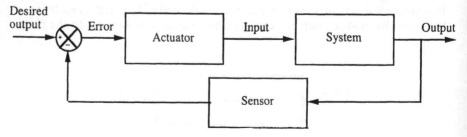

Fig. 8.3 Closed-loop system.

the negative of squared frequencies. The right bottom quadrant also contains only diagonal elements equal to the negative of the product of the frequencies and associated modal damping parameters multiplied by 2. In equation (8.13) $\mathbf{0}$ is also a null matrix. Equation (8.11) is known as the state input equation.

In a closed-loop system, the control forces depend on the system behavior. The behavior can be interpreted as measured outputs. The relation between the output vector y of dimension q and input vector x can be expressed as

$$y = \mathbf{C}x + \mathbf{D}f \qquad (8.14)$$

where $\mathbf{C}$ and $\mathbf{D}$ are $q \times 2n$ and $q \times p$ matrices respectively. Most of the time the output does not depend on the input so that equation (8.14) reduces to

$$y = \mathbf{C}x \qquad (8.15)$$

If the number of sensors measuring y and actuators generating forces f are equal and they are collocated, then $q = p$ and

$$\mathbf{C} = \mathbf{B}^\mathsf{T} \qquad (8.16)$$

A system is completely state controllable if for any initial time, each initial state can be transferred to the final state in a finite time with an unconstrained input vector. Mathematically, a system is completely controllable if the rank of the controllability matrix

$$\mathbf{C} = [\mathbf{B} \colon \mathbf{AB} \colon \mathbf{A}^2\mathbf{B}^1 \colon \cdots \colon \mathbf{A}^{n-1}\mathbf{B}] \qquad (8.17)$$

has n independent columns implying it is of rank n. A system is considered to be observable if every initial state can be exactly determined from the measurements of the output over the final time. Similar to the mathematical definition of controllability, a system is completely observable if the rank of the observability matrix defined as

$$\phi = [\mathbf{C}^\mathsf{T} \colon \mathbf{A}^\mathsf{T}\mathbf{C}^\mathsf{T} \colon (\mathbf{A}^\mathsf{T})^2\mathbf{C}^\mathsf{T} \colon \cdots \colon (\mathbf{A}^\mathsf{T})^{n-1}\mathbf{C}^\mathsf{T}] \qquad (8.18)$$

has n independent columns implying ϕ is of rank n.

The stability of the system is determined by the real part of the eigenvalues of the plant matrix for the open-loop and closed-loop system. The real eigenvalues can be considered as complex eigenvalues with zero imaginary parts. In order for the system to reach stable equilibrium, all eigenvalues must have negative real parts. Even if one eigenvalue has a positive real part, the system would be unstable. A motion in the x directon cannot be controlled by a force in the y direction is a simple example of controllability. Similarly, a motion in the x direction cannot be detected by a sensor in the y direction illustrates the condition of observability.

8.3.3 Optimal control

The objective of optimal feedback control is to determine the control which can transfer the system from initial to final state minimizing the performance index (PI) defined as

$$J = \int_0^\infty (x^\mathsf{T}\mathbf{Q}x + f^\mathsf{T}\mathbf{R}f)\,\mathrm{d}t \qquad (8.19)$$

where $\mathbf{Q}$ and $\mathbf{R}$ are the state and control weighting matrices respectively. The matrix $\mathbf{Q}$ must be positive semidefinite ($x^T\mathbf{Q}x \geqslant 0$) and $\mathbf{R}$ must be positive definite ($f^T\mathbf{R}f > 0$). The dimensions of $\mathbf{Q}$ and $\mathbf{R}$ depend on the size of vectors x and f respectively. The selection of the elements of $\mathbf{Q}$ and $\mathbf{R}$ determine the amount of closed-loop damping and the time required to control the disturbances. The performance index represents a compromise between minimum error and minimum energy criteria. The result of minimizing the performance index and satisfying equation (8.11) gives the state feedback control law

$$f = -\mathbf{G}x \tag{8.20}$$

where $\mathbf{G}$ is the optimum gain matrix given by

$$\mathbf{G} = \mathbf{R}^{-1}\mathbf{B}^T\mathbf{P} \tag{8.21}$$

where $\mathbf{P}$ is a positive definite matrix called the Riccati matrix and is obtained by the solution of the algebraic Riccati equation

$$0 = \mathbf{Q} + \mathbf{P}\mathbf{A} + \mathbf{A}^T\mathbf{P} - \mathbf{P}\mathbf{B}\mathbf{R}^{-1}\mathbf{B}^T\mathbf{P} \tag{8.22}$$

The open-loop system is given by

$$\dot{x} = \mathbf{A}x \tag{8.23}$$

and the closed-loop system is given by

$$\dot{x} = \bar{\mathbf{A}}x \tag{8.24}$$

where

$$\bar{\mathbf{A}} = \mathbf{A} - \mathbf{B}\mathbf{G} \tag{8.25}$$

Equation (8.24) can be obtained by substituting equation (8.20) in equation (8.11). The eigenvalues of the closed-loop matrix $\bar{\mathbf{A}}$ are a set of complex conjugate pairs written as

$$\lambda_i = \tilde{\sigma}_i \pm j\tilde{\omega}_i \qquad i = 1, \ldots, n \tag{8.26}$$

where $j = (-1)^{1/2}$. The sign of $\tilde{\sigma}_i$ must be negative for all i in order to make the system asymptotically stable. The closed-loop damping factors are given by

$$\xi_i = -\frac{\tilde{\sigma}_i}{(\tilde{\sigma}_i^2 + \tilde{\omega}_i^2)^{1/2}} \tag{8.27}$$

The magnitudes of ξ_i associated with each mode determine the extent to which those modes are damped.

For the initial condition $x(0)$, the solutions to equations (8.23) and (8.24) are given by

$$x(t) = \exp(\mathbf{A}t)\,x(0) \tag{8.28}$$

and

$$x(t) = \exp(\bar{\mathbf{A}}t)\,x(0) \tag{8.29}$$

where

$$\exp(\mathbf{A}t) = 1 + \frac{\mathbf{A}t}{1!} + \frac{(\mathbf{A}t)^2}{2!} + \cdots \tag{8.30}$$

and

$$\exp(\bar{\mathbf{A}}t) = 1 + \frac{\bar{\mathbf{A}}t}{1!} + \frac{(\bar{\mathbf{A}}t)^2}{2!} + \cdots \tag{8.31}$$

Equations (8.28) and (8.29) can be used to determine the transient response of the open-loop and closed-loop systems.

The performance index for the given initial condition $x(0)$ is given by

$$PI = x(0)^T p x(0) \tag{8.32}$$

The behavior of the control system depends on the distribution of the closed-loop eigenvalues and the damping parameters. For a structural–control system, these values are functions of the weighting matrices $\mathbf{Q}$ and $\mathbf{R}$ of optimal control and the distribution of structural frequencies in the $\mathbf{A}$ matrix which in turn depends on the distribution of the stiffness of the structure or the cross-sectional areas of the members.

8.3.4 Measure of robustness

The control design is called robust if it can perform in an acceptable manner in the presence of uncertainties. The uncertainties are generally divided into two categories, structural and unstructural. The structural uncertainties are due to the variation in the elements of the plant due to the inaccuracies of the structural frequencies and assumed passive damping. These may be due to the elastic properties of the material used in the formulation of the stiffness matrix or the assumptions made in the derivation of the mass and damping matrices. The unstructured uncertainties are due to the unmodeled dynamics. This is due to the control system being designed for a reduced-order system and higher structural frequencies being ignored in the assembly of the plant matrix. In this chapter, we will discuss one of the measures of the structural uncertainty. A discussion on the effect of unmodeled dynamics on the behavior of the structural and control system is beyond the scope of this chapter.

The perturbed closed-loop state space model can be written as

$$\dot{x} = (\bar{\mathbf{A}} + \varepsilon)x \tag{8.33}$$

where $\bar{\mathbf{A}}$ is a stable matrix and ε is a perturbation matrix. The perturbed system would be stable if the elements ε_{ij} of matrix ε satisfy the relation

$$\varepsilon_{ij} < \frac{1}{\sup_{p \geqslant 0} \rho[|(jp\mathbf{I} - \bar{\mathbf{A}})^{-1}|\mathbf{U}_e]} U_{e_{ij}} = \frac{1}{\rho_s} U_{e_{ij}} \tag{8.34}$$

according to the robustness measure proposed by Qiu and Davison (1986). In Equation (8.34) $|(\cdot)|$ denotes an absolute matrix and $\rho[\cdot]$ denotes the spectral radius of the matrix $[\cdot]$. $U_{e_{ij}}$ are the elements of the perturbation identification matrix $\mathbf{U}_e$. SUP represents the supremum of the matrix over a range of p. The elements of $\mathbf{U}_e$ have assigned values depending on the relative perturbations allowed for those elements of $\bar{\mathbf{A}}$. The spectral radius is equal to the maximum modulus of the complex eigenvalues for a specified operating frequency p. The maximum spectral radius value amongst all possible values of p gives the critical value of p_s. The peaks of the spectral radius for small values of damping occur at the values of p equal to the modulus of the closed-loop frequencies of $\bar{\mathbf{A}}$.

The elements of the perturbation matrix ε are proportional to the elements of the perturbation identification matrix $\mathbf{U}_e$ and inversely proportional to the spectral radius ρ_s. The matrix $\mathbf{U}_e$ can be defined as the normalized closed-loop plant matrix $\bar{\mathbf{A}}$. This gives

$$U_{e_{ij}} = \frac{|\bar{A}_{ij}|}{|\bar{A}_{pq}|} \qquad i > n \tag{8.35}$$

where $|\bar{A}_{pq}|$ is the absolute value of a specific element of $\bar{\mathbf{A}}$. The ratio of ε_{ij} and $|\bar{A}_{ij}|$ defined as ε_r for all the elements is given by

$$\varepsilon_r = \frac{1}{\rho_s |\bar{A}_{pq}|} \tag{8.36}$$

The percent allowable deviation in the elements of $\bar{\mathbf{A}}$ is $100\varepsilon_r$. The percent allowable deviation is independent of the selection of the element $|\bar{A}_{pq}|$.

8.4 SENSITIVITY CALCULATIONS

Sensitivity analysis is the determination of changes in the response functions due to changes in the design variables. This can be achieved either by using the finite difference approach which is generally computationally expensive or by calculating partial derivatives. The sensitivities are needed using most optimization programs. If these gradients cannot be calculated by using analytical expressions, then the finite difference approach can be used. The behavior of the structural and control system depends on the eigenvalues and damping parameters of the closed-loop system defined in equations (8.26) and (8.27). In this section, the expressions for the sensitivities of these functions with respect to the structural design variables are given. The structural design variables are the cross-sectional areas of the members. Any change in the cross-sectional areas affects the mass and stiffness matrices in equation (8.4), and thus the eigenvalues ω_j and eigenvectors ϕ of the homogeneous set of equations. This change alters the plant matrix $\mathbf{A}$ and the input matrix $\mathbf{B}$ in the state equation (equation (8.11)).

The sensitivity of the closed-loop eigenvalues λ_i with respect to the design variables A_l is given by

$$\lambda_{i,l} = \boldsymbol{\beta}_i^{\mathsf{T}} \bar{\mathbf{A}}_{,l} \boldsymbol{\alpha}_i \tag{8.37}$$

The closed-loop matrix given in equation (8.25) can be rewritten as

$$\bar{\mathbf{A}} = \mathbf{A} - \mathbf{X}\mathbf{P} \tag{8.38}$$

where

$$\mathbf{X} = \mathbf{B}\mathbf{R}^{-1}\mathbf{B}^{\mathsf{T}} \tag{8.39}$$

In equation (8.37) $\boldsymbol{\beta}_i^{\mathsf{T}}$ and $\boldsymbol{\alpha}_i$ are the left-hand and right-hand eigenvectors of $\bar{\mathbf{A}}$ defined by the solution to the eigenvalue problems

$$\boldsymbol{\beta}_i^{\mathsf{T}} \bar{\mathbf{A}} = \lambda_i \boldsymbol{\beta}_i^{\mathsf{T}} \tag{8.40}$$

and

$$\bar{\mathbf{A}} \boldsymbol{\alpha}_i = \lambda_i \boldsymbol{\alpha}_i \tag{8.41}$$

respectively. The eigenvectors $\boldsymbol{\beta}_i^{\mathsf{T}}$ and $\boldsymbol{\alpha}_i$ are normalized such that

$$\boldsymbol{\alpha}_i^{\mathsf{T}} \boldsymbol{\alpha}_i = 1 \tag{8.42}$$

and

$$\beta_i^T \alpha_i = 1 \qquad (8.43)$$

These eigenvectors satisfy the following relations

$$\alpha^T \beta = I \qquad (8.44)$$

$$\beta^T \bar{A} \alpha = \lambda \qquad (8.45)$$

$$\beta^T \alpha = I \qquad (8.46)$$

$$\alpha^T = \beta^{-1} \qquad (8.47)$$

where β^T and α are the matrices with each column equal to the eigenvector of the corresponding eigenvalue problem. I is the identity matrix and λ is the square diagonal matrix with elements equal to the eigenvalues.

The partial derivatives of $\bar{A}$ with respect to the structural design variables A_l can be obtained by differentiating equation (8.38) with respect to A_l. This gives

$$\bar{A}_{,l} = A_{,l} - X_{,l}P - XP_{,l} \qquad (8.48)$$

The matrices A and X in this equation are functions of the structural frequencies ω_j and the modal matrix ϕ. The sensitivities of ω_j^2 and ϕ_j are given by

$$\omega_{j,l}^2 = \frac{1}{A_l} \phi_{j,l}^T (k_l - \omega_j^2 m_l) \phi_{j,l} \qquad (8.49)$$

and

$$\phi_{j,l} = \sum_{i=1}^{n} \alpha_{ijl} \phi_i \qquad (8.50)$$

where

$$\alpha_{ijl} = \frac{1}{A_i(\omega_i^2 - \omega_j^2)} \phi_{j,l}^T (k_l - \omega_i^2 m_l) \phi_{i,l} \qquad i \neq j \qquad (8.51)$$

and

$$\alpha_{iil} = -\frac{1}{A_l} \phi_{i,l}^T m_l \phi_{i,l} \qquad (8.52)$$

In equation (8.49), k_l and m_l represent the element stiffness and mass matrix respectively of the *l*th element. In deriving equations (8.49)–(8.52), it is assumed that the element matrices are linear functions of the cross-sectional areas. This is true for the rod elements which are subjected to axial force only. Using equations (8.49)–(8.52), the sensitivities of A and X, required in equation (8.40), can be written. Equation (8.48) also needs the sensitivities of the Riccati matrix P. Differentiating equation (8.22) with respect to the design variables A_l and using equation (8.38) gives

$$\bar{A}^T P_{,l} + P_{,l} \bar{A} = \tilde{B} \qquad (8.53)$$

where

$$\tilde{B} = -A_{,l}^T P - PA_{,l} + PX_{,l}P \qquad (8.54)$$

The sensitivity of the Riccati matrix $P_{,l}$ is given by the solution to the Lyapunov equation (8.53). The solution can be obtained by using available subroutines in control software or using the following procedure.

Premultiplying equation (8.53) by α^T and postmultiplying by α and using the relations in equations (8.44)–(8.47) gives

$$\lambda^T Y + Y\lambda = \tilde{B}^* \tag{8.55}$$

where

$$Y = \alpha^T P_{,i} \alpha \tag{8.56}$$

and

$$\tilde{B}^* = \alpha^T \tilde{B} \alpha \tag{8.57}$$

Premultiplying equation (8.56) by β and postmultiplying by β^T gives

$$P_{,l} = \beta Y \beta^T \tag{8.58}$$

where the elements of matrix Y are given by

$$Y_{ij} = \frac{\tilde{B}^*_{ij}}{\lambda_i + \lambda_j} \tag{8.59}$$

Thus, using equation (8.37), the sensitivities of the complex eigenvalues λ_i can be obtained.

The sensitivities of the spectral radius can be obtained by assuming that the sensitivities are calculated for a specified critical operating frequency p_s and are invariant for small changes of the design variables. This is an approximation. Without this assumption, the analytical partial derivatives cannot be derived since the first derivative of the spectral radius is a discontinuous function of the operating frequency p.

The largest eigenvalue of $|(jp_s I - \bar{A}^{-1})| \cdot U_e$ for the specified critical operating frequency p_s can be written as

$$\lambda_\rho = \lambda_\rho^r + j\lambda_\rho^i \tag{8.60}$$

where λ_ρ^r and λ_ρ^i are the real and imaginary parts of λ_ρ. Then the spectral radius ρ_s can be written as

$$\rho_s = [(\lambda_\rho^r)^2 + (\lambda_\rho^i)^2]^{1/2} \tag{8.61}$$

The sensitivity of ρ_s with respect to the lth design variable can be written as

$$\rho_{s,l} = \frac{\lambda_\rho^r \lambda_{\rho,l}^r + \lambda_\rho^i \lambda_{\rho,l}^i}{\rho_s} \tag{8.62}$$

The sensitivity of the eigenvalues λ_ρ can be written as

$$\lambda_{\rho,l} = \beta_\rho \tilde{\bar{A}}_{\rho,l} \alpha_\rho \tag{8.63}$$

where

$$\tilde{\bar{A}}_\rho = \bar{A}_\rho \cdot U_e \tag{8.64}$$

where

$$\bar{A}_\rho = |(jp_s I - \bar{A})^{-1}| \tag{8.65}$$

In equation (8.63), β_ρ and α_ρ are the left and right eigenvectors of $\tilde{\bar{A}}_\rho$. Differentiating equation (8.64) with respect to the design variable A_l gives

$$\tilde{\bar{A}}_{\rho,l} = \bar{A}_{\rho,l} \cdot U_e + \bar{A}_\rho \cdot U_{e,l} \tag{8.66}$$

The elements of matrix $\bar{A}_\rho$ can be written as

$$\bar{A}_{\rho_{ij}} = [(a_{\rho_{ij}}^r)^2 + (a_{\rho_{ij}}^i)^2]^{1/2} \tag{8.67}$$

where $a_{\rho_{ij}}^r + ja_{\rho_{ij}}^i$ are the elements of the complex matrix $A_\rho = (jp_s I - \bar{A})^{-1}$. The sensitivity of $\bar{A}_{\rho_{ij}}$ can be written as

$$\bar{A}_{\rho_{ij,l}} = \frac{a_{\rho_{ij}}^{\mathrm{r}} a_{\rho_{ij,l}}^{\mathrm{r}} + a_{\rho_{ij}}^{\mathrm{i}} a_{\rho_{ij,l}}^{\mathrm{i}}}{\bar{A}_{\rho_{ij}}} \tag{8.68}$$

The sensitivity of $\mathbf{A}_\rho$ is given by

$$\mathbf{A}_{\rho,l} = - \mathbf{A}_\rho \mathbf{A}_{\rho,l}^{-1} \mathbf{A}_\rho \tag{8.69}$$

where

$$\mathbf{A}_{\rho,l}^{-1} = [jp_s\mathbf{I} - \bar{\mathbf{A}}]_{,l} \tag{8.70}$$

For a specified value of ρ_s

$$\mathbf{A}_{\rho,l}^{-1} = - \bar{\mathbf{A}}_{,l} \tag{8.71}$$

The sensitivity of $\bar{\mathbf{A}}_{,l}$ is given by equations (8.46). Taking into consideration the sign of the elements of $\bar{\mathbf{A}}$ and $\bar{\mathbf{A}}_{ij,l}$, the sensitivity of the elements of $\mathbf{U}_e$ can be written:

$$U_{e_{ij,l}} = \frac{|\bar{A}_{ij}|_{,l}|\bar{A}_{pq}| - |\bar{A}_{ij}||\bar{A}_{pq}|_{,l}}{(|\bar{A}_{pq}|)^2} \quad i > n \tag{8.72}$$

8.5 OPTIMIZATION STEPS

The major steps involved in the simultaneous design of a structure and control system are as follows:

1. For the specified initial cross-section areas of the members, calculate the structural frequencies ω_j and the vibration modes ϕ_j.
2. The plant matrix $\mathbf{A}$ and the input matrix $\mathbf{B}$ are determined.
3. The control problem is solved and the closed-loop eigenvalues and damping parameters are calculated.
4. The sensitivities of the objective function and constraints are calculated.
5. The design variables are modified by using a suitable optimization program such as NEWSUMT-A.
6. With a new values of the design variables, steps 1–5 are repeated until the optimum solution satisfying the constraints are obtained.

Steps 1–5 must be repeated because of the nonlinear nature of the problem.

8.6 NUMERICAL EXAMPLES

The details of two optimization problems are given in this section for illustration of the optimum design of a controlled structure.

EXAMPLE 8.1 TWO-BAR TRUSS

The finite element model of the truss is shown in Fig. 8.4. The elements of the structure are represented by bar elements that allow only axial deformation. The

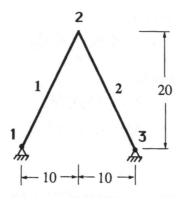

Fig. 8.4 Two-bar truss.

dimensions of structure are defined in unspecified consistent units. The elastic modulus of the members is equal to 1.0 and the density of the structural material ρ is equal to 0.001. A nonstructural mass of two units is attached at node 2. An actuator and a sensor are collocated in element 1 connecting nodes 1 and 2. The weighting matrices $\mathbf{Q}$ and $\mathbf{R}$ in equation (8.19) are assumed to be identity matrices. The design variables are the cross-sectional areas of the two members. The objective is to determine the cross-sectional areas of the members which will give a minimum weight structure and satisfy constraints on the closed-loop eigenvalue $\tilde{\omega}_1$ and damping parameter ξ_1. The areas of members 1 and 2 for the initial design are assumed to be 1000 and 100 units.

The optimization problem for the two-bar truss is specified as follows: minimize the weight

$$W = \sum_{i=1}^{2} \rho_i A_i l_i \tag{8.73}$$

such that

$$\tilde{\omega}_1 \geqslant 1.7393$$
$$\xi_1 \geqslant 0.029\,16 \tag{8.74}$$

where $\tilde{\omega}_1$ and ξ_1 are the imaginary parts of the lowest closed-loop eigenvalue and the associated damping respectively.

The details of the initial design with cross-sectional areas of elements 1 and 2 equal to 1000 and 100 respectively are as follows.

The weight of the initial design is 24.59 units. The mass, stiffness and load distribution matrices are

$$\mathbf{M} = \begin{bmatrix} 2 & 0 \\ 0 & 2 \end{bmatrix} \quad \mathbf{K} = \begin{bmatrix} 9.839 & 16.10 \\ 16.10 & 39.35 \end{bmatrix} \quad \mathbf{D} = \begin{bmatrix} -0.447 \\ -0.894 \end{bmatrix} \tag{8.75}$$

The mass matrix $\mathbf{M}$ is written by taking into consideration only the nonstructural mass. The squares of the structural frequencies for the initial design are

$$\omega_1^2 = 1.378 \quad \omega_2^2 = 23.22 \tag{8.76}$$

The modal matrix normalized with respect to the mass matrix is given by

$$\Phi = \begin{bmatrix} 0.6472 & 0.2847 \\ -0.2847 & 0.6472 \end{bmatrix} \tag{8.77}$$

The plant matrix and the input matrix are

$$A = \begin{bmatrix} 0.0 & 0.0 & 1.0 & 0.0 \\ 0.0 & 0.0 & 0.0 & 1.0 \\ 1.378 & 0.0 & 0.0 & 0.0 \\ 0.0 & -23.22 & 0.0 & 0.0 \end{bmatrix} \tag{8.78}$$

$$B = \begin{bmatrix} 0.0 \\ 0.0 \\ -0.0348 \\ -0.7063 \end{bmatrix} \tag{8.79}$$

The plant matrix A is obtained by assuming that the structural damping parameter ζ is equal to zero in equation (8.12).

The state and control weighting matrices are

$$Q = \begin{bmatrix} 1 & 0 & 0 & 0 \\ 0 & 1 & 0 & 0 \\ 0 & 0 & 1 & 0 \\ 0 & 0 & 0 & 1 \end{bmatrix} \tag{8.80}$$

$$R = [1.0] \tag{8.81}$$

The Riccati matrix obtained by solving equation (8.22) is given by

$$P = \begin{bmatrix} 5.2120 \times 10^1 & -5.6461 \times 10^{-2} & 3.6199 \times 10^{-1} & -8.4474 \times 10^{-2} \\ -5.6461 \times 10^{-2} & 3.3646 \times 10^1 & 1.4253 & 2.1444 \times 10^{-2} \\ 3.6199 \times 10^{-1} & 1.4253 & 3.7862 \times 10^1 & -6.0925 \times 10^{-3} \\ -8.4474 \times 10^{-2} & 2.1444 \times 10^{-2} & -6.0925 \times 10^{-3} & 1.4463 \end{bmatrix} \tag{8.82}$$

The closed-loop plant matrix is given by

$$\bar{A} = \begin{bmatrix} 0.0000 & 0.0000 & 1.0000 & 0.0000 \\ 0.0000 & 0.0000 & 0.0000 & 1.0000 \\ -1.3766 & -2.2523 \times 10^{-3} & -4.5682 \times 10^{-2} & -3.5530 \times 10^{-2} \\ 3.3240 \times 10^{-2} & -2.3264 \times 10^1 & -9.2731 \times 10^{-1} & -7.2124 \times 10^{-1} \end{bmatrix} \tag{8.83}$$

The closed-loop eigenvalues are

$$\begin{aligned} -0.3606 &\pm j4.8062 \\ -0.0228 &\pm j1.1739 \end{aligned} \tag{8.84}$$

The closed-loop damping parameters associated with these eigenvalues are

0.074 82 and 0.019 44 respectively. Comparing the constraints on the optimization problem (equations (8.74)) with the closed-loop eigenvalues and damping parameters of the initial design shows that we seek an optimum design for which $\tilde{\omega}_1$ is same as the initial design but damping increased from 0.019 44 to 0.029 16. The optimum design was obtained by using the NEWSUMT-A program as a black box. It took ten iterations and the weight of the final design was 17.58 with the cross-sectional areas of the two members equal to 684.62 and 101.98.

The details of the optimum design, with cross-sectional areas equal to 684.62 and 101.98, are as follows.

The mass and load distribution matrices would be same as those of the initial design. The stiffness matrix is given by

$$\mathbf{K} = \begin{bmatrix} 7.036 & 10.42 \\ 10.42 & 28.14 \end{bmatrix} \tag{8.85}$$

The squares of the structural frequencies for the final design are

$$\omega_1^2 = 1.378 \qquad \omega_2^2 = 16.21 \tag{8.86}$$

ω_1^2 for the optimum design is same as the initial design owing to the constraint on $\tilde{\omega}_1$. The modal matrix normalized with respect to the mass matrix is given by

$$\boldsymbol{\phi} = \begin{bmatrix} 0.6541 & 0.2686 \\ -0.2686 & 0.6541 \end{bmatrix} \tag{8.87}$$

The plant matrix and the input matrix are

$$\mathbf{A} = \begin{bmatrix} 0.0 & 0.0 & 1.0 & 0.0 \\ 0.0 & 0.0 & 0.0 & 1.0 \\ -1.378 & 0.0 & 0.0 & 0.0 \\ 0.0 & -16.21 & 0.0 & 0.0 \end{bmatrix} \tag{8.88}$$

$$\mathbf{B} = \begin{bmatrix} 0.0 \\ 0.0 \\ -0.0523 \\ -0.7052 \end{bmatrix} \tag{8.89}$$

The weighting matrices $\mathbf{Q}$ and $\mathbf{R}$ should be same as the initial design since they were not functions of the design variables. The Riccati matrix is given by

$$\mathbf{P} = \begin{bmatrix} 3.4725 \times 10^1 & -9.0134 \times 10^{-2} & 3.6102 \times 10^{-1} & -1.2515 \times 10^{-1} \\ -9.0134 \times 10^{-2} & 2.3799 \times 10^1 & 1.4769 & 3.0542 \times 10^{-2} \\ 3.6102 \times 10^{-1} & 1.4769 & 2.5261 \times 10^1 & -1.3560 \times 10^{-2} \\ -1.2515 \times 10^{-1} & 3.0542 \times 10^{-2} & -1.3560 \times 10^{-2} & 1.4618 \end{bmatrix}$$

$$\tag{8.90}$$

The closed-loop plant matrix is given by

$$\bar{\mathbf{A}} = \begin{bmatrix} 0.0000 & 0.0000 & 1.0000 & 0.0000 \\ 0.0000 & 0.0000 & 0.0000 & 1.0000 \\ -1.3747 & -5.1710 \times 10^{-3} & -6.8668 \times 10^{-2} & -5.3902 \times 10^{-2} \\ 4.8909 \times 10^{-2} & -1.6281 \times 10^1 & -9.2537 \times 10^{-1} & -7.2639 \times 10^{-1} \end{bmatrix}$$

$$\tag{8.91}$$

The closed-loop eigenvalues are

$$-0.3633 \pm j4.011$$
$$-0.0343 \pm j1.1739$$

(8.92)

The closed-loop damping parameter associated with the lowest eigenvalue is equal to 0.029 16. The optimum design satisfied both constraints as nearly equality constraints. For the two-bar truss, the optimum design given above is unique in the sense that with any other initial design and the constraints specified in equation (8.74), the optimization procedure would give the same optimum design.

EXAMPLE 8.2 TETRAHEDRAL TRUSS

The twelve-bar truss shown in Fig. 8.5 is optimized with constraints on frequencies and spectral radius. This structure has been used very frequently to illustrate the application of the optimum design of a controlled structure. The coordinates of the node points are given in Table 8.1. A nonstructural mass of 2 units is located at node points 1–4. Young's modulus and material density are the same as those for the two-bar truss. The structure has twelve degrees of freedom. The six

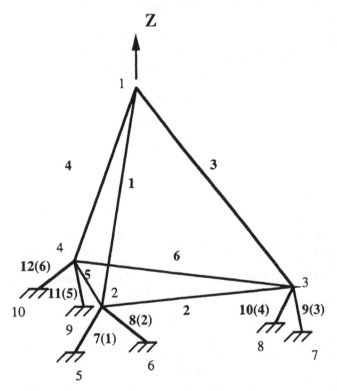

Fig. 8.5 Tetrahedral truss structural model. Actuator numbers are in parentheses.

Table 8.1 Node point coordinate or tetrahedral truss

Node	X	Y	Z
1	0.0	0.0	10.165
2	−5.0	−2.887	2.00
3	5.0	−2.887	2.00
4	0.0	5.7735	2.00
5	−6.0	−1.1547	0.0
6	−4.0	−4.6188	0.0
7	4.0	−4.6188	0.0
8	6.0	−1.1547	0.0
9	−2.0	5.7735	0.0
10	2.0	5.7735	0.0

actuators and sensors are colocated in the elements 7–12. In order to avoid the effect of unmodeled dynamics on the optimum design, all vibration modes are used in the plant matrix. Thus the matrices $\mathbf{A}$ and $\mathbf{B}$ would be 24 × 24 and 24 × 6 respectively. Similarly the weighting matrices $\mathbf{Q}$ and $\mathbf{R}$ would be 24 × 24 and 6 × 6 respectively. The weighting matrices are set equal to $\delta \mathbf{Q}$ and $\gamma \mathbf{R}$. δ and γ are treated as the control design variables. The matrices $\mathbf{Q}$ and $\mathbf{R}$ are assumed to be identity matrices. For the initial design in optimization, δ and γ were set equal to unity.

The cross-sectional areas of the members for the initial design are given in Table 8.2. For this optimization problem, the constraints are imposed on the imaginary parts of the first two lowest frequencies of the closed-loop system and the spectral radius ρ_s. The constraints are

$$\tilde{\omega}_1 - 1.34 \geqslant 0 \tag{8.93}$$

$$\tilde{\omega}_2 - 1.6 \geqslant 0 \tag{8.94}$$

$$5.56 - \rho_s \geqslant 0 \tag{8.95}$$

Table 8.2 Cross-sectional areas of members

Initial design						
Member	1	2	3	4	5	6
Area	1000.0	1000.0	100.0	100.0	1000.0	1000.0
Member	7	8	9	10	11	12
Area	100.0	100.0	100.0	100.0	100.0	100.0
			Weight = 43.69			
Optimum design						
Member	1	2	3	4	5	6
Area	216.4	138.5	165.5	168.0	154.5	468.7
Member	7	8	9	10	11	12
Area	174.6	144.8	155.6	235.4	143.2	170.2
			Weight = 16.01			

The constraint on the spectral radius is based on achieving 10% permissible deviation in all the elements of the closed-loop matrix before the system becomes unstable. The perturbation identification matrix is obtained by normalizing the closed-loop matrix $\bar{A}$ for $\bar{A}_{p,q} = \bar{A}_{13,1}$ (equation (8.35)). The constraint value of ρ_s in equation (8.95) is calculated by using equation (8.36) for $\varepsilon_r = 0.1$ and $|\bar{A}_{p,q}| = |\bar{A}_{13,1}| = 1.34^2$ where 1.34 would be the constraint value of $\tilde{\omega}_1$.

The initial design weighed 43.69 units. The spectral radius ρ_s for this design was 44.35 giving $\varepsilon_r = 0.0126$. Thus the control design would allow 1.26% variation in the closed-loop matrix before becoming unstable according to equation (8.34).

Table 8.3 Iteration history

Iteration number	1	2	3	4	5	6
Weight	43.69	32.01	32.01	25.91	22.12	19.34
Iteration number	7	8	9	10	11	
Weight	16.58	16.11	16.02	16.01	16.01	

Table 8.4 Structural frequencies

Initial design

Mode	1	2	3	4	5	6
ω_j	1.342	1.664	2.890	2.950	3.398	4.204
Mode	7	8	9	10	11	12
ω_j	4.662	4.755	8.539	9.250	10.284	12.905

Optimum design

Mode	1	2	3	4	5	6
ω_j	1.355	1.607	2.865	3.003	3.654	5.175
Mode	7	8	9	10	11	12
ω_j	5.390	5.718	6.147	6.383	6.922	7.884

Table 8.5 Closed-loop frequencies

Initial design

1	$-0.0734 \pm j1.341$	7	$-0.3547 \pm j\ 4.649$
2	$-0.1089 \pm j1.663$	8	$-0.3440 \pm j\ 4.743$
3	$-0.2133 \pm j2.884$	9	$-0.2917 \pm j\ 8.534$
4	$-0.2372 \pm j2.949$	10	$-0.2758 \pm j\ 9.246$
5	$-0.2852 \pm j3.388$	11	$-0.2136 \pm j10.28$
6	$-0.3634 \pm j4.190$	12	$-0.0828 \pm j12.90$

Optimum design

1	$-0.1927 \pm j1.352$	7	$-1.060\ \pm j5.168$
2	$-0.2206 \pm j1.601$	8	$-1.370\ \pm j5.558$
3	$-0.3633 \pm j2.854$	9	$-1.332\ \pm j6.006$
4	$-0.4593 \pm j2.992$	10	$-1.259\ \pm j6.252$
5	$-1.5936 \pm j3.620$	11	$-1.187\ \pm j6.827$
6	$-1.353\ \pm j5.145$	12	$-0.8830 \pm j7.756$

Table 8.6 Closed-loop damping

Initial design						
Mode	1	2	3	4	5	6
ξ_i	0.0546	0.0653	0.0737	0.0801	0.0839	0.0864
Mode	7	8	9	10	11	12
ξ_i	0.0760	0.0723	0.0342	0.0298	0.0207	0.0064
Optimum design						
Mode	1	2	3	4	5	6
ξ_i	0.1411	0.1364	0.1262	0.1517	0.1618	0.2544
Mode	7	8	9	10	11	12
ξ_i	0.2009	0.2365	0.2165	0.1974	0.1713	0.1131

The optimum design weighing 16.01 units was obtained after eleven iterations. The iteration history is given in Table 8.3. The control design parameters δ and γ for the optimum design were 3.89 and 0.257 respectively. The spectral radius ρ_s of the optimum design was 5.560 and ε_r was 0.9999. The cross-sectional areas of the members for the optimum design are given in Table 8.2. The structural frequencies for the two designs are given in Table 8.4. The optimum design has a narrower frequency bandwidth than the initial design. It is well known that in structural optimization with constraints on frequencies, the frequencies come closer. The closed-loop frequencies are listed in Table 8.5. For the optimum design $\tilde{\omega}_1 = 1.352$ and $\tilde{\omega}_2 = 1.601$ indicating that the two constraints on the imaginary parts of the closed-loop frequencies are satisfied. Table 8.6 contains closed-loop damping for the two designs. For the optimum design, the damping associated with all modes is larger than for the initial design.

8.7 OPTIMIZATION PROGRAMS

In addition to the optimization software available in the math libraries of the computer software, there are some public domain programs available for use. The capabilities of some of these programs are discussed below.

8.7.1 NEWSUMT-A (Thareja and Haftka, 1985)

The minimization algorithm used in this program is a sequence of unconstrained minimizations (SUMT) technique. This program can be used for inequality or equality constraints. It uses Newton's method for unconstrained minimization. An extended interior function is used for inequality constraints or an exterior penalty function for equality constraints. The sensitivities can be either calculated by using finite differences or a user-supplied analysis program.

8.7.2 ADS (Vanderplaats, Suginoto and Sprague, 1983)

This is a general-purpose optimization program for solution of nonlinear constrained optimization problems. The program is segmented into three levels, these being strategy, optimizer and one-dimensional search. The available strategies are sequential unconstrained minimization, the augmented Lagrange multiplier method and sequential linear programming. The optimizers include a variable metric method and the method of feasible directions as examples. The feasible direction method amongst all approaches in this software is known to be robust and frequently used. The original feasible direction program is called CONMIN.

8.7.3 IDESIGN (Arora, Thanedar and Tseng, 1985)

This general-purpose optimization program has been under development since the late 1970s. It contains several state-of-the-art nonlinear programming algorithms. It has a wide range of applications with many friendly features. The program can also be used in an interactive mode. Some of the algorithms available in the software are cost function bounding, Pschenichny's linearization method, sequential quadratic programming and conjugate gradient. The gradient information can be user supplied or the finite difference method based on a forward, backward or central scheme can be used.

8.7.4 VMCON (Crane, Hillstrom and Minkoff, 1980)

This program solves a sequence of positive definite quadratic programming subproblems. Each solution determines a direction in which a one-dimensional minimization is performed. It is an iterative algorithm designed to converge to a point that satisfies the first-order necessary condition. These conditions are the Kuhn–Tucker conditions and related conditions on the Lagrange multipliers and the constraints.

The programs discussed above can be used as black boxes. The primary information they need is the objective function, constraints and the sensitivities. Experience has shown that the performance of these programs depend on the problem and the familiarity of the user with them. The control software ORACLE (Armstrong, 1978) is available in the public domain in addition to a number of proprietary programs distributed by various companies.

8.8 CONCLUDING REMARKS

This chapter has reviewed the research work performed by different investigators in developing the algorithms for the simultaneous design of structures and control systems. It was observed that the primary objective of the investigations has been to understand the problem and study the feasibility of integrated techniques. A simple two-bar truss problem was solved to illustrate the application of combined structure and control system optimization. Many of the critical issues in structural and control design such as nonlinearities, dynamic characteristics of sensors and actuators, order of the structure and control modes, use of modern control theories, experimental verification of the simultaneous structure and control design, development of software and efficiencies of the optimization algorithm have still to be investigated.

REFERENCES

Armstrong, E.S. (1978) ORACLS – A system for linear–quadratic–gaussian control law design. *NASA Technical Paper 1106.*

Arora, J.S. (1989) *Introduction to Optimum Design,* McGraw-Hill, New York.

Arora, J.S., Thanedar, P.B. and Tseng, C.H. (1985) User's manual for program IDESIGN. *Technical Report ODL 85.10,* University of Iowa, Iowa City, IA.

Belvin, W.K. and Park, K.C. (1988) Structural tailoring and feedback control synthesis: an interdisciplinary approach. In Proceedings of 29th AIAA/ASME/ASCE/AHS Structures, Structural Dynamics and Materials Conference, Part 1, AIAA-88-2206-CP, Williamsburg, VA, April 18–20, 1988, pp. 1–8.

Bendsoe, M.P., Olhoff, N. and Taylor, J.E. (1987) On the design of structure and controls for optimal performance of actively controlled flexible structures. *Mechanics of Structures and Machines,* 15(3), 265–95.

Bodden, D.S. and Junkins, J.L. (1985) Eigenvalue optimization algorithms for structure/controller design iterations. *Journal of Guidance, Control, and Dynamics,* 8(6), 697–706.

Cheng, F.Y. and Pantelides, C.P. (1988) Combining structural optimization and structural control. *Technical Report NCEER-88-0037,* National Center for Earthquake Engineering Research.

Crane, R.L., Hillstrom, K.E. and Minkoff, M. (1980) Solution of the general nonlinear programming problem with subroutine VMCON. *ANL-80-64,* Argonne National Laboratory, Chicago, IL.

D'Azzo, J.J. and Houpis, C.H. (1989) *Linear Control System Analysis and Design,* McGraw-Hill, New York.

Dracopoulos, T.N. and Öz, H. (1992) Integrated aeroelastic control optimization of laminated composite lifting surfaces. *Journal of Aircraft,* 29(2), 282–8.

Eastep, F., Khot, N.S. and Grandhi, R. (1987) Improving the active vibrational control of large space structures through structural modifications. *Acta Astronautica,* 15(6), 383–9.

Grandhi, R.V. (1989) Structural and control optimization of space structures. *Computers and Structures,* 31(2), 139–50.

Grandhi, R.V. (1990) Optimum design of space structures and active and passive damping, *Engineering with Computers,* 15(6), 177–83.

Grandhi, R.V., Haq, I. and Khot, N.S. (1990) Enhanced robustness in integrated structural/control systems design. In Proceedings of 31st AIAA/ASME/ASCE/AHS/ASC Structures, Structural Dynamics and Materials Conference, Long Beach, CA, April 2–4, pp. 247–57.

Haftka, R.T. (1990) Integrated structure-control optimization of space structures. In Proceedings of AIAA Dynamics Specialist Conference, Long Beach, CA, April 5–6, pp. 1–9.

Haftka, R.T., Gurdal, Z. and Kamat, M.P. (1990) *Structural Optimization,* Kluwer, Dordrecht.

Haftka, R.T., Martinovic, Z.N. and Hallaur, W.L. Jr (1984) Enhanced vibration controllability by minor structural modifications. In Proceedings of 25th AIAA/ASCE/AHS Structures, Structural Dynamics and Materials Conference, Palm Springs, CA, May 14–16, pp. 401–10.

Hale, A.L. (1985) Integrated structural/control synthesis via set-theoretic methods. In Proceedings of 26th AIAA/ASME/ASCE/AHS Structures, Structural Dynamics and Materials Conference, Orlando, FL, April 15–17, pp. 636–41.

Hale, A.L., Lisowski, R.J. and Dahl, W.E. (1985) Optimal simultaneous structural and control design of maneuvering flexible spacecraft. *Journal of Guidance, Control, and Dynamics,* 8(1), 86–93.

Junkins, J.L. and Rew, D.W. (1988) Unified optimization of structures and controls, in *Large*

Space Structures: Dynamics and Control (eds S.N. Atluri and A.K. Amos), Springer, Berlin, pp. 323–53.

Khot, N.S. (1988a) An integrated approach to the minimum weight and optimum control design of space structures, in *Large Space Structures: Dynamics and Control* (eds S.N. Atluri and A.K. Amos), Springer, Berlin, pp. 355–63.

Khot, N.S. (1988b) Structural/control optimization to improve the dynamic response of space structures. *Computational Mechanics*, (3), 179–86.

Khot, N.S. (1990) On optimization of structural and control systems using a reduced order model. *Structural Optimization*, 2(3), 185–91.

Khot, N.S. and Grandhi, R.V. (1988) Structural and control optimization with weight and Frobenius norm as performance functions, in *Structural Optimization* (eds G.I.N. Rozvany and B.L. Karihaloo), Kluwer, Dordrecht, pp. 151–8.

Khot, N.S. and Veley, D.E. (1990) Robustness characteristics of optimum structural/control design. In Proceedings of AIAA Guidance, Navigation and Control Conference, Portland, OR, August 20–22, pp. 394–403.

Khot, N.S. and Veley, D.E. (1991) Use of robustness constraints in the optimum design of space structures. *Journal of Intelligent Materials Substructures*, 2(2), 161–76.

Khot, N.S., Venkayya, V.B. and Eastep, F.E. (1984) Structural modifications of large flexible structures to improve controllability. In Proceedings of AIAA Guidance, Navigation, and Control Conference, Seattle, WA, August 20–22, pp. 420–30.

Khot, N.S., Venkayya, V.B. and Eastep, F.E. (1986) Optimal structural modifications to enhance the active vibration control of flexible structures. *AIAA Journal*, 24(8), 1368–74.

Khot, N.S., Grandhi, R.V. and Venkayya, V.B. (1987) Structural and control optimization of space structures. In Proceedings of 28th AIAA/ASME/ASCE/AHS Structures, Structural Dynamics and Materials Conference, Monterey, CA, April 6–8, pp. 850–60.

Khot, N.S., Öz, H., Grandhi, R.V., *et al.* (1988) Optimal structural design with control gain norm constraint. *AIAA Journal*, 26(5), 604–11.

Kirsch, U. and Moses, F. (1977) Optimization of structures with control forces and displacements. *Engineering Optimization*, 1(3), 37–44.

Lim, K.B. and Junkins, J.L. (1987) Robustness optimization of structural and controller parameter. In Proceedings of 28th AIAA/ASME/ASCE/AHS Structures, Structural Dynamics and Materials Conference, Monterey, CA, April 6–8, pp. 351–61.

Lim, K.B. and Junkins, J.L. (1989) Robustness optimization of structural and controller parameters. *Journal of Guidance, Control and Dynamics*, 12(1), 89–96.

Manning, R.A. and Schmit, L.A. (1990) Control augmented structural synthesis with transient respone constraints. *AIAA Journal*, 28(5), 883–91.

McLaren, M.D. and Slater, G.L. (1990) A covariance approach to integrated control/structure optimization. In Proceedings of AIAA Dynamics Specialist Conference, Long Beach, CA, April 5–6, pp. 189–205.

Meirovitch, L. (1989) *Dynamics and Control of Structures*, Wiley, New York.

Mesquita, L. and Kamat, M.P. (1988) Structural optimization for control of stiffened laminated composite structures. *Journal of Sound and Vibration*, 116(1), 33–48.

Messac, A. and Turner, J. (1984) Dual structural-control optimization of large space structures. In Proceedings of Recent Experience in Multidisciplinary Analysis and Optimization Symposium, NASA Langley Research Center, April 24–26, pp. 775–802.

Miller, D.F. and Shim, J. (1987) Gradient-based combined structural and control optimization. *AIAA Journal*, 10(3), 291–8.

Morris, A.J. (ed.) (1982) *Foundations of Structural Optimization: A Unified Approach*, Wiley, London.

Onoda, J. and Haftka, R.T. (1987) An approach to structure/control simultaneous optimization for large flexible spacecraft. *AIAA Journal*, 25(8), 1133–8.

Onoda, J. and Watanabe, N. (1989) Integrated direct optimization of structure/regulator/observer for large flexible spacecraft. In Proceedings of 30th AIAA/ASME/ASCE/AHS/ASC

Structures, Structural Dynamics and Materials Conference, Mobile, AL, April 3–5, pp. 1336–44.

Padula, S.L., Sandridge, C.A., Walsh, J.L. and Haftka, R.T. (1990) Integrated controls-structures optimization of a large space structure. In Proceedings of 31st AIAA/ASME/ASCE/AHS/ASC Structures, Structural Dynamics and Materials Conference, Long Beach, CA, April 2–4, pp. 258–67.

Qiu, L. and Davison, E.J. (1986) New perturbation bounds for the robust stability of linear state space models. In Proceedings of the 25th Conference on Decision and Control, Athens, pp. 751–5.

Rao, S.S. (1988) Combined structural and control optimization of flexible structures. *Engineering Optimization*, **13**(1), 1–16.

Rao, S.S., Venkayya, V.B. and Khot, N.S. (1988) Game theory approach for the integrated design of structures and controls. *AIAA Journal*, **26**(4), 463–9.

Rao, S.S., Pan, T.S. and Venkayya, V.B. (1990) Robustness improvement of actively controlled structures through structural modifications. *AIAA Journal*, **28**(2), 353–61.

Salama, M., Garba, J., Demsetz, L. and Udwadia, F. (1988) Simultaneous optimization of controlled structures. *Computational Mechanics*, **3**(4), 275–82.

Sepulveda, A.E. and Schmit, L.A., Jr (1990) Optimal placement of actuators and sensors in control augmented structural optimization. In Proceedings of 31st AIAA/ASME/ASCE/AHS/ASC Structures, Structural Dynamics and Materials Conference, Long Beach, CA, April 2–4, pp. 217–40.

Suzuki, S. and Matsuda, S. (1990) Structure/control design synthesis of active flutter suppression system by goal programming. In Proceedings of AIAA Guidance, Navigation and Control Conference, Portland, OR, August 20–22.

Thareja, R. and Haftka, R.T. (1985) NEWSUMT-A. A modified version of NEWSUMT for inequality and equality constraints. VPI Report 148, Aerospace Engineering Department, Virginia Tech., Blacksburg, VA.

Thomas, H.L. and Schmit, L.A., Jr (1989) Control augmented structural synthesis with dynamic stability constraints. In Proceedings of 30th AIAA/ASME/ASCE/AHS/ASC Structures, Structural Dynamics and Materials Conference, Mobile, AL, April 3–5, pp. 521–31.

Thomas, H.L. and Schmit, L.A. (1990) Improved approximations for control augmented structural synthesis. In Proceedings of 31st AIAA/ASME/ASCE/AHS/ASC Structures, Structural Dynamics and Materials Conference, Long Beach, CA, April 2–4, pp. 277–94.

Vanderplaats, G.N., Sugimoto, H. and Sprague, C.M. (1983) ADS-1: a new general-purpose optimization program. In AIAA/ASME/ASCE/AHS 24th Structures, Structural Dynamics and Materials Conference, Lake Tahoe, NV, pp. 117–23.

Zeiler, T.A. and Gilbert, M.G. (1990) Integrated control/structure optimization by multilevel decomposition. In Proceedings of 31st AIAA/ASME/ASCE/AHS/ASC Structures, Structural Dynamics and Materials Conference, Long Beach, CA, April 2–4, pp. 247–57.

9

Database design and management in engineering optimization

JASBIR S. ARORA

9.1 INTRODUCTION

Design of modern structural and mechanical systems requires considerable computer analysis. Also, as more computing power becomes available, the desire to design larger and more complex systems grows. Many of the systems operate in an environment that requires integration of many disciplines in their design process. These systems must be designed to be cost effective and efficient in their operations, requiring use of organized optimization techniques. These techniques are iterative, generating and using massive amounts of data. Multiple programs may be used during the optimization process that require sharing of data. Also, complexity of the design software grows as more disciplines are integrated and more complicated problems need to be solved. These requirements points to the fact that the data and the software must be organized and managed properly. This will allow the design process to be completed in an efficient manner as well as give flexibility in updating and enhancing the design software itself.

This chapter presents basic concepts related to the database design and management in optimization of engineering systems. Most of the concepts are new to the engineering community, so they are explained in some detail. Terminologies used in the database design and management field are defined and explained. Need for the database design and management is discussed. Basic components of a database management system are identified and discussed. The topic of designing a proper database is discussed and a methodology to design a database is given. Examples from the finite element analysis procedure and design of structural systems are used to illustrate various ideas and concepts.

The material presented in this chapter is derived from several references, such as Arora and Mukhopadhyay (1988), Blackburn, Storaasli and Fulton (1982), Bell (1982), Date (1977), Felippa (1979, 1980), Fulton (1987), Martin (1977), Mukhopadhyay and Arora (1987a, b), Pahl (1981), Rajan (1982), Rajan and Bhatti (1983, 1986), SreekantaMurthy and Arora (1985, 1986a, b), SreekantaMurthy et al. (1986) and Vetter and Maddison (1981). These references can be consulted for more details on some of the topics.

9.1.1 Computer-aided design optimization process

To understand the need for use of a database in computer-aided design optimization, one needs to understand the optimization process. Therefore this process is briefly explained here. A first step in optimum design of any system is the precise mathematical statement of the problem (Arora, 1989, 1990). This requires identification of design variables that describe the system, definition of a cost function that needs to be minimized, and identification of constraints that must be satisfied. In design of aircraft components such as stiffened panels and cylinders, the design variables are spacing of the stiffeners, size and shape of stiffeners, and thickness of the skin. In optimization of structural systems such as frames and trusses of fixed configuration, thickness of members, cross-sectional areas of bars, and moment of inertia are the design variables. If shape optimization is the objective, the design variables include parameters related to geometry of the system. In practical applications design variables are usually grouped together to reduce the size of design variable vector.

The constraints for the system are classified into the performance and size constraints. The performance constraints are on stresses, displacements, and local and overall stability requirements in the static case; frequencies and displacements in the dynamic case; flutter velocity and divergence in aeroelastic case, or a combination of these. The size constraints are the minimum and maximum value of the design variables. To evaluate the performance constraints, the system must be analyzed. Finite element and other numerical methods must be used for analysis of large and complex systems. These methods use data such as element number, nodal connectivity, element stiffness matrix, element mass matrix, element load matrix, assembled stiffness, mass and load matrices, displacement vectors, eigenvalues, eigenvectors, buckling modes, decomposed stiffness matrix, and the stress matrix. In general, data used is quite large even when symmetry of the matrices is taken into account. Hypermatrix or other special schemes can be used in dealing with large matrix equations.

In nonlinear programming, the search for the optimum design variables involves iterations. The design variable and other data at the kth iteration are used to compute a search direction. A step size must be calculated in the search direction to move to a new point. This calculation can involve evaluation of problem functions which requires reanalysis of the system. To calculate the search direction, gradients of cost and constraint functions with respect to the design variables are needed. This computation is called design sensitivity analysis which can be quite tedious and time consuming because most functions of the problem are implicitly dependent on the design variables. Therefore, special methods must be developed and used for each class of applications. For design of large structures, efficient design sensitivity analysis is particularly critical. For such structures, substructuring concepts can be effectively integrated into structural analysis, design sensitivity analysis, and optimal design procedures (Haug and Arora, 1979). In this concept, one deals with small-order matrices as the data can be organized substructure-wise (Kamal and Adeli, 1990). It can be seen that the number of matrices and their sizes depend on the number of substructures and their sizes.

In most practical applications, interactive computations and graphics can be profitably employed in design optimization (Arora, 1989; Arora and Tseng, 1988; Al-Saadoun and Arora, 1989; Park and Arora, 1987). At a particular iteration, the designer can study the data of design variables, active constraints, cost function, search direction, sensitivity coefficient etc. Judgements can be made regarding suitability of a particular algorithm, change of system parameters, and perhaps about the problem formulation. The problem conditions can be redefined to achieve convergence to optimal design.

9.1.2 Need for database and its management

It can be seen from the foregoing discussion that sophisticated engineering design optimization methods use large a amount of data and require substantial computer analysis. Numerical simulation techniques, such as the finite element method, must be used. The data generated during the simulation phase must be saved in a database for later use in defining constraints and performing design sensitivity analysis. Once design sensitivity analysis has been completed, a search direction determination subproblem is defined and solved. Note that the size of this subproblem at each iteration depends on the number of active constraints. Therefore, sizes of data sets change from iteration to iteration, making the nature of data to be dynamic. We should be able to create large data sets dynamically, to manipulate them during the iteration, and to delete some of them at the end of the iteration. Useful trend information from each iteration must be saved for processing in later iterations. Note that a row of the history matrix (such as design variable values) is generated at each iteration. However, to use the trend information for a quantity (e.g. a design variable), we need to look at its value at the previous iterations. This implies that we should look at a column of the history matrix, i.e. we should be able to create data in one form and view it in another form. The entire process is frequently interactive where the designer needs to interrupt it, to analyze the data, and to make design decisions. Therefore the designer needs control over the data and the program to guide the iterative process properly towards acceptable designs.

It is concluded that significant improvement in design capability can be achieved with effective management of the design software system and the database. A properly designed database and a database management system when used with interactive computer graphics can be an invaluable tool for the engineer involved in the design process. These capabilities are also essential to allow for proper growth and refinement of the design software system to accommodate new developments in the field.

9.2 TERMINOLOGY AND BASIC DEFINITIONS

The terminology used in database design and management for describing various ideas differs considerably from one group to another and even from one period to another within the same organization. It is therefore necessary to explain various terminologies that are taken from most widely accepted sources (Date, 1977; Martin, 1977; Vetter and Maddison, 1981; Felippa, 1979, 1980; Rajan, 1982; SreekantaMurthy and Arora, 1986a,b). This will also facilitate in the description of various concepts in subsequent sections. They are grouped into three categories – hardware terminology, logical data terminology and physical storage terminology.

9.2.1 Hardware terminology

Auxiliary storage
Storage facilities of large capacity and lower cost but slower access than main memory are called auxiliary storage. They are also referred to as peripheral or secondary storage devices. They are usually assessed via data channels, in which case data is stored and retrieved by physical record blocks. They include magnetic tape and disk units, drums and other devices used to store data.

Cell
This is used as a generic word to mean either track, cylinder, module or other zone delimited by a natural hardware boundary such that the time required to access data increases by a step function when data extends beyond a cell boundary.

Cylinder
An access mechanism may have many reading heads. Each head can read one track. A cylinder refers to a group of tracks that can be read without moving the access mechanism.

Direct access storage device
In this device, access to a position for storage or retrieval of data is not dependent on the position at which data was previously stored or retrieved. It is also called random access device.

Input/output (I/O) device
This is an auxiliary storage device connected to the central processing unit (CPU) by a data channel.

Main memory
This is fast, direct access, electronic memory hardwired to the central processing unit. It holds machine instructions and data that can be accessed in a time of the order of nanoseconds. It is also referred to as core, main storage, or internal memory.

Module
A module of the peripheral storage device is a section of hardware which holds one volume, such as one spindle of disk.

Storage device – logical (logical file, memory device, named space, or logical address space)
This is a subset of the storage space that is treated as a named entity by the operating system for purpose of allocating and releasing storage resources during the execution of a run unit (task). The term is most often applied to auxiliary storage facilities.

Storage facility
This is hardware available to store data at a computer installation.

Track
A track on a direct access device contains data that can be read in a single reading without the head changing its position.

Volume
A volume is normally a single physical unit of any peripheral storage medium such as tapes, disk packs, or cartridges.

9.2.2 Logical data terminology

Arithmetic data
An arithmetic data item has a numeric value with characteristics of base, scale mode and precision, e.g. fixed point data (integer), and floating point data (real and double precision).

Attribute
Properties of an entity are called its attributes. Columns of a two-dimensional table (a relation) are referred to as its attributes. Attributes associate a value from a domain of values for that attribute with each entity in the entity set. For the entity set 'finite elements', NODES for the element, MATERIAL, cross-sectional SHAPE etc. are its attributes. Nodes for an element may be anywhere from 1 to 100, so this defines the domain for the attribute NODES. The attribute MATERIAL may have its domain as several steel and aluminum grades such as (STEEL.1, STEEL.2, ..., ALUM.1, ALUM.2, ...).

Creation
This involves adding new files to the database, initializing the files (i.e. file table definition), data validation, deciding file types etc.

Data aggregate
A data aggregate is a collection of data items within a record. Data aggregates may be vectors or repeating groups.

- A vector is a one-dimensional ordered collection of data items, e.g. node numbers of a structure.
- A repeating group is a collection of data that occurs repeatedly within a data aggregate. For example, degrees of freedom for an element appear in multiples of node numbers:

$$D_{1i} \quad D_{2i} \quad D_{3i}, \qquad i = 1, \dots, n$$

where i is the node number, D_{ki} the kth degree of freedom at node i, and n the number of nodes for the element.

Data definition language (DDL)
This is a set of commands that enables users of a database management system (DBMS) to define data structures to store the data. All data that is to be managed by the DBMS must follow the rules laid down in data definition language. A DBMS must provide a DDL to specify the conceptual scheme and some of the

details regarding the implementation of the conceptual scheme by a physical scheme. It describes relationships among types of entities for a particular data model.

Data independence

This refers to the independence between physical and logical data structures. Physical data structure can change without affecting the user's view of the data when we have data independence. Similarly, logical data structure can change without affecting the physical data structure.

Data item

A data item is the smallest unit of named data. It is also referred to as the data element or field. Each data item has a unique representation. The data item can be any of the following types: arithmetic (integer, real or double precision real), or character string (character, bits).

Data library

This is a named collection of data sets residing on a permanent storage device. It is the most complex data structure upon which a database management system operates.

Data manipulation language (DML)

This is a set of permissible commands that are issued by users or application programs to the DBMS to carry out storage, retrieval or manipulation of data. The DML represents the interface between the application program and the database management system. Thus the data managed by the DBMS can be accessed and processed through the use of DML. It can be an extension of the host language, such as FORTRAN.

Data model

Data model is a representation of the conceptual scheme for the database. Generally a data definition language (a higher-level language) is used to describe the data model. Examples of data model are hierarchical, network and relational.

Data set

This is an ordered collection of logically related data items arranged in a prescribed manner. Each data set has some control information that can be accessed by a programming system.

Data structure

This is a logical arrangement of data as viewed by the users or applications programmers.

Database

A database is a collection of interrelated data stored together without unnecessary redundancy to serve multiple applications. It can also be viewed as a collection of the occurrences of multiple record types, containing relationships between records, data aggregate and data elements. Thus, it is a collection of data files stored on a storage device. The data are stored so that they are independent of programs which use them. A common and controlled approach is used in storing new data and in modifying and retrieving existing data within the database. A database allows a user or application programmer to retrieve or write data without actually making calls to the input/output device; that chore is left to the DBMS.

Database administrator (DBA)
This is the brain of the database management system that provides interfaces between the various parts of the system, does error recovery, and enforces security measures.

Database management system (DBMS)
This is the software that allows one or many persons to use and/or modify the database is called a DBMS. DBMS also deals with security, integrity, synchronization and protection of the database.

Database system
The set of all databases maintained on a computer installation (or computer network) which are administered by a common database manager is called the database system.

Domain
A domain is the set of eligible values for a quantity. For example consider the entity set 'finite elements'; element name, element material type, and length are its attributes. Domains for these attributes can be defined as

$$\text{element name} = (\text{BEAM, TRUSS}, \ldots)$$
$$\text{element material type} = (\text{STEEL, ALUMINUM}, \ldots)$$
$$\text{length} = x: x \geqslant 0 \text{ and } x \leqslant 100$$

Entity
An entity may be 'anything having reality and distinctness of being in fact or in thought', e.g. a finite element, a relation, a pre-processor, or a post-processor for a structural analysis program.

Entity identifier
This uniquely identifies an entity. The identifier is needed by the programmer to record information about a given entity. Also it is needed by by the computer to identify and have means of finding the entity in a storage unit. Entity identifier must be unique, e.g. element number.

Entity key
Entity key is an attribute having different values for each occurring entity and provides unique identification of a tuple (an ordered list or a record). An entity represents a compound key if it corresponds to a group of attributes. It is also called the candidate key.

Entity set
Collection (group) of all similar entities is referred to as an entity set, e.g. a set of finite elements (ELEMENTS) and a set of nodes (NODES) are called entity sets.

Garbage collection
The process of locating all pages of the memory that are no longer in use and adding them to the list of available space is called garbage collection.

Group
This is a data set containing a special 'owner' or 'master' record (the group directory) and a set of member records.

Instances
The current contents of a database is called an instance of the database.

Interrogation
This deals with identification, selection and extraction of data from the database for further processing. It can be divided into two phases:

- the process of selection and identification of needed data and extracting it;
- the processing part which involves computation, display or any other manipulation required including updating parts of the database.

Logical data structure
Data in a particular problem consist of a set of elementary items of data. An item usually consists of single element such as an integer, a bit, a character and a real, or a set of such items. The possible ways in which the data items are structured define different logical data structures. Therefore, it is the data structure as seen by the user of the DBMS without any regard to details of actual storage schemes.

Memory management system (MMS)
This is a system that allocates the available memory to the different entity sets in a program and makes it appear as if more memory is available than what the computer has. It partitions the memory allocated to the DBMS into pages and manages the information contained in them. If the data required is not in the memory it retrieves pages from the secondary storage.

Module
This is a program that performs an identifiable task.

Physical database
This is actual data or information residing in a file in the form of bits, bytes or words. It is the database that actually exists either in the computer memory or on some secondary storage device.

Programming language
This is a language that an application programmer may use, e.g. FORTRAN.

Program library
This is a collection of subroutines that perform primitive functions.

Property
Property is a named characteristic of an entity, e.g. finite element name, and element material type. Properties allow one to identify, characterize, classify and relate entities.

Property value
This is an occurrence of a property of an entity, e.g. 'element name' has property value BEAM.

Primary key
The entity identifier is referred to as the key of the record group or strictly it is a primary key, e.g. element number.

Query language
Query is the process of question and answer that can be accomplished using the query language, i.e. query the database. The commands are generally quite simple and can be used by nonprogramming as well as programming users. These can be interactive commands as well as utilities that can be called from an application program.

Record
A record is a named collection of data elements or data aggregates. When an application program reads data from a database, it may read one complete record at a time.

Relation
A relation is defined as a logical association of related entities or entity sets. It is represented as $R(A1, A2, \ldots, Ai)$, where $A1, A2$, etc. are the attributes of the relation R. Data in a relation can be represented in the form of a table. Each column of the table is called its attribute. Attributes may be integers, reals, characters, fixed length vectors, variable length vectors or matrices.

Schemes
When a database is to be designed, we develop plans for it. Plans consist of an enumeration of the types of entities that the database deals with, the relationships among the entities, and the ways in which the entities and relationships at one level of abstraction are expressed at the next lower level. The term scheme (schema) is used to refer to plans, so we talk about conceptual schemes and physical schemes. The plan for a 'view' is referred to as a subscheme (subschema).

Secondary key
The database may also use a key which does not identify a unique record but identifies all those which have certain properties. This is called the secondary key.

Storage address (address)
This is a label name or number that identifies the place where data is stored in a storage device, or the part of a machine instruction that specifies the allocation of an operand or the destination of a result.

String data
String data are either of the type character or bit. The length of the string data item is equivalent to the number of characters (for a character string) or the number of binary digits (for a bit string) in the item.

Subscheme (view)
The map of a programmer's view of the data is called a subscheme. It is derived from the global logical view of the data – the schema, and external schema. It is an abstract view of a portion of the conceptual database or conceptual scheme. A scheme may have several subschemes. These are defined using the data definition language (DDL).

Systems programmer
This is a person responsible for installation and maintenance of computer programs.

Vectors
These are one-dimensional ordered collections of data items, all of which have identical characteristics. The dimension of a vector is the number of data items contained in it.

Word
This is the standard main storage allocation unit for numeric data. A word consists of a predetermined number of byte characters or bytes, which is addressed and transferred by the computer circuitry as an entity.

9.2.3 Physical data storage terminology

Address
This is a means of assigning data storage locations and subsequently retrieving them on the basis of key for the data.

Bit
This is an abbreviation of binary digit. The term is extended to the actual representation of a binary digit in a storage medium through an encoded two state device.

Byte
This is a generic term to indicate a measurable portion of consecutive binary digits and is the smallest main storage unit addressable by hardware. In machines with character addressing, byte and character are synonymous.

Character
This is a member of a set of elementary symbols that constitute an alphabet interpretable by computer software. It is also a group of consecutive bits that is used to encode one of the above symbols.

File
A file is a named collection of all occurrences of a given type of logical records. It is also a collection of data sets.

Page
This is a basic unit of primary storage; also basic transaction unit between primary and secondary storage.

Paging
In virtual storage systems, the computer memory is made to appear larger than it is by transferring blocks (pages) of data or programs into memory from external storage when they are needed. This is called paging.

Pointer
This is the address of a record (or other data grouping) contained in another record so that a program may access the former record when it has retrieved the latter record. The address can be absolute, relative or symbolic.

Physical data structure
It is important to distinguish explicitly between logical data structures and the ways in which these structures are represented in the memory of a particular computer. This may be dictated by specific hardware and software systems. The way in which a particular logical data structure is represented in the memory or secondary storage of a computer system is known as storage or physical data structure.

Sequential access
A serial access storage device can be characterized as one that relies strictly on physical sequential positioning and accessing of information.

Storage
This is the process of assigning specific areas of storage to specific type of data.

Transaction
An operation performed on a file or physical database is called a transaction, such as create/insert a record, delete a record, update a record, and find a record.

Virtual memory
This is the simulation of large capacity main storage by a multilevel relocation and paging mechanism implemented in the hardware.

9.2.4 Data structures

A structure whose elements are items of data, and whose organization is determined both by the relationship between the data items and by the access functions that are used to store and retrieve them, is called a data structure (Baron and Shapiro, 1980). Examples of data structures are a list, a tree, a vector of fixed length, a vector of variable length, different types of matrices, a relation, etc. Some of the data structures that can be useful for design optimization are discussed below.

Linear structure
A linear structure is one that is stored in the order in which the data are processed. These data can then be used to manipulate information such as insertion and deletion of elements of information. Two such structures are linear list and linked list.

- A **linear list** is a finite ordered list of elements. For example, records in a file containing the design variable value, the upper bound and the lower bound, constitute a linear list. Another example is a vector that is often used to store information whose size (number of elements) is not known *a priori* or one that cannot be created all at once. In this case, the vector is used as a buffer and the buffer is filled with one element at a time. The buffer is emptied once it is full. A disadvantage with a linear list is that if elements in the list must be deleted or inserted then either the entire list or a portion of the list must be modified.

- In a **linked list**, each element contains a pointer to its successor. Only a linked list where each element contains a pointer to the next element (singly linked list) is discussed here. The process of insertion is carried out by modification of one pointer – the pointer preceding the new entry (the pointer value of the new entry now is the same as the old pointer value of the preceding entry). Deletion follows the same process – the pointer value of the preceding entry must be updated.

Trees

A tree is a data structure that is used to represent hierarchical relationships among data items. A tree is described by nodes; the top level node is the root node. Each node may be connected to a node at the lower level that is the child of the parent node. Figure 9.1 shows a tree structure. As an example of data retrieval from node 17, one must go to nodes 4 and 9 first. Trees form a basis for hierarchical data models.

Matrices

These are the most widely used data structures in engineering programs. They are usually stored in memory in contiguous blocks and are accessed by a simple accessing function. A one-dimensional array is called a vector. Arrays can be of two, three or higher dimensions and are stored either in the column order or row order. Matrices can be of different forms, e.g. rectangular, symmetric, banded, upper triangular, lower triangular, sparse, diagonal, etc. Usually a matrix contains

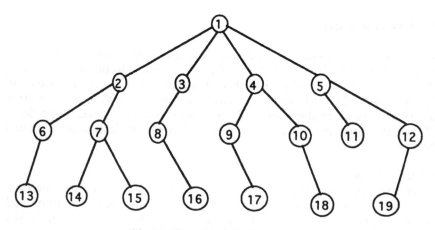

Fig. 9.1 Hierarchical database model.

only one type of data, such as reals or integers. Another similar data structure that is now commonly used is a table where mixed data types can be stored. This forms the basis of the relational data model that is discussed later.

9.3 DATABASE MANAGEMENT CONCEPTS FOR DESIGN OPTIMIZATION

Data management systems in business applications such as accounting, inventory control and task scheduling, are quite sophisticated. However, due to the complex nature of engineering applications, growth of their data management systems has been very slow, although advancements in computer-aided design (graphics) have led to the development of some database systems (DBMS) and concepts (Ulfsby, Steiner and Oian, 1981; Fischer, 1979; Comfort and Erickson, 1978; Fulton and Voigt, 1976; Browne, 1976). Recently the problem has been investigated relative to computer-aided structural analysis and design applications (Rajan, 1982; Rajan and Bhatti, 1983, 1986; SreekantaMurthy and Arora, 1986a, b; Mukhopadhyay and Arora, 1987a, b). This has led to the development of some concepts that are particularly suitable for engineering applications.

9.3.1 Elements of a database

A database can be viewed at many levels of abstraction, such as user's view, conceptual view and physical view. Only the physical database exists on some secondary storage device. The implementation of a physical database is based on a conceptual scheme that is defined in an application program using a data definition language (DDL). The external view represents the data as seen by the interactive terminal users and application programmers. The conceptual view deals with inherent nature of data occurring in the real world and represents a global view of the data. The data organization describing the physical layout is dealt at the internal level. These views – external, conceptual and internal – have been suggested by ANSI/SPARC (American National Standards Institute/Standards Planning and Requirements Committee). A major advantage of using the database capability is that the logical relationships of data are uncoupled from the physical storage scheme; the user need not make calls to files on the computer storage device.

(a) Conceptual data model

Conceptual model (or conceptual scheme) describes the layout of all types of data that need to be stored in a database. It represents real world data independent of any computer constraints and therefore provides a theoretical basis for organizing data of finite element analysis and design optimization problems. A logical approach for conceptualization of data is through information collection and its analysis (Date, 1977). Conceptual model can be derived either by first forming global views and then deriving local views, or by aggregating local views to form a global view. The first approach involves entity identification, relation formation, and name assignment. The other approach is based on combining segregated views of usage and information contents of individual perspectives.

Information collection is through identification of entities based on their properties. To illustrate how this may be done in design optimization, consider the entity PLATE. The entity has properties of two side dimensions, thickness, and material constant E. The associated property values are, say, $100, 50, 0.5$, and 10^7, respectively. These property values belong to a certain domain, for example, domain of property 'dimension' is x, $0 \leqslant x \leqslant 200$. Entity sets are formed by considering entities so identified. Unique names are assigned to the entities and the attributes so formulated. Analysis of information about conceptual objects – entities, attributes and relations – leads to the formation of a conceptual model which replaces the real world information.

(b) Internal model

Logical organization of data to be stored on physical storage media is described by an internal model. It is basically an organization of elementary relations or parts of them, and storing them as a unit to reduce the number of accesses. One approach to define an internal model is by means of relations and use of normalization criteria to obtain relations consistent with the conceptual model (Date, 1977; Martin, 1977). Such a collection of relations reduces redundancy, eliminates undesired anomalies in storage operations, and ensures database integrity. Well-organized procedures need to be used to design the internal data model.

(c) External model

Data structure as seen by an application program or interactive user is called an external data model. Data retrieved from actual physical storage in the database undergoes transformation until it reaches the user. Transformation involves rearrangement of data from internal level to external level into a form acceptable to the application program.

One of the important requirements of a database is to provide a facility for data retrieval by different application programs and users depending on their needs. Different application programs can have different views of a database. To illustrate this consider the needs of two finite element analysis programs. Let one finite element analysis program use a skyline approach to assemble and solve the governing equations. Another program using a hypermatrix approach to perform a similar task needs to use the same database. Basically, the two application programs using some common data such as geometry, material and other finite element idealization data should be able to derive them through an existing database. This aspect of catering to the needs of different applications is possible through an external model.

9.3.2 Elements of a database management system

A database management system (DBMS) is a software system consisting of several programs or subprograms that manages a database created by it. It has capabilities to creat new data sets, delete data sets, modify existing data sets, add or delete data items from data sets and respond to queries about the data in the database. In this section, basic organization and components of a DBMS are presented and explained.

Three types of DBMS can be identified.

- A context-free DBMS is designed to work as a stand-alone package. An application program executes under the control of DBMS, i.e. it acts as the main program. The data manager must be modified or extended when a new application is introduced.
- An application-dependent DBMS is designed for a particular application. Their data definition, data manipulation and query languages use syntax of the application. It is not possible to use the DBMS for other applications unless extensive modifications are done.
- An application-independent DBMS, designed based on the concept of a library of subroutines, is not tied to any particular application. Any application program can call DBMS facilities to define, manipulate and query its database. This type of DBMS is more suitable for design optimization applications.

For more details on different types of DBMS and capabilities of some existing systems, the article by SreekantaMurthy and Arora (1985) should be consulted.

At a higher level of abstraction, a DBMS has two major components as shown in Fig. 9.2: data language interface (DLI), and data storage interface (DSI). These components can have many subcomponents to accomplish their tasks.

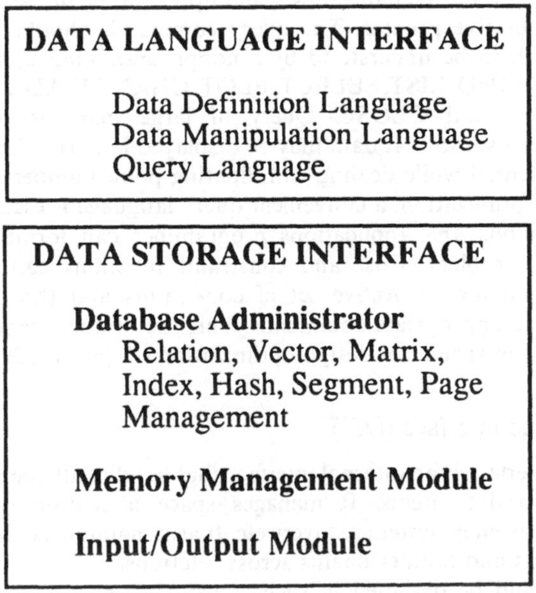

DATA LANGUAGE INTERFACE

Data Definition Language
Data Manipulation Language
Query Language

DATA STORAGE INTERFACE

Database Administrator
Relation, Vector, Matrix,
Index, Hash, Segment, Page
Management

Memory Management Module

Input/Output Module

Fig. 9.2 Components of a database management system.

(a) Data language interface (DLI)

The data language interface is the external interface which can be called directly from a programming language. The high-level data language is embedded within the DLI and is used as the basis for all data definition and manipulation. The language can be in the form of interactive commands, subroutine calls, or some higher-level language constructs. The DLI usually has three subcomponents.

- Data definition language (DDL) is a means to describe data types and logical relations among them. It allows one to assign unique names to data types, to specify the sequence of occurrence, to specify the keys, to assign length of data items, to specify the dimension of a matrix and to specify the password for database security. Since data definition is continuously redefined in an optimization program, DDL must have features to define data dynamically. Also provisions for query of schema must be provided.
- Data manipulation language (DML) contains commands for storing, retrieving and modifying data in the database. These commands should be simple and callable from a higher-level language (such as FORTRAN) as they are frequently used in an application program. They also include utility and schema information commands. Utility commands are used for opening and closing a database, and printing error messages. Schema commands are useful in verifying data definition in situations where they are continuously changing.
- Query language allows a user to interrogate and update the database. Even though some of the design optimization algorithms are automated, it is necessary to provide flexibility and control to the user for modifications during the design process. This control is useful to execute decisions that either cannot be automated or are based on designer's intuition and judgement. For example, a designer may want to change lower and upper limits on the design variables during the iterative process. The query commands should be general and simple enough to be understood by a nonprogramming user. Some typical queries are FIND, LIST, SELECT, PLOT, CHANGE, ADD, DELETE, RENAME, OPEN and CLOSE. Query of large matrices requires special conditional clauses so that data may be displayed in parts. Formatted display of data is essential while dealing with floating point numbers.

  In design optimization, a convenient query language is essential. Using the query commands, the applications programmer can formulate his design optimization problem. Cost and constraint functions can be defined by querying the database. Active set of constraints and their values can be obtained using appropriate commands. This use of query language has been demonstrated in structural design optimization (Rajan, 1982).

(b) Data storage interface (DSI)

Data storage interface is an internal interface that handles all management chores and accesses to data items. It manages space allocation, storage buffers, transaction consistency, system recovery etc. It also maintains indexes on selected fields of relations and pointer chains across relations.

The DSI should be designed in such a way that new data objects or new indexes can be created at any time or existing ones destroyed without exiting or modifications of the system, and without dumping and reloading the data.

One should be able to redefine objects, i.e. change dimensions of matrices or add new fields to relations. Existing programs which execute DSI operations on data aggregates should remain unaffected by the addition of new fields. The DSI has three major subcomponents as shown in Fig. 9.2.

- The database administrator is the brain of the system containing all the administrative information. It has facilities to manage relations, vectors, matrices, pages, indexes etc. Proper algorithms need to be investigated for these tasks for efficient DBMS operations.
- Each DBMS has an internal memory block (often called buffer) to store management information as well as data. This memory needs to be managed judiciously for efficiency of the database operations. The memory management module dynamically controls the allocation of available memory space. The memory is organized into a number of pages, each having the same size. The size of a page is set to a multiple of the physical record. The performance is better with larger page size; however, the space may be wasted if there are too many partially full pages. Small page size leads to increased page replacement activity and maintenance of a larger page table. Variable length pages require more programming effort.
- The input/output module is responsible for all disk operations, i.e. writing and reading of data records on some secondary storage device. For efficiency of operations, system software facilities may be used, making the DBMS system dependent.

9.3.3 Data models

Data in a database can be defined in terms of data sets. Other common approaches are through data models, e.g. hierarchical, network and relational. These data models that represent user's view of the data are described in the sequel with reference to finite element analysis and design optimization data. In all the models, the data definition language is used to define the data structures for the DBMS.

(a) Data set approach

In this approach, the data is organized using uniquely named data sets. Data sets are grouped to form a data library. How the contents of the data sets are managed is completely up to the application programs. Since many of the engineering data are unstructured, the data set offers a simple solution to describe the user's view of the data. Further improvement in this type of organization can be done by defining ordered data sets. For example, row, column or submatrix order may be used to deal with matrix data. The data libraries formed by this approach may be classified according to project or their usage. For example, data of substructures in finite element analysis may be grouped substructure-wise, each in a separate library. This type of data modeling, however, has high redundancy. Also, it is not suitable for interactive use.

(b) Hierarchical model

Hierarchical model organizes data at various levels using a simple tree structure (Date, 1977). This structure appears to fit data of many design problems modeled

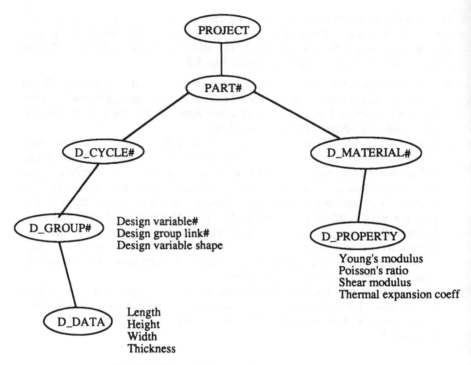

Fig. 9.3 A hierarchical model for design variable data.

using finite elements (Pahl, 1981; Elliott, Kunii and Browne, 1978; Lopez et al., 1978). To illustrate an application of this model, consider the design variable data as shown in Fig. 9.3. The root level in the model represents project name, followed by part number in the immediately next lower level. Part number has two dependent nodes having design cycle (D_CYCLE) and material number (D_MAT) data. At the fourth level design variable group data (D_GROUP) consisting of design variable number, design group link, and design variable shape data are indicated. Material properties data (D_PROPERTY) is also set up at the same level. Lowermost level consists of detailed design data (D_DATA). To retrieve design variable data for a component, one must go through the hierarchy of nodes PART#, D_CYCLE#, D_GROUP#, and D_DATA.

(c) Network model

A collection of arbitrarily connected logical relations is called a network. A data model defined by such a network is called the network model. It is more general than the hierarchical model as it allows many-to-many relations. In the hierarchical model, a parent node can have many child nodes, but a child node must have only one parent. In the network model, a child can also have multiple parent nodes. Disadvantages of the model are in its complexity and the associated data definition language.

Table 9.1 Relation CORD for node coordinate data

NODE#	X	Y	Z
1	720.	0.	0.
2	720.	360.	0.
3	360.	0.	0.
4	360.	360.	0.
5	0.	0.	0.
6	0.	360.	0.

(d) Relational model

Both the hierarchical and network data models have proven to be functionally inefficient. Therefore, the relational model has been developed (Date, 1977; Martin, 1977; Vetter and Maddison, 1981). This model is constructed using a tabular representation of the data which provides a convenient way of representing many engineering data. Each row of the table represents an entity and each column represents an attribute of the relation. Table 9.1 shows an example of a relation CORD for the node coordinate data. This relation has four attributes: NODE#, and X, Y and Z coordinates of the node. There is a distinct relationship between a row entry and the corresponding column entry, and redundant data are not allowed. The mathematical concept of a relation requires the definition of a domain, Cartesian product, and tuples. A set of values in one column is the domain of that attribute, e.g. attribute X for the relation CORD has domain (0., 360., 720.). If there are m domains, then their Cartesian product defines a set of m-tuples (there are as many tuples as there are columns in the table), e.g. each entity (row) of the relation CORD has 4 tuples. In this model, the relational operations, such as JOIN, INTERSECT and PROJECT, can be used on some elementary relations to form new relations.

The relational model is quite appropriate for design optimization applications, since retrieval of data requires smaller preconceived paths. Applications generally require a complete set of related items simultaneously. Retrieving parts of information is not useful. In such a case, the relational model which is set oriented provides a suitable way to organize the design data.

(e) Numerical model

Most of the computations in design optimization involve operations on matrices such as matrix addition, multiplications, solution of simultaneous equations, and eigenvalue calculations. The data models presented earlier are not tailored to handle matrix data effectively. It is necessary to provide a user-friendly facility for defining such numerical data and manipulating a numerical database (Daini, 1982; Rajan and Bhatti, 1983). It is possible to provide such a facility by defining a new data model called numerical model. This numerical model is basically a variation of hierarchical data model having two levels of data representation. At the first level, information pertaining to type, size, and other attributes of the matrix is placed. The second level contains the actual numerical data.

(f) Generalized relational model

Whereas most physical models in business applications can be represented conveniently in the database as relations, engineering applications require both numerical and relational data types. Most large matrices form temporary or semipermanent data private to a program. Relations are either permanent data in public domain used by different users or final results of a program to the end user. Therefore a scheme is needed to represent both relations and matrices in a unified way for integrated engineering applications.

A relation (table) can be imagined as simply a two-dimensional array in which each column has a unique definition. This way, the concept of a matrix is generalized for relations and the resulting model is called the generalized relational model (Mukhopadhyay and Arora, 1987a, b). This scheme is supported by primitive and structured data types. Using these data types users can define their own data models.

The novelty of this model is that the relation is derived from a matrix. In previous attempts, e.g. RIM database management system (Comfort and Erickson, 1978), one tried to extend the relation for matrix data type. That led to clumsy and inefficient handling of numerical data. With the generalized relational model, matrix is the basic data type. The matrix can have elements with composite data structures. Relations are derived from the matrices as vectors of records.

9.3.4 Choice of data model

It is seen that the data models described in previous paragraphs can be used for design optimization applications. The hierarchical data model is suitable where the data to be organized occurs in a truly hierarchical fashion. It has been tried by Lopez (1974) and Pahl (1981) for finite element analysis applications. The model, however, requires complex maneuvers through a chain of pointers to access particular data. Also, it has a fixed structure and offers little flexibility to change to alternate structures. Another drawback of the model is the complexity of database design requiring a tedious process of establishing links between data. If new kinds of data need to be added or new information must be generated from the database, it is necessary to add new links. Generally, this process requires a redesign of the database. Network models have similar problems, although addition of new item is much easier compared with the hierarchical model.

Relational data model provides maximum flexibility of all the three basic data models. Moreover, the model is easier to understand as users find it natural to organize data in the form of tables. A major advantage of the model is the ease with which database can be changed. As the design evolves new attributes and relations can be added, and existing ones deleted easily. The model is more appropriate for design applications, since data storage and retrieval uses a less preconceived path. It is possible to support a simple query structure using this model. Fishwick and Blackburn (1982) have tried this model for finite element analysis and optimization applications with success. Their use of the model was, however, limited to interfacing various programs using an available relational DBMS. A major drawback of the relational model is that it does not provide means to handle large matrix data very efficiently which occurs frequently in engineering applications. A numerical model which is basically a variation of the hierarchical model appears to be quite effective in representing matrix data structures. One can also use the generalized relational model that can treat relations and numerical data in an integrated manner. A prototype DBMS based on this model has been developed and implemented (Mukhopadhyay and Arora, 1987a,b). It has been successfully used in structural optimization applications (Al-Saadoun and Arora, 1989; Park and Arora, 1987; Spires and Arora, 1990; SreekantaMurthy and Arora, 1987). Therefore this model is recommended for engineering applications.

9.3.5 Global and local databases

Computer-aided design of complex structural systems uses several application programs during the design process. Many of these programs require common information such as geometry of the structure, finite element idealization details, material properties, loading conditions, structural stiffness, mass and load distributions, and responses resulting from the analysis runs. Also, it is common that data generated by one program is required for processing in subsequent programs in a certain predetermined pattern. These data do not include transitory information such as intermediate results generated during an analysis run. The transitory information is usually unstructured and its usage pattern is known only to applications that use them. Generally, the transitory information is deleted at the end of a run. Therefore, there is a need for systematic grouping of the data.

A network of databases offers a systematic approach to support data of multiple applications (Jumarie, 1982; Blackburn, Storaasli and Fulton, 1982). A network of databases consists of a global database connected to a number of local databases through a program data interface. Application programs which use them may be thought of as links connecting the databases. A global database contains common information required for all applications whereas a local database contains only application dependent transitory data. Data in a global database is structured and integrity of the database is maintained carefully. Data in a local database, however, is extremely flexible and integrity is not of importance.

The network of databases offers considerable aid in the structural design process. Any changes made to the data in the global databases are immediately available for use in other applications of the system. Any new application program can be added to share the common data. The data views in global databases are clear to all applications and any modified views can be easily incorporated to suit a new application. Local databases are dependent on application programs and are highly efficient in data transfer operations since no overhead is involved in maintaining complicated data structures.

9.4 DATABASE DESIGN TECHNIQUES

The problem is how to organize data in a database, what kind of information is to be stored, what kind of database management system is suitable, and how data is manipulated and used. In this regard, sophisticated techniques are available in business data management area to deal with complex data organization problems. The paper by Koriba (1983) describes several of these approaches and their suitability to computer-aided design (CAD) applications. Buchmann and Dale (1979) and Grabowski and Eigner (1982) have also studied these approaches relative to CAD applications. Most commonly known approaches are ANSI/ SPARC, CODASYL (Committee on Data System Languages, Database Task Group of ACM), relational, hierarchical, and network. Among them, the ANSI/ SPARC approach which recognizes three levels of data views (conceptual, internal and external), provides a generalized framework and basis for a good database design.

Designing a good database is important for successful implementation of finite element analysis and structural design optimization methods (SreekantaMurthy and Arora, 1986a, b). The design procedure should follow well-defined steps. The basic problem is that once all the data items have been identified, how should they be combined to form useful relations. The first step is the extraction of all the characteristics of the information that is to be represented in the database. Analysis of the information to form associations and their integration into one conceptual model is the second step. The conceptual data model obtained by this process is abstract. It is independent of any computer restraint or database management software support. In order for the conceptual model to be useful, it must be expressed in terms that are compatible with a particular DBMS by considering efficiency of storage space and access time. An internal model is developed for this purpose which is compatible with the conceptual data model. Finally, the database design requires accommodation of different users of the database by providing an external data model. The systematic process by which one traverses the different steps of database design and performs the mapping from one level of abstraction to the next is called a database design methodology.

In this section, a methodology based on the ANSI/SPARC approach to design databases for finite element analysis and structural design optimization applications is presented. The methodology considers the following aspects: (i) three views of data – conceptual, internal, and external; (ii) entity set, relationship set, and attributes to form syntactic basic elements of the conceptual model; (iii) relational data model; (iv) matrix data; (v) processing requirements; (vi) normalization of data for relational model. The material for the section is derived from an article by SreekantaMurthy and Arora (1986b).

9.4.1 General concepts

Before presenting the database design methodology a few general concepts that are useful in the design process are described. The idea of an entity–relationship model is explained. Various forms of dependencies between the attributes to form relations are ·explained. The idea of normalization of data is presented and explained.

(a) Entity–relationship

The idea of entities and relationships between them is important in analyzing data and their associations. To explain this concept, consider the entities PRE-

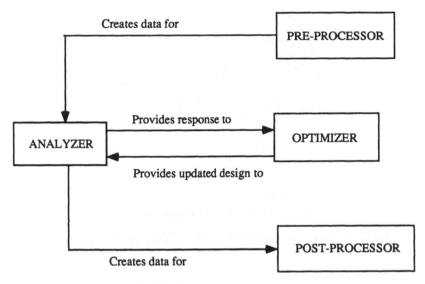

Fig. 9.4 Entity–relationship concept.

PROCESSOR, ANALYZER, POST-PROCESSOR, and OPTIMIZER. Figure 9.4 shows relationships between these entities. PRE-PROCESSOR 'creates data for' the ANALYZER; the ANALYZER determines the structural response and 'provides response to' the OPTIMIZER, and the OPTIMIZER 'provides updated design to' the ANALYZER. The ANALYZER also 'creates data for' the POST-PROCESSOR which provides visual displays for the response quantities. The entity–relationship model for an application is useful in identifying data dependencies and sequence of data usage. This idea of entity–relationship can be also used to identify 'objects' in designing object-oriented design systems.

(b) Functional dependence

An attribute A is functionally dependent on the attribute B of a relation R if at every occurrence of a B value is associated with no more than one A value. This is denoted as $R.B. \rightarrow R.A.$ As an example, consider the relation ELEMENT (EL#, EL-NAME, AREA), where EL# is the element number, EL-NAME is its name and AREA is its cross-sectional area. EL-NAME is functionally dependent on EL#(EL# $\rightarrow$ EL-NAME). AREA is functionally dependent on EL#(EL# $\rightarrow$ AREA). EL# is not functionally dependent on EL-NAME(EL-NAME $\nrightarrow$ EL#), because more than one element could have the same name. Similarly, EL# is not functionally dependent on AREA(AREA $\nrightarrow$ EL#).

An attribute can be functionally dependent on a group of attributes rather than just one attribute. For example, consider the relation CONNECTION for nodal connectivity of triangular finite elements:

$$\text{CONNECTION(EL\#, NODE1\#, NODE2\#, NODE3\#)}$$

Here EL# is functionally dependent on three nodes NODE1#, NODE2#, and

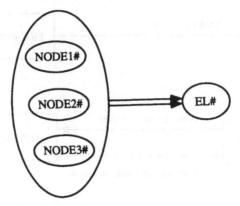

Fig. 9.5 Example of full functional dependency.

NODE3#. Given any one of NODE1#, NODE2#, or NODE3# it is not possible to identify EL#.

(c) Full functional dependency

An attribute or a collection of attributes A of a relation R is said to be fully functionally dependent on another collection of attributes B of R if A is functionally dependent on the whole of B but not on any subset of B. This is written as $R.B. \Rightarrow R.A$. In Fig. 9.5, for example, EL# in the relation CONNECTION of a triangular finite element is fully functionally dependent on concentrated attributes NODE1#, NODE2#, and NODE3# because three nodes combined together define an element. NODE1#, NODE2#, or NODE3# alone does not identify the EL#.

(d) Transitive dependence

Suppose A, B and C are three distinct attributes or attribute collections of a relation R. Suppose the following dependencies always hold: C is functionally dependent on B and B is functionally dependent on A. Then C is functionally dependent on A. If the inverse mapping is nonsimple (i.e. if A is not functionally dependent on B or B is not functionally dependent on C), the C is said to be transitively dependent on A (Fig. 9.6). This is written as

$$R.A \rightarrow R.B, \quad R.B \nrightarrow R.A, \quad R.B \rightarrow R.C$$

Then, we can deduce that

$$R.A \rightarrow R.C, \quad R.C \nrightarrow R.A$$

As an example, consider the relation EL_DISP between element number, element type and degrees of freedom per node:

$$EL\_DISP(EL\#, EL\text{-}TYPE, DOF/NODE)$$

Here

$$EL\# \rightarrow EL\text{-}TYPE, \quad EL\text{-}TYPE \nrightarrow EL\#$$
$$EL\text{-}TYPE \rightarrow DOF/NODE$$

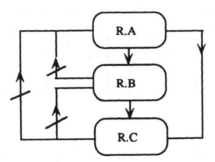

Fig. 9.6 Transitive dependence of *C* on *A*.

Therefore

$$EL\# \rightarrow DOF/NODE \text{ (transitively dependent)}$$
$$DOF/NODE \nrightarrow EL\#$$

9.4.2 Normalization of data

It is seen that data items must be grouped together to form associations. The question is, how to decide what data items should be grouped together? In particular, using a relational model, what relations are needed and what should their attributes be? As a database is changed, older views of data must be preserved so as to avoid having to rewrite the programs using the data. However, certain changes in data associations could force modification of programs which could be extremely disruptive. If grouping of data items and keys is well thought out originally, such disruptions are less likely to occur.

Normalization theory (Date, 1977; Martin, 1977; Vetter and Maddison, 1981) provides certain guidelines to organize data items together to form relations. The theory is built around the concept of normal forms. A relation is said to be in a particular normal form if it satisfies a certain specified set of constraints. Three normal forms – first, second and third – are described below.

(a) First normal form (1NF)

A relation is said to be in the first normal form if and only if it satisfies the constraint of having atomic values. As an example, Fig. 9.7 shows the relation CONNECTION between four attributes: element number (EL#), element name (EL-NAME), node numbers (NODES#) and degrees of freedom per node (DOF/NODE) with domains D_1, D_2, D_3 and D_4, respectively. The relation is first shown not in the 1NF and then in the 1NF.

(b) Second normal form (2NF)

A relation is in second normal form if and only if it is in 1NF and every nonkey attribute is fully functionally dependent on each candidate key. Let us see whether the relation CONNECTION of Fig. 9.7 that is in the 1NF is also in the 2NF.

Domain D_1	Domain D_2	Domain D_3	Domain D_4
EL#	EL-NAME	NODES#	DOF/NODE
1	BEAM3	1 2	6
2	TRUSS3	3 4	3 2
3	PLATE	5	

CONNECTION

Key	Key			Key	Key		
EL#	EL-NAME	NODES #	DOF/NODE	EL#	EL-NAME	NODES#	DOF/NODE
1	BEAM3	1 2	6	1	BEAM3	1	6
2	TRUSS3	3 5	3	1	BEAM3	2	6
3	PLATE	2 3 4 5	2	2	TRUSS3	3	3
				2	TRUSS3	5	3
				3	PLATE	2	2
				3	PLATE	3	2
				3	PLATE	4	2
				3	PLATE	5	2

Not in 1NF   In 1NF

Fig. 9.7 First normal form for the relation CONNECTION.

Consider a nonkey attribute EL-NAME:

$$(EL\#, NODES\#) \rightarrow EL\text{-}NAME$$
$$EL\# \rightarrow EL\text{-}NAME$$
$$NODE\# \nrightarrow EL\text{-}NAME$$

Therefore (EL#, NODES#) $\nrightarrow$ EL-NAME, i.e. EL-NAME is not fully functionally dependent on (EL#, NODES#). Similarly for the nonkey attribute DOF/NODE:

$$(EL\#, NODES\#) \rightarrow DOF/NODE$$
$$EL\# \rightarrow DOF/NODE$$
$$NODE\# \nrightarrow DOF/NODE$$

Therefore (EL#, NODES#) $\nrightarrow$ DOF/NODE. Since neither EL-NAME nor DOF/NODE is fully functionally dependent on candidate key (EL#, NODES#), the relation CONNECTION is not in 2NF.

Conversion of the relation CONNECTION to 2NF consists of replacing it by two of its projections as shown in Fig. 9.8 (note: $\leftarrow$ implies PROJECT operation):

NAM_DOF $\leftarrow$ CONNECTION (EL#, EL-NAME, NODES#, DOF/NODE)
EL_NODE $\leftarrow$ CONNECTION(EL#, EL-NAME, NODES#, DOF/NODE)

Relation EL-NODE does not violate 2NF because its attributes are all keys.

(c) Third Normal Form (3NF)

A relation is in the third normal form if it is in second normal form and its every nonprime attribute is nontransitively dependent on each candidate key of the relation. For example, consider the relation NAM-DOF (Fig. 9.8) to see whether

NAM_DOF

EL#	EL-NAME	DOF/NODE
1	BEAM3	6
2	TRUSS3	3
3	PLATE	2

ELMT_NODE

EL#	NODE#
1	1
1	2
2	3
2	5
3	2
3	3
3	4
3	5

Fig. 9.8 Second normal form for the relation CONNECTION.

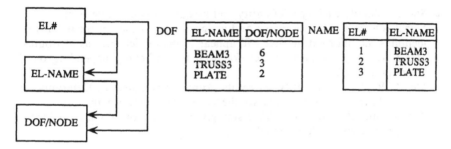

Fig. 9.9 Third normal form for the relation NAM_DOF.

it is in third normal form. It still suffers from a lack of mutual independence among its nonkey attributes. The dependency of DOF/NODE on EL#, though it is functional, is transitive (via EL-NAME). Each EL# value determines an EL-NAME value and in turn determines the DOF/NODE value. This relation is reduced further into relations NAME and DOF in Fig. 9.9. These relations are in the third normal form.

9.4.3 Methodology to develop a conceptual model

Analysis of the data used in finite element analysis and structural design optimization is necessary to develop a conceptual data model. In the analysis, the information in use or needed later is identified, classified, and documented. This forms the basis for a conceptual data model to represent structural design data and the design process as a whole.

The following steps can be used to develop a conceptual data model.

1. Identify all the conceptual data objects of structural analysis and design optimization.
2. Data identified is stored in a number of relations. The data is reduced to elementary relations representing inherent association of data.
3. More elementary relations are derived from the ones formed in step 2. This step uncovers more relationships between the basic data collected in step 1.

4. Redundant and meaningless relations obtained in step 3 are removed to obtain a conceptual data model.

The conceptual model obtained by this process is abstract, representing the inherent nature of structural design data and is independent of any computer restraint or database management software support. These steps are now discussed in detail.

(a) Identification of conceptual data objects

The following steps can be used to identify the conceptual data objects used in structural design. Entity sets and attributes are considered to be the syntactic basic elements of the model. Domain definition can be extended to include vectors and matrices.

- *Step 1.* Identify each type of entity and assign a unique name to it.
- *Step 2.* Determine the domains and assign unique names to them. This step identifies the information that will appear in the model, such as attributes.
- *Step 3.* Identify the primary key for each type of entity depending on the meaning and use.
- *Step 4.* Replace each entity set by its primary key domains. Determine and name relations corresponding to the association between the primary key domain and other domains. This step gives a collection of relations forming a rough conceptual data model.

EXAMPLE 9.1

We consider a sample structural design problem to describe these steps.

- *Step 1.* The following entity sets can be identified for the structure:

    STRUCTURE, BEAM, TRUSS(TR), MEMBRANE-TRI(TRM),
    MEMBRANE-QD(QDM), NODE, ELEMENT(EL)

- *Step 2.* We can identify the following domains:

STRUCTURE#	Structure identification number (integer)
BEAM#	Beam element identification number (integer)
TR#	Truss element identification number (integer)
TRM#	Triangular membrane element identification number (integer)
QDM#	Quadrilateral membrane element identification number (integer)
NODES#	Node number (integer)
EL#	Element number (integer)
EL-TYPE	Element type {BEAM2, BEAM3, TR2, TR3, TRM2, TRM3, QDM2, QDM3}
MATID	Material identification code, for example, {STEEL.1, STEEL.2, ALUM.5, COMP.1}. It also refers to a relation

	or table of material properties; for example, STEEL.1 refers to relation STEEL and material subtype 1
MATPRO	Material property $\{E, \mu, G, \ldots\}$
CSID	Cross-section-type identification code; for example, $\{$THICK.1, THICK.2, RECT.1, CIRC.5, ISEC.6, LSEC.15$\}$. It also refers to a relation of cross-sectional details. For example, RECT.1, refers to a relation RECT and a cross-section subtype 1
CSPRO	Cross-sectional property $\{H, W, T, R, \ldots\}$; $\{$height, width, thichness, radius, $\ldots\}$
DOF#	Degrees of freedom numbers
LOAD-TYP	Load type $\{$CONCENTRATED, DISTRIBUTED, TEMPERATURE, ACCELERATION$\}$
X	X coordinate (real)
Y	Y coordinate (real)
Z	Z coordinate (real)
DESCRIPTION	Description (characters)
VEC	Vectors $\{$integer, real, and double precision vectors$\}$
MATX	Matrices $\{$integer, real, and double precision matrices$\}$
VECID	Vector identification code $\equiv \{x \cdot y | x = $ vector description, $y = $ number$\}$; for example, FORCE.5, LOAD.10
MAXID	Matrix identification code $\equiv \{x \cdot y | x = $ matrix description, $y = $ number$\}$; for example, EL-STIFF.10, EL-MASS.5

- *Step 3.* The following entity keys are identified:

STRUCTURE#	for entity set structure
BEAM#	for entity set beam
TR#	for entity set truss
TRM#	for entity set TRM
QDM#	for entity set QD
EL#	for entity set element

- *Step 4.* In the association between entity sets and domain the entity sets from step 1 are replaced by their primary keys. Attribute names are derived from domain names to provide role identification. The following relation TRM for the triangular membrane element is identified for the entity set TRM:

TRM(TRM#, EL#, EL-TYPE, MATID, E, NODE1#, NODE2#, NODE3#, CSID, T, LOAD-TYP, LOAD#, VECID, VEC, MAXID, MATX)

A triangular membrane element is identified by TRM#. Element number EL# uniquely identifies the finite elements of a structure. Attributes NODE1#, NODE2#, and NODE3# are derived from domain NODES#. Similarly, E is the role name for domain MATPRO. CSID identifies the cross-section property T, the thickness. Vectors and matrices associated with the element are identified through VECID and MAXID, respectively. Similarly, the relations TRUSS, BEAM, QDM are obtained.

(b) Reduction to elementary relations

In the previous section we described a method to identify entities, domains, and relations to produce a rough conceptual model of the structure. Our idea is to develop a conceptual model that contains all the facts and each fact occurring only once. In order to produce a conceptual data model, we transform the rough model into a better model by using a set of elementary relations (Vetter and Maddison, 1981). Using the concept of functional dependencies, full functional dependencies, and transitive dependencies, we can establish rules for reducing a relation to an elementary relation. The following steps are identified to form elementary relations.

- *Step 1*. Replace the original relations by other new relations to eliminate any (nonfull) functional dependencies on candidate keys.
- *Step 2*. Replace the relations obtained in step 1 by other relations to eliminate any transitive dependencies on candidate keys.
- *Step 3*. Go to step 5 if (a) the relation obtained is all keys, or (b) the relation contains a single attribute that is fully functionally dependent on a single candidate key.
- *Step 4*. Determine the primary key for each relation that may be a single attribute or a composite attribute. Take projections of these relations such that each projection contains one primary key and one nonprimary key.
- *Step 5*. Stop when all elementary relations are obtained.

EXAMPLE 9.2

To see how these steps are used, we consider the relation TRM that was given earlier and reduce it to elementary relations:

- *Step 1*.

ER1(TRM#, EL#)	ER2(TRM#, EL-TYPE)
ER3(TRM#, NODE1#)	ER4(TRM#, NODE2#)
ER5(TRM#, NODE3#)	ER6(EL#, TRM#)

- *Step 2*.

ER7(TRM#, MATID)	ER8(MATID, E)
ER9(TRM#, CSID)	ER10(CSID, T)
ER11(TRM#, VECID)	ER12(VECID, VEC)
ER13(TRM#, MAXID)	ER14(MAXID, MATX)

- *Step 3*. The preceding relations contain a single attribute, so go to step 5.
- *Step 4*. Skip.
- *Step 5*. ER1–ER14 are elementary relations.

The steps can be applied to the rest of the relations identified earlier to obtain a set of elementary relations for the sample structural problem.

Table 9.2 Transitive closure for elementary relations

Derived relations	Dependencies	Composition	Semantically meaningful
ER15	EL#→EL-TYPE	EL#→TRM#→EL-TYPE	Yes
ER16	EL#→NODE1#	EL#→TRM#→NODE1#	Yes
ER17	EL#→NODE2#	EL#→TRM#→NODE2#	Yes
ER18	EL#→NODE3#	EL#→TRM#→NODE3#	Yes
ER19	EL#→MATID	EL#→TRM#→MATID	Yes
ER20	EL#→CSID	EL#→TRM#→CSID	Yes
ER21	EL#→LOAD-TYP	EL#→TRM#→LOAD-TYP	Yes
ER22	EL#→MAXID	EL#→TRM#→MAXID	Yes
ER23	TRM#→E	EL#→MAXID→E	No
ER24	TRM#→T	EL#→CS-TYP→T	No
ER25	TRM#→VEC	EL#→VECID→VEC	No
ER26	TRM#→MATX	EL#→MAXID→MATX	No

(c) Determination of transitive closure

While deriving a large number of relations for obtaining a conceptual data model it is prossible that some relations might have been missed. In general, one can derive further elementary relations from any incomplete collection of such relations. To explain in a simple way how such additional relations can be derived, consider two relations ER1(A, B) and ER2(B,C), which imply functional dependencies: A → B and B → C. Taking the product of these functional dependencies, we obtain A → C. Therefore, from suitable pairs of elementary relations representing functional dependencies, further elementary relations can be derived. Deriving all such relations from the initial collection of elementary relations yields a transitively closed collection of elementary relations called the transitive closure (Vetter and Maddison, 1981).

There are problems associated with interpreting relations in transitive closure. For example, consider relations ER1(TRM#, MATID) where TRM# → MATID, and ER7(MATID, E) where MATID → E. Transitive closure for this set yields the relation ER(TRM#, E) which implies TRM# identifies E. This relation, however, does not represent true information since material property E is dependent only on the material number and not on the element number. The relation could be wrongly interpreted. Therefore such semantically meaningless dependencies must be eliminated. It is possible to determine transitive closure by using directed graphs and the connectivity matrix (Vetter and Maddison, 1981).

The transitive closure for the example produces additional dependencies as given in Table 9.2. We have eliminated meaningless dependencies from the list.

(d) Determination of minimal covers

We need to remove redundant elementary relations to provide a minimal set of elementary relations. A minimal cover is the smallest set of elementary relations from which transitive closure can be derived (Vetter and Maddison, 1981). The following points are noted: (i) minimal cover is not unique, (ii) deriving several

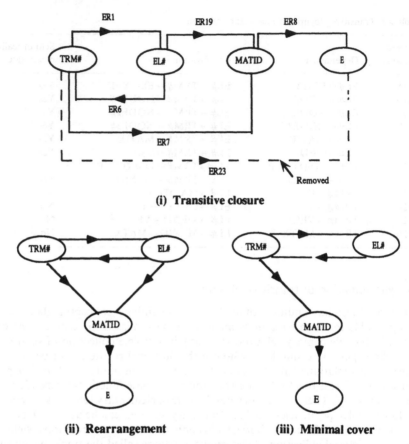

Fig. 9.10 Diagraph representation of minimal cover.

alternative minimal covers from a transitive closure guarantees that every possible minimal cover is found, and (iii) we can select a minimal cover that best fits the structural design process needs.

An example of finding a set of minimal covers from the transitive closure derived in previous sections is given in Fig. 9.10. A set of minimal cover for this transitive closure is {ER1, ER6, ER7, ER8} and {ER1, ER6, ER19, ER8}, out of which one set may be chosen to suit requirements.

The preceding procedure can be applied to other transitive closures derived in the previous section. Thus we can obtain further sets of minimal covers. Each minimal cover is a nonredundant list of elementary relations and is an appropriate conceptual model of the structural design data.

9.4.4 Database design to support an internal model

An internal model deals with the logical organization of data to be stored on physical storage devices. Once a conceptual model has been developed, the internal model needs to be developed so that the data can be stored in files. The methodology to design such a model considers two important aspects: (i) normalization of data, and (ii) processing requirements of data. Normalization of data avoids anomalies in storage and update operations (insert, modify, delete) of data (Date, 1977). The processing requirements are important because they specify how various attributes should be derived from their underlying domains and combined together to from a relation.

An example of element stiffness matrix generation is considered to describe the methodology to design the internal model. Consider the conceptual model given by the following elementary relations:

$$\begin{array}{ll} ER1(TRM\#, EL\text{-}TYPE) & ER2(TRM\#, NODE1\#) \\ ER3(TRM\#, NODE2\#) & ER4(TRM\#, NODE3\#) \\ ER5(TRM\#, MATID) & ER6(MATID, E) \\ ER7(TRM\#, CSID) & ER8(CSID, T) \\ ER9(NODE\#, X) & ER10(NODE\#, Y) \\ ER11(NODE\#, Z) & ER12(TRM\#, MAXID) \\ ER13(MAXID, MATX) & ER14(EL\#, NODES\#) \\ ER15(TRM\#, EL\#) & \end{array}$$

Data needed for the generation of the element stiffness matrix is derived from various domains and represented in a single relation TRM_D as shown in Table 9.3. The domains used are TRM#, MATID, CSID, NODES, X, Y, Z, MAXID, MATX, EL#, EL-TYPE, CSPRO and MATPRO. Our main intention is to obtain all the data required for generation of stiffness matrices for triangular membrane elements in one access or a minimum number of accesses. It is observed that the relation in Table 9.3 is not in the first normal form. Therefore this unnormalized relation should be replaced by a semantically equivalent relation in 1NF as shown in Table 9.4. The advantage of 1NF over the unnormalized relation is that operations required for application programs are less complicated and easy to understand.

To check consistency of this model, first we identify the key attributes. Candidate keys are compound consisting of (EL#, NODES#) and (TRM#, NODES#). The primary key is selected as (TRM#, NODES#). Secondary keys are TRM#, EL#, MATID, CSID, NODES#, and MAXID. These key attributes of the relation are consistent with those in the elementary relation. Second, we need to identify whether all the attributes in the internal model and dependencies between them are consistent with the conceptual model. It can be observed that attributes NODE1#, NODE2#, and NODE3# do not appear in the relation. Therefore these there attributes should be included in the relation. The relation TRM_D is now written as:

$$TRM\_D(TRM\#, EL\#, EL\text{-}TYPE, E, CSID, T, NODE1\#,$$
$$NODE2\#, NODE3\#, X, Y, Z, MAXID, MATX)$$

The functional dependencies reflected by elementary relations ER1 to ER15 are satisfied in the internal model with the values shown in Table 9.4. Therefore at this instant the internal model is consistent with the conceptual model. However, it would be no longer consistent if arbitrary changes in the values of the table are made. Also, note that many values in the relation TRM_D are redundant.

Table 9.3 A tentative internal model TRM_D

TRM#	EL#	EL-TYPE	MATID	E	CSID	T	NODE#	X	Y	Z	MAXID	MATX
1	12	TRM3	STEEL.2	1.0E8	THICK.2	.1	5	1.	6.	8.	STF.1	[•]
							6	2.	5.	9.		
							9	3.	6.	7.		
							15	6.	8.	10.		
2	15	TRM3	ALUM.2	0.9E7	THICK.3	.2	16	4.	7.	9.	STF.2	[•]
							17	5.	9.	2.		

Table 9.4 Relation TRM_D in 1NF

TRM#	EL#	EL-TYPE	MATID	E	CSID	T	NODE#	X	Y	Z	MAXID	MATX
1	12	TRM3	STEEL.2	1.0E8	THICK.2	.1	5	1.	6.	8.	STF.1	[•]
1	12	TRM3	STEEL.2	1.0E8	THICK.2	.1	6	2.	5.	9.	STF.1	[•]
1	12	TRM3	STEEL.2	1.0E8	THICK.2	.1	9	3.	6.	7.	STF.1	[•]
2	15	TRM3	ALUM.2	0.9E7	THICK.3	.2	15	6.	8.	10.	STF.2	[•]
2	15	TRM3	ALUM.2	0.9E7	THICK.3	.2	16	4.	7.	9.	STF.2	[•]
2	15	TRM3	ALUM.2	0.9E7	THICK.3	.2	17	5.	9.	2.	STF.2	[•]

These inconsistencies and redundancies occur because of the anomalies in the 1NF. Thus it is not desirable to use the relation in Table 9.4 to represent the internal model. Modification to 2NF is necessary to avoid the anomalies in the storage operations (Date, 1977). The relation TRM_D should be converted into a set of semantically equivalent relations as follows:

TRM_D1(TRM#, EM#, EL-TYPE, NODE1#, NODE2#, NODE3#,
    MATID, E, CSID, T, MAXID, MATX)
TRM_D2(NODE#, X, Y, Z)
TRM_D3(EL#, NODE#)

The preceding three relations TRM_D1, TRM_D2, and TRM_D3 are all in 2NF because the first two relations do not possess any compound candidate key and the third relation has all keys. Note that by splitting the relation TRM_D no information is lost and the relations are still consistent with the conceptual model. However, the TRM_D1 relation is still not satisfactory since it can lead to anomalies in storage operations. Modification of the relation is necessary to 3NF to avoid anomalies in storage operation. Nonkey attributes must be nontransitively dependent on candidate keys to avoid these anomalies. It can be observed from the relation TRM_D1 of Table 9.5 that attributes E, T, and MATX are transitively dependent on TRM# through MATID, CSID, and MAXID, respectively. Removing these transitive dependencies, we obtain the following relations:

TRM_D4(TRM#, EL#, EL-TYPE, NODE1#, NODE2#, NODE3#,
    MATID, CSID, MAXID)
TRM_D5(MATID, E)
TRM_D6(CSID, T)
TRM_D7(MAXID, MATX)

The preceding three relations together with TRM_D2 and TRM_D3 constitute the internal model for element stiffness matrix generation purpose. This internal model is consistent with the conceptual model identified earlier. Also, note that the number of relations in the internal model is only 6 as compared with 15 elementary relations in the conceptual model.

In summary, the following steps are necessary to derive an internal model that is consistent with the conceptual model. Normalization procedures have to be adopted at each step to reduce redundancy and to eliminate undesired anomalies in storage operation. This ensures integrity of the stored values in the database. At each step unsatisfactory relations are replaced by others.

- *Step 1.* Form relations with attributes derived from a set of domains.
- *Step 2.* Eliminate multiple values at the row–column intersection of the relation table. Vectors and matrices are considered to be single data items for this step.
- *Step 3.* The relations are in the 1NF as a result of step 2. Take projections of 1NF relations to eliminate any nonfull functional dependencies and get relations in the 2NF.
- *Step 4.* Take projection of relations obtained in step 3 to eliminate transitive dependencies to form relations in the 3NF. Thus a set of relations in the 3NF is the internal model.

Table 9.5 Relations in 2 NF

TRM_D1

TRM#	EL#	EL-TYPE	NODE1#	NODE2#	NODE3#	MATID	E	CSID	T	MAXID	MATX
1	12	TRM3	5	6	9	STEEL.2	1.0E8	THICK.2	.1	STF.1	[·]
2	15	TRM3	15	16	17	ALUM.2	0.9E7	THICK.3	.2	STF.2	[·]

TRM_D2

NODE#	X	Y	Z
4	1.	6.	8.
6	2.	5.	9.
9	3.	6.	7.
15	6.	8.	10.
16	4.	7.	9.
17	5.	9.	2.

TRM_D3

EL#	NODE#
12	5
12	6
12	9
15	15
15	16
15	17

9.4.5 Some aspects to accommodate an external model

One of the important requirements of a database is to provide facility for data retrieval by different application programs depending on their needs. Different application programmers can have different views of a database. Transformations are required involving rearrangement of data from the internal level to the external level into a form acceptable to the application program. Some constraints have to be observed while designing an external model. Constraints arise while rearranging data from internal data structure to an external data structure. Any retrieval and storage operations specified on the external model must be correctly transformed into corresponding operations on the internal model, and at the same time, the internal model must be consistent with the conceptual data model. An example of how an external model is derived from an internal model is given subsequently.

Suppose a particular user would like to know the coordinates of nodes of each triangular finite element for generation of element stiffness matrices. This means that the external model

EL_CORD(TRM#, EL#, EL-TYPE, X1, Y1, Z1, X2, Y2, Z2, X3, Y3, Z3)

has to be provided for that particular user. Note that the external view EL-CORD contains data item from two different relations – TRM_D4, TRM_D2. Therefore a procedure is required to transform the internal data model (relations TRM_D4 and TRM_D2) to the external data model (relation EL_CORD). This can be done using JOIN and PROJECT operations (Date, 1977) as follows (note: ← indicates PROJECT; *indicates JOIN):

TRM_A(TRM#, EL#, EL-TYPE, NODE1#) ← TRM_D4
TRM_B(TRM#, EL#, EL-TYPE, NODE2#) ← TRM_D4
TRM_C(TRM#, EM#, EL-TYPE, NODE3#) ← TRM_D4
TRM_D(TRM#, EL#, EL-TYPE, X1, Y1, Z1) = TRM_A*TRM_D2
TRM_E(TRM#, EL#, EL-TYPE, X2, Y2, Z2) = TRM_B*TRM_D2
TRM_F(TRM#, EL#, EL-TYPE, X3, Y3, Z3) = TRM_C*TRM_D2
EL_CORD(TRM#, EL#, EL-TYPE, X1, Y1, Z1, X2, Y2, Z2, X3, Y3, Z3)
   = TRM_D*TRM_E*TRM_F

It can be seen from the algorithm that we did not modify the original relations TRM_D4 and TRM_D2 to retrieve the data required for a particular inquiry. The relations TRM_D4 and TRM_D2 are still consistent with the conceptual model. Therefore pure retrieval operations for rearrangement of data do not cause any inconsistency in data values.

Now, consider the reverse process of transforming external data structure to internal data structure. Suppose a particular user wants to insert the nodal coordinates of a finite element using the external view EL_CORD. Here relation EL_CORD has the only key TRM# and has no reference to the node number to which the element is connected. Insertion is not consistent with the conceptual model because it requires the coordinates of nodes which are dependent on keys NODE#. This restriction is also reflected in the internal model – TRM_D2 that requires NODE3 as key values for insertion. Therefore the transformation of relation EL_CORD into the internal model is not possible. From this example it follows that there are restrictions for rearranging data from the external model to the internal model.

9.4.6 Methodology to incorporate matrix data into a database

In finite element analysis and structural design optimization, we encounter the problem of storage of large-order matrices. This data is unique to the application, so no attempts have been made to design such databases in the business database management area. However, some studies have been reported in the recent engineering literature to deal with large-order matrices. The methodology for representing large matrices can be based on the conceptual, internal and external views.

Conceptually, a matrix is a two-dimensional array of numbers. These numbers appear in a certain pattern; for example, square, sparse, symmetric, diagonal, banded, lower triangular form, upper triangular form, tridiagonal form, hypermatrix form, and skyline form. A matrix is uniquely identified by a name. Rows and columns of the two dimensional array are used for identification of data elements in the matrix. A conceptual view of a matrix can be represented by the following elementary relations:

ER1(NAME, MATRIX TYPE)
ER2(NAME, NUM-OF-ROWS)
ER3(NAME, NUM-OF-COLUMNS)
ER4(NAME, ROW, COLUMN, DATA-ELEMENT-VALUE)
ER5(NAME, NUM-OF-HYPER ROWS)
ER6(NAME, NUM-OF-HYPER COLUMNS)
ER7(NAME, HYPER-ROW, HYPER-COLUMN, ROW, COLUMN,
       DATA-ELEM-VALUE)
ER8(NAME, BAND-WIDTH)
ER10(NAME, SUB-MAT-ROW-SIZE)
ER11(NAME, SUB-MAT-COLUMN-SIZE)
ER12(NAME, VECTOR OF SKYLINE-HEIGHT)
ER13(NAME, HYPER-ROW, HYPER-COLUMN, NULL-OR-NOT)

The attributes of these elementary relations are self-descriptive. These elementary relations completely define a matrix and provide the conceptual structure of the matrix.

An internal (storage) structure for large-order matrices has to be developed that is consistent with the conceptual structure. Storage schemes have to be developed based on efficiency and processing considerations. The special nature of the matrix, that is, sparse, dense, or symmetric, should be used to provide storage efficiency. We can classify various matrix types into two basic types – sparse and dense. Many possible storage schemes are available to store dense and sparse matrices. Conventional storage schemes – row-wise, column-wise, submatrix-wise – are useful for storing dense matrices. Choice among these storage schemes should be based on consideration of several aspects – storage space, processing sequence, matrix operation, page size, flexibility for data modification, ease of transformation to other schemes or user's views, number of addresses required to locate rows or submatrices, and availability of database management system support.

Storage space
The row storage scheme can be used for square, banded, and skyline matrix types. However, this scheme is not appropriate for the hypermatrix. Symmetric,

triangular, and diagonal properties of a square matrix can be used in saving storage space if the variable length of rows is used. Similar schemes can be used for banded and skyline matrices to store data elements that appear in a band or skyline column. Submatrix storage can be used for all matrix types. The submatrix is most appropriate for hypermatrix data. Both schemes have disadvantages when zero elements within a row or submatrix have to be stored.

Processing sequence
Row storage requires that the assembly of matrices, storage, and retrieval be made only row-wise. This becomes inefficient if row-wise processing cannot be done. The submatrix approach is suitable for all types of processing sequences – row-wise, column-wise, or in any arbitrary order.

Matrix operations
Operations such as transposition, addition, multiplications, and solutions of simultaneous equations are frequently carried out at various stages of the structural design process. The row storage scheme is inefficient for matrix transpose when column-wise storage is required. During multiplication of two matrices A and B, a column of B can be obtained only by retrieving all of the rows of B. Therefore, the row storage scheme becomes inappropriate for such an operation. However, the submatrix storage scheme does not impose any such constraints in matrix operation, thus providing a suitable internal storage scheme.

Page size
A page is a unit or block of data stored or retrieved from memory to disk. For a fixed page size, only a number of full rows or a number of full submatrices together with fractional parts of them can be stored or retrieved at a time. It is clear that fragmentation of rows or submatrices takes place depending on the size of rows or submatrices. Large row size will overlap more than one page in memory and cause wastage of space. The submatrix scheme has the advantage of providing flexibility in choosing the submatrix size to minimize fragmentation of pages.

Flexibility for data modification
For modification of rows of a matrix, both row and submatrix storage schemes are suitable. However, the row scheme would be more efficient than the submatrix storage scheme. For modification of a few columns of a matrix, the row storage scheme may require a large number of I/O.

Transformation to other schemes
The submatrix storage scheme requires a minimum number of data accesses to transform to the column-wise or row-wise storage scheme.

Address required
The submatrix storage requires a lower number of addresses to locate data than the row or column storage scheme, provided that submatrices are reasonably large.

For internal storage of large-order matrices in a database, the preceding aspects should be carefully considered. It appears that both submatrix and row storage schemes can be appropriate for various applications.

In order that the internal storage scheme be consistent with the conceptual model, we need to store additional information about the properties of the matrix. That additional information is given by the elementary relations ER1–ER3, ER5, ER6, ER9–ER13. These can be combined and stored in a relation.

So far we have considered schemes for internal organization of large matrices. Since different users may view the same matrix in different forms – banded, skyline, hypermatrix, triangular, or diagonal – it is necessary to provide external views to suit individual needs. The unit of transactions on various views of a matrix may be row-wise, column-wise, submatrix-wise, or data element wise. If the internal scheme is submatrix-wise, the external view need not be submatrix wise. Therefore transformations are necessary to convert the internal matrix data into the form required by a particular user.

Next, we consider the sparse matrix storage scheme. Several storage schemes have been suggested by Pooch and Nieder (1973) and Daini (1982). They are the bit-map scheme, address map scheme, row–column scheme, and threaded list scheme. Out of these, the row–column scheme is simple and easy to use. Also, the row–column scheme can be easily incorporated into the relational model. Therefore, this scheme can be considered for storing sparse matrices encountered in finite element analysis and design applications.

The row-column storage scheme consists of identification of row and column numbers of nonzero elements of a sparse matrix and storing them in a table. This scheme provides flexibility in the modification of data. Any nonzero value generated during the course of a matrix operation can be stored or deleted by simply adding or deleting a row in the stored table. The external view of the row–column storage scheme can be provided through suitable transformation procedures.

Many procedures in structural design, such as element stiffness matrix routines, generate huge amounts of data. Generally, it is not preferable to store such data in a database at the expense of disk space and data transportation time. This inefficiency can be avoided by storing only a minimum amount of data needed to generate the required information (element stiffness matrix). In general, a data model can be replaced by (i) an algorithm that generates the user-requested information, and (ii) a set of (minimum) data that will be used by an algorithm to generate user-requested information.

9.5 CONCLUDING REMARKS

Various concepts of database design and management applicable to engineering design optimization are discussed and illustrated with examples. Conceptualization of design data, and its internal and external views, are discussed. The user's view of data is described through hierarchical, network, numerical and generalized relational models. Even though needs of database management in engineering and business applications are different, there is some commonality between them, and we can take advantage of these developments. One major difference is the need to manage dynamically large amounts of numerical data in engineering and scientific applications. Numerical data models can be used for this purpose. Alternatively, a recently developed generalized relational model can be used to treat various types of data in a unified way (Arora and Mukhopadhyay, 1988).

A methodology to design databases for finite element analysis and structural design is described. The methodology considers several good features of the available database design techniques, such as three views of data organization – conceptual, internal, and external. Tabular and matrix forms of data are included. The relational data model is shown to be quite useful in providing a simple and clear picture of data in the database design. Entities, relations, and attributes are considered to form a conceptual view of data. First, second, and third normal forms of data are suggested to design an internal model. More recently developed normal forms can also be used. Several aspects such as processing, iterative needs, multiple views of data, efficiency of storage and access time, and transitive data are considered in the methodology. Examples used for discussing the methodology are relevant and can be extended to the actual database design for computed-aided analysis and optimization of structural systems.

The need to use organized databases in engineering and scientific applications has increased in recent years. Large and complex systems in multidisciplinary environment need to be designed. Software systems for such applications consisting of perhaps several independent programs can be quite complex. Properly designed databases and the associated database management system are essential in the design process as well as in extending and modifying the software modules. The database capabilities can be used at two levels:

- At a global level, the database can store permanent information about the system being designed, such as the finite element data, element connectivity, node coordinates, material properties, loading conditions, and other such data. These are permanent types of data that may be accessed by different application programs.
- Each application program may generate huge amount of data and save it for later use. These data are usually not needed at the end of the run, so they may be kept in a temporary database and deleted at the end of the run.

Therefore, we see that it is essential for the database management system to have facilities to manage static as well as dynamic data during execution of the program. Concepts of global and local databases provide a means to organize data at various levels in the design optimization environment.

Finally, it is important to note that use of a well-designed database and the associated database management system can speed up the development time for a software system. This has been demonstrated in the literature through the development of a few prototype engineering design systems (Al-Saadoun and Arora, 1989; Park and Arora, 1987; Rajan and Bhatti, 1986; Spires and Arora, 1990; SreekantaMurthy and Arora, 1987).

REFERENCES

Al-Saadoun, S.S. and Arora, J.S. (1989) Interactive design optimization of framed structures. *Journal of Computing in Civil Engineering*, **3** (1), 60–74.

Arora, J.S. (1989) *Introduction to Optimum Design*, McGraw-Hill, New York.

Arora, J.S. (1990) Computational design optimization: a review and future directions. *Structural Safety*, **7**, 131–48.

Arora, J.S. and Mukhopadhyay, S. (1988) An integrated database management system for engineering applications based on an extended relational model. *Engineering with Computers*, **4**, 65–73.

Arora, J.S. and Tseng, C.H. (1988) Interactive design optimization. *Engineering Optimization*, **13** (3), 173–88.

Baron, R.J. and Shapiro, L.G. (1980) *Data Structures and Their Implementation*, Van Nostrand, New York.

Bell, J. (1982) Data modelling of scientific simulation programs. International Conference on Management of Data, 2–4 June, Association for Computing Machinery, Special Interest Group on Management of Data (ACM-SIGMOD), pp. 79–86.

Blackburn, C.L., Storaasli, O.O. and Fulton, R.E. (1982) The role and application of database management in integrated computer-aided design. American Institute of Aeronautics and Astronautics 23rd Structures, Structural Dynamics and Materials Conference, New Orleans, LA, May 10–12, pp. 603–13.

Browne, J.C. (1976) Data definition, structures, and management in scientific computing. Proceedings of Institute for Computer Applications in Science and Engineering (ICASE) Conference on Scientific Computing, pp. 25–56.

Buchmann, A.P. and Dale, A.G. (1979) Evaluation criteria for logical database design methodologies, *Computer-Aided Design*, **11** (3), 121–6.

Comfort, D.L. and Erickson, W.J. (1978) RIM – A prototype for a relational information management system. *NASA Conference Publication 2055*.

Daini, O.A. (1982) Numerical database management system: a model. International Conference on Data Management, Association for Computing Machinery, Special Interest Group on Management of Data (ACM-SIGMOD), pp. 192–9.

Date, C.J. (1977) *An Introduction to Database Systems*, Addison-Wesley, Reading, MA.

Elliott, L., Kunii, H.S. and Browne, J.C. (1978) A data management system for engineering and scientific computing. *NASA Conference Publication 2055*.

Felippa, C.A. (1979) Database management in scientific computing – I. General description. *Computers and Structures*, **10**, 53–61.

Felippa, C.A. (1980) Database management in scientific computing – II. Data structures and program architecture. *Computers and Structures*, **12**, 131–45.

Fischer, W.E. (1979) PIDAS – A database management system for CAD/CAM software. *Computer-Aided Design*, **11** (3), 146–50.

Fishwick, P.A. and Blackburn, C.L. (1982) The integration engineering programs using a relational database scheme, in *Computers in Engineering*. International Computers in Engineering Conference, Vol. 3, ASME, pp. 173–81.

Fulton, R.E. (ed.) (1987) Managing Engineering Data: The Competitive Edge. Proceedings of the Symposium on Engineering Database Management: Critical Issues, American Society of Mechanical Engineers, Computers in Engineering Conference and Exhibition, August 9–13, New York.

Fulton, R.E. and Voigt, S.J. (1976) Computer-aided design and computer science technology. Proceedings of the 3rd ICASE (Institute for Computer Applications in Science and Engineering) Conference on Scientific Computing, pp. 57–82.

Grabowski, H. and Eigner, M. (1982) A data model for a design database. File structures and databases for CAD. Proceedings of the International Federation of Information Processing, pp. 117–44.

Haug, E.J. and Arora, J.S. (1979) *Applied Optimal Design*, Wiley, New York.

Jumarie, G.A. (1982) A decentralized database via micro-computers a preliminary study, in *Computers in Engineering*. International Computers in Engineering Conference, Vol. 4, American Society of Mechanical Engineers, p. 183.

Kamal, O. and Adeli, H. (1990) Automatic partitioning of frame structures for concurrent processing. *Microcomputers in Civil Engineering*, **5** (4), 269–83.

Koriba, M. (1983) Database systems: their applications to CAD software design. *Computer-Aided Design*, **15** (5), 277–88.

Lopez, L.A. (1974) FILES: automated engineering data management system in *Computers in Civil Engineering*. Electronic Computation Conference, American Society of Civil Engineers, pp. 47–71.

Lopez, L.A., Dodds, R.H., Rehak, D.R. and Urzua, J.L. (1978) Application of data management to structures, in *Computing in Civil Engineering*, American Society of Civil Engineers, pp. 477–496.

Martin, J. (1977) *Computer Data-base Organization*, Prentice-Hall, Englewood Cliffs, NJ.

Mukhopadhyay, S. and Arora, J.S. (1987a) Design and implementation issues in an integrated database management system for engineering design environment. *Advances in Engineering Software*, **9** (4), 186–93.

Mukhopadhyay, S. and Arora, J.S. (1987b) Implementation of an efficient run-time support system for engineering design environment. *Advances in Engineering Software*, **9** (4), 178–85.

Pahl, P.J. (1981) Data management in finite element analysis, in *Nonlinear Finite Element Analysis in Structural Mechanics* (eds W. Wunderlich, E. Stein and K.J. Bathe), Springer, Berlin, pp. 714–16.

Park, G.J. and Arora, J.S. (1987) Role of database management in design optimization systems. *Journal of Aircraft*, **24** (11), 745–50.

Pooch, U.W. and Nieder, A. (1973) A survey of indexing techniques for sparse matrices. *Computing Surveys*, **5** (2), 109–33.

Rajan, S.D. (1982) SADDLE: a computer-aided structural analysis and dynamic design language. Ph.D. Dissertation, Civil Engineering, The University of Iowa, Iowa City, IA 52242.

Rajan, S.D. and Bhatti, M.A. (1983) Data management in FEM-based optimization software. *Computers and Structures*, **16** (1–4), 317–25.

Rajan, S.D. and Bhatti, M.A. (1986) SADDLE: a computer-aided structural analysis and dynamic design language. *Computers and Structures*, **22** (2), 185–212.

Spires, D.B. and Arora, J.S. (1990) Optimal design of tall RC-framed tube buildings. *Journal of Structural Engineering*, **116** (4), 877–97.

SreekantaMurthy, T. and Arora, J.S. (1985) A survey of database management in engineering. *Advances in Engineering Software*, **7** (3), 126–33.

SreekantaMurthy, T. and Arora, J.S. (1986a) Database management concepts in design optimization. *Advances in Engineering Software*, **8** (2), 88–97.

SreekantaMurthy, T. and Arora, J.S. (1986b) Database design methodology and database management system for computed-aided structural design optimization. *Engineering with Computers*, **1**, 149–160.

SreekantaMurthy, T. and Arora, J.S. (1987) A structural optimization program using a database management system, in *Managing Engineering Data: The Competitive Edge* (ed. R.E. Fulton). Proceedings of the Symposium on Engineering Database Management: Critical Issues, American Society of Mechanical Engineers, Computers in Engineering Conference and Exhibition, August 9–13, 1987, New York, pp. 59–66.

SreekantaMurthy, T., Shyy, Y.-K. and Arora, J.S. (1986) MIDAS: management of information for design and analysis of systems. *Advances in Engineering Software*, **8** (3), 149–58.

Ulfsby, S., Steiner, S. and Oian, J. (1981) TORNADO: a DBMS for CAD/CAM systems. *Computer-Aided Design*, **13** (4), 193–7.

Vetter, M. and Maddison, R.N. (1981) *Database Design Methodology*, Prentice-Hall, Englewood Cliffs, NJ.

10

Optimization of Topology*

G.I.N. ROZVANY, M. ZHOU and O. SIGMUND

10.1 INTRODUCTION: AIMS AND SIGNIFICANCE OF TOPOLOGY OPTIMIZATION

Topology of a structural system means the spatial sequence or configuration of members and joints or internal boundaries. The two main fields of application of topology optimization are layout optimization of grid-like structures and generalized shape optimization of continua or composites.

A **grid-type structure** has the basic feature that it consists of a system of intersecting members, the cross-sectional dimensions of which are small in comparison with their length, and hence the members can be idealized as one-dimensional continua. Consequences of this feature are that

- the influence of member intersections on strength, stiffness and structural weight can be neglected, and
- the specific cost (e.g. structural weight per unit area or volume) can be expressed as the sum of the costs (weights) of the members running in various directions.

Examples of grid-type structures are trusses, grillages (beam systems), shell-grids and cable nets.

Layout optimization of grid-type structures consists of three simultaneous operations, namely

- topological optimization involving the spatial sequence of members and joints,
- geometrical optimization involving the coordinates of joints, and
- sizing, i.e. optimization of cross-sectional dimensions.

The above concepts are explained on an example in Fig. 10.1, in which all three trusses have the same topology, whilst the trusses in Figs 10.1(b) and 10.1(c) have the same geometry but different cross-sectional dimensions.

Prager and Rozvany (1977b) regarded layout optimization as the most challenging class of problems in structural design because there exists an infinite number of possible topologies which are difficult to classify and quantify; moreover, at each point of the available space potential members may run in an infinite number of directions. At the same time, layout optimization is of

---

*In order to understand this chapter more easily, the reader should study sections 2.6 and 2.7 of Chapter 2.

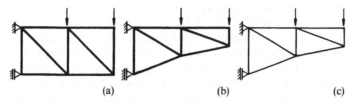

Fig. 10.1 Example illustrating topological, geometrical and cross-sectional properties of a grid-type structure.

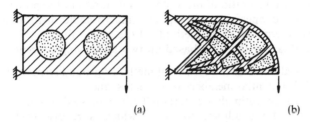

Fig. 10.2 Example illustrating generalized shape optimization.

considerable practical importance, because it results in much greater material savings than pure cross section (sizing) optimization.

The other important field of application of topology optimization is **generalized shape optimization**, in which a simultaneous optimization of both topology and shape of boundaries is required for continua and of interfaces between materials for composites. Figure 10.2(a), for example, shows the initial boundary shape and topology and Fig. 10.2(b) a hypothetical optimal shape and topology for a composite plate in plane stress, where the dotted regions denote the less stiff, weaker and lighter material. For a cellular structure (here, perforated plate), the dotted regions denote cavities (or 'holes').

10.2 OPTIMAL LAYOUT THEORY FOR GRID-TYPE STRUCTURES – BASIC ASPECTS

The theory of optimal structural layouts has been covered extensively in the past, in particular in principal lectures at NATO meetings (Iowa, 1980 (Rozvany, 1981); Troia, 1986 (Rozvany and Ong, 1987); Edinburgh, 1989 (Rozvany, Gollub and Zhou, 1992); Berchtesgaden, 1991 (Rozvany, Zhou and Gollub, 1993); Sesimbra, 1992 (Rozvany, 1993c)), and also in greater detail at a CISM course in Udine in 1990 (Rozvany, 1992b) as well as in books (Rozvany, 1976, 1989) and in book chapters (e.g. Rozvany, 1984). Although the above exposure in the technical literature may appear excessive, the development in this field has also been both rapid and extensive, even during the last few years. For this reason, apart from a brief review of the basic principles and earlier work, mostly new material can be reported in this chapter.

Optimal layout theory, developed in the seventies by Prager and the first author (e.g. Prager and Rozvany, 1977a,b) as a generalization of Michell's (1904) theory for trusses, was originally formulated for grid-like structures on the basis of the simplifying assumptions listed in section 10.1. These simplifications were removed in a more advanced version of layout theory (e.g. Rozvany *et al.*, 1982, 1987), which was applied to structures in which a high proportion of the available space was occupied by material. One could also say that the original, 'classical' layout theory is concerned with structures having a low 'volume fraction' (i.e. material volume/available volume ratio), whilst advanced layout theory deals with structures having a high volume fraction. In the latter, first the microstructure is optimized for given ratio of the stiffnesses or forces in the principal directions and, in a second operation, the optimal macroscopic distribution of micro-structures is determined using methods of the layout theory.

The clasical layout theory is based on two underlying concepts, namely

- the structural universe (in numerical methods: ground structure), which is the union of all potential members or elements, and
- continuum-type optimality criteria (COC), expressed in terms of a fictitious system termed the adjoint structure, which were discussed in Chapter 2 (sections 2.6 and 2.7).

Since the above optimality criteria also provide adjoint strains for vanishing members, their fulfillment for the entire structural universe represents a necessary and sufficient condition of layout optimality if the problem is convex and certain additional requirements (e.g. existence) are satisfied. The above condition of convexity is fulfilled for certain so-called 'self-adjoint' problems, the analytical treatment of which will be discussed in sections 10.3–10.5.

On the basis of optimal layout theory, two basic approaches have been developed, using

- analytical methods for deriving closed form continuum-type solutions representing the exact optimal layout, and
- numerical, discretized iterative methods for deriving approximate (but usually highly accurate) optimal layouts.

Fundamental differences between these two methods are listed in Table 10.1. Layout optimization methods used by the numerical school (e.g. Kirsch, 1989;

Table 10.1 A comparison of analytical and numerical methods based on layout theory

Computational Method	Analytical	Numerical
Structural model	Continuum	Discretized (finite elements)
Procedure	Simultaneous solution of all equations	Iterative solution
Structural universe	Infinite number of members	Finite but large number of members (several thousands)
Prescribed minimum cross-sectional area	Zero	Nonzero but small (10^{-8} to 10^{-12})

Kirsch and Rozvany, 1993) are usually based on the following two-stage procedure:

- first the topology is optimized for a given geometry (i.e. given coordinates of the joints), and then
- for this selected topology the geometry is optimized.

A drawback of this procedure is, of course, that for the new, optimized geometry the old topology may not be optimal any more. Until the introduction of COC–DCOC methods, however, the two-stage procedure was necessary because of the limited optimization capability of other methods, particularly for realistic problems with active stress constraints for a very large number of members in the structural universe.

The new optimality criteria methods (COC–DCOC), which are discussed in Chapter 2, enable us to carry out a simultaneous optimization of topology and geometry, because the number of elements in the structural universe is either infinite (analytical methods) or very large (numerical COC–DCOC methods) and hence topological optimization achieves, in effect, also geometrical optimization.

We shall close this introductory section on layout optimization with some personal notes by the first author.

First, it is often claimed that the layout theory discussed here always results in a 'continuum-type' solution, consisting of a dense grid of members of infinitesimal spacing, whereas numerical layout studies yield 'practical' solutions consisting of a few members only. It will be seen from subsequent sections that this notion is wrong, because for one or several point loads the exact layout consists in general of a few members only (Fig. 10.8, for example), with some notable exceptions (Figs 10.18 and 10.42(a)).

Second, it is generally believed that the interest of the first author's research team is restricted to analytical solutions, mostly in the grillage field. In actual fact, during the last four years the first two authors developed together some of the most efficient numerical methods for both sizing and layout optimization.

Finally, the first author explored recently with relative ease exact optimal topologies of several rather complicated new classes of layout problems, involving grillages and trusses (new solutions under sections 10.4 and 10.5). In this sort of analytical work, one requires a certain intuitive insight as well as an intimate knowledge of optimal layout theory, in order first to guess correctly these new topologies and then to prove their optimality. During the above activity, it occurred to him that, since the tragic deaths of William Prager (Brown University) and Ernest Masur (University of Illinois) and the retirement of William Hemp (Oxford) and Marcel Save (Mons), he is rapidly becoming a sort of a 'last Mohican' of analytical layout optimization. This situation has arisen as a result of the growing popularity of discretized methods in structural optimization, which is fully justified in view of (i) the rapidly increasing computational capability and (ii) the necessity for discretization in most practical problems. Explicit analytical solutions are, however, indispensable in reliably checking the validity and convergence of numerical methods. Too much reliance on automated discretized computer procedures may also affect mentally forthcoming generations of practitioners – one cannot help drawing an analogy with H.G. Wells' 'Time Machine', in which man in some future century loses his ability to think.

10.3 OPTIMAL LAYOUT THEORY: ANALYTICAL SOLUTIONS FOR SOME SIMPLE SELF-ADJOINT PROBLEMS

10.3.1 General formulation

The analytical procedure based on the above layout theory usually consists of the following steps.

- Set up a structural universe consisting of all potential members.
- Determine the specific cost function and continuum-type optimality criteria (COC) for the members.
- Construct an adjoint displacement field satisfying kinematic boundary and continuity conditions.
- Determine on the basis of the optimality criteria and the adjoint strain field the location and direction of nonvanishing (optimal) members.
- Check whether the external loads and the stress resultants along optimal members constitute a statically admissible set.

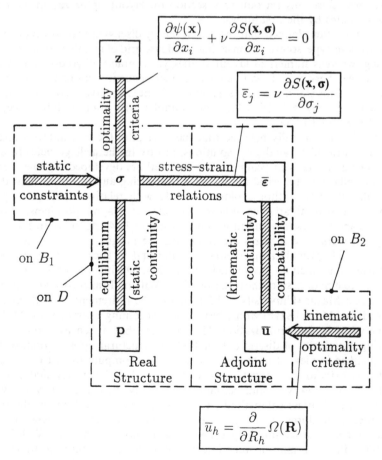

Fig. 10.3 COC algorithm for optimal plastic design.

• Check whether the adjoint strains along vanishing (non-optimal) members satisfy the optimality criteria (usually inequalities) for such members.

A general schematic representation of continuum-based optimality criteria (COC) for linearly elastic structures with stress and displacement constraints was given in Fig. 2.21. We shall discuss two classes of problems in this section: optimal plastic design and optimal elastic design for a compliance constraint.

In the case of optimal plastic design, the real structure is only required to fulfill static admissibility. Since there are only stress constraints but no displacement constraints, the adjoint structure is subject only to prestrains due to active stress constraints, without adjoint loads or generalized stresses. It follows that we only have the kinematic part of the adjoint structure, and hence Fig. 2.21 reduces to the scheme shown in Fig. 10.3. This can be simplified further for one load condition if we express the specific cost directly in terms of the generalized stresses $\psi = \psi(\boldsymbol{\sigma})$, as was done by Prager and Shield (1967) in their pioneering contribution. The corresponding simplified scheme is shown in Fig. 10.4, in which the relations between the real stresses and adjoint strains are termed 'static–kinematic optimality criteria' (Rozvany, 1989).

In the case of optimal elastic design with a so-called compliance constraint,

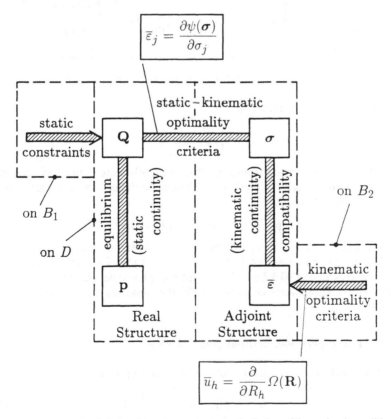

Fig. 10.4 Simplified COC algorithm for optimal plastic design with one load condition.

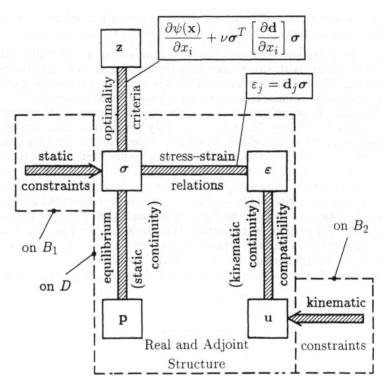

Fig. 10.5 COC algorithm for elastic design with a compliance constraint.

the latter can be represented as

$$\int_D \mathbf{p}\mathbf{u}\,\mathrm{d}\xi \leqslant C \tag{10.1}$$

where C is a given constant and all other symbols were defined in section 2.6. In this case, the 'weighting factors' for the elastic deflections $\mathbf{u}$ are given by the real load $\mathbf{p}$, which implies

$$\mathbf{p} \equiv \bar{\mathbf{p}}, \quad \boldsymbol{\sigma} \equiv \bar{\boldsymbol{\sigma}}, \quad \boldsymbol{\varepsilon} \equiv \bar{\boldsymbol{\varepsilon}}, \quad \mathbf{u} \equiv \bar{\mathbf{u}} \quad \text{(for all } \xi) \tag{10.2}$$

This means that it is only necessary to consider one half of the scheme in Fig. 2.21, as can be seen from Fig. 10.5. It follows that for a certain simple class of problems (sections 10.3.2 and 10.3.3) optimal plastic design and optimal elastic design with a compliance constraint yield the same solution within a constant multiplier.

10.3.2 First application: optimal plastic design

In this subsection, we consider cost functions of the form

$$\psi = k|\sigma|, \quad \Phi = \int_D k|\sigma|\,\mathrm{d}\xi \tag{10.3}$$

where ψ is the member weight per unit length, k is a given constant, σ is a generalized stress (stress resultant), Φ is the total weight, ξ is a spatial coordinate and D is the structural domain (with $\xi \in D$).

For trusses we have (i) $\sigma = N$ where N is the axial member force and (ii) $k = \rho/\sigma_0$ where ρ is the specific weight of the truss material and $\pm \sigma_0$ the yield stress in tension and compression.

For beams of variable width but given depth h, we have (i) $\sigma = M$ where M is the bending moment and (ii) $k = 4\rho/h^2\sigma_0$ where $\pm \sigma_0$ is again the yield stress.

For piecewise differentiable specific cost functions (as in equations (10.3)), the partial derivatives in Fig. 10.4 are replaced by subgradients (e.g. Rozvany, 1989), which means that at slope discontinuities any convex combination of the adjacent slopes can be taken. This implies that the adjoint strains for the specific cost function in relations (10.3) become

$$\bar{\varepsilon} = k \operatorname{sgn} \sigma \quad (\text{for } \sigma \neq 0), \quad |\bar{\varepsilon}| \leqslant k \quad (\text{for } \sigma = 0) \qquad (10.4)$$

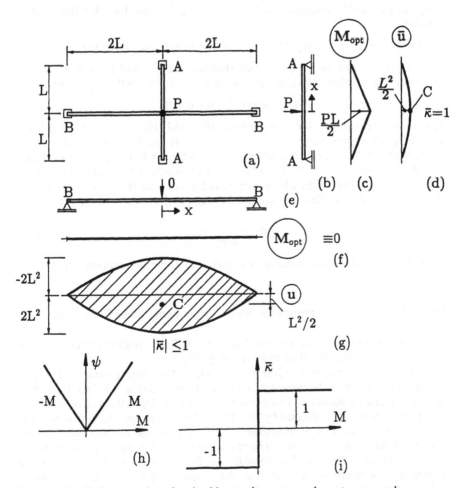

Fig. 10.6 Application of optimal layout theory to an elementary example.

where $\bar\varepsilon$ is the adjoint axial strain in the members. Relations (10.3) and (10.4) are represented graphically for $k = 1$ in Figs 10.6(h) and 10.6(i).

10.3.3 Second application: optimal elastic design for given compliance

We consider the class of problems characterized by

$$\psi = cx, \quad [\mathbf{d}] = 1/rx \tag{10.5}$$

where c and r are given constants, $[\mathbf{d}]$ is the generalized flexibility matrix (section 2.6) and x is a cross-sectional variable.

For trusses, for example, ψ is the member weight per unit length, x is the cross-sectional area and $c = \rho$ is the specific weight. Moreover, $r = E$ is Young's modulus. For beams of constant depth h but variable width x we have $c = h\varrho$ and $r = h^3 E/12$.

By the optimality condition at the top of Fig. 10.5 we have for this class of problems

$$c - v\sigma^2/rx^2 = 0, \quad x = (v/rc)^{1/2}|\sigma|, \quad \varepsilon = \sigma/rx = (c/rv)^{1/2}\,\mathrm{sgn}\,\sigma \quad \text{(for } \sigma \neq 0) \tag{10.6}$$

Moreover, it was shown previously (Rozvany, Zhou and Gollub, 1993) that for vanishing members (with $x = \sigma = 0$) we have the optimality criterion

$$|\varepsilon| \leqslant (c/rv)^{1/2} \quad \text{(for } \sigma = 0) \tag{10.7}$$

The reader is reminded that in equations (10.6) and (10.7) the symbol σ denotes a stress resultant and not a conventional stress. It can be seen from equations (10.4), (10.6) and (10.7) that the adjoint strain fields for both classes of problems considered in sections 10.3.2 and 10.3.3 are identical within a constant multiplier.

It was also shown in the above-mentioned paper that the optimal weights in plastic design (Φ_P) and elastic compliance design (Φ_C) have the following relation:

$$\Phi_C = \frac{c}{k^2 rC}\,\Phi_P^2 \tag{10.8}$$

where C is the prescribed maximum compliance value in equation (10.1).

10.3.4 An elementary example: a beam system consisting of two beams

Although this example has been used before, it will be presented in an extended form here again because of its extreme simplicity.

We consider a structural universe consisting of two simply supported beams, with a point load P at the intersection of the two beams (Fig. 10.6(a)). The rather obvious optimal solution to this problem is given in Figs 10.6(b) and 10.6(e), in which the entire load is carried by the shorter beam AA, and the beam BB, with zero cross-sectional area, is unloaded. The corresponding moment diagrams are shown in Figs 10.6(c) and 10.6(f). The specific cost function and optimality criteria (with $k = 1$) are represented graphically in Figs 10.6(h) and 10.6(i). Since the adjoint displacement must be zero at the simple supports, for the short beam AA we have $\bar u(L) = \bar u(-L) = 0$ and by Fig. 10.6(i) or equation (10.4) with $\sigma = M$,

$\bar{\varepsilon} = \bar{\kappa}$ and $k = 1$ the adjoint curvature $\bar{\kappa} = -\,\mathrm{d}^2\bar{u}/\mathrm{d}x^2 = 1$ for the short beam. The above boundary and curvature conditions imply

$$\bar{u} = -\iint \mathrm{d}x\,\mathrm{d}x + Ax + B, \quad A = 0, \quad B = L^2/2, \quad \bar{u} = L^2/2 - x^2/2 \quad (10.9)$$

as shown in Fig. 10.6(d). For the long beam BB, the bending moment is zero throughout, and hence by equation (10.4) with $\sigma = M$ we have $|\bar{\kappa}| = |-\mathrm{d}^2\bar{u}/\mathrm{d}x^2| \leqslant 1$, giving the nonunique adjoint displacement field $-2L^2 + x^2/2 \leqslant \bar{u} \leqslant 2L^2 - x^2/2$ (shaded area in Fig. 10.6(g)). As the latter does include a central deflection of $L^2/2$ (point C in Fig. 10.6(g)), kinematic admissibility is also satisfied. Since the specific cost function is convex in this problem (Fig. 10.6(h)) and the constraints are linear, necessary and sufficient conditions for optimality have been fulfilled and thus the solution in Figs 10.6(b)–10.6(f) is indeed optimal.

10.3.5 Optimal regions in exact solutions for truss and grillage layouts

In deriving exact optimal layouts for trusses and grillages, the structural universe consists of members in all possible directions at all points of the available space. Since by equation (10.4) the maximum absolute value of the adjoint generalized strain is k, it follows that the direction of any nonvanishing member must coincide with a principal direction of the adjoint strain field, having the principal strain value of $\bar{\varepsilon}_1 = k$. Moreover, the absolute value of the second principal strain value $\bar{\varepsilon}_2$ must not exceed k, that is, $\bar{\varepsilon}_2 \leqslant k$. Adopting $k = 1$, these requirements allow the following optimal regions for trusses and grillages at all points of the available space where loads or non-vanishing members are present:

$$
\begin{aligned}
R^+\colon &\quad \sigma_1 > 0, \quad \sigma_2 = 0, \quad \bar{\varepsilon}_1 = 1, &\quad |\bar{\varepsilon}_2| \leqslant 1 \\
R^-\colon &\quad \sigma_1 < 0, \quad \sigma_2 = 0, \quad \bar{\varepsilon}_1 = -1, &\quad |\bar{\varepsilon}_2| \leqslant 1 \\
S^+\colon &\quad \sigma_1 > 0, \quad \sigma_2 > 0, \quad \bar{\varepsilon}_1 = \bar{\varepsilon}_2 = 1 & (10.10) \\
S^-\colon &\quad \sigma_1 < 0, \quad \sigma_2 < 0, \quad \bar{\varepsilon}_1 = \bar{\varepsilon}_2 = -1 \\
T\colon &\quad \sigma_1 > 0, \quad \sigma_2 < 0, \quad \bar{\varepsilon}_1 = -\bar{\varepsilon}_2 = 1
\end{aligned}
$$

where $\sigma_i = M_i$, $\bar{\varepsilon}_i = \bar{\kappa}_i$ for grillages and $\sigma_i = N_i$, $\bar{\varepsilon}_i = \bar{\varepsilon}_i$ for trusses. Symbols indicating the various types of optimal regions in relations (10.10) are shown in Fig. 10.7.

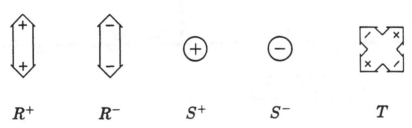

$$R^+ \qquad\qquad R^- \qquad\qquad S^+ \qquad\qquad S^- \qquad\qquad T$$

Fig. 10.7 Symbols used for optimal regions for trusses and grillages.

10.3.6 An example of an exact optimal truss layout

We consider the optimal transmission of a vertical point load by a truss to supports formed by a horizontal and a vertical line (Fig. 10.8(b)). In this problem, the structural universe consists of an infinite number of members (at any given point of the available space ($x \geqslant 0$, $y \leqslant 0$) members may run in any directions).

First we must cover the available space with the optimal regions in relations (10.10) and Fig. 10.7, so that continuity conditions and kinematic boundary conditions are satisfied. The latter consist of:

$$(\text{for } x = 0 \text{ or } y = 0) \quad \bar{u} = \bar{v} = 0 \tag{10.11}$$

where $\bar{u}$ and $\bar{v}$, respectively, are the adjoint displacements in the x and y directions.

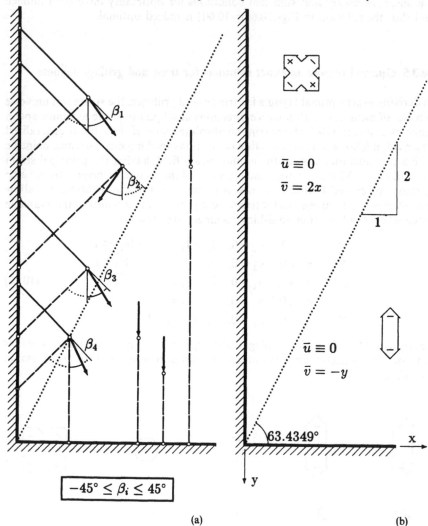

$$-45° \leq \beta_i \leq 45°$$

(a)   (b)

Fig. 10.8 Example of an exact optimal truss layout.

We shall try a topology having an S region at the top and an R^- region at the bottom, with a region boundary having a slope of 2:1 (Fig. 10.8(b)).

In the top region we have

$$\bar{u} \equiv 0, \quad \bar{v} = 2x, \quad \bar{\varepsilon}_x = d\bar{u}/dx = 0, \quad \bar{\varepsilon}_y = d\bar{v}/dy = 0, \quad \bar{\gamma}_{xy} = d\bar{u}/dy + d\bar{v}/dx = 2$$

$$\bar{\varepsilon}_{1,2} = \frac{\bar{\varepsilon}_x + \bar{\varepsilon}_y}{2} \pm \left[\left(\frac{\bar{\varepsilon}_x - \bar{\varepsilon}_y}{2} \right)^2 + \frac{\bar{\gamma}_{xy}^2}{4} \right]^{1/2}, \quad \bar{\varepsilon}_1 = 1, \quad \bar{\varepsilon}_2 = -1$$

$$\alpha = \frac{1}{2} \arctan \frac{\bar{\gamma}_{xy}}{\bar{\varepsilon}_x - \bar{\varepsilon}_y} = \frac{1}{2} \arctan \infty = 45° \tag{10.12}$$

and in the bottom region

$$\bar{u} \equiv 0, \quad \bar{v} = -y, \quad \bar{\varepsilon}_x = d\bar{u}/dx = 0, \quad \bar{\varepsilon}_y = d\bar{v}/dy = -1$$

$$\bar{\gamma}_{xy} = d\bar{u}/dy + d\bar{v}/dx = 0, \quad \bar{\varepsilon}_1 = -1, \quad \bar{\varepsilon}_2 = 0, \quad \alpha = 0 \tag{10.13}$$

where α is the direction of the first principal strain with respect to the vertical.

It can be seen from the above results and relations (10.10) that, in the top region, the optimal bars must run at $\pm 45°$ to the vertical and must have tension and compression, respectively, in the two principal directions. In the bottom region they must be vertical (with a negative, i.e. compressive, force). It can also be checked easily that the displacement fields in Fig. 10.8(b) satisfy (i) the kinematic boundary conditions $\bar{u} = \bar{v} = 0$ along the supports and (ii) continuity conditions along the region boundary. The optimal bar directions and signs of the corresponding forces are shown by arrows in Fig. 10.8(b). Examples of admissible loads for this solution and corresponding optimal truss members are shown in Fig. 10.8(a). Whereas in the top region, the loads may enclose any angle within $\pm 45°$ to the vertical, the forces must be vertical in the bottom region of the adjoint field in Fig. 10.8(b). For loads along the region boundary, the optimal truss may consist of three bars (β_4 in Fig. 10.8(a)). The above solution was obtained in the late 1980s (Rozvany and Gollub, 1990).

The range of admissible load directions in the top region is due to the fact that one of the two bars must be in compression and the other one in tension. Since only bars in the vertical directions are admitted by the adjoint field in the bottom region, the corresponding loads must also be vertical for reasons of equilibrium. All the loads shown in Fig. 10.8(a) may act simultaneously or separately for the given optimal layouts.

Note
In all solutions involving trusses and grillages, continuous thick lines indicate members in tension or in positive moment and broken lines members in compression or negative moment.

10.4 LEAST-WEIGHT GRILLAGES

10.4.1 General aspects

One of the most successful applications of the exact layout theory was the optimization of grillage layouts, as can be seen from the following remark by

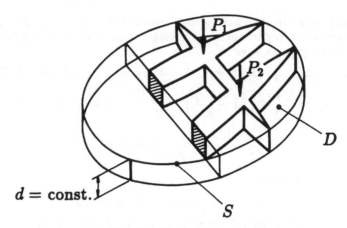

Fig. 10.9 The grillage layout problem.

Prager: "Although the literature on Michell trusses is quite extensive, the mathematically similar theory of grillages of least-weight was only developed during the last decade. Despite its late start, this theory advanced farther than that of optimal trusses. In fact, grillages of least-weight constitute the first class of plane structural systems for which the problem of optimal layout can be solved for almost all loadings and boundary conditions" (Prager and Rozvany, 1977a). The problem of grillage optimization can be described as follows (Fig. 10.9): a structural domain D, bounded by two horizontal planes and some vertical surfaces, is subject to a system of vertical loads which are to be transmitted to given supports by beams of rectangular cross section having a variable width. The beams are to be contained in the structural domain and are to take on a minimum weight (or volume). The beam system is to be designed plastically for a given yield stress or elastically for a given compliance (sections 10.3.2 and 10.3.3).

As can be seen from the quotation above, Prager regarded the grillage optimization problem as particularly important because of the following unique features.

- Grillages constitute the first class of truly two-dimensional structural optimization problems for which closed-form analytical solutions are available for most boundary and loading conditions.
- Optimal grillages are more practical than Michell structures (least-weight trusses), because the latter are subject to instability which is ignored in the formulation.
- The optimal topology of ribs in solid plates and holes in perforated plates has been found to be similar to that of minimum weight grillages (e.g. Cheng and Olhoff, 1981).
- A computer algorithm is available for generating analytically and plotting optimal beam layouts for a wide range of boundary conditions (Rozvany and Hill, 1978; Hill and Rozvany, 1985).
- It has been shown that the same grillage layout is optimal for plastic design and for elastic design with a stress or a compliance or a natural frequency constraint (Rozvany, 1976; Olhoff and Rozvany, 1982).

- The optimal grillage layout is independent of the (nonnegative) load distribution if no internal simple supports are present.
- The adjoint displacement field can be readily generated and it provides an influence surface for any (nonnegative) loading (the total structural weight equals the integral of the product of loads and adjoint deflections).
- A number of additional refinements have been added to the optimal grillage theory, which are reviewed in the next subsection.

10.4.2 Review of earlier developments

Analytical solutions are now available for

- clamped boundaries,
- simply supported boundaries,
- internal simple supports,
- free edges,
- beam supported edges, and
- corners in the boundary.

Earlier extensions of the theory include closed form analytical solutions for plastically designed grillages with

- solutions for up to four loading conditions,
- nonuniform depth,
- partial discretization,
- allowance for cost of supports,
- bending and shear dependent cost,
- upper constraint on the beam density, and
- allowance for selfweight

as well as for elastically designed grillages with

- deflection constraints,
- natural frequency constraints, and
- a combination of stress and deflection constraints.

General theories for clamped and simply supported boundaries were developed already in the early 1970s (e.g. Rozvany, 1973; Rozvany, Hill and Gangadharaiah, 1973). Several comprehensive reviews of earlier work are available (Prager, 1974; Rozvany, 1976; Rozvany and Hill, 1976; Rozvany, 1981, 1984, 1989, Chapter 8, 1992b, 1993a). Quite recent developments are discussed briefly in the next subsection.

10.4.3 Recent developments

(a) Unified constructions for simply supported and clamped boundaries with allowance for cost of supports

The surprisingly simple constructions given in Figs 10.10 and 10.11 were derived quite recently (Rozvany, 1994). As indicated, clamped supports are assumed

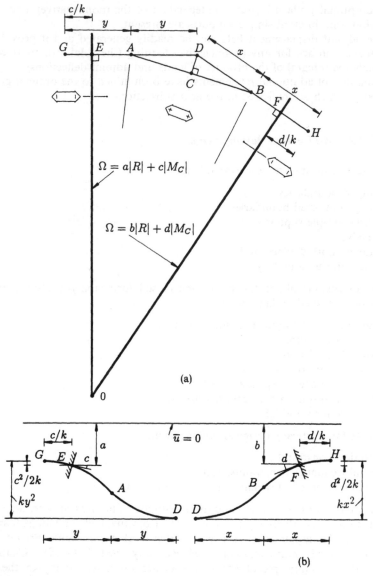

Fig. 10.10 Construction of optimal beam direction for two clamped boundaries with support costs.

to have a cost $\Omega = a|R| + c|M_{\mathrm{C}}|$ or $\Omega = b|R| + d|M_{\mathrm{C}}|$ and simple supports $\Omega = b|R|$ where a, b, c and d are given constants, R is the vertical reaction per unit length and M_{C} the clamping moment per unit length. The distances x and y must also satisfy the following relations:

(Fig. 10.10(a)) $a + ky^2 - c^2/2k = b + kx^2 - d^2/2k$ (10.14)

(Fig. 10.11(a)) $a + ky^2 - c^2/2k = b + kx/2$ (10.15)

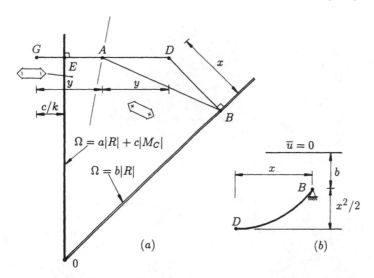

Fig. 10.11 Construction of optimal beam direction for a clamped and a simply supported boundary with support costs.

Naturally, the above unified constructions are also valid for grillages without support cost, in which case we have $a = b = c = d = 0$.

Proof of the constructions in Figs 10.10(a) and 10.11(a) is based on the type of optimal regions in equations (10.10) and satisfaction of the kinematic boundary conditions, as well as continuity and slope continuity conditions along region boundaries.

By Fig. 10.4, the adjoint displacements along supports are given by the partial derivatives of the support cost function Ω with respect to the reaction components. For a reaction cost function $\Omega = a|R| + c|M_C|$, for example, we have the adjoint deflection $\bar{u}$ and slope $\bar{s}$ of

$$\bar{u} = a, \quad \bar{s} = c \quad \text{(for } R > 0, \quad M_C > 0) \tag{10.16}$$

The adjoint deflections satisfying all optimality conditions are shown in Figs 10.10(b) and 10.11(b) (for details, see the paper by Rozvany (1994)).

On the basis of the constructions in Figs. 10.10 and 10.11, a computer program was developed at Essen University by D. Gerdes, for generating analytically and plotting optimal layouts for any combination of analytically defined (straight or circular) simply supported or clamped boundaries. Some examples of these computer-generated analytical solutions are given in Fig. 10.12, in which lines in one direction indicate beams in an R region, lines in two directions at right angles denote beams in a T region and black areas signify S regions (in which all beam directions are equally optimal). The solutions at the top and left bottom have, respectively, two and four internal point supports.

The above-mentioned automated computer algorithm finds points (termed 'centers') for which the type of relations in equations (10.14) and (10.15) are satisfied by more than two boundary points. For example, in Fig. 10.13, the

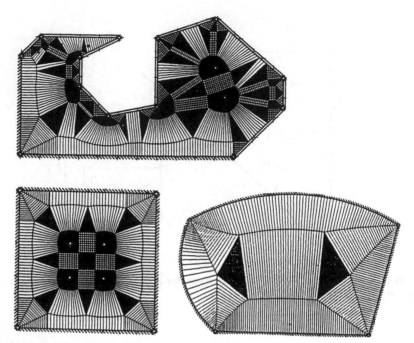

Fig. 10.12 Optimal grillage layouts derived analytically by a computer on the basis of the constructions in Fig. 10.9 and 10.10 (without allowance for support costs).

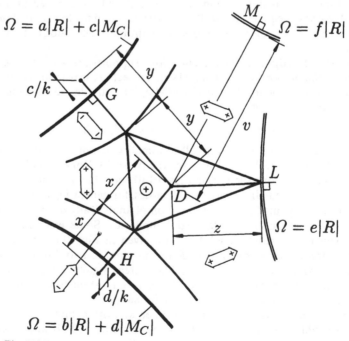

Fig. 10.13 Construction of a 'junction' associated with three boundaries.

center D satisfies the following relation with respect to the boundary points G, H and L:

$$a + ky^2 - c^2/2k = b + kx^2 - d^2/2k = e + kz^2/2 \tag{10.17}$$

where $\overline{DG}$, $\overline{DH}$ and $\overline{DL}$ are normal to the boundary. With respect to each boundary point (G, H, or L) in the above construction, the adjoint deflection values $\bar{u}_D$ are the same, that is $\bar{u}_{DG} = a + ky^2 - c^2/2k$, $\bar{u}_{DM} = b + kx^2 - d^2/2k$ and $\bar{u}_{DL} = e + kz^2$, with $\bar{u}_{DG} = \bar{u}_{DM} = \bar{u}_{DL}$. However, it is an important additional condition of the optimality of the layout in Fig. 10.13 that any other boundary point for which a normal can be drawn from the center D gives a greater $\bar{u}$ value than the one in equation (10.17). For example, for the boundary point M in Fig. 10.13, $\bar{u}_{DM} = f + kv^2/2$ must be greater than $\bar{u}_{DG} = \bar{u}_{DM} = \bar{u}_{DL}$. For each potential center, a large number of such tests must be carried out by the computer if the number of sides is relatively high (top of Fig. 10.12).

For each center, the construction in Fig. 10.13 defines an S^+ region termed a 'junction' (Rozvany, 1976, p. 198). Two such junctions (black triangular areas) are shown in the solution at the bottom right corner of Fig. 10.12.

Junctions are connected with each other and with certain boundary points (usually corners; Fig. 10.12) by so-called 'branches', which are constructed on the basis of Figs. 10.10 and 10.11. Using the above procedure, the computer selects systematically the optimal layout in an analytical representation and plots it. For a layout of such complexity as the one at the top of Fig. 10.12, this requires (e.g. on a HP 9000 work station) only a few minutes of computer time. For a reasonably accurate discretized numerical solution for the same layout, one would require an enormously large number of beam elements in the structural universe, which would make the corresponding computer time and storage requirement prohibitively high.

(b) Grillages with combinations of free, simply supported and clamped edges

For some isolated cases of grillage geometries with free and simply supported edges, solutions were reported some time ago (e.g. Prager and Rozvany, 1977a; Rozvany, 1981). It was found at the time that, in general, the optimal topology along free edges contains a so-called beam-weave, consisting of short beams in negative bending and long beams in positive bending (Fig. 10.14(a)). The general equation for the relation between the distance t along a straight free edge and the angle between the long beams and the edge is (Rozvany and Gerdes, 1994)

$$\frac{t}{a} = \exp\left[\int_\alpha^{\alpha_0} \frac{\sin\gamma \, d\alpha}{\sin(\alpha+\gamma)\cos(2\alpha+2\gamma)\sin\alpha}\right] \tag{10.18}$$

(Fig. 10.14(b)) and the adjoint deflection at a point A (Fig. 10.14(c)) is given by

$$\bar{u} = \left(\frac{\partial\bar{u}}{\partial t}\right)_D (a - t_A) - \int_{t_A}^\alpha \cos(2\alpha)(t - t_A) \, dt \tag{10.19}$$

where

$$\left(\frac{\partial\bar{u}}{\partial t}\right)_D = \frac{a \sin\gamma \sin(2\alpha_0 + \gamma)}{2\sin^2(\alpha+\gamma)} \tag{10.20}$$

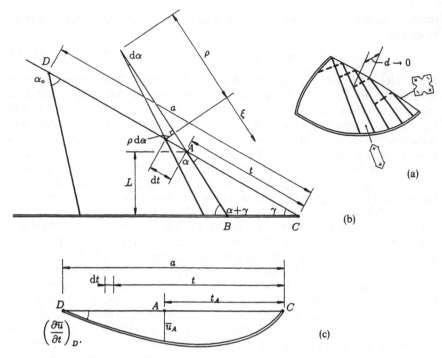

(a)

(b)

(c)

Fig. 10.14 Optimal beam directions in grillages with free and simply supported edges.

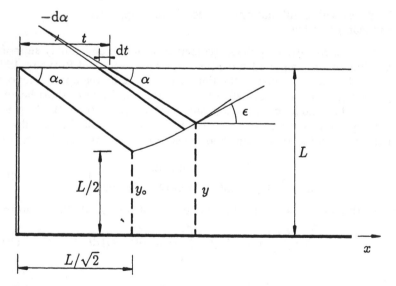

Fig. 10.15 Optimal beam directions in grillages with free, clamped and simply supported straight edges.

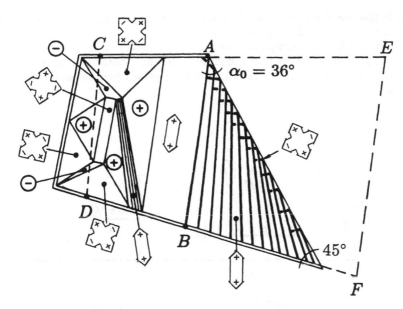

Fig. 10.16 Optimal layout of a grillage with three simply supported edges and one free edge.

Although these general equations can at present be solved only by numerical integration, they do reduce to the known analytical solutions (Rozvany, 1981) for $\gamma = 0°$, 45° and 90°. Moreover, a surprisingly simple geometrical property of the above solutions is that at any point A of the free edge, the adjoint slope in the direction parallel to the simple support is given by the explicit expression

$$\text{slope}_{A,\parallel} = t \sin \gamma \cot(\alpha + \gamma) \tag{10.21}$$

For a straight free edge that is parallel to a straight clamped support, the governing equations have been shown to be (Rozvany and Gerdes, 1994)

$$dt = -\frac{L-y}{\sin^2 \alpha} \frac{1 + \sin^2 \varepsilon}{2 \sin^2(\alpha + \varepsilon) - 1 - \sin^2 \varepsilon} d\alpha, \quad \frac{d^2 \bar{u}}{dt^2} = -\cos(2\alpha)$$

$$y = L - \sin \alpha \left[(L^2 - 2\bar{u})/(1 + \sin^2\alpha)\right]^{1/2}, \quad \tan \varepsilon = dy/dx \tag{10.22}$$

The meaning of the symbols used in equation (10.22) is shown in Fig. 10.15. Equations (10.22) were solved by numerical integration for some examples and compared with discretized solutions (section 10.6.5, Fig. 10.25).

To demonstrate the complexity of grillage layouts for even relatively simple boundary conditions, a least-weight solution for a grillage with three simply supported edges and one free edge is shown in Fig. 10.16. Beams are indicated by thicker lines only in the R^+ region associated with the free edge.

(c) Grillages with partially upward and partially downward loading

It was mentioned earlier that for clamped and simply supported edges the optimal grillage layout is independent of the load distribution if all loads have the same sign. If this condition is violated, then the optimal grillage layout will become (i) load dependent and (ii) much more complicated.

Figure 10.17(a), for example, shows one upward (negative) and three downward point loads. The distances of the loads may appear somewhat artificial, as they were chosen with a view to obtaining simple values for the geometry of the optimal solution.

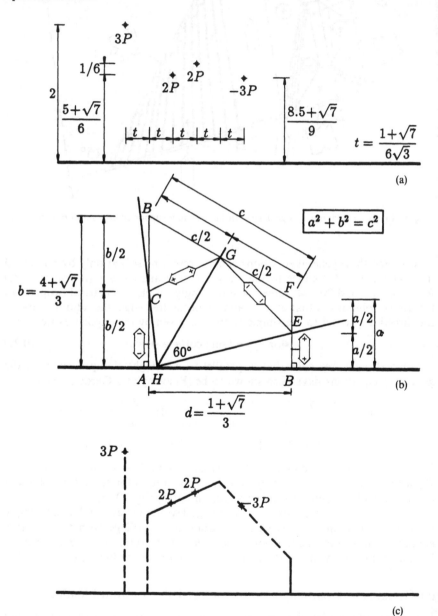

Fig. 10.17 Optimal grillage layout for partially upward and partially downward loads.

As can be seen from Fig. 10.17(b), the solution for the considered class of problems can be obtained by means of introducing a fictitious simple support (HG in Fig. 10.17(b)) having a linearly varying adjoint displacement. Then the construction in Figs 10.10 and 10.11, together with slope continuity across the fictitious support, furnishes the optimal layout (Fig. 10.17(c)). A detailed treatment of the above procedure will be given elsewhere (Rozvany, 1993b). The solution in Fig. 10.17(c) was fully confirmed through numerical solutions by Sigmund.

10.5 LEAST-WEIGHT TRUSSES (MICHELL STRUCTURES)

As mentioned in section 10.2, the first truss solutions of least weight and a general theory for deriving them was published almost 90 years ago (Michell, 1904). The most important publications on this topic during the last 20 years were a book by Hemp (1973) and a paper by Lagache (1981).

The optimal regions for least-weight trusses are given again by relations (10.10), in which we replace σ_i with N_i, i.e. the member force.

Most earlier solutions for least-weight trusses (Fig. 10.18; for a detailed treatment refer to Hemp's book of 1973) consisted of an infinite number of densely spaced members, although the truss was subject to a single point load. This has apparently created the incorrect impression that such 'truss-like continua' (Prager, 1974) constitute the rule rather than the exception.

About three years ago, the first author and his research group started a systematic exploration of optimal layouts for least-weight trusses. It was found (e.g. Rozvany and Gollub, 1990) that Michell structures usually take on such a complicated form only if the supports are statically highly restrictive (e.g. point supports or short line supports), but become relatively simple if the supports consist of longer line segments (for example, Fig. 10.8). This is demonstrated further in Figs 10.19–10.21.

Figure 10.19 shows a quadrilateral support with a point load in two different locations, together with the corresponding optimal truss layouts. Both are very simple, consisting of two bars only. For the optimal bar directions shown, explicit

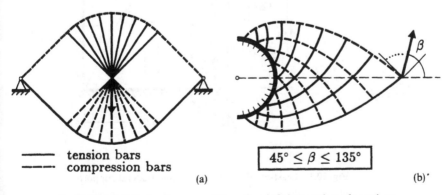

——— tension bars
– – – compression bars

$45° \leq \beta \leq 135°$

(a)   (b)˙

Fig. 10.18 Michell structures consisting of an infinite number of members.

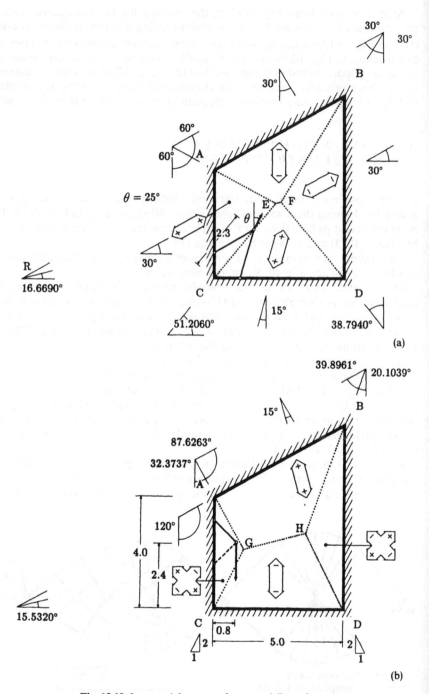

Fig. 10.19 Least-weight trusses for a quadrilateral support.

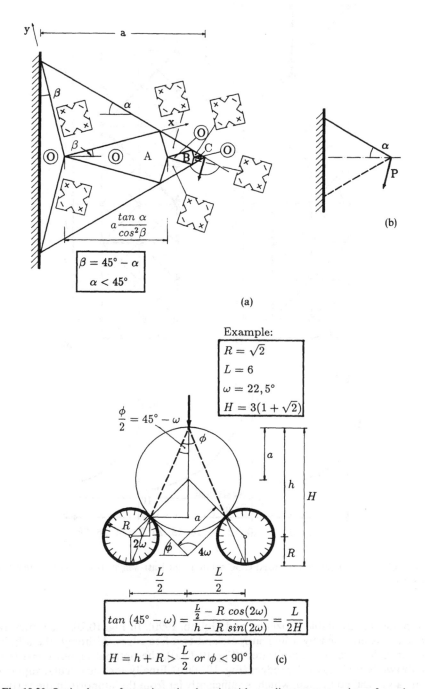

Example:

$$R = \sqrt{2}$$
$$L = 6$$
$$\omega = 22,5°$$
$$H = 3(1 + \sqrt{2})$$

$$\tan(45° - \omega) = \frac{\frac{L}{2} - R\cos(2\omega)}{h - R\sin(2\omega)} = \frac{L}{2H}$$

$$H = h + R > \frac{L}{2} \text{ or } \phi < 90°$$ (c)

Fig. 10.20 Optimal truss for a triangular domain with one line support and two free edges.

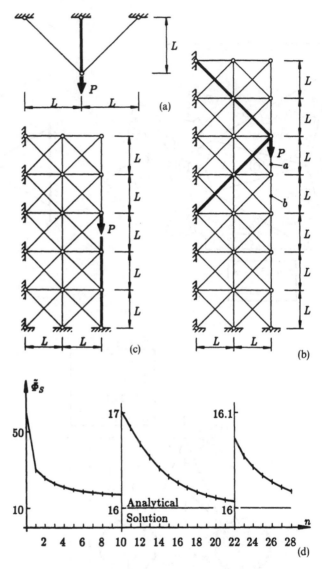

Fig. 10.21 (a)–(c) Iterative discretized COC solutions for least-weight trusses, with (d) iteration history.

formulae are available (Rozvany and Gollub, 1990). In Fig. 10.20, the structural domain is bounded by a line support and two free edges. Although the adjoint strain field is rather complicated (Fig. 10.20(a)), the optimal layout consists of only two bars along the free edges (Fig. 10.20(b)). Finally, for two circular supports and a vertical point load, which is sufficiently far from the supports in the vertical direction, the optimal layout consists of two bars (Fig. 10.20(c)), whose centerline connects the tip of the force with the lowest point of the circles. This solution is valid only if the slope of the bars is greater than 1:1. Further details of the above two solutions are given elsewhere (Rozvany, Gollub and Zhou, 1994).

It should also be mentioned that the theory of least-weight trusses has been extended to least-weight shell grids and cable nets with only axial forces in the members (Rozvany and Prager, 1979), which were later also termed 'Prager structures'. A review of several other papers on this topic was given in a book chapter (Rozvany, 1984).

10.6 DISCRETIZED LAYOUT SOLUTIONS OBTAINED BY ITERATIVE COC–DCOC METHODS

The differences between analytical and numerical COC methods for layout optimization were explained in section 10.2 (Table 10.1). In this section, a number of test examples are presented.

10.6.1 Elementary test examples involving trusses

The first test examples solved by the iterative, discretized COC method (Fig. 10.21) concerned such elementary examples as a three-bar truss or a point load with horizontal and vertical supporting lines (Rozvany *et al.*, 1989). Figure 10.21 shows members of the structural universes in thin line and the optimal members in thick line. It can be seen that the results in Figs 10.21(b) and 10.21(c) fully confirm the analytical solutions in Fig. 10.8.

Using normalized values with $L = P = \rho = C = t = E = 1$ for the problem in Fig. 10.21(b), the analytical solution gives a weight of $\Phi = 16$. With a convergence tolerance value of $T = 10^{-14}$ (Chapter 2, equation (2.82)) and a lower limit on the cross-sectional area of $x\downarrow = 10^{-12}$, the iterative COC–DCOC method (for trusses, the COC and DCOC methods are almost identical) gave a weight value of $\Phi = 16.000\,000\,000\,048$ after 126 iterations, which represents an agreement of twelve significant digits. In the COC solutions, all non-optimal members took on a cross-sectional area of 10^{-12}, except members a and b (Fig. 10.21(b)), which were both under 3×10^{-12}.

10.6.2 Truss-like continuum containing Hencky nets

Another optimal truss layout consisting of an infinite number of members is shown in Fig. 10.22(a). The truss is restricted to the rectangle ABCD and the point load $P = 1$ is to be transmitted to the supporting line AC. In the solution, the triangle ACE contains no members on its interior, whilst the regions AEF and CEG contain straight radial members. The region EFHG consists of a Hencky net with curved members in both principal directions. For obvious reasons, only a finite number of members are indicated. The members along the edges of the truss (AFH and CGH) are 'concentrated' members which have a much bigger cross-sectional area than the others. The above analytical solution was discussed by Hemp (1973, pp. 97–9). The angle μ in Fig. 10.22(a) can be

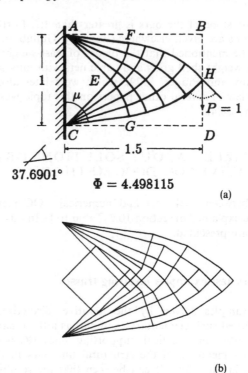

Fig. 10.22 Optimal Michell layout for a rectangular domain with one line support and three free edges: (a) analytical solution (after Hemp, 1973); (b) discretized COC solution.

calculated from Hemp's equation (4.120):

$$1.5 = \frac{1}{2} \int_0^{2\mu} [I_0(t) + I_1(t)] \, dt$$

$$= \frac{1}{2} \left[I_0(2\mu) - 1 + 2 \sum_{n=0}^{\infty} (-1)^n I_{2n+1}(2\mu) \right] \Rightarrow \mu = 82.690\,133° \qquad (10.23)$$

where I_i are modified Bessel functions. Using plastic design (equation (10.3) with $k=1$, or in Hemp's notation $\sqrt{2FR/\sigma}=1$), the total truss weight becomes

$$\Phi = (1 + 2\mu)I_0(2\mu) + 2\mu I_1(2\mu) \qquad (10.24)$$

giving the optimal truss weight

$$\Phi_{\text{opt,plastic}} = 4.498\,115 \qquad (10.25)$$

For elastic design with a compliance constraint, we have by equation (10.8) with $c=k=r=C=1$:

$$\Phi_{\text{opt,compliance}} = 4.498\,115^2 = 20.223\,042 \qquad (10.26)$$

Using an iterative COC–DCOC method with structural universes containing 5055 and 12992 members, respectively, truss weights of $\Phi = 20.540\,807$ and

$\Phi = 20.419\,699$ were obtained. By equation (10.8) these correspond to truss weights of $\Phi = 4.532\,197$ and $\Phi = 4.518\,816$, in plastic design, representing errors of 0.76% and 0.46%, respectively, compared with the analytical solution. In the above calculations a minimum cross-sectional area of $x\downarrow = 10^{-12}$ and a tolerance value of $T = 10^{-8}$ were used. The discretized layout, showing members having a cross-sectional area over $z^e = 0.1$ in compliance design, is shown in Fig. 10.22(b), which exhibits clear similarities with the analytical solution in Fig. 10.22(a).

10.6.3 Optimal grillage layout for a clamped square domain

The optimal layout for this problem was derived originally, in the context of reinforced concrete plates, by Lowe and Melchers (1972–1973) and is shown in Fig. 10.23(a). Sigmund extended the DCOC method for linearly varying elements (Sigmund, Zhou and Rozvany, 1993) and obtained for nine point loads the solution in Figs 10.23(c) and 10.23(d), using a structural universe with 624 beam

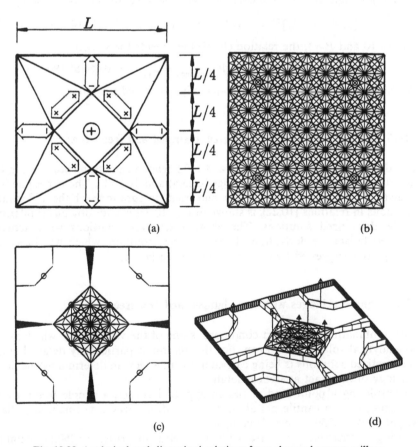

Fig. 10.23 Analytical and discretized solutions for a clamped square grillage.

elements (Fig. 10.23(b)). The analytical and discretized solution gave almost the same nondimensional weight (0.234 619 vs 0.234 620).

10.6.4 Rhombic grillages with two simply supported and two free edges: beam weaves in the solution

For the above problem, the analytical solution (with $d \to 0$) is shown in Fig. 10.24(a). For a finite number of long beams, the structural weight of this layout is given by (Prager and Rozvany, 1977a)

$$\Phi = Pa^2[1 + 3c^2 + (1 - c^2)/n]/2 \qquad (10.27)$$

where n is the number of long beams, whilst $2a$ and $2ca$ with $c > 1$ are the dimensions of the grillage along the two axes of symmetry. Using a structural universe consisting of 620 beam elements, Sigmund obtained the solution in Fig. 10.24(c). Some of the short beams cross more than one long beam because of the nonuniqueness of the optimal solution. On the basis of a side length of 1.0 and an angle at the acute corners of 45°, we have for the above example

$$a = [1/(4 - 2\sqrt{2})]^{1/2} = 0.923\,879\,53, \quad c = \sqrt{2} - 1 = 0.414\,213\,56 \quad (10.28)$$

For $n = 10$ and $P = 1$, the relation (10.27) then yields the weight

$$\Phi_{\text{plastic}} = 0.681\,801\,95, \quad \Phi_{\text{compliance}} = \Phi^2_{\text{plastic}} = 0.464\,853\,90 \quad (10.28a)$$

The numerical solution in Fig. 10.24(c) fully confirmed the latter.

10.6.5 Grillage with simply supported, clamped and free edges

Two discretized solutions, with 466 and 1892 beam elements, respectively, in the structural universe, are shown in Figs 10.25(a) and 10.25(b). The corresponding 'analytical' solution, based on the numerical integrations of the differential equations in relations (10.22), is shown in Fig. 10.25(c), with obvious similarities to the discretized solutions. The above discretized solutions were actually obtained before the derivation of the analytical solution, which was based on the topology suggested by the above discretized layouts.

10.6.6 Square grillages with two clamped and two free edges

This combination of support conditions is one of the very few for which as yet no analytical solution is known. For this reason, a particularly detailed study of discretized solutions is being carried out with a view to determining the likely topology of the exact analytical solution.

Considering a point load at the unsupported corner, a simple beam layout consisting of two cantilevers along the free edges gives a normalized grillage weight of $\Phi = 0.250$ for elastic compliance.

Using a structural universe with 9312 elements (Fig. 10.26(a)), Sigmund obtained the solution in Fig. 10.26(b), with a structural weight of $\Phi = 0.1819$. It

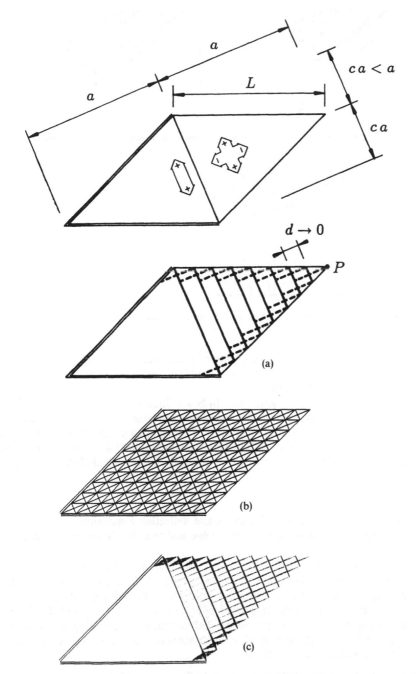

Fig. 10.24 Analytical and discretized solutions for rhombic grillage with two simply supported and two free edges.

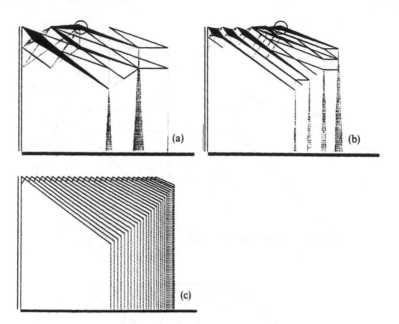

Fig. 10.25 Discretized and analytical solutions for a grillage with simply supported, clamped and free edges.

can be seen that this solution consists of beam weaves over about one half of the free edges, supported by two heavy cantilever beams and balanced by longer beams under negative moment.

Whereas in Figs 10.26(a) and 10.26(b) beam directions are restricted to slopes of 0, 1:1 and 1:2 to the sides, the structural universe in Fig. 10.26(c) includes additional slopes of 1:3 and 2:3 but only 3752 beam elements. The weight of the corresponding solution (Fig. 10.26(d)) is somewhat higher ($\Phi = 0.1906$), but shows better the variation of the direction of the long beams which form part of the beam weave.

10.6.7 A 'practical' solution with stress and deflection constraints: triangular grillage with one free edge and two simply supported edges

The main advantage of discretized layout solutions by COC–DCOC is that they can readily include a combination of a variety of design conditions, such as stress, displacement, natural frequency and system buckling constraints.

The length of the simply supported edge in Fig. 10.27 is 1.0 and its slope to the axis of symmetry is 2:1. The structural universe wth 299 elements is shown in Fig. 10.27(a), and the permissible shear stresses and structural weight in the next three diagrams are as follows:

$$
\begin{array}{lll}
\text{Fig. 10.27(b):} & \tau_a = 17.79, & \Phi = 0.026653 \\
\text{Fig. 10.27(c):} & \tau_a = 10.00, & \Phi = 0.027801 \\
\text{Fig. 10.27(d):} & \tau_a = 6.00, & \Phi = 0.035282
\end{array}
\qquad (10.29)
$$

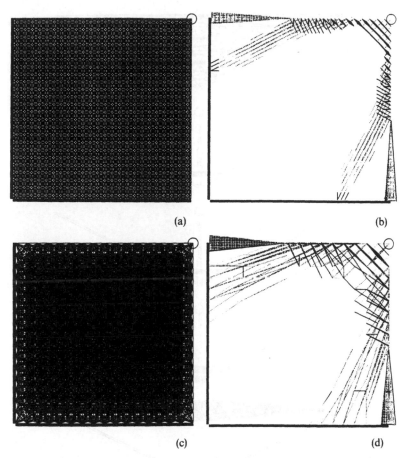

(a) (b)

(c) (d)

Fig. 10.26 Discretized solutions for a square grillage with two clamped and two free edges.

The above results were obtained by Sigmund using a normalized formulation wth $P = \rho = t = 1$. If we restrict the beam elements to the free edge, then we obtain the solution in Fig. 10.27(e), with a weight of $\Phi = 0.161\,626$ ($\tau_{max} = 4.82$ inactive). This latter solution, which was proposed by Lowe and Melchers (1972–1973) as a 'practical optimum', represents over 400% of the optimal weight in this case, although it is still being claimed (Lowe, 1988) that the topology in Fig. 10.27(e) is relatively economical. Figure 10.28(a) shows the analytical solution for the same problem, together with the optimal regions. The short beam across the corner has a theoretically infinitesimal length. For a comparison, the region topology of the Lowe–Melchers solution, which clearly violates the inequality (10.4) in the direction of the axis of symmetry, is given in Fig. 10.28(b). Finally, the exact optimal solution with allowance for support costs of $\Omega = r|R|$ is shown in Fig. 10.28(c), where r is a given constant and R is the reaction. It can be seen that once the cost of reactions is taken into consideration, the solution becomes a transition between the first author's solution (e.g. Rozvany, Hill and Gangadharaiah, 1973)

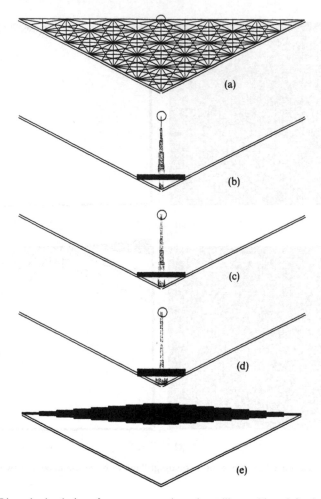

Fig. 10.27 Discretized solutions for a corner region of a grillage with a deflection constraint and shear stress constraints.

and the Lowe and Melchers (1972–1973) solution – a very satisfactory resolution of an old controversy!

10.7 EXTENSION OF THE DCOC ALGORITHM TO COMBINATIONS OF STRESS, DISPLACEMENT AND NATURAL FREQUENCY CONSTRAINTS WITH ALLOWANCE FOR SELFWEIGHT AND STRUCTURAL MASS

10.7.1 Derivation of optimality criteria

We extend in this subsection the derivation of DCOC for stress and displacement constraints and several loading conditions (section 2.3.3) to selfweight and natural

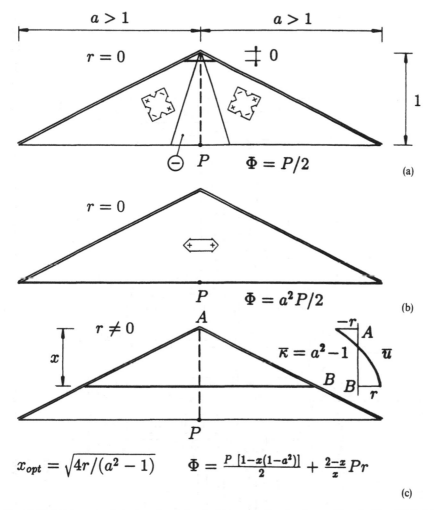

Fig. 10.28 (a) Exact analytical optimal plastic design solution for the problem in Fig. 10.27; (b) region topology of the Lowe and Melchers (1972–1973) solution and (c) solution with allowance for the cost of supports.

frequency constraints. It is assumed that the selfweight of an element equals the element objective function $w^e(x^e)$. The relative initial displacements $\overset{\circ}{\delta}{}^{*e}$ caused by the selfweight are linear functions of w^e, say

$$\{\overset{\circ}{\delta}{}^{*e}\} = \{g^e\}\, w^e \tag{10.30}$$

where the vector $\{g^e\}$ depends on the design variables x^e. The natural frequency constraints can be represented by the corresponding Rayleigh quotients

$$\lambda_g \leqslant \frac{\{\check{U}_g\}^T[K]\{\check{U}_g\}}{\{\check{U}_g\}^T[M + \vec{M}]\{\check{U}_g\}} \quad (g = 1, \ldots, G) \tag{10.31}$$

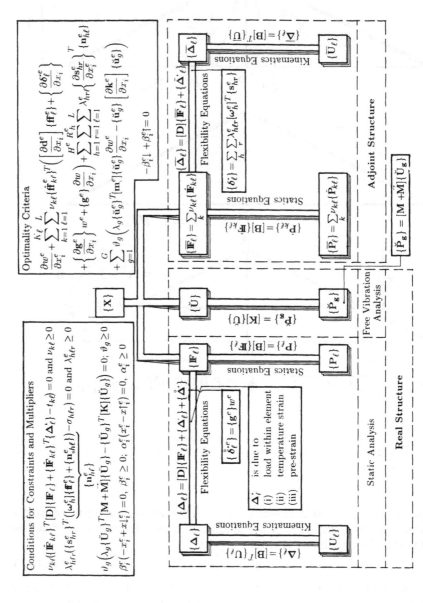

Fig. 10.29 DCOC algorithm for linearly elastic systems with stress, displacement and natural frequency constraints and allowance for selfweight and selfmass – conceptual scheme.

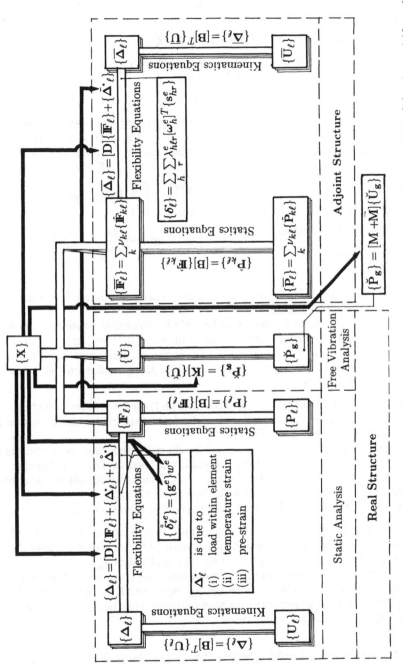

Fig. 10.30 Conceptional diagram showing the dependency of certain relations used in the DCOC algorithm on the design variables $\{X\}$ and on nodal forces $\{F_l\}$.

where λ_g is the lower limit on an eigenvalue (square of natural frequency), $\{\breve{U}_g\}$ is the eigenvector, $[M]$ is the structural mass matrix and $[\vec{M}]$ is the nonstructural mass matrix. The constraint (10.31) can be rewritten in the more convenient form

$$\lambda_g\{\breve{U}_g\}^T[M + \vec{M}]\{\breve{U}_g\} - \{\breve{U}_g\}^T[K]\{\breve{U}_g\} \leqslant 0 \quad (g = 1,...,G) \qquad (10.32)$$

The structural mass matrix $[m^e]$ for each element can be expressed as some linear function of the element weights $w^e(x^e)$,

$$[m^e] = [m_0^e]w^e \qquad (10.33)$$

We use the Lagrange multipliers $\vartheta_g (g = 1,...,G)$ for the natural frequency constraints (10.32), otherwise the notation remains the same as in section 2.3.3.

Necessary conditions for optimality (Kuhn–Tucker conditions) with respect to variation of the design variables $x_i^e (i = 1,...,I; e = 1,...,E)$ then become

$$\frac{\partial w^e}{\partial x_i^e} + \sum_{k=1}^{K_l}\sum_{l=1}^{L} v_{kl}\{\hat{\mathbf{ff}}_{kl}^e\}^T\left(\left[\frac{\partial \mathbf{d}^e}{\partial x_i}\right]\{\mathbf{ff}_l^e\} + \left\{\frac{\partial \pmb{\delta}_l^{*e}}{\partial x_i}\right\} + \left\{\frac{\partial \mathbf{g}^e}{\partial x_i}\right\}w^e + \{\mathbf{g}^e\}\frac{\partial w}{\partial x_i}\right)$$

$$+ \sum_{h=1}^{H^e}\sum_{r=1}^{R_h^e}\sum_{l=1}^{L} \lambda_{hrl}^e \frac{\partial \theta_{hr}^e(\mathbf{n}_{hl}^e)}{\partial x_i^e} + \sum_{g=1}^{G} \vartheta_g\left(\lambda_g\{\breve{\mathbf{u}}_g^e\}^T[m_0^e]\{\breve{\mathbf{u}}_g^e\}\frac{\partial w^e}{\partial x_i} - \{\breve{\mathbf{u}}_g^e\}^T\left[\frac{\partial \mathbf{k}^e}{\partial x_i}\right]\{\mathbf{u}_g^e\}\right)$$

$$- \beta_i^e\downarrow + \beta_i^e\uparrow = 0 \qquad (10.34)$$

Kuhn–Tucker conditions with respect to the real and adjoint forces give the same relations as in section 2.3.3.

A conceptual scheme for the extended DCOC algorithm considered above is given in Fig. 10.29, in which linear stress conditions $\{s_{hr}^e\}^T\{n_{hl}^e\}$ are instead of the nonlinear ones $\theta_{hr}^e(n_{hl}^e)$ in equation (10.34). It can be seen from Fig. 10.29 that any change in the value of the design variables has an effect on the following relations in Fig. 10.29:

- the flexibility matrix $[D]$ (or stiffness matrix $[K]$) used for the real structure;
- the initial displacements $\{\Delta_l^*\}$ caused by loads within the elements, due to changes in member flexibilities;
- the initial displacements $\{\mathring{\Delta}^*\}$ caused by the selfweight, due to (i) changes in the member flexibilities and (ii) changes in the load caused by selfweight;
- the flexibility matrix $[D]$ (or stiffness matrix $[K]$) used for the adjoint structure;
- the stiffness matrix $[K]$ used for the free vibration analysis;
- the structural mass matrix $[M]$ used in the free vibration analysis.

The above feedback effects are represented graphically in Fig. 10.30, which also shows that the adjoint initial displacements $\{\bar{\Delta}^*\}$ are also dependent on the real nodal forces $\{\mathbf{F}_e\}$ (through the Lagrange multipliers λ_{hrl}^e).

For problems with only natural frequency constraints, the optimality conditions (10.34) reduce to those of Berke and Khot (e.g. Berke and Khot, 1987). The detailed computational algorithm for the problems discussed in this section is given elsewhere (Sigmund and Rozvany, 1994).

10.7.2 Test example involving selfweight: square clamped grillage with a central point load

Considering a laminated timber grillage having a depth of 1.0 m, Young's modulus $E = 10^7 \text{ kN m}^{-2}$, specific weight $\rho = 6 \text{ kN m}^{-3}$, minimum width 10^{-5} m,

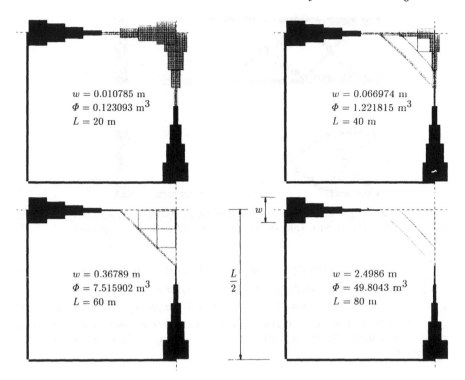

Fig. 10.31 Optimal grillage layouts with allowance for selfweight considering various span lengths (by Sigmund).

prescribed deflection $t = L/300$ where L is the span length, and a point load of 10 kN, Sigmund obtained by the extended DCOC method given in section 10.7.1 the solutions in Fig. 10.31, which also indicates the span length (L), the maximum beam width (w) in m and the total volume Φ in m³. All graphic outputs were scaled such that the maximum width indicated is the same, but the latter for the largest span is in fact over 230 times greater than that for the shortest span. It can be seen that with increasing span length more and more material is moving away from the middle of the grillage to the supports. Naturally, the prescribed minimum width would be completely unrealistic in practice, but the results indicate interesting trends.

10.7.3 Test example: perforated deep cantilever beams optimized for first or second natural frequency

Using the method discussed in section 10.7.1, Sigmund optimized perforated deep beams supported along the left vertical edge. The height of the beam is 1.0 and its length 4.0. At the right end a nonstructural mass is created by fixing the thickness at its maximum value over a length of 0.2. The number of bilinear square elements used is 840, with a minimum thickness of 0.01 and a maximum thickness of 1.0. In the solution in Fig. 10.32(a), the natural frequency was fixed

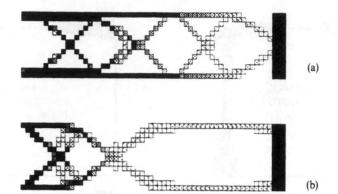

(a)

(b)

Fig. 10.32 Least-weight, deep, perforated cantilever beams designed, respectively, for given first and second natural frequency with nonstructural mass and selfmass (by Sigmund).

at $\sqrt{\lambda_1} = 0.0707$ ($\lambda_1 = 0.005$) and the second natural frequency turned out to be $\sqrt{\lambda_2} = 0.2154$. In Fig. 10.32(b), the second natural frequency was fixed at $\sqrt{\lambda_2} = 0.346$ ($\lambda_2 = 0.12$) and the first natural frequency turned out to be $\sqrt{\lambda_1} = 0.0253$. In Fig. 10.32, black areas indicate the maximum thickness, white areas the minimum thickness, with other areas varying in darkness according to the thickness.

10.8 GENERALIZED SHAPE OPTIMIZATION

10.8.1 Historical background and problem classification

As mentioned in section 10.1, the aim of generalized shape optimization is the simultaneous optimization of both topology and shape of the boundaries of

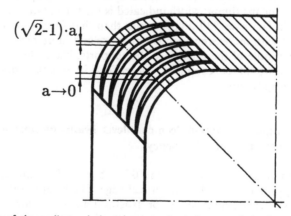

Fig. 10.33 One of the earliest solutions in generalized shape optimization (after Kohn and Strang, 1983).

two- or three-dimensional continua or of the interfaces between different materials in composites.

It was established by Kohn and Strang (1983) in the context of plastic design for torsion of a cross section within a square area (Fig. 10.33) that generalized shape optimization may yield three types of regions in the solution, namely

- solid regions (filled with material),
- empty regions (without material), and
- porous regions (some material, with cavities of infinitesimal size).

Considering elastic perforated plates in bending or in plane stress, it was found by various mathematicians (e.g. Lurie and Cherkaev, 1984; Murat and Tartar, 1985; Kohn and Strang, 1986; Avellaneda, 1987) that one optimal microstructure for a compliance constraint consists of rank-2 laminates, i.e. ribs of first- and second-order infinitesimal width in the two principal directions.

10.8.2 Solid–empty–porous (SEP) solutions

Analytical solutions based on rank-2 laminates for axisymmetric perforated plates, derived by Rozvany *et al.* (1987) as well as Ong, Rozvany and Szeto (1988), have shown that

- a high proportion of the available space in exact optimal solutions consists of porous regions, and
- for low volume fractions (ratio of material volume to available volume) the solution tends to that for least-weight trusses (Michell, 1904) for plane stress and least-weight grillages (e.g. Prager and Rozvany, 1977a) for bending (this conclusion was also confirmed by Allaire and Kohn (1993)).

Following the above developments, Bendsøe (e.g. Bendsøe, 1989) opened up new avenues of research in topology optimization by investigating discretized optimal solutions derived by mathematical programming methods, using either the correct microstructure (rank-2 laminates) or a 'suboptimal' microstructure (square or rectangular holes). The above investigations have also made use of some important analytical relations regarding geometrical properties of optimal topologies (e.g. Pedersen, 1989, 1990). Extensions of this work were reported by several investigators (e.g. Bendsøe and Kikuchi, 1988; Suzuki and Kikuchi, 1991; Diaz and Bendsøe, 1992; Olhoff, Bendsøe and Rasmussen, 1991; Allaire and Kohn, 1993).

Numerical solutions using rank-2 laminates have a considerable theoretical importance because they represent (within a discretization error) an absolute limit on the structural weight for a class of compliance problems. However, other lines of research on topology optimization are justified, because solutions with rank-2 microstructures are somewhat unpractical for the following reasons:

- even an approximate version of rank-2 microstructures with finite rib widths would require very high manufacturing costs;
- rank-2 laminates in perforated structures (as distinct from composite structures) have a zero shear stiffness in one direction, which makes these designs completely unstable if the load direction is changed;

- solutions are only available for a single compliance or natural frequency constraint which are not realistic design problems. This was also demonstrated recently by Sankaranarayanan, Haftka and Kapania (1992).

Moreover, because of nonconvexity, the above solution could represent a local optimum.

Solutions with suboptimal microstructures often turn out to be more practical, because they penalize and thereby suppress porous regions.

The solutions discussed in this subsection often appear in the literature under the term homogenization, which means that an inhomogeneous structural element, containing an infinite number of discontinuities in material or geometrical properties, is replaced by a homogeneous but anisotropic element, whose stiffness is direction but not location dependent within an element. The same idea was used in a very simple form much earlier by others (e.g. Michell, 1904; Prager and Rozvany, 1977b), introducing terms such as 'truss-like continua' or 'grillage-like continua' (Prager, 1974). To engineers this is a rather obvious idealization, but its rigorous mathematical treatment is much more involved (e.g. Bensousson, Lions and Papanicolaou, 1978). It would be incorrect, however, to make the term 'generalized shape optimization' synonymous with 'shape optimization by homogenization'.

10.8.3 Solid–empty (SE) solutions and solid, isotropic microstructure with penalty (SIMP) for intermediate densities

The method described in this subsection has been used extensively by the authors, but it was also tried out earlier by Bendsøe (e.g. Bendsøe, 1989), as an extension of a technique employed by Rossow and Taylor (1973).

From an engineering point of view, it is more useful to aim at solutions in which porous regions are largely suppressed and then a second stage design procedure can produce a 'practical' solution consisting of solid and empty regions only, with smooth boundaries to avoid stress concentrations. This procedure was lucidly demonstrated by Olhoff, Bendsøe and Rasmussen (1991) on an example involving a simply supported beam with a central point load (Figs 10.34(a) and 10.34(b)).

It was suggested by the first author at a meeting in Karlsruhe in 1990 (Rozvany and Zhou, 1991; Zhou and Rozvany, 1991) that porous regions could be suppressed by adding to the material costs, the cost of manufacturing of holes, thereby penalizing porous regions. Once we decide that we want only solid and empty regions in the solution, any range of microstructures that includes the above two as limiting cases can be assumed in the solution process. We can therefore postulate that the specific material cost (e.g. weight) ψ is proportional to the specific stiffness s of perforated regions (Fig. 10.35(a)). This would also be the case if we used a plate of variable thickness (Rossow and Taylor, 1973), in which case we would have to penalize intermediate thicknesses. If the hypothetical microstructure contains holes, however, then the extra manufacturing cost would increase with the number or size of such holes, if we consider a casting or drilling process. However, for an empty region ($s = 0$), the manufacturing costs also become zero. The corresponding specific fabrication cost is shown in

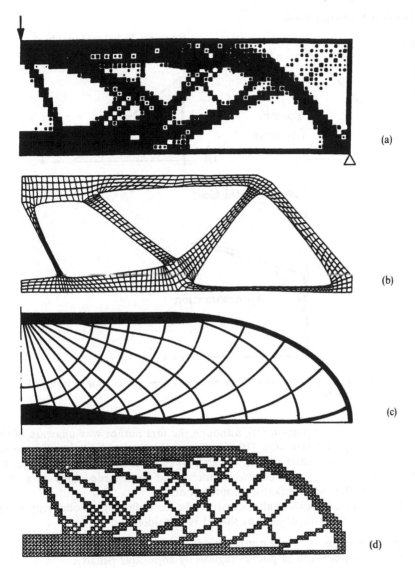

Fig. 10.34 (a) Topology of a perforated simply supported deep beam (one half shown) using a suboptimal microstructure; (b) design after a second stage optimization (after Olhoff, Bendsøe and Rasmussen, 1991); (c) the exact topology suggested by discretized truss solutions (Zhou and Rozvany, 1991); (d) topology derived by using a solid isotropic microstructure with penalty (SIMP) for intermediate densities (Rozvany and Zhou, 1991).

Fig. 10.35(b) and the specific total cost in Fig. 10.35(c), together with an approximation of the type

$$\psi = s^{1/n} \tag{10.35}$$

where *n* is a given constant. The above relation is identical with that suggested

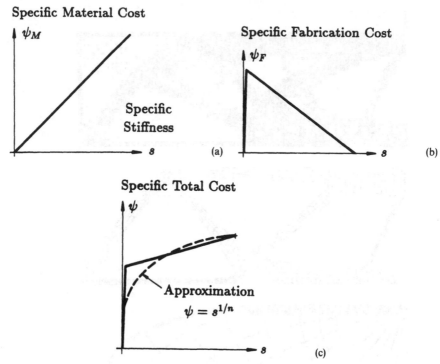

Fig. 10.35 Suppression of porous regions in generalized shape optimization by taking fabrication costs into consideration.

by Bendsøe (1989, equation (7)), although the first author was unaware of this when he proposed equation (10.35). The authors would like to emphasize, however, that this entire line of research owes its existence to the pioneering efforts of Bendsøe, and their own work should be regarded only as a small extension of Bendsøe's milestone contributions.

In selecting a hypothetical microstructure for obtaining a solid–empty (SE) solution, the following objectives should be considered:

- simplicity of analysis and optimization;
- selective suppression of porous regions by adjustable penalty;
- capability of handling a variety of design conditions (e.g. combinations of stress, displacement, natural frequency and system stability constraints for several loading conditions).

It has been found that a solid isotropic microstructure with penalty (SIMP) for intermediate densities, as explained above (Fig. 10.35), largely fulfills these objectives (Rozvany, Zhou and Birker, 1992). For example, the topology obtained by the SIMP model is shown in Fig. 10.34(d). The latter is an excellent approximation of the 'exact' topology for the same problem in Fig. 10.34(c), which was constructed on the basis of the following arguments.

As mentioned earlier, for low volume fractions the optimal topology for perforated plates in plane stress tends to that for least-weight trusses or Michell

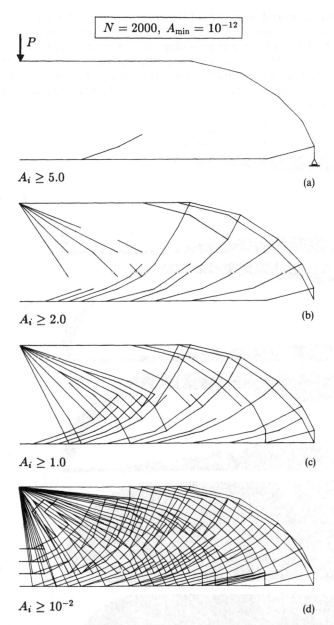

$$N = 2000, \ A_{min} = 10^{-12}$$

P

$A_i \geq 5.0$ (a)

$A_i \geq 2.0$ (b)

$A_i \geq 1.0$ (c)

$A_i \geq 10^{-2}$ (d)

Fig. 10.36 Truss layout obtained by a discretized COC method for the load and support conditions in Fig. 10.34.

structures (Rozvany *et al.*, 1987; Allaire and Kohn, 1993). The layout of a truss, with the same support and loading conditions as the beam in Fig. 10.34, was optimized by the discretized COC procedure and the results are shown in Fig. 10.36, which shows one half of the truss with various ranges of cross-sectional areas in the optimal solution. On the basis of this layout, taking some (analytical) geometrical properties of Hencky nets into consideration, the solution in Fig. 10.34(c) was constructed. In the latter, solid regions appear along the top and bottom chords of the corresponding truss, with the width proportional to the cross-sectional areas of the chords, an empty region in the top right corner and porous regions, with an infinite number of intersecting members in between the chords. The layout in Fig. 10.34(c) was also confirmed analytically (Lewinski, Zhou and Rozvany, 1993).

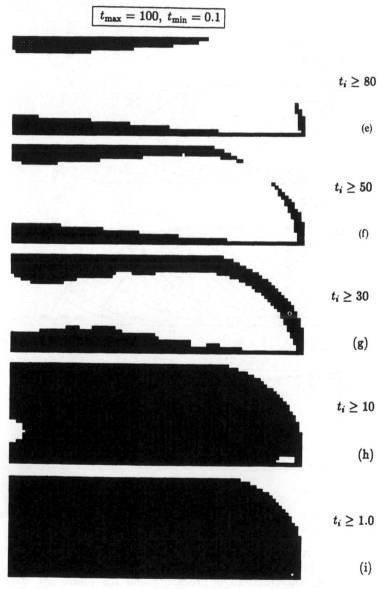

$$t_{max} = 100, \ t_{min} = 0.1$$

$t_i \geq 80$

(e)

$t_i \geq 50$

(f)

$t_i \geq 30$

(g)

$t_i \geq 10$

(h)

$t_i \geq 1.0$

(i)

Fig. 10.37 Solution for a plate of variable thickness obtained by a discretized COC method for the load and support conditions in Fig. 10.34.

The optimal topology in Figs 10.34(c) and 10.34(d) was also made plausible by optimizing a plate of variable thickness, but without penalty for intermediate thicknesses, for the same support and load conditions. The results are shown in Fig. 10.37, which shows that solid isotropic microstructures without penalty tend to give the same average density for larger areas as the SIMP formulation but, instead of ribs, an intermediate density appears in porous regions.

The solution in Fig. 10.34(d) is difficult to compare with a solution containing rank-2 laminates, but it is easier to understand in terms of an alternate optimal microstructure for plates with a compliance constraint, which was derived recently by Vigdergauz (1992). In porous regions, the latter may start at very low volume fractions with a Michell structure or least-weight grillage (Fig. 10.34(c)), then develop roundings at the corners as we increase the volume fraction (Fig. 10.38(a)), finishing up with elliptic holes of decreasing size as the volume fraction approaches unity (Fig. 10.38(b)).

In Fig. 10.39, three types of specific cost functions are compared for a plate in plane stress or bending. The straight line represents the normalized weight per unit area of a plate of variable thickness in plane stress or bending. The next curve shows the weight of a constant thickness perforated plate with rank-2

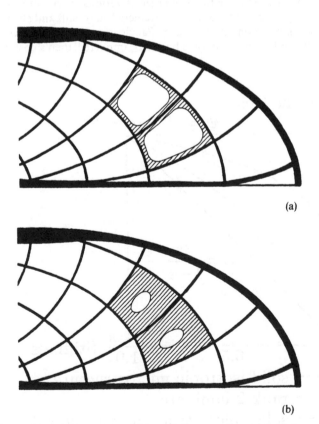

(a)

(b)

Fig. 10.38 Alternative optimal microstructures at higher volume fractions (the spacing should be theoretically infinitesimal).

laminates having an equal stiffness in two directions (Rozvany *et al.*, 1987). Finally, the top curve represents the power type cost function of the SIMP formulation.

When Bendsøe (1989) originally considered this power-type cost function, he used the term 'direct approach' or '0–1 discrete optimization method with a suitable differentiable approximation' using an 'artificial material'. At the same time, Bendsøe expressed preference for optimal or suboptimal anisotropic microstructures, reasoning that the solutions by the direct approach (now SIMP) are (i) highly mesh dependent and (ii) impossible to interpret physically. The authors are more optimistic in these respects, as can be seen from the following.

The SIMP model can be interpreted even for three-dimensional continua with stress and displacement constraints, assuming that both Young's modulus and permissible stress of a fictitious material (or a range of materials) are proportional to its (their) density, which varies between a given maximum value and a very small minimum value. It is then perfectly legitimate for the designer to penalize and thereby to suppress intermediate densities, if he chooses to do so. It is not necessary for such a material to exist, because we are only interested in the end result; however, it is still interesting to remark that the above relations are known to constitute a good estimate for properties of various species of timber, ranging from the very light balsa-wood to the extremely heavy, stiff and strong iron bark (from Australia). Assuming that on some Pacific island only these two tree species grew naturally and all other timbers had to be imported, then a cost optimization

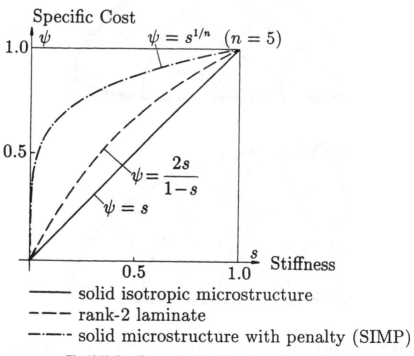

— solid isotropic microstructure
--- rank-2 laminate
—·—·· solid microstructure with penalty (SIMP)

Fig. 10.39 Specific cost functions for various microstructures.

of various timber structures would give exactly the same results as the SIMP model.

It is shown subsequently on test examples that the SIMP formulation yields the correct topology irrespective of the FE mesh used.

10.8.4 Test examples based on the SIMP model

(a) Cantilever beam

The analytical and discretized truss solutions for this problem were shown in Fig. 10.22. A topology obtained by using a suboptimal microstructure (square holes) by Suzuki and Kikuchi (1991) is shown in Fig. 10.40(a). It was shown by the same authors that this topology is largely independent of the number of elements used (for meshes ranging from 32 × 20 to 80 × 50). It can be seen by comparing Figs 10.22(a) and 10.40(a) that in the latter the topology and shape deviate significantly from the optimal ones. It should be taken into consideration that in Fig. 10.40(a) the aspect ratio is higher (1.6 instead of 1.5) and hence the chords should have an almost horizontal tangent at the support. This condition together with other close similarities with Fig. 10.22 were fulfilled better by the SIMP solutions shown in Figs 10.40(b) and 10.40(c), which were obtained, respectively, by M. Zhou (with 10 800 constant strain triangular elements) and T. Birker (1440 isoparametric square elements). For Birker's solutions, two different strategies were used, the iteration histories of which are shown in Fig. 10.41. Whilst in one of them (Fig. 10.41(a)) the initial design consisted of a

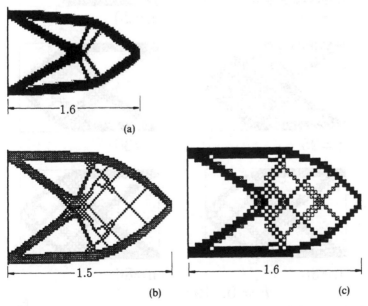

(a)

(b)   (c)

Fig. 10.40 Various discretized solutions for perforated deep cantilever beams with the same loading and support conditions as in Fig. 10.22.

plate of uniform thickness, in the other one (Fig. 10.41(b)), it was a plate of variable thickness optimized without penalty for intermediate densities (as the solutions by Rossow and Taylor (1973)). This latter method for the initial design, suggested by R.V. Kohn, gave a better topology and a slightly lower weight.

(b) Clamped beam

The analytical least-weight truss solution for the above problem is shown in Fig. 10.42(a) and a discretized DCOC solution, obtained by D. Gerdes (Essen University) in Fig. 10.42(c). Finally, a solution by O. Sigmund, who used the

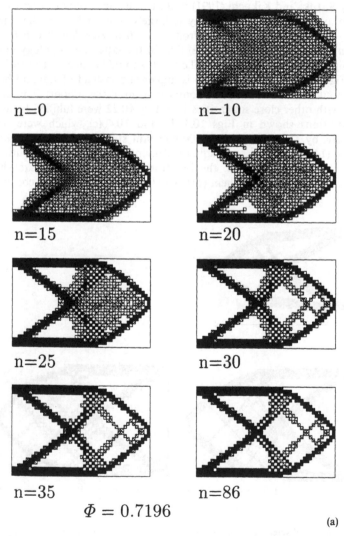

n=0   n=10

n=15   n=20

n=25   n=30

n=35   n=86

$\Phi = 0.7196$

(a)

Fig. 10.41 Iteration histories of the problem in Fig. 10.40 using two different initial solutions.

SIMP model, is given in Fig. 10.42(b). The analytical solution (Fig. 10.42(a)) consists of some circular 'fans' (e.g. ACD, BCE, IGJ, IHK), 'concentrated' straight members (e.g. AD, DG, BE, FH), concentrated curved members (DF, EF, GJ, HK) and Hencky nets consisting of curved members (CDFE). In the discretized solution (Figs 10.42(b) and 10.42(c)), the narrow light fans (e.g. ACD) and heavy members (e.g. AD) are lumped together into a single member, which could only be avoided by using a much finer discretization. However, the discretized solutions in Figs 10.42(b) and 10.42(c) are much more practical, because they consist of a small number of members.

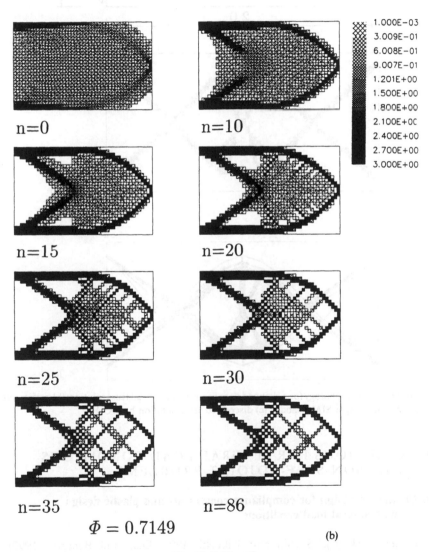

(b)

Fig. 10.41 *(Cont'd)*

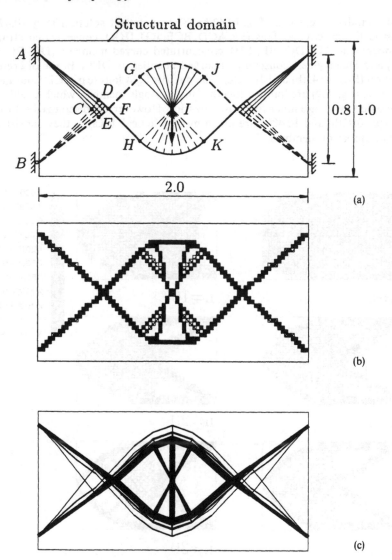

(a)

(b)

(c)

Fig. 10.42 Optimal topology for a clamped deep beam: (a) analytical truss solution; (b) perforated plate solution using a SIMP model; (c) discretized truss solution by DCOC.

10.9 SOLUTIONS FOR SEVERAL LOAD CONDITIONS AND NON-SELF-ADJOINT PROBLEMS

10.9.1 Elastic design for compliance constraints and plastic design with several load conditions

In earlier work (e.g. Suzuki and Kikuchi, 1991; Diaz and Bendsøe, 1992), approximate solutions of the above problems were obtained, in which

- either for each element the maximum compliance value out of all loading conditions was taken and then the sum of such values was calculated for the entire system, or
- a weighted combination of the compliances (or natural frequencies) was taken as objective function.

Both methods simplify the treatment considerably, and the latter is in fact equivalent to a single compliance constraint. The solutions obtained by DCOC are based on the actual multi-load problem.

An analytical layout theory for trusses with several load conditions was developed recently (Rozvany, 1992a). In the notation of Chapter 2 herein, we have the following conditions for layout optimality:

$$\varepsilon_l^e = F_l^e/E^e A^e, \quad \bar{\varepsilon}_l^e = \left(\sum_k v_{lk} F_{lk}^e \right)/E^e A^e$$

$$\frac{E^e}{\varrho^e} \sum_l \varepsilon_l^e \bar{\varepsilon}_l^e \leqslant 1 \quad \text{(for } A^e = 0\text{)}, \quad \frac{E^e}{\varrho^e} \sum_l \varepsilon_l^e \bar{\varepsilon}_l^e = 1 \quad \text{(for } A^e \neq 0\text{)} \qquad (10.36)$$

where l denotes the load condition and k a displacement constraint.

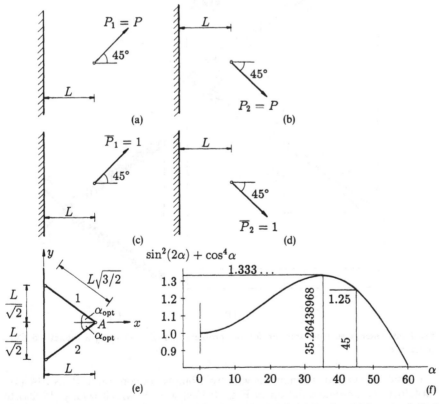

Fig. 10.43 Least weight truss for compliance constraints and two load conditions: analytical solution.

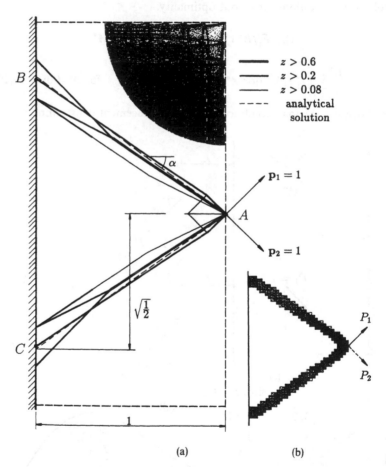

Fig. 10.44 Optimal plastic design for the problem in Fig. 10.43.

Fig. 10.45 Discretized truss solution for the problem in Fig. 10.43 using a discretized COC procedure.

Considering two load conditions with two compliance constraints (Figs 10.43(a)–10.43(d)), the optimal solution in Fig. 10.43(e) was shown (Rozvany, 1992a) to satisfy the conditions (10.36). The actual variation of $\varepsilon_i^e \bar{\varepsilon}_i^e$ is given in Fig. 10.43(f). Since here we are dealing with a convex problem, the above proof establishes global optimality. The corresponding normalized optimal weight is $\Phi = 27/8 = 3.375$.

For a comparison, the optimal plastic design for the two loads in Fig. 10.43 is given in Fig. 10.44 (Zhou and Rozvany, 1991). It can be seen that the latter consists of three members.

Discretized COC solutions for the problem in Fig. 10.43 were obtained (Zhou and Rozvany, 1991) using 7170 and 12 202 members in the structural universe. For the former the discretized solution is shown in Fig. 10.45(a), with part of the structural universe at the top right corner. This structural universe contained only a limited number of member directions and hence it could only approximate the analytical solution (broken line in Fig. 10.45(a)) by using a fairly large number of members, with a weight value of $\Phi = 3.492\,957\,26$ (3.495% error). In the structural universe with 12 202 members, all nodes of an 11×21 grid were connected with all other nodes, and the corresponding discretized solutions consisted basically of only two members, as in the analytical solution. The corresponding weight was $3.375\,668$ (only 0.0198% error).

Using the SIMP procedure, T. Birker obtained the solution in Fig. 10.45(b) for a perforated plate with two load conditions and two compliance constraints. The agreement between both solutions in Fig. 10.45 and that in Fig. 10.43 is rather obvious.

10.9.2 Structural layouts for non-self-adjoint problems

To demonstrate the difficulties arising in the case of non-self-adjoint problems, we consider the layout problem in Fig. 10.46(a), in which we have a point load

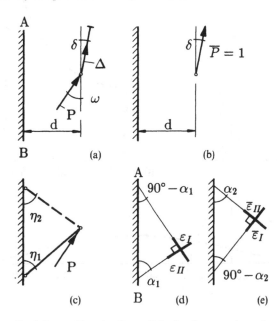

Fig. 10.46 A non-self-adjoint problem: loading, adjoint load, assumed topology, real and adjoint strain fields.

at an angle ω and a prescribed displacement at an angle δ, with $\omega \neq \delta$. The adjoint load for this problem is shown in Fig. 10.46(b). The treatment of this problem was published recently (Rozvany *et al.*, 1993). Figure 10.46(c) shows a possible topology, whilst Figs 10.46(d) and 10.46(e) the corresponding real and adjoint strain fields based on the optimality conditions (10.36). Using the latter, it was shown in the above-mentioned paper that the 'optimal' member directions are symmetrical ($\eta_1 = \eta_2$ in Fig. 10.46(c)). For example, for $\delta = 0°$, $\omega = 20°$ (Fig. 10.47), we obtain $\eta_1 = \eta_2 = 43.845\,305\,2$. Figure 10.47 (top) also shows the variation of the weight for the above displacement constraint and various values of the angle η, with cross-sectional areas optimized for a given angle. It can be seen that the solution obtained by using relation (10.36) is a local optimum (point A), the global optimum being nonstationary (ED in Fig. 10.47) and tending to zero. However, the stationary solution discussed above (point A) can be valid if additional stress conditions are introduced.

Figure 10.47 (bottom) also shows that the optimality conditions (10.36) yield two values for the strain field ε and $\bar{\varepsilon}$, and the corresponding products $\varepsilon\bar{\varepsilon}$ have a maximum at the points where the weight curve has minima or maxima.

From the above example, we may draw the following conclusions.

- Layout optimization of elastic structures with deflection constraints constitutes a non-self-adjoint problem.
- Because of nonconvexity, optimality criteria yield local minima as well as local maxima with respect to some parameters.

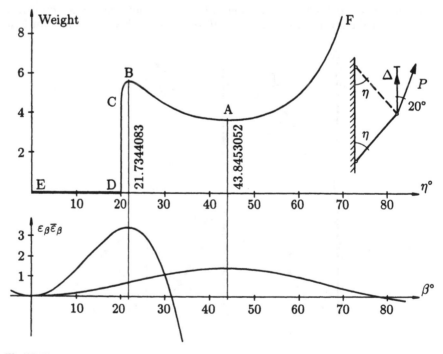

Fig. 10.47 Variation of the weight in dependence of the bar directions and variation of the strain product $\varepsilon\bar{\varepsilon}$.

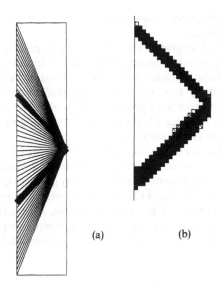

(a) (b)

Fig. 10.48 Discretized truss solution and perforated plate solution for the problem in Fig. 10.46.

- The globally optimal solution for one deflection constraint is, in general, a nonstationary solution whose weight tends to zero.
- For a well-posed problem, stress constraints must also be included in the formulation.

The result in Fig. 10.47 was confirmed by using both DCOC for an optimized truss layout (Fig. 10.48(a), after Gerdes) and a SIMP formulation for a perforated plate (Fig. 10.48(b), after Birker).

10.10 CONCLUDING REMARKS

We can draw the following conclusions from this chapter.

- Topology optimization in structural design has two main applications at present: layout optimization of grid-type structures and generalized shape optimization of cellular continua or composite systems.
- New optimality criteria methods (COC, DCOC) are eminently suitable for topology optimization because of their high optimization capability in terms of the number of variables and the number of active constraints.
- Analytical layout solutions are now available for relatively simple design constraints (a single stress or compliance or natural frequency constraint) and structures with a single variable per cross section (e.g. trusses, grillages of given depth, shell-grids in membrane action), but for these classes of problems the exact optimum can be derived systematically for wide ranges of complicated load and support conditions.
- Discretized layout solutions can be obtained by COC–DCOC methods for many thousand potential members in the structural universe (ground structure) and for combinations of stress, displacement and natural frequency constraints.

- For non-self-adjoint, nonconvex problems, however, the above methods may yield a local optimum.
- Owing to the very fine grid in the structural universe, the proposed methods achieve in effect a simultaneous optimization of topology and geometry.
- In generalized shape optimization (simultaneous optimization of boundary shape and boundary topology), the suggested SIMP formulation (solid, isotropic microstructure with penalty for intermediate densities) seems to be highly efficient in locating SE (solid–empty) topologies which show a good agreement with analytical solutions.
- The test examples used in this chapter are somewhat academic for easy comparison with analytical solutions, but the discretized methods employed are now suitable for large systems with a variety of realistic design constraints.
- Finally, it must be emphasized that topology optimization is a new but rapidly developing field with many unsolved problems. An excellent review of basic features of optimal topologies, some involving severe computational difficulties, was given by Kirsch (1990); for a review see also Kirsch and Rozvany (1993).

ACKNOWLEDGEMENTS

The authors are indebted to the Deutsche Forschungsgemeinschaft for financial support (Project Nos Ro 744/4-1 and Ro 744/6-1), to Sabine Liebermann (text processing), Peter Moche (drafting) and Susann Rozvany (editing and some text processing).

REFERENCES

Allaire, G. and Kohn, R.V. (1993) Topology optimization and optimal shape design using homogenization, in *Topology Design of Structures*. Proceedings of NATO ARW, Sesimbra, Kluwer, Dordrecht, pp. 207–18.

Avellaneda, M. (1987) Optimal bounds and microgeometries for elastic two-phase composites. *SIAM Journal of Applied Mathematics*, **47**, 1216–28.

Bendsøe, M.P. (1989) Optimal shape design as a material distribution problem. *Structural Optimization*, **1**, 193–202.

Bendsøe, M.P. and Kikuchi, N. (1988) Generating optimal topologies in structural design using a homogenization method. *Computer Methods in Applied Mechanical Engineering*, **71**, 197–224.

Bensousson, A., Lions, J.-L. and Papanicolaou, G. (1978) *Asymptotic Analysis for Periodic Structures*, North-Holland, Amsterdam.

Berke, L. and Khot, N. (1987) Structural optimization using optimality criteria, in *Computer Aided Optimal Design: Structural and Mechanical Systems* (ed. C.A. Mota Soares), Springer, Berlin, pp. 271–312.

Cheng, K.T. and Olhoff, N. (1981) An investigation concerning optimal design of solid elastic plates. *International Journal of Solids and Structures*, **17**, 305–23.

Diaz, A.R. and Bendsøe, M.P. (1992) Shape optimization of multipurpose structures by a homogenization method. *Structural Optimization*, **4**, 17–22.

Hemp, W.S. (1973) *Optimum Structures*, Clarendon, Oxford.

Hill, R.H. and Rozvany, G.I.N. (1985) Prager's layout theory: a non-numeric computer alogrithm for generating optimal structural configurations and weight influence surfaces. *Computer Methods in Applied Mechanical Engineering*, **49**, 131–48.

Kirsch, U. (1989) Optimal topologies of structures. *Applied Mechanics Review*, **42**, 223–38.

Kirsch, U. (1990) On singular topologies in structural design. *Structural Optimization*, **2**, 133–42.

Kirsch, U. and Rozvany, G.I.N. (1993) Design considerations in the optimization of structural topologies, in *Optimization of Large Structural Systems* (ed. G.I.N. Rozvany), Proceedings of NATO ASI, Berchtesgaden, 1991, Kluwer, Dordrecht, pp. 121–41.

Kohn, R.V. and Strang, G. (1983) Optimal design for torsional rigidity, in *Hybrid and Mixed Finite Element Methods* (eds S.N. Atluri, R.H. Gallagher, and O.C. Zienkiewicz), Wiley, Chichester, pp. 281–8.

Kohn, R.V. and Strang, G. (1986) Optimal design and relaxation of variational problems, I, II, and III. *Communications in Pure and Applied Mathematics*, **39**, 113–37, 139–82, 353–77.

Lagache, J.-M. (1981) Developments in Michell theory, in *Proceedings of International Symposium on Optimal Structure Design, Tucson, 1981* (eds E. Atrek and R.H. Gallagher), University of Arizona, Tucson, AZ, pp. 4.9–4.16.

Lewiński, T., Zhou, M. and Rozvany, G.I.N. (1993) Exact least-weight truss layouts for rectangular domains with various support conditions. *Structural Optimization*, **6**, 65–67.

Lowe, P.G. (1988) Optimization of systems in bending conjecture, bounds and estimates relating to moment volume and shape, in *Proceedings of IUTAM Symposium on Structural Optimization, Melbourne, 1988* (eds G.I.N. Rozvany and B.L. Karihaloo), Kluwer, Dordrecht, pp. 169–76.

Lowe, P.G. and Melchers, R.E. (1972–1973) On the theory of optimal constant thickness fibre-reinforced plates, I, II, III. *International Journal of Mechanical Science*, **14**, 311–24; **15**, 157–70, 711–26.

Lurie, K.A. and Cherkaev, A.V. (1984) G-closure for some particular set of admissible material characteristics for the problem of bending of thin elastic plates. *Journal of Optimization Theory and Applications*, **42**, 305–16.

Michell, A.G.M. (1904) The limits of economy of material in frame-structures. *Philosophical Magazine*, **8**, 589–97.

Murat, F. and Tartar, L. (1985) Calcul des variations et homogénéisation, in *Les Méthodes de l'Homogénéisation: Théorie et Applications en Physique*, Eyrolles, Paris, pp. 319–70.

Olhoff, N., Bendsøe, M.P. and Rasmussen, J. (1991) On CAD-integrated structural topology and design optimization. *Computer Methods in Applied Mechanical Engineering*, **89**, 259–79.

Olhoff, N. and Rozvany, G.I.N. (1982) Optimal grillage layouts for given natural frequency. *Journal of Engineering Mechanics of ASCE*, **108**, 971–5.

Ong, T.G., Rozvany, G.I.N. and Szeto, W.T. (1988) Least-weight design of perforated elastic plates for given compliance: non-zero Poisson's ratio. *Computer Methods in Applied Mechanical Engineering*, **66**, 301–22.

Pedersen, P. (1989) On optimal orientation of orthotropic materials. *Structural Optimization*, **1**, 101–6.

Pedersen, P. (1990) Bounds on elastic energy in solids of orthotropic materials. *Structural Optimization*, **2**, 55–63.

Prager, W. (1974) *Introduction to Structural Optimization*, Course held in International Centre for Mechanical Science, Udine, CISM 212, Springer, Vienna.

Prager, W. and Rozvany, G.I.N. (1977a) Optimal layout of grillages. *Journal of Structural Mechanics*, **5**(1) 1–18.

Prager, W. and Rozvany, G.I.N. (1977b) Optimization of structural geometry, in *Dynamical Systems* (eds A.R. Bednarek and L. Cesari), Academic Press, New York, pp. 265–93.

Prager, W. and Shield, R.T. (1967) A general theory of optimal plastic design. *Journal of Applied Mechanics*, **34**(1), 184–6.

Rossow, M.P. and Taylor, J.E. (1973) A finite element method for the optimal design of variable thickness sheets. *AIAA Journal*, **11**, 1566–9.

Rozvany, G.I.N. (1973) Optimal force transmission of flexure-clamped boundaries. *Journal of Structural Mechanics*, **2**, 57–82.

Rozvany, G.I.N. (1976) *Optimal Design of Flexural Systems*, Pergamon, Oxford.

Rozvany, G.I.N. (1981) Optimality criteria for grids, shells and arches, in *Optimization of Distributed Parameter Structures* (eds E.J. Haug and J. Cea), Proceedings of NATO ASI, Iowa City, 1980, Sijthoff and Noordhoff, Alphen aan den Rijn, pp. 112–51.

Rozvany, G.I.N. (1984) Structural layout theory: the present state of knowledge, in *New Directions in Optimum Structural Design* (eds E. Atrek, R.H. Gallagher, K.M. Ragsdell and O.C. Zienkiewicz), Wiley, Chichester, pp. 167–95.

Rozvany, G.I.N. (1989) *Structural Design via Optimality Criteria*, Kluwer, Dordrecht.

Rozvany, G.I.N. (1992a) Optimal layout theory: analytical solutions for elastic structures with several deflection constraints and load conditions. *Structural Optimization*, 4, 247–9.

Rozvany, G.I.N. (1992b) Optimal layout theory, in *Shape and Layout Optimization of Structural Systems and Optimality Criteria Methods* (ed. G.I.N. Rozvany), CISM Course, Udine, 1990, Springer, Vienna, pp. 75–163.

Rozvany, G.I.N. (1993a) Topological optimization of grillages: past controversies and new directions. *International Journal of Mechanical Science* (in press).

Rozvany, G.I.N. (1993b) Optimal grillage layouts for partially upward and partially downward loading. *Structural Optimization* (to be submitted).

Rozvany, G.I.N. (1993c) Layout theory for grid-type structures, in *Topology Design of Structures* (eds M.P. Bendsøe and C.A. Mota Soares), Proceedings of NATO ARW, Sesimbra, 1992, Kluwer, Dordrecht, pp. 251–72.

Rozvany, G.I.N. and Gerdes, D. (1994) Optimal layout of grillages with free, simply supported and clamped edges. *Structural Optimization* (to be submitted).

Rozvany, G.I.N. and Gollub, W. (1990) Michell layouts for various combinations of line supports, part I. *International Journal of Mechanical Science*, 32(12) 1021–43.

Rozvany, G.I.N. (1994) Optimal layout theory – allowance for the cost of supports and optimization of support locations. *Mechanics and Structures of Machines*, 22 (in press).

Rozvany, G.I.N. and Hill, R.H. (1976) General theory of optimal load transmission by flexure. *Advances in Applied Mechanics*, 16, 184–308.

Rozvany, G.I.N. and Hill, R.H. (1978) A computer algorithm for deriving analytically and plotting optimal structural layout. *Computers and Structures*, 10, 295–300.

Rozvany, G.I.N. and Ong, T.G. (1987) Minimum-weight plate design via Prager's layout theory (Prager memorial lecture), in *Computer Aided Optimal Design: Structural and Mechanical Systems* (ed. C.A. Mota Soares), Proceedings of NATO ASI, Troia, 1986, Springer, Berlin, pp. 165–79.

Rozvany, G.I.N. and Prager, W. (1979) A new class of structural optimization problems: optimal archgrids. *Computer Methods in Applied Mechanical Engineering*, 19, 127–50.

Rozvany, G.I.N. and Zhou, M. (1991) Applications of the COC method in layout optimization. in *Proceedings of International Conference on Engineering Optimization in Design Processes* (eds H. Eschenauer, C. Matteck and N. Olhoff), Karlsruhe, 1990, Springer, Berlin, pp. 59–70.

Rozvany, G.I.N., Gollub, W. and Zhou, M. (1992) Layout optimization in structural design, in *Proceedings of NATO ASI, Optimization and Decision Support Systems in Civil Engineering* (ed. B.H.V. Topping), Edinburgh, June 25–July 7, 1989, Kluwer, Dordrecht.

Rozvany, G.I.N., Gollub, W. and Zhou, M. (1994) Michell layouts for various combinations of line supports, part II. *International Journal of Mechanical Science* (to be submitted).

Rozvany, G.I.N., Hill, R.H. and Gangadharaiah, C. (1973) Grillages of least weight – simply supported boundaries. *International Journal of Mechanical Science*, 15, 665–77.

Rozvany, G.I.N., Olhoff, N., Cheng, K.-T. and Taylor, J.E. (1982) On the solid plate paradox in structural optimization. *Journal of Structural Mechanics*, 10, 1–32.

Rozvany, G.I.N., Olhoff, N., Bendsøe, M.P. *et al.* (1987) Least-weight design of perforated elastic plates I, II. *International Journal of Solids and Structures*, 23, 521–36, 537–50.

Rozvany, G.I.N., Sigmund, O., Lewinski, T., Gerdes, D. and Birker, T. (1993) Exact optimal structural layouts for non-self adjoint problems. *Structural Optimization*, 5, 204–6.

Rozvany, G.I.N., Zhou, M. and Birker, T. (1992) Generalized shape optimization without homogenization. *Structural Optimization*, 4, 250–2.

Rozvany, G.I.N., Zhou, M. and Gollub, W. (1993) Layout optimization by COC methods: analytical solutions, in *Optimization of Large Structural Systems* (ed. G.I.N. Rozvany), Proceedings of NATO ASI, Berchtesgaden, 1991, Kluwer, Dordrecht, pp. 77–102.

Rozvany, G.I.N., Zhou, M., Rotthaus, M., Gollub, W. and Spengemann, F. (1989) Continuum-type optimality criteria methods for large structural systems with a displacement constraint – part I. *Structural Optimization*, 1, 47–72.

Sankaranarayanan, S., Haftka, R.T. and Kapania, R.K. (1992) Truss topology optimization with simultaneous analysis and design. Proceedings of 33rd AIAA/ASME/ASCE/AHS/ASC Structural Dynamics Materials Conference, Dallas, TX, AIAA, Washington, DC, pp. 2576–85.

Sigmund, O. and Rozvany, G.I.N. (1994) Extension of discretized OC methods to combined stress, displacement and natural frequency constraints including selfweight – part I: beams and grillages. *Structural Optimization* (to be submitted).

Sigmund, O., Zhou, M. and Rozvany, G.I.N. (1993) Layout optimization of large FE-systems by new optimality criteria methods: applications to beam systems. in *Proceedings of NATO ASI Concurrent Engineering Tools and Technologies for Mechanical System Design* (ed. E.H. Haug), Iowa, 1992, Springer, Berlin.

Suzuki, K. and Kikuchi, N. (1991) A homogenization method for shape and topology optimization. *Computer Methods in Applied Mechanical Engineering*, 93, 291–318.

Vigdergauz, S. (1992) Two-dimensional grained composites of extreme rigidity. 18th International Congress on Theoretical Applied Mechanics, Haifa.

Zhou, M. and Rozvany, G.I.N. (1991) The COC algorithm, part II: topological, geometrical and generalized shape optimization. *Computer Methods in Applied Engineering*, 89, 309–36.

11

Practical shape optimization for mechanical structures

NOBORU KIKUCHI and KAZUTO HORIMATSU

11.1 INTRODUCTION

Reduction of the duration of design and manufacturing processes becomes much more important than reducing the cost of raw material of a product while the required functionality is fulfilled, to reduce the overall cost and to lead to success of a new product. In order to reflect this situation to research and development in design methodology, we must reconstruct the notion of structural and mechanical design optimization. In the past most structural and mechanical design optimization techniques tried to minimize the cost of raw material under certain constraints which are implied from mechanics and manufacturing requirement. However, now, the most important matter is how easily certain design can be improved with minimal effort by design engineers rather than just considering minimization of the cost (i.e. weight) of raw material. In other words it becomes very important to examine how the optimal design can be achieved for solving the minimum cost (weight) problem. In this chapter we shall describe a method to find the optimum with less effort without constructing an excessively large system of design optimization that integrates geometric modeling, finite element analysis, sensitivity analysis, and optimization algorithms.

The standard procedure of mechanical–structural design is that, after defining the shape and topology of a structure based on the conceptual or existing design, a discrete model is developed for stress–thermal–flow analysis. Then, defining appropriate design variables such as representative physical dimensions of a structure, sensitivity of the objective and constraint functions for design optimization are computed to provide necessary information for optimization algorithms based on mathematical programming methods such as sequential linear programming (SLP), sequential quadratic programming (SQP), and other methods. In short, the procedure of design optimization consists of four modules: solid modeling for the initial design, finite element analysis, sensitivity analysis, and optimization methods.

Nowadays, it is easy to find appropriate general-purpose programs for solid modeling, finite element analysis, and optimization methods. Commercial and public domain software is available to most of structural design engineers. For example, I-DEAS for SDRC does have all of these capabilities including some capability of sensitivity analysis as an advanced engineering option, PATRAN from PDA Engineering can provide decent capability of finite element modeling, MSC/ NASTRAN from MacNeal-Schwendler Corporation is the most widely used finite element analysis program in which sensitivity analysis for limited design variables and optimization algorithms are also available as an advanced option, while ANSYS from Swanson Analysis Systems, NISA II from EMRC, and MECANICA from RASNA claim handy design optimization capability. There are also many other software packages for finite element analysis which are applicable to design optimization study. For optimization algorithms, we can find, for example, ADS and DOT from VMA, IDESIGN from University of Iowa, VMCON from Argonne National Laboratory, and others.

Most of the commercially available software is developed as an extended variation from finite element analysis programs, in particular in the United States. On the other hand, structural optimization software is developed differently in Europe. Most European systems are constructed based on sensitivity analysis or optimization algorithms, and finite element analysis is regarded as a black box type of capability in the design system. They simply call an existing finite element program such as SAP IV, ASKA, SAMCEF, MSC/NASTRAN, and others. For example, ELFINI from Dassault, OPTI/SAMCEF from University of Liege,. OPTSYS from SAAB-SCANIA, OASIS from the Royal Institute of Technology, Stockholm, LAGRANGE from MBB, STARS from RAE and others have been developed for structural optimization in Europe (Hörnlein, 1987). Although all of these systems use finite element methods, the essential portion is optimization algorithms as well as sensitivity analysis. This difference between the United States and Europe seems to be created by the nature of the developing process of this software. In Europe, structural optimization systems are generated by researchers in structural optimization, while finite element software houses in the United States added design optimization capability to their existing commercially available finite element programs. Therefore, a large difference can be expected among these structural optimization software packages, in particular, in basic philosophy.

A limitation found both in European and United States' systems is that they are closed. More precisely, the state equations we can deal with in these systems are limited to the ones which are originally designed in the finite element analysis codes. Furthermore, modeling for finite element analysis is not fully integrated in most of design systems. The majority of existing design optimization is for stress analysis, and other analysis areas such as heat conduction, fluid flow, metal forming, and others are at this moment excluded from design optimization. However, as noted earlier, these wide ranging state equations must be solved in mechanical design. Thus it is required in a design system that we can deal with any kind of state equation using different analysis software i.e. a design system must be open. If other discrete methods such as boundary element and finite difference methods are applied to solve state equations, we should still be able to incorporate these different types of analysis capabilities in a design optimization system.

It is noted that structural optimization was studied mainly by structural engineers in the aeronautical and aerospace industry in the 1970s and 1980s. Most European structural optimization software packages are thus developed in the large-scale aerospace industry, and they target primarily application to aircraft or space structures. Design variables are sizes of the cross section of members of a frame structure, thickness of a plate or shell, and the orientation of fibers and the thickness of laminae of a fiber-reinforced laminate in these systems. Stress and thermal analyses are thus within linear range, i.e. they are based on the theory of linear elasticity and linear thermoelasticity. On the other hand, mechanical design requires optimal design of not only the sizes of a structure but also its shape, and it mostly deals with design of components or parts of a structure rather than a large-scale whole body of complex structure such as ships, aircraft, and other large-scale space structures. Thus the number of design variables is much smaller than that of aerospace type structures, but the shape design becomes far more important in mechanical design. Thus, analysis model development involving shape geometry is the most important in mechanical design. No matter how capable the sensitivity analysis and optimization algorithms that are developed in a design system, if an appropriate finite element model cannot be generated automatically in every design step during execution of the system, the optimum cannot be easily obtained. This implies that a design system must involve a flexible modeling capability that can reflect any geometrical design change by an optimization algorithm. In other words, a design system may be constructed based on modeling software rather than finite element or sensitivity–optimization software by adding other necessary modules for design optimization.

Although design optimization capability was added into finite element analysis programs in the United States in 1980s, shape optimization capability is still very limited even in MSC/NASTRAN, ANSYS, MECANICA, and NISA II. It is thus required to study and develop a simple methodology that can deal with both sizing and shape optimization as well as both linear and nonlinear state equations involving possibly large plastic deformation, heat conduction, and fluid flow. Nonlinearity and time dependency yield a significant change in the concept how design sensitivity is computed. Since most design problems in aerospace engineering are limited to elastic design, sensitivity analysis is rather straight-forward even by using the analytical and semianalytical methods.

In this chapter we shall describe a method to construct a very small scale but sufficiently flexible optimization system that is mostly written in UNIX C-shell scripts by combining existing modeling, finite element analysis, and optimization algorithm capability using the concept of open-ended software modules in PATRAN from PDA Engineering. More precisely, we shall develop an optimization system using PATRAN's PCL (Patran Control Language) that allows communication between PATRAN and the UNIX OS so that other flexible operations can be implemented outside of PATRAN without terminating execution of PATRAN itself. It is noteworthy that the concept that we shall introduce to develop an optimization system should be general in the sense that the present approach is applicable to any software that has control flow commands to execute a variety of modules and communication commands to the operating system. These are used as a high-level computer language to write a program of design optimization. Since PATRAN is primarily for developing a finite element model by using automatic mesh generation, the present system allows us to deal with sizing and shape optimization in the same manner, while the nature of static/ dynamic, or linear/nonlinear problems for analysis does not affect the system itself since the analysis is assumed to be independent of the system. In this sense, this can provide a much more flexible and powerful capability than that existing in commercially available codes for design optimization, and yet the system is far smaller than any existing programs.

11.2 STRUCTURAL OPTIMIZATION PROCEDURE

The standard procedure of structural optimization consists of the following four major modules:

1. developing a model for analysis;
2. finite element analysis;
3. design sensitivity analysis;
4. application of optimization algorithms.

Development of a finite element analysis model is the same with the usual preprocessing for finite element analysis in sizing optimization problems, and then it can be completely independent of the design optimization system for such problems. However, it must be a module inside the design system for shape optimization, since geometric modeling itself is subject to design change. In general, finite element methods are applied to solve the state equation, but also other analysis methods such as boundary element and finite difference methods should be applicable. In other words, the module for analysis need not be restricted to finite element methods. It is noteworthy that design sensitivity calculation is closely related to analysis. There are several methods available such as finite difference, semianalytical, and analytical methods together with application of direct and adjoint methods. An appropriate method of sensitivity analysis may be determined by the characteristics of the state equation and a choice of method in the analysis module. The last part of the design optimization is the set of optimization algorithms to solve the optimal design problem that is formulated as a constrained nonlinear programming problem.

If the expected applicability range of a design system is sufficiently small, it is not difficult to develop a FORTRAN or C program that involves all of these four modules. For example, if the cross-sectional size of members of frame structures is designed for both stress and eigenvalue analyses, it is rather easy to develop a complete design optimization FORTRAN program since it is free from the difficulty of mesh generation for a finite element model, and since its state equations are linear and definite so that sensitivity can be easily computed even by an analytical method. Development of such a program can be strightforward for graduate students who have necessary background of finite element methods, sensitivity analysis, and some optimization algorithms. In the 1970s when the availability and capability of commercial or public domain finite element codes and optimization methods were limited, this approach of complete development of a design optimization system was very popular for a restricted class of structural optimization problem. However, there are many well-developed sophisticated software packages for finite element analysis and optimization at present, and they are used daily in design practice. Since many of these software packages are already available in most of industry and academia, it is better to develop a design system that can utilize these as intact existing modules so that it does not require additional effort to be familiar with analysis and optimization modules. Furthermore, since each organization has its own preference and rather extensive experience in use of a particular finite element code, the design optimization system should be able to integrate this specific one as well as other choices for other users. This means that the optimization system should be able to choose any of them according to the nature of the state equation and to the preference of a user. Since most commercially available design optimization codes are developed as an enhancement of their original finite element analysis programs, they do not possess this flexibility and openness. As mentioned earlier, state equations in structural design are not just an equilibrium of linearly elastic structures. They are sometimes heat conduction equations with convection and radiation, motion of non-Newtonian fluids, equilibrium of large deformation elastoplastic bodies, and others. Thus a design system should not be restricted by a particular finite element analysis capability. It should be designed with new concept based on the nature of mechanical design that has much larger scope than analysis of the state equation.

Since there are many available commercial and public domain packages of optimization algorithms based on the theory of mathematical programming methods for both linear and nonlinear programming problems, it is much simpler to adopt an appropriate one from existing software. In structural optimization, ADS based on the sequential linear programming and IDESIGN with sequential quadratic programming are widely used in the United States. ADS is now enhanced as DOT and is marketed by VMA.

11.3 SENSITIVITY ANALYSIS BY THE FINITE DIFFERENCE METHOD

In order to utilize an optimization algorithm, it is at least required to compute the values and the first derivatives of performance functions (that is the objective and constraint functions) g with respect to a design variable d, i.e. $g(u, d)$ and $Dg/Dd(u, d)$, where u is the state variable used to define the state equation (that is the equilibrium equation for stress analysis, the equation of motion is dynamics, the heat conduction equation for thermal analysis, ...) of a physical system to be optimized.

If the state equation is linear and time independent, and if it is expressed by

$$\mathbf{K}\mathbf{u} = \mathbf{f} \tag{11.1}$$

then differentiation of this in a design variable d yields

$$\frac{\partial \mathbf{K}}{\partial d}\mathbf{u} + \mathbf{K}\frac{\partial \mathbf{u}}{\partial d} = \frac{\partial \mathbf{f}}{\partial d} \Rightarrow \frac{\partial \mathbf{u}}{\partial d} = \mathbf{K}^{-1}\left(\frac{\partial \mathbf{f}}{\partial d} - \frac{\partial \mathbf{K}}{\partial d}\mathbf{u}\right) \tag{11.2}$$

where $\mathbf{K}^{-1}$ is the inverse of a linear operator (e.g. the global stiffness matrix in stress analysis by the finite element method). Thus the sensitivity of the performance function g, i.e. the first derivative of g with respect to a design variable d, is given by

$$\frac{Dg}{Dd} = \frac{\partial g}{\partial d} + \frac{\partial g}{\partial \mathbf{u}}\frac{\partial \mathbf{u}}{\partial d} = \frac{\partial g}{\partial d} + \frac{\partial g}{\partial \mathbf{u}}\left[\mathbf{K}^{-1}\left(\frac{\partial \mathbf{f}}{\partial d} - \frac{\partial \mathbf{K}}{\partial d}\mathbf{u}\right)\right] \tag{11.3}$$

This form of sensitivity requires explicit representation of a performance function in terms of the state variable u and a design variable d.

If stress analysis is considered for a three-dimensional solid structure, and if a performance function g is identified with the maximum principal stress, it is certainly a function of the displacement u, but its function form is not described in an explicit form, since the principal stresses are identified with the eigenvalues of the symmetric 3×3 stress tensor. Thus application of the above analytically derived sensitivity might not be useful for performance functions which are implicit functions of the state variable u. Even in the case that performance functions are explicitly expressed in the state and design variables, utilization of the above form of sensitivity might not be practical since we have to compute the first derivative of the linear operator (stiffness matrix in stress analysis) $\mathbf{K}$ of the state equation. For example, in a three-dimensional shell structure, the analytical partial derivative of the stiffness matrix $\mathbf{K}$ with respect to a nodal coordinate of a node may not be obtained in a simple form. For a shell formulation in the finite element method we use a local coordinate system attached to each finite element that is defined by the nodal coordinates of the four corner nodes of the element. The element stiffness matrix $\mathbf{K}^e_{\text{local}}$ is transformed to the one $\mathbf{K}^e_{\text{global}}$ in the global coordinate system and it is assembled to form the stiffness matrix, i.e.

$$\mathbf{K}^e_{\text{global}} = \mathbf{R}^\mathrm{T} \mathbf{K}^e_{\text{local}} \mathbf{R} \tag{11.4}$$

where $\mathbf{R}$ is the transformation matrix from the local to the global coordinate system. Thus the first derivative of the stiffness matrix is computed by

$$\frac{\partial \mathbf{K}^e_{\text{global}}}{\partial d} = \left(\frac{\partial \mathbf{R}}{\partial d}\right)^\mathrm{T} \mathbf{K}^e_{\text{local}} \mathbf{R} + \mathbf{R}^\mathrm{T} \frac{\partial \mathbf{K}^e_{\text{local}}}{\partial d} \mathbf{R} + \mathbf{R}^\mathrm{T} \mathbf{K}^e_{\text{local}} \frac{\partial \mathbf{R}}{\partial d} \tag{11.5}$$

this is, the first derivatives of both coordinate transformation and element stiffness matrices must be analytically obtained. It is thus not practical to compute the analytical first derivative of the stiffness matrix. This leads to a finite difference scheme to compute the first derivative of the operator $\mathbf{K}$ of the state equation:

$$\frac{\partial K}{\partial d} \approx \frac{K|_{d+\Delta d} - K|_{d-\Delta d}}{2\,\Delta d} \tag{11.6}$$

by evaluating $\mathbf{K}$ at the two perturbed designs $d + \Delta d$ and $d - \Delta d$, where Δd is a sufficiently small design change at the current design d. This approximation yields the semianalytical method to compute the design sensitivity

$$\frac{Dg}{Dd} = \frac{\partial g}{\partial d} + \frac{\partial g}{\partial \mathbf{u}}\frac{\partial \mathbf{u}}{\partial d} \approx \frac{\partial g}{\partial d} + \frac{\partial g}{\partial \mathbf{u}}\left[\mathbf{K}^{-1}\left(\frac{\mathbf{f}|_{d+\Delta d} - \mathbf{f}|_{d+\Delta d}}{2\,\Delta d} - \frac{K|_{d+\Delta d} - K|_{d-\Delta d}}{2\,\Delta d}\mathbf{u}\right)\right] \tag{11.7}$$

The semianalytical method does not require us to know the exact explicit form of function in the state variable u. If the operator can be evaluated at two different designs in the analysis procedure (that is, if we can output the stiffness matrix and the load vector at two different design stages), we can compute the design sensitivity by using the semianalytical method. In other words, we need direct access to the source code of analysis to develop an efficient sensitivity analysis program using the semianalytical method.

Therefore, for general problems, it might be much simpler to apply the finite difference approximation to compute the sensitivity of a performance function g:

$$\frac{Dg}{Dd} \approx \frac{g|_{d+\Delta d} - g|_{d-\Delta d}}{2\,\Delta d} \tag{11.8}$$

In this way, the explicit function form of g in d and u is not required, and even the first derivative of the linear operator of the state equation need not be calculated. This means that the finite difference approximation is applicable, both for linear and for nonlinear state equations, while the time dependency of the state equation does not affect at all the computation of the design sensitivity. The disadvantage of this method is the requirement of two analyses per design variable to compute the sensitivity. If the state equation is solved at the current design, three analyses are required. Thus, if m performance functions and n design variables are involved in a design problem, $(2n + 1)m$ analyses are required to compute the sensitivity and the value of performance functions.

If the number of design variables is large as in sizing problems for aerospace structures, the finite difference method is not practical at all. However, if the number of design variables is rather small, it becomes powerful. As mentioned earlier, in most mechanical design problems the number of design variables is very small, because of the requirement from cost effective manufacturing. It is nearly impossible to have varying thickness everywhere in a structure in mechanical design, while several different thicknesses may be introduced in parts. Similarly, the introduction of very arbitrary shapes of holes to reduce the total weight is unrealistic. The usual practice is to make circular holes. Such a design problem is then described to find the best locations and radii of a given number of holes. Furthermore, mechanical design requires consideration of dynamics and impact as well as static strength of a structure. This means that the state equation is far more complicated than that for most of aeronautical–aerospace structures. This nature of the mechanical design yields the consequence that analytical, or even semianalytical, methods are not practical to compute the sensitivity.

It is also noted that no special development is required for sensitivity analysis if the finite difference method is applied, while analytical and semianalytical methods need considerable code-developing effort to be implemented. If shape design of a structure is considered in a multidisciplinary setting, e.g. if the shape of a structure is optimized by minimizing the drag generated by fluid flow outside the structure as well as the maximum stress of the structure, we must calculate sensitivity in two different disciplines, i.e. two different state equations. Since analysis programs for stress and fluids are in general not integrated in a single software system by a single software house, it is not realistic for most analysis programs to request the provision of a sensitivity analysis capability based on the analytical or semianalytical methods, at least at present. The finite difference method is, however, applicable together with already existing analysis programs for different kinds of state equations to compute sensitivity without any modification or enhancement of these at all. The flexibility of the finite difference method for sensitivity is truly enormous.

Furthermore, in a distributed computing environment, it is possible to use several workstations at once to compute design sensitivity. This implies the possibility that two or three workstations are used at the same time to make the analysis in a parallel manner. More precisely, a workstation analyzes the state equation at the design $d + \Delta d$, while another workstation analyzes it at the design $d - \Delta d$. In this way, we can compute sensitivity without waiting for the long time required by sequential calculation using a single workstation. If the analyzer is using an iterative method to solve the state equation, it is also possible to take advantage of this by using the fact that the state variable at the designs $d + \Delta d$ and $d - \Delta d$ should be quite close to the one at the current design d. Then solutions at the designs $d + \Delta d$ and $d - \Delta d$ can be easily computed by any iterative methods by starting from the solution at the current design d. This indicates that research into the reduction of the computing time for the finite difference method may have more merit than developing a general-purpose program of design sensitivity analysis based on the analytical or semianalytical method.

In summary, it is concluded that the finite difference method is the most appropriate method to compute sensitivity for most practical optimization problems in structural design, since

- calculation time may not be excessive because the number of design variables is not so large, say 5–20 variables at most,
- a design optimization system can be very flexible because sensitivity is calculated using any kind of analyzer (not only an FEM analyzer) for any kind of state equation,
- the size of the software to be developed for the design system is very small and easy to maintain, and
- a workstation that forms and modifies the design model need not be the same one used for sensitivity calculation in a distributed computing environment.

Sensitivity can be calculated using the finite difference method using the following procedure:

   LOOP 1: For each design variable
        Analysis of $g|_{d+\Delta d}$
        Analysis of $g|_{d-\Delta d}$
        LOOP 2: For performance function
            Calculation of Sensitivity $\dfrac{g|_{d+\Delta d} - g|_{d-\Delta d}}{2\,\Delta d}$

   NEXT 2
   NEXT 1

Here the central finite difference approximation is applied instead of forward or backward finite difference schemes, although it requires more computation time. It is noted that the central difference method is more accurate than the others. Indeed, if a performance function g is sufficiently smooth, the approximation errors are given as follows: central difference,

$$\frac{g|_{d+\Delta d} - g|_{d-\Delta d}}{2\,\Delta d} - \frac{\mathrm{D}g}{\mathrm{D}d}\bigg|_{d} \approx \frac{1}{3}(\Delta d)^2 \frac{\mathrm{D}^3 g}{\mathrm{D}d^3}\bigg|_{d} \tag{11.9}$$

forward difference,

$$\frac{g|_{d+\Delta d} - g|_{d}}{\Delta d} - \frac{\mathrm{D}g}{\mathrm{D}d}\bigg|_{d} \approx \frac{1}{2}(\Delta d) \frac{\mathrm{D}^2 g}{\mathrm{D}d^2}\bigg|_{d} \tag{11.10}$$

backward difference,

$$\frac{g|_{d} - g|_{d-\Delta d}}{\Delta d} - \frac{\mathrm{D}g}{\mathrm{D}d}\bigg|_{d} \approx -\frac{1}{2}(\Delta d) \frac{\mathrm{D}^2 g}{\mathrm{D}d^2}\bigg|_{d} \tag{11.11}$$

Thus the central difference scheme is a second-order accurate approximation. This means that it is not necessary to assume too small a design perturbation Δd to calculate the sensitivity. If the central difference scheme is applied, errors in the forward and backward difference schemes can be estimated by

$$\frac{1}{2}(\Delta d)\frac{D^2 g}{Dd^2}\bigg|_d \approx \frac{1}{2}\frac{g|_{d+\Delta d} - 2g|_d + g|_{d-\Delta d}}{\Delta d} \tag{11.12}$$

This may be applicable to modify the size of design perturbation Δd by specifying the allowable tolerance ε for the approximation error in the forward or backward difference scheme. Indeed, we first estimate the second derivative of the performance function using a trial design perturbation Δd_{trial}, and then we check whether the estimated error

$$E_{\text{estimated}} = \left|\frac{1}{2}(\Delta d_{\text{trial}})\frac{g|_{d+\Delta d} - 2g|_d + g|_{d-\Delta d}}{\Delta d_{\text{trial}}^2}\right| \tag{11.13}$$

is smaller or larger than a given tolerance ε. If this estimated error is larger than ε, then we define a new perturbation $\Delta d_{\text{desired}}$ by

$$\Delta d_{\text{desired}} \leqslant \frac{2\varepsilon\,\Delta d_{\text{trial}}^2}{|g|_{d+\Delta d} - 2g|_d + g|_{d-\Delta d}|} \tag{11.14}$$

For the sensitivity, however, we shall apply the central difference scheme to ensure at least one order higher accuracy than this estimation based on the forward or backward difference approximation. Since it is possible to estimate the upper bound of the amount of the approximation error, the finite difference method is now applicable with more confidence to compute design sensitivity.

We shall provide an example of error estimation of the finite difference method for sensitivity analysis related to the shape of the body. To do this, the bending

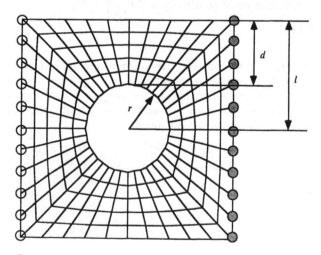

○ Applied Transverse Load
○ Fixed Nodes

Fig. 11.1 Finite element mesh and definition of design variable.

problem of a plate with a circular hole is considered shown in Fig. 11.1. The nodes on the left edge are fixed, and a transverse force is applied uniformly on the right edge. If the performance function is the maximum principal stress $\bar{\sigma}$ of the structure, and the design variable is the radius of the circular hole r, the finite difference approximation error is estimated by

$$E_{\text{estimated}} = \left| \frac{1}{2} \frac{\bar{\sigma}|_{r+\Delta r} - 2\bar{\sigma}|_r + \bar{\sigma}|_{r-\Delta r}}{\Delta r_{\text{trial}}} \right| \tag{11.15}$$

where the size of perturbation Δr is determined by $\Delta r = 10^{\alpha} r (\alpha = -1, -2, -3)$. Figure 11.2 shows the estimated error of the sensitivity of the maximum principal stress with respect to the radius of the circular hole. The estimated error ratio is plotted with the design variable r changed from initial value 2.0 to the upper limit 4.0. It follows from the plotting of the estimated error in Fig. 11.2 that the lowest error is obtained at $\alpha = -2$ while much worse results are obtained for $\alpha = -1$ and -3. In the case that $\alpha = -1$, we may say that too large a Δr is assumed. However, for $\alpha = -3$, the roundoff error becomes much bigger than approximation error. In the finite difference approximation, the roundoff error occurs in the calculation of $g|_{d+\Delta d} - g|_{d-\Delta d}$, and thus the error is quantified by

$$E_{\text{roundoff}} = \frac{g|_{d+\Delta d} - g|_{d-\Delta d}}{g|_d} \tag{11.16}$$

In the present example, the roundoff error is $O(\Delta d) \approx 10^{-3}$ at $\alpha = -3$. This implies we would lose 3 digits in the calculation. This is the reason why the error declines when the value of performance function and the perturbation size Δr become large.

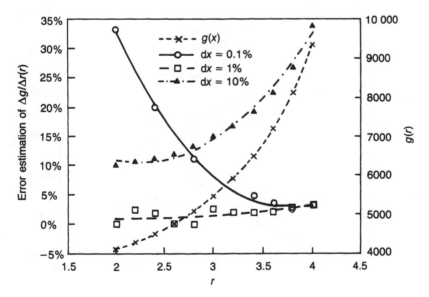

Fig. 11.2 Estimated error in the original definition of design variable ($dx = (\Delta r/r) \times 100\%$).

In this problem, we can improve the error by the finite difference approximation, and also the distribution can be much more constant, by modifying the definition of the design variable by

$$d = l - r$$

$$\Delta d = 10^{-\alpha}(l - r) \tag{11.17}$$

Using this new definition of the design variable, the approximation error is estimated by

$$E_{\text{estimated}} = \left| \frac{1}{2} \frac{\bar{\sigma}|_{d+\Delta d} - 2\bar{\sigma}|_d + \bar{\sigma}|_{d-\Delta d}}{\Delta d_{\text{trial}}} \right| \tag{11.18}$$

and is shown in Fig. 11.3. Since the perturbation size Δd becomes smaller as the value of performance function is larger, both approximation and roundoff error decline.

From this error analysis, the definition of the design variables is rather important to control the amount of approximation error, but the finite difference approximation is practically accurate enough to perform the optimization. In other words, the error is not too large (less than 2% at $\alpha = -2$) even in the forward or backward finite difference approximation that is a one order accurate approximation than the central difference scheme applied in our optimization study. It is also noted that the perturbation size does not need to be too small. The estimated error is less than 15% even for the case that the perturbation size

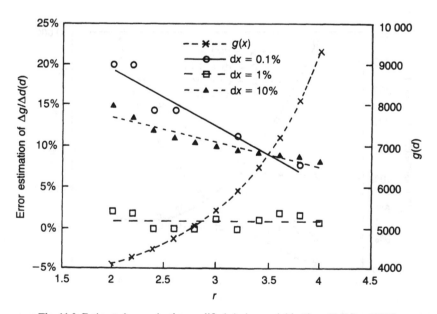

Fig. 11.3 Estimated error in the modified design variable $(\text{d}x = (\Delta d/d) \times 100\%)$.

is 10% of design variable. However, it is important that the perturbation size should not be too small because roundoff error becomes large.

11.4 SHAPE OPTIMIZATION IN STRUCTURAL DESIGN

Since shape optimization problems are more important in mechanical structure design, an optimization system should easily be able to deal with the shape of a structure. The characteristic of shape design is that the geometric finite element model must be modified during the optimization procedure, while sizing problems can be solved with a fixed geometrical finite element model. This nature of the adaptation of the finite element mesh has been the major difficulty in shape optimization. Many researchers have been challenged to solve this problem and it is not completely solved yet, especially for three-dimensional structures. In the early days, the direction of mesh movement was specified according to the shape change of a structure, while the topology of the finite element connectivity was fixed as the initial one. This approach failed to yield the optimum in many problems by crashing elements or destroying convexity of finite elements unless we can specify an appropriate direction of mesh movement by predicting the final shape in advance. Thus, we have reached to the notion that finite element meshes must be regenerated completely only by using boundary information that defines the shape of a structure for fully automated shape optimization methods (Bennett and Botkin, 1985; Botkin and Bennett, 1985; Botkin, Yang and Bennett, 1986; Belegundu and Zhang, 1992; Manicka and Belegundu, 1991). This means that a design optimization system must contain the capability of finite element mesh generation which can regenerate a model completely without terminating an optimization program. This suggests that the model-developing (i.e. pre-processing including automatic mesh generation) capability is the key for success in shape optimization in mechanical design, and it may be the most important module for development of a design system. Design boundary segments–patches may be expressed by appropriate splines, and the coordinate of control points of spline curves and surfaces may be the design variable in shape optimization, while this information should provide necessary information for generating finite element meshes at every design stage. This is a major difference of the mechanical design from structural optimization studied for example in aeronautical–aerospace engineering. The importance of the modeling module in a mechanical design system is thus very high. We shall now describe details of the development of shape optimization of an elastic structure. An extensive literature on this subject can be found, for example, in Haftka and Gandhi (1985, 1986), Haftka, Gurdal and Kamat (1990) and also Bennett and Botkin (1986).

The notion of shape optimization was introduced by Zienkiewicz and Campbell (1973) for design of the outside shape of a dam structure by using sensitivity analysis with respect to the coordinates of the nodes of a finite element model on the design boundary and the sequential linear programming method. Sensitivity is calculated by the semianalytical approach with the direct method. After this work, most shape optimization of a structure follows the notion that nodal coordinates of a finite element model are discrete design variables. Since they consider a rather minor shape modification of the dam profile, adaptation of the finite element model according to a design change is made only in the horizontal node movement without any sophistication. If a more general case of shape optimization is concerned, finite element model adaptation methods should be clearly described.

In the existing literature of shape optimization, we can find two major methods. The first way, shown in Fig. 11.4, is to specify the direction of node (either nodes of a finite element model or control points if the shape is defined by a spline function) movement that yields noncrashing of the mesh as much as possible, e.g. Fleury (1987). Setting of the direction of nodal adaptation requires a fair amount of advance information on the final shape of the structure. Otherwise the specification sometimes yields excessive finite element distortion or crashing of elements. The second widely applied method is the one Schnack (1979) introduced. Nodes on the design boundary are moved toward the normal direction as shown in Fig. 11.5. In this case, finite element distortion by change of boundary shape is not so large, while the possibility of contacting nodes on the design boundary may not be completely avoided. The second approach is thus far more robust than the first one for a large modification of the boundary shape.

After adapting the nodes on the design boundary, the location of the rest of nodes must be appropriately modified to maintain good quality of the finite element model. A method for the adaptation of interior nodes is the so-called mesh smoothing scheme. To relocate a node, all the finite elements connected to it are first searched, and then the nodes connected to it are identified. Using these nodal locations, we shall relocate the node to the center of 'mass' using the equation

$$(\mathbf{x}_n)_{\text{new}} = (1 - \alpha)(\mathbf{x}_n)_{\text{old}} + \alpha \frac{1}{n_{\max}} \sum_{i=1}^{n_{\max}} \mathbf{x}_{ni} \tag{11.19}$$

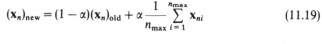

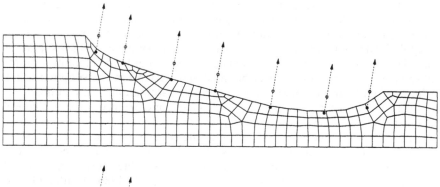

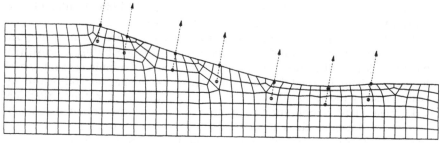

Fig. 11.4 Adaptation to the specified direction.

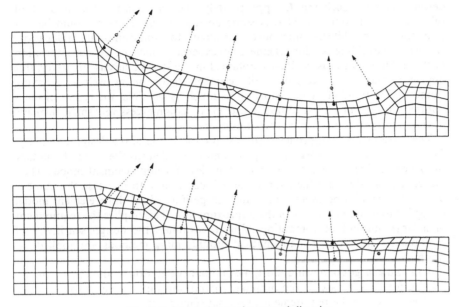

Fig. 11.5 Adaptation to the normal direction.

where node n is relocated with a specified parameter α such that $0 < \alpha \leqslant 1, n_{max}$ is the number of nodes connected to node n, and x_{ni} are the coordinates of node ni (Fig. 11.6). Another relocation method, introduced by Choi (1986), may be obtained by solving

$$\mathbf{K} \, \Delta \mathbf{x} = \mathbf{f}_x \qquad (11.20)$$

where $\mathbf{K}$ is the global stiffness matrix for stress analysis, $\mathbf{f}_x$ is a specified generalized load that controls mesh density, and $\Delta \mathbf{x}$ is the amount of relocation of the nodal coordinates $\mathbf{x}$. Here $\Delta \mathbf{x}$ is specified for the nodes on the design boundary at each

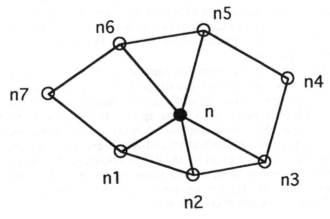

Fig. 11.6 Node relocation scheme.

design change. Specifying $\mathbf{f_x}$ appropriately, we can control the amount of relocation in the domain. That is some portion of the domain should have a large relocation, while the other portion is subject to a small amount of relocation. It is noted that this relocation scheme is not expensive, since the LU decomposition of the stiffness matrix $\mathbf{K}$ is already obtained in the stress analysis. However, this approach may yield excessive finite element distortion as well as crashing elements. To avoid element distortion as much as possible, the center of mass approach is recommended, although it tends to assign excessively small elements in the vicinity of the concave outer boundary.

There are also two approaches to describe the shape of the design boundary. The majority of early work on shape optimization defined the design boundary by piecewise linear segments connecting nodes of a finite element model. Then the design variables are the coordinates of such nodes. In this case the number of design variables becomes significantly large if a refined finite element model is applied to ensure accuracy of the approximation. This implies the requirement of a large amount of computing effort to evaluate the design sensitivity of performance functions if mathematical programming methods are applied for optimization.

To avoid a large computing effort, the optimality criterion method is, in general, applied for such a case. If the optimality condition for the shape optimization problem involves the requirement of the satisfaction of

$$g(\mathbf{x}) = 1 \tag{11.21}$$

on the design boundary, the new boundary location (defined by the nodes) is then resized by

$$\mathbf{x}_{\text{new}} = \mathbf{x}_{\text{old}} [g(\mathbf{x}_{\text{old}})]^{\alpha} \tag{11.22}$$

for an appropriately given iteration factor α close to 1. If the optimality condition is described by the Mises equivalent stress, the function g may be defined by the ratio of the Mises stress at an arbitrary point and its average value (or the specified maximum allowable Mises stress) on the design variable. If the optimality condition is given by the mutual strain energy density, g is accordingly defined by using the mutual energy. The optimality criterion method does not require formal calculation of the design sensitivity of a set of performance functions. Thus if the optimization problem is written with several constraints, construction of an effective resizing algorithm is not obvious, and furthermore it may not yield the true optimum.

It is also noted that, if the optimality condition involves stress and strain, these must be accurately evaluated on the design boundary. In most finite element methods, very accurate stress and strain can be obtained at the one-order-lower Gaussian quadrature points than those applied to compute element stiffness matrices. However, these points are located inside finite elements. Thus we must introduce an extrapolation method to evaluate stress and strain on the nodes or along the element boundaries on the design boundary. A commonly applied extrapolation method is based on the least-squares method (Zienkiewicz and Taylor, 1989). Interpolating a component q of stress–strain by the same shape functions $\mathbf{N}$ for displacement components

$$q = \mathbf{N}^{\mathsf{T}} \mathbf{Q} \tag{11.23}$$

where $\mathbf{Q}$ is the vector of the nodal values of q, we find $\mathbf{Q}$ as the solution of the least-squares problem

$$\min_{\mathbf{Q}} \frac{1}{2} \sum_{e=1}^{N_e} \int_{\Omega_e} (q - q_h)^2 \, \mathrm{d}\Omega_e \qquad (11.24)$$

where q_h is the finite element approximation of a component of stress–strain defined by the finite element approximation $\mathbf{u}_h$ of the displacement in equilibrium. Here N_e represents the total number of finite elements, and Ω_e is the domain of an arbitrary finite element. If the size and orientation of finite elements moderately vary in a finite element model, this least-squares extrapolation provides quite reliable stress–strain values on the boundary. However, when a finite element model contains large and small finite elements alternately, this simple least-squares method may yield very inaccurate stress–strain values. Accurate finite element approximations of stress and strain on the boundary can also be obtained by applying the so-called mixed finite element approximation whose potential is examined by Haber (1987) in shape optimization of an elastic structure. In this case, stress and strain are independently approximated from displacement. Despite a considerable increase in the total number of degrees of freedom in a finite element model, an effective method to solve a mixed problem is studied in Zienkiewicz and Taylor (1989) using an iteration method.

Another disadvantage of the approach by defining the design boundary by a set of nodal points is that mere application of a resizing algorithm of the nodal location may yield a physically unrealistic oscillatory optimum shape, because of 'instability' of stress and strain on the boundary, as shown in Braibant and Fleury (1984). The total number of degrees of freedom of displacement should be larger than those for stress and strain in the displacement finite element approximation. However, if the least-squares method is applied to compute stress and strain on the boundary, those degrees of freedom exceed those for the displacement. This leads to instability in the approximation. Stresses and strains obtained by least squares in general possess a slightly oscillatory distribution, and then this slight oscillation is magnified during the iterations for shape optimization. Thus some smoothing algorithms for resizing or adaptive schemes to have an accurate finite element approximation of stress and strain must be introduced to provide a smooth optimum shape, (Kikuchi *et al.*, 1986).

To overcome the defect due to the definition of the boundary by nodal points, Braibant and Fleury (1984) introduced the method of *B* splines to describe the shape of design boundaries. After decomposing the design boundary into a set of design segments consisting of lines, arcs, and curves, each boundary segment is represented by an appropriate spline function. Then a finite element model is constructed independently of the number of control points of the splines. In most shape optimization problems, design boundary segments are defined by a few control points. If a cubic curve is expected, four control points are required. If an arc is desired, the location of the center and the radius are regarded as the control points. No matter how refined the finite element model introduced is the number of control points stays the same, that is, the total number of the discrete design variables can be fixed and few. Furthermore, as Braibant and Fleury (1984) showed, a smooth optimum shape can be obtained without introducing special techniques of smoothing and adaptive finite element methods. Therefore, the majority of researchers and structural engineers are now using this second approach to define the design boundary and the discrete design variables in shape optimization.

Sensitivity analysis for shape optimization has been studied by Choi and Haug (1983), Haug *et al.* (1983), and others. Since extensive treatment of sensitivity analysis can be found in Haug, Choi and Komkov (1985), we shall only describe a brief explanation of shape sensitivity. If the design boundary segment is described by the spline expression

$$\mathbf{x} = \sum_{i=1}^{N_c} \mathbf{x}_{ci}\phi_i(s) \quad 0 \leqslant s \leqslant 1 \tag{11.25}$$

where N_c is the number of control points, $x_{ci}, i = 1,\ldots,N_c$ are the coordinates of the control points, s is the parametric coordinate, ϕ_i are the basis functions for spline expression. If either the analytical or the semianalytical method is applied to compute the design sensitivity, we must calculate the sensitivity

$$\frac{\mathbf{D}g}{\mathbf{D}\mathbf{x}_{cj}}(\mathbf{u}) = \frac{\partial g}{\partial \mathbf{u}}\frac{\partial \mathbf{u}}{\partial \mathbf{x}_{cj}} = \frac{\partial g}{\partial \mathbf{u}}\frac{\partial \mathbf{u}}{\partial \mathbf{x}(s)}\frac{\partial \mathbf{x}(s)}{\partial \mathbf{x}_{cj}} = \frac{\partial g}{\partial \mathbf{u}}\frac{\partial \mathbf{u}}{\partial \mathbf{x}(s)}\frac{\partial}{\partial \mathbf{x}_{cj}}\sum_{i=1}^{N_c}\mathbf{x}_{ci}\phi_i(s) = \frac{\partial g}{\partial \mathbf{u}}\frac{\partial \mathbf{u}}{\partial \mathbf{x}(s)}\phi_i(s) \tag{11.26}$$

for a performance function g of only the displacement $\mathbf{u}$. If N_{smax} nodes are placed on this spline, sensitivity of g is thus computed by

$$\frac{\mathbf{D}g}{\mathbf{D}\mathbf{x}_{cj}}(\mathbf{u}) = \frac{\partial g}{\partial \mathbf{u}}\frac{\partial \mathbf{u}}{\partial \mathbf{x}_{cj}} = \frac{\partial g}{\partial \mathbf{u}}\sum_{i=1}^{N_{smax}}\frac{\partial \mathbf{u}}{\partial \mathbf{x}_i}\phi_j(s_i) \tag{11.27}$$

where $\mathbf{x}_i, i = 1,\ldots,N_{smax}$ are the nodal coordinates of the finite element model on the design boundary segment, and s_i are the parametric coordinates corresponding to $\mathbf{x}_i$. Therefore, we must compute the sensitivity of the displacement $\mathbf{u}$ with respect to all of the nodal coordinates on the design boundary segment of a finite element model. If a refined finite element model is used for stress analysis, computing sensitivity might become fairly large despite the small number of control points for the spline expression in the case that the analytical or semianalytical method is applied. It thus follows from this fact that design sensitivity for shape optimization should be effectively computed by finite difference methods by taking advantage of a few control points of the spline expression. Since a fairly small design perturbation is taken in the finite difference approximation for the sensitivity calculation, we need not regenerate the whole finite element model. We simply relocate the nodes on the design boundary segment to the curve perturbed, while the rest of the nodes in the model are fixed. Since the element connectivity and the total number of degrees of freedom of a finite element model are also fixed, sensitivity calculation by the finite difference method can be well justified.

Another issue we must discuss in shape optimization is the possible need for remeshing (or rezoning) of a finite element model, especially for the case of a large difference in the initial and optimum shape configurations. If the design change from the initial to the final is moderate, it is sufficient to adapt the finite element mesh by moving the nodes on the design boundary in the normal or specified direction, as desired above. However, if the design change becomes large, element distortion might not be reduced sufficiently by application of mesh smoothing schemes only. Excessively refined or coarse finite elements are generated in certain domains of the model, and then the quality of the analysis may deteriorate significantly. To avoid this, complete remeshing may be required; that is the whole finite element model must be completely regenerated by the current shape design. This remeshing concept is introduced by Bennett and Botkin (1985) as mentioned earlier.

11.5 IMPLEMENTATION OF STRUCTURAL OPTIMIZATION

On the basis of the concept described above, a prototype system has been implemented to solve sizing and shape optimization problems by integrating a generalized geometric modeler, an FEM analysis code, and an optimizer in the form small-size software driver. This system is flexible, easy to maintain, and can distribute the calculation to different types of computers, one to a graphic workstation with which the users are accustomed, and the other to faster computers with possibly a vector/parallel processor to carry out the analysis, as mentioned previously.

The system consists of (1) **PATRAN** as a geometric modeler and an automatic mesh generator, (2) the driver and file format converter written by PCL (Patran Command Language), (3) a general-purpose FEM analysis code, (4) a set of UNIX C-shell scripts to calculate sensitivity, (5) ADS for an optimizer, and (6) Mathematica to calculate the objective function value and derivatives. It is implemented on an APOLLO DN4000 (for man–machine interface and for system flow control) and a SONY NWS3710 (for finite element analysis) in the present study. The flowchart of this system is given in Fig. 11.7.

This prototype system can optimize (i.e. minimize) the volume or weight of a three-dimensional structure under constraints on the maximum principal stress, maximum strain energy density, maximum Mises equivalent stress, or maximum deformation, while physical dimensions of shells as well as their shape are the set of design variables. For example, an optimal design problem is formulated by

$$\text{minimize} \quad \text{weight}(\mathbf{d})$$

$$\text{subject to} \quad \frac{\text{stress}_{\text{maximum}}(\mathbf{d})}{\text{stress}_{\text{allowable}}} - 1 \leqslant 0$$

$$\underline{d}_i \leqslant d_i \leqslant \bar{d}_i, i = 1, 2, \ldots, n \qquad (11.28)$$

where $\underline{d}_i$ and $\bar{d}_i$ are the given lower and upper limit of each design variable. In this example, the weight is minimized in dimensions (sizes) of a structure, while the maximum deformation is restricted in the allowable range. In general, because the objective function and constraints are nonlinear functions of the design variable, the optimization problem may be solved by the sequential linear programming method. In this case, the derivatives of the constraints with respect to the design variables must be given to the system in the linearizing process to solve the problem.

Since the major portion of the optimization system consists of already existing (commercially available or public) codes, the only substantial portion of the present system is the main control flow routine written in UNIX C-shell scripts for sensitivity analysis and the optimization algorithm, while it calls the geometric

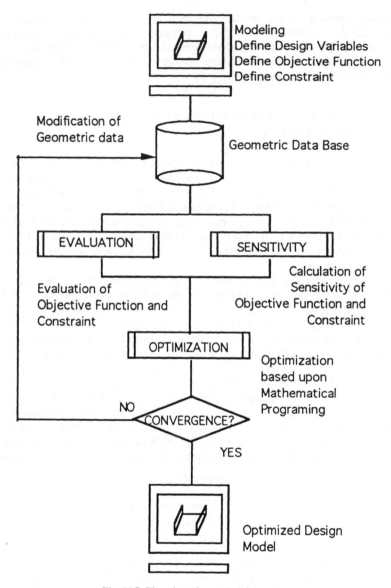

Fig. 11.7 Flowchart for optimizing system.

modeler and finite element analysis code. Because of the choice of sensitivity calculation using the central finite difference scheme, whose size of perturbation can be controlled by the error analysis described in the previous section, the main routine becomes very simple. A sample program is given in Appendix 11A.

We simply call the geometric modeling module with automatic finite element mesh generation module, a finite element analysis code, and a program for the numerical optimization algorithm. If it is intended to use different geometric modules, analysis codes, and optimizers are in the system of optimization, we simply call the desired modules. Most of the rest of the program of the optimization is unchanged.

It is clear that if the total number of design variables is reasonably small, and if computing speed is not so critical, this simple and small design optimization system written in UNIX C-shell scripts can solve quite a large number of design optimization problems in both size and shape design.

11.6 EXAMPLES OF OPTIMIZATION BY THE PRESENT DESIGN SYSTEM

We shall give several examples of optimum structures using the above design system together with PATRAN's PCL and PATRAN to control geometric quantities for size and shape optimization and to develop finite element models automatically in the optimization system. As mentioned in the previous section, ADS, a public domain software package, is applied for optimization. More precisely, the method of feasible direction in ADS for constrained minimization is used as the optimizer, in which the golden method is used as the one-dimensional search.

11.6.1 Case of a folded plate

The geometry of a folded plate is given in Fig. 11.8. The objective of this example is to reduce the volume of the structure, while the maximum principal stress should not exceed the allowable value of the maximum principal stress by

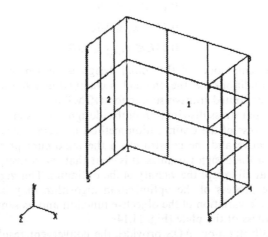

INTERFACE? 1. NEUTRAL 2. NASTRAN 3. ANSYS 4. ABAQUS 5. IGES 6. END
PLOT
INTERFACE? 1. NEUTRAL 2. NASTRAN 3. ANSYS 4. ABAQUS 5. IGES 6. END
RUN, DECK

Fig. 11.8 Original model (patches in PATRAN for mesh generation).

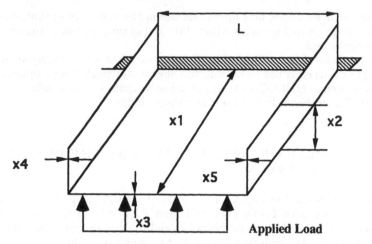

Fig. 11.9 Definition of design variables.

reducing the weight. Design variables are shown in Fig. 11.9. While the width of the plate $L(1.0\,\text{m})$ is fixed, design variables are the height of two folded portions, the length of the plate, and the thickness of the folded plate, and they are expressed by x_i, $i = 1, 2, \ldots, 5$. The upper and lower bounds of the design variables are also specified. This problem is mathematically formulated by

$$\text{minimize} \quad \text{volume}(\mathbf{x}) = Lx_1x_3 + 2x_1x_2(x_4 + x_5)$$

$$\text{subject to:} \quad \bar{\sigma}(r, x) \equiv \text{maximum principal stress}$$

$$\bar{\sigma}(r, x) \leqslant \bar{\sigma}_{\text{Allowable}}$$

$$0.1\,L \leqslant x_1 \leqslant 0.9\,L$$

$$0.4\,L \leqslant x_2 \leqslant 0.6L$$

$$0.03\,L \leqslant x_{3,4,5} \leqslant 0.07L \tag{11.29}$$

A finite element model of the initial design is shown in Fig. 11.10 with the boundary conditions, i.e. the one end is clamped and the other is subject to a uniformly distributed transverse traction (10 MPa).

The principal stress distribution of this design state is given in Fig. 11.11. The iteration process of the optimization method in ADS is described in Fig. 11.12 for the objective and the constraint on the maximum principal stress and in Fig. 11.13 for the design variables. It is clear that the convergence process is not monotonic at all except the vicinity of the optimum. This zigzag behavior of the convergence process of the optimization algorithm may be explained by the nonmonotonic variation of the objective function and its sensitivity with respect to the thickness of the plate (Fig. 11.14).

At the 13th iteration, ADS provides the convergent result with the optimal values of the design variables as shown in Table 11.1. The initial volume is reduced (by 20%) by changing the length of the folded plate to the lower limit while the height of the folded portions is also reduced to 0.48 m from 0.50 m without exceeding of the limit value of the maximum principal stress of the

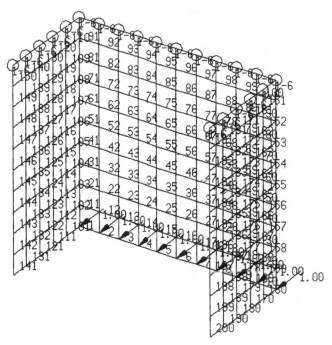

TYPE, DATA, SID, INCLUDE-LIST, CID-LIST

TYPE, DATA, SID, INCLUDE-LIST, CID-LIST

Fig. 11.10 Mesh and boundary condition.

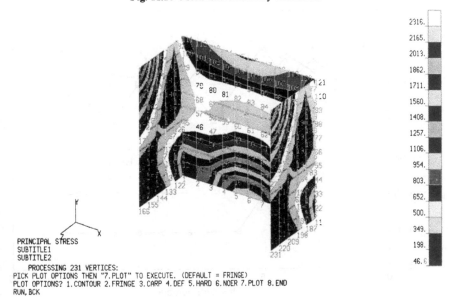

PRINCIPAL STRESS
SUBTITLE1
SUBTITLE2
  PROCESSING 231 VERTICES:
PICK PLOT OPTIONS THEN "7.PLOT" TO EXECUTE. (DEFAULT = FRINGE)
PLOT OPTIONS? 1.CONTOUR 2.FRINGE 3.CARP 4.DEF 5.HARD 6.NOER 7.PLOT 8.END
RUN,BCK

Fig. 11.11 Initial distribution of principal stress.

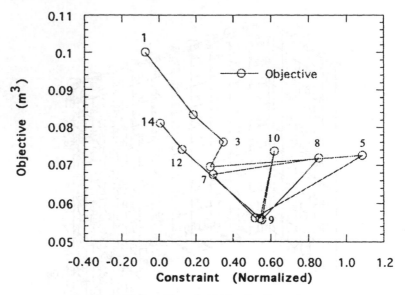

Fig. 11.12 Transition of objective function value.

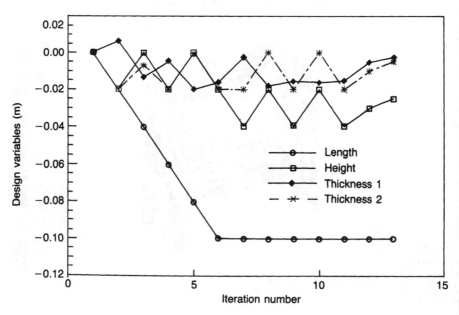

Fig. 11.13 Transition of design variables.

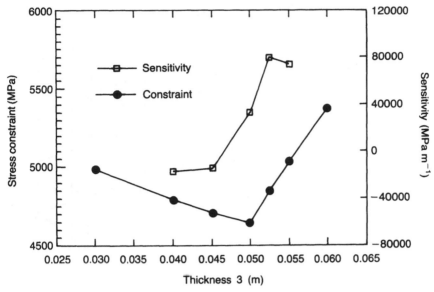

Fig. 11.14 Transition and sensitivity of stress constraint.

Table 11.1 Result of optimization example 1

		Initial	Lower	Upper	Optimum
Volume (m³)		0.100			0.080
Constraint		−0.070			0.010
Design Var(m)	1	1.000	0.900	1.100	0.900
	2	0.500	0.400	0.600	0.480
	3	0.050	0.030	0.070	0.050
	4	0.050	0.030	0.070	0.040
	5	0.050	0.030	0.070	0.040

structure. The thickness of the folded portions is reduced to 0.04 m from 0.05 m, and the thickness of the base plate is unchanged although it changes in a zigzag manner in the optimization process. Figures 11.15 and 11.16 show the optimal model. Figure 11.17 shows the distribution of the maximum principal stress at the optimum. It is clear that the pattern of the distribution and the maximum value are unchanged from those of the initial design.

Since the height of the folded portions is a design variable, this problem possesses the nature of shape optimization as well as that of sizing optimization by modifying the thickness and the length of a plate. While the height of the folded portion is modified, the finite element must be regenerated. As shown in Figs. 11.10 and 11.16, the finite element model of the initial design is different from that of the optimum design; the number of elements has been changed from 200 and 180. In this case, PATRAN regenerates the finite element model with a fixed size of elements which we specified.

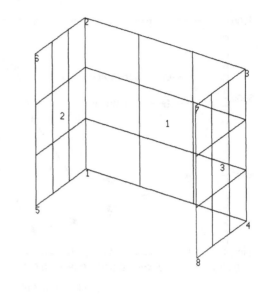

INPUT DIRECTIVE OR "END"
PLOT
INPUT DIRECTIVE OR "END"
RUN, BCK

Fig. 11.15 Optimal model (patches in PATRAN for.mesh generation).

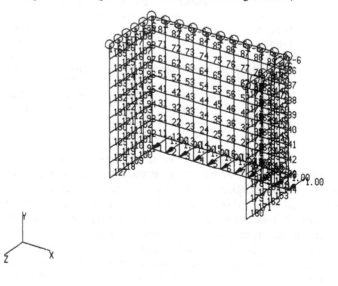

INPUT DIRECTIVE OR "END"
DISP, 1, PLOT
INPUT DIRECTIVE OR "END"
RUN, BCK

Fig. 11.16 Optimal finite element model.

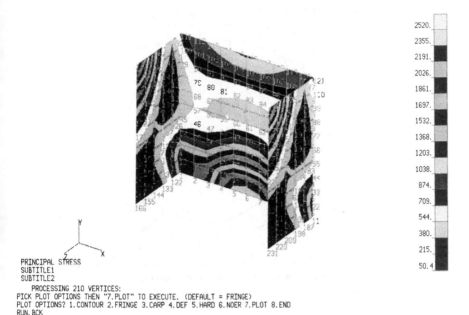

PRINCIPAL STRESS
SUBTITLE1
SUBTITLE2
    PROCESSING 210 VERTICES:
PICK PLOT OPTIONS THEN "7.PLOT" TO EXECUTE. (DEFAULT = FRINGE)
PLOT OPTIONS? 1.CONTOUR 2.FRINGE 3.CARP 4.DEF 5.HARD 6.NOER 7.PLOT 8.END
RUN, BCK

Fig. 11.17 Stress distribution of optimal model.

11.6.2 Plate with a circular hole

We shall consider a shape design problem of a plate that is fixed along one edge and is subject to a uniform transverse traction along the opposite edge. The original geometry of the plate model is shown in Fig. 11.18. A circular hole is generated. In this example, the weight of the plate is minimized without increasing the maximum principal stress by changing the location of the center and the radius of the circular hole. Although the original shape of the circular hole is not modified in the optimization, the finite element model geometry must be changed because of modification to the location and radius of the hole. Thus, this must be classified as a shape design problem despite the definition of the design variables, the location and the size of a hole (Fig. 11.19). In order to maintain the symmetry of the structure, we shall place the hole along the center (horizontal) line. This problem is mathematically formulated as follows:

$$\text{minimize} \quad (L^2 - \pi r^2)t$$

$$\text{subject to} \quad \bar{\sigma}(r, x) \equiv \text{maximum principal stress}$$

$$\bar{\sigma}(r, x) \leqslant \bar{\sigma}_{\text{Initial}}$$

$$x + r \leqslant 0.9L$$

$$x - r \leqslant 0.1L \qquad\qquad (11.30)$$

$$0 \leqslant x \leqslant L$$

$$0 \leqslant r \leqslant 0.5L$$

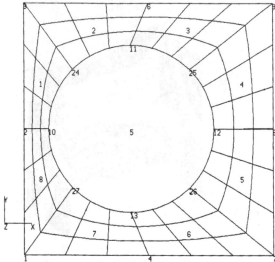

MODE? 1.GEOMETRY 2.ANALYSIS MODEL 3.ANALYZE 4.RESULTS 5.INTERFACE 6.STOP
PLOT
MODE? 1.GEOMETRY 2.ANALYSIS MODEL 3.ANALYZE 4.RESULTS 5.INTERFACE 6.STOP
RUN, BCK

Fig. 11.18 Original model (patches in PATRAN for mesh generation).

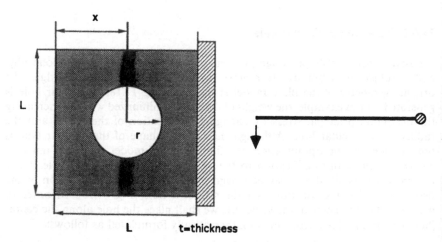

Fig. 11.19 Definition of design variables.

where L is the size of a square plate, t is the plate thickness, r is the radius of the hole, and x is the (horizontal) location of the center of the hole. Here L and t are assumed to be fixed ($L = 1.0$ m, $t = 0.01$ m), i.e. only two design variables are defined in this shape design problem. The finite element model of the initial design state is shown in Fig. 11.20 with the boundary conditions. A small circle along the right edge indicates the fixed condition of a node, while the value '1.00'

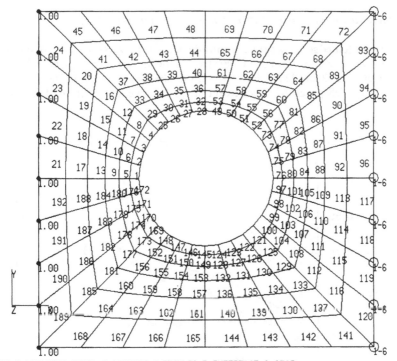

:TRY 2.ANALYSIS MODEL 3.ANALYZE 4.RESULTS 5.INTERFACE 6.STOP

:TRY 2.ANALYSIS MODEL 3.ANALYZE 4.RESULTS 5.INTERFACE 6.STOP

Fig. 11.20 Mesh and boundary conditions.

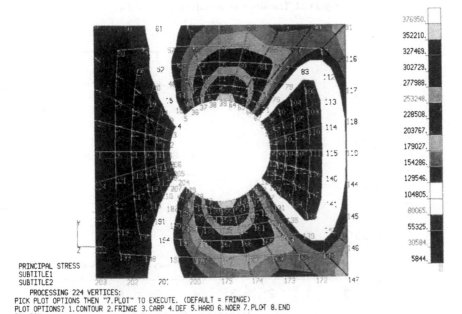

PRINCIPAL STRESS
SUBTITLE1
SUBTITLE2
PROCESSING 224 VERTICES:
PICK PLOT OPTIONS THEN "7.PLOT" TO EXECUTE. (DEFAULT = FRINGE)
PLOT OPTIONS? 1.CONTOUR 2.FRINGE 3.CARP 4.DEF 5.HARD 6.NOER 7.PLOT 8.END
RUN,BCK

Fig. 11.21 Initial distribution of principal stress.

along the left edge shows the magnitude of the transverse force (1.00 Ma) applied on a node. The maximum principal stress of the initial design is given in Fig. 11.21.

Figures 11.22 and 11.23 show the iteration history of the optimization algorithm form the ADS program. Since lower stresses are generated in the left side of the

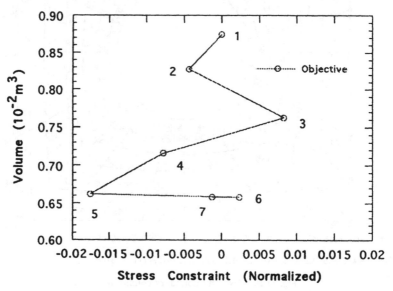

Fig. 11.22 Transition of objective function value.

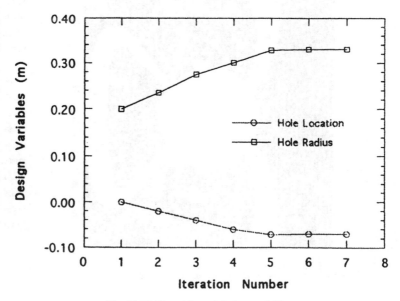

Fig. 11.23 Transition of design variables.

Table 11.2 Result of optimization example 2

	Initial	Lower	Upper	Optimum
Volume (10^{-2} m^3)	0.8743			0.6582
Stress constraint	0.0000			-0.0013
Edge constraint	-2.0000			-0.0008
Hole location (m)	0.5000	0.1000	0.9000	0.4299
Hole radius (m)	0.2000	0.1000	0.5000	0.3298

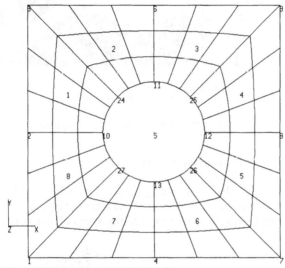

```
INTERFACE? 1.NEUTRAL 2.NASTRAN 3.ANSYS 4.ABAQUS 5.IGES 6.END
6
MODE? 1.GEOMETRY 2.ANALYSIS MODEL 3.ANALYZE 4.RESULTS 5.INTERFACE 6.STOP
RUN, BCK
```

Fig. 11.24 Optimal model (patches in PATRAN for mesh generation).

plate, the hole location should be shifted to the left until it hits the limit $x - r \leqslant 0.1\,L$ (Fig. 11.23). As shown in Fig. 11.22, an unfeasible design is generated at the third and sixth iterations, but most of designs in the optimization process are feasible, i.e. do not violate the constraints we have set. Convergence is very monotonic. The volume is reduced to 0.6582 from 0.8743 by the optimum radius $r = 0.3298$ and the optimum location $x = 0.4299$ (Table 11.2). The geometric model of the optimum design is given in Fig. 11.24, and the distribution of the maximum principal stress is shown in Fig. 11.25. The overall pattern of the stress distribution is unchanged, although the stress in the left side of the hole becomes much larger than the original one.

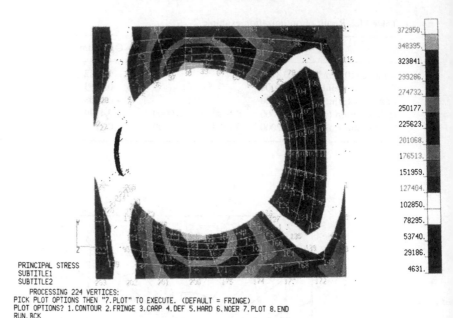

PRINCIPAL STRESS
SUBTITLE1
SUBTITLE2
PROCESSING 224 VERTICES:
PICK PLOT OPTIONS THEN "7.PLOT" TO EXECUTE. (DEFAULT = FRINGE)
PLOT OPTIONS? 1.CONTOUR 2.FRINGE 3.CARP 4.DEF 5.HARD 6.NOER 7.PLOT 8.END
RUN, BCK

Fig. 11.25 Stress distribution of optimal model.

11.6.3 Plate with rib reinforcement

Let a square plate be reinforced by ribs as shown in Fig. 11.26. We shall consider a sizing optimization problem that minimizes the maximum deflection of the rib-reinforced plate by modifying the thickness of ribs without changing the rib

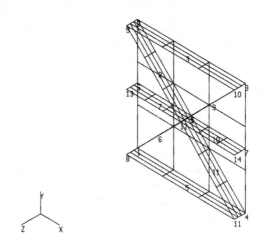

5.INTERFACE 6.STOP 7.(RESERVED) 8.(USER MENU)
RUN, BCK
CREATING IMAGE BACKUP FILE:
"patran.rst.3"

Fig. 11.26 Original model (patches in PATRAN for mesh generation).

height. The total volume of the structure is also bounded to the one at the initial design. This problem is then mathematically defined by

minimize $\bar{u}(\mathbf{x}) \equiv$ maximum deformation

subject to $V = L^2 t + hL\left[x_1 + x_2 + \dfrac{x_4 + x_7 + \sqrt{2}(x_3 + x_5 + x_6 + x_8)}{2} \right]$

$V \leqslant V_{\text{Initial}}$

$0.001\,L \leqslant x \leqslant 0.1\,L$ (11.31)

where L is the size of a square plate, t is the thickness of the plate, h is the rib

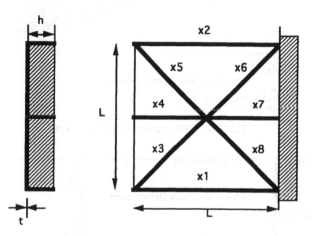

Fig. 11.27 Definition of design variables.

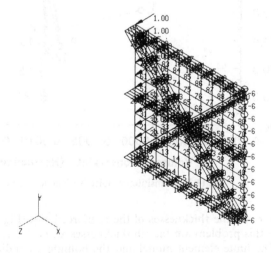

```
     5. INTERFACE 6. STOP 7. (RESERVED) 8. (USER MENU)
RUN, BCK
CREATING IMAGE BACKUP FILE:
"patran.rst.4"
```

Fig. 11.28 Mesh and boundary condition.

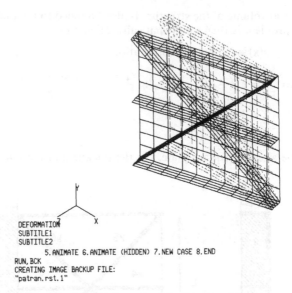

DEFORMATION
SUBTITLE1
SUBTITLE2
    5.ANIMATE 6.ANIMATE (HIDDEN) 7.NEW CASE 8.END
RUN,BCK
CREATING IMAGE BACKUP FILE:
"patran.rst.1"

Fig. 11.29 Deformation of original model.

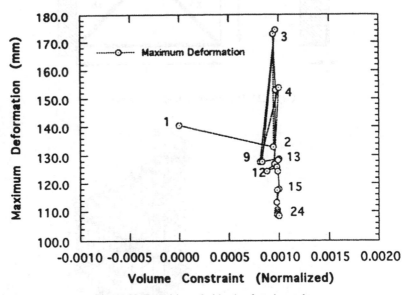

Fig. 11.30 Transition of objective function value.

height, and x_i are the thicknesses of the reinforced ribs (Fig. 11.27). The design variables in this problem are the rib thicknesses x_i, $i = 1, \ldots, 8$, while L, h, and t are fixed. The finite element model and the boundary conditions are shown in Fig. 11.28. Reinforced ribs are modeled as plates, and are discretized by 4-node quadrilateral flat shell elements which are similar to QUAD4 elements in MSC/NASTRAN. The right edge of the plate is fixed, while uniform transverse loads

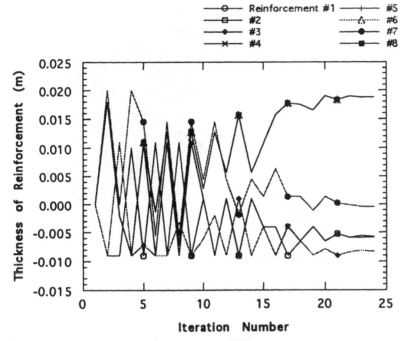

Fig. 11.31 Transition of design variables.

Table 11.3 Result of optimization example 3

		Initial	Lower	Upper	Optimum
Deformation (cm)		1.404			1.085
Volume		0.000			0.001
Thickness (m)	1	0.010	0.001	0.100	0.004
	2	0.010	0.001	0.100	0.004
	3	0.010	0.001	0.100	0.002
	4	0.010	0.001	0.100	0.004
	5	0.010	0.001	0.100	0.002
	6	0.010	0.001	0.100	0.029
	7	0.010	0.001	0.100	0.010
	8	0.010	0.001	0.100	0.029

are applied at every node along the left edge of the square plate. It is noted that the ribs are not subject to boundary conditions, i.e. the fixed condition is not imposed on the nodes of ribs. The deformed shape of the plate at the initial design is given in Fig. 11.29.

The iteration process of the optimization algorithm is shown in Figs. 11.30 and 11.31. The volume constraint is slightly violated, say about 1%, which is specified as allowable tolerance in ADS optimization code. As shown in Fig. 11.30, the volume constraint is always within the feasible range with a given allowable

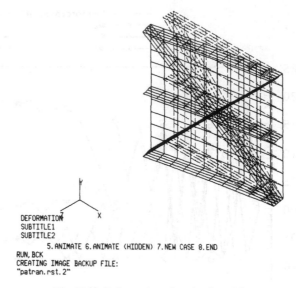

DEFORMATION
SUBTITLE1
SUBTITLE2
    5.ANIMATE 6.ANIMATE (HIDDEN) 7.NEW CASE 8.END
RUN, BCK
CREATING IMAGE BACKUP FILE:
"patran.rst.2"

Fig. 11.32 Deformation of optimal model.

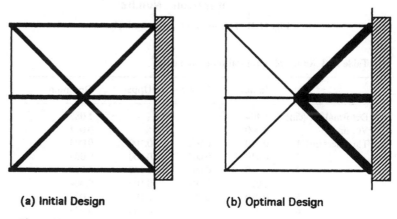

(a) Initial Design   **(b) Optimal Design**

Fig. 11.33 Optimization of reinforcement: (a) initial design; (b) optimal design.

tolerance, while the objective function, the maximum deflection, is converging to the optimum in a zigzag manner. The behavior of the design variables is also very wild in most of the iteration history except in the vicinity of the optimum as shown in Fig. 11.31. The values of the maximum deflection and the design variables as given in Table 11.3. The deformed configuration of the rib-reinforced plate at the optimum is shown in Fig. 11.32. It is clear that both deformation and design maintain symmetry as shown in Fig. 11.33, which schematically represents the thickness of the ribs at both the initial and the optimum design. Thick ribs are formed in the vicinity of the fixed edge of the plate.

11.6.4 Optimum shell configuration

We shall present a capability of the design optimization system that can manage the shape change of a structure defined by a free-form Bezier surface in three-dimensional space. In this case we modify the shape of the Bezier surface by moving the control points. To do this we shall form the best shell configuration that minimizes the maximum principal stress under the volume constraint of the shell structure.

The original shape of a plate is defined by 16 control points and its patches are given in Fig. 11.34. As shown in Fig. 11.35, the size of the projected surface of a plate-shell structure onto the $x-y$ plane is fixed to be L, the thickness of the thin structure is also fixed to be t, while the z coordinates of the four interior control points are varying in design. The objective is to reduce the maximum principal stress of the structure by increasing by 10% the original volume of the flat shell (i.e. plate) structure. This problem is mathematically represented by

$$\text{minimize} \quad \bar{\sigma}(\mathbf{x}) \equiv \text{maximum principal stress}$$

$$\text{subject to} \quad V = t \iint_{\Omega} \left| \frac{\partial B(\mathbf{x}, u, v)}{\partial u} \times \frac{\partial B(\mathbf{x}, u, v)}{\partial v} \right| du \, dv$$

$$V \leqslant 1.1 \, V_{\text{Initial}}$$

$$-\frac{L}{3} \leqslant x \leqslant \frac{L}{3} \tag{11.32}$$

Here the lower and upper bound of the z coordinates of the four control points of the Bezier surface are also specified to be within 1/3 of the size of the projected plane. The finite element model generated by the patch model in Fig. 11.34 in

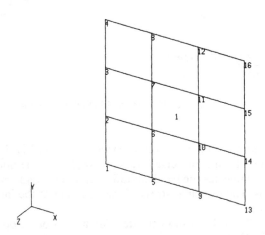

5. INTERFACE 6. STOP 7. (RESERVED) 8. (USER MENU)
RUN, BCK
CREATING IMAGE BACKUP FILE:
"patran.rst.4"

Fig. 11.34 Original model (patches in PATRAN for mesh generation).

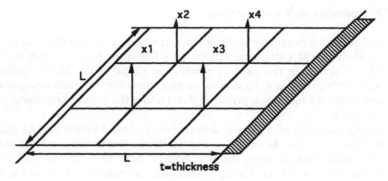

Fig. 11.35 Definition of design variables.

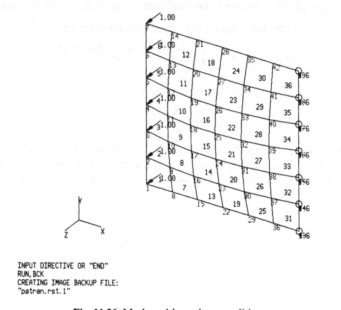

INPUT DIRECTIVE OR "END"
RUN, BCK
CREATING IMAGE BACKUP FILE:
"patran.rst.1"

Fig. 11.36 Mesh and boundary condition.

PATRAN and the boundary conditions are given in Fig. 11.36. The right edge is fixed, and uniform transverse forces are applied at the nodes on the left edge. A small circle indicates the fixed condition at a node, and small arrows represent applied loads. The principal stress of the structure at the initial design is shown in Fig. 11.37.

Figures 11.38 and 11.39 show the iteration history of the optimization algorithm in ADS code. Convergence is very smooth and monotonic in this problem. The maximum principal stress is monotonically reduced by increasing the z coordinates of the control points. Because of the symmetry of this problem, four design variables have the same value in each design process (Table 11.4). Figure 11.40 shows the PATRAN patch model at the optimum. The four interior control

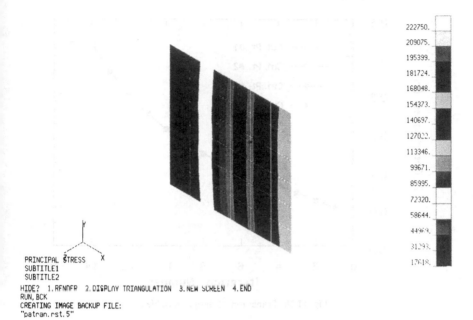

222750.
209075.
195399.
181724.
168048.
154373.
140697.
127022.
113346.
99671.
85995.
72320.
58644.
44969.
31293.
17618.

PRINCIPAL STRESS   X
  SUBTITLE1
  SUBTITLE2
HIDE? 1.RENDER 2.DISPLAY TRIANGULATION 3.NEW SCREEN 4.END
RUN, BCK
CREATING IMAGE BACKUP FILE:
"patran.rst.5"

Fig. 11.37 Initial distribution of principal stress.

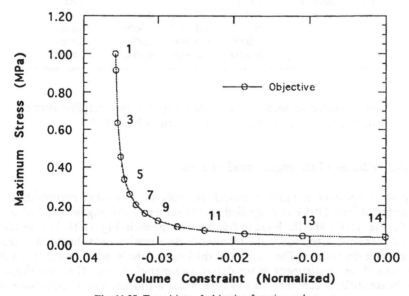

Fig. 11.38 Transition of objective function value.

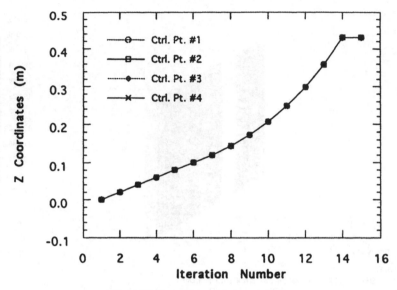

Fig. 11.39 Transition of design variables.

Table 11.4 Result of optimization example 4

	Initial	Lower	Upper	Optimum
Maxmum stress (MPa)	1.0000			0.0340
Volume constraint	−0.0357			−0.0001
Control point (m) 1	0.0000	−0.5000	0.5000	0.4300
2	0.0000	−0.5000	0.5000	0.4300
3	0.0000	−0.5000	0.5000	0.4300
4	0.0000	−0.5000	0.5000	0.4300

points are displaced to form a shell defined by the Bezier spline surface. The stress distribution at the optimum shell is given in Fig. 11.41.

11.6.5 Stiffener of the engine hood of a car

We shall consider a more practical example of structural optimization. A distributed load (10 kPa) is applied to the center of the engine hood of a car body (Fig. 11.42), and the hood is deformed as shown in Fig. 11.43. The objective of this example is to reduce the amount of deformation over the hood without increasing the volume. The shape and thickness of the hood cannot be modified because of the requirement of the design and manufacturing. Only the shape of the beam stiffener can vary. In this case, a beam stiffener with hat-shaped cross section is used (Fig. 11.44). Although any dimensions in the cross section can be

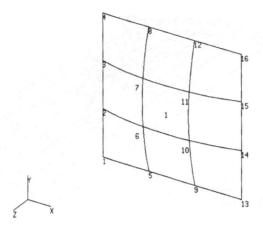

5. INTERFACE 6. STOP 7. (RESERVED) 8. (USER MENU)
RUN, BCK
CREATING IMAGE BACKUP FILE:
"patran.rst.3"

Fig. 11.40 Optimal model.

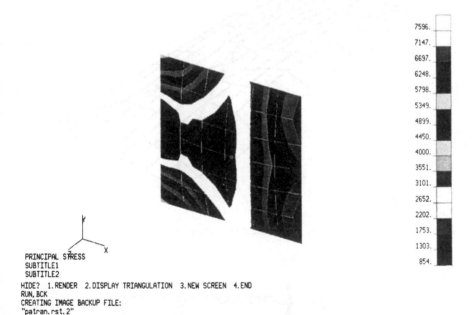

PRINCIPAL STRESS
SUBTITLE1
SUBTITLE2
HIDE? 1.RENDER 2.DISPLAY TRIANGULATION 3.NEW SCREEN 4.END
RUN, BCK
CREATING IMAGE BACKUP FILE:
"patran.rst.2"

Fig. 11.41 Stress distribution of optimal model.

design variables in optimization, we shall fix the sizes of f and h for simplicity. From the symmetry of the structure, the beams on the left-hand side and the right-hand side should have the same cross section, and then w and t of the front and rear beam are considered to be design variables. In most finite element

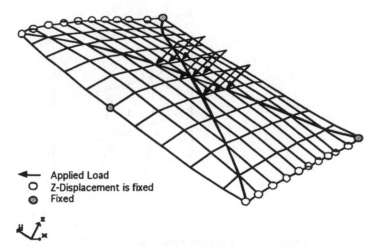

Applied Load
○ Z-Displacement is fixed
◉ Fixed

Fig. 11.42 Discrete model and boundary conditions.

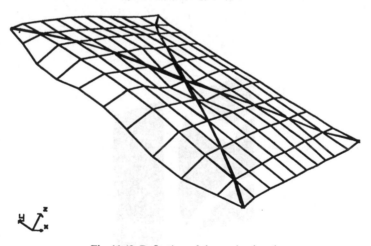

Fig. 11.43 Deflection of the engine hood.

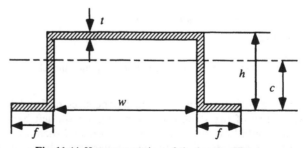

Fig. 11.44 Hat cross section of the beam stiffener.

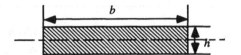

Fig. 11.45 A rectangular cross section.

analysis codes, the moment of inertia of the cross section of the beam must be given as the element property. The neutral line is calculated from

$$\int_0^c dA = \int_c^h dA \Leftrightarrow c = \frac{w + 2h + 2t - 2f}{4} \qquad (11.33)$$

The moment of inertia is calculated using c from

$$I_{\text{hat}} = \int_{-c}^{h-c} y^2 \, dA = \int_{-c}^{-c+t} 2fy^2 \, dy + \int_{-c+t}^{h-c} 2ty^2 \, dy + \int_{h-c-t}^{h-c} wy^2 \, dy \quad (11.34)$$

In this example, we use NIKE3D code from Lawrence Livermore National Laboratory to analyze the deformation of such a three-dimensional shell structure. NIKE3D does not have an input card to specify the moment of inertia but has two kinds of beam elements: one is for rectangular cross sections while the other is for pipe cross sections. If the effect due to the shear strain can be ignored, and if only pure bending of the beam is considered, the hat-type cross section beam can be described as a rectanglar beam (Fig. 11.45). The moment of inertia of the rectangular beam is given by

$$I_{\text{rect}} = \frac{bh^3}{12} \qquad (11.35)$$

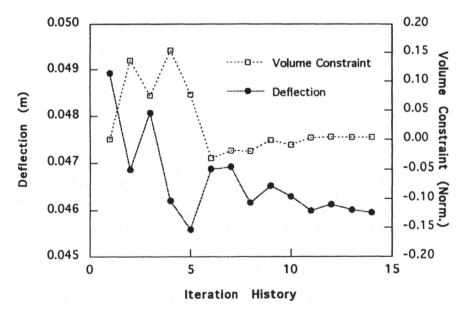

Fig. 11.46 Iteration history of performance functions.

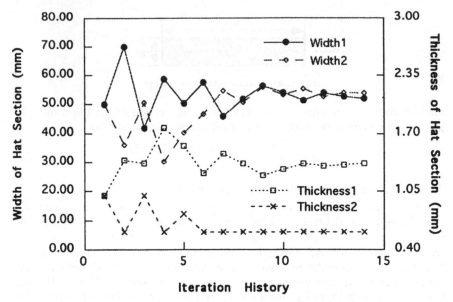

Fig. 11.47 Iteration history of the dimensions of hat cross section.

Table 11.5 Result of example of the engine hood

	Initial	Lower	Upper	Optimal
w (front (mm)	50.00	10.00	100.00	52.05
t	1.00	0.50	10.00	1.37
w (rear)	50.00	10.00	100.00	53.94
t	1.00	0.50	10.00	0.60
Deflection (m)	4.89	–	–	4.60
Volume constraint [1]	0.00	–	–	0.00

If the value h for the rectangular beam is fixed, and if the shear strain effect is ignored, then b may be a function of w and t:

$$b \approx \frac{12}{h^3} I_{hat}(w, t) \tag{11.36}$$

Using this, the shell structure is analyzed and the cross section of beam stiffener for reinforcement of the shell is optimized. The iteration history of the performance function is given in Fig. 11.46 and dimensions of the cross section are given in Fig. 11.47. As the result of this optimization, the deflection over the structure is reduced to 4.60 from 4.89 without increasing the volume (Table 11.5).

11.7 REMARKS AND CONCLUSION

As shown above a very small system mostly written in UNIX C-shell scripts with PATRAN's PCL can solve design optimization problems which involve both size and shape of a structure by calling a finite element analysis program, optimization code, and symbolic manipulation software from the system. If the speed of the process is not too important, the present system can handle most of structural optimization problems. Furthermore, we have great flexibility in the choice of analysis and optimization codes, while commercially available design optimization systems are very sophisticated, but possess substantial restriction on these modules. In other words, our system can deal with a much wider class of design optimization problems than any other existing ones.

The key of the system is the capability of automatic mesh generation of the analysis model, since both sizing and shape of a structure are involved in most mechanical design problems. Thus the optimization system should be constructed by emphasizing the importance of the model development capability. Otherwise it will not be practical at all, no matter how sophisticated a sensitivity analysis is implemented. In spite of the restriction of PATRAN's automatic mesh generation capability, it can handle a fairly large class of problems. If a more capable full-scale automatic mesh generation capability that can control mesh density with simple commands is introduced, we can enhance the capability of the present design optimization system developed in this study. The disadvantage of the present method is the necessity of a large amount of processing time, because of the central difference method used to compute sensitivity of performance functions. It is certain that we have used other workstations to carry out the finite element analysis to reduce the overall processing time. However, still this may be insufficient if the number of design variables becomes large. At this moment, we must restrict the number of design variables to at most 20–30.

ACKNOWLEDGEMENTS

During the present work, the first author was partially supported by Suzuki Corporation and RTB Corporation, while the second author was supported by Sony Corporation (through Dr K. Arita, Mr H. Harimaya and Mr K.C. Imaeda). They express sincere appreciation for this support.

Development of the optimization system was performed by using a computer system of Sony Corporation of America and the facilities of the Computer Aided Engineering Network, the College of Engineering, the University of Michigan, and using software supported by Mr J. Andusiak. Suggestions from Mr S. Ida, Mr M. Okamoto, Mr Fukushima and Mr Y. Park were very useful for implementing the prototype system.

APPENDIX 11A SAMPLE PROGRAM

```
# !/bin/csh
#
# Optimization Procedure using UNIX C Shell Script
#
# set default file name and command aliases

set   PRM_FILE = parameters.dat
```

```
set   INFO_FILE = info.dat
set   BC = "bc -1"
set   ANALYZE = analysis

# Initialization of optimization parameters
# (Optimizer is called) and status is saved
set INFO = `optinit`

# set number of design variables/constraints
set NDV = `head -2 $PRM_FILE | tail -1 | awk '{print $1}'`
set NCON = `head -2 $PRM_FILE | tail -1 | awk '{print $2}'`
# Start iteration
set ino = 1
while ($INFO > 0)
  # Save current design variables into a file
  optdv > current.dv

  switch ($INFO)
    # When the values of performance functions are required
    case 1:
      # Generate geometric model
      model.sh `cat current.dv`

      Generate FE model with automatic mesh generator
      omesh > /dev/null

      # Convert FEM Model to the INPUT format of analyzer
      convert.sh meshgen.fem2

      # Analyze current design
      ${ANALYZE}

      # Calculate performance functions at the initial design
      #     Volume (Objective)
      volume.sh < current.dv > current.val

      #     Max Principal Stress (Constraint #1)
      #         (Calculate the value from OUTPUT file of analyzer)
      if  ($ino = = 1) then
          awk -f max_ps.awk OUTFEM > normalizer
          set normal = `cat normalizer`
          echo "0.0" >> current.val
      else
          echo `awk -f max_ps.awk OUTFEM` / $norm1 – 1.0| \
              $BC >> current.val
      endif

      #     Edge Constraint (Constraint #2)
      edge.sh < current.dv >> current.val

      # Reconstructing Parameter file
      optval current.val
      breaksw
```

```
# When the gradients of performance functions are required
case 2:
    # Calculate Perturbation Size
    set pertub = 0.001
    set ptmp0 = `perturb.sh current.dv`
    echo > temp.dat
    foreach p ($ptmp0)
        echo $p \* $pertub | $BC >> temp.dat
    end
    set psize = `cat temp.dat`
    echo > temp.dat
    foreach p ($psize)
        echo $p \* 2.0 | $BC >> temp.dat
    end
    set psize2 = `cat temp.dat`

    # Calculate Gradient for each design variables
    set i = 1
    while ($i < = $NDV)

        # Create perturbation model for design var #i
        sed $(i), \$d current.dv > temp.dat
        sed -n ${i}p current.dv |\
        awk '{print $1 + '$psize[$i]'}' >> temp.dat
        sed 1, ${i}d current.dv >> temp.dat

        # Calculate Performances at the Plus Perturbed Design
        model.sh `cat temp.dat`
        omesh > /dev/null
        convert.sh meshgen.fem2

        # Analyze perturbed model
        ${ANALYZE}

        # Calculate performances at the perturbed design
        volume.sh < temp.dat > obj + ${i}.val
        echo `awk -f max_ps.awk OUTFEM`/$norm1 |\
        $BC >> obj + ${i}.val
        edge.sh < temp.dat >> obj + ${i}.val

        # Create perturbation model for design var #i
        sed ${i}, \$d current.dv > temp.dat
        sed -n ${i}p current.dv |\
        awk '{print $1-'$psize[1]'}' >> temp.dat
        sed 1, ${i}d current.dv >> temp.dat
        # Calculate Performances at the Minus Perturbed Design
        model.sh `cat temp.dat`
        omesh > /dev/null
        convert.sh meshgen.fem2

        # Analyze perturbed model
        $ (ANALYZE}
```

```
            # Calculate Objectives at the perturbed design
            volume.sh < temp.dat > obj-${i}.val
            echo `awk -f max_ps.awk OUTFEM`/$norm1|\
               $BC ≫ obj-${i}.val
            edge.sh ⟨ temp.dat ≫ obj-${i}.val

            # Calculate Sensitivity
            set PERF_P = `cat obj + ${i}.val`
            set PERF_M = `cat obj − ${i}.val`
            set j = 1
            @ ONCON = $NCON + 1
            while ($j < = $ONCON)
               echo "(${PERF_P[$j]} − ${PERF_M[$j]}]/$psize2[$i]"|\
               $BC ≫ current.grad
               @ j+ +
            end
            @ i+ +
      end

      # Reconstruct parameter file
      optgrad < current.grad

      breaksw
   endsw

   # Calculate the next trial design variables
   #    (Optimizer is called) and save status
   optimize
   set INFO = `cat $INFO_-FILE`
   @ ino+ +
end

# Output the result of optimization

optout
```

REFERENCES

Belegundu, A.D. and Zhang, S. (1992) Mesh distortion control in shape optimization. AIAA Proceedings of 33rd Structures, Structural Dynamics and Material Conference, Dallas, TX, April 13–15, 1992.

Bennet, J.A. and Botkin, M.E. (1985) Structural shape optimization with geometric problem description and adaptive mesh refinement. *AIIA Journal*, **23**(3), 458–64.

Bennet, J.A. and Botkin, M.E. (eds) (1986) *The Optimum Shape: Automated Structural Design*, GM Symposia Series, Plenum, New York and London.

Botkin, M.E. and Bennet, J.A. (1985) Shape optimization of three-dimensional folded plate structures. *AIAA Journal*, **23**(11), 1804–10.

Botkin, M.E., Yang, R.J. and Bennet, J.A. (1986) Shape optimization of three-dimensional stamped and solid automotive components, in *The Optimum Shape: Atuomated Structural Deisgn* (eds J.A. Bennet and M.E. Botkin), GM Symposia Series, Plenum, New York and London, pp. 235–61.

Braibant, V. and Fleury, C. (1984) Shape optimal design using B-splines. *Computer Methods in Applied Mechanical Engineering*, **44**, 247–67.

Choi, K.K. (1986) Shape design sensitivity analysis and optimal design of structural systems, in *Computer Aided Optimal Design: Structural and Mechanical Systems* (ed. C.A. Mota Soares), NATO ASI Series, Springer, pp. 439–92.

Choi, K.K. and Haug, E.J. (1983) Shape design sensitivity analysis of elastic structures. *Journal of Structural Mechanics*, **11**(2), 231–69.

Fleury, C. (1987) Computer aided optimal design of elsatic structures, in *Computer Aided Optimal Design: Structural and Mechanical Systems* (ed. C.A. Mota Soares), NATO ASI Series, Springer, pp. 831–900.

Haber, R.B. (1987) A new variational approach to structural shape design sensitivity analysis, in *Computer Aided Optimal Design: Structural and Mechanical Systems* (ed. C.A. Mota Soares), NATO ASI Series, Springer, pp. 573–88.

Haftka, R.T. and Grandhi, R.V. (1985) Structural shape optimization – a survey. The 26th AIAA SDM Conference, CP No. 85-0772, pp. 617–28.

Haftka, R.T. and Grandhi, R.V. (1986) Structural shape optimization – survey. *Computer Methods in Applied Mechanics and Engineering*, **57**(1), 91–106.

Haftka, R.T., Gurdal, Z. and Kamat, M.P. (1990) *Elements of Structural Optimization*, 2nd edition, Kluwer.

Haug, E.J., Choi, K.K., Hou, K.W. and Yoo, Y.M. (1983) A variational method for shape optimal design of elastic structures, in *New Directions in Optimum Structural Design* (eds E. Atrek, R.H. Gallagher, K.M. Ragsdell and O.C. Zienkiewicz), Wiley, New York, pp. 105–37.

Haug, E.J., Choi, K.K. and Komkov, V. (1985) *Design Sensitivity Analysis of Structural Systems*, Academic Press, New York.

Hörnlein, H.R.E.M. (1987) Take-off in optimum structural design, in *Computer Aided Optimal Design: Structural and Mechanical Systems* (ed C.A. Mota Soares), NATO ASI Series, Springer, 1987, pp. 901–19.

Kikuchi, N., Chung, K.Y., Torigaki, T. and Taylor, J.E. (1986) Adaptive finite element methods for shape optimization of linearly elastic structures, in *The Optimum Shape: Automated Structural Design* (eds. J.A. Bennet and M.E. Botkin), GM Symposia Series, Plenum, pp. 139–69.

Manicka, Y. and Belegundu, A.D. (1991) Analytical shape sensitivity by implicit differentiation for general velocity fields. *The First US National Congress on Computational Mechanics*, Chicago, IL, July 21–24.

Schnack, E. (1979) An optimization procedure for stress concentration by finite element technique. *International Journal of Numerical Methods in Engineering*, **14**, 115–24.

Zienkiewicz, O.C. and Campbell, J.S. (1973) Shape optimization and sequential linear programming, in *Optimal Structural Design* (eds R.H. Gallagher and O.C. Zienkiewicz), Wiley, London, Chapter 7.

Zienkiewicz, O.C. and Taylor, R.L. (1989) Mixed formulation and constraints – complete field methods, in *The Finite Element Method*, 4th edition, Volume 1, McGraw-Hill, London, Chapter 12.

SOFTWARE BIBLIOGRAPHY

ADS

Vanderplaats, G.N. (1985) *ADS: A FORTRAN Program for Automated Design Synthesis*, Engineering Design Optimization, Inc. Santa Bardara, CA.

Vanderplaats, G.N. (1987) *ADS Version 2.01 Manual.*

ANSYS

ANSYS User's Manual, Swanson Analysis Systems Inc., Houston, PA.

DOT

DOT User's Manual Version 2.04, VMA Engineering, Goleta, CA.

ELFINI

Lecina, G. and Petiau, C. (1987) Advances in optimal design with composite materials, in *Computer Aided Optimal Design: Structural and Mechanical Systems* (ed. C.A. Mota Soares), NATO ASI Series, Springer, pp. 943–54.

I-DEAS

L-DEAS User's Guide Vol. I/II, SDRC.
L-DEAS Model Solution and Optimization User's Guide, SDRC.
Ward, P. and Cobb, W.G.C. (1989) Application of I-DEAS optimization for the static and dynamic optimization of engineering structures, in *Computer Aided Optimum Design of Structures: Applications* (eds C.A. Brebbia and S. Hernandez), Computational Mechanics Publications, Springer, pp. 33–50.

IDESIGN

Arora, J.S. and Haug, E.J. (1977) Methods of design sensitivity analysis in structural optimization. *AIAA Journal*, **17**(9), 970–4.
Arora, J.S. and Tseng, C.H. (1987) *User Manual for IDESIGN: Version 3.5*, Optimal Design Laboratory, College of Engineering, The University of Iowa, Iowa City, IA.
Haug, E.J. and Arora, J.S. (1979) *Applied Optimal Design*, Wiley, New York.
Lim, O.K. and Arora, J.S. (1986) An active set RQP algorithm for optimal design. *Computer Methods in Applied Mechanics and Engineering*.
Wu, C.C. and Arora, J.S. (1987) Design sensitivity analysis and of nonlinear structures, in *Computer Aided Optimal Design: Structural and Mechanical Systems* (ed. C.A. Mota Soares), NATO ASI Series, Springer, pp. 589–603.

MATHEMATICA

Wolfram, S. (1991) *Mathematica*, 2nd edition, Addison-Wesley, Reading, MA.

MECHANICA

MECHANICA User's Manual, Rasna Corp., San Jose, CA.

NASTRAN

MSC/NASTRAN User's and Application Manuals, The MacNeal-Schwendler Corporation, 1984.
Fleury, C. and Liefooghe, D. (1986) Shape optimal design on an engineering workstation. *MSC/NASTRAN User's Conference*, University City, CA, March 20–21, 1986.
Nagendra, G.K. and Fleury, C. (1986) Sensitivity and optimization of composite structures using

MSC/NASTRAN. *NASA/VPI Symposium on Sensitivity Analysis in Engineering*, NASA Langley Research Center, Hampton, VA.

Vanderplaats, G.N., Miura, H., Nagendra, G. and Wallerstein, D. (1989) Optimization of large scale structures using MSC/NASTRAN, in *Computer Aided Optimum Design of Structures: Applications* (eds C.A. Brebbia and S. Hernandez), Computational Mechanics Publications, Springer, 1989, pp. 51–68.

NISA II

NISA II User's Manual, EMRC, Troy, MI, 1988.

NISAOPT

NISAOPT User's Manual, EMRC, Troy, MI, 1988.

PATRAN

PATRAN Plus User Manual, PATRAN Division/PDA Engineering, July 1988.

PCL

PATRAN Command Language Guide, PATRAN Division/PDA Engineering, September 1989.

OASIS

Esping, B.J.D. (1985) A CAD approach to the minimum weight design problem. *International Journal of Numerical Methods in Engineering*, **21**, 1049–66.

Esping, B.J.P. (1986) The OASIS structural optimization system. *Computers and Structures*, **23**(3), 365–77.

Esping, B.J.D. and Holm, D. (1987) A CAD approach to structural optimization, in *Computer Aided Optimal Design: Structural and Mechanical Systems* (ed. C.A. Mota Soares), NATO ASI Series, Springer, pp. 987–1001.ʼ

OPTSYS

Bråmå, T. (1989) Applications of structural optimization software in the design process, in *Computer Aided Optimum Design of Structures: Applications* (eds C.A. Brebbia and S. Hernandez), Computational Mechanics Publications, Springer, pp. 13–21.

SAMCEF

SAMCEF. Systeme d'Analyse des Milieux Continus per Elements Finis, University of Liege, Belgium.

See also Fleury (1987) in the References.

SHAPE

Atrek, E. (1989) SHAPE: a program for shape optimization of continuum structures, in *Computer Aided Optimum Design of Structures: Applications* (eds C.A. Brebbia and S. Hernandez), Computational Mechanics Publications, Springer, pp. 135–44.

STROPT

Hariran, M., Paeng, J.K. and Belsare, S. (1988) STROPT – the structural optimization system. Proceedings of the 7th International Conference on Vehicle Structural Mechanics, Detroit, MI, April 11–13, 1988, SAE, pp. 37–47.

STARS

Bartholomew, P. and Morris, A.J. (1984) STARS: a software package for structural optimization, in *New Directions in Optimum Structural Design* (eds E. Atrek *et al.*), Wiley, New York, 1984.

VMCON

Crane, R.L., Hillstrom, K.E. and Minkoff, M. (1980) Solution of the general nonlinear programming problem with subroutine VMCON. *Report ANL-80-64*, Argonne National Laboratory, Argonne, II.

12

Practical optimization of structural steel frameworks

DONALD E. GRIERSON

12.1 INTRODUCTION

The means by which a structural design evolves from some initial form to a final form can be described as an optimization process whereby attempts are made to minimize features or to maximize benefits while satisfying design criteria reflecting performance and fabrication requirements. Many researchers over many years have pursued the quest to produce optimal designs. As early as the 1600s, Leonardo da Vinci and Galileo conducted design optimization through planned trials of models or actual full-scale structures. In the 1700s and 1800s, researchers such as Newton, Lagrange and Euler produced optimal designs using numeric calculations based on an optimality criterion that specified the strength of the structure to be uniform throughout all its parts. Maxwell in the late 1900s and Michell in the early part of the twentieth century altered structural form to achieve optimum designs having minimum structure weight. Since the advent of the computer in the last half-century there has been a veritable explosion of studies concerned with the optimal design of structural systems. One recent survey estimates that the modern literature has some 150 books and monographs and over 2500 research papers that deal with the topic of structure design optimization (Cohn, 1991). While some may argue that this long and extensive history of research has resulted in disappointingly few applications of optimization theory in professional practise, one notable exception in this regard concerns the design of structural steel building frameworks. In fact, today, there are a number of commercial software packages based on member-by-member (Computers and Structures, 1989; Research Engineers, 1987) and/or whole-structure (Grierson and Cameron, 1991) optimization that the professional designer may use to produce economical designs of such structures.

This chapter presents the details of a design synthesis strategy that employs member-by-member and whole-structure optimization in tandem to conduct computer-automated design of least-weight structural steel frameworks, where members are automatically sized using commercial standard steel sections in full conformance with governing steel design code provisions for strength–stability and stiffness. A number of design examples from professional practice are presented to illustrate the scope and effectiveness of the optimization technique.

The underlying features and functions of the corresponding computer software essential for its use in professional practice are not dealt with herein, but may be briefly summarized as follows: capability to conduct first-order and second-order analysis of planar and space frameworks; account for the strength–stability of individual members and the stiffness of the structure as a whole; clause-by-clause verification of the design in conformance with all of the relevant provisions of the governing steel design standard (e.g. American Institute of Steel Construction, 1989; Canadian Standards Association, 1989); automatic sizing of members using commercially available standard steel sections; account for a broad range of commercially available section shapes (e.g. W, T, hollow-box, single and double angle, etc.); determination of local-buckling classifications of member sections; calculation of effective length factors of members; calculation of the unbraced compression flange length of flexible members; account for fixed, pinned, spring and roller supports, as well as for pinned releases at connections and bolted connections; account for various gusset-plate thicknesses for back-to-back double-angle sections; account for strong-axis and weak-axis bending of member sections; capability to specify common section properties for groups of members, to impose limitations on section depths, and to fix the section properties of selected members; account for a variety of nodal and member load types, as well as for external effects due to temperature change and support settlement.

12.2 DESIGN FORMULATION

Design is herein defined as the detailed proportioning of the members of a structural steel framework to satisfy strength, stiffness and fabrication criteria. This definition assumes that the structural layout and loadings have been finalized and it only remains to size the structural components. Within this context the design process may be reduced to the following sequence of tasks: assume an initial trial design; analyze the trial structure to determine behavioral responses; compare these responses with design code specifications; select a new trial design either to remedy any code violations or to improve on the economy of the previous design (herein, structure weight is taken as the measure of design economy); reanalyze and continue the process until the design is deemed satisfactory. This iterative procedure has been traditionally carried out using trial and error, which often leads to a less-than-optimal final design after a tedious analysis–design process that owes its termination only to the skilled intuition and patience of the designer. Conducted in this way, the process becomes quite cumbersome because of the many facets of the design that must be coordinated. The initial design must be selected carefully because its quality has significant bearing on how quickly the design evolves to a satisfactory state. Design code verification is an exhaustive and time-consuming activity that requires selective application of critical code clauses at various stages of the synthesis process. The larger and more complicated the structure, the more difficult it becomes to have simultaneous concern for both design economy and design performance. On the other hand, contrary to the trial-and-error approach, computer-based optimization techniques can be effectively applied to overcome most if not all of these difficulties.

The design optimization seeks a least-weight structure while satisfying all strength and stiffness performance requirements simultaneously under static loading. For a structure having $i = 1, 2, ..., n$ members, and taking the cross-sectional area a_i as the sizing variable for each member i, the general form of the design problem may be stated as

$$\text{minimize} \quad \sum_{i=1}^{n} w_i a_i \tag{12.1}$$

$$\text{subject to} \quad \underline{\delta}_j \leqslant \delta_j \leqslant \bar{\delta}_j \quad (j = 1, 2, ..., d) \tag{12.2}$$

$$\underline{\sigma}_k \leqslant \sigma_k \leqslant \bar{\sigma}_k \quad (k = 1, 2, ..., s) \tag{12.3}$$

$$a_i \in A_i \quad (i = 1, 2, ..., n) \tag{12.4}$$

Equation (12.1) defines the weight of the structure (w_i is the weight coefficient for member i given by material density × member length); equations (12.2) define d constraints on displacements δ_j (under- and overscored quantities denote specified lower and upper bounds, respectively); equations (12.3) define s constraints on stresses σ_k; equations (12.4) require each cross-sectional area a_i to belong to the discrete set of areas $A_i \equiv \{a_1, a_2, ...\}_i$ prevailing for the section shape specified for member i (e.g. W, T, hollow-box, double-angle, etc.).

In their present form, the displacement and stress constraints (12.2) and (12.3) are implicit nonlinear functions of the cross-sectional areas a_i. To facilitate computer solution, it is necessary to express these constraints as explicit functions of the sizing variables a_i. In fact, while introducing some approximations into the formulation, it is computationally advantageous to express these constraints as linear functions of the design variables. This suggests, then, to approximate each displacement and stress by its first-order Taylor series expansion. Moreover, since displacement and stresses in elastic structures under static loads generally vary inversely with member sizes a_i, it follows that better quality explicit linear constraint approximations result if the Taylor series expansions are expressed in terms of the reciprocal-sizing variables

$$x_i = \frac{1}{a_i} \quad (i = 1, 2, ..., n) \tag{12.5}$$

This is because displacements δ_j and stresses σ_k generally vary directly with the x_i variables and, therefore, the nonlinear relationships in x_i space between δ_j and x_i, and between σ_k and x_i, are shallower functions than are the corresponding δ_j–a_i and σ_k–a_i relationships in a_i space, i.e. the tangent-plane Taylor series expansions approximate the actual nonlinear functions more closely in x_i space than in a_i space.

The first-order Taylor series approximation of each displacement δ_j in terms of reciprocal sizing variables x_i is

$$\delta_j = \delta_j^0 + \sum_{i=1}^{n} \left(\frac{\partial \delta_j}{\partial x_i} \right)^0 (x_i - x_i^0) \tag{12.6}$$

while that for each stress σ_k is

$$\sigma_k = \sigma_k^0 + \sum_{i=1}^{n} \left(\frac{\partial \sigma_k}{\partial x_i} \right)^0 (x_i - x_i^0) \tag{12.7}$$

where the superscript (0) indicates known quantities for the current design stage (e.g. the initial 'trial' design). The quantities $(\partial \delta_j / \partial x_i)^0$ and $(\partial \sigma_k / \partial x_i)^0$ are displacement and stress gradients referenced to the current design, while the x_i are the sizing variables for the next weight optimization. It is readily shown that

$$\delta_j^0 = \sum_{i=1}^{n} \left(\frac{\partial \delta_j}{\partial x_i}\right)^0 x_i^0 \tag{12.8}$$

$$\sigma_k^0 = \sum_{i=1}^{n} \left(\frac{\partial \sigma_k}{\partial x_i}\right)^0 x_i^0 \tag{12.9}$$

Therefore, from equations (12.1)–(12.9), the statement of the weight optimization problem becomes

$$\text{minimize} \quad \sum_{i=1}^{n} w_i / x_i \tag{12.10}$$

$$\text{subject to} \quad \underline{\delta}_j \leqslant \sum_{i=1}^{n} \left(\frac{\partial \delta_j}{\partial x_i}\right)^0 x_i \leqslant \bar{\delta}_j \quad (j = 1, 2, \ldots, d) \tag{12.11}$$

$$\underline{\sigma}_k \leqslant \sum_{i=1}^{n} \left(\frac{\partial \sigma_k}{\partial x_i}\right)^0 x_i \leqslant \bar{\sigma}_k \quad (k = 1, 2, \ldots, s) \tag{12.12}$$

$$x_i \in X_i \quad (i = 1, 2, \ldots, n) \tag{12.13}$$

where the elements of each discrete set $X_i = \{x_1, x_2, \ldots\}$ in equations (12.13) are the reciprocals of the cross-sectional areas constituting the corresponding discrete set A_i in equations (12.4). The reciprocal areas x_i are selected from the database of standard steel sections having the shape specified for member i. Table 12.1 presents the range of standard section shapes produced by North American steel

Table 12.1 Standard steel sections

Section shape	Section description	Number of sections	
		USA	Canada
WWF	Welded wide flange	–	83
W	Wide flange	187	196
HP	Bearing piles	15	18
M	Miscellaneous beams and columns	8	7
S	Standard beams	31	31
C	Standard channels	29	31
MC	Miscellaneous channels	40	36
RHS	Rectangular hollow sections (tubing)	127	65
SHS	Square hollow sections (tubing)	51	61
CHS	Circular hollow sections (pipe)	37	63
WWT	Tees cut from WWF sections	–	41
WT	Tees cut from W sections	187	190
EL1L	Equal leg angles	51	50
UL1L	Unequal leg angles	80	56
EL2L	Two equal leg angles back to back	40	50
LL2L	Two unequal leg angles: long leg back to back	48	56
SL2L	Two unequal leg angles: short leg back to back	48	56

mills, including the approximate number of sections available for each shape. Theoretically, the number of elements x_i in each set X_i is equal to the number of sections listed in Table 12.1 for the section shape prevailing for member i. Practically, however, this number is taken to be considerably smaller, for reasons that are discussed later.

The displacement and stress gradients in equations (12.11) and (12.12) may be evaluated using one of a number of different sensitivity analysis techniques. For example, a finite difference technique based on, say, a central difference scheme may be employed to approximate the gradients as

$$\frac{\partial \delta_j}{\partial x_i} = \frac{\delta_j(x_i + \Delta x_i) - \delta_j(x_i - \Delta x_i)}{2 \Delta x_i} \tag{12.14}$$

$$\frac{\partial \sigma_k}{\partial x_i} = \frac{\sigma_k(x_i + \Delta x_i) - \sigma_k(x_i - \Delta x_i)}{2 \Delta x_i} \tag{12.15}$$

Despite their apparent simplicity, however, equations (12.14) and (12.15) are computationally expensive to evaluate because they involve $2n$ analyses of the structure, where n is the number of variables x_i undergoing specified small perturbations Δx_i. Alternatively, an analytical technique may be employed to establish the displacement and stress gradients. To this end, displacement gradients are first considered in the following.

Each individual nodal displacement δ_j of concern to the design is related to the overall vector of nodal displacements $\mathbf{u}$ for the structure as

$$\delta_j = \mathbf{b}_j^T \mathbf{u} \tag{12.16}$$

where $\mathbf{b}_j$ is a Boolean vector that identifies δ_j from among all nodal displacements (e.g. if $\mathbf{b}_j^T = [1, 0 \ldots 0]$ then $\delta_j = u_1$) and, depending on the requirements of the design, the displacements $\mathbf{u}$ are found from either first-order or second-order analysis of the current structure by solving the equilibrium conditions

$$\mathbf{Ku} = \mathbf{P} \tag{12.17}$$

for each vector $\mathbf{P}$ of applied loads, where $\mathbf{K}$ is the structure stiffness matrix, to obtain

$$\mathbf{u} = \mathbf{K}^{-1} \mathbf{P} \tag{12.18}$$

Now, differentiate equation (12.16) with respect to each reciprocal-sizing variable x_i to give

$$\frac{\partial \delta_j}{\partial x_i} = \mathbf{b}_j^T \frac{\partial \mathbf{u}}{\partial x_i} + \mathbf{u}^T \frac{\partial \mathbf{b}_j}{\partial x_i} = \mathbf{b}_j^T \frac{\partial \mathbf{u}}{\partial x_i} \tag{12.19}$$

which recognizes that $\partial \mathbf{b}_j / \partial x_i = 0$. Next, differentiate equation (12.17) w.r.t. x_i to give

$$\mathbf{K} \frac{\partial \mathbf{u}}{\partial x_j} + \frac{\partial \mathbf{K}}{\partial x_i} \mathbf{u} = \frac{\partial \mathbf{P}}{\partial x_i} = 0 \tag{12.20}$$

which recognizes for the design of steel building frameworks that $\partial \mathbf{P}/\partial x_i = 0$ since applied loads $\mathbf{P}$ are generally taken independent of changes in member sizes x_i. Solve equation (12.20) to find

$$\frac{\partial \mathbf{u}}{\partial x_i} = -\mathbf{K}^{-1}\frac{\partial \mathbf{K}}{\partial x_i}\mathbf{u} \tag{12.21}$$

The stiffness gradient $\partial\mathbf{K}/\partial x_i$ in equation (12.21) can be calculated using a finite difference approximation. Alternatively, an analytical approach can be taken to find $\partial\mathbf{K}/\partial x_i$ for steel building frameworks since for each member i it is generally possible to express the member stiffness matrix $\mathbf{K}_i$ as a function of solely the reciprocal-sizing variable x_i, either exactly or instantaneous-approximately, such that the structure stiffness matrix can be expressed as

$$\mathbf{K} = \sum_{i=1}^{n}\mathbf{K}_i = \sum_{i=1}^{n}\frac{1}{x_i}\mathbf{K}_i^* \tag{12.22}$$

where $\mathbf{K}_i^*$ is a constant matrix. Evidently, equation (12.22) is exact for pin-jointed truss structures that experience axial forces alone but only approximate for structures that also experience flexure, torsion and shear. In the latter case, the matrix $\mathbf{K}_i^*$ is updated from stage to stage of the design process to account for the changing relationships between the section area a_i and the other section inertia parameters concerned with flexural, torsional and shear behavior. (Appendix 12A illustrates these relationships for several wide-flange sections that are typically used in steel building frameworks.)

Now, differentiate equation (12.22) w.r.t. x_i to obtain the stiffness gradient as

$$\frac{\partial \mathbf{K}}{\partial x_i} = \frac{\partial}{\partial x_i}\left(\sum_{i=1}^{n}\frac{1}{x_i}\mathbf{K}_i^*\right) = \frac{\partial}{\partial x_i}\left(\frac{1}{x_i}\mathbf{K}_i^*\right) = -\frac{\mathbf{K}_i^*}{x_i^2} = -\frac{\mathbf{K}_i}{x_i} \tag{12.23}$$

which recognizes that $\partial\mathbf{K}_r/\partial x_i = \mathbf{0}$ when $r \neq i$. Having $\partial\mathbf{K}/\partial x_i$ from equation (12.23), it is now possible to calculate the displacement gradient $\partial\delta_j/\partial x_i$ through equation (12.19) using either a pseudo-load or virtual-load method. The pseudo-load method proceeds by observing that the gradient vector $\partial\mathbf{u}/\partial x_i$ in equation (12.21) is actually the solution of the system of equilibrium equations for the structure subject to the vector of pseudo-loads $-(\partial\mathbf{K}/\partial x_i)\mathbf{u}$. The resulting vector $\partial\mathbf{u}/\partial x_i$ is then substituted into equation (12.19) along with the vector $\mathbf{b}_j$ to find the displacement gradient $\partial\delta_j/\partial x_i$. For $\partial\delta_j/\partial x_i$ to be found for all $i = 1, 2, \ldots, n$ sizing variables x_i, which is typically the case, this method involves analyzing the structure through equation (12.21) for n different pseudo-load cases. Alternatively, the virtual-load method substitutes *a priori* for $\partial\mathbf{u}/\partial x_i$ from equation (12.21) into equation (12.19) to give

$$\frac{\partial\delta_j}{\partial x_i} = -\mathbf{b}_j^T\mathbf{K}^{-1}\frac{\partial\mathbf{K}}{\partial x_i}\mathbf{u} \tag{12.24}$$

which becomes from equation (12.23)

$$\frac{\partial\delta_j}{\partial x_i} = \frac{1}{x_i}(\mathbf{b}_j^T\mathbf{K}^{-1}\mathbf{K}_i\mathbf{u}) \tag{12.25}$$

It is then observed that if a unit virtual load is associated with the real displacement δ_j in equation (12.16), by the principle of virtual work equivalence the vector $\mathbf{b}_j$ may be viewed as a vector of virtual loads associated with the vector of real displacements **u** in equation (12.16). The vector of virtual nodal displacements **u**$_j$ corresponding to the virtual loads **b**$_j$ is found as

$$\mathbf{u}_j = \mathbf{K}^{-1}\mathbf{b}_j \qquad (12.26)$$

and then substituted into equation (12.25) to find the displacement gradient as

$$\frac{\partial \delta_j}{\partial x_i} = \frac{1}{x_i}(\mathbf{u}_j^T \mathbf{K}_i \mathbf{u}) \qquad (12.27)$$

Contrary to the pseudo-load method, the virtual-load method only involves a single analysis of the structure for the virtual loads **b**$_j$ to find the gradient $\partial \delta_j / \partial x_i$ for all $i = 1, 2, \ldots, n$ sizing variables x_i. On the other hand, if the gradients of $j = 1, 2, \ldots, m$ different displacements δ_j are to be found, which is typically the case, then the virtual-load method will involve m analyses of the structure while the pseudo-load method will still only involve n analyses regardless of the number of displacements of concern to the design of the structure. Therefore, assuming that the method requiring the fewest analyses of the structure is the most efficient, the pseudo-load method should be applied if the number n of sizing variables is less than the number m of response gradients. Otherwise, the virtual-load method should be applied if $n > m$. In actual fact, either method is quite efficient since each analysis of the structure to establish response gradients is really nothing more than a back-substitution operation for different pseudo- or virtual loads since the inverse or factorized stiffness matrix $\mathbf{K}^{-1}$ is already available from the basic analysis equation (12.18) that determines the displacements **u** due to the actual loading **P**. The virtual-load method (12.27) is adopted to calculate displacement gradients for the design examples presented later in the chapter.

Stress gradients are similarly calculated analytically using the virtual-load method described in the foregoing for displacement gradients. Here, each individual member stress σ_k of concern to the design is related to the overall vector of nodal displacements **u** for the structure as

$$\sigma_k = \mathbf{t}_k^T \mathbf{u} \qquad (12.28)$$

where the vector **t**$_k$ is row k of the global-axis stress matrix for the member associated with the stress σ_k. The vector **t**$_k$ is invariant for truss structures composed of axial members alone. For flexural structures, however, the vector **t**$_k$ is a function of the neutral-axis position for the member section associated with the stress σ_k, and therefore it varies as the section design changes over the synthesis history. Upon viewing **t**$_k$ as a vector of virtual loads applied to the structure to cause virtual nodal displacements

$$\mathbf{u}_k = \mathbf{K}^{-1}\mathbf{t}_k \qquad (12.29)$$

the gradient of the stress σ_k with respect to change in each reciprocal-sizing variable x_i is found as

$$\frac{\partial \sigma_k}{\partial x_i} = \frac{1}{x_i}(\mathbf{u}_k^T \mathbf{K}_i \mathbf{u}) \qquad (12.30)$$

Equation (12.30) is based on the assumption that the vector **t**$_k$ is invariant with changes in x_i for the current design stage, for both truss and flexural structures. The error inherent in this assumption for flexural structures becomes negligible as the synthesis process converges to a final design state where no changes take place in the sizing variable x_i for a number of successive design stages.

An illustration of the calculation of displacement and stress gradients using the virtual-load method is presented in Appendix 12B for a simple truss framework.

12.3 SYNTHESIS PROCESS

By virtue of the approximate nature of the performance constraints equations (12.11) and (12.12), the design synthesis is conducted through an iterative process that involves solving the weight optimization problem (12.10)–(12.13) during each design cycle. Attention is focused in the following on presenting the details of the iterative synthesis process that specifically pertain to the design of practical steel frameworks using standard steel sections in conformance with the provisions of a governing steel code (e.g. American Institute of Steel Construction, 1989; Canadian Standards Association, 1989).

The synthesis process begins by selecting an initial 'trial' design of the structure. There are several ways in which this can be done. One method is to employ some heuristics based on designer experience to derive an acceptable preliminary design. The advantage of this approach is that it provides an initial design that is reasonably well proportioned. A disadvantage, particularly for complex structures, is that it may not be readily possible to find an acceptable initial design using simple heuristics alone. Another approach is to identify the standard section shape for each member (e.g. W, T, etc.) and then to select the largest section from the corresponding database of available sections (i.e. as commercially supplied by steel mills). The advantage of this approach is that it is simple to apply to establish an initial design. Also, as the largest available sections are being used, this approach clearly signals that the design is probably ill-posed if the initial design is found to be infeasible for the performance requirements of the governing steel design code. A disadvantage of this approach is that the initial design is generally not very well proportioned. However, a reasonably proportioned design is generally found after one cycle of the synthesis process by solving the weight optimization problem for continuous-valued sizing variables by temporarily replacing the discrete sizing constraints (12.13) with the simple bounding conditions

$$\underline{x}_i \leqslant x_i \leqslant \bar{x}_i \quad (i = 1, 2, \ldots, n) \tag{12.31}$$

where $\underline{x}_i = 1/\bar{a}_i$ and $\bar{x}_i = 1/\underline{a}_i$, in which $\bar{a}_i$ and $\underline{a}_i$ are the largest and smallest section areas available for member i, respectively. This latter approach to establish an initial design is adopted for the design examples presented later in the chapter.

Having a reasonably well-proportioned initial design, structural analysis is conducted to determine the corresponding elastic displacements and stresses for each load case. Either first-order or second-order analysis is performed depending on the requirements of the governing steel design code. If first-order analysis is the basis for the design calculations then effective length factors greater than unity are adopted for flexural members; if second-order analysis is employed then all member effective length factors are typically set equal to unity.

Having the stresses and displacements from structural analysis, the database of standard steel sections is searched, member by member, to find the least-weight sections that satisfy the stress-related strength–stability requirements of the governing steel design code while simultaneously ensuring that all displacement-related stiffness constraints are satisfied. This searching activity, hereafter referred to as member-by-member optimization, has two distinct aspects. The concern for strength–stability involves selecting standard sections taking into account axial tension and compression, bending, shear and torsion effects, and combinations of these effects, for individual members considered in isolation from the structure. This aspect of the search is somewhat more straightforward than that when displacements are of concern where member sections must be economically selected while at the same time ensuring that the stiffness of the assembled structure is satisfactory. To this latter end, an effective technique adopted herein is based on calculating displacement gradients (e.g. Appendix 12B) so as to establish the relative influence that individual members have on displacements, which then allows member sections to be selected while having concern for both displacements and economy (Cameron, 1989; Cameron *et al.*, 1992).

Having the member sections resulting from the member-by-member optimization procedure, structural analysis is again conducted and the strength–stability and stiffness requirements for the design are checked for the resulting stresses and displacements. If no design violations are detected, the synthesis process continues on to the next stage involving formal optimization. Otherwise, the design is infeasible and member-by-member optimization is conducted again.

The formal optimization stage of each cycle of the synthesis process involves formulating and solving the weight optimization problem posed by equations (12.10)–(12.13). With a view to computational efficiency, it is desirable to delete temporarily relatively inactive displacement and stress constraints from the formulation. For steel frameworks, this generally results in a significant reduction in the size of the active constraint set and thus allows either for faster computation for a certain size structure, or for the design of a larger structure than would be otherwise possible on a computer with fixed memory capacity. Deleted constraints are continuously monitored and added to the constraint set if and when they become active for a subsequent stage of the synthesis process. Moreover, to overcome possible design history oscillations that impede or thwart convergence of the synthesis process, it is recommended to retain continuously a constraint in the active set even if it subsequently becomes passive to the design (Grierson and Lee, 1984).

The weight coefficients w_i in the objective function (12.10) remain constant throughout the synthesis process. This is also true for the displacement bounds $\bar{\delta}_j$ and $\underline{\delta}_j$ in equations (12.11). On the other hand, the allowable tensile and compressive stress bounds in equations (12.12) must be updated for each design cycle because corresponding axial, flexural, shear and torsional capacities are dependent on unbraced member lengths and the geometry of member sections, both of which vary over the synthesis history. For example, the lengths of unbraced compression flanges depend on the bending moment distribution, which changes with modifications in the stiffness distribution for the structure. Alternatively, the width–thickness ratios of section plate elements change as different standard sections are selected for a member from stage to stage of the synthesis process. The tensile stress bound $\bar{\sigma}_k$ accounts for the type of member connection (bolted, etc.), while the compressive stress bound $\underline{\sigma}_k$ depends on the stress state (axial, flexural, combined, etc.) and accounts for both local section buckling and overall member buckling.

For the current standard section sizes for the members of the structure, sensitivity analysis is conducted to establish the gradients $(\partial \delta_j/\partial x_i)^0$ and $(\partial \sigma_k/\partial x_i)^0$ of the active displacement and stress constraints in equations (12.11) and (12.12). It is important to note that the computational effort required to find these gradients is quite nominal since direct use is made of the inverse or factorized form of the structure stiffness matrix that has been previously established by the conventional analysis conducted for the current cycle of the synthesis process (e.g. Appendix 12B).

To complete the formulation of the weight optimization problem of equations (12.10)–(12.13) for the current design stage, it remains to update each set of discrete sizing variables X_i in equations (12.13) from the corresponding database of standard steel sections to reflect only those sections having adequate strength–stability and stiffness properties for the current set of active stress and displacement constraints. In fact, the total number of sections reflected in each set X_i is limited so as to restrict the extent of movement in the design space during solution of the weight optimization problem, thereby preserving the integrity of the first-order displacement and stress gradients in equations (12.11) and (12.12). The updated subset of standard sections for each member (or fabrication group of members) is arranged such as to bracket the current design section, thereby allowing for the selection of either a smaller or a larger section as the outcome of the weight optimization that solves equations (12.10)–(12.13).

12.4 OPTIMIZATION ALGORITHM

The solution of the least-weight design problem (12.10)–(12.13), hereafter referred to as the formal optimization stage of each design cycle, can be found using one of a number of different algorithms. An optimality criteria (OC) method involves temporarily replacing the discrete sizing constraints (12.13) with the simple bounding conditions (12.31) and deriving an optimality criterion based on the stationary conditions for the Lagrangian function involving equations (12.10)–(12.12). The Lagrange variables, which identify the active performance constraints for the design, are then determined along with the sizing variables x_i through a recursive algorithm that converges after a number of cycles to a nonstandard design (i.e. a set of x_i values that do not correspond to standard steel sections). A pseudo-discrete algorithm based on the optimality criterion is then applied to assign progressively a standard section to each of the members. This OC method is particularly well suited to the design of structures for which the number of performance constraints is small compared with the number of sizing variables, such as for the lateral stiffness design of tall steel frameworks subject to inter-story drift constraints (Chan, 1993).

Another optimization algorithm for solving the design problem (12.10)–(12.13) is a generalized optimality criteria (GOC) technique that is often referred to in the literature as the dual method (Schmit and Fleury, 1979; Fleury, 1979). The details of the GOC method, which is used for the design examples presented later in this chapter, are briefly developed in the following. To simplify the associated discussion it is convenient to express the optimization problem (12.10)–(12.13) in the following form:

$$\text{minimize} \quad \sum_{i=1}^{n} w_i/x_i \tag{12.32}$$

$$\text{subject to} \quad \sum_{i=1}^{n} c_{ir}x_i \leqslant \bar{g}_r \quad (r = 1, 2, \ldots, m) \tag{12.33}$$

$$x_i \in X_i \quad (i = 1, 2, \ldots, n) \tag{12.34}$$

where equations (12.33) are a concise expression of the displacement and stress constraints (12.11) and (12.12): the gradients $c_{ir} \equiv \pm (\partial \delta_j/\partial x_i)^0, \pm (\partial \sigma_k/\partial x_i)^0$; the bounds $\bar{g}_r \equiv \underline{\delta}_j, \bar{\delta}_j, \underline{\sigma}_k, \bar{\sigma}_k$; the total number of constraints $m = d + s$.

To commence the development of the GOC method, temporarily ignore the discrete sizing constraints (12.34) and formulate the Lagrangian function

$$\mathcal{L}(\mathbf{x}, \lambda) = \sum_{i=1}^{n} \frac{w_i}{x_i} + \sum_{r=1}^{m} \lambda_r \left(\sum_{i=1}^{n} c_{ir}x_i - \bar{g}_r \right) \tag{12.35}$$

where the Lagrange variables are sign restricted as

$$\lambda_r \geqslant 0 \quad (r = 1, 2, \ldots, m) \tag{12.36}$$

The GOC method proceeds by finding a saddle-point $(\mathbf{x}^*, \lambda^*)$ of equation (12.35) while accounting for the discrete sizing constraints (12.34), which in turn provides a solution $\mathbf{x}^*$ to the original optimization problem (12.32)–(12.34) (Zangwill, 1969). A saddle-point of equation (12.35) is given by a solution of the min–max problem

$$(\mathbf{x}^*, \lambda^*) = \max_{\lambda} [\min_{\mathbf{x}} \mathcal{L}(\mathbf{x}, \lambda)] \tag{12.37}$$

The problem posed by equation (12.37) is solved by first minimizing the Lagrangian function over the $\mathbf{x}$ variables to establish a function in terms of λ variables alone, i.e.

$$\mathcal{L}_m(\lambda) = \min_{\mathbf{x}} \mathcal{L}(\mathbf{x}, \lambda) = \min_{\mathbf{x}} \sum_{i=1}^{n} L_i(x_i, \lambda) \tag{12.38}$$

where, from equation (12.35), the functions

$$L_i(x_i, \lambda) = \frac{w_i}{x_i} + \sum_{r=1}^{m} \lambda_r (c_{ir}x_i - \bar{g}_r) \quad (i = 1, 2, \ldots, n) \tag{12.39}$$

are each expressed in terms of a single sizing variable x_i.

Equation (12.38), which is called the dual function, is readily established by separately minimizing each function $L_i(x_i, \lambda)$ in equations (12.39) over the corresponding discrete set of values X_i for sizing variable x_i and then summing, i.e.

$$\mathcal{L}_m(\lambda) = \sum_{i=1}^{n} \min_{x_i \in X_i} L_i(x_i, \lambda) \tag{12.40}$$

As the λ variables vary in value, the value of the sizing variable x_i for which the corresponding function $L_i(x_i, \lambda)$ is minimum may shift from one discrete value x_i^k to the next discrete value x_i^{k+1} in the set X_i. When this happens, continuity of the dual function $\mathcal{L}_m(\lambda)$ is maintained by requiring that

$$L_i(x_i^k, \lambda) = L_i(x_i^{k+1}, \lambda) \tag{12.41}$$

which reduces to, from equations (12.39),

$$\sum_{r=1}^{m} \lambda_r c_{ir} = \frac{w_i}{x_i^k x_i^{k+1}} \tag{12.42}$$

From equation (12.42), the specific discrete value of each sizing variable x_i for which the corresponding function $L_i(x_i, \lambda)$ is minimum is given by

$$x_i^0 = x_i^k \tag{12.43}$$

if

$$\frac{w_i}{x_i^{k-1} x_i^k} < \sum_{r=1}^{m} \lambda_r^0 c_{ir} < \frac{w_i}{x_i^k x_i^{k+1}} \tag{12.44}$$

where the values of the Lagrange variables λ_r^0 are initially assumed (e.g. if constraint m of equations (12.33) is the most active for the initial design of the structure, set $\lambda_m^0 = 1$ and $\lambda_r^0 = 0 \, (r = 1, 2, \ldots, m-1)$); thereafter, the λ_r^0 values are known from the previous iteration step of the maximization process conducted to solve equation (12.37), as outlined in the following.

Having established the dual function $\mathcal{L}_m(\lambda)$, the GOC method completes the solution of equation (12.37) by solving the maximization problem, from equations (12.36), (12.39) and (12.40),

$$\text{maximize} \quad \mathcal{L}_m(\lambda) = \sum_{i=1}^{n} \frac{w_i}{x_i^0} + \sum_{r=1}^{m} \lambda_r \left(\sum_{i=1}^{n} c_{ir} x_i^0 - \bar{g}_r \right) \tag{12.45}$$

$$\text{subject to} \quad \lambda_r \geqslant 0 \quad (r = 1, 2, \ldots, m) \tag{12.46}$$

where the x_i^0 values are known from equations (12.43) and (12.44). The problem posed by equations (12.45) and (12.46) is solved using a gradient projection method (Zangwill, 1969; Haftka and Gürdal, 1992) to update progressively the Lagrange variables as

$$\lambda = \lambda^0 + \alpha s \tag{12.47}$$

where s is the search direction, α the step length and λ^0 the variable values found at the end of the previous search step. Note from equations (12.43) and (12.44) that the x_i^0 values in equation (12.45) are subject to change as the λ_r^0 values change. The gradient projection search (12.47) progressively accounts for the bounds on the λ_r variables, equations (12.46), and the discontinuities in the x_i variables, equations (12.44), to find the solution point $(\mathbf{x}^*, \lambda^*)$ for which equation (12.45) is maximized. The point $\mathbf{x}^*$ is the solution of the design optimization problem (12.32)–(12.34).

The GOC method for discrete optimization is illustrated in Appendix 12C for the design of a simple truss structure.

12.5 EXAMPLE APPLICATIONS

The least-weight designs of five different steel frameworks are presented in the following. The designs are conducted using a software system based on the optimization technology described in the foregoing (Grierson and Cameron, 1991). The first design example has been selected more for its illustrative value than for its practicality and is presented in some detail. The remaining four design examples are presented somewhat more briefly and are representative of actual designs carried out in professional practice by structural engineering firms. (The names of the individuals and firms contributing particular design examples are acknowledged at the end of the chapter.)

EXAMPLE 12.1

The three-bay by two-bay by two-story steel framework in Fig. 12.1 is to be designed in accordance with the strength–stability and stiffness provisions of the American Allowable Stress Design steel code AISC-89 (American Institute of Steel Construction, 1989). The bay spacings and story heights in the X, Z and Y axis directions are 20, 16.4 and 10 feet, respectively.

First-order effects alone are to be accounted for, and sidesway is to be permitted. The in-plane and out-of-plane effective length factors K_x and K_y for the continuously connected X direction girders and the Y direction columns are to be calculated on the basis of their end stiffnesses (Johnston, 1976). The effective length factors for the pinned-end Z direction girders and the diagonal bracing

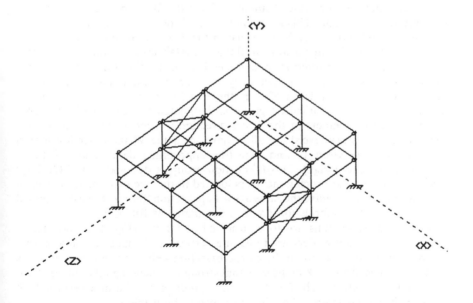

FABRICATION GROUPS

GX1-1 = First Story X-Girders   IC1-2 = Interior Columns

GX2-2 = Second Story X-Girders   EC1-2 = Exterior Columns

GZ1-1 = First Story Z-Girders   BRACE = Diagonal Braces

GZ2-2 = Second Story Z-Girders

Fig. 12.1 Steel building framework.

members are specified to be $K_x = K_y = 1.0$. The maximum allowable slenderness ratios for each member are specified to be $KL/r = 200$ in compression and $KL/r = 300$ in tension (where L is the member length and r the section radius of gyration).

Member sections are to be selected from available AISC standard steel sections (American Institute of Steel Construction, 1989). All column and girder members are to have wide-flange (W) shapes, while all diagonal bracing members are to have square hollow section (SHS) shapes. For these specified section types, the Young's modulus for the steel is $29\,000\,\mathrm{klbf\,in^{-2}}$, while the shear modulus is $11\,200\,\mathrm{klbf\,in^{-2}}$, the yield stress is $36\,\mathrm{klbf\,in^{-2}}$ and $50\,\mathrm{klbf\,in^{-2}}$ for the W and SHS sections, respectively, and the ultimate stress is $58\,\mathrm{klbf\,in^{-2}}$ and $62\,\mathrm{klbf\,in^{-2}}$ for the W and SHS sections, respectively.

To satisfy practical fabrication requirements, the members of the structure are to be grouped as follows (Fig. 12.1). (a) The eight continuously connected girder members located in the X axis direction at story level 1 (i.e. the second floor) are all to have the same W-shape section (the depth and size to be determined by the design). This fabrication group is designated as GX1-1. (b) The eight continuously connected girder members located in the X axis direction at story level 2 (i.e. the roof) are all to have the same W-shape section. This fabrication group is designated as GX2-2. (c) The nine pinned-end girder members located in the Z axis direction at story level 1 are all to have the W-shape section W10 × 33 (i.e. fixed for the design). This fabrication group is designated as GZ1-1. (d) The nine pinned-end girder members located in the Z axis direction at story level 2 are all to have the W-shape section W10 × 33. This fabrication group is designated as GZ2-2. (e) The four interior column members in the Y axis direction (i.e. two interior columns for each of the two story levels) are all to have the same W-shape section with a nominal depth of 12 in (the actual section size to be determined by the design). This fabrication group is designated as ICI-2. (f) The twenty exterior column members in the Y axis direction (i.e. ten exterior columns for each of the two story levels) are all to have the same W-shape section with a nominal depth between 10 and 14 in (the appropriate depth and size to be determined by the design). This fabrication group is designated as EC1-2. (g) The eight pinned-end bracing members located in the two exterior Z–Y faces of the structure are all to have the same SHS-shape section (the size to be determined by the design). This fabrication group is designated as BRACE.

Member cross-sections are to be oriented as follows (Fig. 12.1): (a) section webs for X direction girders are to lie in a vertical X–Y plane such that strong axis bending takes place about the Z axis; (b) section webs for Y direction columns are to lie in a vertical X–Y plane such that strong-axis bending takes place about the Z axis; (c) section webs for Z direction girders are to lie in a vertical Y–Z plane such that strong-axis bending takes place about the X axis.

The service loading on the structure is (Fig. 12.1) as follows. (a) Wind loading acts horizontally in the positive X axis direction of magnitude $11.0\,\mathrm{klbf\,in^{-2}}$ at each exterior node of the Z–Y face of the structure for which the X coordinate is 0. This loading is designated as WIND:X. (b) Wind loading acts horizontally in the positive Z axis direction of magnitude $9.0\,\mathrm{klbf\,in^{-2}}$ at each exterior node of the X–Y face of the structure for which the Z coordinate is 0. This loading is designated as WIND:Z. (c) Gravity uniformly distributed loading acts vertically in the negative Y axis direction of magnitude $1.4\,\mathrm{klbf\,in^{-2}\,ft.^{-1}}$ on each of the X

axis girders in the second floor and the roof. This loading is designated as GRAVITY:X. (d) Gravity point loading acts vertically in the negative Y axis direction of magnitude 14.5 klbf in^{-2} at midspan of each of the Z axis girders in the second floor and the roof. This loading is designated as GRAVITY:Z.

Two load combinations are considered for the design of the structure, as follows:

GRAVITY:X + GRAVITY:Z + WIND:Z

GRAVITY:X + GRAVITY:Z + WIND:X

Both load combinations define service loading schemes for which the stress related provisions of the AISC-ASD design code are to be satisfied (American Institute of Steel Construction, 1989). The second load combination also defines the service loading scheme for which the lateral sway in the X axis direction is limited to 0.8 in at the roof level (i.e. $h/300$).

The least-weight design of the framework is conducted using the synthesis process involving formal and member-by-member optimization described earlier in this chapter. The iterative synthesis history and final design for the steel framework are shown in Fig. 12.2. The AISC standard steel section shapes and designations given in Fig. 12.2 apply for the seven fabrication groups defined in Fig. 12.1. The initial structure weight given in Fig. 12.2 refers to an arbitrarily

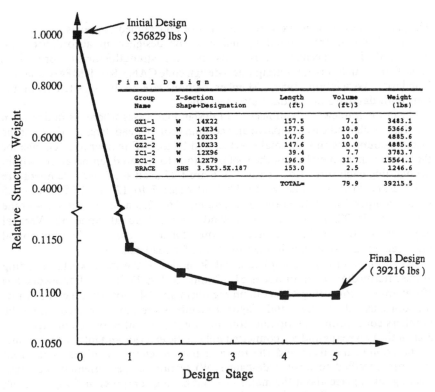

Fig. 12.2 Example 12.1: synthesis history and final design.

selected conservative design (i.e. largest section sizes satisfying fabrication requirements). As shown in Fig. 12.2, the optimization process converged after five
iterations to a least-weight final design satisfying the imposed X axis sway
limitation and all relevant strength/stability requirements of the AISC–ASD steel
design code (American Institute of Steel Construction, 1989; Cameron, Xu and
Grierson, 1991).

The design optimization involved 7 sizing variables, 4 displacement constraints
and 364 stress constraints, and required approximately 90 minutes execution
time on a 386-based microcomputer. In fact, if formal optimization is excluded
and member-by-member optimization is alone applied, a very similar final design
is found after only approximately 5 minutes of computer time. However, this
result is peculiar to this example as the use of formal optimization generally
yields lower-weight designs compared with those produced by member-by-
member optimization alone. In fact, as is demonstrated for the following example,
the two optimization strategies generally work in tandem to produce a design
that is an improvement on that produced by either strategy alone.

EXAMPLE 12.2

The planar steel framework with trussed roof and pinned supports in Fig. 12.3
is part of a mill crane building and is to be designed in accordance with
user-specified displacement constraints and the strength/stability provisions
of the Canadian limit-states design steel design code CAN3-S16.1-M84 (Canadian
Standards Association, 1989). Member sections are to be selected from the
Canadian database of standard steel sections.

The crane framework is subject to 14 design load cases where, as indicated in
Fig. 12.3, each load case is a particular combination of dead, live, wind and crane
loading, coupled with a thermal effect caused by a 50 °C temperature change for
the bottom chord members of the roof truss (due to elevated temperatures within
the building enclosure). Load cases 1 to 6 are 1.25 dead + 1.5 live + temperature
+ 1.5 crane load at nodes 27 to 32; load cases 7 to 12 are 1.25 dead + 1.5
wind + temperature + 1.5 crane load at nodes 27 to 32; load case 13 is live + crane
load at node 32; load case 14 is wind + crane load at node 32. Vertical
displacement at node 32 is limited to 50 mm for load case 13, while horizontal
displacement at node 26 is limited to 20 mm for load case 14.

Out-of-plane bracing is applied at all 56 nodes of the framework, including
at the intermediate column nodes 51, 52 and 53 (Fig. 12.3). The framework has
103 members, consisting of 9 column members and 94 roof truss members. For
the roof truss, the vertical and diagonal members are pin connected, while the
members constituting the top and bottom chords are continuous connected over
each of the two spans and pin connected only at the exterior and interior columns.

The column members and the top and bottom chord members in the roof
truss are specified to have wide-flange (W) section shapes oriented such that
bending takes place about the major axis and having compression flange bracing
only at members ends. The vertical and diagonal members in the roof truss are

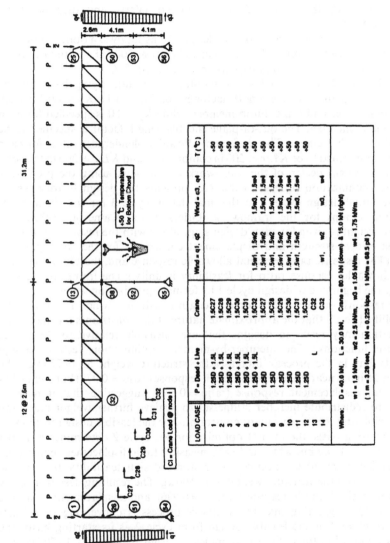

LOAD CASE	P = Dead + Live	Crane	Wind = q1, q2	Wind = q3, q4	T(°C)
1	1.25D + 1.5L	1.5C27			+50
2	1.25D + 1.5L	1.5C28			+50
3	1.25D + 1.5L	1.5C29			+50
4	1.25D + 1.5L	1.5C30			+50
5	1.25D + 1.5L	1.5C31			+50
6	1.25D + 1.5L	1.5C32			+50
7	1.25D	1.5C27	1.5w1, 1.5w2	1.5w3, 1.5w4	+50
8	1.25D	1.5C28	1.5w1, 1.5w2	1.5w3, 1.5w4	+50
9	1.25D	1.5C29	1.5w1, 1.5w2	1.5w3, 1.5w4	+50
10	1.25D	1.5C30	1.5w1, 1.5w2	1.5w3, 1.5w4	+50
11	1.25D	1.5C31	1.5w1, 1.5w2	1.5w3, 1.5w4	+50
12	1.25D	1.5C32	1.5w1, 1.5w2	1.5w3, 1.5w4	+50
13		C32	w1, w2	w3, w4	
14	L	C32	w2	w4	

Where: D = 40.0 kN, L = 30.0 kN, Crane = 80.0 kN (down) & 15.0 kN (right)

w1 = 1.5 kN/m, w2 = 2.5 kN/m, w3 = 1.05 kN/m, w4 = 1.75 kN/m

(1 m = 3.28 feet, 1 kN = 0.225 kips, 1 kN/m = 68.5 plf)

Fig. 12.3 Mill crane building framework.

specified to have double-angle section shapes with long legs back to back (⌐L) separated by a 10 mm thick gusset plate.

To satisfy fabrication requirements, six groups of members are identified for which all members in each group are specified to have common section properties. The six fabrication groups are exterior column members (COLext), interior column members (COLint), top chord truss members (CHtop), bottom chord truss members (CHbot), vertical truss members (Vert), and diagonal truss members (Diag).

The steel material properties for the design are as follows: Young's modulus, 200 000 MPa; shear modulus, 77 000 MPa; yield stress, 300 MPa; ultimate stress, 450 MPa; coefficient of thermal expansion, $0.117 \times 10^{-4}\,^\circ C^{-1}$.

Second-order (P–delta) analysis is adopted as the underlying basis for the design. The in-plane effective length factor is specified to be $K_x = 1.0$ for each roof truss member and top column member, while $K_x = 2.0$ is specified for each lower column member. The out-of-plane effective length factor is specified to be $K_y = 1.0$ for each member. The maximum allowable slenderness ratios for each member are specified to be $KL/r = 200$ in compression and $KL/r = 300$ in tension.

The least-weight design of the framework is conducted using the previously described optimization process. Table 12.2 presents the design history results from the initial design cycle 1 to the final design cycle 3 for the six member fabrication groups for the framework (Grierson and Cameron, 1989). The structure weight is given for each design stage along with the response ratios for the most critical strength and displacement constraints for the corresponding design. (The response ratio = actual/allowable response and, therefore, the maximum allowable response ratio for feasibility is unity.) The framework design found at the beginning of design cycle 1 by formal optimization weighs 19 665 kg but the response ratio is 1.292 for a strength constraint, which implies 29.2% infeasibility. (Recall that the approximate nature of the constraints for the formal optimization may result in designs that are infeasible relative to the actual constraint functions.) The member-by-member optimization restores design feasibility but, in the process, increases the structure weight to 19 725 kg. The resulting design is nearly active for strength response (ratio = 0.978) but somewhat inactive for displacement response (ratio = 0.638), suggesting that it may be possible to redistribute member stiffnesses to achieve further weight reduction without violating the displacement constraint. For the feasible design from cycle 1 as its starting basis, the formal optimization for cycle 2 determines a design that weighs 17 110 kg but which is 13.8% infeasible for a displacement constraint. The member-by-member optimization restores design feasibility but, in the process, increases the structure weight to 17 500 kg. The resulting design is nearly active for both strength response (ratio = 0.996) and displacement response (ratio = 0.910), suggesting that the synthesis process is converging to a least-weight structure. For the feasible design from cycle 2 as its starting basis, the formal optimization for cycle 3 determines a design that weighs 16 795 kg but which is 4.8% infeasible for a strength constraint. The member-by-member optimization determines a heavier design weighing 17 190 kg, but, now, the critical strength constraint is identically satisfied (ratio = 1.000) and the critical displacement constraint is nearly active (ratio = 0.969). The tight degree of satisfaction of both the strength and displacement constraints indicates that the synthesis process has converged to a least-weight feasible design of the framework, as given

Table 12.2 Design history results for mill crane building framework

Design cycle	1 Section designation		2 Section designation		3 Section designation	
	Formal optimization	Member-by-member optimization	Formal optimization	Member-by-member optimization	Formal optimization	Member-by-member optimization
Member group						
COLext	W610 × 174	W610 × 174	W610 × 113	W610 × 140	W610 × 125	W610 × 125
COLint	W610 × 195	W610 × 195	W530 × 123	W610 × 125	W610 × 125	W610 × 125
CHtop	W360 × 57	W250 × 49	W250 × 49	W250 × 49	W250 × 49	W250 × 49
CHbot	W410 × 85	W310 × 97	W250 × 89	W310 × 86	W250 × 80	W310 × 86
Vert	JL 100 × 90 × 10	JL 125 × 90 × 10	JL 125 × 90 × 10	JL 125 × 90 × 10	JL 125 × 90 × 10	JL 125 × 90 × 10
Diag	JL 150 × 100 × 10	JL 100 × 75 × 13	JL 125 × 90 × 10	JL 125 × 90 × 10	JL 125 × 90 × 10	JL 125 × 90 × 10
Structure weight (kg)	19 665	19 725	17 110	17 500	16 795	17 190
Critical strength response ratio	1.292	0.978	0.994	0.596	1.048	1.000
Critical displacement response ratio	0.632	0.638	1.138	0.910	0.971	0.969

Note: W $d \times m \equiv$ wide-flange section; depth d (mm) and mass m (kg m^{-1}).
JL $l_1 \times l_2 \times t \equiv$ double-angle section with long legs back to back; long leg length l_1 (mm), short leg length l_2 (mm), and thickness t (mm).

in the last column of Table 12.2. The iterative design process involved 2886 strength and displacement constraints and required approximately 120 minutes execution time on an IBM PS/2 Model 30 microcomputer.

As an illustration that formal optimization and member-by-member optimization together produce the lightest design, the design in the first column in Table 12.2 was taken as the starting basis for member-by-member optimization applied alone and the iterative synthesis process converged after three cycles to a feasible design weighing 18 070 kg. This design is 18 070–17 190 kg = 880 kg or 5.1% heavier that the final design in Table 12.2 found using formal and member-by-member optimization in tandem.

EXAMPLE 12.3

This example concerns the design of an individual component part of the materials handling plant shown in Fig. 12.4. Material is input from the ground by conveyor

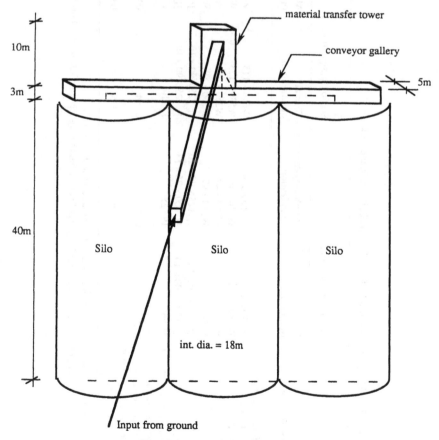

Fig. 12.4 Materials handling plant.

to a transfer tower located above the central silo, where it is then distributed to the three silos through a conveyor gallery. The supporting roof system for the central silo shown in Fig. 12.5 consists of eleven stringers supported in span by two girders. As shown in Fig. 12.6, the roof girders are supported on the outer silo wall by short connector beams splice-connected at the girder ends.

It is required to design the silo roof girder and end-connectors shown . in Fig. 12.6 in accordance with the provisions of the CAN3-S16.1-M89 steel design code (Canadian Standards Association, 1989). Both the girder and connectors are to have welded wide-flange (WWF) sections oriented such that bending takes place about the major axis. The sections are to be selected from the database of Canadian standard WWF steel sections. The girder is required to have a nominal depth between 1000 mm and 1200 mm, while each connector is required to have a nominal depth of 500 mm.

The steel material properties for the design of the girder and connectors are as follows: Youngs modulus, 200 000 MPa; shear modulus, 77 000 MPa; yield stress, 300 MPa; and ultimate stress, 450 MPa.

Both the girder and connectors are subject to uniformly distributed selfweight loading, while the girder is additionally subject to the selfweight of the stringers applied as point loads at the stringer locations (this loading is designated as

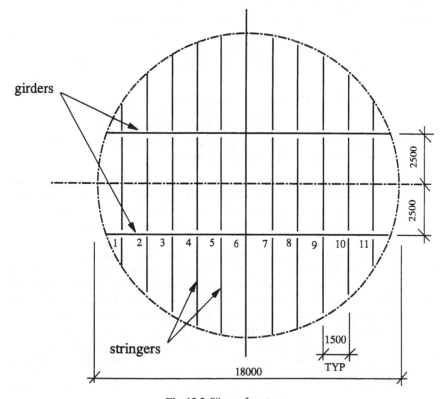

Fig. 12.5 Silo roof system.

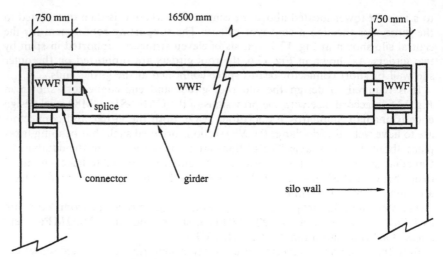

Fig. 12.6 Roof girder and end-connectors.

DEAD). The girder is subject to live loading and snow loading applied as point loads at the stringer locations (these two loadings are designated as LIVE and SNOW). The girder is also subject to conveyor loading and wind loading applied, in each case, as two point loads at stringer locations 4 and 8 in Fig. 12.5 (these two loadings are designated as CONVEYOR and WIND). The roof girder and end-connectors are to be designed accounting for the following five load combinations:

> 1.25*DEAD + 1.5* (LIVE + CONVEYOR)
> 1.25* DEAD + 1.05* (LIVE + CONVEYOR + SNOW)
> 1.25* DEAD + 1.05* (LIVE + CONVEYOR + WIND)
> 1.25* DEAD + 0.9* (LIVE + CONVEYOR + SNOW + WIND)
> LIVE + CONVEYOR + SNOW + WIND

The strength–stability provisions of the CAN3-S16.1-M89 steel design code (Canadian Standards Association, 1989) are to be satisfied for the girder and connectors for the first four factored load combinations, while the vertical deflection at midspan of the girder is not to exceed 50 mm for the final unfactored service load combination.

The in-plane and out-of-plane effective length factors are specified to be $K_x = K_y = 1.0$ for the girder and connectors. The girder is specified to be fully braced against lateral-torsional buckling throughout its length, while the compression flanges of the end-connectors are considered to be unbraced. The maximum allowable slenderness ratios for both girder and connectors are specified to be $KL/r = 200$ in compression and $KL/r = 300$ in tension.

The design of the roof girder and end-connectors is conducted using the optimization process described earlier in this chapter. The iterative synthesis history and optimal WWF section designations are given in Fig. 12.7. The initial design weight given in Fig. 12.7 corresponds to the largest WWF sections satisfying the depth limitations specified for the girder and connectors. The

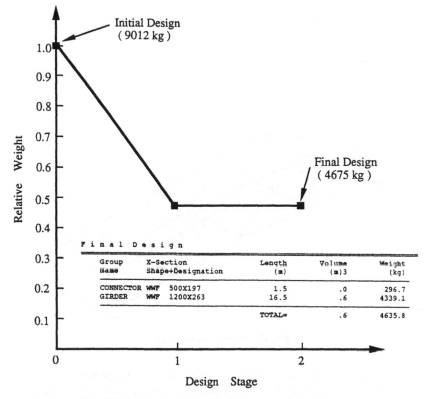

Fig. 12.7 Example 12.3: synthesis history and final design.

optimization process converged after two iterations to a least-weight final design satisfying all of the relevant strength–stability provisions of the CAN3-S16.1-M89 steel code (Canadian Standards Association, 1989) as well as the imposed limitation on the midspan deflection of the girder under service loads (in fact, this latter stiffness condition was the controlling design constraint). The design optimization required approximately 1 minute execution time on a 386-based microcomputer.

EXAMPLE 12.4

The Warren pony truss structure shown in Fig. 12.8 is a pedestrian footbridge with a 5% camber that is to be designed in accordance with the provisions of the CAN3-S16.1-M89 steel design code (Canadian Standards Association, 1989), and the highway bridge code for Ontario, Canada (Ontario Ministry of Transport and Communications, 1983). The bridge has a span of 9.14 m (30 ft) and a width of 1.83 m (6 ft). The member joints are all moment-resisting rigid connections.

To satisfy practical construction requirements, the 47 members of the bridge

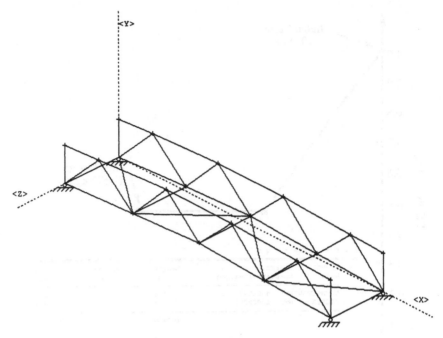

Fig. 12.8 Pedestrian footbridge.

structure are collected into six fabrication groups: top chord members (T_CHORD); bottom chord members (B_CHORD); diagonal bracing members in bridge side walls (DIAG); diagonal bracing members in bridge floor (BRACE); parallel floor beams (BEAM); vertical end-members in side walls (VERT).

The in-plane and out-of-plane effective length factors for the T_CHORD and VERT members are taken to be $K_x = 2.0$ and $K_y = 1.0$, respectively, while those for the B_CHORD and BEAM members are taken to be $K_x = K_y = 1.0$. The effective length factors for the DIAG members are taken to be $K_x = 1.2$ and $K_y = 1.0$, while those for the BRACE members are taken to be $K_x = K_y = 0.2$. The maximum allowable slenderness ratios for each members are specified to be $KL/r = 200$ in compression and $KL/r = 300$ in tension.

The members of the bridge are to have rectangular hollow section (RHS) shapes, with the exception that the floor bracing members are to have square hollow section (SHS) shapes. Member sections are to be selected from the Canadian database of standard RHS and SHS steel sections: Young's modulus, 200 000 MPa; shear modulus, 77 000 MPa; yield stress, 350 MPa; ultimate stress, 450 MPa. Member sections are to be oriented such as to undergo strong-axis bending, with the exception that weak-axis bending is to be experienced by the sections of the beam members in the bridge floor.

The bridge is subject to gravity dead and live loading in the Y axis direction (Fig. 12.8), wind loading in the Y axis and Z axis directions, rail loading on the top chord members in the Z axis direction and temperature loading on all members. The bridge is to be designed for eight different factored combinations of these loads.

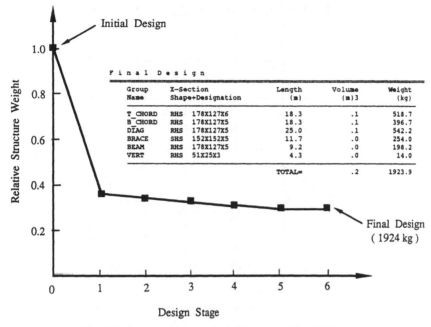

Fig. 12.9 Example 12.4: synthesis history and final design.

The design of the pedestrian footbridge is conducted using the previously described optimization procedure. The iterative synthesis history and final design for the structure are shown in Fig. 12.9. The optimization process converged after six iterations to a least-weight final design satisfying all relevant strength–stability requirements of the governing design code (displacement constraints were not imposed for the design), and required approximately 8 minutes execution time on a 386-based microcomputer. In fact, an almost optimal design weighing only 2% heavier than the final design is found after completion of the first design stage in approximately 1.5 minutes execution time.

EXAMPLE 12.5

The framework shown in Fig. 12.10 is part of a thermal power station boiler house structure that measures 138 ft × 167 ft in plan and requires approximately 2000 tons of structural steel. The boiler itself measures 75 ft × 79 ft × 187 ft high, weighs approximately 6000 tons and is suspended from large plate girders supported by twelve major columns. Other columns support the various components of the boiler house, such as elevators and stairs, access platforms, piping, air heaters and flue gas ductwork. The boiler house is required to withstand high seismic forces that are transmitted to the structure through bumpers at specific locations, and then to the foundations via a complex network of horizontal diaphragms and vertical bracing systems. The structure is designed

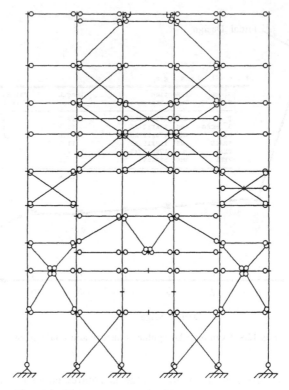

Fig. 12.10 Thermal power station boiler house structure.

Table 12.3 Example 12.5: final design

Group name	X-section shape + designation	Length (ft)	Volume (ft³)	Weight (lb)
1	W 18 × 65	51.7	6.9	3367.9
2	W 18 × 50	42.5	4.3	2128.1
3	W 18 × 50	46.4	4.7	2322.5
4	W 18 × 50	48.2	4.9	2415.7
5	W 14 × 90	103.5	19.0	9345.5
6	W 18 × 97	95.0	18.8	9225.2
7	W 18 × 76	82.7	12.8	6285.1
8	W 18 × 76	96.4	14.9	7329.3
9	W 18 × 76	51.7	8.0	3932.2
10	W 14 × 159	77.8	25.2	12374.2
11	W 18 × 50	36.3	3.7	1820.5
12	W 18 × 50	48.2	4.9	2415.7
13	W 14 × 61	51.7	6.4	3156.3
14	W 14 × 61	47.7	5.9	2910.5
15	W 14 × 61	51.7	6.4	3156.3
16	W 14 × 34	47.7	3.3	1626.0

Table 12.3 (*Continued*)

Group name	X-section shape + designation	Length (ft)	Volume (ft³)	Weight (lb)
17	W 10 × 39	107.8	8.6	4226.6
18	W 12 × 30	202.0	12.3	6051.2
19	W 12 × 65	25.3	3.3	1643.6
20	W 16 × 31	75.8	4.8	2354.4
21	W 14 × 30	25.3	1.6	761.6
22	W 10 × 49	80.3	8.0	3940.3
23	W 14 × 38	69.8	5.4	2666.0
24	W 10 × 39	82.3	6.6	3225.3
25	W 10 × 39	74.8	6.0	2933.1
26	W 12 × 30	63.5	3.9	1902.3
27	W 10 × 39	77.0	6.1	3017.2
28	W 10 × 45	21.8	2.0	990.0
29	W 8 × 40	21.9	1.8	871.9
30	W 10 × 45	106.2	9.8	4812.9
31	W 10 × 39	55.0	4.4	2157.4
32	W 12 × 30	27.5	1.7	824.7
33	W 14 × 43	34.4	3.0	1478.2
34	W 12 × 30	17.6	1.1	527.6
35	W 12 × 30	78.2	4.8	2343.8
36	W 12 × 30	48.3	3.0	1448.3
37	W 12 × 30	36.3	2.2	1088.6
38	W 10 × 39	58.3	4.7	2285.3
39	W 12 × 30	45.9	2.8	1373.8
40	W 14 × 34	54.0	3.8	1840.3
41	W 10 × 39	27.0	2.2	1058.2
42	W 10 × 39	54.0	4.3	2116.4
43	W 10 × 45	135.0	12.5	6119.1
44	W 14 × 43	27.0	2.4	1159.4
45	W 10 × 39	27.0	2.2	1058.2
46	W 12 × 30	22.8	1.4	681.5
47	W 12 × 30	22.8	1.4	681.5
48	W 12 × 30	22.8	1.4	681.5
49	W 12 × 30	45.5	2.8	1363.0
50	W 24 × 55	45.5	5.1	2512.0
51	W 24 × 55	68.3	7.7	3768.1
52	W 12 × 30	22.8	1.4	681.5
53	W 12 × 30	22.8	1.4	681.5
54	W 12 × 30	18.9	1.2	565.4
55	W 12 × 30	18.9	1.2	565.4
56	W 8 × 31	18.9	1.2	587.3
57	W 14 × 193	44.0	17.4	8517.3
58	W 36 × 150	44.0	13.5	6627.9
59	W 24 × 162	44.0	14.6	7152.7
60	W 14 × 34	44.0	3.1	1499.5
61	W 14 × 34	44.0	3.1	1499.5
62	W 18 × 50	27.0	2.8	1352.6
Total			365.8	179505.0

as a series of planar frameworks, such as in Fig. 12.10, connected together by horizontal diaphragms.

The framework in Fig. 12.10 is to be designed in accordance with the strength–stability requirements of the AISC-89 steel design code (American Institute of Steel Construction, 1989), and user-specified horizontal displacement limitations are imposed at various heights above the foundation level. The various member joints are either moment-resisting welded connections or pin-jointed bolted connections (denoted by O in Fig. 12.10). The 181 members of the framework are collected into 62 fabrication groups, each one of which is specified to have a W-flange (W) section shape to be selected from the American database of standard W-shape steel sections. Maximum depth limitations are imposed on each of the 62 sections to satisfy practical construction considerations. The maximum allowable slenderness ratios for the members are specified to be 200 in compression and 300 in tension: Young's modulus, $29\,000\,\text{klbf in}^{-2}$; shear modulus, $11\,200\,\text{klbf in}^{-2}$; yield stress, $50\,\text{klbf in}^{-2}$; and ultimate stress, $65\,\text{klbf in}^{-2}$. The framework is subject to gravity dead and live loading, lateral wind loading and horizontal seismic loading (modeled as equivalent static

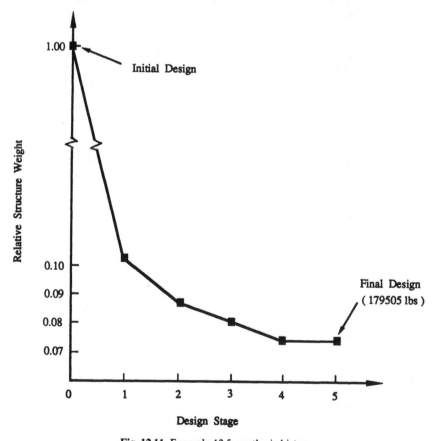

Fig. 12.11 Example 12.5: synthesis history.

loading). The design is conducted accounting for ten different combinations of these loads.

The design optimization of the framework was carried out in two phases. Initially, the previously described optimization procedure was applied to establish a preliminary design that provided an estimate of the required steel tonnage and the foundation loads. Here, the effective length factors for all members were taken to be $K_x = K_y = 1.0$, and all beams were assumed to be fully laterally braced. As well, all loads were applied as nodal loading alone (minimum section sizes were imposed for the pin-jointed beam members to compensate for the absence of member loads). The framework was assumed to be sidesway prevented and second-order (P–delta) effects were not accounted for.

From the results of the preliminary design phase, member effective length factors and lateral bracing parameters were adjusted to reflect actual conditions. Actual nodal and member loads were then applied on the framework, and the optimization was conducted again to determine the final design listed in Table 12.3. The iterative synthesis history for the design is shown in Fig. 12.11. The optimization process converged after five iterations to a least-weight structure satisfying the imposed displacement constraints and all relevant strength–stability requirements of the AISC-89 steel design code (American Institute of Steel Construction, 1989) and required approximately 2.5 hours execution time on a 386-based microcomputer. A subsequent second-order analysis of the final design in Table 12.3 verified that all strength–stability and stiffness requirements are also satisfied when P–delta effects are accounted for.

12.6 CONCLUDING REMARKS

The practical design of structural steel frameworks is very effectively carried out using optimization techniques, such as those described in this chapter. Member-by-member and formal optimization employed in tandem permits the simultaneous consideration of both strength and stiffness aspects of the design, thereby eliminating much of the guesswork from the synthesis process. Ths use of optimization techniques results in reduced designer time and more economical designs compared to that when conventional trial-and-error techniques are employed. In fact, an optimal structural steel design typically represents a 3–10% saving in steel weight compared with conventional design, which for a large-scale structure can result in a significant monetary saving.

Much research is being carried out in many areas to improve further the applicability of computer-based optimization techniques to practical structural design. The body of work is too broad and varied to cite here, but listed in the following are some representative studies currently underway by the writer and his research associates at the University of Waterloo. The optimization strategy presented in this chapter is being extended to account for the cost and design of optimal member connections (Xu, 1993; Xu and Grierson, 1992). Optimality criteria techniques are being developed for the lateral-resistant design of tall three-dimensional steel building frameworks (Chan, 1993; Chan and Grierson, 1993), and for the strength and stiffness design of reinforced concrete building structures (Moharrami, 1993; Moharrami and Grierson, 1993). Genetic algorithms are being developed to establish not only optimal section sizes but also the optimal topological layout of the members for structural steel frameworks (Pak, 1993; Grierson and Pak, 1992). Finally, new programming paradigms are being employed to develop enhanced computer-based capabilities for structural design (Biedermann, 1993; Biedermann and Grierson, 1993).

ACKNOWLEDGEMENTS

The writer wishes to thank Lei Xu, Chun-Man Chan and Nancy Simpson of the University of Waterloo for their help in preparing the text manuscript for this chapter. Gordon Cameron of Waterloo Engineering Software supplied the details for design example 12.2, as did freelance structural engineer Edward Greig for design examples 12.3 and 12.4. The writer is indebted to them, and to the anonymous-by-choice individual and firm that supplied the details for design example 12.5.

APPENDIX 12A SECTION PROPERTY RELATIONSHIPS

Consider a structural steel member subject to a three-dimensional force field involving axial force, biaxial shear forces, torsional moment and biaxial bending moments, as indicated in Fig. 12A.1. The geometrical properties of the member cross section are A_x, A_y, A_z the axial and shear areas, and I_x, I_y, I_z, the torsional and bending moments of inertia.

For commercial standard steel sections, each of the properties A_y, A_z, I_x, I_y and I_z may be expressed in terms of the axial area A_x as follows (Chan, 1993):

$$1/A_Y = C_{AY}(1/A_X) + C'_{AY}$$
$$1/A_Z = C_{AZ}(1/A_X) + C'_{AZ}$$
$$1/I_X = C_{IX}(1/A_X) + C'_{IX} \qquad \text{(12A.1)}$$
$$1/I_Y = C_{IY}(1/A_X) + C'_{IY}$$
$$1/I_Z = C_{IZ}(1/A_X) + C'_{IZ}$$

where the coefficients C and C' are determined by linear regression analysis and have different values depending on the type and size of the section.

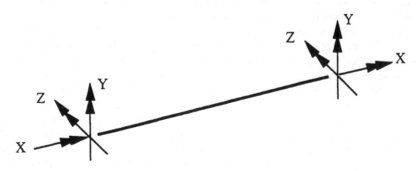

Fig. 12A.1 Three-dimensional member force field.

Table 12A.1 Relationships between cross section area A_X and shear areas A_Y and A_Z for AISC-LRFD W14 and W24 sections

Sections (Number of sections)	$\dfrac{1}{A_Y} = \dfrac{C_{AY}}{A_X} + C'_{AY}$		Maximum error (%)	$\dfrac{1}{A_Z} = \dfrac{C_{AZ}}{A_X} + C'_{AZ}$		Maximum error (%)
	C_{AY}	C'_{AY}		C_{AZ}	C'_{AZ}	
W14 × 22–26 (2)	1.435 326	0.095 276	0.00	2.561 815	−0.096 225	0.00
W14 × 30–38 (3)	1.634 329	0.085 120	0.98	2.084 494	−0.044 015	0.92
W14 × 43–53 (3)	3.004 888	0.001 081	0.26	1.572 869	−0.006 856	0.02
W14 × 61–82 (4)	3.814 517	−0.019 917	0.98	1.383 969	0.000 091	0.39
W14 × 90–132 (5)	4.725 603	−0.015 799	0.86	1.293 418	−0.000 474	0.34
W14 × 145–176 (3)	4.944 198	−0.016 278	0.01	1.262 608	−0.000 032	0.12
W14 × 193–257 (4)	4.641 405	−0.009 561	0.60	1.276 141	−0.000 330	0.16
W14 × 283–426 (6)	4.419 709	−0.006 828	0.24	1.279 435	−0.000 335	0.33
W14 × 455–730 (6)	4.108 586	−0.004 616	0.24	1.263 193	−0.000 185	0.14
W24 × 55–62 (2)	1.393 000	0.021 422	0.00	3.090 550	−0.049 433	0.00
W24 × 68–84 (3)	1.428 414	0.030 837	0.43	2.517 252	−0.030 232	0.43
W24 × 94–103 (2)	1.857 359	0.012 822	0.00	2.049 032	−0.010 935	0.00
W24 × 104–131 (3)	2.328 042	0.007 155	0.15	1.755 494	−0.005 079	0.01
W24 × 146–192 (4)	2.486 482	0.004 511	0.42	1.657 739	−0.003 055	0.21
W24 × 207–306 (5)	2.897 758	−0.003 009	0.24	1.518 109	−0.000 560	0.15
W24 × 335–492 (5)	2.810 400	−0.002 391	0.58	1.512 870	−0.000 536	0.49

Table 12.A2 Relationships between cross section area A_X and moments of inertia I_Z, I_Y and I_X for AISC-LRFD W12 and W24 sections

Sections (number of sections)	$\dfrac{1}{I_Z} = \dfrac{C_{IZ}}{A_X} + C'_{IX}$		Maximum error (%)	$\dfrac{1}{I_Y} = \dfrac{C_{IY}}{A_X} + C_{IY}$		Maximum error (%)	$\dfrac{1}{I_X} = \dfrac{C_{IX}}{A_X} + C'_{IX}$		Maximum error (%)
	C_{IZ}	C'_{IZ}		C_{IY}	C'_{IY}		C_{IX}	C'_{IX}	
W14 × 22–26 (2)	0.039 240	−0.001 021	0.00	1.273 642	−0.053 390	0.00	82.520 01	−7.953 04	0.00
W14 × 30–38 (3)	0.035 327	−0.000 573	0.63	0.570 689	−0.013 801	0.81	56.961 14	−3.878 09	3.50
W14 × 43–53 (3)	0.031 973	−0.000 203	0.08	0.314 017	−0.002 804	0.03	28.398 04	−1.312 02	1.40
W14 × 61–82 (4)	0.029 882	−0.000 110	0.31	0.181 081	−0.000 790	0.50	17.591 99	−0.540 04	3.65
W14 × 90–132 (5)	0.029 025	−0.000 097	0.42	0.078 045	−0.000 193	0.38	13.145 35	−0.261 87	5.69
W14 × 145–176 (3)	0.028 522	−0.000 084	0.11	0.068 920	−0.000 138	0.07	6.705 84	−0.092 28	1.58
W14 × 193–257 (4)	0.028 078	−0.000 077	0.10	0.067 919	−0.000 123	0.14	3.616 11	−0.035 52	2.68
W14 × 283–426 (6)	0.027 137	−0.000 065	0.31	0.067 382	−0.000 115	0.13	1.566 43	−0.009 66	5.12
W14 × 455–730 (6)	0.024 444	−0.000 044	0.62	0.063 449	−0.000 084	0.30	0.611 77	−0.002 20	7.09
W24 × 55–62 (2)	0.014 090	−0.000 129	0.00	0.792 936	−0.014 582	0.00	38.721 68	−1.542 77	0.00
W24 × 68–84 (3)	0.013 433	−0.000 123	0.18	0.389 350	−0.005 212	0.40	28.162 79	−0.875 74	2.20
W24 × 94–103 (2)	0.011 956	−0.000 061	0.00	0.248 872	0.000 190	0.00	15.968 55	−0.386 37	0.00
W24 × 104–131 (3)	0.011 006	−0.000 037	0.07	0.137 181	−0.000 621	0.02	15.678 37	−0.303 99	1.96
W24 × 146–192 (4)	0.010 668	−0.000 030	0.16	0.120 064	−0.000 247	0.58	7.577 72	−0.103 49	3.69
W24 × 207–306 (5)	0.009 939	−0.000 017	0.18	0.120 004	−0.000 249	0.15	3.129 03	−0.026 87	5.37
W24 × 335–492 (5)	0.009 775	−0.000 016	0.41	0.114 721	−0.000 201	0.67	1.268 54	−0.006 74	5.90

Equations (12A.1) are illustrated in Tables 12A.1 and 12A.2 for W14 and W24 standard steel sections from the AISC-LRFD steel design code (American Institute of Steel Construction, 1986). Each set of coefficients C and C' applies for a group of from two to six sections for which the section properties A_Y, A_Z, I_X, I_Y and I_Z predicted by equations (12A.1), for given area A_X, approximate the actual section properties within small percentage error. Perhaps an exception is the prediction of the torsional moment of inertia I_X for larger size sections, for which percentage errors up to 7% are noted to occur in Table 12A.2. However, this is not a significant shortcoming as torsional capacity requirements rarely govern the design of steel members of building frameworks. Similar results to those in Tables 12A.1 and 12A.2 may be derived for a broad range of commercially available standard steel sections.

Equations (12A.1) permit the stiffness matrix for a structural steel member to be instantaneously approximated in terms of its cross section area A_X alone, thereby permitting the analytical derivation of stress and displacement gradients for the assembled framework of members (e.g. Appendix 12B).

APPENDIX 12B DISPLACEMENT AND STRESS GRADIENTS

Consider the pin-jointed truss loaded as shown in Fig. 12B.1. The truss has six members of cross section area $a_i (i = 1, 2, \ldots, 6)$ and five independent nodal displacements $u_j (j = 1, 2, \ldots, 5)$. It is required to find the gradient (sensitivity) of the lateral displacement u_3 to changes in the reciprocal cross section areas $x_i = 1/a_i$ from (equation (12.27))

$$\frac{\partial u_3}{\partial x_i} = \frac{1}{x_i}(u_3^T K_i u) \quad (i = 1, 2, \ldots, 6) \tag{12B.1}$$

Similarly, it is required to find the gradient of the axial stress σ_5 in member 5

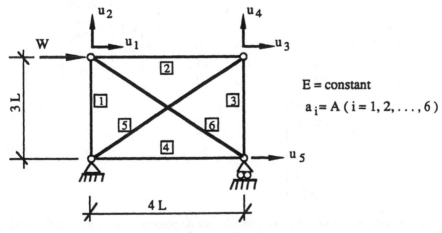

Fig. 12B.1 Pin-jointed truss.

to changes in the reciprocal areas x_i from (equation (12.30))

$$\frac{\partial \sigma_5}{\partial x_i} = \frac{1}{x_i}(\mathbf{u}_5^T \mathbf{K}_i \mathbf{u}) \quad (i = 1, 2, \ldots, 6) \tag{12B.2}$$

To evaluate equations (12B.1) and (12B.2), the nodal displacements are found from

$$\mathbf{u} = \mathbf{K}^{-1}\mathbf{P} \tag{12B.3}$$

which becomes for the truss in Fig. 12B.1,

$$
\begin{bmatrix} u_1 \\ u_2 \\ u_3 \\ u_4 \\ u_5 \end{bmatrix} = \frac{L}{EA}
\begin{bmatrix}
6.750 & & & & \\
1.125 & 2.813 & & \text{SYM} & \\
4.750 & 0.792 & 6.157 & & \\
-1.125 & -0.187 & -1.458 & 2.813 & \\
2.000 & -0.333 & 1.407 & -0.333 & 3.407
\end{bmatrix}
\begin{bmatrix} W \\ 0 \\ 0 \\ 0 \\ 0 \end{bmatrix}
$$

$$
= \frac{WL}{EA}
\begin{bmatrix}
6.750 \\ 1.125 \\ 4.750 \\ -1.125 \\ 2.000
\end{bmatrix} \tag{12B.4}
$$

where E is Young's material modulus and, as indicated in Figure 12B.1, L and A are member length and section area parameters, respectively.

Having displacements $\mathbf{u}$ from equation (12B.4), the member axial stresses are found from

$$\sigma = \mathbf{T}\mathbf{u} \tag{12B.5}$$

where the transformation matrix $\mathbf{T}$ accounts for member stress–strain relations and the structure topology. Equation (12B.5) becomes for the truss in Fig. 12B.1

$$
\begin{bmatrix} \sigma_1 \\ \sigma_2 \\ \sigma_3 \\ \sigma_4 \\ \sigma_5 \\ \sigma_6 \end{bmatrix} = \frac{E}{L}
\begin{bmatrix}
0 & 1/3 & 0 & 0 & 0 \\
-1/4 & 0 & 1/4 & 0 & 0 \\
0 & 0 & 0 & 1/3 & 0 \\
0 & 0 & 0 & 0 & 1/4 \\
0 & 0 & 0.8/5 & 0.6/5 & 0 \\
-0.8/5 & 0.6/5 & 0 & 0 & 0.8/5
\end{bmatrix}
\frac{WL}{EA}
\begin{bmatrix}
6.750 \\ 1.125 \\ 4.750 \\ -1.125 \\ 2.000
\end{bmatrix}
$$

$$
= \frac{W}{A}
\begin{bmatrix}
0.375 \\ -0.500 \\ -0.375 \\ 0.500 \\ 0.625 \\ -0.625
\end{bmatrix} \tag{12B.6}
$$

The vector of virtual loads associated with displacement u_3 is $\mathbf{b}_3^T = [0\ 0\ 1\ 0\ 0]$, as shown in Fig. 12B.2, and the corresponding vector of virtual nodal displace-

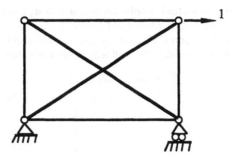

Fig. 12B.2 Virtual loading b_3.

ments is found as

$$\mathbf{u}_3 = \mathbf{K}^{-1}\mathbf{b}_3 = \mathbf{K}^{-1}\begin{bmatrix} 0 \\ 0 \\ 1 \\ 0 \\ 0 \end{bmatrix} = \frac{L}{EA}\begin{bmatrix} 4.750 \\ 0.792 \\ 6.157 \\ -1.458 \\ 1.407 \end{bmatrix} \qquad (12B.7)$$

where the inverse stiffness matrix $\mathbf{K}^{-1}$ is as given in equation (12B.4).

The fifth row of the transformation matrix in equation (12B.6) identifies the vector of virtual loads associated with member stress σ_5 to be $\mathbf{t}_5^T = [0\ 0\ 0.8\ 0.6\ 0]E/5L$, as shown in Fig. 12B.3, and the corresponding vector of virtual nodal displacements is found as

$$\mathbf{u}_5 = \mathbf{K}^{-1}\mathbf{t}_5 = \mathbf{K}^{-1}\frac{E}{5L}\begin{bmatrix} 0 \\ 0 \\ 0.8 \\ 0.6 \\ 0 \end{bmatrix} = \frac{1}{A}\begin{bmatrix} 0.625 \\ 0.104 \\ 0.810 \\ 0.104 \\ 0.185 \end{bmatrix} \qquad (12B.8)$$

where, again, matrix $\mathbf{K}^{-1}$ is as given in equation (12B.4).

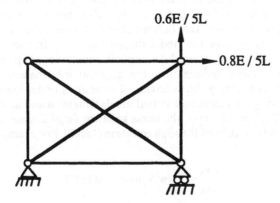

Fig. 12B.3 Virtual loading t_5.

Having the actual and virtual displacements **u** and **u**$_3$ from equations (12B.4) and (12B.7), respectively, the displacement gradient $\partial u_3/\partial x_i$ is readily calculated through equation (12B.1) upon identifying the (global-axis) stiffness matrix **K**$_i$ for each member i. For example, the stiffness matrix for member $i = 6$ is

$$\mathbf{K}_6 = \frac{EA}{5L} \begin{bmatrix} 0.64 & -0.48 & 0 & 0 & -0.64 \\ -0.48 & 0.36 & 0 & 0 & 0.48 \\ 0 & 0 & 0 & 0 & 0 \\ 0 & 0 & 0 & 0 & 0 \\ -0.64 & 0.48 & 0 & 0 & 0.64 \end{bmatrix} \tag{12B.9}$$

and, from Fig. 12B.1, the reciprocal area for member 6 is $x_6 = 1/a_6 = 1/A$. Therefore, from equations (12B.1), (12B.4), (12B.7) and (12B.9), the gradient of the displacement u_3 with respect to changes in the reciprocal-area x_6 for member 6 is

$$\frac{\partial u_3}{\partial x_6} = \frac{1}{x_6}(\mathbf{u}_3^T \mathbf{K}_6 \mathbf{u}) = +1.375\frac{WL}{A^2}$$

where the positive sign indicates that the nodal displacement u_3 varies directly with x_6 (which means that, as expected, the displacement u_3 varies inversely with changes in the actual section area $a_6 = 1/x_6$ for, member 6). The displacement gradient $\partial u_3/\partial x_i$ is similarly found for the remaining five members $i = 1, 2, \ldots, 5$, such that the vector of gradients for all six members is

$$\left[\frac{\partial u_3}{\partial \mathbf{x}}\right]^T = [0.297 \ -0.704 \ 0.547 \ 0.704 \ 2.532 \ 1.375]WL/A^2$$

The negative sign for the gradient corresponding to member 2 indicates that the displacement u_3 varies inversely with the reciprocal area x_2 and, therefore, directly with the actual area a_2 for member 2 (i.e. u_3 increases as a_2 increases, and vice versa). This counterintuitive result is explained by the fact that since member 2 is in axial compression, for the loading shown in Fig. 12B.1, any increase in a_2 will tend to decrease the member axial shortening and, thereby, increase the nodal displacement u_3. Another interesting result can be derived from the fact that members 2 and 4 correspond to gradients that have equal magnitude but opposite sign. Namely, for the loading shown in Fig. 12B.1, any symmetrical change in the section areas of the top and bottom chord members 2 and 4 will have no effect on the magnitude of the nodal displacement u_3. In fact, in addition to providing the means to formulate explicit constraint approximations for computer-based design optimization, the gradient information itself provides valuable insight concerning the behavior of structures under loads.

Similarly, having the actual and virtual displacements **u** and **u**$_5$ from equations (12B.4) and (12B.8), respectively, the stress gradient $\partial \sigma_5/\partial x_i$ associated with each member i is readily calculated through equation (12B.2). For example, for member $i = 6$,

$$\frac{\partial \sigma_5}{\partial x_6} = \frac{1}{x_6}(\mathbf{u}_5^T \mathbf{K}_6 \mathbf{u}) = +0.181W$$

where the positive sign indicates that the axial stress σ_5 in member 5 varies directly with x_6, or that σ_5 varies inversely with changes in the actual section area $a_6 = 1/x_6$ for member 6. The stress gradient $\partial \sigma_5/\partial x_i$ is similarly found for the remaining five members $i = 1, 2, \ldots, 5$, such that the vector of gradients for all six members is

$$\left[\frac{\partial \sigma_5}{\partial \mathbf{x}}\right]^{\mathrm{T}} = [0.039 \ -0.093 \ -0.039 \ 0.093 \ 0.444 \ 0.181]W$$

The negative signs for the gradients corresponding to members 2 and 3 indicate that the stress σ_5 in member 5 varies directly with changes in the section areas a_2 and a_3 for members 2 and 3. The fact that member pairs 1,3 and 2,4 each correspond to gradients that have equal magnitude but opposite sign means that any symmetrical change in the section areas of the vertical members 1,3 and/or the chord members 2,4 for the truss will have no effect on the magnitude of the axial stress σ_5 in member 5.

APPENDIX 12C GOC METHOD FOR DISCRETE OPTIMIZATION

Consider the pin-jointed truss loaded as shown in Fig. 12C.1 (Schmit and Fleury, 1979; Haftka and Gürdal, 1992). The truss has two members of cross section area a_1 and a_2, and two independent nodal displacements u_1 and u_2. It is required to find the least-weight truss by selecting each of the section areas a_1 and a_2 from the discrete set of areas

$$A = \left\{1, \frac{3}{2}, 2\right\} \tag{12C.1}$$

while satisfying the displacement constraints

$$u_1 \leqslant 0.75PL/E; \quad u_2 \leqslant 0.25PL/E \tag{12C.2}$$

As the truss is statically determinate, the displacements are readily found to be

$$u_1 = \frac{PL}{2E}\left(\frac{1}{a_1} + \frac{1}{a_2}\right); \quad u_2 = \frac{PL}{2E}\left(\frac{1}{a_1} - \frac{1}{a_2}\right) \tag{12C.3}$$

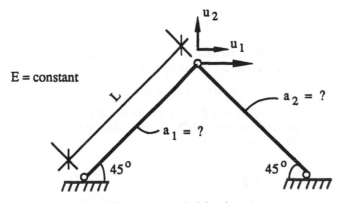

Fig. 12C.1 Two-bar pin-jointed truss.

The weight of the truss is given by

$$W = \rho L(a_1 + a_2) \tag{12C.4}$$

where L is the length of each members and ρ is the material density.

From equations (12C.1)–(12C.4), the discrete optimization problem expressed in terms of reciprocal sizing variables $x_1 = 1/a_1$ and $x_2 = 1/a_2$ is

$$\text{minimize} \quad \frac{1}{x_1} + \frac{1}{x_2} \tag{12C.5}$$

$$\text{subject to} \quad x_1 + x_2 \leqslant 1.5 \tag{12C.6}$$

$$x_1 - x_2 \leqslant 0.5 \tag{12C.7}$$

$$x_1, x_2 \in \left\{ \frac{1}{2}, \frac{2}{3}, 1 \right\} \tag{12C.8}$$

The generalized optimality criteria (GOC) method is applied to solve the problem posed by equations (12C.5)–(12C.8). To that end, the Lagrangian function is formed as

$$\mathscr{L}(\mathbf{x}, \lambda) = \frac{1}{x_1} + \frac{1}{x_2} + \lambda_1(x_1 + x_2 - 1.5) + \lambda_2(x_1 - x_2 - 0.5) \tag{12C.9}$$

where the Lagrange variables are sign restricted as

$$\lambda_1 \geqslant 0; \quad \lambda_2 \geqslant 0 \tag{12C.10}$$

A saddle-point $(\mathbf{x}^*, \lambda^*)$ of equation (12C.9) is given by a solution of the min–max problem

$$(\mathbf{x}^*, \lambda^*) = \max_\lambda [\min_\mathbf{x} \mathscr{L}(\mathbf{x}, \lambda)] = \max_\lambda [\mathscr{L}_m(\lambda)] \tag{12C.11}$$

where the dual function is

$$\mathscr{L}_m(\lambda) = \sum_{i=1}^{2} \min_{x_i} L_i(x_i, \lambda) \tag{12C.12}$$

in which the individual functions of the variables x_i are, from equation (12C.9),

$$L_1(x_1, \lambda) = \frac{1}{x_1} + (\lambda_1 + \lambda_2)x_1 - 1.5\lambda_1 - 0.5\lambda_2 \tag{12C.13}$$

$$L_2(x_2, \lambda) = \frac{1}{x_2} + (\lambda_1 - \lambda_2)x_2 \tag{12C.14}$$

When the dual function is maximized over the λ variables in equation (12C.11), its continuity is maintained by requiring that

$$L_i(x_i^k, \lambda) = L_i(x_i^{k+1}, \lambda) \quad (i = 1, 2) \tag{12C.15}$$

where x_i^k and x_i^{k+1} are successive discrete values of each variable x_i from equation (12C.8). Therefore, from equations (12C.8), (12C.13) and (12C.15), the boundaries for changes in the values of x_1 are

$$\frac{1}{1/2} + \frac{1}{2}(\lambda_1 + \lambda_2) = \frac{1}{2/3} + \frac{2}{3}(\lambda_1 + \lambda_2)$$

$$\frac{1}{2/3} + \frac{2}{3}(\lambda_1 + \lambda_2) = \frac{1}{1} + (\lambda_1 + \lambda_2)$$

which reduce to

$$\lambda_1 + \lambda_2 = 3; \quad \lambda_1 - \lambda_2 = 1.5 \tag{12C.16}$$

Similarly, from equations (12C.8), (12C.14) and (12C.15), the boundaries for changes in x_2 are

$$\lambda_1 - \lambda_2 = 3; \quad \lambda_1 - \lambda_2 = 1.5 \tag{12C.17}$$

As equation (12C.10) requires variables λ_1 and λ_2 to be nonnegative, equations (12C.16) and (12C.17) divide the positive quadrant of the (λ_1, λ_2) plane into the six regions shown in Fig. 12C.2, where given in parentheses for each region are the fixed values of the variables x_1, x_2 prevailing for the dual function $\mathscr{L}_m(\lambda)$. For example, suppose that the search to maximize $\mathscr{L}_m(\lambda)$ in equation (12C.11) commences at the origin $\lambda = (0, 0)$. From equation (12C.9), the Lagrangian function $\mathscr{L}(\mathbf{x}, \lambda) = 1/x_1 + 1/x_2$, which is minimized for the discrete values $x_1 = x_2 = 1$, as indicated in Fig. 12C.2, such that the dual function is, from equations (12C.12)–(12C.14),

$$\mathscr{L}_m(\lambda) = 2 + 0.5\lambda_1 - 0.5\lambda_2 \tag{12C.18}$$

Since λ_2 cannot be reduced below zero, the right-hand side of equation (12C.18) is maximized by increasing λ_1 to the boundary of the region at $\lambda = (1.5, 0)$, at which point $\mathscr{L}_m(\lambda) = 2.75$. From Fig. 12C.2, it is now possible to move into one of the two regions for which $\mathbf{x} = (2/3, 1)$ or $\mathbf{x} = (2/3, 2/3)$ in a further attempt to maximize the dual function. In the former region, from equations (12C.12)–(12C.14), the dual function is

$$\mathscr{L}_m(\lambda) = 2.5 + \lambda_1/6 - 5\lambda_2/6 \tag{12C.19}$$

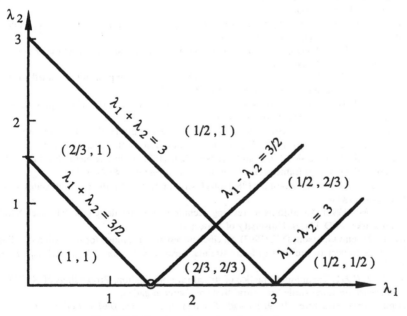

Fig. 12C.2 Constant (x_1, x_2) regions in λ space.

while in the latter region it is

$$\mathscr{L}_m(X) = 3 - \lambda_1/6 - \lambda_2/2 \qquad (12C.20)$$

It is not possible to maximize further either of the dual functions (12C.19) and (12C.20) for nonnegative λ values different from $\lambda^* = (1.5, 0)$, which means that the solution of the min–max problem (12C.11) occurs for any one of three pairs of x_i values (1,1), (2/3, 1) or (2/3, 2/3). However, only $x^* = (2/3, 2/3)$ satisfies the constraints (12C.6) and (12C.7), for which the optimal cross section areas are $a_1 = a_2 = 3/2$ and the value of the objective function (12C.5) is 3.

A final comment is in order concerning the fact that the minimum value of the objective function (12C.5) is 3 while the maximum value of the corresponding dual function (12C.12) is 2.75. This difference serves as a reminder that the discrete optimization problem is not convex and, therefore, that there is no guarantee that the solution found is optimal and/or unique (as it happens, the solution point $x^* = (2/3, 2/3)$ found for the present example is one of two optima, the other being $x^* = (1/2, 1)$). The consequence of this for the design of structural steel frameworks is that the set of discrete standard sections found by the GOC method may correspond to but one of several near-optimal designs for the structure. Moreover, while all such designs are likely to be quite similar to each other from the viewpoint of both structure weight and standard section selection, it is theoretically possible that there may exist designs of similar weight but with radically different standard sections for the members.

REFERENCES

American Institute of Steel Construction (1986) *Manual of Steel Construction: Load and Resistance Factor Design.*

American Institute of Steel Construction (1989) Specification for structural steel buildings: allowable stress design and plastic design.

Biedermann, J.D. (1992) Object-oriented computer-automated design of structures. PhD Thesis, Civil Engineering, University of Waterloo.

Biedermann, J.D. and Grierson, D.E. (1993) An object-oriented approach to detailed structural design. *Microcomputers in Civil Engineering*, **8**, 225–31.

Cameron, G.E. (1989) A knowledge-based expert system for structural steel design. PhD Thesis; Civil Engineering, University of Waterloo.

Cameron, G.E., Xu, L. and Grierson, D.E. (1991) Discrete optimal design of 3D frameworks. ASCE Structural Congress/10th Electronic Computation Conference, Indianapolis, IN.

Cameron, G.E., Chan, C.-M., Xu, L. and Grierson, D.E. (1992) Alternate methods for the optimal design of slender steel frameworks. *Computers and Structures*, **44**(4), 735–41.

Canadian Standards Association (1989) Steel structures for buildings (limit states design). *CAN3-S16.1M89.*

Chan, C.-M. (1992) An automatic resizing technique for the design of tall steel building frameworks. PhD Thesis, University of Waterloo.

Chan, C.-M. and Grierson, D.E. (1993) An efficient resizing technique for the design of tall steel buildings subject to multiple drift constraints. *Structural Design of Tall Buildings Journal*, **2**, 17–32.

Cohn, M.Z. (1991) Theory and practice of structural optimization. Proceedings of NATO–ASI Conference on Large-scale Structural Systems, Berchtesgaden.

Computers and Structures (1989) *ETABS, Building Analysis and Design*, Berkeley, CA.

Fleury, C. (1979) Structural weight optimization by dual methods of convex programming. *International Journal of Numerical Methods in Engineering*, **14**, 1761–83.

Grierson, D.E. and Cameron, G.E. (1989) Microcomputer-based optimization of steel structures in professional practice. *Microcomputers in Civil Engineering*, **4**(4), 289–96.

Grierson, D.E. and Cameron, G.E. (1991) *SODA: Structural Optimization Design and Analysis*, User Manual, Waterloo Engineering Software, Waterloo.

Grierson, D.E. and Lee, W.-H. (1984) Optimal synthesis of steel frameworks using standard sections. *Journal of Structural Mechanics*, **12**(3), 335–70.

Grierson, D.E. and Pak, W.H. (1992) Discrete optimal design using a genetic algorithm. Proceedings of NATO–ARW Conference on Topology Design of Structures, Sesimbra.

Haftka, R.T. and Gürdal, Z. (1992) *Elements of Structural Optimization*, 3rd edition, Kluwer, Dordrecht.

Johnston, B.G. (ed.) (1976) *Guide to Stability Design Criteria for Metal Structures*, 3rd edition, Wiley, New York.

Moharrami, H. (1993) Design optimization of reinforced concrete structures. PhD Thesis, Civil Engineering, University of Waterloo.

Moharrami, H. and Grierson, D.E. (1993) Computer-automated design of reinforced concrete frameworks. *ASCE Journal of Structural Engineering*, **119**(7) 2036–58.

Ontario Ministry of Transportation and Communications (1983) Ontario highway bridge design code.

Pak, W.H. (1993) Genetic algorithms in structural engineering. MASc Thesis, Civil Engineering, University of Waterloo.

Research Engineers (1987) *STAAD-III, Integrated Structural Design System*, New Jersey.

Schmit, L.A., Jr, and Fleury, C. (1979) An improved analysis/synthesis capability based on dual methods – ACCESS 3. Proceedings of AIAA/ASME/AHS 20th Structures, Structural Dynamics and Materials Conference, St. Louis, MO.

Xu, L. (1993) Optimal design of steel frameworks with semi-rigid connections. PhD Thesis, Civil Engineering, University of Waterloo.

Xu, L. and Grierson, D.E. (1992) Computer-automated design of semi-rigid steel frameworks. *ASCE Journal of Structural Engineering*, **119**(6) 1740–60

Zangwill, W.I. (1969) *Nonlinear Programming: A Unified Approach*, Prentice-Hall, Englewood Cliffs, NJ.

13

Reliability-based structural optimization

DAN M. FRANGOPOL and FRED MOSES

13.1 INTRODUCTION

The aim of structural optimization is to achieve the best possible design. Using deterministic concepts this has been interpreted in the following manner (Moses, 1973). The objective function, load conditions, design requirements, failure modes and design variables are all treated in a nonstatistical fashion. Furthermore, it is assumed that design codes provide adequate safety factors to cover any likelihood of failure. By using automated optimization procedures, member sizes are found so that the objective function (e.g. cost, weight) is minimized without violating any design requirements. In some instances deterministic optimization promotes structures with less redundancy and smaller ultimate overload margins than obtained with more conventional and conservative design procedures (Feng and Moses, 1986b; Frangopol and Klisinski, 1989a). Consequently, deterministic optimized structures will usually have higher failure probabilities than unoptimized structures. The reason for this is that the safety factors specified in design codes usually refer to the design of elements, such as beams and colums, and do not allow for a simultaneous occurrence of many failure modes in an optimized structure designed to its limits (Moses, 1973). Therefore, a balance must be developed between the safety needs of the structure and the aims of reducing its cost. Clearly, this requires the use of reliability-based design concepts in structural optimization.

Uncertainties are unavoidable in the design and evaluation of engineering structures. Structural reliability theory has been developed during the past three decades to handle problems involving such uncertainties. This theory has had considerable impact in recent years on structural design. For example, partial safety factors based on load and strength uncertainties have been proposed in the United States, Canada and Europe for building, bridge and offshore platform design codes. Today, the need for designing the best structures by balancing the conflicting goals of safety and economy provides the motivation for the application of structural reliability theory in structural optimization.

Reliability is a quantitative measure of confidence or belief that a system will successfully serve its intended purpose. Implicit in such a definition is the assumption that success and failure can be uniquely defined. With only two possible outcomes, reliability can be defined by the probability of success or the probability of failure, where

$$P_{failure} = 1.0 - P_{success} = 1 - \text{reliability} \qquad (13.1)$$

In most practical engineering undertakings, the probability of success is so close to 1.0 that it is customary to reference reliability by the probability of failure. The probability of failure can be interpreted as the number of failures that would occur given a large set of independent trials, i.e. 'a one in a thousand chance'. (In some instances, e.g. meeting structural serviceability criteria, there is no clear-cut distinction as to whether a system has passed or failed the requirements. In other words, the observation of the event even after the event has occurred is in dispute. A branch of probability known as fuzzy-set theory has recently arisen to deal with such problems in a decision mode. For the present chapter, the assumption is made of a crisp observation such that there is agreement regarding an outcome that is either a survival or a failure event.)

The aim of this chapter, which is partly based on previous work of the authors and their coworkers, is to present a brief state-of-the-art review of reliability-based structural optimization (RBSO) with applications. This review is divided into eight sections treating basic concepts and problem types, basic reliability modeling, element reliability-based optimization, design code optimization, analysis of system reliability, system reliability-based optimization, multiobjective reliability-based optimization, and residual and damage-oriented reliability-based optimization.

13.2 BASIC CONCEPTS AND PROBLEM TYPES

13.2.1 Concept development

Forsell (1924) is apparently the first who formulated the structural optimization problem as minimization of total expected cost. The term 'total cost' implies the initial cost (sum of all costs associated with erection and operation of the structure during its projected design life without failure) and the expected cost of failure (sum of all costs associated with the probability of failure, e.g. expected cost of repairs, expected cost of disruption of normal use, expected cost of loss of human life). This total cost criterion governs nearly all of the work that flourished during the 1950s and is recorded in Johnson (1953), Ferry-Borges (1954), Freudenthal (1956), and Paez and Torroja (1959), among others.

During the next decade (1960–1970) the importance of the Bayesian decision theory in structural optimization was recognized, and several devices aimed at improving the value of the optimum reliability-based solutions were introduced by Benjamin (1968), Cornell (1969a) and Turkstra (1967, 1970), among others. In the same decade, the need to reduce structural weight without compromising structural reliability, particularly in aerospace applications, was also recognized. The first major efforts along this line were led by Hilton and Feigen (1960). Subsequently, other efforts in this category were carried out by Kalaba (1962), Switzky (1964), and Moses and Kinser (1967), among others. The first survey of the optimum reliability-based design field has been written by Moses (1969); it was stated in this study, which focused on the relationship between reliability and optimization, that an optimization procedure which uses overall structural failure probability as the behavior constraint should produce more balanced designs, consistent with the development of rational safety. Furthermore, it was also stated that a truly optimum design should consider the behavior of the structure for various types of loading conditions as well as possible strength deteriorations.

By 1970 it had become apparent that the concepts and methods of probability are the bases for the development of optimum criteria for structural design (Turkstra, 1970). As a result, during the past two decades many investigators focused their efforts on developing the following:

- strategies for identifying failure modes for structural systems which incorporate brittle or ductile behavior, or both (Moses, 1977, 1982; Thoft-Christensen and Murotsu, 1986; Ishikawa and Iizuka, 1987; Zimmerman, Corotis and Ellis, 1991; among others);
- more sophisticated methods, taking failure mode correlation into account, for determining overall system reliability of both brittle and ductile structures (Vanmarcke, 1971; Ang and Ma, 1981; Chou, McIntosh and Corotis, 1983; Ditlevsen, 1979; Grimmelt and Schuëller, 1983; among others).
- random process and field structural reliability models (Vanmarcke, 1984; Madsen, Krenk and Lind, 1986; Melchers, 1987; Wen and Chen, 1989; among others), and simulation techniques (Deak, 1980; Bjerager, 1988; Bucher, 1988; Verma, Fu and Moses, 1990; among others).
- rational automatable structural design optimization techniques (Kirsch, 1981; Osyczka, 1984; Vanderplaats, 1984, 1986; Frangopol and Klisinski, 1989a; among others).
- automated RBSO procedures for minimum material cost with failure probability constraint, for maximum safety based on fixed weight or total expected cost, and for minimum total expected cost (Vanmarcke, 1971; Mau and Sexsmith, 1972; Moses, 1973; Frangopol, 1985b, c, 1986a; Feng and Moses, 1986a; Thoft-Christensen and Murotsu, 1986; Nakib and Frangopol, 1990a, b; among others).
- multiconstraint reliability-based optimal procedures for structural systems subject to probability requirements imposed at both serviceability and ultimate limit states (Cohn and Parimi, 1972; Parimi and Cohn, 1978; Frangopol, 1985b, 1986a, b, 1987; Frangopol and Fu, 1990; among others).
- damage-tolerant design and optimization of nondeterministic systems (Feng and Moses, 1986b; Frangopol and Fu, 1989; Frangopol, Klisinski and Iizuka, 1991; among others).
- reliability-based optimization of inspection and repair strategies (Sørensen and Thoft-Christensen, 1987; Fu, Liu and Moses, 1991; among others).
- time variant RBSO (Kim and Wen, 1987; Iizuka, 1991; among others).
- reliability-based shape optimization of structural and material systems (Thoft-Christensen, 1987, 1991; Sørensen and Enevoldsen, 1989; Enevoldsen, Sørensen and Sigurdsson, 1990; Murotsu and Shao, 1990; Shao, 1991; among others).
- multiobjective structural optimization using reliability-based philosophy (Frangopol and Fu, 1989, 1990; Frangopol and Iizuka, 1991a, b; Fu and Frangopol, 1990a, b; Iizuka, 1991; among others).
- sensitivity studies and techniques to establish response changes of the RBSO

solutions to change in problem parameters (Moses, 1970; Frangopol, 1985a; Sørensen and Enevoldsen, 1989; among others).

The preceding developments allow reliability-based design concepts, which offer a fundamentally new approach to optimum design, to be used by structural engineers to a wide spectrum of design problems including trusses, frames, offshore platforms, highway bridges, shells and pressure vessels.

13.2.2 Problem types

Optimizing structural members and systems using reliability-based concepts and methods raises questions as to the meaning of an optimum design solution (Moses, 1969). Many objective functions have been proposed (Hilton and Feigen, 1960; Kalaba, 1962; Switzky, 1964; Moses and Kinser, 1967; Moses, 1969, 1970, 1973, 1977; Moses and Stevenson, 1970; Turkstra, 1970; Rosenblueth and Mendoza, 1971; Vanmarcke, 1971; Cohn and Parimi, 1972; Mau and Sexsmith, 1972; Frangopol and Rondal, 1976; Parimi and Cohn, 1978; Surahman and Rojiani, 1983; Rojiani and Bailey, 1984; Frangopol, 1984a,b, 1985b,c, 1986b, 1987; Feng and Moses, 1986b; Rosenblueth, 1986; Thoft-Christensen and Murotsu, 1986; Frangopol and Nakib, 1987; Ishikawa and Iizuka, 1987; Kim and Wen, 1987, 1990; Rackwitz and Cuntze, 1987; Sørensen, 1987, 1988; Sørensen and Thoft-Christensen, 1987; Thoft-Christensen, 1987, 1991; Sørensen and Enevoldsen, 1989; Frangopol and Fu, 1989, 1990; Enevoldsen, Sørensen and Sigurdsson, 1990; Murotsu and Shao, 1990; Nakib and Frangopol, 1990a,b; Fu and Frangopol 1990a,b; Frangopol and Iizuka, 1991a,b; Iizuka, 1991; Mahadevan and Haldar, 1991; Shao, 1991; among others).

These include cost and utility functions that should be minimized and maximized, respectively, as follows.

1. Minimization of the total expected cost of the structure $C = C(\mathbf{X})$ expressed as

$$C = C_0 + C_f P_f \tag{13.2}$$

where C_0 is the initial cost, which is a function of the vector of design variables $\mathbf{X}$, C_f the cost of failure, and P_f the probability of failure, which is also a function of $\mathbf{X}$.

2. Maximization of the total expected utility function, U, expressed in a monetary form as

$$U = B - C_0 - L \tag{13.3}$$

where B is the benefit derived from the existence of the structure, and L the expected loss due to failure. As shown by Rosenblueth (1986), in some cases B can be a function of the design variables $\mathbf{X}$. All the quantities in equation (13.3) are expected present values. Conversion of future into present values is done by means of an exponential function (Rosenblueth, 1986).

Because of the difficulties in assuming monetary values to all failure consequences (e.g. placing monetary value on human life, environmental effects), the preceding (i.e. equations (13.2) and (13.3)) reliability-based unconstrained optimization formulations are of limited interest for practical purposes. As an alternative to placing a monetary value on failure consequences, Hilton and Feigen (1960) were the first to propose a reliability-based weight minimization formulation as follows.

3. Find the design variable vetcor

$$\mathbf{X} = (X_1 \, X_2 \cdots X_n)^t \tag{13.4}$$

that will minimize the objective function

$$W = W(\mathbf{X}) \tag{13.5}$$

subject to

$$P_f = P_f(\mathbf{X}) \leqslant P_f^0 \tag{13.6}$$

in which W is the total structural weight, P_f the probability of failure of the structure, and P_f^0 the allowable probability of failure. Side constraints of the form

$$X_i^l \leqslant X_i \leqslant X_i^u \qquad i = 1, 2, \ldots, n \tag{13.7}$$

can also be imposed, in which the superscripts l and u denote lower and upper bounds, respectively. In the above constrained optimization formulation (i.e. equations (13.4)–(13.7)) it is assumed that the measure of performance is given by the weight of the structure.

For the particular case of weakest-link structures (e.g. statically determinate systems) with independent failure modes under a single loading condition, Switzky (1964) showed that at the optimum (i.e. $P_f = P_f^0$) the following approximate linear relationship exists

$$\frac{W_i}{W} = \frac{P_{fi}}{P_f^0} \tag{13.8}$$

where W_i and P_{fi} are the weight and the probability of failure of member i, respectively. Other assumptions were used in the development of equation (13.8) including that the failure probability and weight functions are both linear, $P_f = \sum P_{fi}$ and $W = \sum W_i$, and that a change in P_f^0 does not affect the ratio W_i/W.

Another formulation of the optimum design problem in a reliability-based context is to minimize the probability of failure of a structure for a fixed structural weight (initial cost). In mathematical terms the formulation of this problem is as follows.

4. Find the design variable vector $\mathbf{X}$, that will minimize the objective function P_f, subject to

$$W(\mathbf{X}) \leqslant W^0 \tag{13.9}$$

where W^0 is the allowable total structural weight (initial cost).

The above four formulations are single-objective single-limit state RBSO problems. In practice there are many situations in which the designer wants to optimize a structure with regard to the occurrence of two or more limit states. In general, the RBSO formulation with collapse and loss of serviceability as the failure criteria may be stated in mathematical form as follows:

$$W = W(\mathbf{X}) \leqslant W^0 \tag{13.10}$$

$$P_{fcol} = P_{fcol}(\mathbf{X}) \leqslant P_{fcol}^0 \qquad (13.11)$$

$$P_{funs} = P_{funs}(\mathbf{X}) \leqslant P_{funs}^0 \qquad (13.12)$$

in which P_{fcol} is the probability of plastic collapse, P_{funs} the probability of unserviceability, and the superscript ^{0} denotes prescribed (allowable) values. Three single-objective biconstraint formulations could be used for finding the optimum solution as follows.

Find the design variable vector $\mathbf{X}$ which minimizes

5.

$$W(\mathbf{X}) \qquad (13.13)$$

subject to

$$P_{fcol}(\mathbf{X}) \leqslant P_{fcol}^0 \qquad (13.14)$$

$$P_{funs}(\mathbf{X}) \leqslant P_{funs}^0 \qquad (13.15)$$

6.

$$P_{fcol}(\mathbf{X}) \qquad (13.16)$$

subject to

$$W(\mathbf{X}) \leqslant W^0 \qquad (13.17)$$

$$P_{funs}(\mathbf{X}) \leqslant P_{funs}^0 \qquad (13.18)$$

7.

$$P_{funs}(\mathbf{X}) \qquad (13.19)$$

subject to

$$W(\mathbf{X}) \leqslant W^0 \qquad (13.20)$$

$$P_{fcol}(\mathbf{X}) \leqslant P_{fcol}^0 \qquad (13.21)$$

The formulation 5 is useful when the optimization consists in finding the minimum weight (or initial cost) structure, given the maximum specifed values of the risk levels with regard to occurrence of both collapse and unserviceability. The formulations 6 and 7 are useful when the most reliable structure against the occurrence of collapse and unserviceability is, respectively, sought, given both the maximum specified total weight (or initial cost) of the structure and the preassigned maximum risk level against the occurrence of the other limit state of interest. For the particular case in which the cost is of no concern (i.e. W^0 is not specified) in formulations 6 and 7, the most reliable structure against the occurrence of collapse and unserviceability is, respectively, sought, given the allowable risk level against the occurrence of the other limit state of interest.

If m limit states are of concern and the total weight is the objective to be minimized, the RBSO formulation may be stated in mathematical form as follows.

8. Find the design variable vector $\mathbf{X}$ that will minimize the total weight $W(\mathbf{X})$ subject to

$$P_{fj} = P_{fj}(\mathbf{X}) \leqslant P_{fj}^0 \qquad j = 1, 2, \ldots, m \qquad (13.22)$$

where P_{fj} is the probability of occurrence of limit state j.

In RBSO, the selection of the objective function, the limit states and their allowable probability levels are extremely important considerations (Parimi and

Cohn, 1978; Frangopol, 1985c; Kim and Wen, 1987). The limit states in equation (13.22) may be at the member level or system level, or both, and the allowable probability levels should be chosen accordingly. Much effort has been devoted in (a) developing reliability-based element (member)-level code formats (Ellingwood *et al.*, 1980; Ellingwood and Galambos, 1982; Corotis, 1985; among others), (b) improving the reliability bases of structural safety and design at both element and system level (Ferry-Borges, 1954; Freudenthal, 1956; Cornell, 1967, 1969b; Benjamin, 1968; Ang and Cornell, 1974; Moses, 1974, 1982, 1990; Ditlevsen, 1979; Ang and Ma, 1981; Chou, McIntosh and Corotis, 1983; Shinozuka, 1983; Vanmarcke, 1984; Madsen, Krenk and Lind, 1986; Thoft-Christensen and Murotsu, 1986; Melchers, 1987; Bjerager, 1988; Bucher, 1988; Frangopol, 1989; Wen and Chen, 1989; Frangopol and Corotis, 1990; Verma, Fu and Moses, 1990; among others), (c) choosing an optimum level of the prescribed failure probabilities P_{fj}^0 (Turkstra, 1967, 1970; Rosenblueth, 1986; among others), and (d) developing computational optimization algorithms (Rosenbrock, 1960; Schmit, 1969; Kirsch, 1981; Duckstein, 1984; Koski, 1984; Osyczka, 1984; Vanderplaats, 1984, 1986; Arora, 1989; Frangopol and Klisinski, 1989a; among others).

13.2.3 Hierarchy

The *n* design variables in the vector **X** (equation (13.4)) are varied during the RBSO process in order to improve the design until the optimum is reached. These variables can be divided into five groups (Schmit, 1969; Kirsch, 1981; Thoft-Christensen, 1991):

- sizing design variables (e.g. the cross-sectional dimensions of structural members);
- shape (configuration, geometric layout) design variables (e.g. the coordinates of joints in a framed structure, the length of spans in a continuous bridge);
- material design variables (e.g. the types of mechanical and/or physical properties of the materials to be used);
- topological (general arrangement) design variables (e.g. the number of spans in a bridge);
- structural system types (e.g. truss structure, framed structure, plated structure).

Most RBSO studies deal exclusively with the first and lowest category in this hierarchy of the design variables (i.e. the selection of cross-sectional sizes) because of the relative simplicity of this problem. However, it should be noted that extending optimization capabilities upward into the design variable hierarchy (e.g. consideration of both shape and cross-sectional design variables) may appreciably improve the optimum solution (Murotsu and Shao, 1990; Shao, 1991; Thoft-Christensen, 1991). Some computational experience with the optimum fiber orientation of a multiaxial fiber reinforced laminate under nondeterministic conditions has been recently reported by Shao (1991). It is apparent that much work remains to be done until the stage arrives when the designer will be able to obtain for a specific structure a practical total optimum reliability-based solution in which the type of structure, topology, material shape and member sizes are all treated simultaneously as design variables.

13.3 BASIC RELIABILITY MODELING

13.3.1 Introduction

Probability deals with the future outcome of real or imagined experiments. If the outcome of the experiment was predetermined, then we are dealing with a deterministic and not a random phenomenon. It is the natural or uncontrolled variations as well as the lack of professional knowledge of all the important variables that lead to uncertainty in the result of an experiment.

The use of structural reliability concepts for structural safety requires descriptions of random phenomena and calculation of failure probability and risk. Because insufficient observational data exists for applying the methods of classical statistics, the tools of probability theory must be used. These can be applied in several ways to predict events.

- Fit existing probability functions such as normal, log-normal, or Weibull distributions to the phenomena (load, resistance, etc.) being described.
- Derive distribution functions from an underlying physical process such as a sum, product, repeated trial. For example, the maximum event in, say, 25 years may be derived from distribution of maximum annual events by a process which leads to an extremal distribution.
- Compute distributions derived from functions of more basic random variables as, for example, the wind-induced structure response being a function of the random variables associated with wind speed, drag coefficient, dynamic behavior, etc.

Considerable background material on probability applications in structures may be found in the books by Benjamin and Cornell (1970), and Ang and Tang (1975), or the more recent texts on structural reliability by Thoft-Christensen and Baker (1982), Ang and Tang (1984), Yao (1985), Madsen, Krenk and Lind (1986), Thoft-Christensen and Murotsu (1986), and Melchers (1987), among others. Several journals now deal extensively with the subject, including *Structural Safety* and *Probabilistic Engineering Mechanics*. Specialized conferences on structural reliability include the ASCE Speciality Conference and the international conferences ICASP (International Conference on Applications of Statistics and Probability in Civil Engineering) and ICOSSAR (International Conference on Structural Safety and Reliability), which are periodically held every four years in different countries.

13.3.2 Fundamental reliability problem

A basic design decision problem is now considered. In structural engineering this problem is called a fundamental reliability problem which considers the case where all the uncertainty in strength is associated with one capacity variable, R, and all the load variability with one demand variable, S. The fundamental structural reliability or $R-S$ problem has several aspects:

- calculation of failure probability, P_f, or probability that S is greater than R from the distributions of R and S;

- the sensitivity of P_f to assumptions in the model, distribution function, probability laws and parameters such as means, standard deviations, etc.;
- a subsequent decision problem is to optimize a component by selecting its safety margin for either a target failure probability or a minimization of total expected cost.

A precise calculation for risk may be illustrated by writing the exact integral for calculating failure probability, P_f, as

$$P_f = P[R < S] = \int_0^\infty F_R(t) f_S(t) \, dt \qquad (13.23)$$

where, by definition, the distribution function of resistance, R, can be written as

$$F_R(t) = P[R < t] \qquad (13.24)$$

$f_S(t)$ is the probability density of the demand random variable, S. As an alternative form, the risk may be written as

$$P_f = 1 - \int_0^\infty F_S(t) f_R(t) \, dt \qquad (13.25)$$

Figure 13.1 shows a pictorial representation of these distributions.

For an exact solution, such integrals must usually be evaluated numerically. Simplifications result if one of the two uncertainties dominate, i.e. has much larger dispersion. If the coefficient of variation of resistance, V_R, greatly exceeds the coefficient of variation of load, V_S, then

$$P_f \cong F_R(\bar{S}) \qquad (13.26)$$

where $\bar{S}$ is the mean value of S.

Some closed-form solutions of equation (13.23) are possible in certain cases.

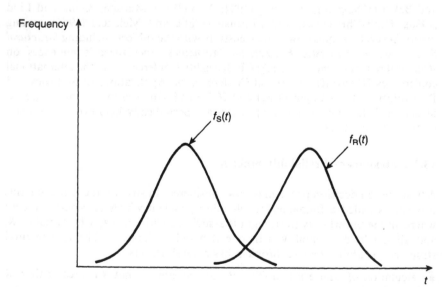

Fig. 13.1 Fundamental reliability illustration showing distribution of load effect, S, and resistance, R.

Consider the case where both load and strength are normal (Gaussian) distributed variables.

Let $Z = R - S$, where Z can be viewed as the margin of safety. Z is also normal because R and S are normal. The statistical moments of Z are the mean

$$\bar{Z} = \bar{R} - \bar{S} \tag{13.27}$$

and the variance (for independent R and S)

$$\sigma_Z^2 = \sigma_R^2 + \sigma_S^2 \tag{13.28}$$

The failure probability is then

$$P_f = P[Z < 0] = F_U\left(\frac{0 - \bar{Z}}{\sigma_Z}\right) = F_U(-1/V_Z) \tag{13.29}$$

where F_U is the standard normal form, and V_Z the coefficient of variation (COV) of Z.

In terms of the central or mean safety factor, $\bar{n} = \bar{R}/\bar{S}$

$$P_f = 1 - F_U\left[\frac{\bar{R} - \bar{S}}{(\sigma_R^2 + \sigma_S^2)^{1/2}}\right] = 1 - F_U\left[\frac{\bar{n} - 1}{(\bar{n}^2 V_R^2 + V_S^2)^{1/2}}\right] \tag{13.30}$$

A similar closed-form expression for P_f occurs when both R and S are log-normal variables. In that case, we obtain for values of V_R and V_S up to 30%

$$P_f = 1 - F_U[\ln(\bar{R}/\bar{S})/(V_R^2 + V_S^2)^{1/2}] \tag{13.31}$$

13.3.3 Simplified reliability design formats

The approach described above for calculating failure probabilities was developed by Freudenthal (1956) and Ang and Cornell (1974). Given a knowledge of the frequency distributions of load and strength, equation (13.23) can be used when sufficient data is available to determine the distribution functions for R and S. It is thus a powerful analytical tool for predicting and controlling safety. It has severe limitations, however, both physical and psychological, which restrict its general application and acceptance. For example, the load random variable is not just a measurable environmental quantity, but also represents uncertainties in modeling load effects within the structure as well as in some of the basic limitations in the theory of mechanics. Similarly, the strength distribution depends not only on measurable material values, but also on assumptions of theoretical behavior of materials, quality of workmanship and fabrication control.

The major obstacles in accepting a fully probabilistic analytical treatment of safety directly in design can be summarized as follows.

- There are limitations in available data, particularly in a consistent form. This is combined with the low failure probabilities usually demanded of structures. A statistician, for example, would have to place a low confidence level on the results of most reliability predictions.
- Calculation of P_f generally requires computer analysis programs that are independent of the usual design analysis programs. This makes it difficult and

inefficient to use during the design process since an iteration among programs would be required.

• Even if engineers felt it were possible to compute failure probabilities, the legal and professional responsibilities for establishing acceptable risks by consultants, owners, regulators, or even code writers, would present a severe problem.

• Safety factors in design are not always meant to protect against the phenomena being checked but rather to consider other unforeseen events such as possible future changes in function of the structure, blunders and accidents, serviceability requirements and possible rather gross approximate analysis and design assumptions.

Some of these objections to probabilistic analysis in structural engineering may not be as valid for such 'industrialized' structure types such as transmission lines or cranes. In cases where many structures of a basically similar design are to be constructed, this justifies a large engineering design and analysis cost to establish risk levels.

To overcome objections to explicit probability-based design procedures, there have been a number of attempts to introduce simplified formats. This work has been done by Cornell (1969b), Rosenblueth and Esteva (1972), Lind (1973) and Ang and Cornell (1974). It tries to incorporate as much as possible of the probabilistic basis while making the result appear in a format close to current design procedures.

For example, in the 'second moment format', the essential feature is that only mean and variance are used to describe a random variable. In order to have an explicit expression for the risk, an assumed probability law must be used. Alternatively, in place of a specified P_f level, it is required that the mean safety margin must be β standard deviations above zero, i.e.

$$\bar{Z} \geqslant \beta \sigma_Z \tag{13.32}$$

where β is the **safety index** (also called **reliability index**). In terms of load and strength values, the safety criterion is met if

$$\bar{R} \geqslant \bar{S} + \beta(\sigma_R^2 + \sigma_S^2)^{1/2} \tag{13.33}$$

The safety index, β, may be related to risk levels if the fundamental reliability problem is used with a normal distribution of load and strength, as follows:

β	P_f
0	0.5
1	0.158655
2	0.022750
2.5	0.006210
3	0.001350
4	0.316712×10^{-4}

Alternatively, a log-normal assumption for R and S leads to the safety expression

$$\bar{R} \geqslant \bar{S} \exp[\beta(V_R^2 + V_S^2)^{1/2}] \tag{13.34}$$

Both equations (13.33) and (13.34) can easily be used to derive safety factors for use in design checking. Such factors depend on the desired safety index, β, and the uncertainty level expressed by V_R and V_S.

13.3.4 Advanced reliability design formats

The first step in component reliability modeling is to define the li¬it state condition. This is usually written in the form of a limit state requirement:

$$g(X_1, X_2, X_3,...) = 0 \qquad (13.35)$$

The X_i are the random variables (e.g. load, resistance, dimensional, material). g is defined such that a realization of the performance function g which is positive means the component has not failed.

The function g is basically deterministic in nature and expresses the mechanics of the strength check. A performance function (also called state or failure function) must be clear in the sense that if values of all the random variables were precisely known, then the engineer would know whether the structure failed (i.e. $g < 0$) or not ($g > 0$). In this approach, analysis or equation approximations can be treated as additional random variables. However, if after knowing the magnitudes of each random variable, the engineer is still not certain regarding failure or survival, it means the mechanics are not well understood. Such problems are obviously not ripe enough for reliability analysis when even the deterministic solutions are not known.

The first-order second moment approximation to reliability fits a mean and variance to the function g. The best approximation to the mean, $\bar{g}$, is

$$\bar{g} = g(\bar{X}_1, \bar{X}_2, \bar{X}_3,...) \qquad (13.36)$$

i.e. set all the variables equal to the mean value and compute the corresponding g.

Approximating the performance function g by a first-order Taylor series, expanded at the mean value of the variables, X_i, gives for the variance of g, σ_g^2, the following expression:

$$\sigma_g^2 = \sum (\partial g / \partial X_i)^2 \sigma_{X_i}^2 \qquad (13.37)$$

provided that the random variables, X_i, are uncorrelated. Knowing the mean and variance of g allows the safety index to be computed:

$$\beta = \bar{g} / \sigma_g \qquad (13.38)$$

If g were in fact normal, β could provide an exact value for the failure probability P_f from a normal table. By the central limit theorem, g would be normal if the performance function were linear and all the X_i were also normal. This first-order second moment expression for β looks attractive since β can easily be found using finite difference approximation for the derivative of g. A major difficulty arose with the expression, however, soon after it was proposed, which is called the lack of invariance. That is, one would like the same result regardless of mathematical form of g as long as the physical models were the same. This turns out not to be the case because of the approximation of the derivative at the mean value.

This 'lack of invariance' was solved by the Hasofer–Lind (Hasofer and Lind, 1978) reliability criterion which recognized that the linearized approximation to the variance of g should not be done at the mean, $\bar{g}$, but rather at a point on the limit state surface (i.e. $g = 0$). The 'best' point recommended to do this linearization would be the point (known as the **design point**) on the limit state surface $g = 0$ which is closest to the mean value. The reason for selecting this closest point is that it is likely to have the highest probability density. This design point can be found in several ways by iteration (algorithms of Rackwitz and Fiessler (1978), Baker (1977) or Wirshing (1985)) or by minimization (Shinozuka, 1983). The design point is the point on the limit state surface closest to the mean value (i.e. shortest distance to the mean). Mathematically, we now expand g about a point on the limit state surface (also called failure surface) rather than the mean. The problem is finding the point, say $(X_1^*, X_2^*, \ldots)$ identically satisfying $g(\mathbf{X}^*) = 0$. This leads to an iterative calculation for the safety index.

This iteration provides the safety index for a particular design. The inverse design problem is to find the safety factors so that a target safety index is reached. It can be shown that if the function g is linear in $\mathbf{X}$ and the variables X_i are all normal, β correlates to an exact probability of failure, P_f, using a normal table.

The AFOSM or advanced first-order second moment method goes another step further by approximating the 'true' nonnormal distribution of a variable X_i by replacing it by a normal distribution (with appropriate mean and variance) obtained by matching at the design point X_i^*. Thus, the equivalent or approximating normal distribution will have (a) the same density and (b) the same cumulative probability as the actual nonnormal distribution at the point X_i^*. In fact the method can go further and treat correlated random variables by first performing an uncoupling that produces an independent set of variables. The transformation of the performance function to these new uncorrelated variables must also be carried out. Computer programs for automatically performing such safety index calculations for general performance functions of correlated variables with nonnormal distributions are generally available. In this approach, the safety index is uniquely defined as the shortest distance from the origin (normalized mean value) to a point on the limit state surface (Hasofer–Lind definition of safety or reliability index (Hasofer and Lind, 1978)). These AFOSM methods are all related to finding the safety index, or the probability of failure of individual components provided that a performance function and distributions for the random variables are both given. Only one failure mode is usually considered.

13.4 ELEMENT RELIABILITY-BASED OPTIMIZATION

13.4.1 General

Reliability-based specifications and design procedures are usually aimed at design of individual elements such as beams, columns and connections (Moses, 1974). This reflects the activities of designers who look on design as satisfying individual component safety checks. Thus, there evolved reliability-oriented prediction programs to estimate the probability of failure of single components. To apply such probabilistic results, design methods were developed so that the designers could be satisfied that a target element reliability index was realized (Moses, 1990).

13.4.2 Optimization formulations

The reliability-based optimization problem at the element level may be formulated in various ways, as follows.

1. Minimization of the total expected cost of the element $C_{el} = C_{el}(X)$ expressed as a function of the probability of occurrence of the dominant element limit state (e.g. collapse):

$$C_{el} = C_{o,el} + C_{fd,el} P_{fd,el} = C_{o,el} + C_{Fd,el} \qquad (13.39)$$

where $C_{o,el}$ is the element initial cost, which is a function of the vector of design variables $X = (X_1 \, X_2 \cdots X_n)^t$, $C_{fd,el}$ the cost of element failure due to the occurrence of the dominant limit state, and $P_{fd,el}$ the probability of failure (i.e. occurrence of the dominant limit state of interest) of the element, which is also a function of X.

    An illustration of the variation of total, initial and failure costs as a function of the risk $P_{fd,el}$ is shown in Fig. 13.2 (Moses, 1977). Increased initial cost causes a decreased risk and a reduction in the (expected) failure cost $C_{Fd,el}$. An optimum design (i.e. minimum expected total cost) is reached when an increase of initial cost $C_{o,el}$ is balanced by a reduction in the (expected) failure cost $C_{Fd,el}$.

2. Minimization of the total expected cost of the element $C_{el} = C_{el}(X)$ expressed as a function of the probabilities of occurrence of dominant ultimate (e.g. collapse) and serviceability (e.g. excessive elastic deformation) limit states $P_{f1,el}$ and $P_{f2,el}$ respectively:

$$C_{el} = C_{o,el} + C_{f1,el} P_{f1,el} + C_{f2,el} P_{f2,el} \qquad (13.40)$$

where $C_{f1,el}$ and $C_{f2,el}$ are the costs of failure of the element by occurrence of dominant ultimate and serviceability limit states, respectively. Both $P_{f1,el}$ and $P_{f2,el}$ are functions of the design variable vector X.

3. Minimization of the total expected cost of the element $C_{el} = C_{el}(X)$ taking

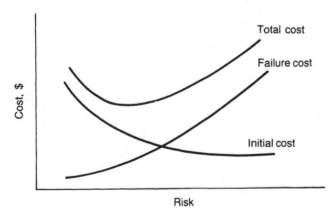

Fig. 13.2 Total cost, failure cost and initial cost versus risk (Moses, 1977).

into account all possible ultimate and serviceability limit states:

$$C_{el} = C_{o,el} + \sum_{j=1}^{m} C_{fj,el} P_{fj,el} \quad j = 1, 2, \ldots, m \tag{13.41}$$

where m is the total number of limit states of the element.

4. Minimization of the total weight of the element:

$$W_{el} = W_{el}(\mathbf{X}) \tag{13.42}$$

subject to

$$P_{fj,el}(\mathbf{X}) \leqslant P^o_{fj,el} \quad j = 1, 2, \ldots, m \tag{13.43a}$$

or, alternatively,

$$\beta_{j,el}(\mathbf{X}) \geqslant \beta^o_{j,el} \quad j = 1, 2, \ldots, m \tag{13.43b}$$

where

$$\beta_{j,el}(\mathbf{X}) = \Phi^{-1}[1 - P_{fj,el}(\mathbf{X})] \tag{13.44}$$

is the reliability index with respect to the occurrence of limit state j, $\Phi^{-1}(a)$ is the value of the standard nominal variate at the probability level a, $P^o_{fj,el}$ and $\beta^o_{j,el}$ are allowable values of $P_{fj,el}$ and $\beta_{j,el}$, respectively. Fabrication or other side constraints can also be imposed in the form given by equation (13.7).

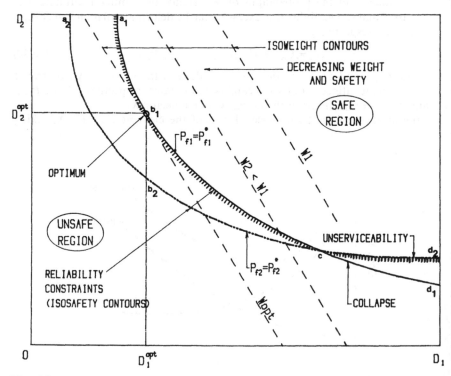

Fig. 13.3 Design space with two reliability constraints: collapse critical at the optimum (Frangopol, 1986a).

5. Minimization of the probability of occurrence of a specified limit state k:

$$P_{fk,el} = P_{fk,el}(\mathbf{X}) \tag{13.45a}$$

subject to

$$W_{el}(\mathbf{X}) \leqslant W^{\circ} \tag{13.46}$$

$$P_{fj,el}(\mathbf{X}) \leqslant P^{\circ}_{fj,el} \quad j = 1, 2, \ldots, m \quad j \neq k \tag{13.47a}$$

Alternatively, using reliability indices described in the previous section instead of failure probabilities the above formulation could be viewed as the maximization of the reliability index with respect to the occurrence of a specified limit state k:

$$\beta_{k,el} = \beta_{k,el}(\mathbf{X}) \tag{13.45b}$$

subject to the constraints given by both equation (13.46) and the following $m-1$ inequalities:

$$\beta_{j,el}(\mathbf{X}) \geqslant \beta^{\circ}_{j,el} \quad j = 1, 2, \ldots, m \quad j \neq k \tag{13.47b}$$

Of course, side constraints can be added to this formulation in the form given by equation (13.7).

Figures 13.3, 13.4 and 13.5 (Frangopol, 1986a) are useful in understanding the solution of the formulation 4 for the particular case of the reliability-based

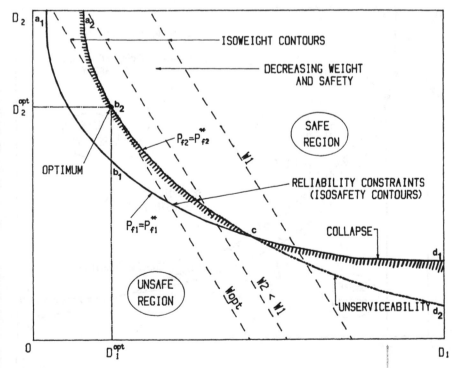

Fig. 13.4 Design space with two reliability constraints: unserviceability critical at the optimum (Frangopol, 1986a).

optimization of an element with two limit states: collapse (e.g. formation of one mechanism) and unserviceability (e.g. first plastic hinge occurrence). In Figs 13.3–13.5 the design variable vector $\mathbf{X}$ has two components ($X_1 = D_1$ and $X_2 = D_2$) and the active collapse and loss of serviceability constraints are denoted as $P_{f1} = P_{f1}^*$ and $P_{f2} = P_{f2}^*$, respectively, where $P_{f1}^* = P_{f1}^o$ and $P_{f2}^* = P_{f2}^o$ are allowable failure probabilities.

The boundary of the acceptable design space (which contains points with acceptable probabilities of plastic collapse and first plastic hinge occurrences) depends on the relative magnitudes of the acceptable risk levels P_{f1}^* and P_{f2}^*. In Fig. 13.3 the collapse constraint is the only one which is critical at the optimum (point b_1 in Fig. 13.3), while in Fig. 13.4 the unserviceability constraint is the only one which is critical at the optimum (point b_2 in Fig. 13.4).

It is interesting to note that (a) the points lying on some regions of the isosafety contours with regard to loss of serviceability limit state (e.g. regions a_2b_2c in Fig. 13.3, cd_2 in Fig. 13.4, and bc_2d in Fig. 13.5) have unacceptable reliabilities with regard to plastic collapse ($P_{f1} > P_{f1}^*$), (b) the points lying on some regions of the isosafety contours with regard to collapse (e.g. regions cd_1 in Fig. 13.3, a_1b_1c in Fig. 13.4, and a_1b and de_1 in Fig. 13.5) have unacceptable reliabilities with regard to loss of serviceability ($P_{f2} > P_{f2}^*$), and (c) both reliability constraints are active (i.e. satisfied as equalities) at one (e.g. point c in Figs 13.3 and 13.4)

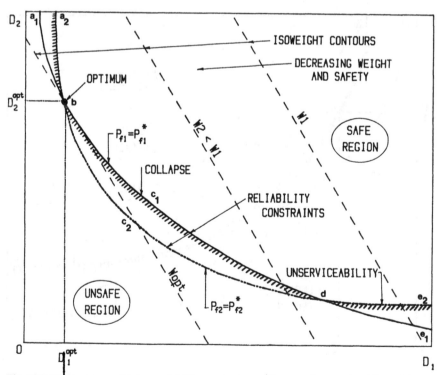

Fig. 13.5 Design space with two reliability constraints: collapse and unserviceability critical at the optimum (Frangopol, 1986a).

or two (e.g. points b and d in Fig. 13.5) design points. It is also relevant to note that only one constraint (e.g. reliability against collapse) may govern the boundary of the acceptable design space; consequently, in this case the two reliability constraints have no common points. Another interesting remark is that one or both reliability constraints may be satisfied as equalities at the optimum. For example, the minimum weight solution is governed by the collapse constraint in Fig. 13.3 (optimum point b_1), by the unserviceability constraint in Fig. 13.4 (optimum point b_2), and by both constraints satisfied as equalities in Fig. 13.5 (optimum point b).

13.4.3 Applications

From the above formulations it is clear that an important step in element reliability-based optimization is the evaluation of risk associated with various element failure modes. If proper failure modes are not selected, the solution will not be useful. Another step in achieving the optimum solution is to use a reliable and efficient optimization algorithm. The illustrative examples presented in this section show only some of the possible applications of reliability-based optimization methods in elements design.

EXAMPLE 13.1 STEEL BEAM-COLUMN (Iizuka, 1991)

A new approach for reliability analysis of steel beam-columns was recently presented by Iizuka (1991) using the theorem of total probability. In this approach both beam-column instability and material yielding are considered as limit states and both loading and material properties are considered as random variables. Figure 13.6 (Iizuka, 1991) displays the variations of the reliability indices against both material yielding and instability with the change in the cross-sectional area of the beam-column. A discrete constrained minimization programming method, considering standard steel beam-column selection based on minimum area design (equations (13.42) and (13.43b)), will permit optimum selection of nondeterministic beam-columns.

EXAMPLE 13.2 STEEL BRIDGE GIRDER DESIGN FOR FATIGUE (Moses, 1977)

In highway bridges the design fatigue stress is a function of truck volume (load cycles), specific weld category life and shape of the random load histogram. The applied stress (load effect) is a function of truck bending moment, section modulus, impact factor and girder load distribution. These random variables have been studied in great detail in a number of field measurement, laboratory and analytical studies. These studies have evaluated the means and variances of these random

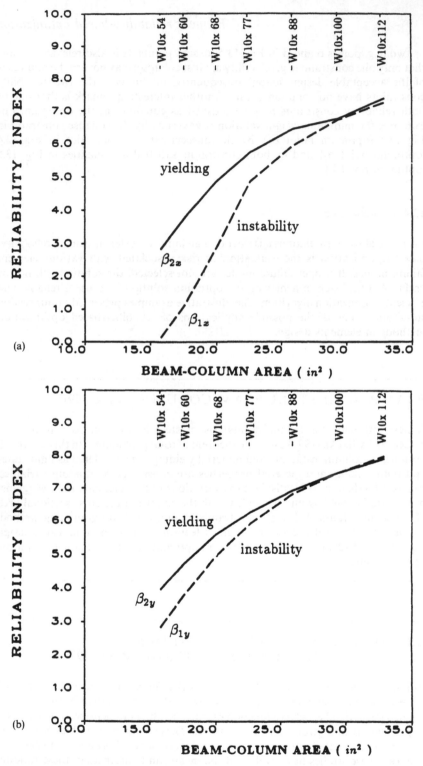

Fig. 13.6 Beam-column reliability index: (a) instability can only occur about x axis; (b) instability can only occur about y axis (Iizuka, 1991).

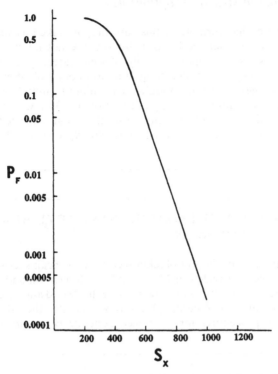

Fig. 13.7 Probability of fatigue failure, P_f, vs girder section modulus, S_x (Moses, 1977).

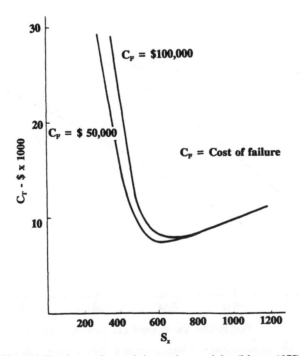

Fig. 13.8 Total cost, C_T, vs girder section modulus (Moses, 1977).

variables. The design variable in this case may be taken as simply the section modulus S_x since the fatigue life distributions are not known as a function of say girder thickness or depth but only of weld category. Figure 13.7 (Moses, 1977) shows for a specific girder design the probability of fatigue failure (during the bridge lifetime) vs section modulus. By adding to initial cost $C_0 = C_{0,el}$ the expected cost of failure $C_F = C_{Fd,el}$ (equation (13.39)) we obtain the curves in Fig. 13.8 (Moses, 1977) which show the total expected cost $C_T = C_{el}$ (equation (13.39)) vs the girder section modulus S_x for two values of the cost of failure $C_F = C_{fd,el}$.

EXAMPLE 13.3 REINFORCED CONCRETE BEAM
(Moses, 1973)

Means and standard deviations of reinforced concrete beams resistances in several failure modes were provided in Moses (1973), Ellingwood and Ang (1974), and Surahman and Rojiani (1983), among others. Ellingwood and Ang (1974), working within the reliability-based design philosophy, evaluated the risks associated with existing design procedures for reinforced concrete beams in both flexure and

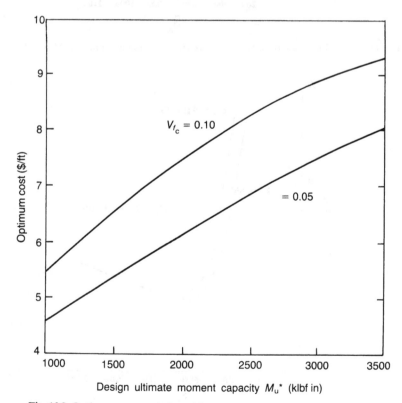

Fig. 13.9 Optimum cost vs design ultimate moment capacity (Moses, 1973).

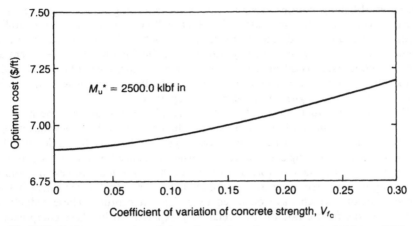

Fig. 13.10 Optimum cost vs coefficient of variation of concrete strength (Moses, 1973).

shear. They were able to calculate the probability of failure in flexure and shear of simply supported beams. Load, strength and model prediction uncertainties were taken into account. Although Ellingwood and Ang's (1974) study does not include optimization, it is important because the risk of unfavorable performance of reinforced concrete beams commensurate with the level of uncertainty provides a consistent measure of design safety and thus furnishes a systematic measure for comparing design alternatives.

Some examples of reliability-based under-reinforced concrete beam optimization results are presented in Figs 13.9 and 13.10 (Moses, 1973), showing the variation of the optimum cost with design ultimate moment capacity, M_u^*, and coefficient of variation of concrete strength, V_{f_c}. The total (expected) cost includes concrete, steel and forming costs. The design problem is to find the mean values of the steel area, depth and width of the rectangular cross section, steel bar strength, and concrete strength to minimize the total cost subject to five constraints including both strength and side requirements.

Other optimum element reliability-based design examples under both time invariant and time varying loads have been presented (Iizuka, 1991; among others). In all reliability-based optimization applications, it is necessary to subject the results to sensitivity analysis to determine the influence of input parameters, including distribution functions and coefficients of variation, on optimum solutions (Moses, 1970, 1973; Frangopol, 1985a; among others).

13.5 DESIGN CODE OPTIMIZATION

13.5.1 Introduction

The implementation of reliability concepts into engineering practice may occur in several ways, for example, as an aid to evaluation of alternative design concepts or by establishing routine design criteria for checking structures. The most broadly acceptable implementation, however, will be by introducing a reliability basis into documented design checking codes such as in the steel, concrete or other code practices. Such industry-wide design checking formulae represent an industry consensus based on theory, practice and experience. Once it is in the book (blue, white, red, or whatever the corresponding cover), it will have to be considered by all designers in their routine operations.

Consequently, much of recent structural reliability research has been in the area of code implementation, that is, to provide the methodology by which design codes can introduce the benefits and advantages of reliability-based design. This is not to say that existing codes were not earlier influenced by reliability concepts since, in fact, codes have historically been modified in a quasi-Bayesian manner. This is apparent when one examines, for example, the reductions in safety factors for steel beams which have occurred over the last century. These reductions reflect the low failure rate or high reliability experienced by these components. A lower safety factor presumably provides a more economical design decision. Similarly, the recent code increases in seismic factors reflect the large damages that did occur during some earthquakes and the need to increase reliability of seismic resistant designs.

The difference between a historical or quasi-Bayesian evolution of codes and the present reliability developments is the latter formulation of the safety decision problem in a comprehensive statistical decision-based framework. These include the various types of load intensities such as dead, live, wind and seismic, analysis uncertainties, material properties, strength formula and failure consequences. The aim is to treat all the uncertainties consistently and the corresponding limit states and to derive design checking criteria which will lead to acceptable and economical reliability levels.

Despite some slow initial progress in introducing changes, the development of reliability-based checking formats has advanced rapidly in recent years. The new codes are often called limit-state formats but in reality they are independent of both the particular methodology for analyzing the force distribution within the structure and even the maximum acceptable strength definition for failure. Conventional linear structural analyses are in fact mostly used in code applications described as limit state design. The approximations inherent in an elastic force distribution are only one of the uncertainties in predicting the load effect and resistance.

13.5.2 Design checking formats

Traditional design checking procedures are based on allowable or working stress design (WSD) methods. In this format, the maximum design stress due to applied loads is restricted to be equal to the nominal yield divided by a safety factor. That is,

$$R/SF = D + L + W + \cdots \qquad (13.48)$$

where R is the nominal resistance or component strength, D the dead load effect, L the live load effect, W the environmental load effect, and SF the safety factor. This approach, if precisely followed, would lead to obvious contradictions. For example, environmental load is more uncertain than live load which, in turn, is usually more uncertain than dead load. Also, the combination of different loading effects makes it less likely that the full design value will be reached compared with situations in which only one load effect is present. Similarly, some components, e.g. members in compression or connections, may often require higher safety factors than members in tension because of their failure consequences.

Over the years, however, many of the obvious disadvantages of the WSD format have been indirectly eliminated. For example, the one-third increase in allowable stresses for some load combinations or the deliberate conservative reserves in connection formulae do, in fact, accomplish some of the goals of reliability-based formats. The limitations of WSD, however, become apparent in several distinct areas.

- It is difficult to distinguish the overall safety margin in WSD since safety is enhanced by the safety factor as well as deliberate but unquantified bias in strength formula, analysis procedures, etc.
- It is difficult to develop consistent safety levels when extending the provisions to new material applications or environmental exposures.
- It lacks flexibility in approaching nontraditional applications such as evaluating remedial construction, repair, rehabilitation, or short-term load exposures.
- Allowable stresses are awkward when dealing with nonlinear structure behavior problems such as seismic response, foundation flexibility, or dynamic response to ocean waves.

Reliability procedures have been shown to offer a comprehensive guideline for code writing. The first step is a code checking format followed by optimization of the checking factors.

13.5.3 Load and resistance factor design (LRFD) calibration

The terminology denoted as 'load and resistance factor design' has been generally adopted to implement reliability-based design practices in the United States. It is known in Canada and the United Kingdom as limit-state design while in other parts of Europe it is called partial safety factor design. Similarly, in concrete design practice it has been called ultimate strength design, while in highway bridge design it is called load factor design. The basic format is repeated here as:

$$\phi_i R = \gamma_D D + \gamma_L L + \gamma_W W + \cdots \tag{13.49}$$

where R, D, L and W retain the same definitions as in WSD, ϕ_i is the resistance factor on the ith component type, e.g. beam bending, shear, γ_D the load factor applied to dead load effect, γ_L the load factor applied to live load effect, and γ_W the load factor applied to environmental load effect.

In checking, there may be more than one load combination used for the right-hand side in equation (13.49), for example, gravity alone, gravity plus wind, gravity plus seismic, overturning, etc. This is similar to the formulation in a WSD code. Each partial safety factor in equation (13.49), i.e. ϕ on the resistance and γ on the load, is associated with its corresponding quantity.

The load and resistance factors depend on several items.

- The bias of that variable, i.e. ratio of mean value to nominal code recommended formula.
- Corresponding uncertainty of that variable.
- An overall calibration exercise which varies the partial load and resistance factors in order to come as close as possible to the target safety index over a range or possible parameters. The latter exercise is known as code calibration or code optimization and is described in more detail in the next section.

The LRFD format in equation (13.49) is not unique, since additional factors may also be used to augment the ϕ and γ shown. For example, a system factor may reflect consequences of component failure such that

$$\phi_i = \phi_{i,R}\phi_S \tag{13.50}$$

where ϕ_S is the system factor and $\phi_{i,R}$ the resistance factor, for a component in the absence of consideration of failure consequences. The system factor can depend on redundancy, ductility, and consequences of component failure. Similarly the γ factor on load effects can be made a function of analysis procedures, such as

$$\gamma_L = \gamma_I\gamma_A \tag{13.51}$$

where γ_A is a factor based on analysis uncertainty including modeling details and any field performance proof testing of the analysis method, γ_I a factor based on environmental load intensity uncertainty, and γ_L the overall load factor.

Additional partial safety factors can lead to a finer 'tuning' of the code safety margin to the specific application that the user is designing. This approach is similar to European practices such as in the CEB code which contain several additional partial factors such as separate factors for workmanship and quality control procedures. Many of these latter applications are more appropriate for concrete structures, rather than steel structures, but there are instances where such factors apply also in steel.

There are obvious advantages to having more factors in attaining uniform reliability over different applications. The disadvantages, however, include the added effort required by the designer in applying more factors. In particular, there may be subjective interpretations in using such factors as workmanship, ductility, or failure consequences. There is the further limitation that adding more factors implies that the statistical data is sufficiently refined to reflect a breakdown of uncertainties into a greater number of variables. One possible area that benefits from more factors is the application to reevaluation of structures. For example, a recent code developed for evaluating existing bridges considers a variety of load and resistance factors based on the level of effort (i) to inspect the existing structure, (ii) to improve maintenance to avoid further deterioration, (iii) to perform accurate rather than routine approximate analysis, and (iv) to limit or enforce the external operational loading (Verma and Moses, 1989).

13.5.4 Code optimization

The process by which load and resistance factors are derived for codes has been denoted as **code calibration**. A direct reliability approach would be to analyze each component using the reliability and safety index calculation. Rather than applying equation (13.49), the member checking criterion would be that a design component is acceptable if the safety index exceeds the required target value. This approach is sometimes called a level II approach. A level III approach would use an integration or Monte Carlo simulation to evaluate the risk directly to avoid any approximations that are inherent in the AFOSM safety index method (section 13.3.4). The so-called level I procedure is generally used in US–LRFD codes and requires the partial factors to be specified by the code writers in such a way that they produce, on the average, the desired target safety index. For designers, the checking procedures are carried out in a deterministic manner.

In the code calibration, the following steps are needed.

1. Assemble a representative population of components that will be checked by the code. These components should be generally acceptable to industry and code writers. That is, they were checked by existing (WSD) codes or are known historically to have acceptable performance.
2. Develop a reliability formulation including a failure function (g), and statistical parameters and distributions for each of the random variables, including load, analysis, and strength.
3. Compute safety indices for each component by the AFOSM method.
4. Average these values and produce a target safety index, β^{T}.
5. Find the load and resistance factors (γ and ϕ) appearing in the LRFD format (equation (13.49)) which minimize the following functions and satisfy the constraints:

$$\text{find} \qquad \gamma_i, \phi_i$$

$$\text{minimize} \quad \sum_i W_i(\beta_i - \beta^{\mathrm{T}})^2 \tag{13.52}$$

$$\text{subject to} \quad \sum_i W_i = 1.0 \tag{13.53}$$

where W_i is the weighting factor assigned to design component i in the population, β_i the computed reliability for component i using parameters γ_i and ϕ_y, and β^{T} the target reliability.

In addition, another constraint in the optimization may be imposed to ensure that the average risk for all components equals the risk target, P_f^{T}, or

$$\sum_i W_i P_{fi} = P_f^{\mathrm{T}} \tag{13.54}$$

where P_{fi} is the risk of failure or inverse function of safety index β_i and P_f^{T} the probability of failure equivalent to target reliability, β^{T}.

In general, most studies have found that the safety indices found with WSD will have unexplainable variation. This is a result of the fact that design margins were not historically derived with any reliability models and further the presence of only a single safety factor in the format is too restrictive to achieve uniform indices (β_i). The LRFD format is more flexible simply by virtue of having more factors and hence can reduce the scatter in β and produce components with more uniform reliability. This type of code optimization is well researched and documented. Standard minimization algorithms may be used to solve equations (13.52)–(13.54).

13.6 ANALYSIS OF SYSTEM RELIABILITY

13.6.1 Introduction

What is meant by the conventional design of structures is the design checking of individual components or members (beams, columns, piles, connections, etc.). Codes do not usually check the overall structure capacity. In some types of structures, several components must simultaneously fail in sequence before there is overall structure damage or failure. On the other hand, in other types of structures there may be a multitude of critical members; the failure of any single one of the members may lead to catastrophic conditions.

This discussion suggests that there is an area of investigation related to reliability of the overall structure system as distinct from the reliability of components. This is important since what is meant by reliability in an owner's or the public's mind is not related to component, but rather to the system reliability with respect to some specific consequence of failure.

Another limitation in code checking is that the target β values discussed in section 13.5 refer to notional probabilities of component failure due to occurrence of overload and understrength. Most reported structural failures are due to accidents, fabrication mistakes, or human error which are not always influenced by code-specified safety margins. Rather, these types of failures must be controlled by tighter design review and inspection procedures which reveal such gross errors before they cause failure. System reliability can, in many cases, identify whether there are critical locations from which failure may initiate and assist in the best utilization of resources to promote better design accuracy, ductility, redundancy, or inspection. In-service assessments of existing structures are another area where comparative system reliability estimates can be used to recommend levels of in-depth inspection, analysis, testing, or repair.

It is important in some instances for engineers to understand both the limitations of present reliability-based code procedures as well as the potential usefulness of system reliability application. Examples include remedial studies of existing structures, comparison of new alternative structure concepts and geometries, and calibration of actual system reserve levels. The considerable research in recent years in structure system reliability is due to interest in nuclear plants, bridges, long span roofs, and offshore platforms where ultimate capacity rather than component serviceability is important and where identification of critical members for inspection, review, and strengthening is an option. In the future, more attention is expected to focus on system reliability, both in code work and in specific structural investigations for criteria selection, concept evaluation, or remedial re-evaluation.

13.6.2 Basic system models

The basic problem of system reliability is to estimate the system reliability from the reliability of the components. Several approaches have been proposed including direct methods (e.g. Monte Carlo simulation, point estimate, and response surface) and failure mode analysis.

The direct methods which are simple to apply have been generally thought to be too time consuming and offer little insight into the contributing factors in the system behavior. Recent developments, however, in simulation, such as importance sampling or directional simulation, have offered improvements in computation efficiencies while removing some of the limitations in other analysis approaches (Hasofer, 1987; Moses, Fu and Verma, 1989). Expected further cost reductions in computer time suggest that in the future there will be further advances in the direct methods.

The failure mode analysis will be emphasized here because of its wider usage in the structure application and the fact that it more easily lends itself to an optimization formulation. Three characteristics of a structure model greatly affecting its reliability performance are (i) the material behavior, (ii) the geometry, and (iii) the statistical correlation (Moses, 1982).

A structural system is an assemblage of components. Both the reliability and behavior of the components affect system reliability. Various types of material models are used for component behavior – these include elastic brittle, elastic–plastic (ductile), strain hardening in which loads increase after component yielding, and semibrittle in which loads decrease after yielding. Ductile components continue to carry load after failure. Brittle components shed all of their load to remaining components upon failure. Because of the load redistribution that occurs after component failure, the system reliability must be closely connected to the structural analysis.

The second system characteristic of importance is the geometry. Series and parallel geometries are idealized models that represent the two extremes of system behavior. However, most structures contain some combination, and it is instructive initially to consider these two limiting cases.

A **series system** fails if any component fails, and is analogous to a 'weakest link' chain which fails if any link fails. Statically determinate structures, and independent subsystems such as superstructure, and foundation, are examples of series systems. The reliability model of series systems is the same for both ductile and brittle components.

The resistance, R, of ideal series systems can be found from the n component resistances:

$$R = \min_{i=1}^{n} (R_i) \qquad (13.55)$$

where R_i is the resistance of the ith component.

If the component resistances are independent, the distribution function, $F_R(r)$, of a series system can be written:

$$F_R(r) = 1 - \prod_{i=1}^{n} [1 - F_{Ri}(r)] \qquad (13.56)$$

Series systems may have very poor reliability characteristics. The reliability of a series system will always be equal to or less than the reliability of the least reliable element. The more elements are added to the series system, the worse its reliability becomes.

The third characteristic of a system is the statistical correlation between component failure events. Correlation between component resistances indicates that there is some dependence between their strengths, so that if one element is above its mean strength, there is a high probability that the other strengths also exceed their mean. Such correlation may arise as a result of common material sources, similar fabrication, inspection, and control methods, and perhaps a uniform interpretation of the strength formulae by the designer. Independence of element resistance means that the strengths are not statistically correlated. More importantly, in cases of environmental loadings, correlation of component failure events results from the fact that the same external loading affects the reliability of each component.

For a series system, increasing the correlation between element resistances decreases the probability of failure (or increases the reliability) of the system.

Independence of element strengths increases the probability that one of the elements (the 'weakest link') will fail, and hence the system will fail. Increasing statistical correlation improves the reliability of a series system, i.e. common or identical material supply, fabrication, inspection, testing, etc. increases reliability.

Unlike a series system, a **parallel system** allows for redistribution of load following an element failure. As the failure of a single element is not tantamount to failure of the system, parallel systems are often referred to as 'fail-safe' systems. In a true parallel system, all elements must fail for the system to fail.

The reliability characteristics of parallel systems depend upon the element failure mode. Parallel systems can be considered for both ductile elements and brittle elements.

Ductile parallel systems have good reliability characteristics. The reliability is at least as good as the most reliable element. Increasing the number of parallel elements increases the reliability of the system.

The coefficient of variation of system resistance R is increased by any positive correlation between elements. Thus, increasing the correlation between component resistances increases the uncertainty in the system resistance, and increases the probability of failure (or decreases reliability). Increasing correlation reduces the reliability of a parallel system.

Brittle elements that fail carry little or no load thereafter. In parallel structures, the load that was carried by the failed element must be distributed to the remaining elements. This greatly complicates the reliability analysis. The results depend on the safety factor levels used. If a member adjacent to a failed element has sufficient reserve capacity, it can pick up the redistributed load. Otherwise, it will fail and a cascading effect of failed members may result.

Several factors suggest that except for highly redundant systems (e.g. yarn-type systems), the reliability of brittle parallel systems can be approximated by a 'weakest link' analysis. Unless strength variability or reserve margins are large, failure of one element will usually trigger consecutive failures in other elements following load redistribution. This failure pattern is often termed 'progressive collapse'.

13.6.3 Failure modes analysis

A failure mode is any distinct sequence of component failures that causes the system to fail. For example, in a simple frame, plastic hinges can form leading to beam, sway, and combined types of plastic mechanisms. In typical structural frames or trusses, there is a very large number of possible failure modes (some are more probable than others). Failure mode analysis methods seek the modes that are most likely to occur, and use these modes to estimate the system reliability. The individual failure modes are considerably less complex to analyze than the failure of the system.

The modal failure probability, $P(g_i)$, for mode i can be written in terms of the modal safety margin, g_i:

$$P(g_i) = \text{prob}(g_i \leqslant 0) \tag{13.57}$$

and can be calculated from a knowledge of the distribution of load and component resistances. For a particular failure mode to occur, all of the components in that mode must fail. Thus, a failure mode can be modeled as a parallel system. The structure fails if any failure mode occurs. The overall structure can be modeled as a system with the failure modes in series, and the structure can be thought of as a series–parallel system.

The reliability of a structural system can be improved by increasing the length of the failure path, i.e. increasing the number of components that must fail before the structure fails (increasing the number of components in parallel). The length of the failure path can be increased by increasing the structural redundancy (i.e. the degree of static indeterminacy). However, increasing the structural redundancy by increasing the number of members also increases the number of possible failure modes. Increasing the number of possible failure modes (a series system) decreases system reliability. Some initial calculations by Gorman (1985) show that the greatest marginal benefit is achieved in going from zero redundancy (i.e. statically determinate) to two and three degrees of redundancy.

For the system, we then have

$$P_f(\text{system}) = \text{prob}(\text{any } g_i \leqslant 0) \tag{13.58}$$

(a) Correlation between modes

In general, there will be correlation between the failure modes, i.e. given that one mode has a high safety margin there is a greater probability that another mode will also have a high safety margin. The modal correlation arises from the following: (a) correlation between the component strengths due to common material, design methods, fabrication, etc.; (b) a single component may appear in several different modes and create correlation between the modes; (c) most importantly, there is a common source of loading that affects the safety of each mode.

The correlation plays an important role on the system reliability. If, for example, load coefficient of variation (COV) greatly exceeds strength COV, there may be high modal correlation. In this case, the most critical mode (highest probability of occurrence) may be close to the overall system reliability. If strength COV exceeds load COV, the correlation between modes may be small. In this case, it is necessary to consider in the system model all modes with significant probabilities of occurrence and the system reliability may act like a chain or weakest link model.

(b) Large systems

For application to large-scale realistic structures, the system analysis should be divided into two parts: (a) engineering modeling, which means identification, description and enumeration of the statistically significant collapse modes, and (b) probabilistic calculations, to determine individual mode failure probabilities (or safety indices) and then combine them into an overall system assessment. For simple structures, such as one-story frames (plastic analysis), the identification and enumeration of collapse modes in terms of loads and component resistances is straightforward. For large structural frameworks, the search for critical modes is difficult and it is not possible to assure that all significant failure modes have been found (Moses and Rashedi, 1983).

Collapse mode events are correlated through loading and resistances, so an exact solution to equation (13.58) is usually impossible. Several investigations considered this combination problem by finding either approximate solutions or bounds (Nordal, Cornell and Karamchandani, 1987). Recently, there have been considerable improvements in system reliability bounds. These upper and lower bounds on system reliability include correlation. Monte Carlo simulation is another precise method to combine the statistics of each collapse mode to estimate the system reliability (Fu and Moses, 1988). Improvements in importance sampling techniques have made such approaches highly efficient.

13.6.4 A practical expression of system reliability

A conclusion drawn from the idealized behavior studies (e.g. series, parallel, ductile, brittle, correlated, independent) is that there is a multiplicity of possible modes which are interconnected in geometry, material behavior and correlation assumptions. This suggests that detailed failure mode analyses are needed for practical systems which can be studied to define realistic reliability ranges within the parametric values likely to be encountered.

Gorman and Moses (1979) had initiated studies to develop an incremental loading method. This was extended by Moses and Stahl (1979) to offshore platform frameworks. The method leads to a failure mode equation and second-order modeling (mean and variance) of structure systems appropriate for computing safety indices.

For general member behavior, the enumeration of significant modes is difficult and various fault tree search procedures are necessary. The object is to express the reserve margins, g, for each failure mode in terms of basic load and element resistance variables. This simplifies the subsequent computation of modal safety indices. The combination of these failure modes into a system reliability index is also complex.

To investigate a general system reliability approach, consider the incremental analysis technique. It identifies a failure mode by following a load path from initial component failure to system collapse and leads to a linear failure expression. The basic steps of the incremental loading method are quite simple. A much greater problem is the detailed 'book-keeping' that is involved in generating and keeping track of each significant possible failure alternative in a large realistically modeled system.

Gorman (1979) used heuristic techniques and Monte Carlo simulation to cause different failure sequences to appear. The latter were encouraged by using artificially high strength coefficients of variation. Linear programming and nonlinear programming techniques have also been used for finding ductile collapse modes for frames first analyzed for basic plastic mechanisms (Rashedi and Moses, 1986). More general approaches for systems with different member behavior have used safety index criteria to enumerate significant failure modes. It is important for large frameworks to reduce the number of structural reanalysis.

The incremental procedure begins with critical components and the possible existence of redundant load paths. For design and quality assurance, it also provides the overall importance of a member to the load carrying capacity of the system under a particular loading condition. The steps are outlined as follows:

1. Analyze the structural system intact with no members failed.
2. Identify critical components. This may be based on (a) mean strength as done above, (b) a safety index, or (c) other measures of importance based on professional judgment.
3. Allow a critical component to fail and recognize its ductile, brittle or semi-brittle behavior.
4. Examine the changes in component utilization ratios as candidates for a failure sequence.
5. For each component identified in step 4, extrapolate the load factor to find its failure value. Ignore sequences which have high load factors.
6. Remove in succession components identified in steps 4 and 5 as remaining important.
7. The procedure is continued until either the load needed to cause additional failures is larger and hence the path can be ignored or else several components have failed with little change in load factor suggesting that the load is flattening out.
8. Continue the analysis (steps 3–7) with other critical members.

This strategy may be viewed with a failure tree (Moses and Rashedi, 1983). Each branch leads to a failure mode expression. Some paths may be similar so failure mode correlation in assessing system reliability is important. It is often found that only two or three failed components may provide a sufficiently accurate reliability. Further, changes in utilization ratios for typical frameworks suggest that a component failure significantly affects only members located adjacent to it or in a parallel load path arrangement.

The identification strategy can be automated for analyzing large frameworks (Bjerager, Karamchandani and Cornell, 1987). Moses and Rashedi (1983) extended the basic incremental loading model to include brittle and semibrittle components, multiple loads, significant mode identification and even partial failure models. Applications included both evaluation of design concepts (e.g. structure topology and redundancy) and inspection criteria. The latter utilize the tree search to identify which members are critical to the system and should require additional quality review in design, construction and maintenance. The difficult task of combining the failure mode expressions into an overall system risk was also studied by Gorman (1979) who used Monte Carlo simulation. Fu and Moses (1988) found a new approach in importance sampling that both provided good accuracy and high efficiency. The major recent extensions of the failure tree system modeling were done by Cornell and his associates (Bjerager, Karamchandani and Cornell, 1987; Nordal, Cornell and Karamchandani, 1987). They have developed programs for practical modeling of offshore platform frameworks. Among the advances were improved assessments and comparisons of alternate structure geometries and topologies, effects of redundancy, simulated damage conditions and detailed investigation of structure redistribution capabilities, improved reanalysis, more accurate modal combination probabilities and more realistic post-failure behavior models. Further work was identified to include more detailed and realistic post-failure element models including combined loading cases and more intelligent generation of failure paths.

13.7 SYSTEM RELIABILITY-BASED OPTIMIZATION

13.7.1 General

In the previous section, various methods and techniques for reliability analysis of structural systems have been considered. Depending on the type and topology of the structure, material, configuration, sizing of members, and types of connections, as well as on the loading acting on the structure, the reliability of the system can be vastly different. The results of structural system reliability analysis may be used to assess the adequacy and relative merits of alternative system designs with respect to established objectives and to find the optimum solution. This section is devoted to this subject.

13.7.2 Unconstrained and single-constrained formulations

A summary of the classical (i.e. single-objective) system reliability-based optimization formulations was given by Parimi and Cohn (1978) and Frangopol (1987). The context of these classical formulations is provided by (a) the philosophy of limit states design for both steel and reinforced concrete structures, and (b) the structural system reliability and optimization approaches. The classical system reliability-based optimization formulations may be classified into three groups, namely

1. minimizing the total expected cost, which includes (a) the initial cost as a function of design variables, and (b) the expected loss in case of failure (equation (13.2)),
2. minimizing the initial cost (or structure weight), with system failure probability specified as a constraint, and
3. minimizing the system failure probability, with initial cost or weight specified as a constraint.

As mentioned previously, in the unconstrained formulation 1 some of the expected loss such as human life may not be easily converted into monetary value. Therefore, such a formulation is difficult to implement. Contributions to this formulation and to its applications have been made by Turkstra (1970), Rosenblueth and Mendoza (1971), Mau and Sexsmith (1972), Frangopol and Rondal (1976), Surahman and Rojiani (1983), Rojiani and Bailey (1984), and Rosenblueth (1986). Formulation 2 implicitly assumes a lifetime expected cost translated to the allowable value of the system failure probability.

Among the most used single-objective single-constraint system reliability-based optimization formulations pertaining to groups 2 and 3 above it is worthwhile mentioning the following. For group 2,

1. minimizing the initial cost (weight) W, when the allowable probability of collapse of the system P^o_{fcol} is prescribed (the probability of loss of system serviceability P_{funs} is of no concern):

$$\min W(\mathbf{X}) \tag{13.59}$$

$$P_{fcol}(\mathbf{X}) \leqslant P^o_{fcol} \tag{13.60}$$

where $\mathbf{X} = (X_1 \, X_2 \cdots X_n)^t$ is the design variable vector;

2. minimizing the initial cost (weight) W, when the allowable probability of unserviceability of the system P^o_{funs} is prescribed (the probability of system collapse P_{fcol} is of no concern):

$$\min W(\mathbf{X}) \tag{13.61}$$

$$P_{funs}(\mathbf{X}) \leqslant P^o_{funs} \tag{13.62}$$

For group 3,

1. minimizing the probability of collapse of the system P_{fcol}, when the allowable initial cost (weight) W^o is prescribed (the probability of system unserviceability is of no concern):

$$\min P_{fcol}(\mathbf{X}) \tag{13.63}$$

$$W(\mathbf{X}) \leqslant W^o \tag{13.64}$$

2. minimizing the probability of system unserviceability, P_{funs}, when the allowable initial cost (weight) W^o is prescribed (the probability of system collapse is of no concern):

$$\min P_{funs}(\mathbf{X}) \tag{13.65}$$

$$W(\mathbf{X}) \leqslant W^o \tag{13.66}$$

Fabrication or other side constraints can also be added to the above four formulations in the form given by equation (13.7).

13.7.3 Multiconstrained formulations

In practice there are many situations in which the designer wants to optimize a structure with regard to the occurrence of two or more limit states. In general, the problem of optimization under uncertainty with three limit states (i, j, k) as the failure criteria may be stated in mathematical programming form as follows (Frangopol, 1987):

$$W = W(\mathbf{X}) \leqslant W^o \quad \text{cost (weight) constraint} \tag{13.67}$$

$$P_{fi} = P_{fi}(\mathbf{X}) \leqslant P^0_{fi} \quad \text{reliability constraint with respect to limit state } i \tag{13.68}$$

$$P_{fj} = P_{fj}(\mathbf{X}) \leqslant P^0_{fj} \quad \text{reliability constraint with respect to limit state } j \tag{13.69}$$

$$P_{fk} = P_{fk}(\mathbf{X}) \leqslant P^0_{fk} \quad \text{reliability constraint with respect to limit state } k \tag{13.70}$$

in which $\mathbf{X}$ is the vector of design variables, W the initial cost (or weight) of the structure, P_{fi} the probability of occurrence of limit state i (e.g. excessive deformation), P_{fj} the probability of occurrence of limit state j (e.g. instability), P_{fk} the probability of occurrence of limit state k (e.g. plastic collapse), and the superscript ^{0} denotes prescribed (allowable) values.

Sixteen different single-objective multiconstrained optimum design problems could be obtained from the combinations of equations (13.67)–(13.70) as follows:

$$\min W, \text{ subject to constraints } (13.68), (13.69) \text{ and } (13.70) \qquad (13.71)$$

$$\min P_{fi}, \text{ subject to constraints } (13.67), (13.69) \text{ and } (13.70) \qquad (13.72)$$

$$\min P_{fj}, \text{ subject to constraints } (13.67), (13.68) \text{ and } (13.70) \qquad (13.73)$$

$$\min P_{fk}, \text{ subject to constraints } (13.67), (13.68) \text{ and } (13.69) \qquad (13.74)$$

$$\min W, \text{ subject to constraints } (13.68) \text{ and } (13.69) \qquad (13.75)$$

$$\min W, \text{ subject to constraints } (13.68) \text{ and } (13.70) \qquad (13.76)$$

$$\min W, \text{ subject to constraints } (13.69) \text{ and } (13.70) \qquad (13.77)$$

$$\min P_{fi}, \text{ subject to constraints } (13.67) \text{ and } (13.69) \qquad (13.78)$$

$$\min P_{fi}, \text{ subject to constraints } (13.67) \text{ and } (13.70) \qquad (13.79)$$

$$\min P_{fi}, \text{ subject to constraints } (13.69) \text{ and } (13.70) \qquad (13.80)$$

$$\min P_{fj}, \text{ subject to constraints } (13.67) \text{ and } (13.68) \qquad (13.81)$$

$$\min P_{fj}, \text{ subject to constraints } (13.67) \text{ and } (13.70) \qquad (13.82)$$

$$\min P_{fj}, \text{ subject to constraints } (13.68) \text{ and } (13.70) \qquad (13.83)$$

$$\min P_{fk}, \text{ subject to constraints } (13.67) \text{ and } (13.68) \qquad (13.84)$$

$$\min P_{fk}, \text{ subject to constraints } (13.67) \text{ and } (13.69) \qquad (13.85)$$

$$\min P_{fk}, \text{ subject to constraints } (13.68) \text{ and } (13.69) \qquad (13.86)$$

The formulations given by equations (13.71)–(13.86) are multiconstraint system RBSO problems with three (equations (13.71)–(13.74) or two (equations (13.75)–(13.86)) constraints. The formulations (13.71), (13.75)–(13.77) are useful when the optimization consists of finding the minimum weight structure, given the allowable risk levels with regard to the occurrence of three (formulation (13.71)) or two (formulations (13.75)–(13.77)) limit states. The multiconstraint formulations given by equations (13.72)–(13.74) and (13.78)–(13.86) are useful when the most reliable structure with respect to the occurrence of a limit state is sought, given (a) the allowable cost (formulations (13.72)–(13.74), (13.78), (13.79), (13.81), (13.82), (13.84) and (13.85) and the preassigned values of the risk levels with respect to the occurrence of the other limit state(s) of concern, or (b) the preassigned values of the risk levels with respect to the occurrence of two other limit states concern (formulations (13.80), (13.83) and (13.86)).

It is interesting to note that when only two of the four equations (13.67)–(13.70) are chosen in the system reliability-based optimization process, twelve different formulations are possible, as follows:

$$\min W, \text{ subject to constraint } (13.68) \qquad (13.87)$$

$$\min W, \text{ subject to constraint } (13.69) \qquad (13.88)$$

min W, subject to constraint (13.70)   (13.89)

min P_{fi}, subject to constraint (13.67)   (13.90)

min P_{fi}, subject to constraint (13.69)   (13.91)

min P_{fi}, subject to constraint (13.70)   (13.92)

min P_{fj}, subject to constraint (13.67)   (13.93)

min P_{fj}, subject to constraint (13.68)   (13.94)

min P_{fj}, subject to constraint (13.70)   (13.95)

min P_{fk}, subject to constraint (13.67)   (13.96)

min P_{fk}, subject to constraint (13.68)   (13.97)

min P_{fk}, subject to constraint (13.69)   (13.98)

The above single-constrained formulations are useful when the system optimization consists of determining (a) the minimum weight structure, subject to a reliability constraint with respect to occurrence of a single limit state (formulations (13.87)–(13.89)) (also the formulations of equations (13.59) and (13.60) and of equation (13.61) and (13.62)), or (b) the most reliable structure with respect to the occurrence of a single limit state, given constraints on weight (formulations (13.90), (13.93) and (13.96)) or on risk of occurrence of another limit state (formulations (13.91), (13.92), (13.94), (13.95), (13.97) and (13.98)).

The selection of the objective function, allowable cost (weight) level, limit states and associated risk levels are crucial considerations in system RBSO. As shown previously, a variety of objectives can be chosen in the optimization process. Consequently, an optimum design solution is always relative. The selection of allowable cost and/or risk levels in RBSO is a problem beyond the scope of this chapter. According to limit states design specifications (e.g. Ellingwood *et al.*, 1980; Ellingwood and Galambos, 1982; Corotis, 1985) a moderate (say, $10^{-2} \geqslant P_f^0 \geqslant 10^{-3}$) and a low (say, $10^{-4} \geqslant P_f^0 \geqslant 10^{-6}$) risk level has to be required for serviceability and ultimate limit states, respectively.

As outlined in section 13.4, present reliability-based limit states design codes are almost exclusively aimed towards an element-level reliability format. This is achieved through the use of load and resistance factor design checking equations at the element level. Design according to these equations may not satisfy the required system reliability level. Conversely, an optimum design based on system reliability requirements only may not guarantee the element-level reliability requirements in the current codes. Consequently, RBSO with consideration of both system-level and element-level reliability requirements may at present produce an attractive balanced (system–element) risk consistent design approach. Efforts in this direction were made by Kim and Wen (1987), Nakib and Frangopol (1990a), and Mahadevan and Haldar (1991), among others.

13.7.4 Applications

The examples used herein to illustrate the applications of the system reliability-based optimization approach cover both steel and reinforced concrete structures. In addition to these examples, numerous additional applications are described by Cohn and Parimi (1972), Frangopol (1984b, 1986b, c), Ishikawa and Iizuka (1987), Kim and Wen (1987), Mahadevan and Haldar (1991), Moses (1969, 1970), Moses and Kinser (1967), Nakib and Frangopol (1990a), Parimi and Cohn (1978), Rackwitz and Cuntze (1987), Rojiani and Bailey (1984), Sørensen (1987, 1988), Stevenson and Moses (1970), Surahman and Rojiani (1983), Thoft-Christensen (1987), Thoft-Christensen and Murotsu (1986), and Vanmarcke (1971). The following single-limit state (Examples 13.4 to 13.6) and multilimit state (Example 13.7) system reliability-based optimization examples are only briefly presented. The interested reader will find the detailed data and explanation of these examples in the original references from which they were excerpted.

EXAMPLE 13.4 TWO-STORY TWO-BAY FRAME
(Moses and Stevenson, 1970; Moses, 1973)

The nondeterministic two-story two-bay steel frame under the mean random loads shown in Fig. 13.11 was optimized by Moses and Stevenson (1970) for plastic collapse failure using the system reliability-based optimization formulation of equations (13.59) and (13.60). There are six mean moment capacity design variables, including three for beams and three for columns, and as many as fifty-three collapse modes. A deterministic collapse analysis using linear programming would indicate at least six simultaneous active collapse mechanisms at the optimum. Table 13.1 (Moses, 1973) shows some examples of system RBSO optima. Moses and Stevenson (1970) pointed out that only one collapse mode

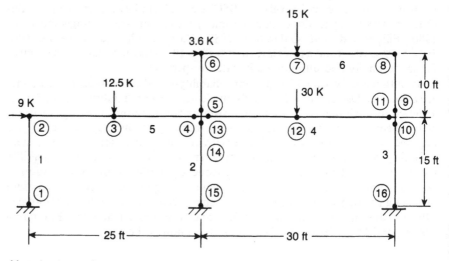

Mean loads are shown

1   Member number

①   Critical joint number

Fig. 13.11 Two-story two-bay frame (Moses and Stevenson, 1970).

Table 13.1 Optimum-design results of two-story two-bay frame shown in Fig. 13.11 (Moses, 1973)

Example number	Optimum mean moment capacities (klbf ft)						COV		Allowable P_f	Weight function	Frequency distribution
	M_1	M_2	M_3	M_4	M_5	M_6	Moment capacity	Load			
1	29.2	95.8	84.4	175.0	73.2	74.4	0.10	0.20	7.78(−2)*	312.47	Normal
2	27.8	96.3	84.4	173.8	72.0	77.9	0.10	0.20	7.80(−2)	312.89	Log-normal
Monte Carlo value of P_f (9500 trials)									7.59(−2)		
3	28.0	78.7	71.0	170.9	69.4	74.9	0.20	0.10	7.72(−2)	297.26	Normal
4	27.3	78.3	71.3	166.4	65.1	74.9	0.20	0.10	7.16(−2)	293.53	Log-normal
5	29.1	87.8	72.3	170.3	68.0	74.1	0.15	0.15	7.52(−2)	300.56	Normal
Monte Carlo value of P_f (7000 trials)									7.50(−2)		

*Exponents of failure probability are shown in parentheses (m) and should be read as 10^m.

dominates, in contributing most of the system failure probability to the allowable value. A check on the system reliability analysis was obtained for two cases in Table 13.1 by comparison with a Monte Carlo simulation of the system failure probability. It is interesting to note that the central-safety-factor checks alone, as used in conventional deterministic optimization methods, were not good indications of system failure probability. In fact, it was shown that some collapse modes of the frame with higher safety factors also had higher occurrence probabilities.

EXAMPLE 13.5 TWO-STORY ONE-BAY FRAME
(Frangopol, 1986b)

The constrained optimization formulation aimed at minimizing the probability of system collapse $P_f = P_{fcol}$ when the allowable weight $W = W^0$ is prescribed (equations (13.63) and (13.64)) is illustrated in Fig. 13.12(b) for the two-story one-bay nondeterministic steel plastic frame in Fig. 13.12(a) (Frangopol, 1986b). Figure 13.12(b) shows isosafety contours $P_f = P_{fcol}$ of the frame with respect to plastic collapse, obtained using sixty possible collapse modes, for two values of the prescribed weight function $W = 7000\,\text{kN m}^2$ and $W = 6500\,\text{kN m}^2$. The load and strength variabilities and correlations are included in the design. The design variables are the three mean values of the beams and columns plastic moments $\bar{M}_1$, $\bar{M}_2$ and $\bar{M}_3$ indicated in Fig. 13.12(a). As expected, a higher allowable weight means a lower optimum probability of collapse.

Both the two-story two-bay frame (Moses and Stevenson, 1970) and the two-story one-bay frame (Frangopol, 1986b) showed that the optimum proportioning of a nondeterministic structure with many collapse modes for minimum weight (Moses and Stevenson, 1970) or for minimum collapse failure probability (Frangopol, 1986b) is a complex interplay of members participating in a different collapse mode. The computer-based optimization is needed in both the system reliability analysis and the mathematical programming methods for finding the minimum weight or the maximum safety design.

EXAMPLE 13.6 STEEL PORTAL FRAMES (Frangopol, 1985a;
Nakib and Frangopol, 1990b)

- The nondeterministic steel plastic frame in Fig. 13.13(a) was optimized for minimum weight, with a prescribed collapse failure probability, $P_f = 10^{-5}$ (equations (13.59) and (13.60)). The two loads and plastic moments at the seven critical sections are assumed random variables with prescribed variabilities and correlations (Frangopol, 1985a).
   The purpose here is to study the sensitivity of the optimum solution of the frame shown in Fig. 13.13(a) to various types of plastic moment correlation

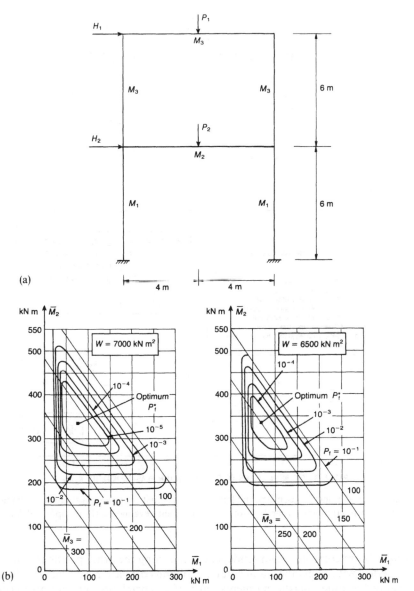

Fig. 13.12 Two-story one-bay frame: (a) geometry and loading; (b) minimization of the probability of collapse failure for a specified total weight, W (Frangopol, 1986b).

and various methods for determining the overall probability of plastic collapse of the structure. In order to study this sensitivity five different cases of resistance correlation are considered as follows (Frangopol, 1985a): (a) independence among all plastic moments; (b) perfect within columns and column–column correlation, and independence among the other plastic moments; (c) perfect

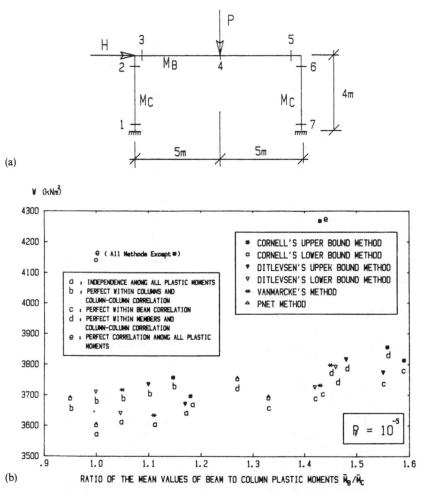

Fig. 13.13 Portal frame: (a) geometry and loading; (b) sensitivity of optimum solution to changes in strength correlation and in method for system reliability evaluation (Frangopol, 1985a).

within beam correlation, and independence among the other plastic moments; (d) perfect within members and column–column correlation, and independence among the other plastic moments; (e) perfect correlation among all plastic moments.

For each one of the five cases of correlation among plastic moments just mentioned, six different methods for determining the overall probability of plastic collapse failure of the frame are used: Cornell's lower and upper bounds method (Cornell, 1967); Ditlevsen's lower and upper bounds method (Ditlevsen, 1979); Vanmarcke's method (Vamarcke, 1971); PNET method, choosing the value 0.8 for the demarcating correlation coefficient (Ang and Ma, 1981). Figure 13.13(b) provides information on the effect of varying both the statistical correlation of the plastic moments and the method for determining the overall

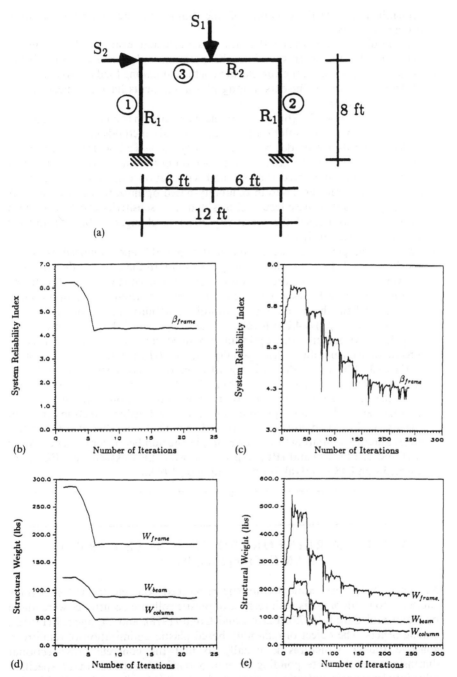

Fig. 13.14 Portal frame: (a) geometry and loading; (b) reliability index iteration history (method of feasible directions); (c) reliability index iteration history (interior penalty method); (d) weight iteration history (method of feasible directions); (e) weight iteration history (interior penalty method) (Nakib and Frangopol, 1990b).

probability of plastic collapse of the frame on the optimum solution (Frangopol, 1985a).

It should be emphasized that for this particular example the dominant failure modes that correspond to the optimum solution are always beam modes. For this reason the correlation coefficient among loads as well as some of the correlation coefficients among plastic moments have no effect on the optimum solution.

As shown in Fig. 13.13(b), plastic moment correlation can have a considerable effect on the optimum solution for each of the six methods considered.

The information provided by this sensitivity analysis allows for identification of those factors with the greatest influence on optimum solutions. It appears from the results presented that strength correlation can play an important role in the accuracy achieved in reliability-based optimization calculations. A further point is that these results shows that their sensitivity effect to strength correlation may exceed the sensitivity effect to changes in the method for system reliability evaluation.

- Another example of a nondeterministic steel portal frame optimized by Nakib and Frangopol (1990b) is shown in Fig. 13.14(a). In the optimization process, the weight of the structure is minimized with a constraint on the frame reliability index with respect to plastic collapse occurrence. The optimum is obtained when the minimum allowable reliability index of the frame $\beta^0_{col} = \Phi^{-1}(1 - P^0_{fcol}) = 4.265$ is reached (i.e. $P^0_{fcol} = 10^{-5}$). Using an interactive graphics environment, the optimum obtained by using both the feasible directions (FD) and the interior penalty function (IPF) methods are compared in Figs 13.14(b) to 13.14(e) with regard to convergence histories of both the system reliability index (Figs 13.14(b) and 13.14(c)) and weight (Figs. 13.14(d) and 13.14(e)). Despite the fact that both optimization methods achieve approximately the same optimum solution, the feasible directions method offers a significant advantage with regard to both the number of convergence cycles and the computing time. For example, the number of convergence cycles is 21 and 236 for FD and IPF, respectively, while their associated CPU times are 5.83 s and 48.18 s (Nakib and Frangopol, 1990b).

EXAMPLE 13.6 REINFORCED CONCRETE PORTAL FRAME (Frangopol, 1985d)

Attempts to optimize nondeterministic reinforced concrete framed structures subjected to random loads with respect to plastic collapse occurrence were made by Cohn and Parimi (1972), Parimi and Cohn (1978), and Frangopol (1985d), among others. The object of reliability-based plastic optimization of reinforced concrete framed structures is usually to find the concrete cross-sectional dimensions and the corresponding amount of reinforcing steel so that a specified reliability level against plastic collapse can be provided and an adopted objective be minimized. In general, it is assumed that the minimization of the total cost of concrete and longitudinal steel is the design objective. If the concrete section sizes are assigned from reliability-based elastic design computations, the reliability-

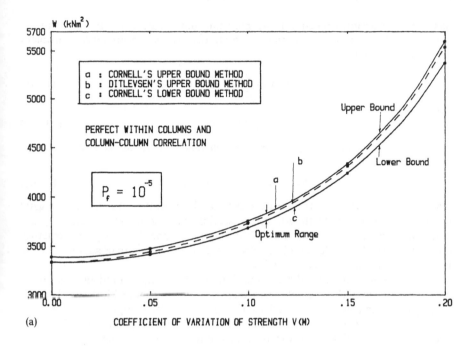

(a)　COEFFICIENT OF VARIATION OF STRENGTH V (M)

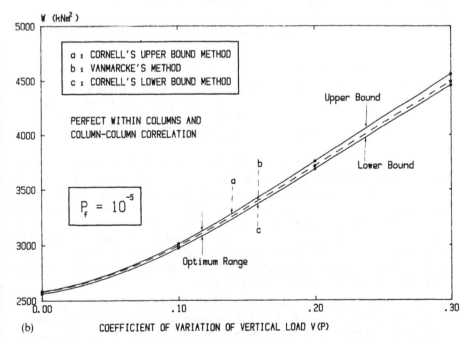

(b)　COEFFICIENT OF VARIATION OF VERTICAL LOAD V (P)

Fig. 13.15 Sensitivity of optimum solution of a reinforced concrete portal frame to changes in the coefficients of variation of (a) plastic moment (strength) and (b) vertical load (Frangopol, 1985d).

based plastic design objective is simpler: the minimization of the longitudinal steel weight (or volume). Using this assumption, the plastic optimization problem can be formulated as follows:

$$\min W = \min \sum_{j=1}^{s} (l_j^+ \bar{M}_j^+ + l_j^- \bar{M}_j^-) \tag{13.99}$$

subject to

$$P_f(\bar{\mathbf{M}}) \leqslant P_f^0 \tag{13.100}$$

in which P_f^0 is the allowable probability of plastic collapse, and $\bar{\mathbf{M}}$ is the optimum design vector defined by

$$\bar{\mathbf{M}} = \{\bar{M}_1^+ \ \bar{M}_1^- \cdots \bar{M}_s^+ \ \bar{M}_s^-\}^t \tag{13.101}$$

where $\bar{M}_j^+$ and $\bar{M}_j^-$ are the mean values of the positive and negative plastic moments associated with the jth critical section. The moments M_j^+ and M_j^- are assumed to govern over the prescribed lengths l_j^+ and l_j^- along the structure and, consequently, the longitudinal steel areas A_j^+ and A_j^- are constant over the lengths associated with the critical section j.

The optimum reliability-based solution of the nondeterministic reinforced concrete frame with the geometry and loading shown in Fig. 13.13(a) was obtained using the assumptions presented in Frangopol (1985d). Figures 13.15(a) and 13.15(b) show the sensitivity of the optimum objective (equation (13.99)) to changes in the coefficients of variation of strength (plastic moment) and vertical load, respectively. As expected, the optimum weight increases with increasing uncertainties in strength and/or loading.

EXAMPLE 13.7 MULTILIMIT STATE RELIABILITY ANALYSIS AND OPTIMIZATION OF A STEEL PORTAL FRAME (Frangopol and Nakib, 1987; Frangopol, 1985c)

Practical design is much more complex than the single-level system reliability approach illustrated in the previous examples. Usually, a structure is proportioned for conformance with a set of limit states including both serviceability and ultimate requirements. Therefore, to evaluate the reliability of a structure a multilevel system reliability approach has to be used. In order to accomplish this, an incremental mean loading technique was formulated for evaluating the risks associated with various limit states of a structural sytem (Moses, 1982; Frangopol and Nakib, 1986). This technique was applied in the following multilimit state reliability analysis and optimization examples.

- For a nondeterministic steel portal frame under two random loads (i.e. lateral load H and gravity load P), Fig. 13.16 (Frangopol and Nakib, 1987) indicates isosafety mean loading functions for four limit states including excessive deflection, first yield, first plastic hinge, and plastic collapse. All points on the three serviceability limit state curves (i.e. deflection, first yield, first plastic hinge) represent different combinations of mean loads leading to the same

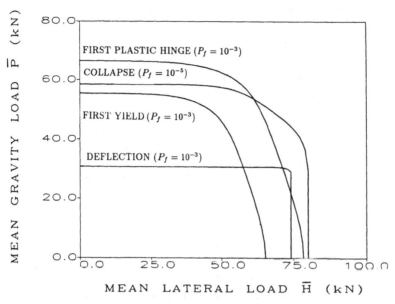

Fig. 13.16 Isosafety mean loading functions with respect to four limit states of a portal frame (Frangopol and Nakib, 1987).

value of the overall probability of frame unreliability (i.e. $P_f = 10^{-3}$, or $\beta = 3.09$), and all points on the ultimate limit state curve (i.e. plastic collapse) represent different combinations of mean loads leading to the same value of the overall probability of plastic collapse (i.e. $P_f = 10^{-5}$, or $\beta = 4.265$).

Figures 13.17 and 13.18 (Frangopol and Nakib, 1987) show the system reliability levels (i.e. system safety index β) with respect to three limit states (i.e. first yield, first plastic hinge, and plastic collapse in Fig. 13.17, and excessive deflection, first yield and first plastic hinge in Fig. 13.18) corresponding to all possible combinations of mean loads represented by the isosafety mean loading functions with respect to excessive deflection and plastic collapse, respectively. It is important to note that using the isosafety loading function against excessive deflection as the only serviceability limit state (Fig. 13.17) does not protect against unreliability by occurrence of other limit states for large values of the mean load ratio $\bar{H}/\bar{P}$ (i.e. say >2.0). Also, using the isosafety loading function against plastic collapse (Fig. 13.18) does not always protect against unreliability by loss of serviceability. Consequently, the serviceability and ultimate limit states have to be considered simultaneously in the reliability-based analysis and optimization processes. This latter case is demonstrated in the following example.

● This example is the rectangular fix-ended single-story rigid portal frame with deterministic geometry and random strengths (M_B is the plastic moment of the critical sections of the beam and M_C the plastic moment of the critical sections of the columns) shown in Fig. 13.13(a). The frame is subjected to two random concentrated loads H and P. The statistical information is given in Frangopol (1985c). The total weight of the frame is approximated by the linear

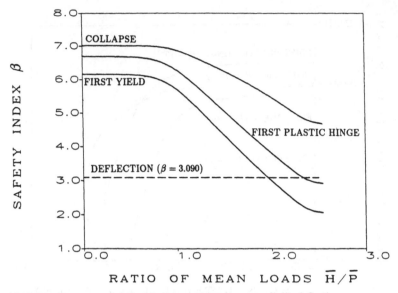

Fig. 13.17 Isosafety mean loading function with respect to limit deflection occurrence and associated reliability levels of three other limit states (Frangopol and Nakib, 1987).

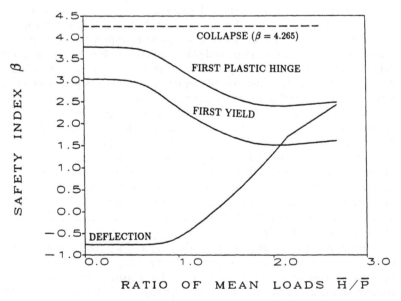

Fig. 13.18 Isosafety mean loading function with respect to plastic collapse occurrence and associated reliability levels of three other limit states (Frangopol and Nakib, 1987).

function $W = l\bar{M}_B + 2h\bar{M}_C$, in which $l = 10$ m (beam span) and $h = 4$ m (column height).

Figure 13.19 (Frangopol, 1985c) contains the design space with the two isoreliability contours with respect to plastic collapse and loss of serviceability: $P_{f1}^0 = 10^{-5}$ and $P_{f2}^0 = 10^{-2}$, respectively. All points on the curves $P_{f1}^0 = 10^{-5}$ and $P_{f2}^0 = 10^{-2}$ have the probabilities of plastic collapse and unserviceability equal to the prescribed values 10^{-5} and 10^{-2}, respectively. Also shown is the linear objective function, W, that should be minimized. The boundary of the feasible design space (which contains points with acceptable probabilities of failure) depends on the optimization formulation as follows: boundary $a_2bc_1de_2$ corresponds to the biconstraint (collapse and unserviceability) formulation given by equation (13.75) and boundaries $a_1bc_1de_1$ and $a_2bc_2de_2$ correspond to the single-constraint optimization formulations with reliability against collapse only (equations (13.59) and (13.60)) and unserviceability only (equations (13.61) and (13.62)), respectively. Consequently, if the biconstraint formulation given by equation (13.75) or the single-constraint formulation given by equations (13.59) and (13.60) is chosen, the optimum solution (the point at which W is tangent to the P_f constraint) is the design point c_1 (collapse

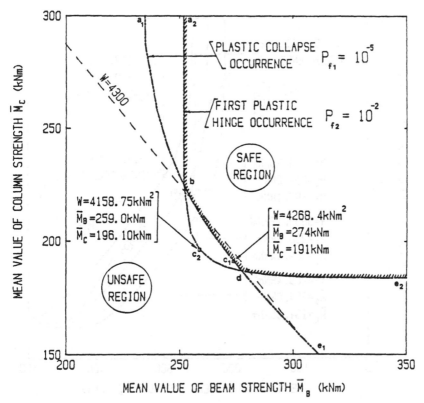

Fig. 13.19 Design space with two system reliability constraints used in minimum weight optimization of a steel portal frame (Frangopol, 1985c).

constraint is critical at the optimum). On the other hand, if the single-constraint formulation given by equations (13.59) and (13.60) is chosen, the optimum solution is the design point c_2. It is interesting to note that (a) both reliability constraints are satisfied as equalities at two design points (b and d), (b) the points lying on the curve bc_2d have unacceptable reliabilities with regard to plastic collapse ($P_{f1} > 10^{-5}$), and (c) the points lying on the curves a_1b and de_1 have unacceptable reliabilities with regard to loss of serviceability ($P_{f2} > 10^{-2}$).

Figures 13.20 and 13.21 (Frangopol, 1985c) show minimum weight solutions for various sets of specified risk levels against plastic collapse and unserviceability, respectively. Figure 13.20 presents solutions for different values of P_{f1}^0 when P_{f2}^0 is kept constant, while Fig. 13.21 presents solutions for different values of P_{f2}^0 when P_{f1}^0 is kept constant.

The results in Figs 13.20 and 13.21 indicate the possible fallacy in a single-limit-state optimization approach. For example, when a biconstraint optimization formulation is used the solutions d_1 in Fig. 13.20 and c_2 and d_2 in Fig. 13.21 have unacceptable reliabilities with regard to unserviceability and plastic collapse, respectively.

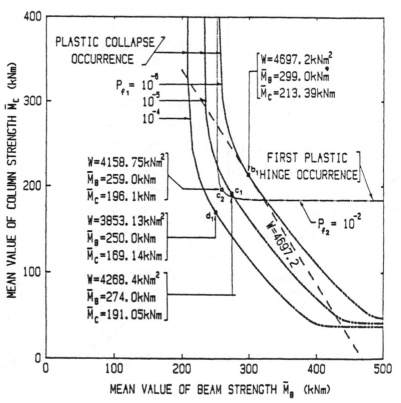

Fig. 13.20 Effect of prescribed risk level with respect to plastic collapse, P_{f1}^0, on optimum solution (Frangopol, 1985c).

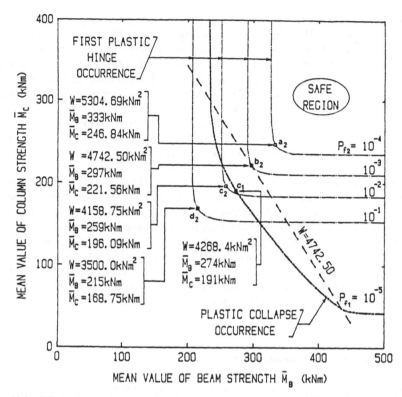

Fig. 13.21 Effect of prescribed risk level with respect to unserviceability, $P_{f_2}^0$, on optimum solution (Frangopol, 1985c).

13.8 MULTIOBJECTIVE RELIABILITY-BASED OPTIMIZATION

13.8.1 General

Single-objective optimization has been the basic approach in most previous work on the design of structural systems. The purpose was to seek optimal values of design variables which minimize or maximize a specific single quantity termed the objective function, while satisfying a variety of behavioral and geometrical conditions, termed constraints. In this definition of structural optimization, the quality of a structural system is evaluated using a single criterion (e.g. total expected cost, total weight, system reliability). For most design problems, however, engineers are confronted with alternatives that are conflicting in nature. For these problems, where the quality of a design is evaluated using several competing criteria simultaneously, multiobjective optimization (also called vector, multicriterion or Pareto optimization) should be used.

Multiobjective optimization of structural systems is an important idea that has only recently been brought to the attention of the structural optimization community by Duckstein (1984), Koski (1984), Osyczka (1984), Frangopol and Klisinski (1989b, 1991), and Fu and Frangopol (1990a, b), among others. It was shown that there are many structural design situations in which several conflicting objectives should be considered. For example, a structural system is expected to be designed such that both its total weight and maximum displacements be minimized. In such a situation, the designer's goal is to minimize not a single objective but several objectives simultaneously.

The main difference between single- and multiobjective optimization is that the latter almost always is characterized not by a single solution but by a set of solutions. These solutions are called Pareto optimum, noninferior or nondominated solutions. If a point belongs to a Pareto set there is no way of improving any objective without worsening at least another one. The advantage of the multiobjective optimization is that it allows one to choose between different results and to find possibly the best compromise. It requires, however, a considerable computation effort. In this section which is highly based on the results obtained during the past four years at the University of Colorado (Frangopol and Fu, 1989, 1990; Frangopol and Iizuka, 1991a, b; Frangopol and Klisinski, 1989b, 1991; Fu and Frangopol, 1990a, b), the mathematical formulation and the solution methods used in multiobjective structural optimization are briefly reviewed and examples of solutions for both nondeterministic truss and frame systems are presented.

13.8.2 Mathematical formulation

A multiobjective optimization problem can be formulated in the following way (Duckstein, 1984; Koski, 1984):

$$\min \mathbf{F(X)} \tag{13.102}$$

where $\mathbf{X} \in \Omega$ and $\mathbf{F} : \Omega \to R^m$ is a vector objective function given by

$$\mathbf{F(X)} = [F_1(\mathbf{X}) \, F_2(\mathbf{X}) \cdots F_m(\mathbf{X})]^t \tag{13.103}$$

Each component of this vector describes a single objective

$$F_i : \Omega \to R \qquad i = 1, 2, \ldots, m \tag{13.104}$$

The design variable vector $\mathbf{X}$ belongs to the feasible set Ω defined by equality and inequality constraints as follows:

$$\Omega = (\mathbf{X} \in R^n : \mathbf{H(X)} = \mathbf{0}, \mathbf{G(X)} \leqslant \mathbf{0}) \tag{13.105}$$

The image of the feasible set Ω in the objective function space is denoted by $\mathbf{F}(\Omega)$.

In a multiobjective RBSO problem, the components of the objective vector $\mathbf{F(X)}$ are certain characteristic properties of the structural system to be designed, such as failure cost, material volume or weight, system probabilities of ultimate collapse, first yielding, or plastic hinge occurrences, among others. The components of the design variable vector $\mathbf{X}$ are parameters to be detemined in the optimization process, such as cross-sectional dimensions or mean plastic moments. Some of the objective vectors used in RBSO were specified by Frangopol and Fu (1989, 1990), Frangopol and Iizuka (1991a, b), Fu and Frangopol (1990a, b), and

Iizuka (1991), as follows:

$$\mathbf{F}_1(\mathbf{X}) = (V(\mathbf{X}) \quad P_{fcol}(\mathbf{X}) \quad P_{fyld}(\mathbf{X}) \quad P_{fdfm}(\mathbf{X}))^t \tag{13.106}$$

$$\mathbf{F}_2(\mathbf{X}) = (V(\mathbf{X}) \quad P_{fint}(\mathbf{X}) \quad P_{fdmg}(\mathbf{X}))^t \tag{13.107}$$

$$\mathbf{F}_3(\mathbf{X}) = (V(\mathbf{X}) \quad P_{fint}(\mathbf{X}) \quad -R_1(\mathbf{X}))^t \tag{13.108}$$

$$\mathbf{F}_4(\mathbf{X}) = (V(\mathbf{X}) \quad -\beta_{yld}(\mathbf{X}) \quad -R_2(\mathbf{X}))^t \tag{13.109}$$

in which V is material volume of the structure, P_{fcol} the system probability of plastic collapse, P_{fyld} the system probability of first yielding, P_{fdfm} the system probability of excessive elastic deformation, P_{fint} the system probability of collapse initiated from the intact state (i.e. $P_{fint} = P_{fcol}$), P_{fdmg} the system probability of collapse initiated from a given damaged state (e.g. the system probability of collapse given that one element is completely removed), R_1 the redundancy factor of the system given by the ratio (Fu and Frangopol, 1990a)

$$R_1 = (P_{fdmg} - P_{fint})/P_{fint} \tag{13.110}$$

and R_2 the redundancy factor of the system given by the difference (Frangopol and Iizuka, 1991b)

$$R_2 = \beta_{col} - \beta_{yld} \tag{13.111}$$

in which β_{col} is the system reliability index with respect to plastic collapse (i.e. $\Phi^{-1}(1 - P_{fcol})$) and β_{yld} the system reliability index with respect to first yielding (i.e. $\Phi^{-1}(1 - P_{fyld})$). Other definitions of redundancy factors are given in Fu, Liu and Moses (1991), and Iizuka (1991), among others.

If the components of the objective vector $\mathbf{F}(\mathbf{X})$ are not fully independent there is not usually a unique point at which all the objective functions reach their minima simultaneously. As previously mentioned, the solution of a multi-objective optimization problem is called Pareto optimum or nondominated solution. Two types of nondominated solutions can be defined as follows (Duckstein, 1984; Koski, 1984):

- a point $\mathbf{X}_0 \in \Omega$ is weakly nondominated solution if and only if there is no $\mathbf{X} \in \Omega$ such that

$$\mathbf{F}(\mathbf{X}) < \mathbf{F}(\mathbf{X}_0) \tag{13.112}$$

- a point $\mathbf{X}_0 \in \Omega$ is a strongly nondominated solution if there is no $\mathbf{X} \in \Omega$ such that

$$F(\mathbf{X}) \leqslant F(\mathbf{X}_0) \tag{13.113}$$

and for at least one component i

$$F_i(\mathbf{X}) \leqslant F_i(\mathbf{X}_0) \qquad i = 1, 2, \ldots, m \tag{13.114}$$

If the solution is strongly nondominated it is also weakly nondominated. Such a strongly nondominated solution is also called Pareto optimum. In other words, the above definitions state that if the solution is strongly nondominated, no one of the objective functions can be decreased without causing a simultaneous increase in at least one other objective.

The main task of multiobjective optimization is to find the set of strongly nondominated solutions (also called Pareto optimal objectives or minimal curve) in the objective function space and the corresponding values of the design

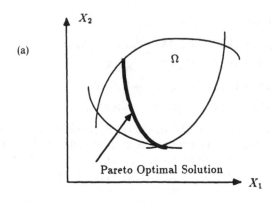

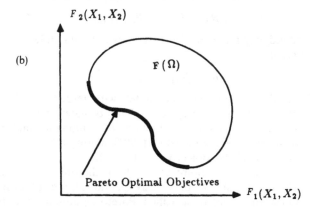

Fig. 13.22 Multiobjective optimization: (a) design space; (b) objective space (Fu and Frangopol, 1990b).

variables (Pareto optimal solution) in the feasible region space. Figure 13.22 (Fu and Frangopol, 1990b) qualitatively displays the feasible space Ω, the Pareto optimal solution, and its corresponding Pareto optimal objectives in a biobjective optimization problem with two design variables.

13.8.3 Solution techniques

There are a number of multiobjective optimization solution techniques described in the literature (Duckstein, 1984; Koski, 1984; Osyczka, 1984; among others). Not all of them are, however, suitable for reliability-based structural optimization. Three solution methods which may be applied to multiobjective reliability-based structural optimization are briefly described herein: the weighting method, the minimax method and the constraint method. For a more complete description of these methods the interested reader is referred to Duckstein (1984), Koski (1984), and Osyczka (1984), among others.

- The basic idea of the **weighting method** is to define the objective function F as a scalar product of the weight vector $\mathbf{W} = [W_1 \ W_2 \cdots W_m]^t$ and the objective vector $\mathbf{F}$, as follows:

$$F = \mathbf{W} \cdot \mathbf{F} \tag{13.115}$$

Without loss of generality the vector $\mathbf{W}$ can be nomalized. The Pareto optimal objective set can be theoretically obtained by varying the weight vector $\mathbf{W}$. In the objective function space the single objective function F is linear. For a fixed vector $\mathbf{W}$, the optimization process results in a point at which a hyperplane representing the single objective function is tangential to the image of the feasible set $\mathbf{F}(\Omega)$. Only if the set $\mathbf{F}(\Omega)$ is convex is the weighting method able to generate all the strongly nondominated solutions. The second drawback of this technique is the difficulty involved in choosing the proper weight factors. Since the shape of the image of the feasible set $\mathbf{F}(\Omega)$ is generally unknown it is almost impossible to predict where the solution will be located. Sometimes the problem is also very sensitive to the variation of weights. In such a case the weighting approach can prove to be unsatisfactory.

- Another method for solving vector optimization problems is the **minimax method** described in Koski (1984), among others. This method introduces distance norms into the objective function space. In this method the reference point from which distances are measured is the so-called ideal solution. This solution can be described by the vector

$$\mathbf{F}^* = [F_1^* \ F_2^* \cdots F_m^*]^t \tag{13.116}$$

where all its components are obtained as the solutions of m independent single-objective optimization problems:

$$F_i^* = \min_{\mathbf{X} \in \Omega} F_i(\mathbf{X}) \tag{13.117}$$

Generally, the ideal solution is not feasible, so it does not belong to the set $\mathbf{F}(\Omega)$.

In the minimax method the norm is defined as follows:

$$^-\max_i W_i(F_i - F_i^*) \qquad i = 1, 2, \ldots, m \tag{13.118}$$

where the components W_i and F_i of the vectors $\mathbf{W}$ and $\mathbf{F}$, respectively, were previously defined. For a given vector $\mathbf{W}$, the norm associated with equation (13.118) corresponds to the search in a direction of some line starting from the ideal solution. If the other norms are used, this approach generates the entire family of nondominated solutions.

The minimax approach eliminates one drawback of the weighting method, because it is also suitable for nonconvex problems. It may also, however, be sensitive to the values of the weight factors. The prediction where the solution will be located is improved. To use this method it is necessary to know first the ideal solution, which calls for solving m scalar optimization problems: it is the price to pay for using this method.

- Another alternative to the previous methods is the **constraint method**. The main idea of this method is to convert $m - 1$ objective functions into constraints. This can be obtained by the assumption that the values below some prescribed levels for all these functions are satisfactory. Without loss of generality it may be assumed that the components $F_2, F_3, \ldots, F_m$ of the objective vector will be constrained and only F_1 will be minimized.

In mathematical terms, the constraint method is formulated according to

Duckstein (1984) and Koski (1984) as follows. The original minimization problem (equations (13.102–13.105)) is replaced by

$$\min_{\Omega \cap \Omega_0} F_1(\mathbf{X}) \tag{13.119}$$

where

$$\mathbf{X} \in \Omega \cap \Omega_0 \text{ and } \Omega_0 = (\mathbf{X} : F_i(\mathbf{X}) \leqslant E_i, i = 2, \ldots, m) \tag{13.120}$$

and

$$\mathbf{E} = [E_2 \ E_3 \cdots E_m]^t \tag{13.121}$$

The entire Pareto set can be obtained by varying the **E** vector. The constraint method applies to nonconvex problems and does not require any additional computations.

The computational experience gained at the University of Colorado has indicated that the constraint method is the most appropriate technique for solving both deterministic (Frangopol and Klisinski, 1989b, 1991) and reliability-based (Fu and Frangopol, 1990b; Frangopol and Iizuka, 1991a, b; Iizuka, 1991) multiobjective structural optimization problems. For this reason, the constraint method may be treated as a basic numerical technique in multi-objective RBSO.

There also many other multiobjective optimization techniques (e.g. Duckstein, 1984; Koski, 1984; Osyczka, 1984), but they are less or not suitable for multi-objective RBSO.

13.8.4 Applications

The solution of a multiobjective RBSO formulation provides alternatives to the decision maker which are optimal in the Pareto (noninferior) sense. As the numerical result of this optimization process, a space of Pareto optimal objectives (decision support space) can be offered to the designers (Fig. 13.22(b)). This space contains optimal combinations of the individual objective functions. These combinations are optimal in the sense that none of the objectives can be further reduced without increasing at least one of the others. According to specific conditions encountered in design practice, the decision makers can choose from the decision support space one of the alternatives as the final preferred solution. For example, system reliabilities with respect to different limit states may be given different preference levels in order to reach a balanced design. Trade-off techniques for determining the preferred solution from a decision support space have been discussed in the literature by Osyczka (1984), among others. In the following numerical examples decision support spaces are provided for both nondeterministic truss and frame systems. The interested reader will find the detailed data and explanation of these examples in the original references from which they were excerpted.

EXAMPLE 13.8 FIVE-BAR TRUSS (Fu and Frangopol, 1990a, b)

Consider the simple steel truss structure shown in Fig. 13.23 (Fu and Frangopol, 1990b). This structural system consists of five uniaxial components and is acted on by the load S. The deterministic cross-sectional areas of the truss members constitute the design variable vector $\mathbf{A} = \mathbf{X}$. The three cross-sectional areas to be determined are A_1, A_2, and A_3 for vertical, horizontal, and diagonal members, respectively.

All members are assumed to exhibit ductile (elastic–perfectly plastic) behavior. The random variables considered in this problem are the three uniaxial capacities (C_1, C_2, and C_3) and the load S. The component capacities are related to component areas by $C_i = A_i \sigma_{yi}$ ($i = 1, 2, 3$), where σ_{yi} is the random yielding stress of component i; it contributes randomness to the component capacity C_i. The mean values of σ_{yi} for tension and compression are $25\,\mathrm{kN\,cm^{-2}}$ and $12.5\,\mathrm{kN\,cm^{-2}}$, respectively. The mean value of S is $23.8\,\mathrm{kN}$ and the coefficients of variation of σ_{yi} and S are 0.10 and 0.20, respectively. In the following, the five-bar truss is optimized successively considering the objective vectors $\mathbf{F}_2(\mathbf{X})$ and $\mathbf{F}_3(\mathbf{X})$ given by equations (13.107) and (13.108), respectively.

- *Objective vector* $\mathbf{F}_2(\mathbf{X})$(*Fu and Frangopol, 1990b*). In this case, the truss in Fig. 13.23 is optimized to satisfy the conditions

$$\min[V(\mathbf{A}), P_{\mathrm{fint}}(\mathbf{A}), P_{\mathrm{fdmg}}(\mathbf{A})] \tag{13.122}$$

subject to

$$\mathbf{A} \in \Omega = (\mathbf{A} \mid A_i \geqslant 2\,\mathrm{cm^2}) \qquad i = 1, 2, 3 \tag{13.123}$$

where $V(\mathbf{A})$ is the total material volume, $P_{\mathrm{fint}}(\mathbf{A})$ the probability of system collapse initiated from intact state, and $P_{\mathrm{fdmg}}(\mathbf{A})$ the probability of system collapse initiated from damaged state in which any one and only one of the five members of the truss is removed. The system failure probabilities P_{fint} and P_{fdmg} are evaluated by Ditlevsen's upper bound method (Ditlevsen, 1979).

Figure 13.24 (Fu and Frangopol, 1990b) displays the decision support space (Pareto optimal objectives) for the five-bar truss shown in Fig. 13.23, which is obtained by conducting the multiobjective optimization given by equations (13.122) and (13.123) with the constraint method described earlier. This space indicates trade-offs among the three objectives, V, P_{fint} and P_{fdmg}. A point on the surface shown in Fig. 13.24 represents a combination of the objectives

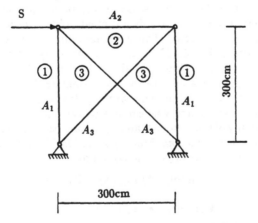

Fig. 13.23 Five-bar truss: geometry and loading (Fu and Frangopol, 1990b).

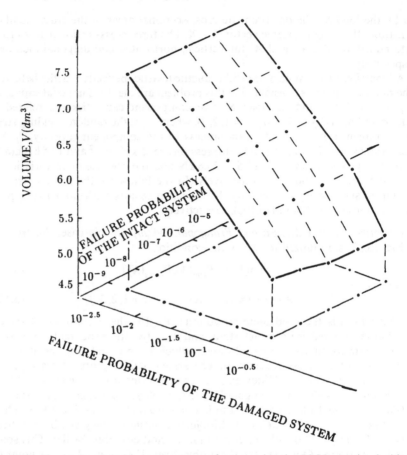

Fig. 13.24 Five-bar truss: decision support space for three-objective RBSO associated with the objective vector $F_2(X)$ (Fu and Frangopol, 1990b).

which is optimal in the Pareto sense. These points provide alternatives to decision makers. V and P_{fdmg} are the dominant objectives for this problem in most of the region considered. This is because one component removed from the five-bar truss in Fig. 13.23 results in a statically determinate structural system which has a much lower reliability level than the intact system. This low level of residual reliability of the indeterminate five-bar system is taken into account in the optimal design searching. It is important to note that for high redundant truss systems, where removal of any member will not drastically affect the reliability of the system, V and P_{fint} could be the dominant objectives.

- *Objective vector* $F_3(X)$(*Fu and Frangopol, 1990a*). In this case, the truss in Fig. 13.23 is optimized to satisfy the conditions

$$\min[V(A), P_{fint}(A), - R_1(A)] \qquad (13.124)$$

subject to

$$A \in \Omega = (A \,|\, A_i \geqslant 2\,\text{cm}^2) \qquad i = 1, 2, 3 \qquad (13.125)$$

where the redundancy factor R_1 is given by equation (13.110).

The trade-offs among the three objectives in equation (13.124) are described by the decision support space shown in Fig. 13.25 (Fu and Frangopol, 1990a). It is important to notice that considering only the system reliability in structural optimization does not always result in a rational design in the sense of redundancy gaining. For example, Figure 13.25 shows $\ln(R_1) = 8$ when V is minimized to $5\,\text{dm}^3$ under the constraint $P_{\text{fint}} \leqslant 10^{-8}$. However, in this case, the decision space in Fig. 13.25 indicates that by consuming about the same amount of material under the same system reliability requirement the system redundancy can be substantially increased to $\ln(R_1) = 11$. This is reached by including both reliability and redundancy in the optimization process.

EXAMPLE 13.9 ONE-STORY TWO-BAY FRAME
(Fu and Frangopol, 1990b)

A nondeterministic steel frame under three random loads $(S_i,\ i = 1, 2, 3)$ is considered in this example as shown in Fig. 13.26 (Fu and Frangopol, 1990b). This single-story two-bay structure has five random bending components $(M_i,\ i = 1, 2, \ldots, 5)$. The design variables are the deterministic cross-sectional areas $\mathbf{A} = \mathbf{X}$ of those components. Three cross-sectional areas are to be determined, namely left-hand beam A_2, right-hand beam A_3 and columns A_1.

The frame in Fig. 13.26 is optimized considering the objective vector $\mathbf{F}_1(\mathbf{X})$ given by equation (13.106). The formulation is as follows:

$$\min[V(\mathbf{A}), P_{\text{fcol}}(\mathbf{A}), P_{\text{fyld}}(\mathbf{A}), P_{\text{fdfm}}(\mathbf{A})] \tag{13.126}$$

subject to

$$\mathbf{A} \in \Omega = (\mathbf{A} \,|\, A_i \geqslant 2.5\,\text{cm}^2) \qquad i = 1, 2, 3 \tag{13.127}$$

Twenty-eight failure modes are considered for computing P_{fcol}. The evaluation of P_{fyld} involves consideration of the elastic moments at twelve critical sections (six column sections and six beam sections) of the frame shown in Fig. 13.26. They are either at a connection or at a load acting section. The system failure due to yielding occurrence is evaluated by considering that at least one of those twelve moments may exceed the elastic moment capacity of the corresponding critical section. The linear elastic deformations at three sections are considered for the evaluation of P_{fdfm}, namely section 7 (in the horizontal direction), and sections 3 and 6 (both in the vertical direction). They are regarded as excessive if 4 cm, 4 cm, and 6 cm are exceeded, respectively.

Figure 13.27 (Fu and Frangopol, 1990b) presents the decision support space (i.e. Pareto optimal objectives) for the four-objective vector optimization problem associated with the frame in Fig. 13.26. It exhibits a group of constant volume surfaces subject to the three failure probabilities of collapse, first yielding and deformation. These surfaces display the interaction among the four objectives in equation (13.126). A point on these isovolume surfaces is defined by its coordinates in the three-dimensional unreliability decision space. It can be seen in Fig. 13.27 that the collapse probability is not interactive in most cases. Its effect increases when this probability is very low (close to 10^{-9}) and the other two failure probabilities are relatively high (say, $\geqslant 10^{-3}$). This can be observed in the lower

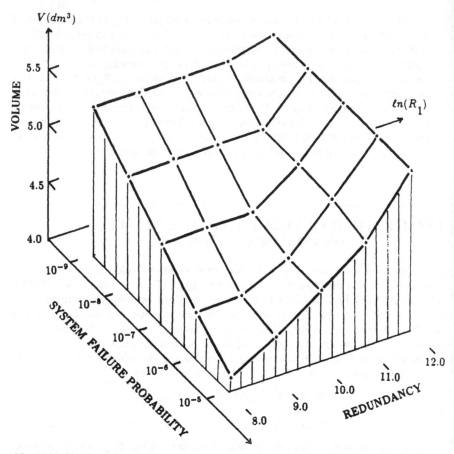

Fig. 13.25 Five-bar truss: decision support space for three-objective RBSO associated with the objective vector $F_3(X)$ (Fu and Frangopol, 1990a).

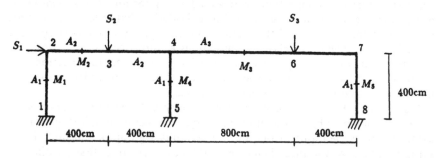

Fig. 13.26 One-story two-bay frame: geometry and loading (Fu and Frangopol, 1990b).

right-hand part of the decision space shown in Fig. 13.27. It is then possible to consider only the other two failure probabilities in certain regions for final decision making. This observation also indicates that P_{fcol} may not be the dominant factor to be taken into account in design. Considering only the collapse risk in the

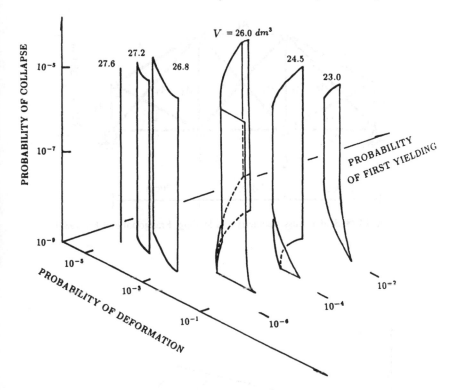

Fig. 13.27 One-story two-bay frame: decision support space for four-objective RBSO (Fu and Frangopol, 1990b).

optimization process may implicitly sacrifice reliability with respect to excessive deformation or to other limit states of concern.

EXAMPLE 13.10 TRUSS BRIDGE (Frangopol and Fu, 1989)

The multiobjective optimization of the nondeterministic truss bridge in Fig. 13.28 (Frangopol and Fu, 1989) is presented here for illustration. The material to be used for uniaxial components is assumed ductile (elastic–perfectly plastic). The design variables are chosen as the cross-sectional areas of these uniaxial components, namely $\mathbf{A} = \mathbf{X}$.

Considering the symmetric vertical (vehicle) load S as shown in Fig. 13.28, symmetric components are required. The members are classified in four deterministic design variables (i.e. cross-sectional areas) as follows: vertical members (A_1), lower chord members (A_2), diagonal members (A_3), and upper chord members (A_4). They are indicated also in Fig. 13.28.

The statistical parameters of the random variables (i.e. load S and resistances C_1, C_2, C_3, C_4 of the four types of members) are given in Frangopol and Fu

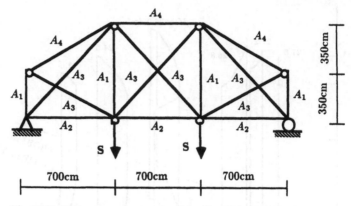

Fig. 13.28 Bridge truss: geometry and loading (Frangopol and Fu, 1989).

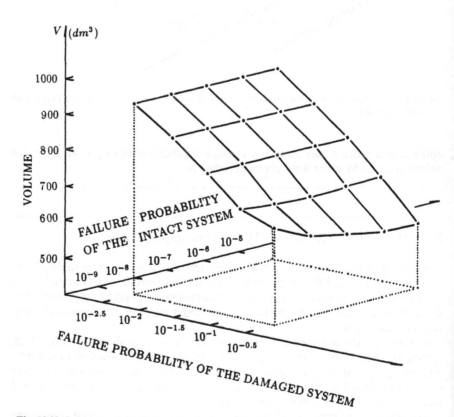

Fig. 13.29 Bridge truss: decision support space for three-objective RBSO associated with the objective vector $\mathbf{F}_2(\mathbf{X})$ (Frangopol and Fu, 1989).

(1989). It is noted that the compression capacities of the truss members take into account buckling effects indirectly. Consequently, they are different from the tension capacities.

The truss bridge is optimized considering the objective vector $\mathbf{F}_2(\mathbf{X})$ given by equation (13.107). The formulation is as follows:

$$\min[V(\mathbf{A}), P_{\text{fint}}(\mathbf{A}), P_{\text{fdmg}}(\mathbf{A})] \tag{13.128}$$

subject to

$$\mathbf{A} \in \Omega = (\mathbf{A} \,|\, A_i \geqslant 25\,\text{cm}^2) \qquad i = 1, 2, 3, 4 \tag{13.129}$$

In equation (13.128), P_{fint} and P_{fdmg} have the same meanings as in the five-bar truss example (equation (13.122)) optimized previously.

Figure 13.29 (Frangopol and Fu, 1989) shows the decision support space obtained by solving the problem defined by equations (13.128) and (13.129). It provides Pareto optimal objectives to the decision maker for the final decision. It is easy to recognize that V and P_{fdmg} are dominant figures for the final decision making. Consequently, the final design can be determined based only on the trade-offs between V and P_{fdmg}, and P_{fint} will be automatically kept at a satisfactorily low level.

EXAMPLE 13.11 TWO-STORY FOUR-BAY FRAME (Frangopol and Iizuka, 1991a, b)

Consider the nondeterministic two-story four-bay steel frame shown in Fig. 13.30 (Frangopol and Iizuka, 1991a). The geometry of the frame ($L_1 = 3\,\text{m}$, $L_2 = 4\,\text{m}$, $H_1 = 6\,\text{m}$, $H_2 = 4.8\,\text{m}$) and the cross-sectional area design vector $\mathbf{X} = [A_1$ (bottom columns) A_2 (bottom beams) A_3 (top columns) A_4 (top beams)$]^t$ are deterministic. The loads and the yield stress at each section are random variables with distributions, mean values, coefficients of variation, and correlations given in Iizuka (1991).

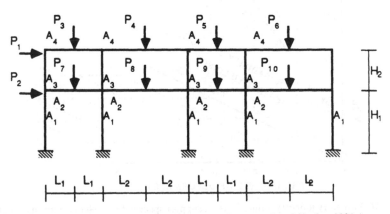

Fig. 13.30 Two-story four-bay steel frame (Frangopol and Iizuka, 1991a).

The two-story four-bay frame is optimized successively considering the objective vectors $\mathbf{F}_1(\mathbf{X})$ and $\mathbf{F}_4(\mathbf{X})$ given by equations (13.106) and (13.109), respectively,

- *Objective vector $\mathbf{F}_1(\mathbf{X})$ (Frangopol and Iizuka, 1991a).* In this case, the frame in Fig. 13.30 is optimized to satisfy the conditions

$$\min[V(\mathbf{A}), P_{fcol}(\mathbf{A}), P_{fyld}(\mathbf{A}), P_{fdfm}(\mathbf{A})] \qquad (13.130)$$

subject to

$$\mathbf{A} \in \Omega = (\mathbf{A} \mid A_i \geqslant 0) \qquad i = 1, 2, 3, 4 \qquad (13.131)$$

The system probabilities of collapse and unserviceability are computed using the method proposed by Ang and Ma (1981) and modified by Ishikawa and Iizuka (1987). The allowable vertical and horizontal elastic displacements of the frame in Fig. 13.30 are given in Iizuka (1991). The probabilities P_{fyld} and P_{fdef} are calculated using forty-four critical sections (i.e. top and bottom for each column, and two joints and the midspan for each beam) and ten reference sections (i.e., one for each load), respectively.

Figures 13.31 and 13.32 (Iizuka, 1991) show different representations of decision support spaces for the Pareto optimal objectives associated with the frame in Fig. 13.30. For example, in Fig. 13.31 the isoprobability of collapse surfaces are drawn in the three-dimensional space defined by the structural volume and probabilities of deformation and yielding. This space is more convenient than that of Fig. 13.32, when the designer's purpose is to balance cost and serviceability requirements given the collapse reliability level. Consequently,

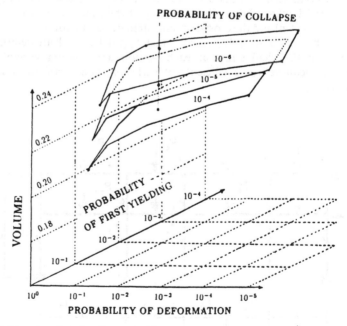

Fig. 13.31 Two-story four-bay frame: decision support space with isoreliability surfaces against collapse for four-objective RBSO associated with the objective vector $\mathbf{F}_1(\mathbf{X})$ (Iizuka, 1991).

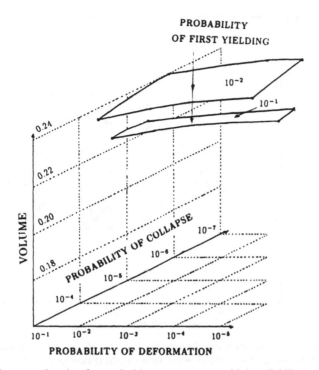

PROBABILITY
OF FIRST YIELDING

Fig. 13.32 Two-story four-bay frame: decision support space with isoreliability surfaces against first yielding for four-objective RBSO associated with the objective vector $\mathbf{F}_1(\mathbf{X})$ (Iizuka, 1991).

appropriate decision support spaces must be chosen according to specified purposes.

- *Objective vector* $\mathbf{F}_4(\mathbf{X})$ *(Frangopol and Iizuka, 1991b)*. In this case, the frame in Fig. 13.30 is optimized to satisfy the conditions

$$\min[V(\mathbf{A}), -\beta_{\text{yld}}(\mathbf{A}), -R_2(\mathbf{A})] \tag{13.132}$$

subject to

$$\mathbf{A} \in \Omega = (\mathbf{A} \mid A_i \geqslant 0) \qquad i = 1, 2, 3, 4 \tag{13.133}$$

where the reliability index with respect to first yielding is $\beta_{\text{yld}} = \Phi^{-1}(1 - P_{\text{fyld}})$, and the redundancy factor R_2 is given by equation (13.111). Consequently, the purpose is to obtain Pareto optimum solutions by considering the weight (material volume) of the structure, the system reliability with respect to first yielding occurrence (i.e. system serviceability requirement), and system redundancy as three equally important goals of the design process.

Figure 13.33 (Frangopol and Iizuka, 1991b) displays the Pareto optimal objectives in the space of system yielding reliability index β_{yld} and system redundancy factor R_2. In this space five isovolume contours and the two borders which enclose the domain of strongly nondominated solutions are indicated. The variations of Pareto optimal objectives and solutions along the isovolume contour A–B in Fig. 13.33 are shown in Figs 13.34(a) and 13.34(b), respectively. Several trade-offs can be observed.

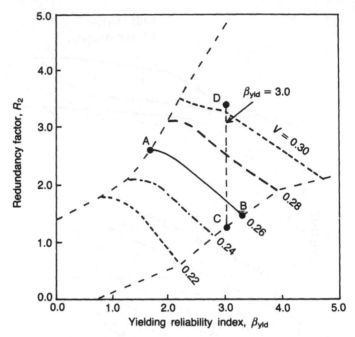

Fig. 13.33 Two-story four-bay frame: Pareto objective set with isovolume contours for three-objective RBSO associated with the objective vector $\mathbf{F}_4(\mathbf{X})$ (Frangopol and Iizuka, 1991b).

Figures 13.35(a) and 13.35(b) show the variations of Pareto optimal objectives and solutions, respectively, along the isoreliability contour C–D shown in Fig. 13.33. From Fig. 13.35(a) it can be observed that a small increase in volume causes great improvement of redundancy under a constant reliability index with respect to first yielding. The conclusion is obvious: it is surely worthwhile to add so little material to improve system performance so much as a result of higher redundancy.

13.9 RESIDUAL AND DAMAGE-ORIENTED RELIABILITY-BASED OPTIMIZATION

13.9.1 Introduction

A major limitation in many structural optimization programs is the use of deterministic constraints on stresses and performance. Overall safety should consider the statistical properties of loads, materials and analysis variables. System reliability analysis has improved as tools for predicting the overall safety under extreme load conditions. This section presents the optimization of structures with system reliability constraints imposed on both the performance of the original intact structure (reserve reliability) as well as the structure response after specified accident or damage scenarios (residual reliability). Several examples (in

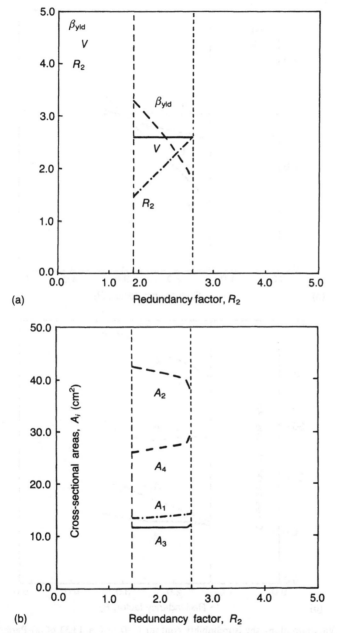

Fig. 13.34 Variation along the isovolume contour $A-B$ in Fig. 13.33 of (a) Pareto objective set and (b) Pareto solution set.

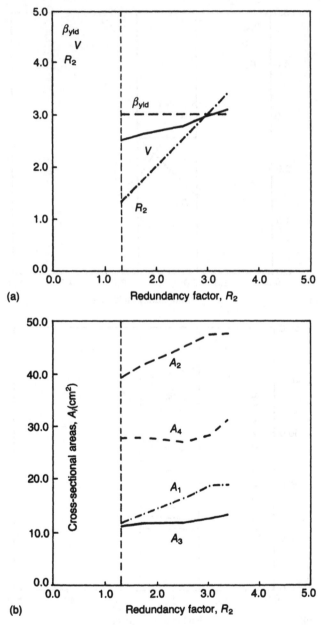

Fig. 13.35 Variation along the isoreliability contour $C-D$ in Fig. 13.33 of (a) Pareto objective set and (b) Pareto solution set.

addition to those presented in Section 13.8) show the benefits of such optimization for selecting both member sizes as well as structure topologies. Different behavior models include ductile and brittle behavior and degrees of strength correlation. The methods allow the designers to specify different accident scenarios and their corresponding target reliability levels as well as the target reliability for the intact geometry.

During the expected lifetime of a structure, various types of deterioration or damage may occur as a result of corrosion, fatigue and/or accidental loss of structural member capacity. Therefore, if the design only optimized the system reliability for the initial intact structure, this still does not guarantee that the structure is safe over its lifetime.

The method considers system reliability constraints for system reserve (for the initial installed structure) and system residual reliability corresponding to a damaged structural condition. These methods are illustrated for a truss structural optimization example.

In order to select a practical system reliability analysis, several methods including bounding and simulation were compared. The study found that the average value of the well-known Ditlevsen bounds on system reliability is an adequate approach owing to its accuracy, efficiency and especially its stable sensitivity analysis. The optimization algorithm uses a modified method of feasible directions for minimization.

Several examples are included to illustrate the minimum weight of structures subject to overall system reliability constraints both for the intact structure and for life cycle performance including damaged conditions.

13.9.2 System reliability formulations

Using the incremental loading methods of section 13.6, linear failure mode expressions (FME) are derived in the form of safety margins as follows:

$$g_j = \sum_j c_{ji} R_i - \sum_k d_{jk} S_k \quad (j = 1, 2, \ldots, M) \tag{13.134}$$

where g_j is the jth failure mode margin of safety, R_i is the component strength i in mode j, S_k are the loads acting in mode j, and c_{ji} and d_{jk} are constant coefficients representing the participation of strength and load variables in a failure mode. System failure occurs if any g_j is less than zero. M is the total number of significant failure modes identified. If the load and material statistical data is given, then the number of modes, M, that must be examined is fixed by the accuracy requirements. During the structure optimization, M may be changed as a result of the relative contributions of different collapse modes to the system reliability.

(a) Reliability bounding methods

The simplest system reliability bounds are of the following form (Cornell, 1967):

$$P_f(\text{sys}) \begin{cases} \leqslant \sum\limits_{m=1}^{M} P_{f,j} \\ \\ \geqslant \max\limits_{j} P_{f,j} \end{cases} \tag{13.135}$$

where $P_f(\text{sys})$ is the system failure probability and $P_{f,i}$ the failure probability of any single mode of collapse. Alternate and more exact system reliability bounds are available (Ditlevsen, 1979):

$$P_f(\text{sys}) \begin{cases} \leqslant \sum_{j=1}^{M} P_{f,j} - \sum_{j=2}^{M} \max_{i<j} P_{f,ij} \\ \geqslant P_{f,1} + \sum_{j=2}^{M} \max\left(0; P_{f,j} - \sum_{i=1}^{j-1} P_{f,ij} \right) \end{cases} \tag{13.136}$$

where the doubled subscripts of P_f indicate simultaneous occurrence of two collapse modes, and $P_{f,1}$ is the probability of occurrence of the dominant failure mode.

Several examples have demonstrated that the use of the bounds provides accurate and stable calculation of system reliability applicable to structures which typically have a multitude of significant collapse modes. The studies included brittle and ductile components, correlated and independent strength members and a range of strength and load variables.

13.9.3 Optimization formulation for truss structures

(a) Case 1: deterministic optimization for intact truss structures

The structural optimization considered is restricted to weight minimization of trusses with fixed geometry and material properties. The cross-sectional areas of structural members are the design variables, A_i. Consider the 15-bar truss (shown in Fig. 13.36) (strength correlation coefficient $\rho = 0.5$). The deterministic fully

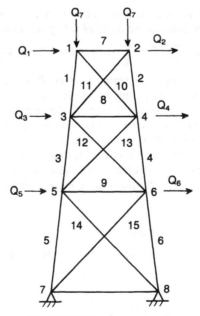

Fig. 13.36 15-bar truss example.

stressed design (FSD) for this truss structure, $\sigma = 16\,000\,\text{kN}\,\text{m}^{-2}$, corresponds to a total truss weight $W = 52.83$.

The reliability analysis for the FSD truss gave the following results for the system reliability index, β_s: for ductile material behavior,

$$\text{FSD truss} \quad \beta_s = 2.86$$
$$(\text{SF} = 2.0)$$

for brittle material behavior,

$$\text{FSD truss} \quad \beta_s = 1.72$$
$$(\text{SF} = 2.0)$$

It was found that the weight of structure is evidently reduced from that of the original design, by using a deterministic optimization program, but the safety level is also evidently reduced and therefore the failure probability is increased. It was also seen that while the original design has similar system reliability β_s for ductile and brittle behavior (3.79 and 3.41 respectively), the FSD β_s is much lower for brittle compared with ductile behavior. Thus, structures designed by deterministic optimization may not be optimal structures in a safety sense and, in some instances, these structures can be unsafe structures.

(b) Case 2: reliability optimization for intact truss structures

In order to overcome shortcomings in deterministic structure optimization, the reliability optimization for intact truss structures is compared. Find member areas (A), to

$$\text{min weight}(A)$$
$$\text{subject to } \beta_s(A) \geqslant \beta_s^* \tag{13.137}$$

where β_s^* is the target structural system reliability index required by the designer and $\beta_s(A)$ is the truss structural reliability index when design variables are A (the value used is β_M, the average of the Ditlevsen bounds).

Using this optimization formulation, the truss structure weight can be minimized while providing the safety levels required for the original intact structures. For example, with this formulation for the 15-bar truss and $\beta_s^* = 4.0$, the following results are found for the ductile case:

$$\text{original} \quad W = 86.43, \beta_s = 3.79$$
$$\text{optimum} \quad W = 76.16, \beta_s = 4.00$$

Thus, in this optimal formulation, the truss weight is actually reduced from 86.43 to 76.16 while at the same time the structure safety level is increased. Other optimization results were presented showing optimal weights for different target system β_s.

(c) Case 3: truss optimization under reserve and residual reliability constraints

During the expected lifetime of a structure, various types of deterioration or damage often occur as a result of corrosion, fatigue, fracture, and/or accidental loss of structural members. Therefore, sufficient reserve based on the system reliability constraints for the intact structure cannot ensure that the optimal structure is also safe during its expected or damaged lifetime.

After optimization, the weight of the 15-bar truss in Fig. 13.36 is 76.16 with the reliability (reserve or intact system) index $\beta_s = 4.0$. If the top horizontal bar (bar 7) is suddenly lost because of damage, then the system reliability β_s of the damaged structure is reduced to 1.15. This corresponds to a failure probability increased to 0.1251 from the intact value of 0.3167×10^{-4} or a ratio of 3950. The occurrence of such a lost member is not an uncommon event, say in marine offshore structures, due to fatigue, boat collision or dropped objects during operations.

Thus, all possible damage situations or accident scenarios during the expected lifetime of the structure should be considered. According to this safety philosophy, the following formulation is used for considering both system reserve and system residual reliability requirements simultaneously.

Find the vector A, denoted by (A), such that:

$$\min W(A)$$

subject to:

$$\beta_s(A) \geq \beta_s^*$$
$$\beta_k(A) \geq \beta_k^* \quad (k = 1, 2, \ldots, P) \tag{13.138}$$

where β_s^* is the target structural reliability index required for intact structure, β_k^* is the target structural reliability index required for the kth residual structure (possible damage scenarios must be defined), $\beta_s(A)$ is the intact truss structure reliability index, when design variables are A, $\beta_k(A)$ is the kth residual truss structure reliability index, also when design variables are A, and P is the number of residual or damage cases considered.

This optimal formulation can ensure a tolerable safety level for a structure during its expected lifetime for all possible damage cases. For example, with this optimal formulation consider the 15-bar truss structures with intact and lost bars again. Set $\beta_s^* = 4.0$ for the intact system and $\beta_7^* = 3.5$ for the scenario in which truss bar 7 accidentally fails. For this case, the new optimal structure weight is increased to 76.81 from $W = 76.16$ which is the optimal result for intact structure with $\beta_s^* = 4.0$. At the same time, $\beta_s = 4.0$ and β_7 is increased to 3.5 from $\beta = 1.15$. Thus, only a slight weight change makes the structure satisfactory.

As another example, consider the 13-bar truss shown in Fig. 13.37. Again, set $\beta_s^* = 4.0$ for the intact truss and $\beta_7^* = 3.5$ for a constraint for the truss without bar 7. Using only the intact system constraint ($\beta_s^* = 4.0$) the structure weight is 77.23 kN but $\beta_7 = -1.57$. That is, this truss is almost certain to fail if an accident damages member 7. This low β value in this case is due to the less redundant geometry in the 13-member truss compared with the 15-member truss considered above. Using both intact $\beta_s^* = 4.0$ and residual $\beta_7^* = 3.5$ constraints, the structure weight is increased to 85.21 from $W = 77.23$. This represents a weight increase of about 10%. Similarly, looking at other potential accident scenarios would also have the effect of raising the weight.

Comparisons of this nature also help to compare overall topologies such as the 15-bar vs 13-bar designs. In general, deterministic optimization will often lead to nonredundant deterministic structures. The approach herein accounts for both intact and possible damage scenarios in fixing member sizes.

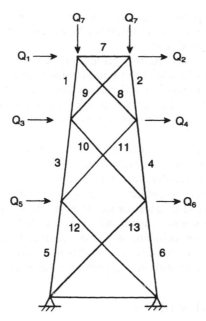

Fig. 13.37 13-bar truss example.

13.9.4 System reliability with a damage distribution

In some applications, a partial damage such as corrosion loss of strength should be considered in the optimization. The results above can be extended to a distribution of member damage with a random variable D.

The system probability of failure P_{RSD} can be found by a conditional integration over D as follows:

$$P_{RSD} = \int_0^1 P_{f_{(D=x)}} f_D(x)\, dx \qquad (13.139)$$

This formulation was given in Liu and Moses (1991) for different member damage formulations, and residual reliability targets. Results show weight increases with expected damage level. If damages in more than one member simultaneously are considered, then the optimal formulation can be expanded. Some illustrations were given in Liu and Moses (1991) for different member-group damages and residual reliability levels. In actual practice, uniform damage in all members is likely to be due to corrosion and is likely to be small. Large damage in single members will probably be repaired. In general, the damage distributions will have to be correlated to inspection intervals and maintenance practices.

13.10 CONCLUSIONS

- In order to obtain a consistent approach to safety in an optimization, the criteria should be based on a probabilistic formulation of failure events.

- Reliability-based optimization permits specification of either load or risk constraints and illustrates the interplay beteween cost allocation to specific components and overall risk.
- System reliability is needed to account properly for the large number of potential failure modes and their complex interaction in terms of common loadings, strength correlation and relative occurrence risks.
- System optimization can be cast in a risk formulation which has evolved to a stage which allows multiple collapse modes to be identified and overall risk to be computed.
- Reliability-based design has reached a stage where both serviceability and ultimate limit states constraints can be incorporated in the same probabilistic framework. This is especially important in certain structures such as reinforced concrete frames where serviceability limits may control the optimization.
- Formulation of system constraints for large structures has been accomplished using incremental loading schemes with fault tree logic. Realistic applications to weight and cost optimization with risk constraints have been reported.
- Further work is needed to make the models more accessible to designers and to provide a consistent database acceptable for making risk–benefit trade-offs.

ACKNOWLEDGEMENTS

Several postdoctoral researchers at both the University of Colorado and Case Western Reserve University including G. Fu, M. Klisinski, Y. Liu and D. Verma, and a number of former graduate students including M. Iizuka, M. Gorman, Y.-H. Lee, R. Nakib, J. Stevenson, R. Rashedi and K. Yoshida contributed to the results presented in this chapter. Their contributions and assistance are greatly appreciated. Support for part of the reliability-based structural optimization research at the University of Colorado has mainly come from the National Science Foundation under grants MSM-8618108, MSM-8800882 and MSM-9013017, and at Case Western Reserve University, also from the National Science Foundation grants and also research contracts with Transportation Research Board, Federal Highway Administration and Ohio Department of Transportation. This support is gratefully acknowledged.

REFERENCES

Ang, A. H.-S. and Cornell, C.A. (1974) Reliability bases of structural safety and design. *Journal of the Structural Division, ASCE*, **100**(ST9), 1755–69.

Ang, A. H.-S. and Ma, H.-F. (1981) On the reliability of structural systems, in Proceedings of the Third International Conference on Structural Safety and Reliability: ICOSSAR'81, Trondheim, pp. 295–314.

Ang. A. H.-S. and Tang, W.H. (1975) *Probability Concepts in Engineering Planning and Design*, Vol. I, Wiley, New York.

Ang, A. H.-S. and Tang, W.H. (1984) *Probability Concepts in Engineering Planning and Design*, Vol. II, Wiley, New York.

Arora, J.S. (1989) *Indroduction to Optimum Design*, McGraw-Hill, New York.

Baker, M. (1977) Rationalization of safety and serviceability factors in structural codes. *CIRIA Report No. 63*, Construction Industry Research and Information Association, London.

Benjamin, J.R. (1968) Probabilistic structural analysis and design. *Journal of the Structural Division, ASCE,* **94**(ST7), 1665–79.

Benjamin, J.R. and Cornell, C.A. (1970) *Probability, Statistics, and Decision for Civil Engineers,* McGraw-Hill, New York.

Bjerager, P. (1988) Prbability integration by numerical simulation. *Journal of Engineering Mechanics, ASCE,* **114**(8), 1285–302.

Bjerager, P., Karamchandani, A. and Cornell, C.A. (1987) Failure tree analysis in structural system reliability, in *Reliability and Risk Analysis in Civil Engineering* (ed. N.C. Lind), Vol. 2, Proceedings of ICASP 5, Vancouver, pp. 985–96.

Bucher, C.G. (1988) Adaptive sampling: an iterative fast Monte-Carlo procedure. *Structural Safety,* **5**(2), 119–26.

Chou, K.C., McIntosh, C. and Corotis, R.B. (1983) Observations on structural system reliability and the role of modal correlations. *Structural Safety,* **1**, 189–98.

Cohn, M.Z. and Parimi, S.R. (1972) Multi-criteria probabilistic design of reinforced concrete structures, in *Inelasticity and Non-linearity in Structural Concrete* (ed. M.Z. Cohn), SM Study No. 8, University of Waterloo, Waterloo, Ontario, pp. 471–92.

Cornell, C.A. (1967) Bounds on the reliability of structural systems. *Journal of the Structural Division, ASCE,* **93**(ST1), 171–200.

Cornell, C.A. (1969a) Bayesian statistical decision theory and reliability-based design, in Proceedings of the International Conference on Structural Safety and Reliability, Washington, DC, pp. 47–66.

Cornell, C.A. (1969b) A probability-based structural code. *Journal of the American Concrete Institute, ACI,* **66**(12), 974–85.

Corotis, R.B. (1985) Probability-based design codes. *Concrete International,* **7**, 42–9.

Deak, I. (1980) Three digit accurate multiple normal probabilities. *Numerische Mathematik,* **35**, 369–80.

Ditlevsen, O. (1979) Narrow reliability bounds for structural systems. *Journal of Structural Mechanics,* **7**(4), 453–72.

Duckstein, L. (1984) Multiobjective optimization in structural design: the model choice problem, in *New Directions in Optimum Structural Design* (eds. E. Atrek, R.H. Gallagher, K.M. Ragsdell and O.C. Zienkiewicz), Wiley, Chichester, pp. 459–81.

Ellingwood, B. and Ang, A. H.-S. (1974) Reliability bases of structural safety and design. *Journal of the Structural Division, ASCE,* **100**(ST9), 1771–88.

Ellingwood, B. and Galambos, T.V. (1982) Probability based criteria for structural design. *Structural Safety,* **1**, 15–26.

Ellingwood, B., Galambos, T.V., MacGregor, J.G. and Cornell, C.A. (1980) Development of a probability based load criterion for American National Standard A58. *NBS Special Publication 577,* Washington, DC.

Enevoldsen, I., Sørensen, J.D. and Sigurdsson, G. (1990) Reliability-based shape optimization using stochastic finite element methods. *Structural Reliability Theory Paper No. 73,* Institute of Building Technology and Structural Engineering, Aalborg University, Aalborg.

Feng, Y.S. and Moses, F. (1986a) A method of structural optimization based on structural system reliability. *Journal of Structural Mechanics,* **14**(4), 437–53.

Feng, Y.S. and Moses, F. (1986b) Optimum design, redundancy and reliability of structural systems. *Computers and Structures,* **24**(2), 239–51.

Ferry-Borges, J. (1954) *O Dimensionamento de Estruturas,* Ministry of Public Works, National Laboratory of Civil Engineering, **54**, Lisbon.

Forssell, C. (1924) Ekonomi och byggnadsvasen (economy and construction), *Sunt Fornoft,* **4**, 74–7 (in Swedish). (Translated in excerpts in Lind, S.M. (1970) *Structural Reliability and Codified Design,* SM study No. 3, Solid Mechanics Division, University of Waterloo, Waterloo.)

Frangopol, D.M. (1984a) A reliability-based optimization technique for automatic plastic design. *Computer Methods in Applied Mechanics and Engineering,* **44**, 105–17.

Frangopol, D.M. (1984b) Interactive reliability-based structural optimization. *Computers and Structures*, **19**(4), 559–63.

Frangopol, D.M. (1985a) Sensitivity of reliability-based optimum design. *Journal of Structural Engineering, ASCE*, **111**(8), 1703–21.

Frangopol, D.M. (1985b) Multicriteria reliability-based optimum design. *Structural Safety*, **3**(1), 23–8.

Frangopol, D.M. (1985c) Structural optimization using reliability concepts. *Journal of Structural Engineering, ASCE*, **111**(11), 2288–301.

Frangopol, D.M. (1985d) Towards reliability-based computer aided optimization of reinforced concrete structures. *Engineering Optimization*, **8**(4), 301–13.

Frangopol, D.M. (1986a) Computer-automated design of structural systems under reliability-based performance constraints. *Engineering Computations*, **3**(2), 109–15.

Frangopol, D.M. (1986b) Structural optimization under conditions of uncertainty, with reference to serviceability and ultimate limit states, in *Recent Developments in Structural Optimization* (ed. F.Y. Cheng), ASCE, New York, pp. 54–71.

Frangopol, D.M. (1986c) Computer-automated sensitivity analysis in reliability-based plastic design. *Computers and Structures*, **22**(1), 63–75.

Frangopol, D.M. (1987) Unified approach to reliability-based structural optimization, in *Dynamics of Structures* (ed. J.M. Roesset), ASCE, New York, pp. 156–67.

Frangopol, D.M. (ed.) (1989) *New Directions in Structural System Reliability*, University of Colorado, Boulder, CO.

Frangopol, D.M. and Corotis, R.B. (eds) (1990) System reliability in structural analysis, design and optimization. *Structural Safety*, **7**(2–4), 83–312.

Frangopol, D.M. and Fu, G. (1989) Optimization of structural systems under reserve and residual reliability requirements, in *Lecture Notes in Engineering* (eds C.A. Brebbia and S.A. Orszag), Vol. 48 (ed. P. Thoft-Christensen), Springer, Berlin, pp. 135–45.

Frangopol, D.M. and Fu, G. (1990) Limit states reliability interaction in optimum design of structural systems, in *Structural Safety and Reliability* (eds. A. H-S. Ang, M. Shinozuka and G.I. Schuëller), Vol. III, ASCE, New York, pp. 1879–86.

Frangopol, D.M. and Iizuka, M. (1991a) Multiobjective decision support spaces for optimum design of nondeterministic structural systems, in *Probabilistic Safety Assessment and Management* (ed. G. Apostolakis), Vol. 2, Elsevier, Amsterdam, pp. 977–82.

Frangopol, D.M. and Iizuka, M. (1991b) Pareto optimum solutions for nondeterministic systems, in Proceedings of the Sixth International Conference on Applications of Statistics and Probability in Civil Engineering, ICASP6 (eds L. Esteva and S.E. Ruis), Mexico City, Vol. 1, pp. 216–23.

Frangopol, D.M. and Klisinski, M. (1989a) Material behavior and optimum design of structural systems. *Journal of Structural Engineering, ASCE*, **115**(5), 1054–75.

Frangopol, D.M. and Klisinski, M. (1989b) Vector optimization of structural systems, in *Computer Utilization in Structural Engineering* (ed. J.K. Nelson), ASCE, New York, pp. 490–9.

Frangopol, D.M. and Klisinski, M. (1991) Computational experience with vector optimization techniques for structural systems, in *Lecture Notes in Engineering* (eds C.A. Brebbia and S.A. Orszag), Vol. 61 (eds A. Der Kiureghian and P. Thoft-Christensen), Springer, Berlin, pp. 99–111.

Frangopol, D.M. and Nakib, R. (1986) Isosafety loading functions in system reliability analysis. *Computers and Structures*, **24**(3), 425–36.

Frangopol, D.M. and Nakib, R. (1987) Reliability of structural systems with multiple limit states, in *Materials and Member Behavior* (ed. D.S. Ellifritt), ASCE, New York, 1987, pp. 638–46.

Frangopol, D.M. and Rondal, J. (1976) Considerations on optimum combination of safety and economy, in Final Report, Tenth Congress of the International Association for Bridge and Structural Engineering, Tokyo, pp. 45–8.

Frangopol, D.M., Klisinski, M. and Iizuka, M. (1991) Computational experience with damage-tolerant optimization of structural systems, in Proceedings of the 1st International Conference on Computational Stochastic Mechanics (eds P.D. Spanos and C.A. Brebbia),

Computational Mechanics Publications, Southampton, and Elsevier Applied Science, London, pp. 199–210.

Freudenthal, A.M. (1956) Safety and the probability of structural failure. *Transactions, ASCE,* **121**, 1337–75.

Fu, G. and Frangopol, D.M. (1990a) Balancing weight, system reliability and redundancy in a multiobjective optimization framework. *Structural Safety,* **7**(2–4), 165–75.

Fu, G. and Frangopol, D.M. (1990b) Reliability-based vector optimization of structural systems. *Journal of Structural Engineering, ASCE,* **116**(8), 2141–61.

Fu, G. and Moses, F. (1988) Importance sampling in structural system reliability, in *Probabilistic Methods in Civil Engineering* (ed. P.D. Spanos). Proceedings of the 5th ASCE Specialty Conference on Probabilistic Methods, Blacksburg, VA, pp. 340–3.

Fu, G., Liu, Y. and Moses, F. (1991) Management of structural system reliability, in *Lecture Notes in Engineering* (eds C.A. Brebbia and S.A. Orszag), Vol. 61 (eds A. Der Kiureghian and P. Thoft-Christensen), Springer, Berlin, pp. 113–28.

Gorman, M.R. (1979) Reliability of structural systems. *Report CE 79-2,* Civil Engineering Department, Case Western Reserve University, Cleveland, OH, May 1979.

Gorman, M.R. (1985) Resistance modeling. *ASCE Short Course Notes on Structural Reliability Analysis of Offshore Platforms,* May 1985.

Gorman, M.R. and Moses, F. (1979) Direct estimate of structural system reliability. ASCE Seventh Electronic Computation Conference, St. Louis, MO.

Grimmelt, M.J. and Schuëller, G.I. (1983) Benchmark study on methods to determine collapse failure probabilities of redundant structures. *Structural Safety,* **1**(2), 93–106.

Hasofer, A.M. (1987) Directional simulation with applications to outcrossings of Gaussian processes, in *Reliability and Risk Analysis in Civil Engineering* (ed. N.C. Lind), Vol. 1, Proceedings ICASP 5, Vancouver, pp. 35–45.

Hasofer, A.M. and Lind, N.C. (1978) Exact and invariant second-moment code format. *Journal of the Engineering Mechanics Division, ASCE,* **100**(EM1), 829–44.

Hilton, H.H. and Feigen, M. (1960) Minimum weight analysis based on structural reliability. *Journal of the Aerospace Sciences,* **27**, 641–53.

Iizuka, M. (1991) Time invariant and time variant reliability analysis and optimization of structural systems. Ph.D. Thesis, Department of Civil Engineering, University of Colorado, Boulder, CO.

Ishikawa, N. and Iizuka, M. (1987) Optimum reliability-based design of large framed structures. *Engineering Optimization,* **10**(4), 245–61.

Johnson, A.I. (1953) *Strength, Safety and Economical Dimensions of Structures,* Division of Building Statistics and Structural Engineering, Royal Institute of Technology, **12**, Stockholm.

Kalaba, R.E. (1962) Design of minimum weight structures given reliability and cost. *Journal of the Aerospace Sciences,* **29**, 355–6.

Kim, S.H. and Wen, Y.K. (1987) Reliability-based structural optimization under stochastic time varying loads. *Civil Engineering Studies, Structural Research Series No. 533,* University of Illinois, Urbana, IL.

Kim, S.H. and Wen, Y.K. (1990) Optimization of structures under stochastic loads. *Structural Safety,* **7**(2–4), 177–90.

Kirsch, U. (1981) *Optimum Structural Design,* McGraw-Hill, New York.

Koski, J. (1984) Multicriterion optimization in structural design, in *New Directions in Optimum Structural Design* (eds E. Atrek, R.H. Ragsdell and O.C. Zienkiewicz), Wiley, Chicester, pp. 483–503.

Lind, N.C. (1973) The design of structural design norms. *Journal of Structural Mechanics,* **1**(3), 357–70.

Liu, Y. and Moses, F. (1991) Bridge design with reserve and residual reliability constraints. *Structural Safety,* **11**(1), 29–42.

Madsen, H.O., Krenk, S. and Lind, N.C. (1986) *Methods of Structural Safety,* Prentice-Hall, Englewood Cliffs, NJ.

Mahadevan, S. and Haldar, A. (1991) Reliability-based optimization using SFEM, In *Lecture*

Notes in Engineering (eds C.A. Brebbia and S.A. Orszag), Vol. 61 (eds A. Der Kiureghian and P. Thoft-Christensen), Springer, Berlin, pp. 241–50.

Mau, S.-T. and Sexsmith, R.G. (1972) Minimum expected cost optimization. *Journal of the Structural Division, ASCE*, **98**(ST9), 2043–58.

Melchers, R.E. (1987) *Structural Reliability Analysis and Prediction*, Ellis Horwood, Chichester.

Moses, F. (1969) Approaches to structural reliability and optimization, in *An Introduction to Structural Optimization* (ed. M.Z. Cohn), SM Study No. 1, Solid Mechanics Division, University of Waterloo, Waterloo, Ontario, pp. 81–120.

Moses, F. (1970) Sensitivity studies in structural reliability, in *Structural Reliability and Codified Design* (ed. N.C. Lind), SM Study No. 3, Solid Mechanics Division, University of Waterloo, Waterloo, Ontario, pp. 1–18.

Moses, F. (1973) Design for reliability – concepts and applications, in *Optimum Structural Design* (eds R.H. Gallagher and O.C. Zienkiewicz), Wiley, New York, pp. 241–65.

Moses, F. (1974) Reliability of structural systems. *Journal of the Structural Division, ASCE*, **100**(ST9), 1813–20.

Moses, F. (1977) Structural system reliability and optimization. *Computers and Structures*, **7**, 283–90.

Moses, F. (1982) System reliability developments in structural engineering. *Structural Safety*, **1**(1), 3–13.

Moses, F. (1990) New directions and research needs in system reliability research. *Structural Safety*, **7**(2–4), 93–100.

Moses, F. and Kinser, D.E. (1967) Optimum structural design with failure probability constraints. *AIAA Journal*, **5**(6), 1152–8.

Moses, F. and Rashedi, M.R. (1983) The application of system reliability to structural safety, in *Applications of Statistics and Probability in Soil and Structural Engineering* (eds G. Augusti, A. Borri and G. Vannucchi), Proceedings ICASP 4, Vol. 1, Florence, pp. 573–84.

Moses, F. and Stahl, B. (1979) Reliability analysis format for offshore structures. *Journal of Petroleum Technology*, (March), 347–54.

Moses, F. and Stevenson, J.D. (1970) Reliability-based structural design. *Journal of the Structural Division, ASCE*, **96**(ST2), 221–44.

Moses, F., Fu, G. and Verma, D. (1989) Advanced simulation methods in system reliability, in *Computational Mechanics of Reliability Analysis* (eds W.K. Liu and T. Belytschko), Elme, Lausanne.

Murotsu, Y. and Shao, S. (1990) Optimum shape design of truss structures based on reliability. *Structural Optimization*, **2**(2), 65–76.

Nakib, R. and Frangopol, D.M. (1990a) RSBA and RSBA-OPT: two computer programs for structural system reliability analysis and optimization. *Computers and Structures*, **36**(1), 13–27.

Nakib, R. and Frangopol, D.M. (1990b) Reliability-based structural optimization using interactive graphics. *Computers and Structures*, **37**(1), 27–34.

Nordal, H., Cornell, C.A. and Karamchandani, A. (1987) A structural system reliability case study of an eight leg steel jacket offshore production platform, in Proceedings of the Marine Structural Reliability Symposium, SNAME, Arlington, VA, October 1987.

Osyczka, A. (1984) *Multicriterion Optimization in Engineering*, Ellis Horwood, Chichester.

Paez, A. and Torroja, E. (1959) *La Determination del Coefficiente de Seguridad en las Distintas Obras*, Instituto Technico de la Construccion y del Cemento, Madrid.

Parimi, S.R. and Cohn, M.Z. (1978) Optimum solutions in probabilistic structural design. *Journal of Applied Mechanics*, **2**(1), 47–92.

Rackwitz, R. and Cuntze, R. (1987) Formulations of reliability-oriented optimization. *Engineering Optimization*, **11**(1,2), 69–76.

Rackwitz, R. and Fiessler, B. (1978) Structural reliability under combined random load sequences. *Computers and Structures*, **9**, 489–94.

Rashedi, R. and Moses, F. (1986) Application of linear programming to structural system reliability. *Computers and Structures*, **24**(3), 375–84.

Rojiani, K.B. and Bailey, G.L. (1984) Reliability-based optimum design of steel structures, in *New Directions in Optimum Structural Design* (eds E. Atrek, R.H. Gallagher, K.M. Ragsdell and O.C. Zienkiewicz), Wiley, Chichester, pp. 443–57.

Rosenblueth, E. (1986) Optimum reliabilities and optimum design. *Structural Saftey*, 3(1), 69–83.

Rosenblueth, E. and Esteva, L. (1972) Reliability basis for some mexican codes, in *Probabilistic Design of Reinforced Concrete Buildings*, ACI Publication SP-31, pp. 1–41.

Rosenbleuth, E. and Mendoza, E. (1971) Reliability optimization in isostatic structures. *Journal of the Engineering Mechanics Division, ASCE*, 97(EM6), 1625–40.

Rosenbrock, H.H. (1960) An automatic method for finding the greatest or least value of a function. *Computer Journal*, 3, 175–84.

Schmit, L.A. (1969) Problem formulation, methods and solutions in the optimum design of structures, in *An Introduction to Structural Optimization* (ed. M.Z. Cohn), SM Study No. 1, Solid Mechanics Division, University of Waterloo, Waterloo, Ontario, pp. 19–46.

Shao, S. (1991) *Reliability-based shape optimization of structural and material systems*. Ph.D. Thesis, Division of Engineering, University of Osaka Prefecture, Osaka.

Shinozuka, M. (1983) Basic analysis of structural safety. *Journal of Structural Engineering, ASCE*, 109(3), 721–40.

Sørensen, J.D. (1987) Reliability-based optimization of structural systems. *Structural Reliability Theory Paper No. 32*, Institute of Building Technology and Structural Engineering, Aalborg University, Aalborg.

Sørensen, J.D. (1988) Optimal design with reliability constraints. *Structural Reliability Theory Paper No. 45*, Institute of Building Technology and Structural Engineering. Aalborg University, Aalborg.

Sørensen, J.D. and Enevoldsen, I. (1989) Sensitivity analysis in reliability-based shape optimization. *Structural Reliability Theory Paper No. 69*, Institute of Building Technology and Structural Engineering, Aalborg University, Aalborg.

Sørensen, J.D. and Thoft-Christensen, P. (1987) Integrated reliability-based optimal design of structures. *Structural Reliability Theory Paper No. 29*, Institute of Building Technology and Structural Engineering, Aalborg University, Aalborg.

Stevenson, J. and Moses, F. (1970) Reliability analysis of frame structures. *Journal of the Structural Division, ASCE*, 96(ST11), 2409–27.

Surahman, A. and Rojiani, K.B. (1983) Reliability-based optimum design of concrete frames. *Journal of Structural Engineering, ASCE*, 109(3), 741–57.

Switzky, H. (1964) Minimum weight design with structural reliability, in Proceedings of the Fifth Annual Structures and Materials Conference, pp. 316–22.

Thoft-Christensen, P. (1987) Application of optimization methods in structural systems reliability theory. *Structural Reliability Theory Paper No. 33*, Institute of Building Technology and Structural Engineering, Aalborg University, Aalborg.

Thoft-Christensen, P. (1991) On reliability-based structural optimization, in *Lecture Notes in Engineering* (eds C.A. Brebbia and S.A. Orszag), Vol. 61 (eds A. Der Kiureghian and P. Thoft-Christensen), Springer, Berlin, pp. 387–402.

Thoft-Christensen, P. and Baker, M.J. (1982) *Structural Reliability Theory and Its Applications*, Springer, Berlin.

Thoft-Christensen, P. and Murotsu, Y. (1986) *Applications of Structural Systems Relaibility Theory*, Springer, Berlin.

Turkstra, C.J. (1967) Choice of failure probabilities. *Journal of the Structural Division, ASCE*, 93(ST6), 189–200.

Turkstra, C.J. (1970) *Theory of Structural Design Decisions* (ed. N.C. Lind), SM Study No. 2, Solid Mechanics Division, University of Waterloo, Waterloo, Ontario.

Vanderplaats, G.N. (1984) *Numerical Optimization Techniques for Engineering Design: With Applications*, McGraw-Hill, New York.

Vanderplaats, G.N. (1986) *ADS – A Fortran Program for Automated Design Synthesis*, Version 1.10. Engineering Design Optimization, Inc. Santa Barbara, CA.

Vanmarcke, E. (1971) Matrix formulation of reliability analysis and reliability-based design. *Computers and Structures*, **3**, 757–70.

Vanmarcke, E. (1984) *Random Fields: Analysis and Synthesis*. The MIT Press, Cambridge, MA.

Verma, D., Fu, G. and Moses, F. (1990) Efficient structural system reliability assessment by Monte-Carlo methods, in *Structural Safety and Reliability* (eds A. H-S. Ang, M. Shinozuka and G.I. Schuëller), Vol. II, ASCE, New York, pp. 895–901.

Verma, D. and Moses, F. (1989) Calibration of bridge-strength evaluation code. *Journal of Structural Engineering, ASCE*, **115**(8), 1538–54.

Wen, Y.K. and Chen, H-C. (1989) System reliability under time varying loads: I. *Journal of Engineering Mechanics, ASCE*, **115**(4), 808–23.

Wirshing, P. (1985) Reliability methods. *ASCE Short Course Notes on Structural Reliability Analysis of Offshore Platforms*, May, 1985.

Yao, J.T.P. (1985) *Safety and Reliability of Existing Structures*, Pitman, Boston.

Zimmerman, J.J., Corotis, R.B. and Ellis, J.H. (1991) Stochastic programs for identifying significant collapse modes in structural systems, in *Lecture Notes in Engineering* (eds C.A. Brebbia and S.A. Orszag), Vol. 61 (eds A. Der Kiureghian and P. Thoft-Christensen), Springer, Berlin, pp. 359–67.

Index

Milton Keynes UK
Ingram Content Group UK Ltd.
UKHW021930071024
449327UK00022B/1750

9 780367 864026

Transport Processes in Concrete

Robert Černy and Pavla Rovnaníková

CRC Press
Taylor & Francis Group
Boca Raton London New York

CRC Press is an imprint of the
Taylor & Francis Group, an **informa** business
A SPON PRESS BOOK

CRC Press
Taylor & Francis Group
6000 Broken Sound Parkway NW, Suite 300
Boca Raton, FL 33487-2742

First issued in paperback 2019

© 2002 by Taylor & Francis Group, LLC
CRC Press is an imprint of Taylor & Francis Group, an Informa business

No claim to original U.S. Government works

ISBN-13: 978-0-415-24264-6 (hbk)
ISBN-13: 978-0-367-44704-5 (pbk)

This book contains information obtained from authentic and highly regarded sources. Reasonable efforts have been made to publish reliable data and information, but the author and publisher cannot assume responsibility for the validity of all materials or the consequences of their use. The authors and publishers have attempted to trace the copyright holders of all material reproduced in this publication and apologize to copyright holders if permission to publish in this form has not been obtained. If any copyright material has not been acknowledged please write and let us know so we may rectify in any future reprint.

Except as permitted under U.S. Copyright Law, no part of this book may be reprinted, reproduced, transmitted, or utilized in any form by any electronic, mechanical, or other means, now known or hereafter invented, including photocopying, microfilming, and recording, or in any information storage or retrieval system, without written permission from the publishers.

For permission to photocopy or use material electronically from this work, please access www.copyright .com (http://www.copyright.com/) or contact the Copyright Clearance Center, Inc. (CCC), 222 Rosewood Drive, Danvers, MA 01923, 978-750-8400. CCC is a not-for-profit organization that provides licenses and registration for a variety of users. For organizations that have been granted a photocopy license by the CCC, a separate system of payment has been arranged.

Trademark Notice: Product or corporate names may be trademarks or registered trademarks, and are used only for identification and explanation without intent to infringe.

Publisher's Note
This book has been prepared from camera-ready copy provided by the authors.

British Library Cataloguing in Publication Data
A catalogue record for this book is available from the British Library

Library of Congress Cataloging in Publication Data
A catalog record for this book has been requested

**Visit the Taylor & Francis Web site at
http://www.taylorandfrancis.com**

**and the CRC Press Web site at
http://www.crcpress.com**

Contents

Preface

Heat, moisture and chemical compounds transport in concrete are among the most fundamental phenomena determining the practical performance of concrete. For instance, mechanical properties of concrete depend on temperature, moisture and salt content in a significant way, and these parameters can only be determined by an application of advanced modelling of transport processes. Service-life prediction models cannot give reasonable results without appropriate modelling of transport processes. Any calculation in building physics cannot be done without modelling the transport phenomena in the proper way.

The main purpose of this book is to provide a complex physical-chemical analysis of transport processes in concrete aimed at direct application in the computational assessment of concrete performance in a variety of situations that occur on building sites. In the first part of the book, a survey of physical and chemical processes and transport mechanisms in concrete is made, and their significance for affecting heat, moisture and chemical compounds transport is analysed. Then, fundamentals of modelling transport phenomena in chemically reacting mixtures are given. In the principal part of the book, a critical analysis is undertaken of current models commonly employed for describing transport phenomena both generally in porous materials and particularly in concrete. As a result of this analysis, the most promising models suitable for advanced calculations of transport processes in concrete are recommended. An unavoidable condition for functionality of every mathematical model is an exact knowledge of its input parameters. Therefore, an overview of methods for experimental determination of both field variables and transport and storage parameters characterizing heat, moisture and chemical compounds transport in concrete is given. Finally, the possibilities of advanced models of transport processes in concrete are illustrated in typical examples of practical calculations, covering such topics as building-physics related assessment of the envelope, hydration-heat related problems in large-scale concrete structures, the behaviour of concrete in fire conditions, embedded steel corrosion induced by salt transport, and service life prediction.

The description of physical and chemical processes in concrete includes both the physical aspects, described in a similar way to general books on fluid transport in porous media, and chemical aspects, described in the manner of general books on properties of concrete. As a result, an integrated physical and chemical view on transport phenomena in concrete is provided.

The fundamentals of transport phenomena in materials are explained at the standard level for continuum physics, but contrary to the most purely theoretical books on this topic, the authors are aware of the fact that they write a book for engineers. Therefore the book is written in a way that can be understood easily, without need for derivations, i.e. there are no mathematical proofs and no lemmas, but there are physical discussions in detail.

The models of heat, moisture and chemical compounds transport are described in a very detailed way. In order to bring the topic near to the reader, the authors derive all the models themselves from the beginning to the end in much more detail than was done in the original papers, and indicate the strengths and weaknesses of each particular model.

The authors believe that the problem of modelling transport processes in concrete should not be treated as a calculation problem only, as is often done in the practice. They understand the modelling process as a complex system consisting of (i) model formulation, (ii) measuring all necessary input parameters of the model, and (iii) model verification, which can only be done by measuring exactly the same field variables as those generated by the model under the same conditions as those employed in the model. Therefore, not only models are formulated, but also methods for determination of hygric and thermal properties of concrete are surveyed in a detailed way, as well as methods for determination of basic field variables of heat, moisture and chemical-compound transport. In this way, the reader should be able to implement the above-mentioned modelling procedure in practice, not just theory.

The authors would like to express sincere thanks to Dr. John Grunewald, the developer of the DELPHIN 4.1 computer code, who provided us with complete documentation of the code, and to Dr. Dariusz Gawin, the developer of the computer code for modelling transport processes in concrete at high temperatures, who provided the necessary material in Section 6.3. Special thanks are also due to our families, particularly to Robert's wife Marie and his children Maruška and Robík, for their understanding during the writing of this book, which was often done at the expense of spending time with them.

Prague and Brno, July 2001   Robert Černý and Pavla Rovnaníková

## CHAPTER ONE

Physical and Chemical Processes in Concrete

1.1 THE COMPOSITION OF PORTLAND CEMENT

Cement is a powdered hydraulic binder, which is characterized by the fact that the products of its hydration are air-resistant, damp-proof and stable even under water. Based upon composition, cements are divided into silicate, aluminous and special cements. Portland clinker provides the basis of silicate cements, with a setting regulator added and some hydraulic active or pozzolana additives. Aluminous cement is made of a mixture of bauxite and limestone, but this cement has not been used for structural concrete for more than 40 years because the conversion of hydration products leads to mechanical failures. Since the first half of 20th century, Portland cement has been the most frequently used cement.

The sections to follow will solely deal with cement that is based on silicates, i.e. based on Portland clinker.

Cement is made by burning the raw material mix of suitable kinds of limestone containing 76–78% $CaCO_3$ and aluminosilicates contained in limestone or added externally in a form of clays. In case of lack of some constituents in the raw material mix, supplementary and correction components are added.

Raw materials may contain further components that may affect the quality of clinker. The presence of Na_2O and K_2O affect the temperature of melting, high alkali content leads to a potential alkali aggregate reaction if a reactive aggregate is present. Calcium fluoride CaF_2 accelerates the decomposition of calcite and leads to the decrease of temperature of alite formation. The content of chlorides in the raw material must be kept low, because there is a risk of steel corrosion in reinforced concrete and also the creation of kiln rings and preheater deposits. Magnesium oxide MgO causes magnesium expansion of cement by forming $Mg(OH)_2$ with greater molar volume.

Silicon dioxide, aluminium oxide and iron trioxide are added to the raw material mixture in a form of clay; the most suitable clays are those of kaolinite type, such as kaolinite, halloysite, allophane. Steel dust or siderite are used as correction components in the composition of the raw material mixture for increasing the content of Fe_2O_3, bauxite for Al_2O_3 and quartz sand or kieselguhr for SiO_2.

The proportion of components in the raw material mixture must be such that the entire CaO may react with the clinker minerals to prevent the pres-

Table 1.1 The mean chemical composition of Portland cement (Lea, 1971)

Constituent	Content [%]
CaO	60–66
SiO_2	20–24
Al_2O_3	4–8
Fe_2O_3	1.5–4
MgO	0.5–2
$Na_2O + K_2O$	0.8–1.5
SO_3	1–3.5

ence of free lime in clinker. Free CaO in clinker is dead burnt, and through delayed hydration causes the failures of cement binder.

From the relation for LSF (lime saturation factor),

$$\text{LSF} = CaO/(2.8SiO_2 + 1.18Al_2O_3 + 0.65Fe_2O_3),$$

it follows that if the LSF values equal 100, the whole CaO contained in the raw material mixture has reacted with the clinker minerals. The LSF of common types of cement is within the range of 87–92. Higher values are attained in cements of a marked content of C_3S and C_3A, which is conditioned by a very reactive raw material, otherwise free CaO originates (Šauman, 1993). The mixture is burnt in a rotary kiln at a temperature about 1450°C. The resulting product is Portland clinker, the composition of which depends upon the composition of the raw material mixture and the manner of burning.

The cooling of clinker is an important step in the production of cement. The rate of cooling affects the ratio between the crystalline and amorphous phases of clinker and also influences the size of crystals of clinker minerals. A slower rate of cooling leads to the formation of crystals, which is manifested in more difficult grinding of clinker and a lower degree of hydration. Fast cooling leads to the formation of C_3A predominantly in the vitreous phase, which is manifested in a higher resistance to the action of sulphates.

After being cooled and matured, clinker is ground together with gypsum, which regulates the rate of forming solid structures of the cement binder in its initial stage of hydration. In mixed cements, additives such as blast furnace granulated slag, fly ash, natural pozzolana, and others are added.

The chemical compositions of clinker and cement are expressed as the presence of particular oxides in percentage terms. The limits of concentrations of Portland cement are given in Table 1.1.

The content of the main oxides, i.e. of CaO, SiO_2, Al_2O_3, Fe_2O_3, usually amounts to 96–98%. Further components are in the minority, and in cement are specified based on their content in the raw materials used. The mineralogical composition and properties of cement follow from the chemical

Table 1.2 The average content of clinker minerals in clinker

Constituent	Content [%]
Tricalcium silicate (alite)	63
Dicalcium silicate (belite)	20
Tricalcium aluminate	8
Calcium aluminate ferite	7
Free CaO	< 1

composition. The kinds of cement that show a higher initial strength, a higher hydration heat and a lower resistance against the action of aggressive matters, have a higher content of CaO. The content of Al_2O_3 affects the properties in the same way. By contrast, the increase of SiO_2 and Fe_2O_3 content represents the decrease of the initial strength but an increase of long-term strength, lower hydration heat and higher resistance against the action of aggressive matters. The content of MgO is usually limited to 6% because of the possibility it can cause volume changes to the cement binder by the formation of $Mg(OH)_2$.

The content of SO_3 in cements is substantially increased by adding gypsum used for the regulation of setting. The concentration of SO_3 in cement does not usually exceed 3.5% of the materials.

The amount of alkalis depends on the technology of cement production. Common concentrations of alkalis are up to 1.5%, the amount of potassium oxide is two-fold to three-fold that of the content of sodium oxide. A higher content of alkalis is dangerous due to the possibility of their reaction with the aggregate containing amorphous silicon dioxide.

The compounds present in cement are formed by the mutual reaction of the components of the raw material mixture during burning, i.e. by calcium oxide, silicon dioxide, aluminium oxide and ferric oxide together with minor admixtures such as alkalis and magnesium oxide. The compounds formed, which are also called clinker minerals, are usually described by a combination of formulas of two or more oxides. The summary formulas of more compound clinker products would be intricate. For instance, calcium orthosilicate may be described by formula $2CaO.SiO_2$ or in a summary way as the salt of orthosilicic acid Ca_2SiO_4. Here, in order to write formulas in a simpler way, short forms will be used, where $C = CaO$, $S = SiO_2$, $A = Al_2O_3$, $F = Fe_2O_3$, $H = H_2O$, e.g. tricalcium silicate will be written as C_3S.

The properties of clinker, and subsequently those of cement, are determined by their mineralogical composition. Basic minerals are as follows: alite, belite, the aluminate phase, the ferrite phase. The average content of clinker minerals in Portland clinker is presented in Table 1.2.

Alite. The compound $3CaO.SiO_2$ (C_3S), which is formed at a temperature higher than 1350°C, is a main part of alite. A typical content of oxidic impurities in alite is from 3 to 4% (Taylor, 1992). These are chiefly MgO, Al_2O_3

and Fe_2O_3 forming solid solutions in its structure. Tricalcium silicate is a polymorphous compound forming crystals in triclinic, monoclinic and rhombohedral systems depending on the temperature. The pure compound, when cooled to the laboratory temperature, forms crystals in the triclinic system. In clinker, where foreign ions are present, this occurs in the monoclinic system or in the mixture of monoclinic and triclinic crystals after having been cooled. MgO substitutes CaO in the structure of C_3S, Al^{3+} ions replace Si^{4+}. The presence of MgO stabilizes the monoclinic alite (Šauman, 1993).

Belite. The compound $2CaO.SiO_2$, occurring in four modifications α, α', β and γ is a nature of belite. The structure of polymorphs α, α', β is very similar, but γ-C_2S differs in having a low density. The form β-C_2S is frequently found in ordinary Portland cement clinkers. Form α, occurring rarely during the fast cooling of clinker, has a noticeable higher strength than β-form. The presence of stabilizing ions, whose content in the form of oxides is 4-6%, prevents the transformation of β-C_2S into the powdered γ-form. The oxides are mostly MgO, Al_2O_3, and Fe_2O_3.

Aluminate phase. Pure tricalcium aluminate crystallizes in the cubic system. The structure is formed by Ca^{2+} ions and AlO_4 tetraeders. C_3A in clinker crystallizes either in the cubic or orthorhombic system and forms a dark interstitial material. Iron oxides, magnesium oxide, silicon dioxide, sodium oxide and potassium oxide are present in the structure of the aluminate phase in the form of a solid solution, the amount being about 13% for the cubic form, and 20% for the orthorhombic form (Taylor, 1992).

Aluminoferrite phase. The composition of solid solutions of the aluminoferrite phase is between C_6AF_2 and C_6A_2F, and it is usually described by the formula C_4AF. Sometimes it is marked as $C_2(A,F)$, since this is a solid solution of hypothetical constituents C_2A and C_2F. Magnesium oxide, silicon dioxide, titanium oxide, manganese oxide and sodium oxide represent 10% of impurities in the structure of the aluminoferrite phase.

Free CaO. In good clinker, the content of free CaO does not exceed 1%. With respect to the burning temperature of clinker, a higher content of CaO may cause volume changes due to delayed hydration of CaO, and by the formation of a more voluminous $Ca(OH)_2$.

It is generally known that finer particles of clinker lead to a faster growth of strength and also to higher final strengths. A faster growth of strength is caused by the size of the surface, on which hydration reactions proceed. Final higher strengths of cements with a higher proportion of smaller particles are attained because the particles react completely with water, i.e. these form products of hydration from all the present clinker materials that are capable of reactions. In finely ground cements, the residues of grains that have not been hydrated do not occur. The chemical and mineralogical composition of proportions of particular sizes of fractions is not regular. The results presented in Lea (1971) show greater differences in the content of alite and belite (51–64% and 13–39%) in particular fractions than other clinker minerals. The size of particles of ordinary ground clinker ranges from 2×10^{-6} to 9×10^{-5} m.

The size of the smallest particles determined in clinker was 3×10^{-8} m (van Breugel, 1991).

The specific surface of cement is defined as the surface area per unit mass. In cements, it is expressed in units of m^2/kg. Ordinary Portland cements have a specific surface of about 300 m^2/kg. Generally, cements are produced with surfaces ranging from 250 up to 500 m^2/kg. With special extra quick–binding cements, the specific surface is larger than 500 m^2/kg.

The specific surface is a function of the distribution of the particle sizes and their shapes, which means that at the same distribution of particles, the specific surface may be different. It is evident from microscopic observation that the cement particles are not of spherical shape.

To reduce the power demand of cement and to improve the properties of concrete from the viewpoint of water impermeability, corrosion resistance and mechanical properties, silicate clinker additives affecting the properties of the final product are added. Inorganic additives are divided into pozzolana active substances with a high content of amorphous SiO_2 in the reactive state, and often also Al_2O_3 but these have a very low content of CaO, and hydraulically active substances with a higher content of CaO. Pozzolana active substances, when mixed with lime, form hydrated calcium silicates as early as at a normal temperature. If Al_2O_3 is present, then calcium aluminate hydrates are formed. Latently hydraulic substances such as the granulated blast furnace slag react with calcium ions of low concentrations also, and products of marked strength are formed.

The additives used are either waste materials produced in large quantities in other branches of industry or extracted natural materials, which are sometimes treated. These include granulated blast furnace slag, fly ash, natural pozzolana, and heat-treated clayey.

Blast furnace slag is a by-product of iron production, which originates due to the reactions of attendant constituents of iron ore (quartz, clays and other) with a slag-forming addition (limestone) in the blast furnace as a melt at temperatures of 1350 to 1550°C. The slag is formed in the roentgenoamorphous state by fast cooling the slag below 800°C by means of wetting with water. Alkaline slag contains a number of minerals, especially gehlenite $2CaO.Al_2O_3.SiO_2$, akermanite $2CaO.MgO.2SiO_2$, belite $2CaO.SiO_2$, merwinite $3CaO.MgO.2SiO_2$, wollastonite β-$CaO.SiO_2$, and others. The usability of slag as a latently hydraulic binder is given by the proportion of acid oxides $H_f = (CaO + MgO)/(SiO_2 + Al_2O_3)$. If the proportion is > 1, the slag is suitable for mixed cements. The condition for using blast furnace slag is its resistance against silicate disintegration, when it originates from β-C_2S by modifier conversion γ- C_2S of a higher molar volume, and further against ferrite disintegration caused by FeS oxidation into sulphate, iron hydroxide and sulphuric acid are formed by the successive hydrolysis. The slag, which is cooled slowly, crystallizes and has no hydraulic properties (Šauman, 1993).

Fly ash is a pozzolana active substance formed as a waste from coal burning. The chemical composition of fly ash shows that this contains 40–50%

of SiO_2, 15–35% of Al_2O_3, 4–10% of Fe_2O_3, 2–8% of CaO. Lignite fly ash contains as much as 20% of CaO. The share of the vitreous phase, ranging from 60 to 70%, and the content of SiO_2 in this phase, are important for the binding properties of fly ash. Quartz SiO_2, mullite $3Al_2O_3.2SiO_2$, hematite Fe_2O_3, rutile and anatas TiO_2 and carbon have been found in fly ash.

In some countries, natural pozzolana, which is characterized by a high content of amorphous reactive SiO_2, is mined. The majority of the natural pozzolana is of volcanic origin, in particular tuffs and trasses, that also contain inert minerals, e.g. quartz and feldspars, besides basic reactive constituents. Sedimentation materials, such as diatomite formed by shells of prehistoric animals or spongilite also rank among pozzolanas. Pozzolana contains 50–70% of SiO_2, 10–20% of Al_2O_3, and smaller amounts of Fe_2O_3, Na_2O and K_2O, CaO and MgO.

Some clays with a high content of clayey minerals react with lime as early as at a normal temperature. This is used for the stabilization of soils, but with respect to physical properties of products (swelling, shrinkage during drying up), these cannot be used as an addition to concrete. When clayey minerals are heated to temperatures above 600°C, imperfectly crystallic or amorphous dehydrated aluminosilicates are formed. These aluminosilicates are pozzolana active. Kaolinite $Al_2O_3.2SiO_2.2H_2O$, which at the temperature of approximately 600°C passes to metakaolin, a pozzolana active substance, may serve as an example (Coleman and McWhinnie, 2000).

1.2 HYDRATION PROCESSES OF CEMENT

The reaction of clinker minerals with water leads to the formation of a solid structure of the cement stone. Their reactivity with water is mutually very different. However, it is possible to notice a difference in the reactivity of the same clinker mineral depending upon the structure, polymorphous kind, amount and composition of solid solutions, and the content of minor components. The rate of reaction of basic clinker minerals decreases in the sequence

$$C_3A > C_3S > \text{aluminoferrite phase} > \beta\text{-}C_2S.$$

The greatest difference in the degree of hydration of clinker minerals is at the beginning of the hydration process. While the reaction of C_3S after 28 days is 70%, the reaction of C_2S with water is considerably slower. It should be pointed out that the hydration rate depends on the size of hydrating grains. If cement grains are big, then the residues of nonhydrated clinker minerals are found in concrete that may be several tens of years old.

The hydration of clinker minerals leads to the formation of compounds that form a strong structure of cement. These chemical reactions are revealed externally by the mechanical properties of the final product, i.e. concrete.

Knowledge of the conditions of the course of hydration leads to control of this process, to the possibility of influencing the process by adding chemical substances, and creating the most suitable physical conditions, and also to the explanation of some anomalous phenomena that may finally lead to the failure of concrete.

A fast initial hydration occurs in the reaction of tricalcium silicate with water. The ions of Ca^{2+} pass to the water phase and form a supersaturated solution. A layer of calcium hydrosilicates is formed on the surface of grains. This process lasts a very short period of several minutes only and then a rest time follows, which lasts up to several hours, when the reaction proceeds with a very low rate. When the rest time is over, fast hydration is restored. The rate reaches its maximum after about 10 hours, and then gradually decreases.

The hydration of C_3S may be expressed in a somewhat simplified way by the following formula:

$$2 \ (3CaO.SiO_2) + 6 \ H_2O \rightarrow 3CaO.2SiO_2.3H_2O + 3 \ Ca(OH)_2.$$

This, however, does not fully express the complexity of the process. The reaction cannot be expressed precisely in the stoichiometric way, since the calcium silicate gel, marked CSH, has a variable composition based upon the conditions under which the hydration takes place. The composition is influenced by hydration temperature, the time of hydration, the amount of water present, and by the potential presence of additives. The intermediate product of hydration coats the surface of the grains of C_3S from the very beginning. In the final product, the ratio C:S is from 1.4 to 1.6 (Šauman, 1993). This ratio increases with the decreasing water/cement ratio. The amount of chemically bound water ranges from 0.204 to 0.230 g per 1 g of C_3S. The product formed is a substance that can be identified with problems due to the almost roentgenoamorphous character and a considerable variability of its composition. The CSH phase is the main bearer of strength in a hardened product. The second product of hydration is $Ca(OH)_2$ which, due to its low solubility (160 mg in 100 g of water at 20°C), appears in the hydrated product in a crystalline form. The hydroxide ions form an alkaline environment in the porous solution that is important for the protection of the steel reinforcement against corrosion.

β-$CaO.SiO_2$ is a substantial constituent of belite, and with water it reacts in a similar way to alite under conditions of the formation of CSH gel and $Ca(OH)_2$. This reaction is very slow due to the hydration of other clinker minerals. Only after several weeks will the surface of grains be covered with amorphous hydrated calcium silicates. A substantial part of belite reacts in a period of time longer than one month. Even after several years, the residues of non hydrated belite are found in the hardened cement binder.

The hydration reaction of dicalcium silicate with water may be described by the following general equation:

$$2 (2CaO.SiO_2) + 4 H_2O \rightarrow 3CaO.2SiO_2.3H_2O + Ca(OH)_2.$$

As with the hydration of alite, the product of hydration is the CSH phase. The content of $Ca(OH)_2$ in hydrated products is lower.

Compared with clinker minerals, C_3A reacts with water in the fastest way. The hydration without the gypsum presence first leads to the formation of hexagonal intermediate products of the plate character, C_4AH_{19} and C_2AH_8. C_4AH_{19} is stable under water only, and in the air it converts to C_4AH_{13} quickly. Both products of hydration are unstable, and later on, they convert to a less soluble and thermodynamically more stable C_3AH_6. The conversion of the above compounds to C_3AH_6 is fast at temperatures above $30°C$ (Šauman, 1993). At a temperature of about $20°C$, the intermediate products may exist even 24 hours before they convert. It has been proved that the final product C_3AH_6 is directly formed at a temperature of $80°C$. Taking into consideration the rate of hydration of C_3A, for practical uses of cement, the reaction should be reduced by adding gypsum. The hydration with the presence of gypsum $CaSO_4.2H_2O$ leads in the first phase to the formation of ettringite $3CaO.Al_2O_3.3CaSO_4.32H_2O$:

$$3CaO.Al_2O_3 + 3 (CaSO_4.2H_2O) + 26 H_2O \rightarrow$$
$$3CaO.Al_2O_3.3CaSO_4.32H_2O.$$

When the entire sulphate is consumed, ettringite reacts with other amounts of C_3A in the formation of monosulphate $3CaO.Al_2O_3.CaSO_4.12H_2O$:

$$3CaO.Al_2O_3.3CaSO_4.32H_2O + 2 (3CaO.Al_2O_3) + 4 H_2O \rightarrow$$
$$3 (3CaO.Al_2O_3.CaSO_4.12H_2O).$$

Ettringite crystallizes in the form of long, needle-like crystals; monosulphate is tabular or leaf-like. The principle of the reaction reduction is based on the formation of a layer of ettringite on the surface of C_3A particles, which considerably reduces the diffusion of water molecules to the unreacted surface of the grain.

The hydration of calcium aluminoferrites occurs parallel to the hydration of C_3A, but is much slower. The reaction rate increases with the increasing A/F ratio in the compound. During hydration, hexagonal crystals $C_4(A,F)H_{19}$ and $C_4(A,F)H_{13}$ are the first to form, then converting to the stable $C_3(A,F)H_6$. In the final phase, hydration with the presence of lime leads to the formation of C_3AH_6, α-Fe_2O_3 and $Ca(OH)_2$.

Ettringite with the substituted Fe^{3+} ions that after consuming all the gypsum converts to a monosulphate, is first formed during hydration in the presence of gypsum. With respect to a fast reaction of C_3A with gypsum, the share of ettringite and a monosulphate formed from the ferrite phase is low.

Almost every kind of cement contains free lime amounting to 1% that is not bound to silicate and aluminate compounds. Free lime reacts with water during the formation of hydroxide:

$$CaO + H_2O \rightarrow Ca(OH)_2.$$

Due to the high temperature of the burning of clinker, CaO is largely unreactive, and delayed hydration may cause the volume instability of cement and concrete through increase in volume.

MgO, which is in large measure bound in the structure of alite and other clinker minerals, reacts in a similar way. A small amount in the form of periclase reacts very slowly during the formation of a more voluminous $Mg(OH)_2$ (brucite), the formation of which may lead to the degradation of concrete. For this reason, the content of both components in cement is restricted.

The hydration of clinker minerals may be affected by adding some electrolytes, and also by nonelectrolytes. Hydration is accelerated by inorganic salts represented, above all, by chlorides, and sodium, potassium, calcium and lithium nitrates. Cations of metals forming less soluble hydroxides and complex salts compared to the solubility of $Ca(OH)_2$, e.g. Zn^{2+} or Pb^{2+}, act as retarders. The influence of the presence of ZnO on the hydration of Portland cement is shown in Fig. 1.1 (Rovnaníková and Rovnaník, 1997). Also some anions, forming insoluble calcium salts such as carbonates, phosphates, fluorides or sulphates, act as retarders. Hydration is considerably retarded by the presence of some organic substances, especially carbohydrates, which in higher concentrations completely stop hydration reactions. $CaCl_2$ reacts with C_3A in the formation of Friedel's salt $3CaO.Al_2O_3.CaCl_2.10H_2O$, which is formed of the surface of particles, and in this way the reaction is retarded but the hydration rate of C_3S and β-C_2S is increased.

The hydration of Portland cement is an intricate complex of reactions between clinker minerals, calcium sulphate and water. The moment when cement is mixed with water starts a number of reactions that finally lead to the formation of a truly solid structure of the cement binder, which changes in the course of time. The pores are gradually reduced through their filling with the products of reaction, and at the same time, the strength that reflects the course of chemical reactions increases. The hydration of cement may be divided into four consecutive periods. In the first minutes after the mixture of cement and water is formed, the cement grains are wetted, $CaSO_4$ is dissolved and C_3A is hydrated. The aluminate phase reacts with gypsum, and on the surface of the grains, ettringite $3CaO.Al_2O_3.3CaSO_4.32H_2O$ is formed, which prevents further hydration for a certain period of time. Simultaneously, a coat of CSH gels is formed on the surface of C_3S grains. In the second period, which lasts from 1 to 6 hours depending on the conditions of hydration, hydration reactions take place at a low rate. At the end of this period, the concentration of Ca^{2+} ions reaches the state of an oversaturated solution and the crystal nuclei of $Ca(OH)_2$ and CSH gels begin to form. In the third pe-

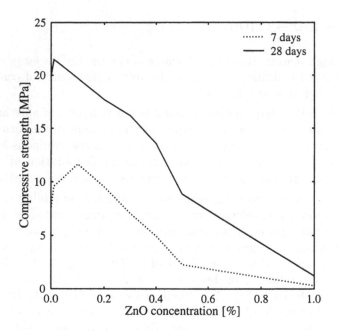

Figure 1.1 The development of compressive strength with the presence of ZnO

riod, which may take 8 to 14 hours, C_3S reacts quickly at the formation of CSH gels and $Ca(OH)_2$, which is precipitated in the form of crystals of port-landite. In this period, a basic solid structure of the binder is created, which is formed by CSH gels, crystalline phases of aluminates and ferrites, and by ettringite. At the same time, the porosity is reduced and the plastic mixture changes into a solid one. The beginning and the end of setting take place in this period. The concentration of SO_4^{2-} ions decreases by consuming $CaSO_4$ in cement and ettringite begins to react with unreacted C_3A to form mono-sulphate $3CaO.Al_2O_3.CaSO_4.12H_2O$. After all the ettringite has transformed to monosulphate, no $CaSO_4$ is available for the reaction with the aluminofer-rite phase. Other types of hydrates of these clinker minerals, that react with monosulphate to the products with the content of Fe_2O_3 in the crystal lattice, are formed. In the fourth period, days and months, the role of belite, that also forms CSH gels and minor shares of $Ca(OH)_2$, increases. In this phase, a stable solid structure with permanently decreasing volume of pores is formed.

The hydration rate reaches its maximum several hours (usually about 10 to 14 hours) after the start of the hydration, and then it gradually decreases with time. The cause of these changes is the fact that the diffusion of water through products of hydration to the nonhydrated residues of cement grains is becoming the regulating action. After 28 days a considerable proportion of clinker minerals (70 to 90%) is hydrated, except for belite, whose hydration still takes place for a long time.

The rate of hydration after 28 days is relatively low, and the nonhydrated residues of cement grains hydrate depending on the availability of water and fineness of cement grinding, and therefore depending on the size of the original grains. Big grains have such a layer of products of hydration on their surface, that the diffusion of water molecules through this layer is made more difficult.

Hydration is also affected by the surroundings, i.e. by temperature and by the relative humidity of air. The increased temperature accelerates hydration reactions, first of all, in the initial periods of hydration, and thus also the rate of the development of the solid structure of the binder. The influence of temperature in the fourth period is already relatively low. The cement mix with water, cured at 38°C, shows, after the fast initial increase of strength in 3 days, a decrease after 7 days, and negligible increase after 28 days.

The effect of low temperatures on cement hydration is quite different. The hydration reactions in the three initial periods are slow and so is the increase of strength. In the fourth period, a marked increase in rate and growth of strength occur. The strength reaches higher values after 2 to 3 months, compared with cements cured at higher temperatures.

The effect of the increased temperature upon the degree of hydration related to 28 days hydration of cement paste is shown in Fig. 1.2 (Rovnaníková, 1981). The degree of hydration of the same cement cured at 20°C in 1 day was 0.52, in 3 days 0.62, and in 7 days 0.79 compared to the 28-day hydration. At a temperature of 60°C, the same proportion of clinker minerals is hydrated in 8 hours, as at the temperature of 20°C in 7 days.

Such knowledge of the course of hydration leads to the possibility of control of this process by adding chemical substances, and by creating the most favourable physical conditions. Also, it can explain some anomalous phenomena that may lead to the failure of concrete.

1.3 HYDRATION HEAT

Hydration heat is released during the reaction of cement with water. From the viewpoint of the production of concrete, hydration heat is a very important property. In structures built of large concrete blocks, e.g. foundations for reactors, dams, etc., considerable temperature differences between the centre and the external layer of the concrete block occur due to the effect of the development of the hydration heat of cement, and with respect to the relatively low thermal conductivity of concrete. The reaction takes place internally in an adiabatic regime, which means that the temperature increases due to the

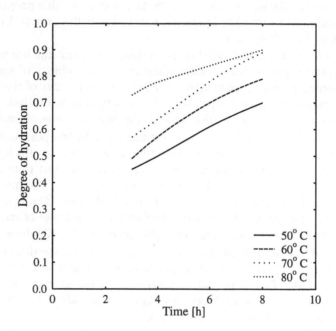

Figure 1.2 The effect of temperature on the degree of hydration of
the cement paste

released heat, and hydration takes place thereafter at a higher temperature.
On the surface, the heat is lost to the environment, so the hydration takes
place under approximately isothermal conditions.

Hydration heat of cement is a result of the processes that comprise the
actions connected with the release or absorption of heat. These are as follows:
wetting of cement in water; dissolution of clinker minerals; formation of hydrated silicates and calcium aluminates; crystallization and recrystallization
of silicates, aluminates and calcium ferrites, and water adsorption on products
of hydration.

The development of hydration heat is a function of time. The initial stages
of the development of heat are shown in Fig. 1.3. The course corresponds to
the reactions of particular clinker minerals to water.

The total course of the heat development is shown in Fig. 1.4. Ordinary
cements release about 50% of the heat in the course of the first 3 days at a
temperature of 20°C, from 60 to 75% in 7 days. Every clinker mineral releases
different amount of hydration heat in reaction with water.

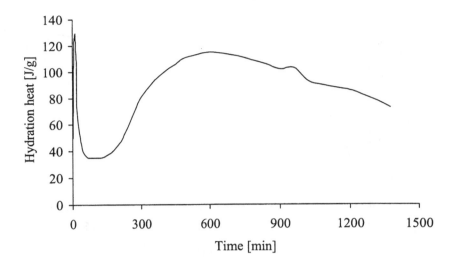

Figure 1.3 The course of the development of cement hydration heat
in initial stages

As follows from Table 1.3, the development of hydration heat depends on
the mineralogical composition of cement. Cements with high content of alite
and C_3A develop a high hydration heat.

These cements are suitable for laying concrete at low temperatures, when
after starting the reaction, for instance by heating the batch water or mix,
the forming heat makes the acceleration of hydration reactions possible. Conversely, for making large concrete blocks, it is advantageous to use cements
with a low content of alite and C_3A, and a high content of belite.

The addition of gypsum to clinker affects the development of hydration
heat. It has been discovered experimentally (Swiete and Roth, 1973) that the

Table 1.3 Hydration heat of clinker minerals

Clinker minerals	Hydration heat [J/g]
C_3S	520
β-C_2S	230
C_3A	860
C_4AF	465
CaO	1161

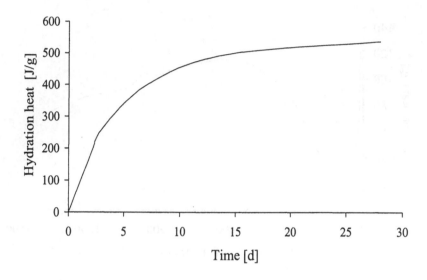

Figure 1.4 Total course of the development of cement hydration heat

first maximum in the development of hydration heat in clinker is higher by 9.5°C and the second maximum by 5.5°C, than the corresponding maximum of Portland cement. The addition of gypsum reduces the hydration heat, especially in the initial stage of hydration.

Hydration heat is developed depending on the size of its specific surface. The rate of the hydration heat development increases with the increasing surface, but the same final values need not to be reached. The cause lies in the fact that in large cement grains the whole grain does not hydrate because there is such a thick layer of hydration products on its surface that water diffusion is stopped. The influence of the specific surface on the development of hydration heat is shown in Table 1.4.

When clinker is ground together with additives, the total specific surface is dependent on the grindability of particular constituents of the mixture. For example, when clinker is ground together with gypsum, clinker acts as an aid to grinding because it is harder. Thus the size of gypsum grains is considerably smaller than that of clinker.

The rate of hydration, like any other chemical reaction, is affected by temperature. By increasing temperature, the hydration rate increases, and thus the rate of the development of hydration heat increases as well. The values of 313 and 390 J/g (Tínková, 1977) were determined after 7 days of hydration by measuring the hydration heat of cement cured at 20°C and 50°C. The acceleration of hydration is inversely proportional to the increased

Table 1.4 Hydration heat of cement (in J/g) with different specific surfaces
(in m^2/kg)

Specific surface	204	306	406	504	600
Hydration heat 1 day	21	42	52	77	84
2 days	40	79	100	134	143
3 days	54	107	130	165	173
4 days	75	125	148	184	192
5 days	84	140	161	198	205
6 days	92	150	173	211	215
7 days	100	159	184	217	222

temperature. The effect of temperature markedly manifests itself above 40°C. This fact is caused by the existence of several mutually connected phases of the reaction, the kinetics of which is differently affected by temperature.

The process of cement hydration was observed to depend on temperature by using the isothermal microcalorimeter (Courtault, 1974). The temperatures, at which the hydration took place were in the temperature range from 20 to 70°C. The development of hydration heat is given in Fig. 1.5.

The increase in the rate of hydration heat development is prominent at the beginning of hydration. After a longer period >180 days, the differences in hydration heat are no longer noticeable.

The setting and hardening of cement is a summary description of physical and chemical processes in the formed dispersion system. All substances affecting any part of these complex processes may affect them. In this respect, the electrolytes that can increase or decrease the solubility of particular constituents of the cement clinker may reveal their presence, and may affect the formation of a larger number of the colloidal particles or affect the crystallization, etc. So the addition of electrolytes affects even the development of hydration heat. The hydration heat of cement is reduced by adding additives because the silicate additives, such as blast furnace granulated slag, natural pozzolana, or fly ash, release less heat than cement in reaction with water, when cement is present. The courses of the development of hydration heat of Portland cement containing different amounts of blast furnace granulated slag measured by the adiabatic calorimeter are shown in Fig. 1.6.

1.4 PROPERTIES OF AGGREGATES

Aggregates in concrete are most frequently of inorganic origin. However, organic components may also be used. The kind of aggregates in concrete is

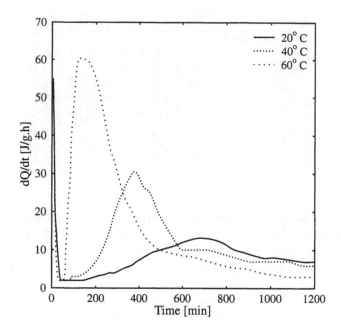

Figure 1.5 The development of hydration heat at various temperatures under
isothermal conditions

selected based on the manner of utilizing the structural element or concrete
structure, and on the local availability. They may either be natural or made
of secondary raw materials. Wood in various shapes and grain sizes is the
most frequently used of all organic fillers. Special kinds of fillers in concrete
are reactive powders, by means of which a high strength of concrete and fibres
having special functions in concrete may be secured.

Based on the consistency, aggregates used for concrete may be classified
as compact aggregates and porous (light) aggregates. Compact aggregates
comprise a mixture of uncrushed and crushed grains of natural substances of
a consistent structure. Porous aggregates represent a number of natural expanded materials such as perlite or sintered natural materials or technogenous
materials made, e.g. of clays or power plant fly ash. Various waste materials
such as blast furnace or steel-making slag, fly ash, foundry sands, etc. may
fulfil the function of a filler agent like aggregates.

Sand and gravel may contain foreign particles, which are grains of a different origin and composition than the aggregates themselves (e.g. lumps of

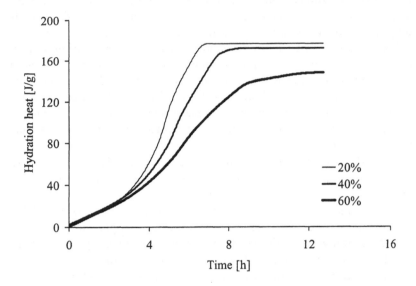

Figure 1.6 Influence of blast furnace slag on evolution of hydration heat

coal, wood). The most dangerous harmful substances contained in aggregates and sand are the compounds of sulphur and organic, most frequently humic substances.

The aggregates for concrete must fulfil quality requirements to prevent negative effects on concrete properties. These especially refer to the size, shape and quality of the grain surface, granulometric curve, physical properties, petrographical and mineralogical composition and chemical properties of aggregates.

1.4.1 Natural Aggregates

Natural dense aggregates for concrete that are extracted from the fluvial deposits or in the deposits of disintegrated rocks in quarries are characterized by the rounded edges. Crushed aggregates, which are characterized by their irregular shapes of grains and rough surface, are obtained by means of crushing the blasted massif in quarries. The aggregates for structural concrete are prepared by sorting, and in some cases by washing, in order to remove clayey and loamy constituents.

From the viewpoint of the particles size, aggregates are divided into sand (grain size 0/4 mm) and gravel (most frequent grain size 4/32 mm). Sand grains (0/4 mm) are usually composed of particular minerals. From the mineralogical point of view, gravel stones are heterogeneous. In sand, there is quartz supplemented by sodium, potassium and calcium feldspars, and mica lamellas; black particles of iron compounds and a number of other minerals represent minor abundance. Different amounts of clayey minerals, whose content is limited by standards, are the constituents of sand. The presence of organic compounds, e.g. of humic acids, is also unacceptable, because these compounds negatively affect the hydration reaction of cement. This is reflected in a slow rate of reactions and in the formation of products that manifest themselves by reducing strength.

Finely ground powders of the same mineralogical composition without any content of clays and loams, of grain size to 0.125 mm, are used as aggregates to improve the workability and to ease the pumping of fresh concrete. In case where the stone powder is pozzolana active, i.e. reacts with $Ca(OH)_2$ from the pore solution with hydrosilicates at a normal temperature, then the concrete strength and water-tightness usually increases. This especially refers to some kinds of rocks containing SiO_2 in the amorphous form.

Commonly used fractions of gravel are 4/8, 8/16, 16/22 and 22/32. The aggregates of grain size up to 125 mm are used for concretes for very massive structures. The granulometric curve of the aggregates should be selected, so that the minimum possible porosity of the aggregates may be achieved. Therefore the aggregates are composed of several fractions, with two fractions as a minimum. It is advantageous for the aggregates mixture to be coarse-grained if possible, since in this way the need for the binder that must precoat the particles to make the compact mass is reduced. Additives of fine grained reactive fillers, as mentioned above, are used to achieve a higher density and strength of concrete. The maximum size of the grains of the aggregates must be 1/4, exceptionally 1/3 of the minimum thickness of the structure to create a uniform composition of the aggregates and the binder within the whole volume of the structural element.

The shape and the shape index of grains defined as a ratio of the largest grain size to the smallest one, provided that the grain is placed in a prism, are important. A higher value of the shape index (> 3) leads to an increase in the mixture porosity, because the grains are mutually wedged in during the compaction.

The quality of the grain surface affects the mix workability and determines the coherence with the cement binder. It results from the rock structure, from the properties of grains or crystals of particular minerals, from the manner of the extraction and processing of aggregates.

The surface is influenced by a potential disintegration of the surface parts of the rock and by its pollution with the products of disintegration, or by its coating with the cemented particles of loams, clays, iron oxides, gypsum and carbonates. Washed aggregates free of adhered parts are used for extra high quality structural concrete. The surface may be smooth to glassy or rough to porous. Large porosity leads to the deterioration of the mix workability.

Aggregates are weathered by the long-term action of external factors. This action is mechanical (pressure of ice in cracks), physical (changes of temperature and humidity), chemical (the action of water containing dissolved aggressive substances) and biological (microorganisms, plant roots). The weathering of feldspars (albite, orthoclase, plagioclase) leading to the formation of clayey minerals of kaolinite type is very frequent.

The quality of the surface and the composition of the aggregates are significant for the formation of the interfacial transition zone (ITZ). The formation of a strong or less strong junction between the aggregates and the binder depends on the composition of the aggregates. The following three mechanisms of bonding between aggregates and the cement matrix have been described (Zimbelman, 1985, Yuan and Cuo, 1987): mechanical anchoring of hydration products in the uneven surface of aggregates; growth of hydration products on the surface of aggregates; and formation of a chemical bond of reactive aggregates with the cement matrix. The last of the above-mentioned mechanisms is advantageous, unless the quality of the contact between the aggregates and the binder is affected and the more voluminous compounds are formed, as in the case, for example, of the reaction of alkalis with the active silica. It has been found that ITZ in limestone aggregates is of porous character in the initial stages of hardening (Tasong et al., 1999a, Tasong et al., 1999b).

Light, usually porous aggregates, are used for concretes of bulk density below 1600 kg/m^3. Lightweight concretes are made, above all, because of their excellent thermal insulation properties. For their production, natural porous aggregates are used, and are either of volcanic or sedimentary origin. The disadvantage of natural porous aggregates comprising tuffs, spongilites, trasses, kieselguhr, and pumice is the fluctuation in their quality. The open pores formed by crushing reduce the total effect of concrete lightening. Pumice, which originated in the release of gases during the hardening of lava, is the most frequently used natural aggregate. In view of the chemical composition of the above raw materials, there is a danger of the alkali–silica reaction with the increased concentration of alkalis. This especially applies to sintered clays, claystones and slates. The artificial porous aggregates made of clays are formed by the expansion of clays of suitable composition due to heating. Besides aluminosilicates, the loam must also be composed of flux agents (Fe_2O_3, CaO, MgO, alkalis) that cause the smelting and subsequent sintering of the grains surface at a relatively low temperature. The raw material mixture must also contain substances capable of releasing gas at a higher temperature. Then the expansion proceeds by the action of gases formed in the softened mass on the closed surface, which prevents free escape of gases. The bulk density ranges from 500 to 900 kg/m^3, the absorption capacity is up to 25% of the mass.

The above porous aggregates behave like pozzolana active substances. In the surface layer, these react with $Ca(OH)_2$ and the calcium hydrosilicates are formed, which results in the increase of the ITZ strength between the aggregates and the binder.

Fine portions of light aggregates may be made of perlite, which is hydrated volcanic glass and expands due to heating. The expanded perlite is made for a grain size of 4 mm, and therefore this may replace sand in lightweight concrete. Perlite represents small beads that, due to heating, expand under the influence of released water vapour. The bulk density is within the range of 50 to 200 kg/m^3, the absorption capacity is very high, reaching almost 400% of the mass. The behaviour of vermiculite, which belongs to the group of clayey micas, is similar. Heating to about 1150^0C results in a significant volume increase of the tiny particles of vermiculite.

1.4.2 Aggregates Made of Secondary Raw Materials

Among the secondary raw materials, crushed crystalline blast furnace slag slowly cooled in the air is used as aggregate for concrete. The slag should contain 29% of SiO_2 at least, and 43% of CaO as maximum, because with a higher content, calcium disintegration may occur, i.e. β-C_2S may change to γ-C_2S. The change is connected with the volume enlargement by 9 to 10% and with the disintegration of the slag. The content of iron in the oxidation state Fe^{2+} is also not suitable because iron may be oxidized to Fe^{3+}, and more voluminous compounds leading to concrete degradation may be formed.

Coal slag represents the sintered residues after the firing of coal. The composition of coal slag depends on the minerals present in coal, and on the manner of burning that determines the amount of combustible residue. Coal slag must be matured free in the air so as to enable the access of CO_2 and air humidity or rainfall. The present oxides CaO and MgO formed by disintegration of carbonates during the process of burning are relatively unreactive, which may lead to a reaction with water and the formation of hydroxides as late as in the solid structure of concrete. The molar volume of hydroxides is larger than that of the original oxides. This may lead to failure of the concrete structure. Water and CO_2 react with CaO and MgO, and harmless compounds such as hydroxide or carbonate are formed before coal slag is used for concrete.

The content of S^{2-} and the total amount of sulphur that reacts with $Ca(OH)_2$ from the cement binder while gypsum is being formed is also important. Sulphates may lead to the formation of ettringite, if favourite conditions for its formation are created. Besides this, the sulphate content is a cause of the steel reinforcement corrosion.

The organic compounds that remain unburnt in case of unsuitable burning equipment may negatively affect the hydration reactions of cement and thus cause not only the slower development of strength but also the reduction of final strengths. In view of the diversity of properties, coal slag as aggregate is used for less valuable plain concretes.

Concrete from demolished structures is also used as aggregate. This concrete is crushed to a required grain size. The grains of the crushed concrete have a suitable shape index, a lower bulk density and a higher sorption capacity compared to compact aggregates. In aggregates made from the recycled concrete, it is necessary to observe potential presence of foreign substances. This especially refers to concretes from bridges, roads and road facilities, because these concretes are often polluted by the presence of chlorides. The reinforced concrete might then contain an internal source of chlorides leading to the reinforcement corrosion.

The sand from the sand moulds may be used for concrete depending on the kind of binder by means of which the grains have been bonded to make a strong mass. Sands containing organic binders are not suitable because these are dangerous from the viewpoint of the formation of a solid structure of the cement binder. The sands bonded by water glass increase the content of alkalis in the mixture through the sodium carbonate present. These sands are recycled after being washed or are used for concretes that are not threatened by the alkali-silica reaction.

Crushed bricks may also be utilized as a substitute aggregate. However, the content of sulphates, which are often present in the ceramic raw material in a form of $MgSO_4$, $CaSO_4$ or Na_2SO_4, must be ascertained for the bricks. The critical content of sulphates in conversion to SO_3 is 1% of mass. A higher content of sulphates leads to reactions with $Ca(OH)_2$ and with hydrated calcium aluminates.

Waste materials are also utilized for the production of porous aggregates. Fly ash produced in power plants is processed to sintered pellets or extruded granules. The porous mass inside is surrounded by a closed surface. The bulk density ranges from 1100 to 1400 kg/m^3, the sorption capacity is 15–39% of mass. Blast furnace slag, which must fulfil the assumption that it will not disintegrate, is used for the production of slag pumice. The slag foaming is carried out in special equipment by the fast cooling of water. Slag pumice does not have a compact surface, the bulk density is 700 to 1200 kg/m^3 and the sorption capacity up to 20% of mass.

1.4.3 Other Components

Materials of organic origin, namely the <u>wooden mass</u> in the form of splinters or chips, are also used as special fillers for concrete. Compounds such as tanning materials, sugars, salts of organic acids and dyes are leached from small pieces of wood or chips by cold water. In the alkaline environment, gums are dissolved and hemicelluloses are extracted. All the above mentioned compounds negatively affect the cement hydration reactions, which means that their course is being slowed down, and the result is reflected in low strengths of concrete.

Adsorption and chemical sorption of water vapour and capillary condensation are the cause of wood moisture content. Through water absorption, moisture in wood increases and wood expands its volume. During desiccation, wood reduces its volume again.

The above facts clearly show that the use of wood as a filler for dense concrete is not suitable. Wood splinters are utilized for the production of structural elements serving as thermal insulation. For these purposes, wood particles are treated, for example, by a water glass solution that infiltrates into the surface layer, and by the action of CO_2 forms the $SiO_2.nH_2O$ gel, preventing the transport of water into the wood and of the organic compounds from wood into the mixing water.

Reactive inorganic powders are used for reducing the porosity of the cement binder in concrete. These are pozzolana active pulverized fuel ashes (PFA) containing amorphous SiO_2 that reacts with $Ca(OH)_2$ together with the formation of CSH gels. On average, the size of the ash grains is comparable to the grains of ordinary cement. The rate of the pozzolana reaction is lower than the hydration reaction of cement. Therefore, the strength development in concretes containing ashes is slower.

Silica fumes containing (based on the origin) as much as 99% of amorphous SiO_2 with a mean grain size between 0.2 to 10 μm significantly participate in the reduction of the binder porosity. The products of reaction fill the pores between the hydration products of cement, and the binder becomes compact. This leads to a high water impermeability of concrete and to an increase in strength.

Through their function in concrete, fibres are on the border between the special filler and the reinforcement. Concretes reinforced with fibres are called fibre reinforced cement composites, where fibres of both inorganic and organic origin are used. Typical properties of the fibres used in the composites with the cement matrix are given in Table 1.5, see Bentur and Mindes (1990).

The most frequently used fibres for dense concrete are steel fibres, alkali-resistant glass fibres and polypropylene fibres. The reason for using the above fibres rests in the increase of flexural strength of concrete and the prevention of crack formation formed during the autogenous shrinkage of the cement binder during its hydration. The presence of the fibres increases the residual strength when concrete is exposed to high temperatures. Other kinds of fibres are used for the production of insulating boards.

The influence of fibres on the concrete infrastructure is given by the transition zone, in which the character of the cement binder differs greatly from the character of the binder remote from the fibre. It has been found that the matrix close to the fibre is porous. At the beginning of hydration when the transition zone is water-filled, substantially voluminous crystals of $Ca(OH)_2$ deposited on the surface of the fibres in large cavities are formed (Bentur and Mindes, 1990). The quality of the transition zone also depends, on whether the reaction of cement with water proceeds near the surface of the monofilament fibre (e.g. steel fibres) or of the bundled filaments (e.g. glass fibres). While the monofilament fibres are surrounded by the cement binder on the

Table 1.5 Properties of fibres used in composites with cement matrix

Fiber	Density [kg/m^3]	Diameter [μm]	Modulus of elasticity [GPa]
Steel	7840	5–500	200
Glass	2600	9–15	70–80
Fibrillated polypropylene	900	20–200	5–77
Aramid	1450	10	65–133
Carbon (high strength)	1900	9	230
Cellulose	1200	–	10
Acrylic	1180	18	14–19.5
Polyethylene	950	–	0.3
Sisal	1500	10–50	–
Cement matrix (for comparison)	2500	–	10–45

Table 1.5 Properties of fibres used in composites with cement matrix (continued)

Fiber	Tensile strength [GPa]	Elongation at break [%]
Steel	0.5–2.0	
Glass	2–4	2–3.5
Fibrillated polypropylene	0.5–0.75	8.0
Aramid	3.6	2.1–4.0
Carbon (high strength)	2.6	1.0
Cellulose	0.3–0.5	–
Acrylic	0.4–1.0	3
Polyethylene	0.7×10^{-3}	10
Sisal	0.8	3.0
Cement matrix (for comparison)	3.7×10^{-3}	0.02

whole surface, the products of hydration in the bundled fibres are gradually growing between single fibres.

The adhesion of the cement matrix to the fibre, which results from the character of the chemical bonds in compounds forming the fibre and the quality of its surface, are other important factors affecting the microstructure of a fibre composite. The compounds with a polar covalent bond will have a larger adhesion to the cement matrix than fibres with a nonpolar bond. The anchorage of fibres with smooth surface is worse than that with surface unevenness.

The resistance of fibres to the alkaline environment of the cement binder is an important factor. This especially refers to ordinary glass fibres that are dissolved in the alkaline environment, and also to some kinds of natural fibres, for example, sisal fibres, where hemicelluloses and lignin are dissolved and disintegrated. Thus the fibres are divided into small cells, which leads to the loss of the reinforcing efficiency.

The orientation, the length of fibres and their amount in the matrix are other factors affecting the microstructure of the composite and propagation of cracks. Two-dimensional distribution of fibres increases resistance to the formation of cracks by 50%; in three-dimensional distribution, there is a high probability cracks will not arise. The resistance to the formation and propagation of cracks improves with the length of fibres. However, the homogeneous distribution of fibres in the cement matrix in longer fibres is problematic.

1.5 STRUCTURE OF CONCRETE

1.5.1 Basic Structure

Concrete is a composite material consisting of basic constituents formed by fillers and binding agents that may be supplemented with additives. Fillers in concrete are gravels and sands. Sometimes, special kinds of fillers, especially those with a very small size of particles, are added to the concrete mix. These fillers may behave actively, which means they react with products of hydration of cement, especially with calcium hydroxide. Cement hydrates and creates the hydrated products that are of amorphous character, while calcium hydroxide is crystalline.

Concrete is a porous material. Generally, a porous medium can be characterized as a solid containing other phases in its volume. The parts of the porous space that contain the liquid and gaseous phases are known as pores. The pores are interconnected, so a number of continuous paths from one side of the medium to the another exist.

A more exact definition of a porous medium was formulated by Bear et al. (1968):

(a) A porous medium is a portion of space occupied by heterogeneous or multiphase matter. At least one of the phases comprising this matter is not solid. There may be gaseous and/or liquid phases. The solid phase is called the solid matrix. The space within the porous medium domain that is not part of the solid matrix, is referred to as void space or pore space.

(b) The solid phase should be distributed throughout the porous medium within the domain occupied by the porous medium; solid must be present inside each representative elementary volume (REV). An essential characteristic of a porous medium is that the specific surface of the solid matrix is relatively high. This affects the behaviour of fluids in porous media in a significant way. Another basic feature of a porous medium is that the various openings comprising the void space are relatively narrow.

(c) At least some of the pores comprising the void space should be interconnected. The interconnected pore space is sometimes termed the effective pore space. As far as flow through porous media is concerned, unconnected pores may be considered as part of the solid matrix. Certain portions of the interconnected space may, in fact, also be ineffective as far as flow through the medium is concerned. For example, pores may be dead-end pores or blind pores, i.e. pores or channels with only a narrow single connection to the interconnected pore space, so that almost no flow occurs through them.

Similarly as in other porous materials, the basic structure of concrete is formed by the solid phase, i.e. by aggregates and the products of cement hydration, by the liquid phase, which is a pore solution, and by the gaseous phase, which is air with gas admixtures. In the cement paste, gel, capillary and technological pores are formed. The diameter of gel pores that are formed in products of hydration, is 10^{-9} m and these pores are filled with pore solution. The diameter of capillary pores is in the range of 10^{-6} to 10^{-5} m and these are formed in the cement binder, depending on the w/c ratio. When the w/c ratio increases, the more capillary pores originate. Capillary pores ($> 5 \times 10^{-6}$ m) control permeability and durability of concrete, smaller pores create the stress during drying (Jawed et al., 1983). Technological pores are formed in processing fresh concrete, when air is enclosed in its mass. Their size ranges from 10^{-4} to 10^{-3} m.

Small cracks occur in the concrete, when its composition and curing is unsuitable. These cracks share the permeability of the cement binder in concrete. To eliminate the formation of cracks, steel or nonmetallic fibres are added as a smeared reinforcement. When fibres are used, the internal structure of concrete depends upon the homogeneity of their smeareding.

A thin layer of the filler on the aggregates and fibres grains (if these are present in concrete) is a very important factor in the evaluation of the quality of concrete. This layer, called the interfacial transition zone (ITZ), represents the connection between the surface of the aggregates or a fibre and the cement binder, i.e. the adhesion of the cement matrix on the grains of aggregates or fibres. A typical thickness of the ITZ is within the range of 1×10^{-5} to 3×10^{-5} m (Garboczi and Berryman, 2000). Synthetic polymer fibres do not make a sufficient adhesion of the cement binder to the fibre possible, since the character of chemical bonds is different. While in polymers the bonds are prevailingly covalent nonpolar, in the silicate matrix polar bonds are predominant.

The character of the cement paste in concrete depends upon the addition of some additives. Plasticizers and superplasticizers reduce the amount of mixing water, while the workability of fresh concrete is maintained. Water in concrete is consumed in the hydration of cement, and any surplus evaporates from concrete. During the evaporation of water, a smaller number of capillary pores are formed than in case, the plasticizer would not be used.

Air entraining agents cause the formation of a system of air bubbles of diameter 10^{-5} to 10^{-3} m, which create a space for potential crystallization of ice or of salts. The efficiency of an air-entraining agent depends upon a uniform distribution of bubbles in concrete, on their number in relation to the w/c ratio, and on the manner of processing fresh concrete. Sealing additives for waterproofing concrete either reduce the volume and cross-section of capillary pores in the cement paste or hydrophobize the surface of pores. In either case, transport processes in concrete are affected.

1.5.2 Models of the Porous Structure

The models of porous structure are usually constructed with the primary aim of characterizing a porous medium by its pore distribution curve. Using these models, it is in principle possible to express the water transport material parameters, such as moisture diffusivity, hydraulic conductivity or water permeability, as functions of the capillary structure of the particular porous material. The simplest models (see, e.g., Burdine, 1953) use the concept of cylindrical capillaries. In this case, the volume flux (in m^3/s) in a single capillary with a radius r can be expressed by the Hagen–Poisseuille law,

$$\dot{V}(r) = -\frac{\pi r^4}{8\eta}\frac{\partial p}{\partial x},\tag{1.1}$$

where η is the dynamic viscosity of water, p the pressure in the capillary. If water flow in the whole porous body is to be modelled, it is necessary to use the measured pore distribution curve $f(r)$,

$$\int_{R_{min}}^{R_{max}} f(r)\,\mathrm{d}r = 1,\tag{1.2}$$

where R_{min} and R_{max} are minimum and maximum pore radii of the particular porous medium, respectively, and to take into account the tortuosity factor τ, $0 < \tau < 1$, which expresses the fact that the porous body consists of a complicated system of curved pores, and water molecules have to follow a significantly longer path than they would have to in a straight capillary tube.

Using the common relation between the mass flux j (in kg/m²s) and volume flux $\dot{V}$,

$$j = \frac{\dot{V}}{S}\rho_w, \tag{1.3}$$

where S is the cross-sectional area of a capillary tube and ρ_w the density of water, and equation (1.2), we obtain

$$j(R) = -\rho_w \tau \int_{R_{min}}^{R} \frac{\dot{V}}{S} f(r) \, dr \cdot \frac{\partial p}{\partial x}. \tag{1.4}$$

Comparing (1.4) with the definition relation for water permeability K (in m²),

$$j = -\rho_w \frac{K}{\eta} \frac{\partial p}{\partial x}, \tag{1.5}$$

and hydraulic conductivity k (in m/s)

$$j = -\frac{k}{g} \frac{\partial p}{\partial x}, \tag{1.6}$$

where g is the gravity acceleration, we obtain

$$K(R) = \frac{1}{8}\tau \int_{R_{min}}^{R} r^2 f(r) \, dr \tag{1.7}$$

and

$$k(R) = \frac{\rho_w g}{8\eta}\tau \int_{R_{min}}^{R} r^2 f(r) \, dr. \tag{1.8}$$

In equations (1.7), (1.8), K and k are expressed as functions of the pore radius R, which is not very useful for practical applications. However, the moisture content can also be expressed as a function of the pore radius,

$$w(R) = w_{sat} \int_{R_{min}}^{R} f(r) \, dr, \tag{1.9}$$

where w is the moisture content by volume (in m³/m³), w_{sat} the saturated moisture content. Therefore, calculating a particular value of K or k for a specified pore radius R from (1.7), (1.8), this value can be immediately assigned to the respective moisture content $w(R)$ using equation (1.9).

The moisture diffusivity κ is defined by the relation

$$j = -\rho_w \kappa \frac{\partial w}{\partial x}, \tag{1.10}$$

where w is the moisture content by volume. If equation (1.10) for moisture flux is to be matched with (1.4), it is necessary to express first a relation between the pressure p and moisture content w. Assuming that the pressure

p can be considered as capillary pressure (see Section 1.6.2.3), we can write for a capillary with the radius r the following relation:

$$p = -\frac{2\sigma}{r}\cos\varphi, \tag{1.11}$$

where σ is the surface tension of water, φ the contact angle. Then, for the radius R we have

$$\frac{\partial p}{\partial x} = \frac{\partial p}{\partial R}\frac{\partial R}{\partial x} = \frac{\partial p}{\partial R}\frac{\partial R}{\partial w}\frac{\partial w}{\partial x}. \tag{1.12}$$

From equation (1.9) it follows that

$$dw = w_{\text{sat}}f(R)\,dR, \tag{1.13}$$

and therefore

$$\frac{dR}{dw} = \frac{1}{f(R)w_{\text{sat}}}. \tag{1.14}$$

Substituting (1.14) into (1.12) and using (1.11) for calculation of $\frac{\partial p}{\partial R}$, we obtain

$$\frac{\partial p}{\partial x} = \frac{2\sigma\cos\varphi}{R^2}\frac{1}{f(R)w_{\text{sat}}}\frac{\partial w}{\partial x}, \tag{1.15}$$

and substituting (1.15) into (1.4) we arrive at

$$j(R) = -\rho_{\text{w}}\tau\frac{\sigma\cos\varphi}{4\eta w_{\text{sat}}}\frac{1}{R^2 f(R)}\int_{R_{\min}}^{R} r^2 f(r)\,dr \cdot \frac{\partial w}{\partial x}. \tag{1.16}$$

Comparing (1.16) and (1.10) we finally obtain

$$\kappa(R) = \tau\frac{\sigma\cos\varphi}{4\eta w_{\text{sat}}}\frac{1}{R^2 f(R)}\int_{R_{\min}}^{R} r^2 f(r)\,dr. \tag{1.17}$$

Again, if a dependence of κ on the moisture content w is required, $w(R)$ can be calculated using (1.9).

In equations (1.7), (1.8), (1.17), the tortuosity τ may be a function of the pore radius R, and therefore also of the moisture content w. This means that for a direct calculation of K, k, κ from equations (1.7), (1.8), (1.17), the $\tau(R)$ function would have to be measured in some way. The most common way to get such an expression is to introduce the concept of relative transport parameters, so that for instance the relative hydraulic conductivity k_{r} (dimensionless) is defined as

$$k_{\text{r}} = \frac{k}{k_{\text{sat}}}, \tag{1.18}$$

where k_{sat} is the saturated value of hydraulic conductivity.

From equation (1.8) it follows that

$$k_{sat} = k(R_{max}) = \frac{\rho_w g}{8\eta} \tau(R_{max}) \int_{R_{min}}^{R_{max}} r^2 f(r) \, dr. \tag{1.19}$$

Using (1.8) and (1.19) we obtain

$$k_r(R) = \frac{\tau(R)}{\tau(R_{max})} \frac{\int_{R_{min}}^{R} r^2 f(r) \, dr}{\int_{R_{min}}^{R_{max}} r^2 f(r) \, dr}. \tag{1.20}$$

Assuming (Mualem, 1976) that

$$\frac{\tau(R)}{\tau(R_{max})} = \left(\frac{w}{w_{sat}}\right)^n, \tag{1.21}$$

and substituting (1.21) into (1.20) we get

$$k_r(w) = \left(\frac{w}{w_{sat}}\right)^n \frac{\int_{R_{min}}^{R} r^2 f(r) \, dr}{\int_{R_{min}}^{R_{max}} r^2 f(r) \, dr}, \tag{1.22}$$

where the exponent n is the only unknown quantity, which should be determined experimentally.

In an analogous way, we get a similar relation for the relative water permeability $K_r = K/K_{sat}$,

$$K_r(w) = \left(\frac{w}{w_{sat}}\right)^n \frac{\int_{R_{min}}^{R} r^2 f(r) \, dr}{\int_{R_{min}}^{R_{max}} r^2 f(r) \, dr}. \tag{1.23}$$

In formulating a similar relation for the relative moisture diffusivity $\kappa_r = \kappa/\kappa_{max}$, where $\kappa_{max} = \kappa(R_{max})$, the Mualem assumption (1.21) can also be applied, but we arrive at a slightly more complicated expression:

$$\kappa_r(w) = \left(\frac{w}{w_{sat}}\right)^n \frac{R_{max}^2 f(R_{max})}{R^2 f(R)} \frac{\int_{R_{min}}^{R} r^2 f(r) \, dr}{\int_{R_{min}}^{R_{max}} r^2 f(r) \, dr}. \tag{1.24}$$

It should be noted that from a purely theoretical point of view, κ_{max} in fact does not mean the saturated value of moisture diffusivity but some maximum value defined by equation (1.17) under assumption $R = R_{max}$. In the saturated state, we have $w = w_{sat} = const.$ and according to (1.10) κ theoretically does not have a finite value. Therefore, water transport cannot be realized by a moisture gradient but only by some other thermodynamic force, for instance a gradient of external pressure in this case.

Comparing relations (1.22), (1.23) and (1.24) from the point of view of their practical applicability, we realize that (1.22) and (1.23) can be employed providing K_{sat} or k_{sat} and n are known. Measuring the saturated values of water permeability is quite a common procedure (see Section 5.3.2.5 for details), and the value of n can be obtained either experimentally or by a fitting procedure using measured $K(w)$ or $k(w)$ data (see Mualem, 1976, for details). On the other hand, the application of equation (1.24) is not so straightforward as that of (1.22) and (1.23), because the value of κ_{sat} does not exist, and an additional assumption concerning κ_{max} is to be adopted. If we assume that κ_{max} is the maximum value of κ, which is obtained using inverse modelling of water suction curves (see Section 5.3.2.3 for details), we can treat equation (1.24) in the same way as (1.22) and (1.23). However, it should be noted that from the theoretical point of view this treatment is not so pure as that employed for (1.22) and (1.23).

Until now, we assumed the simplest model of the porous structure, where the basic element consists of just one capillary of the pore class r (i.e. the pore with the radius r). This is basically the model of Burdine (1953). Later, models with more complicated basic elements were constructed. Mualem (1976) formulated a model with the basic element consisting of two capillaries placed in a series and having different lengths and radii. The relative hydraulic conductivity was then expressed in the form

$$k_{\mathrm{r}}(w) = \left(\frac{w}{w_{\mathrm{sat}}} \right)^n \left(\frac{\int_{R_{\min}}^{R} r f(r) \, \mathrm{d}r}{\int_{R_{\min}}^{R_{\max}} r f(r) \, \mathrm{d}r} \right)^2 . \tag{1.25}$$

A whole series of different models was formulated by Neiss (1982). The basic elements consisted here of a system of cylindrical capillaries with various lengths and radii, combined arbitrarily both in series and parallel way. The corresponding relations for the relative hydraulic conductivity were naturally much more complicated (see Neiss, 1982, for details).

All the above models can generally be classified as continuous models, where the only information on the porous structure comes from the pore distribution curve determined experimentally. In recent years, computer modelling techniques representing the microstructure of concrete and its components using discrete elements have been employed (see, e.g., Garboczi and Bentz, 1992, Bentz et al., 1997, Garboczi and Bentz, 1998) with an increasing frequency. These models (their advanced versions are usually called multi-scale models – see Bentz et al., 1998) can estimate the hygric transport parameters or salt transport parameters in concrete using the following input parameters (see Bentz et al., 1995): mixture proportions (aggregate content, water to cement ratio, air content); the particle size distribution of cement and aggregates; the degree of hydration of the cement paste in concrete. The basic task solved by this class of models in the process of predicting hygric and salt transport properties is the conversion of a digital image model into a random conductor network. Then, the diffusivity or conductivity are calculated using a combination of specific algorithms, developed mostly for simple lattice problems. An example of such a treatment can be found for instance in Garboczi and Bentz (1992).

The <u>multi-scale models</u> have a great potential for future use, but currently their application to the determination of hygric or salt transport properties can still be considered as a novel technique, which is under continuous development. Therefore, we will not go into details here and refer the interested reader to the original work dealing with this kind of models. For a basic reference, the electronic monograph by Garboczi et al. (2001) can be mentioned in this respect, where all the basic aspects of the theory and a series of practical applications are given.

1.6 BASIC PHYSICAL TRANSPORT MECHANISMS

1.6.1 The Continuum Approach to Porous Media

The internal geometry of a porous body is very complicated. If we want to try to model the fluid flow in a porous medium by a classical continuum approach, which is common in fluid mechanics, i.e. to define the domain of fluid flow and to formulate balance equations and boundary conditions, we would arrive at practically unresolvable difficulties due to the necessity to describe accurately the geometry of the internal solid surfaces demarcating the space where the fluid can flow. In addition, a classical description of transport phenomena by methods of continuum physics would not be appropriate from the physical point of view for all types of pores. In large pores, the description of fluid transport by continuum methods would probably be a relatively good representation of the physical reality. However, in the porous space may also appear pores having diameters in the nanometre range, which are almost comparable with the distance between the molecules, and here the continuum approach is no longer valid, because the interactive forces between the molecules should be taken into account.

From the theoretical point of view, it would certainly be better to employ methods of classical mechanics to describe transport processes in small pores, and to describe the given system of molecules as a mass-point system. For this kind of description, however, we have to know not only the initial position and momentum of all the points but also all acting forces, which is always doubtful at a molecular level. In addition, an exact formulation of such a problem would lead to a so-called many-body problem, because for instance in 1 cm^3 of a gas, there are approximately 10^{19} molecules, in 1 cm^3 of a liquid or solid material we have $10^{22} - 10^{23}$ molecules. Such a problem is generally unsolvable. From classical mechanics it is known that even a solution of a three-body problem is difficult to solve at a general level.

A way out from the difficulties of rigorous application of classical and continuum mechanics approaches may consist in using methods of statistical mechanics, which are designed for such a description of large systems of particles. The statistical approach is macroscopic in principle, but its application makes it possible to describe a fluid as a continuum. This kind of description consists in choosing certain coarser or finer levels of space averaging in the fluid filling out the pore space (which is, on the other hand, a microscopic treatment) and replacing the point values of the particular quantities by their average values over the given space domain. In this way, the molecular structure of the fluid is smoothed and replaced by an artificially formed continuum.

A starting point for such a treatment is the concept of a particle (Bear, 1988). A particle is an ensemble of many molecules contained in a small volume. Its size is much larger than the mean free path of a single molecule but much smaller than the space domain containing the fluid, so that meaningful values of fluid properties can be obtained by an averaging procedure. These values are then related to some mathematical point, which might be the centre of gravity of the space domain under averaging. At every mathematical point we thus have a particle having definite dynamic and kinematic properties.

From the point of view of model concepts of classical physics, such a treatment is denoted as quasi-continuous (Madelung, 1964), and it is applicable to practically the whole continuum mechanics. In a description of fluid transport in a porous medium, this concept is emphasized by the fact that the dimensions of the particles are substantially larger than for instance in the case of water flow in pipes, due to the nonhomogeneity of the porous space.

The necessary dimension of the particle or elementary volume, which is then considered as a point and termed the physical or material point, can be determined using the definition of bulk density or specific mass as given by Prandtl and Tietjens (1934).

Bulk density is the ratio between the mass of a given amount of matter Δm and the volume ΔV occupied by this matter. In determining the minimum necessary volume of a material point ΔV_i ensuring that the bulk density $\rho_i = \frac{\Delta m_i}{\Delta V_i}$ represents the matter in the vicinity of a point P, we choose at first a sufficiently large volume ΔV_1 (it can even be the whole domain, for instance). Then, we determine ρ_i for a sequence of volumes $\Delta V_1 > \Delta V_2 > \Delta V_3 > \ldots > \Delta V_n$ and construct a function $\rho = \rho(\Delta V)$. The minimum necessary volume of a material point ΔV_0 we determine as a lower limit of the volume domain, where the density is practically constant, i.e. no significant changes of density for small changes of volume can be observed. If the volume decreases to less than the value of ΔV_0, marked fluctuations appear in the $\rho(\Delta V)$ function, which are caused by the nonhomogeneity of matter within the volume ΔV. The minimum volume ΔV_0, which makes it possible still to consider the matter as a continuum, is called the representative elementary volume (REV – see Bear, 1988).

To determine the REV, the volume porosity $n = V_p/V$ (the volume of pores per the whole volume of the porous body) is sometimes employed instead of the bulk density, because it is certainly a quantity more characteristic of a porous medium. The principle of REV determination is the same: it is the minimum volume in the domain where porosity is practically constant. In general, REV separates the volume domain, where the porous medium can be considered as a continuum, from the domain where microscopic effects prevail.

1.6.2 Coupled Water Vapour and Water Transport in the Porous System

In solid materials, water can be present in two basic forms, namely as free water or bound water. However, there is no sharp transition between these two forms. Usually we consider as <u>free water</u> that part of moisture in the material, whose physical properties approach those of pure water. As <u>bound water</u> we denote that part of moisture that is bonded by sorption forces in the material and whose properties are significantly different from those of pure water. For removal of bound water from the material, it is necessary to supply the latent heat of evaporation, and the sorption heat as well. Water of hydration and water of crystallization are also considered as bound water.

The bonding of water to a material can be chemical, physical-chemical or physical. The <u>chemical bond</u> is the most stable. A molecule of water bonded by chemical forces directly in the molecule of a substance can be removed from the material only by its destruction, accompanied by changes of chemical structure and other properties (e.g. water of hydration and water of crystallization). <u>Physical-chemical</u> bonds are characteristic of water bonding to a material surface. The strength of the bonds is affected by the structure of the surface in a significant way. Hydrogen bonds or van der Waals bonds are the most typical representatives of physical-chemical bonding. <u>Physically bound</u> water exhibits the weakest bond and the smallest interaction with the molecules of the solid material. Capillary forces (due to surface tension forces) are typical for this kind of bonding.

Inorganic building materials contain pores with radii from 10^{-9} m to 10^{-3} m. The dimensions of the pores are the decisive factor for the determination of the phase of water that can be present in a particular pore, and consequently also for the mode of moisture transfer here (see, e.g., Kaviany, 1995).

1.6.2.1 Water Vapour Transport and Knudsen Diffusion

The basic criterion for the assessment, whether water in a given pore occurs in the form of water vapour or only as isolated molecules, is the Knudsen number Kn:

$$Kn = \frac{\lambda}{d}, \tag{1.26}$$

where λ is the mean free path of molecules of water vapour and d is the diameter of the pore. For the mean free path the following relation is valid (see, e.g., Hildebrand, 1963):

$$\lambda = \frac{k_{\mathrm{B}} T}{2^{5/2} \pi R_{\mathrm{m}}^2 p},$$ (1.27)

where k_{B} is the Boltzmann constant, $k_{\mathrm{B}} = 1.38 \times 10^{-23}$ J/K, πR_{m}^2 the collision cross section of the molecule (for water vapour $2R_{\mathrm{m}} = 4.6 \times 10^{-10}$ m), and p the pressure.

Thus, for water vapour under normal conditions, $T = 293$ K, we have

$$\lambda = \frac{1.38 \cdot 10^{-23} \cdot 293}{2^{5/2} \pi (2.3 \cdot 10^{-10})^2 \cdot 10^5} \text{ m} = 4.3 \times 10^{-8} \text{ m}.$$ (1.28)

For $Kn \gg 1$ (a general criterion of $Kn > 10$ is adopted in most cases) we cannot talk about any definite phase of water in the pores, and the transport is realized in the form of isolated water molecules. This type of transport is termed effusion or <u>Knudsen diffusion</u>.

On the other hand, for $Kn \ll 1$ the mode of water vapour transport in the pores is similar as in the case of flow in pipes and can be approximated by the Hagen–Poisseuille law

$$j = \frac{Q_{\mathrm{V}} \cdot \rho}{S} = -\frac{R^2}{8\eta} \cdot \frac{pM}{R_{\mathrm{g}} T} \cdot \frac{\mathrm{d}p}{\mathrm{d}x},$$ (1.29)

where Q_{V} is the volume flux in m^3/s, S the cross section of the pore, R the pore radius, η the dynamic viscosity, M the molar mass of water vapour, R_{g} the universal gas constant.

This mode of moisture transfer is called viscous flow, and Knudsen determined that $Kn = 0.14$ is its limiting value (see Cunningham and Williams, 1980).

For Knudsen numbers between 0.1 and 10, we talk about a so-called transition region, which is characterized by the molecular slip on the pore walls. Contrary to the Hagen–Poisseuille law, where the condition $v = 0$ is assumed on the pore wall, here we have

$$\tau = -\eta \frac{\mathrm{d}v}{\mathrm{d}r} = \alpha v,$$ (1.30)

where α is the slip friction coefficient.

In deriving the relation for water vapour flux, we start in the same way as with the Hagen–Poisseuille law. On an elementary cylinder of the fluid, the resistance force of internal friction acts in the direction against the motion,

$$\mathrm{d}F_{\mathrm{r}} = 2\pi r \cdot \mathrm{d}x \cdot \tau = -2\pi r \cdot \mathrm{d}x \cdot \eta \frac{\mathrm{d}v}{\mathrm{d}r}.$$ (1.31)

The elementary pressure force between the bases of the cylinder with the elementary pressure difference $\mathrm{d}p$ is

$$\mathrm{d}F_{\mathrm{p}} = \pi r^2 \mathrm{d}p.$$ (1.32)

These two forces have to be in an equilibrium. Therefore,

$$-2\pi r\eta\frac{\mathrm{d}v}{\mathrm{d}r}\mathrm{d}x = \pi r^2 \mathrm{d}p \tag{1.33}$$

and

$$\mathrm{d}v = -\frac{1}{2\eta}\frac{\mathrm{d}p}{\mathrm{d}x}r\,\mathrm{d}r. \tag{1.34}$$

The integration of (1.34) then leads to

$$v = -\int \frac{1}{2\eta}\frac{\mathrm{d}p}{\mathrm{d}x}r\,\mathrm{d}r = -\frac{1}{2\eta}\frac{\mathrm{d}p}{\mathrm{d}x}\frac{r^2}{2} + K. \tag{1.35}$$

The integration constant K can be determined from the boundary conditions. We note that for the classical Hagen–Poisseuille law we had $v(R) = 0$, and therefore

$$K = \frac{1}{2\eta}\frac{\mathrm{d}p}{\mathrm{d}x}\frac{R^2}{2} \tag{1.36}$$

and

$$v = \frac{1}{4\eta}\frac{\mathrm{d}p}{\mathrm{d}x}\left(R^2 - r^2\right). \tag{1.37}$$

In our case we have the condition

$$\tau(R) = \alpha v(R) = -\eta\left(\frac{\mathrm{d}v}{\mathrm{d}r}\right)_{r=R}. \tag{1.38}$$

For an application of this condition we can utilize the relation (1.34), which immediately leads to

$$\eta\left(\frac{\mathrm{d}v}{\mathrm{d}r}\right)_{r=R} = -\frac{1}{2}\frac{\mathrm{d}p}{\mathrm{d}x}R. \tag{1.39}$$

Substituting into (1.38) we get

$$\frac{1}{2}\frac{\mathrm{d}p}{\mathrm{d}r}R = \alpha v(R), \tag{1.40}$$

and therefore

$$v(R) = \frac{1}{2\alpha}\frac{\mathrm{d}p}{\mathrm{d}x}R. \tag{1.41}$$

From the general solution (1.35) it follows that

$$v(R) = -\frac{1}{2\eta}\frac{\mathrm{d}p}{\mathrm{d}x}\frac{R^2}{2} + K. \tag{1.42}$$

Comparing (1.41) and (1.42) we obtain the integration constant K,

$$K = \frac{R^2}{4\eta}\frac{\mathrm{d}p}{\mathrm{d}x} + \frac{R}{2\alpha}\frac{\mathrm{d}p}{\mathrm{d}x}, \tag{1.43}$$

and the final relation for the velocity of flow reads

$$v = \frac{1}{4\eta} \frac{dp}{dx} \left(R^2 - r^2 + \frac{2R\eta}{\alpha} \right).$$
(1.44)

The volume flux in a pipe of a radius R can be calculated as follows:

$$Q_{V} = \int v dS = \int\limits_{0}^{R} \frac{1}{4\eta} \frac{dp}{dx} \left(R^2 - r^2 + \frac{2R\eta}{\alpha} \right) 2\pi r \, dr$$

$$= \frac{\pi}{2\eta} \frac{dp}{dx} \int\limits_{0}^{R} \left(R^2 r - r^3 + \frac{2Rr\eta}{\alpha} \right) dr$$

$$= \frac{\pi}{2\eta} \frac{dp}{dx} \left[R^2 \frac{r^2}{2} - \frac{r^4}{4} + Rr^2 \frac{\eta}{\alpha} \right]_{0}^{R}$$

$$= \pi \frac{dp}{dx} \left(\frac{R^4}{8\eta} + \frac{R^3}{2\alpha} \right).$$
(1.45)

The mass flux can then be determined as

$$j = \frac{Q_{V} \cdot \rho}{S} = \frac{Q_{V} \cdot \rho}{\pi R^2} = \rho \frac{dp}{dx} \left(\frac{R^2}{8\eta} + \frac{R}{2\alpha} \right).$$
(1.46)

The density of water vapour ρ can be expressed using the equation of state for an ideal gas

$$pV = \frac{m}{M} R_{g} T,$$
(1.47)

which leads to

$$\rho = \frac{pM}{R_{g} T},$$
(1.48)

and therefore

$$j = \left(\frac{R^2}{8\eta} + \frac{R}{2\alpha} \right) \frac{pM}{R_{g} T} \frac{dp}{dx}.$$
(1.49)

Eventually, it is necessary to note that for the sake of simplification, the derivation was done in a scalar form. Taking into account the vector character of the quantities j, v, $\frac{dp}{dx}$, which were considered in the form of their absolute values until now, the mass flux in accordance to the laws of irreversible thermodynamics is positive in the direction of negative gradient of a particular quantity, which acts as a generalized thermodynamic force. Such a quantity in our case is the pressure p and therefore the generalized force is its gradient, $\text{grad } p = \frac{dp}{dx}$. Therefore, the following sign convection is correct in (1.49):

$$j = -\left(\frac{R^2}{8\eta} + \frac{R}{2\alpha}\right)\frac{pM}{R_g T}\frac{dp}{dx}. \tag{1.50}$$

In the sense of schemes of linear irreversible thermodynamics, the expression

$$\left(\frac{R^2}{8\eta} + \frac{R}{2\alpha}\right)\frac{pM}{R_g T}$$

is a phenomenological coefficient defined as the proportionality factor between the thermodynamic flux j and the thermodynamic force $\text{grad } p$.

As follows from a comparison with the classical relation for viscous flow given by the Hagen–Poisseuille law

$$j = -\frac{R^2}{8\eta}\frac{pM}{R_g T}\frac{dp}{dx}, \tag{1.51}$$

the molecular slip leads to an increase in the water vapour flux, which is a logical conclusion taking into account that the velocity of the flow has nonzero values on the wall. The slip friction coefficient α has to be determined on the basis of experiment.

We note that for the limiting case $\alpha \to \infty$, we have $v \to 0$, because τ must have a finite value. This is, however, exactly the Hagen–Poisseuille law.

For $Kn \gg 1$, typically $Kn > 10$, it is not possible to derive a relation for mass flux by a modification of the viscous regime and Hagen-Poisseuille law, because the deviations would be too large. Knudsen introduced for this case a semi-empirical relation in the form

$$j = -\left(a_1 p + D_K\frac{1 + a_2 p}{1 + a_3 p}\right)\frac{M}{R_g T}\frac{dp}{dx}, \tag{1.52}$$

where D_K is the Knudsen diffusivity (m^2/s), which is defined in such a way that for $p \to 0$ the following relation is valid:

$$j = -D_K\frac{d\rho}{dx} = -D_K\frac{M}{R_g T}\frac{dp}{dx}. \tag{1.53}$$

The constants a_1, a_2, a_3, D_K depend on the type of gas and on the properties of the material forming the porous space and have to be determined experimentally.

Comparing the relations (1.50) and (1.52), we can see that the coefficient α is in certain range of Kn proportional to the pressure p. Again, it is a logical conclusion, because with the increase of pressure the situation eventually points towards $\alpha \to \infty$, and therefore to a viscous regime.

Another semi-empirical relation was introduced by Weber (see Cunningham and Williams, 1980),

$$j = -\frac{K}{\eta}\frac{pM}{R_g T}\frac{dp}{dx} - \frac{4}{3}\left(1 - \frac{Kn}{1 + Kn}\right)D_{\text{slip}}\frac{M}{RgT}\frac{dp}{dx}$$

$$-\frac{Kn}{Kn+1}D_{\mathrm{K}}\frac{M}{R_{\mathrm{g}}T}\frac{\mathrm{d}p}{\mathrm{d}x},\tag{1.54}$$

where the first term is for the viscous flow, the second for the slip flow and the third for the Knudsen flow, and D_{slip} and D_{K} are determined by the kinetic theory of gases.

In the description of mechanisms of water transport in porous materials, we employed in principle hydraulic or more precisely convection models only, where the basic driving force was the pressure gradient and the basic transport parameter the velocity of the flowing water vapour.

However, water vapour transport in a porous body can also be described by purely phenomenological relations following from the laws of linear irreversible thermodynamics. For the transport of gases with sufficiently high pressure it was determined experimentally that the velocity of diffusion, or in other words, the mass flux in absence of the total pressure gradient, is proportional to the concentration gradient or to the gradient of partial density of the given component. This relation is known as the Fick's law,

$$\vec{j}_i = \rho_i \vec{v}_i = -D_{\mathrm{m}i}\,\mathrm{grad}\,\rho_i, \qquad \mathrm{grad}\,p = 0,\tag{1.55}$$

and the proportionality constant between the flux and gradient of partial density is denoted as the diffusion coefficient $D_{\mathrm{m}i}$. The index i denotes a component of the mixture, because the diffusion can take place in a mixture of two or more components. The Fick's law was originally formulated for a mixture of only two gases but later experiments showed that it can be also employed to the phenomenological description of gas transport through a porous medium, where in fact the porous matrix can be treated as the second component in the model concept. Models of this type are usually termed diffusion models.

This description cannot be applied for very low pressures, because for the Knudsen number we have $Kn \gg 1$, and the presence of other molecules of water vapour does not affect the diffusion process any more. In such case the mutual collisions of molecules are no longer the limiting factor of the transport; rather, the collisions of the water vapour molecules with the surface of the porous matrix play the decisive role. The transport of isolated molecules through the porous matrix can be then expressed by Knudsen diffusion,

$$\vec{j}_i = \rho_i \vec{v}_i = -D_{\mathrm{K}i}\,\mathrm{grad}\,\rho_i,\tag{1.56}$$

where $D_{\mathrm{K}i}$ is the Knudsen diffusion coefficient of the component i, which is independent of both ρ_i and the presence of other components – contrary to Fick's law, where the diffusion coefficient depends both on the partial density of a component ρ_i and on the partial densities of other components.

1.6.2.2 Transport of the Surface Phase of Water

During water vapour transport through a porous medium, some of the molecules are adsorbed on the surface of the porous matrix due to the interaction between the molecules of the gas and the solid surface. This process is accompanied by the release of adsorption heat ΔE_a (in J/kg). The molar heat of adsorption $\Delta E_a \cdot M$ is within the range of $5 \times 10^2 - 5 \times 10^5$ J/mol for various types of gases and surfaces. The adsorbed molecules can then diffuse along the solid surface, which is another mechanism of water transport in a porous medium.

The adsorbed water vapour on the surface of the porous matrix is a further phase of water in the porous material. We call it the surface phase, and its properties are significantly different from the properties of water in the volume phase or water vapour. Between the molecules of water and of the solid surface, van der Waals bonds are formed, which are stronger than hydrogen bonds characteristic of the volume phase of water. The surface phase of water is therefore not a classical liquid.

Transport of the surface phase of water can be expressed either using the hydraulic analogy, with the pressure gradient as driving force, or using the analogy with Fick's law. In the first case we have (see, e.g., Kaviany, 1995)

$$\vec{j} = -\frac{R_g\sqrt{T}}{R_{ad}A_0 M} \frac{\rho_{ad}^2}{p} \exp\left(-\frac{\Delta E_a M}{R_g T}\right) \text{ grad } p, \tag{1.57}$$

where R_{ad} is the adsorption resistance, which depends on the material parameters of the gas and of the solid surface, ρ_{ad} the density of the adsorbed phase of water, and A_0 the specific surface (in m^{-1}), where $A_0 = \frac{A_{fs}}{V}$ and A_{fs} is the area of the free surface of the pores. For the density of the adsorbed phase we have

$$\rho_{ad} = \frac{A_0 p M}{R_g T} \frac{d\phi}{dp}, \tag{1.58}$$

where ϕ is the so-called two-dimensional pressure or surface pressure (in N/m).

Fick's diffusion for the surface phase of water can be formulated as

$$\vec{j} = -D_{ad} \text{ grad } \rho_{ad}, \tag{1.59}$$

where the diffusion coefficient D_{ad} is based on applied empirical relations. For instance Cunningham and Williams (1980) introduce the following relation for D_{ad} (in m^2/s):

$$D_{ad} = 1.6 \cdot 10^{-6} \exp\left(-\frac{\Delta E_a M}{R_g T}\right). \tag{1.60}$$

1.6.2.3 Liquid Water Transport

In the porous medium, transport of liquid water can also occur. The system of pores of various dimensions can be considered as a system of capillaries. If the surface of the body comes in direct contact with water, the capillary forces begin to act, and the porous system begins to draw in water. The mechanism of this phenomenon can be best explained on the example of one capillary.

The mechanical equilibrium on the liquid/gas interface is expressed by the Laplace equation

$$p_v - p_l = \sigma \left(\frac{1}{r_1} + \frac{1}{r_2} \right), \tag{1.61}$$

where p_v is the pressure in the gaseous phase, p_l the pressure in the liquid phase, σ the surface tension on the interface, and r_1, r_2 the radii of curvature in two perpendicular directions.

In practical applications, a simplified form of this equation is mostly used, which assumes a spherical interface with $r_1 = r_2 = r$, and therefore

$$p_v - p_l = \frac{2\sigma}{r}. \tag{1.62}$$

In a capillary having walls of a solid material, it is necessary also to include the influence of the solid surface, namely of the adhesive forces between the molecules of the liquid and the solid. This effect is expressed by the contact angle Φ, so that the resulting form of the mechanical equilibrium for a cylindrical capillary is given by the Young–Laplace equation

$$p_v - p_l = \frac{2\sigma}{r} \cos \Phi \tag{1.63}$$

or

$$p_c = -\frac{2\sigma}{r} \cos \Phi, \tag{1.64}$$

where

$$p_c = p_l - p_v \tag{1.65}$$

is the capillary pressure. The Young–Laplace equation expresses the basic principle of capillary sorption of water in a porous medium. Taking into account that in contact between water, air and the most inorganic porous substances $\Phi \in (0, 90^0)$, so the capillary pressure is negative and the porous material draws in water.

In a real porous material the surface of a pore is normally not a smooth plane as required by the Young–Laplace equation. In addition, the contact angle is mostly very small, typically $\Phi < 10^0$. Therefore, equation (1.64) is simplified by assuming $\Phi \sim 0$ and

$$p_c = -\frac{2\sigma}{r}. \tag{1.66}$$

1.6.2.4 Phase Change Processes in the Porous System

A further phenomenon significantly affecting the transport of water and water vapour in a porous body is phase change processes between water and water vapour. On penetration of water vapour into the porous material, capillary condensation can occur. This phenomenon can be explained by the laws of reversible thermodynamics, on the basis of the assumption that on the interface between water and water vapour a thermodynamic equilibrium is established.

We start with the Gibbs–Duhem equation

$$S dT - V dp + n d\mu = 0, \tag{1.67}$$

where S is the entropy, T the thermodynamic temperature, V the volume, p the pressure, n the number of moles of a given substance, and μ the chemical potential. Assuming isothermal conditions, $dT = 0$, for the gaseous and liquid phases we can write, respectively

$$-V_v dp_v + n_v d\mu_v = 0 \tag{1.68}$$

$$-V_l dp_l + n_l d\mu_l = 0. \tag{1.69}$$

In thermodynamic equilibrium, the following relation is valid:

$$\mu_v = \mu_l, \tag{1.70}$$

therefore also

$$d\mu_v = d\mu_l, \tag{1.71}$$

and

$$\frac{V_v}{n_v} dp_v = \frac{V_l}{n_l} dp_l \tag{1.72}$$

or

$$v_v \, dp_v = v_l \, dp_l, \tag{1.73}$$

where v_v, v_l are molar volumes of the gaseous and liquid phases, respectively (i.e. the volumes of 1 mole of a given substance).

In further modifications of equation (1.73), we will first try to include the capillary pressure. We formally subtract the expression $v_l \, dp_v$ from both sides of this equation,

$$v_l d(p_l - p_v) = (v_v - v_l) dp_v \tag{1.74}$$

and

$$d(p_l - p_v) = \frac{v_v - v_l}{v_l} dp_v. \tag{1.75}$$

We know that

$$p_l - p_v = p_c = -\frac{2\sigma}{r}. \tag{1.76}$$

Substituting (1.76) into (1.75) we get

$$\mathrm{d}\left(\frac{2\sigma}{r}\right) = \frac{v_l - v_v}{v_l}\mathrm{d}p_v, \tag{1.77}$$

and by modifying the left-hand side

$$-\frac{2\sigma}{r^2}\mathrm{d}r = \frac{v_l - v_v}{v_l}\mathrm{d}p_v. \tag{1.78}$$

As $v_v \gg v_l$, we can write

$$\frac{2\sigma}{r^2}\mathrm{d}r = \frac{v_v}{v_l}\mathrm{d}p_v. \tag{1.79}$$

The molar volume v_v can be expressed using the equation of state for an ideal gas,

$$p_v V_v = n_v R_g T, \tag{1.80}$$

so that

$$v_v = \frac{V_v}{n_v} = \frac{R_g T}{p_v}, \tag{1.81}$$

and therefore

$$\frac{2\sigma}{r^2}\mathrm{d}r = \frac{R_g T}{v_l}\frac{1}{p_v}\mathrm{d}p_v. \tag{1.82}$$

Integrating now over the limits from r to ∞ for the radius and from any p_v to the saturation value of partial pressure of water vapour p_{vs} (for the radius of curvature $r \to \infty$ we have $p_v = p_{vs}$),

$$2\sigma \int_r^\infty \frac{1}{r^2}\mathrm{d}r = \frac{R_g T}{v_l}\int_{p_v}^{p_{vs}} \frac{1}{p_v}\mathrm{d}p_v, \tag{1.83}$$

we get

$$-2\sigma \left(\frac{1}{r}\right)_r^\infty = \frac{R_g T}{v_l}[\ln p_v]_{p_v}^{p_{vs}}, \tag{1.84}$$

and therefore

$$\ln \frac{p_v}{p_{vs}} = -\frac{2\sigma v_l}{R_g T}\frac{1}{r}. \tag{1.85}$$

The equation (1.85) is called the <u>Kelvin equation</u>.

Using now the definition of the relative humidity φ,

$$\varphi = \frac{p_v}{p_{vs}}, \tag{1.86}$$

and expressing the molar volume of the liquid phase as

$$v_l = \frac{M}{\rho_l}, \tag{1.87}$$

the Kelvin equation (1.85) can be modified into the form

$$\varphi = \exp\left(-\frac{2\sigma M}{\rho_l R_g T}\frac{1}{r}\right). \tag{1.88}$$

From a practical point of view, the Kelvin equation determines the limiting value of relative humidity φ, which can be achieved in a capillary of radius r. For this value of φ, water and water vapour in a capillary are in thermodynamic equilibrium. If the amount of water vapour in the capillary exceeds this value of φ, condensation begins; this phenomenon is termed capillary condensation. From physical point of view, the Kelvin equation expresses the fact that partial pressure of saturated water vapour decreases with increasing surface curvature, i.e. with decreasing radius of the capillary.

We substitute now in the Kelvin equation the numerical values in order to find out for what pores the capillary condensation has practical importance. We have

$$\varphi = \exp\left(-\frac{2 \cdot 0.073 \cdot 0.018}{1000 \cdot 8.314 \cdot 298}\frac{1}{r}\right) = \exp\left(-1.06 \cdot 10^{-9}\frac{1}{r}\right), \tag{1.89}$$

and therefore for instance for $r = 1$ nm we get $\varphi = 0.35$, for $r = 2$ nm $\varphi = 0.59$, and for $r = 50$ nm $\varphi = 0.98$. Accordingly, the region of capillary condensation begins somewhere at 1 nm or 2 nm and ends at 50 nm pores. For smaller pores, the surface adsorption is the dominant mechanism of water sorption. In larger pores, only water vapour transport can take place. Liquid water can appear there only due to capillary suction via direct contact of the porous body with water.

1.6.2.5 Sorption Isotherms

Practical determination of water sorption in a porous material has to be performed experimentally, and a sorption isotherm has to be measured. The sorption isotherm expresses the dependence between the relative humidity φ in the system of pores and water content w in the porous body. One point of the sorption isotherm can be determined by putting a porous body in an environment with constant temperature and constant relative humidity and leaving it there until partial pressures of water vapour in the body and in the surrounding air reach equilibrium. The water content is then measured by weighing the porous body. Repeating this treatment with various values of relative humidity, we can get a complete sorption isotherm as a function expressed in point-wise form.

The shape of a sorption isotherm depends on the type of porous material. A generally recognized classification was introduced by Brunauer et al. (1938). They divided sorption isotherms (called BET isotherms) into five groups. The most common are groups I and II.

In porous building materials including concrete, almost exclusively the sorption isotherms are of type II according to the original Brunauer et al. (1938) classification, i.e. with an S-shape. We will now describe this isotherm type in more detail, and interpret particular parts of it. At low relative humidity, the molecules of water are bound in one layer to the surface of pores by hydrogen bonds or van der Waals forces. This phase of sorption expresses the relatively fast moisture increase in the initial regions of the curve. Once all surfaces of pores are covered by one layer of molecules, further layers begin to be formed, typically 2–3 at maximum. This phase is characterized by the linear part of the isotherm. The final phase is the phase of capillary condensation, which can be expressed for a single capillary by the Kelvin equation. For very high relative humidity, i.e. $\varphi \geq 98\%$, it is practically impossible to determine the sorption isotherm, because such high relative humidity cannot be maintained constant for a sufficiently long time.

The thermodynamic equilibrium between the relative humidity (or capillary pressure) and the amount of absorbed water expressed by the sorption isotherm is not the only one that can be achieved under various conditions. At desorption we mostly get a different dependence of water content on the relative humidity content than at sorption. The desorption isotherm expressed in the from of a $w = w(\varphi)$ function always lies above the sorption isotherm.

This effect is called hysteresis and is caused by two main mechanisms. The first is the hysteresis of the contact angle, which at sorption is significantly smaller than at desorption. The probable causes of this fact are water contamination on the solid surface, surface roughness or immobility of the surface layer of water (Adamson, 1982).

The capillary hysteresis, which is the second main mechanism in this respect, is a consequence of the appearance of pores with alternating wide and narrow parts. In the narrow pore necks, evaporation is significantly slower than in wider parts because (see Kelvin equation (1.85)) the partial pressure of saturated water vapour is lower there. The phenomenon, which accelerates capillary condensation, thus slows down the drying process at desorption.

Theoretically, the hysteresis can also be induced by a third factor, namely by the irreversible processes taking place at water sorption, but it is assumed that in most cases this factor is much less significant than the previous two (Everett, 1967).

Alternating of sorption and desorption processes can lead to a variety of further equilibrium states (Hansen, 1986).

The position and shape of a sorption isotherm also strongly depend on temperature. At higher temperatures, the transport of water molecules is faster, the bonds can be released more easily, and therefore both sorption and desorption isotherms (expressed as $w = w(\varphi)$ functions) corresponding to higher temperatures are lower than those corresponding to lower temperatures (some water molecules already bound on the solid surface can be released, polymolecular layers on the pore walls are thinner). Looking at the Kelvin equation (1.85), we can see that with increasing T the exponent is decreasing and φ is increasing. Therefore, the sorption isotherm is shifted to the right, i.e. capillary condensation occurs (all other conditions being the same) at higher relative humidity.

1.6.3 Salt Transport

In porous media, salts may be present either in the form of electrolytic aqueous solutions or in the form of crystals. Therefore, salt transport in a porous space can take place only if salts are dissolved in water, and any salt transport is always related to water transport. However, capillary forces are affected by the presence of salts in water, because surface tension generally depends on salt concentration. The viscosity of the fluid also depends on the amount of salts in the solution, and therefore the transport coefficients are different than in the case of pure liquid water.

In addition to the main salt transport mechanism, where the salt solution is driven by the capillary forces as a whole and ions are transported by advection with the liquid, diffusion of dissolved salts in the electrolyte can take place as well. This is driven by the salt concentration gradient, or, more generally, by the chemical potential gradient.

The most salts are highly dissociated in water, which leads to a fast sorption of the particular ions on the pore walls. Due to the ion sorption, an electrical double layer is formed on the solid surface (see, e.g., Adamson, 1990, Zhang and Gjørv, 1997). The surface potential is determined by the ions rigidly adsorbed on the pore wall, which are sometimes called potential determining ions. The potential decreases in the direction away from the solid surface depending on the properties of the solution. Therefore, in the so-called Stern layer the solvated ions are immobile and can be considered as bound to the surface. The movement of ions and flow of solvent can only take place outside the Stern layer. The electrical double layer interferes with the ionic clouds in the solution, and affects the movement of both ions and water. This effect is naturally more pronounced in smaller pores, where the effective thickness of the double layer is comparable with the pore radius.

The electrical double layer on the surface of pore walls builds up a repulsion potential barrier. For a given solid, the attraction potential due to the van der Waals forces is constant. Therefore, the potential barrier is primarily determined by the properties of the electrolytic solution and the thickness of the double layer. When the ionic clouds have the same sign as that of the electrical double layer on the pore wall, the increased repulsion barrier may prevent ions from entering the particular pore. According to the colloid stability theory (see Adamson, 1990), the diffusing force must be large enough to overcome the potential barrier in order to move the species closer to each other. Thus, in a dilute solution the repulsion is overwhelming, while in more concentrated solutions, the repulsion potential decreases. As a consequence, for a given ionic concentration, the ions can enter only pores above a certain size, or, in other words, for a given pore diameter, there is a minimum concentration of ions that are able to diffuse in.

Therefore, the surface phase plays in transport of aqueous solutions a much more important role than in pure water transport.

In modelling salt transport (or, more precisely, ion transport) in a porous medium, the two basic phases, namely the free phase and the bound phase, should be distinguished. Therefore, salt (or ion) binding isotherms have to be determined, which express the equilibrium relation between the amount of free ions in the solution and the amount of bound ions (by both physisorption and chemisorption) on the pore walls in the porous medium. A simple nonlinear equilibrium ion binding isotherm was suggested by Freundlich (1926),

$$C_b = K \cdot C_f^m, \tag{1.90}$$

where C_b is the bound ion concentration, C_f the free ion concentration, and K and m are empirical parameters determined experimentally. Another simple two-parametric function was proposed by Langmuir,

$$\frac{1}{C_b} = \frac{1}{kC_{bm}} \cdot \frac{1}{C_f} + \frac{1}{C_{bm}}, \tag{1.91}$$

where k is the adsorption constant, and C_{bm} the bound chloride content at saturated monolayer adsorption.

Taking into account the above modes of ion transport in porous media, the basic ion transport equation under isothermal conditions can be expressed as (see Bear and Bachmat, 1990):

$$\frac{\partial}{\partial t}(wC_f) = \text{div} (wD \text{ grad } C_f) - \text{div} (wC_f \vec{v}) - \frac{\partial C_b}{\partial t}, \tag{1.92}$$

where w is the relative moisture content by volume, D the diffusion coefficient of the particular ion in water (in some references it is called hydrodynamic dispersion coefficient), and v the velocity of the liquid phase.

1.6.4 Gas Transport

The transport of gases through a porous medium can be modelled in a very similar way as water vapour transport. The basic criterion for the assessment, whether a gas in a given pore occurs in the form of a real gas or only as isolated molecules, is again the Knudsen number Kn. If the Knudsen number is large (typically greater than 10), i.e. the mean free path of the gas is much greater than the pore diameter, Knudsen flow is the main mode of gas transport. For the pores having much greater diameter than is the mean free path, i.e. typically $Kn < 0.1$, a viscous flow regime is the dominant mode of gas transport, and the flow is very similar to the fluid flow in pipes. In the transition region between these two extreme cases, molecular slip on the pore walls is the effect characteristic for gas transport, resulting in non-zero velocities of the gas at the walls. Contrary to water vapour transport, the phase changes of the most common gases do not take place during the transport in porous medium, which makes the measurement of transport parameters easier. Also, many common gases are inert to the material of the porous matrix, so the bonds between the gas molecules and the pore walls are not established when a gas molecule enters the porous space, and therefore surface diffusion does not take place as an important transport mechanisms, contrary to the case of water transport.

1.7 INTERACTION OF CHEMICAL COMPOUNDS WITH THE SOLID MATRIX

Concrete as a composite material contains, both in the binding agent and in the aggregates, compounds which are susceptible to reaction with the compounds that are in mutual contact. These are gaseous substances in the above-ground parts of a construction, or solutions that are in contact with the concrete of hydraulic structures and underground parts of constructions.

1.7.1 Interaction with the Cement Binder

Hydrated cement contains compounds that are, to a certain degree, susceptible to reaction with compounds contained in the environment surrounding concrete. In particular, these include water, aqueous solution and atmospheric gases.

Calcium hydroxide is the most soluble of all products of cement hydration; in pores it is present in the form of saturated solutions. The solubility of the hydroxide makes the reaction with compounds surrounding concrete possible. Its concentration in the pore solution is supplemented from the solid phase until all the solid $Ca(OH)_2$ is dissolved. Other reactions leading to the reduction of the hydroxide concentration in the pore solution reduce the pH value. The solubility of other products of cement hydration is very low. Nevertheless, under certain conditions, they are liable to reactions leading to their transformation into soluble, nonbinding or voluminous compounds.

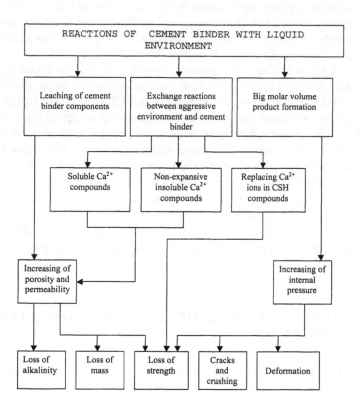

Figure 1.7 Reactions of cement binder with liquid environment

Concrete structures are most frequently in contact with surrounding atmosphere containing a number of compounds that react with the cement binder components. The concrete of hydraulic structures is washed by surface or ground water that, depending on the content of dissolved compounds, affects its constituents forming products of reaction. These products may be the cause of the change in the pore structure of concrete, which manifests itself in change in its physical properties. The solutions of acids, hydroxides or salts that are in contact with concrete structures of industrial buildings are the cause of reactions that may lead to the total destruction of concrete. A schema for the action effect of aggressive environment is given in Fig. 1.7.

Chemical reactions between the constituents of the hydrated cement binder with the liquid environment, i.e. with water and solutions of acids, hydroxides and salts, may be divided into reactions with:

- water with low content of salts (soft water)

- water containing aggressive carbon dioxide (CO_{2aggr})

- water containing humic substances

- solutions of acids

- solutions of hydroxides

- solutions of magnesium ions

- solutions of ammonium ions

- solutions of sulphate ions

- solutions of chlorides

- organic compounds

- other compounds.

The reaction of the constituents of the hydrated cement binder with gases constitutes an independent group. In particular, this concerns the reactions with atmospheric CO_2 called carbonation, and with other acid gases in the atmosphere.

1.7.1.1 Water

Water that is in contact with concrete structures may be of natural, industrial or agricultural origin. Natural water, which comprises surface water and groundwater, contains solutes of both inorganic and organic origin.

Rain water passing through the atmosphere dissolves solids, liquids and also gases dispersed in this atmosphere. The most important is carbon dioxide CO_2, which is a permanent constituent of air. Its solubility in water at temperature of 20°C and pressure of 100 kPa is 1707 mg/l (Malý and Malá, 1996). CO_2 contained in rain water, in concentrations from 0.1 up to 0.3 mmol/l, considerably increases the aggressive effect on rock formations and building materials. Sulphur dioxide SO_2, which enriches rain water through the formation of sulphurous acid, is formed by burning fossil fuels containing sulphur. A proportion of the SO_2 oxidizes to sulphur trioxide SO_3 from which, through a subsequent reaction, sulphuric acid is formed. Nitrogen oxides, generally denoted by NO_x, are formed in the operation of combustion engines. In contact with water, these oxides form nitric acid in the final phase. The breakdown of chemical industrial plant may release other gaseous components into the atmosphere. For example, HCl forming together with the rain water the solution of hydrochloric acid reacts intensely with the products of hydration in concrete. While measures for a marked reduction in emission of sulphur dioxide and nitrogen oxide in the atmosphere have been taken,

Table 1.6 Solubility of calcium hydroxide

Temperature [°C]	Solubility of $Ca(OH)_2$ [g in 100 g of water]
0	0.18
20	0.16
60	0.12
100	0.08

carbon dioxide is not usually monitored, although originates in all processes of combustion and in production technologies such as the production of lime and cement, the processes of fermentation, etc.

Ground waters contain substances dissolved during the passage of water through rock formations. They contain high concentrations of dissolved inorganic ions, especially Na^+, K^+, Ca^{2+}, and Mg^{2+}, which are in an ion balance with anions HCO_3^-, Cl^-, SO_4^{2-} and NO_3^-. Waters containing > 1000 mg of solutes in 1 litre of water are called mineral waters.

Surface water, which as well as the above-mentioned inorganic ions also contains organic compounds, is formed by underground water rising to the surface together with rain water. As far as the organic compounds are concerned, these are usually humic acids, which together with CO_2 are the most frequent cause of surface water acidity.

Rain water and water in unpolluted water courses is called soft water, with a low content of calcium and magnesium hydrogencarbonate and other compounds tending to dissolve and dilute some constituents of hydrated cement in concrete. Above all it is calcium hydroxide $Ca(OH)_2$ that is most soluble and forms a saturated solution in the pores of the binder. Moreover, in the hardened binder, it is present in a solid crystalline form, portlandite. After 28 days of cement hydration, from 12 to 14% of the hydroxide is present in the hardened cement binder; after three months it is about 20% of cement mass. The solubility of calcium hydroxide decreases with increasing temperature, as shown in Table 1.6.

The solubility of hydroxide is affected by the presence of ions in water, while the same ions, i.e. Ca^{2+} and OH^-, decrease its solubility. By contrast, different ions, e.g. Na^+, K^+, Cl^- or SO_4^{2-} increase their solubility. 2.82 g of $Ca(OH)_2$ will dissolve in one litre of 1% solution of Na_2SO_4, in 2% solution it is as much as 3.96 g. On the contrary, in one litre of NaOH solution of 5 g/l concentration, only 0.124 g of $Ca(OH)_2$ will dissolve (Dobrý and Palek, 1988).

The products of hydration of clinker minerals are stable, if their particular balanced concentrations are on the phase boundary in contact with hydroxide solution. The reduction of $Ca(OH)_2$ content in the pore solution leads to the instability of products of hydration, which results in the formation of hydrates of lower alkalinity. When the concentration of $Ca(OH)_2$ drops below 1.1 g of CaO in 1 litre, highly alkaline hydrosilicates and hydroaluminates of $3CaO.2SiO_2.3H_2O$ and $4CaO.Al_2O_3.13H_2O$ types begin to decompose. With further reduction in $Ca(OH)_2$ concentration in the solution, other types of aluminohydrates and hydroferrites hydrolyze. In the final phase, the hydrolytic decomposition may lead to the formation of hydrated oxides SiO_2, Al_2O_3 and Fe_2O_3 (Moskvin, 1980).

The intensity of the dissolution and extraction of $Ca(OH)_2$ is determined both by the composition of concrete and especially of the hydrated cement binder, and by the transport of water into the concrete structure. The rate of dissolution and extraction depends on the character of the water surrounding the concrete – whether it is stagnant water, running water or infiltration water. In stagnant water, hydroxide from the concrete surface layers is dissolved only after reaching saturation at a given temperature, and further amounts of hydroxide do not dissolve any more. Water on the phase boundary does not reach the concentration of saturated solution of calcium hydroxide. However, it may extract higher shares of hydroxide, especially in cases when physical and chemical processes are connected with the mechanical processes during which particles of concrete are washed off.

Natural waters in areas with humus or peaty soils contain humic acids. These are high-molecular cyclic compounds that belong to a group of polyphenols with side chains. They contain hydrophilic group, particularly carboxyl and hydroxyl groups that cause their acidity. Their dissociation constants are 10^{-4} to 10^{-5}. The presence of humic substances is very dangerous during the hydration reactions of cement, because even in low concentrations they function as active retarders. In hardened concrete, the humic acids react with calcium oxide leading to the formation of insoluble calcium humates. These compounds form an insoluble layer on the surface of concrete that prevents further reaction. If the water that is close to the concrete surface is free flowing then these products will be washed off and the reaction may continue.

Carbon dioxide present in natural waters may originate from two sources, namely magma, and chemical and biological processes. CO_2 dissolved in water is called free, and is dissolved in the molecular form of $CO_2.nH_2O$, of which about 1% only reacts to form carbonic acid H_2CO_3. The carbonic acid is diluted and dissociated in two stages,

$$H_2CO_3 \rightarrow H^+ + HCO_3^-$$

$$HCO_3^- \rightarrow H^+ + CO_3^{2-}$$

with dissociation constants $K_1 = 10^{-6.4}$ and $K_2 = 10^{-10.3}$. The ionic forms of carbon dioxide occurrence in water are HCO_3^- and CO_3^{2-}. The equilibrium between CO_2, HCO_3^- and CO_3^{2-} is established depending on pH, for which see Table 1.7.

Table 1.7 Forms of CO_2 in dependence on pH

CO_2 form	pH
CO_2 molecular	< 4.5
CO_2, HCO_3^-	4.5–8.3
HCO_3^-, CO_3^{2-}	8.3–10
CO_3^{2-}	> 10

The pH of the majority of natural waters ranges from 4.5 to 8.3, where CO_2 and HCO_3^- are present side by side. At pH 8.3, HCO_3^- markedly prevails. Above this value, HCO_3^- and CO_3^{2-} are in the equilibrium, and for pH > 10 when there is a negligible amount of OH^- ions in the solution, only CO_3^{2-} occurs. The equilibrium between CO_2 and HCO_3^- is established depending on pH, and also on CO_3^{2-} and Ca^{2+} concentrations. This equilibrium is not disturbed if $CaCO_3$ is added to the aqueous solution. For aqueous solutions of $CaCO_3$, the following relation is valid:

$$K_s = c(Ca^{2+}) \cdot c(CO_3^{2-}) = 10^{-8.22} \text{ (at } 25°C).$$

If the concentration of CO_2 is higher than that corresponding to the above equilibrium, the excess carbon dioxide reacts with $CaCO_3$ according to equation

$$CaCO_3 + CO_{2aggr} + H_2O \leftrightarrow Ca^{2+} + 2\,HCO_3^-.$$

In this way, it substantially increases its solubility. This carbon dioxide is called aggressive (CO_{2aggr}) for concrete.

Soon after concrete has been made, a carbonate layer created by the reaction of the air CO_2 with $Ca(OH)_2$ is formed on the concrete surface. Free carbon dioxide contained in natural waters reacts with calcium hydroxide in the cement binder, and forms very slightly soluble calcium carbonate $CaCO_3$,

$$Ca(OH)_2 + CO_2 \rightarrow CaCO_3 + H_2O.$$

In waters with CO_{2aggr} content, $CaCO_3$ formed on the concrete surface is changed into a soluble form of calcium hydrogencarbonate, which is washed out from the surface. The process of the carbonate formation and consequently that of hydrogencarbonate, is then repeated.

In concrete submerged in water with CO_{2aggr}, $Ca(OH)_2$ diffuses from the internal part of the cement binder to the concrete–water inter-phases, and changes the conditions of the equilibrium leading to the transformation of the equivalent amount of HCO_3^- into $CaCO_3$. The reaction between $CaCO_3$

and CO_{2aggr} takes place faster than the diffusion of $Ca(OH)_2$ from the pores. Then, the carbonate layer of $CaCO_3$, which was formed on the concrete surface before its submersion to the aggressive water, is gradually degraded. This reaction takes place either until the rate of diffusion becomes equal to the reaction of $CaCO_3$ and CO_{2aggr}, or until the direct reaction with hydrated clinker minerals begins. With continuous supply of $Ca(OH)_2$ to the concrete surface, $CaCO_3$ is formed and consequently dissolved to hydrogencarbonate again. While concrete in contact with air has a carbonated layer, depending upon the age of concrete, from one to several tens of millimetres thick, concrete surrounded by water containing aggressive CO_{2aggr} has a layer of only a few tenths of millimetre. In stagnant waters or in slowly running waters, equilibrium is established after a certain period of time. Running water continuously supplies the surface with further amounts of water containing CO_{2aggr}. Therefore, equilibrium will not be established, and the reaction will take place until the total decomposition of the cement binder in concrete is completed. The reaction rate depends, among other things, on the concentration of CO_{2aggr} in water.

With filtration of water containing CO_{2aggr}, which affects concrete under the hydrostatic pressure, dissolution of the surface of the carbonated layer takes place. The water passing through is gradually saturated with hydrogencarbonate, which during further passage reacts with $Ca(OH)_2$,

$$Ca(HCO_3)_2 + Ca(OH)_2 \rightarrow 2\,CaCO_3 + 2\,H_2O.$$

This reaction results in the formation of insoluble $CaCO_3$, which by its deposition in pores changes the porous structure of the cement binder in concrete. Then the CO_{2aggr} free water passing through dissolves $Ca(OH)_2$, which is washed off on the exposed side of the concrete structure, where it carbonates through atmospheric CO_2. In horizontal structures, small dripstones are formed on their surface. In vertical structures, the surface is covered with a crust. The progressive movement of the $CaCO_3$ created in the concrete layer towards the exposed side and the dissolution of $Ca(OH)_2$ supplies lead to increase of concrete corrodability in the final phase.

Observation of the dependence of disturbances to concrete containing carbonate and silicate aggregates by water containing CO_{2aggr} led to the conclusion that irrespective of the kind of aggregates used the cement binder is selectively disturbed by aggressive water. The cement binder contains $CaCO_3$ formed through the reaction of the aggressive CO_{2aggr} with $Ca(OH)_2$, the structure of which is less compact than in limestone aggregate. Therefore, the rate of the reaction of CO_{2aggr} with limestone aggregate is lower, because $CaCO_3$ is firmly built into the calcite crystal lattice (Pištěková, 1999).

In some cases, water affects concrete through hydrostatic pressure. If concrete is water permeable, filtration of water through concrete occurs while calcium hydroxide and other constituents of the hardened cement binder are

Table 1.8 pH value of some solutions

Solution	pH value
NaOH, c = 1 mol/l	14
NaOH, c = 0.1 mol/l	13
Ca(OH)$_2$, saturated at 25°C	12.45
Neutral water	7
HCl, c = 0.001 mol/l	3
HCl, c = 0,1 mol/l	1
HCl, c = 1 mol/l	0

being dissolved. At a low rate of filtration, a saturated solution of Ca(OH)$_2$ is formed, and the amount extracted is determined by the amount of water filtered through concrete. The filtration of water does not cause any changes in the pore structure if the amount is small and the rate of filtration is low. The change of concentration of Ca(OH)$_2$ in concrete occurs at the moment when water is capable of dissolving large amounts of hydroxide. Concrete is permeable to water and the volume of filtered water is high. The dissolution of portlandite, and consequently of hydrated clinker minerals, leads to the increase in porosity of the hardened cement binder in concrete. The rate of change of concrete porosity is affected by the mineralogical composition and size of particles of the cement used, by the presence of hydraulic and pozzolana active additives, and last but not least, by the compaction and curing of concrete during its processing. Pozzolana active additives bind calcium hydroxide to the less soluble hydrated calcium silicates. Very effective are silica fumes, which, owing to their very small grain size of 0.2 μm, fill in the space between the products of cement hydration. This leads to the increase in density and water tightness of concrete.

Running water may cause mechanical failures of surface layers of concrete through erosion or cavitation. The cement binder is washed off from the surface of the concrete, and the aggregate grains are then exposed to the external environment.

1.7.1.2 Acids and Hydroxides

Acids and hydroxides are in contact with concrete structures in various areas of human activity. The acidity or alkalinity of the environment is characterized by a pH value determining the concentration of hydrogen ions in a solution. The pH values of some solutions are presented in Table 1.8.

In industrial areas, the rainfall is acidic, with pH values reaching about 3. When the rain passes through the polluted atmosphere, the reaction of acid-forming oxides with water proceeds, and solutions of acids are formed (acid rains). The most frequent constituents of acid rains are carbon dioxide, sulphurous acid, sulphuric acid and nitric acid. The rate of reaction of the above acids with products of hydration depends on the concentration of acids and their dissociation. By decreasing the pH of solutions, i.e. increasing the concentration of acids, the rate of the reaction increases. Industrial wastewater, especially that from chemical industry, contains higher concentrations of strong mineral acids. The acids dissociated for the most part in the water solution react with the cement binder very quickly, which may lead to the total disintegration of concrete. The mineral acids occurring most frequently in wastewaters are sulphuric acid, nitric acid, phosphoric acid, hydrofluoric acid and hydrochloric acid. Impurities in industrial wastewaters are removed in industrial wastewater treatment plants, where some parts of the processing equipment are made of concrete or are built on concrete foundation.

Organic acids are produced in agricultural activities via the enzymatic decomposition of green plants and excrement of farm animals, and also in the food processing and chemical industries. The most frequently occuring organic acids in these various areas are formic acid, acetic acid, lactic acid, citric acid, tartaric acid, butyric acid, oleic acid, palmitic acid, and stearic acid.

Acids neutralize calcium hydroxide and react with calcium hydrosilicates and calcium hydroaluminates during the formation of calcium salts that have no binding properties. The reaction takes place along the whole contact place of the acid with the concrete surface. The rate of reaction depends on the solubility of calcium salts formed here, and also on the qualities of the layer of reaction products created there. The more soluble the calcium salts formed are and the faster they are washed out by the surrounding water, the faster is the rate of reaction. In reactions during which insoluble products are formed, initially the rate of reaction is high but then gradually decreases. In this case, slightly soluble or insoluble product of reaction adheres on the surface, and the rate of reaction is controlled by the diffusion of the solution through the layer of these products of solution. The diffusion is determined by the concentration gradient, by the thickness of the layer and by its diffusion permeability. If concrete is affected by strongly diluted solutions of acids whose concentration of hydrogen ions corresponds to pH = 4, the differences are evident in the rate of reaction of Portland cement. In that case the rate of reaction is the highest, while in mixed cement with hydraulic additives it decreases. If the concentration of H^+ ions corresponds to pH < 4, then the influence of hydraulic additives in mixed cements is not great. According to the solubility of calcium salts (highly, moderately and barely soluble salts), the action of acids may be divided into three groups, as set out in Table 1.9.

As well as calcium salts, SiO_2, which creates a gel layer on the surface grains, is also formed in reactions of acids with the products of cement hydration. The soluble calcium salts are washed out from this layer. In the case of insoluble salts, the gel layer has a coating of increased density.

Table 1.9 Classification of acids according to the solubility of calcium salts
(g in 100 g of water at 20°C)

Group	Acid	Ca^{2+} salt	Salt solubility
I Highly soluble	HCl	$CaCl_2$	73.9
	HNO_3	$Ca(NO_3)_2$	127.0
II Moderately soluble	H_2SO_4	$CaSO_4.2H_2O$	0.256
	H_2SO_3	$CaSO_3.0.5H_2O$	0.0043
	H_3PO_4	$Ca_3(PO_4)_2$	0.002
III Barely soluble	HF	CaF_2	0.0016

Among inorganic acids, hydrochloric acid and nitric acid are highly re-active. These acids react effectively with the hardened cement binder even in low concentrations ($\sim$ 1%). Highly soluble calcium nitrate and calcium chloride are taken into the solution, and the concrete surface is exposed to further reaction with an acid. The reaction of these acids leads to the to-tal decomposition of concrete; the rate of this decomposition depends on the concentration of the acid. Sulphuric acid reacts in two ways. Hydrogen ions neutralize OH^- ions in the pore solution; sulphate ions react during the for-mation of $CaSO_4$, which together with the aluminate component of the binder forms ettringite $3CaO.Al_2O_3.3CaSO_4.32H_2O$. Phosphoric acid is less aggres-sive, because through the reaction, little soluble calcium phosphate is formed. The increase in concentration of acids manifests itself in the increase in the rate of reaction.

Solutions of acids, which through their hydrolysis give rise to the relevant acid in the solution, behave in a similar way. For example, solution of iron trichloride causes the formation of an insoluble hydroxide and hydrochloric acid dissociated in the solution.

Organic acids are little dissociated in aqueous solutions. The influence of the solubility of calcium salts on the rate of reaction of organic acids with products of cement hydration is like that of mineral acids. Classification of some organic acids into three groups according to the solubility of their calcium salts is given in Table 1.10.

While solutions of acids in Group I completely decompose concrete after two months, acids of Group II reduce the compression strength for the same period of time by approximately 10%. In solutions of oxalic acid, strength actually slightly increases, because the insoluble oxalate forms an impermeable coating on the concrete surface that fills the pores and prevents contact of the acid with concrete. Of the above organic acids, lactic acid and acetic acid are the most reactive. Formic acid and citric acid react with the products of cement hydration at a slower rate, but their final effect is destructive.

The solutions of alkaline hydroxides react with the cement binder in con-crete at concentrations higher than 50 g/l, especially at higher temperatures.

Table 1.10 Solubility of calcium salts of organic acids
(g in 100 g of water at 20°C)

Group	Acid	Ca^{2+} salts solubility
I Highly soluble	CH_3COOH (acetic acid)	52.0
	HCOOH (formic acid)	16.1
	$CH_3CHOHCOOH$ (lactic acid)	10.5
II Moderately soluble	$(CH_2)_2COH(COOH)_3$ citric acid	0.085
	$(CH)_2(OH)_2(COOH)_2$ (tartaric acid)	0.037
III Barely soluble	$(COOH)_2$ oxalic acid	0.00067

In a strong alkaline environment (e.g. 50% solution of NaOH), both silicate and aluminate products of hydration of clinker minerals dissolve. Soluble alkaline silicates, which on the reduction of the concentration of OH^- ions through the air carbonation cause the formation of $SiO_2.nH_2O$ gel, are formed from the silicate compounds. Aluminates are transformed to the soluble form of alkaline aluminates, which by reducing pH lead to the formation of the amorphous form of Al_2O_3. In the final phase, the action of the concentrated solutions of alkaline hydroxides may be the cause of the total dissolution of the cement binder and the destruction of concrete.

The rate of the reaction of diluted solutions of alkaline hydroxides with the products of hydration of clinker minerals is insignificant with respect to the change of concrete quality.

1.7.1.3 Salt Solutions

Solutions of salts are in contact with concrete both in the natural environment and also in industry and agriculture. Natural waters, as mentioned earlier, are in most cases solutions of salts. There are also some solutions in industry and agriculture that contain constituents which significantly affect the character of the pore structure of the cement binder in concrete.

The ions that react with the hydrated cement binder in concrete in an especially significant way are magnesium ions, ammonium ions, sulphate ions, and chloride ions. Particular kinds of ions in the solution are equivalent to ions of the opposite charge. In the reactions of the above ions with the cement binder, in many cases the ions of opposite charges are decisive.

Solutions of salts may act in a physical way only, when no chemical reaction with the constituents of the binder proceeds. Solutions of salts diffuse into the concrete pore structure, while water is simultaneously evaporated. The concentration of dissolved salts in the solution present in the pores increases until it reaches the state of the saturated solution at a given temperature. When the saturation threshold is exceeded, salts begin to crystallize. After the pores have been filled with the salt crystals, pressures are generated that during the formation of crystallohydrates may lead to the destruction of concrete. Crystallization pressures of salts can reach tens of MPa.

When soluble <u>magnesium salts</u>, e.g. $MgCl_2$, $Mg(NO_3)_2$, come in contact with concrete, substitution reactions with calcium hydroxide proceed and magnesium hydroxide $Mg(OH)_2$ (brucite) is formed according to the equation

$$Ca(OH)_2 + MgCl_2 \rightarrow Mg(OH)_2 + CaCl_2$$

or

$$Ca(OH)_2 + Mg(NO_3)_2 \rightarrow Mg(OH)_2 + Ca(NO_3)_2.$$

Magnesium hydroxide has a very low solubility (1.8 mg in 100 g of water at 20°C), and therefore the whole calcium hydroxide present in the pore solution is converted to highly soluble salts that are leached from concrete. Magnesium hydroxide forms a hard skin of the concrete surface, and since it has no binding properties, it does not influence the strength of the cement binder. The reduction in the concentration of calcium hydroxide leads to the decrease in pH of the pore solution, and thus creates conditions for the hydrolytic decomposition of calcium-rich hydrosilicates (Taylor, 1992)

$$xCaO.SiO_2.(aq) + z\ H_2O \rightarrow y\ Ca^{2+} + 2y\ OH^- + (x\text{-}y)CaO.SiO_2.(aq).$$

<u>Magnesium sulphate</u> $MgSO_4$, whose solubility is 59.6 g in 100 g of water at 20°C, reacts during the formation of $Mg(OH)_2$ and gypsum $CaSO_4.2H_2O$ according to the equation

$$Ca(OH)_2 + MgSO_4 + 2\ H_2O \rightarrow Mg(OH)_2 + CaSO_4.2H_2O.$$

$CaSO_4.2H_2O$ crystallizes in pores and reacts with aluminate constituents of the cement binder in concrete, forming ettringite during this process. This is decomposed in the presence of $MgSO_4$ to $CaSO_4.2H_2O$, $Al(OH)_3$ and $Mg(OH)_2$ (Lea, 1971).

Assessment of the reaction of ammonium ions with the hydrated cement binder depends on the anion to which these ions are bonded. Ammonia NH_3

together with water forms ammonium hydroxide, which is a weak base. If an ammonium cation is bound to a strong acid anion, hydrolysis occurs in the aqueous solution, and fully dissociated acid and slightly dissociated ammonium hydroxide are formed. Then strong acids react with the cement binder in concrete and calcium salts are being formed. Gaseous ammonia, which shifts the chemical equilibrium to the benefit of the products, calcium salts, is released in the presence of strong calcium hydroxide:

$$Ca(OH)_2 + 2\ NH_4Cl \rightarrow 2\ NH_3 + 2\ H_2O + CaCl_2.$$

Ammonium chloride NH_4Cl and ammonium nitrate NH_4NO_3 are ammonium salts of strong acids, through reactions of which with $Ca(OH)_2$ soluble calcium salts are formed. Their reaction with the cement binder leads to the total decomposition of concrete.

Ammonium hydrogencarbonate reacts with calcium hydroxide and gaseous NH_3 is released. Hydrogencarbonate ions form with Ca^{2+} soluble calcium hydrogencarbonate according to the following equation:

$$Ca(OH)_2 + 2\ NH_4HCO_3 \rightarrow 2\ NH_3 + 2\ H_2O + Ca(HCO_3)_2.$$

Calcium hydrogencarbonate participates in the carbonate equilibrium, and may also be leached from the cement binder.

The action of ammonium sulphate $(NH_4)_2SO_4$ leads to the formation of slightly soluble $CaSO_4.2H_2O$, which crystallizes in the pores of the binder and participates in the formation of ettringite:

$$Ca(OH)_2 + (NH_4)_2SO_4 \rightarrow 2\ NH_3 + CaSO_4.2H_2O.$$

$(NH_4)_2CO_3$, $(NH_4)_2(COO)_2$ and NH_4F, which together with $Ca(OH)_2$ form very slightly soluble calcium salts, are less dangerous. These products of reaction reduce the porosity of the concrete surface layer in the first phase.

The reactions of sulphates with concrete belong to a group of reactions leading to the formation of compounds whose molar volume is larger than the molar volume of the original constituents. The formation of voluminous compounds is accompanied by pressure acting on the walls of capillary pores, which in some cases may exceed the concrete strength and lead to its destruction.

Sulphates may be found most frequently in natural waters in the form of Na_2SO_4, K_2SO_4, $MgSO_4$, $CaSO_4$ and $FeSO_4$. These compounds are transported into groundwater through dissolving the minerals present in the permeable layers of the earth's crust. There are vast areas of the earth where there is a high content of sulphates in groundwater, up to several grams per litre.

In some areas, notably in the Arabian Peninsula, gypsum $CaSO_4.2H_2O$ is a part of the sand used as aggregate to concrete. In this case, there exists an internal source of sulphur ions that can cause damage in concrete.

The compounds of sulphur may also occur in the aggregates in the form of sulphides and sulphates. Sulphates are most frequently present in the form of gypsum. In sand, these may also be present as sodium sulphate or magnesium sulphate, in the form of single grains; they may also be a part of the aggregate grains. The above mentioned sulphates are soluble, and the result of their action on the cement binder is damage leading to decomposition. Sulphides are bound to insoluble compounds, and if no oxidizing substances that would oxidize them to sulphates are present, these are not harmful. In an alkaline environment, and through the action of bacteria, the compounds of sulphides are oxidized to sulphates, which in that case represent an internal source.

Sulphur dioxide originating during the burning of brown coal containing sulphur is partly oxidized to sulphur trioxide in the atmosphere. This trioxide together with water produces sulphuric acid, which in the form of acid rain falls to the earth's surface, and by penetrating permeable layers it reacts to form salts that consequently enrich the groundwater with sulphates.

Sulphates are also the product of the bacterial decomposition of organic substances by sulphur bacteria of *Thiobacillus thioparus*, *Thiobacillus thioox-idans*, *Thiobacillus denitrificans* and other types, leading to the formation of hydrogen sulphide H_2S. This product, under aerobic conditions, is consequently oxidized to sulphates (Wasserbauer, 2000). Some oxidizing reactions of thiobacilli proceed according to the equations

$$2\,H_2S + O_2 \rightarrow 2\,S + 2\,H_2O$$

$$H_2S + 2\,O_2 \rightarrow SO_4^{2-} + 2\,H^+$$

$$5\,S^{2-} + 8\,NO_3^- + 8\,H^+ \rightarrow 5\,SO_4^{2-} + 4\,N_2 + 4\,H_2O.$$

Very often, bacterial processes take place in sewers where, in the final phase, sulphuric acid is formed. This acid damages concrete in two ways: as an acid and by sulphur ions.

The reaction of sulphur ions with the products of cement hydration principally leads to the following products: calcium sulphate dihydrate $CaSO_4.2H_2O$ and ettringite $3CaO.Al_2O_3.3CaSO_4.32H_2O$. While the formation of calcium sulphate is conditioned by the presence of Ca^{2+} ions in the pore solution, ettringite is formed by the reaction of sulphates with aluminate compounds. In both cases, this leads to the expansion of the products (Tian and Cohen, 2000). It has been proved that the presence of sulphate ions may also lead to the formation of thaumasite, which can be described by the formula $CaSiO_3.CaCO_3.CaSO_4.15H_2O$ (Hobbs and Taylor, 2000).

Calcium sulphate can be formed by the reaction of sulphate ions, originating from the liquid environment surrounding concrete, with Ca^{2+} ions of the cement binder, according to the equation

$$Ca(OH)_2 + Na_2SO_4 + 2 H_2O \rightarrow CaSO_4.2H_2O + 2 NaOH.$$

In a presence of water, two molecules of water are bound to a molecule of $CaSO_4$, and $CaSO_4.2H_2O$ is formed. This compound is slightly soluble, the solubility amounting to 256 mg in 100 g of water at 20°C. Sulphates also react with C-S-H gels in the cement binder, while $CaSO_4.2H_2O$ is being formed. The calcium sulphate formed reacts with highly alkaline products of the hydration of calcium aluminates, which affect the formation of ettringite. The reaction may be expressed by the chemical equation

$$3CaO.Al_2O_3.6H_2O + 3 CaSO_4.2H_2O + 20 H_2O \rightarrow$$
$$3CaO.Al_2O_3.3CaSO_4.32H_2O.$$

Ettringite $3CaO.Al_2O_3.3CaSO_4.32H_2O$ is irregularly soluble in water. If the concentration of $Ca(OH)_2$ in the pore solution is low, then ettringite does not originate in a solid phase but in a solution. With increasing concentration of hydroxide, the solubility of ettringite decreases. In a saturated solution of $Ca(OH)_2$, ettringite is formed in a solid phase. The molar volume of ettringite is 2.65 times larger than the sum of molar volumes of compounds that are forming it. The pressure exerted on the walls of pores subsequently leads to cohesion failure of concrete. The solubility of ettringite increases with temperature and in presence of some salts. For example, in 2–3% solution of NaCl, conditions are created such that ettringite in a solid phase does not form (Moskvin, 1980).

During the reaction of $MgSO_4$ solution, brucite $Mg(OH)_2$ is formed, a very slightly soluble compound that reacts with OH^- ions from the pore solution and promotes further dissolution of solid $Ca(OH)_2$. A low pH value leads to the decomposition of ettringite (Lea, 1971).

The sulphate resistance of concrete depends on the mineralogical composition of the cement used. The hydration of cements of a lower content of C_3S, and those of a higher content of β-C_2S leads to the formation of a smaller amount of $Ca(OH)_2$, which creates worse conditions for the formation of ettringite in the solid phase. A higher content of C_3S guarantees a high concentration of $Ca(OH)_2$ in the pore solution, which leads to the formation of insoluble ettringite. The increased content of C_3A in cement increases the possibility of reaction with sulphate ions. Therefore, one of the measures to prevent the decomposition of concrete in a sulphate environment is to use cement with a reduced content of C_3A, which is commonly called sulphate resistant. The use of silica fumes, which react with calcium hydroxide and thus reduce, to a certain degree, the basicity of the pore solution in the cement binder, is also advantageous.

Chlorides come into contact with concrete in transport structures such as roads and road facilities. During winter maintenance, bridges must not be cured with sodium chloride. However, a considerable amount of NaCl is supplied to bridge structures on the wheels of passing vehicles.

Chlorides are monitored in concrete particularly from the viewpoint of their effects on the steel reinforcement. The presence of chlorides causes the corrosion of the reinforcement leading to the formation of corrosion products, whose molar volume is almost seven times larger than the volume of the original iron. This change exerts pressure on the concrete covering layer, where cracks are formed, and in the final phase this layer peels off.

The chloride threshold is a widely discussed question. In earlier publications the chloride threshold was derived from the ratio of concentrations of chlorides and hydroxides, $c(Cl^-)/c(OH^-)$ (Tutti, 1982). The recommended threshold value was 0.6. However, the concentration of chlorides causing corrosion of steel reinforcement is also dependent on other parameters, such as concrete humidity, w/c ratio, and content of cement in concrete (Neville, 1995, Delagrave et al, 1996, Thangavel and Rengaswamy, 1998).

Concrete itself can be damaged by sodium chloride through physical action. Capillary pores of concrete absorb the solution of NaCl, which crystallizes after the water has been evaporated. At low temperatures, NaCl crystallizes as dihydrate, $NaCl.2H_2O$. At temperature $+0.15°C$, this is stable in a saturated solution of NaCl (Remy, 1961). The crystallization pressure of NaCl is 55.4 MPa. Unless the concrete is aerated, this pressure causes the disintegration of the surface layers of the concrete.

The chemical action of chlorides on the cement binder is related to the presence of the relevant cations. Chlorides of alkaline metals do not react with the cement binder, because alkaline metals form soluble hydroxides. The chlorides of metals, which form insoluble hydroxides, are harmful to the cement binder. Their solutions are slightly acid due to the withdrawal of OH^- ions to insoluble compounds. Cations like Mg^{2+}, Al^{3+}, and Fe^{3+} react with $Ca(OH)_2$ to form slightly soluble hydroxides. Then free Ca^{2+} ions together with Cl^- form highly soluble calcium chloride $CaCl_2$, which is easily diluted from the cement binder.

Chlorides are bound to Friedel's salt $3CaO.Al_2O_3.CaCl_2.12H_2O$, to a small extent. This compound is unstable in water, in solutions of $CaSO_4$, $Ca(OH)_2$ and in alkaline solutions; the stability increases in a solution of $CaCl_2$ (Lea, 1971).

1.7.1.4 Gases

The majority of concrete structures are in contact with atmospheric gases. Some of them, such as oxygen and nitrogen, do not react with the cement binder. Carbon dioxide CO_2, the concentration of which reaches almost 0.04% by volume, is the most significant of the gases normally present in the atmosphere for reaction with the cement binder in concrete. In places where the sources of additional CO_2 occur, for example in the production of lime and cement, during the burning of fuels or in fermentation processes, its content in the air increases.

Other gases that react with the cement binder occur in the atmosphere in negligible amounts; their concentration increases in places where these gases are produced. This applies to sulphur dioxide SO_2, from nondesulphurized or imperfectly desulphurized flue gases of brown coal containing sulphur. Other reactive gases originate in industry, in particular chemical production.

From the chemical point of view, reactive gases may be divided into acid, alkaline and neutral. The effect of CO_2 on the cement binder, generally called carbonation (because the main product of the reaction are carbonates), has been given the maximum attention. The main reason for the study of the carbonation of concrete is the fact that carbon dioxide is a gas of acid character, reacting in the presence of water with alkaline products of cement hydration. The pH value of the pore solution is moved to the neutral range, and thus the stability of iron oxides and hydroxides forming the protective layer on the surface of the steel reinforcement and protecting it against corrosion is reduced.

The course and rate of carbonation depend on many factors. Especially important is the content of $Ca(OH)_2$ in the cement binder, which depends on the amount and kind of cement in the concrete. Concretes made of Portland cement and using super plasticizers form a structure in the binder which has a minimum amount of capillary pores and a high content of $Ca(OH)_2$. Other factors affecting the rate of carbonation are the humidity of the environment, partial pressure of CO_2 in the ambient atmosphere, temperature, and porosity of the cement binder in the concrete. Capillary pores formed during cement hydration make diffusion of CO_2 deep into the concrete possible. Gel pores, which under standard conditions are filled with water or solutions, are impermeable to CO_2.

Carbon dioxide dissolves in water depending on the temperature and its partial pressure above the liquid level. About 1% reacts to form carbonic acid H_2CO_3 (see section 1.7.1.1). Pores completely filled with water prevent the penetration of CO_2 deep into concrete. On the other hand, the presence of water is a necessary condition of carbonation. In fully dried concrete, the reaction of CO_2 or of other gases with the cement binder does not occur. The dependence of the rate of carbonation on the relative humidity of air has been measured. It is clear from Fig. 1.8 (Matoušek and Drochytka, 1998), that the fastest carbonation takes place within the range of relative humidity of 50 to 96% RH. Lower relative humidity of air slows the rate of carbonation. At the other extreme, when the relative humidity is over 96%, its rate drops very fast.

In the first phase of carbonation, CO_2 reacts with calcium hydroxide that originated during the hydration of C_3S and β-C_2S, and forms a saturated solution on the walls of capillary pores of the cement binder. In addition,

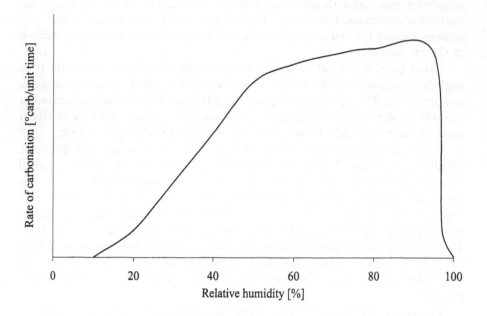

Figure 1.8 Rate of carbonation versus relative humidity of air

$Ca(OH)_2$ is present in the binder in a solid phase, in a form of portlandite. Calcium carbonate $CaCO_3$ is formed through the reaction

$$Ca(OH)_2 + CO_2 \rightarrow CaCO_3 + H_2O.$$

Due to the fact that the $CaCO_3$ that is formed has a larger molar volume than the original $Ca(OH)_2$, it fills the capillary pores and reduces their permeability. Based on the diameter of the capillary pores, the compactness of concrete surface layers may be increased, and further processes of carbonation may be slowed. The reduction of concentration of $Ca(OH)_2$ in the pore solution promotes the dissolution of its other fractions that are carbonated. Thus the content of hydroxide that is necessary for maintaining a constantly high value of pH in the pore solution, which in fresh concrete reaches 12.45, is reduced. After the major part of the $Ca(OH)_2$ has been consumed, and consequently the pH has dropped, the conditions of the stability of products of cement hydration are also changed, and CO_2 begins to react with them (Matoušek, 1972). This complex set of reactions may be described by the simplified equations

$$xCaO.SiO_2.zH_2O + x\, CO_2 + n\, H_2O \rightarrow x\, CaCO_3 + SiO_2.(z+n)H_2O$$

$$xCaO.Al_2O_3.zH_2O + x\ CO_2 + n\ H_2O \rightarrow$$
$$x\ CaCO_3 + 2\ Al(OH)_3 + (z+n-3)H_2O.$$

The form of the calcium carbonate produced is important for understanding the state of the carbonated concrete. Fine-grained crystals of $CaCO_3$ together with the gel of silicic acid on the surface of products of hydration are formed at the beginning of the carbonation of calcium hydrosilicates and hydroaluminates. In the later stage of carbonation, at a sufficient relative humidity of air, large crystals of aragonite and calcite are formed through recrystallization, which in exceptional cases may lead to the disintegration of the cement binder in concrete (Matoušek and Drochytka, 1998).

The course of concrete carbonation has been mathematically modelled, based on Fick's law for diffusion,

$$x_c = A\sqrt{t}, \tag{1.93}$$

where x_c is the depth of carbonation at time t, and A an empirical constant. After the material and physical parameters have been included, the equation takes on, according to Papadakis et al. (1991, 1992), the following form:

$$x_c = \sqrt{\frac{2\ D^c_{e,CO_2}[CO_2]^0}{[Ca(OH)_2]^0 + 3\ [CSH]^0 + 3\ [C_3S]^0 + 2\ [C_2S]^0}t}, \tag{1.94}$$

where D^c_{e,CO_2} is the effective diffusivity of CO_2 for fully carbonated concrete in m^2/s, $[CO_2]^0$ the concentration of CO_2 in the concrete surface layer in mol/m^3, $[Ca(OH)_2]^0$, $[CSH]^0$, $[C_3S]^0$ and $[C_2S]^0$ the initial concentrations at time $t = 0$ in mol/m^3 of concrete, and t time in years.

The denominator in equation (1.94) represents the total molar concentration of CaO that may be carbonated. The effective diffusivity in the numerator fully corresponds to the carbonated concrete, and is a function of porosity ϵ in time t and the degree of pore saturation f, which depends on the ambient relative humidity and the size and distribution of pores in concrete.

Another model (Bob and Afana, 1993) determining the depth of carbonation is expressed by equation

$$x_c(X,t) = \frac{150\ Ckd}{f_c}\sqrt{t}, \tag{1.95}$$

where C is the coefficient characterizing the type of cement, d the coefficient characterizing the carbonation depth, f_c concrete strength in MPa, and k the coefficient of the environment humidity. Besides the coefficients of an empirical character, the equation considers the compression strength of concrete, which includes the effect of the content and kind of cement, the water/cement ratio, the kind and amount of aggregates, and the manner of concrete processing. This characteristic quantity may easily be found for all types of concrete.

Table 1.11 Input variables

Variable	Description	Mean value	Std. dev.	Probability distr.
C	Cement type coefficient	1.2	0.12	Normal
d	Carbonation depth	1.5	0.15	Normal
f_c	Compression strength of concrete	32.3	2.1	Two parametric log-normal
k	Environmental humidity coefficient	0.6	0.06	Normal
Φ	Uncertainty factor of model	1.0	0.1	Normal

Due to the random variability of input values, the model used a probabilistic approach, in which the parameters are described as distribution functions (Keršner et al., 1996), see Table 1.11.

Thus the carbonation depth $x_c(X, t)$ is a function of random input values, and statistical parameters of x_c may be estimated by means of the statistical analysis, e.g. by the mean value or by the standard deviation. Simulation of the Monte Carlo type or numerical simulation via Latin Hypercube Sampling (LHS), which uses a small number of simulations to gain a good estimate of statistic parameters, may be carried out to advantage.

As input, these probability simulations utilize the variables together with statistics gained from earlier measurements, and data from the literature or from an expert estimate. These may not be quite accurate. The mathematical model of the advance of the carbonation front cannot be regarded as an exact one. The so-called primary information, which depends on the attainable input data and the model used, is the result of calculation. Information for an arbitrarily chosen time may be obtained by means of the numerical simulation.

The prediction of the carbonation depth may be improved by combining the primary statistics (mean values and standard deviations gained by numerical simulation) and the statistics from measuring the carbonation depth on a structure. This prediction utilizes Bayes' theorem, and may be combined with the LHS method. Typical results are shown in Fig. 1.9 (Keršner et al., 1996).

By improving the estimate of the carbonation depth versus time, or the time taken for the carbonation front to reach the reinforcement, the estimate of the reinforced concrete structure service life may be perfected.

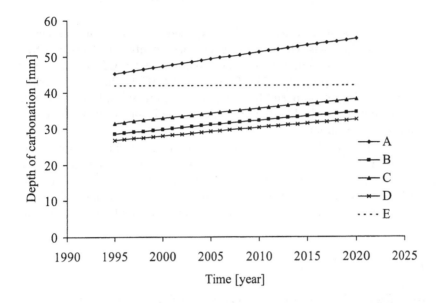

Figure 1.9 Carbonation depth versus time (A: Mean value + standard deviation: Prior, B: Mean value − standard deviation: Prior, C: Mean value + standard deviation: Bayesian updating, D: Mean value − standard deviation: Bayesian updating, E: Mean value of cover)

Far less attention has been given to the action of sulphur dioxide, called sulphatization, in spite of the fact that SO_2 is a gas of acid character and occurs in the atmosphere. SO_2 reacts with water to form a medium strong acid ($pK_1 = 1.76$, $pK_2 = 7.21$). The humidity and SO_2 concentration in the ambient environment determine the kind and amount of reaction intermediates and products. Sulphur dioxide reacts, first of all, with $Ca(OH)_2$, and after that has been consumed, reacts with highly alkaline calcium hydrosilicates and hydroaluminates. Calcium sulphite hemihydrate $CaSO_3.0.5H_2O$ is formed first which in aqueous solution easily oxidizes to sulphate $CaSO_4.0.5H_2O$. If the relative humidity of air is > 75%, a considerable proportion of dihydrate $CaSO_4.2H_2O$ is created (Matoušek and Drochytka, 1998).

In the beginning, the products fill capillary pores, which leads to an increase in strength of the concrete surface layers. In the final stage, large-volume gypsum crystals develop by recrystallization in the humid environment. This leads to the formation of cracks and finally to the loss of strength.

At a very low porosity of concrete, i.e. when the cement binder is very consistent, the reaction is confined to the concrete surface layer only, and results in efflorescence:

$$2\ Ca(OH)_2 + 2\ SO_2 + 2\ H_2O \rightarrow 2\ CaSO_3.0.5H_2O + 3\ H_2O.$$

In practice, concrete structures are often exposed to a simultaneous action of CO_2 and SO_2. In areas where there is increased content of CO_2 and SO_2 in the ambient environment, the complete reaction of $Ca(OH)_2$ takes place in concretes of higher porosity. The fine-grained $CaCO_3$ that is formed by the reaction is easily decomposed by SO_2, and calcium sulphite and calcium sulphates are then formed. The reaction rate is conditioned by the humidity of the ambient environment.

Hydrogen sulphide H_2S is a product of a number of chemical processes and of the anaerobic decomposition of organic substances containing sulphur. It is a gas with a marked smell, which reacts with water to form a weak acid ($pK_1=7.07$, $pK_2=12.20$). Hydrogen sulphide is oxidized in microbiological way to form sulphuric acid, which reacts with the cement binder (Biczók, 1972),

$$H_2S + 2\ O_2 \rightarrow H_2SO_4.$$

The direct effect of H_2S on concrete may be described by equations

$$Ca(OH)_2 + H_2S \rightarrow CaS + 2\ H_2O$$

$$2\ CaCO_3 + H_2S \rightarrow Ca(HCO_3)_2 + CaS$$

$$CaS + H_2S \rightarrow Ca(HS)_2.$$

Calcium hydrogencarbonate and calcium hydrogensulphide are soluble in water and may be easily leached from concrete. Thus the equilibrium of the reactions is shifted in the direction of the reaction products. H_2S is more harmful for steel reinforcement, which may be completely destroyed by it.

Nitrogen oxides originate as secondary products in the processes occurring in combustion engines and in the manufacture of various chemical products. The redox system of nitrogen oxides is very unstable, so through reacting with oxygen in the air, NO readily oxidizes to NO_2, which reacts with water to form nitric acid HNO_3:

$$NO + 0.5\ O_2 \rightarrow NO_2$$

$$3\ NO_2 + H_2O \rightarrow 2\ HNO_3 + NO$$

or

$$2\ NO_2 + H_2O \rightarrow HNO_3 + HNO_2.$$

The reaction rate depends on the oxide concentration and on the relative humidity in the ambient environment. Nitric acid is a strong acid, which reacts very quickly with both $Ca(OH)_2$ and other products of cement hydration, forming highly soluble $Ca(NO_3)_2$.

Ammonia NH_3 together with water forms a weak hydroxide NH_4OH (with dissociation constant $pK_b = 4.75$). Neither gaseous ammonia nor ammonium hydroxide reacts with the cement binder. Ammonia ions are damaging to concrete only in connection with anions such as NO_3^- or Cl^-.

Gaseous chlorine Cl_2 originates in various chemical processes; it can be produced by the electrolysis of NaCl melt. Gaseous chlorine reacts with water, and hypochlorous acid (dissociation constant $pK_a = 7.53$) and hydrochloric acid are formed,

$$Cl_2 + H_2O \leftrightarrow HClO + HCl,$$

and then the unstable HClO is decomposed,

$$HClO \rightarrow 2\,HCl + O_2.$$

If the gaseous chlorine reacts with water, it affects the cement binder and highly soluble salts are formed. In the case gaseous chlorine reacts directly with the solution of $Ca(OH)_2$ in the pore solution, calcium chloride-hypochlorite $Ca(OCl)Cl$ is formed. This easily eliminates oxygen, and the resulting $CaCl_2$ is leached from the binder or in a presence of $Ca(OH)_2$ crystallizes as a binary compound $CaCl_2.3Ca(OH)_2.12H_2O$ in the form of long thin needle crystals (Remy, 1961).

Oxygen and nitrogen, which are the two largest components in the ambient air, do not react with the cement binder. However, the presence of oxygen is necessary for the electrochemical corrosion of the steel reinforcement in concrete.

1.7.2 Interaction with Aggregates

Aggregates contain compounds that react with the compounds surrounding concrete, which may result in its degradation. The schema for the reactions of aggregates with aggressive environments is given in Fig. 1.10.

Aggregates often contain pyrite FeS_2, which is oxidized in the alkali environment of the cement binder according to equation

$$FeS_2 + H_2O + 3.5\,O_2 \rightarrow FeSO_4 + H_2SO_4.$$

The iron (II) sulphate formed oxidizes to iron (III) sulphate. In humid envi-

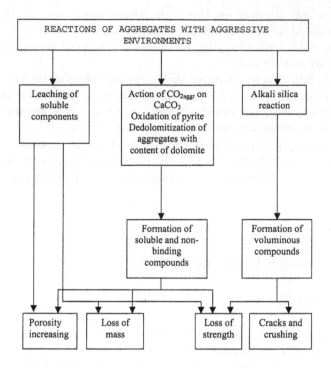

Figure 1.10 Reactions of aggregates with aggressive environments

ronments this hydrolyses, and ferric hydroxide and sulphuric acid are formed. Sulphuric acid reacts with calcium hydroxide and aluminate components of cement. In the final phase, the reactions of pyrite lead to the formation of ettringite, and ferric hydroxide manifests itself by rusty spots on the concrete surface.

Humic substances occurring in sand or in waters in areas covered with peat soils are formed during the biochemical disintegration of dead animal or plant organisms. These are brown-coloured high-molecular cyclic compounds comprising a complex of polyphenols. Humic substances having carboxyl and hydroxyl groups in the side chains cause the acidity and sorptive and ion exchanging properties. These behave like organic acids with a dissociation constant in the range of $pK_a = 3$–4. Their salts mostly form complex compounds, and their formation is one of the causes of leaching of metal elements (Ca^{2+}) from the cement matrix. Even in small amounts, the presence of humic substances prevents cement from setting, which is the reason for detecting their presence in the aggregates used for making concrete.

The use of limestone aggregates is problematic in an environment with a high content of aggressive CO_{2aggr}. Karst effects occur when $CaCO_3$ reacts, and when the soluble hydrogencarbonate that is leached from the concrete, is formed. The reaction takes place according to the equation

$$CaCO_3 + CO_{2aggr} + H_2O \leftrightarrow Ca^{2+} + 2\,HCO_3^-.$$

However, when concrete is degraded by the action of CO_{2aggr}, an important role is played by the cement matrix, which reacts with CO_2 irrespective of the kind of aggregates used (Pištěková, 1999).

Reactions that may in their ultimate consequences lead to the degradation of concrete may proceed between the aggregates and some constituents of cement. This especially concerns the reaction of alkalis from cement with the aggregates containing amorphous silica.

The alkali–silica reaction is conditioned by a sufficiently high concentration of alkali in cement, by the presence of amorphous silica and by humidity. In cement, alkalis are present in a form of sulphates and oxides. Sulphates react with aluminates, and slightly soluble ettringite and equivalent amounts of OH^-, Na^+ and K^+ ions that are present in the pore solution are formed. Alkaline oxides also react with water when the above ions are formed. Alkaline hydroxides in the pore solution reduce the solubility of $Ca(OH)_2$ present in the cement binder, and the pH value of the pore solution becomes higher than 12.45, which is the pH attained by a saturated solution of $Ca(OH)_2$ at 25°C. Alkalis may also originate from external sources, such as concrete additives, de-icing salts, and groundwater, or may be a part of the gravel and sand used. Alkali reaction is not expected in cements with low alkali content and also when external sources of alkalis are not present.

Silica is subject to the alkali reaction especially in an amorphous form, even if the reaction may occur also in crystalline forms, notably if these are of microcrystalline character. Aggregates containing opals, chalcedonies, flints, cherts, tuffs, and other amorphous or imperfectly crystalline forms of SiO_2 are subject to the alkali–silica reaction. Opals are most susceptible to the reaction due to their less compact structure. The reaction proceeds on the grain surface, where the oxygen bridge between two atoms of silicon is broken (Bažant and Steffens, 2000),

$$Si-O-Si + H_2O \rightarrow Si-OH + Si-OH.$$

This reaction commonly proceeds only on the grain surface of the amorphous or imperfectly crystalline SiO_2, not to any depth. In the environment with a high concentration of OH^- ions, a reaction between silanol and hydroxide groups occurs,

$$Si-OH + OH^- \rightarrow SiO^- + H_2O,$$

and a gel-like layer begins to form on the grains surface. Sodium, potassium

and calcium ions penetrate through the gel-like layer to the non-reacting surface. The calcium ions form CSH gels, which are not expansive. With a high
concentration of alkalis, calcium is not able to penetrate through the gel-like
layer sufficiently fast, so an alkali gel is formed, which with a sufficient amount
of water can exert large pressures leading to the failure through the formation of branched cracks. If the alkali concentration is low or if the grains are
very small, there is a prerequisite that the expansive gel will not originate
or the amount of this will be small, and concrete will not fail. The risk of
the alkali–silica reaction may be mitigated if lithium salts are added to the
concrete mix. The gel that is formed in the presence of lithium salts is not
expansive.

The dolomite aggregates also react with alkalis. The change of dolomite
to calcium (de-dolomitization) is another reaction connected with volume increase. This reaction, like the alkali–silica reaction, is conditioned by a higher
content of alkalis, humidity and dolomite aggregates. This reaction may be
described in a simplified way by the equation

$$CaMg(CO_3)_2 + 2\ (Na, K, Li)OH \rightarrow Mg(OH)_2 + CaCO_3$$
$$+\ (Na, K, Li)_2CO_3.$$

The alkali carbonate formed reacts with $Ca(OH)_2$ and another alkali hydroxide is formed. This alkali hydroxide again attacks another piece of dolomite,
thus securing a continuing reaction. Dolomite limestone containing 40–60%
of dolomite and 10–20% of clays are susceptible to this reaction.

References

Adamson, A.W., 1990, *Physical Chemistry of Surfaces*, 5th edition. (New
York: Wiley).

Bažant, Z.P., Steffens, A., 2000, Mathematical model for kinetics of alkali-
silica reaction in concrete. *Cement and Concrete Research*, **30**, pp.
419–428.

Bear, J., 1988, *Dynamics of Fluids in Porous Media*. (New York: Dover
Publications).

Bear, J., Bachmat, Y., 1990, *Introduction to Modelling of Transport Phenomena in Porous Media*. (Dordrecht: Kluwer).

Bear, J., Zaslavsky, D., Irmay, S., 1968, *Physical Principles of Water Perco-
lation and Seepage*. (Paris: UNESCO).

Bentur, A., Mindess, S., 1990, *Fibre Reinforced Cementitious Composites*.
(London: Elsevier).

Bentz, D.P., Detwiler, R.J., Garboczi, E.J., Halamickova, P., Schwartz, L.M., 1997, Multi-scale modelling of the diffusivity of mortar and concrete. In: *Proceedings of the RILEM International Workshop Chloride Penetration into Concrete*, St-Remy-les-Chevreuse, edited by Nilsson, L.O., Ollivier, J.P. (St-Remy-les-Chevreuse: RILEM), pp. 85–94.

Bentz, D.P., Garboczi, E.J., Lagergren, E.S., 1998, Multi-scale microstructural modelling of concrete diffusivity: identification of significant variables. *Cement, Concrete and Aggregates* **20**, pp. 129–139.

Biczók, I., 1972, *Concrete Corrosion, Concrete Protection*. (Budapest: Akadémiai Kiadó).

Bob, C., Afana, E., 1993, On-site assessment of concrete carbonation. In: *Proceedings of the Conference on Failure of Concrete Structure*. (Štrbské Pleso: Expertcentrum), pp. 84–87.

Brunauer, S., Emmet, P.H., Teller, E., 1938, Adsorption of gases in multi-molecular layers. *Journal of the American Chemical Society*, **60**, pp. 309–319.

Burdine, N.T., 1953, Relative permeability calculations from pore-size distribution data. *Trans. AIME*, **198**, pp. 71–78.

Coleman, N.J., McWhinnie, W.R., 2000, Solid state chemistry of metakaolin-blended ordinary Portland cement. *Journal of Materials Science*, **35**, pp. 2701–2710.

Courtault, B., 1974, Influence de la température sur l'évolution thermique des pates de ciments. *Revue des matériaux de construction*, No. 687, pp. 117–122.

Cunningham, R.E., Williams, R.J., 1980, *Diffusion in Gases and Porous Media*. (New York: Plenum).

Delagrave, A. et al, 1996, Influence of chloride ions and level on the durability of high performance cement pastes, Part II. *Cement and Concrete Research*, **26**, pp. 749–760.

Dobrý, O., Palek, L., 1988, *Concrete Corrosion in Civil Engineering*. (Prague: SNTL), (in Czech).

Everett, D.H., 1967, Adsorption hysteresis. In: *The Solid–Gas Interface*, Vol. 2, edited by E.A. Flood. (New York: Dekker), Chapter 36.

Freundlich, C.G.L., 1926, *Colloid and Capillary Chemistry*. (London: Metheun).

Garboczi, E.J., Bentz, D.P., 1992, Computer simulation of the diffusivity of cement-based materials. *Journal of Materials Science*, **27**, pp. 2083–2092.

Garboczi, E.J., Bentz, D.P., 1998, Multi-scale analytical/numerical theory of the diffusivity of concrete. *Journal of Advanced Cement-Based Materials*, **8**, pp. 77–88.

Garboczi, E.J., Bentz, D.P., Snyder, K.A., Martys, N.S., Stutzman, P.E., Ferraris, C.F., 2001, *Modelling and Measuring the Structure and Properties of Cement-Based Materials.* NIST, Gaithersurg, http:/ciks.cbt.nist.gov/garbocz/monograph/

Garboczi, E.J., Berryman, J.G., 2000, New effective medium theory for the diffusivity or conductivity of multi-scale concrete microstructure model. *Concrete Science and Engineering*, **2**, pp. 88–96.

Hansen, K.K., 1986, *Sorption Isotherms*, Technical Report 163/86. (Lyngby: The Technical University of Denmark).

Hildebrand, J.H., 1963, *An Introduction to Kinetic Theory.* (New York: Reinhold).

Hobbs, D.W., Taylor, M.G., 2000, Nature of thaumasite sulphate attack mechanism in field concrete. *Cement and Concrete Research*, **30**, pp. 529–533.

Jawed, I., Skalny, J., Young, J. F., 1983, Hydration of Portland cement. In: *Structure and Performance of Cements*, edited by Barnes, P. (London and New York: Applied Science Publishers), p. 311.

Kaviany, M., 1995, *Principles of Heat Transfer in Porous Media*, 2nd edition. (New York: Springer Verlag).

Keršner, Z., Rovnaníková, P., Suza, I., Novák, D., Teplý, B., 1996, When is it necessary to repair reinforced concrete structures? *Beton a zdivo (Concrete and Masonry)*, **4**, pp. 2–4, (in Czech).

Lea, F.M, 1972, *The Chemistry of Cement and Concrete.* (New York: Chemical Publishing Company, Inc).

Madelung, E., 1964, *Die wissentschaftlichen Hilfsmittel des Physikers*, 7th edition. (Berlin: Springer Verlag).

Malý, J., Malá, J. , 1996, *The Chemistry and Technology of Water.* (Brno: NOEL 2000), (in Czech).

Matoušek, M., 1972, Some problems of stability of calciumhydrosilicates and their conversion by carbon dioxide in concrete construction. In: *Proceedings of the 2nd Symposium SILICHEM.* (Brno: DT Press), pp. 166–174, (in Czech).

Matoušek, M., 1980, The action of selected atmospheric factors on building structures, DSc Thesis. (Brno: FCE BUT), (in Czech).

Matoušek, M. and Drochytka, R., 1998, *Atmospheric Corrosion of Concrete.* (Prague: IKAS), (in Czech).

Moskvin, V.M et al, 1980, *Korozija betona i železobetona i metody ich zaščity.* (Moscow: Strojizdat), (in Russian).

Mualem, Y., 1976, A new model for predicting the hydraulic conductivity of unsaturated soils. *Water Resources Research*, **12**, pp. 513–522.

Neiss, J., 1982, *Numerische Simulation des Wärme und Feuchtetransports und der Eisbildung in Böden.* Fortschritt Berichte der VDI-Zeitschriften, Reihe 3, Nr. 73, VDI Verlag.

Neville, A., 1995, Chloride attack of reinforced concrete: an overview. *Materials and Structures*, **28**, pp. 63–70.

Papadakis, G.V., Fardis, M.N. and Vayenas, C.G., 1992, Effect of composition, environmental factors and cement-lime coating on concrete carbonation. *Materials and Structures*, **25**, pp. 293–304.

Papadakis, G.V., Vayenas, C.G. and Fardis, M.N., 1991, Fundamental modelling and experimental investigation of concrete carbonation. *ACI Materials Journal*, **88**, pp. 363–373.

Pištěková, M., 1999, The evaluation of aggressive water effects on quality of concrete. In: *Proceedings of the 11th Int. Conf. FCE BUT.* (Brno: FCE BUT), (in Czech).

Prandtl, L., Tietjens, O.G., 1934, *Fundamentals of Hydro- and Aeromechanics.* (New York: McGraw-Hill).

Remy, H., 1961, *Inorganic Chemistry.* (Prague: SNTL), (in Czech)

Rovnaníková, P. and Rovnaník, P., 1997, Cement solidification waste with ZnO content. In: *Proceedings of the IXth SILICHEM*, edited by Hoffmann, O. (Brno: Academic Publishing CERM), pp. 35–36.

Rovnaníková, P., 1981, Utilizing of hydration heat during acceleration of concrete solidification. PhD Thesis. (Brno: FCE TU Brno), (in Czech).

Šauman, J., 1993, *Maltoviny I.* (Brno: PC DIR), (in Czech).

Swiete, H.E., Roth, W., 1973, *Beitrag zur Hydratationswärme von Portland Zement und deren Bestimmungmethoden.* (Leverkusen: Westdeutsch Verlag).

Tasong, W.A., Lynsdale, C. J. and Cripps, J. C., 1999a, Aggregate–cement paste interface. Part I. Influence of aggregate geochemistry. *Cement and Concrete Research*, **29**, pp. 1019–1025.

Tasong, W.A., Lynsdale, C. J. and Cripps, J. C., 1999b, The influence of aggregates on interfacial bonding mechanisms in concrete. *Proceedings of International Conference on A Vision for Next Millenium*, edited by Swamy, R.N. (Sheffield: Academic Press), pp. 381–392.

Taylor, H.F.W., 1992, *Cement Chemistry*. (London: Academic Press).

Thangavel, K. and Rengaswamy, N. S., 1998, Relation between chloride/hydroxide ratio and corrosion rate of steel in concrete. *Cement and Concrete Composites*, **20**, pp. 283–292.

Tian, B., Cohen, M.D., 2000, Does gypsum formation during sulphate attack on concrete lead to expansion? *Cement and Concrete Research*, **30**, pp. 117–123.

Tínková, J., 1977, Hydration heat of cements. PhD Thesis. (Brno: FCU BTU Brno).

Tuutti, K, 1982, *Corrosion Steel in Concrete*. (Stockholm: Cement and Concrete Research Institute).

Van Breugel, K., 1991, *Simulation of Hydration and Formation of Structure in Hardening Cement-Based Materials*. (Delft: Technische Universitteit).

Wasserbauer, R., 2000, *Biological Degradation of Buildings*. (Prague: ABF), (in Czech).

Yuan, C. Z. and Cuo, W. J., 1987, Bond between marble and cement paste. *Cement and Concrete Research*, **17**, pp. 544–552.

Zhang, T., Gjørv, O.E., 1997, Diffusion behavior of chloride ions in concrete. In: *Proceedings of the RILEM International Workshop Chloride Penetration into Concrete*, St-Remy-les-Chevreuse, edited by Nilsson, L.O., Ollivier, J.P.. (St-Remy-les-Chevreuse: RILEM), pp. 53–63.

Zimbelman, R., 1985, Contribution to the problem of cement-aggregate bond. *Cement and Concrete Research*, **15**, pp. 801–808.

CHAPTER TWO

Fundamentals of Transport Phenomena in Chemically Reacting Mixtures

2.1 BASIC CONCEPTS OF CONTINUUM MECHANICS

In mechanics, the concept of a continuous medium (see, e.g., Malvern, 1969) is based on a simplification of physical reality, which consists in disregarding the molecular structure of matter and picturing it as being without gaps and empty spaces. Further, it is supposed that all the mathematical functions in the theory are continuous functions, except for a finite number of interior surfaces (denoted as discontinuity surfaces) separating the continuity regions. This concept makes it possible to define the physical quantities such as density, velocity, stress, etc. at a mathematical point, which is the most important factor for modelling purposes. Consequently, the mathematical methods of differential and integral calculus immediately become available for the study of physical phenomena, and mathematical models can be constructed on the basis of physical concepts using common mathematical tools.

The point-wise defined physical quantities in a continuum together constitute a field; for instance, density is a function of both space coordinates and time. Therefore, they are often called field quantities. The field quantities naturally have their global counterparts, which are analogous to the quantities commonly employed in discrete theories working with the mass-point concept. Thus, for instance, density as a field quantity corresponds to the global quantity of mass of certain continuous space domain.

A general relation between a global quantity G and its corresponding field quantity F can be expressed as

$$F_V = \lim_{\Delta V \to 0} \frac{\Delta G}{\Delta V} = \frac{\mathrm{d}G}{\mathrm{d}V} \tag{2.1}$$

or

$$F_M = \lim_{\Delta M \to 0} \frac{\Delta G}{\Delta M} = \frac{\mathrm{d}G}{\mathrm{d}M}, \tag{2.2}$$

where V is the volume, M is the mass.

These definition relations express the fact that a field quantity can be obtained by relating the given global quantity to a particular volume or mass, or more precisely by performing a limit transition to zero for the volume or mass elements. If a global quantity is related to the volume, the respective field quantity is termed a density of a quantity, if is it related to the mass, it is called a specific quantity. For instance, specific energy or energy density are field quantities related to the global quantity of energy.

In practical analysis of processes in a continuous body, two basic points of view are adopted. The first of them works with so-called spatial points of a continuous domain, i.e. fixed points in a body. This concept would be quite sufficient for a description of processes in a rigid solid body, where the distances between the particular geometrical points in the domain are not changed, and the motion of the body can be characterized by a single value of velocity as a function of time only.

However, the majority of continuous bodies are deformable. In other words, the volume and the surface of a porous body are changed during the process. Consequently, the concept of fixed points cannot be used as the only mode of treatment in such cases, because some mathematical points disappear and some newly appear, according to changes in the spatial configuration of the body. In this case, the concept of a material point appears very useful. A material point is simply a macroscopic particle of the continuum, which means a very small volume of the material having constant mass (i.e. the number of elementary particles in this macroscopic particle is constant). The volume of the material point is a function of time and also of its position. Therefore, the finite number of material points constitutes the material body, and their mutual movements characterize the deformation processes in the continuous domain as a whole.

As we will see in what follows, both spatial point and material point concepts are employed in parallel in continuum mechanics, and their application can be conveniently combined.

2.1.1 Coordinates and Configuration

We assume a continuous body B consisting of n material points P. In an arbitrary time t, the body occupies a volume V of the physical space and it is bounded by a surface S. We establish at time $t = 0$ a one-to-one correspondence between each material point P and a set of geometrical points (X_1, X_2, X_3), where X_1, X_2, X_3 are the coordinates of a point in a chosen coordinate system. In this way, we obtain the initial configuration of the body, which is in fact its spatial layout. In most cases, we consider this initial state of the body at time $t = 0$ as the reference state, which is undistorted, and the coordinates of the initial state we denote as material or Lagrangian coordinates. A material point P_0 can be in this initial state located by its position vector $\vec{X}$ in the given coordinate system $OX_1X_2X_3$, which for the sake of simplicity can be considered Cartesian,

$$\vec{X} = X_1\vec{I}_1 + X_2\vec{I}_2 + X_3\vec{I}_3 = X_K\vec{I}_K, \tag{2.3}$$

where $\vec{I}_1, \vec{I}_2, \vec{I}_3$ are the unit vectors in the direction of the coordinate axes X_1, X_2, X_3, and the expression $X_K \vec{I}_K$ denotes the so-called Einstein summation, expressed in the following convention: if in any mathematical expression an index appears two times, this index is considered as a summation index, and the whole expression as a sum (mostly from 1 to 3, if classical three-dimensional space is considered).

The body can also possess other configurations. We assume that in some other (possibly deformed) state our material point P_0 is converted into point P, and its position can be expressed in another (possibly also Cartesian) coordinate system $ox_1x_2x_3$ using the position vector

$$\vec{x} = x_1\vec{e}_1 + x_2\vec{e}_2 + x_3\vec{e}_3 = x_k\vec{e}_k, \tag{2.4}$$

where $\vec{e}_1, \vec{e}_2, \vec{e}_3$ are the unit vectors in the direction of the coordinate axes x_1, x_2, x_3. The coordinates x_1, x_2, x_3 of this another (possibly deformed) state giving the instant position of the material point are called spatial or <u>Eulerian coordinates</u>.

As both X_K and x_k are the coordinates of the same set of material points, the following one-to-one mapping must exist:

$$x_k = x_k(X_K) \leftrightarrow X_K = X_K(x_k), \tag{2.5}$$

so that the Jacobian

$$J = \det(\partial X_K / \partial x_k) \neq 0 \tag{2.6}$$

at all points of the body B.

In the case of orthogonal Cartesian systems, the mutual orientation of the material axes $OX_1X_2X_3$ and spatial axes $ox_1x_2x_3$ can be characterized by the direction cosines $\alpha_{kK} = \cos(x_k, X_K)$, $\alpha_{Kk} = \cos(X_K, x_k)$,

$$\vec{e}_k \cdot \vec{I}_K = \vec{I}_K \cdot \vec{e}_k = \alpha_{kK} = \alpha_{Kk}. \tag{2.7}$$

For more complicated systems, such as those with curvilinear coordinates, other appropriate transformation relations can be formulated, the details of which can be found for instance in Eringen (1967).

2.1.2 Motion and Deformation

The motion of a continuum is characterized by the time evolution of the position of every material point. Mathematically, it can be described by a set of one-parameter family of configurations

$$x_k = x_k(\vec{X}, t), \quad k = 1, 2, 3. \tag{2.8}$$

Accordingly, a material point X_K (or briefly $\vec{X}$) in the domain B is carried to a spatial place x_k (or briefly $\vec{x}$). We assume that the mapping is single-valued and possesses continuous first-order partial derivatives with respect to their arguments, except possibly a finite number of singular surfaces, lines and points. Moreover, the unique inverse of (2.8) exists in the neighbourhood of the spatial point $\vec{x}$,

$$X_K = X_K(\vec{x}, t), \quad K = 1, 2, 3, \tag{2.9}$$

at time t. Thus the mappings (2.8) and (2.9) are one-to-one near a material point P at time t, except for some singular surfaces, lines and points.

This important assumption is known as the axiom of continuity (see, e.g., Eringen, 1967). It assumes that matter is indestructible (no domain of finite volume is deformed into a zero or infinite volume) and impenetrable (every volume is converted into another volume, every surface into a surface and every curve into a curve, so no one portion of matter can penetrate into another).

The ultimate goal of continuum mechanics is to determine the forms of the three functions (2.8) or (2.9), if the geometry of the body is known at one time and the boundary and initial conditions are specified. If the motion is specified, the positions of all points of the body (and therefore also the shape of the body) and internal field parameters can, in principle, be calculated. This information is essential in engineering design and technological applications.

The term motion is more general in continuum mechanics than in the discrete mechanics of mass-points. It includes two basic modes, namely deformation and convection. The term deformation denotes the change of shape of a continuum from some initial (undistorted) configuration to another (deformed) one. In the analysis of deformation, we take into account the initial and final configurations only, and disregard any intermediate states. On the other hand, the term convection is employed to denote the state of continuous movement of a continuum.

In the description of the deformation of a continuous body, the displacement vector $\vec{u}$ plays a crucial role. It is defined as the vector that extends from a material point P_0 in the initial state to its corresponding point P in the deformed state. We can express it in both spatial coordinates

$$\vec{u} = u_k \vec{e}_k, \tag{2.10}$$

and material coordinates

$$\vec{U} = U_K \vec{I}_K, \tag{2.11}$$

where for the components of the displacement vectors U_K and u_k (under the assumption that we are working with Cartesian coordinate systems) the following transformation relation is valid:

$$U_K = \alpha_{Kk} u_k. \tag{2.12}$$

Denoting as $\vec{b}$ the vector that extends from the origin of material coordinates to the origin of the spatial coordinates, we can write for the displacement vector

$$\vec{u} = \vec{b} + \vec{x} - \vec{X}. \tag{2.13}$$

The problem can be simplified by a choice of common origins for the coordinate systems $OX_1X_2X_3$ and $ox_1x_2x_3$, which leads to $\vec{b} = 0$. We then have

$$\vec{u} = \vec{x} - \vec{X}. \tag{2.14}$$

For the Cartesian coordinates $ox_1x_2x_3$ we can write

$$u_k = x_k - \alpha_{kK}X_K. \tag{2.15}$$

Finally, if also the axes of the coordinate systems $OX_1X_2X_3$ and $ox_1x_2x_3$ coincide, we get

$$u_k = x_k - X_k. \tag{2.16}$$

In the solution of deformation problems, we always have to distinguish which type of deformation we have. In the general case, we work with so-called finite deformations (or large deformations), where the displacement vector $\vec{u}$ has a finite value. The formulation of such problems leads to geometrically nonlinear relations, whose solution is relatively difficult. The reader can find details on the large deformation problems for instance in Eringen (1967).

We will deal only with so-called small deformation (or infinitesimal deformation) problems in what follows, where we aim to express not the displacement vector $\vec{u}$ as in the case of large deformations but its differential $d\vec{u}$.

In comparison with finite deformations, the solution to small deformations is significantly simpler, because for very small changes we can assume also that the change of the coordinate system is very small (infinitesimal) and the coordinate systems $OX_1X_2X_3$ and $ox_1x_2x_3$ coincide.

The infinitesimal change of the displacement vector $d\vec{u}$ can be expressed in the coordinate system $ox_1x_2x_3$ by the relation for the total differential,

$$du_i = \frac{\partial u_i}{\partial x_j}dx_j, \tag{2.17}$$

or in the vector form

$$d\vec{u} = \operatorname{grad} \vec{u}^{\mathrm{T}} \cdot d\vec{x} = d\vec{x} \cdot \operatorname{grad} \vec{u}. \tag{2.18}$$

The expression (2.17) can be modified as

$$du_i = \left[\frac{1}{2}\left(\frac{\partial u_i}{\partial x_j} + \frac{\partial u_j}{\partial x_i}\right) + \frac{1}{2}\left(\frac{\partial u_i}{\partial x_j} - \frac{\partial u_j}{\partial x_i}\right)\right]dx_j$$

$$= [(\operatorname{grad} \vec{u})_{ij}^{\mathrm{S}} - (\operatorname{grad} \vec{u})_{ij}^{\mathrm{A}}]dx_j. \tag{2.19}$$

The first term in the square brackets is the Euler infinitesimal strain tensor (briefly strain tensor) **U**,

$$U_{ij} = \frac{1}{2}\left(\frac{\partial u_i}{\partial x_j} + \frac{\partial u_j}{\partial x_i}\right) = (\operatorname{grad} \vec{u})_{ij}^{\mathrm{S}}, \tag{2.20}$$

where $(\text{grad } \vec{u})_{ij}^S$ is a component of the symmetric part of the grad $\vec{u}$ tensor. The second term is (except for the sign) the Euler infinitesimal <u>rotation tensor</u> (briefly rotation tensor) **R**,

$$-R_{ij} = \frac{1}{2}\left(\frac{\partial u_i}{\partial x_j} - \frac{\partial u_j}{\partial x_i}\right) = -(\text{grad } \vec{u})_{ij}^A, \tag{2.21}$$

where $(\text{grad } \vec{u})_{ij}^A$ is a component of the antisymmetric part of the grad $\vec{u}$ tensor.

The expression (2.19) can then be rewritten in the form

$$d\vec{u} = \mathbf{U} \cdot d\vec{x} - \mathbf{R} \cdot d\vec{x} = (\mathbf{U} - \mathbf{R}) \cdot d\vec{x}$$

$$= d\vec{x} \cdot (\mathbf{U} + \mathbf{R}) = d\vec{x} \cdot \text{grad } \vec{u}. \tag{2.22}$$

We will be mainly concerned in what follows with strain only. Details on the rotation problems can be found for instance in Eringen (1967) or Malvern (1969).

The strain tensor **U** can for illustrative purposes be rewritten in a matrix form,

$$\mathbf{U} = \begin{pmatrix} U_{11} & U_{12} & U_{13} \\ U_{21} & U_{22} & U_{23} \\ U_{31} & U_{32} & U_{33} \end{pmatrix}. \tag{2.23}$$

The diagonal terms are called normal components of the strain tensor (or normal strains), and they express the <u>extensions</u> along the respective axes,

$$U_{ii} = \frac{dx_i - dX_i}{dX_i}, \tag{2.24}$$

while the nondiagonal terms are called <u>shear strains</u> and they are equal to one half of the shear angle γ_{ij},

$$U_{ij} = \frac{1}{2}\gamma_{ij}. \tag{2.25}$$

Using the components of the strain tensor, the <u>relative volume change</u> ξ can also be expressed,

$$\xi = \frac{dv - dV}{dV} = U_{ii} = \frac{\partial u_i}{\partial x_i} = \text{div } \vec{u}, \tag{2.26}$$

where dV is the elementary volume before deformation and dv the elementary volume after deformation.

The strain tensor **U** can be written as the sum of two tensors, one representing a spherical state of strain, in which each normal strain has the same numerical value U, and all shear strains are equal to zero, and the second called the strain deviator $^d\mathbf{U}$,

$$\mathbf{U} = \delta \cdot U + {}^d\mathbf{U}, \tag{2.27}$$

where

$$U = \frac{1}{3}U_{ii} = \frac{1}{3} \text{ div } \vec{u} \tag{2.28}$$

$$^{\text{d}}\mathbf{U} = (\text{grad } \vec{u})^{\text{S}} - \frac{1}{3}\delta \text{ div } \vec{u}, \tag{2.29}$$

and δ is the Kronecker delta, i.e. the unit tensor of second order.

According to relation (2.26), the sum of the diagonal parts of the strain tensor is equal to the relative volume change of the material. Therefore, the scalar part of the strain tensor expresses the volume change and the deviatoric part the change of shape of a material.

2.1.3 Kinematics of a Continuous Medium

There are four basic types of description of motion (see Truesdell, 1965):

1. Material description, whose independent variables are the particle X and the time t.

2. Referential description, whose independent variables are the position $\vec{X}$ of the particle in an arbitrarily chosen reference configuration, and the time t. If the reference configuration is chosen to be the actual initial configuration at $t = 0$, the referential description is often called the Lagrangian description, though many authors call it the material description (e.g., Eringen, 1967) using the reference position $\vec{X}$ as a label for the material particle X of the first type of description listed above.

3. Spatial description, whose independent variables are the present position $\vec{x}$ of the particle at time t and the present time t. The spatial description fixes attention on a given region of space instead of a given body of matter, and is often called Eulerian description.

4. Relative description, whose independent variables are the present position $\vec{x}$ of the particle and a variable time τ. The variable time τ is the time when the particle occupied another position $\vec{x}'$, and the motion is described with $\vec{x}'$ as dependent variable $\vec{x}' = \vec{x}'(\vec{x}, \tau)$, while the configuration at time t is the reference configuration.

We will use in what follows the Lagrangian and Eulerian descriptions only. The Lagrangian description deals with the motion of a material point. We take a particle (material point) X, expressed by the position vector $\vec{X}$ in the reference (initial) coordinate system, and identify the place $\vec{x}$, which is occupied by the particle X at time t. This can be written symbolically as

$$\vec{x} = \vec{x}(\vec{X}, t), \tag{2.30}$$

where $\vec{X}$ are material coordinates, i.e. the coordinates of the initial undistorted state of the particular material point, $\vec{x}$ are the spatial coordinates of the same material point at a particular time t.

The Eulerian description considers the state of a fixed spatial point. We take a fixed spatial point and investigate the properties of the continuum at this point. In other words, we observe how the particular material points move into and out from this fixed point of space. Our aim is to find the function

$$\vec{X} = \vec{X}(\vec{x}, t), \tag{2.31}$$

where $\vec{x}$ are again the coordinates of the fixed spatial point, $\vec{X}$ are the coordinates of the particular material points in the initial state. The Euler description can be understood as a treatment that makes it possible to identify an initial position of a particular material point that occupies the position $\vec{x}$ at present. The equation of motion (2.30) describes the motion of the same material point. Therefore, the mapping $\vec{x} = \vec{x}(\vec{X}, t)$ is one-to-one, i.e. to every point X_i must correspond one and only one point x_i in any instant of time t. Under this assumption, an inverse mapping $\vec{X} = \vec{X}(\vec{x}, t)$ having the same properties must exist. It is, however, identical with the Eulerian equation of motion (2.31). Thus, it can be concluded that the Lagrangian and Eulerian descriptions are equivalent, and represent equally good descriptions of the motion of a continuum.

Following the Lagrangian and Eulerian descriptions, the time rates of the particular field quantities in a continuum can be expressed both in material and spatial coordinate systems, respectively, i.e. the time derivatives can be related to either a material point $\vec{X}$ or to a fixed spatial point $\vec{x}$.

For the <u>material time derivative</u> of an arbitrary quantity A (which can be a scalar, a vector or a tensor) we can write

$$\frac{\mathrm{d}A}{\mathrm{d}t} = \left(\frac{\partial A}{\partial t} \right)_{\vec{X}}, \tag{2.32}$$

where the subscript $\vec{X}$ means that $\vec{X}$ is held constant in the differentiation.

In an analogous way, the <u>spatial time derivative</u> is defined in a fixed spatial point $\vec{x}$,

$$\frac{\partial A}{\partial t} = \left(\frac{\partial A}{\partial t} \right)_{\vec{x}}. \tag{2.33}$$

In solving practical problems of continuum mechanics, we can usually get relatively simple and clear formulations by using the material derivative. However, this treatment is mostly impractical, because in many cases we are interested in an analysis of changes in a given locality, i.e. in a fixed spatial point. Therefore, we also need a relation between the material derivative and the spatial derivative. Using the chain rule of calculus we obtain

$$\left(\frac{\partial A}{\partial t} \right)_{\vec{X}} = \left(\frac{\partial A}{\partial t} \right)_{\vec{x}} + \left(\frac{\partial A}{\partial \vec{x}} \right)_{t} \cdot \left(\frac{\partial \vec{x}}{\partial t} \right)_{\vec{X}}. \tag{2.34}$$

The last term in equation (2.34) represents the definition of the velocity,

$$\vec{v} = \frac{\mathrm{d}\vec{x}}{\mathrm{d}t} = \left(\frac{\partial \vec{x}}{\partial t}\right)_{\vec{X}}, \tag{2.35}$$

the last but one is the gradient of a field quantity A. Thus, we finally get the following relation:

$$\frac{\mathrm{d}A}{\mathrm{d}t} = \frac{\partial A}{\partial t} + \vec{v} \cdot \operatorname{grad} A. \tag{2.36}$$

The last term in equation (2.36) is called the convection derivative.

Among the basic kinematic quantities, we have already defined the displacement vector and the velocity. Now we are going to continue the definitions.

The acceleration is given by the relation

$$\vec{a} = \frac{\mathrm{d}\vec{v}}{\mathrm{d}t} = \frac{\partial \vec{v}}{\partial t} + \vec{v} \cdot \operatorname{grad} \vec{v}. \tag{2.37}$$

The material time derivatives of the strain tensor $\mathbf{U}$ and rotation tensor $\mathbf{R}$ are called the rate of deformation tensor $\mathbf{D}$ and spin tensor (vorticity tensor) $\mathbf{W}$, respectively,

$$\mathbf{D} = \frac{\mathrm{d}\mathbf{U}}{\mathrm{d}t} = \frac{\mathrm{d}}{\mathrm{d}t}(\operatorname{grad}\vec{u})^{\mathrm{S}} = \left(\operatorname{grad}\frac{\mathrm{d}\vec{u}}{\mathrm{d}t}\right)^{\mathrm{S}} = (\operatorname{grad}\vec{v})^{\mathrm{S}} \tag{2.38}$$

and

$$\mathbf{W} = \frac{\mathrm{d}\mathbf{R}}{\mathrm{d}t} = \frac{\mathrm{d}}{\mathrm{d}t}(\operatorname{grad}\vec{u})^{\mathrm{A}} = \left(\operatorname{grad}\frac{\mathrm{d}\vec{u}}{\mathrm{d}t}\right)^{\mathrm{A}} = (\operatorname{grad}\vec{v})^{\mathrm{A}}. \tag{2.39}$$

We can easily see that in an analogous way to the displacements, from (2.38) and (2.39) it follows that

$$\mathbf{D} + \mathbf{W} = \operatorname{grad}\vec{v}. \tag{2.40}$$

Finally, we have to express the material time derivatives of the square of the arc length $(\mathrm{d}s)^2$, the surface element $\mathrm{d}\vec{S}$ and the volume element $\mathrm{d}V$.

For the square of the arc length we can write

$$\frac{\mathrm{d}}{\mathrm{d}t}[(\mathrm{d}s)^2] = \frac{\mathrm{d}}{\mathrm{d}t}(\mathrm{d}\vec{x} \cdot \mathrm{d}\vec{x}) = 2 \cdot \mathrm{d}\vec{x} \cdot \frac{\mathrm{d}}{\mathrm{d}t}(\mathrm{d}\vec{x}). \tag{2.41}$$

According to the definition of velocity (2.35)

$$\mathrm{d}\vec{x} = \vec{v} \cdot \mathrm{d}t, \tag{2.42}$$

and therefore

$$\frac{\mathrm{d}}{\mathrm{d}t}(\mathrm{d}\vec{x}) = \mathrm{d}\vec{v}. \tag{2.43}$$

The velocity differential $d\vec{v}$ can be written formally as

$$d\vec{v} = d\vec{x} \cdot \text{grad } \vec{v} = (\text{grad } \vec{v})^T \cdot d\vec{x}. \qquad (2.44)$$

Substituting (2.44) and (2.40) into (2.41) we get

$$\frac{d}{dt}[(ds)^2] = 2 \cdot d\vec{x} \cdot \mathbf{D} \cdot \vec{x} + 2 \cdot d\vec{x} \cdot \mathbf{W} \cdot \vec{x}. \qquad (2.45)$$

As the spin tensor $\mathbf{W}$ is antisymmetric, the last term on the right-hand side of equation (2.45) is equal to zero, and finally we have

$$\frac{d}{dt}[(ds)^2] = 2 \cdot d\vec{x} \cdot \mathbf{D} \cdot \vec{x}. \qquad (2.46)$$

For the surface element we can write (see, e.g., Eringen, 1967, for an exact derivation)

$$\frac{d}{dt}(d\vec{S}) = \text{div } \vec{v} \cdot d\vec{S} - \text{grad } \vec{v} \cdot d\vec{S}. \qquad (2.47)$$

The volume element in the initial undistorted state dV_0 can be expressed

$$dV_0 = dX_1 dX_2 dX_3, \qquad (2.48)$$

and the volume element dV in time t

$$dV = dx_1 dx_2 dx_3. \qquad (2.49)$$

The transformation relation between these two elements is given by

$$dV = J \cdot dV_0, \qquad (2.50)$$

where J is the Jacobian,

$$J = \left| \frac{\partial x_i}{\partial X_j} \right|. \qquad (2.51)$$

For the material time derivative of the volume element dV we then have

$$\frac{d}{dt}(dV) = \frac{d}{dt}(J \cdot dV_0) = \frac{dJ}{dt} \cdot dV_0. \qquad (2.52)$$

From the definition of the Jacobian it follows (see for instance Mase, 1970, for details) that

$$\frac{dJ}{dt} = J \cdot \frac{\partial v_i}{\partial x_i} = J \cdot \text{div } \vec{v}. \qquad (2.53)$$

Substituting (2.53) and (2.50) into (2.52) we get

$$\frac{d}{dt}(dV) = \text{div } \vec{v} \cdot dV. \qquad (2.54)$$

Finally, the material time derivative of the relative volume change can be expressed using (2.26) and by interchanging the order of differentiation as

$$\frac{d\xi}{dt} = \frac{dU_{ii}}{dt} = \frac{d}{dt}(\text{div }\vec{u}) = \text{div }\vec{v}. \tag{2.55}$$

Until now, we were concerned with the kinematics of field quantities of a continuum only. The kinematics of global, integral quantities is equally important in the continuum mechanics. Therefore, we will now derive the material time derivatives of line, surface and volume integrals.

For the line integral of an arbitrary field quantity A over a material line C we have

$$\frac{d}{dt}\int_C A \, d\vec{x} = \int_C \frac{dA}{dt} \, d\vec{x} + \int_C A \frac{d}{dt}(d\vec{x}). \tag{2.56}$$

From (2.43) and (2.44) it follows that

$$\frac{d}{dt}(d\vec{x}) = (\text{grad }\vec{v})^T \cdot d\vec{x}. \tag{2.57}$$

Substituting (2.57) into (2.56) we get

$$\frac{d}{dt}\int_C A \, d\vec{x} = \int_C \frac{dA}{dt} \, d\vec{x} + \int_C A \cdot (\text{grad }\vec{v})^T \cdot d\vec{x}. \tag{2.58}$$

The counterpart of (2.58) for a spatially fixed line c is

$$\frac{\partial}{\partial t}\int_c A \, d\vec{x} = \int_c \frac{\partial A}{\partial t} \, d\vec{x}. \tag{2.59}$$

For the surface integral over a material surface S we obtain using (2.47)

$$\frac{d}{dt}\int_S A \, d\vec{S} = \int_S \frac{dA}{dt} \, d\vec{S} + \int_S A \frac{d}{dt}(d\vec{S})$$

$$= \int_S \left[\frac{dA}{dt} + A \text{ div }\vec{v}\right] d\vec{S} - \int_S A \cdot \text{grad }\vec{v} \cdot d\vec{S}. \tag{2.60}$$

The counterpart of (2.60) for a spatially fixed surface s is

$$\frac{\partial}{\partial t}\int_s A \, d\vec{S} = \int_s \frac{\partial A}{\partial t} \, d\vec{S}. \tag{2.61}$$

If in (2.60) we select the vector field $\vec{A}$, we obtain

$$\frac{d}{dt}\int_S \vec{A} \cdot d\vec{S} = \int_S \left[\frac{d\vec{A}}{dt} + \vec{A} \cdot \text{div }\vec{v} - \vec{A} \cdot \text{grad }\vec{v}\right] \cdot d\vec{S}. \tag{2.62}$$

In a modification of the right-hand side of (2.62) we can employ the relations

$$\frac{d\vec{A}}{dt} = \frac{\partial \vec{A}}{\partial t} + \vec{v} \cdot \text{grad }\vec{A} \tag{2.63}$$

and

$$\text{curl}\,(\vec{A} \times \vec{v}) = \vec{A} \cdot \text{div}\,\vec{v} - \vec{v} \cdot \text{div}\,\vec{A} + \vec{v} \cdot \text{grad}\,\vec{A} - \vec{A} \cdot \text{grad}\,\vec{v}. \quad (2.64)$$

Substituting (2.63) and (2.64) in (2.62) we obtain

$$\frac{\mathrm{d}}{\mathrm{d}t} \int_S \vec{A} \cdot \mathrm{d}\vec{S} = \int_S \left[\frac{\partial \vec{A}}{\partial t} + \text{curl}\,(\vec{A} \times \vec{v}) + \vec{v} \cdot \text{div}\,\vec{A} \right] \cdot \mathrm{d}\vec{S}. \quad (2.65)$$

As shown by Eringen (1967), equation (2.65) is also valid for an arbitrary spatial surface $s(t)$ bounded by a closed curve $c(t)$, which is moving with the velocity $\vec{v}$,

$$\frac{\mathrm{d}}{\mathrm{d}t} \int_{s(t)} \vec{A} \cdot \mathrm{d}\vec{S} = \int_{s(t)} \left[\frac{\partial \vec{A}}{\partial t} + \text{curl}\,(\vec{A} \times \vec{v}) + \vec{v} \cdot \text{div}\,\vec{A} \right] \cdot \mathrm{d}\vec{S}. \quad (2.66)$$

Using the Stokes theorem for a conversion between surface and line integrals,

$$\int_{s(t)} \text{curl}\,(\vec{A} \times \vec{v}) \cdot \mathrm{d}\vec{S} = \int_{c(t)} (\vec{A} \times \vec{v}) \cdot \mathrm{d}\vec{x}, \quad (2.67)$$

we convert the second term of the right-hand side of equation (2.66) and get

$$\frac{\mathrm{d}}{\mathrm{d}t} \int_{s(t)} \vec{A} \cdot \mathrm{d}\vec{S} = \int_{s(t)} \left(\frac{\partial \vec{A}}{\partial t} + \vec{v} \cdot \text{div}\,\vec{A} \right) \cdot \mathrm{d}\vec{S}$$

$$+ \int_{c(t)} (\vec{A} \times \vec{v}) \cdot \mathrm{d}\vec{x}. \quad (2.68)$$

For the volume integral over a material volume V, using (2.54) we get

$$\frac{\mathrm{d}}{\mathrm{d}t} \int_V A \, \mathrm{d}V = \int_V \frac{\mathrm{d}A}{\mathrm{d}t} \, \mathrm{d}V + \int_V A \frac{\mathrm{d}}{\mathrm{d}t} (\mathrm{d}V)$$

$$= \int_V \left(\frac{\mathrm{d}A}{\mathrm{d}t} + A \, \text{div}\,\vec{v} \right) \mathrm{d}V, \quad (2.69)$$

which can be further modified using the conversion relation between the material and spatial derivatives (2.36)

$$\int_V \left(\frac{\mathrm{d}A}{\mathrm{d}t} + A \, \text{div}\,\vec{v} \right) \mathrm{d}V = \int_V \left(\frac{\partial A}{\partial t} + \vec{v} \cdot \text{grad}\,A + A \, \text{div}\,\vec{v} \right) \mathrm{d}V$$

$$= \int_V \left(\frac{\partial A}{\partial t} + \text{div}\,(A\vec{v}) \right) \mathrm{d}V, \quad (2.70)$$

so that we can write

$$\frac{\mathrm{d}}{\mathrm{d}t} \int_V A \, \mathrm{d}V = \int_V \left(\frac{\partial A}{\partial t} + \text{div}\,(A\vec{v}) \right) \mathrm{d}V. \quad (2.71)$$

The last term of equation (2.71) can be modified using the Gauss theorem, which for a conversion between the volume and surface integrals can be written as

$$\int_V \operatorname{div} \vec{A} \, dV = \int_S \vec{A} \cdot d\vec{S}. \tag{2.72}$$

Thus, we have

$$\frac{d}{dt} \int_V A \, dV = \int_V \frac{\partial A}{\partial t} \, dV + \int_S A\vec{v} \cdot d\vec{S}. \tag{2.73}$$

If we select V and S to coincide instantaneously with a spatial volume v and its bounding surface s, equation (2.73) can also be written as

$$\frac{d}{dt} \int_V A \, dV = \int_v \frac{\partial A}{\partial t} \, dV + \int_s A\vec{v} \cdot d\vec{S}. \tag{2.74}$$

Equation (2.73) can be extended (see Eringen, 1967, for a detailed derivation and a proof) to an arbitrary spatial volume $v(t)$ bounded by a closed surface $s(t)$, which is moving with the velocity $\vec{v}$,

$$\frac{d}{dt} \int_{v(t)} A \, dV = \int_{v(t)} \frac{\partial A}{\partial t} \, dV + \int_{s(t)} A\vec{v} \cdot d\vec{S}. \tag{2.75}$$

We now extend the results (2.65) and (2.73) to material surfaces and volumes containing moving discontinuity lines and surfaces.

First, we will consider the case of a material volume V intersected by a discontinuity surface $\sigma(t)$ moving with a velocity $\vec{v}$. Thus, we have two volumes V^+ and V^- instead of the original volume V, which are bounded by $S^+ + \sigma^+$ and $S^- + \sigma^-$, respectively. Consequently, we have to employ (2.73) for the two volumes V^+ and V^-. We obtain

$$\frac{d}{dt} \int_{V^+} A \, dV = \int_{V^+} \frac{\partial A}{\partial t} \, dV + \int_{S^+} A\vec{v} \cdot d\vec{S} - \int_{\sigma^+} A\vec{v} \cdot d\vec{S} \tag{2.76}$$

and

$$\frac{d}{dt} \int_{V^-} A \, dV = \int_{V^-} \frac{\partial A}{\partial t} \, dV + \int_{S^-} A\vec{v} \cdot d\vec{S} + \int_{\sigma^-} A\vec{v} \cdot d\vec{S}. \tag{2.77}$$

We note that the last terms on the right-hand sides of equations (2.76) and (2.77) have opposite signs. This is due to the fact that $d\vec{S}$ always has the direction of the outward normal vector of surface $\vec{S}$, but $\vec{v}$ is a velocity vector, which is independent of this notation. Thus, it has to appear once with the positive sign, once with the negative sign. Here we adopted the convention, that the vector $\vec{v}$ acts in the direction into the volume V^+.

Adding equations (2.76) and (2.77) and letting σ^+ and σ^- approach σ we obtain

$$\frac{d}{dt} \int_{V-\sigma} A \, dV = \int_{V-\sigma} \frac{\partial A}{\partial t} \, dV + \int_{S-\sigma} A\vec{v} \cdot d\vec{S} - \int_\sigma [\![A\vec{v}]\!] \cdot d\vec{S}, \tag{2.78}$$

where $[\![A]\!] = A^+ - A^-$, A^+ and A^- are the values of A on σ as approached from the positive and negative side of the normal $\vec{n}$ of σ.

Substituting the Gauss theorem, which in our case has the form

$$\int_{S-\sigma} A\vec{v} \cdot d\vec{S} = \int_{V-\sigma} \text{div}\,(A\vec{v})\, dV + \int_{\sigma} [\![A\vec{v}]\!] \cdot d\vec{S}, \qquad (2.79)$$

we finally get

$$\frac{d}{dt}\int_{V-\sigma} A\, dV = \int_{V-\sigma} \left[\frac{\partial A}{\partial t} + \text{div}\,(A\vec{v})\right] dV$$

$$+ \int_{\sigma} [\![A(\vec{v}-\vec{\nu})]\!] \cdot d\vec{S}. \qquad (2.80)$$

In a similar way we will work with the material surface S intersected by a discontinuity line $\gamma(t)$ moving with a velocity ν. In this case, we have two surfaces S^+ and S^- instead of the original surface S, which are bounded by $C^+ + \gamma^+$ and $C^- + \gamma^-$, respectively. We employ (2.68) for the two surfaces S^+ and S^- and get

$$\frac{d}{dt}\int_{S+} \vec{A} \cdot d\vec{S} = \int_{S+} \left(\frac{\partial \vec{A}}{\partial t} + \vec{v} \cdot \text{div}\,\vec{A}\right) \cdot d\vec{S}$$

$$+ \int_{C+} (\vec{A} \times \vec{v}) \cdot d\vec{x} - \int_{\gamma+} (\vec{A} \times \vec{\nu}) \cdot d\vec{x} \qquad (2.81)$$

and

$$\frac{d}{dt}\int_{S-} \vec{A} \cdot d\vec{S} = \int_{S-} \left(\frac{\partial \vec{A}}{\partial t} + \vec{v} \cdot \text{div}\,\vec{A}\right) \cdot d\vec{S}$$

$$+ \int_{C-} (\vec{A} \times \vec{v}) \cdot d\vec{x} + \int_{\gamma-} (\vec{A} \times \vec{\nu}) \cdot d\vec{x}. \qquad (2.82)$$

Adding (2.81) and (2.82) and letting γ^+ and γ^- approach γ we obtain

$$\frac{d}{dt}\int_{S-\gamma} \vec{A} \cdot d\vec{S} = \int_{S-\gamma} \left(\frac{\partial \vec{A}}{\partial t} + \vec{v} \cdot \text{div}\,\vec{A}\right) \cdot d\vec{S}$$

$$+ \int_{C-\gamma} (\vec{A} \times \vec{v}) \cdot d\vec{x} - \int_{\gamma} [\![\vec{A} \times \vec{\nu}]\!] \cdot d\vec{x}. \qquad (2.83)$$

Using the Stokes theorem in the form

$$\int_{C-\gamma} (\vec{A} \times \vec{v}) \cdot d\vec{c} = \int_{S-\gamma} \text{curl}\,(\vec{A} \times \vec{v}) \cdot d\vec{S} + \int_{\gamma} [\![\vec{A} \times \vec{\nu}]\!] \cdot d\vec{x} \qquad (2.84)$$

we finally get

$$\frac{d}{dt} \int_{S-\gamma} \vec{A} \cdot d\vec{S} = \int_{S-\gamma} \left(\frac{\partial \vec{A}}{\partial t} + \mathrm{curl}\,(\vec{A} \times \vec{v}) + \vec{v} \cdot \mathrm{div}\,\vec{A} \right) \cdot d\vec{S}$$

$$+ \int_{\gamma} [\![\vec{A} \times (\vec{v} - \vec{\nu})]\!] \cdot d\vec{x}. \tag{2.85}$$

2.1.4 Forces and Stress

Forces can be classified (see, e.g., Malvern, 1969) as <u>external forces</u> acting on a body or <u>internal forces</u> acting between two parts of the body. However, this classification is a matter of convention only. By a suitable choice of a free body imagined to be cut out of the original body, any force which was internal in the original body, may become an external force on the isolated free body. The term free body is not limited to a solid body, but may be used to denote a portion of fluid instantaneously bounded by an arbitrary closed surface. This closed surface may consist in part of an actual free surface of a liquid or an actual container wall bounding the liquid, it may consist of a part of an actual surface of a solid body, but it may also be wholly or in part an imaginary surface within the liquid or the solid.

The external forces acting at any instant on a chosen free body are classified in continuum mechanics into two kinds, body forces and surface forces.

<u>Body forces</u> act on the elements of volume or mass inside the body. Typical representatives of body forces are real forces such as gravity or electromagnetic forces, but also fictitious forces such as inertia or the Coriolis force. Body forces are sometimes called action-at-a-distance forces. Within the framework of classical mechanics, body forces act immediately on the whole volume of a free body, i.e. they are propagating in space in a field form by an infinite velocity.

<u>Surface forces</u> are contact forces acting on the free body at its bounding surface. They propagate from the surface of the body inwards by mutual interactions of the molecules of the body. Therefore, the propagation velocity of the surface forces is finite. Typical representatives of surface forces are friction or pressure.

Body forces are reckoned per unit mass or per unit volume. The force density $\vec{f}$ is defined as

$$\vec{f} = \lim_{\Delta V \to 0} \frac{\Delta \vec{F}}{\Delta V} = \frac{d\vec{F}}{dV}, \tag{2.86}$$

and the specific force (also called the force intensity) is defined as

$$\vec{b} = \lim_{\Delta M \to 0} \frac{\Delta \vec{F}}{\Delta M} = \frac{d\vec{F}}{dM}, \tag{2.87}$$

where $\Delta \vec{F}$ is the total body force acting on the volume ΔV. From (2.86) and (2.87) it follows that

$$\vec{f} = \rho \vec{b}, \tag{2.88}$$

where ρ is the density,

$$\rho = \frac{\mathrm{d}M}{\mathrm{d}V}. \tag{2.89}$$

The total force acting on a finite volume V can be determined by an integration of (2.86) and (2.87),

$$\vec{F} = \int_V \vec{f} \, \mathrm{d}V = \int_V \rho \vec{b} \, \mathrm{d}V. \tag{2.90}$$

Surface forces are reckoned per unit area of the surface, across which they act. The traction (sometimes it is called stress vector) $\vec{t}$ is defined as

$$\vec{t} = \lim_{\Delta S \to 0} \frac{\Delta \vec{F}_S}{\Delta S} = \frac{\mathrm{d}\vec{F}_S}{\mathrm{d}S}, \tag{2.91}$$

where $\Delta \vec{F}_S$ is the total surface force acting on the surface ΔS. The total force acting on a finite surface S can be written

$$\vec{F}_S = \int_S \vec{t} \, \mathrm{d}S. \tag{2.92}$$

The moment density $\vec{m}$ of a force density $\vec{f}$ about a fixed point A is defined as

$$\vec{m} = \vec{r} \times \vec{f}, \tag{2.93}$$

where $\vec{r}$ is the vector from the moment centre A to any point P on the line of action of the force,

$$\vec{r} = \vec{x} - \vec{x}_A, \tag{2.94}$$

$\vec{x}$ is the position vector of the point P, and $\vec{x}_A$ is the position vector of the point A. The total moment of all body forces acting on a finite volume V is given by

$$\vec{M} = \int_V (\vec{r} \times \vec{f}) \, \mathrm{d}V = \int_V (\vec{r} \times \vec{b}) \rho \, \mathrm{d}V. \tag{2.95}$$

The moment surface density $\vec{m}_S$ of surface forces $\vec{t}$ can be expressed in a similar fashion,

$$\vec{m}_S = \vec{r} \times \vec{t}, \tag{2.96}$$

and the total moment of the distributed force on a finite surface S can be obtained by means of a surface integral of (2.96),

$$\vec{M}_S = \int_S (\vec{r} \times \vec{t}) \, \mathrm{d}S. \tag{2.97}$$

Now, we will reconsider the surface forces defined by equation (2.91) and will analyse their action in more details. At first, we will take into account the orientation of the surface element ΔS, which is given by its outward unit normal vector $\vec{n}$, and denote the stress vector more precisely as $\vec{t}^{(\vec{n})}$. This notation expresses the fact that the stress vector at a given point P depends on the orientation of the chosen surface element ΔS. If we consider at point P another surface element, say $\Delta S'$, given by the normal vector $\vec{n}'$, the stress vector at point P would have another value and orientation.

If there exists in the given body a state of force equilibrium, then at any point on an arbitrarily chosen surface the external surface force is equal (except for the sign) to the internal surface force as follows from Newton's third law. Thus, if at a point P there acts an external surface force given by the stress vector $\vec{t}^{(\vec{n})}$ on the material inside the volume V, then at the same point P there also acts an internal surface force (the reaction force to the external force) given by the stress vector $-\vec{t}^{(\vec{n})}$ on the material outside the volume V. All possible stress vectors $\vec{t}^{(\vec{n})}$, corresponding to the normal vectors $\vec{n}$, at point P together determine the state of stress at this point. There is an infinite number of such couples $(\vec{n}, \vec{t}^{(\vec{n})})$ but for a full description of the state of stress at a given point, three values of the stress vector $\vec{t}^{(\vec{n})}$ corresponding to three normal vectors of three surface elements perpendicular one another at the point P are sufficient.

Each of the three stress vectors can be rewritten as

$$\vec{t}^{(\vec{e}_i)} = t_1^{(\vec{e}_i)}\vec{e}_1 + t_2^{(\vec{e}_i)}\vec{e}_2 + t_3^{(\vec{e}_i)}\vec{e}_3 = t_j^{(\vec{e}_i)}\vec{e}_j, \quad i = 1, 2, 3. \tag{2.98}$$

Thus, we have nine components of the three stress vectors, which together constitute a second-order tensor $\mathbf{T}$ that is called the <u>stress tensor</u>. A component T_{ij} of this tensor can be expressed as

$$T_{ij} = t_j^{(\vec{e}_i)}. \tag{2.99}$$

We can see that in numerical terms, the component T_{ij} is equal to the force acting in the direction of the j-th axis on a unit surface element having a normal vector perpendicular to the i-th axis. The T_{ii} components are termed <u>normal stresses</u>, the $T_{ij}(i \neq j)$ components <u>shear stresses</u>. The relation between the stress vector $\vec{t}^{(\vec{n})}$ and the stress tensor $\mathbf{T}$ follows directly from relations (2.98) and (2.99),

$$t_i^{(\vec{n})} = n_j T_{ji} \tag{2.100}$$

or in a vector notation

$$\vec{t}^{(\vec{n})} = \vec{n} \cdot \mathbf{T}. \tag{2.101}$$

As with the strain tensor $\mathbf{U}$, the stress tensor $\mathbf{T}$ can also be written as the sum of two tensors, one representing a spherical (or hydrostatic) state of stress in which each normal stress has the same numerical value T and all shear stresses are equal to zero, and the second called the stress deviator $^{\mathrm{d}}\mathbf{T}$,

$$\mathbf{T} = \delta \cdot T + {}^{\mathrm{d}}\mathbf{T}, \tag{2.102}$$

where

$$T = \frac{1}{3}T_{ii} = -p \tag{2.103}$$

$$^{\mathrm{d}}\mathbf{T} = \mathbf{T} - \frac{1}{3}\delta T, \tag{2.104}$$

and δ is the Kronecker delta, i.e. the unit tensor of second order, and p is the pressure.

2.1.5 Work and Power

In mass-point mechanics, <u>work</u> (the mechanical work) expresses the path effect of a force,

$$A = \int_{l} \vec{F} \cdot \mathrm{d}\vec{l}. \tag{2.105}$$

In the continuum mechanics, as well as movement of the whole body there is also deformation. Just as during deformation processes the particular parts of the body change their mutual positions, so too here a work is also done. The work can be performed by all forces that are external to the given system. Such forces are called <u>deformation forces</u>.

As with all other forces in continuum mechanics, deformation forces can be classified into volume forces and surface forces. The work of volume forces can be obtained directly from the definition relations (2.90) and (2.105), though the path vector $\vec{l}$ should be replaced by the displacement vector $\vec{u}$,

$$A_{\mathrm{V}} = \int_{V} \left(\int_{\vec{u}} \vec{f} \cdot \mathrm{d}\vec{u} \right) \mathrm{d}V. \tag{2.106}$$

For the surface forces we employ relations (2.92) and (2.101),

$$\vec{F}_{\mathrm{S}} = \int_{S} \vec{n} \cdot \mathbf{T} \, \mathrm{d}S = \int_{S} \vec{t} \, \mathrm{d}S. \tag{2.107}$$

Substituting (2.107) in (2.105) we get

$$A_{\mathrm{S}} = \int_{\vec{u}} \vec{F}_{\mathrm{S}} \cdot \mathrm{d}\vec{u} = \int_{S} \left(\int_{\vec{u}} \vec{t} \cdot \mathrm{d}\vec{u} \right) \mathrm{d}S. \tag{2.108}$$

The <u>power</u> of volume and surface forces we can get by differentiating the relations (2.106) and (2.108) with respect to time,

$$P_{\mathrm{V}} = \frac{\mathrm{d}A_{\mathrm{V}}}{\mathrm{d}t} = \int_{V} \vec{f} \cdot \vec{v} \, \mathrm{d}V \tag{2.109}$$

$$P_S = \frac{dA_S}{dt} = \int_S \vec{t} \cdot \vec{v} \, dS = \int_S (\vec{v} \cdot \mathbf{T}) \cdot d\vec{S}$$

$$= \int_V \text{div} \, (\vec{v} \cdot \mathbf{T}) \, dV. \tag{2.110}$$

2.2 BALANCE LAWS FOR A SINGLE CONTINUUM CONTAINING MOVING DISCONTINUITIES

In continuum physics, the following eight balance laws are deemed to be valid, irrespective of material constitution and geometry (see, e.g., Eringen, 1967):

1. Conservation of mass

2. Balance of momentum

3. Balance of moment of momentum

4. Conservation of energy

5. Entropy inequality

6. Conservation of charge

7. Faraday's law

8. Ampere's law.

For mechanical media in which electromagnetic effects are not important, the last three laws can be disregarded. Because in the analysis of transport processes in concrete electromagnetic effects play an important role only exceptionally, we will confine ourselves to thermomechanical effects and will posit the first five laws only.

We note that the domain of applicability of the balance laws that are deemed to apply here is subject to restrictions arising from the relativistic speeds and dimensions and quantum mechanical phenomena. For material velocities near to the velocity of light, these laws require modifications. For atomic and nuclear phenomena, where quantum effects are important, again major modifications of these laws are necessary.

In this section, we will concentrate on the formulation of balance laws as independent axioms. We will not for instance explain the thermodynamic considerations concerning the links between the energy and the entropy and will not analyse entropy production and related phenomena. This will be done in the subsequent section devoted to the basic concerns of continuum thermodynamics and constitutive equations.

The global balance laws of continuum mechanics can be formulated on the basis of an analogy with the mechanics of discrete systems, except that the term mass-point has to be replaced by the term the material volume V in the sense of continuum mechanics. Thus we have:

1. Conservation of mass M

$$\frac{\mathrm{d}M}{\mathrm{d}t} = \frac{\mathrm{d}}{\mathrm{d}t} \int_V \rho \, \mathrm{d}V = 0. \tag{2.111}$$

2. Balance of momentum $\vec{H}$

$$\frac{\mathrm{d}\vec{H}}{\mathrm{d}t} = \frac{\mathrm{d}}{\mathrm{d}t} \int_V \rho\vec{v} \, \mathrm{d}V = \vec{F}_\mathrm{e}, \tag{2.112}$$

where $\vec{F}_\mathrm{e}$ are the external forces,

$$\vec{F}_\mathrm{e} = \int_S \mathbf{T} \cdot \mathrm{d}\vec{S} + \int_V \vec{f} \, \mathrm{d}V + \sum_i \vec{F}_i, \tag{2.113}$$

$\vec{F}_i$ are concentrated forces acting at isolated points $\vec{x}_i$ of the body, and the remaining symbols were explained in the Section 2.1.4.

3. Balance of moment of momentum $\vec{N}$

$$\frac{\mathrm{d}\vec{N}}{\mathrm{d}t} = \frac{\mathrm{d}}{\mathrm{d}t} \int_V \rho \cdot (\vec{r} \times \vec{v}) \, \mathrm{d}V = \vec{M}_\mathrm{e}, \tag{2.114}$$

where $\vec{M}_\mathrm{e}$ is the moment of external forces,

$$\vec{M}_\mathrm{e} = \int_S [\vec{r} \times (\vec{n} \cdot \mathbf{T})] \, \mathrm{d}S + \int_V (\vec{r} \times \vec{f}) \, \mathrm{d}V + \sum_i \vec{r}_i \times \vec{F}_i + \sum_i \vec{M}_i \tag{2.115}$$

and $\vec{M}_i$ represent couples that may act at isolated points.

4. Conservation of energy E

$$\frac{\mathrm{d}E}{\mathrm{d}t} = \frac{\mathrm{d}}{\mathrm{d}t} \int_V \rho e \, \mathrm{d}V = \frac{\mathrm{d}A_\mathrm{e}}{\mathrm{d}t} + \frac{\mathrm{d}Q}{\mathrm{d}t}, \tag{2.116}$$

where e is the specific total energy, which in continuum thermodynamics is usually understood as the sum of specific internal and specific kinetic energy, namely

$$e = u + \frac{1}{2}\vec{v} \cdot \vec{v}, \tag{2.117}$$

A_e is the work of external forces, which can be expressed using the relations (2.106) and (2.108) derived earlier. The power of external forces is given by (2.109), (2.110), therefore we can write

$$\frac{\mathrm{d}A_\mathrm{e}}{\mathrm{d}t} = \int_V \vec{f} \cdot \vec{v} \, \mathrm{d}V + \int_S (\vec{v} \cdot \mathbf{T}) \cdot \mathrm{d}\vec{S} + \sum_i \vec{F}_i \cdot \vec{v}_i + \sum_i \vec{M}_i \cdot \vec{\omega}_i, \tag{2.118}$$

where $\vec{\omega}$ is the angular velocity, Q the heat supplied to the system,

$$\frac{dQ}{dt} = -\int_S \vec{j}_Q \cdot d\vec{S} + \int_V I_Q \, dV + \sum_i H_i, \tag{2.119}$$

$\vec{j}_Q$ the heat flux through the surface S, taken as positive in the direction of the outward normal vector to the surface S (therefore the minus sign), I_Q the internal heat source density, and H_i isolated heat sources at points $\vec{x}_i$.

5. Entropy S inequality

$$\Gamma = \frac{dS}{dt} - B_V - B_S \geq 0, \tag{2.120}$$

where Γ is the total entropy production rate, and the particular terms in (2.120) can be expressed as follows:

$$\frac{dS}{dt} = \frac{d}{dt} \int_V \rho s \, dV \tag{2.121}$$

is the time rate of change of the total entropy S, s the specific entropy,

$$B_S = -\int_S \vec{j}_s \cdot d\vec{S}, \tag{2.122}$$

the time rate of change of entropy due to the influx of entropy $\vec{j}_s$ through the surface S of the body,

$$B_V = \int_V \rho b_s \, dV + \sum_i B_i \tag{2.123}$$

the time rate of change of entropy supplied by body sources, b_s the specific entropy source, and B_i isolated entropy sources.

The foregoing laws are deemed to hold for all bodies irrespective of the nature of the body, its geometry and constitution. From the point of view of the physical theory they are considered as axioms, and therefore they are basic to discussion of all physical phenomena. However, from the point of view of practical solutions of problems of continuum mechanics, they cannot be employed directly to obtain values of field quantities, which are of major concern in this respect. Therefore, local balance laws have to be derived from the global laws. The requirements of clarity and direct practical applicability lead to the necessity of expressing these laws in terms of fixed spatial points.

For the derivation of <u>local balance laws</u>, we first formally express the global balance equations (2.111), (2.112), (2.114), (2.116), (2.120) in a unified form. Neglecting (for the sake of simplicity) the isolated effects, and on the other hand taking into account the <u>discontinuity surfaces</u> σ (which definitely have an important role to play in local balances) within the material volume V, we can write for a scalar quantity A

$$\frac{d}{dt} \int_{V-\sigma} A \, dV = \int_{S-\sigma} \vec{A}_S \cdot d\vec{S} + \int_V A_I \, dV \tag{2.124}$$

and for a vector quantity $\vec{A}$

$$\frac{\mathrm{d}}{\mathrm{d}t} \int_{V-\sigma} \vec{A}\,\mathrm{d}V = \int_{S-\sigma} \mathbf{A}_\mathrm{S} \cdot \mathrm{d}\vec{S} + \int_V \vec{A}_\mathrm{I}\,\mathrm{d}V, \tag{2.125}$$

where $\vec{A}_\mathrm{S}$ and $\mathbf{A}_\mathrm{S}$ are the surface fluxes of the quantities A and $\vec{A}$, respectively, A_I and $\vec{A}_\mathrm{I}$ the volume sources of A and $\vec{A}$. The equations (2.124) or (2.125) are called <u>master balance laws</u>. They are valid for the first four balance laws, for the entropy inequality the sign $=$ has to be replaced by $\geq$.

The left-hand sides of equations (2.124), (2.125) can be expressed using the relation for material time derivative (2.80), which for a scalar quantity A has the form

$$\frac{\mathrm{d}}{\mathrm{d}t} \int_{V-\sigma} A\,\mathrm{d}V = \int_{V-\sigma} \left[\frac{\partial A}{\partial t} + \operatorname{div}(A\vec{v}) \right] \mathrm{d}V$$

$$+ \int_\sigma [\![A(\vec{v} - \vec{v})]\!] \cdot \mathrm{d}\vec{S}. \tag{2.126}$$

The first terms on the right-hand sides of equations (2.124), (2.125) can be modified using the modified Gauss theorem (2.79), so that we get for a scalar quantity A

$$\int_{S-\sigma} \vec{A}_\mathrm{S} \cdot \mathrm{d}\vec{S} = \int_{V-\sigma} \operatorname{div}(\vec{A}_\mathrm{S})\,\mathrm{d}V + \int_\sigma [\![\vec{A}_\mathrm{S}]\!] \cdot \mathrm{d}\vec{S}. \tag{2.127}$$

Substituting (2.126) and (2.127) into the original balance equation (2.124), we obtain

$$\int_{V-\sigma} \left[\frac{\partial A}{\partial t} + \operatorname{div}(A\vec{v}) - \operatorname{div}\vec{A}_\mathrm{S} - A_\mathrm{I} \right] \mathrm{d}V$$

$$+ \int_\sigma [\![A(\vec{v} - \vec{v}) - \vec{A}_\mathrm{S}]\!] \cdot \mathrm{d}\vec{S} = 0. \tag{2.128}$$

In an analogous way, for a vector quantity $\vec{A}$ we get

$$\int_{V-\sigma} \left[\frac{\partial \vec{A}}{\partial t} + \operatorname{div}(\vec{A} \otimes \vec{v}) - \operatorname{div}\mathbf{A}_\mathrm{S} - \vec{A}_\mathrm{I} \right] \mathrm{d}V$$

$$+ \int_\sigma [\![\vec{A} \otimes (\vec{v} - \vec{v}) - \mathbf{A}_\mathrm{S}]\!] \cdot \mathrm{d}\vec{S} = 0, \tag{2.129}$$

where the symbol $\vec{A} \otimes \vec{B}$ denotes the tensor product of two vectors.

The master balance laws (2.128) and (2.129) were derived for an arbitrary volume V. Therefore, these are valid for every part of a body, and the volume V may be replaced by any small volume v within the body, an arbitrary locality, i.e. at the limit an arbitrary fixed space point. This postulate is called the postulate of locality, and the process is called localization.

Application of the localization process to (2.128) and (2.129) leads immediately to the master equations for the local balance laws,

$$\frac{\partial A}{\partial t} + \operatorname{div}(A\vec{v}) - \operatorname{div}\vec{A}_S - A_I = 0 \qquad \text{in } V - \sigma \tag{2.130}$$

$$[\![A(\vec{v} - \vec{v}) - \vec{A}_S]\!] \cdot \vec{n} = 0 \qquad \text{on } \sigma \tag{2.131}$$

and

$$\frac{\partial \vec{A}}{\partial t} + \operatorname{div}(\vec{A} \otimes \vec{v}) - \operatorname{div}\mathbf{A}_S - \vec{A}_I = 0 \qquad \text{in } V - \sigma \tag{2.132}$$

$$[\![\vec{A} \otimes (\vec{v} - \vec{v}) - \mathbf{A}_S]\!] \cdot \vec{n} = 0 \qquad \text{on } \sigma. \tag{2.133}$$

The equations (2.130) and (2.132) are the field equations, (2.131) and (2.133) are their accompanying jump conditions. The individual local balance laws can all be derived using the master laws (2.130)–(2.133). This will be done in the later subsections.

2.2.1 Mass Balance and the Continuity Equation

The global equation of mass conservation is expressed by (2.111), i.e. taking into account the discontinuity surfaces as well we have

$$\frac{\mathrm{d}}{\mathrm{d}t} \int_{V-\sigma} \rho \, \mathrm{d}V = 0. \tag{2.134}$$

Comparing (2.134) with (2.124) we identify

$$A = \rho \tag{2.135}$$

$$\vec{A}_S = 0 \tag{2.136}$$

$$A_I = 0. \tag{2.137}$$

Substituting (2.135)–(2.137) into (2.130) and (2.131) we obtain

$$\frac{\partial \rho}{\partial t} + \operatorname{div}(\rho\vec{v}) = 0 \qquad \text{in } V - \sigma \tag{2.138}$$

$$[\![\rho(\vec{v} - \vec{v})]\!] \cdot \vec{n} = 0 \qquad \text{on } \sigma. \tag{2.139}$$

The local balance equations of mass at a fixed spatial point of both the volume V and the discontinuity surface σ (2.138) and (2.139) show that the mass is locally conserved both in V and on σ.

The first term in equation (2.138) expresses the time rate of the increase of mass in a unit fixed volume, the second the mass outflow from this volume. Equation (2.138) is often called continuity equation, and the term

$$\vec{j}_\text{M} = \rho \cdot \vec{v} \tag{2.140}$$

is called the mass flux.

2.2.2 Momentum Principles and Equations of Motion

The global balance law of momentum is given by (2.112). Neglecting the forces at isolated points $\vec{F}_i$ and taking into account the discontinuity surfaces σ we get

$$\frac{\text{d}}{\text{d}t} \int_{V-\sigma} \rho \vec{v} \, \text{d}V = \int_{S-\sigma} \mathbf{T} \cdot \text{d}\vec{S} + \int_{V-\sigma} \vec{f} \, \text{d}V. \tag{2.141}$$

Comparing (2.141) with the master law (2.125) we obtain

$$\vec{A} = \rho \vec{v} \tag{2.142}$$

$$\mathbf{A}_S = \mathbf{T} \tag{2.143}$$

$$\vec{A}_\text{I} = \vec{f}. \tag{2.144}$$

Substituting (2.142)–(2.144) into (2.132) and (2.133) we get

$$\frac{\partial}{\partial t}(\rho \vec{v}) + \text{div}\,(\rho \vec{v} \otimes \vec{v}) - \text{div}\,\mathbf{T} - \vec{f} = 0 \qquad \text{in } V - \sigma \tag{2.145}$$

$$[\![\rho \cdot \vec{v} \otimes (\vec{v} - \vec{\nu}) - \mathbf{T}]\!] \cdot \vec{n} = 0 \qquad \text{on } \sigma. \tag{2.146}$$

The first term in equation (2.145) expresses the time rate of the increase of momentum in a unit fixed volume, the second the outflow of momentum from this volume due to convective transport, the third the outflow of momentum from this volume due to molecular transport. The last term expresses the source of momentum, which is due to the volume forces. Therefore, by contrast to the mass, the momentum is not conserved in a fixed volume, but it is balanced by the volume forces.

The equation (2.146) shows that on the discontinuity surface σ, the momentum is conserved.

The momentum balance equation (2.145) is often expressed in the form of an equation of motion, which in mass-point mechanics was expressed by the Newton's second law,

$$\vec{F} = m\frac{\text{d}\vec{v}}{\text{d}t} = m\vec{a}, \tag{2.147}$$

where for the acceleration we use the formula

$$\vec{a} = \frac{d\vec{v}}{dt}. \tag{2.148}$$

For that purpose, we have to modify the first two terms of equation (2.145), namely

$$\frac{\partial}{\partial t}(\rho\vec{v}) + \text{div} \ (\rho\vec{v} \otimes \vec{v}) = \rho\frac{\partial\vec{v}}{\partial t} + \vec{v}\frac{\partial\rho}{\partial t} + \rho\vec{v} \cdot \text{grad} \ \vec{v} + \vec{v} \cdot \text{div} \ \rho\vec{v}. \tag{2.149}$$

The sum of the second and the fourth term on the right-hand side of (2.149) is equal to zero, taking into account the mass conservation equation (2.138). The sum of the first and the third term is clearly the material derivative of $\vec{v}$ multiplied by ρ (see (2.36)). Thus, the equation of motion can be written as

$$\rho\frac{d\vec{v}}{dt} = \vec{f} + \text{div} \ \mathbf{T}. \tag{2.150}$$

The algebraic modifications that we have made to the first two terms of the momentum balance equation can be generalized, so we can write for an arbitrary vector quantity $\vec{A}$

$$\rho\frac{d\vec{A}}{dt} = \frac{\partial}{\partial t}(\rho\vec{A}) + \text{div} \ (\rho\vec{A} \otimes \vec{v}). \tag{2.151}$$

In an analogous way we can write a similar relation for a scalar quantity A,

$$\rho\frac{dA}{dt} = \frac{\partial}{\partial t}(\rho A) + \text{div} \ (\rho A \cdot \vec{v}). \tag{2.152}$$

The global balance of moment of momentum (2.114) over a body V swept by a discontinuity surface σ can under simplifying assumptions $\vec{F_i} = 0$, $\vec{M_i} = 0$ be written as

$$\frac{d}{dt}\int_{V-\sigma} \rho \cdot (\vec{r} \times \vec{v}) \ dV = \int_{S-\sigma} [\vec{r} \times (\vec{n} \cdot \mathbf{T})] \ dS$$

$$+ \int_{V-\sigma} (\vec{r} \times \vec{f}) \ dV. \tag{2.153}$$

Contrary to the previous balance laws, in this case the correspondence between the master law in the simple form (2.125) and the global balance of moment of momentum (2.153) is not so obvious because of the appearance of vector products in (2.153). Comparing (2.153) with the master law (2.125) we get

$$\vec{A} = \rho\vec{r} \times \vec{v} \tag{2.154}$$

$$\mathbf{A}_S \cdot \vec{n} = \vec{r} \times (\vec{n} \cdot \mathbf{T}) \tag{2.155}$$

$$\vec{A}_{\mathrm{I}} = \vec{r} \times \vec{f}. \tag{2.156}$$

We can see immediately that (2.155) in particular is not very suitable for direct substitution into (2.132), but needs further modifications. Therefore, we will perform the derivation of the local balances step by step.

We begin with the term on the left-hand side of equation (2.153) using the conversion relation (2.80), which for a vector quantity $\vec{A}$ has the form

$$\frac{\mathrm{d}}{\mathrm{d}t} \int_{V-\sigma} \vec{A}\, \mathrm{d}V = \int_{V-\sigma} \left[\frac{\partial \vec{A}}{\partial t} + \mathrm{div}\,(\vec{A} \otimes \vec{v}) \right] \mathrm{d}V$$

$$+ \int_{\sigma} [\vec{A} \otimes (\vec{v} - \vec{\nu})] \cdot \mathrm{d}\vec{S}. \tag{2.157}$$

Substituting (2.154) into (2.157) we obtain

$$\frac{\mathrm{d}}{\mathrm{d}t} \int_{V-\sigma} \rho(\vec{r} \times \vec{v})\, \mathrm{d}V = \int_{V-\sigma} \left[\frac{\partial}{\partial t}[\rho(\vec{r} \times \vec{v}] + \mathrm{div}\,[\rho(\vec{r} \times \vec{v}) \otimes \vec{v})] \right] \mathrm{d}V$$

$$+ \int_{\sigma} [\rho(\vec{r} \times \vec{v}) \otimes (\vec{v} - \vec{\nu})] \cdot \mathrm{d}\vec{S}. \tag{2.158}$$

In the first term of the right-hand side of equation (2.158) we employ the fact that the vector product of two identical vectors is equal to zero, so we have

$$\frac{\partial}{\partial t}[\rho(\vec{r} \times \vec{v}] = \vec{r} \times \frac{\partial}{\partial t}(\rho\vec{v}). \tag{2.159}$$

The second term can be modified using the component notation,

$$\mathrm{div}\,[\rho(\vec{r} \times \vec{v}) \otimes \vec{v})] = \vec{e}_i \times \vec{e}_j \frac{\partial}{\partial x_k}(\rho x_i v_j v_k)$$

$$= \vec{e}_i \times \vec{e}_j x_i \frac{\partial}{\partial x_k}(\rho v_j v_k) + \vec{e}_i \times \vec{e}_j \frac{\partial x_i}{\partial x_k}\rho v_j v_k$$

$$= \vec{e}_i x_i \times \vec{e}_j \frac{\partial}{\partial x_k}(\rho v_j v_k) + e_i \times \vec{e}_j \rho v_i v_j$$

$$= \vec{r} \times \mathrm{div}\,(\rho\vec{v} \otimes \vec{v}) + \rho\vec{v} \times \vec{v} = \vec{r} \times \mathrm{div}\,(\rho\vec{v} \otimes \vec{v}). \tag{2.160}$$

The first term on the right-hand side of equation (2.153) we will write again in components. We have

$$\vec{r} = \vec{e}_j x_j \tag{2.161}$$

$$\vec{n} \cdot \mathbf{T} = \vec{e}_i n_k T_{ki}, \tag{2.162}$$

and therefore

$$\vec{r} \times (\vec{n} \cdot \mathbf{T}) = \vec{e}_j x_j \times \vec{e}_i n_k T_{ki} = \epsilon_{jim} \vec{e}_m n_k x_j T_{ki}, \tag{2.163}$$

where ϵ is the Levi–Civita tensor (i.e. the antisymmetric tensor of third order). Thus, we can write

$$\int_{S-\sigma} [\vec{r} \times (\vec{n} \cdot \mathbf{T})] \, dS = \int_{S-\sigma} \epsilon_{jim} \vec{e}_m n_k x_j T_{ki} \, dS. \tag{2.164}$$

Using the Gauss theorem and the rule on the derivative of a product we obtain

$$\int_{S-\sigma} [\vec{r} \times (\vec{n} \cdot \mathbf{T})] \, dS = \int_{S-\sigma} \epsilon_{jim} \vec{e}_m n_k x_j T_{ki} \, dS$$

$$= \int_{V-\sigma} \epsilon_{jim} \vec{e}_m \frac{\partial (x_j T_{ki})}{\partial x_k} \, dV + \int_\sigma [\vec{r} \times (\vec{n} \cdot \mathbf{T})] \, dS$$

$$= \int_{V-\sigma} \vec{e}_m \epsilon_{jim} T_{ji} \, dV$$

$$+ \int_{V-\sigma} (\vec{e}_j \times \vec{e}_i) x_j \frac{\partial T_{ki}}{\partial x_k} \, dV + \int_\sigma [\vec{r} \times (\vec{n} \cdot \mathbf{T})] \, dS. \tag{2.165}$$

Coming back to the vector and tensor notation we get

$$\int_{S-\sigma} [\vec{r} \times (\vec{n} \cdot \mathbf{T})] \, dS = \int_{V-\sigma} \vec{e}_m \epsilon_{jim} T_{ji} \, dV + \int_{V-\sigma} (\vec{r} \times \operatorname{div} \mathbf{T}) \, dV$$

$$+ \int_\sigma [\![\vec{r} \times (\vec{n} \cdot \mathbf{T})]\!] \, dS. \tag{2.166}$$

Substituting now (2.159), (2.160), (2.166) into (2.153), we arrive at

$$\int_{V-\sigma} \left[\vec{r} \times \left(\rho \frac{d\vec{v}}{dt} - \vec{f} - \operatorname{div} \mathbf{T} \right) \right] \, dV = \int_{V-\sigma} \vec{e}_m \epsilon_{jim} T_{ji} \, dV \tag{2.167}$$

and

$$\int_\sigma [\![(\vec{r} \times \rho \vec{v}) \otimes (\vec{v} - \vec{\nu})]\!] \cdot d\vec{S} = \int_\sigma [\![\vec{r} \times (\vec{n} \cdot \mathbf{T})]\!] \, dS. \tag{2.168}$$

The term on the left-hand side of equation (2.167) is equal to zero, as it follows from the equation of motion (2.150). The resulting global balance equation of moment of momentum can then be written as

$$\int_{V-\sigma} \vec{e}_m \epsilon_{jim} T_{ji} \, dV = 0, \tag{2.169}$$

which leads to the local balance equation

$$\ddot{e}_m \epsilon_{jim} T_{ji} = 0. \tag{2.170}$$

This balance equation is significantly different from those derived before. It simply states that a certain vector is equal to zero. Consequently, all the components of this vector are equal to zero,

$$\epsilon_{jim} T_{ji} = 0, \quad m = 1, 2, 3. \tag{2.171}$$

From the properties of the Levi–Civita tensor it immediately follows that

$$T_{ij} = T_{ji}, \tag{2.172}$$

which is the final form of the local balance equation of the moment of momentum. The only new fact coming from this equation is that the stress tensor **T** is symmetric.

Assuming the position vector $\vec{r}$ to be continuous across the discontinuity surface σ, which is quite logical, from (2.168) it follows that

$$\vec{r} \times [\![(\rho \vec{v} \otimes (\vec{v} - \vec{\nu}) - \mathbf{T})]\!] \cdot \vec{n} = 0, \tag{2.173}$$

which is satisfied identically because of (2.146).

2.2.3 Energy Balance

The global balance law of energy is given by equation (2.116). Disregarding the isolated sources but on the other hand taking into account the discontinuity surfaces σ we get

$$\frac{\mathrm{d}}{\mathrm{d}t} \int_{V-\sigma} \left(\rho u + \frac{1}{2} \rho \vec{v} \cdot \vec{v} \right) \mathrm{d}V = \int_{V-\sigma} \vec{f} \cdot \vec{v} \, \mathrm{d}V + \int_{S-\sigma} (\vec{v} \cdot \mathbf{T}) \cdot \mathrm{d}\vec{S}$$

$$- \int_{S-\sigma} \vec{j}_Q \cdot \mathrm{d}\vec{S} + \int_{V-\sigma} I_Q \, \mathrm{d}V. \tag{2.174}$$

Comparing (2.174) with the master law (2.124) we obtain

$$A = \rho u + \frac{1}{2} \rho \vec{v} \cdot \vec{v} = \rho u + \frac{1}{2} \rho v^2 \tag{2.175}$$

$$\vec{A}_S = \vec{v} \cdot \mathbf{T} - \vec{j}_Q \tag{2.176}$$

$$A_I = \vec{f} \cdot \vec{v} + I_Q. \tag{2.177}$$

Substituting (2.175)–(2.177) in (2.130) and (2.131) we get

$$\frac{\partial}{\partial t} \left(\rho u + \frac{1}{2} \rho v^2 \right) + \mathrm{div} \left[\left(\rho u + \frac{1}{2} \rho v^2 \right) \cdot \vec{v} \right] - \mathrm{div} \left(\vec{v} \cdot \mathbf{T} \right) + \mathrm{div} \, \vec{j}_Q$$

$$- \vec{f} \cdot \vec{v} - I_Q = 0 \qquad \text{in } V - \sigma \tag{2.178}$$

$$[\![(\rho u + \frac{1}{2} \rho \vec{v}^2) \cdot (\vec{v} - \vec{\dot{v}}) - \vec{v} \cdot \mathbf{T} + \vec{j}_Q]\!] \cdot \vec{n} = 0 \qquad \text{on } \sigma. \tag{2.179}$$

We assumed the total energy E consisted of the kinetic energy and the internal energy. In some applications, it is useful to consider either the kinetic energy or the internal energy separately. Therefore, we will derive their balance equations in what follows.

The balance equation of kinetic energy we can obtain by simple reasoning. According to the master law (2.130), it has to contain the term

$$\frac{\partial}{\partial t} \left(\frac{1}{2} \rho v^2 \right),$$

expressing the time rate of kinetic energy in unit fixed volume. We can immediately see that the easiest way to obtain such a term is to multiply the first term of the momentum balance (2.145) by the velocity vector $\vec{v}$. Therefore, we begin with the multiplication of the momentum balance equation (2.145) by $\vec{v}$,

$$\vec{v} \cdot \frac{\partial}{\partial t} (\rho \vec{v}) + \vec{v} \cdot \operatorname{div} (\rho \vec{v} \otimes \vec{v})$$

$$- \vec{v} \cdot \operatorname{div} \mathbf{T} - \vec{v} \cdot \vec{f} = 0 \qquad \text{in } V - \sigma. \tag{2.180}$$

In equation (2.180), we have to modify the first two terms. For the first term we immediately have

$$\vec{v} \cdot \frac{\partial}{\partial t} (\rho \vec{v}) = \frac{\partial}{\partial t} \left(\frac{1}{2} \rho v^2 \right) + \frac{1}{2} \frac{\partial \rho}{\partial t} v^2. \tag{2.181}$$

Substituting the mass balance equation (2.138) into (2.181) we get

$$\vec{v} \cdot \frac{\partial}{\partial t} (\rho \vec{v}) = \frac{\partial}{\partial t} \left(\frac{1}{2} \rho v^2 \right) - \frac{1}{2} v^2 \operatorname{div} \rho \vec{v}. \tag{2.182}$$

The algebraic modifications of the second term in equation (2.180) are a little more complicated,

$$\vec{v} \cdot \operatorname{div} (\rho \vec{v} \otimes \vec{v}) = \vec{v} \cdot (\vec{v} \cdot \operatorname{grad} \rho \vec{v} + \rho \vec{v} \cdot \operatorname{div} \vec{v})$$

$$= \vec{v} \cdot \left[\operatorname{grad} \left(\frac{1}{2} \rho v^2 \right) + \frac{1}{2} v^2 \operatorname{grad} \rho + \rho \vec{v} \cdot \operatorname{div} \vec{v} \right]$$

$$= \vec{v} \cdot \operatorname{grad} \left(\frac{1}{2} \rho v^2 \right) + \frac{1}{2} v^2 \cdot \vec{v} \cdot \operatorname{grad} \rho + \rho v^2 \operatorname{div} \vec{v}$$

$$= \vec{v} \cdot \text{grad} \left(\frac{1}{2} \rho v^2 \right) + \frac{1}{2} \rho v^2 \text{ div } \vec{v}$$

$$+ \frac{1}{2} v^2 \cdot \vec{v} \cdot \text{ grad } \rho + \frac{1}{2} \rho v^2 \text{ div } \vec{v}$$

$$= \text{div} \left(\frac{1}{2} \rho v^2 \cdot \vec{v} \right) + \frac{1}{2} v^2 \text{ div } \rho \vec{v}. \tag{2.183}$$

From (2.182) and (2.183) we obtain

$$\vec{v} \cdot \frac{\partial}{\partial t} (\rho \vec{v}) + \vec{v} \cdot \text{ div } (\rho \vec{v} \otimes \vec{v}) = \frac{\partial}{\partial t} \left(\frac{1}{2} \rho v^2 \right) + \text{ div } \left(\frac{1}{2} \rho v^2 \cdot \vec{v} \right). \tag{2.184}$$

The third term in equation (2.180) can be modified using the rule on differentiation of a product,

$$\vec{v} \cdot \text{ div } \mathbf{T} = \text{ div } (\vec{v} \cdot \mathbf{T}) - \mathbf{T} : \text{ grad } \vec{v}, \tag{2.185}$$

where the symbol $\mathbf{A} : \mathbf{B}$ denotes a scalar product of two tensors. The last term in (2.185) can be further modified using the fact that the stress tensor $\mathbf{T}$ is symmetric and the definition relation of the rate of deformation tensor $\mathbf{D}$ (2.38),

$$\mathbf{D} = (\text{grad } \vec{v})^{\text{S}}. \tag{2.186}$$

We obtain

$$\vec{v} \cdot \text{ div } \mathbf{T} = \text{ div } (\vec{v} \cdot \mathbf{T}) - \mathbf{T} : \mathbf{D}. \tag{2.187}$$

Substituting (2.184) and (2.187) into (2.180), we get the kinetic energy balance equation in the form

$$\frac{\partial}{\partial t} \left(\frac{1}{2} \rho v^2 \right) + \text{ div } \left(\frac{1}{2} \rho v^2 \cdot \vec{v} \right) - \text{ div } (\vec{v} \cdot \mathbf{T})$$

$$+ \mathbf{T} : \mathbf{D} - \vec{v} \cdot \vec{f} = 0 \qquad \text{in } V - \sigma. \tag{2.188}$$

Comparing (2.188) with the master law (2.124) we get

$$A = \frac{1}{2} \rho v^2 \tag{2.189}$$

$$\vec{A}_{\text{S}} = \vec{v} \cdot \mathbf{T} \tag{2.190}$$

$$A_{\text{I}} = \vec{f} \cdot \vec{v} - \mathbf{T} : \mathbf{D}, \tag{2.191}$$

so that we can state that equation (2.188) is really the balance equation of kinetic energy. The only term in (2.188) that appears for the first time in our formulas is $\mathbf{T} : \mathbf{D}$. It is called <u>stress power</u> (see, e.g., Malvern, 1969) and expresses the time rate of the deformation work.

The balance equation of kinetic energy on the discontinuity surface σ we obtain by a direct application of the master law (2.131) together with (2.189)–(2.191),

$$[\tfrac{1}{2}\rho v^2 \cdot (\vec{v} - \vec{\nu}) - \vec{v} \cdot \mathbf{T}] \cdot \vec{n} = 0 \qquad \text{on } \sigma. \tag{2.192}$$

As we defined the total energy as the sum of the kinetic energy and the internal energy, the balance equation of <u>internal energy</u> can be obtained immediately by subtracting the balance equations of kinetic energy (2.188) and (2.192) from the respective balance equations of total energy (2.178), (2.179). We can write

$$\frac{\partial}{\partial t}(\rho u) + \text{div }(\rho u \vec{v}) = -\text{div }\vec{j}_Q + I_Q + \mathbf{T} : \mathbf{D} \qquad \text{in } V - \sigma. \tag{2.193}$$

$$[\![\rho u \cdot (\vec{v} - \vec{\nu}) + \vec{j}_Q]\!] \cdot \vec{n} = 0 \qquad \text{on } \sigma. \tag{2.194}$$

In classical mechanics of mass points, as well as the kinetic energy we also worked with the <u>potential energy</u>, which has not been mentioned so far. Therefore, we now will derive the balance equation of potential energy of volume forces in continuum mechanics. As in mass-points mechanics we will assume that the volume forces are conservative and possess a scalar potential φ, defined by the relation

$$dE_p = \varphi(\vec{x})\, dm = \rho\varphi(\vec{x})\, dV, \tag{2.195}$$

where E_p is the potential energy.

The conservative volume forces have to obey the same two basic relations as in mass-point mechanics, namely

$$\vec{b} = -\text{grad }\varphi \tag{2.196}$$

$$\frac{\partial \varphi}{\partial t} = 0, \tag{2.197}$$

where $\vec{b}$ is the specific force.

From the master law (2.130) it follows that we should identify the term

$$\frac{\partial}{\partial t}(\rho\varphi).$$

In the modifications of this term we employ the mass balance (2.138) and the rule on product differentiation,

$$\frac{\partial(\rho\varphi)}{\partial t} = \rho\frac{\partial\varphi}{\partial t} + \varphi\frac{\partial\rho}{\partial t} = -\varphi\,\text{div }(\rho\vec{v}) = -\text{div }(\rho\varphi\vec{v}) + \rho\vec{v}\cdot\text{grad }\varphi$$

$$= -\operatorname{div}(\rho\varphi\vec{v}) - \rho\vec{v}\cdot\vec{b} = -\operatorname{div}(\rho\varphi\vec{v}) - \vec{v}\cdot\vec{f}, \tag{2.198}$$

so that we finally get

$$\frac{\partial(\rho\varphi)}{\partial t} + \operatorname{div}(\rho\varphi\vec{v}) + \vec{v}\cdot\vec{f} = 0 \qquad \text{in } V - \sigma. \tag{2.199}$$

Comparing (2.199) with the master law (2.124), we obtain

$$A = \rho\varphi \tag{2.200}$$

$$\vec{A}_S = 0 \tag{2.201}$$

$$A_I = -\vec{v}\cdot\vec{f}. \tag{2.202}$$

Thus, equation (2.199) is really the balance equation of potential energy of volume forces.

The balance equation of potential energy on the discontinuity surface σ can be obtained by a substitution of (2.200)–(2.202) into the master law (2.131),

$$[\![\rho\varphi\cdot(\vec{v}-\vec{\nu})]\!]\cdot\vec{n} = 0 \qquad \text{on } \sigma. \tag{2.203}$$

Now we have the balance equations of kinetic energy and potential energy (2.188), (2.192), (2.199), (2.203), so we can formulate the respective balance equations of <u>mechanical energy</u> by summing them,

$$\frac{\partial}{\partial t}\left(\rho\varphi + \frac{1}{2}\rho v^2\right) + \operatorname{div}\left(\rho\varphi\cdot\vec{v} + \frac{1}{2}\rho v^2\cdot\vec{v}\right)$$

$$- \operatorname{div}(\vec{v}\cdot\mathbf{T}) + \mathbf{T}:\mathbf{D} = 0 \qquad \text{in } V - \sigma. \tag{2.204}$$

$$[\![\left(\rho\varphi + \frac{1}{2}\rho v^2\right)\cdot(\vec{v}-\vec{\nu}) - \vec{v}\cdot\mathbf{T}]\!]\cdot\vec{n} = 0 \qquad \text{on } \sigma. \tag{2.205}$$

The total energy e can be alternatively defined as the sum of the kinetic energy, the potential energy and the internal energy,

$$e = \frac{1}{2}v^2 + \varphi + u. \tag{2.206}$$

In that case, the balance equations of the total energy (2.178) and (2.179) would have to be modified as

$$\frac{\partial}{\partial t}\left(\rho u + \frac{1}{2}\rho v^2 + \rho\varphi\right) + \operatorname{div}\left[\left(\rho u + \frac{1}{2}\rho v^2 + \rho\varphi\right)\cdot\vec{v}\right]$$

$$- \operatorname{div}(\vec{v}\cdot\mathbf{T}) + \operatorname{div}\vec{j}_Q - I_Q = 0 \qquad \text{in } V - \sigma \tag{2.207}$$

$$[\![\left(\rho u + \frac{1}{2}\rho\vec{v}^2 + \rho\varphi\right)\cdot(\vec{v}-\vec{\nu}) - \vec{v}\cdot\mathbf{T} + \vec{j}_Q]\!]\cdot\vec{n} = 0 \qquad \text{on } \sigma. \tag{2.208}$$

2.2.4 Entropy Inequality

The global entropy inequality (2.120) over $V - \sigma$ swept by discontinuity surface σ can be expressed (disregarding again the isolated entropy sources, $B_i = 0$) in the form

$$\Gamma = \frac{\mathrm{d}}{\mathrm{d}t} \int_{V-\sigma} \rho s \, \mathrm{d}V + \int_{S-\sigma} \vec{j}_\mathrm{s} \cdot \mathrm{d}\vec{S} - \int_{V-\sigma} \rho b_\mathrm{s} \, \mathrm{d}V \geq 0. \tag{2.209}$$

Comparing (2.209) with (2.124) we identify

$$A = \rho s \tag{2.210}$$

$$\vec{A}_\mathrm{S} = -\vec{j}_\mathrm{s} \tag{2.211}$$

$$A_\mathrm{I} = \rho b_\mathrm{s}. \tag{2.212}$$

Substituting (2.210)–(2.212) in (2.130) and (2.131), we obtain

$$\rho \gamma = \frac{\partial(\rho s)}{\partial t} + \operatorname{div}(\rho s \vec{v}) + \operatorname{div} \vec{j}_\mathrm{s} - \rho b_\mathrm{s} \geq 0 \qquad \text{in } V - \sigma, \tag{2.213}$$

where the specific entropy production rate γ is defined as

$$\Gamma = \int_{V-\sigma} \rho \gamma \, \mathrm{d}V, \tag{2.214}$$

and

$$[\![\rho s(\vec{v} - \vec{v}) + \vec{j}_\mathrm{s}]\!] \cdot \vec{n} \geq 0 \qquad \text{on } \sigma. \tag{2.215}$$

2.3 CONSTITUTIVE EQUATIONS FOR A SINGLE CONTINUUM

The balance laws formulated in the previous section are the basic principles of continuum mechanics. These are valid for all material groups and for various external conditions. However, these axioms are not sufficient for a complex description of a specific material and a specific process, because they clearly contain more unknown field quantities than there are equations formulating the relations between them. Therefore, the system of balance laws must be completed by constitutive equations, which formulate the remaining relations necessary to achieve the solvability of the system of equations for field variables.

The constitutive equations never achieve the generality of the balance laws. These are always specific to a particular material and a particular type of the process. Plainly, the constitutive relations cannot be formulated in an arbitrary way, for there are thermodynamic restrictions on the response of bodies arising from the energy balance and the entropy inequality. Without these restrictions, mathematically characterized ideal bodies may not represent physical bodies, and their predicted response may not be realistic. We will now provide a short overview of thermodynamic concepts and principles necessary for the formulation of constitutive equations.

2.3.1 Basic Concepts

Classical thermodynamics is concerned (see, e.g., Eringen, 1967) with the study of the thermal behaviour of bodies consisting of a finite number of parts at rest or in uniform motion. A body as a whole at each point in time is assigned a positive number T called temperature. This number measures how hot or cold is the body at a given time. Experiments show that no body can be cooled below a certain limit. If this limit is set at zero, then the temperature is called absolute. Therefore, we have

$$T > 0 \tag{2.216}$$

$$\inf T = 0. \tag{2.217}$$

In continuum mechanics, this concept motivates the introduction of a function $T(\vec{X}, t)$ at a material point $\vec{X}$ at time t as the absolute temperature of $\vec{X}$ at time t. Thus, a thermomechanical process is defined as a pair of functions

$$T = T(\vec{X}, t) \tag{2.218}$$

$$\vec{x} = \vec{x}(\vec{X}, t). \tag{2.219}$$

If the thermomechanical process is known, the motions and temperatures of all points of the body can be determined, in principle. However, as we already mentioned before, there are too many unknown field quantities in the balance laws. Some of them were defined, but then they were left without clear relations to other field parameters and to the process in general, namely u, $\vec{j}_Q$, γ, $\vec{j}_s$ and $\mathbf{T}$. This problem is addressed within the framework of continuum thermodynamics by introducing the constitutive equations relating these quantities to the process. In the most general form, which is valid for very large class of materials, we can write a set of constitutive equations as follows:

$$u(\vec{X}, t) = u'[\vec{x}(\vec{X}', t'), T(\vec{X}', t')] \tag{2.220}$$

$$\vec{j}_Q(\vec{X}, t) = \vec{j}_Q'[\vec{x}(\vec{X}', t'), T(\vec{X}', t')] \tag{2.221}$$

$$\gamma(\vec{X}, t) = \gamma'[\vec{x}(\vec{X}', t'), T(\vec{X}', t')] \tag{2.222}$$

$$\vec{j}_s(\vec{X}, t) = \vec{j}_s'[\vec{x}(\vec{X}', t'), T(\vec{X}', t')] \tag{2.223}$$

$$\mathbf{T}(\vec{X}, t) = \mathbf{T}'[\vec{x}(\vec{X}', t'), T(\vec{X}', t')]. \tag{2.224}$$

It is clear that if the constitutive equations are known, by substituting them into the balance laws we obtain adequate number of equations to determine, in principle, $\vec{x}(\vec{X}, t)$ and $T(\vec{X}, t)$.

Every thermomechanical process $\vec{x}(\vec{X}, t)$ and $T(\vec{X}, t)$ (see, e.g., Eringen, 1967) that satisfies the local balance laws and constitutive equations is called a thermodynamic process. Every thermodynamic process is further subject to the law of entropy inequality, which puts further restrictions on all thermomechanical processes. A thermodynamic process is called admissible if it does not violate the entropy inequality.

Practically this means that we have to restrict the constitutive equations, i.e. the functionals u', $\vec{j}_Q'$, γ', $\vec{j}_s'$, $\mathbf{T}'$, by the admissibility condition derived from the entropy inequality. Such a condition rules out certain thermodynamic processes that can take place in a body with more general response functions than those restricted by the entropy inequality. In this sense, we in fact place restrictions on the nature of the body rather than on the nature of the thermodynamic process.

Thermodynamics has not yet developed to a stage where it can include general constitutive equations such as (2.220)–(2.224). From a practical point of view, for local theories it is not necessary to consider the problem in such a broad way. First, the principle of determinism in classical physics dictates that the past history of motions and temperatures up to the present time determines the present response. Therefore, the response functionals contain only the histories up to the present time, i.e.

$$u(\vec{X}, t) = u'[\vec{x}(\vec{X}', t - \tau'), T(\vec{X}', t - \tau')] \tag{2.225}$$

$$\vec{j}_Q(\vec{X}, t) = \vec{j}_Q'[\vec{x}(\vec{X}', t - \tau'), T(\vec{X}', t - \tau')] \tag{2.226}$$

$$\gamma(\vec{X}, t) = \gamma'[\vec{x}(\vec{X}', t - \tau'), T(\vec{X}', t - \tau')] \tag{2.227}$$

$$\vec{j}_s(\vec{X}, t) = \vec{j}_s'[\vec{x}(\vec{X}', t - \tau'), T(\vec{X}', t - \tau')] \tag{2.228}$$

$$\mathbf{T}(\vec{X}, t) = \mathbf{T}'[\vec{x}(\vec{X}', t - \tau'), T(\vec{X}', t - \tau')]. \tag{2.229}$$

Moreover, local theories and simple materials, which are discussed in this book, require consideration of only the motions and temperature of the infinitesimal neighbourhood of a material point $\vec{X}$. This means we can replace the argument functions by just first two terms in the Taylor series expansion.

Taking into account small (infinitesimal) deformations only, we can finally write

$$u(\vec{x},t) = u'[\vec{x}(t-\tau'), \text{ grad } \vec{x}(t-\tau'), T(t-\tau'), \text{ grad } T(t-\tau')] \quad (2.230)$$

$$\vec{j}_Q(\vec{x},t) = \vec{j}_Q'[\vec{x}(t-\tau'), \text{ grad } \vec{x}(t-\tau'), T(t-\tau'), \text{ grad } T(t-\tau')] \quad (2.231)$$

$$\gamma(\vec{x},t) = \gamma'[\vec{x}(t-\tau'), \text{ grad } \vec{x}(t-\tau'), T(t-\tau'), \text{ grad } T(t-\tau')] \quad (2.232)$$

$$\vec{j}_s(\vec{x},t) = \vec{j}_s'[\vec{x}(t-\tau'), \text{ grad } \vec{x}(t-\tau'), T(t-\tau'), \text{ grad } T(t-\tau')] \quad (2.233)$$

$$\mathbf{T}(\vec{x},t) = \mathbf{T}'[\vec{x}(t-\tau'), \text{ grad } \vec{x}(t-\tau'), T(t-\tau'), \text{ grad } T(t-\tau')]. \quad (2.234)$$

We note that all simple memory-dependent materials are included in the constitutive equations (2.230)–(2.234), for instance elastic solids, viscous fluids, viscoelastic materials of all kinds, etc.

It also should be noted that the constitutive equations (2.230)–(2.234) are still very general, and in this general form they are usually applied within the framework of rational thermodynamics (see, e.g., Truesdell, 1969, Noll, 1974). Classical irreversible thermodynamics works with a much simpler form of linear constitutive equations, where so-called generalized thermodynamic fluxes J_i are assumed to be linear functions of generalized thermodynamic forces X_j, and also of unspecified functions of all the thermodynamic state variables such as temperature T or pressure p,

$$J_i = J_i(X_j, T, p, ...). \qquad (2.235)$$

The generalized fluxes may be for example the heat flux $\vec{j}_Q$, the stress tensor $\mathbf{T}$, and the generalized forces the temperature gradient grad T or pressure gradient grad p, etc.

In what follows we will analyse the identification of thermodynamic forces and fluxes, which is a principal task of classical irreversible thermodynamics, and the practical construction of constitutive equations. Only the basic ideas and basic methods in this area will be discussed, as it is not the purpose of this book to deal with theoretical aspects of irreversible thermodynamics. The reader can find details in a number of well known basic textbooks on thermodynamics, such as Prigogine (1961), de Groot and Mazur (1962), Fitts (1962), Haase (1969), and Lavenda (1978).

In practical construction of a constitutive equation, the type of the thermodynamic process is of great importance. Thermodynamic processes are generally of two types, reversible and irreversible. A process is called reversible if the system and all elements of its environment can be completely restored to their respective initial states after the process has occurred, i.e. the system passes through the same sequence of states (only in reverse order) and the same path in the direct process and in the reversed process. If the same (only of reversed order) sequence of states in the reversed process as in the direct process cannot be followed, the process is called irreversible.

The basic axioms of classical thermodynamics are the first and the second laws of thermodynamics. The first law is the law of conservation of internal energy,

$$dU = \delta Q - \delta A, \tag{2.236}$$

where Q is the net heat received by the system during a change of state and A is the net work done by the system during a change of state. In equation (2.236) only dU is a total differential, δQ and δA have to be understood as referring to some differential changes only. The second law defines the total differential of entropy S for reversible processes as

$$dS = \frac{\delta Q}{T}, \tag{2.237}$$

while for the irreversible processes we have

$$dS > \frac{\delta Q}{T}. \tag{2.238}$$

Therefore, generally we can write for all type of processes

$$dS \geq \frac{\delta Q}{T}, \tag{2.239}$$

which is called the principle of the increase in entropy. It should be noted, that this principle is valid only in isolated systems.

The first and second law for a reversible process can be combined with the result

$$dU = TdS - \delta A, \tag{2.240}$$

and if we suppose that δA is the deformation work, which for a reversible process can be written as

$$\delta A = - \int_V \mathbf{T^R} : d\mathbf{U^R} \, dV, \tag{2.241}$$

where $\mathbf{T^R}$ is the reversible part of the stress tensor and $\mathbf{U^R}$ is the reversible part of the strain tensor, we finally get

$$dU = TdS + \int_V \mathbf{T^R} : d\mathbf{U^R} \, dV. \tag{2.242}$$

Equation (2.242) is called Gibbs equation.

In continuum thermodynamics we work with specific quantities rather than with global quantities. Thus, the Gibbs equation (2.242) can be reformulated as

$$\rho \, du = \rho T \, ds + \mathbf{T^R} : d\mathbf{U^R}. \tag{2.243}$$

As we can see, equation (2.243) expresses the relation between the internal energy and the entropy. In the thermodynamics of irreversible processes this equation is of central importance. In combination with the balance equation of internal energy and the entropy inequality, it makes it possible to identify the specific entropy production rate γ, and therefore the extent of irreversible processes in a material. As follows from the second law of thermodynamics, for the reversible processes γ vanishes.

We will derive the specific entropy production rate γ for our simple case of a one-component material. We will employ the balance equation of internal energy

$$\frac{\partial}{\partial t}(\rho u) + \text{div}\,(\rho u \vec{v}) = -\,\text{div}\,\vec{j}_Q + I_Q + \mathbf{T} : \mathbf{D} \tag{2.244}$$

and the entropy inequality

$$\rho\gamma = \frac{\partial(\rho s)}{\partial t} + \text{div}\,(\rho s \vec{v}) + \text{div}\,\vec{j}_s - \rho b_s \geq 0. \tag{2.245}$$

For practical reasons, we will also treat the entropy inequality as a balance equation (without any loss of generality and still taking into account that $\gamma \geq 0$), i.e. we will rewrite it in the form

$$\frac{\partial(\rho s)}{\partial t} + \text{div}\,(\rho s \vec{v}) + \text{div}\,\vec{j}_s - \rho b_s - \rho\gamma = 0. \tag{2.246}$$

In order to get comparable terms in the three mentioned equations, we will first differentiate the Gibbs equation (2.243) with respect to time,

$$\rho\frac{du}{dt} = \rho T\frac{ds}{dt} + \mathbf{T}^{\mathbf{R}} : \mathbf{D}^{\mathbf{R}}. \tag{2.247}$$

Then, we use the conversion relation (2.152) to modify equations (2.244) and (2.246),

$$\rho\frac{du}{dt} = -\,\text{div}\,\vec{j}_Q + I_Q + \mathbf{T} : \mathbf{D} \tag{2.248}$$

$$\rho\frac{ds}{dt} + \text{div}\,\vec{j}_s - \rho b_s - \rho\gamma = 0. \tag{2.249}$$

The last term on the right-hand side of equation (2.248) can be formally rewritten as

$$\mathbf{T} : \mathbf{D} = (\mathbf{T}^{\mathbf{R}} + \mathbf{T}^{\mathbf{D}}) : (\mathbf{D}^{\mathbf{R}} + \mathbf{D}^{\mathbf{D}}) = \mathbf{T}^{\mathbf{R}} : \mathbf{D}^{\mathbf{R}} + \mathbf{T}^{\mathbf{D}} : \mathbf{D}^{\mathbf{R}}$$

$$+ \mathbf{T}^{\mathbf{R}} : \mathbf{D}^{\mathbf{D}} + \mathbf{T}^{\mathbf{D}} : \mathbf{D}^{\mathbf{D}}, \tag{2.250}$$

where the superscript D denotes the dissipative part of a quantity. Substituting (2.248) with (2.250) into the Gibbs equation (2.247) we get

$$\rho T\frac{ds}{dt} + \mathbf{T}^{\mathbf{R}} : \mathbf{D}^{\mathbf{R}} = -\,\text{div}\,\vec{j}_Q + I_Q + \mathbf{T}^{\mathbf{R}} : \mathbf{D}^{\mathbf{R}} + \mathbf{T}^{\mathbf{D}} : \mathbf{D}^{\mathbf{R}}$$

$$+ \mathbf{T^R} : \mathbf{D^D} + \mathbf{T^D} : \mathbf{D^D}, \tag{2.251}$$

and after simple modifications

$$\rho \frac{ds}{dt} + \frac{1}{T} \operatorname{div} \vec{j}_Q - \frac{1}{T} I_Q - \frac{1}{T} \mathbf{T^D} : \mathbf{D^R}$$

$$- \frac{1}{T} \mathbf{T^R} : \mathbf{D^D} - \frac{1}{T} \mathbf{T^D} : \mathbf{D^D} = 0. \tag{2.252}$$

Clearly, equation (2.252) is an alternative expression for the entropy balance equation (2.249). Comparing the respective terms in (2.252), and taking into account that the second term of equation (2.252) can be modified as

$$\frac{1}{T} \operatorname{div} \vec{j}_Q = \operatorname{div} \frac{\vec{j}_Q}{T} + \frac{\vec{j}_Q}{T^2} \operatorname{grad} T, \tag{2.253}$$

we arrive at

$$\vec{j}_s = \frac{\vec{j}_Q}{T} \tag{2.254}$$

$$\rho b_s = \frac{1}{T} I_Q \tag{2.255}$$

$$\rho \gamma = -\frac{\vec{j}_Q}{T^2} \operatorname{grad} T + \frac{1}{T} \mathbf{T^D} : \mathbf{D^R} + \frac{1}{T} \mathbf{T^R} : \mathbf{D^D} + \frac{1}{T} \mathbf{T^D} : \mathbf{D^D}. \tag{2.256}$$

Equation (2.254) shows that the entropy flux is caused by reduced heat flux $\frac{\vec{j}_Q}{T}$. Equation (2.255) demonstrates that internal heat sources of unspecified origin (for instance due to radiation) can lead to entropy production inside the body in the form of the term b_s.

Equation (2.256) is the most important result of our algebraic operations. It is fundamental for the identification of thermodynamic forces and fluxes, and therefore also for the formulation of phenomenological equations. Clearly, the first term on the right-hand side arises from heat conduction, the second term is connected to the gradient of velocity field giving rise to viscous flow, the third term is related to plastic deformation, and the final term to a combination of viscous and plastic effects.

The structure of the expression for γ is of a bilinear form (see, e.g., de Groot and Mazur, 1962) and consists of a sum of products of two factors. One of these factors in each term is a flow quantity (heat flux $\vec{j}_Q$, reversible or dissipative parts of the stress tensor $\mathbf{T}$), which was already introduced in the balance laws. The other factor in each term is related to the gradient of a field variable (temperature or velocity in our case). These quantities, which multiply the fluxes in the expression for the entropy production, are called thermodynamic forces. Therefore, equation (2.256) can be schematically written in the form

$$\rho\gamma = \sum_i J_i x_i, \tag{2.257}$$

where J_i are the thermodynamic fluxes and x_i the thermodynamic forces (more precisely, the force densities in the sense of continuum theory).

One of the basic tasks performed by irreversible thermodynamics is the formulation of relations between the thermodynamic forces and thermodynamic fluxes. These relations are called phenomenological equations (see, e.g., de Groot and Mazur, 1962). They represent constitutive equations characterizing the irreversible processes.

On the other hand, reversible processes (or, more exactly the processes that are close to reversibility, for as is known from classical thermodynamics, a reversible process is in practice only an idealization) can also take place in a continuum. Constitutive equations characterizing the reversible processes are called equations of state. These express the relations between state variables in a state of equilibrium, where all fluxes vanish simultaneously with the thermodynamic forces.

2.3.2 Construction of Equations of State and Phenomenological Equations

Before a formulation of any equation of state, first it is necessary to know the number of independent macroscopic parameters (state variables) characterizing the macroscopic state of the particular system. This number depends on external conditions and it has to be determined in an empirical way.

If the number of independent state parameters is known, the number of independent relations between the state parameters is known as well. These relations can be formulated for an arbitrary number of state parameters, which is equal to or lower than the number of independent parameters. However, again they have to be determined empirically. The most important among these relations is the equation of state (another one is for example the dependence of specific heat capacity on temperature), which is a relation between all independent state variables. Any equation of state can be formulated only in a state of thermodynamic equilibrium. If any nonzero thermodynamic fluxes appear in the system, no equation of state can be formulated.

In continuum thermodynamics, where the systems (i.e. the bodies) can be formally divided into a finite number of subsystems, called spatial or material volumes, the hypothesis of local equilibrium is often employed. According to this hypothesis, the relaxation times (i.e. the times that the system or subsystem takes to achieve under constant external conditions the state of equilibrium from a chosen initial state) of these volumes are an order of magnitude lower than the relaxation time of the system as a whole. Consequently, the state of thermodynamic equilibrium in the particular localities is established much sooner than in the whole system. In a model solution, we can then assume that while in the particular localities a state of equilibrium is established or equilibrium (reversible) processes take place, in the whole system there are nonequilibrium (irreversible) processes. This means, for instance, that in a specific material or spatial volume we have a constant temperature, but in two neighbouring volumes the temperatures are different, so that between these volumes a temperature gradient exists, resulting in nonzero heat flux.

For the irreversible processes between the subsystems, we have to formulate the phenomenological equations. If the gradients of particular macroscopic quantities (the thermodynamic forces) are not very high, the macroscopic (thermodynamic) fluxes can be expressed as linear functions of these gradients. The part of thermodynamics that is based on this assumption, is called the linear irreversible thermodynamics.

There are many processes where the assumption of linearity can be applied, for instance Fourier's law of thermal conductivity, Ohm's law in the theory of electricity, the Navier–Stokes law of viscosity, etc. However, these examples are valid only for so-called pure phenomena, i.e. pure thermal conductivity, pure electrical conductivity, pure viscous effects, etc. In reality, existence of pure processes, where only one thermodynamic force is acting, is exceptional. Mostly, the effects of particular thermodynamic forces are combined, i.e. within the framework of the linear theory superposed. Thus, for instance, the temperature gradient results not only in heat flux but also in electric current. Consequently, every macroscopic (thermodynamic) flux can be expressed as a linear function (superposition) of the gradients of the particular macroscopic quantities (of the thermodynamic forces). The main aim of the phenomenological theory of irreversible processes is the identification and formulation of relations between the particular fluxes and forces. We will now show the methods that are employed in this identification process.

We assume that a thermodynamic system is characterized by n state parameters, and denote $\psi_1, ..., \psi_n$ the deviations of state parameters from their equilibrium values. Thus, the quantities $\psi_1, ..., \psi_n$ characterize the level of irreversibility of the given system. For the sake of simplicity, we will assume that the system is closed, i.e. $\vec{j}_s = 0$, and also that the internal heat sources I_Q are equal to zero, which leads to $b_s = 0$.

The entropy S is a function of parameters ψ_i. As ψ_i are functions of time, we can write for the entropy production rate Γ defined by (2.120) the following:

$$\Gamma = \frac{dS}{dt} = \frac{\partial S}{\partial \psi_i} \frac{d\psi_i}{dt}, \tag{2.258}$$

where the same indices i denote again the Einstein summation.

In the sense of our discussion from the previous subsection, the negative time rate $-\frac{d\psi_i}{dt}$ is related to a transfer of some thermodynamic quantity, and therefore it is the i-th thermodynamic flux J_i,

$$J_i = -\frac{\mathrm{d}\psi_i}{\mathrm{d}t}. \tag{2.259}$$

The derivative $-\frac{\partial S}{\partial \psi_i}$ is then a thermodynamic force X_i corresponding to the parameter ψ_i,

$$X_i = -\frac{\partial S}{\partial \psi_i}. \tag{2.260}$$

Substituting (2.259) and (2.260) in (2.258) we obtain

$$\Gamma = J_i X_i. \tag{2.261}$$

In further derivation, we will assume that the system of quantities ψ_i fully characterizes the approaching of the system to the state of thermodynamic equilibrium and that in the course of this process no other deviations than ψ_i occur. In that case the time rate $\frac{\mathrm{d}\psi_i}{\mathrm{d}t}$ of the quantities ψ_i is a function of parameters ψ_i in that instant of time,

$$\frac{\mathrm{d}\psi_i}{\mathrm{d}t} = -J_i = -J_i(\psi_1, \psi_2, ..., \psi_n). \tag{2.262}$$

For small deviations from the equilibrium, i.e. $\psi_i \to 0$, we can employ a linear approximation in the form

$$-\frac{\mathrm{d}\psi_i}{\mathrm{d}t} = J_i = \beta_{ik}\psi_k, \quad i = 1, 2, ..., n, \tag{2.263}$$

where

$$\beta_{ik} = \left(\frac{\partial J_i}{\partial \psi_k}\right)_0 \tag{2.264}$$

are constant coefficients. We note that equation (2.263) cannot contain terms of zero order, because for

$$\psi_1 = \psi_2 = ... = \psi_n = 0,$$

i.e. for the state of equilibrium, we have

$$\frac{\mathrm{d}\psi_i}{\mathrm{d}t} = 0.$$

Our next aim will be the incorporation of the generalized thermodynamic forces X_i into the relation (2.263) for $\frac{\mathrm{d}\psi_i}{\mathrm{d}t}$. In the state of thermodynamic equilibrium all forces X_i have to be equal to zero. For small deviations from the equilibrium we can employ again the expansion with only linear terms,

$$X_i(\psi_1, \psi_2, ..., \psi_n) = \phi_{ik}\psi_k, \tag{2.265}$$

where

$$\phi_{ik} = \left(\frac{\partial X_i}{\partial \psi_k}\right)_0 = \left(\frac{\partial^2 S}{\partial \psi_i \psi_k}\right)_0. \tag{2.266}$$

From the system of equations (2.265) we can express $\psi_i = \psi_i(X_1, X_2, ..., X_n)$,

$$\psi_i(X_1, X_2, ..., X_n) = \chi_{ik} X_k. \tag{2.267}$$

Substituting (2.267) into (2.263) we obtain

$$\frac{d\psi_i}{dt} = -J_i = -L_{ik} X_k, \tag{2.268}$$

where the coefficients L_{ik},

$$L_{ik} = \beta_{ij} \chi_{jk}, \tag{2.269}$$

are called the kinetic coefficients, which obey the well-known Onsager reciprocity relation,

$$L_{ik} = L_{ki}. \tag{2.270}$$

Taking into account the definition of the kinetic coefficients, we can see that the Onsager relations (2.270) bind together the i-th and k-th components of a thermodynamic flux.

Substituting (2.268) into (2.261) we get the final relation for the entropy production of a closed system,

$$\Gamma = L_{ik} X_i X_k. \tag{2.271}$$

2.3.3 Examples of Constitutive Equations

The advanced continuum theories of constitutive equations, which have been developed within the last 30–40 years, begin with very general functional constitutive equations. They aim to determine the limits imposed on the forms of the equations by certain general principles, and specialize the equations as late as possible and as little as possible. This approach has the advantage of not overlooking coupling effects between different kind of behaviour (e.g. thermal, mechanical and electrical) and also provides many general results applicable to all of the possible specializations. These general theories are very useful in formulating new constitutive equations on the basis of observed macroscopic properties of a material or from physical theories of the molecular behaviour. They can help considerably after the pertinent constitutive variables have been selected on the basis of experience and intuition. The theories can restrict the possible forms of the functional dependence on the assumed independent constitutive variables. This reduces the scope of the required experimental exploitation of the material properties and may also reveal some of the limitations on the applicability of a formulation.

We will not give details of the general theories of constitutive equations here, for that is beyond the scope of this book, which is aimed at transport processes in concrete only. Rather, we will give some examples of simple constitutive equations describing various kinds of ideal material response.

2.3.3.1 Heat Conduction in a One-Component Rigid Material

Heat conduction in a one-component rigid material can be expressed by one of the most simple constitutive equations, Fourier's law,

$$\vec{j}_Q = -\lambda \operatorname{grad} T, \tag{2.272}$$

where $\vec{j}_Q$ is the heat flux, T the temperature and λ the thermal conductivity.

2.3.3.2 Elastic Material in Isothermal Conditions

A material is called ideally elastic if a body formed of the material recovers its original form completely upon removal of forces causing the deformation, and there is a one-to-one relationship between the state of stress and the state of strain, for a given temperature. Therefore, no creep at constant load or stress relaxation at constant strain are included in the classical theory of elasticity.

In the case where strains are sufficiently small and the geometry change is negligible, we can assume a linear relation between the reversible parts of the stress tensor $\mathbf{T^R}$ and the strain tensor $\mathbf{U^R}$,

$$\mathbf{T^R} = \mathbf{C} : \mathbf{U^R}, \tag{2.273}$$

or in the component form

$$T_{ij}^R = C_{ijkm} U_{km}^R. \tag{2.274}$$

The proportionality factor $\mathbf{C}$ is a fourth-order tensor, which has 81 components and is mostly called the modulus of elasticity tensor. The relations (2.273) or (2.274) are often called the generalized Hooke's law. These nine equations expressing the stress components as linear homogeneous functions of the nine strain components are the classical elastic constitutive equations.

Not all of the 81 components of the tensor $\mathbf{C}$ are independent. Taking into account the symmetry of the tensors $\mathbf{T^R}$, $\mathbf{U^R}$ and $\mathbf{C}$ (see, e.g., Eringen, 1967), we obtain

$$C_{ijkm} = C_{jikm} = C_{ijmk} = C_{kmij}, \tag{2.275}$$

and therefore only 21 components of the $\mathbf{C}$ tensor are independent.

For the practical applications, the number of 21 independent constants describing the elastic behaviour of a material is still too high. Therefore, the principles of material symmetry are further applied. If the material is elastically isotropic, i.e. if there are no preferred directions in the material, it can be shown that among the 81 components of the **C** tensor 60 are equal to zero, and the remaining 21 can be expressed using just two independent constants λ and μ, which are called <u>Lamé elastic constants</u>. Then, we can write for the components of the **C** tensor

$$C_{ijkm} = \lambda \delta_{ij}\delta_{km} + \mu(\delta_{jk}\delta_{im} + \delta_{ik}\delta_{jm}) \tag{2.276}$$

or

$$C_{iiii} = \lambda + 2\mu, \quad i = 1, 2, 3 \tag{2.277}$$

$$C_{iijj} = \lambda, \quad i, j = 1, 2, 3 \tag{2.278}$$

$$C_{ijij} = C_{ijji} = \mu, \quad i, j = 1, 2, 3. \tag{2.279}$$

We note that in equations (2.277)–(2.279) the same indices exceptionally do not express a summation but a fact, that the same equation is valid for all indices i, j from 1 to 3.

Substituting (2.276) into Hooke's law (2.274) we obtain

$$T_{ij}^{\mathbf{R}} = \lambda \delta_{ij} U_{kk}^{\mathbf{R}} + 2\mu U_{ij}^{\mathbf{R}}, \tag{2.280}$$

where again k is the summation index according to the Einstein summation.

The second Lamé constant μ is often called the <u>shear modulus</u> and denoted as G. Besides the Lamé elastic constants, three other more familiar constants are commonly employed in technical practice, namely the <u>bulk modulus</u> K, <u>Young's modulus</u> E and the <u>Poisson's ratio</u> σ. These can be expressed as follows:

$$K = \lambda + \frac{2}{3}\mu \tag{2.281}$$

$$E = \frac{9K\mu}{3K + \mu} \tag{2.282}$$

$$\sigma = \frac{3K - 2\mu}{2(3K + \mu)}. \tag{2.283}$$

Using these relations, Hooke's law can also be expressed by other pairs of the elastic constants than (λ, μ), namely (K, G) and (E, σ), which in technical practice is a more common case. We can write

$$T_{ij}^{\mathbf{R}} = K\delta_{ij}U_{kk}^{\mathbf{R}} + 2G\left(U_{ij}^{\mathbf{R}} - \frac{1}{3}\delta_{ij}U_{kk}^{\mathbf{R}}\right) \tag{2.284}$$

$$T_{ij}^{\mathbf{R}} = \frac{E}{1+\sigma}\left(U_{ij}^{\mathbf{R}} + \delta_{ij}\frac{\sigma}{1-2\sigma}U_{kk}^{\mathbf{R}}\right). \tag{2.285}$$

The equation (2.284) can be formally rewritten using the decomposition of stress and strain tensors into their scalar and deviatoric parts. We obtain

$$T^{\mathbf{R}} = 3K\,U^{\mathbf{R}} \tag{2.286}$$

for the scalar parts and

$$^{\mathrm{d}}\mathbf{T}^{\mathbf{R}} = 2G\,^{\mathrm{d}}\mathbf{U}^{\mathbf{R}} \tag{2.287}$$

for the deviatoric parts.

2.3.3.3 Viscoelastic Material in Isothermal Conditions

For classical elastic materials, the stress tensor was proportional to the strain tensor and the Hooke's law was valid for all elastic materials. In the case of viscoelastic materials, the elastic and viscous characteristics are combined, and this combination can be done in several different ways.

The most simple is the Kelvin–Voigt model, which assumes that the stress tensor can be divided into the reversible and irreversible (dissipative) parts,

$$\mathbf{T} = \mathbf{T}^{\mathbf{R}} + \mathbf{T}^{\mathbf{D}}. \tag{2.288}$$

For the reversible part of the stress tensor $\mathbf{T}^{\mathbf{R}}$, we can apply Hooke's law in the same form as for classical elastic materials. The dissipative part $\mathbf{T}^{\mathbf{D}}$ can be expressed as function of the rate of deformation tensor,

$$\mathbf{D}^{\mathbf{R}} = \frac{\mathrm{d}\mathbf{U}^{R}}{\mathrm{d}t}. \tag{2.289}$$

In the case of small velocities, this function can be approximated by a linear relation,

$$\mathbf{T}^{\mathbf{D}} = \mathbf{C}^{\mathrm{v}}\frac{\mathrm{d}\mathbf{U}^{\mathbf{R}}}{\mathrm{d}t} \tag{2.290}$$

or in components

$$T_{ij}^{\mathbf{D}} = C_{ijkm}^{\mathrm{v}}\frac{\mathrm{d}U_{km}^{\mathbf{R}}}{\mathrm{d}t}. \tag{2.291}$$

The proportionality constant $\mathbf{C}^{\mathrm{v}}$ is again a fourth-order tensor, which is called the viscosity tensor. Equation (2.290) is called the Navier–Stokes law or Newton's law of viscosity. The viscosity tensor $\mathbf{C}^{\mathrm{v}}$ has with regard to symmetry the same properties as the modulus of elasticity tensor $\mathbf{C}$, so that for isotropic materials its 81 components can be reduced to just two constants, λ^{v}, μ^{v}, called the first and the second coefficient of viscosity,

$$C_{ijkm}^{\mathrm{v}} = \lambda^{\mathrm{v}}\delta_{ij}\delta_{km} + \mu^{\mathrm{v}}(\delta_{jk}\delta_{im} + \delta_{ik}\delta_{jm}). \tag{2.292}$$

In practical applications, the pair of coefficients (λ^v, μ^v) is not very common. More often is the application of the bulk viscosity κ and the dynamic viscosity η, which can be expressed as

$$\kappa = \lambda^v + \frac{2}{3}\mu^v \tag{2.293}$$

$$\eta = \mu^v. \tag{2.294}$$

Substituting (2.292) into the Navier–Stokes law (2.291) we obtain

$$T_{ij}^D = \lambda^v \delta_{ij} \frac{dU_{kk}^R}{dt} + 2\mu^v \frac{dU_{ij}^R}{dt}. \tag{2.295}$$

For the pair of constants (κ, η) we have

$$T_{ij}^D = \kappa \delta_{ij} \frac{dU_{kk}^R}{dt} + 2\eta \left(\frac{dU_{ij}^R}{dt} - \frac{1}{3}\delta_{ij} \frac{dU_{kk}^R}{dt} \right). \tag{2.296}$$

In equation (2.296), the last term on the right-hand side in parenthesis is the deviator of the rate of deformation tensor $\mathbf{D^R}$, the first term includes the scalar part of $\mathbf{D^R}$. Therefore, we can formally write for the scalar parts

$$T^D = 3\kappa \, D^R \tag{2.297}$$

and for the deviatoric parts

$$^d\mathbf{T^D} = 2\eta \, ^d\mathbf{D^R}. \tag{2.298}$$

The Kelvin–Voigt model of a viscoelastic material can then be expressed by adding the equations (2.286), (2.297) for the scalar parts and (2.287), (2.298) for the deviatoric parts,

$$T = 3K \, U^R + 3\kappa \, D^R \tag{2.299}$$

$$^d\mathbf{T} = 2G \, ^d\mathbf{U^R} + 2\eta \, ^d\mathbf{D^R}. \tag{2.300}$$

We will now analyse the deviatoric part of the Kelvin–Voigt constitutive equation. We introduce the retardation or relaxation time τ as

$$\tau = \frac{\eta}{G}. \tag{2.301}$$

Substituting (2.301) into (2.300) we obtain

$$^d\mathbf{T} = 2G \, (^d\mathbf{U^R} + \tau \, ^d\mathbf{D^R}). \tag{2.302}$$

Using the definition relation for the rate of deformation tensor (2.38) in (2.302), we get the following solution for the deviator of the reversible part of the strain tensor $\mathbf{U^R}$:

$$^d\mathbf{U^R} = \frac{^d\mathbf{T_0}}{2G}(1 - e^{-t/\tau}), \tag{2.303}$$

which is called the Kelvin–Voigt retarded elasticity. The meaning of the expression retarded elasticity immediately follows from equation (2.303). It shows that in the Kelvin–Voigt model, there is no instantaneous elastic response (glassy response), but if stress $^d\mathbf{T}_0$ is applied at $t = 0$ and is held constant, the equilibrium strain is approached in an asymptotic way.

The Kelvin–Voigt model is one of so called spring and dashpot models, where $2G$ acts as a spring and 2η as a dashpot. The Kelvin–Voigt element presents a parallel arrangement of a spring and a dashpot. Therefore, the displacement is the same for the spring and the dashpot, and the total axial force (total stress) is the sum of the spring force and the dashpot force, as expressed by the relation (2.288).

Second basic spring and dashpot model is the Maxwell model, where the spring and the dashpot are arranged as series. Here, the force is the same for both spring and dashpot, but the displacements or velocities are added. For the deviatoric parts we have

$$^d\mathbf{D} = {}^d\mathbf{D}^\mathbf{R} + {}^d\mathbf{D}^\mathbf{D} = \frac{1}{2G}\frac{d^d\mathbf{T}^\mathbf{R}}{dt} + \frac{1}{2\eta}{}^d\mathbf{T}^\mathbf{R}, \tag{2.304}$$

or using the relaxation time

$$2G\,{}^d\mathbf{D} = \frac{d^d\mathbf{T}^\mathbf{R}}{dt} + \frac{1}{\tau}\,{}^d\mathbf{T}^\mathbf{R}. \tag{2.305}$$

The integration of equation (2.304) leads to

$$^d\mathbf{U} = \frac{1}{2G}{}^d\mathbf{T} + \frac{1}{2\eta}{}^d\mathbf{T}\cdot t, \tag{2.306}$$

which means that under constant stress $^d\mathbf{T}$, after an instantaneous glassy response, steady creep takes place, where strain is a linear function of time.

The solution of equation (2.305) for the deviator of the stress tensor $\mathbf{T}$ leads to the following expression:

$$^d\mathbf{T}^\mathbf{R} = 2G\,{}^d\mathbf{U}_0 \cdot e^{-t/\tau}, \tag{2.307}$$

where $^d\mathbf{U}_0$ is the strain suddenly applied at $t = 0$. Equation (2.307) expresses the Maxwell relaxation, where zero stress is approached asymptotically.

2.3.3.4 Plastic Material in Isothermal Conditions

Contrary to the elastic and viscous models, the constitutive equations for plasticity are not well established yet, although extensive work in the field has been done, particularly in the last few decades. Several different idealizations have been proposed but all of them fail to account completely for the phenomena observed experimentally.

The basic idealized stress–strain relations, where σ denotes the stress and ϵ the strain in a one-dimensional case, are the following:

1. Rigid, perfectly plastic material

$$\sigma = Y, \tag{2.308}$$

where Y is the yield stress, i.e. the elastic limit stress or in other words the minimum value of stress permitting inelastic response of a material.

2. Elastic, perfectly plastic material

$$\sigma = E\epsilon \qquad \epsilon < \epsilon_e \tag{2.309}$$

$$\sigma = Y \qquad \epsilon \geq \epsilon_e, \tag{2.310}$$

where ϵ_e is the maximum elastic strain, E is the elasticity modulus.

3. Rigid, work-hardening material

$$\sigma = Y + E_2\epsilon, \tag{2.311}$$

where E_2 is the elasticity modulus in the work-hardening plastic range, i.e. in the post-yield conditions, where a permanent strain ϵ_p remains after unloading and where stress increases much more slowly with strain than in the elastic range, $E_2 << E$.

4. Elastic, work-hardening material

$$\sigma = E\epsilon \qquad \epsilon < \epsilon_e \tag{2.312}$$

$$\sigma = Y + E_2\epsilon \qquad \epsilon \geq \epsilon_e. \tag{2.313}$$

As follows from this basic survey, three features can be observed on the stress–strain curves. An initial steeply rising linear elastic range is followed by yield and then a work-hardening plastic range.

Any constitutive theory of plasticity must specify the yield condition under combined stresses and the post-yield behavior. We will discuss the simplest models only to illustrate the basic ways of solving the plasticity problems.

The Levy–Mises perfectly plastic constitutive equations read as follows:

$$^d\mathbf{T} = \frac{k}{\sqrt{II_{\mathbf{D}^\mathbf{D}}}}\,{}^d\mathbf{D}^\mathbf{D}, \tag{2.314}$$

where k is a constant and

$$II_{\mathbf{D}^\mathbf{D}} = \frac{1}{2}(D_{ij}^\mathbf{D}D_{ij}^\mathbf{D} - D_{ii}^\mathbf{D}D_{jj}^\mathbf{D}). \tag{2.315}$$

Further, plastic incompressibility is assumed, i.e.

$$D_{ii}^\mathbf{D} = 0, \tag{2.316}$$

and therefore

$$D_{ij}^\mathbf{D} = {}^d D_{ij}^\mathbf{D}. \tag{2.317}$$

The Mises yield condition immediately following from (2.314) is formulated as

$$\frac{1}{2}{}^{d}T_{ij}{}^{d}T_{ij} = k^2.$$

$$(2.318)$$

The rates of deformations D_{ij}^{D} are interpreted as the rates of change of the plastic (dissipative) part of the strains, since the elastic strains are equal to zero, as follows from (2.314). As k is assumed to be constant, the yield condition does not change with plastic deformation in this perfectly plastic theory.

Prandtl–Reuss elastoplastic constitutive equations can be formulated as

$$dU_{ij}^{D} = {}^{d}T_{ij} \cdot d\lambda,$$

$$(2.319)$$

where

$$d\lambda = \frac{\sqrt{II_{\mathbf{U^D}}}}{k},$$

$$(2.320)$$

$$II_{\mathbf{U^D}} = \frac{1}{2}(U_{ij}^{D}U_{ij}^{D} - U_{ii}^{D}U_{jj}^{D}),$$

$$(2.321)$$

$d\mathbf{U^D}$ is the increment of the dissipative (plastic) part of the total strain $d\mathbf{U}$, $d\mathbf{U} = d\mathbf{U^R} + d\mathbf{U^D}$, and the increment of the elastic part of the total strain $d\mathbf{U^R}$ is determined by the inverse Hooke's law. The Mises yield condition (2.318) is adopted in the Prandtl–Reuss model as well as in the Levy–Mises model.

The constitutive equation (2.319) is expressed in incremental form and cannot be directly integrated. Therefore, it is more convenient to write it in terms of rates of deformation, which follows from the time differentiation of (2.319),

$$D_{ij}^{D} = \frac{d\lambda}{dt}{}^{d}T_{ij},$$

$$(2.322)$$

where

$$\frac{d\lambda}{dt} = \frac{\sqrt{II_{\mathbf{D^D}}}}{k}.$$

$$(2.323)$$

Prager viscoplastic constitutive equations can be written as

$${}^{d}T_{ij} = \left(2\eta + \frac{k}{\sqrt{II_{\mathbf{D^D}}}}\right) D_{ij}^{D},$$

$$(2.324)$$

where

$$II_{\mathbf{D^D}} = \frac{1}{2}(D_{ij}^{D}D_{ij}^{D} - D_{ii}^{D}D_{jj}^{D})$$

$$(2.325)$$

and again the Mises yield condition is assumed to be valid.

2.4 BALANCE LAWS FOR A LINEAR CHEMICALLY REACT-ING MIXTURE WITH DISCONTINUITIES

2.4.1 Fundamentals of the Linear Theory of Mixtures

A mixture is a system of $n \geq 2$ deformable materials, each of which is regarded as a continuum; the i-th component is denoted by A_i. The continuum theory of mixtures (see, e.g., Atkin and Craine, 1975, 1976) is based on the fundamental assumption that the mixture can be viewed as a superposition of n single continua, each following its own motion, and that at any time t each place $\vec{x}$ in the mixture is occupied simultaneously by several different particles $\vec{X}^i$, one from each component A_i. Therefore, the equation of motion of the mixture is described by the n equations

$$\vec{x}^i = \vec{x}^i(\vec{X}^i, t). \tag{2.326}$$

We define the partial density of a constituent i as

$$\rho_i = \frac{\mathrm{d}M_i}{\mathrm{d}V}, \tag{2.327}$$

where M_i is the mass of the component i, V the total volume. The total density of the mixture ρ can then be expressed as

$$\rho = \sum_i \rho_i. \tag{2.328}$$

Similarly, the velocity of the i-th component is given by

$$\vec{v}^i = \frac{\mathrm{d}\vec{x}^i}{\mathrm{d}t}. \tag{2.329}$$

It is convenient to introduce a mean or barycentric velocity $\vec{v}$ of the mixture. It is defined by the requirement that the total mass flow is the sum of the individual mass flows, so that

$$\rho\vec{v} = \sum_i \rho_i\vec{v}^i. \tag{2.330}$$

The velocity of a component A_i related to the barycentric velocity is called the diffusion velocity $\vec{u}^i$,

$$\vec{u}^i = \vec{v}^i - \vec{v}, \tag{2.331}$$

the quantity

$$\vec{j}_\mathrm{D}^i = \rho_i\vec{u}^i \tag{2.332}$$

is the diffusion flux of a component i.

It should be noted that the n diffusion fluxes are not independent. Substituting (2.331) into (2.330) and using (2.328) we get

$$\sum_i \rho_i \vec{u}^i = 0. \tag{2.333}$$

The material time derivatives are defined as

$$\frac{\mathrm{d}^i A}{\mathrm{d}t} = \frac{\partial A}{\partial t} + \vec{v}^i \cdot \operatorname{grad} A \tag{2.334}$$

$$\frac{\mathrm{d}A}{\mathrm{d}t} = \frac{\partial A}{\partial t} + \vec{v} \cdot \operatorname{grad} A, \tag{2.335}$$

where the derivative $\frac{\mathrm{d}A}{\mathrm{d}t}$ follows the mean motion of the mixture whilst $\frac{\mathrm{d}^i A}{\mathrm{d}t}$ follows the motion of the component i. From (2.334), (2.335) and (2.331) we get

$$\frac{\mathrm{d}^i A}{\mathrm{d}t} = \frac{\partial A}{\partial t} + \vec{u}^i \cdot \operatorname{grad} A. \tag{2.336}$$

Summing (2.336) over all i and using (2.328) and (2.333) we obtain

$$\sum_i \rho_i \frac{\mathrm{d}^i A}{\mathrm{d}t} = \rho \frac{\mathrm{d}A}{\mathrm{d}t}. \tag{2.337}$$

We note that all additive quantities follow the same superposition rule (2.328) as the densities, for instance the partial pressures, partial densities of internal energy and enthalpy, etc.

2.4.2 Mass Balance

The occurrence of chemical reactions and/or phase change processes between the components can change the mass of any component of a mixture. Therefore, the mass of an individual component A_i need not to be conserved and for its balance we can write

$$\frac{\mathrm{d}M_i}{\mathrm{d}t} = \frac{\mathrm{d}}{\mathrm{d}t} \int_{V-\sigma} \rho_i \, \mathrm{d}V = \int_{V-\sigma} m_i \, \mathrm{d}V, \tag{2.338}$$

where m_i is the rate of mass production of a component i in kg/m^{3}s. Assuming that j chemical reactions appear in the system, we can write

$$m_i = \sum_i \theta_{ij} J_j, \tag{2.339}$$

where J_j is the chemical reaction rate of the j-th reaction, the coefficients θ_{ij} are proportional to the stoichiometric coefficients with which component

i appears in the j-th chemical reaction, and to the molar mass of the i-th component.

In deriving the local balance equations for the volume $V - \sigma$ and the discontinuity surface σ, we employ slightly modified forms of the unified balance law (2.124) and the master balance laws (2.130) and (2.131), where A has to be replaced by A_i, $\vec{v}$ by $\vec{v}^i$, and the additional source terms A_{Ri}, A_{SRi}, appearing due to the interactions between the components, have to be included in the volume $V - \sigma$ and on the discontinuity surface σ, respectively. We can write these modified equations in the form

$$\frac{d}{dt} \int_{V-\sigma} A_i \, dV = \int_{S-\sigma} \vec{A}_{Si} \cdot d\vec{S} + \int_V A_{Ii} \, dV + \int_{V-\sigma} A_{Ri} \, dV \quad (2.340)$$

$$\frac{\partial A_i}{\partial t} + \text{div} \, (A_i \vec{v}^i) - \text{div} \, \vec{A}_S^i - A_{Ii} = A_{Ri} \qquad \text{in } V - \sigma \qquad (2.341)$$

$$[\![A_i(\vec{v}^i - \vec{v}) - \vec{A}_S^i]\!] \cdot \vec{n} = A_{SR}^i \qquad \text{on } \sigma. \qquad (2.342)$$

Comparing (2.338) with (2.340) we identify

$$A_i = \rho_i \qquad (2.343)$$

$$\vec{A}_S^i = 0 \qquad (2.344)$$

$$A_{Ii} = 0 \qquad (2.345)$$

$$A_{Ri} = m_i, \qquad (2.346)$$

and using (2.342) we get

$$A_{SRi} = m_{Si}, \qquad (2.347)$$

where m_{Si} is the surface mass source of the i-th component.

Substituting (2.343)–(2.347) into (2.341) and (2.342) we obtain

$$\frac{\partial \rho_i}{\partial t} + \text{div} \, (\rho_i \vec{v}^i) = m_i \qquad \text{in } V - \sigma \qquad (2.348)$$

$$[\![\rho_i(\vec{v}^i - \vec{v})]\!] \cdot \vec{n} = m_{Si} \qquad \text{on } \sigma. \qquad (2.349)$$

Summing (2.348) and (2.349) over all i components, we obtain the total mass balance of the mixture,

$$\sum_i \frac{\partial \rho_i}{\partial t} + \text{div} \sum_i \rho_i \vec{v}^i = \sum_i m_i \qquad \text{in } V - \sigma \qquad (2.350)$$

$$\sum_i [\![\rho_i(\vec{v}^i - \vec{v})]\!] \cdot \vec{n} = \sum_i m_{Si} \qquad \text{on } \sigma. \tag{2.351}$$

Using (2.328) and (2.330) then

$$\frac{\partial \rho}{\partial t} + \text{div} \, (\rho \vec{v}) = \sum_i m_i \qquad \text{in } V - \sigma \tag{2.352}$$

$$[\![\rho(\vec{v} - \vec{v})]\!] \cdot \vec{n} = \sum_i m_{Si} \qquad \text{on } \sigma. \tag{2.353}$$

For the mixture as a whole, the conservation of mass must be valid. In other words, whilst we allow for interchange of mass between the components, the creation of additional mass in the mixture is forbidden. Therefore,

$$\sum_i m_i = 0 \tag{2.354}$$

$$\sum_i m_{Si} = 0, \tag{2.355}$$

and the balance equations (2.352) and (2.353) are identical with the balance equations of mass of a single continuum (2.138) and (2.139).

2.4.3 Momentum Balances

Before stating the balance law of momentum for A_i, we will consider the forces acting on this component within the volume $V - \sigma$ and outside this volume. By analogy to the balance equation for a single continuum, we introduce the volume force density $\vec{f}^i$ acting on A_i, and the partial stress tensor $\mathbf{T}^i$, which plays a similar role in the theory of mixtures as the stress tensor for a single continuum. Two additional effects not appearing in the balance equation for a single continuum have to be introduced. The first is the momentum supplied to A_i due to chemical reactions with the other components, and the second is the transfer of momentum due to the other interaction effects, for example the relative motion of the components.

Taking into account the above remarks, we can write the momentum balance for A_i in the form

$$\frac{\mathrm{d}}{\mathrm{d}t} \int_{V-\sigma} \rho_i \vec{v}^i \, \mathrm{d}V = \int_{S-\sigma} \mathbf{T}^i \cdot \mathrm{d}\vec{S} + \int_{V-\sigma} \vec{f}^i \, \mathrm{d}V$$

$$+ \int_{V-\sigma} m_i \vec{J}^i \, \mathrm{d}V + \int_{V-\sigma} \vec{p}^i \, \mathrm{d}V, \tag{2.356}$$

where $\vec{J}^i$ has the dimension of velocity, and $\vec{p}^i$ is the diffusive force exerted on A_i by the other components.

The local balance equations for the volume $V - \sigma$ and the discontinuity surface σ will be derived by an application of the unified balance law (2.125) and the master balance laws (2.132) and (2.133) modified in a similar way as the corresponding equations for the scalar quantities in the previous subsection. Again, $\vec{A}$ has to be replaced by $\vec{A}^i$, $\vec{v}$ by $\vec{v}^i$, and the additional source terms $\vec{A}^i_R$, $\vec{A}^i_{SR}$, appearing due to the interactions between the components, have to be included in the volume $V - \sigma$ and on the discontinuity surface σ, respectively. We can write the following:

$$\frac{d}{dt} \int_{V-\sigma} \vec{A}^i \, dV = \int_{S-\sigma} \mathbf{A}^i_S \cdot d\vec{S} + \int_V \vec{A}^i_I \, dV + \int_{V-\sigma} \vec{A}^i_R \, dV \qquad (2.357)$$

$$\frac{\partial \vec{A}^i}{\partial t} + \text{div} \, (\vec{A}^i \otimes \vec{v}^i) - \text{div} \, \mathbf{A}^i_S - \vec{A}^i_I = \vec{A}^i_R \qquad \text{in } V - \sigma \qquad (2.358)$$

$$[\![\vec{A}^i \otimes (\vec{v}^i - \vec{v}) - \mathbf{A}^i_S]\!] \cdot \vec{n} = \vec{A}^i_{SR} \qquad \text{on } \sigma. \qquad (2.359)$$

Comparing (2.356) with the master law (2.357), we get

$$\vec{A}^i = \rho_i \vec{v}^i \qquad (2.360)$$

$$\mathbf{A}^i_S = \mathbf{T}^i \qquad (2.361)$$

$$\vec{A}^i_I = \vec{f}^i. \qquad (2.362)$$

$$\vec{A}^i_R = m_i \vec{J}^i + \vec{p}^i, \qquad (2.363)$$

and using (2.359) we obtain

$$\vec{A}^i_{SR} = m_{Si} \vec{J}^i_S + \vec{p}^i_S. \qquad (2.364)$$

Substituting (2.360)–(2.364) into (2.358) and (2.359), we get

$$\frac{\partial}{\partial t} (\rho_i \vec{v}^i) + \text{div} \, (\rho_i \vec{v}^i \otimes \vec{v}^i) - \text{div} \, \mathbf{T}^i - \vec{f}^i$$

$$= m_i \vec{J}^i + \vec{p}^i \qquad \text{in } V - \sigma \qquad (2.365)$$

$$[\![\rho_i \cdot \vec{v}^i \otimes (\vec{v}^i - \vec{v}) - \mathbf{T}^i]\!] \cdot \vec{n} = m_{Si} \vec{J}^i_S + \vec{p}^i_S \qquad \text{on } \sigma. \qquad (2.366)$$

The momentum balance equation of a component i (2.365) can be expressed in the form of an equation of motion. We follow a similar treatment as in the case of a single continuum and modify the first two terms of equation (2.365), namely

$$\frac{\partial}{\partial t}(\rho_i \vec{v}^i) + \text{div}\,(\rho_i \vec{v}^i \otimes \vec{v}^i) = \rho_i \frac{\partial \vec{v}^i}{\partial t} + \vec{v}^i \frac{\partial \rho_i}{\partial t}$$

$$+ \rho_i \vec{v}^i \cdot \text{grad}\,\vec{v}^i + \vec{v}^i \cdot \text{div}\,\rho_i \vec{v}^i. \tag{2.367}$$

The sum of the second and the fourth term on the right-hand side of (2.367) is equal to $m_i \vec{v}^i$, taking into account the mass balance equation (2.348). The sum of the first and the third term is clearly the material derivative of $\vec{v}^i$ multiplied by ρ^i. Thus, the equation of motion can be written as

$$\rho_i \frac{\mathrm{d}^i \vec{v}^i}{\mathrm{d}t} = \vec{f}^i + \text{div}\,\mathbf{T}^i + m_i \vec{J}^i + \vec{p}^i - m_i \vec{v}^i. \tag{2.368}$$

The total momentum balances of the mixture we obtain by summing (2.365) and (2.366) over all components i,

$$\frac{\partial}{\partial t}\left(\sum_i \rho_i \vec{v}^i\right) + \text{div}\,\left(\sum_i \rho_i \vec{v}^i \otimes \vec{v}^i\right) - \text{div}\,\sum_i \mathbf{T}^i - \sum_i \vec{f}^i$$

$$= \sum_i (m_i \vec{J}^i + \vec{p}^i) \qquad \text{in } V - \sigma \tag{2.369}$$

$$\left[\!\left[\sum_i [\rho_i \cdot \vec{v}^i \otimes (\vec{v}^i - \vec{v})] - \sum_i \mathbf{T}^i\right]\!\right] \cdot \vec{n} = \sum_i (m_{\mathrm{S}i} \vec{J}_{\mathrm{S}}^i + \vec{p}_{\mathrm{S}}^i) \qquad \text{on } \sigma. \tag{2.370}$$

Using (2.328), (2.330) and (2.333) we obtain

$$\frac{\partial}{\partial t}(\rho \vec{v}) + \text{div}\,(\rho \vec{v} \otimes \vec{v}) + \sum_i \text{div}\,(\rho_i \vec{u}^i \otimes \vec{u}^i) - \text{div}\,\sum_i \mathbf{T}^i - \sum_i \vec{f}^i$$

$$= \sum_i (m_i \vec{J}^i + \vec{p}^i) \qquad \text{in } V - \sigma \tag{2.371}$$

$$\left[\!\left[\rho \cdot \vec{v} \otimes (\vec{v} - \vec{v}) + \sum_i \rho_i \vec{u}^i \otimes \vec{u}^i - \sum_i \mathbf{T}^i\right]\!\right] \cdot \vec{n}$$

$$= \sum_i (m_{\mathrm{S}i} \vec{J}_{\mathrm{S}}^i + \vec{p}_{\mathrm{S}}^i) \qquad \text{on } \sigma. \tag{2.372}$$

The forms of (2.371) and (2.372) suggest (see Eringen and Ingram, 1965) the introduction of the total stress tensor in the form

$$\mathbf{T} = \sum_i (\mathbf{T}^i - \rho_i \vec{u}^i \otimes \vec{u}^i). \tag{2.373}$$

The force densities acting on the individual components of the mixture are additive, therefore we can write

$$\vec{f} = \sum_i \vec{f^i}. \tag{2.374}$$

As we have assumed overall mass conservation in the mixture, and introduced the diffusive force $\vec{p}^i$ as an internal effect, which naturally does not affect the mixture as a whole, we have

$$\sum_i (m_i \vec{J}^i + \vec{p}^i) = 0 \tag{2.375}$$

and

$$\sum_i (m_{\mathrm{S}i} \vec{J}^i_{\mathrm{S}} + \vec{p}^i_{\mathrm{S}}) = 0. \tag{2.376}$$

Substituting (2.373)–(2.376) into (2.371) and (2.372), we obtain the final balance equations of momentum of the mixture

$$\frac{\partial}{\partial t}(\rho \vec{v}) + \operatorname{div}(\rho \vec{v} \otimes \vec{v}) - \operatorname{div}\mathbf{T} - \vec{f} = 0 \qquad \text{in } V - \sigma \tag{2.377}$$

$$[\![\rho \cdot \vec{v} \otimes (\vec{v} - \vec{v}') - \mathbf{T}]\!] \cdot \vec{n} = 0 \qquad \text{on } \sigma, \tag{2.378}$$

which are formally identical with the balance equations of momentum for a single continuum (2.145) and (2.146).

The global balance of moment of momentum for A_i can be written, taking into account the effects considered in the derivation of the momentum balance, as follows

$$\frac{\mathrm{d}}{\mathrm{d}t} \int_{V-\sigma} \rho_i \cdot (\vec{r} \times \vec{v}^i) \, \mathrm{d}V = \int_{S-\sigma} [\vec{r} \times (\vec{n} \cdot \mathbf{T}^i)] \, \mathrm{d}S + \int_{V-\sigma} (\vec{r} \times \vec{f}^i) \, \mathrm{d}V$$

$$+ \int_{V-\sigma} m_i \vec{r} \times \vec{J}^i \, \mathrm{d}V + \int_{V-\sigma} \vec{r} \times \vec{p}^i \, \mathrm{d}V + \int_{V-\sigma} \epsilon : \lambda^i \, \mathrm{d}V, \tag{2.379}$$

where λ^i is an antisymmetric second-order tensor representing an internal body couple acting on A_i, ϵ is the Levi–Civita tensor, and $\epsilon : \lambda^i$ means the vector with the components $\epsilon_{ijk}\lambda^i_{jk}$. The couple accounts for the change of moment of momentum between A_i and the other components, which does not arise from the interaction terms included in the balance of momentum. This additional change is a purely internal effect and does not affect the global balance equation of the mixture.

Again as in the case of the balance of moment of momentum of a single continuum, the correspondence between the master law in the simple form (2.357) and the global balance of moment of momentum (2.379) is not obvious because of the appearance of vector products in (2.379). Comparing (2.379) with the master law (2.357) we get

$$\vec{A}^i = \rho_i \vec{r} \times \vec{v}^i \tag{2.380}$$

$$\mathbf{A}_S^i \cdot \vec{n} = \vec{r} \times (\vec{n} \cdot \mathbf{T}^i) \tag{2.381}$$

$$\vec{A}_I^i = \vec{r} \times \vec{f}^i \tag{2.382}$$

$$\vec{A}_R^i = \vec{r} \times (m_i \vec{J}^i + \vec{p}^i) + \epsilon : \lambda^i, \tag{2.383}$$

and using (2.359) we obtain

$$\vec{A}_{SR}^i = \vec{r} \times (m_{Si} \vec{J}_S^i + \vec{p}_S^i) + \epsilon : \lambda_S^i. \tag{2.384}$$

The further derivation is quite similar to that performed for the moment of momentum balance for a single component. Therefore, we will not carry it out in such a detailed way as before. The term on the left hand side of equation (2.379) becomes

$$\frac{\mathrm{d}}{\mathrm{d}t} \int_{V-\sigma} \rho_i (\vec{r} \times \vec{v}^i) \, \mathrm{d}V$$

$$= \int_{V-\sigma} \left[\frac{\partial}{\partial t} [\rho_i (\vec{r} \times \vec{v}^i)] + \mathrm{div} \, [\rho_i (\vec{r} \times \vec{v}^i) \otimes \vec{v}^i)] \right] \, \mathrm{d}V$$

$$+ \int_{\sigma} [\rho_i (\vec{r} \times \vec{v}^i) \otimes (\vec{v}^i - \vec{v})] \cdot \mathrm{d}\vec{S}. \tag{2.385}$$

In the first term of the right-hand side of equation (2.385) we use the identity

$$\frac{\partial}{\partial t} [\rho (\vec{r} \times \vec{v}^i] = \vec{r} \times \frac{\partial}{\partial t} (\rho \vec{v}^i), \tag{2.386}$$

the second term becomes

$$\mathrm{div} \, [\rho_i (\vec{r} \times \vec{v}^i) \otimes \vec{v}^i)] = \vec{r} \times \, \mathrm{div} \, (\rho_i \vec{v}^i \otimes \vec{v}^i). \tag{2.387}$$

The first term on the right-hand side of equation (2.379) is modified to the form

$$\int_{S-\sigma} [\vec{r} \times (\vec{n} \cdot \mathbf{T}^i)] \, \mathrm{d}S = \int_{V-\sigma} \vec{e}_m \epsilon_{jkm} T^i_{jk} \, \mathrm{d}V + \int_{V-\sigma} (\vec{r} \times \, \mathrm{div} \, \mathbf{T}^i) \, \mathrm{d}V$$

$$+ \int_{\sigma} [\vec{r} \times (\vec{n} \cdot \mathbf{T}^i)] \, \mathrm{d}S. \tag{2.388}$$

Substituting (2.386), (2.387), (2.388) into (2.379) we get

$$\int_{V-\sigma} \left[\vec{r} \times \left(\rho_i \frac{d^i \vec{v}^i}{dt} - \vec{f}^i - \text{div } \mathbf{T}^i - m_i \vec{J}^i - \vec{p}^i + m_i \vec{v}^i \right) \right] dV$$

$$= \int_{V-\sigma} (\vec{e}_m \epsilon_{jkm} T^i_{jk} + \vec{e}_k \epsilon_{kjm} \lambda^i_{jm}) \, dV \qquad (2.389)$$

and taking into account (2.384) then

$$\int_\sigma [(\vec{r} \times \rho_i \vec{v}^i) \otimes (\vec{v}^i - \vec{v})] \cdot d\vec{S} = \int_\sigma [\vec{r} \times (\vec{n} \cdot \mathbf{T}^i)] \, dS$$

$$+ \int_\sigma (\vec{r} \times (m_{Si} \vec{J}^i_S + \vec{p}^i_S) + \epsilon : \lambda^i_S) \, dS. \qquad (2.390)$$

The term on the left-hand side of equation (2.389) is equal to zero, as follows from the equation of motion (2.368). The resulting global balance equation of moment of momentum can then be written as

$$\int_{V-\sigma} (\vec{e}_m \epsilon_{jkm} T^i_{jk} + \vec{e}_k \epsilon_{kjm} \lambda^i_{jm}) \, dV = 0, \qquad (2.391)$$

which leads to the local balance equation

$$\vec{e}_m \epsilon_{jim} T^i_{ji} + \vec{e}_i \epsilon_{ijm} \lambda^i_{jm} = 0. \qquad (2.392)$$

From the properties of the Levi–Civita tensor it finally follows that

$$\mathbf{T}^{i,A} + \lambda^{i,A} = 0, \qquad (2.393)$$

where the superscript A means the antisymmetric part of a tensor.

The balance equation of the moment of momentum on the discontinuity surface σ (2.390) can be modified under the assumption that the position vector $\vec{r}$ is continuous across the discontinuity surface σ to the form

$$\vec{r} \times [(\rho_i \vec{v}^i \otimes (\vec{v}^i - \vec{v}) - \mathbf{T}^i] \cdot \vec{n} = \vec{r} \times (m_{Si} \vec{J}^i_S + \vec{p}^i_S) + \epsilon : \lambda^i_S. \qquad (2.394)$$

The total balance equation of the moment of momentum of the mixture in $V - \sigma$ can be obtained by summing equation (2.393),

$$\sum_i (\mathbf{T}^{i,A} + \lambda^{i,A}) = 0. \qquad (2.395)$$

As λ^i is a purely internal effect, we have

$$\sum_i \lambda^{i,A} = 0, \qquad (2.396)$$

which after substitution into (2.395) leads to

$$\sum_i \mathbf{T}^{i,A} = 0, \qquad (2.397)$$

and finally to

$$\mathbf{T}^{\mathrm{A}} = 0, \tag{2.398}$$

because the condition

$$\sum_i (\rho_i \vec{u}^i \otimes \vec{u}^i)^{\mathrm{A}} = 0 \tag{2.399}$$

is satisfied automatically. The equation (2.398) is clearly identical with the equation (2.172) representing the balance equation of moment of momentum of a single continuum.

The total balance equation of moment of momentum on the discontinuity surface σ is obtained by summing equation (2.394),

$$\sum_i \vec{r} \times \llbracket (\rho_i \vec{v}^i \otimes (\vec{v}^i - \vec{v}) - \mathbf{T}^i \rrbracket \cdot \vec{n}$$

$$= \sum_i \vec{r} \times (m_{\mathrm{S}i} \vec{J}_{\mathrm{S}}^i + \vec{p}_{\mathrm{S}}^i) + \epsilon : \lambda_{\mathrm{S}}^i. \tag{2.400}$$

The right-hand side of equation (2.400) is equal to zero, because it contains internal effects only. For the modification of the left-hand side, we employ the definition of the total stress tensor (2.373), and obtain finally the relation

$$\vec{r} \times \llbracket (\rho_i \vec{v} \otimes (\vec{v} - \vec{v}) - \mathbf{T} \rrbracket \cdot \vec{n} = 0, \tag{2.401}$$

which is the same as the moment of momentum balance of a single continuum and is satisfied identically.

2.4.4 Energy Balance

Considering the additional effects identified in the balance equations of mass and momentum, the global balance law of energy for A_i can be written in the form

$$\frac{\mathrm{d}}{\mathrm{d}t} \int_{V-\sigma} \left(\rho_i u_i + \frac{1}{2} \rho_i (\vec{v}^i)^2 \right) \mathrm{d}V = \int_{V-\sigma} \vec{f}^i \cdot \vec{v}^i \, \mathrm{d}V$$

$$+ \int_{S-\sigma} (\vec{v}^i \cdot \mathbf{T}^i) \cdot \mathrm{d}\vec{S}$$

$$- \int_{S-\sigma} \vec{j}_{\mathrm{Q}}^i \cdot \mathrm{d}\vec{S} + \int_{V-\sigma} I_{\mathrm{Q}i} \, \mathrm{d}V$$

$$+ \int_{V-\sigma} m_i G_i \, \mathrm{d}V + \int_{V-\sigma} \vec{p}^i \cdot \vec{v}^i \, \mathrm{d}V + \int_{V-\sigma} Q_i, \tag{2.402}$$

where G_i is the specific internal energy of the mass produced by chemical reaction (in J/kg), Q_i the energy source due to the interaction with other components.

Comparing (2.402) with the master law (2.340) we get

$$A_i = \rho_i u_i + \frac{1}{2}\rho(\vec{v}^i)^2 \tag{2.403}$$

$$\vec{A}_S^i = \vec{v}^i \cdot \mathbf{T}^i - \vec{j}_Q^i \tag{2.404}$$

$$A_{Ii} = \vec{f}^i \cdot \vec{v}^i + I_{Qi} \tag{2.405}$$

$$A_{Ri} = m_i G_i + \vec{p}^i \cdot \vec{v}^i + Q_i, \tag{2.406}$$

and using (2.342) we obtain

$$A_{SRi} = m_{Si} G_{Si} + \vec{p}_S^i \cdot \vec{v}^i + Q_{Si}. \tag{2.407}$$

Substituting (2.403)–(2.407) into (2.341) and (2.342) we get

$$\frac{\partial}{\partial t}\left(\rho_i u_i + \frac{1}{2}\rho_i(\vec{v}^i)^2\right) + \text{div}\left[\left(\rho_i u_i + \frac{1}{2}\rho(\vec{v}^i)^2\right)\cdot\vec{v}^i\right]$$

$$- \text{div}\,(\vec{v}^i \cdot \mathbf{T}^i) + \text{div}\,\vec{j}_Q^i - \vec{f}^i \cdot \vec{v}^i - I_{Qi}$$

$$= m_i G_i + \vec{p}^i \cdot \vec{v}^i + Q_i \qquad \text{in } V - \sigma \tag{2.408}$$

$$[\![(\rho_i u_i + \frac{1}{2}\rho_i(\vec{v}^i)^2)\cdot(\vec{v}^i - \vec{v}) - \vec{v}^i \cdot \mathbf{T}^i + \vec{j}_Q^i]\!] \cdot \vec{n}$$

$$= m_{Si} G_{Si} + \vec{p}_S^i \cdot \vec{v}^i + Q_{Si} \qquad \text{on } \sigma. \tag{2.409}$$

The total energy balances can be obtained by summing (2.408) and (2.409) over all i components,

$$\frac{\partial}{\partial t}\left(\sum_i \rho_i u_i + \frac{1}{2}\rho_i(\vec{v}^i)^2\right) + \text{div}\left[\sum_i\left(\rho_i u_i + \frac{1}{2}\rho_i(\vec{v}^i)^2\right)\cdot\vec{v}^i\right]$$

$$- \text{div}\,(\sum_i \vec{v}^i \cdot \mathbf{T}^i) + \text{div}\,\sum_i \vec{j}_Q^i$$

$$- \sum_i \vec{f}^i \cdot \vec{v}^i - \sum_i I_{Qi}$$

$$= \sum_i (m_i G^i + \vec{p}^i \cdot \vec{v}^i + Q_i) \qquad \text{in } V - \sigma \qquad (2.410)$$

$$[\sum_i (\rho_i u_i + \frac{1}{2}\rho_i(\vec{v}^i)^2) \cdot (\vec{v}^i - \vec{v}) - \sum_i \vec{v}^i \cdot \mathbf{T}^i + \sum_i \vec{j}_Q^i] \cdot \vec{n}$$

$$= \sum_i (m_{Si} G_{Si} + \vec{p}_S^i \cdot \vec{v}^i + Q_{Si}) \qquad \text{on } \sigma. \qquad (2.411)$$

The sums in equations (2.410) and (2.411) are a little more complicated than we experienced before. Therefore, we will go into some more details in this special case. In the algebraic modifications, we employ equations (2.328), (2.330) and (2.333). In the first term in (2.410) we have

$$\sum_i \rho_i u_i + \sum_i \frac{1}{2}\rho_i(\vec{v}^i)^2 = \sum_i \rho_i u_i + \sum_i \frac{1}{2}\rho_i(\vec{v})^2 + \sum_i \frac{1}{2}\rho_i(\vec{u}^i)^2, \quad (2.412)$$

in the second term

$$\sum_i \rho_i u_i \cdot \vec{v}^i + \sum_i \frac{1}{2}\rho_i(\vec{v}^i)^2 \cdot \vec{v}^i = \sum_i \rho_i u_i \cdot \vec{v} + \sum_i \left(\rho_i u_i \cdot \vec{u}^i \right.$$

$$\left. + \frac{1}{2}\rho_i(\vec{v})^2 \cdot \vec{v} + \frac{3}{2}\rho_i(\vec{u}^i)^2 \cdot \vec{v} + \frac{1}{2}\rho_i(\vec{u}^i)^2 \cdot \vec{u}^i \right), \qquad (2.413)$$

in the third then

$$\sum_i \vec{v}^i \cdot \mathbf{T}^i = \sum_i \vec{v} \cdot \mathbf{T}^i + \sum_i \vec{u}^i \cdot \mathbf{T}^i, \qquad (2.414)$$

and in the fourth

$$\sum_i \vec{f}^i \cdot \vec{v}^i = \sum_i \vec{f}^i \cdot \vec{v} + \sum_i f^i \cdot \vec{u}^i. \qquad (2.415)$$

Using expressions (2.412)–(2.415), we can define several new quantities, which will help us in simplifying the final energy balance of the mixture (see Eringen and Ingram, 1965). The <u>intrinsic energy</u> $\bar{u}$ of the mixture is given as a sum of the individual internal energies and the kinetic energies of diffusion,

$$\rho\bar{u} = \sum \left(\rho_i u_i + \frac{1}{2}\rho_i(\vec{u}^i)^2 \right), \qquad (2.416)$$

the total heat flux vector $\vec{j}_Q^t$ includes the remaining parts of the first two terms in (2.410), and the diffusion part of the the third term,

$$\vec{j}_Q^t = \sum_i \vec{j}_Q^i + \sum_i \left[\rho_i u_i \vec{u}^i + \frac{1}{2}\rho_i(\vec{u}^i)^2 \cdot \vec{u}^i + \rho_i(\vec{u}^i)^2 \cdot \vec{v} \right]$$

$$-\sum_i \mathbf{T}^i \cdot \vec{u}^i, \tag{2.417}$$

the total source term I_Q^t includes the diffusive part of the fourth term,

$$I_Q^t = \sum_i I_{Qi} + \sum_i \vec{f}^i \cdot \vec{u}^i. \tag{2.418}$$

The right-hand sides of equations (2.410) and (2.411) include internal effects only, therefore

$$\sum_i (m_i G^i + \vec{p}^i \cdot \vec{v}^i + Q_i) = 0 \tag{2.419}$$

$$\sum_i (m_{Si} G_{Si} + \vec{p}_S^i \cdot \vec{v}^i + Q_{Si}) = 0. \tag{2.420}$$

Substituting (2.416)–(2.420) into (2.410) and (2.411), we get final balance equations of the energy of the mixture

$$\frac{\partial}{\partial t} \left(\rho \bar{u} + \frac{1}{2} \rho v^2 \right) + \operatorname{div} \left[\left(\rho \bar{u} + \frac{1}{2} \rho v^2 \right) \cdot \vec{v} \right] - \operatorname{div} (\vec{v} \cdot \mathbf{T}) + \operatorname{div} \vec{j}_Q^t$$

$$- \vec{f} \cdot \vec{v} - I_Q^t = 0 \qquad \text{in } V - \sigma \tag{2.421}$$

$$[\![(\rho \bar{u} + \frac{1}{2} \rho \vec{v}^2) \cdot (\vec{v} - \vec{v}) - \vec{v} \cdot \mathbf{T} + \vec{j}_Q^t]\!] \cdot \vec{n} = 0 \qquad \text{on } \sigma, \tag{2.422}$$

which are formally identical with the balance equations of energy for a single continuum (2.178) and (2.179).

The balance equation of the kinetic energy of a component A_i can be obtained by multiplication of the momentum balance equation (2.365) by $\vec{v}^i$,

$$\vec{v}^i \cdot \frac{\partial}{\partial t} (\rho_i \vec{v}^i) + \vec{v}^i \cdot \operatorname{div} (\rho_i \vec{v}^i \otimes \vec{v}^i) - \vec{v}^i \cdot \operatorname{div} \mathbf{T}^i$$

$$- \vec{v}^i \cdot \vec{f}^i = m_i \vec{v}^i \cdot \vec{J}^i + \vec{v}^i \cdot \vec{p}^i \qquad \text{in } V - \sigma. \tag{2.423}$$

The modifications of the first terms we will do in a similar way as in the case of the respective equation for the single continuum. Applying the results obtained in Section 2.2 and taking into account, that mass of a component i is not conserved, but it is given by equation (2.348), we obtain

$$\vec{v}^i \cdot \frac{\partial}{\partial t} (\rho_i \vec{v}^i) = \frac{\partial}{\partial t} \left(\frac{1}{2} \rho_i (\vec{v}^i)^2 \right) - \frac{1}{2} (\vec{v}^i)^2 \operatorname{div} \rho_i \vec{v}^i + \frac{1}{2} m_i (\vec{v}^i)^2 \tag{2.424}$$

$$\vec{v}^i \cdot \operatorname{div} (\rho_i \vec{v}^i \otimes \vec{v}^i)$$

$$= \text{div}\left(\frac{1}{2}\rho_i(\vec{v}^i)^2 \cdot \vec{v}^i\right) + \frac{1}{2}(\vec{v}^i)^2 \, \text{div} \, \rho_i\vec{v}^i \tag{2.425}$$

$$\vec{v}^i \cdot \text{div} \, \mathbf{T}^i = \text{div}\,(\vec{v}^i \cdot \mathbf{T}^i) - \mathbf{T}^i : \mathbf{D}^i. \tag{2.426}$$

Substituting (2.424), (2.425) and (2.426) into (2.423), we get the kinetic energy balance equation for a component A_i as

$$\frac{\partial}{\partial t}\left(\frac{1}{2}\rho_i(\vec{v}^i)^2\right) + \text{div}\left(\frac{1}{2}\rho_i(\vec{v}^i)^2 \cdot \vec{v}^i\right)$$

$$- \text{div}\,(\vec{v}^i \cdot \mathbf{T}^i) + \mathbf{T}^i : \mathbf{D}^i - \vec{v}^i \cdot \vec{f}^i$$

$$= -\frac{1}{2}m_i(\vec{v}^i)^2 + m_i\vec{v}^i \cdot \vec{J}^i + \vec{v}^i \cdot \vec{p}^i. \tag{2.427}$$

Comparing (2.427) with the master law (2.340) we get

$$A_i = \frac{1}{2}\rho_i(\vec{v}^i)^2 \tag{2.428}$$

$$\vec{A}_\mathrm{S}^i = \vec{v}^i \cdot \mathbf{T}^i \tag{2.429}$$

$$A_{\mathrm{I}i} = \vec{f}^i \cdot \vec{v}^i - \mathbf{T}^i : \mathbf{D}^i, \tag{2.430}$$

$$A_{\mathrm{R}i} = -\frac{1}{2}m_i(\vec{v}^i)^2 + m_i\vec{v}^i \cdot \vec{J}^i + \vec{v}^i \cdot \vec{p}^i. \tag{2.431}$$

The balance law of kinetic energy for A_i on the discontinuity surface σ follows directly from the master law (2.342) using (2.428)–(2.431),

$$[\![\frac{1}{2}\rho_i(\vec{v}^i)^2 \cdot (\vec{v}^i - \vec{v}) - \vec{v}^i \cdot \mathbf{T}^i]\!] \cdot \vec{n} = m_{\mathrm{S}i}\vec{v}^i \cdot \vec{J}_\mathrm{S}^i + \vec{v}^i \cdot \vec{p}_\mathrm{S}^i \quad \text{on } \sigma. \tag{2.432}$$

In formulating the balance equations of the kinetic energy of the mixture, we employ the relations (2.412)–(2.415), derived for the total energy. In addition, we express

$$\mathbf{T}^i : \mathbf{D}^i = \mathbf{T}^i : \mathbf{D} + \mathbf{T}^i : \text{grad} \, \vec{u}^i. \tag{2.433}$$

Summing (2.427) over all i and using (2.412)–(2.415), (2.433) we get

$$\frac{\partial}{\partial t}\left(\frac{1}{2}\rho(\vec{v})^2\right) + \text{div}\left(\frac{1}{2}\rho(\vec{v})^2 \cdot \vec{v}\right) - \text{div}\,(\vec{v} \cdot \mathbf{T})$$

$$+ \mathbf{T} : \mathbf{D} - \vec{v} \cdot \vec{f} + \frac{\partial}{\partial t}\left(\sum_i \frac{1}{2}\rho_i(\vec{u}^i)^2\right)$$

$$+ \text{ div } \sum_i \left(\frac{1}{2} \rho_i (\vec{u}^i)^2 + \frac{3}{2} \rho_i (\vec{u}^i)^2 \cdot \vec{v} + \frac{1}{2} \rho_i (\vec{u}^i)^2 \cdot \vec{u}^i \right)$$

$$- \text{ div } \sum_i \vec{u}^i \cdot \mathbf{T}^i + \sum_i \mathbf{T}^i : \text{ grad } \vec{u}^i - \sum_i \vec{f}^i \cdot \vec{u}^i$$

$$= \sum_i \left(-\frac{1}{2} m_i (\vec{v}^i)^2 + m_i \vec{v}^i \cdot \vec{J}^i + \vec{v}^i \cdot \vec{p}^i \right). \tag{2.434}$$

The right-hand side of equation (2.434) can be modified using (2.331), (2.354) and (2.375), so that we get

$$\sum_i \left(-\frac{1}{2} m_i (\vec{v}^i)^2 + m_i \vec{v}^i \cdot \vec{J}^i + \vec{v}^i \cdot \vec{p}^i \right)$$

$$= \sum_i \left(-\frac{1}{2} m_i (\vec{u}^i)^2 + m_i \vec{u}^i \cdot \vec{J}^i + \vec{u}^i \cdot \vec{p}^i \right). \tag{2.435}$$

For the balance law of kinetic energy of the mixture on the discontinuity surface σ we get by summing (2.432)

$$\left[\frac{1}{2} \rho(\vec{v})^2 \cdot (\vec{v} - \vec{\nu}) - \vec{v} \cdot \mathbf{T} + \sum_i \left(\frac{1}{2} \rho_i (\vec{u}^i)^2 \right. \right.$$

$$\left. \left. + \frac{3}{2} \rho_i (\vec{u}^i)^2 \cdot \vec{v} + \frac{1}{2} \rho_i (\vec{u}^i)^2 \cdot \vec{u}^i - \sum_i \vec{u}^i \cdot \mathbf{T}^i \right) \right] \cdot \vec{n}$$

$$= \sum_i \left(m_{\text{S}i} \vec{u}^i \cdot \vec{J}^i_{\text{S}} + \vec{u}^i \cdot \vec{p}^i_{\text{S}} \right) \qquad \text{on } \sigma. \tag{2.436}$$

The balance laws of internal energy of A_i can be obtained again by subtracting the kinetic energy balances (2.427), (2.432) from the total energy balances (2.408), (2.409). We have

$$\frac{\partial}{\partial t} (\rho_i u_i) + \text{ div } (\rho_i u_i \vec{v}^i) = - \text{ div } \vec{j}^i_{\text{Q}} + I_{\text{Q}i} + \mathbf{T}^i : \mathbf{D}^i + \frac{1}{2} m_i (\vec{v}^i)^2$$

$$- m_i \vec{v}^i \cdot \vec{J}^i + m_i G_i + Q_i \qquad \text{in } V - \sigma. \tag{2.437}$$

$$\left[\rho_i u_i \cdot (\vec{v}^i - \vec{\nu}) + \vec{j}^i_{\text{Q}} \right] \cdot \vec{n}$$

$$= - m_{\text{S}i} \vec{v}^i \cdot \vec{J}^i_{\text{S}i} + m_{\text{S}i} G_{\text{S}i} + Q_{\text{S}i} \qquad \text{on } \sigma. \tag{2.438}$$

Summing (2.437) we get the balance equation of internal energy of the mixture,

$$\frac{\partial}{\partial t}(\rho u) + \operatorname{div}(\rho u \vec{v}) = -\operatorname{div} \vec{j}_Q^{\,g} + I_Q^g + \mathbf{T} : \mathbf{D}$$

$$+ \Psi + \sum_i Q_i \qquad \text{in } V - \sigma, \tag{2.439}$$

where we assume

$$\rho u = \sum_i \rho_i u_i, \tag{2.440}$$

for the generalized heat flux vector we have

$$\vec{j}_Q^{\,g} = \sum_i \vec{j}_Q^{\,i} + \sum_i \rho_i u_i \vec{u}^i, \tag{2.441}$$

for the generalized heat source term

$$I_Q^g = \sum_i I_{Qi} + \sum_i \mathbf{T}^i : \operatorname{grad} \vec{u}^i, \tag{2.442}$$

and for the generalized chemical reaction term

$$\Psi = \sum_i \Psi_i = -\sum_i m_i \vec{v}^i \cdot \vec{J}^i + \sum_i m_i G_i + \sum_i \frac{1}{2} m_i (\vec{v}^i)^2$$

$$= \sum_i m_i \left[G_i - \vec{v}^i \cdot \left(\vec{J}^i - \frac{1}{2} \vec{v}^i \right) \right]. \tag{2.443}$$

The balance equation of internal energy of the mixture on the discontinuity surface we again get by summing (2.438),

$$[\![\rho u \cdot (\vec{v} - \vec{\nu}) + \vec{j}_Q^{\,g}]\!] \cdot \vec{n}$$

$$= -\sum_i \left(m_{Si} \vec{u}^i \cdot \vec{J}_{Si}^i + m_{Si} G_{Si} + Q_{Si} \right) \qquad \text{on } \sigma. \tag{2.444}$$

The balance equations of potential energy for A_i can be obtained in a similar way, as we have shown in the case of a single continuum,

$$\frac{\partial(\rho_i \varphi_i)}{\partial t} + \operatorname{div}(\rho_i \varphi_i \vec{v}^i) + \vec{v}^i \cdot \vec{f}^i = 0 \qquad \text{in } V - \sigma, \tag{2.445}$$

where

$$\vec{f}^i = -\rho_i \operatorname{grad} \varphi_i. \tag{2.446}$$

Comparing (2.445) with the master law (2.340), we get

$$A_i = \rho_i \varphi_i \tag{2.447}$$

$$\vec{A}_S^i = 0 \tag{2.448}$$

$$A_{Ii} = -\vec{v}^i \cdot \vec{f}^i. \tag{2.449}$$

Using (2.447)–(2.449), we have for the discontinuity surface σ

$$[\![\rho_i \varphi_i \cdot (\vec{v}^i - \vec{\nu})]\!] \cdot \vec{n} = 0 \qquad \text{on } \sigma. \tag{2.450}$$

The balance equations of potential energy of the mixture we obtain by summing (2.445) and (2.450) over all i, so that we get

$$\frac{\partial(\rho\varphi)}{\partial t} + \text{div}\,(\rho\varphi\vec{v}) + \text{div}\,\left(\sum_i \rho_i\varphi_i\vec{u}^i\right) + \vec{v}\cdot\vec{f}$$

$$+ \sum_i \vec{u}^i \cdot \vec{f}^i = 0 \qquad \text{in } V - \sigma \tag{2.451}$$

$$[\![\rho\varphi\cdot(\vec{v}-\vec{\nu}) + \sum_i \rho_i\varphi_i\vec{u}^i]\!]\cdot\vec{n} = 0 \qquad \text{on } \sigma, \tag{2.452}$$

where

$$\rho\varphi = \sum_i \rho_i\varphi_i. \tag{2.453}$$

The balance equations of mechanical energy we obtain by summing the respective balance equations of kinetic and potential energy, so that we have for the component A_i

$$\frac{\partial}{\partial t}\left(\rho_i\varphi_i + \frac{1}{2}\rho_i(\vec{v}^i)^2\right) + \text{div}\,\left[\left(\rho_i\varphi_i + \frac{1}{2}\rho_i(\vec{v}^i)^2\right)\cdot\vec{v}^i\right]$$

$$- \text{div}\,(\vec{v}^i\cdot\mathbf{T}^i) + \mathbf{T}^i : \mathbf{D}^i$$

$$= -\frac{1}{2}m_i(\vec{v}^i)^2 + m_i\vec{v}^i\cdot\vec{J}^i + \vec{v}^i\cdot\vec{p}^i \qquad \text{in } V - \sigma. \tag{2.454}$$

$$[\![\left(\rho_i\varphi_i + \frac{1}{2}\rho_i(\vec{v}^i)^2\right)\cdot(\vec{v}^i - \vec{\nu}) - \vec{v}^i\cdot\mathbf{T}^i]\!]\cdot\vec{n}$$

$$= m_{Si}\vec{v}^i\cdot\vec{J}^i_S + \vec{v}^i\cdot\vec{p}^i_S \qquad \text{on } \sigma. \tag{2.455}$$

For the mixture we can write the balance equations of mechanical energy in the form

$$\frac{\partial}{\partial t}\left(\rho\varphi + \frac{1}{2}\rho(\vec{v})^2\right) + \operatorname{div}\left[\left(\rho\varphi + \frac{1}{2}\rho(\vec{v})^2\right)\cdot\vec{v}\right] - \operatorname{div}(\vec{v}\cdot\mathbf{T})$$

$$+\,\mathbf{T}:\mathbf{D} + \operatorname{div}\left(\sum_i \rho_i\varphi_i\vec{u}^i\right) + \frac{\partial}{\partial t}\left(\sum_i \frac{1}{2}\rho_i(\vec{u}^i)^2\right)$$

$$+\operatorname{div}\sum_i\left(\frac{1}{2}\rho_i(\vec{u}^i)^2 + \frac{3}{2}\rho_i(\vec{u}^i)^2\cdot\vec{v} + \frac{1}{2}\rho_i(\vec{u}^i)^2\cdot\vec{u}^i\right)$$

$$-\operatorname{div}\sum_i \vec{u}^i\cdot\mathbf{T}^i + \sum_i \mathbf{T}^i : \operatorname{grad}\vec{u}^i$$

$$=\sum_i\left(-\frac{1}{2}m_i(\vec{u}^i)^2 + m_i\vec{u}^i\cdot\vec{J}^i + \vec{u}^i\cdot\vec{p}^i\right)\qquad \text{in } V-\sigma \quad (2.456)$$

$$\left[\!\left(\rho\varphi + \frac{1}{2}\rho(\vec{v})^2\right)\cdot(\vec{v}-\vec{v}) - \vec{v}\cdot\mathbf{T} + \sum_i \rho_i\varphi_i\vec{u}^i + \sum_i\left(\frac{1}{2}\rho_i(\vec{u}^i)^2\right.\right.$$

$$\left.\left.+\frac{3}{2}\rho_i(\vec{u}^i)^2\cdot\vec{v} + \frac{1}{2}\rho_i(\vec{u}^i)^2\cdot\vec{u}^i - \sum_i \vec{u}^i\cdot\mathbf{T}^i\right)\!\right]\cdot\vec{n}$$

$$=\sum_i\left(m_{\mathrm{S}i}\vec{u}^i\cdot\vec{J}^i_{\mathrm{S}} + \vec{u}^i\cdot\vec{p}^i_{\mathrm{S}}\right)\qquad \text{on } \sigma. \qquad (2.457)$$

For the total energy defined as the sum of the kinetic, the potential and the internal energy we can write for A_i by summing the respective balances of mechanical and internal energy

$$\frac{\partial}{\partial t}\left(\rho_i u_i + \rho_i\varphi_i + \frac{1}{2}\rho_i(\vec{v}^i)^2\right) + \operatorname{div}\left[\left(\rho_i u_i + \rho_i\varphi_i\frac{1}{2}\rho(\vec{v}^i)^2\right)\cdot\vec{v}^i\right]$$

$$-\operatorname{div}(\vec{v}^i\cdot\mathbf{T}^i) + \operatorname{div}\vec{j}^i_{\mathrm{Q}} - I_{\mathrm{Q}i}$$

$$=m_i G_i + \vec{p}^i\cdot\vec{v}^i + Q_i \qquad \text{in } V-\sigma \qquad (2.458)$$

$$\left[(\rho_i u_i + \rho_i\varphi_i\frac{1}{2}\rho_i(\vec{v}^i)^2)\cdot(\vec{v}^i - \vec{v}) - \vec{v}^i\cdot\mathbf{T}^i + \vec{j}^i_{\mathrm{Q}}\right]\cdot\vec{n}$$

$$= m_{\text{S}i}G_{\text{S}i} + \vec{p}_{\text{S}}^{i} \cdot \vec{v}^i + Q_{\text{S}i} \qquad \text{on } \sigma. \tag{2.459}$$

For the mixture we have

$$\frac{\partial}{\partial t}\left(\rho\bar{u} + \rho\varphi + \frac{1}{2}\rho v^2\right) + \text{div}\left[\left(\rho\bar{u} + \rho\varphi + \frac{1}{2}\rho v^2\right)\cdot\vec{v}\right]$$

$$- \text{div}\,(\vec{v}\cdot\mathbf{T}) + \text{div}\,\vec{j}_{\text{Q}}^{\,t} - I_Q = 0 \qquad \text{in } V - \sigma \tag{2.460}$$

$$[(\rho\bar{u} + \rho\varphi + \frac{1}{2}\rho\vec{v}^2)\cdot(\vec{v} - \vec{v}) - \vec{v}\cdot\mathbf{T} + \vec{j}_{\text{Q}}^{\,t}]\cdot\vec{n} = 0 \qquad \text{on } \sigma, \tag{2.461}$$

where

$$\vec{j}_{\text{Q}}^{\,t} = \sum_i \vec{j}_{\text{Q}i} + \sum_i \rho_i\varphi_i\vec{u}^i$$

$$+ \sum_i\left[\rho_i u_i\vec{u}^i + \frac{1}{2}\rho_i(\vec{u}^i)^2\cdot\vec{u}^i + \rho_i(\vec{u}^i)^2\cdot\vec{v}\right] - \sum_i\mathbf{T}^i\cdot\vec{u}^i, \tag{2.462}$$

$$I_Q = \sum_i I_{\text{Q}i}. \tag{2.463}$$

2.4.5 Entropy Inequality

The entropy production rate of a component A_i of the mixture can be expressed as

$$\Gamma_i = \frac{\mathrm{d}}{\mathrm{d}t}\int_{V-\sigma}\rho_i s_i\,\mathrm{d}V + \int_{S-\sigma}\vec{j}_{\text{s}}^{\,i}\cdot\mathrm{d}\vec{S} - \int_{V-\sigma}\rho_i b_{\text{s}i}\,\mathrm{d}V$$

$$- \int_{V-\sigma}(m_i R_i + Z_i)\,\mathrm{d}V, \tag{2.464}$$

where $m_i R_i$ is the entropy production rate due to the chemical reactions, Z_i the rate of entropy transfer into the i-th component from the other components. Comparing formally (2.464) with (2.340) (we note that in equation (2.464), Γ_i can be either positive or negative) we identify

$$A_i = \rho_i s_i \tag{2.465}$$

$$\vec{A}_{\text{S}}^{i} = -\vec{j}_{\text{s}}^{\,i} \tag{2.466}$$

$$A_{\text{I}i} = \rho_i b_{\text{s}i}. \tag{2.467}$$

$$A_{\mathrm{R}i} = m_i R_i + Z_i, \tag{2.468}$$

and using (2.342) we get

$$A_{\mathrm{SR}i} = m_{\mathrm{S}i} R_{\mathrm{S}i} + Z_{\mathrm{S}i}. \tag{2.469}$$

Substituting (2.465)–(2.467) into (2.341) and (2.342), we obtain

$$\rho_i \gamma_i = \frac{\partial(\rho_i s_i)}{\partial t} + \mathrm{div}\,(\rho_i s_i \vec{v}^i)$$

$$+ \mathrm{div}\,\vec{j}_{\mathrm{s}}^{\,i} - \rho_i b_{\mathrm{s}i} - m_i R_i - Z_i \qquad \mathrm{in}\ V - \sigma, \tag{2.470}$$

where the specific entropy production rate γ_i is defined as

$$\Gamma_i = \int_{V-\sigma} \rho_i \gamma_i\, \mathrm{d}V, \tag{2.471}$$

and

$$\rho_i \gamma_{\mathrm{S}i} = [\![\rho_i s_i(\vec{v}^i - \vec{v}) + \vec{j}_{\mathrm{s}}^{\,i}]\!] \cdot \vec{n} - m_{\mathrm{S}i} R_{\mathrm{S}i} - Z_{\mathrm{S}i} \qquad \mathrm{on}\ \sigma. \tag{2.472}$$

Summing the production rates (2.470) and (2.472) over all i, we obtain the entropy production rates of the mixture, which have to remain non-negative in accordance with the second law of thermodynamics,

$$\sum_i \rho_i \gamma_i = \frac{\partial}{\partial t} \sum_i \rho_i s_i + \mathrm{div}\, \sum_i \rho_i s_i \vec{v}^i + \mathrm{div}\, \sum_i \vec{j}_{\mathrm{s}}^{\,i}$$

$$- \sum_i \rho_i b_{\mathrm{s}i} - \sum_i (m_i R_i + Z_i) \geq 0 \qquad \mathrm{in}\ V - \sigma, \tag{2.473}$$

$$\sum_i \rho_i \gamma_{\mathrm{S}i} = [\![\sum_i \rho_i s_i(\vec{v}^i - \vec{v}) + \sum_i \vec{j}_{\mathrm{s}}^{\,i}]\!] \cdot \vec{n}$$

$$- \sum_i (m_{\mathrm{S}i} R_{\mathrm{S}i} + Z_{\mathrm{S}i}) \geq 0 \qquad \mathrm{on}\ \sigma. \tag{2.474}$$

Using the summing rules

$$\rho\gamma = \sum_i \rho_i \gamma_i \tag{2.475}$$

$$\rho s = \sum_i \rho_i s_i \tag{2.476}$$

$$\vec{j}_{\mathrm{s}} = \sum_i \vec{j}_{\mathrm{s}}^{\,i} \tag{2.477}$$

$$\rho b_{\mathrm{s}} = \sum_i \rho_i b_{\mathrm{s}i}, \tag{2.478}$$

we finally get

$$\rho\gamma = \frac{\partial}{\partial t}(\rho s) + \operatorname{div}(\rho s\vec{v}) + \operatorname{div} \vec{j}_{\mathrm{s}}^{\mathrm{g}}$$

$$- \rho b_{\mathrm{s}} - \sum_i (m_i R_i + Z_i) \geq 0 \qquad \text{in } V - \sigma, \tag{2.479}$$

$$\rho\gamma_{\mathrm{S}} = [\![\rho s(\vec{v} - \vec{\nu}) + \vec{j}_{\mathrm{s}}^{\mathrm{g}}]\!] \cdot \vec{n}$$

$$- \sum_i (m_{\mathrm{S}i} R_{\mathrm{S}i} + Z_{\mathrm{S}i}) \geq 0 \qquad \text{on } \sigma, \tag{2.480}$$

where for the generalized entropy flux we have

$$\vec{j}_{\mathrm{s}}^{\mathrm{g}} = \sum_i \vec{j}_{\mathrm{s}}^i + \sum_i \rho_i s_i \vec{u}^i. \tag{2.481}$$

2.5 CONSTITUTIVE EQUATIONS FOR A LINEAR MIXTURE

In the constitutive theory of mixtures, basically the same principles as for the single continuum are employed. We only have to take into account the effects of interaction processes between the components, as shown in the previous section. We will not give a general theory of constitutive equations of a mixture here, for this is outside the scope of this book. We will concentrate instead on just a few details, which are useful in the construction of phenomenological equations.

One of the principal tasks of the constitutive theory is the derivation of the relation for entropy production rate γ. We will use a similar treatment to that in the case of a single continuum, and employ the balance equation of internal energy of the mixture in the form

$$\frac{\partial}{\partial t}(\rho u) + \operatorname{div}(\rho u\vec{v}) = - \operatorname{div} \vec{j}_{\mathrm{Q}}^{\mathrm{g}} + I_{\mathrm{Q}}^{\mathrm{g}} + \mathbf{T} : \mathbf{D}$$

$$+ \sum_i \Psi_i + \sum_i Q_i \qquad \text{in } V - \sigma, \tag{2.482}$$

and the entropy balance equation obtained from the entropy inequality,

$$\frac{\partial}{\partial t}(\rho s) + \operatorname{div}(\rho s\vec{v}) + \operatorname{div} \vec{j}_{\mathrm{s}}^{\mathrm{g}}$$

$$- \rho b_s - \sum_i (m_i R_i + Z_i) - \rho \gamma = 0 \qquad \text{in } V - \sigma. \qquad (2.483)$$

The Gibbs equation for the system, where the number of particles is changed, can be written as

$$\rho \, du = \rho T \, ds + \mathbf{T^R} : d\mathbf{U^R} + \sum_i \mu_i \, d\rho_i, \qquad (2.484)$$

where μ_i is the specific chemical potential (in J/kg). Differentiation of the Gibbs equation with respect to time leads to

$$\rho \frac{du}{dt} = \rho T \frac{ds}{dt} + \mathbf{T^R} : \mathbf{D^R} + \sum_i \mu_i \frac{d\rho_i}{dt}. \qquad (2.485)$$

In order to get comparable terms in the Gibbs equation, internal energy balance and entropy balance, we use the conversion relation (2.152) to modify equations (2.482) and (2.483),

$$\rho \frac{du}{dt} = - \operatorname{div} \vec{j}_Q^g + I_Q^g + \mathbf{T} : \mathbf{D}$$

$$+ \sum_i \Psi_i + \sum_i Q_i \qquad \text{in } V - \sigma, \qquad (2.486)$$

$$\rho \frac{ds}{dt} = - \operatorname{div} \vec{j}_s^g$$

$$+ \rho b_s + \sum_i (m_i R_i + Z_i) + \rho \gamma = 0 \qquad \text{in } V - \sigma. \qquad (2.487)$$

For $\mathbf{T} : \mathbf{D}$ we use the common formula

$$\mathbf{T} : \mathbf{D} = (\mathbf{T^R} + \mathbf{T^D}) : (\mathbf{D^R} + \mathbf{D^D}) = \mathbf{T^R} : \mathbf{D^R} + \mathbf{T^D} : \mathbf{D^R}$$

$$+ \mathbf{T^R} : \mathbf{D^D} + \mathbf{T^D} : \mathbf{D^D}, \qquad (2.488)$$

substitute it into (2.486), and get

$$\rho \frac{du}{dt} = - \operatorname{div} \vec{j}_Q^g + I_Q^g$$

$$+ \mathbf{T^R} : \mathbf{D^R} + \mathbf{T^D} : \mathbf{D^R} + \mathbf{T^R} : \mathbf{D^D} + \mathbf{T^D} : \mathbf{D^D}$$

$$+ \sum_i \Psi_i + \sum_i Q_i \qquad \text{in } V - \sigma. \qquad (2.489)$$

The last term in the Gibbs equation (2.485) can be modified using the balance equation of mass of a component i (2.348) of the mixture, which can be rewritten using the material derivatives. According to (2.335) and (2.334) we have

$$\frac{d^i \rho_i}{dt} = \frac{\partial \rho_i}{\partial t} + \vec{v}^i \cdot \text{grad } \rho_i = \frac{d\rho_i}{dt} + \vec{u}^i \cdot \text{grad } \rho_i. \tag{2.490}$$

Using (2.490) the left-hand side of (2.348) can be written in the form

$$\frac{\partial \rho_i}{\partial t} + \text{div } (\rho_i \vec{v}^i) = \frac{\partial \rho_i}{\partial t} + \vec{v}^i \cdot \text{grad } \rho_i + \rho_i \text{ div } \vec{v}^i$$

$$= \frac{d\rho_i}{dt} + \rho_i \text{ div } \vec{v}^i. \tag{2.491}$$

Substituting (2.491) into (2.348) and using (2.332) we finally get

$$\frac{d\rho_i}{dt} + \text{div } \vec{j}_D^i + \rho_i \text{ div } \vec{v} = m_i. \tag{2.492}$$

Substitution of (2.492) into the Gibbs equation (2.493) leads then to

$$\rho \frac{du}{dt} = \rho T \frac{ds}{dt} + \mathbf{T^R} : \mathbf{D^R} + \sum_i m_i \mu_i$$

$$- \sum_i \mu_i \text{ div } \vec{j}_D^i - \sum_i \mu_i \rho_i \text{ div } \vec{v}. \tag{2.493}$$

Equations (2.489) and (2.493) both contain the time rate of specific internal energy, but only (2.493) contains the time rate of specific entropy. Therefore, an analogue to the entropy balance can be obtained by a combination of these two equations. After some simple algebraic operations we obtain

$$\rho \frac{ds}{dt} = -\frac{1}{T} \text{ div } \vec{j}_Q^g + \frac{1}{T} I_Q^g + \frac{1}{T} \mathbf{T^D} : \mathbf{D^R} + \frac{1}{T} \mathbf{T^R} : \mathbf{D^D} + \frac{1}{T} \mathbf{T^D} : \mathbf{D^D}$$

$$- \frac{1}{T} \sum_i m_i \mu_i + \frac{1}{T} \sum_i \mu_i \text{ div } \vec{j}_D^i - \frac{1}{T} \sum_i \mu_i \rho_i \text{ div } \vec{v}$$

$$+ \frac{1}{T} \sum_i \Psi_i + \frac{1}{T} \sum_i Q_i \qquad \text{in } V - \sigma. \tag{2.494}$$

In order to get comparable terms in entropy balance equations (2.483) and (2.494), some algebraic adjustments are still needed. We can write

$$\frac{1}{T} \text{ div } \vec{j}_Q^g = \text{div } \frac{\vec{j}_Q^g}{T} + \frac{\vec{j}_Q^g}{T^2} \text{ grad } T \tag{2.495}$$

$$\frac{1}{T} \mu_i \text{ div } \vec{j}_D^i = \text{div } \left(\frac{\mu_i}{T} \vec{j}_D^i\right) - \frac{\vec{j}_D^i}{T} \text{ grad } \mu_i + \frac{\mu_i \vec{j}_D^i}{T^2} \text{ grad } T, \tag{2.496}$$

and substituting into (2.494) then

$$\rho \frac{ds}{dt} = - \operatorname{div} \frac{\vec{j}_Q^g}{T} - \frac{\vec{j}_Q^g - \sum_i \mu_i \vec{j}_D^i}{T^2} \operatorname{grad} T$$

$$+ \sum_i \operatorname{div} \left(\frac{\mu_i}{T} \vec{j}_D^i \right) - \sum_i \frac{\vec{j}_D^i}{T} \operatorname{grad} \mu_i + \frac{1}{T} I_Q^g + \frac{1}{T} \mathbf{T}^D : \mathbf{D}^R$$

$$+ \frac{1}{T} \mathbf{T}^R : \mathbf{D}^D + \frac{1}{T} \mathbf{T}^D : \mathbf{D}^D - \frac{1}{T} \sum_i m_i \mu_i + \frac{1}{T} \sum_i \mu_i \rho_i \operatorname{div} \vec{v}$$

$$+ \frac{1}{T} \sum_i \Psi_i + \frac{1}{T} \sum_i Q_i \qquad \text{in } V - \sigma. \tag{2.497}$$

Comparing (2.483) and (2.497), we can identify the unknown entropy related terms in (2.483), so that we have

$$\vec{j}_s^g = \frac{\vec{j}_Q^g - \sum_i \mu_i \vec{j}_D^i}{T} \tag{2.498}$$

$$\rho b_s = \frac{I_Q^g}{T} \tag{2.499}$$

$$\sum_i Z_i = \sum_i \frac{Q_i}{T} \tag{2.500}$$

$$\sum_i m_i R_i = \sum_i \frac{\Psi_i}{T} + \sum_i \frac{\mu_i}{T} \rho_i \operatorname{div} \vec{v}, \tag{2.501}$$

and finally for the entropy production rate

$$\rho \gamma = - \frac{\vec{j}_Q^g - \sum_i \mu_i \vec{j}_D^i}{T^2} \operatorname{grad} T - \sum_i \frac{\vec{j}_D^i}{T} \operatorname{grad} \mu_i - \frac{1}{T} \sum_i m_i \mu_i$$

$$+ \frac{1}{T} \mathbf{T}^D : \mathbf{D}^R + \frac{1}{T} \mathbf{T}^R : \mathbf{D}^D + \frac{1}{T} \mathbf{T}^D : \mathbf{D}^D. \tag{2.502}$$

The chemical reaction term in the entropy production can be modified into a more common form using expression (2.339) for the rate of mass production. We can write

$$\sum_i m_i \mu_i = \sum_i \sum_j \mu_i \theta_{ij} J_j^c = \sum_j \left(\sum_i \theta_{ij} \mu_i \right) J_j^c = \sum_j A_j J_j^c, \tag{2.503}$$

where A_j is the chemical affinity of the j-th reaction.

Equation (2.502), as in the case of a single continuum, has a bilinear form, but it contains some additional terms. This bilinear form consists of a sum of products of the particular thermodynamic forces x_i and fluxes J_i, which can be either scalars, vectors or tensors. Schematically we can write

$$\sigma = \sum_i J_i x_i \tag{2.504}$$

and considering (2.502) we can identify the fluxes and forces as follows:

$$J_1 = \vec{j}_Q^g - \sum_i \mu_i \vec{j}_D^i \tag{2.505}$$

$$x_1 = \frac{\operatorname{grad} T}{T^2} \tag{2.506}$$

$$J_k = \vec{j}_D^k, \qquad k = 2, \ldots, n \tag{2.507}$$

$$x_k = \frac{\operatorname{grad} \mu_k}{T}, \qquad k = 2, \ldots, n \tag{2.508}$$

$$J_{n+j} = J_j^c, \qquad j = 1, \ldots, n_j \tag{2.509}$$

$$x_{n+j} = \frac{A_j}{T}, \qquad j = 1, \ldots, n_j, \tag{2.510}$$

n_j is the number of chemical reactions taking place in the system,

$$J_{n+n_j+1} = -\mathbf{T}^D \tag{2.511}$$

$$x_{n+n_j+1} = \frac{\mathbf{D}^R}{T} \tag{2.512}$$

$$J_{n+n_j+2} = -\mathbf{T}^R \tag{2.513}$$

$$x_{n+n_j+2} = \frac{\mathbf{D}^D}{T}. \tag{2.514}$$

The procedure for constructing equations of state and phenomenological equations for a linear mixture is quite similar to that described for a single continuum. The thermodynamic forces and fluxes identified in the expression for the entropy production are organized into a system using kinetic coefficients L_{ik},

$$J_i = \sum_k L_{ik} x_k, \tag{2.515}$$

and applying some thermodynamic restrictions such as the Onsager reciprocity relation or the Curie symmetry principle. The details of this procedure can be found in any of the specialized textbooks on thermodynamics, for example Prigogine (1961), de Groot and Mazur (1962), Fitts (1962), Haase (1969), and Lavenda (1978).

References

Atkin, R.J., Craine, R.E., 1975, Continuum theories of mixtures: basic theory and historical development. *Q. J. Mech. Appl. Math.*, **29**, pp. 209–244.

Atkin, R.J., Craine, R.E., 1976, Continuum theories of mixtures: applications. *J. Inst. Maths Applics*, **17**, pp. 153–207.

Eringen, A. C., 1967, *Mechanics of Continua, Vols. I,II,III.* (New York: John Wiley and Sons).

Eringen, A.C., Ingram, D., 1965, A continuum theory of chemically reacting media-I. *Int. J. Engng. Sci.*, **3**, pp. 197–212.

Fitts, D.D., 1962, *Non-Equilibrium Thermodynamics.* (New York: McGraw-Hill).

de Groot, S.R., Mazur, P., 1962, *Non-Equilibrium Thermodynamics.* (Amsterdam: North-Holland).

Haase, R., 1969, *Thermodynamics of Irreversible Processes.* (Reading: Addison-Wesley).

Keenan, J.H., 1946, *Thermodynamics*, 4th edition. (New York: John Wiley and Sons).

Kvasnica, J., 1965, *Thermodynamics.* (Praha: SNTL), (in Czech).

Lavenda, B.H., 1978, *Thermodynamics of Irreversible Processes.* (London: Macmillan).

Malvern, L.E., 1969, *Introduction to the Mechanics of a Continuous Medium.* (Englewood Cliffs: Prentice-Hall).

Mase, G.E., 1970, *Theory and Problems of Continuum Mechanics.* (New York: McGraw-Hill).

Noll, W., 1974, *The Foundations of Mechanics and Thermodynamics.* (Berlin: Springer).

Prigogine, I., 1961, *Introduction to Thermodynamics of Irreversible Processes.* (New York: Interscience).

Truesdell, C., 1965, *The Elements of Continuum Mechanics.* (New York: Springer Verlag).

Truesdell, C., 1969, *Rational Thermodynamics.* (New York: McGraw-Hill).

## CHAPTER THREE

Models of Heat, Moisture and Chemical Compounds Transport in Porous Materials

3.1 MODELS OF WATER AND WATER VAPOUR TRANSPORT

According to the main transport mechanisms and their corresponding mathematical description, three basic groups of water and water vapour transport models can be distinguished, namely convection models, diffusion models and hybrid models.

3.1.1 Convection Models

Modelling water transport in porous materials on the way water in a porous system flows through particular channels similar, for instance, to pipes, where convection is the main transfer mechanism and the gradient of pressure acts as the main driving force, is probably the most natural way of describing this phenomenon. Therefore, this type of model appeared historically as the first.

The simplest case is the flow in an inert material fully saturated by water, where it can be assumed that water flows in all pores. Practical importance is accorded only to the steady state, when the rate of flow Q (in m^3/s) is constant and can be expressed by a simple relation

$$Q = \frac{V}{t},\qquad(3.1)$$

where V is the volume of flowing water, t the time. The macroscopic velocity v is then determined by the relation

$$v = \frac{V}{At},\qquad(3.2)$$

where A is the cross-section of the porous material perpendicular to the direction of the flow. The velocity v is termed the filtration velocity, Darcy velocity or apparent velocity. In fact, it is the average velocity of the flow taken for the whole porous body.

As well as the filtration velocity, the mean pore velocity v_p can also be introduced, defined as the velocity of water flow in the pores,

$$v_p = \frac{v}{\Pi},\qquad(3.3)$$

where v is the Darcy velocity, Π the porosity,

$$\Pi = \frac{V_p}{V},\tag{3.4}$$

V_p is the pore volume, V the volume of the porous body. Taking into account that $\Pi < 1$, the following condition is valid: $v_p > v$.

The most important application of the convection model is water flow in soils induced by gravity force. For this case, Darcy (1856) formulated his empirical relation, which historically was the first describing water transport in porous materials. Darcy investigated water flow in a vertically oriented sand column and found a linear relationship between the velocity and the hydraulic gradient s, which is the ratio between the height of the water column ΔH and the length of the porous material in the direction of the flow L,

$$s = \frac{\Delta H}{L}.\tag{3.5}$$

Darcy's equation can be mathematically expressed as

$$v = K_S \frac{\Delta H}{L},\tag{3.6}$$

where the coefficient K_S (in m/s) is called the hydraulic conductivity or filtration coefficient.

ΔH is in principle the difference in water level before the water inlet into and after the water outlet from the porous body, but these levels are not identical with the boundaries of the porous body in the vertical direction, i.e. $\Delta H \neq L$.

The hydraulic gradient can be determined using two piezometers, which measure water pressure in the sand at two different elevations z_1, z_2 as the height of water ascension above the mouth of the piezometer into the pipe with sand H_1, H_2. Then we have

$$v = K_S \frac{\Delta H'}{L'},\tag{3.7}$$

where $\Delta H' = H_2 - H_1$, $L = z_2 - z_1$. Introducing the total head h as the sum of the elevation head z and hydraulic head (sometimes called pressure head or water potential) H, $h = z + H$, and taking into account that water flow is in the direction of negative z−axis, we obtain

$$v = -K_S \frac{h_2 - h_1}{z_2 - z_1},\tag{3.8}$$

and at the limit $L = z_2 - z_1 \to 0$ then

$$v = -K_S \frac{dh}{dz},\tag{3.9}$$

or in general, for an arbitrary flow direction,

$$\vec{v} = -K_S \operatorname{grad} h.\tag{3.10}$$

For unsaturated flow, an analogous relation to that for saturated flow can be formulated. We have only to take into account that a part of the pores is filled with the air and the amount of water in the porous body can vary during the flow process. The unsaturated flow was first described by Buckingham (1907) using the relation

$$\vec{v} = -K(H) \text{ grad } h, \tag{3.11}$$

which is called Darcy–Buckingham equation. The coefficient K is the unsaturated hydraulic conductivity, which depends on the drainage stage of the material, i.e. on the average moisture content by volume w or on the average hydraulic head H.

The Darcy–Buckingham equation can be directly applied for unsaturated steady-state flow. However, in practice there often appear situations when moisture content varies with time, i.e. $\frac{dw}{dt} \neq 0$, and therefore transient unsaturated flow takes place. We then have to solve simultaneously the flow velocity and the charging or discharging of water in pores.

The flow velocity is determined by the Darcy–Buckingham equation. The rate of pore charge and discharge can be solved by the continuity equation, i.e. the mass balance of water in the porous body. Thus, we have two basic equations

$$\vec{v} = -K(H) \text{ grad } h, \tag{3.12}$$

$$\frac{\partial \rho_v}{\partial t} + \text{div } \vec{j} = 0, \tag{3.13}$$

where $\vec{j}$ is the water flux in the porous body, ρ_v is the partial moisture density, i.e. the mass of water per unit volume of the porous body,

$$\rho_v = \frac{m_v}{V}. \tag{3.14}$$

The relation between the partial moisture density and the moisture content by volume, w,

$$w = \frac{V_w}{V}, \tag{3.15}$$

V_w is the volume of water in the porous body, is as follows:

$$\rho_v = \rho_w w, \tag{3.16}$$

where ρ_w is the density of water.

The Darcy velocity $\vec{v}$ is defined as the volume rate of flow Q (in m^3/s) per $1\ m^2$ of the cross-section of the porous body (see equation (3.2)). The mass rate of flow Q_M (in kg/s) is defined as

$$Q_M = \rho_w Q, \tag{3.17}$$

therefore the water flux j (in kg/m²s) can be expressed as

$$j = \frac{Q_M}{A} = \frac{V_w \rho_w}{At} = \frac{m_w}{At}. \tag{3.18}$$

Using the definition relation for the Darcy velocity (3.2) and the just-derived expression for water flux j (3.18), we have

$$\vec{j} = \rho_w \vec{v}. \tag{3.19}$$

Substituting (3.16) and (3.19) under assumption $\rho_w = const.$ into the continuity equation (3.13) we obtain

$$\frac{\partial w}{\partial t} + \text{div } \vec{v} = 0. \tag{3.20}$$

Substituting the Darcy–Buckingham equation into the continuity equation we get the equation

$$\frac{\partial w}{\partial t} = \text{div} \left(K(H) \text{ grad } h \right), \tag{3.21}$$

which was first derived by Richards (1931) and therefore it is often called the Richards equation. For a practical application it must be further modified. Using the definition relation for the total head h,

$$h = H + z, \tag{3.22}$$

and the dependence of volumetric moisture content on the hydraulic head (moisture retention curve or water retention curve) $w = w(H)$, which can be measured by common methods (see Chapter 5), we obtain

$$\frac{\partial w}{\partial H} \frac{\partial H}{\partial t} = \text{div} \left(K(H) \text{ grad } H \right) + \text{div} \left(K(H) \text{ grad } z \right). \tag{3.23}$$

Defining the water capacity c_w as

$$c_w = \frac{\partial w}{\partial H} \tag{3.24}$$

and taking into account that

$$\text{div} \left(K(H) \text{ grad } z \right) = \frac{\partial}{\partial x} \left(K(H) \frac{\partial z}{\partial x} \right) + \frac{\partial}{\partial y} \left(K(H) \frac{\partial z}{\partial y} \right)$$

$$+ \frac{\partial}{\partial z} \left(K(H) \frac{\partial z}{\partial z} \right) = \frac{\partial K(H)}{\partial z}, \tag{3.25}$$

we have

$$c_w \frac{\partial H}{\partial t} = \text{div} \left(K(H) \text{ grad } H \right) + \frac{\partial K(H)}{\partial z}, \tag{3.26}$$

which is the diffusion equation for the unknown H.

A similar equation can also be derived for the unknown volumetric moisture content w. We again employ the Richards equation (3.21), the known dependence $w = w(H)$ and the definition relation for the total head $h = H + z$. Then

$$\text{grad } H = \frac{\partial H}{\partial w} \text{ grad } w, \tag{3.27}$$

and $K(w)$ can be expressed using $K(H)$ and $w(H)$, so that

$$\frac{\partial w}{\partial t} = \text{div} \left(K(w) \frac{\partial H}{\partial w} \text{ grad } w \right) + \text{div} \left(K(w) \text{ grad } z \right). \tag{3.28}$$

Defining the water diffusivity D as

$$D(w) = K(w) \frac{\partial H}{\partial w}, \tag{3.29}$$

and modifying the second term on the right-hand side of equation (3.28) in a similar way as before, i.e.

$$\text{div} \left(K(w) \text{ grad } z \right) = \frac{\partial K(w)}{\partial z} = \frac{\partial K(w)}{\partial w} \frac{\partial w}{\partial z}, \tag{3.30}$$

we obtain

$$\frac{\partial w}{\partial t} = \text{div} \left(D(w) \text{ grad } w \right) + \frac{\partial K(w)}{\partial w} \frac{\partial w}{\partial z}, \tag{3.31}$$

which is the diffusion equation for the volumetric moisture content. This form of diffusion equation was first derived by Klute (1952). The last term on the right-hand side expresses the effect of gravity. Neglecting the effect of gravity in equation (3.31), we obtain the classical diffusion equation for moisture content,

$$\frac{\partial w}{\partial t} = \text{div} \left(D(w) \text{ grad } w \right). \tag{3.32}$$

Convection models mostly describe the flow of liquid water only, which in modelling water flow in soils is the most important mode of moisture transport. In cases where it is also necessary to include the flow of water vapour, diffusion models are mostly employed (see, e.g., Philip, 1957), as described in the subsequent section.

3.1.2 Diffusion Models

The diffusion theory of moisture transfer in porous materials is based on the assumption that moisture in the liquid or gaseous phase penetrates through the porous system by a diffusion mechanism in a similar way, for instance, to in a system of two gases. This concept was introduced for the first time in the description of drying processes of porous materials practically simultaneously and independently by Krischer and Lykov. Krischer (1942) formulated for the transfer of liquid moisture the relation

$$\dot{m}_{\mathrm{w}} = \rho_{\mathrm{w}} \kappa \frac{\partial w}{\partial x}, \tag{3.33}$$

where $\dot{m}_{\mathrm{w}}$ is the moisture flux, ρ_{w} the density of water, w the moisture content by volume,

$$w = \frac{V_{\mathrm{w}}}{V}, \tag{3.34}$$

V_{w} the volume of water, V the volume of the porous body, κ the moisture diffusivity. For the transport of water vapour he introduced

$$\dot{m}_{\mathrm{d}} = \frac{1}{\mu} \frac{D}{R_{\mathrm{v}}T} \frac{\partial p_{\mathrm{d}}}{\partial x}, \tag{3.35}$$

where $\dot{m}_{\mathrm{d}}$ is the flux of water vapour, p_{d} the partial pressure of water vapour in the air, D the diffusion coefficient of water vapour in the air, μ the water vapour diffusion resistance factor, which expresses how many times is the porous material less permeable for water vapour than the air, R_{v} the gas constant of water vapour, and T the temperature in K.

Let us pause now for a moment and analyse how Krischer's relations are constructed. In the relation for liquid moisture transport, we immediately see that

$$\rho_{\mathrm{w}} \cdot w = \rho_v, \tag{3.36}$$

which is the mass of water per unit volume of the porous body (the partial density of water). This can easily be proven by substituting the definition relations for density of water,

$$\rho_{\mathrm{w}} = \frac{m_{\mathrm{w}}}{V_{\mathrm{w}}}, \tag{3.37}$$

and for the moisture content by volume (3.34) into (3.36), so that

$$\rho_v = \rho_{\mathrm{w}} w = \frac{m_{\mathrm{w}}}{V_{\mathrm{w}}} \frac{V_{\mathrm{w}}}{V} = \frac{m_{\mathrm{w}}}{V}. \tag{3.38}$$

Clearly, the last expression in (3.38) is the partial density of water in the porous body. We have in fact

$$\dot{m}_{\mathrm{w}} = \kappa \frac{\partial \rho_v}{\partial x}, \tag{3.39}$$

which is (except for the sign) analogous to Fick's law of diffusion,

$$j = -D \frac{\partial c}{\partial x}, \tag{3.40}$$

where D is the diffusion coefficient (in m^2/s), c is the concentration (in kg/m^3).

In the relation for water vapour diffusion, the state equation of an ideal gas is hidden. This has the form

$$pV = nRT. \tag{3.41}$$

For an amount of substance n (the number of moles) we can write

$$n = \frac{m}{M}, \tag{3.42}$$

and therefore

$$p = \rho \frac{RT}{M}. \tag{3.43}$$

Thus, for water vapour we have

$$p_\mathrm{d} = \rho_\mathrm{p} R_\mathrm{v} T, \tag{3.44}$$

where

$$R_\mathrm{v} = \frac{R}{M}, \tag{3.45}$$

ρ_p is the partial density of water vapour, $M = 0.018$ kg/mol for water, the remaining symbols are the same as used before. Substituting for p_d into (3.35) we get

$$\dot{m}_\mathrm{d} = \frac{D}{\mu} \frac{1}{R_\mathrm{v} T} \frac{\partial}{\partial x} (\rho_\mathrm{p} R_\mathrm{v} T), \tag{3.46}$$

and under assumption $T = \text{const.}$ then

$$\dot{m}_\mathrm{d} = \frac{D}{\mu} \frac{\partial \rho_\mathrm{p}}{\partial x}. \tag{3.47}$$

Denoting

$$D_\mathrm{p} = \frac{D}{\mu} \tag{3.48}$$

as the diffusion coefficient of water vapour in porous material, we again obtain a relation that is analogous (except for the sign) to Fick's law,

$$\dot{m}_\mathrm{d} = D_\mathrm{p} \frac{\partial \rho_\mathrm{p}}{\partial x}. \tag{3.49}$$

The choice of an opposite sign convention in the material relations for water and water vapour transport to the which is common in irreversible thermodynamics, is probably related to the fact that these relations were formulated in 1930s, when the applications of irreversible thermodynamics were still rare, particularly in technical sciences. Krischer's material relations were formulated in more or less an intuitive way, and in their original form they did not have any direct relation to the thermodynamic theories.

Lykov (1954) introduced for water and water vapour transfer the relations

$$j_i = -a_{mi}\rho_0 \frac{\partial u}{\partial x}, \tag{3.50}$$

where the index $i = 1$ denotes water vapour, $i = 2$ water, ρ_0 is the partial density of the porous matrix, u the <u>moisture content by mass</u>,

$$u = u_1 + u_2, \tag{3.51}$$

j_i the flux, and a_{mi} the diffusion coefficients of water vapour and water in the porous body.

There is only one principal difference between Lykov's and Krischer's formulations. Lykov assumes the gradient of total moisture content is the driving force for both water and water vapour transport, while Krischer takes water and water vapour transport separately, i.e. the driving force for water vapour transport is the gradient of its partial pressure in the air, the driving force for water transport is the gradient of liquid moisture content. Both formulations have their advantages and disadvantages, as we will see later in more complex cases of transport phenomena in porous materials.

For a practical solution to the problem of moisture transport, we have to substitute Krischer's and Lykov's material relations into the mass balance equation, which can be written in one dimension schematically as

$$\frac{\partial \rho_i}{\partial t} = -\frac{\partial j_i}{\partial x}, \tag{3.52}$$

where ρ_i is the partial density of moisture (in kg/m^3), j_i the moisture flux (in kg/m^{2}s).

For Lykov's formulations we have

$$\rho_1 = \rho_0 u_1 \tag{3.53}$$

$$\rho_2 = \rho_0 u_2, \tag{3.54}$$

and therefore

$$\frac{\partial u_1}{\partial t} = \frac{\partial}{\partial x}\left(a_{m1}\frac{\partial u}{\partial x}\right) \tag{3.55}$$

$$\frac{\partial u_2}{\partial t} = \frac{\partial}{\partial x}\left(a_{m2}\frac{\partial u}{\partial x}\right). \tag{3.56}$$

Adding these two equations, we obtain the relation for total moisture transport,

$$\frac{\partial u}{\partial t} = \frac{\partial}{\partial x}\left(a_m\frac{\partial u}{\partial x}\right), \tag{3.57}$$

where $a_\mathrm{m} = a_\mathrm{m1} + a_\mathrm{m2}$.

In Krischer's model, we have for the liquid moisture

$$\rho_2 = \rho_\mathrm{w} w \tag{3.58}$$

$$j_2 = -\dot{m}_\mathrm{w}, \tag{3.59}$$

and therefore

$$\frac{\partial w}{\partial t} = \frac{\partial}{\partial x}\left(\kappa \frac{\partial w}{\partial x}\right), \tag{3.60}$$

for water vapour

$$\rho_1 = p_1 \frac{1}{R_\mathrm{v} T} \tag{3.61}$$

$$j_1 = -\dot{m}_\mathrm{d}, \tag{3.62}$$

and therefore under assumption $T = \mathrm{const.}$

$$\frac{\partial \rho_1}{\partial t} = \frac{\partial}{\partial x}\left(\frac{D}{\mu}\frac{\partial \rho_1}{\partial x}\right). \tag{3.63}$$

The formulation of the total mass balance equation for liquid and gaseous moisture is for Krischer's model a little more complicated than for Lykov's model. The necessary first step is to multiply the equation for liquid water transport (3.60) by ρ_w and the equation for water vapour transport (3.63) by $\frac{1}{R_\mathrm{v} T}$ in order to get the same physical dimensions in both equations,

$$\rho_\mathrm{w}\frac{\partial w}{\partial t} = \rho_\mathrm{w}\frac{\partial}{\partial x}\left(\kappa \frac{\partial w}{\partial x}\right), \tag{3.64}$$

$$\frac{1}{R_\mathrm{v} T}\frac{\partial p_1}{\partial t} = \frac{1}{R_\mathrm{v} T}\frac{\partial}{\partial x}\left(\frac{D}{\mu}\frac{\partial p_1}{\partial x}\right). \tag{3.65}$$

Krischer further introduces two additional corrections, which Lykov does not take into account. The first is the reduction of the accumulation term for water vapour on the left-hand side of equation (3.65) by the multiplication factor $(\Pi - w)$ (Π is the porosity, $\Pi = V_\mathrm{p}/V$, V_p is the volume of the pores, V the volume of the whole porous body), which expresses the fact that water vapour can appear only in that part of the porous space where water in the liquid state is not present, and its accumulation is therefore limited to only that space. On the right-hand side this correction is not necessary, because in fact it is included already in the diffusion resistance factor μ, which itself reduces the diffusion flux on the one hand to the porous space (this is the reduction by the factor Π), and on the other μ depends on the water content w, which itself is a correction to w.

The second Krischer correction is inclusion of <u>Stefan diffusion</u>. The principle of Stefan diffusion is in our case as follows: both water and water vapour are present at the same time in the specimen, and therefore their phase change processes take place continuously. Due to water evaporation, the amount of water vapour in the basic gas system increases. Therefore, the absolute pressure in the vicinity of the interface between the gaseous and liquid phase increases as well, and its gradient leads to the removal of water vapour from the interface. The real flux of water vapour is then higher than it would correspond to the gradient of its partial pressure (volume increase due to the evaporation of a part of liquid water is negligible because the density of water vapour is much lower than that of liquid water). The physical situation can be simplified by introducing the assumption that the process is so slow that the absolute pressure equilibrates in the whole volume, i.e.

$$p = p_1 + p_a = \text{const.} \tag{3.66}$$

Another logical assumption is that in the vicinity of the surface of the liquid, the total air flux (i.e. the sum of the diffusion and convection fluxes) must be equal to zero. If this assumption were not valid, the amount of dry air near the interface would increase continuously (water vapour flows in the direction away from the interface, air flows to the interface due to classical diffusion, in the direction of the negative gradient of its concentration), and the total pressure near the interface would increase. As $p = \text{const.}$, convective transport of the water vapour–air mixture in the direction from the interface must exist.

Therefore, we can write for the water vapour flux

$$j_1 = -D_{1a}\frac{\partial \rho_1}{\partial x} + \rho_1 v, \tag{3.67}$$

and for the air flux

$$j_a = -D_{a1}\frac{\partial \rho_a}{\partial x} + \rho_a v. \tag{3.68}$$

Under assumptions $p = \text{const.}$, $T = \text{const.}$, it follows from the state equation of an ideal gas that

$$\rho = p\frac{1}{R_v T} = \text{const.}, \tag{3.69}$$

and therefore

$$\rho_1 + \rho_a = \rho = \text{const.} \tag{3.70}$$

From (3.70) it immediately follows that

$$\frac{\partial \rho_1}{\partial x} = -\frac{\partial \rho_a}{\partial x}. \tag{3.71}$$

For the diffusion coefficients of water vapour in the air we use the common relation $D_{1a} = D_{a1}$, which is valid for interdiffusion.

From equation (3.68) we can determine the velocity of the mixture v,

$$v = \frac{D_{a1}}{\rho_a} \frac{\partial \rho_a}{\partial x} = -\frac{D_{1a}}{\rho - \rho_1} \frac{\partial \rho_1}{\partial x}. \tag{3.72}$$

Substituting (3.72) into (3.67) we get

$$j_1 = -D_{1a} \frac{\partial \rho_1}{\partial x} - D_{1a} \frac{\rho_1}{\rho - \rho_1} \frac{\partial \rho_1}{\partial x}$$

$$= -D_{1a} \frac{\partial \rho_1}{\partial x} \left(1 + \frac{\rho_1}{\rho - \rho_1} \right) = -D_{1a} \frac{\rho}{\rho - \rho_1} \frac{\partial \rho_1}{\partial x}. \tag{3.73}$$

By introducing

$$p_1 = \rho_1 \frac{1}{R_v T} \tag{3.74}$$

for water vapour and

$$p = \rho \frac{1}{R_s T} \tag{3.75}$$

for water vapour–air mixture and under assumption $R_v \approx R_s$ we obtain

$$j_1 = -D_{1a} \frac{1}{R_v T} \frac{p}{p - p_1} \frac{\partial p_1}{\partial x}. \tag{3.76}$$

The last equation is valid for diffusion of water vapour in the air. If diffusion of water vapour in a porous body is to be described, we have to use Krischer's diffusion resistance factor μ, which results in

$$j_1 = -\frac{1}{\mu} \frac{D}{R_v T} \frac{p}{p - p_1} \frac{\partial p_1}{\partial x}, \tag{3.77}$$

where Stefan diffusion is expressed by the multiplication factor $p/(p - p_1)$.

Now, we can return to the total mass balance of Krischer's model. For water we have

$$\rho_w \frac{\partial w}{\partial t} = \rho_w \frac{\partial}{\partial x} \left(\kappa \frac{\partial w}{\partial x} \right), \tag{3.78}$$

and for water vapour

$$\frac{\Pi - w}{R_v T} \frac{\partial p_1}{\partial t} = \frac{1}{R_v T} \frac{\partial}{\partial x} \left(\frac{D}{\mu} \frac{p}{p - p_1} \frac{\partial p_1}{\partial x} \right). \tag{3.79}$$

The total mass balance of moisture therefore reads:

$$\rho_w \frac{\partial w}{\partial t} + \frac{\Pi - w}{R_v T} \frac{\partial p_1}{\partial t} = \rho_w \frac{\partial}{\partial x} \left(\kappa \frac{\partial w}{\partial x} \right)$$

$$+ \frac{1}{R_v T} \frac{\partial}{\partial x} \left(\frac{D}{\mu} \frac{p}{p - p_1} \frac{\partial p_1}{\partial x} \right). \tag{3.80}$$

The afore-given derivation was performed in one dimension, for the sake of simplicity, but we assumed the material parameters as functions of the primary field variables. In Krischer's and Lykov's original formulations it was assumed that all material parameters are constant. The total mass balances of moisture were therefore a little simpler in both models,

$$\frac{\partial u}{\partial t} = a_m \frac{\partial^2 u}{\partial x^2} \tag{3.81}$$

in Lykov's model, and

$$\rho_w \frac{\partial w}{\partial t} + \frac{\Pi - w}{R_v T} \frac{\partial p_1}{\partial t} = \rho_w \kappa \frac{\partial^2 w}{\partial x^2} + \frac{D}{\mu R_v T} \frac{p}{p - p_1} \frac{\partial^2 p_1}{\partial x^2} \tag{3.82}$$

in Krischer's model. On the other hand, an extension to a three-dimensional formulation under the common assumption that the material is isotropic and the material parameters remain scalar quantities is quite straightforward,

$$\frac{\partial u}{\partial t} = \text{div } (a_m \text{ grad } u) \tag{3.83}$$

for Lykov's model, and

$$\rho_w \frac{\partial w}{\partial t} + \frac{\Pi - w}{R_v T} \frac{\partial p_1}{\partial t} = \rho_w \text{ div } (\kappa \text{ grad } w)$$

$$+ \frac{1}{R_v T} \text{ div } \left(\frac{D}{\mu} \frac{p}{p - p_1} \text{ grad } p_1 \right) \tag{3.84}$$

for Krischer's model.

3.1.2.1 Diffusion Models with Sharp or Steep Interface

In some cases, the diffusion theory of moisture transport in porous materials has led to disagreement with experimental measurements, particularly if direct contact of material with water was imposed. While the experimentally determined curves showed a steep or even sharp interface between the moist and dry parts of the specimen, the classical diffusion models led to gradual transition and a diffuse, unsharp interface between these two parts.

A logical possibility to deal with the so-called diffusion paradox, which is characteristic for any diffusion equation, namely that a part of the substance (although very small) propagates in the medium with a basically independent velocity, is the formulation of a model with a sharp interface between the moist and dry part of the specimen, in other words the application of a Stefan problem.

The <u>Stefan problem</u> was originally formulated by Stefan (1889) for phase change processes between water and ice. From a mathematical point of view, it consisted of simultaneous determination of temperature field and position of the unknown interface between the solid and liquid parts of the space interval. Later, further applications arose, called Stefan-like problems, which were from the physical point of view very diverse but mathematically had a basic feature that is common to all Stefan problems, namely the solution of the problem (most often diffusion) on a time-varying space interval.

The application of Stefan problem to the transport of liquid moisture in porous materials was first carried out by Černý (1987), and consisted basically in the formulation of conditions on the interface between the moist and dry parts of the specimen. Assuming simple moisture transport without any additional effects, i.e. the moisture gradient is the only driving force, we obtain

$$u(x = s(t), t) = u_x \tag{3.85}$$

$$u_x \frac{ds}{dt} = -\kappa(u_x) \left(\frac{\partial u}{\partial x} \right)_{x=s(t)}, \tag{3.86}$$

where κ is the moisture diffusivity.

The first condition means that a constant moisture content u_x is maintained on the sharp interface $x = s(t)$ during the whole process. Physically it means, for instance, that some of the water is bonded by adhesion forces onto the pore walls, i.e. the presence of surface phase of water, which can be transferred in a limited way only within the porous medium after once being bonded.

The second condition is the mass balance of moisture on the interface $x = s(t)$. Multiplying this condition formally by the partial density of the porous matrix ρ_0 (u is the moisture content by mass), the interpretation of the particular terms is as follows: on the left-hand side we have $\rho_0 u_x \frac{ds}{dt}$, which is the mass of water per unit area of the interface that is picked up at the interface per unit time and in this way in fact excluded from taking further part in the diffusion process. The term on the right-hand side is the water flux in the direction towards the interface.

The basic diffusion equation remains unchanged,

$$\frac{\partial u}{\partial t} = \frac{\partial}{\partial x} \left(\kappa \frac{\partial u}{\partial x} \right), \tag{3.87}$$

we only have to take into account that it is valid on the space interval $\langle 0, s(t) \rangle$, for $x > s(t)$ we have $u = 0$.

The Stefan-like formulation with a sharp interface is not the only way to deal with the disagreement between theoretical description and experimental moisture suction curves. Another possibility is a formulation of a solution that would lead to a sufficiently steep interface between the moist and dry parts of the space interval. This type of solution was formulated by Häupl and Stopp (1980). They applied Krischer's formulation of liquid moisture transport in the form

$$\rho_w \frac{\partial w}{\partial t} = \rho_w \frac{\partial}{\partial x}\left(\kappa(w)\frac{\partial w}{\partial x}\right),$$ (3.88)

and assumed a special type of the $\kappa(w)$ function,

$$\kappa(w) = \kappa_0 \left[(n+1)\left(\frac{w}{w_s}\right)^{\frac{1}{n}} - n\left(\frac{w}{w_s}\right)^{\frac{2}{n}}\right],$$ (3.89)

where w_s is the saturation moisture content of the material, $\kappa_0 = \kappa(w_s)$. In equation (3.89), the value of w_s is known, but κ_0 and n are not known in advance. Therefore, these have to be determined experimentally.

The analytical solution of the diffusion equation (3.88) with the $\kappa(w)$ function in the form of (3.89) can be formulated in the form

$$w(x,t) = w_s \left(1 - \sqrt{\frac{1}{2n(n+1)\kappa_0}}\frac{x}{\sqrt{t}}\right)^n.$$ (3.90)

Clearly, this solution has a real interpretation only when the term in parentheses on the right-hand side of equation (3.90) is greater than or equal to zero. For the limiting case $w = 0$ we obtain

$$x_{\max} = \sqrt{2n(n+1)\kappa_0} \cdot \sqrt{t},$$ (3.91)

which is in fact the position of the interface between the moist and dry part of the specimen. Measuring this position as function of time, i.e. determining

$$x_{\max} = B\sqrt{t},$$ (3.92)

we obtain the first experimental quantity necessary for the determination of the unknown quantities κ_0 and n, because a comparison of (3.91) and (3.92) gives

$$B = \sqrt{2n(n+1)\kappa_0}.$$ (3.93)

The second experimental quantity, which is relatively easy to measure, is the total amount of water in the material as function of time, which can be expressed in the form

$$m_w(t) = A \cdot S \cdot \sqrt{t},$$ (3.94)

where S is the cross-sectional area of the specimen. The only unknown parameter A can be determined from the measured mass of the sample at specified time intervals.

A quantity corresponding to the value of A can also be derived from the analytical solution (3.90). We know that

$$m_w(t) = S \int_0^t j \, dt,$$ (3.95)

where j is the moisture flux at the contact surface of the material with water, i.e. $x = 0$,

$$j = -\rho_w \kappa(w_s) \left(\frac{\partial w}{\partial x}\right)_{x=0}.$$ (3.96)

From (3.89) we have

$$\kappa(w_s) = \kappa_0,$$ (3.97)

and from (3.90)

$$\left(\frac{\partial w}{\partial x}\right)_{x=0} = w_s \left[n \left(1 - \sqrt{\frac{1}{2n(n+1)\kappa_0}} \frac{x}{\sqrt{t}}\right)^{n-1} \right.$$

$$\left. \times \left(-\sqrt{\frac{1}{2n(n+1)\kappa_0}} \frac{1}{\sqrt{t}}\right)\right]_{x=0}$$

$$= -w_s n \sqrt{\frac{1}{2n(n+1)\kappa_0}} \frac{1}{\sqrt{t}} = -w_s \sqrt{\frac{n}{2(n+1)\kappa_0}} \frac{1}{\sqrt{t}}.$$ (3.98)

Therefore,

$$j = \rho_w w_s \kappa_0 \sqrt{\frac{n}{2(n+1)\kappa_0}} \frac{1}{\sqrt{t}} = \rho_w w_s \sqrt{\frac{\kappa_0 n}{2(n+1)}} \frac{1}{\sqrt{t}}.$$ (3.99)

Substituting (3.99) into (3.95) then

$$m_w(t) = S \rho_w w_s \sqrt{\frac{\kappa_0 n}{2(n+1)}} \int_0^t \frac{1}{\sqrt{t}}\, dt$$

$$= S \rho_w w_s \sqrt{\frac{\kappa_0 n}{2(n+1)}} \frac{t^{1/2}}{\frac{1}{2}} = S \rho_w w_s \sqrt{\frac{2\kappa_0 n}{n+1}} \sqrt{t}$$ (3.100)

and

$$A = \rho_w w_s \sqrt{\frac{2\kappa_0 n}{n+1}}.$$ (3.101)

From equation (3.93) we easily obtain

$$\kappa_0 = \frac{B^2}{2n(n+1)}.$$ (3.102)

Substituting (3.102) into (3.101) we get

$$A = \rho_w w_s \sqrt{\frac{2n}{n+1}} \frac{B}{\sqrt{2n(n+1)}},$$ (3.103)

after elementary modifications

$$A = \rho_{\text{w}} w_{\text{s}} \frac{B}{n+1},$$ (3.104)

and finally

$$n = \frac{B}{A} \rho_{\text{w}} w_{\text{s}} - 1.$$ (3.105)

Knowing now κ_0 and n, we can determine $\kappa(w)$ from the relation (3.89) and $w(x,t)$ from (3.90).

3.1.3 Hybrid Models

Diffusion and convection models were for more or less historical reasons developed separately, and communication between these basic schools was minimal. Nevertheless, both models have their advantages and disadvantages and could conveniently be combined.

One of the possibilities for such synthesis is formulation of moisture transport using the linear theory of mixtures, which makes it possible to include both diffusion and convection fluxes. One of first attempts in this direction was by Černý (1987).

This formulation is in fact a straightforward application of the continuum theory of mixtures, as introduced in detail by for instance Atkin and Craine (1975, 1976). It consists of the basic assumption that the particular phases of the porous medium, i.e. the solid matrix and water in the simplest case, can be considered as partial continua, which differ from each other by their field parameters (velocity, density, etc.), and the whole moist material is represented by another continuum, namely the mixture of both components.

Within the framework of this theory, the partial moisture density can be expressed as

$$\rho_{\text{v}} = \frac{\text{d}m_{\text{w}}}{\text{d}V},$$ (3.106)

where m_{w} is the mass of water, V the volume of the porous body, and for the partial density of the porous matrix we have

$$\rho_{\text{s}} = \frac{\text{d}m_{\text{s}}}{\text{d}V}.$$ (3.107)

Using the definition relation for the moisture content by mass

$$u = \frac{\text{d}m_{\text{w}}}{\text{d}m_{\text{s}}},$$ (3.108)

we get

$$\rho_{\text{v}} = \rho_{\text{s}} u.$$ (3.109)

The balance equation of mass of the moisture component is expressed within the framework of the theory of mixtures in the form

$$\frac{\partial \rho_v}{\partial t} + \text{div } \vec{j} = 0, \tag{3.110}$$

where $\vec{j}$ is the total moisture flux,

$$\vec{j} = \rho_v \vec{v}_v, \tag{3.111}$$

and $\vec{v}_v$ the velocity of the moisture component. The total moisture flux can be expressed as a sum of diffusion and convection fluxes,

$$\vec{j} = \vec{j}_d + \vec{j}_c, \tag{3.112}$$

where

$$\vec{j}_c = \rho_v \vec{v}, \tag{3.113}$$

$$\vec{j}_d = \rho_v (\vec{v}_v - \vec{v}), \tag{3.114}$$

$\vec{v}$ is the barycentric velocity (i.e. the velocity of the mixture), defined by the relation

$$(\rho_s + \rho_v) \cdot \vec{v} = \rho_s \vec{v}_s + \rho_v \vec{v}_v, \tag{3.115}$$

where $\vec{v}_s$ is the velocity of the porous matrix. Assuming that the porous matrix is a nondeformable solid body, we can put

$$\vec{v}_s = 0, \tag{3.116}$$

and therefore

$$(\rho_s + \rho_v) \vec{v} = \rho_v \vec{v}_v. \tag{3.117}$$

According to the laws of linear irreversible thermodynamics, the diffusion moisture flux $\vec{j}_d$ can be expressed by the phenomenological relation

$$\vec{j}_d = -\kappa_D \text{ grad } \rho_v, \tag{3.118}$$

so that for the total moisture flux we have

$$\vec{j} = -\kappa_D \text{ grad } \rho_v + \rho_v \vec{v}. \tag{3.119}$$

Substituting (3.119) into the mass balance of the component of moisture (3.110) we get

$$\frac{\partial \rho_v}{\partial t} = \text{div } (\kappa_D \text{ grad } \rho_v) - \text{div } (\rho_v \vec{v}). \tag{3.120}$$

In the general case, the velocity of the mixture $\vec{v}$ has to be determined from the momentum balance of the mixture. In our simple case of a two-component mixture, where one of its components has zero velocity, the relations that we already introduced will be sufficient.

From equation (3.117) it follows that

$$\vec{v}_v = \left(\frac{\rho_s}{\rho_v} + 1\right)\vec{v} = \frac{u+1}{u}\vec{v}. \tag{3.121}$$

A comparison of the two relations for the diffusion fluxes (3.114) and (3.118) gives

$$\rho_v(\vec{v}_v - \vec{v}) = -\kappa_D \operatorname{grad} \rho_v, \tag{3.122}$$

substituting for $\vec{v}_v$ from (3.121) we get

$$\rho_v\left(\frac{u+1}{u} - 1\right)\vec{v} = -\kappa_D \operatorname{grad} \rho_v, \tag{3.123}$$

and after simple algebraic operations then

$$\frac{\rho_v}{u}\vec{v} = -\kappa_D \operatorname{grad} \rho_v, \tag{3.124}$$

which leads to

$$\vec{v} = -\frac{\kappa_D}{\rho_s} \operatorname{grad} \rho_v. \tag{3.125}$$

Substituting into the mass balance of moisture (3.120) under assumption

$$\rho_s = const., \tag{3.126}$$

we then get

$$\frac{\partial u}{\partial t} = \operatorname{div}(\kappa_D \operatorname{grad} u) + \operatorname{div}(u\kappa_D \operatorname{grad} u), \tag{3.127}$$

and after a modification of the last term on the right-hand side finally

$$\frac{\partial u}{\partial t} = (1+u)\operatorname{div}(\kappa_D \operatorname{grad} u) + \kappa_D(\operatorname{grad} u)^2. \tag{3.128}$$

3.2 COUPLED HEAT AND MOISTURE TRANSPORT

The models from the previous section assumed moisture transfer in isothermal conditions. This is a very rare case in practice, which can only be employed in some specific situations when temperature differences are very small.

A typical example is groundwater flow. At a sufficient depth under surface level, thermal conditions are more or less stable, and for not very thick soil or rock layers, the condition $T = const.$ is a reasonable approximation. On the other hand, in modelling moisture transport in envelope parts of building structures the assumption of constant temperature is plainly wrong. For instance in winter period we can have $+20°C$ on the internal side and $-20°C$ on the external side, in the summer period then the temperature on the facade can increase up to $60 - 70°C$.

The models of pure moisture transport in isothermal conditions were formulated mostly in an intuitive way, and employed the analogy with known empirical relations, for instance for diffusion in gases or water flow in pipes. The situation with models of coupled heat and moisture transport is somewhat more complicated, and the application of laws of irreversible thermodynamics is unavoidable.

To demonstrate a model application of thermodynamic laws to transport phenomena of heat and moisture in porous materials, we will first choose the simplest model to present the basic ideas and features of this application. This is the model of diffusive heat and liquid moisture transport without taking into account phase change processes. Its derivation will be performed on the basis of an analogous case of coupled heat conduction and diffusion introduced by Kvasnica (1965).

3.2.1 Elementary Thermodynamic Model

We will consider the porous body as a mixture consisting of two components, namely the porous matrix and water. This system will neither move nor deform, so that for its convective velocity we will have $\vec{v} = 0$. Under these conditions, only two balance equations are sufficient, namely for the internal energy of the mixture and for the mass of the water component, which can be written in the form

$$\frac{\partial}{\partial t}(\rho u) + \text{div } \vec{j}_Q = 0 \tag{3.129}$$

$$\frac{\partial}{\partial t}(\rho c) + \text{div } \vec{j}_c = 0, \tag{3.130}$$

where ρ is the total density of the mixture, u the specific internal energy, c the mass concentration of moisture,

$$c = \frac{\rho_v}{\rho}, \tag{3.131}$$

ρ_v the partial density of moisture, $\vec{j}_Q$ the heat flux, $\vec{j}_c$ the moisture flux.

The relations between the moisture concentration c and previously introduced moisture content by mass u (we will denote it as u_m in this section in order to avoid confusion with the specific internal energy, which is also traditionally denoted in thermodynamics as u) and volumetric moisture content w can be easily expressed using the partial moisture density ρ_v,

$$\rho_v = \rho c = \rho_w w = \rho_s u_m. \tag{3.132}$$

Therefore, the derivation, which will be done with moisture concentration c here, could easily be performed for volume or mass moisture content in an analogous way, with only formal adjustments. The main aim of our thermodynamic derivation will be the identification of the heat flux $\vec{j}_Q$ and moisture flux $\vec{j}_c$.

Considering a closed system with no additional internal heat sources, i.e.

$$\vec{j}_s = 0 \tag{3.133}$$

$$b_s = 0 \tag{3.134}$$

in the entropy inequality (see Chapter 2 for details)

$$\Gamma = \frac{\mathrm{d}}{\mathrm{d}t} \int_V \rho s \, \mathrm{d}V + \int_V \rho b_s \, \mathrm{d}V + \sum_i B_i - \int_S \vec{j}_s \cdot \mathrm{d}\vec{S} \geq 0, \tag{3.135}$$

we can write for entropy production

$$\Gamma = \frac{\partial S}{\partial t} = \frac{\partial}{\partial t} \int_V \rho s \, \mathrm{d}V, \tag{3.136}$$

where s is the specific entropy.

If n_1, n_2 are numbers of particles of the particular components in a unit mass of the mixture, then for concentrations we have

$$c_i = n_i m_i, \quad i = 1, 2, \tag{3.137}$$

where m_i is the mass of one particle, and therefore

$$c_1 + c_2 = n_1 m_1 + n_2 m_2 = 1. \tag{3.138}$$

For systems with a change of number of particles and for the specific quantities, the first law of thermodynamics reads

$$\mathrm{d}u = T\mathrm{d}s - p\mathrm{d}v + \mu_1 \mathrm{d}n_1 + \mu_2 \mathrm{d}n_2, \tag{3.139}$$

where p is the pressure, v the specific volume defined as

$$v = \frac{1}{\rho}, \tag{3.140}$$

with the differential in the form

$$\mathrm{d}v = \mathrm{d}\left(\frac{1}{\rho}\right) = -\frac{1}{\rho^2}\mathrm{d}\rho, \tag{3.141}$$

and μ_1, μ_2 are the chemical potentials of the components 1, 2.

The last two terms in equation (3.139) can be modified using (3.137) and (3.138). For the sake of simplicity, we will denote $c = c_1$ in what follows, so that we can write

$$\mu_1 \mathrm{d}n_1 = \mu_1 \mathrm{d}\left(\frac{c}{m_1}\right) = \frac{\mu_1}{m_1}\mathrm{d}c \tag{3.142}$$

$$\mu_2 dn_2 = \mu_2 d \left(\frac{1-c}{m_2} \right) = -\frac{\mu_2}{m_2} dc. \tag{3.143}$$

Now, we will define the specific chemical potential μ (i.e. the chemical potential per unit mass) by the relation

$$\mu = \frac{\mu_1}{m_1} - \frac{\mu_2}{m_2}. \tag{3.144}$$

Substituting (3.141)–(3.144) into the first law of thermodynamics (3.139) we get

$$du = T ds + \frac{p}{\rho^2} d\rho + \mu dc. \tag{3.145}$$

We will further define the specific Gibbs potential g as

$$g = u - Ts + \frac{p}{\rho} = \mu c. \tag{3.146}$$

For its differential dg we have

$$dg = du - T ds - s dT + pd \left(\frac{1}{\rho} \right) + \frac{1}{\rho} dp. \tag{3.147}$$

Substituting (3.145) into (3.147) we obtain

$$dg = -s dT + \frac{1}{\rho} dp + \mu dc. \tag{3.148}$$

Now we have all the necessary relations and can begin the derivation of phenomenological relations for heat and moisture transport. The most important task in this respect is derivation of the entropy production. According to (3.136) we have to express the term

$$\frac{\partial}{\partial t}(\rho s) = \rho \frac{\partial s}{\partial t} + s \frac{\partial \rho}{\partial t}. \tag{3.149}$$

From the first law of thermodynamics (3.145) it follows that

$$\frac{\partial u}{\partial t} = T \frac{\partial s}{\partial t} + \frac{p}{\rho^2} \frac{\partial \rho}{\partial t} + \mu \frac{\partial c}{\partial t} \tag{3.150}$$

or

$$\frac{\partial s}{\partial t} = \frac{1}{T} \left(\frac{\partial u}{\partial t} - \frac{p}{\rho^2} \frac{\partial \rho}{\partial t} - \mu \frac{\partial c}{\partial t} \right). \tag{3.151}$$

The balance equations for the internal energy of the mixture and mass of water component (3.129) and (3.130) can be rewritten in the form

$$\rho \frac{\partial u}{\partial t} = - \operatorname{div} \vec{j}_Q - u \frac{\partial \rho}{\partial t} \tag{3.152}$$

$$\rho \frac{\partial c}{\partial t} = - \operatorname{div} \vec{j}_c - c \frac{\partial \rho}{\partial t}. \tag{3.153}$$

Substituting now (3.151)–(3.153) into (3.149), we obtain

$$\frac{\partial}{\partial t}(\rho s) = \frac{\rho}{T}\frac{\partial u}{\partial t} - \frac{p}{\rho T}\frac{\partial \rho}{\partial t} - \frac{\mu \rho}{T}\frac{\partial c}{\partial t} + s\frac{\partial \rho}{\partial t}$$

$$= -\frac{1}{T}\operatorname{div}\vec{j}_Q - \frac{u}{T}\frac{\partial \rho}{\partial t} + \frac{\mu}{T}\operatorname{div}\vec{j}_c + \frac{\mu c}{T}\frac{\partial \rho}{\partial t} - \frac{p}{\rho T}\frac{\partial \rho}{\partial t} + s\frac{\partial \rho}{\partial t}$$

$$= -\frac{1}{T}\operatorname{div}\vec{j}_Q + \frac{\mu}{T}\operatorname{div}\vec{j}_c - \frac{1}{T}\left(u + \frac{p}{\rho} - Ts - \mu c\right)\frac{\partial \rho}{\partial t}. \qquad (3.154)$$

Applying the relation (3.146) for the specific Gibbs potential, we observe that the last term in equation (3.154) is equal to zero, and therefore

$$\frac{\partial}{\partial t}(\rho s) = -\frac{1}{T}\operatorname{div}\vec{j}_Q + \frac{\mu}{T}\operatorname{div}\vec{j}_c. \qquad (3.155)$$

Modifying the last term of equation (3.155) using the common rule of differentiation of a product and substituting (3.155) into the relation of entropy production (3.136), we obtain

$$\Gamma = \frac{\partial S}{\partial t} = -\int_V \frac{1}{T}\operatorname{div}(\vec{j}_Q - \mu\vec{j}_c)\,dV - \int_V \frac{\vec{j}_c}{T}\operatorname{grad}\mu\,dV. \qquad (3.156)$$

The relation (3.156) has to be converted into a form in which there are gradients of certain thermodynamic quantities, which could serve as thermodynamic forces X_i. To achieve that, we have to apply the product differentiation rule one more time, now to the integrand in the first term of the right-hand side of equation (3.156), namely

$$\Gamma = \frac{\partial S}{\partial t} = -\int_V \operatorname{div}\left[\frac{1}{T}(\vec{j}_Q - \mu\vec{j}_c)\right]dV$$

$$+ \int_V (\vec{j}_Q - \mu\vec{j}_c)\operatorname{grad}\left(\frac{1}{T}\right)dV - \int_V \frac{\vec{j}_c}{T}\operatorname{grad}\mu\,dV. \qquad (3.157)$$

Expressing

$$\operatorname{grad}\left(\frac{1}{T}\right) = -\frac{1}{T^2}\operatorname{grad}T, \qquad (3.158)$$

and applying the Gauss formula to the first term of the right-hand side of equation (3.157), we obtain

$$\Gamma = \frac{\partial S}{\partial t} = -\int_V \frac{1}{T^2}(\vec{j}_Q - \mu\vec{j}_c)\operatorname{grad}T\,dV - \int_V \frac{\vec{j}_c}{T}\operatorname{grad}\mu\,dV$$

$$- \int_S \vec{n}\cdot\frac{1}{T}(\vec{j}_Q - \mu\vec{j}_c)\,dS. \qquad (3.159)$$

Under the assumption that the system is closed, the energy flux across the surface is equal to zero,

$$\vec{j}_Q - \mu \vec{j}_c = 0, \tag{3.160}$$

and therefore the last term on the right-hand side of equation (3.159) is equal to zero.

The resulting form of entropy production,

$$\Gamma = \frac{\partial S}{\partial t} = -\int_V \frac{1}{T^2}(\vec{j}_Q - \mu\vec{j}_c)\,\text{grad}\,T\,dV - \int \frac{\vec{j}_c}{T}\,\text{grad}\,\mu\,dV, \tag{3.161}$$

is already suitable for the identification of thermodynamic forces and fluxes. We can write

$$\vec{J}_1 = \vec{j}_c \tag{3.162}$$

$$\vec{J}_2 = \vec{j}_Q - \mu\vec{j}_c \tag{3.163}$$

$$\vec{X}_1 = -\frac{\text{grad}\,\mu}{T} \tag{3.164}$$

$$\vec{X}_2 = -\frac{\text{grad}\,T}{T^2}. \tag{3.165}$$

On the basis of the linearity law, the following relations can be written:

$$\vec{J}_1 = L_{11}\vec{X}_1 + L_{12}\vec{X}_2 \tag{3.166}$$

$$\vec{J}_2 = L_{21}\vec{X}_1 + L_{22}\vec{X}_2. \tag{3.167}$$

Putting now

$$L_{11} = \alpha T \tag{3.168}$$

$$L_{12} = \beta T^2 \tag{3.169}$$

$$L_{21} = \zeta T \tag{3.170}$$

$$L_{22} = \gamma T^2, \tag{3.171}$$

and using the Onsager reciprocity relation, so that

$$\zeta = \beta T, \tag{3.172}$$

we obtain for the particular fluxes

$$\vec{j}_c = -\alpha\,\text{grad}\,\mu - \beta\,\text{grad}\,T \tag{3.173}$$

$$\vec{j}_Q - \mu \vec{j}_c = -\beta T \operatorname{grad} \mu - \gamma \operatorname{grad} T. \tag{3.174}$$

Eliminating grad μ from equations (3.173)–(3.174), we get

$$\vec{j}_Q - \left(\mu + \frac{\beta T}{\alpha}\right) \vec{j}_c = -\left(\gamma - \frac{\beta^2 T}{\alpha}\right) \operatorname{grad} T. \tag{3.175}$$

In equation (3.175) $\vec{j}_c$ determines the mass flux. If $\vec{j}_c = 0$, Fourier's law of heat conduction must be valid, and for the thermal conductivity λ we obtain

$$\lambda = \gamma - \frac{\beta^2 T}{\alpha}. \tag{3.176}$$

Chemical potential is a quantity that is mostly not very suitable for practical applications. Therefore it should be eliminated from equation (3.173). Because $\mu = \mu(c, p, T)$, we can write

$$\operatorname{grad} \mu = \left(\frac{\partial \mu}{\partial c}\right)_{p,T} \operatorname{grad} c + \left(\frac{\partial \mu}{\partial p}\right)_{c,T} \operatorname{grad} p$$

$$+ \left(\frac{\partial \mu}{\partial T}\right)_{c,p} \operatorname{grad} T. \tag{3.177}$$

Substituting (3.177) into (3.173), we get

$$\vec{j}_c = -\alpha \left(\frac{\partial \mu}{\partial c}\right)_{p,T} \operatorname{grad} c - \alpha \left(\frac{\partial \mu}{\partial p}\right)_{c,T} \operatorname{grad} p$$

$$- \left[\beta + \alpha \left(\frac{\partial \mu}{\partial T}\right)_{c,p}\right] \operatorname{grad} T. \tag{3.178}$$

Introducing the definitions

$$D = \frac{\alpha}{\rho} \left(\frac{\partial \mu}{\partial c}\right)_{p,T} \tag{3.179}$$

$$\delta = \frac{D\Theta}{T} = \frac{1}{\rho} \left[\beta + \alpha \left(\frac{\partial \mu}{\partial T}\right)_{c,p}\right] \tag{3.180}$$

$$\varphi = \frac{D\pi}{p} = \frac{\alpha}{\rho} = \left(\frac{\partial \mu}{\partial p}\right)_{c,T}, \tag{3.181}$$

where D is the diffusion coefficient, δ the thermodiffusion coefficient (Soret coefficient), φ the barodiffusion coefficient, Θ the thermodiffusion ratio, π the barodiffusion ratio, and substituting (3.179)–(3.181) into (3.178), we obtain

$$\vec{j}_c = -\rho(D \operatorname{grad} c + \delta \operatorname{grad} T + \varphi \operatorname{grad} p). \tag{3.182}$$

The mass flux $\vec{j}_c$ is now identified in a sufficient way, and we can return to the heat flux. In equation (3.175) it is necessary to express $\frac{\beta T}{\alpha}$. From the definition relation (3.180) it follows that

$$\beta = \frac{\rho D\Theta}{T} - \alpha \left(\frac{\partial \mu}{\partial T}\right)_{c,p}. \tag{3.183}$$

Multiplying equation (3.183) by the expression $\frac{T}{\alpha}$ and using relation (3.179), we obtain

$$\frac{\beta T}{\alpha} = \Theta \left(\frac{\partial \mu}{\partial c}\right)_{p,T} - T \left(\frac{\partial \mu}{\partial T}\right)_{c,p}. \tag{3.184}$$

Substituting (3.184) into (3.175) then

$$\vec{j}_Q = \left[\Theta \left(\frac{\partial \mu}{\partial c}\right)_{p,T} - T \left(\frac{\partial \mu}{\partial T}\right)_{c,p} + \mu\right] \vec{j}_c - \lambda \operatorname{grad} T. \tag{3.185}$$

Denoting

$$K_\mu = \Theta \left(\frac{\partial \mu}{\partial c}\right)_{p,T} - T \left(\frac{\partial \mu}{\partial T}\right)_{c,p} \tag{3.186}$$

and substituting (3.186) and (3.182) into (3.185), we get

$$\vec{j}_Q - \mu\vec{j}_c = -\lambda^* \operatorname{grad} T - k_c \operatorname{grad} c - k_p \operatorname{grad} p, \tag{3.187}$$

where

$$\lambda^* = K_\mu \rho \delta + \lambda \tag{3.188}$$

$$k_c = K_\mu \rho D \tag{3.189}$$

$$k_p = K_\mu \rho \varphi, \tag{3.190}$$

λ^* is the generalized thermal conductivity, k_c the diffusion-thermo coefficient (Dufour coefficient), k_p the diffusion-baro coefficient.

Equations (3.182) and (3.187) for mass flux and heat flux can now be substituted into the appropriate balance equations for the mass of a component of the mixture and for the internal energy of the mixture. For the component mass balance we use equation (3.153), in formulating the energy balance we use relations (3.136) and (3.156). Substituting (3.136) into (3.156) and multiplying the resulting equation by T we get (after conversion into a local form):

$$T\frac{\partial(\rho s)}{\partial t} + \operatorname{div}(\vec{j}_Q - \mu\vec{j}_c) + \vec{j}_c \operatorname{grad} \mu = 0. \tag{3.191}$$

For demonstration purposes, the problem just formulated would still be too complicated. Therefore, we will introduce further simplifications. We will consider isobaric conditions,

$$p = const., \qquad \text{grad } p = 0, \tag{3.192}$$

and further

$$\rho = const. \tag{3.193}$$

For an isobaric process we have

$$dQ = mc_p dT, \tag{3.194}$$

where c_p is the specific heat capacity at constant pressure. According to the second law of thermodynamics (assuming the local equilibrium hypothesis) we can write

$$T ds = \frac{dQ}{m}. \tag{3.195}$$

Substituting (3.194) into (3.195), we obtain

$$T ds = c_p dT. \tag{3.196}$$

In the most of practical applications

$$\vec{j}_c \text{ grad } \mu << \text{ div } (\vec{j}_Q - \mu \vec{j}_c), \tag{3.197}$$

and therefore the term $(\vec{j}_c \text{ grad } \mu)$ in the balance equation (3.191) can be neglected. Substituting now (3.182), (3.187), (3.192)–(3.193) and (3.196) into the balance equations (3.153) and (3.191), we obtain the resulting system of equations for coupled heat and mass transfer in a porous medium in the form

$$\frac{\partial c}{\partial t} = \text{ div } (D \text{ grad } c + \delta \text{ grad } T) \tag{3.198}$$

$$\rho c_p \frac{\partial T}{\partial t} = \text{ div } (\lambda^* \text{ grad } T + k_c \text{ grad } c). \tag{3.199}$$

Neglecting the cross effects and putting

$$k_c = 0, \qquad \delta = 0 \tag{3.200}$$

in equations (3.198), (3.199), we arrive at pure diffusion and pure heat conduction,

$$\frac{\partial c}{\partial t} = \text{ div } (D \text{ grad } c) \tag{3.201}$$

$$\rho c_p \frac{\partial T}{\partial t} = \text{ div } (\lambda \text{ grad } T). \tag{3.202}$$

3.2.2 Convection Models

The most recognized convection model of coupled heat and moisture transport (which, however, in its final form represents a diffusion formulation) is that of Philip and de Vries (1957).

In their description of liquid water transport in porous materials, Philip and de Vries started with the Darcy's equation in the form

$$\vec{j}_{lc} = -\rho_1 K_1 \operatorname{grad} \psi, \tag{3.203}$$

where ρ_1 is the density of water, K_1 the hydraulic conductivity, ψ the pressure head, $\vec{j}_{lc}$ the water flux, and assumed that pressure head is function of moisture content by volume and temperature,

$$\psi = \psi(w, T). \tag{3.204}$$

Then,

$$\operatorname{grad} \psi = \frac{\partial \psi}{\partial w} \operatorname{grad} w + \frac{\partial \psi}{\partial T} \operatorname{grad} T,$$

and after substitution into the Darcy's equation

$$\vec{j}_{lc} = -\rho_1 \left(K_1 \frac{\partial \psi}{\partial w} \operatorname{grad} w + \frac{\partial \psi}{\partial T} \operatorname{grad} T \right)$$

$$= -\rho_1 (D_{wl} \operatorname{grad} w + D_{Tl} \operatorname{grad} T), \tag{3.205}$$

where D_{wl} and D_{Tl} are the coefficients of water diffusion in the porous body induced by moisture gradient and temperature gradient. These coefficients can be expressed in terms of classical convective quantities from the Darcy's equation as follows. Pressure head can be calculated from the known relation for the height of rise of a liquid in capillary of radius r,

$$\pi r^2 \psi \rho_1 g = -2\pi r \sigma \cos \varphi, \tag{3.206}$$

and therefore

$$\psi = -\frac{2\sigma \cos \varphi}{\rho_1 g r}, \tag{3.207}$$

where g is the gravity acceleration, σ the surface tension, φ the contact angle, which is for water $\sim 8°$, therefore $\cos \varphi \sim 1$, and

$$\psi = -\frac{2\sigma(T)}{\rho_1 g r(w)}. \tag{3.208}$$

This relation is called the Laplace equation, and its more general form was given in Chapter 1. The Laplace equation can be applied to the calculation of the thermodiffusion coefficient D_{Tl},

$$D_{Tl} = K_1 \frac{\partial \psi}{\partial T} = -K_1 \frac{2}{\rho_1 g r(w)} \frac{\partial \sigma}{\partial T} = K_1 \psi \frac{1}{\sigma} \frac{\partial \sigma}{\partial T}. \tag{3.209}$$

The relation for D_{wl} can be derived in an analogous way,

$$D_{\text{wl}} = K_1 \frac{\partial \psi}{\partial w} = K_1 \left(-\frac{2\sigma}{\rho_{\text{l}} g}\right) \left(-\frac{1}{r^2}\right) \frac{\partial r}{\partial w} = -K_1 \frac{\psi}{r} \frac{\partial r}{\partial w}. \tag{3.210}$$

Besides the gradient of capillary pressure (or pressure head), Philip and de Vries took into account gravity as another driving force in water transport,

$$\vec{j}_{\text{lg}} = -\rho_{\text{l}} K_1 \vec{k}, \tag{3.211}$$

where $\vec{k}$ is a unit vector in the vertical direction, with the positive orientation upwards.

For the water vapour transport they applied Fick's law, extended for Stefan diffusion (similarly as Krischer, see Section 3.1),

$$\vec{j}_{\text{v}} = -\alpha a D \frac{p}{p - p_{\text{v}}} \text{ grad } \rho_{\text{v}}, \tag{3.212}$$

where α is the tortuosity, a geometrical factor ($\alpha \ll 1$), which expresses the extension of the path length of a molecule in the system of curved pores in comparison to its path in the air, a the relative volume ratio of the air in partially saturated material,

$$a = \Pi - w, \tag{3.213}$$

Π the porosity, D the diffusion coefficient of water vapour in the air, ρ_{v} the partial density of water vapour in the porous body, p the total pressure, p_{v} the partial pressure of water vapour in the air. For the partial pressure of water vapour they used the common relation using relative humidity φ,

$$p_{\text{v}} = \varphi p_{\text{vs}}(T), \tag{3.214}$$

where p_{vs} is the saturated water vapour pressure at temperature T, for the relation between p_{v} and ρ_{v} then the equation of state of an ideal gas,

$$p_{\text{v}} = \frac{RT}{M} \rho_{\text{v}}, \tag{3.215}$$

and assumed $\rho_{\text{v}} = \rho_{\text{v}}(w, T)$, therefore

$$\text{grad } \rho_{\text{v}} = \frac{\partial \rho_{\text{v}}}{\partial w} \text{ grad } w + \frac{\partial \rho_{\text{v}}}{\partial T} \text{ grad } T, \tag{3.216}$$

so that

$$\vec{j}_{\text{v}} = -\alpha a D \frac{p}{p - p_{\text{v}}} \left(\frac{\partial \rho_{\text{v}}}{\partial w} \text{ grad } w + \frac{\partial \rho_{\text{v}}}{\partial T} \text{ grad } T\right)$$

$$= -\rho_{\text{l}}(D_{\text{wv}} \text{ grad } w + D_{\text{Tv}} \text{ grad } T). \tag{3.217}$$

In order to express the new coefficients D_{wv}, D_{Tv} in terms of classical diffusion quantities, it is necessary to determine $\frac{\partial \rho_v}{\partial w}, \frac{\partial \rho_v}{\partial T}$. From the equation of state it follows

$$\rho_v = \frac{M}{RT} p_v = \frac{M}{RT} \varphi p_{vs}(T). \tag{3.218}$$

Therefore

$$\frac{\partial \rho_v}{\partial w} = \frac{\partial \rho_v}{\partial \varphi} \frac{\partial \varphi}{\partial w} = \frac{M}{RT} p_{vs} \frac{\partial \varphi}{\partial w} \tag{3.219}$$

$$\frac{\partial \rho_v}{\partial T} = -\frac{M}{RT^2} \varphi p_{vs}(T) + \frac{M}{RT} p_{vs} \frac{\partial \varphi}{\partial T} + \frac{M}{RT} \varphi \frac{dp_{vs}}{dT}$$

$$= \frac{M}{RT} \left(-\varphi \frac{p_{vs}}{T} + \varphi \frac{dp_{vs}}{dT} + p_{vs} \frac{\partial \varphi}{\partial T} \right), \tag{3.220}$$

where $\frac{\partial \varphi}{\partial w}, \frac{\partial \varphi}{\partial T}$ can be obtained from the moisture storage function $\varphi = \varphi(w, T)$, which for $T = T_0$ is in fact the sorption isotherm, $\varphi = \varphi(w, T_0)$.

Substituting (3.219) and (3.220) into (3.217) we obtain

$$D_{wv} = \frac{1}{\rho_l} \alpha a D \frac{p}{p - p_v} \frac{M}{RT} p_{vs} \frac{\partial \varphi}{\partial w} \tag{3.221}$$

$$D_{Tv} = \frac{1}{\rho_l} \alpha a D \frac{p}{p - p_v} \frac{M}{RT} \left(-\varphi \frac{p_{vs}}{T} + \varphi \frac{dp_{vs}}{dT} + p_{vs} \frac{\partial \varphi}{\partial T} \right). \tag{3.222}$$

The balance equations in the original Philip and de Vries model were considered in a simplified form, assuming for instance $\rho_l w \gg \rho_v$. The moisture balance had the following form:

$$\frac{\partial}{\partial t} (\rho_l w) = - \text{div} (\vec{j}_{lc} + \vec{j}_{lg} + \vec{j}_v), \tag{3.223}$$

and the heat balance:

$$C \frac{\partial T}{\partial t} = - \text{div} \, \vec{j}_q - L \, \text{div} \, (D_{wv} \, \text{grad} \, w), \tag{3.224}$$

where C is the volumetric specific heat capacity of the system (in J/m^{3}K), L the latent heat of evaporation of water, and for the heat flux we have

$$\vec{j}_q = -\lambda \, \text{grad} \, T. \tag{3.225}$$

After substitution for the particular fluxes and under assumption $\rho_l = \text{const.}$ the final balance equations of the Philip and de Vries model had the following form:

$$\frac{\partial w}{\partial t} = \text{div} \, (D_w \, \text{grad} \, w) + \text{div} \, (D_T \, \text{grad} \, T) + \frac{\partial K_l}{\partial z} \tag{3.226}$$

$$C\frac{\partial T}{\partial t} = \text{div} (\lambda \text{ grad } T) - L \text{ div} (D_{wv} \text{ grad } w), \tag{3.227}$$

where $D_w = D_{wl} + D_{wv}$, $D_T = D_{Tw} + D_{Tv}$.

The model of Philip and de Vries is a closed system of two equations for two unknown field variables w, T.

In the later, refined de Vries (1987) model, the heat flux is considered in a more complicated form,

$$\vec{j}_q = -\lambda \text{ grad } T + h_l(\vec{j}_{cl} + \vec{j}_{lg}) + h_v \vec{j}_v, \tag{3.228}$$

where the first term is for heat conduction, the second for convection of the liquid phase, the third for convection of the gaseous phase and phase transition of water to water vapour. For the particular specific enthalpies we have

$$h_l = c_l(T - T_0) \tag{3.229}$$

$$h_v = c_v(T - T_0) + L_v, \tag{3.230}$$

where c_l, c_v are the specific heat capacities of the liquid and the gaseous phase of water, respectively, L_v the latent heat of evaporation, and T_0 a reference temperature.

The moisture balance equation in this innovated model looked as follows:

$$\frac{\partial}{\partial t}(\rho_l w + a\rho_v) = -\text{div}(\vec{j}_{lc} + \vec{j}_{lg} + \vec{j}_v), \tag{3.231}$$

and the heat balance:

$$\frac{\partial}{\partial t}(\rho_s h_s + \rho_l h_l w + a\rho_v h_v) = -\text{div}\vec{j}_q, \tag{3.232}$$

where ρ_s is the partial density of the porous matrix, and h_s the specific enthalpy of the porous matrix,

$$h_s = c_s(T - T_0). \tag{3.233}$$

The system of two equations (3.231), (3.232) for three unknowns w, ρ_v, T was closed using the moisture storage function $\rho_v = \rho_v(w, T)$.

After substituting for the particular fluxes into (3.231), (3.232) de Vries obtained the following moisture and heat balance equations

$$\frac{\partial}{\partial t}(\rho_l w + a\rho_v) = \text{div} [\rho_l(D_{wl} + D_{wv}) \text{ grad } w]$$

$$+ \text{div} [\rho_l(D_{Tl} + D_{Tv}) \text{ grad } T] + \frac{\partial}{\partial z}(\rho_l K_l) \tag{3.234}$$

$$\frac{\partial}{\partial t}(\rho_s h_s + \rho_l h_l w + a\rho_v h_v) = \text{div} (\lambda \text{ grad } T)$$

$$- \operatorname{div} [h_1 \rho_1 (D_{wl} \operatorname{grad} w + D_{Tv} \operatorname{grad} T)]$$

$$- \operatorname{div} [h_v \rho_1 (D_{wv} \operatorname{grad} w + D_{Tv} \operatorname{grad} T)] - \frac{\partial}{\partial z} (h_1 \rho_1 K_1), \quad (3.235)$$

and under assumptions $\rho_1 = \mathrm{const.}$, $\rho_s = \mathrm{const.}$, $c_s = \mathrm{const.}$, $c_1 = \mathrm{const.}$, $c_v = \mathrm{const.}$ the following final form was obtained

$$\frac{\partial w}{\partial t} = \operatorname{div} (D_w \operatorname{grad} w) + \operatorname{div} (D_T \operatorname{grad} T)$$

$$+ \frac{\partial K_1}{\partial z} - \frac{1}{\rho_1} \frac{\partial}{\partial t} (a \rho_v) \quad (3.236)$$

$$\rho_s c^* \frac{\partial T}{\partial t} = \operatorname{div} (\lambda \operatorname{grad} T) - \operatorname{div} [\rho_1 T (c_1 D_{wl} + c_v D_{wv}) \operatorname{grad} w]$$

$$- \operatorname{div} [\rho_1 T (c_1 D_{Tl} + c_v D_{Tv}) \operatorname{grad} T]$$

$$- \operatorname{div} [\rho_1 L_v (D_{wv} \operatorname{grad} w + D_{Tv} \operatorname{grad} T)]$$

$$- \frac{\partial}{\partial t} (a \rho_v L_v) - \frac{\partial}{\partial z} (\rho_1 c_1 K_1 T) - \rho_1 c_1 T \frac{\partial w}{\partial t} - c_v T \frac{\partial}{\partial t} (a \rho_v), \quad (3.237)$$

where $D_w = D_{wl} + D_{wv}$, $D_T = D_{Tl} + D_{Tv}$,

$$c^* = c_s + \frac{\rho_1}{\rho_s} c_1 w + \frac{\rho_v}{\rho_s} a c_v. \quad (3.238)$$

The refined de Vries model in its final form is again a system of two equations for the two unknowns w, T. The partial density of water vapour ρ_v is determined on the basis of the moisture storage function $\rho_v = \rho_v(w, T)$, and from the point of view of practical calculation, it is treated in fact as a material parameter, similarly for instance to $\lambda = \lambda(w, T)$.

The original Philip and de Vries model was developed in somewhat different sense by Milly (1982). The balance equations in Milly's model were formulated as follows:

$$\frac{\partial}{\partial t} [\rho_s (u_1 + u_v)] = - \operatorname{div} (\vec{j}_{lc} + \vec{j}_{lg} + \vec{j}_v) \quad (3.239)$$

$$\frac{\partial}{\partial t} [\rho_s (h_s + h_1 u_1 + h_v u_v)] = - \operatorname{div} \vec{j}_q, \quad (3.240)$$

where u_1 is the liquid moisture content by mass, $u_1 = w \rho_1 / \rho_s$, u_v the gaseous moisture content by mass, $u_v = a \rho_v / \rho_s$, the other symbols have the same meaning as in the Philip and de Vries model.

For the particular fluxes Milly used the following relations. For liquid water transport induced by capillary forces Darcy's equation without any adjustments,

$$\vec{j}_{lc} = -\rho_l K_l \,\text{grad}\,\psi,\tag{3.241}$$

and for water transport due to gravity the same relation as Philip and de Vries,

$$\vec{j}_{lg} = -\rho_l K_l \vec{k}.\tag{3.242}$$

The phenomenological relation for heat flux was also the same,

$$\vec{j}_q = -\lambda\,\text{grad}\,T + h_l(\vec{j}_{lc} + \vec{j}_{lg}) + h_v\vec{j}_v.\tag{3.243}$$

In the formulation of the relation for water vapour flux Milly started from Fick's law

$$\vec{j}_v = -\alpha a D \frac{p}{p - p_v}\,\text{grad}\,\rho_v,\tag{3.244}$$

and employed the relations

$$p_v = \varphi p_{vs}(T)\tag{3.245}$$

$$p_v = \frac{RT}{M}\rho_v.\tag{3.246}$$

In contradiction to Philip and de Vries, however, he also used the Kelvin equation between the pressure head ψ and the relative humidity φ in the form

$$\psi = \frac{RT}{gM}\ln\varphi.\tag{3.247}$$

Then, he assumed $\rho_v = \rho_v(\psi, T)$, and therefore

$$\text{grad}\,\rho_v = \frac{\partial\rho_v}{\partial\psi}\,\text{grad}\,\psi + \frac{\partial\rho_v}{\partial T}\,\text{grad}\,T\tag{3.248}$$

and

$$\vec{j}_v = -\alpha a D \frac{p}{p - p_v}\left(\frac{\partial\rho_v}{\partial\psi}\,\text{grad}\,\psi + \frac{\partial\rho_v}{\partial T}\,\text{grad}\,T\right)$$

$$= -\rho_l(K_v\,\text{grad}\,\psi + D_{Tv}\,\text{grad}\,T).\tag{3.249}$$

In order to derive the definition formulas for the coefficients K_v, D_{Tv} it is necessary to determine $\frac{\partial\rho_v}{\partial\psi}$, $\frac{\partial\rho_v}{\partial T}$. From the equation of state of an ideal gas we get

$$\rho_v = \frac{M}{RT}p_v = \frac{M}{RT}\varphi p_{vs}(T).\tag{3.250}$$

Therefore

$$\frac{\partial \rho_v}{\partial \psi} = \frac{M}{RT} p_{vs}(T) \frac{\partial \varphi}{\partial \psi}$$

(3.251)

$$\frac{\partial \rho_v}{\partial T} = -\frac{M}{RT^2} \varphi p_{vs} + \frac{M}{RT} p_{vs} \frac{\partial \varphi}{\partial T} + \frac{M}{RT} \varphi \frac{dp_{vs}}{dT}.$$

(3.252)

For the determination of $\frac{\partial \varphi}{\partial \psi}$, $\frac{\partial \varphi}{\partial T}$, we use the Kelvin equation and an algebraic modification using $\ln \varphi$

$$\frac{\partial \varphi}{\partial \psi} = \varphi \frac{\partial \ln \varphi}{\partial \psi}$$

(3.253)

$$\frac{\partial \varphi}{\partial T} = \varphi \frac{\partial \ln \varphi}{\partial T},$$

(3.254)

so that

$$d\psi = \frac{RT}{gM} d\ln \varphi$$

(3.255)

and

$$\frac{\partial \ln \varphi}{\partial \psi} = \frac{gM}{RT}.$$

(3.256)

As

$$\frac{1}{T} = \frac{R}{gM\psi} \ln \varphi,$$

(3.257)

then

$$d\left(\frac{1}{T}\right) = \frac{R}{gM\psi} d\ln \varphi$$

(3.258)

and

$$-\frac{1}{T^2} dT = \frac{R}{gM\psi} d\ln \varphi.$$

(3.259)

Consequently,

$$\frac{\partial \ln \varphi}{\partial T} = -\frac{gM\psi}{RT^2},$$

(3.260)

and finally

$$\frac{\partial \varphi}{\partial T} = -\frac{gM\psi\varphi}{RT^2}.$$

(3.261)

Therefore we have

$$\frac{\partial \rho_v}{\partial \psi} = \left(\frac{M}{RT}\right)^2 g\varphi p_{vs}$$

(3.262)

$$\frac{\partial \rho_v}{\partial T} = -\frac{M}{RT^2}\varphi p_{vs} - \frac{M}{RT^2}\frac{M}{RT}p_{vs}\varphi g\psi + \frac{M}{RT}\varphi\frac{dp_{vs}}{dT}$$

$$= \frac{M}{RT}\varphi\left(-\frac{p_{vs}}{T} - \frac{M}{RT^2}p_{vs}g\psi + \frac{dp_{vs}}{dT}\right) \tag{3.263}$$

and

$$K_v = \frac{1}{\rho_l}\alpha a D\frac{p}{p-p_v}\left(\frac{M}{RT}\right)^2 g\varphi p_{vs} \tag{3.264}$$

$$D_{Tv} = \frac{1}{\rho_l}\alpha a D\frac{p}{p-p_v}\frac{M}{RT}\varphi\left(-\frac{p_{vs}}{T} + \frac{dp_{vs}}{dT} - \frac{M}{RT^2}g\psi p_{vs}\right). \tag{3.265}$$

In the two Milly's balance equations, there appear four unknown field variables, u_l, u_v, ψ and T. Therefore, two more relations between these variables have to be formulated. The first of them is the relation between liquid moisture content, pressure head and temperature, namely

$$u_l = u_l(\psi(t), T), \tag{3.266}$$

which is the water retention curve determined experimentally in the usual way (see Chapter 5).

The second necessary relation (between u_v, ψ and T) can be derived using the Kelvin equation

$$\psi = \frac{RT}{gM}\ln\varphi, \tag{3.267}$$

modified into the form

$$\varphi = e^{\frac{\psi gM}{RT}}, \tag{3.268}$$

together with the definition relation for the relative humidity φ

$$\varphi = \frac{p_v}{p_{vs}(T)}, \tag{3.269}$$

and the equation of state of an ideal gas

$$p_v = \frac{RT}{M}\rho_v. \tag{3.270}$$

Substituting (3.270) into (3.269) we get

$$\varphi = \frac{RT\rho_v}{Mp_{vs}}, \tag{3.271}$$

using the definition relation for u_v,

$$u_v = a\frac{\rho_v}{\rho_s}, \tag{3.272}$$

which leads to

$$\rho_v = u_v \frac{\rho_s}{a},$$ (3.273)

and substituting into (3.271) we obtain

$$\varphi = \frac{RT u_v \rho_s}{M p_{vs} a}.$$ (3.274)

Therefore, for u_v we have

$$u_v = \varphi p_{vs} \frac{a}{\rho_s} \frac{M}{RT}.$$ (3.275)

As $a = \Pi - w$ and $\rho_l w = \rho_s u_l$, we can write

$$a = \Pi - \frac{\rho_s}{\rho_l} u_l.$$ (3.276)

Substituting (3.276) into (3.275) we get

$$u_v = \varphi p_{vs} \left(\frac{\Pi}{\rho_s} - \frac{u_l}{\rho_l} \right) \frac{M}{RT},$$ (3.277)

and using the Kelvin equation (3.268) then

$$u_v = p_{vs} \frac{M}{RT} \left(\frac{\Pi}{\rho_s} - \frac{u_l}{\rho_l} \right) \cdot e^{\frac{\psi g M}{RT}},$$ (3.278)

which is the final relation necessary to make the system solvable.

Equation (3.278) makes it possible to eliminate u_v from the equation system for heat and moisture transfer (3.239), (3.240). To perform this elimination, it is necessary to express $\frac{\partial u_v}{\partial t}$, which appears in both these equations. We can write

$$\frac{\partial u_v}{\partial t} = \frac{\partial u_v}{\partial u_l} \frac{\partial u_l}{\partial t} + \frac{\partial u_v}{\partial \psi} \frac{\partial \psi}{\partial t} + \frac{\partial u_v}{\partial T} \frac{\partial T}{\partial t} = -A \frac{\partial u_l}{\partial t} + B \frac{\partial \psi}{\partial t} + C \frac{\partial T}{\partial t}.$$ (3.279)

From (3.278) it follows that

$$-A = \frac{\partial u_v}{\partial u_l} = -\frac{1}{\rho_l} \varphi p_{vs} \frac{M}{RT}$$ (3.280)

$$B = \frac{\partial u_v}{\partial \psi} = p_{vs} \frac{a}{\rho_s} \frac{M}{RT} e^{\frac{\psi g M}{RT}} \frac{g M}{RT} = \varphi p_{vs} \frac{a}{\rho_s} g \left(\frac{M}{RT} \right)^2$$ (3.281)

$$C = \frac{\partial u_v}{\partial T} = \frac{\partial \varphi}{\partial T} p_{vs} \frac{a}{\rho_s} \frac{M}{RT} + \varphi \frac{dp_{vs}}{dT} \frac{a}{\rho_s} \frac{M}{RT} + \varphi p_{vs} \frac{a}{\rho_s} \frac{M}{R} \frac{d}{dT} \left(\frac{1}{T} \right)$$

$$= -\frac{\psi g M}{RT^2} \varphi p_{vs} \frac{a}{\rho_s} \frac{M}{RT} + \varphi \frac{dp_{vs}}{dT} \frac{a}{\rho_s} \frac{M}{RT} - \varphi p_{vs} \frac{a}{\rho_s} \frac{M}{RT^2}$$

$$= \frac{1}{\rho_s} \frac{M}{RT} a p_{vs} \varphi \left(-\frac{\psi g M}{RT^2} + \frac{1}{p_{vs}} \frac{dp_{vs}}{dT} - \frac{1}{T} \right). \tag{3.282}$$

Substituting now the respective relations for $\vec{j}_{lc}$, $\vec{j}_{lg}$, $\vec{j}_v$, $\vec{j}_q$ and $\frac{\partial u_v}{\partial t}$ into the balance equations (3.239) and (3.240) under the assumption that $\rho_s = $ const., $\rho_l = $ const., we obtain the final balances of the Milly's model in the form

$$(1 - A)\frac{\partial u_l}{\partial t} + B\frac{\partial \psi}{\partial t} + C\frac{\partial T}{\partial t}$$

$$= \frac{\rho_l}{\rho_s} \left\{ \operatorname{div} \left[(K_l + K_v) \operatorname{grad} \psi + D_{Tv} \operatorname{grad} T \right] + \frac{\partial K_l}{\partial z} \right\} \tag{3.283}$$

$$(h_l - Ah_v)\frac{\partial u_l}{\partial t} + Bh_v\frac{\partial \psi}{\partial t} + (c_s + c_l u_l + c_v u_v + h_v C)\frac{\partial T}{\partial t}$$

$$= \frac{1}{\rho_s} \operatorname{div} \left[\lambda \operatorname{grad} T - h_l \rho_l K_l \operatorname{grad} \psi \right.$$

$$\left. - h_v \rho_l (K_v \operatorname{grad} \psi + D_{Tv} \operatorname{grad} T) \right] - \frac{\rho_l}{\rho_s} \frac{\partial}{\partial z}(h_l K_l), \tag{3.284}$$

which have to be completed by the measured water retention curve

$$u_l = u_l(\psi(t), T). \tag{3.285}$$

3.2.3 Diffusion Models

As with the pure moisture transport, the first diffusion models of coupled heat and moisture transport were formulated by Krischer and Lykov.

Krischer's model (Krischer, 1963) was also formulated in an intuitive way and its original version can be expressed as follows:

$$\rho_w \frac{\partial w}{\partial t} + \frac{\Pi - w}{R_v T} \frac{\partial p_1}{\partial t} = \rho_w \kappa \frac{\partial^2 w}{\partial x^2} + \frac{D}{\mu R_v T} \frac{p}{p - p_1} \frac{\partial^2 p_1}{\partial x^2} \tag{3.286}$$

$$\rho c \frac{\partial T}{\partial t} = \lambda \frac{\partial^2 T}{\partial x^2} + L_v \left(\frac{D}{\mu R_v T} \frac{p}{p - p_1} \frac{\partial^2 p_1}{\partial x^2} - \frac{\Pi - w}{R_v T} \frac{\partial p_1}{\partial t} \right), \tag{3.287}$$

where c is the specific heat capacity, ρ the partial density of the porous body, L_v the latent heat of evaporation of water, λ the thermal conductivity, and the other symbols have the same meaning as in Section 3.1.

We can see that in Krischer's model the phase change processes between water and water vapour were taken into account (the second term on the right-hand side) but a comparison with our elementary thermodynamic model (see Section 3.2.1) shows, that cross effects between the heat and moisture transport (Soret and Dufour effects) are missing. Later, this fact influenced in a negative way the whole German school, which was developed on the basis of the Krischer's concept.

In Krischer's two equations for heat and moisture transport, there appear three unknown quantities, namely w, p_1, T. Therefore, one more relation between these quantities needs to be formulated to make the system of equations solvable. Krischer employed the sorption isotherm, i.e. the dependence of the volumetric moisture content w on the relative humidity φ or more generally on the partial pressure of water vapour p_1 under equilibrium conditions, $\varphi = \varphi(w)$ or $p_1 = p_1(w)$.

The relative humidity is defined as

$$\varphi = \frac{p_1}{p_s}, \tag{3.288}$$

where p_s is the partial pressure of saturated water vapour in the air. Therefore, in the conditions of $T = \text{const.}$, the relations $\varphi = \varphi(w)$, $p_1 = p_1(w)$ are identical.

The assumption that equilibrium conditions between the liquid and vapour phases of water exist within the porous body, which is expressed by the sorption isotherm, is probably the most logical solution of the problem of simultaneous water and water vapour transport. On the other hand, it is a relatively strong assumption, which particularly in faster transport processes can be questionable. Therefore, Krischer's model is suitable for common moisture transport processes, for instance in envelope parts of building structures, but for fast processes taking place at higher temperatures its applicability may be limited.

The Lykov's model of coupled heat and moisture transport (Lykov, 1956, Lykov, 1972) was from the thermodynamic point of view more advanced then the Krischer's model. Lykov formulated balance equations for heat and moisture in the following way (we note, that the condition $\vec{v} = 0$, which is general for all diffusion models, is adopted in the model). The balance equation for water vapour was

$$\rho_0 \frac{\partial u_1}{\partial t} = - \operatorname{div} \vec{j}_1 + I_1, \tag{3.289}$$

and the water balance equation

$$\rho_0 \frac{\partial u_2}{\partial t} = - \operatorname{div} \vec{j}_2 + I_2, \tag{3.290}$$

where ρ_0 is the partial density of the porous matrix, $\vec{j}_1, \vec{j}_2$ the water vapour and water fluxes, respectively, I_1, I_2 the source terms of water vapour and water, respectively, where

$$I_1 = -I_2, \tag{3.291}$$

so that only their mutual transition is assumed.

The total balance of moisture Lykov obtained by summing up equations (3.289) and (3.290) is

$$\rho_0 \frac{\partial u}{\partial t} = - \operatorname{div} \vec{j}_1 - \operatorname{div} \vec{j}_2. \tag{3.292}$$

In the relations for $\vec{j}_1$, $\vec{j}_2$ Lykov had also already introduced the temperature gradient as the second generalized thermodynamic force besides moisture gradient,

$$\vec{j}_i = -a_{mi}\rho_0 \operatorname{grad} u - a_{mi}^T \rho_0 \operatorname{grad} T$$

$$= -a_{mi}\rho_0(\operatorname{grad} u + \delta_i \operatorname{grad} T), \qquad i = 1, 2, \tag{3.293}$$

where a_{mi} is the diffusion coefficient, a_{mi}^T the thermodiffusion coefficient, and δ_i the relative thermodiffusion coefficient. Substituting (3.293) into (3.292) under assumptions $\rho_0 = \text{const.}$, $a_{mi} = \text{const.}$, $a_{mi}^T = \text{const.}$, he arrived at

$$\frac{\partial u}{\partial t} = a_m \operatorname{div} \operatorname{grad} u + a_m^T \operatorname{div} \operatorname{grad} T. \tag{3.294}$$

The heat balance in the Lykov's model was formulated as

$$\frac{\partial}{\partial t}\left(h_0\rho_0 + \sum_i h_i\rho_0 u_i\right) = - \operatorname{div}\left(\vec{j}_Q + \sum_i h_i\vec{j}_i\right), \tag{3.295}$$

where h_i is the partial specific enthalpy of the component i, $i = 1$ denotes water vapour again, $i = 2$ water and $i = 0$ the porous matrix, the expression $\sum_i h_i\vec{j}_i$ is the energy flux due to diffusion in a multi-component system.

Using the relation

$$c_i = \frac{\mathrm{d}h_i}{\mathrm{d}T}, \tag{3.296}$$

where c_i is the specific heat capacity of a component i, and under assumption $\rho_0 = \text{const.}$, we get

$$\left(\rho_0 c_0 + \sum_i \rho_0 c_i u_i\right)\frac{\partial T}{\partial t} + \rho_0 \sum_i h_i \frac{\partial u_i}{\partial t}$$

$$= - \operatorname{div} \vec{j}_Q - \sum_i c_i \vec{j}_i \operatorname{grad} T - \sum_i h_i \operatorname{div} \vec{j}_i. \tag{3.297}$$

Multiplying equations (3.289) and (3.290) by h_i,

$$\rho_0 h_i \frac{\partial u_i}{\partial t} = -h_i \operatorname{div} \vec{j}_i + h_i I_i, \qquad i = 1, 2, \tag{3.298}$$

summing them and substituting into (3.297), we get

$$(\rho_0 c_0 + \sum_i \rho_0 c_i u_i)\frac{\partial T}{\partial t} = - \text{div } \vec{j}_Q - \sum_i c_i \vec{j}_i \text{ grad } T - \sum_i h_i I_i. \quad (3.299)$$

Denoting

$$c = c_0 + \sum_i c_i u_i \quad (3.300)$$

as the total specific heat capacity of the system and taking into account that for $i = 1, 2$

$$-h_1 I_1 - h_2 I_2 = -I_1(h_1 - h_2) = -I_1 L_v, \quad (3.301)$$

where L_v is the latent heat of evaporation of water, we obtain the heat balance in the form

$$\rho_0 c \frac{\partial T}{\partial t} = - \text{div } \vec{j}_Q - \sum_i c_i \vec{j}_i \text{ grad } T - I_1 L_v. \quad (3.302)$$

The heat flux $\vec{j}_Q$ in Lykov's model is assumed in the basic Fourier's form,

$$\vec{j}_Q = -\lambda \text{ grad } T, \quad (3.303)$$

the moisture source I_1 is determined from the balance equation of mass of water vapour (3.289) under the assumption

$$\frac{\partial u_1}{\partial t} = 0, \quad (3.304)$$

so that

$$I_1 = \text{ div } \vec{j}_1. \quad (3.305)$$

After substitution for

$$\vec{j}_1 = -\rho_0 a_{m1} \text{ grad } u - \rho_0 a_{m1}^T \text{ grad } T \quad (3.306)$$

(all material parameters are assumed to be constant in the original Lykov's model) then

$$\rho_0 c \frac{\partial T}{\partial t} = \lambda \text{ div grad } T + \rho_0[(c_1 a_{m1} + c_2 a_{m2}) \text{ grad } u$$

$$+ (c_1 a_{m1}^T + c_2 a_{m2}^T) \text{ grad } T] \cdot \text{ grad } T$$

$$+ L_v \rho_0 (a_{m1} \text{ div grad } u + a_{m1}^T \text{ div grad } T). \quad (3.307)$$

Denoting

$$a = \frac{\lambda}{\rho_0 c} \quad (3.308)$$

the thermal diffusivity, the final heat balance equation reads

$$\frac{\partial T}{\partial t} = \left(a + a_{m1} \frac{L_v}{c} \right) \text{ div grad } T + a_{m1} \frac{L_v}{c} \text{ div grad } u$$

$$+ \frac{1}{c} [(c_1 a_{m1} + c_2 a_{m2}) \text{ grad } u$$

$$+ (c_1 a_{m1}^T + c_2 a_{m2}^T) \text{ grad } T] \cdot \text{ grad } T. \tag{3.309}$$

The Lykov's model is constructed in a less intuitive way than the Krischer's model. It utilizes more knowledge in the field of continuum physics and irreversible thermodynamics. A comparison with our elementary thermodynamic model shows that Lykov includes Soret's effect, but its cross counterpart, Dufour's effect, is not expressed in an exact way. Nevertheless, the heat transfer induced by mass flux (which is the effect described by Dufour's coefficient) is taken into account by the term $- \text{ div } \sum_i h_i \vec{j}_i$. Therefore, it can be concluded, that both cross effects of heat and moisture transport are included in Lykov's model.

Lykov's formulation of material relations for moisture transfer, where the gradient of total moisture acts as a generalized thermodynamic force, but the gradients of liquid and gaseous moisture no longer appear, leads to a simpler description than in the Krischer's model. But on the other hand, it leads to problems with determination of the amount of evaporated water in the heat balance. Lykov's solution with the assumption

$$\frac{\partial u_1}{\partial t} = 0 \tag{3.310}$$

or

$$\left| \frac{\partial u_1}{\partial t} \right| << \left| \text{ div } \vec{j}_1 \right| = |I_1| \tag{3.311}$$

in practice means that all the evaporated water can be removed immediately from where it was generated, and the amount of water vapour is constant in time. This is in fact a second extreme to Krischer's assumption, that in the porous space, there is an equilibrium state between the liquid and vapour phase. Therefore, Lykov's model better expresses situations where faster moisture transport appears, such as when the transport processes take place at higher temperatures. On the other hand, for a slower moisture transport, such as in usual conditions in envelope parts of building structures, will Lykov's model lead to substantial errors.

The basic Krischer's and Lykov's models were further developed and improved by many of their successors. On the basis of the Krischer's model, so-called German school was built. We will mention its main representatives here.

In the description of the amount of water vapour Kiessl (1983) used its partial density ρ_p instead of the partial pressure, and expressed the water vapour flux in the form

$$\dot{m}_D = D \frac{\partial \rho_p}{\partial x}. \tag{3.312}$$

The relative humidity φ was defined using the partial densities

$$\varphi = \frac{\rho_p}{\rho_{p,s}}, \tag{3.313}$$

where $\rho_{p,s}$ is the partial density of saturated water vapour. Substituting (3.313) into (3.312), we get

$$\dot{m}_D = D\rho_{p,s}\frac{\partial \varphi}{\partial x} + D\varphi\frac{\partial \rho_{p,s}}{\partial T}\frac{\partial T}{\partial x}$$

$$= \rho_w \text{FDP}\frac{\partial \varphi}{\partial x} + \rho_w \text{FDT}\frac{\partial T}{\partial x}, \tag{3.314}$$

where FDP is the coefficient of diffusion moisture transport due to grad φ, and FDT the coefficient of diffusion moisture transport due to grad T.

For the liquid moisture transport Kiessl used the same relation as Krischer,

$$\dot{m}_w = \rho_w \text{FKU}\frac{\partial w}{\partial x}, \tag{3.315}$$

where FKU is the coefficient of capillary moisture transport.

The moisture balance then had the following form:

$$\frac{\partial}{\partial t}\left[\rho_w w + (\Pi - w)\varphi\rho_{p,s}\right]$$

$$= \rho_w \frac{\partial}{\partial x}\left(\text{FKU}\frac{\partial w}{\partial x} + \text{FDP}\frac{\partial \varphi}{\partial x} + \text{FDT}\frac{\partial T}{\partial x}\right). \tag{3.316}$$

The heat balance was similar to that in the Krischer's model,

$$\frac{\partial}{\partial t}(\rho_0 c_0 T + \rho_w c_w w T) = \frac{\partial}{\partial x}\left(\lambda\frac{\partial}{\partial x}\right)$$

$$+ L_v \left\{\rho_w \frac{\partial}{\partial x}\left(\text{FDP}\frac{\partial \varphi}{\partial x} + \text{FDT}\frac{\partial T}{\partial x}\right)\right.$$

$$\left. - \frac{\partial}{\partial t}\left[(\Pi - w)\varphi\rho_{p,s}\right]\right\}. \tag{3.317}$$

Material parameters in Kiessl's model were considered as functions of field variables, namely temperature and moisture content. The problem of simultaneous water and water vapour transport Kiessl solved in the same way as Krischer using the sorption isotherm $\varphi = \varphi(w)$.

In addition, Kiessl attempted to unify the description of moisture transport in the hygroscopic and overhygroscopic range. The characteristic feature of the hygroscopic range is that the moisture content in the porous material within this range can be achieved by moistening in the air, i.e. by the water vapour contained in the air. It is limited on the upper side by the value of the hygroscopic moisture content w_h, $w \leq w_h$ that corresponds to the value of relative humidity $\varphi \to 1$ (in practical conditions $\varphi \sim 0.95$). In this range, the sorption isotherm is valid.

The moisture range $w_h < w \leq w_s$ is termed overhygroscopic, and it can only be achieved by direct contact of the material with water. Here w_s is the saturation moisture content. In this range, moisture transfer is realized in the liquid phase only, and we assume $\varphi = 1$.

For the unification of the mathematical description of moisture transport in these two ranges, Kiessl introduced so-called moisture potential Φ. In the hygroscopic range, he put $\Phi = \varphi$ (for $\varphi < 90\%$), in the overhygroscopic range then $\Phi = 1.7 + 0.1 \log r$, where r is the equivalent radius of pores in the given material.

The Kiessl's idea of introducing the moisture potential as the only quantity describing moisture transport was very useful from a practical point of view, because it has led to a very simple expression for the moisture transport across the interface between two different layers of building elements consisting of more layers. Kiessl assumed that on the interface the moisture potential is continuous, i.e.

$$[\![\Phi]\!] = 0, \tag{3.318}$$

where the symbol $[\![.]\!]$ means a jump of a quantity across an interface. On the other hand, the definition of moisture potential in the overhygroscopic range was too artificial, and Kiessl introduced it without any theoretical justification. The function of the model in the overhygroscopic range was therefore questionable from the beginnings.

The way of introducing the temperature gradient into the material relation for the water vapour transport is also somewhat questionable. By comparison with our elementary thermodynamic model, the gradient of the partial density of water vapour ρ_p remains the only thermodynamic force affecting the water vapour transport in the Kiessl's model anyway. The conversion of grad ρ_p to grad φ and grad T is only a formal mathematical adjustment. So nor Kiessl's model takes into account the Soret's and Dufour's effects.

Another model on the basis of Krischer's concept was formulated by Künzel (1995). The basic moisture balance equation was formulated in the form

$$\frac{\partial \rho_v}{\partial t} = \text{div} \left(D_w \, \text{grad} \, \rho_v + \delta_p \, \text{grad} \, p_1 \right), \tag{3.319}$$

where ρ_v is the partial moisture density, i.e. the mass of water per unit volume of the porous body, D_w the capillary water transport coefficient, δ_p the water vapour permeability, and p_1 the partial pressure of water vapour.

For the description of simultaneous water and water vapour transport Künzel chose the relative humidity φ as the only moisture potential for both hygroscopic and overhygroscopic range. In the hygroscopic range, he used the standard sorption isotherm $w = w(\varphi)$ as a relation between the amounts of liquid and gaseous moisture. As $\rho_v = \rho_w w$, $\rho_w \sim$ const., from $w = w(\varphi)$ directly follows the relation $\rho_v = \rho_v(\varphi)$, which Künzel called the moisture storage function, and $\frac{d\rho_v}{d\varphi}$ can be determined unambiguously. Further, the definition relation for the relative humidity $p_1 = \varphi p_s$, where p_s is the partial pressure of saturated water vapour in the air, was also applied in the derivation of the computational version of the model.

In the range of capillary water, which was considered by Künzel as the range with $\varphi > 0.95$, where φ is no longer measurable by classical methods, he used the Kelvin equation in the form

$$\varphi = \exp\left(-\frac{P_K}{\rho_w R_v T}\right), \tag{3.320}$$

where P_K is the capillary pressure (suction tension),

$$P_K = \frac{2\sigma \cos \Theta}{r}, \tag{3.321}$$

σ the surface tension, Θ the contact angle, and r the radius of the pore.

In the so-called oversaturated range, where $\varphi = 1$ and all pores are filled with water (which can be achieved for instance by water saturation under external pressure), Künzel assumed on the basis of Kelvin equation that $P_K = 0$, and therefore no liquid phase transport is realized in this range.

In comparison with Kiessl's model, Künzel introduces certain modifications of the moisture potential concept, for instance the division of the overhygroscopic range into two subranges. Also, the extension of the sorption isotherm to the range $\varphi > 0.95$ and its replacement by a more general moisture storage function is a substantial improvement compared to the Kiessl's model.

Substituting the relations for the moisture storage function $\rho_v = \rho_v(\varphi)$ and for the partial pressure of water vapour $p_1 = \varphi p_s$ into the balance equation (3.319) Künzel then obtained the moisture balance equation, where the relative humidity φ is the only unknown field variable,

$$\frac{d\rho_v}{d\varphi} \frac{\partial \varphi}{\partial t} = \operatorname{div}\left[D_\varphi \operatorname{grad} \varphi + \delta_p \operatorname{grad}(\varphi p_s)\right], \tag{3.322}$$

and where

$$D_\varphi = D_w \frac{d\rho_v}{d\varphi} \tag{3.323}$$

is the liquid water transport coefficient.

The heat balance in Künzel's model was formulated in a more or less classical way,

$$\frac{dH}{dT} \frac{\partial T}{\partial t} = \operatorname{div}(\lambda \operatorname{grad} T) + L_v \operatorname{div}[\delta_p \operatorname{grad}(\varphi p_s)], \tag{3.324}$$

where H is the enthalpy density (in J/m^3).

Generally, in comparison with Kiessl's or Krischer's model Künzel's model brings certain simplifications. In both moisture and heat balance equations Künzel assumes that the time variation of the amount of gaseous moisture is negligible, i.e. the terms

$$[(\Pi - w)\varphi\rho_{p,s})]$$

or

$$\frac{\partial}{\partial t}\left(\frac{\Pi - w}{R_V T}\frac{\partial p_1}{\partial t}\right)$$

are equal to zero. Also, his definition of moisture potential is more easily applicable in a computational implementation of the model. On the other hand, as well as in Krischer's and Kiessl's model, in Künzel's model Soret's and Dufour's effects are also missing.

The Häupl and Stopp (1987) model is apparently the most advanced within the classical German school. For moisture transport, Häupl and Stopp introduced the material relation

$$\vec{j}_v = -\rho_w a_w \operatorname{grad} w - \delta \operatorname{grad} p_d + \rho_d \vec{v} + \rho_w a_g \vec{g}, \tag{3.325}$$

where the first term is for the capillary water transport, the second term for water vapour diffusion, the third term for the convection of water vapour in a mixture with the air, and the last term for the water flow due to gravity forces.

Among the coefficients, which were not yet defined, a_w is the capillary water conductivity, a_g the conductivity of water due to gravity, and $\vec{v}$ the velocity of the water vapour–air mixture.

For the energy transfer they employed the relation

$$\vec{j}_Q = -\lambda \operatorname{grad} T - \rho_w h_w a_w \operatorname{grad} w - h_d \delta \operatorname{grad} p_d$$

$$+ \rho_w h_w a_g \vec{g} + \rho_d h_d \vec{v} + \rho_a h_a \vec{v}, \tag{3.326}$$

where the individual terms express consecutively the heat conduction, specific enthalpy fluxes due to the capillary water transport, water vapour diffusion, water transport due to gravity, and convective water vapour and air transport.

In equation (3.326), h denotes the specific enthalpy, and the indices w, d, a are for water, water vapour and air, respectively. On the basis of the definition of specific heat capacity, we can write linearized relations in the form

$$h_i = c_i T, \quad i = w, d, a. \tag{3.327}$$

The air transport is expressed by the relation

$$\vec{j}_a = \rho_a \vec{v}. \tag{3.328}$$

The velocity of the water vapour–air mixture Häupl and Stopp expressed using the gradient of the total pressure, $p = p_d + p_a$,

$$\vec{v} = -l \text{ grad } p, \tag{3.329}$$

where l is the permeability of the water vapour–air mixture (in $m^3 s/kg$).

The source term in the heat balance due to the water vapour evaporation was given in the form

$$I_Q = \text{div } (L_v \delta \text{ grad } p_d) - \text{div } (L_v \rho_d \vec{v}) - \frac{\partial}{\partial t}[\rho_d L_v(w_s - w)], \tag{3.330}$$

where ρ_d is the density of water vapour, w_s the saturation moisture content.

For the moisture accumulation term Häupl and Stopp introduced:

$$A_V = \frac{\partial}{\partial t}\left(\frac{dm_d}{dV} + \frac{dm_w}{dV}\right) = \frac{\partial}{\partial t}[\rho_d(w_s - w) + \rho_w w], \tag{3.331}$$

where ρ_d is the density of water vapour, the particular terms correspond to water vapour and water, respectively.

The energy accumulation term they formulated as

$$A_U = \frac{\partial}{\partial t}\left(\frac{dU}{dV}\right) = \frac{\partial}{\partial t}[\rho_s c_s T + \rho_w c_w T w$$

$$+ \rho_d c_d T(w_s - w) + \rho_a c_a T(w_s - w)], \tag{3.332}$$

where the individual terms correspond consecutively to the porous matrix, water, water vapour and the air, U is the internal energy,

$$\frac{dU}{dV} = \rho h - p \sim \rho h \sim \rho c T. \tag{3.333}$$

The air mass accumulation term was in the form

$$A_a = \frac{\partial}{\partial t}\left(\frac{dm_a}{dV}\right) = \frac{\partial}{\partial t}[\rho_a(w_s - w)]. \tag{3.334}$$

Using the quantities defined above, the balance equations can be formulated schematically in the following way:

$$A_V = - \text{div } \vec{j}_v + I_V \tag{3.335}$$

$$A_U = - \text{div } \vec{j}_Q + I_Q \tag{3.336}$$

$$A_a = - \text{div } \vec{j}_a + I_a, \tag{3.337}$$

where the moisture source term I_V can be for instance the rain falling on the facade, and an air source term I_a can appear, for instance due to the wind.

In equations (3.335)–(3.337) we have four unknown field quantities, w, T, p_d, p_a. The necessary fourth equation in the Häupl and Stopp model is again the sorption isotherm $w = w(\varphi, T)$. In the formal adjustment of equations (3.335)–(3.337) to a form suitable for a computational solution, Häupl and Stopp used for partial pressure of water vapour classical relations,

$$p_\mathrm{d} = \varphi(w, T)p_\mathrm{s}(T), \qquad 0 \le w \le w_\mathrm{h} \tag{3.338}$$

$$p_\mathrm{d} = p_\mathrm{s}(T), \qquad w_\mathrm{h} \le w \le w_\mathrm{s}, \tag{3.339}$$

and chose w as the basic variable for the description of moisture transport, so that they obtained

$$\operatorname{grad} p_\mathrm{d} = \operatorname{grad}\left[\varphi(w, T)p_\mathrm{s}(T)\right] = p_\mathrm{s}(T)\operatorname{grad}\varphi(w, T)$$

$$+ \varphi(w, T)\operatorname{grad} p_\mathrm{s}(T) = p_\mathrm{s}(T)\left(\frac{\partial\varphi}{\partial w}\operatorname{grad} w + \frac{\partial\varphi}{\partial T}\operatorname{grad} T\right)$$

$$+ \varphi(w, T)\frac{\mathrm{d}p_\mathrm{s}}{\mathrm{d}T}\operatorname{grad} T, \tag{3.340}$$

where the sorption isotherm $\varphi(w, T)$ was treated as a material parameter, i.e. a known function of w and T.

The resulting system of balance equations had the following form. The moisture balance was formulated as

$$\frac{\partial}{\partial t}[\rho_\mathrm{d}(w_\mathrm{s} - w) + \rho_\mathrm{w}w] = \operatorname{div}\left[\left(\rho_\mathrm{w}a_\mathrm{w} + \delta p_\mathrm{s}(T)\frac{\partial\varphi}{\partial w}\right)\operatorname{grad} w\right]$$

$$+ \operatorname{div}\left[\delta\left(p_\mathrm{s}(T)\frac{\partial\varphi}{\partial T} + \varphi(w, T)\frac{\mathrm{d}p_\mathrm{s}}{\mathrm{d}T}\right)\operatorname{grad} T\right]$$

$$- \operatorname{div}(\rho_\mathrm{w}a_\mathrm{g}\vec{g}) - \operatorname{div}(\rho_\mathrm{d}\vec{v}) + I_\mathrm{V}, \tag{3.341}$$

the energy balance

$$\frac{\partial}{\partial t}[\rho_\mathrm{s}c_\mathrm{s}T + \rho_\mathrm{w}c_\mathrm{w}Tw + \rho_\mathrm{d}c_\mathrm{d}T(w_\mathrm{s} - w) + \rho_\mathrm{a}c_\mathrm{a}T(w_\mathrm{s} - w)]$$

$$= \operatorname{div}\left\{\left[h_\mathrm{w}\rho_\mathrm{w}a_\mathrm{w} + (h_\mathrm{d} + L_\mathrm{v})\delta p_\mathrm{s}\frac{\partial\varphi}{\partial w}\right]\operatorname{grad} w\right\}$$

$$+ \operatorname{div}\left\{\left[\lambda + (h_\mathrm{d} + L_\mathrm{v})\delta\left(p_\mathrm{s}\frac{\partial\varphi}{\partial T} + \varphi\frac{\mathrm{d}p_\mathrm{s}}{\mathrm{d}T}\right)\right]\operatorname{grad} T\right\}$$

$$- \operatorname{div}(h_\mathrm{w}\rho_\mathrm{w}a_\mathrm{g}\vec{g}) - \operatorname{div}[(\rho_\mathrm{d}h_\mathrm{d} + \rho_\mathrm{a}h_\mathrm{a} + \rho_\mathrm{d}L_\mathrm{v})\vec{v}]$$

$$- \frac{\partial}{\partial t}[\rho_\mathrm{d}L_\mathrm{v}(w_\mathrm{s} - w)], \tag{3.342}$$

and the air mass balance

$$\frac{\partial}{\partial t}[\rho_a(w_s - w)] = -\operatorname{div}(\rho_a \vec{v}).$$

(3.343)

In equations (3.341)–(3.343), for the sake of simplicity $w, T, \vec{v}$ were used as the unknown field variables. Using equations (3.328) and (3.340), the problem can be reformulated for the unknowns w, T, p_a.

The velocity of the water vapour–air mixture $\vec{v}$ can be then expressed in the form

$$\vec{v} = -l \operatorname{grad} p = -l \operatorname{grad} p_a - l \operatorname{grad} p_d$$

$$= -l \operatorname{grad} p_a - l p_s \frac{\partial \varphi}{\partial w} \operatorname{grad} w$$

$$- l \left(p_s \frac{\partial \varphi}{\partial T} + \varphi \frac{dp_s}{dT} \right) \operatorname{grad} T$$

(3.344)

and substituted into the appropriate balance equations of moisture, energy and air.

It should be noted that nor Häupl and Stopp consider the cross effects in quite the same way as in the basic schemes of irreversible thermodynamics (see our elementary thermodynamic model). The effect of temperature gradient on liquid moisture transport was not taken into account in the Häupl and Stopp model at all (apparently the influence of Krischer), for water vapour they used a similar treatment as Kiessl or Künzel, and introduced the effect of temperature gradient indirectly as a consequence of the effect of the gradient of partial pressure of water vapour (which of course is not identical to Soret's effect). On the other hand, the influence of moisture flux on heat transfer was taken into account. It was modelled in an analogous way to Lykov's model, namely by the additional enthalpy fluxes, but the application of thermodynamic schemes was more straightforward than in Lykov's model. Consequently, Dufour's coefficient can be identified in the Häupl and Stopp (1987) model exactly as the term

$$B = h_w \rho_w a_w + h_d \delta p_s \frac{\partial \varphi}{\partial w}.$$

(3.345)

3.2.4 Hybrid Models

In constructing a hybrid model of coupled heat and moisture transport, an analogous treatment to the case of moisture transport can be employed. Again, the linear theory of mixtures, which makes it possible to include both diffusion and convection fluxes, is applied to couple the particular phases of the system, namely the solid matrix, water and water vapour. Thus, we have for partial densities

$$\rho = \sum_{i=1}^{3} \rho_i, \tag{3.346}$$

where index 1 denotes the solid matrix, 2 is for water and 3 for water vapour. For the velocities (assuming $\vec{v}_1 = 0$, $\rho_1 = const.$), we introduce the relation

$$\rho \vec{v} = \rho_2 \vec{v}_2 + \rho_3 \vec{v}_3, \tag{3.347}$$

for the convective fluxes

$$\vec{j}_{c,i} = \rho_i \vec{v}, \quad i = 2, 3, \tag{3.348}$$

for the diffusion fluxes

$$\vec{j}_{d,i} = \rho_i (\vec{v}_i - \vec{v}) = -\kappa_i \, \text{grad} \, \rho_i - \rho \delta_i \, \text{grad} \, T + M_i \vec{g}, \quad i = 2, 3, \tag{3.349}$$

where κ_i is the diffusivity, δ_i the thermodiffusion coefficient, M_i the mobility of the i-th component, and $\vec{g}$ the acceleration due to gravity.

The balance equations of mass of the particular components can be written as

$$\frac{\partial \rho_i}{\partial t} + \text{div} \, \vec{j}_{c,i} + \text{div} \, \vec{j}_i = R_i, \quad i = 2, 3, \tag{3.350}$$

where R_i is the source term of the component i,

$$\sum_{i=2,3} R_i = 0, \tag{3.351}$$

which is due to either water evaporation or water vapour condensation. Under the assumption that any changes of the volumetric relative content of water and temperature lead to the evaporation or condensation of an amount of water vapour corresponding to the change given by the sorption isotherm $\varphi_0(w, T)$, the source term of water vapour can be expressed in the form

$$R_3 = \frac{\partial \rho_{vo}}{\partial t}, \tag{3.352}$$

where

$$\rho_{vo} = \varphi_0(w, T) \cdot \rho_{vs}(w_m - w), \tag{3.353}$$

where ρ_{vs} is the partial density of saturated water vapour in the air, $\rho_{vs} = \rho_{vs}(T)$, w_m the maximum relative content of water (equal to the relative volume content of pores), and $\rho_{vs}(w_m - w)$ the mass of saturated water vapour per unit volume of the porous medium. Substituting for the fluxes into the balance equation of mass (3.350), we get

$$\frac{\partial \rho_i}{\partial t} = \text{div} \, (\kappa_i \, \text{grad} \, \rho_i + \rho \delta_i \, \text{grad} \, T - M_i \vec{g})$$

$$- \operatorname{div}(\rho_i \vec{v}) + R_i, \qquad i = 2, 3. \tag{3.354}$$

For the mass balance of the mixture we have

$$\frac{\partial \rho}{\partial t} + \operatorname{div}(\rho \vec{v}) = 0. \tag{3.355}$$

Momentum balance of the mixture reads

$$\rho \left(\frac{\partial \vec{v}}{\partial t} + \vec{v} \cdot \operatorname{grad} \vec{v} \right) = - \operatorname{grad} p + \vec{f}, \tag{3.356}$$

where p is the pressure of the mixture, $\vec{f}$ the force per unit volume,

$$\vec{f} = \vec{f}_g + \vec{f}_r, \tag{3.357}$$

$\vec{f}_g$ the gravitational force per unit volume,

$$\vec{f}_g = \rho \vec{g}, \tag{3.358}$$

$\vec{f}_r$ the resistance force of the porous medium per unit volume, which can be expressed for low velocities by a linear relation to the velocity,

$$\vec{f}_r = -A\vec{v}, \tag{3.359}$$

where A is an empirical constant. Balance equations of the momentum of the components 2, 3 can be written as

$$\rho_i \left(\frac{\partial \vec{v}_i}{\partial t} + \vec{v}_i \cdot \operatorname{grad} \vec{v} \right) = - \operatorname{grad} p_i + \vec{f}_i, \qquad i = 2, 3, \tag{3.360}$$

where

$$\vec{f}_i = \vec{f}_{g,i} + \vec{f}_{r,i}, \qquad i = 2, 3, \tag{3.361}$$

$$\vec{f}_{g,i} = \rho_i \vec{g}, \qquad i = 2, 3 \tag{3.362}$$

$$\vec{f}_{r,i} = -AC_i \vec{v}_i, \qquad i = 2, 3 \tag{3.363}$$

$$p = \sum_{i=1}^{3} p_i, \tag{3.364}$$

p_i is the partial pressure of the component i, $\vec{f}_i$ the volume force acting on the component i, $C_i = \rho_i / \rho$.

The balance equation for the internal energy of the mixture can be formulated in the form

$$\rho \left(\frac{\partial u}{\partial t} + \vec{v} \cdot \operatorname{grad} u \right) + \operatorname{div} \left(\vec{j}_q + \sum_{i=2,3} \rho_i u_i \vec{v}_i \right) = R_u, \tag{3.365}$$

where u is the specific internal energy, R_u the source of internal energy, $\vec{j}_q$ the heat flux,

$$\vec{j}_q = -\lambda \ \mathrm{grad} \ T - \sum_{i=2,3} \beta_i \ \mathrm{grad} \ \rho_i + \sum_{i=2,3} \frac{\beta_i M_i}{\rho_i \kappa_i} \vec{g}, \tag{3.366}$$

λ the thermal conductivity, β_i the diffusion-thermo coefficient (Dufour's coefficient) of the component i.

In modifying the internal energy balance equation, we employ the Gibbs relation

$$\rho \ du = \rho c_p \ dT + \frac{p}{\rho} \ d\rho + \sum_{i=2,3} H_i \ d\rho_i, \tag{3.367}$$

where c_p is the specific heat capacity at constant pressure of the mixture,

$$c_p = \sum_{i=1}^{3} C_i c_{p,i}, \tag{3.368}$$

H_i the partial specific enthalpy,

$$H_i = h_i - h_1, \tag{3.369}$$

h_i the specific enthalpy of the component i,

$$\rho_i h_i = \rho_i u_i + p_i. \tag{3.370}$$

Using (3.367)–(3.370), (3.349), and (3.366), we obtain the following:

$$\sum_{i=1}^{3} \rho_i c_{p,i} \left(\frac{\partial T}{\partial t} + \vec{v} \cdot \mathrm{grad} \ T \right) + \sum_{i=2,3} \left(H_i + \frac{p}{\rho} \right) \left(\frac{\partial \rho_i}{\partial t} + \vec{v} \cdot \mathrm{grad} \ \rho_i \right)$$

$$= \mathrm{div} \left(\lambda \ \mathrm{grad} \ T + \sum_{i=2,3} \beta_i \ \mathrm{grad} \ \rho_i - \sum_{i=2,3} \frac{\beta_i M_i}{\rho_i \kappa_i} \vec{g} \right)$$

$$+ \mathrm{div} \left[\sum_{i=2,3} h_i (\kappa_i \ \mathrm{grad} \ \rho_i + \rho \delta_i \ \mathrm{grad} \ T - M_i \vec{g}) \right]$$

$$- \mathrm{div} \left(\sum_{i=2,3} \rho_i h_i \vec{v} \right) + \mathrm{div} \left(\sum_{i=2,3} p_i \vec{v}_i \right) + R_u. \tag{3.371}$$

Assuming now, that

$$\rho_i h_i \gg p_i \tag{3.372}$$

in most cases, we can simplify equation (3.371) as follows:

$$\sum_{i=1}^{3} \rho_i c_{\mathrm{p},i} \left(\frac{\partial T}{\partial t} + \vec{v} \cdot \operatorname{grad} T \right) + \sum_{i=2,3} H_i \left(\frac{\partial \rho_i}{\partial t} + \vec{v} \cdot \operatorname{grad} \rho_i \right)$$

$$= \operatorname{div} \left(\lambda \operatorname{grad} T + \sum_{i=2,3} \beta_i \operatorname{grad} \rho_i - \sum_{i=2,3} \frac{\beta_i M_i}{\rho_i \kappa_i} \vec{g} \right)$$

$$+ \operatorname{div} \left[\sum_{i=2,3} h_i (\kappa_i \operatorname{grad} \rho_i + \rho \delta_i \operatorname{grad} T - M_i \vec{g}) \right]$$

$$- \operatorname{div} \left(\sum_{i=2,3} \rho_i h_i \vec{v} \right) + R_{\mathrm{u}}. \tag{3.373}$$

For the source term of internal energy we have

$$R_{\mathrm{u}} = -\frac{\partial \rho_{\mathrm{vo}}}{\partial t} L_{\mathrm{v}}, \tag{3.374}$$

where L_{v} is the latent heat of evaporation of water.

We now have a system of ten independent scalar equations (3.354), (3.355), (3.360), (3.373) for the twelve unknown scalar quantities ρ_2, ρ_3, $\vec{v}_2$, $\vec{v}_3$, p, p_2, p_3, T. Therefore, two additional equations need to be formulated, which are the equations of state of water and water vapour

$$\rho_2 = \rho_{2\mathrm{o}}[1 - \beta_{\mathrm{T}}(T - T_\mathrm{o}) + \beta_{\mathrm{p}}(p_2 - p_{2\mathrm{o}})], \tag{3.375}$$

$$p_3 = \rho_3 T \frac{R}{M}, \tag{3.376}$$

where β_{T} is the volume thermal expansion coefficient, β_{p} the volume compression coefficient, $\rho_{2\mathrm{o}}$, T_o, $p_{2\mathrm{o}}$ are the reference values of the respective quantities, R the universal gas constant, M the molar mass of the mixture, and T is the temperature in K.

This system of twelve scalar equations for the twelve unknown scalar quantities we obtained can now be solved in an unambiguous way.

3.2.5 Heat and Moisture Transport Across the Interfaces

Most classical models of heat and moisture transport in porous materials do not deal with moisture transport across interfaces between two porous materials in an explicit way. In practical calculations, the simplest solution is commonly used. It is assumed that there is ideal contact between the two materials, and therefore we have continuous heat and moisture flux across the interface,

$$[j_Q] = 0, \tag{3.377}$$

$$[j_w] = 0, \tag{3.378}$$

where the symbol $[.]$ denotes a jump of a quantity across an interface. For instance in the case of Lykov's model and moisture transport it means that

$$a_{m+}(\text{grad } u)_+ + a_{m+}^T (\text{grad } T)_+$$

$$= a_{m-}(\text{grad } u)_- + a_{m-}^T (\text{grad } T)_-, \tag{3.379}$$

where the indices $+, -$ denote the limits of the particular quantities on the right and the left, respectively.

The simplest complementary condition for ideal contact is an assumption of equality of temperatures and moistures, taken as limits on the right and the left to the interface,

$$u_+ = u_- \tag{3.380}$$

$$T_+ = T_-. \tag{3.381}$$

This requirement can sometimes lead to dramatic but unclear moisture profiles in the regions close to the interface, particularly if the bordering materials have considerably different transfer coefficients.

Some models of the German school (Künzel, 1995, Kiessl, 1983) solve this problem by replacing moisture content as the unknown field variable by moisture potential Φ (Kiessl, 1983), or by relative humidity φ (Künzel, 1995). They introduce at the interface the condition

$$[\Phi] = 0, \tag{3.382}$$

(Kiessl) or

$$[\varphi] = 0 \tag{3.383}$$

(Künzel).

However, as already stated, this solution is partial, and it is valid in the hygroscopic range only, where a sorption isotherm can be employed.

A systematic analysis of transport phenomena on an interface between two porous materials was done by Brocken (1998). He started with the classical definition of a hydraulic contact given by Bear and Bachmat (1990), namely that the macroscopic capillary pressure must be continuous across the interface,

$$[\![p_c]\!] = 0. \tag{3.384}$$

For two identical porous materials this condition permits application of classical models, because these two materials can be considered as a monolithic system. At the interface between two different materials, this condition leads to a moisture jump across the interface, which however can be determined explicitly on the basis of the measured functions $p_c = p_c(w)$ for both materials.

The water retention curves (sometimes denoted as suction curves) $p_c = p_c(w)$ can be determined according to the American ASTM D3152 - 72 standard using a pressure membrane apparatus. Among the models we introduced before, the water retention curves are explicitly applied by Milly (1982).

The condition of continuity of moisture flux is also valid for an ideal hydraulic contact,

$$[\![j]\!] = 0. \tag{3.385}$$

If there is non-ideal hydraulic contact on the interface, Brocken (1998) characterizes it by interface permeability, K_{if}, a macroscopic parameter defined by the relation

$$K_{if} = -\frac{j_{if}}{p_{c-}(w_-) - p_{c+}(w_+)}, \tag{3.386}$$

where j_{if} is the moisture flux across the interface. For the capillary pressure we therefore have

$$[\![p_c]\!] = -\frac{j_{if}}{K_{if}}, \tag{3.387}$$

which means that a jump of capillary pressure appears at the interface. Physical reasons for such a discontinuity can be, for instance, a discontinuity in the porous structure of both materials on the interface (i.e. the path for water molecules is broken, because the pores of both materials do not link up) or an existence of air gap between the two porous materials. For the resulting moisture flux across the interface we have

$$j_+ = j_- = j_{if}, \tag{3.388}$$

which in the simplest case of isothermal transport means

$$\rho_l D_{w+}(\text{grad } w)_+ = \rho_l D_{w-}(\text{grad } w)_- = K_{if}(p_{c-} - p_{c+}). \tag{3.389}$$

3.3 TRANSPORT OF CHEMICAL COMPOUNDS

Transport of chemical compounds in porous materials is usually considered as a diffusion process. The simplest models assume that it can be described by Fick's law,

$$\vec{j}_i = -D_i \operatorname{grad} \rho_i, \tag{3.390}$$

where ρ_i is the partial density of the chemical compound in the porous material, D_i the diffusion coefficient, and $\vec{j}_i$ the mass flux of the compound. Substitution of (3.390) into the mass balance equation

$$\frac{\partial \rho_i}{\partial t} + \operatorname{div} \vec{j}_i = 0 \tag{3.391}$$

leads then to Fick's second law,

$$\frac{\partial \rho_i}{\partial t} = \operatorname{div} (D_i \operatorname{grad} \rho_i). \tag{3.392}$$

Sometimes, the mass concentration C_i is preferred instead of the partial density ρ_i,

$$C_i = \frac{\rho_i}{\rho}, \tag{3.393}$$

where ρ is the total density of the porous material. Under assumption

$$\rho = const. \tag{3.394}$$

equation (3.392) then reads

$$\frac{\partial C_i}{\partial t} = \operatorname{div} (D_i \operatorname{grad} C_i). \tag{3.395}$$

In the chemical literature, the molar concentration C_{mi} (in mol/m^3) is often employed, defined as

$$C_{mi} = \frac{\rho_i}{M_i}, \tag{3.396}$$

where M_i is the molar mass of the compound in kg/mol. Again, substituting (3.396) into (3.392) leads to the usual form of Fick's second law,

$$\frac{\partial C_{mi}}{\partial t} = \operatorname{div} (D_i \operatorname{grad} C_{mi}). \tag{3.397}$$

The transport equations based on Fick's law assume the concentration gradient is the only driving force affecting the transport of chemical compounds. However, taken strictly, Fick's law and its variants are valid for non-ionic diffusants only, as analysed for instance by Chatterji (1994a–c), (1995a–b). For ionic diffusants, the effect of an electric field should be taken into account, expressed by the Nernst–Planck or Nernst–Einstein equations.

The Nernst–Einstein diffusion equation can be derived as follows. The diffusion force acting on one particle of the i-th component can be expressed as

$$\vec{F}_i = -\frac{1}{N_A} \operatorname{grad} \mu_i, \tag{3.398}$$

where N_A is the Avogadro constant, i.e. the number of particles per mole, $N_A = 6.023 \cdot 10^{23}$ mol^{-1}, μ_i the chemical potential (in J/mol). The force $\vec{F}_i$ acting on a ion of the i-th component possessing electric charge Q_i results in the motion of the ion with velocity $\vec{v}_i$,

$$\vec{v}_i = u_i \vec{E} = \frac{\vec{F}_i}{Q_i} u_i, \tag{3.399}$$

where $\vec{E}$ is the electric field intensity, u_i the mobility of ions of the i-th component. Substituting (3.398) into (3.399) we obtain

$$\vec{v}_i = -\frac{u_i}{N_A Q_i} \operatorname{grad} \mu_i, \tag{3.400}$$

where $N_A Q_i$ is the charge of a mole of ions of the i-th component, which can be expressed using the Faraday constant, $F = 96487$ C/mol,

$$N_A Q_i = F z_i, \tag{3.401}$$

z_i is the valence of ions of the i-th component. Substituting (3.401) into (3.400) we get

$$\vec{v}_i = -\frac{u_i}{F z_i} \operatorname{grad} \mu_i. \tag{3.402}$$

The chemical potential of the i-th component can be expressed in the form

$$\mu_i = \mu_i^0 + RT \ln C_{mi}, \tag{3.403}$$

where C_{mi} is the molar concentration (in mol/m^3). Therefore, we have

$$\operatorname{grad} \mu_i = \frac{RT}{C_{mi}} \operatorname{grad} C_{mi}. \tag{3.404}$$

Using the conversion relation (3.396) between the molar concentration and the partial density ρ_i in (3.404) we get

$$\operatorname{grad} \mu_i = \frac{RT}{\rho_i} \operatorname{grad} \rho_i. \tag{3.405}$$

The substitution of (3.405) into (3.402) then leads to

$$\vec{v}_i = -\frac{u_i}{F z_i} \frac{RT}{\rho_i} \operatorname{grad} \rho_i. \tag{3.406}$$

For the mass flux of the i-th component (in kg/m^{2}s) we can write

$$\vec{j}_i = \rho_i \vec{v}_i. \tag{3.407}$$

Using (3.406) in (3.407), we arrive at the relation

$$\vec{j}_i = -\frac{u_i}{F z_i} RT \text{ grad } \rho_i, \tag{3.408}$$

which is formally identical to Fick's law. A comparison with equation (3.390) immediately gives

$$D_i = \frac{u_i}{F z_i} RT, \tag{3.409}$$

which is the Nernst–Einstein equation for diffusion in an electric field. It can be further modified using the following relation for the molar conductivity Λ_i ($\Lambda_i = \sigma_i / C_{\text{m}i}$, where σ_i is the electric conductivity):

$$\Lambda_i = z_i F u_i. \tag{3.410}$$

Substituting (3.410) into (3.409), we obtain

$$D_i = \frac{\Lambda_i}{F^2 z_i^2} RT, \tag{3.411}$$

which is the most common relation for the Nernst–Einstein diffusion coefficient.

The <u>Nernst–Planck equation</u> follows from a straightforward application of Faraday's first law for mass transfer in electrolytes. The original equation

$$\Delta m = \frac{M}{zF} I \Delta t, \tag{3.412}$$

where Δm is the mass transported during the time interval Δt due to the electric current I, can easily be modified relating the mass change to unit time and unit area, so that we get for the mass flux $\vec{j}_i$ of the i-th component the following:

$$\vec{j}_i = \frac{M_i}{z_i F} \vec{j}_{\text{el},i}, \tag{3.413}$$

where $\vec{j}_{\text{el},i}$ is the electric current density (in A/m^2). The electric current density can be expressed by Ohm's law in the form

$$\vec{j}_{\text{el},i} = \sigma_i \vec{E}. \tag{3.414}$$

For the electric conductivity of the i-th component we have

$$\sigma_i = n_i z_i u_i e, \tag{3.415}$$

where e is the elementary charge, and n_i the number of particles per unit volume, which can be expressed as

$$n_i = \frac{\rho_i}{M_i} N_A. \tag{3.416}$$

Substituting (3.416) into (3.415) we obtain

$$\sigma_i = \frac{\rho_i}{M_i} N_A z_i u_i e = \frac{\rho_i}{M_i} z_i F u_i, \tag{3.417}$$

because $F = N_A e$. After substitution of (3.417) into (3.414) and subsequent substitution of (3.414) into (3.413), we get

$$\vec{j}_i = \frac{M_i}{z_i F} \vec{E} \frac{\rho_i}{M_i} z_i F u_i = \rho_i u_i \vec{E}. \tag{3.418}$$

From the Nernst–Einstein equation (3.409) it follows that

$$u_i = D_i \frac{z_i F}{RT}. \tag{3.419}$$

Substituting (3.419) into (3.418), we already obtain the final form of the Nernst–Planck equation,

$$\vec{j}_i = \frac{z_i F}{RT} D_i \rho_i \vec{E}, \tag{3.420}$$

where the diffusion coefficient D_i is sometimes called migration coefficient.

In formulating a relation for mass flux of a chemical compound i due to both the concentration gradient and electric field intensity (gradient of electric potential), the superposition principle can be applied. This is a common way of formulating constitutive equations within the framework of classical linear irreversible thermodynamics. We obtain

$$\vec{j}_i = -D_i \operatorname{grad} \rho_i + \frac{z_i F}{RT} D_i \rho_i \vec{E}. \tag{3.421}$$

Equation (3.421) is often employed in evaluating the migration coefficients (e.g., Andrade et al., 1997, Arsenault et al., 1997, Tang, 1997, etc.), Chatterji (1995a,b) recommends it for transport of chemical compounds in concrete in general.

The chemical compounds in porous media are mostly present in the form of water solution. Therefore, they move together with water through the porous space, and in describing their transport it is necessary to take into account the convective effects as well. According to the linear theory of mixtures, we can divide the total flux of a chemical compound i into diffusion and convection parts,

$$\vec{j}_i = \vec{j}_{\mathrm{dif},i} + \vec{j}_{\mathrm{con},i} = \rho_i(\vec{v}_i - \vec{v}) + \rho_i \vec{v} = \rho_i \vec{u}_i + \rho_i \vec{v}, \tag{3.422}$$

where $\vec{v}$ is the barycentric velocity of the liquid phase consisting of water and dissolved chemical compounds, $\vec{v}_i$ the velocity of the i-th compound, and $\vec{u}_i$ the relative velocity of the chemical compound, which expresses the diffusion effects. For the diffusion flux $\vec{j}_{\mathrm{dif},i}$ we already have equation (3.421). Therefore, we can write for the total flux

$$\vec{j}_i = -D_i \operatorname{grad} \rho_i + \frac{z_i F}{RT} D_i \rho_i \vec{E} + \rho_i \vec{v}, \tag{3.423}$$

and after substitution into the mass balance equation (3.391) we get

$$\frac{\partial \rho_i}{\partial t} = \text{div}\,(D_i\,\text{grad}\,\rho_i) - \text{div}\,\left(\frac{z_i F}{RT} D_i \rho_i \vec{E}\right) - \text{div}\,(\rho_i \vec{v}). \qquad (3.424)$$

Clearly, equation (3.424) has to be solved together with the balance equations describing water transport in the porous medium, because the barycentric velocity of the liquid phase, which acts as a drift velocity of the chemical compound, has to be identified. This problem will be solved in the next section dealing with the complex models.

Up to now, we were concerned with the transport of chemical compounds only, and did not analyse the way they appear in the porous body. Basically, there are two possible ways a chemical compound can enter porous space. The first is an external flux from the surroundings, which can be expressed by a boundary condition of Neumann type,

$$\vec{j}_i(\Gamma) = \vec{j}_{p,i}, \qquad (3.425)$$

where Γ is the boundary of the porous material, $\vec{j}_{p,i}$ the prescribed external flux.

The second possibility is a chemical reaction, either with the porous matrix or with some other components that are present in the porous space. In this case a source term $\dot{m}_i$ has to be included in equation (3.424),

$$\frac{\partial \rho_i}{\partial t} = \text{div}\,(D_i\,\text{grad}\,\rho_i) - \text{div}\,\left(\frac{z_i F}{RT} D_i \rho_i \vec{E}\right) - \text{div}\,(\rho_i \vec{v}) + \dot{m}_i, \quad (3.426)$$

which represents the mass production rate of the i-th component in kg/m³s.

The mass production rate can be expressed using the rates of the particular chemical reactions in the system, in which the i-th component takes part. For the rate of the j-th reaction J_j, we can write (see, e.g., Moore, 1972)

$$J_j = \frac{1}{\nu_{ij}}\left(\frac{\partial C_{mi}}{\partial t}\right)_j, \qquad (3.427)$$

where J_j is in mol/m³s, ν_{ij} is the stoichiometric factor of the i-th component in the j-th reaction ($\nu_{ij} > 0$ for the products, $\nu_{ij} < 0$ for the reactants), and C_{mi} the molar concentration of the i-th component in mol/m³. Using the partial density ρ_i in equation (3.427) instead of the molar concentration (see the conversion relation (3.396)), we obtain

$$J_j = \frac{1}{\nu_{ij} M_i}\left(\frac{\partial \rho_i}{\partial t}\right)_j, \qquad (3.428)$$

where $\left(\frac{\partial \rho_i}{\partial t}\right)_j$ is the mass production rate of the i-th component due to the j-th chemical reaction. From (3.428) it directly follows that

$$\dot{m}_i = M_i \sum_j \nu_{ij} J_j. \qquad (3.429)$$

We note that the reaction rate J_j is always positive, as for reactants $\nu_{ij} < 0$ and $\left(\frac{\partial \rho_i}{\partial t}\right)_j < 0$, while for the products $\nu_{ij} > 0$ and $\left(\frac{\partial \rho_i}{\partial t}\right)_j > 0$. Therefore, the j-th part of $\dot{m}_i$ is positive, if the i-th component is a reaction product ($\nu_{ij} > 0$), and negative, if it is a reactant ($\nu_{ij} < 0$).

3.4 COMPLEX MODELS

3.4.1 Coupled Heat, Moisture and Chemical Compounds Transport

Models of coupled heat, moisture and chemical compounds transport are a logical extension of heat and moisture transport models. This arose from the need to monitor the salt transport phenomena in building materials as well.

The most straightforward way to formulate such a model is an application of a hierarchical-mixture treatment to the system of porous solid skeleton, water, water vapour, salts dissolved in water, air and gaseous admixtures. The principal assumption of this model is that the system (ABC) consists basically of three components, water and dissolved admixtures (A), water vapour, air and gaseous admixtures (B), and solid porous skeleton (C), and that no chemical reactions occur in the system.

In accordance with the linear theory of mixtures, for densities and concentrations in the ABC system we can write

$$\rho = \sum_{i=A,B,C} \rho_i \tag{3.430}$$

$$C_i = \frac{\rho_i}{\rho}, \qquad i = A, B, C, \tag{3.431}$$

and for the diffusion fluxes

$$\vec{j}_i = \rho_i(\vec{v}_i - \vec{v}), \qquad i = A, B, C, \tag{3.432}$$

where $\vec{v}$ is the velocity of the mixture, and $\vec{v}_i$ the velocities of the particular components,

$$\rho\vec{v} = \sum_{A,B,C} \rho_i\vec{v}_i. \tag{3.433}$$

Assuming now that the solid phase does not move,

$$\vec{v}_C = 0, \tag{3.434}$$

we have

$$\rho\vec{v} = \sum_{i=A,B} \rho_i\vec{v}_i. \tag{3.435}$$

Phenomenological relations for mass fluxes of the components A, B we can write in the form

$$\vec{j}_i = -\kappa_i \ \text{grad} \ \rho_i - \rho\delta_i \ \text{grad} \ T + M_i\vec{g}, \qquad i = \text{A}, \text{B}, \tag{3.436}$$

where κ_i, δ_i, and M_i are the diffusivity, the thermodiffusion coefficient, and mobility, respectively, of the component i in the porous material, T the temperature, p the pressure, and g the gravity acceleration.

Balance equations for the mass of the components A,B can generally be written in the form

$$\frac{\partial \rho_i}{\partial t} + \text{div} \ \vec{j}_{c,i} + \text{div} \ \vec{j}_i = R_i, \qquad i = \text{A}, \text{B}, \tag{3.437}$$

where $\vec{j}_{c,i}$ is the convective flux of the component i,

$$\vec{j}_{c,i} = \rho_i\vec{v}, \qquad i = \text{A}, \text{B}, \tag{3.438}$$

R_i the source term of the component i,

$$\sum_{i=\text{A},\text{B}} R_i = 0, \tag{3.439}$$

which is due to either water evaporation or water vapour condensation, and will be identified precisely below.

Substituting equations (3.436), (3.438) into (3.437), we get

$$\frac{\partial \rho_i}{\partial t} = \text{div} \ (\kappa_i \ \text{grad} \ \rho_i + \rho\delta_i \ \text{grad} \ T - M_i\vec{g})$$

$$- \text{div} \ (\rho_i\vec{v}) + R_i, \qquad i = \text{A}, \text{B}. \tag{3.440}$$

The mass balance of the ABC mixture can be written as

$$\frac{\partial \rho}{\partial t} + \text{div} \ (\rho\vec{v}) = 0, \tag{3.441}$$

and assuming

$$\rho_{\text{C}} = const. \tag{3.442}$$

then

$$\frac{\partial \rho_\text{A}}{\partial t} + \frac{\partial \rho_\text{B}}{\partial t} + \text{div} \ (\rho_\text{A}\vec{v}) + \text{div} \ (\rho_\text{B}\vec{v}) + \rho_\text{C} \ \text{div} \ \vec{v} = 0. \tag{3.443}$$

Combining (3.440), (3.443), we obtain

$$-\rho_\text{C} \ \text{div} \ \vec{v} = \sum_{i=\text{A},\text{B}} \text{div} \ (\kappa_i \ \text{grad} \ \rho_i + \rho\delta_i \ \text{grad} \ T - M_i\vec{g}). \tag{3.444}$$

Momentum balance of the ABC mixture in the porous medium can be expressed as

$$\rho \left(\frac{\partial \vec{v}}{\partial t} + \vec{v} \cdot \text{grad} \ \vec{v} \right) = - \ \text{grad} \ p + \vec{f}, \tag{3.445}$$

where p is the pressure of the ABC mixture, $\vec{f}$ the force per unit volume,

$$\vec{f} = \vec{f_g} + \vec{f_r}, \tag{3.446}$$

$\vec{f_g}$ the gravitational force per unit volume,

$$\vec{f_g} = \rho \vec{g}, \tag{3.447}$$

$\vec{g}$ the gravity acceleration, and $\vec{f_r}$ the resistance force of the porous medium per unit volume. This can be expressed for low velocities by a linear relation to the velocity,

$$\vec{f_r} = -A\vec{v}, \tag{3.448}$$

where A is an empirical constant. For higher velocities we would have for instance

$$\vec{f_r} = -A\vec{v} - \vec{B} \cdot (\vec{v} \otimes \vec{v}), \tag{3.449}$$

where $\vec{B}$ is a vector empirical constant.

The corresponding momentum balances of the components i read

$$\rho_i \left(\frac{\partial \vec{v_i}}{\partial t} + \vec{v_i} \cdot \operatorname{grad} \vec{v} \right) = - \operatorname{grad} p_i + \vec{f_i}, \qquad i = A, B, \tag{3.450}$$

where

$$\vec{f_i} = \vec{f_{g,i}} + \vec{f_{r,i}}, \qquad i = A, B \tag{3.451}$$

$$\vec{f_{g,i}} = \rho_i \vec{g}, \qquad i = A, B \tag{3.452}$$

$$\vec{f_{r,i}} = -AC_i \vec{v_i}, \qquad i = A, B \tag{3.453}$$

$$p = \sum_{i=A,B,C} p_i, \tag{3.454}$$

p_i is the partial pressure of the component i, $\vec{f_i}$ the volume force acting on the component i.

In formulating the internal energy balance of the ABC mixture we will neglect the radiation terms, and using (3.434) we obtain

$$\rho \left(\frac{\partial u}{\partial t} + \vec{v} \cdot \operatorname{grad} u \right) + \operatorname{div} \left(\vec{j_q} + \sum_{i=A,B} \rho_i u_i \vec{v_i} \right) = R_u, \tag{3.455}$$

where u is the specific internal energy, R_u the source of internal energy, $\vec{j_q}$ the heat flux,

$$\vec{j_q} = -\lambda \operatorname{grad} T - \sum_{i=A,B} \beta_i \operatorname{grad} \rho_i + \sum_{i=A,B} \frac{\beta_i M_i}{\rho_i \kappa_i} \vec{g}, \tag{3.456}$$

λ the thermal conductivity, and β_i the diffusion-thermo coefficient (Dufour's coefficient) of the component i.

In modifying the internal energy balance equation, we employ the Gibbs relation

$$\rho \, du = \rho c_\mathrm{p} \, dT + \frac{p}{\rho} \, d\rho + \sum_{i=\mathrm{A,B}} H_i \, d\rho_i, \qquad (3.457)$$

where c_p is the specific heat capacity at constant pressure of the mixture,

$$c_\mathrm{p} = \sum_{i=\mathrm{A,B,C}} C_i c_{\mathrm{p},i}, \qquad (3.458)$$

H_i the partial specific enthalpy,

$$H_i = h_i - h_\mathrm{C}, \qquad (3.459)$$

h_i the specific enthalpy of the component i,

$$\rho_i h_i = \rho_i u_i + p_i. \qquad (3.460)$$

Using (3.457)–(3.460), (3.432), (3.434) and (3.456), we obtain the following:

$$\sum_{i=\mathrm{A,B,C}} \rho_i c_{\mathrm{p},i} \left(\frac{\partial T}{\partial t} + \vec{v} \cdot \operatorname{grad} T \right)$$

$$+ \sum_{i=\mathrm{A,B}} \left(H_i + \frac{p}{\rho} \right) \left(\frac{\partial \rho_i}{\partial t} + \vec{v} \cdot \operatorname{grad} \rho_i \right)$$

$$= \operatorname{div} \left(\lambda \operatorname{grad} T + \sum_{i=\mathrm{A,B}} \beta_i \operatorname{grad} \rho_i - \sum_{i=\mathrm{A,B}} \frac{\beta_i M_i}{\rho_i \kappa_i} \vec{g} \right)$$

$$+ \operatorname{div} \left[\sum_{i=\mathrm{A,B}} h_i (\kappa_i \operatorname{grad} \rho_i + \rho \delta_i \operatorname{grad} T - M_i \vec{g}) \right]$$

$$- \operatorname{div} \left(\sum_{i=\mathrm{A,B}} \rho_i h_i \vec{v} \right) + \operatorname{div} \left(\sum_{i=\mathrm{A,B}} p_i \vec{v}_i \right) + R_\mathrm{u}. \qquad (3.461)$$

Assuming now that

$$\rho_i h_i \gg p_i \qquad (3.462)$$

in most cases, we can simplify equation (3.461) as follows:

$$\sum_{i=\mathrm{A,B,C}} \rho_i c_{\mathrm{p},i} \left(\frac{\partial T}{\partial t} + \vec{v} \cdot \operatorname{grad} T \right) + \sum_{i=\mathrm{A,B}} H_i \left(\frac{\partial \rho_i}{\partial t} + \vec{v} \cdot \operatorname{grad} \rho_i \right)$$

$$= \operatorname{div} \left(\lambda \operatorname{grad} T + \sum_{i=A,B} \beta_i \operatorname{grad} \rho_i - \sum_{i=A,B} \frac{\beta_i M_i}{\rho_i \kappa_i} \vec{g} \right)$$

$$+ \operatorname{div} \left[\sum_{i=A,B} h_i (\kappa_i \operatorname{grad} \rho_i + \rho \delta_i \operatorname{grad} T - M_i \vec{g}) \right]$$

$$- \operatorname{div} \left(\sum_{i=A,B} \rho_i h_i \vec{v} \right) + R_u. \tag{3.463}$$

In the second part of the solution, we have to take into account the fact that the components A,B of the ABC mixture are themselves mixtures. Component A consists of water (w) and dissolved salts (pollutants) (p), component B of water vapour (v), air (a) and gaseous admixtures (g).

First we will deal with component A. The standard application of the theory of mixtures leads to the relations

$$\rho_A = \sum_{i=w,p} \rho_i \tag{3.464}$$

$$\vec{j}_i = \rho_i (\vec{v}_i - \vec{v}_A), \qquad i = w, p \tag{3.465}$$

$$\rho_A \vec{v}_A = \sum_{i=w,p} \rho_i \vec{v}_i, \tag{3.466}$$

while the phenomenological relations read

$$\vec{j}_w = -\kappa_w \operatorname{grad} \rho_w - \rho_A \delta_w \operatorname{grad} T + M_w \vec{g} \tag{3.467}$$

$$\vec{j}_p = -D_p \operatorname{grad} \rho_p - \rho_A \delta_p \operatorname{grad} T + M_p \vec{g}, \tag{3.468}$$

where D_p is the diffusion coefficient of the salts in water.

The balance equations for the particular components w, p can be written as follows:

$$\frac{\partial \rho_p}{\partial t} + \operatorname{div} \vec{j}_{c,p} + \operatorname{div} \vec{j}_p = 0 \tag{3.469}$$

$$\frac{\partial \rho_w}{\partial t} + \operatorname{div} \vec{j}_{c,w} + \operatorname{div} \vec{j}_w = R_w, \tag{3.470}$$

where the convective fluxes can be expressed as

$$\vec{j}_{c,i} = \rho_i \vec{v}_A, \qquad i = w, p, \tag{3.471}$$

and for the water source term R_w we can write

$$R_\text{w} = R_\text{A}. \tag{3.472}$$

Substituting for the respective fluxes in (3.469), (3.470), we obtain

$$\frac{\partial \rho_\text{p}}{\partial t} = \text{div} \left(D_\text{p} \ \text{grad} \ \rho_\text{p} + \rho_\text{A} \delta_\text{p} \ \text{grad} \ T - M_\text{p} \vec{g} \right) - \text{div} \left(\rho_\text{p} \vec{v}_\text{A} \right) \tag{3.473}$$

$$\frac{\partial \rho_\text{w}}{\partial t} = \text{div} \left(\kappa_\text{w} \ \text{grad} \ \rho_\text{w} + \rho_\text{A} \delta_\text{w} \ \text{grad} \ T - M_\text{w} \vec{g} \right)$$

$$- \text{div} \left(\rho_\text{w} \vec{v}_\text{A} \right) + R_\text{w}. \tag{3.474}$$

In formulating the analogous relations for the component B of the ABC mixture, we begin with the definitions of densities and diffusion fluxes as before,

$$\rho_\text{B} = \sum_{i=\text{v,g,a}} \rho_i \tag{3.475}$$

$$\vec{j}_i = \rho_i (\vec{v}_i - \vec{v}_\text{B}), \qquad i = \text{v,g,a} \tag{3.476}$$

$$\rho_\text{B} \vec{v}_\text{B} = \sum_{i=\text{v,g,a}} \rho_i \vec{v}_i. \tag{3.477}$$

In further derivation, we have to take into account the phase transition processes between water and water vapour, and to quantify the extent of these transitions. We utilize the moisture storage functions

$$\varphi = \varphi_0(w, T), \tag{3.478}$$

where w is the volumetric water content (with dissolved salts),

$$w = \frac{V_\text{A}}{V} = \frac{\frac{m_\text{A}}{V}}{\frac{m_\text{A}}{V_\text{A}}} = \frac{\rho_\text{A}}{\rho_\text{dw}}, \tag{3.479}$$

ρ_dw the density of water (with dissolved salts), φ the relative air humidity,

$$\varphi = \frac{\rho_\text{v}}{\rho_\text{vs}(w_\text{m} - w)}, \tag{3.480}$$

ρ_vs the density of saturated water vapour, $\rho_\text{vs} = \rho_\text{vs}(T)$, w_m the maximum relative content of water (equal to the relative volume content of pores), and $\rho_\text{vs}(w_\text{m} - w)$ the mass of saturated water vapour per unit volume of the porous medium (the ABC mixture).

The moisture storage functions $\varphi_0(w, T)$ describe the equilibrium state between water and water vapour within the porous body. Therefore, the content of water vapour in the porous medium should be directed to this equilibrium state. In formulating the extent of phase transitions between water and water vapour, we will assume that any changes of the volumetric water content and temperature lead to the evaporation or condensation of the amount of water vapour, corresponding to the change given by the sorption isotherm. Thus, the source term for water vapour can be expressed in the form

$$R_{\mathrm{v}} = \frac{\partial \rho_{\mathrm{vo}}}{\partial t}, \tag{3.481}$$

where

$$\rho_{\mathrm{vo}} = \varphi_0(w, T) \cdot \rho_{\mathrm{vs}}(w_{\mathrm{m}} - w). \tag{3.482}$$

The flux of water vapour reads

$$\vec{j}_v = -D_{\mathrm{v}} \ \mathrm{grad} \ \rho_{\mathrm{v}} - \rho_{\mathrm{B}}\delta_{\mathrm{v}} \ \mathrm{grad} \ T + M_{\mathrm{v}}\vec{g}, \tag{3.483}$$

and the mass balance for water vapour can be written in the form

$$\frac{\partial \rho_{\mathrm{v}}}{\partial t} = \mathrm{div} \ (D_{\mathrm{v}} \ \mathrm{grad} \ \rho_{\mathrm{v}} + \rho_{\mathrm{B}}\delta_{\mathrm{v}} \ \mathrm{grad} \ T - M_{\mathrm{v}}\vec{g})$$

$$- \mathrm{div} \ (\rho_{\mathrm{v}}\vec{v}_{\mathrm{B}}) + \frac{\partial \rho_{\mathrm{vo}}}{\partial t}. \tag{3.484}$$

As

$$R_{\mathrm{B}} = R_{\mathrm{v}} \tag{3.485}$$

and

$$R_{\mathrm{A}} = -R_{\mathrm{B}}, \tag{3.486}$$

we also can write

$$R_{\mathrm{w}} = -\frac{\partial \rho_{\mathrm{vo}}}{\partial t}, \tag{3.487}$$

and for the internal energy source term we have

$$R_{\mathrm{u}} = -\frac{\partial \rho_{\mathrm{vo}}}{\partial t} L_{\mathrm{v}}, \tag{3.488}$$

where L_{v} is the latent heat of evaporation of water.

The phenomenological relation and balance equation of mass of the gaseous admixtures can be formulated as follows:

$$\vec{j}_g = -D_{\mathrm{g}} \ \mathrm{grad} \ \rho_{\mathrm{g}} - \rho_{\mathrm{B}}\delta_{\mathrm{g}} \ \mathrm{grad} \ T + M_{\mathrm{g}}\vec{g} \tag{3.489}$$

$$\frac{\partial \rho_{\mathrm{g}}}{\partial t} = \mathrm{div} \ (D_{\mathrm{g}} \ \mathrm{grad} \ \rho_{\mathrm{g}} + \rho_{\mathrm{B}}\delta_{\mathrm{g}} \ \mathrm{grad} \ T - M_{\mathrm{g}}\vec{g}) - \mathrm{div} \ (\rho_{\mathrm{g}}\vec{v}_{\mathrm{B}}). \tag{3.490}$$

In order to get a closed system of equations, we have to formulate the equations of state of the components A, B, i.e.

$$p_A = f_A(T, \rho_A) \tag{3.491}$$

$$p_B = f_B(T, \rho_B). \tag{3.492}$$

For the liquid mixture A we can have, for instance,

$$\rho_A = \rho_{Ao}[1 - \beta_T(T - T_o) + \beta_p(p_A - p_{Ao})], \tag{3.493}$$

where β_T is the volume thermal expansion coefficient, β_p the volume compression coefficient, ρ_{Ao}, T_o, p_{Ao} the reference values of the respective quantities. For the gaseous mixture B we can write the equation of state of an ideal gas,

$$p_B = \rho_B T \frac{R}{M}, \tag{3.494}$$

where R is the universal gas constant, M the molar mass of the mixture, T the temperature in K.

Finally, we should note that both ρ_v and ρ_{vo} are subject to the limitation given by their maximum allowable value, $\rho_{vs}(w_m - w)$. Therefore,

$$\varphi_o(w, T) = 1, \qquad w > w_h, \tag{3.495}$$

where w_h is the maximum hygroscopic relative moisture content, and

$$\rho_v \leq \rho_{vs}(w_m - w). \tag{3.496}$$

For ρ_w we have a limitation given by the maximum volume content of the pores, so that

$$\rho_w \leq w_m \rho_{dw}, \tag{3.497}$$

where ρ_{dw} is the density of water (ρ_w is the mass of water per unit volume of the porous body). The conditions (3.496) and (3.497) are enforced by the chosen boundary conditions.

Summarizing the results of the above derivation of transport equations of the hierarchical mixture, we have equations (3.440), (3.473), (3.484), (3.490), (3.444), (3.445), (3.450), (3.463), (3.491), (3.492) for the basic unknowns ρ_A, ρ_B, ρ_p, ρ_v, ρ_g, $\vec{v}$, $\vec{v}_A$, $\vec{v}_B$, p, p_A, p_B, T, and (3.474), (3.475) for the additional unknown variables ρ_w, ρ_a. These equations form themselves a closed system, which makes it possible to determine all variables unambiguously.

In finding the practical solution of the system of equations just formulated, we can utilize the fact that the equations are not fully coupled, and solve the system in several consecutive steps.

The first step is to find the solution of the basic system of equations of the ABC mixture, i.e. equations (3.440), (3.444), (3.445), (3.463) for the unknowns ρ_A, ρ_B, $\vec{v}$, p, T. As the meaning of the total pressure p of the mixture is not fully clear (what is the pressure of the solid phase?), we use equation (3.454) and the assumption

$$p_C = const.,\tag{3.498}$$

and substitute for p in the momentum balance (3.445). Also, the pressure of the gaseous component B can be expressed by the equation of state (3.492), and the only pressure then remaining in the momentum balance of the mixture is the pressure p_A of the liquid component A, which is already acceptable. The momentum balance then reads

$$\rho\left(\frac{\partial \vec{v}}{\partial t} + \vec{v}\cdot \text{grad } \vec{v}\right) = -\text{grad } p_A - \frac{RT}{M}\text{grad } \rho_B$$

$$-\frac{\rho_B R}{M}\text{grad } T + \vec{f},\tag{3.499}$$

and we have p_A as the unknown variable instead of p and equation (3.499) instead of (3.445), all other equations remain unchanged.

The second step is the determination of velocities of components A, B, $\vec{v}_A$, $\vec{v}_B$. This results from a simple comparison of equations (3.432), (3.436), i.e.

$$\vec{v}_i = -\frac{1}{\rho_i}(\kappa_i \text{ grad } \rho_i + \rho\delta_i \text{ grad } T - M_i\vec{g}) + \vec{v}, \qquad i = \text{A}, \text{B}.\tag{3.500}$$

The last step is the calculation of the densities of the particular components p, w of the A component of the ABC mixture and components v, g, a of the B component of the ABC mixture. For this calculation, we have equations (3.473), (3.474), (3.484), (3.490) and (3.475).

The formulation of transport equations of water, water vapour and dissolved and gaseous pollutants in porous medium, which was shown above, is an exact derivation within the framework of linear irreversible thermodynamics and linear theory of mixtures. In this formulation, diffusion fluxes are strictly separated from the convective fluxes, and as a consequence there emerge relatively complicated system of equations, where both densities and velocities are unknown. Also, the diffusion material parameters are defined in a relatively complicated way. This is usually not employed in the practice, as it results in difficulties in measuring these parameters and solving complicated advection-diffusion inverse problems.

For the sake of simplification of the general model, we can employ the formulation where the diffusion parameters of main components A, B include both diffusive and convective fluxes, i.e.

$$\vec{j}_i = \rho_i\vec{v}_i, \qquad i = \text{A}, \text{B}.\tag{3.501}$$

The formulation of phenomenological relations (3.436), (3.456) will be a little more complicated in this case, because one more thermodynamic force has to be taken into account. This is the pressure gradient, which in the previous general formulation appeared only in the momentum balance as the driving force of the convective motion. We will have

$$\vec{j}_i = -\kappa_i \ \mathrm{grad} \ \rho_i - \rho\delta_i \ \mathrm{grad} \ T$$

$$- \rho\psi_i \ \mathrm{grad} \ p + M_i\vec{g}, \qquad i = A, B, \tag{3.502}$$

where ψ is the barodiffusion coefficient,

$$\vec{j}_\mathrm{q} = -\lambda \ \mathrm{grad} \ T - \sum_{i=A,B} \beta_i \ \mathrm{grad} \ \rho_i - \sum_{i=A,B} \gamma_i \ \mathrm{grad} \ p$$

$$+ \sum_{i=A,B} \frac{\beta_i M_i}{\rho_i \kappa_i} \vec{g}, \tag{3.503}$$

where γ is the diffusion-baro coefficient.

The mass balance equations of the components A, B then read

$$\frac{\partial \rho_A}{\partial t} = \mathrm{div} \ (\kappa_A \ \mathrm{grad} \ \rho_A + \rho\delta_A \ \mathrm{grad} \ T$$

$$+ \rho\psi_A \ \mathrm{grad} \ p - M_A\vec{g}) - \frac{\partial \rho_{vo}}{\partial t}, \tag{3.504}$$

$$\frac{\partial \rho_B}{\partial t} = \mathrm{div} \ (\kappa_B \ \mathrm{grad} \ \rho_B + \rho\delta_B \ \mathrm{grad} \ T$$

$$+ \rho\psi_B \ \mathrm{grad} \ p - M_B\vec{g}) + \frac{\partial \rho_{vo}}{\partial t}, \tag{3.505}$$

and using (3.502) we get the balance equations of the components p, v, g in the form

$$\frac{\partial \rho_\mathrm{p}}{\partial t} = \mathrm{div} \ (D_\mathrm{p} \ \mathrm{grad} \ \rho_\mathrm{p} + \rho_A\delta_\mathrm{p} \ \mathrm{grad} \ T + \rho_A\psi_\mathrm{p} \ \mathrm{grad} \ p_A - M_\mathrm{p}\vec{g})$$

$$+ \ \mathrm{div} \ \left[\frac{\rho_\mathrm{p}}{\rho_A} (\kappa_A \ \mathrm{grad} \ \rho_A + \rho\delta_A \ \mathrm{grad} \ T \right.$$

$$\left. + \rho\psi_A \ \mathrm{grad} \ p - M_A\vec{g}) \right] \tag{3.506}$$

$$\frac{\partial \rho_\mathrm{v}}{\partial t} = \mathrm{div} \ (D_\mathrm{v} \ \mathrm{grad} \ \rho_\mathrm{v} + \rho_B\delta_\mathrm{v} \ \mathrm{grad} \ T + \rho_B\psi_\mathrm{v} \ \mathrm{grad} \ p_B - M_\mathrm{v}\vec{g})$$

$$+ \ \mathrm{div} \ \left[\frac{\rho_\mathrm{v}}{\rho_B} (\kappa_B \ \mathrm{grad} \ \rho_B + \rho\delta_B \ \mathrm{grad} \ T + \rho\psi_B \ \mathrm{grad} \ p \right.$$

$$-M_{\mathrm{B}}\vec{g})\Big] + \frac{\partial \rho_{\mathrm{vo}}}{\partial t} \tag{3.507}$$

$$\frac{\partial \rho_{\mathrm{g}}}{\partial t} = \mathrm{div}\,(D_{\mathrm{g}}\ \mathrm{grad}\ \rho_{\mathrm{g}} + \rho_{\mathrm{B}}\delta_{\mathrm{g}}\ \mathrm{grad}\ T + \rho_{\mathrm{B}}\psi_{\mathrm{g}}\ \mathrm{grad}\ p_{\mathrm{B}} - M_{\mathrm{g}}\vec{g})$$

$$+\ \mathrm{div}\ \left[\frac{\rho_{\mathrm{g}}}{\rho_{\mathrm{B}}}\,(\kappa_{\mathrm{B}}\ \mathrm{grad}\ \rho_{\mathrm{B}} + \rho\delta_{\mathrm{B}}\ \mathrm{grad}\ T \right.$$

$$\left. +\rho\psi_{\mathrm{B}}\ \mathrm{grad}\ p - M_{\mathrm{B}}\vec{g})\right]. \tag{3.508}$$

The internal energy balance equation (3.463) is simplified using (3.501), (3.502) as follows:

$$\sum_{i=\mathrm{A,B,C}} \rho_i c_{\mathrm{p},i}\left(\frac{\partial T}{\partial t} + \vec{v}\cdot\ \mathrm{grad}\ T\right) + \sum_{i=\mathrm{A,B}} H_i\left(\frac{\partial \rho_i}{\partial t} + \vec{v}\cdot\ \mathrm{grad}\ \rho_i\right)$$

$$= \mathrm{div}\ \left(\lambda\ \mathrm{grad}\ T + \sum_{i=\mathrm{A,B}} \beta_i\ \mathrm{grad}\ \rho_i\right.$$

$$+ \sum_{i=\mathrm{A,B}} \gamma_i\ \mathrm{grad}\ p - \sum_{i=\mathrm{A,B}} \frac{\beta_i M_i}{\rho_i \kappa_i}\vec{g}\Bigg)$$

$$+\ \mathrm{div}\ \left[\sum_{i=\mathrm{A,B}} h_i(\kappa_i\ \mathrm{grad}\ \rho_i + \rho\delta_i\ \mathrm{grad}\ T\right.$$

$$\left. + \rho\psi_i\ \mathrm{grad}\ p - M_i\vec{g})\right] - L_{\mathrm{v}}\frac{\partial \rho_{\mathrm{vo}}}{\partial t}. \tag{3.509}$$

The velocity of the mixture $\vec{v}$ can be expressed using (3.435), (3.501), (3.502),

$$\vec{v} = -\frac{1}{\rho}\sum_{i=\mathrm{A,B}} (\kappa_i\ \mathrm{grad}\ \rho_i + \rho\delta_i\ \mathrm{grad}\ T + \rho\psi_i\ \mathrm{grad}\ p - M_i\vec{g}). \tag{3.510}$$

Substituting (3.504), (3.505), (3.510) into (3.509), we get

$$\sum_{i=\mathrm{A,B,C}} \rho_i c_{\mathrm{p},i}\frac{\partial T}{\partial t} = \mathrm{div}\ \left(\lambda\ \mathrm{grad}\ T + \sum_{i=\mathrm{A,B}} \beta_i\ \mathrm{grad}\ \rho_i\right.$$

$$+ \sum_{i=\mathrm{A,B}} \gamma_i\ \mathrm{grad}\ p - \sum_{i=\mathrm{A,B}} \frac{\beta_i M_i}{\rho_i \kappa_i}\vec{g}\Bigg)$$

$$+ h_C \ \text{div} \ \left(\sum_{i=A,B} \kappa_i \ \text{grad} \ \rho_i + \rho \delta_i \ \text{grad} \ T \right.$$

$$\left. + \rho \psi_i \ \text{grad} \ p - M_i \vec{g} \right)$$

$$+ \frac{1}{\rho} \sum_{i=A,B,C} \rho_i c_{p,i} \sum_{i=A,B} (\kappa_i \ \text{grad} \ \rho_i + \rho \delta_i \ \text{grad} \ T$$

$$+ \rho \psi_i \ \text{grad} \ p - M_i \vec{g}) \cdot \text{grad} \ T$$

$$+ \sum_{i=A,B} \frac{H_i}{\rho} (\kappa_i \ \text{grad} \ \rho_i + \rho \delta_i \ \text{grad} \ T$$

$$+ \rho \psi_i \ \text{grad} \ p - M_i \vec{g}) \cdot \text{grad} \ \rho_i$$

$$+ \sum_{i=A,B} (\kappa_i \ \text{grad} \ \rho_i + \rho \delta_i \ \text{grad} \ T$$

$$+ \rho \psi_i \ \text{grad} \ p - M_i \vec{g}) \cdot \text{grad} \ h_i - L_v \frac{\partial \rho_{vo}}{\partial t}. \tag{3.511}$$

The system of equations (3.504), (3.505), (3.506), (3.507), (3.508), (3.511) can now be employed for determining the unknows ρ_A, ρ_B, ρ_p, ρ_v, ρ_g, T.

The momentum balance equations are not necessary in the diffusion approximation, for calculating pressure gradients we employ the equations of state (3.491), (3.492), equation (3.454) and the condition

$$p_C = const. \tag{3.512}$$

The remaining unknown fields $\vec{v}$, $\vec{v}_A$, $\vec{v}_B$, ρ_w, ρ_a can be calculated a posteriori using equations (3.510), (3.501), (3.502), (3.474), (3.475) and the previously calculated variables ρ_A, ρ_B, ρ_p, ρ_v, ρ_g, T.

Another possible simplification consists in neglecting the diffusion fluxes,

$$\vec{j}_i = 0, \qquad i = A, B, \tag{3.513}$$

and consequently

$$\vec{v}_i = \vec{v}, \qquad i = A, B. \tag{3.514}$$

However, with regard to (3.433) this leads to

$$\vec{v}_C = \vec{v}, \tag{3.515}$$

which is a logical impossibility.

Therefore, the convective approximation can be better introduced with the assumption of a motionless porous matrix (3.434). In this case we obtain the mass balance equations of components A, B in the form

$$\frac{\partial \rho_A}{\partial t} + \text{div}\,(\rho_A \vec{v}_A) = -\frac{\partial \rho_{vo}}{\partial t} \tag{3.516}$$

$$\frac{\partial \rho_B}{\partial t} + \text{div}\,(\rho_B \vec{v}_B) = \frac{\partial \rho_{vo}}{\partial t}. \tag{3.517}$$

In an analogous way we get the respective balances for the components p, v, g. Here the convective approximation

$$\vec{j}_i = 0, \qquad i = w, p, v, g, a \tag{3.518}$$

does not lead to similar illogical consequences to those for the ABC mixture and can be applied directly, giving the results

$$\vec{v}_w = \vec{v}_p = \vec{v}_A \tag{3.519}$$

$$\vec{v}_v = \vec{v}_g = \vec{v}_a = \vec{v}_B. \tag{3.520}$$

In this way we obtain

$$\frac{\partial \rho_p}{\partial t} + \text{div}\,(\rho_p \vec{v}_A) = 0 \tag{3.521}$$

$$\frac{\partial \rho_v}{\partial t} + \text{div}\,(\rho_v \vec{v}_B) = \frac{\partial \rho_{vo}}{\partial t} \tag{3.522}$$

$$\frac{\partial \rho_g}{\partial t} + \text{div}\,(\rho_g \vec{v}_B) = 0. \tag{3.523}$$

The internal energy balance reads

$$\sum_{i=A,B,C} \rho_i c_{p,i} \left(\frac{\partial T}{\partial t} + \vec{v} \cdot \text{grad}\,T \right) + \sum_{i=A,B} H_i \left(\frac{\partial \rho_i}{\partial t} + \vec{v} \cdot \text{grad}\,\rho_i \right)$$

$$= \text{div}\,(\lambda\ \text{grad}\,T) - \text{div}\left(\sum_{i=A,B} \rho_i h_i \vec{v}_i \right) - L_v \frac{\partial \rho_{vo}}{\partial t}. \tag{3.524}$$

We now have equations (3.516), (3.517), (3.521), (3.522), (3.523), (3.524), (3.441), (3.445), (3.450), (3.491), (3.492) for the unknowns ρ_A, ρ_B, ρ_p, ρ_v, ρ_g, T, $\vec{v}$, $\vec{v}_A$, $\vec{v}_B$, p, p_A, p_B.

In the case of saturated flow of water (with dissolved salts) through the porous medium we have

$$w = w_m, \tag{3.525}$$

and therefore

$$\rho_B = 0 \tag{3.526}$$

$$R_w = 0 \tag{3.527}$$

$$R_v = 0 \tag{3.528}$$

$$R_u = 0 \tag{3.529}$$

$$\rho_A = \rho_{A,m} = const., \tag{3.530}$$

where

$$\rho_{A,m} = \rho_C \frac{w_m}{1 - w_m}. \tag{3.531}$$

The mass balance equation for the ABC mixture (3.441) now reads

$$\text{div } \vec{v} = 0, \tag{3.532}$$

and the mass balance for component A is reduced to the form

$$\text{div } \vec{v}_A = 0, \tag{3.533}$$

the relation (3.435) becomes

$$\rho \vec{v} = \rho_A \vec{v}_A. \tag{3.534}$$

The internal energy balance is simplified as follows:

$$\sum_{i=A,C} \rho_i c_{p,i} \left(\frac{\partial T}{\partial t} + \vec{v} \cdot \text{grad } T \right)$$

$$= \text{div } (\lambda \text{ grad } T) - \rho \vec{v} \cdot \text{grad } h_A. \tag{3.535}$$

We now have equations (3.532), (3.445), (3.535) for the unknowns $\vec{v}$, p, T. The remaining quantities $\vec{v}_A$, p_A can be determined from (3.450), (3.534).

Assuming one-dimensional saturated flow in a porous material, from (3.532) it follows that

$$v = const. \tag{3.536}$$

and (3.445) is simplified significantly,

$$\frac{\partial p}{\partial z} = -\rho g - Av, \tag{3.537}$$

which makes it possible to express the velocity in the form

$$v = -\frac{1}{A} \frac{\partial p}{\partial z} - \frac{\rho g}{A}. \tag{3.538}$$

Denoting

$$p = \rho g h, \tag{3.539}$$

where h is the pressure head, and

$$K_\mathrm{s} = \frac{\rho g}{A}, \tag{3.540}$$

where K_s is the hydraulic conductivity, we obtain

$$v = -K_\mathrm{s} \ \mathrm{grad} \ \Phi, \tag{3.541}$$

where

$$\Phi = h + z \tag{3.542}$$

is the total head as the sum of the pressure head and the elevation head. Equation (3.541) is Darcy's equation, which is commonly used in hydraulics for modelling groundwater flow. We note that Darcy's equation could be derived as a simplification of the momentum balance equation for one-dimensional saturated flow only.

Of course, the hybrid model formulated above and its simplifications are not the only possible ways to construct complex models of heat, moisture and chemical compounds transport. One of the most successful attempts to formulate such complex model was by Grunewald (1997). His model is also a kind of hybrid model and takes into account both diffusion and convection effects.

The balance equation for liquid and gaseous moisture was formulated in Grunewald (1997) model as follows:

$$\frac{\partial}{\partial t}(\rho_\mathrm{w} w_\mathrm{l} + \rho_\mathrm{v} w_\mathrm{g}) = - \ \mathrm{div} \ [(\rho_\mathrm{w} \vec{v}^{ml} - \vec{j}_\mathrm{dif}^{ms}) w_\mathrm{l}$$

$$+ (\rho_\mathrm{v} \vec{v}^{mg} + \vec{j}_\mathrm{dif}^{mv}) w_\mathrm{g}], \tag{3.543}$$

where w_l, w_g are the moisture contents by volume in the liquid and gaseous phase, respectively (more accurately, the volume fractions of the whole liquid and gaseous phase in the system), ρ_w the density of water, ρ_v the partial density of water vapour, $\vec{v}^{ml}$, $\vec{v}^{mg}$ the barycentric velocities of water and water vapour, respectively, $\vec{j}_\mathrm{dif}^{ms}$ the diffusion flux of salts, and $\vec{j}_\mathrm{dif}^{mv}$ the diffusion flux of water vapour.

For the particular convective fluxes in equation (3.543) we have

$$\rho_\mathrm{l} \vec{v}^{ml} = -K_\mathrm{l}(\mathrm{grad} \ p_\mathrm{l} + \rho_\mathrm{l} \vec{g}) \tag{3.544}$$

$$\rho_\mathrm{g} \vec{v}^{mg} = -K_\mathrm{g}(\mathrm{grad} \ p_\mathrm{g} + \rho_\mathrm{g} \vec{g}), \tag{3.545}$$

where ρ_l, ρ_g are the densities of the liquid and gaseous phases, respectively, K_l, K_g the convective coefficients in the liquid and gaseous phase, respectively (with the physical dimension of s), p_l, p_g are the pressures in the liquid and gaseous phases, respectively.

For the diffusive fluxes we can write:

$$\vec{j}_{\mathrm{dif}}^{\mathrm{ms}} = -D^s \operatorname{grad} C_s - D_p^s \operatorname{grad} p_l \tag{3.546}$$

$$\vec{j}_{\mathrm{dif}}^{\mathrm{mv}} = -D^v \operatorname{grad} C_v - D_p^v \operatorname{grad} p_g, \tag{3.547}$$

where ρ_s is the partial density of salts in the liquid phase, D^s the diffusion coefficient of salts in the liquid, D^v the diffusion coefficient of water vapour (both in kg/ms), C_s, C_v the concentrations of salts and water vapour, respectively, D_p^s, D_p^v the barodiffusion coefficients of salts and water vapour, respectively,

$$C_v = \frac{\rho_v}{\rho_g} = \frac{\rho_v}{\rho_v + \rho_a} \tag{3.548}$$

$$C_s = \frac{\rho_s}{\rho_l} = \frac{\rho_s}{\rho_w + \rho_s}. \tag{3.549}$$

The air mass balance reads:

$$\frac{\partial}{\partial t}(\rho_a w_g) = -\operatorname{div}\left[(\rho_a \vec{v}^{mg} - \vec{j}_{\mathrm{dif}}^{\mathrm{mv}})w_g\right], \tag{3.550}$$

where ρ_a is the partial density of air.

The salts mass balance is written as

$$\frac{\partial}{\partial t}(\rho_s w_l + \rho_p w_p) = -\operatorname{div}\left[(\rho_s \vec{v}^{ml} + \vec{j}_{\mathrm{dif}}^{\mathrm{ms}})w_l\right], \tag{3.551}$$

where ρ_p is the partial density of the crystalline salt, w_p the volume fraction of the crystalline salt, ρ_s the partial density of dissolved salts.

The heat balance is formulated in the following way:

$$\frac{\partial}{\partial t}\left[\rho_m u_m + \rho_p u_p w_p + \rho_l u_l w_l + (\rho_v u_v + \rho_a u_a)w_g\right]$$

$$= -\operatorname{div}\left[\rho_l u_l \vec{v}^{ml} w_l + (\rho_v u_v + \rho_a u_a)\vec{v}^{mg} w_g\right]$$

$$-\operatorname{div}\left[\vec{j}_{\mathrm{dif}}^{Q} + (h_s - h_w)\vec{j}_{\mathrm{dif}}^{\mathrm{ms}} w_l + (h_v - h_a)\vec{j}_{\mathrm{dif}}^{\mathrm{mv}} w_g\right], \tag{3.552}$$

where u with the particular indices are specific internal energies, the index m means the porous matrix,

$$\vec{j}_{\mathrm{dif}}^{Q} = -\lambda \operatorname{grad} T, \tag{3.553}$$

where λ is the thermal conductivity, $(h_s - h_w)$ the difference in partial specific enthalpies of salts and water, and $(h_v - h_a)$ the difference in partial specific enthalpies of water vapour and air.

In the four balance equations, there appear nine unknown field variables w_l, w_g, p_l, p_g, w_p, T, ρ_v, ρ_a, ρ_s. Therefore, the system of equations has to be completed by five additional relations between these variables.

For an easier description of salt transport Grunewald introduced the salt potential ψ_s,

$$\psi_s = \frac{\rho_s}{\rho_{s,max}} + \frac{w_p}{\Phi}, \tag{3.554}$$

where Φ is the porosity. He assumed that salt crystallization can be initiated only if $\rho_s = \rho_{s,max}$, for $\rho_s < \rho_{s,max}$ $w_p = 0$.

Therefore, according to the equation (3.554)

$$\rho_s = \begin{cases} \psi_s \rho_{s,max} & 0 \le \psi_s \le 1 \\ \rho_{s,max} & 1 < \psi_s \le 2 \end{cases} \tag{3.555}$$

$$\frac{\partial \rho_s}{\partial \psi_s} = \begin{cases} \rho_{s,max} & 0 \le \psi_s \le 1 \\ 0 & 1 < \psi_s \le 2 \end{cases} \tag{3.556}$$

$$w_p = \begin{cases} 0 & 0 \le \psi_s \le 1 \\ (\psi_s - 1)\Phi & 1 < \psi_s \le 2 \end{cases} \tag{3.557}$$

$$\frac{\partial w_p}{\partial \psi_s} = \begin{cases} 0 & 0 \le \psi_s \le 1 \\ \Phi & 1 < \psi_s \le 2 \end{cases}. \tag{3.558}$$

The relation between the partial pressure of water vapour p_v and liquid moisture content w_l is given again by the sorption isotherm, so that

$$p_v = \varphi(w_l, \psi_s, T) \cdot p_{sat}(T), \tag{3.559}$$

the relations between the pressures and partial densities of water vapour and air can be expressed by the equation of state of an ideal gas,

$$p_v = \frac{\rho_v}{M_v} RT \tag{3.560}$$

$$p_a = \frac{\rho_a}{M_a} RT, \tag{3.561}$$

and according to the basic principles of the theory of mixtures

$$p_g = p_v + p_a. \tag{3.562}$$

The pressure of the liquid phase Grunewald expressed by the capillary pressure,

$$p_c = p_l - p_g, \tag{3.563}$$

and the relation between capillary pressure and liquid moisture content by the water retention curve,

$$p_c = p_c(w_l). \tag{3.564}$$

Taking into account that according to the Kelvin equation the following relation between p_c and φ is valid,

$$p_c = \rho_l \frac{R}{M} T \ln \varphi, \tag{3.565}$$

the sorption isotherm and water retention curve can be unified in just one <u>moisture potential</u> function describing the relation between the liquid and gaseous moisture contents. This can be formally written as

$$\varphi = \varphi(w_l, \psi_s, T), \qquad 0 \leq w_l \leq w_{\text{sat}}, \tag{3.566}$$

where for $\varphi < 95 - 98\%$ the sorption isotherm is valid, while for $\varphi > 95 - 98\%$ the water retention curve with capillary pressure recalculated to relative humidity using the Kelvin equation is applied.

In order to express the relation between the volume fractions of the particular phases, we use

$$\Phi = w_l + w_g + w_p, \tag{3.567}$$

and therefore

$$\frac{\partial w_p}{\partial w_l} = -1 \tag{3.568}$$

$$\frac{\partial w_g}{\partial w_l} = -1. \tag{3.569}$$

Now, choosing w_l, ψ_s, p_a and T as the four basic field variables, the remaining ones can be determined using the relations given above. Therefore, the original set of balance equations (3.543), (3.550), (3.551), (3.552) can be modified into a form that can be directly solved as a system of four equations for four unknown field quantities.

The left-hand sides of the balance equations (3.543), (3.550), (3.551), (3.552) containing the accumulation terms can be modified as follows. In the moisture balance we have

$$\frac{\partial}{\partial t}(\rho_w w_l + \rho_v w_g) = \rho_w \frac{\partial w_l}{\partial t} + w_l \frac{\partial \rho_w}{\partial t} + \rho_v \frac{\partial w_g}{\partial t} + w_g \frac{\partial \rho_v}{\partial t}. \tag{3.570}$$

We use relations

$$\rho_v = p_v \frac{M_v}{RT} = \varphi(w_l, \psi_s, T) p_{\text{sat}}(T) \frac{M_v}{RT} \tag{3.571}$$

$$\Phi = w_l + w_g + w_p \tag{3.572}$$

$$\frac{\partial \rho_v}{\partial t} = \frac{M_v}{R} \frac{\partial}{\partial t} \left(\frac{\varphi p_{sat}}{T} \right)$$ (3.573)

$$\frac{\partial}{\partial t} \left(\frac{\varphi p_{sat}}{T} \right) = \frac{\partial \varphi}{\partial t} \frac{p_{sat}}{T} + \frac{\varphi}{T} \frac{dp_{sat}}{dT} - \frac{1}{T^2} \frac{\partial T}{\partial t} \varphi p_{sat}$$ (3.574)

$$\frac{\partial \varphi}{\partial t} = \frac{\partial \varphi}{\partial w_l} \frac{\partial w_l}{\partial t} + \frac{\partial \varphi}{\partial T} \frac{\partial T}{\partial t} + \frac{\partial \varphi}{\partial \psi_s} \frac{\partial \psi_s}{\partial t}$$ (3.575)

$$\frac{dp_{sat}}{dt} = \frac{dp_{sat}}{dT} \frac{\partial T}{\partial t}$$ (3.576)

$$\frac{\partial \rho_w}{\partial t} = \frac{\partial \rho_w}{\partial T} \frac{\partial T}{\partial t}$$ (3.577)

$$\frac{\partial w_g}{\partial t} = \frac{\partial w_g}{\partial w_l} \frac{\partial w_l}{\partial t} + \frac{\partial w_g}{\partial w_p} \frac{\partial w_p}{\partial t} = -\frac{\partial w_l}{\partial t} - \frac{\partial w_p}{\partial \psi_s} \frac{\partial \psi_s}{\partial t}.$$ (3.578)

Substituting into (3.570), we get

$$\frac{\partial}{\partial t}(\rho_w w_l + \rho_v w_g) = \left(\rho_w - \rho_v + w_g \frac{M_v}{R} \frac{p_{sat}}{T} \frac{\partial \varphi}{\partial w_l} \right) \frac{\partial w_l}{\partial t}$$

$$+ \left(-\rho_v \frac{\partial w_p}{\partial \psi_s} + w_g \frac{p_{sat}}{T} \frac{M_v}{R} \frac{\partial \varphi}{\partial \psi_s} \right) \frac{\partial \psi_s}{\partial t}$$

$$+ \left[w_l \frac{\partial \rho_w}{\partial T} + \frac{w_g M_v}{RT} \left(\varphi \frac{dp_{sat}}{dT} - \frac{p_v}{T} + p_{sat} \frac{\partial \varphi}{\partial T} \right) \right] \frac{\partial T}{\partial t}.$$ (3.579)

In the air mass balance we can write

$$\frac{\partial}{\partial t}(\rho_a w_g) = w_g \frac{\partial \rho_a}{\partial t} + \rho_a \frac{\partial w_g}{\partial t}.$$ (3.580)

We use

$$\rho_a = p_a \frac{M_a}{RT},$$ (3.581)

and therefore

$$\frac{\partial \rho_a}{\partial t} = \frac{M_a}{RT} \frac{\partial p_a}{\partial t} - p_a \frac{M_a}{RT^2} \frac{\partial T}{\partial t},$$ (3.582)

for $\frac{\partial w_g}{\partial t}$ we employ (3.578), and finally have

$$\frac{\partial}{\partial t}(\rho_a w_g) = -\rho_a \frac{\partial w_l}{\partial t} + w_g \frac{M_a}{RT} \frac{\partial p_a}{\partial t}$$

$$- \rho_a \frac{\partial w_p}{\partial \psi_s} \frac{\partial \psi_s}{\partial t} - w_g \frac{p_a M_a}{RT^2} \frac{\partial T}{\partial t}. \tag{3.583}$$

The left-hand side of the salts mass balance is first rewritten as

$$\frac{\partial}{\partial t}(\rho_s w_l + \rho_p w_p) = w_l \frac{\partial \rho_s}{\partial t} + \rho_s \frac{\partial w_l}{\partial t} + \rho_p \frac{\partial w_p}{\partial t} + w_p \frac{\partial \rho_p}{\partial t}. \tag{3.584}$$

We use

$$\frac{\partial \rho_s}{\partial t} = \frac{\partial \rho_s}{\partial \psi_s} \frac{\partial \psi_s}{\partial t} \tag{3.585}$$

$$\frac{\partial \rho_p}{\partial t} = \frac{\partial \rho_p}{\partial \psi_s} \frac{\partial \psi_s}{\partial t} = 0 \tag{3.586}$$

$$\frac{\partial w_p}{\partial t} = \frac{\partial w_p}{\partial \psi_s} \frac{\partial \psi_s}{\partial t}, \tag{3.587}$$

and obtain

$$\frac{\partial}{\partial t}(\rho_s w_l + \rho_p w_p) = \rho_s \frac{\partial w_l}{\partial t} + \left(w_l \frac{\partial \rho_s}{\partial \psi_s} + \rho_p \frac{\partial w_p}{\partial \psi_s} \right) \frac{\partial \psi_s}{\partial t}. \tag{3.588}$$

Finally, in the heat balance we have

$$\frac{\partial}{\partial t}[\rho_m u_m + \rho_p u_p w_p + \rho_l u_l w_l + (\rho_v u_v + \rho_a u_a) w_g]$$

$$= \rho_m \frac{\partial u_m}{\partial t} + u_m \frac{\partial \rho_m}{\partial t} + \frac{\partial \rho_p}{\partial t} u_p w_p + \rho_p \frac{\partial u_p}{\partial t} w_p + \rho_p u_p \frac{\partial w_p}{\partial t}$$

$$+ \frac{\partial \rho_l}{\partial t} u_l w_l + \rho_l \frac{\partial u_l}{\partial t} w_l + \rho_l u_l \frac{\partial w_l}{\partial t} + \frac{\partial \rho_v}{\partial t} u_v w_g + \rho_v \frac{\partial u_v}{\partial t} w_g$$

$$+ \rho_v u_v \frac{\partial w_g}{\partial t} + \frac{\partial \rho_a}{\partial t} u_a w_g + \rho_a \frac{\partial u_a}{\partial t} w_g + \rho_a u_a \frac{\partial w_g}{\partial t}. \tag{3.589}$$

We use relations

$$u_i = u_i(T), \quad i = m, p, l, v, a, \tag{3.590}$$

therefore

$$\frac{\partial u_i}{\partial t} = \frac{\partial u_i}{\partial T} \frac{\partial T}{\partial t}, \tag{3.591}$$

further

$$\frac{\partial \rho_m}{\partial t} = 0, \tag{3.592}$$

$$\frac{\partial \rho_p}{\partial t} = 0 \tag{3.593}$$

$$\frac{\partial \rho_l}{\partial t} = \frac{\partial \rho_l}{\partial T}\frac{\partial T}{\partial t} \tag{3.594}$$

$$\frac{\partial w_p}{\partial t} = \frac{\partial w_p}{\partial \psi_s}\frac{\partial \psi_s}{\partial t}, \tag{3.595}$$

for $\frac{\partial \rho_v}{\partial t}$ we employ (3.573)–(3.574), so that

$$\frac{\partial \rho_v}{\partial t} = \frac{M_v}{R}\frac{p_{sat}}{T}\frac{\partial \varphi}{\partial w_l}\frac{\partial w_l}{\partial t} + \frac{M_v}{R}\frac{p_{sat}}{T}\frac{\partial \varphi}{\partial \psi_s}\frac{\partial \psi_s}{\partial t}$$

$$+ \frac{M_v}{RT}\left(\varphi\frac{dp_{sat}}{dT} - \frac{p_v}{T} + p_{sat}\frac{\partial \varphi}{\partial T}\right)\frac{\partial T}{\partial t}, \tag{3.596}$$

for $\frac{\partial w_g}{\partial t}$ we have equation (3.578), for $\frac{\partial \rho_a}{\partial t}$ (3.582).
Substituting into (3.589) we get

$$\frac{\partial}{\partial t}\left[\rho_m u_m + \rho_p u_p w_p + \rho_l u_l w_l + (\rho_v u_v + \rho_a u_a)w_g\right]$$

$$= \left(\rho_l u_l + u_v w_g\frac{M_v}{R}\frac{p_{sat}}{T}\frac{\partial \varphi}{\partial w_l} - \rho_v u_v - \rho_a u_a\right)\frac{\partial w_l}{\partial t}$$

$$+ u_a w_g\frac{M_a}{RT}\frac{\partial p_a}{\partial t}$$

$$+ \left[\rho_p u_p\frac{\partial w_p}{\partial \psi_s} + u_v w_g\frac{M_v}{R}\frac{p_{sat}}{T}\frac{\partial \varphi}{\partial \psi_s} - (\rho_v u_v + \rho_a u_a)\frac{\partial w_p}{\partial \psi_s}\right]\frac{\partial \psi_s}{\partial t}$$

$$+ \left[\rho_m\frac{\partial u_m}{\partial T} + \rho_p w_p\frac{\partial u_p}{\partial T} + u_l w_l\frac{\partial \rho_l}{\partial T} + \rho_l w_l\frac{\partial u_l}{\partial T}\right.$$

$$+ u_v w_g\frac{M_v}{RT}\left(\varphi\frac{dp_{sat}}{dT} - \frac{p_v}{T} + p_{sat}\frac{\partial \varphi}{\partial T}\right) + \rho_v w_g\frac{\partial u_v}{\partial T}$$

$$\left. - u_a w_g\frac{p_a M_a}{RT^2} + \rho_a w_g\frac{\partial u_a}{\partial T}\right]\frac{\partial T}{\partial t}. \tag{3.597}$$

In an analogous way we modify the right-hand sides of equations (3.543), (3.550), (3.551), (3.552), primarily the expressions for the particular fluxes. Grunewald (1997) concentrated on the problem of coupling the liquid and gaseous phases of water, which is not yet a fully resolved problem. The fluxes describing heat and salt transport can be considered in the form given above, namely (3.546), (3.553).

The basic material relations for the fluxes of liquid and gaseous phases read as follows. In the liquid phase we have

$$\vec{j}^{ml} = \rho_l \vec{v}^{ml} = -K_l(\text{grad } p_l + \rho_l \vec{g}), \tag{3.598}$$

in the gaseous phase

$$\vec{j}^{mv} = \rho_v \vec{v}^{mg} + \vec{j}^{mv}_{dif} = -\frac{\rho_v}{\rho_g} K_g(\text{grad } p_g + \rho_g \vec{g})$$

$$- D^v \text{ grad } C_v - D^v_p \text{ grad } p_g. \tag{3.599}$$

The material relation for the liquid phase can be modified using the definition equation for capillary pressure p_c,

$$p_l = p_c + p_g, \tag{3.600}$$

which makes it possible to eliminate the gradient of pressure of the liquid phase, namely

$$\vec{j}^{ml} = \rho_l \vec{v}^{ml} = -K_l(\text{grad } p_c + \rho_l \vec{g}) - K_l \text{ grad } p_g. \tag{3.601}$$

In the material relation for the gaseous phase, we have to express first grad C_v. According to the definition

$$C_v = \frac{\rho_v}{\rho_a + \rho_v}, \tag{3.602}$$

and using the equation of state of ideal gas,

$$\rho = p\frac{M}{RT}, \tag{3.603}$$

we have

$$C_v = \frac{p_v M_v}{p_v M_v + p_a M_a}, \tag{3.604}$$

and

$$\text{grad } C_v = \frac{\partial C_v}{\partial p_v} \text{ grad } p_v + \frac{\partial C_v}{\partial p_a} \text{ grad } p_a. \tag{3.605}$$

The particular derivatives from (3.605) can be expressed as

$$\frac{\partial C_v}{\partial p_v} = \frac{M_v(p_v M_v + p_a M_a) - p_v M_v M_v}{(p_v M_v + p_a M_a)^2}$$

$$= \frac{p_a M_a M_v}{(p_v M_v + p_a M_a)^2} = \frac{p_a}{p_x^2} \tag{3.606}$$

$$\frac{\partial C_v}{\partial p_a} = -\frac{p_v M_v M_a}{(p_v M_v + p_a M_a)^2} = -\frac{p_v}{p_x^2}. \qquad (3.607)$$

Substituting into (3.605), we get

$$\text{grad } C_v = \frac{p_a}{p_x^2} \text{ grad } p_v - \frac{p_v}{p_x^2} \text{ grad } p_a. \qquad (3.608)$$

Taking into account that in the material relation for the gaseous phase, there appears grad p_g, it will be more convenient to modify the expression for grad C_v, so that it would include grad p_g instead of grad p_a. This can be easily done by adding and subtracting the term

$$\frac{p_v}{p_x^2} \text{ grad } p_v$$

to and from the terms on the right-hand side of equation (3.608). Then, taking into account

$$p_g = p_v + p_a, \qquad (3.609)$$

we have

$$-\frac{p_v}{p_x^2} \text{ grad } p_a - \frac{p_v}{p_x^2} \text{ grad } p_v = -\frac{p_v}{p_x^2} \text{ grad } p_g \qquad (3.610)$$

$$\frac{p_a}{p_x^2} \text{ grad } p_v + \frac{p_v}{p_x^2} \text{ grad } p_v = \frac{p_g}{p_x^2} \text{ grad } p_v \qquad (3.611)$$

and

$$\text{grad } C_v = \frac{p_g}{p_x^2} \text{ grad } p_v - \frac{p_v}{p_x^2} \text{ grad } p_g. \qquad (3.612)$$

Substituting now into (3.599), we get

$$\vec{j}^{mv} = -\frac{p_g}{p_x^2} D^v \text{ grad } p_v - \frac{\rho_v}{\rho_g} K_g \text{ grad } p_g$$

$$-\left(D_p^v - D^v \frac{p_v}{p_x^2} \right) \text{ grad } p_g - K_g \rho_v \vec{g}. \qquad (3.613)$$

The relation between partial pressure of water vapour p_v and capillary pressure p_c gives the Kelvin equation,

$$p_c = \rho_l \frac{R}{M_v} T \ln \frac{p_v}{p_{sat}}. \qquad (3.614)$$

Therefore, in equation (3.613) we can express

$$\text{grad } p_v = \frac{\partial p_v}{\partial p_c} \text{ grad } p_c + \frac{\partial p_v}{\partial T} \text{ grad } T, \qquad (3.615)$$

and according to (3.614)

$$p_v = p_{sat} \exp\left(\frac{p_c M_v}{\rho_l RT}\right).$$ (3.616)

Then,

$$\frac{\partial p_v}{\partial p_c} = p_{sat} \frac{M_v}{\rho_l RT} \exp\left(\frac{p_c M_v}{\rho_l RT}\right) = p_v \frac{M_v}{\rho_l RT},$$ (3.617)

and using the equation of state of ideal gas,

$$p_v = \frac{\rho_v}{M_v} RT,$$ (3.618)

we get

$$\frac{\partial p_v}{\partial p_c} = \frac{\rho_v}{\rho_l}.$$ (3.619)

The second expression in $\operatorname{grad} p_v$ is a little more complicated,

$$\frac{\partial p_v}{\partial T} = p_{sat} \exp\left(\frac{p_c M_v}{\rho_l RT}\right) \frac{M_v p_c}{\rho_l R}\left(-\frac{1}{T^2}\right) + \frac{dp_{sat}}{dT}\exp\left(\frac{p_c M_v}{\rho_l RT}\right)$$

$$= -p_v \frac{p_c M_v}{\rho_l RT^2} + \frac{dp_{sat}}{dT}\frac{p_v}{p_{sat}} = -\frac{\rho_v}{\rho_l}\frac{p_c}{T} + \frac{dp_{sat}}{dT}\varphi.$$ (3.620)

Substituting the derived expressions for the derivatives of p_v in (3.615), we have

$$\operatorname{grad} p_v = \frac{\rho_v}{\rho_l}\operatorname{grad} p_c + \left(\varphi\frac{dp_{sat}}{dT} - \frac{\rho_v}{\rho_l}\frac{p_c}{T}\right)\operatorname{grad} T,$$ (3.621)

and using (3.621) in (3.613) we obtain the final material relation for water vapour transport

$$\vec{j}^{mv} = -\frac{p_g}{p_x^2}\frac{\rho_v}{\rho_l}D^v \operatorname{grad} p_c - \frac{\rho_v}{\rho_g}K_g \operatorname{grad} p_g - \left(D_p^v - D^v\frac{p_v}{p_x^2}\right)\operatorname{grad} p_g$$

$$- K_g \rho_v \vec{g} - \frac{p_g}{p_x^2}D^v\left(\varphi\frac{dp_{sat}}{dT} - \frac{\rho_v}{\rho_l}\frac{p_c}{T}\right)\operatorname{grad} T.$$ (3.622)

The total moisture flux is defined as a combination of water and water vapour fluxes, taking into account their volume fractions,

$$\vec{j}^{mw} = \vec{j}^{ml} w_l + \vec{j}^{mv} w_g.$$ (3.623)

Substituting (3.601) and (3.622) in (3.623), we arrive at

$$\vec{j}^{mw} = -\left(K_l w_l + \frac{p_g}{p_x^2}\frac{\rho_v}{\rho_l}D^v w_g\right)\operatorname{grad} p_c$$

$$- \left[K_1 w_1 + \left(D_p^v - D^v \frac{p_v}{p_x^2} \right) w_g \right] \operatorname{grad} p_g - \frac{\rho_v}{\rho_g} K_g w_g \operatorname{grad} p_g$$

$$- K_1 w_1 \rho_1 \vec{g} - K_g w_g \rho_v \vec{g}$$

$$- \frac{p_g}{p_x^2} D^v w_g \left(\varphi \frac{\mathrm{d} p_{sat}}{\mathrm{d} T} - \frac{\rho_v}{\rho_1} \frac{p_c}{T} \right) \operatorname{grad} T, \tag{3.624}$$

where

$$L_c = K_1 w_1 + \frac{p_g}{p_x^2} \frac{\rho_v}{\rho_1} D^v w_g \tag{3.625}$$

is the capillary moisture conductivity, and

$$L_p = K_1 w_1 + \left(D_p^v - D^v \frac{p_v}{p_x^2} \right) w_g \tag{3.626}$$

the pressure moisture conductivity (remember that p_g is the total pressure in the gaseous phase).

Equation (3.624) can be reformulated in such a way that the gradient of partial pressure of water vapour would be the basic thermodynamic driving force instead of the gradient of capillary pressure. We employ the Kelvin equation in the form of (3.614), and express $\operatorname{grad} p_c$ using $\operatorname{grad} p_v$ and $\operatorname{grad} T$,

$$\operatorname{grad} p_c = \frac{\partial p_c}{\partial p_v} \operatorname{grad} p_v + \frac{\partial p_c}{\partial T} \operatorname{grad} T. \tag{3.627}$$

The particular derivatives are as follows:

$$\frac{\partial p_c}{\partial p_v} = \rho_1 \frac{R}{M_v} \frac{T}{p_v} = \frac{\rho_1}{\rho_v} \tag{3.628}$$

$$\frac{\partial p_c}{\partial T} = \rho_1 \frac{R}{M_v} \ln \frac{p_v}{p_{sat}} + \rho_1 \frac{R}{M_v} \frac{T}{\frac{p_v}{p_{sat}}} \left(-\frac{p_v}{p_{sat}^2} \right) \frac{\mathrm{d} p_{sat}}{\mathrm{d} T}$$

$$= \rho_1 \frac{R}{M_v} \ln \frac{p_v}{p_{sat}} - \rho_1 \frac{R}{M_v} \frac{T}{p_{sat}} \frac{\mathrm{d} p_{sat}}{\mathrm{d} T}, \tag{3.629}$$

and therefore

$$\operatorname{grad} p_c = \frac{\rho_1}{\rho_v} \operatorname{grad} p_v$$

$$+ \left(\rho_1 \frac{R}{M_v} \ln \frac{p_v}{p_{sat}} - \rho_1 \frac{R}{M_v} \frac{T}{p_{sat}} \frac{\mathrm{d} p_{sat}}{\mathrm{d} T} \right) \operatorname{grad} T. \tag{3.630}$$

Substituting (3.630) into (3.624), we get

$$\vec{j}^{\,mw} = -\left(K_1 w_1 \frac{\rho_1}{\rho_v} + \frac{p_g}{p_x^2} D^v w_g \right) \operatorname{grad} p_v$$

$$-\left[K_1 w_1 + \left(D_p^v - D^v \frac{p_v}{p_x^2} \right) w_g \right] \operatorname{grad} p_g - \frac{\rho_v}{\rho_g} K_g w_g \operatorname{grad} p_g$$

$$- K_1 w_1 \rho_1 \vec{g} - K_g w_g \rho_v \vec{g} - L_T \operatorname{grad} T, \tag{3.631}$$

where

$$L_v = K_1 w_1 \frac{\rho_1}{\rho_v} + \frac{p_g}{p_x^2} D^v w_g \tag{3.632}$$

is the hygroscopic moisture conductivity, an analogue to the capillary moisture conductivity. Comparing (3.625) with (3.632), we obtain

$$L_v = \frac{\rho_1}{\rho_v} L_c. \tag{3.633}$$

For the thermodiffusion coefficient L_T, we have

$$L_T = \left(K_1 w_1 + \frac{p_g}{p_x^2} \frac{\rho_v}{\rho_1} D^v w_g \right) \left(\rho_1 \frac{R}{M_v} \ln \frac{p_v}{p_{sat}} - \rho_1 \frac{R}{M_v} \frac{T}{p_{sat}} \frac{dp_{sat}}{dT} \right)$$

$$+ \frac{p_g}{p_x^2} D^v w_g \left(\varphi \frac{dp_{sat}}{dT} - \frac{\rho_v}{\rho_1} \frac{p_c}{T} \right). \tag{3.634}$$

This definition formula is a little too complicated, therefore we will modify it further,

$$L_T = K_1 w_1 \rho_1 \frac{R}{M_v} \ln \varphi + \frac{p_g}{p_x^2} \frac{\rho_v}{\rho_1} D^v w_g \rho_1 \frac{R}{M_v} \ln \varphi$$

$$- K_1 w_1 \rho_1 \frac{R}{M_v} \frac{T}{p_{sat}} \frac{dp_{sat}}{dT} - \frac{p_g}{p_x^2} \frac{\rho_v}{\rho_1} D^v w_g \rho_1 \frac{R}{M_v} \frac{T}{p_{sat}}$$

$$+ \frac{p_g}{p_x^2} D^v w_g \varphi \frac{dp_{sat}}{dT} - \frac{p_g}{p_x^2} D^v w_g \frac{\rho_v}{\rho_1} \frac{p_c}{T}$$

$$= K_1 w_1 \frac{p_c}{T} - K_1 w_1 \rho_1 \frac{R}{M_v} \frac{T}{p_{sat}} \frac{dp_{sat}}{dT}$$

$$+ \frac{p_g}{p_x^2} \frac{\rho_v}{\rho_1} D^v w_g \left(\rho_1 \frac{R}{M_v} \ln \varphi - \frac{p_c}{T} \right)$$

$$+ \frac{p_g}{p_x^2} D^v w_g \frac{dp_{sat}}{dT} \left(-\rho_v \frac{RT}{M_v} \frac{1}{p_{sat}} + \varphi \right). \tag{3.635}$$

From the Kelvin equation (3.614) it follows that the last two terms in (3.635) are equal to zero, therefore

$$L_T = K_l w_l \frac{p_c}{T} - K_l w_l \rho_l \frac{R}{M_v} \frac{T}{p_{sat}} \frac{dp_{sat}}{dT}, \tag{3.636}$$

and using the equation of state in the form

$$\frac{RT}{M_v} = \frac{p_v}{\rho_v}, \tag{3.637}$$

we finally get

$$L_T = K_l w_l \frac{p_c}{T} - K_l w_l \frac{\rho_l}{\rho_v} \varphi \frac{dp_{sat}}{dT}. \tag{3.638}$$

We will now take a closer view to the just-defined coefficients L_c, L_p, L_v. All of them are functions of liquid moisture content w_l and temperature T. The coefficients L_c and L_v can be conveniently used in a complementary way, i.e. L_c in the overhygroscopic range and L_v in the hygroscopic range.

Further, in the overhygroscopic range the capillary moisture transport is dominant, and therefore $L_c \sim L_p$, in the case $w_l = w_{sat}$ then $L_c = L_p$. In the hygroscopic range, the condition $L_c \sim L_p$ is still valid, at least in an order of magnitude sense, and therefore from (3.633) it follows that $L_p \ll L_v$. As grad $p_g \sim$ grad p_v, we can conclude that in the hygroscopic range the moisture transport induced by the gradient of total pressure (excluding the convective motion of the gaseous phase, of course) is much less important than the transport induced by the gradient of the partial pressure of water vapour.

In order to solve the system of balance equations (3.543), (3.550), (3.551), (3.552) with the included constitutive relations, it is necessary to express the transport coefficients K_l, D^v, D_p^v by the moisture conductivities L_c, L_v. This is possible only by separating the liquid and gaseous phases of moisture.

Remember that in equations (3.624) and (3.631), moisture fluxes induced by the gradients of capillary pressure and partial water vapour pressure include parts corresponding to both liquid water and water vapour. If the total moisture flux is to be divided into the parts corresponding to water and water vapour, a logical solution is a phase dividing function,

$$f^{lv} = f^{lv}(w_l), \tag{3.639}$$

with the following basic properties:

$$f^{lv}(0) = 0 \tag{3.640}$$

$$f^{lv}(w_{sat}) = 1. \tag{3.641}$$

Then, the particular parts of the total moisture flux,

$$\vec{j}^{mw} = \vec{j}^{ml} w_l + \vec{j}^{mv} w_g, \tag{3.642}$$

can be expressed using the definition relations (3.598), (3.599) and modified equations (3.601) and (3.622) as follows:

$$\vec{j}^{ml} w_l = -[f^{lv}(w_l) L_c](\text{grad } p_c + \text{ grad } p_g + \rho_l \vec{g}) \tag{3.643}$$

$$\vec{j}^{mv} w_g = -\left\{ [1 - f^{lv}(w_l)] L_c \right\} (\text{grad } p_c + \text{ grad } p_g + \rho_l \vec{g})$$

$$- \frac{\rho_v}{\rho_g} K_g w_g \text{ grad } p_g - \left[\frac{p_g}{p_x^2} D^v \left(\varphi \frac{dp_{sat}}{dT} - \frac{\rho_v}{\rho_l} \frac{p_c}{T} \right) w_g \right] \text{grad } T$$

$$- \left(K^g - D^v \frac{p_v}{p_x^2} \right) w_g \rho_v \vec{g}. \tag{3.644}$$

Substituting (3.630) into (3.644) and comparing equations (3.643) and (3.644) with the original relations for the same fluxes (3.598), (3.599), we obtain

$$K_l w_l = f^{lv}(w_l) L_c = \frac{\rho_v}{\rho_l} f^{lv}(w_l) L_v \tag{3.645}$$

$$D^v w_g = \frac{\rho_l}{\rho_v} [1 - f^{lv}(w_l)] \frac{p_x^2}{p_g} L_c = [1 - f^{lv}(w_l)] \frac{p_x^2}{p_g} L_v \tag{3.646}$$

$$D_p^v w_g = [1 - f^{lv}(w_l)] L_c = \frac{\rho_v}{\rho_l} [1 - f^{lv}(w_l)] L_v. \tag{3.647}$$

Now, we have three basic material parameters for measurements, namely L_c in the overhygroscopic range, L_v in the hygroscopic range and f^{lv} in the whole range of moistures. Grunewald introduced the following form of the phase dividing function. For $0 \leq w_l \leq w_{hyg}$ he defined

$$f^{lv}(w_l) = \left(\frac{w_l}{w_{hyg}} \right)^{k_v} \cdot f^{lv}_{hyg}, \tag{3.648}$$

and for $w_{hyg} \leq w_l \leq w_{sat}$

$$f^{lv}(w_l) = 1 - \left(\frac{w_{sat} - w_l}{w_{sat} - w_{hyg}} \right)^{k_c} \cdot (1 - f^{lv}_{hyg}), \tag{3.649}$$

where the unknown parameters k_v in the hygroscopic range and k_c in the overhygroscopic range have to be determined experimentally.

3.4.2 Hygro-Thermo-Mechanical Models

Most models of heat, moisture and chemical compound transport in porous materials are based on the assumption of incompressibility of the porous matrix. Practically this means that the porous space remains unchanged during the transport process, and the investigation of transport processes is restricted to the fluid phase only.

This assumption looks like a good approximation of physical reality for many practical problems. For instance, for most building materials under normal conditions it can probably be employed without any significant loss of generality in describing transport processes. However, there are physical situations where the inclusion of the solid matrix in the mathematical models is unavoidable. Transport processes in soils is the most common example of such a situation. It is quite obvious that for most soils both external and internal pressure can lead to significant volume changes in the solid matrix, and therefore to the variations of both spatial configuration and volume of the porous space. Another example is modelling of transport processes in building materials at high temperatures. Here the chemical reactions in the solid matrix can change both its structure and its mechanical properties, so the volume of the porous space can be changed, for instance due to the release of some volatile compounds. Also, the stiffness of the material may decrease in a significant way, which may in practice preclude the incompressibility assumption.

Over the last decade or two, there have appeared with an increasing frequency so-called hygro-thermo-mechanical models, which deal with both heat and mass transfer in the porous space and with the deformation processes in the solid matrix. How to couple the hygrothermal and mechanical processes however remains a major problem, and it can be stated that none of the current models solves it in a satisfactory way. This remains a challenge for research work in the near future.

The simplest hygro-thermo-mechanical models separate the hygrothermal and mechanical analyses, and the coupling of both effects is done a posteriori, i.e. coupling the results of both models, which affect the input parameters of each other. An example of such a model is that of Majorana et al. (1998). They assumed the heat flux $\vec{q}$ and the moisture flux $\vec{J}$ in the simplest form,

$$\vec{q} = -a_{\mathrm{TT}} \ \mathrm{grad} \ T, \tag{3.650}$$

$$\vec{J} = -c \ \mathrm{grad} \ h, \tag{3.651}$$

where a_{TT} is the thermal conductivity, h the relative humidity, and the coefficient c they called permeability. The relation between relative humidity and water content w was

$$\mathrm{d}h = k \ \mathrm{d}w + K \ \mathrm{d}T + \mathrm{d}h_{\mathrm{s}}, \tag{3.652}$$

where

$$K = \frac{\partial h}{\partial T} \qquad (3.653)$$

is the hygrothermic coefficient representing the change in relative humidity due to a one-degree change in temperature at constant water content,

$$k = \frac{\partial h}{\partial w} \qquad (3.654)$$

is the cotangent of the slope of the isotherm $w = w(h)$, and dh_s the self-desiccation.

The balance equations for mass of water and heat they formulated as

$$\frac{\partial w}{\partial t} = -\operatorname{div} \vec{J} + \frac{\partial w_d}{\partial t} \qquad (3.655)$$

$$\rho C \frac{\partial T}{\partial t} - C_a \frac{\partial w}{\partial t} - C_w \vec{J} \cdot \operatorname{grad} T = -\operatorname{div} \vec{q}, \qquad (3.656)$$

where the water content w is taken as the mass of all free water per unit volume of the material, w_d represents the total mass of free water released in the process of drying, ρ is the mass density, C the isobaric specific heat capacity including chemically bound water but excluding free water, C_a the heat of sorption of free water (per unit mass of free water), C_w the isobaric specific heat capacity of liquid water, and the term $C_w \vec{J} \cdot \operatorname{grad} T$ is the rate of heat supply due to water convection.

The momentum balance equation for the whole multiphase medium was taken as

$$\operatorname{div} \sigma + \rho_m \vec{g} = 0, \qquad (3.657)$$

where σ is the total stress, ρ_m the density of the multiphase medium, and $\vec{g}$ the gravity acceleration.

The mechanical behaviour was described using a viscoelastic model, where coupling of creep and damage was considered. As the application of standard mechanical models was quite straightforward, we will not repeat this derivation here but instead refer the reader to the original paper Majorana et al. (1998).

Another simple hygro-thermo-mechanical model based on the application of a common thermoelastic model used in continuum mechanics (see, e.g., Malvern, 1969) was formulated by Hrstka et al. (1999). They assumed heat conduction equation in the following form:

$$\rho c \frac{\partial T}{\partial t} = \frac{\partial}{\partial x} \left(\lambda \frac{\partial T}{\partial x} \right) + \frac{\partial}{\partial y} \left(\lambda \frac{\partial T}{\partial y} \right) + \frac{\partial}{\partial z} \left(\lambda \frac{\partial T}{\partial z} \right) + I(x, y, z, t), \qquad (3.658)$$

where the term $I(x, y, z, t)$ represents the internal heat source, for instance due to the hydration heat.

The moisture balance equation was assumed in a quite simplified form suitable for the initial phase of the hydration process,

$$\frac{\partial w}{\partial t} = J(x, y, z, t), \tag{3.659}$$

where w is the moisture content, and $J(x, y, z, t)$ the moisture sink term specified on the basis of the differential function of the hydration heat production $I(x, y, z, t)$, such that the initial value of the moisture content is determined from the known water/cement ratio of the concrete mixture, and the final value is equal to the maximum hygroscopic moisture content.

The momentum balance equation was formulated as

$$(\Lambda + \mu) \operatorname{grad} \operatorname{div} \vec{u} + \mu \nabla^2 \vec{u} - (3\Lambda + 2\mu) \alpha_T \operatorname{grad} T$$

$$- (3\Lambda + 2\mu) \alpha_w \operatorname{grad} w = 0, \tag{3.660}$$

where $\vec{u}$ stands for the displacement vector, α_T is the linear thermal expansion coefficient, α_w the linear hygric expansion coefficient. The material properties Λ and μ are Lamé's constants, which are related to more frequently used parameters E and ν (Young's modulus and Poisson number):

$$\Lambda = \frac{E\nu}{(1 + \nu)(1 - 2\nu)}, \tag{3.661}$$

$$\mu = \frac{E}{2(1 + \nu)}. \tag{3.662}$$

In calculating stress fields from the displacement values determined on the basis of equation (3.660), Hrstka et al. (1999) employed Hooke's general theorem, which defines the relation between the stress tensor σ and the deformation tensor ϵ:

$$\sigma_{ik} = \left[\Lambda \sum m \, \epsilon_{mm} - \alpha_T (3\Lambda + 2\mu)(T - T_0) \right.$$

$$\left. - \alpha_w (3\Lambda + 2\mu)(w - w_0) \right] \delta_{ik} + 2\mu \epsilon_{ik}. \tag{3.663}$$

Probably the most advanced hygro-thermo-mechanical model so far employed for practical computations is that of Lewis and Schrefler (1998), which was developed for modelling transport processes in soils.

Lewis and Schrefler formulated the mass balance equation for the solid phase in the following form:

$$\frac{1 - n}{\rho^s} \frac{d^s \rho^s}{dt} - \frac{d^s n}{dt} + (1 - n) \operatorname{div} \vec{v}^s = 0, \tag{3.664}$$

where n is the porosity,

$$n = \frac{dv^w + dv^g}{dv}, \tag{3.665}$$

v^{w} is the volume of the liquid phase, v^{g} the volume of the gaseous phase, v the total volume of the porous body, ρ^{s} the phase-averaged solid density (i.e. the partial density of the solid phase), and $\vec{v}^{\mathrm{s}}$ the mass-averaged solid velocity, and for the material derivative in the solid phase we have

$$\frac{\mathrm{d}^{\mathrm{s}}\rho^{\mathrm{s}}}{\mathrm{d}t} = \frac{\partial\rho^{\mathrm{s}}}{\partial t} + \vec{v}^{\mathrm{s}} \cdot \operatorname{grad}\rho^{\mathrm{s}}. \tag{3.666}$$

For liquid water with incompressible solid grains:

$$\frac{nS_{\mathrm{w}}}{K_{\mathrm{w}}}\frac{\mathrm{d}^{\mathrm{s}}p^{\mathrm{w}}}{\mathrm{d}t} + S_{\mathrm{w}}\operatorname{div}\vec{v}^{\mathrm{s}} - \beta_{\mathrm{sw}}\frac{\mathrm{d}^{\mathrm{s}}T}{\mathrm{d}t} + n\frac{\mathrm{d}^{\mathrm{s}}S_{\mathrm{w}}}{\mathrm{d}t}$$

$$+ \frac{1}{\rho^{\mathrm{w}}}\operatorname{div}\left\{\rho^{\mathrm{w}}\frac{kk^{\mathrm{rw}}}{\mu^{\mathrm{w}}}\left[-\operatorname{grad}p^{\mathrm{w}}\right.\right.$$

$$\left.\left. + \rho^{\mathrm{w}}(\vec{g} - \vec{a}^{\mathrm{s}} - \vec{a}^{\mathrm{ws}})\right]\right\} = -\frac{\dot{m}}{\rho^{\mathrm{w}}}, \tag{3.667}$$

where for the degree of water saturation S_{w} we have

$$S_{\mathrm{w}} = \frac{\mathrm{d}v^{\mathrm{w}}}{\mathrm{d}v^{\mathrm{w}} + \mathrm{d}v^{\mathrm{g}}}, \tag{3.668}$$

K_{w} is the bulk modulus of water, p^{w} the pressure of the liquid phase,

$$\beta_{\mathrm{sw}} = S_{\mathrm{w}}[(\alpha - n)\beta_{\mathrm{s}} + n\beta_{\mathrm{w}}], \tag{3.669}$$

α is Biot's constant,

$$1 - \alpha = \frac{K_{\mathrm{T}}}{K_{\mathrm{s}}}, \tag{3.670}$$

K_{T} the bulk modulus of the porous matrix, K_{s} the bulk modulus of the solid grain material contained in the fluid phase, β_{s} the thermal expansion coefficient for the solid, β_{w} the thermal expansion coefficient for the liquid, $\mathbf{k}$ the permeability tensor of the multiphase medium, k^{rw} the relative permeability of the liquid phase (a dimensionless parameter varying from zero to one), μ^{w} the dynamic viscosity, $\vec{g}$ the gravity acceleration, ρ^{w} the phase-averaged liquid density. The acceleration of the solid phase $\vec{a}^{\mathrm{s}}$ is given as

$$\vec{a}^{\mathrm{s}} = \frac{\partial\vec{v}^{\mathrm{s}}}{\partial t} + \vec{v}^{\mathrm{s}} \cdot \operatorname{grad}\vec{v}^{\mathrm{s}}, \tag{3.671}$$

the relative acceleration $\vec{a}^{\mathrm{ws}}$ then

$$\vec{a}^{\mathrm{ws}} = \frac{\partial\vec{v}^{\mathrm{ws}}}{\partial t} + \vec{v}^{\mathrm{s}} \cdot \operatorname{grad}\vec{v}^{\mathrm{ws}}, \tag{3.672}$$

$$\vec{v}^{\mathrm{ws}} = \vec{v}^{\mathrm{w}} - \vec{v}^{\mathrm{s}}, \tag{3.673}$$

and finally $-\dot{m}$ is the quantity of water lost through evaporation per unit time and unit volume.

For liquid water with compressible solid grains Lewis and Schrefler obtained

$$\left(\frac{\alpha - n}{K_s}S_w^2 + \frac{nS_w}{K_w}\right)\frac{d^s p^w}{dt} + \frac{\alpha - n}{K_s}S_w S_g \frac{d^s p^g}{dt} + \alpha S_w \operatorname{div} \vec{v}^s$$

$$- \beta_{sw}\frac{d^s T}{dt} + \left(\frac{\alpha - n}{K_s}S_w p^w - \frac{\alpha - n}{K_s}S_w p^g + n\right)\frac{d^s S_w}{dt}$$

$$+ \frac{1}{\rho^w}\operatorname{div}\left\{\rho^w \frac{kk^{rw}}{\mu^w}\left[-\operatorname{grad} p^w\right.\right.$$

$$+\rho^w(\vec{g} - \vec{a}^s - \vec{a}^{ws})]\} = -\frac{\dot{m}}{\rho^w}, \tag{3.674}$$

where for the degree of gas saturation S_g we have

$$S_g = \frac{dv^g}{dv^w + dv^g}, \tag{3.675}$$

and the other quantities are the same as in the balance equation (3.667). For gas with incompressible solid grains:

$$S_g \operatorname{div} \vec{v}^s - \beta_s(1-n)S_g\frac{d^s T}{dt} - n\frac{d^s S_w}{dt}$$

$$+ \frac{nS_g}{\rho^g}\frac{d^s}{dt}\left[\frac{1}{\Theta R}(p^{ga}M_a + p^{gw}M_w)\right]$$

$$+ \frac{1}{\rho^g}\operatorname{div}\left\{\rho^g\frac{kk^{rg}}{\mu^g}\left[-\operatorname{grad} p^g + \rho^g(\vec{g} - \vec{a}^s - \vec{a}^{gs})\right]\right\} = \frac{\dot{m}}{\rho^g}, \tag{3.676}$$

where ρ^g is the phase-averaged gas density, Θ the absolute temperature, R the universal gas constant, p^{ga} the partial pressure of dry air,

$$p^{ga} = \rho^{ga}\frac{\Theta R}{M_a}, \tag{3.677}$$

p^{gw} the partial pressure of the water vapour,

$$p^{gw} = \rho^{gw}\frac{\Theta R}{M_w}, \tag{3.678}$$

M_a the molar mass of air, M_w the molar mass of water vapour, ρ^{ga} the partial density of dry air, ρ^{gw} the partial density of water vapour,

$$\rho^g = \rho^{ga} + \rho^{gw}, \tag{3.679}$$

p^{g} the total pressure of moist air,

$$p^{\text{g}} = p^{\text{ga}} + p^{\text{gw}}. \tag{3.680}$$

For the molar mass of the mixture of dry air and water vapour we can write

$$M_{\text{g}} = \frac{1}{\dfrac{\rho^{\text{gw}}}{\rho^{\text{g}}}\dfrac{1}{M_{\text{w}}} + \dfrac{\rho^{\text{ga}}}{\rho^{\text{g}}}\dfrac{1}{M_{\text{a}}}}, \tag{3.681}$$

where k^{rg} is the relative permeability of the gaseous phase (a dimensionless parameter varying from zero to one), μ^{g} the dynamic viscosity of the gaseous phase, and the relative acceleration $\vec{a}^{\text{gs}}$ is defined as

$$\vec{a}^{\text{gs}} = \frac{\partial \vec{v}^{\text{gs}}}{\partial t} + \vec{v}^{\text{s}} \cdot \text{grad}\ \vec{v}^{\text{gs}}, \tag{3.682}$$

$$\vec{v}^{\text{gs}} = \vec{v}^{\text{g}} - \vec{v}^{\vec{s}}. \tag{3.683}$$

For gas with compressible solid grains:

$$\frac{\alpha - n}{K_{\text{s}}} S_{\text{w}} S_{\text{g}} \frac{\mathrm{d}^{\text{s}} p^{\text{w}}}{\mathrm{d}t} + \frac{\alpha - n}{K_{\text{s}}} S_{\text{g}}^2 \frac{\mathrm{d}^{\text{s}} p^{\text{g}}}{\mathrm{d}t} - \left(n + \frac{\alpha - n}{K_{\text{s}}} p^{\text{c}} S_{\text{g}}\right) \frac{\mathrm{d}^{\text{s}} S_{\text{w}}}{\mathrm{d}t}$$

$$+ \alpha S_{\text{g}}\ \text{div}\ \vec{v}^{\text{s}} - \beta_{\text{s}}(\alpha - n) S_{\text{g}} \frac{\mathrm{d}^{\text{s}} T}{\mathrm{d}t}$$

$$+ \frac{n S_{\text{g}}}{\rho^{\text{g}}} \frac{\mathrm{d}^{\text{s}}}{\mathrm{d}t} \left[\frac{1}{\Theta R}(p^{\text{ga}} M_{\text{a}} + p^{\text{gw}} M_{\text{w}})\right]$$

$$+ \frac{1}{\rho^{\text{g}}}\ \text{div}\ \left\{\rho^{\text{g}} \frac{\mathbf{k} k^{\text{rg}}}{\mu^{\text{g}}}[-\ \text{grad}\ p^{\text{g}} + \rho^{\text{g}}(\vec{g} - \vec{a}^{\text{s}} - \vec{a}^{\text{gs}})]\right\} = \frac{\dot{m}}{\rho^{\text{g}}}, \tag{3.684}$$

where p^{c} is the capillary pressure,

$$p^{\text{c}} = p^{\text{g}} - p^{\text{w}}. \tag{3.685}$$

The mass balance for water species (liquid water and water vapour) with incompressible solid grains Lewis and Schrefler formulated in the form:

$$(\rho^{\text{w}} S_{\text{w}} + \rho^{\text{gw}} S_{\text{g}})\ \text{div}\ \vec{v}^{\text{s}} - \beta_{\text{swg}} \frac{\mathrm{d}^{\text{s}} T}{\mathrm{d}t} + n(\rho^{\text{w}} - \rho^{\text{gw}}) \frac{\mathrm{d}^{\text{s}} S_{\text{w}}}{\mathrm{d}t}$$

$$+ \frac{n \rho^{\text{w}} S_{\text{w}}}{K_{\text{w}}} \frac{\mathrm{d}^{\text{s}} p^{\text{w}}}{\mathrm{d}t} + S_{\text{g}} n \frac{\mathrm{d}^{\text{s}}}{\mathrm{d}t}\left(\frac{M_{\text{w}}}{\Theta R} p^{\text{gw}}\right)$$

$$- \text{div}\ \left[\rho^{\text{g}} \frac{M_{\text{a}} M_{\text{w}}}{M_{\text{g}}^2} \mathbf{D}_{\text{g}}\ \text{grad}\ \left(\frac{p^{\text{gw}}}{p^{\text{g}}}\right)\right]$$

$$+ \ \mathrm{div} \ \left\{ \rho^{\mathrm{gw}} \frac{kk^{\mathrm{rg}}}{\mu^{\mathrm{g}}} \left[- \ \mathrm{grad} \ p^{\mathrm{g}} + \rho^{\mathrm{g}} (\vec{g} - \vec{a}^{\mathrm{s}} - \vec{a}^{\mathrm{gs}}) \right] \right\}$$

$$+ \ \mathrm{div} \ \left\{ \rho^{\mathrm{w}} \frac{kk^{\mathrm{rw}}}{\mu^{\mathrm{w}}} \left[- \ \mathrm{grad} \ p^{\mathrm{w}} + \rho^{\mathrm{w}} (\vec{g} - \vec{a}^{\mathrm{s}} - \vec{a}^{\mathrm{ws}}) \right] \right\} = 0, \quad (3.686)$$

where $\mathbf{D_g}$ is the effective dispersion tensor of the gaseous phase,

$$\beta_{\mathrm{swg}} = \beta_{\mathrm{s}}(1 - n)(S_{\mathrm{g}}\rho^{\mathrm{gw}} + \rho^{\mathrm{w}} S_{\mathrm{w}}) + n\beta_{\mathrm{w}}\rho^{\mathrm{w}} S_{\mathrm{w}}. \quad (3.687)$$

For water species (liquid water and water vapour) with compressible solid grains they have written:

$$\left[\frac{\alpha - n}{K_{\mathrm{s}}} S_{\mathrm{w}}(\rho^{\mathrm{gw}} S_{\mathrm{g}} + \rho^{\mathrm{w}} S_{\mathrm{w}}) + \rho^{\mathrm{w}} \frac{nS_{\mathrm{w}}}{K_{\mathrm{w}}} \right] \frac{\mathrm{d}^{\mathrm{s}} p^{\mathrm{w}}}{\mathrm{d}t}$$

$$+ \frac{\alpha - n}{K_{\mathrm{s}}} S_{\mathrm{g}}(\rho^{\mathrm{gw}} S_{\mathrm{g}} + \rho^{\mathrm{w}} S_{\mathrm{w}}) \frac{\mathrm{d}^{\mathrm{s}} p^{\mathrm{g}}}{\mathrm{d}t}$$

$$+ \left[\frac{\alpha - n}{K_{\mathrm{s}}} (\rho^{\mathrm{w}} S_{\mathrm{w}} p^{\mathrm{w}} + \rho^{\mathrm{gw}} S_{\mathrm{g}} p^{\mathrm{c}} - \rho^{\mathrm{w}} S_{\mathrm{w}} p^{\mathrm{c}}) + n(\rho^{\mathrm{w}} - \rho^{\mathrm{gw}}) \right] \frac{\mathrm{d}^{\mathrm{s}} S_{\mathrm{w}}}{\mathrm{d}t}$$

$$+ \alpha(\rho^{\mathrm{w}} S_{\mathrm{w}} + \rho^{gw} S_{\mathrm{g}}) \ \mathrm{div} \ \vec{v}^{\mathrm{s}}$$

$$- \beta_{\mathrm{swg}} \frac{\mathrm{d}^{\mathrm{s}} T}{\mathrm{d}t} + S_{\mathrm{g}} n \frac{\mathrm{d}^{\mathrm{s}}}{\mathrm{d}t} \left(\frac{M_{\mathrm{w}}}{\Theta R} p^{\mathrm{gw}} \right)$$

$$- \ \mathrm{div} \ \left[\rho^{\mathrm{g}} \frac{M_{\mathrm{a}} M_{\mathrm{w}}}{M_{\mathrm{g}}^2} \mathbf{D_g} \ \mathrm{grad} \ \left(\frac{p^{\mathrm{gw}}}{p^{\mathrm{g}}} \right) \right]$$

$$+ \ \mathrm{div} \ \left\{ \rho^{\mathrm{gw}} \frac{kk^{\mathrm{rg}}}{\mu^{\mathrm{g}}} \left[- \ \mathrm{grad} \ p^{\mathrm{g}} + \rho^{\mathrm{g}} (\vec{g} - \vec{a}^{\mathrm{s}} - \vec{a}^{\mathrm{gs}}) \right] \right\}$$

$$+ \ \mathrm{div} \ \left\{ \rho^{\mathrm{w}} \frac{kk^{\mathrm{rw}}}{\mu^{\mathrm{w}}} \left[- \ \mathrm{grad} \ p^{\mathrm{w}} + \rho^{\mathrm{w}} (\vec{g} - \vec{a}^{\mathrm{s}} - \vec{a}^{\mathrm{ws}}) \right] \right\} = 0, \quad (3.688)$$

where

$$\beta_{\mathrm{swg}} = \beta_{\mathrm{s}}(\alpha - n)(S_{\mathrm{g}}\rho^{\mathrm{gw}} + \rho^{\mathrm{w}} S_{\mathrm{w}}) + n\beta_{\mathrm{w}}\rho^{\mathrm{w}} S_{\mathrm{w}}. \quad (3.689)$$

Momentum balance equations for fluids Lewis and Schrefler formulated as follows:

$$\eta^{\pi} \vec{v}^{\pi s} = \frac{kk^{\mathrm{r}\pi}}{\mu^{\pi}} \left[- \ \mathrm{grad} \ p^{\pi} + \rho^{\pi} (\vec{g} - \vec{a}^{\mathrm{s}} - \vec{a}^{\pi s}) \right], \quad (3.690)$$

where the superscript π stands for the particular fluid phase, and η^π is the volume fraction of the phase π,

$$\eta^\pi = nS_\pi. \tag{3.691}$$

For dry air and water vapour the body force term is usually neglected.

The momentum balance equation for the multiphase medium reads:

$$-\rho\vec{a}^s - nS_w\rho^w(\vec{a}^{ws} + \vec{v}^{ws} \cdot \text{grad } \vec{v}^w)$$

$$- nS_g\rho^g(\vec{a}^{gs} + \vec{v}^{gs} \cdot \text{grad } \vec{v}^g) + \text{div } \sigma + \rho\vec{g} = 0, \tag{3.692}$$

where for the total density of the system we have

$$\rho = (1-n)\rho^s + nS_w\rho^w + nS_g\rho^g, \tag{3.693}$$

and for the total stress tensor σ

$$\sigma = (1-n)\sigma^s + n(S_w\sigma^w + S_g\sigma^g). \tag{3.694}$$

For the fluid phases the stress tensor is given as:

$$\sigma^\pi = \tau^\pi - \mathbf{I}p^\pi, \tag{3.695}$$

where τ is the shear stress, which for an ideal fluid can be neglected, $\mathbf{I}$ the unit tensor of the second order, p the pressure. Denoting

$$p^s = S_g p^g + S_w p^w, \tag{3.696}$$

and substituting (3.695) and (3.696) into (3.694), we obtain

$$\sigma = (1-n)\sigma^s - n\mathbf{I}p^s = \sigma' - \mathbf{I}p^s, \tag{3.697}$$

where the stress tensor is split into two components, the pore pressure effect, and the component that deforms the solid skeleton, namely the effective stress tensor of the solid phase σ',

$$\sigma' = (1-n)(\sigma^s + \mathbf{I}p^s) = \sigma + \mathbf{I}(S_w p^w + S_g p^g), \tag{3.698}$$

which can be further modified by accounting for the deformability of the grains using the Biot's coefficient α,

$$\sigma'' = \sigma + \mathbf{I}\alpha(S_w p^w + S_g p^g). \tag{3.699}$$

The mechanical constitutive equations can be then constructed using the effective stress tensor. Lewis and Schrefler (1998) introduced elastic and elastoplastic models following the basic constitutive theory given for instance in Malvern (1969).

The enthalpy balance for the multiphase medium was formulated in Lewis and Schrefler (1998) as follows:

$$(\rho C_p)_{\text{eff}}\frac{\partial T}{\partial t} + (\rho_w C_p^w \vec{v}^w + \rho_g C_p^g \vec{v}^g) \cdot \text{grad } T$$

$$- \operatorname{div}\left(\chi_{\text{eff}} \operatorname{grad} T\right) = -\dot{m}\Delta H_{\text{vap}}, \tag{3.700}$$

where the subscript eff means the effective value of a quantity, χ is the thermal conductivity,

$$\chi_{\text{eff}} = \chi^{\text{s}} + \chi^{\text{w}} + \chi^{\text{g}} \tag{3.701}$$

$$(\rho C_{\text{p}})_{\text{eff}} = \rho^{\text{s}} C_{\text{p}}^{\text{s}} + \rho^{\text{w}} C_{\text{p}}^{\text{w}} + \rho^{\text{g}} C_{\text{p}}^{\text{g}}, \tag{3.702}$$

ΔH_{vap} is the latent heat of vaporization,

$$\Delta H_{\text{vap}} = H^{\text{gw}} - H^{\text{w}}. \tag{3.703}$$

A purely theoretical analysis of hygro-thermo-mechanical behaviour of a porous medium was done by Coussy (1995). A comprehensive analysis of various mechanical models of deformable porous media was performed for a saturated porous material. The extension to a reactive partially saturated porous medium in the last chapter of the Coussy's book was done in a less detailed way but it can be anticipated that further work in this direction can lead to very prospective results.

We will not give details of mechanical constitutive equations applied in porous media here, because it is outside the scope of this book, which is aimed rather to heat and mass transport phenomena than to the mechanistic effects. The reader can find theoretical details directly in Coussy (1995) and in the numerous references cited by Coussy.

3.4.3 Transport Processes at High Temperatures

Models of heat, moisture and salt transport described so far can generally be applied within a limited temperature range, where phase-change processes in water are realized only within the framework of capillary condensation or evaporation, which are in fact given by the sorption isotherm. Therefore, these models are applicable in the temperature range of about 5°C to 90°C. For temperature ranges above 100°C, it is necessary to use modified models that take into account, for instance, changes in the porous system and employ a more complex phase diagram of water.

One of the first models including the phase transition of water to water vapour in the temperature range above 100°C was published by Lykov (1956).

For the description of phase changes of water Lykov used the liquid-vapour phase change coefficient ϵ, defined by the formula

$$\epsilon = \frac{\mathrm{d}_i u}{\mathrm{d}u} = \frac{\mathrm{d}_i u}{\mathrm{d}_i u + \mathrm{d}_e u}, \tag{3.704}$$

where $\mathrm{d}u$ is the total change of moisture, $u = u_1 + u_2$, index 1 denotes water vapour, index 2 water, $\mathrm{d}_e u$ is the change of moisture due to its transport, and

$d_i u$ the change of moisture due to the phase change of water to water vapour. In the formulation of the moisture balance he put

$$u_1 = 0 \tag{3.705}$$

$$u = u_1 + u_2 = u_2, \tag{3.706}$$

and the source term of moisture $I_2 = -I_1$ (see Lykov's model of heat and moisture transport in Section 3.2.3) then had the following form:

$$I_2 = \epsilon \frac{\partial u}{\partial t}. \tag{3.707}$$

Balance equations of moisture were formulated accordingly,

$$\frac{\partial u}{\partial t} = a_{m2} \text{ div grad } u + a_{m2}\delta \text{ div grad } T + \epsilon \frac{\partial u}{\partial t} \tag{3.708}$$

$$\frac{\partial T}{\partial t} = a \text{ div grad } T + \epsilon \frac{L_{21}}{c} \frac{\partial u}{\partial t}. \tag{3.709}$$

Modifying the moisture balance (3.708), we obtain

$$\frac{\partial u}{\partial t} = \frac{a_{m2}}{1 - \epsilon} (\text{div grad } u + \delta_2 \text{ div grad } T). \tag{3.710}$$

Comparing this balance equation with the general moisture balance of the Lykov's model from Section 3.2.3 (3.294), namely

$$\frac{\partial u}{\partial t} = a_m (\text{div grad } u + \delta \text{ div grad } T), \tag{3.711}$$

we get

$$a_m = \frac{a_{m2}}{1 - \epsilon} \tag{3.712}$$

$$\delta = \delta_2, \tag{3.713}$$

and therefore

$$a_m(1 - \epsilon) = a_{m2} \tag{3.714}$$

$$a_m = a_{m2} + a_m\epsilon. \tag{3.715}$$

According to the definition we have

$$a_m = a_{m1} + a_{m2}. \tag{3.716}$$

Thus, comparing (3.715) and (3.716), we get

$$a_{m1} = a_m\epsilon, \tag{3.717}$$

and finally

$$\epsilon = \frac{a_{m1}}{a_m} = \frac{a_{m1}}{a_{m1} + a_{m2}}. \tag{3.718}$$

Substituting (3.710) in (3.709), we obtain the heat balance in the form

$$\frac{\partial T}{\partial t} = \left(a + \frac{a_{m2}\delta_2}{1-\epsilon}\epsilon\frac{L_{21}}{c}\right) \text{ div grad } T + \frac{a_{m2}}{1-\epsilon}\epsilon\frac{L_{21}}{c} \text{ div grad } u. \tag{3.719}$$

The balance equations (3.710), (3.719) can we written schematically as

$$\frac{\partial u}{\partial t} = K_{11} \text{ div grad } u + K_{12} \text{ div grad } T \tag{3.720}$$

$$\frac{\partial T}{\partial t} = K_{21} \text{ div grad } u + K_{22} \text{ div grad } T, \tag{3.721}$$

where the coefficients $K_{ij}, i = 1, 2; j = 1, 2$ are defined as

$$K_{11} = \frac{a_{m2}}{1-\epsilon} = a_m \tag{3.722}$$

$$K_{12} = \frac{a_{m2}\delta_2}{1-\epsilon} = a_m\delta \tag{3.723}$$

$$K_{21} = \epsilon\frac{L_{21}}{c}\frac{a_{m2}}{1-\epsilon} = \epsilon\frac{L_{21}}{c}a_m \tag{3.724}$$

$$K_{22} = a + \epsilon\frac{L_{21}}{c}\frac{a_{m2}\delta_2}{1-\epsilon} = a + \epsilon\frac{L_{21}}{c}a_m\delta. \tag{3.725}$$

Lykov's idea about the coefficient ϵ is very simple and straightforward; at first glance its definition looks plausible. The difficulties begin to emerge in the moment when it is necessary to quantify it, i.e. to determine it experimentally. A closer view of the coefficient ϵ shows that it is not an objectively measured quantity at all. This can be illustrated by the following simple analysis. In a difference form, the Lykov's coefficient ϵ can be defined as

$$\epsilon(u, T) = \frac{\Delta u_i}{\Delta u_i + \Delta u_e}. \tag{3.726}$$

The value of Δu_i can be estimated by measuring the drying velocity of a specimen at various temperatures under conditions whereby water transport is minimized. This can be achieved, for instance, by a choice of appropriate shape of the specimens under test. The specimens could have a plate form with lateral faces water-proofed, while the thickness of the plate b is as small as possible, i.e. $b << \sqrt{A}$, where A is the area of the board face.

From the measured curves $u_i = u_i(t)$ we can obtain $\frac{\partial u_i}{\partial t}$ by a numerical derivation, and therefore

$$\Delta u_i = \left| \frac{\partial u_i}{\partial t} \right| \Delta t. \tag{3.727}$$

The relation for Δu_e we can get by the following analysis. The amount of water removed during time Δt through area A from a given place x_o can be expressed as

$$\Delta m_{\rm v} = j \cdot \Delta t \cdot A, \tag{3.728}$$

where

$$j = -\rho_{\rm s} \left(a_{\rm m} \ {\rm grad} \ u + a_{\rm m}\delta \ {\rm grad} \ T \right). \tag{3.729}$$

According to the definition of moisture content by mass we can write

$$\Delta u_e = \frac{\Delta m_{\rm v}}{m_{\rm s}} = \frac{|j \cdot \Delta t \cdot A|}{V \cdot \rho_{\rm s}}, \tag{3.730}$$

where

$$V = A \cdot b \tag{3.731}$$

is the volume of the specimen. Substituting (3.729) and (3.731) in (3.730) we get

$$\Delta u_e = \frac{|j \cdot \Delta t|}{\rho_{\rm s} \cdot b} = |a_{\rm m} \ {\rm grad} \ u + a_{\rm m}\delta \ {\rm grad} \ T| \frac{\Delta t}{b}, \tag{3.732}$$

and by further substitution of (3.727) and (3.732) in (3.726) eventually

$$\epsilon = \frac{\left| b\frac{\partial u}{\partial t} \right|}{\left| b\frac{\partial u}{\partial t} \right| + |a_{\rm m} \ {\rm grad} \ u + a_{\rm m}\delta \ {\rm grad} \ T|}. \tag{3.733}$$

From (3.733) it follows that the coefficient ϵ depends on moisture and temperature gradients. Therefore, it is not an unambiguously defined material parameter. Rather, its values depend on the particular experimental setup.

A possible way to replace the Lykov's coefficient ϵ by another coefficient, which would be defined in an unambiguous way, is by application of drying curves $u_i = u_i(T, t)$ and defining the evaporation coefficient ψ in the form (Černý and Venzmer, 1991)

$$\psi(u, T, \varphi) = \frac{\partial u_i(T, t, \varphi)}{\partial t}, \tag{3.734}$$

where φ is the relative humidity.

From balance equations (3.708) and (3.709) for moisture and heat transport it follows directly that

$$\frac{\partial u}{\partial t} = a_{\rm m} \ {\rm div} \ {\rm grad} \ u + a_{\rm m}\delta \ {\rm div} \ {\rm grad} \ T + \psi(u, T, \varphi) \tag{3.735}$$

$$\frac{\partial T}{\partial t} = a \ \text{div grad} \ T + \frac{L_\text{v}}{c_\text{p}} \psi(u, T, \varphi). \tag{3.736}$$

Compared to Lykov's original model, this modification with the evaporation coefficient ψ is simpler and also the coefficient ψ is more easily measurable than Lykov's coefficient ϵ. On the other hand, the assumption that during measurement of the evaporation coefficient ψ the moisture transport is realized by water evaporation only, is very strong. It would be fully valid only on the condition that the thickness of the specimen approached zero. This requirement is naturally confronted with practical problems in producing specimens. Another problem is the determination of relative humidity in the specimen.

Another possibility for an expression for the phase change processes of water at higher temperatures is an application of the discontinuity surface theory (Černý, 1989). The starting point of this model is a division of a particular element of the structure into two domains. Domain A is on the side of the structure exposed to higher temperatures. It does not contain any water in liquid state, and only heat and water vapour transport take place there. Domain B is exposed to lower temperatures, so heat, water and water vapour transport occur here. The domains A and B are separated by the discontinuity surface Γ, where jump conditions of heat and liquid water content apply. Taking into account that the evaporation velocity strongly depends on temperature, we can assume that for the case of fast heat transport (such as in the conditions of a fire), the phase transition of water to water vapour occurs at boiling temperature only, and the evaporation processes at lower temperatures can be neglected.

As a consequence of the above assumptions, the process of heat penetration into the structure can be monitored as a movement of the discontinuity surface Γ, which is in fact the isotherm of temperature equal to the boiling point of water. Assuming further, that in both domains A and B no condensation of water vapour occurs, we can write in the domain A

$$\frac{\partial u_1}{\partial t} = \text{div} \ (a_{\text{m}1} \ \text{grad} \ u_1 + a_{\text{m}1}\delta_1 \ \text{grad} \ T) \tag{3.737}$$

$$c_\text{A}\rho_\text{s}\frac{\partial T}{\partial t} = \text{div} \ (\lambda_\text{A} \ \text{grad} \ T)$$

$$+ \ c_1 \ (a_{\text{m}1} \ \text{grad} \ u_1 + a_{\text{m}1}\delta_1 \ \text{grad} \ T) \cdot \text{grad} \ T, \tag{3.738}$$

and in the domain B

$$\frac{\partial u_i}{\partial t} = \text{div} \ (a_{\text{m}i} \ \text{grad} \ u_i + a_{\text{m}i}\delta_i \ \text{grad} \ T) \quad i = 1, 2 \tag{3.739}$$

$$c_\text{B}\rho_\text{s}\frac{\partial T}{\partial t} = \text{div} \ (\lambda_\text{B} \ \text{grad} \ T)$$

$$+ \sum_{i=1}^{2} c_i \left(a_{mi} \text{ grad } u_i + a_{mi} \delta_i \text{ grad } T \right) \cdot \text{grad } T, \tag{3.740}$$

where as in the classical Lykov's models index 1 denotes water vapour, index 2 water, etc.

On the discontinuity surface Γ, we have the following thermodynamic equilibrium conditions:

$$T_A = T_B = T_v \tag{3.741}$$

$$u_{2,B} = u_{2,x} \tag{3.742}$$

$$u_{2,A} = 0, \tag{3.743}$$

where T_v is the boiling temperature of water.

Finally, the heat balance on the discontinuity surface Γ has to be formulated. For the jump of heat fluxes on Γ in the normal direction $[\![j_{Q,N}]\!]$, we can write

$$[\![j_{Q,N}]\!] = L_v u_{2,x} \rho_s v_{N,Q}, \tag{3.744}$$

where $v_{N,Q}$ is the velocity of the discontinuity surface in the normal direction. Also, we have

$$[\![j_{Q,N}]\!] = -\lambda_A \vec{n} \cdot (\text{grad } T)_A + \lambda_B \vec{n} \cdot (\text{grad } T)_B, \tag{3.745}$$

where $\vec{n}$ is the unit normal vector to the surface Γ.
For the moisture flux at evaporation in the direction of the normal to the surface Γ we have

$$j_{u_1,N} = -\rho_s u_{2,x} v_{N,Q} = -[\![j_{Q,N}]\!] \cdot \frac{1}{L_v}. \tag{3.746}$$

Using now the balance of mass of liquid water on the surface Γ, we get

$$u_{2,x} \rho_s v_N^s = [\![j_{u_2,N}]\!], \tag{3.747}$$

where

$$[\![j_{u_2,N}]\!] = -\rho_s \left(a_{m2} \vec{n} \cdot \text{grad } u_2 + a_{m2} \delta_2 \vec{n} \cdot \text{grad } T \right) - j_{u_1,N}. \tag{3.748}$$

Substituting (3.745), (3.746) and (3.748) in (3.747), we obtain the final equation for the velocity of the discontinuity surface Γ in the form

$$v_N^s = -\frac{1}{u_{2,x}} \left(a_{m2} \vec{n} \cdot \text{grad } u_2 + a_{m2} \delta_2 \vec{n} \cdot \text{grad } T \right)_B$$

$$- \frac{\lambda_A}{L_v \rho_s u_{2,x}} \vec{n} \cdot (\text{grad } T)_A + \frac{\lambda_B}{L_v \rho_s u_{2,x}} \vec{n} \cdot (\text{grad } T)_B. \tag{3.749}$$

What remains is the definition of boundary conditions for water vapour on the discontinuity surface Γ, in other words the decision about how the flux of water vapour $j_{u_1,N}$, formed on the discontinuity surface, is divided into the domains A, B. We get the following:

$$\rho_s \left(a_{m1}\vec{n} \cdot \operatorname{grad} u_1 + a_{m1}\delta_1\vec{n} \cdot \operatorname{grad} T\right)_A = -X_A j_{u_1,N} \qquad (3.750)$$

$$\rho_s \left(a_{m1}\vec{n} \cdot \operatorname{grad} u_1 + a_{m1}\delta_1\vec{n} \cdot \operatorname{grad} T\right)_B = X_B j_{u_1,N} \qquad (3.751)$$

$$X_A + X_B = 1, \qquad (3.752)$$

where X_A, X_B are the functions characterizing the ratio, in which the flux $j_{u_1,N}$ is divided into the domains A, B, which have to be determined experimentally.

Simple models of heat and moisture transport at higher temperatures based on Lykov's model are certainly useful tools for the calculations in specialized cases, for which they were in fact derived. One such characteristic example is one-sided heating of an envelope structure, occurring at a fire. However, for more complicated cases these models fail logically, because the necessary material parameters describing the water–water vapour phase transition are not defined in an unambiguous way.

A significantly more complicated and more realistic model of heat and moisture transport in high temperature conditions was formulated by Bažant and Thonguthai (1978, 1979). They started with the general thermodynamic relations for moisture and heat fluxes,

$$\vec{j} = -a_{wv} \operatorname{grad} w - a_{wT} \operatorname{grad} T \qquad (3.753)$$

$$\vec{q} = -a_{Tv} \operatorname{grad} w - a_{TT} \operatorname{grad} T, \qquad (3.754)$$

where w is the partial density of water (in kg/m³). As $w = w(p,T)$, where p is the pore water pressure, we can write

$$\operatorname{grad} w = \frac{\partial w}{\partial p} \operatorname{grad} p + \frac{\partial w}{\partial T} \operatorname{grad} T, \qquad (3.755)$$

and substitute into (3.753) and (3.754),

$$\vec{j} = -a_{ww} \frac{\partial w}{\partial p} \operatorname{grad} p - \left(a_{wT} + a_{ww}\frac{\partial w}{\partial T}\right) \operatorname{grad} T \qquad (3.756)$$

$$\vec{q} = -a_{Tw} \frac{\partial w}{\partial p} \operatorname{grad} p - \left(a_{TT} + a_{Tw}\frac{\partial w}{\partial T}\right) \operatorname{grad} T, \qquad (3.757)$$

where the coefficient $a = a_{ww}\frac{\partial w}{\partial p}$ is the permeability (in s, $a_p = a.g$ is the Darcy's coefficient of hydraulic conductivity in m/s).

In further derivation, Bažant and Thonguthai followed common procedures and neglected the cross effects, so that the phenomenological relations were trivial,

$$\vec{j} = -a \operatorname{grad} p, \tag{3.758}$$

$$\vec{q} = -b \operatorname{grad} T, \tag{3.759}$$

where b is the thermal conductivity, $b = a_{TT}$.

The balance equation for moisture was formulated in the form

$$\frac{\partial w}{\partial t} = -\operatorname{div} \vec{j} + \frac{\partial w_d}{\partial t}, \tag{3.760}$$

where w_d is the partial density of chemically bound water, which is released into the porous space as a result of heating of the porous matrix. In concrete, $\frac{\partial w_d}{\partial t}$ is positive for temperatures higher than 120°C, for instance due to the dehydration of the cement binder. However, for lower temperatures it is negative due to the hydration processes, because it expresses the amount of free water which hydrated, i.e. was incorporated into the porous matrix.

The heat balance equation was formulated in the form

$$\rho c \frac{\partial T}{\partial t} - c_a \frac{\partial w}{\partial t} = -\operatorname{div} \vec{q} + c_w \vec{j} \cdot \operatorname{grad} T, \tag{3.761}$$

where c_a is the sorption heat of free water (per kilogram of free water), ρ and c are the density and the specific heat capacity of the porous matrix including the chemically bound water, and c_w the specific heat capacity of water. The term $(c_w \vec{j} \cdot \operatorname{grad} T)$ expresses the heat transport due to water flow.

In the balance equations for moisture and heat (3.760) and (3.761), there appear three unknown field quantities: the pore pressure p, the partial moisture density w and the temperature T. Therefore, the system of equations has to be completed by one constitutive relation including these parameters.

For temperatures below the critical point of water (374.15°C) it is necessary to distinguish between the saturated and unsaturated state of the porous material. For higher temperatures this is not necessary, because the liquid phase of water does not exist for any pressure.

In the unsaturated state the pore pressure $p < p_s(T)$ and the Kelvin equation is valid, which Bažant and Thonguthai use in a classical form,

$$\ln \frac{p}{p_s(T)} = \frac{2M}{R_g} \frac{1}{r} \frac{\sigma(T)}{\rho_w T}. \tag{3.762}$$

Under the assumption that the pore structure is not changed during the process, we can write

$$\frac{\rho_w T}{\sigma(T)} \ln \frac{p}{p_s(T)} = \frac{2M}{R_g} \frac{1}{r} = const. \tag{3.763}$$

Denoting

$$m = \frac{\frac{\rho_{wo} T_o}{\sigma_o}}{\frac{\rho_w T}{\sigma}}, \tag{3.764}$$

where the index o denotes the values of the particular quantities at the reference temperature 25°C, and

$$\varphi = \frac{p}{p_s(T)}, \tag{3.765}$$

where φ is the relative humidity, we obtain

$$\frac{1}{m} \ln \varphi = const., \tag{3.766}$$

and finally

$$\varphi^{\frac{1}{m}} = const. \tag{3.767}$$

Bažant and Thonguthai then assumed, on the basis of experimental data, that for temperature 25°C

$$\frac{w}{w_s} \sim \varphi, \tag{3.768}$$

and therefore (m is a temperature related factor),

$$\frac{w}{w_s} = \varphi^{\frac{1}{m}}. \tag{3.769}$$

For concrete this relation leads to the pressures p, which are too high in a comparison with an experiment. Therefore, they employed finally an empirical formula of the form

$$\frac{w}{c} = \left(\frac{w_1}{c}\varphi\right)^{\frac{1}{m(T)}}, \quad \varphi = \frac{p}{p_s(T)}, \quad \varphi \le 0.96, \tag{3.770}$$

where

$$m(T) = 1.04 - \frac{T'}{22.34 + T'}, \quad T' = \left(\frac{T + 10}{T_o + 10}\right)^2,$$

$$T \text{ in } °C, \quad T_0 = 25°C \tag{3.771}$$

$$w_1 = w_s(25°C), \tag{3.772}$$

and c is the mass of cement per 1 m³ of concrete.

The relation (3.770) is naturally valid for concrete only, it cannot be applied for other materials. We also note that for $\varphi = 1$, the relation (3.770) leads to

$$w_s = c \left(\frac{w_1}{c}\right)^{\frac{1}{m(T)}}, \tag{3.773}$$

and therefore w_s decreases with increasing temperature.

In the saturated state, i.e. $w = w_s$, a simple calculation of the pore pressure p from the values of w and T using the thermodynamic tables and with a consideration of porosity n as the volume being available for the pore water, is not possible any more. The reason is that for temperatures above 100°C, considerable changes in the porous structure occur, and the porous space that is available for the pore water increases significantly, due to the overall increase in the total porous space and to the decrease in the amount of adsorbed water in the total moisture content w. The effect of porosity change and of elastic volume expansion is described by Bažant and Thonguthai using the thermodynamic relation

$$w = (1 + 3\epsilon^v)\frac{n}{v}, \quad \varphi = \frac{p}{p_s} \geq 1.04, \tag{3.774}$$

where n is the porosity, v the specific volume of water, $v = v(T, p)$,

$$d\epsilon^v = \frac{d\sigma^v}{3K} + \alpha dT, \tag{3.775}$$

$\sigma^v = n.p$, ϵ^v is the linear volumetric strain, α the thermal expansion coefficient, and K the bulk modulus.

The porosity of many materials increases with temperature. For concrete this fact is caused primarily by partial dehydration at higher temperatures, i.e. by the release of some molecules of water, which were chemically bound under normal temperature conditions. The amount of water released by the dehydration w_d is known, and therefore

$$n = \left[n_o + \frac{w_d - w_{do}}{\rho_o} \right] \cdot P(\varphi) \quad \text{for} \quad \varphi \geq 1.04, \tag{3.776}$$

where $\rho_o = 1000 \text{ kg/m}^3$, $w_d - w_{do}$ is the decrease in the amount of chemically bound water comparing to the temperature 25°C, $n_o = n(25°C)$, and $P(\varphi)$ is an empirical correction function,

$$P(\varphi) = 1 + 0.12(\varphi - 1.04), \quad \varphi = \frac{p}{p_s(t)}. \tag{3.777}$$

In the transition range between $\varphi = 0.96$ and $\varphi = 1.04$, Bažant and Thonguthai assumed a linear course between points corresponding to $w = w_{96}$ from the relation

$$\frac{w}{c} = \left(\frac{w_1}{c}\varphi\right)^{\frac{1}{m(T)}} \tag{3.778}$$

and $w = w_{104}$ from the relation

$$w = (1 + 3\epsilon^v)\frac{n}{v}, \tag{3.779}$$

where both values were determined for the same temperature.

References

Andrade, C., Castellote, M., Cervigon, D., Alonso, C., 1997, Fundamentals of migration experiments. In: *Proceedings of the RILEM International Workshop Chloride Penetration into Concrete*, St-Remy-les-Chevreuse, edited by Nilsson, L.O., Ollivier, J.P. (St-Remy-les-Chevreuse: RILEM), pp. 95–114.

Arsenault, J., Bigas, J.-P., Ollivier, J.-P., 1997, Determination of chloride diffusion coefficient using two different steady-state methods: influence of concentration gradient. In: *Proceedings of the RILEM International Workshop Chloride Penetration into Concrete*, St-Remy-les-Chevreuse, edited by Nilsson, L.O., Ollivier, J.P. (St-Remy-les-Chevreuse: RILEM), pp. 150–161.

Atkin, R.J., Craine, R.E., 1975, Continuum theories of mixtures: basic theory and historical development. *Q. J. Mech. Appl. Math.*, **29**, pp. 209–244.

Atkin, R.J., Craine, R.E., 1976, Continuum theories of mixtures: applications. *J. Inst. Maths Applics*, **17**, pp. 153–207.

Bažant, Z.P., Thonguthai, W., 1978, Pore pressure and drying of concrete at high temperatures. *Journal of Engineering Mechanics Division, Proceedings of the American Society of Civil Engineers*, **104**, pp. 1059–1079.

Bažant, Z.P., Thonguthai, W., 1979, Pore pressure in heated concrete walls. *Magazine of Concrete Research*, **31**, pp. 67–76.

Bear, J., Bachmat, Y., 1990, *Introduction to Modeling of Transport Phenomena in Porous Media*, Vol. 4. (Dordrecht: Kluwer).

Brdička, R., Dvořák, J., 1977, *Fundamentals of Physical Chemistry*. (Praha: Academia), (in Czech).

Brocken, H.J.P., 1998, *Moisture Transport in Brick Masonry: The Grey Area between Bricks*, PhD Thesis. (Eindhoven: TU Eindhoven).

Buckingham, E., 1907, Studies on the movement of soil moisture. (Washington: U.S. Dept. Agr. Bur. Soils, Bull. 38).

Černý, R., 1987, *Application of Stefan Problem to the Moisture Transport in Porous Materials*, PhD Thesis. (Praha: ČVUT), (in Czech).

Černý, R., 1989, Heat and moisture transport during a fire. A model with moving boundary. *Stavebnícky časopis*, **37**, pp. 89–99 (in Czech).

Černý, R., Venzmer, H., 1991, Ein modifiziertes Lykow-Modell zum Wärme-
und Feuchtetransports in kapillarporösen Stoffen. *Gesundheitsingenieur*,
112, pp. 41–44.

Chatterji, S., 1994a, Transportation of ions through cement based materials. Part I: Fundamental equations and basic measurement techniques.
Cement and Concrete Research, **24**, pp. 907–912.

Chatterji, S., 1994b, Transportation of ions through cement based materials. Part II: Adaptation of the fundamental equations and relevant
comments. *Cement and Concrete Research*, **24**, pp. 1010–1014.

Chatterji, S., 1994c, Transportation of ions through cement based materials.
Part III: Experimental evidence for the basic equations and some important deductions. *Cement and Concrete Research*, **24**, pp. 1229–1236.

Chatterji, S., 1995a, A critical review of the factors affecting ion migration
through cement based materials. *Il Cemento*, **92**, pp. 139–150.

Chatterji, S., 1995b, On the applicability of the Fick's second law to chloride
ion migration through Portland cement concrete. *Cement and Concrete
Research*, **25**, pp. 299–303.

Coussy, O., 1995, *Mechanics of Porous Continua*. (Chichester: John Wiley
and Sons).

Darcy, H., 1856, *Les Fontaines Publiques de la Ville de Dijon*. (Paris: Victor
Dalmont).

Grunewald, J., 1997, *Diffusiver und konvektiver Stoff- und Energietransport
in kapillarporösen Baustoffen*, PhD Thesis. (Dresden: TU Dresden).

Häupl, P., Stopp, H., 1980, Ein Beitrag zum Feuchtigkeitstransport in Bauw-
erksteilen. In: *Proceedings of 3. Bauklimatisches Symposium*. (Dresden:
TU Dresden), pp. 93–102.

Häupl, P., Stopp, H., 1987, *Feuchtetransport in Baustoffen und Bauwerk-
steilen*, Dissertation B. (Dresden: TU Dresden).

Hrstka, O., Černý, R., Rovnaníková, 1999, Hygrothermal stress induced
problems in large scale sprayed concrete structures. In: *Specialist Techniques and Materials for Concrete Construction*, R.K. Dhir, N.A. Henderson (eds.), (London: Thomas Telford), pp. 103–109.

Kiessl, K., 1983, *Kapillarer und dampfförmiger Feuchtetransport in mehrschichtigen Bauteilen*, PhD Thesis. (Essen: Universität Essen).

Klute, A., 1952, A numerical method for solving the flow equation for water
in unsaturated material. *Soil Science*, **73**, pp. 105–116.

Krischer, O., 1942, Der Wärme- und Stoffaustausch im Trocknungsgut. *VDI-Forschungsheft*, **415**, pp. 1–22.

Krischer, O., 1963, *Die wissentschaftliche Grundlagen der Trocknungstechnik*, 2. Auflage. (Berlin: Springer Verlag).

Kvasnica, J., 1965, *Thermodynamics*. (Praha: SNTL), (in Czech).

Künzel, H.M., 1995, *Simultaneous Heat and Moisture Transport in Building Components*, PhD Thesis. (Stuttgart: IRB Verlag).

Lewis, R.W., Schrefler, B.A., 1998, *The Finite Element Method in the Static and Dynamic Deformation and Consolidation of Porous Media*, 2nd Edition. (Chichester: John Wiley and Sons).

Lykov, A.V., 1954, *Javlenija perenosa v kapilljarnoporistych telach*. (Moskva: Gosenergoizdat), (in Russian).

Lykov, A.V., 1956, *Teplo i massoobmen v processach sushki*. (Moskva: Gosenergoizdat), (in Russian).

Lykov, A.V., 1972, *Teplomassoobmen*. (Moskva: Energia), (in Russian).

Majorana, C.E., Salomoni, V., Schrefler, B.A., 1998, Hygrothermal and mechanical model of concrete at high temperature. *Materials and Structures*, **31**, pp. 378–386.

Malvern, L.E., 1969, *Introduction to the Mechanics of a Continuous Medium*. (Englewood Cliffs: Prentice-Hall).

Milly, P.C.D., 1982, Moisture and heat transport in hysteretic, inhomogeneous porous media: a matrix head-based formulation and a numerical model. *Water Resources Research*, **18**, pp. 489–498.

Moore, W.J., 1972, *Physical Chemistry*, 4th Edition. (Englewood Cliffs: Prentice-Hall).

Philip, J.R., 1957, The theory of infiltration. *Soil Science*, **83**, pp. 345–357, 435–448, 160–178, 257–264, 329–339.

Philip, J.R., de Vries, D.A., 1957, Moisture movement in porous materials under temperature gradients. *Transactions of the American Geophysical Union*, **38**, pp. 222–232.

Richards, L.A., 1931, Capillary conduction of liquids in porous medium. *Physics*, **1**, pp. 318–333.

Stefan, J., 1889, Über einige probleme der Theorie der Wärmeleitung. *Sitzber. Wien. Akad. Mat. Naturw.*, **98**, pp. 473–484.

Tang, L., 1997, On chloride diffusion coefficients obtained by using the electrically accelerated methods. In: *Proceedings of the RILEM International Workshop Chloride Penetration into Concrete*, St-Remy-les-Chevreuse, edited by Nilsson, L.O., Ollivier, J.P. (St-Remy-les-Chevreuse: RILEM), pp. 126–134.

de Vries, D.A., 1987, The theory of heat and moisture transfer in porous media revisited. *International Journal of Heat and Mass Transfer*, **30**, pp. 1343–1350.

CHAPTER FOUR

Modelling Transport Processes in Concrete

4.1 COMMONLY EMPLOYED MODELS

The first models of moisture transport in concrete were initiated by the necessity to describe the drying shrinkage processes. Therefore, the researchers were primarily concerned with the description of concrete drying. The models were mostly inspired by Fick's law, and employed the linear diffusion equation for relative humidity φ,

$$\frac{\partial \varphi}{\partial t} = D_\varphi \text{ div grad } \varphi, \tag{4.1}$$

or water content w,

$$\frac{\partial w}{\partial t} = D_w \text{ div grad } w, \tag{4.2}$$

where D_φ, D_w are the respective diffusion coefficients (diffusivities). Examples of such applications can be found, e.g., in Carlson (1937), Pickett (1942), Hancox (1967), Hilsdorf (1967), and many other papers up to the 1960s. However, these models gave a very poor fit to experimental data over long periods of time. Already Carlson (1937) had observed that drying becomes much slower than given by an extrapolation of the initial drying curve, which a linear diffusion theory would predict. Therefore, Pickett (1942) proposed considering the diffusivity as function of time. This, however, was a very crude approximation, which did not give a satisfactory fit to data for various specimen thicknesses and shapes.

Bažant and Najjar (1972) proposed the diffusivity in concrete drying models should be considered as dependent on pore humidity, temperature and other variables, and arrived at

$$\frac{\partial \varphi}{\partial t} = \text{div} \left(D \text{ grad } \varphi \right) + \frac{\partial \varphi_s}{\partial t} + K_T \frac{\partial T}{\partial t}, \tag{4.3}$$

where $\frac{\partial \varphi_s}{\partial t}$ is the self-desiccation, which expresses the decrease of relative humidity due to the hydration process, $\varphi_s(t)$ the pore humidity, which would be established in time t in a sealed, initially wet specimen, and $K_T = \left(\frac{\partial \varphi}{\partial T} \right)_w$ the hygrothermic coefficient expressing the change of relative humidity due to the change of temperature. Alternatively, Bažant and Najjar (1972) proposed a nonlinear diffusion equation for water content also,

$$\frac{\partial w_e}{\partial t} = \text{div} \left(D \text{ grad } w_e \right) - \frac{\partial w_n}{\partial t}, \tag{4.4}$$

where w_e is the evaporable water content, w_n the non-evaporable water content, and the last term expresses the rate of loss of the evaporable water content due to hydration. However, in actual applications they preferred the formulation of nonlinear diffusion equation involving relative humidity, mainly for practical reasons (the drop in relative humidity due to self-desiccation is rather small, the usual formulation of boundary conditions is in terms of φ, etc.).

In the model by Bažant and Najjar (1972) water transport in the saturated state was considered in the form of Darcy's equation,

$$j = -\rho_w C_{sat} \text{ grad } \frac{p}{\rho_w g}, \tag{4.5}$$

where j is the water flux (in $kgm^{-2}s^{-1}$), C_{sat} the hydraulic conductivity in saturated state in m/s (in the model by Bažant and Najjar (1972) it was called permeability, which is not correct in terms of the commonly used notation in soil physics, see, e.g., Bear, 1988, Bear and Bachmat, 1990), and p the pore pressure.

From the point of view of later findings (see Chapter 3), the model of Bažant and Najjar (1972) suffers from a major problem. No difference between the maximum hygroscopic moisture content and saturated moisture content is assumed. Consequently, a discontinuity in the diffusion coefficient appears in the transition point between the unsaturated and saturated state at $\varphi = 1$, which leads to ambiguous solutions. Nevertheless, the model can be considered as very advanced, taking into account the time it was developed.

Recently, Baroughel-Bouny et al. (1999) introduced a more complex model describing isothermal concrete drying. The mass balance equations for liquid water l, water vapour v, and the air a, respectively, were formulated in the form

$$\frac{\partial}{\partial t}(\Phi \rho_l S_l) = - \text{div} \left(\Phi \rho_l S_l \vec{v}_l \right) - \mu_{l \to v} \tag{4.6}$$

$$\frac{\partial}{\partial t}[\Phi \rho_v (1 - S_l)] = - \text{div} \left[\Phi \rho_v (1 - S_l) \vec{v}_v \right] + \mu_{l \to v} \tag{4.7}$$

$$\frac{\partial}{\partial t}[\Phi \rho_a (1 - S_l)] = - \text{div} \left[\Phi \rho_a (1 - S_l) \vec{v}_a \right], \tag{4.8}$$

where Φ is the total porosity, S_l the liquid water saturation,

$$S_l = \frac{\Phi_l}{\Phi}, \tag{4.9}$$

Φ_l the volumetric content of liquid water, ρ_i the densities, $\mu_{l\to v}$ denotes the rate of mass change of liquid water into water vapour per unit of overall material volume, and $\vec{v}_i$ are the velocities of the particular fluids.

The transport of liquid water and of the gaseous phase were expressed using Darcy's equation,

$$\Phi_i \vec{v}_i = -\frac{K}{\eta_i} k_{ri}(S_l) \operatorname{grad} p_i, \tag{4.10}$$

where K is the permeability of the porous material (in m^2), which is an intrinsic property of the material independent of the saturation process, η_i the dynamic viscosity of the fluid i, and $k_{ri}(S_l)$ the relative permeability (dimensionless) of the fluid i. In the gaseous phase it was assumed that

$$p_g = p_v + p_a, \tag{4.11}$$

$$\vec{v}_g = C_v \vec{v}_v + C_a \vec{v}_a, \tag{4.12}$$

where C_i is the molar ratio of the constituent i,

$$C_i = \frac{p_i}{p_g}. \tag{4.13}$$

The relative diffusion process of water vapour and dry air phases relative to the gaseous mixture was assumed to be governed by Fick's law,

$$\Phi_g(\vec{v}_j - \vec{v}_g) = -F(S_l) \frac{D_{va}(T)}{p_g C_j} \operatorname{grad} C_j, \tag{4.14}$$

where j is either v or a,

$$D_{va}(T) = 0.217 \cdot 10^{-4} p_{atm} \left(\frac{T}{T_0}\right)^{1.88}, \tag{4.15}$$

$p_{atm} = 101325$ Pa, $T_0 = 273$ K, and the resistance factor $F(S_l)$ accounts for the tortuosity effects and the reduction of space offered to gas diffusion.

The set of equations (4.6)–(4.8), (4.10), (4.14) has to be completed by the equations of state. These were formulated in the form

$$p_j M_j = RT \rho_j \tag{4.16}$$

$$p_g - p_l = p_c(S_l), \tag{4.17}$$

where j denotes either v or a, i.e. the gaseous substances were assumed to be ideal gases, M_j is the molar mass of the constituent j, and p_c the capillary pressure.

The phase change process between water and water vapour was assumed to remain permanently in local thermodynamic equilibrium, therefore the isothermal Clapeyron's law could be used,

$$\frac{dp_v}{\rho_v} - \frac{dp_l}{\rho_l} = 0. \tag{4.18}$$

The material parameters necessary for obtaining the numerical data from the model by Baroughel-Bouny et al. (1999) are: the moisture storage function (called capillary curve by the authors) $p_c = p_c(S_l)$, the intrinsic permeability K, the viscosities η_i, the relative permeabilities $k_{ri}(S_l)$, and the resistance factor $F(S_l)$. Taking into account how the authors treat these material data, the main weakness of the model by Baroughel-Bouny et al. (1999) is the reliability of the input data. The intrinsic permeability K is determined either by a fitting procedure or by application of theoretical models. Also the relative permeabilities and the resistance factor are determined using theoretical models. However, this problem can be solved relatively easily by measuring the usual moisture storage and moisture transport material parameters (see Chapter 5), which can be related to the coefficients defined by Baroughel-Bouny et al. (1999) without serious problems. For instance, the relative permeabilities and the intrinsic permeability can be expressed using the measured values of hydraulic conductivity or a combination of the moisture storage function and the moisture diffusivity. Similarly, the resistance factor $F(S_l)$ can be expressed using the commonly measured water vapour diffusion resistance factor μ. Therefore, the model can be generally considered as well applicable for practical calculations.

Ulm et al. (1999) employed the nonlinear diffusion model by Bažant and Najjar (1972) in the form

$$\frac{\partial C}{\partial t} = \text{div}\,(D(C)\,\text{grad}\,C), \tag{4.19}$$

with

$$D(C) = a\exp(bC), \tag{4.20}$$

for the modelling of <u>hygric stresses</u> induced by drying of the concrete element of a nuclear cooling tower combined with <u>thermal stresses</u> induced by heating. The stress increments were calculated according to the relation

$$d\sigma = \frac{E}{1-\nu}d\mathbf{e} + \frac{E}{3(1-2\nu)}(d\epsilon - 3\kappa_s\,dC - 3\alpha\,dT)\mathbf{I}, \tag{4.21}$$

where E is the Young modulus, ν the Poisson ratio, $\mathbf{e} = \epsilon - 1/3\epsilon_s\mathbf{I}$ the deviator of the strain tensor ϵ, ϵ_s the scalar part of the strain tensor ϵ, $\epsilon_s = \text{tr}\,\epsilon$, which expresses the isotropic volume changes (see Chapter 2), α the thermal dilatation coefficient, and κ_s the hygric dilatation coefficient.

Modelling of <u>water penetration</u> into concrete is another very frequent application of the theory of transport processes. In the last ten or twenty years, it was inspired mainly by the necessity to describe degradation processes in concrete due to the intrusion of harmful substances, such as chlorides or sulphates where water transport plays a dominant role.

The simplest way to describe uptake of water by unsaturated, hardened concrete is in terms of sorptivity. The <u>sorptivity</u> S (in m/s$^{1/2}$) is defined (see, e.g., Hall, 1989, Ho and Lewis, 1987, McCarter et al., 1996, etc.) as

$$I = S \cdot t^{1/2}, \tag{4.22}$$

where I is the cumulative mass of water imbibed (in m).

Equation (4.22) is a simplification of the general expression for the cumulative mass of water in terms of the square root of time-rule commonly employed in the theory of transport processes in porous media (see, e.g., Krischer, 1962), which is obtained by dividing the original equation,

$$i = A \cdot t^{1/2}, \tag{4.23}$$

by the density of water, ρ_w. In equation (4.23), i is the real cumulative mass of water (in kg/m^2), and A the <u>water absorption coefficient</u> (in kg/m^{2}s$^{1/2}$), $A = S \cdot \rho_w$.

Water sorptivity can be related to the diffusivity by solving the diffusion equation with the boundary conditions corresponding to the sorptivity test, i.e. the sample is initially air-dry and the surface is in contact with free water, in other words, it is saturated. Mathematically, the problem can be described as

$$\frac{\partial \theta}{\partial t} = \frac{\partial}{\partial x}\left(D(\theta)\frac{\partial \theta}{\partial x}\right), \quad 0 < x < \infty \tag{4.24}$$

$$\theta(0, t) = 1 \tag{4.25}$$

$$\theta(x, 0) = 0 \tag{4.26}$$

$$\theta(\infty, t) = 0, \tag{4.27}$$

where θ is the reduced water content,

$$\theta = \frac{\Theta - \Theta_i}{\Theta_s - \Theta_i}, \tag{4.28}$$

Θ the volumetric water content (volume of water per bulk volume of concrete), Θ_s the saturated water content, and Θ_i the initial water content.

Using the Boltzmann variable $\eta = x/t^{1/2}$, the boundary conditions in the sorptivity test are transformed as: $\theta = 1$ for $\eta = 0$ and $\theta = 0$ for $\eta \to \infty$. The sorptivity can then be expressed in the form (Hall, 1989):

$$S = \int_{\Theta_i}^{\Theta_s} \eta \, d\Theta = (\Theta_s - \Theta_i) \int_0^1 \eta \, d\theta, \tag{4.29}$$

and the solution to the nonlinear problem (4.24)–(4.27) is (Parlange et al., 1992)

$$2\int_0^1 \frac{D(a)}{a} \, da = s\eta + \frac{B}{2}\eta^2, \tag{4.30}$$

where

$$s = \frac{S}{\Theta_s - \Theta_i} = \left(\int_0^1 (1 + \theta) D(\theta) \, d\theta \right)^{1/2} \tag{4.31}$$

$$B = 2 - \frac{s^2}{\int_0^1 D \, d\theta}. \tag{4.32}$$

Using this solution, Lockington et al. (1999) derived for the commonly used type of the $D(\theta)$ function,

$$D = D_0 \cdot e^{n\theta}, \tag{4.33}$$

the following relation:

$$2D_0[E_i(n) - E_i(n\theta)] = s\eta + \frac{B}{2}\eta^2, \tag{4.34}$$

where

$$E_i(x) = -\int_{-x}^\infty \frac{e^{-t}}{t} \, dt \tag{4.35}$$

and

$$s^2 = D_0[e^n(2n^{-1} - n^{-2}) - n^{-1} + n^{-2}] \tag{4.36}$$

$$B = \frac{e^n n^{-1} - 1 - n^{-1}}{e^n - 1}. \tag{4.37}$$

Nevertheless, measurement of sorptivity alone is not sufficient to identify the two unknown parameters D_0 and n in equation (4.33), and one additional equation is necessary. Lockington et al. (1999) on the basis of previous measurements of diffusivities of mortars and concretes put $n = 6$ as a standard value for concrete, and then calculated the unknown D_0 from equation (4.36). Lockington et al. (1999) also introduced the solution of equation (4.30) for the diffusivity function of type

$$D = D_0 \cdot \theta^n. \tag{4.38}$$

They obtained

$$2D_0 \frac{1 - \theta^n}{n} = s\eta + \frac{B}{2}\eta^2, \tag{4.39}$$

$$s^2 = D_0 \frac{3 + 2n}{(1 + n)(2 + n)} \tag{4.40}$$

$$B = 2 - \frac{(1 + n)s^2}{D_0}. \tag{4.41}$$

Putting the universal exponent for concrete $n = 4$, D_0 was calculated from (4.40).

Modelling solely <u>heat transport</u> in concrete (without moisture transport) is very rare at present. Practically the only field where it is commonly employed is in modelling the evolution of heat in large concrete blocks during the hydration process. Hrstka et al. (1999) modelled the temperature fields in a large scale sprayed concrete structure. They assumed heat conduction equation in the following form:

$$\rho c \frac{\partial T}{\partial t} = \frac{\partial}{\partial x}\left(\lambda\frac{\partial T}{\partial x}\right) + \frac{\partial}{\partial y}\left(\lambda\frac{\partial T}{\partial y}\right) + \frac{\partial}{\partial z}\left(\lambda\frac{\partial T}{\partial z}\right) + I(x, y, z, t), \quad (4.42)$$

where the term $I(x, y, z, t)$ represents the internal heat source due to the hydration heat production (in W/m^3). This term can be obtained as the time derivative of the measured cumulative hydration heat production curve (which is in J/m^3). Similar calculations on 1×1 m concrete column were done by Khan et al. (1998).

<u>Coupled heat and moisture transport</u> in concrete structures is usually done for the sake of analysis of hygrothermal performance of concrete envelopes. For modelling temperature and moisture fields in a concrete envelope Kiessl (1983) employed the following model (for details, see Chapter 3):

$$\frac{\partial}{\partial t}\left[\rho_{\mathrm{w}}w + (\Pi - w)\varphi\rho_{\mathrm{p,s}}\right]$$

$$= \rho_{\mathrm{w}}\frac{\partial}{\partial x}\left(\mathrm{FKU}\frac{\partial w}{\partial x} + \mathrm{FDP}\frac{\partial \varphi}{\partial x} + \mathrm{FDT}\frac{\partial T}{\partial x}\right) \quad (4.43)$$

$$\frac{\partial}{\partial t}(\rho_0 c_0 T + \rho_{\mathrm{w}}c_{\mathrm{w}}wT) = \frac{\partial}{\partial x}\left(\lambda\frac{\partial}{\partial x}\right)$$

$$+ L_{\mathrm{v}}\left\{\rho_{\mathrm{w}}\frac{\partial}{\partial x}\left(\mathrm{FDP}\frac{\partial \varphi}{\partial x} + \mathrm{FDT}\frac{\partial T}{\partial x}\right)\right.$$

$$\left. - \frac{\partial}{\partial t}\left[(\Pi - w)\varphi\rho_{\mathrm{p,s}}\right]\right\}, \quad (4.44)$$

where ρ_{w} is the density of water, w the volumetric water content, $\rho_{\mathrm{p,s}}$ the partial density of saturated water vapour, φ the relative humidity, FDP the coefficient of diffusion moisture transport due to grad φ, FDT the coefficient of diffusion moisture transport due to grad T, FKU the coefficient of capillary moisture transport, ρ_0 the partial density of the solid matrix, c_0 the specific heat capacity of the solid matrix, c_{w} the specific heat capacity of water, and L_{v} the latent heat of evaporation of water. The coupling of water and water vapour transport was achieved in the model by Kiessl (1983) by the sorption isotherm.

A more advanced model of coupled heat and moisture transport including air transport also was employed by Häupl et al. (1997) for the analysis of the hygrothermal behaviour of a concrete wall with polystyrene foam outside insulation. They formulated the moisture, air and internal energy balance equations, respectively, in the form

$$\frac{\partial}{\partial t}(\rho_l w_l + \rho_v w_g) = -\,\text{div}\,\vec{j}^{\,\text{mw}} + \sigma^{\text{mw}} \tag{4.45}$$

$$\frac{\partial}{\partial t}(\rho_a w_g) = -\,\text{div}\,\vec{j}^{\,\text{ma}} + \sigma^{\text{ma}} \tag{4.46}$$

$$\frac{\partial}{\partial t}(\rho u) = -\,\text{div}\,\vec{j}^{\,\text{u}} + \sigma^{\text{u}}, \tag{4.47}$$

where w_i is the volumetric fraction of the i-th phase, ρ_i the partial density of the i-th phase, $\vec{j}^{\,\text{mw}}$ the total moisture flux, $\vec{j}^{\,\text{ma}}$ the total air flux, $\vec{j}^{\,\text{u}}$ the total flux of internal energy, and σ^{mw}, σ^{ma}, σ^{u} the productions terms for moisture, air and internal energy, respectively. The subscripts denote the following phases: l the liquid, g the gaseous phase, v the vapour, a the air.

The fluxes of particular constituents were in the Häupl et al. (1997) model formulated as follows.
Moisture flux equations:
    convective flux of the liquid phase

$$\vec{j}^{\,\text{ml}}_{\text{con}} = \rho_l \vec{v}^{\,\text{ml}} = -\rho_l K_l(\text{grad}\,p_l - \rho_l \vec{g}) \tag{4.48}$$

    convective flux of the vapour phase

$$\vec{j}^{\,\text{mv}}_{\text{con}} = \rho_v \vec{v}^{\,\text{mg}} = -\rho_v K_g(\text{grad}\,p_g - \rho_g \vec{g}) \tag{4.49}$$

    diffusive flux of the vapour phase

$$\vec{j}^{\,\text{mv}}_{\text{dif}} = -\rho_v D^v \,\text{grad}\,C_v - D^v_p \,\text{grad}\,p_g \tag{4.50}$$

    total moisture flux

$$\vec{j}^{\,\text{mw}} = \vec{j}^{\,\text{ml}} w_l + (\vec{j}^{\,\text{mv}}_{\text{con}} + \vec{j}^{\,\text{mv}}_{\text{dif}}) w_g. \tag{4.51}$$

Air flux equations:
    convective air flux

$$\vec{j}^{\,\text{ma}}_{\text{con}} = \rho_a \vec{v}^{\,\text{ma}} = -\rho_a K_g(\text{grad}\,p_g - \rho_g \vec{g}) \tag{4.52}$$

    diffusive air flux

$$\vec{j}^{\,\text{ma}}_{\text{dif}} = -\vec{j}^{\,\text{mv}}_{\text{dif}} \tag{4.53}$$

    total air flux

$$\vec{j}^{\,\text{ma}} = (\vec{j}^{\,\text{ma}}_{\text{con}} + \vec{j}^{\,\text{ma}}_{\text{dif}}) w_g. \tag{4.54}$$

Internal energy flux equations:
  total heat conduction

$$\vec{j}^q = -\lambda \operatorname{grad} T \tag{4.55}$$

liquid enthalpy flux

$$\vec{j}^{hl} = \vec{j}^{ml}_{con} h_l \tag{4.56}$$

vapour enthalpy flux

$$\vec{j}^{hv} = (\vec{j}^{mv}_{con} + \vec{j}^{mv}_{dif}) h_v \tag{4.57}$$

air enthalpy flux

$$\vec{j}^{ha} = (\vec{j}^{ma}_{con} + \vec{j}^{ma}_{dif}) h_a \tag{4.58}$$

total flux of internal energy

$$\vec{j}^u = \vec{j}^q + \vec{j}^{hl} w_l + (\vec{j}^{hv} + \vec{j}^{ha}) w_g. \tag{4.59}$$

In the moisture, air and internal energy flux equations, K_i is the laminar convection coefficient of the i-th phase, D^v the diffusion coefficient of the vapour phase, D^v_p the barycentric pressure diffusion coefficient of the vapour phase, λ the total thermal conductivity, and C_v the vapour mass concentration, $C_v = \rho_v / \rho_g$.

After substituting the relations for moisture, air and internal energy fluxes into the system of balance equations (4.45)–(4.47), a system of three equations for five variables p_l, p_g, w_l, T, C_v is obtained. Therefore, two additional relations are necessary, for instance

$$C_v = C_v(w_l, p_g, T) \tag{4.60}$$

$$p_l = p_l(w_l, p_g, T). \tag{4.61}$$

The functions (4.60), (4.61) can be formulated using two material relations, which can be determined experimentally, namely the inverse sorption isotherm

$$\varphi = \varphi(w_l, T) \tag{4.62}$$

and the water retention curve

$$p_c = p_c(w_l, T), \tag{4.63}$$

where φ is the relative humidity,

$$\varphi(w_l, T) = \frac{p_v(w_l, T)}{p_{sat}(T)}, \tag{4.64}$$

$p_{sat}(T)$ the saturated water vapour pressure at temperature T, and p_c the capillary pressure,

$$p_c(w_l, T) = p_l(w_l, p_g, T) - p_g. \tag{4.65}$$

Using (4.62), (4.63), and utilizing the equations of state for the water vapour and the air, respectively, which are assumed to be ideal gases,

$$\rho_v = \frac{p_v M_v}{RT}$$

(4.66)

$$\rho_a = \frac{p_a M_a}{RT} = \frac{(p_g - p_v) M_a}{RT},$$

(4.67)

the quantities p_l and C_v can be eliminated from the system of equations (4.45)–(4.47), so that a system of three equations can be obtained for the three unknowns w_l, p_g, T, which is a closed solvable system.

Models of chemical compounds transport in concrete are commonly employed in service life predictions. In particular, modelling of chloride diffusion, which makes it possible to predict the service life of concrete subject to reinforcement corrosion, is frequently used in practice. Most of the applied models are very simple, and employ Fick's diffusion equation for the chloride concentration with a constant diffusion coefficient and constant initial and boundary conditions,

$$\frac{\partial C}{\partial t} = D \text{ div grad } C$$

(4.68)

$$C(0, t) = C_0$$

(4.69)

$$C(\infty, t) = 0$$

(4.70)

$$C(x, 0) = 0.$$

(4.71)

The problem (4.68)–(4.71) has a very simple mathematical solution (see, e.g., Carslaw and Jaeger, 1959):

$$C(x, t) = C_0 \left[1 - \text{erf} \left(\frac{x}{2\sqrt{Dt}} \right) \right],$$

(4.72)

and this is the main reason of its high frequency of application. This model was employed for instance by Tuutti (1982), Funahashi (1990), Cady and Weyers (1992), Weyers (1998), Zemajtis et al. (1998), Costa and Appleton (1999) and many others. Some of the authors, for example Zemajtis et al. (1998), assume the surface concentration as a function of the square root of time, or in more general form (e.g., Costa and Appleton, 1999) as a power function of time,

$$C_0(t) = C_1 \cdot t^n,$$

(4.73)

where C_1 is the surface chloride concentration after one year, t is expressed in years, and n is an empirical coefficient.

The main weaknesses of these models are the assumption of the constant diffusion coefficient and the fact that they neglect the influence of water transport on the transport of chemical compounds. As a consequence, a single value of the diffusion coefficient can never be obtained from the measured concentration profiles, particularly if the measurements are performed over longer time periods. The diffusion coefficient then appears as a function of time. For example, Mangat and Molloy (1992) proposed for the diffusion coefficient of chlorides in concrete a power function,

$$D(t) = D_1 \cdot t^{-m}, \tag{4.74}$$

where D_1 is the diffusion coefficient at one year, t is expressed in years and m is an empirical coefficient.

The tendency of many authors to work with very simple mathematical models is quite understandable, but sometimes it may lead to serious errors. As an example, the erroneous treatment of Costa and Appleton (1999) may be cited. They have taken the analytical solution (4.72) of the problem (4.68)–(4.71), and substituted into this solution the surface concentration from (4.73) instead of the constant C_0 and the diffusion coefficient from (4.74) instead of the constant D. This is simply not possible. The diffusion coefficient (4.74) has to be substituted into the nonlinear form of equation (4.68),

$$\frac{\partial C}{\partial t} = \text{div} \, (D \, \text{grad} \, C), \tag{4.75}$$

instead of D, and the boundary condition (4.73) has to be used instead of (4.69). Then, the newly formulated problem has to be solved either analytically or numerically, and the new solution will naturally be quite different from that which follows from the erroneous treatment of Costa and Appleton (1999).

A more advanced model of chloride transport in concrete was introduced by Boddy et al. (1999). They formulated the chloride transport equation in the form

$$\frac{\partial C}{\partial t} = D \frac{\partial^2 C}{\partial x^2} - v \frac{\partial C}{\partial x} + \frac{\rho}{n} \frac{\partial C_s}{\partial t}, \tag{4.76}$$

where D is the diffusion coefficient, ρ the density of concrete, C_s the concentration of bound chlorides, v the average linear velocity of water,

$$v = \frac{Q}{nA} = -\frac{k}{n} \frac{\partial h}{\partial x}, \tag{4.77}$$

Q the water flow rate (in m^3/s), n the porosity, A the cross sectional area, k the hydraulic conductivity, and h the hydraulic head. Water transport was formulated according to the model of Buenfeld et al. (1997),

$$\frac{\partial w}{\partial t} = \frac{\partial}{\partial x} \left(D_w(w) \frac{\partial w}{\partial x} \right) - \frac{S_p^2}{2x_i} \frac{\partial w}{\partial x}, \tag{4.78}$$

where w is the water content, D_w the water vapour diffusivity, x the depth from the wet face, x_i the depth from the wet/dry interface, S_p the sorptivity.

The basic idea of including water transport in the chloride transport equation in the model by Boddy et al. (1999) was certainly good but the practical applicability of the model in the published form remained questionable. Boddy et al. (1999) still assumed a linear chloride diffusion problem with the diffusion coefficient as a function of time and temperature. The problem of coupling water transport, expressed by the material relation (4.77), and the chloride transport was not solved at all. The application of equation (4.78) for the description of water transport was not a very good idea. In this equation, water transport and water vapour transport are not coupled, which normally can be done very easily using the sorption isotherm (see Section 1.6 and Chapter 3). In addition, this equation containing both water content w and the position of the wet/dry interface x_i as independent variables cannot be solved without an additional condition for x_i (it is in fact a Stefan-like problem, see Section 3.1.2.1 for details), which was, however, not formulated. Therefore, despite the fact that the effort of the authors was in the right direction, the model in its current form cannot be recommended for practical applications.

Johannesson (1999) formulated a model of the diffusion of a mixture of cations and anions (dissolved in water) in concrete. The mass balance for a constituent i was written in the form

$$\frac{\partial n_i}{\partial t} = -\operatorname{div}(n_i \vec{u}_i), \quad i = 1, ..., m, \tag{4.79}$$

where n_i (in mol/m^3) denotes the molar concentration of the i-th constituent, and $\vec{u}_i$ is the diffusion velocity.

The constitutive relations were formulated in several ways. In the first approach he assumed

$$n_i \vec{u}_i = -D_i \operatorname{grad} n_i - E_i \nu_i \operatorname{grad} \Phi, \tag{4.80}$$

where D_i is the separate diffusion constant for the i-th constituent, E_i (in mol/s/m) a material constant describing the tendency of the i-th ion diffusing towards opposite charged domains or being repelled from like-charged domains, and ν_i the valence number. The state variable Φ was defined as

$$\Phi = \sum_{i=1}^{m} n_i \nu_i, \tag{4.81}$$

i.e. it determined the charge imbalance at a certain material point in the domain.

In the second approach Johannesson (1999) assumed that the m cations and anions can potentially form k acids or salts, and the charge will be balanced at every material point,

$$\Phi = \sum_{i=1}^{m} n_i \nu_i = 0. \tag{4.82}$$

The constitutive relation was then formulated as follows:

$$n_j \vec{u}_j = -D_j \operatorname{grad} n_j, \quad j = 1, ..., k, \tag{4.83}$$

where n_j is a neutrally charged package of an anion and a cation forming the neutrally charged substance j, D_j a weighted value of the separate diffusion constants for the cation and the anion forming the neutrally charged substance j,

$$D_j = 2 \left(\frac{1}{D^+} + \frac{1}{D^-} \right)^{-1}. \tag{4.84}$$

The third approach was similar to the first one, only the influence of the electric field was determined using the electrostatic potential ϕ, so that the constitutive relations had the form

$$n_i \vec{u}_i = -D_i \operatorname{grad} n_i - A_i \nu_i n_i \operatorname{grad} \phi, \tag{4.85}$$

where A_i (in m^2/s/V) is the ionic mobility.

Bažant and Thonguthai (1978), (1979) formulated a model of the coupled heat and moisture transport in heated concrete describing the fire response of a concrete structure, and solved a hot-spot problem on a concrete wall in cylindrical symmetry. They formulated the moisture and heat balance equations, respectively, in the form

$$\frac{\partial w}{\partial t} = -\operatorname{div} \vec{j} + \frac{\partial w_d}{\partial t}, \tag{4.86}$$

$$\rho c \frac{\partial T}{\partial t} - c_a \frac{\partial w}{\partial t} = -\operatorname{div} \vec{q} + c_w \vec{j} \cdot \operatorname{grad} T, \tag{4.87}$$

where w is the free water content (in kg/m^3), w_d the mass of free water per unit volume that has been released into the porous space as a result of heating of the porous matrix, $\vec{j}$ the water flux, c_a the heat capacity of free water, including the adsorption heat of adsorbed water layers on pore walls (in J per kilogram of free water), ρ and c the density and specific heat capacity of the porous matrix including the chemically bound water, respectively, c_w the specific heat capacity of water, and $\vec{q}$ the heat flux. The term $(c_w \vec{j} \cdot \operatorname{grad} T)$ expresses the heat transport due to the water flow. In equations (4.86) and (4.87), the following constitutive relations were used for the moisture and heat fluxes:

$$\vec{j} = -\frac{a}{g} \operatorname{grad} p \tag{4.88}$$

$$\vec{q} = -b \operatorname{grad} T, \tag{4.89}$$

where p is the pore pressure of water vapour, a the permeability (in m/s, therefore in fact the hydraulic conductivity in the notation common in soil physics, see, e.g., Bear, 1988, Bear and Bachmat, 1990), g the gravity acceleration, and b the thermal conductivity. The heat capacity terms in equation (4.87) were expressed as

$$\rho c \frac{\partial T}{\partial t} = \rho_s c_s \frac{\partial T}{\partial t} - c_d \frac{\partial w_d}{\partial t}, \tag{4.90}$$

$$c_a \frac{\partial w}{\partial t} = \frac{\partial}{\partial t}(w_c H) - c_{ad} \frac{\partial w_{ad}}{\partial t}, \tag{4.91}$$

where ρ_s, c_s are the mass density and the specific heat capacity of solid microstructure excluding hydrate water, c_d the heat of dehydration, c_{ad} the heat of adsorption on pore walls, w_c the amount of capillary water, $w_c = w - w_{ad}$, w_{ad} the amount of water adsorbed on pore walls, and H the enthalpy of water. The heat of vaporization of water does not figure explicitly, though it may be included under the term $\frac{\partial}{\partial t}(w_c H)$.

Finally, the relation between the water content w and the pore pressure p was given by the sorption relation

$$w = w(p, T). \tag{4.92}$$

Substituting (4.92) together with (4.88) and (4.89) into the moisture and heat balance equations (4.86) and (4.87), Bažant and Thonguthai (1978), (1979) finally obtained for a two-dimensional axisymmetric case the following set of equations for p and T:

$$\frac{\partial}{\partial r}\left(a\frac{\partial p}{\partial r}\right) + \frac{a}{r}\frac{\partial p}{\partial r} + \frac{\partial}{\partial z}\left(a\frac{\partial p}{\partial z}\right) + A_1\frac{\partial p}{\partial t} + A_2\frac{\partial T}{\partial t} + A_3 = 0 \tag{4.93}$$

$$\frac{\partial}{\partial r}\left(b\frac{\partial T}{\partial r}\right) + \frac{b}{r}\frac{\partial T}{\partial r} + \frac{\partial}{\partial z}\left(b\frac{\partial T}{\partial z}\right)$$

$$+ A_4\frac{\partial T}{\partial r} + A_5\frac{\partial T}{\partial z} + A_6\frac{\partial T}{\partial t} + A_7\frac{\partial p}{\partial t} = 0, \tag{4.94}$$

where

$$A_1 = -\frac{\partial w}{\partial p} \tag{4.95}$$

$$A_2 = -\frac{\partial w}{\partial T} \tag{4.96}$$

$$A_3 = \frac{\partial w_d}{\partial t} \tag{4.97}$$

$$A_4 = -a c_w \frac{\partial p}{\partial r} \tag{4.98}$$

$$A_5 = -ac_w \frac{\partial p}{\partial z} \tag{4.99}$$

$$A_6 = c_a \frac{\partial w}{\partial T} - \rho c \tag{4.100}$$

$$A_7 = c_a \frac{\partial w}{\partial p}. \tag{4.101}$$

Majorana et al. (1998) employed a combination of the models by Bažant and Najjar (1972) and Bažant and Thonguthai (1979) to the modelling of hygrothermal and mechanical behaviour of concrete structures at high temperatures. They formulated the moisture and heat balance equations, respectively, in the form

$$\frac{\partial \varphi}{\partial t} - \mathrm{div}\,(D \,\mathrm{grad}\,\varphi) - \frac{\partial \varphi_s}{\partial t} - K \frac{\partial T}{\partial t} = 0 \tag{4.102}$$

$$\rho c \frac{\partial T}{\partial t} - \mathrm{div}\,(\lambda \,\mathrm{grad}\,T) - \frac{\partial Q_h}{\partial t} = 0, \tag{4.103}$$

where φ is the relative humidity, T the temperature, D the diffusivity, $K = \left(\frac{\partial \varphi}{\partial T}\right)_w$ the hygrothermic coefficient, $\frac{\partial \varphi_s}{\partial t}$ expresses the change of relative humidity due to self-desiccation, ρ is the density of concrete, c the isobaric specific heat capacity of concrete, λ the thermal conductivity, and the term $\frac{\partial Q_h}{\partial t}$ represents the heat power density sources (in W/m^3) due to the water sorption on the walls of the porous system and to the water movement induced heat convection.

The linear momentum balance equation for the whole multiphase medium was first written in the form

$$\mathrm{div}\,\sigma + \rho_m \vec{g} = 0, \tag{4.104}$$

where σ is the total stress tensor, ρ_m the density of the multiphase medium, and $\vec{g}$ the acceleration due to gravity, but for practical calculations, the virtual work principle was employed,

$$\int \delta\epsilon : \sigma \,\mathrm{d}V = \delta\vec{u} \cdot \vec{F}, \tag{4.105}$$

where $\vec{u}$ is the displacement vector, $\vec{F}$ the force vector.

In the constitutive equations of the mechanistic analysis, superposition of creep and damage effects was assumed, which for constant strain can be schematically expressed as

$$\sigma(t) = [1 - D(\epsilon)]R(t, t')[\epsilon(t') - \epsilon_0(t')], \tag{4.106}$$

where D is the damage factor ($0 \leq D \leq 1$) as a parameter measuring the reduction of the resistant area due to cracks beginning and spreading, $R(t, t')$

the relaxation function, ϵ the strain tensor, and ϵ_0 the tensor of strain due to the variations of temperature and pore pressure, which also includes the irreversible part of the thermal and hygric strains.

The system of hygrothermal and mechanical balance equations was not directly coupled in the model by Majorana et al. (1998), and the hygrothermal and mechanical analyses were done separately. Nevertheless, their model is among the most advanced in the field of description of transport processes in concrete to date.

A similar model to that of Majorana et al. (1998) was developed by Gawin et al. (1999) but the heat and moisture transport equations were formulated on the basis of the general thermo-hygro-mechanical model of Lewis and Schrefler (1998). The balance equations of mass of the dry air, mass of water and energy, respectively, were written in the form:

$$\phi \frac{\partial}{\partial t}[(1-S)\rho_{ga}] + (1-S)\rho_{ga}\frac{\partial \phi_{hydr}}{\partial t} + \alpha(1-S)\rho_{ga}\frac{\partial}{\partial t}(\text{div } \vec{u})$$

$$+ \text{ div } (\rho_{ga}\vec{v}_g) + \text{ div } (\rho_g\vec{v}_{ga}^d) = 0 \tag{4.107}$$

$$\phi \frac{\partial}{\partial t}[(1-S)\rho_{gw}] + (1-S)\rho_{gw}\frac{\partial \phi_{hydr}}{\partial t} + \alpha(1-S)\rho_{gw}\frac{\partial}{\partial t}(\text{div } \vec{u})$$

$$+ \text{ div } (\rho_{gw}\vec{v}_g) + \text{ div } (\rho_g\vec{v}_{gw}^d)$$

$$= -\phi \frac{\partial}{\partial t}(S\rho_w) - S\rho_w\frac{\partial \phi_{hydr}}{\partial t} - \alpha S\rho_w\frac{\partial}{\partial t}(\text{div } \vec{u})$$

$$- \text{ div } (\rho_w\vec{v}_l) - \frac{\partial}{\partial t}(\Delta m_{hydr}) \tag{4.108}$$

$$\rho c_p \frac{\partial T}{\partial t} + (c_{pw}\rho_w\vec{v}_l + c_{pg}\rho_g\vec{v}_g)\text{ grad } T - \text{ div } (\lambda_{eff}\text{ grad } T)$$

$$= \Delta h_{phase}\left[\phi \frac{\partial}{\partial t}(S\rho_w) + S\rho_w\frac{\partial \phi_{hydr}}{\partial t} + \alpha S\rho_w\frac{\partial}{\partial t}(\text{div } \vec{u})\right.$$

$$\left. + \text{ div } (\rho_w\vec{v}_l)\right] + \Delta h_{hydr}\frac{\partial}{\partial t}(\Delta m_{hydr}), \tag{4.109}$$

where ϕ is the total porosity (pore volume by total volume), S the liquid phase volumetric saturation (liquid volume by pore volume), ρ_{ga} the mass concentration of the dry air in the gas phase, ϕ_{hydr} the part of porosity resulting from the dehydration of concrete, which is given by the relation

$$\frac{\partial \phi_{hydr}}{\partial t} = -\frac{\partial}{\partial t}\left(\frac{\Delta m_{hydr}}{\rho_s}\right), \tag{4.110}$$

Δm_{hydr} the mass source related to hydration/dehydration process, ρ_s the solid state density, α the Biot's constant,

$$\alpha = 1 - \frac{K_T}{K_M}, \tag{4.111}$$

K_T the bulk modulus of the porous medium, K_M the bulk modulus of the solid phase, $\vec{u}$ the displacement vector of the solid matrix, $\vec{v}_g$ the velocity of the gaseous phase, $\vec{v}_{ga}^d$ the relative average diffusion velocity of dry air, ρ_{gw} the mass concentration of water vapour in the gas phase (in kg/m³), $\vec{v}_{gw}^d$ the relative average diffusion velocity of water vapour, $\vec{v}_l$ the velocity of the liquid phase, ρ the apparent density of the porous medium, c_p the effective specific heat capacity of the porous medium, ρ_w the density of liquid phase, ρ_g the density of gaseous phase, c_{pw} the specific heat capacity of the liquid phase, c_{pg} the specific heat capacity of the gaseous phase, λ_{eff} the effective thermal conductivity, Δh_{phase} the latent heat of the phase change, for $S \leq S_{\text{ssp}}$ (S_{ssp} is the solid saturation point) we have $\Delta h_{\text{phase}} = \Delta h_{\text{adsorp}}$, Δh_{adsorp} is the latent heat of adsorption, for $S > S_{\text{ssp}}$ $\Delta h_{\text{phase}} = \Delta h_{\text{vap}}$, Δh_{vap} is the latent heat of vaporization, and Δh_{hydr} the amount of heat per unit mass produced by hydration.

The linear momentum balance equation of the whole mixture was written as

$$\text{div}\left(\frac{\partial \sigma}{\partial t}\right) + \frac{\partial \rho}{\partial t}\vec{b} = 0, \tag{4.112}$$

where σ is the total stress tensor, ρ the averaged density of the multi-phase medium,

$$\rho = (1 - \phi)\rho_s + \phi S \rho_w + \phi(1 - S)\rho_g, \tag{4.113}$$

and $\vec{b}$ the specific force (in J/kg) vector.

In the balance equations of energy and linear momentum, the solid state density is present, which is not a constant because it depends on the extent of the hydration process. Therefore, the solid mass conservation equation also has to be written,

$$\frac{\partial}{\partial t}[(1 - \phi)\rho_s] + \text{div}\left[(1 - \phi)\rho_s\vec{v}_s\right] = \frac{\partial}{\partial t}(\Delta m_{\text{hydr}}), \tag{4.114}$$

where $\vec{v}_s$ is the velocity of the solid phase.

The hygric constitutive equations were in the model of Gawin et al. (1999) written in the following form.
For capillary water:

$$\vec{v}_l = -\frac{K \cdot K_{rw}}{\mu_w}(\text{grad } p_g - \text{grad } p_c - \rho_w\vec{b}), \tag{4.115}$$

where p_c is the capillary pressure (defined in such a way that it has negative values, i.e. it expresses the suction of water),

$$p_c = p_g - p_l, \tag{4.116}$$

p_g the pressure of the gaseous phase, p_l the pressure of the liquid phase, K the absolute permeability (in m^2), K_{rw} the relative permeability of the liquid phase, μ_w the dynamic viscosity of the liquid phase.
For the gaseous phase:

$$\vec{v}_g = -\frac{K \cdot K_{rg}}{\mu_g} \text{ grad } p_g, \qquad (4.117)$$

where K_{rg} the relative permeability of the gaseous phase, μ_w the dynamic viscosity of the gaseous phase.
For the bound water:

$$\vec{v}_b = -D_b \text{ grad } S_b, \qquad (4.118)$$

where D_b is the bound water diffusion coefficient, and the degree of saturation of the bound water, S_b, is given by:

$$S_b = S \quad \text{for } S \leq S_{ssp} \qquad (4.119)$$

and

$$S_b = S_{ssp} \quad \text{for } S > S_{ssp}. \qquad (4.120)$$

For the description of the diffusion process of the binary gas mixture of dry air and water vapour, Fick's law was applied,

$$\vec{v}_{ga} = -\frac{M_a M_w}{M_g^2} D_{eff} \text{ grad } \left(\frac{p_{ga}}{p_g}\right) =$$

$$= \frac{M_a M_w}{M_g^2} D_{eff} \text{ grad } \left(\frac{p_{gw}}{p_g}\right) = -\vec{v}_{gw}, \qquad (4.121)$$

where M_a is the molar mass of dry air, M_w the molar mass of water, M_g the molar mass of the binary mixture,

$$\frac{1}{M_g} = \frac{\rho_{gw}}{\rho_g} \frac{1}{M_w} + \frac{\rho_{ga}}{\rho_g} \frac{1}{M_a}, \qquad (4.122)$$

p_{ga} the partial pressure of dry air, p_{gw} the partial pressure of water vapour, and D_{eff} the effective diffusivity of the gas mixture. For the gaseous constituents, the equation of state of ideal gas was adopted, so that

$$p_{ga} = \rho_{ga} \frac{RT}{M_a} \qquad (4.123)$$

$$p_{gw} = \rho_{gw} \frac{RT}{M_w} \qquad (4.124)$$

$$p_g = \rho_g \frac{RT}{M_g}, \qquad (4.125)$$

and Dalton's law was assumed to be valid,

$$p_g = p_{ga} + p_{gw}.$$ (4.126)

In the mechanical constitutive equations, the concept of Bishop's stress σ' (i.e. the effective stress) responsible for all deformations in concrete was adopted,

$$\sigma' = \sigma + \alpha p \mathbf{I},$$ (4.127)

where $\mathbf{I}$ is the unit tensor, p is the average pressure of the mixture of fluids filling the voids,

$$p = p_g - p_{atm} \quad \text{for } S \leq S_{ssp}$$ (4.128)

$$p = (S - S_{ssp})p_l + [1 - (S - S_{ssp})]p_g - p_{atm} \quad \text{for } S > S_{ssp}.$$ (4.129)

The constitutive equation for the solid skeleton was then assumed in the form

$$d\sigma' = K_M(d\epsilon - d\epsilon_T - d\epsilon_0),$$ (4.130)

where K_M is the bulk modulus of the concrete matrix,

$$d\epsilon_T = \mathbf{I}\frac{\beta_s}{3}\,dT,$$ (4.131)

the strain increment caused by thermoelastic expansion, β_s the volume thermal expansion coefficient of the solid, $d\epsilon_0$ represents the autogenous strain increments related to variation of capillary pressure and the irreversible part of the thermal strains.

In the model by Gawin et al. (1999) the effect of damage was considered in the form of the damage parameter D, $0 \leq D \leq 1$, defined as

$$\bar{\sigma} = \sigma\frac{A}{\bar{A}} = \frac{\sigma}{1 - D},$$ (4.132)

where A is the resistant area of the uncracked material, $\bar{A}$ the resistant area of the damaged material. The meaning of the modified effective stress in (4.132) is different from that of equation (4.127). The effective stress in (4.127) has to be introduced into (4.132) and subsequent equations. Since the damaging mechanisms are different in a uniaxial traction experiment compared to a compression one, the final equation for the stress becomes

$$\sigma_i = \left\{ (1 - A_i)K_0 + \frac{A_i\epsilon}{\exp\left[B_i(\epsilon - K_0)\right]} \right\} E, \quad i = t, c,$$ (4.133)

where the indices t and c denote the traction and the compression, respectively, A_i, B_i are material characteristics, which can be determined experimentally, E the elastic modulus, K_0 the initial value of the hardening/softening parameter $K(D)$, which satisfies the principle of maximum of the Saint-Venant strain (loading function)

$$f(\epsilon, K_0) = \bar{\epsilon} - K(D), \tag{4.134}$$

the equivalent strain $\bar{\epsilon}$ is defined as

$$\bar{\epsilon} = \sqrt{\sum_i ((\langle \epsilon_i \rangle_+)^2}, \tag{4.135}$$

$\langle x \rangle = (|x| + x)/2$, ϵ_i are the principal strains.

4.2 APPLICABILITY LIMITS OF GENERAL MODELS OF TRANSPORT PHENOMENA IN POROUS MATERIALS

The majority of general models of transport phenomena in porous materials, which were analysed in Chapter 3, were not formulated for specific applications to concrete. Diffusion models were developed primarily for applications in building physics, convection models for applications in soil physics. However, for most common calculations of transport phenomena in concrete structures these general models can be employed without any change. It is only necessary to measure the transport and storage parameters of concrete, which have to be used as input parameters in the particular models. For some more specific applications, such as modelling the transport processes in hydrating concrete or in concrete at high temperatures, most general models could not be employed in their current form, but special adjustments would have to be made. Nevertheless, these adjustments could be included without serious problems, because they mostly consist in addition of source/sink terms in the particular balance equations, as it has been analysed before.

Chapter 3 provided a historical overview of models of transport phenomena in porous materials with a detailed derivation of balance equations and an analysis of constitutive relations employed in the particular models. Here, we will discuss only typical representatives of different classes of models that are currently employed for modelling transport processes in porous materials in general, and which would be suitable without any substantial changes for application to transport processes in mature concrete. We will confine ourselves to the models capable of easily including the transport phenomena that are specific to concrete alone.

Among the diffusion models, commonly employed in building physics related calculations, the computer code WUFI developed on the basis of the model by Künzel (1995) belongs to those that are suitable for modelling moisture and heat transport in mature concrete without additional modifications. The basic transport equations of the model are as follows:

$$\frac{d\rho_v}{d\varphi} \frac{\partial \varphi}{\partial t} = \text{div} \left[D_\varphi \text{ grad } \varphi + \delta_p \text{ grad } (\varphi p_s) \right] \tag{4.136}$$

$$\frac{\mathrm{d}H}{\mathrm{d}T}\frac{\partial T}{\partial t} = \mathrm{div}\,(\lambda\,\mathrm{grad}\,T) + L_{\mathrm{v}}\,\mathrm{div}\,[\delta_{\mathrm{p}}\,\mathrm{grad}\,(\varphi p_{\mathrm{s}})], \qquad (4.137)$$

where ρ_{v} is the partial moisture density, i.e. the mass of water per unit volume of the porous body, δ_{p} the water vapour permeability, φ the relative humidity, $\rho_{\mathrm{v}} = \rho_{\mathrm{v}}(\varphi)$ the moisture storage function, p_{s} the saturated water vapour pressure,

$$D_{\varphi} = D_{\mathrm{w}}\frac{\mathrm{d}\rho_{\mathrm{v}}}{\mathrm{d}\varphi} \qquad (4.138)$$

the liquid water transport coefficient, D_{w} the capillary water transport coefficient, H the enthalpy density, T the temperature, λ the thermal conductivity, and L_{v} the latent heat of vaporization of water.

Among the currently employed convection models of simultaneous heat and moisture transport, the computational model by Milly (1982) developed on the basis of Philip and de Vries (1957) model can be considered as suitable for application to mature concrete. The model assumes knowledge of the water retention curve

$$u_{\mathrm{l}} = u_{\mathrm{l}}(\psi(t), T), \qquad (4.139)$$

where u_{l} is the liquid moisture content by mass, ψ the pressure head, and the basic transport equations can be written in the form

$$(1 - A)\frac{\partial u_{\mathrm{l}}}{\partial t} + B\frac{\partial \psi}{\partial t} + C\frac{\partial T}{\partial t}$$

$$= \frac{\rho_{\mathrm{l}}}{\rho_{\mathrm{s}}}\left\{\mathrm{div}\,[(K_{\mathrm{l}} + K_{\mathrm{v}})\,\mathrm{grad}\,\psi + D_{\mathrm{Tv}}\,\mathrm{grad}\,T] + \frac{\partial K_{\mathrm{l}}}{\partial z}\right\} \qquad (4.140)$$

$$(h_{\mathrm{l}} - Ah_{\mathrm{v}})\frac{\partial u_{\mathrm{l}}}{\partial t} + Bh_{\mathrm{v}}\frac{\partial \psi}{\partial t} + (c_{\mathrm{s}} + c_{\mathrm{l}}u_{\mathrm{l}} + c_{\mathrm{v}}u_{\mathrm{v}} + h_{\mathrm{v}}C)\frac{\partial T}{\partial t}$$

$$= \frac{1}{\rho_{\mathrm{s}}}\,\mathrm{div}\,[\lambda\,\mathrm{grad}\,T - h_{\mathrm{l}}\rho_{\mathrm{l}}K_{\mathrm{l}}\,\mathrm{grad}\,\psi$$

$$- h_{\mathrm{v}}\rho_{\mathrm{l}}(K_{\mathrm{v}}\,\mathrm{grad}\,\psi + D_{\mathrm{Tv}}\,\mathrm{grad}\,T)] - \frac{\rho_{\mathrm{l}}}{\rho_{\mathrm{s}}}\frac{\partial}{\partial z}(h_{\mathrm{l}}K_{\mathrm{l}}), \qquad (4.141)$$

where

$$-A = \frac{\partial u_{\mathrm{v}}}{\partial u_{\mathrm{l}}} = -\frac{1}{\rho_{\mathrm{l}}}\varphi p_{\mathrm{vs}}\frac{M}{RT} \qquad (4.142)$$

$$B = \frac{\partial u_{\mathrm{v}}}{\partial \psi} = p_{\mathrm{vs}}\frac{a}{\rho_{\mathrm{s}}}\frac{M}{RT}e^{\frac{\psi_{g}M}{RT}}\frac{gM}{RT} = \varphi p_{\mathrm{vs}}\frac{a}{\rho_{\mathrm{s}}}g\left(\frac{M}{RT}\right)^{2} \qquad (4.143)$$

$$C = \frac{\partial u_v}{\partial T} = \frac{1}{\rho_s} \frac{M}{RT} a p_{vs} \varphi \left(-\frac{\psi g M}{RT^2} + \frac{1}{p_{vs}} \frac{dp_{vs}}{dT} - \frac{1}{T} \right), \tag{4.144}$$

T is the temperature, ρ_s the partial density of the solid matrix, ρ_l the density of water, K_l, K_v the hydraulic conductivities of liquid water and water vapour, respectively, D_{Tv} the thermodiffusion coefficient of water vapour, h_l, h_v the specific enthalpies of liquid water and water vapour, respectively, the gaseous moisture content by mass, $u_v = (\Pi - w)\rho_v/\rho_s$, is expressed as

$$u_v = p_{vs} \frac{M}{RT} \left(\frac{\Pi}{\rho_s} - \frac{u_l}{\rho_l} \right) \cdot e^{\frac{\psi g M}{RT}}, \tag{4.145}$$

p_{vs} is the saturated water vapour pressure, R the universal gas constant, M the molar mass of water vapour, c_s, c_l, c_v the specific heat capacities of solid matrix, liquid water and water vapour, respectively, λ the thermal conductivity.

Among the more complex models of transport phenomena in porous materials, the hygro-thermo-mechanical model of Lewis and Schrefler (1998) has the potential to describe properly all transport processes in concrete. The general transport equations of the model can be expressed as follows:

$$\frac{1-n}{\rho^s} \frac{d^s \rho^s}{dt} - \frac{d^s n}{dt} + (1-n) \operatorname{div} \vec{v}^s = 0 \tag{4.146}$$

$$\frac{n S_w}{K_w} \frac{d^s p^w}{dt} + S_w \operatorname{div} \vec{v}^s - \beta_{sw} \frac{d^s T}{dt} + n \frac{d^s S_w}{dt}$$

$$+ \frac{1}{\rho^w} \operatorname{div} \left\{ \rho^w \frac{k k^{rw}}{\mu^w} \left[-\operatorname{grad} p^w \right. \right.$$

$$\left. \left. + \rho^w (\vec{g} - \vec{a}^s - \vec{a}^{ws}) \right] \right\} = -\frac{\dot{m}}{\rho^w} \tag{4.147}$$

$$S_g \operatorname{div} \vec{v}^s - \beta_s (1-n) S_g \frac{d^s T}{dt} - n \frac{d^s S_w}{dt}$$

$$+ \frac{n S_g}{\rho^g} \frac{d^s}{dt} \left[\frac{1}{\Theta R} (p^{ga} M_a + p^{gw} M_w) \right]$$

$$+ \frac{1}{\rho^g} \operatorname{div} \left\{ \rho^g \frac{k k^{rg}}{\mu^g} \left[-\operatorname{grad} p^g \right. \right.$$

$$\left. \left. + \rho^g (\vec{g} - \vec{a}^s - \vec{a}^{gs}) \right] \right\} = \frac{\dot{m}}{\rho^g} \tag{4.148}$$

$$\eta^\pi \vec{v}^{\pi s} = \frac{\mathbf{k} k^{r\pi}}{\mu^\pi} \left[-\operatorname{grad} p^\pi + \rho^\pi (\vec{g} - \vec{a}^s - \vec{a}^{\pi s}) \right] \tag{4.149}$$

$$-\rho \vec{a}^s - n S_w \rho^w (\vec{a}^{ws} + \vec{v}^{ws} \cdot \operatorname{grad} \vec{v}^w)$$

$$- n S_g \rho^g (\vec{a}^{gs} + \vec{v}^{gs} \cdot \operatorname{grad} \vec{v}^g) + \operatorname{div} \sigma + \rho \vec{g} = 0 \tag{4.150}$$

$$(\rho C_p)_{\text{eff}} \frac{\partial T}{\partial t} + (\rho_w C_p^w \vec{v}^w + \rho_g C_p^g \vec{v}^g) \cdot \operatorname{grad} T$$

$$- \operatorname{div} (\chi_{\text{eff}} \operatorname{grad} T) = -\dot{m} \Delta H_{\text{vap}}, \tag{4.151}$$

where n is the porosity, ρ^s the the partial density of the solid phase, $\vec{v}^s$ the mass-averaged solid velocity, S_w the degree of water saturation, K_w the bulk modulus of water, p^w the pressure of liquid phase, β_{sw} the volume thermal expansion coefficient of the liquid phase, T the temperature, $\mathbf{k}$ the permeability tensor of the multiphase medium, k^{rw} the relative permeability of the liquid phase, μ^w the dynamic viscosity, $\vec{g}$ the gravity acceleration, ρ^w the phase-averaged liquid density, $\vec{a}^s$ the acceleration of the solid phase, $\vec{a}^{ws}$ the relative acceleration of the liquid phase, $\dot{m}$ the mass production term, S_g the degree of gas saturation, β_s the volume thermal expansion coefficient of the solid phase, ρ^g the phase- averaged gas density, Θ the absolute temperature, R the universal gas constant, p^{ga} the partial pressure of dry air, p^{gw} the partial pressure of the water vapour, M_a the molar mass of air, M_w the molar mass of water vapour, k^{rg} the relative permeability of the gaseous phase, μ^g the dynamic viscosity of the gaseous phase, $\vec{a}^{gs}$ the relative acceleration of the gaseous phase, the superscript π stands for the particular fluid phase, η^π is the volume fraction of the phase π, ρ the total density of the system, $\vec{v}_w$ the velocity of the liquid phase, $\vec{v}_g$ the velocity of the gaseous phase, $\vec{v}_{ws}$ the relative velocity of the liquid phase, $\vec{v}_{gs}$ the relative velocity of the gaseous phase, σ the total stress tensor, the subscript eff means the effective value of a quantity, χ is the thermal conductivity, C_p the specific heat capacity of the system at constant pressure, C_p^w the specific heat capacity of the liquid phase at constant pressure, C_p^g the specific heat capacity of the gaseous phase at constant pressure, and ΔH_{vap} the latent heat of vaporization of water.

As it follows both from the analysis of models developed specifically for describing transport processes in concrete given in Section 4.1, and from the above analysis of general models for description of transport processes in porous materials, there are numerous computational models available that are capable of simulating either coupled heat and moisture transport or just salt transport in mature concrete, but models including coupled heat, moisture and salt transport are still rare. It is not such a big problem to write the necessary transport equations and constitutive relations, but more difficult is the practical realization, which has to include material parameters depending not only on temperature and moisture content but also on the amount of the particular salts.

Among the currently employed computational models of coupled heat, air, moisture and salt transport in porous materials, the computer code DELPHIN 4.1 (Grunewald, 2000), based on the theoretical analysis of heat, air, moisture and salt transport in Grunewald (1997), belongs to the most advanced. The model is capable of simulating not only transport processes in mature concrete but also in hydrating concrete, because it includes all necessary source/sink terms.

We will now provide a brief overview of the basic transport equations and constitutive relations of the Grunewald (1997) model, exactly in the form they are employed in DELPHIN 4.1. A complete theoretical analysis including the variations in defining the storage and transport material parameters can be found in Section 3.4.1.

The basic transport equations of moisture, air, salt and internal energy are formulated as follows:

$$s_{11}\frac{\partial w_l}{\partial t} + s_{12}\frac{\partial p_g}{\partial t} + s_{13}\frac{\partial c_s}{\partial t} + s_{14}\frac{\partial T}{\partial t}$$

$$= -\operatorname{div}\left[(\rho_w\vec{v}^{ml} - \vec{j}_{dif}^{ms} - \vec{j}_{disp}^{ms})w_l + (\rho_v\vec{v}^{mg} + \vec{j}_{dif}^{mv})w_g\right] \qquad (4.152)$$

$$s_{21}\frac{\partial w_l}{\partial t} + s_{22}\frac{\partial p_g}{\partial t} + s_{23}\frac{\partial c_s}{\partial t} + s_{24}\frac{\partial T}{\partial t}$$

$$= -\operatorname{div}\left[(\rho_a\vec{v}^{mg} - \vec{j}_{dif}^{mv})w_g\right] \qquad (4.153)$$

$$s_{31}\frac{\partial w_l}{\partial t} + s_{32}\frac{\partial p_g}{\partial t} + s_{33}\frac{\partial c_s}{\partial t} + s_{34}\frac{\partial T}{\partial t}$$

$$= -\operatorname{div}\left[(\rho_s\vec{v}^{ml} + \vec{j}_{dif}^{ms} + \vec{j}_{disp}^{ms})w_l\right] \qquad (4.154)$$

$$s_{41}\frac{\partial w_l}{\partial t} + s_{42}\frac{\partial p_g}{\partial t} + s_{43}\frac{\partial c_s}{\partial t} + s_{44}\frac{\partial T}{\partial t}$$

$$= -\operatorname{div}\left[\rho_l u_l\vec{v}^{ml}w_l + (\rho_v u_v + \rho_a u_a)\vec{v}^{mg}w_g\right] - \operatorname{div}\vec{j}_{dif}^{Q}$$

$$- \operatorname{div}\left[(h_s - h_w)(\vec{j}_{dif}^{ms} + \vec{j}_{disp}^{ms})w_l\right] - \operatorname{div}\left[(h_v - h_a)\vec{j}_{dif}^{mv}w_g\right], \qquad (4.155)$$

where w_1, w_g are the moisture contents by volume in the liquid and gaseous phase, respectively (more accurately, the volume fractions of the whole liquid and gaseous phase in the system), ρ_w the density of water, ρ_v the partial density of water vapour, $\vec{v}^{ml}$, $\vec{v}^{mg}$ the barycentric velocities of water and water vapour, respectively, $\vec{j}_{dif}^{ms}$ the diffusion flux of salts, $\vec{j}_{disp}^{ms}$ the dispersion flux of salts, $\vec{j}_{dif}^{mv}$ the diffusion flux of water, ρ_a the partial density of air, c_s the salt concentration by mass,

$$c_s = \frac{\rho_s}{\rho_l} = \frac{\rho_s}{\rho_w + \rho_s}, \tag{4.156}$$

ρ_s the partial density of dissolved salts, T temperature, p_g the pressure of the gaseous phase, u with the particular indices are specific internal energies, $\vec{j}_{dif}^{Q}$ the heat flux, $(h_s - h_w)$ the difference in partial specific enthalpies of salts and water, and $(h_v - h_a)$ the difference in partial specific enthalpies of water vapour and air.

The basic diffusion and convection fluxes are expressed in the following way:

$$\vec{j}_{con}^{ml} = \rho_l \vec{v}^{ml} = -K_l(\operatorname{grad} p_c + \rho_l \vec{g}) - K_l \operatorname{grad} p_g \tag{4.157}$$

$$\vec{j}_{con}^{mv} = \rho_g \vec{v}^{mg} = -K_g(\operatorname{grad} p_g + \rho_g \vec{g}) \tag{4.158}$$

$$\vec{j}_{dif}^{mv} = -\rho_g D^v \operatorname{grad} c_v - D_p^v \frac{M_g}{RT} \operatorname{grad} p_g \tag{4.159}$$

$$\vec{j}_{dif}^{ms} = -\rho_l D^s \operatorname{grad} c_s - \rho_l \kappa_l D_p^s(\operatorname{grad} p_c + \operatorname{grad} p_g) \tag{4.160}$$

$$\vec{j}_{disp}^{ms} = -\frac{|\vec{v}^{ml}|}{w_l} D_d^s \operatorname{grad} c_s \tag{4.161}$$

$$\vec{j}_{dif}^{Q} = -\lambda \operatorname{grad} T, \tag{4.162}$$

where ρ_l, ρ_g are the densities of the liquid and gaseous phases, respectively, K_l, K_g the convective coefficients in the liquid and gaseous phase, respectively (with the physical dimension of s), p_c the capillary pressure,

$$p_c = p_l - p_g, \tag{4.163}$$

p_l, p_g the pressures in the liquid and gaseous phases, respectively, D^s the diffusion coefficient of salts in the liquid, D^v the diffusion coefficient of water vapour, λ the thermal conductivity, c_v the concentration of water vapour,

$$c_v = \frac{\rho_v}{\rho_g} = \frac{\rho_v}{\rho_v + \rho_a}, \tag{4.164}$$

D_p^s, D_p^v the barodiffusion coefficients of salts and water vapour, respectively, D_d^s the dispersion coefficient of salt in water, κ_l a correction coefficient (in Pa^{-1}),

$$\kappa_l = \frac{\partial w_l}{\partial p_l}. \tag{4.165}$$

The storage parameters $s_{11} - s_{44}$ have the following form:

$$s_{11} = \rho_w - \rho_v + w_g \frac{M_v}{RT} p_{sat} \frac{\partial \varphi}{\partial w_l} \tag{4.166}$$

$$s_{12} = 0 \tag{4.167}$$

$$s_{13} = -\rho_v \frac{\partial w_p}{\partial c_s} + w_l \frac{\partial \rho_w}{\partial c_s} + w_g p_{sat} \frac{M_v}{RT} \frac{\partial \varphi}{\partial c_s} \tag{4.168}$$

$$s_{14} = w_l \frac{\partial \rho_w}{\partial T} + w_g \frac{M_v}{RT} \left(\varphi \frac{dp_{sat}}{dT} - \frac{p_v}{T} + p_{sat} \frac{\partial \varphi}{\partial T} \right) - \rho_v \frac{\partial w_p}{\partial T} \tag{4.169}$$

$$s_{21} = -\rho_a - w_g \frac{M_a}{RT} p_{sat} \frac{\partial \varphi}{\partial w_l} \tag{4.170}$$

$$s_{22} = w_g \frac{M_a}{RT} \tag{4.171}$$

$$s_{23} = -\rho_a \frac{\partial w_p}{\partial c_s} - w_g \frac{M_a}{RT} p_{sat} \frac{\partial \varphi}{\partial c_s} \tag{4.172}$$

$$s_{24} = -w_g \frac{M_a}{RT} \left(\varphi \frac{dp_{sat}}{dT} + \frac{p_a}{T} + p_{sat} \frac{\partial \varphi}{\partial T} \right) - \rho_a \frac{\partial w_p}{\partial T} \tag{4.173}$$

$$s_{31} = c_s \rho_l \tag{4.174}$$

$$s_{32} = 0 \tag{4.175}$$

$$s_{33} = w_l \rho_l + \rho_p \frac{\partial w_p}{\partial c_s} + w_l c_s \frac{\partial \rho_l}{\partial c_s} \tag{4.176}$$

$$s_{34} = \rho_p \frac{\partial w_p}{\partial T} + w_l c_s \frac{\partial \rho_l}{\partial T} \tag{4.177}$$

$$s_{41} = \rho_l u_l + u_v w_g \frac{M_v}{RT} p_{sat} \frac{\partial \varphi}{\partial w_l}$$

$$- u_a w_g \frac{M_a}{RT} p_{sat} \frac{\partial \varphi}{\partial w_l} - \rho_v u_v - \rho_a u_a \tag{4.178}$$

$$s_{42} = u_a w_g \frac{M_a}{RT} \tag{4.179}$$

$$s_{43} = (\rho_p u_p - \rho_v u_v - \rho_a u_a) \frac{\partial w_p}{\partial c_s} + \rho_l w_l \frac{\partial u_l}{\partial c_s} + u_l w_l \frac{\partial \rho_l}{\partial c_s}$$

$$+ \left(u_v w_g \frac{M_v}{RT} - u_a w_g \frac{M_a}{RT} \right) p_{sat} \frac{\partial \varphi}{\partial c_s} \tag{4.180}$$

$$s_{44} = \rho_m \frac{\partial u_m}{\partial T} + \rho_p w_p \frac{\partial u_p}{\partial T} + \rho_l w_l \frac{\partial u_l}{\partial T} + u_l w_l \frac{\partial \rho_l}{\partial T}$$

$$+ u_v w_g \frac{M_v}{RT} \left(\varphi \frac{dp_{sat}}{dT} - \frac{p_v}{T} + p_{sat} \frac{\partial \varphi}{\partial T} \right)$$

$$- u_a w_g \frac{M_a}{RT} \left(\varphi \frac{dp_{sat}}{dT} + \frac{p_a}{T} + p_{sat} \frac{\partial \varphi}{\partial T} \right)$$

$$+ \left(\rho_v \frac{\partial u_v}{\partial T} + \rho_a \frac{\partial u_a}{\partial T} \right) w_g + (\rho_p u_p - \rho_v u_v - \rho_a u_a) \frac{\partial w_p}{\partial T}, \tag{4.181}$$

where φ is the relative humidity, p_{sat} the saturated water vapour pressure, ρ_p the partial density of crystalline salt, w_p the volume fraction of the crystalline salt, R the universal gas constant, M_v, M_a and M_g the molar masses of water vapour, air and the gaseous mixture, respectively, and the index m is for the porous matrix.

4.3 RECOMMENDED MODEL OF TRANSPORT PROCESSES IN CONCRETE IN USUAL SERVICE CONDITIONS

As it has been analysed in the previous section, the model by Grunewald (1997) in the computer form DELPHIN 4.1 (Grunewald, 2000) can be considered as very suitable for modelling transport processes in concrete in usual service conditions, because it includes not only heat and moisture transport but also salt and air transport, and all these phenomena are coupled. Therefore, we can recommend it for the most practical applications.

We will introduce basic features of DELPHIN 4.1, as described in the user manual (Grunewald, 2000). DELPHIN 4.1 is available free of charge for academic use, and it can be downloaded from ftp://abks07.arch.tu-dresden.de. DELPHIN 4.1 is subject to continuous improvements. Complete up-to-date information can be obtained directly from the code developer Dr.-Ing. John Grunewald:

tel. +49 0351 463 6186, fax +49 0351 463 2627,

e-mail: grunewald@abkfs2.arch.tu-dresden.de.

The numerical simulation program DELPHIN 4.1 has been developed at the Institute of Building Climatology of the Technical University of Dresden in order to support the investigation of the coupled heat, air, salt and moisture transport in porous building materials. The simulation of the thermal and hygric behaviour of constructive building details is possible for 1D, 2D and axial-symmetric 3D problems. The program can be used to simulate transient mass and energy transport processes for arbitrary standard and natural climatic boundary conditions (temperature, relative humidity, driving rain, wind speed, wind direction, short and long wave radiation).

The modelling of transient transport processes leads to a system of nonlinear partial differential equations. DELPHIN 4.1 solves the resulting system of coupled partial differential equations by numerical integration over time. A large number of variables (moisture contents, air pressures, salt concentrations, temperatures, diffusive and convective fluxes of liquid water, water vapour, air, salt, heat and enthalpy, etc.), which characterize the hygrothermal state of building constructions, can be obtained as a function of space and time. A particular advantage of the numerical simulation program is the possibility of investigation of variants concerning different constructions, different materials and different climatic loads. Construction details of buildings and building materials can be optimised using the numerical simulation, and the reliability of constructions for different given indoor and outdoor climates can be judged.

The program is written in C/C++ (ANSI-standard) in order to provide as much portability as possible. It includes a solver package for integration of the resulting system of ordinary differential equations: the CVODE-package (Cohen and Hindmarsh, 1996).

DELPHIN 4.1 has been installed and is available for the majority of the known platforms and operating systems (IBM RS 6000 SP2: AIX, DEC Alpha: DecOS, HP-Workstations: HPUX, Sun-Workstations: Solaris, Apple Macintosh: System 7-8, PC: Windows 95/98/NT/2000). The memory allocation demand depends on the problem size. Computers with about 64 MB RAM and 200 MByte free disk space are recommended.

In order to facilitate the installation of DELPHIN 4.1, a setup routine is available for Windows operating systems. The DELPHIN 4.1 setup copies the executable program files and the configuration files into the user selected installation directory. At the same time, several example projects, climatic data files, material data files, the DELPHIN 4.1 documentation and the description of the theoretical background are installed by the setup routine.

The DELPHIN 4.1 setup automatically creates program symbols, sets key variables in your Windows registry and adds an entry in the Windows start menu. To remove the DELPHIN 4.1 installation, the automatic software deinstallation tool in the Windows system control should be used. Do not delete the DELPHIN 4.1 installation directory manually. After a successful installation, the delphin.exe file can be started for calculation of the example projects. The path can be relative or absolute. Projects are arbitrarily located in the directory system, so the user may create a project directory outside the installation directory.

Normally, no path settings are required in the users local autoexec.bat file. For working with projects outside the installation directory, an environment variable PROJPATH can be set. The DELPHIN 4.1 program takes the PROJPATH variable into account only if the argument ProjectPath is a relative path, otherwise the given (absolute) path is used. If a batch file out of the installation directory is used for delphin.exe calls, an environment variable DELPHIN41PATH indicating the installation directory must be set in the autoexec.bat file. For working with climatic data files out of the installation directory, an environment variable CLIMAPATH can be set. The DELPHIN 4.1 program takes the CLIMAPATH into account only if the path specifications of climatic data files are relative:

set DELPHIN41PATH=InstallDir, if a batch file out of the InstallDir is used,
set PROJPATH=YourProjectPath for projects out of the InstallDir (optional),
set CLIMAPATH=YourClimaPath for climatic data files out of the InstallDir (optional).

Computer code DELPHIN 4.1 consists of several programs for data input, initialisation and output, numerical calculation and data representation, as shown in the program structure in Fig. 4.1.

The user interface program PreTool has been developed for PC, Windows 95/98/NT/2000, to facilitate the input data handling. Using PreTool, the input files can be edited and the user has access to the databases of materials, climatic data and construction details. In order to enable changes in the input files, the information about projects is stored in a user friendly, human readable format.

After finishing the input work, the simulation can be started by PreTool. A manual start is necessary in cases when PreTool is not available (Workstations,

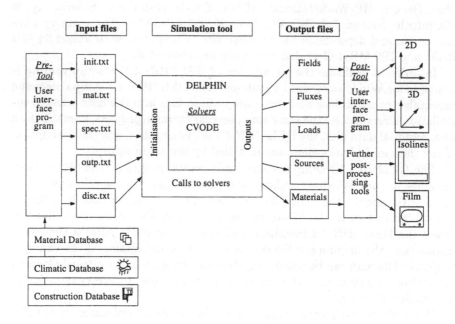

Figure 4.1 Program structure of DELPHIN 4.1

Supercomputers). The time required for simulations depends strongly on several factors: the system size (number of unknowns), the climatic boundary conditions and the material properties. The simulation time can vary between a couple of seconds (steady state, 1D problems) and several months (complex 2D/3D calculations using the entire set of the natural climatic data).

The DELPHIN 4.1 simulation tool reads the input files, carries out the numerical simulation and writes ASCII output files. The outputs are distinguished into *fields*, *fluxes*, *loads* and *sources* according to their assignments to the geometry of the construction (elements, element sides, boundaries). The output directories are created by DELPHIN 4.1 after starting a simulation.

For the graphical representation of simulation results, a postprocessing tool can be used. The usage of external programs is also possible since the output files are easily readable. DELPHIN 4.1 allows the definition of different output file formats, so the user may select the most convenient form.

The simulation problems in DELPHIN 4.1 are handled as projects. The project structure of DELPHIN 4.1 is directory organized, i.e. projects may contain subprojects, subprojects may contain subsubprojects and the projects may be arbitrarily located in the directory system. In many cases, variants of projects with relatively small changes in the input files have to be investigated. So, common input files can be used.

Working with projects is possible in DELPHIN 4.1 either in the traditional way (DELPHIN 4.1 is called with the project path, which is used as input/output directory) or by a project file (the project file containing the input file location specifications is given as calling argument). In the latter case, the location of the project file is used as output directory.

The specification of the properties and conditions of simulation problems very often requires handling of physical quantities. In order to facilitate this work, DELPHIN 4.1 allows free definition and use of units. All defined units are contained in a unit table file units.txt (installed by the setup program), which is used for initialisation of the available physical units at the simulation start. This table is a user editable ASCII file. Additional units may be inserted, if required, which are valid for input and output for the next DELPHIN 4.1 call.

The input file init.txt contains keywords, flags and parameters for the selection of the balance equations, the physical models, the limits of the state variables and the solver. With respect to the solver, specific settings, optional inputs and outputs have to be given. We will introduce and explain the basic variables of the init.txt file in order to demonstrate the capabilities of DELPHIN 4.1.

The variables *MstEqu*, *AirEqu*, *SalEqu* and *EneEqu* are switches for the balance equations for moisture mass, air mass, salt mass and internal energy. Zero means that the balance equation is switched off, one that the balance equation is switched on.

The set of unknown state variables depends on the selected balance equations. The moisture balance selects the moisture content as the unknown variable. The gas pressure, the total salt concentration in the liquid phase and the temperature are the unknown variables, when respectively the air mass balance, the salt mass balance and the energy balance are switched on. Expressing the storage of the considered quantities by the temporal change of the unknown state variables, the balance equation system can be rewritten as a system of parabolic differential equations. After a semidiscretization of the spatial coordinates, the number of unknowns in the resulting system of ordinary differential equations equals the number of elements times the number of selected balance equations.

TranspModel is the variable that selects the appropriate driving force of the liquid water transport: the liquid water pressure gradient or the water content gradient. The choice of the transport model also influences the requirements on material properties. Depending on the transport model selected, the liquid water conductivity or the liquid water diffusivity is primarily used.

StorageModel is the variable that selects the storage model of the system of parabolic differential equations. The choice of storage model (for instance only main storage effects or full storage effects) influences the calculation speed and the accuracy of the results. Therefore, some experimentation with the storage models is recommended.

FieldCoeffAvg selects the spatial averaging procedure of the transport coefficients in the homogeneous material. *BoundCoeffAvg* selects the spatial averaging procedure of the transport coefficients at the material layer boundary interfaces. The variables used are the averaged coefficient (to be multiplied with a difference, not a gradient), the left- and right-hand side coefficients, and the left- and right-hand side spatial steps. An additional resistance can be assigned to the boundary between material layers. This can simulate for instance the interface permeability (see Section 3.2.5).

The user may impose general limit values to the water content, the gas pressure, the salt concentration and the temperature. DELPHIN 4.1 checks the initial conditions during the initialisation and the solver calls the subroutines using these limits. If the ranges are exceeded during the initialisation, the calculation is stopped by an error message. During the simulation, an exceed causes warnings, which are printed to the file err-log.txt. If there are warnings, DELPHIN 4.1 gives an indication by printing of an E at the left hand side of the last line of the screen header output. The size of the file err-log.txt is limited to 100 warnings.

In order to establish a connection between the moisture storage functions, i.e. the sorption isotherm and the water retention characteristic, the Kelvin equation is implemented (see Section 1.6). In DELPHIN 4.1, the standard potential of the water retention characteristic is the pressure head. Alternative potentials for the water retention characteristic may be chosen: the logarithmic pressure head, the logarithmic capillary pressure and the capillary pressure.

The mass densities of the liquid phase, liquid water, water vapour, dry air and the gaseous phase are necessary to formulate physical storage and transport quantities. The gas constant of moist air and modified gas constants for water vapour and moist air are required for the moisture transport coefficient expressions. The implementation of the specific enthalpy/internal energy functions of the liquid phase, solid material, precipitated salt, saturated liquid phase, liquid water, water vapour and dry air is done using linear relations with respect to temperature.

The transport coefficients and the storage coefficients of the system of parabolic partial differential equations for coupled heat, air, salt and moisture transport are dependent on the material properties and the variables of state. The material properties are described by material functions (analytical functions with parameters and/or interpolation and approximation of measured data). The available material specifications are contained in the material database files in ASCII format. The material specifications can be selected manually or with the preprocessing tool PreTool, copied from the material database files and stored in the input file mat.txt.

A material input file mat.txt contains a list of material specifications related to a project or a group of projects. A project related storage of (possibly changed) function types and parameters offers advantages. Simulations with a variation of the material parameters can be carried out, and the influence of the variations on the results can be investigated.

The required material properties for a simulation depend on the equation coupling (see description of the input parameters in init.txt) and the moisture handling (liquid water conductivity or diffusivity, implicit or explicit phase dividing function). DELPHIN 4.1 searches for the required material properties according to the current settings. PreTool performs graphical and numerical outputs of the material functions.

A material specification consists of a header part and a part for specification of material properties. The defined material properties of DELPHIN 4.1 are shown in Table 4.1. These properties relate to the current state of modelling of coupled heat, air, salt and moisture transport. A further completion of the list of defined material properties depends on the progress of research in this field. DELPHIN 4.1 searches for the defined names of the material properties in the material file and initialises accordingly. Undefined material properties are ignored.

The set of required material properties for a simulation problem depends on the balance equations to be solved and on the handling of the moisture transport. With the moisture mass balance switched on, DELPHIN 4.1 offers different possibilities to describe moisture transport properties.

MSTCOND: The primary moisture transport properties are the (hygroscopic and capillary) moisture conductivities *MstCond* and the phase dividing function (see Section 3.4.1 for the definition relations and exact derivation of these parameters) *PhaseDiv*. The moisture conductivities describe the total moisture flux due to a water vapour pressure gradient or a capillary pressure gradient. Then, the total moisture flux is separated into a liquid water and a water vapour flux using the phase dividing function. If a measured phase dividing function and moisture conductivities are available, the *MSTCOND* case is preferable.

WATCOND or WATDIFF: The primary moisture transport properties are the liquid water conductivity *WatCond* or the liquid water diffusivity *WatDiff* and the water vapour diffusivity *VapDiff*. A phase separation of the total moisture flux is given by these properties. An implicitly given phase dividing function can be calculated and is not required on input.

POREMOD: a pore model is used to calculate the shape of the moisture conductivities from the moisture retention characteristic. The saturated capillary moisture conductivity, the dry hygroscopic moisture conductivity and the phase dividing function have to be given for that case.

The material property *Storage* denotes the storage capacity of the material in respect to heat and moisture. The storage coefficients of the system of parabolic partial differential equations are dependent on the material parameters. The parameters required for calculation of the storage coefficients are the density, the specific heat capacity and the open porosity of the material.

The material property *MstRetC* denotes the moisture retention characteristic of the material expressed by the sorption isotherm in the hygroscopic moisture range and the water retention characteristic in the capillary moisture range.

Table 4.1 Defined material properties of DELPHIN 4.1

MatProp	Explanation
Storage	Storage properties: specific heat capacity, density and porosity
MstRetC	Moisture retention characteristic: sorption isotherm in the hygroscopic moisture range and water retention characteristic in the capillary moisture range
PhaseDiv	Phase dividing function
Tortuosity	Relative tortuosity function: using a pore model in order to calculate the shape of the moisture conductivities from the moisture retention characteristic
MstCond	Moisture conductivities: hygroscopic and capillary moisture conductivity used in the hygroscopic and capillary moisture range, respectively
WatCond	Liquid water conductivity, assigned to the capillary pressure gradient as driving force of the liquid water flux
WatDiff	Liquid water diffusivity, assigned to the water content gradient as driving force of the liquid water flux
VapDiff	Water vapour diffusivity, assigned to the water vapour pressure gradient as driving force of the vapour diffusion
AirPerm	Air permeability
SalRetC	Salt retention characteristic
SalDifN	Normal diffusion coefficient for salt
SalDifP	Pressure diffusion coefficient for salt
SalDisp	Dispersion coefficient for salt
TheCond	Thermal conductivity

The material property *PhaseDiv* denotes the phase dividing function of the material. The phase dividing function determines the percentages of the liquid water flux and the water vapour diffusion as a function of the moisture content for isothermal and isobaric conditions. The phase dividing function is required for the calculation of the transport coefficients for liquid water and water vapour from the moisture conductivities.

The relative moisture conductivities can be calculated by integration of the moisture retention functions using pore models and tortuosity functions. The available pore models are denoted in respect to the authors (Burdine, 1953; Mualem, 1976; Neiss, 1982). Except for the closed form equation, the integration of the pore model function is carried out numerically. The resulting function depends on the moisture content (the upper border of the integral) and is expressed by a cubic spline interpolation.

The material property *MstCond* denotes the moisture conductivities of the material expressed by the hygroscopic moisture conductivity in the hygroscopic moisture range and the capillary moisture conductivity in the capillary moisture range. The moisture flux (liquid water + water vapour flux) caused by the water vapour pressure gradient is described by the hygroscopic moisture conductivity, and the moisture flux caused by the capillary pressure gradient is described by the capillary moisture conductivity (see Section 3.4.1 for exact definitions). The hygroscopic moisture conductivity is implemented in the hygroscopic moisture range and the capillary moisture conductivity in the capillary moisture range. Both functions are defined over the entire moisture range. Therefore, they call each other if the limit between them, the maximum hygroscopic moisture content, is exceeded.

The material property *WatCond* denotes the liquid water conductivity of the material. The liquid water conductivity is defined over the entire moisture range and describes the liquid water flux caused by the liquid water pressure gradient.

The material property *WatDiff* denotes the liquid water diffusivity of the material over the entire moisture range. It describes the liquid water flux caused by the moisture content gradient.

The material property *VapDiff* denotes the water vapour diffusivities of the material defined over the entire moisture range. The total water vapour diffusion includes the normal diffusion (due to the water vapour mass concentration gradient) and the pressure diffusion (due to the gas pressure gradient). For isobaric conditions, there is no pressure diffusion and the mass concentration gradient can be replaced by the partial pressure gradient.

The material property *AirPerm* is used to select different function types for the description of the air permeability of materials. The advective flux of the gaseous phase is described by the product of the air permeability and gas pressure gradient.

The material property *SalRetC* is used to select different function types of the salt retention characteristic (ion binding isotherm) of materials, which is defined as a relation between the dissolved salt concentration and the precip-

Table 4.2 Condition types of DELPHIN 4.1

Condition	Explanation
Climatic conditions	Internal and external time dependent loads: can be assigned to the boundary conditions
Boundary conditions	Boundary conditions: can be assigned to the boundary domain of the problem
Initial conditions	Initial conditions: can be assigned to the field domain of the problem
Field conditions	Internal sources and sinks: can be assigned to the field domain of the problem
Contact conditions	Additional transport resistances: can be assigned to the material boundary interfaces

itated salt. The relative salt concentration is calculated using the solubility.

The material property *SalDifN* denotes the normal diffusion coefficient for salt, which is defined over the entire moisture range. It describes the salt diffusion in the liquid phase due to a salt mass concentration gradient. The normal salt diffusion is regarded as diffusion in each other of both components of the liquid phase: (pure) liquid water and dissolved salt.

The material property *SalDifP* denotes the pressure diffusion coefficient for salt defined over the entire moisture range. It describes the salt diffusion due to a pressure gradient in the liquid phase. The pressure salt diffusion is regarded as diffusion in each other of both components of the liquid phase: (pure) liquid water and dissolved salt.

The material property *SalDisp* denotes the dispersion coefficient for salt defined over the entire moisture range. It describes the salt dispersion due to a salt mass concentration gradient in the liquid phase. Salt dispersion is regarded as a diffusive effect caused by the macroscopic averaging of the microscopic inhomogeneous velocity field of the liquid phase.

The material property *TheCond* is used to select different function types of the thermal conductivity of materials. The reduced heat flux of the porous material is described by the product of the thermal conductivity and the temperature gradient.

The different conditions for a simulation problem are described by specification data records contained in the input file spec.txt. The defined condition types are listed in Table 4.2. The available climatic, boundary, initial, field and contact conditions are predefined in the program header file del-enums.h, which is automatically installed by the DELPHIN 4.1 setup.

While the specified climatic conditions can be assigned to the boundary conditions in the input file spec.txt, the boundary, initial, field and contact conditions can be assigned to the spatial domain of the problem in the input file disc.txt.

DELPHIN 4.1 searches in the input file spec.txt for the *SearchString* of the condition types during the initialisation. According to the available condition types, lists of specification data records follow the *SearchString*. If *SearchString* cannot be found in the input file, then DELPHIN 4.1 assumes, that there is no specification data record list of the respective condition type. Only those specification data records that are used for assignments have to be contained in the input file spec.txt. Other unused specification data records can be contained there. Sometimes, it is advantageous to use a common specification file containing the specification data records for a group of similar problems. A specification record describes a property of the problem in a certain spatial and/or temporal range. The specification data records are distinguished by record names given by the user. The required parameters of the specification data records depend on the condition type and the selected keywords.

The climatic conditions describe time dependent loads corresponding to boundary conditions or field conditions. The evolution of the climatic conditions with time can be expressed by periodic analytical functions or by numerical values initialised from climatic data files. The defined climatic conditions are given in Table 4.3.

DELPHIN 4.1 initialises the climatic data records, if *ClimaConditions* is found in the input file spec.txt. Otherwise, it is assumed that there is no need for specification of climatic conditions for the problem.

The climatic data records are assigned to the boundary conditions. Constant climatic conditions can be given directly as boundary parameters. For problems with constant climatic conditions only, there is no need for specification of climatic data records. For climatic data records using climatic data files, the file names have to be specified. The path can be an absolute or a relative path. DELPHIN 4.1 takes the environment variable CLIMAPATH into account (which can be set in the file autoexec.bat), if the given path is a relative path. A climatic data record contains the defined name of the climatic condition, its record name, the keyword for *Course* and a set of parameters. The time evolution of climatic conditions is controlled by keywords. The current values of the climatic conditions are calculated from periodic functions or from data sets.

The numerical data initialised from climatic data files may be a set of measured data on a reference object, a test reference year (TRY) or another arbitrary climatic data. The climatic data files should contain a header block with information about the climatic condition, its input/output unit, the given time format and its usage. The usage can be cyclic or noncyclic. In case of cyclic use, the simulation time may exceed the given time period of the data file. For noncyclic use of climatic data, the given time period cannot be exceeded. The comment block is optional and ignored by DELPHIN 4.1 during the initialisation.

The boundary conditions describe the surface properties of problems exposed to the external climatic loads. The defined boundary conditions and

Table 4.3 Predefined climatic conditions

Clim. Cond.	SI Unit	Explanation
WatCont	m^3/m^3	Contact water content for the boundary condition water contact
PreHead	m	Pressure head of environmental water (e.g. ground water with a positive water pressure)
NorRain	kg/m^2s	Rain flow density normal to the surface area of the wall
VerRain	kg/m^2s	Vertical rain flow density, normal to the ground
WindDir	Rad	Wind direction (0° denotes north, 90° east, 180° south and 270° west)
WindVel	m/s	Wind speed
RelHum	—	Environmental relative humidity
Temper	K	Environmental temperature
SaltConc	—	Relative salt concentration
HeatFlux	W/m^2	Imposed heat flux
VapPres	Pa	Environmental water vapour pressure
ShWRad	W/m^2	Short wave radiation normal to the wall surface
DirRad	W/m^2	Direct short wave radiation on a horizontal surface
DifRad	W/m^2	Diffuse short wave radiation on a horizontal surface
LoWRad	W/m^2	Long wave radiation normal to the wall surface
LWEmis	W/m^2	Long wave radiation emitted by a horizontal surface
SkyRad	W/m^2	Long wave radiation reflected on a horizontal surface by the clouds and particles in the atmosphere
GasFlux	kg/m^2s	Imposed gaseous mass flux (dry air + water vapour)
GasPres	Pa	Environmental air pressure (dry air + water vapour)
WatProd	kg/m^3s	Water mass production density
AirProd	kg/m^3s	Air mass production density
EneProd	W/m^3	Energy production density
XWat Flux	kg/m^2s	Internal water flux through cracks or gaps in x-direction
YWat Flux	kg/m^2s	Internal water flux through cracks or gaps in y-direction

Table 4.4 Predefined boundary conditions

Balance eq.	Boundary condition	KeyWord	Boundary fluxes
Moisture mass	Water contact	WatContc	WatFlux kg/m^{2}s
			WatEnth J/m^{2}s
			SalFlux kg/m^{2}s
	Driving rain	Rain	WatFlux kg/m^{2}s
			WatEnth J/m^{2}s
	Water vapour diffusion	VapDiff	VapDiff kg/m^{2}s
			VapEnth J/m^{2}s
Air mass	Air convection	AirPress	GasFlux kg/m^{2}s
	and diffusion		AirFlux kg/m^{2}s
			AirEnth J/m^{2}s
			VapFlux kg/m^{2}s
			VapEnth J/m^{2}s
Salt mass	Salt diffusion	SaltDiff	SalDiff kg/m^{2}s
Energy	Heat conduction	HeatCond	HeatFlux W/m^2
	Short wave radiation	ShWRad	ShWRad W/m^2
	Long wave radiation	LoWRad	LoWRad W/m^2

their related balance equations and boundary fluxes are shown in Table 4.4. The boundary data records are assigned to the spatial boundary domain of the problem. The assignment of boundary data records causes the calculation of boundary fluxes. If no boundary condition is assigned, then the surface is assumed to be impermeably tight in respect to the concerning quantity. DELPHIN 4.1 takes into account only those boundary conditions that are relevant for the used balance equations. For example, if a driving rain flux and heat conduction are imposed as boundary conditions and only the moisture mass balance is switched on, the driving rain flux is taken into account but the heat conduction is ignored.

The boundary data records are applied to the problem for their specified time limits. If the simulation time exceeds the specified time limit of a record, then the application of the respective boundary record is removed and the related fluxes are set back to zero. This offers the possibility of changing the application of boundary data records during the simulation time by assignment of several records with different time limits to the same spatial range of the problem. In case of overlapping time limits, the preceding record (in specification order) is taken into account as the currently assigned record. The models used for the calculation of the boundary fluxes and the required parameters depend on the kind of the boundary condition selected. DELPHIN 4.1 recognizes the boundary condition kinds shown in Table 4.5.

The boundary condition *Water contact* imposes a water flux on the boundary domain of the problem. In case of a switched on salt mass balance, an

Table 4.5 Kinds of boundary conditions

Kind	Explanation
ZERO	The boundary flux is zero. This is identical to the case where no boundary condition is assigned
SURFVAL	A surface value is prescribed, for example a moisture content or a temperature value (may be dependent on time)
IMPFLUX	An imposed flux is prescribed, for example a rain flux density, short wave or long wave radiation (may be dependent on time). Absorption and emission coefficients are input parameters
EXCHANGE	The boundary flux is calculated using exchange coefficients. The exchange coefficients are input parameters
EXTENDED	An extended model is used for calculation of the boundary flux. Generally, this requires more parameters

advective salt flux is taken into account according to the salt concentration of the contact water. Possible applications of the boundary condition *Water contact* are: direct water contact of underground building constructions and the simulation of water adsorption/imbibition experiments. The water contact can have a positive or a negative water pressure. Ground water normally has a positive water pressure, but unsaturated soils have negative water pressures. Laboratory experiments can also be carried out with positive and negative boundary water pressures. The required climatic condition for *Water contact* is either an environmental water content or a pressure head of the water contact. Both climatic conditions are alternatively applicable. In case of consideration of the salt advection, the relative salt concentration of the environmental water is taken into account as climatic condition.

The boundary condition *Rain* assigns a driving rain to the boundary domain of the problem. *Rain* cannot be cumulated with the boundary condition *Water contact*, but both boundary conditions can replace each other during the simulation. If *Rain* and *Water contact* are assigned to the same spatial and temporal range, then DELPHIN 4.1 takes only *Water contact* into account. The rain flux density normal to the wall surface can be directly imposed with Kind= IMPFLUX or can be calculated from the vertical rain flux density with Kind= EXTENDED taking into account the wall direction and the climatic conditions *Wind direction* and *Wind speed*. The formula behind Kind= EXTENDED is limited to vertical walls. The wall directions are defined as

North=0°, East=90°, South=180° and West=270°. In both cases DELPHIN 4.1 compares the rain flux density normal to the wall surface with a maximum water flux that can be taken up by the wall surface. If the normal rain flux exceeds the maximum water flux, then it is limited to the maximum water flux. The maximum water flux is calculated depending on the surface water content. In the case of surface saturation, the maximum water flux is zero.

The boundary condition *Vapour diffusion* assigns a water vapour diffusion flux to the boundary domain of the problem. The climatic conditions required for *Vapour diffusion* are the temperature and relative humidity (alternatively the partial water vapour pressure). For the Kind= EXCHANGE, the water vapour diffusion flux is calculated with a constant exchange coefficient. The extended model with Kind= EXTENDED uses an exchange coefficient, that depends on the wind direction and wind speed as well as on the wall direction. All directions are taken into account as North=0°, East=90°, South=180° and West=270°.

The boundary condition *Air pressure* imposes a gas mass flux to the boundary domain of the problem. The gaseous phase is regarded as a mixture of dry air and water vapour. Therefore, the boundary condition *Air pressure* causes an advective air flux and an advective water vapour flux. The advective air flux is taken into account, if the air mass balance is switched on, and the advective water vapour flux is taken into account if the moisture mass balance is switched on. The climatic conditions required for the boundary condition *Air pressure* are the environmental temperature, the relative humidity (alternatively the partial water vapour pressure) and the gas pressure (environmental air pressure inclusive the partial water vapour pressure). In case of an imposed air flux (Kind= IMPFLUX), the gaseous mass flux should be given. Otherwise (in case of Kind= EXCHANGE), an exchange coefficient is required and the volumetric air flow is calculated from the gas pressure difference between the surface and the environment.

The boundary condition *Salt diffusion* imposes a salt diffusion flux to the boundary domain of the problem. Salt diffusion can only take place when the surface is contacted with salted water. The climatic conditions required for the *Salt diffusion* are the relative salt concentrations of the contacted water. For the exchange model, the salt diffusion flux is calculated with a constant exchange coefficient.

The boundary condition *Heat conduction* imposes a heat flux on the boundary domain of the problem. The climatic conditions required for the *Heat conduction* are the environmental temperature or the heat flux. With Kind= SURFVAL, a time-dependent surface temperature can be assigned as a surface value. No further parameter is required in this case. For the imposed flux case, a heat flux is applied. For the exchange model, the heat flux is calculated with a constant exchange coefficient. The extended model uses an exchange coefficient that depends on the wall direction and on the wind direction and wind speed. All directions are taken into account as North=0°, East=90°, South=180° and West=270°.

The boundary condition *Short wave radiation* incorporates the short wave radiation balance at the outside surface of walls. The short wave radiation flux normal to the wall surface can be directly applied (imposed flux) or it can be calculated from the direct and the diffuse radiation taking the wall direction and the wall inclination into account (extended model). The directions are defined as North=0°, East=90°, South=180° and West=270°, as well as Horizontal=0° and Vertical=90°, respectively. The parameters required for the boundary condition *Short wave radiation* are the absorption coefficient and the environmental albedo (the reflection coefficient of the ground surroundings for short wave radiation). Additionally, the geographical latitude is a necessary parameter for the extended short wave radiation model.

The boundary condition *Long wave radiation* describes the long wave radiation balance on the boundary domain of the problem. The long wave radiation flux normal to the wall surface can be directly imposed (Kind= IMPFLUX) or it can be calculated from the long wave emission and the sky radiation (Kind= EXTENDED). The required parameter is the emission coefficient for long wave radiation. The climatic data of the test reference years (TRYs) generally do already include an emission coefficient of 0.9 for long wave emission and sky radiation, and it must not be taken into account twice.

The initial conditions describe the initial state of the problem. They have to be set concerning the quantities moisture content, gas pressure, salt concentration and temperature. The relative humidity can alternatively be used instead of the moisture content. If the assigned balance equation of an initial condition is switched off, then the initial condition remains unchanged during simulation. DELPHIN 4.1 initialises the initial data records, if *InitConditions* is found in the input file spec.txt. Otherwise, the default initial conditions set in init.txt are used.

DELPHIN 4.1 supports the following initial condition kinds. A field of single values can be given as initial condition. There are two possibilities. The field can consist of one value or of more than one value. Constant initial condition: In case of only one value, DELPHIN 4.1 takes it into account as a constant initial value. When the respective initial data record is assigned to an element selection in disc.txt, then the elements are initialised with the given constant initial value. Initial spatial distribution: In case of more than one value, DELPHIN 4.1 takes them into account as a spatial distribution. When the respective initial data record is assigned to an element selection in disc.txt, then the elements are initialised with the given field of initial values. The number of initial values must match the number of selected elements in this case. Note that the order of assignment of values to the elements is determined by the selection order of the element range. For reuse of precalculated spatial distributions of state variables as initial condition, select the COMPRESSED output format and copy the values from the output file.

The field conditions describe the internal sources and sinks of the problem concerning moisture, air and energy. This feature of DELPHIN 4.1 can be utilized in modelling transport processes in hydrating concrete. If no field condition is assigned, the production densities of the respective quantities are assumed to be zero. The evolution in time of the sources or sinks are expressed by the internal production densities for moisture mass, air mass and energy. They are taken into account as climatic conditions. Additionally, sources and sinks may depend on the variables of state. In order to describe problem-specific sources and sinks, normally the introduction of further kinds of field conditions is necessary. DELPHIN 4.1 initialises the field data records if the keyword *FieldConditions* is found in the input file spec.txt. Otherwise, no sources and sinks are taken into account. DELPHIN 4.1 supports the field condition kinds as follows.

VOLPROD: A time-dependent production density is used. The evolution in time of the production density must be known for the whole time period of application. This means that the production density must not depend on the local state variables.

WATPENE: A source/sink for a specific problem. A rainfall water flux density is applied to a roof construction and a part of the rainfall water penetrates through gaps behind the insulation. The water flux in the outflow gap behind the insulation is taken into account as a water source and an energy sink.

WATFLUX: A source/sink for a specific problem. Cold water flows through cracks in the material and causes a cooling effect on the construction. The cooling effect is taken into account as an energy sink.

The field condition *Water source* imposes an internal water production density on the field domain of the problem. The required climatic conditions for *Water source* depend on the selected keyword for Kind.

The field condition *Air source* imposes an internal air production density on the field domain of the problem. The required climatic condition for *Air source* is the air production density.

The field condition *Energy source* imposes an internal energy production density on the field domain of the problem. The required climatic conditions for *Energy source* depend on the selected keyword for Kind.

The contact conditions describe the contact properties of materials. This means that an additional resistance can be assigned to the interface layer between different materials. For example, a small air gap between two materials may cause a high resistance for liquid water suction or a vapour barrier can be handled without discretization of the thin layer. If no contact condition is assigned, then an ideal contact between the materials in respect to the transported quantities is assumed. Table 4.6 shows the defined contact conditions, SI units and explanations. DELPHIN 4.1 initialises the contact data records, if the keyword *KontcConditions* is found in the input file spec.txt. Otherwise, it is assumed, that there are no specifications of contact conditions for the problem.

The input file outp.txt is used to specify the time management and the desired output of the problem. While the time management can be described by a small number of variables, the output is specified by output data records containing the output quantities, their number of outputs during the simu-

Table 4.6 Predefined contact conditions

Contact condition	SI Unit	Explanation
WatCond	m/s	Additional resistance in respect to liquid water transport
VapDiff	s/m	Additional resistance in respect to water vapour diffusion
AirPerm	m/s	Additional resistance in respect to advective air transport
SalDiff	m/s	Additional resistance in respect to salt diffusion
SalDisp	m^2/kg	Additional resistance in respect to salt dispersion
TheCond	$m^2 K/W$	Additional resistance in respect to heat conduction

lation, keywords for spatial and temporal integration and output file format specifications. The output data records can be assigned in disc.txt to the spatial domain of the problem. While an output data record describes how the respective quantity has to be output, the assignment of the record in disc.txt determines the spatial range, where the output has to be performed.

For calculation of the total simulation time and its division into time intervals, the time management variables are required. The output time intervals are set globally, while the number of outputs per interval are set individually for each output quantity. This allows the use of different output time scales in respect to each quantity. This is particularly advantageous when some variables are requested at small time intervals on a very fine output scale, while other variables can be put out on a rough scale.

CalcStartDate: The start date of the calculation is required for the climatic data files obtained from Test Reference Years (TRY) or other time-dependent climatic data. A simulation may start at any arbitrary time point in respect to the climatic time scale.

TimeScale: For output, a simulation time scale or a climatic time scale can be selected. The simulation time scale starts at the initial time equal to zero, while the climatic time scale starts at the specified calculation start date. Screen output gives both the climatic time (in DATE format) and the simulation time (in FULL format). The output to data files uses the specified time scale and the specified time formats of the output quantities.

IntervLength: For specification of the total simulation time, a field of time intervals is used. The number of time intervals is arbitrary. The total simulation time is calculated from the sum of all time intervals. The number of outputs per time interval must be set for each output specification record.

IntgTimeStep: The temporal integration procedure for time dependent output quantities requires an internal integration time step. The variables are linearly interpolated and their integrals are calculated using a trapezoidal rule with the specified time step. The temporal integrals and mean values are individually calculated for each output time step of the respective quantities.

The settings *ScrMainOutp* and *DatMainOutp* in the file outp.txt are the main output switches controlling the output streams to screen and to data file. If the main switch for screen or for data file output is set to zero, then all output to this channel is suppressed. Otherwise, the channel is free for output. The number of values per line of the screen output is controlled by the variable *ScrLineLength*.

The fields of moisture content, gas pressure, salt concentration and temperature are derived from the solution of the balance equation system. Further output quantities can be calculated from these primary variables: other fields (vapour pressure, relative humidity, overhygroscopic moisture content, etc.), the fluxes of the considered extensive quantities and other time-dependent variables. Therefore, classes of output variables (control classes) are distinguished according to their assignment to the spatial domain of the problem. The following four available control classes are defined:

Fields: Field quantities, assigned to the centroids of elements;
Fluxes: Flux quantities, assigned to element sides;
Loads: Climatic loads, assigned to boundary element sides;
Sources: Source quantities, assigned to the centroids of elements.

For each control class, there exists a list of predefined control variables. Control variables are output quantities that can be selected. The predefined control variables of DELPHIN 4.1 are listed in Tables 4.7–4.10.

The output to be performed by DELPHIN 4.1 is specified by output data records in the input file outp.txt. The program searches for the keywords *FieldsOutput*, *FluxesOutput*, *LoadsOutput* and *SourcesOutput*. DELPHIN 4.1 expects a list of controls with output data records if the respective control class keyword was found. If a control class keyword cannot be found, then DELPHIN 4.1 assumes that no output of the respective control class is desired. The format of output data records is as follows. The description of a control output starts with the keyword *ConClass=*, which can be *Field=*, *Flux=*, *Load=* or *Source=*. Then the name of the control *ControlName* and an output specification record follow. After specification, the data records can be assigned to the spatial domain of the problem in the input file disc.txt. The variables of the output data records to be set are explained in Table 4.11.

Normally, the current values of time-dependent output variables are requested for output. But it may be the case that current values are disadvantageous or yield inaccurate results in respect to the requested task. For example, if a simulation under transient climatic conditions starts on 1 January at 00:00 and the time scale for output of the boundary heat flux is set to once a day, the result will be totally different when compared to the same simulation, but started on 1 January at 12:00. The reason is the considerable daily variation of the heat flux. In the first case, the current heat flux at

Table 4.7 Field controls for output

Field	Description	SI Unit
WatCont	Volumetric water content	m^3/m^3
WatPres	Liquid water pressure	Pa
PreHead	Pressure head	m
pFValue	Logarithmic pressure head	logcm
VapPres	Water vapour pressure	Pa
SatPres	Water vapour saturation pressure	Pa
RelHum	Relative humidity in the gaseous phase	—
OvhWatC	Overhygroscopic volumetric water content, calculated as water content minus hygroscopic moisture content	m^3/m^3
WatMass	Water mass content	kg
OvhWatM	Overhygroscopic water mass content	kg
GasPres	Pressure of the gaseous phase	Pa
AirPres	Air pressure of the gaseous phase, calculated as gas pressure minus vapour pressure	Pa
DifPres	Difference of the gas pressure to 100000 Pa	Pa
SalConc	Relative salt concentration in the liquid phase	—
SalMConc	Salt mass in the liquid phase related to solid mass	kg/kg
SalPrec	Precipitated salt content	m^3/m^3
Temper	Temperature, obtained from the solution vector	K
EntCont	Enthalpy content, calculated as sum of the enthalpy contents of the solid, liquid and gaseous phase	J

midnight is obtained, and in the latter case, the current heat flux at mid-day. Alternatively, the output of the heat flux on a finer time scale and the postcalculation of a mean value would unnecessarily increase the size of the data files, take more effort and yield a more or less inaccurate mean value. To avoid these difficulties and to reduce the required work for postprocessing, DELPHIN 4.1 offers the possibility of temporal integration of time-dependent variables. The temporal integration is carried out by linear interpolation in time and value-time-area calculation using the internal time step set by the input variable *IntgTimeStep*.

Many variables are spatial distributions defined over a geometrical range. Beside single values, spatial integrals or weighed sums of the variables may be of interest. For example, this can be the total moisture content of a material layer or of the entire construction (a selection of elements) or a heat flux through a surface area (a selection of element sides). For such cases,

Table 4.8 Flux controls for output

Flux	Description	SI Unit
WatFlux	Advective mass flux of the liquid phase	kg/m^2s
WatEnth	Liquid water enthalpy flux	W/m^2
VapFlux	Advective water vapour mass flux	kg/m^2s
VapDiff	Water vapour mass diffusion flux	kg/m^2s
VapEnth	Water vapour enthalpy flux	W/m^2
GasFlux	Advective flux of the gaseous phase	kg/m^2s
AirFlux	Advective dry air mass flux	kg/m^2s
AirEnth	Air enthalpy flux	W/m^2
SalFlux	Advective salt flux of the liquid phase	kg/m^2s
SalDiff	Salt mass diffusion	kg/m^2s
HeatFlux	Reduced heat flux	W/m^2
ShWRad	Boundary short wave radiation	W/m^2
LoWRad	Boundary long wave radiation	W/m^2

Table 4.9 Load controls for output

Load	Description	SI Unit
GasPres	Gas pressure	Pa
Temper	Temperature	K
VapPres	vapour pressure	Pa
AirPres	Air pressure	Pa
RelHum	Relative humidity	—
VerRain	Vertical rain	kg/m^2s
NorRain	Normal rain	kg/m^2s
WindDir	Wind direction	Rad
WindVel	Wind velocity	m/s
LoWRad	Long wave radiation	W/m^2
ShWRad	Short wave radiation	W/m^2
DirRad	Direct solar radiation	W/m^2
DifRad	Diffuse radiation	W/m^2
LWEmis	Long wave emission	W/m^2
SkyRad	Sky radiation	W/m^2

Table 4.10 Source controls for output

Source	Description	SI Unit
WatProd	Water mass production density	kg/m³s
AirProd	Air mass production density	kg/m³s
EneProd	Internal energy production density	W/m³

Table 4.11 Variables of output data records

Variable	Explanation
Record	A user given name in order to distinguish the data records of a control
NumOutp	Number of outputs per time interval.
ScrOutp	Output to screen (0=no output, 1=output)
DatOutp	Output to data file (0=no output, 1=output)
IO Unit	Output unit (must be translatable into the defined SI unit of the control)
TimeMode	Temporal integration method
TimeFormat	Time output format
IntgMode	Spatial integration method
DatFormat	Data output format

DELPHIN 4.1 offers the possibility of spatial integration.

The output quantities can be written to output files using different formats for data output. This is performed by DELPHIN 4.1 in order to enable easy data handling and postprocessing. The output format depends on the selected geometrical range, the spatial integration method, the data type (SCALAR, VECTOR), and the selection of a keyword for data output format. DELPHIN 4.1 sets the presentation type (2D: single variable depending on time, 3D: one-dimensional field depending on time and one spatial coordinate, 4D: two-dimensional field depending on time and two spatial coordinates) in respect to the selected geometrical range and to the spatial integration method.

The numerical simulation of mass and energy transport processes with DELPHIN 4.1 requires the description of the geometry of the problem and its spatial discretization. The input file disc.txt specifies the discretization and the assignment of properties and conditions to the spatial domain of the simulation problem.

DELPHIN 4.1 allows the simulation of 1D, 2D and axial symmetric 3D problems. The variables *Xsteps*, *Ysteps*, *Xgeometry* and *Ygeometry* of the block *Discretization* describe a rectangular grid of volume elements used for discretization of the construction. The *Dimension* of the problem is determined by DELPHIN 4.1 from the number of values of the variables *Xsteps* and *Ysteps*. If the number of values given for *Xsteps* exceeds 1 and one value is given for *Ysteps*, then the problem is taken into account as a one-dimensional problem with flux in the x-direction. More than one value for both variables, *Xsteps* and *Ysteps*, describe a two-dimensional problem. The variable *Zsteps* has always only one value. For problems with a plane geometry, it denotes the size of the construction in the z-direction, in the axial symmetric cases, it denotes the angle of the cylindrical section. Integrated output quantities are calculated taking *Zsteps* into account, but the choice of *Zsteps* has no influence on the calculation of the state variables of the problem.

It is recommended that discretization for moisture transport calculations be set up with 1 mm steps at the boundaries and at the material interface layers. In most cases, the steps can be increased in the homogeneous material. The exception are suction experiments with direct water contact, where an equidistant discretization should be preferred.

The numerical error in time integration of the system of ordinary differential equations is controlled by the solver, but this does not apply for the numerical error caused by discretization. The user is responsible for a proper discretization of the problem. To get an idea of the discretization error, the calculation should be redone with a finer mesh until the results stabilize.

The spatial discretization of a construction results in a network of volume elements occupying a rectangular spatial domain. The calculation domain can be arbitrarily shaped, but it has to be generated entirely by rectangular elements. The calculation domain is generated by the numbering of the volume elements. All elements with a number greater than 0 belong to the calculation domain, the elements getting the element number 0 generate void space. In this way, an arbitrarily shaped two-dimensional domain can be obtained which may consist of several nonconnected subdomains. The numbering of elements starts from 1 and must be continuously increasing. Doubled or missing element numbers are not allowed. The order of element numbering is arbitrary but it influences the calculation speed. The highest calculation speed can be achieved by minimizing the maximum difference in element numbers of neighbouring volume elements.

The properties and conditions of the problem are assigned to the construction in the input file disc.txt. There is no need for assignment of all specified data records to the construction, but all assigned records must have been specified. The assignments refer to spatial field domains (elements) or to spatial boundary domains (element sides). While the elements have been numbered during the initialisation of the input file disc.txt as described above, the element sides are internally numbered by DELPHIN 4.1. So, each property or condition can be assigned to an arbitrary range of the calculation domain. During the initialization of disc.txt, DELPHIN 4.1 searches for the following keywords: *BoundCondAss* (optional), *MaterialAss* (Materials must be assigned), *KontCondAss* (optional), *InitCondAss* (optional), *FieldCondAss* (optional), *FieldsOutpAss* (optional), *FluxesOutpAss* (optional), *LoadsOutpAss* (optional), *SourcesOutpAss* (optional). The assignment of properties, conditions and output specifications to the construction is optional, except the material assignments must be found. If the respective keyword is found, DELPHIN 4.1 reads the following assignment list.

The handling of multiple assignments of records of the same specification type to identical or overlapping spatial ranges depends on the specification type. For materials, initial conditions and contact conditions, a consecutive assignment replaces a previous assignment. For the specification types having a time limit, i.e. boundary conditions, field conditions and contact conditions, the record assignments are changed during the simulation according to their time limits. The user should take care with overlapping time limits. The preceding record (in specification order) is set as currently assigned, if the consecutive record overlaps its time limit.

4.4 RECOMMENDED MODEL FOR CONCRETE IN SEVERE CONDITIONS

Modelling transport processes in concrete in severe conditions often means a description of concrete behaviour after its high temperature exposure. Among the models of transport processes in concrete exposed to high temperatures, we can recommend the model by Gawin et al. (1999). This model was already introduced in detail in Section 4.1. Therefore, we will not repeat the transport equations and constitutive relations here, and will confine ourselves only to the main features of the model and the reasons why we recommend it for common applications.

The model is based on the general model of heat, moisture, air and momentum transport by Lewis and Schrefler (1998). Concerning moisture transport, it belongs to the class of convection models, and transport of both liquid and gaseous phase is expressed by the permeability. Diffusion is assumed only between the particular components of the gaseous phase and for the bound water. The solid density and the porosity are assumed to be variable and to depend on the hydration/dehydration processes. Dehydration in high temperature conditions is modelled using the concept of Bažant and Thonguthai (1978), (1979), which was described in detail in Sections 3.4.3 and 4.1. The model does not take into account salt transport but for modelling transport processes in concrete at high temperatures, salt transport certainly does not belong to the most important transport mechanisms.

Generally, models of transport processes in concrete at high temperatures are not so frequently employed as models developed for description of concrete in usual service conditions. Therefore, only few models are available, as it has been analysed in Section 3.4.3. If we exclude the simplest models, which are suitable for the calculations of very specialized cases only, such as the models by Lykov (1956) or by Černý and Venzmer (1991), from our considerations, there remains in practice only one sophisticated model, which was employed for calculation of transport processes in concrete at high temperatures for many years, namely the model by Bažant and Thonguthai (1978), (1979).

The model of Gawin et al. (1999) is a logical extension of this model. It retains the strongest part of the Bažant and Thonguthai (1978), (1979) model, which is the description of dehydration processes, and introduces a more sophisticated way of modelling moisture transport according to the widely recognized model by Lewis and Schrefler (1998). Therefore, the model by Gawin et al. (1999) should be recommended for common applications to transport processes in concrete at high temperatures. The computational version of the model was developed by the authors of the Gawin et al. (1999) paper for their own calculations but as far as we know, the computer code is not freely available.

4.5 REQUIREMENTS FOR THE MATERIAL PARAMETERS EMPLOYED IN THE MODELS

Computational models for simulation of transport processes in concrete make it possible to calculate fields of basic <u>state variables</u> (sometimes called as <u>field variables</u>, because they appear as unknowns in the field equations), which are necessary for an evaluation of the hygric, thermal, chemical or mechanical state of concrete. This evaluation can result in a prediction of further performance of the particular concrete structure and/or in an assessment of its service life, which are the primary aims of most such calculations. In this section, we will overview the main physical quantities appearing in the most often used models of transport processes in concrete. We will not repeat their definitions here; the reader can find all necessary information in Chapter 3 and the foregoing sections of this chapter.

The number of state variables must always be the same as the number of solved transport equations and complementary conditions, in order to achieve solvability of the system of equations. For a numerical solution of a particular problem, the complementary conditions have to be included in the system of transport equations, so that the number of primary state variables is reduced in such a way that one state variable corresponds to one transport equation.

The basic state variables calculated by the computational models may differ in accordance with the type of the particular physical model employed. The widest diversity appears in the choice of the primary state variable corresponding to the moisture balance equation, which may be partial moisture density, moisture content by mass, moisture content by volume, relative humidity, partial pressure of water vapour, partial density of water vapour, water pressure, capillary pressure, pressure head. As it has been analysed before (see Chapter 3), this diversity is due to the parallel development of different classes of models of moisture transport by different schools established in the particular research fields. However, recalculation of any such variable to any other one is a relatively easy task, because it requires only knowledge of moisture storage functions for the particular material.

On the other hand, the primary variable corresponding to the heat or internal energy transport equation is always temperature, and molar or mass concentration is always the state variable corresponding to the transport equation of mass of a chemical compound. This is due to the fact that practically all models logically have a common background in thermodynamics and physical chemistry.

In the transport equations, there appear numerous parameters that are specific for every particular material. Therefore, these are generally termed material parameters. Two basic types of such parameters can be recognized, namely the transport parameters and the storage parameters. Transport parameters are the phenomenological coefficients, which appear in the constitutive equations as proportionality factors between the generalized thermodynamic forces and fluxes.

The transport equations are generally formulated for the particular mass and energy densities, and in their original form they do not contain the basic state variables, which appear (in the form of their gradients) in the constitutive equations. So it is necessary to express the differentials of the mass and energy densities using the differentials of the basic state variables. For this reason, we define storage parameters, which make it possible to calculate the partial derivatives of the particular mass or energy densities with respect to the basic state variables.

As with the state variables, the widest diversity in defining material parameters appears in describing moisture transport, because this definition must be in accordance with the chosen basic thermodynamic forces. Most of the models use two hygric transport parameters, one for water vapour transport, which usually also includes Knudsen diffusion and surface diffusion, and the second for liquid water transport. For water vapour transport, diffusion coefficient of water vapour (in m^2/s), water vapour diffusion permeability (in s), water vapour resistance factor (dimensionless), or water vapour permeability (in m^2) are mostly employed. For liquid water transport, we have either water (moisture) diffusivity (in m^2/s) or hydraulic conductivity (in m/s) or water permeability (in m^2).

Every model of moisture transport has to include minimally one hygric storage parameter, depending on the class of moisture transport model, but definition of two parameters, one for hygroscopic range, and second for the overhygroscopic range, is physically more correct. In the hygroscopic range, the sorption isotherm is the main parameter used in most models, while in the overhygroscopic range, the water retention curve is employed. As has been mentioned before, these two moisture storage parameters can be unified in a single relation using the Kelvin equation.

The basic transport and storage parameters corresponding to heat transport are quite unified, as with the state variable. Thermal conductivity appears as the main heat transport parameter, and specific heat capacity is employed as the main heat storage parameter.

In describing salt transport, the main transport parameter employed is the salt diffusion coefficient. The main salt storage parameter is the ion binding isotherm (in the unsaturated range) or the salt retention characteristic (in the whole concentration range).

It should be noted that we have surveyed only the basic transport and storage parameters that are employed in most models describing moisture, heat and salt transport in concrete, and therefore are commonly measured. Some of the other parameters that appear in a thorough thermodynamic analysis of coupled heat, moisture and salt transport can be found for instance in the Grunewald (1997), (2000) model. However, these additional parameters, which are sometimes called second order parameters because they express second order effects, are not employed in most models and not commonly measured.

The material parameters mentioned before are indispensable input parameters of the computational models similar to the initial and boundary conditions. Without exact knowledge of them it is not possible to obtain reasonable values of state variables from any numerical simulation. As the numerical schemes of the currently used models are mostly advanced and numerically stable, the accuracy of material parameters is the most important factor determining the accuracy of obtained results. Therefore, the material parameters should always be determined as functions of all state variables used in a particular model. In the next chapter, we will introduce basic methods for determining transport and storage parameters appearing in the computational models of transport processes in concrete. As the experimental methods for determining these parameters must always employ measurements of state variables, we will give a basic overview of methods for measuring the state variables of heat, moisture and chemical compounds transport as well.

References

Baroghel-Bouny, V., Mainguy, M., Lassabatere, T., Coussy, O., 1999, Characterization and identification of equilibrium and transfer moisture properties for ordinary and high-performance cementitious materials. *Cement and Concrete Research*, **29**, pp. 1225–1238.

Bažant, Z.P., Najjar, L.J., 1972, Nonlinear water diffusion in nonsaturated concrete. *Materials and Structures*, **5**, pp. 3–20.

Bažant, Z.P., Thonguthai, W., 1978, Pore pressure and drying of concrete at high temperatures. *Journal of Engineering Mechanics Division, Proceedings of the American Society of Civil Engineers*, **104**, pp. 1059–1079.

Bažant, Z.P., Thonguthai, W., 1979, Pore pressure in heated concrete walls, *Magazine of Concrete Research*, **31**, pp. 67–76.

Bear, J., 1988, *Dynamics of Fluids in Porous Media.* (New York: Dover Publications).

Bear, J., Bachmat, Y., 1990, *Introduction to Modeling of Transport Phenomena in Porous Media*, Vol. 4. (Dordrecht: Kluwer).

Boddy, A., Bentz, E., Thomas, M.D.A., Hooton, R.D., 1999, An overview and sensitivity study of a multimechanistic chloride transport model. *Cement and Concrete Research*, **29**, pp. 827–837.

Buenfeld, N.R., Shurafa-Daoudi, M.T., McLoughlin, I.M., 1997, Chloride transport due to the wick action in concrete. In: *Proceedings of the RILEM International Workshop Chloride Penetration into Concrete*, St-Remy-les-Chevreuse, edited by Nilsson, L.O., Ollivier, J.P. (St-Remy-les-Chevreuse: RILEM), pp. 315–324.

Burdine, N. T., 1953. Relative permeability calculations from pore-size distribution data. *Trans. AIME*, **198**, pp. 71–78.

Cady, P.D., Weyers, R.E., 1992, Predicting service life of concrete bridge decks subject to reinforcement corrosion. In: *Corrosion Forms and Control for Infrastructure*, ASTM STP 1137. (Philadelphia: American Society for Testing and Materials), pp. 328–338.

Carlson, R.W., 1937, Drying shrinkage of large concrete members. *American Concrete Institute Journal*, **33**, pp. 327–335.

Carslaw, H.S., Jaeger, J.C., 1959, *Conduction of Heat in Solids.* (Oxford: Clarendon Press).

Cohen, S. D., Hindmarsh, A. C., 1996. CVODE, A Stiff/Nonstiff ODE Solver in C. *Computers in Physics*, **10**, pp. 138–143.

Černý, R., Venzmer, H., 1991, Ein modifiziertes Lykow-Modell zum Wärme- und Feuchtetransports in kapillarporösen Stoffen. *Gesundheitsingenieur*, **112**, pp. 41–44.

Costa, A., Appleton, J., 1999, Chloride penetration into concrete in marine environment – Part II: Prediction of long term chloride penetration. *Materials and Structures*, **32**, pp. 354–359.

Funahashi, M., 1990, Predicting corrosion-free service life of a concrete structure in a chloride environment. *ACI Materials Journal*, **87**, pp. 581–587.

Gawin, D., Majorana, C.F., Schrefler, B.A., 1999, Numerical analysis of hygro-thermal behaviour and damage of concrete at high temperature. *Mechanics of Cohesive Frictional Materials*, **4**, pp. 37–74.

Grunewald, J., 1997, *Diffusiver und konvektiver Stoff- und Energietransport in kapillarporösen Baustoffen*, PhD Thesis. (Dresden: TU Dresden).

Grunewald, J., 2000, *DELPHIN 4.1 – Documentation, Theoretical Fundamentals*. (Dresden: TU Dresden).

Hall, C., 1989, Water sorptivity of mortars and concretes: a review. *Magazine of Concrete Research*, **41**, pp. 51–61.

Hancox, N.L., 1967, A note on the form of the rate of drying curve for cement paste and its use in analyzing the drying behavior of this material. *RILEM Bulletin*, **36**, pp. 197–201.

Häupl, P., Grunewald, J., Fechner, H., Stopp, H., 1997, Coupled heat, air and moisture transfer in building structures. *Int. J. Heat Mass Transfer*, **40**, pp. 1633–1642.

Hilsdorf, H.K., 1967, A method to estimate the water content of concrete shields. *Nuclear Engineering and Design*, **6**, pp. 251–263.

Ho, D.W., Lewis, R.K., 1987, The water sorptivity of concretes: the influence of constituents under continuous curing. *Durability of Building Materials*, **4**, pp. 241–252.

Hrstka, O., Černý, R., Rovnaníková, 1999, Hygrothermal Stress Induced Problems in Large Scale Sprayed Concrete Structures. In: *Specialist Techniques and Materials for Concrete Construction*, R.K. Dhir, N.A. Henderson (eds.). (London: Thomas Telford), pp. 103–109.

Johannesson, B.F., 1999, Diffusion of a mixture of cations and anions dissolved in water. *Cement and Concrete Research*, **29**, pp. 1261–1270.

Kiessl, K., 1983, *Kapillarer und dampfförmiger Feuchtetransport in mehrschichtigen Bauteilen*, PhD Thesis. (Essen: Universität Essen).

Khan, A.A., Cook, W.D., Mitchell, D., 1998, Thermal properties and transient thermal analysis of structural members during hydration. *ACI Materials Journal*, **95**, pp. 293–302.

Krischer, O., 1963, *Die wissenschaftliche Grundlagen der Trocknungstechnik*, 2. Auflage. (Berlin: Springer Verlag).

Künzel, H.M., 1995, *Simultaneous Heat and Moisture Transport in Building Components*, PhD Thesis. (Stuttgart: IRB Verlag).

Lewis, R.W., Schrefler, B.A., 1998, *The Finite Element Method in the Static and Dynamic Deformation and Consolidation of Porous Media*, 2nd Edition. (Chichester: John Wiley and Sons).

Lockington, D., Parlange, J.Y., Dux, P., 1999, Sorptivity and the estimation of water penetration into unsaturated concrete. *Materials and Structures*, **32**, pp. 342–347.

Lykov, A.V., 1956, *Teplo i massoobmen v processach sushki.* (Moskva: Gosenergoizdat), (in Russian).

Majorana, C.E., Salomoni, V., Schrefler, B.A., 1998, Hygrothermal and mechanical model of concrete at high temperature. *Materials and Structures*, **31**, pp. 378–386.

Milly, P.C.D., 1982, Moisture and heat transport in hysteretic, inhomogeneous porous media: a matrix head-based formulation and a numerical model. *Water Resources Research*, **18**, pp. 489–498.

McCarter, W.J., Ezirim, H., Emerson, M., 1996, Properties of concrete in the cover zone: water penetration, sorptivity and ionic ingress. *Magazine of Concrete Research*, **48**, pp. 149–156.

Mualem, Y., 1976. A new model for predicting the hydraulic conductivity of unsaturated porous media. *Water Resources Research*, **12**, pp. 513–522.

Neiss, J., 1982. *Numerische Simulation des Wärme- und Feuchtetransports und der Eisbildung in Böden.* Fortschritt-Berichte der VDI-Zeitschriften, Reihe 3, Nr. 73, VDI-Verlag.

Parlange, M.B., Prasad, S.N., Parlange, J.Y., Romkens, M.J., 1992, Extension of the Heaslet-Alksne technique to arbitrary soil-water diffusivities. *Water Resources Research*, **28**, pp. 2793–2797.

Pickett, G., 1942, The effect of change in moisture content on the creep of concrete under a sustained load. *American Concrete Institute Journal*, **36**, pp. 333–355.

Philip, J.R., de Vries, D.A., 1957, Moisture movement in porous materials under temperature gradients. *Transactions of the American Geophysical Union*, **38**, pp. 222–232.

Tuuti, K., 1982, *Corrosion of Steel in Concrete.* (Stockholm, Swedish Cement and Concrete Research Institute).

Ulm, F.J., Rossi, P., Schaller, I., Chauvel, D., 1999, Durability scaling of cracking in HPC structures subject to hygromechanical gradients. *Journal of Structural Engineering*, **125**, pp. 693–702.

Weyers, R.E., 1998, Service life model for concrete structures in chloride laden environments. *ACI Materials Journal*, **95**, pp. 445–453.

Zemajtis, J., Weyers, R.E., Sprinkel, M.M., 1998, Corrosion protection service life of low-permeable concretes and low-permeable concrete with a corrosion inhibitor. In: *Transportation Research Report 1642*. (Washington: National Research Council), pp. 51–59.

## CHAPTER FIVE

Experimental Methods for Determination of Field Variables and Material Properties of Concrete

5.1 FIELD VARIABLES OF HEAT, MOISTURE AND CHEMICAL COMPOUNDS TRANSPORT

5.1.1 Moisture Content

Water in all its phases possesses many anomalous properties, which also affect the properties of a porous material. Therefore, there are various methods for determining moisture content in porous materials and various moisture meters. The measuring methods can be classified into two basic groups:

- Absolute methods (or direct methods), which determine the actual water content in the material after its removal from the specimen (by drying, extraction).

- Relative methods (or indirect methods), which determine the amount of water in the specimen on the basis of measuring another physical quantity (permittivity, electric conductivity, absorption of radiation energy, etc.), which has a clear relation to the amount of water in the material.

We will first survey the particular kinds of methods for measuring the moisture content in porous materials, in order to become familiar with the applied measuring principles (see, e.g., Fexa and Široký, 1983). Then we will look at the details.

1. Direct methods

   (a) Gravimetric method
   Water is removed from the specimen by drying in exactly defined conditions, and its amount is determined from the loss of mass after drying or by titration. Water contained in a material can also be removed by annealing in dry air flow and trapping in a tube filled

by a drying agent. This method is fundamental to many standard treatments for determining moisture content in materials and is commonly employed to calibrate other methods.

(b) Extraction method
Water is extracted from the material into a proper solvent where its amount is determined by titration.

(c) Azeotropic distillation
Water is distilled with an organic solvent into a calibrated receiver.

2. Indirect methods

(a) Resistance methods
Electric resistance of porous materials varies with the amount of water they contain, which can be utilized in determining moisture content. The method cannot be employed for materials containing a high amount of salts, because then the ionic conductivity would play a dominant role.

(b) Dielectric methods
The real (and in some applications also the imaginary) part of the complex permittivity is measured as a function of the amount of water.

(c) Radiometric methods
The methods are based on measuring the absorption of β or γ radiation or inhibition or scattering of neutrons.

(d) Nuclear magnetic resonance method
The absorption of high-frequency energy in a material exposed to the magnetic field is measured as a function of water content. The method can distinguish between the free and bound water.

(e) Absorption and reflection of infrared radiation
By measuring the reflection of infrared radiation, the surface water content can be determined.

(f) Chemical methods
These methods are based on the chemical reaction of water contained in the material with specific agents, such as CaC_2.

(g) Ultrasonic methods
These measure the ultrasound velocity, which varies with the water content.

We will now describe particular methods of measuring the moisture content of porous materials in more detail. The methods will be introduced in order of frequency of their application in practice.

5.1.1.1 Gravimetric Method

This method is the most frequently used in the practice. Its application requires a basic knowledge on the bond energy of water to the porous matrix. The basic classification of water in porous materials into free and bound water is not an exact expression of the physical-chemical reality. Strictly speaking, a capillary porous material cannot contain free water, because water is always bonded to the matrix by certain forces. These forces are most intense in the beginning of the sorption isotherm at low relative humidities, when a monomolecular layer is formed, and they gradually weaken with increasing moisture content in a material (see Section 1.6 for details).

The bond energy of water in a porous material can be defined as the work necessary for the conversion of a certain amount of water from the bound state in the porous material into the free (unbound) state. Leroy (1954) found that this specific energy e (in J/kg) is given by an analogous relation to the osmotic pressure,

$$e = \frac{RT}{M} \ln \frac{p_s}{p}, \tag{5.1}$$

where p_s is the saturated water vapour pressure at temperature T (which is given in kelvins), p the partial pressure of water vapour at temperature T, M the molar mass of water, and R the universal gas constant.

In considering this relation, it should be recalled (see, e.g., Moore, 1972), that the osmotic pressure is defined as the pressure induced by the penetration of the solvent into the solution, i.e. by the spontaneous dilution of the solution. It can be measured, for instance, by using a membrane between the solution and the solvent, which lets pass the solvent but not the dissolved substance. The increase in pressure in the solution can be determined, for instance, by using the equilibrium with the hydrostatic pressure. The osmotic pressure Π can also be calculated using the vapour pressures above the solvent and the solution,

$$\Pi = \frac{RT}{V_A} \ln \frac{p_A^*}{p_A}, \tag{5.2}$$

where V_A is the partial molar volume of the solvent A (in m^3/mol), p_A^* the partial vapour pressure above the pure substance A (the solvent), and p_A the partial vapour pressure above the solution of the substance B in the solvent A.

The specific energy e corresponding to the osmotic pressure Π can be obtained as

$$e = \frac{\Pi}{\rho_A} = \frac{RT}{\rho_A V_A} \ln \frac{p_A^*}{p_A} = \frac{RT}{M_A} \ln \frac{p_A^*}{p_A}, \tag{5.3}$$

where ρ_A is the density of the solvent, M_A the molar mass of the solvent.

In an analogy between the specific bond energy and the osmotic pressure, Leroy (1954) assumed that water is substance A in the above notation (i.e. the solvent), while substance B is the solid phase matrix. Then, p_A^* corresponds to p_s in equation (5.1), and p_A to p. In justification of this analogy, we will use the following argument. According to the definition relation (5.2), the osmotic pressure is the external pressure that would have to act on the solution, so that the vapour pressure of solvent A above this solution would increase to the value of the vapour pressure above pure solvent A. In our case the bond energy is the energy necessary to remove the water from the material. According to Leroy's analogy, it is the energy necessary to increase the partial pressure of water vapour to the value of the saturated pressure at the given temperature. "The solvent" is therefore water A in the material, and the "dissolved substance" the solid phase matrix B. In other words, water has to be removed from the material or, viewed the other way round, water penetrates into the "solution", i.e. into the mixture of water and the solid phase matrix.

Coming back to relation (5.1), we immediately see that for $p = 0$ we have $e \to \infty$, and therefore it is impossible to remove all water contained in the material. In further analysis, we will use Raoult's law, which for ideal solutions can be written in the form

$$x_A = \frac{p_A}{p_A^*}, \tag{5.4}$$

where x_A is the molar concentration of substance A,

$$x_A = \frac{n_A}{n_A + n_B}, \tag{5.5}$$

n_A, n_B are the numbers of moles of substances A and B. Using the symbols in equation (5.1), we can write

$$x_A = \frac{p}{p_s}. \tag{5.6}$$

Substituting equation (5.6) into (5.1), we get

$$e = \frac{RT}{M_A} \ln \frac{p_s}{p} = \frac{RT}{M_A} \ln \frac{1}{x_A} = -\frac{RT}{M_A} \ln x_A = -\frac{RT}{M_A} \ln(1 - x_B). \tag{5.7}$$

Using the Taylor expansion for the last term of equation (5.7),

$$-\ln(1 - x_B) = x_B + \frac{1}{2}x_B^2 + \ldots \sim x_B, \tag{5.8}$$

we finally obtain

$$e \sim \frac{RT}{M_A} x_B, \tag{5.9}$$

which means that in the first approximation of equation (5.7), the energy is directly proportional to the molar concentration of the porous matrix. In other words, with increasing water content in the porous material the bond energy decreases.

If the molar concentration of the porous matrix is not known (which is a very often case in the practice), it can be approximated by the relative mass ratio. We will define the dryness $w_{r,s}$ by the relation

$$w_{r,s} = \frac{m_d}{m_{w,b}}, \qquad (5.10)$$

where m_d is the mass of the dry material, $m_{w,b}$ the mass of bound water. For high water concentrations, i.e. $x_B \to 0$, and $n_A \gg n_B$, the molar concentration of the porous matrix can be expressed as

$$x_B = \frac{n_B}{n_A + n_B} \sim \frac{n_B}{n_A}, \qquad (5.11)$$

and therefore for the specific energy the following approximation relation can be written:

$$e = a \cdot w_{r,s} + b, \qquad (5.12)$$

where a, b are constants.

Equations (5.1) and (5.12) can be utilized to the determination of the total amount of water in the porous material in a gravimetric experiment. We use the following procedure:

1. A set of gravimetric experiments is done with various temperatures in the dryer, and the total loss of water Δm_T is determined as a function of temperature. Using equation (5.1), we can then calculate bond energy e as a function of temperature.

2. We assume that after drying to the highest temperature T_{max}, some amount of residual water still remained bonded in the porous material, and estimate this amount of water by the upper and lower limits Δm_1, Δm_2. Using equation (5.10) we calculate for all temperatures of the gravimetric experiment the dryness $w_{r,s}$, where for $T = T_{max}$ we put

$$m_{w,b} = \Delta m_1, \qquad (5.13)$$

respectively

$$m_{w,b} = \Delta m_2, \qquad (5.14)$$

for $T_x < T_{max}$ then

$$m_{w,b} = \Delta m_1 + \Delta m_{T_{max}} - \Delta m_{T_x}, \qquad (5.15)$$

respectively

$$m_{w,b} = \Delta m_2 + \Delta m_{T_{max}} - \Delta m_{T_x}. \qquad (5.16)$$

3. We calculate e using equation (5.1) for all temperatures of the gravimetric experiment, and construct for both Δm_1 and Δm_2 the functions
$e = e(w_{r,s})$, where every point $[w_{r,s,i}, e_i]$ corresponds to the temperature T_i. According to (5.12), this relation should be linear. If Δm_1 and
Δm_2 were really the upper and lower limits, the curvatures of functions
$e(w_{r,s})$ corresponding to Δm_1 and Δm_2 have different signs.

The correct value of Δm apparently lies between Δm_1 and Δm_2, and it can
be determined by a simple iterative treatment following the above procedure
again with different upper or lower limits. In practice, it is very difficult to
maintain the water vapour pressure in the dryer at a defined value. Therefore,
in calculations we commonly use the water vapour pressure in the laboratory.
In the dryer this pressure is usually lower, therefore also the amount of residual
water in the porous material is slightly lower than that, which was calculated.

The above method for the determination of the amount of residual water
in a porous material is very time consuming. Therefore, in most practical
measurements this correction is not performed, and such drying conditions
are chosen instead that the amount of residual water in the material is very
low, i.e. negligible compared to the measured water loss during the drying
process. According to Leroy's equation (5.1), we have only two possible ways
to achieve that. The first is to increase the saturated water vapour pressure p_s,
i.e. to increase the drying temperature. The second is to decrease the partial
pressure of water vapour in the dryer p, i.e. using some drying agent such as
P_2O_5 in the dryer. Usual conditions for drying porous building materials are
at temperatures of $105 - 110°C$ without a drying agent. Drying usually takes
several hours or days, depending on the dimensions of the specimens. Some
building materials, particularly those containing organic substances, may be
decomposed at these temperatures. In that case we can choose, for instance,
drying at temperatures of about $60 - 70°C$ under lowered pressure in the
dryer (mostly it is sufficient to lower only to about 15 kPa), which can be
achieved by a simple vacuum pump. It may happen that some materials do
not allow heating even to $60°C$, although that would be an extraordinary
case for building materials. The moisture content can then be determined by
drying in a desiccator at room temperature conditions using a drying agent.
This kind of drying can take weeks or months. The drying rate can only be
increased by choosing proper drying agents. The most important feature of a
drying agent is the dryability expressed usually in milligrams of residual water
per one liter of the air. The best drying agent in this respect is P_2O_5, with
0.000025 mg/l. Two more important features of drying agents are the surface
area and the regeneration of the active surface after water absorption. In
this respect, the best drying agent is BaO, which absorbs water very fast and
forms $Ba(OH)_2$ with up to 16 molecules of water per 1 molecule of $Ba(OH)_2$.

The accuracy of the determination of moisture content by the gravimetric
method is very important, because it is often employed as a standard method
for the calibration of other moisture meters. Besides the neglect of the amount

of residual water, which was mentioned before in more detail, weighing the specimens outside the dryer can also lead to substantial errors, because the sorption of moisture from the surrounding air is very fast, particularly for hot specimens just taken from the dryer. This problem can be solved either by weighing the specimen in the dryer or by cooling the specimen in the desiccator with drying agents before weighing.

5.1.1.2 Resistance Methods

The change in the amount of water in a material is accompanied by a change of its resistivity, which can be as high as several orders of magnitude. A typical value for the resistivity of a dry porous material is within the range of $10^8 - 10^{13}$ Ωm. The presence of water in this material can decrease its resistivity to 10^{-4} Ωm. Therefore, the moist material cannot be considered as a simple mixture of the dry material and water. In this respect, the way water is bonded to the material and the presence of ions, which dissociate in water, are the most important factors. The dependence of the electric resistance R on moisture content u at constant temperature usually has the form

$$R = a \cdot u^{-b}, \tag{5.17}$$

where a, b are empirical constants. However, this relation is valid over a limited range of moistures only, typically for hygroscopic moisture content corresponding to a relative humidity of 30–90%. For higher moistures, the resistance decreases more slowly than according to the given exponential relation, because the importance of the volume phase of water in larger pores is increasing. Ionic conductivity does not play here such an important role as in the surface phase. Its effect is similar as in free water. With increasing temperature the resistance of a moist material decreases, which is due to the ionic conductivity.

In practical measurements of moisture content by resistance methods (see, e.g., Pande, 1974), needle electrodes are most often used, which are installed in the measured material. Resistance measurements are performed mostly by using the Wheatstone bridge.

Resistance moisture meters always have to be calibrated for the given material and the given moisture range. Determination of moisture content by these moisture meters is very simple, and they can be applied over a wide range of materials. Therefore, their application in practice is very frequent. However, they have some substantial disadvantages that have to be taken into account in particular applications.

The most important problem with resistance moisture meters is that the obtained data depend on the amount of salts in water. For materials with a high amount of salts, the resistance moisture meters are practically inapplicable, because the errors rapidly increase with the increasing moisture content. Another problem is the low accuracy in the determination of large resistances, which results in lower precision of resistance moisture meters in the range of low moistures. The measurements can be negatively affected also by the polarization of electrodes due to electrochemical processes on their surface, which lead to the formation of an electrochemical cell. This cell is polarized counter to the applied voltage, so that the measured resistances are higher and the measured moistures lower than the real values.

5.1.1.3 Dielectric Methods

Dielectric methods for the determination of moisture content consist in an analysis of the behaviour of dielectrics in a time-varying electric field. The current density flowing through a capacitor can be expressed by the relation

$$\vec{j} = \sigma \vec{E} + \hat{\epsilon}\frac{\mathrm{d}\vec{E}}{\mathrm{d}t}, \tag{5.18}$$

where $\vec{E}$ is the electric field intensity, σ the electric conductivity, and $\hat{\epsilon}$ the complex permittivity,

$$\hat{\epsilon} = \epsilon' - i\epsilon''. \tag{5.19}$$

Therefore, the properties of a dielectric can be characterized by the quantities σ, ϵ', ϵ'', which generally depend on the composition of the dielectric, its structure, the frequency of the electric field, temperature and pressure. For an isotropic medium the permittivity is a scalar quantity expressing the influence of the medium on the electric field intensity (see, e.g., Smythe, 1968). Therefore, the dielectric properties of materials are mostly expressed by the relative permittivity,

$$\epsilon'_r = \frac{\epsilon'}{\epsilon_0} \tag{5.20}$$

$$\epsilon''_r = \frac{\epsilon''}{\epsilon_0}, \tag{5.21}$$

where $\epsilon_0 = 8.854 \cdot 10^{-12} \mathrm{F/m}$ is the permittivity of vacuum. If the alternating electric field is harmonic, we can write

$$\vec{E} = \vec{E}_0 \cdot e^{i\omega t}, \tag{5.22}$$

where $\omega = 2\pi f$ is the angular frequency of the field, f the frequency, and therefore we have

$$\vec{j} = (\sigma + i\omega\hat{\epsilon})\,\vec{E}_0 \cdot e^{i\omega t} = (\sigma + \omega\epsilon'' + i\omega\epsilon')\,\vec{E}_0 \cdot e^{i\omega t}. \tag{5.23}$$

If the imaginary part of $\hat{\epsilon}$ is equal to zero and the conductivity is also zero, the phase shift between $\vec{j}$ and $\vec{E}$ is 90°. In reality, the phase shift differs from the value of 90° by the loss angle δ. This can be caused by the electron or hole conductivity of the dielectric or by the dipole losses due to the relaxation phenomena (orientation or dipole polarization), the dielectric can also be inhomogeneous and electric charges can appear on the phase interfaces. Using equation (5.23), we can obtain for the loss angle the following relation:

$$\tan \delta = \frac{\sigma + \omega \epsilon''}{\omega \epsilon'} = \frac{\sigma}{\omega \epsilon'} + \frac{\epsilon''}{\epsilon'}. \tag{5.24}$$

In the high-frequency range of 10–100 GHz (i.e. in the range of cm and mm waves) the dielectric properties can be determined by optical methods, for instance by using the complex refraction index $\hat{n}$,

$$\hat{n} = n(1 - ik), \tag{5.25}$$

where n is the refraction index, k the extinction coefficient expressing the absorption properties of the medium. The analysis of Maxwell equations (see, e.g., Born and Wolf, 1991) leads to the relations

$$\epsilon_r' = n^2(1 - k^2) \tag{5.26}$$

$$\epsilon_r'' = 2n^2 k. \tag{5.27}$$

The determination of moisture content using the permittivity measurements is based on the fact that the static relative permittivity of pure water is equal to approximately 80 at 20°C, while for most dry building materials it ranges from 2 to 6.

The permittivity of materials is strongly affected by the orientability of molecules in the electric field. This characteristic is high for water in gaseous and liquid phase, but is significantly lower for water bound to a material by various sorption forces, which makes the orientation of water molecules more difficult. This feature makes it possible to distinguish between the particular types of bond of water to the material using the permittivity but on the other hand, it results in the dependence of the sensitivity of moisture measurements on the amount of water in the material. The relative permittivity of water bonded in a monomolecular layer is approximately 2.5, but for further layers it increases relatively fast. Therefore, the dependence of relative permittivity on the moisture content is generally characterized by an abrupt change at the transition from a monomolecular to a polymolecular layer. Consequently, the methods of moisture measurements based on the determination of changes of relative permittivity have low sensitivity in the range of low moistures and their application is rather limited.

Based on the frequency of the applied electric field, the dielectric methods can be divided into two groups:

1. Capacitance methods

2. Microwave methods.

Capacitance methods are employed in the range of lower frequencies, typically from 100 kHz to 100 MHz. The permittivity is determined using a capacitor, usually with the measured material as its dielectric. The measuring capacitor usually has either a simple plate form (for measuring solid materials) or consists of two coaxial cylinders (for measuring loose materials). For the capacity of a plate capacitor, we have the standard relation (see, e.g., Smythe, 1968)

$$C = \epsilon_0 \epsilon_r \frac{S}{d}, \tag{5.28}$$

where S is the area of the capacitor plates, and d the distance between the plates.

The admittance of a plate capacitor can be expressed using equation (5.23) in the form

$$\hat{Y} = G + i\omega C = (\sigma + i\omega\epsilon_0\hat{\epsilon}_r)\frac{S}{d} = \hat{y}\frac{S}{d}, \tag{5.29}$$

where G is the electric conductance, Y the admittance, y the admittivity, and ω the angular frequency of the generator of alternating voltage. Denoting

$$C_0 = \epsilon_0 \frac{S}{d} \tag{5.30}$$

the capacity of the empty capacitor, we can write for the admittance

$$\hat{Y} = G + i\omega\hat{\epsilon}_r \cdot C_0 = G + \omega\epsilon_r'' \cdot C_0 + i\omega\epsilon_r' \cdot C_0, \tag{5.31}$$

and for the loss angle we obtain

$$\tan\delta = \frac{Y_r}{Y_{im}} = \frac{G + \omega\epsilon_r'' \cdot C_0}{\omega\epsilon_r' \cdot C_0} = \frac{G}{\omega\epsilon_r' \cdot C_0} + \frac{\epsilon_r''}{\epsilon_r'}. \tag{5.32}$$

The relations (5.29), (5.31), (5.32) can be employed for the determination of complex relative permittivity. The measuring capacitor with the capacity C_0 is filled by the measured material, and the changes in its capacity and in the loss angle are determined by an appropriate method (see, e.g., Pande, 1974). The most frequently used methods are bridge methods and resonance methods. In the case of bridge methods, an alternating voltage supply of an angular frequency ω is placed into the measuring circuit, and the impedance of the capacitor with the measured material is determined. The resonance methods are based on the determination of the change of resonance frequency of a series or parallel LC circuit, where the capacitor with the measured material as a dielectric is placed. This frequency is in both cases expressed by the Thompson relation

$$f = \frac{1}{2\pi} \frac{1}{\sqrt{LC}}, \tag{5.33}$$

where L is the inductance of the coil. It can be measured, for instance, by using the inductance coupling (see Kašpar, 1984) of an active and a passive resonance circuit. In this measuring method, the capacitor with the measured material is placed in the passive circuit, and the capacity of the active circuit is changed by a rotary capacitor until the resonance with the passive circuit is achieved. The resonance is indicated by a voltage decrease in the active circuit.

Recently, a capacitance method working on a frequency of 350 kHz was employed for monitoring the moisture content in early hydration stages of cement mortar (Tydlitát et al., 2000) and cement paste (Černý et al., 2001a).

Microwave methods differ from the capacitance methods principally only in the applied frequency of the power supply. In the spectrum of electromagnetic waves microwaves occupy the frequency range of 1 GHz to 100 GHz, i.e. the range of dm to mm waves. Their wavelength is comparable with the dimensions of the electric elements such as capacitors or resistors, and so capacitance methods cannot be employed. Also, for the signal transmission coaxial cables or waveguides are employed instead of the usual wires. Moisture measurements are performed in a relatively narrow range of microwaves between 2 GHz and 12 GHz. The main reason is that in this range the microwave technology is most advanced, as most common radars work in this frequency range.

Determination of moisture content using the microwave methods is performed on the basis of measuring the change of relative permittivity due to the change of moisture, because microwave methods also belong to the dielectric methods. However, the methods for measuring ϵ_r are quite different from those that are used at lower frequencies. The most often used techniques of ϵ_r determination employ waveguides, cavity resonators, or the measurements are performed directly in a free space. A survey of various experimental set-ups can be found in McIntyre (1986).

The measurements in a waveguide are based on the fact that if a part of the waveguide is filled by a dielectric, the velocity of propagation of electromagnetic waves is changed, and the wave is reflected by all air-dielectric and dielectric-metal interfaces. The phase shift induced by these effects can be detected. The disadvantage of this method is that the specimens of the measured material have to fit exactly into the internal dimensions of the waveguide, which sometimes might be a difficult requirement.

The measurements in resonators are based on the determination of the change in the resonance frequency induced by inserting the dielectric. In practical experimental set-ups, almost exclusively used are cylindrical cavity resonators with dimensions corresponding to the formation of transverse electric (TE) or transverse magnetic (TM) waves. It should be noted here that TE and TM waves (see, e.g., Smythe, 1964) are linearly polarized waves, so that in the case of TE the vector of electric field intensity $\vec{E}$ is perpendicular to the plane of incidence, while in the case of TM the vector of magnetic field intensity $\vec{H}$ is perpendicular to the plane of incidence. The resonator can be tuned to certain limits by a sliding piston. The dielectric must be placed exactly along its axis, because the electromagnetic field has to be symmetric. The frequency change is determined by the change of the position of the piston necessary to achieve resonance after the dielectric has been inserted. Resonance systems are only rarely used for building materials, though an example of an application to tricalcium silicate can be found in Reboul (1978).

The measurements in free space are based on the determination of either microwave attenuation after transmission through a material sample or microwave reflection from its surface. In the GHz range, most dielectrics can be considered as absorbing media for electromagnetic waves. Therefore, the amplitudes of the vectors $\vec{E}$ and $\vec{H}$ decrease with the thickness x of the dielectric according to the relations

$$\vec{E} = \vec{E}_0 \cdot e^{-\alpha x} \tag{5.34}$$

$$\vec{H} = \vec{H}_0 \cdot e^{-\alpha x}, \tag{5.35}$$

where α is the absorption coefficient. From the theory of electromagnetic waves (see, e.g., Fexa and Široký, 1983), the following relation for α can be derived:

$$\alpha = \frac{2\pi}{\lambda_0} \sqrt{\frac{\epsilon_r'}{2} \sqrt{1 + \tan^2 \delta} - 1}, \tag{5.36}$$

where

$$\tan \delta = \frac{\epsilon_r''}{\epsilon_r'} \tag{5.37}$$

$$\hat{\epsilon}_r = \epsilon_r' - i\epsilon_r''. \tag{5.38}$$

As power P increases with E^2 (see, e.g., Smythe, 1968), the power attenuation of electromagnetic waves in free space can be expressed in the form

$$P(x) = P(0) \cdot e^{-2\alpha x}, \tag{5.39}$$

and for the attenuation A we have

$$A = 10 \log \frac{P(0)}{P(x)} = 10 \log(e^{2\alpha x}) = 20\,\alpha x \log e = 8.7\,\alpha x. \tag{5.40}$$

A microwave moisture meter based on measuring the attenuation consists of the transmitting and the receiving parts. The signal is transmitted from a supply to the specimen using a horn antenna, and on the other side of the specimen it is detected by another horn antenna and evaluated. The differences in the power of the microwaves going in and coming out of the specimen make it possible to determine the attenuation directly from relation (5.40). Practical moisture measurements are performed by comparing the attenuations measured on the dry and the moist samples. Examples of applications of these methods to concrete can be found in Wittmann and Schlude (1975), Leschnik et al. (1995).

Similarly to the attenuation, the reflection of microwaves from the surface of a measured object is also a function of the moisture content. The basic experimental set-up is analogous to measuring the attenuation, the only difference being that the reflected wave is detected instead of the transmitted wave. The reflected wave has to be evaluated on the same side of the specimen, where the power supply is, and the transmitting antenna is employed as a receiver as well. Therefore, a directional couple has to be included in the measuring system, making it possible to detect only the energy reflected from the material specimen. The moisture measurements are again performed by a comparison of reflected powers measured on the dry and the moist specimens. Taking into account that the calculation of permittivity directly from the reflectivity is relatively complicated, calibration is used instead in most practical applications, and the dependence of the detected voltage signal on the moisture content is directly determined.

Among the practical applications of reflection systems to measurements on concrete, the open-ended coaxial probe/waveguide method is among the most frequently used over the last decade (see Thompson, 1997, Mouhasseb et al., 1995, Rhim and Büyüköztürk, 1998). Also, an "infinite sample" method based on the analysis of the interference pattern produced by the incident and reflected waves from a sample put into a waveguide was employed (Moukwa et al., 1991).

Some of the basic microwave methods can be conveniently combined in one experimental set-up. A logical combination seems to be reflection-transmission methods as this was demonstrated for measurements on concrete by Gorur et al. (1982).

The aforementioned methods work with continuous microwaves. Another possibility for determining the moisture content using microwaves is to measure the travel time and absorption of a short ($\sim$ 100 ps) microwave impulse through a material specimen of a known thickness (Maierhofer and Wöstmann, 1997). The advantage of such a method is that the measurements are completely nondestructive and large surfaces can be scanned rapidly. Recently, this method was applied to measurements on hydrating cement paste (Černý et al., 2001b).

A specific methodology among the microwave impulse techniques is the time-domain reflectometry (TDR). A device based on the TDR principle (see Nissen and Moldrup, 1995) launches electromagnetic waves and then measures the amplitudes of the reflected waves together with the time intervals between launching the waves and detecting the reflections. The fundamental element in any TDR equipment used for the determination of moisture content in porous materials is a metallic cable tester. This usually consists of four main components: a step-pulse generator, a coaxial cable, a sampler and an oscilloscope.

The step-pulse generator produces the electromagnetic waves. The electric part of the electromagnetic waves consists of sine waves covering a large frequency range, but which frequencies the step-pulse generator produces is not arbitrary. If a sine wave is superimposed on harmonic sine waves, where the highest frequency tends towards infinity, the result will be a perfect periodic square wave. This is what happens in the step-pulse generator. The periodic square wave is commonly called "voltage step" (see Nissen and Moldrup, 1995). In the cable tester Tektronix 1502B, which is often used in soil science, the voltage steps are produced by superimposing a sine wave with a ground frequency of 16.6 kHz on harmonic sine waves up to 1.75 GHz. One voltage step is transmitted over a period of 10 μs, and then there is a pause in the transmission lasting 50 μs. This pause ensures, that standing waves die out before new waves are launched from the cable tester. The rise time of the voltage steps, which depends on the highest frequency of the sine waves (for an infinite frequency it would be equal to zero), is for this particular case about 200 ps.

The coaxial cable connects the step-pulse generator and the sampler. The shield of the coaxial cable is connected to earth and its electric potential is 0 V. The electromagnetic waves produced by the step-pulse generator are launched into the conductor in the coaxial cable with a voltage drop of several tenths of a volt (for Tektronix 1502B it is 0.225 V) between the conductor and the shield.

The TDR probe itself is conductively connected to the coaxial cable in such a way that the cable is open ended and the probe forms this open end. In principle, the coaxial cable and the probe differ only in the type of dielectric. While the cable has usually polyethylene as a dielectric, the dielectric of the probe is the measured porous material.

The sampler detects the electromagnetic waves launched by the step-pulse generator and transmitted by the coaxial cable-TDR probe system. It generally consists of two main components, a high precision timing device and a high precision voltmeter. When the electromagnetic waves launched by the generator are detected by the sampler, the sampler starts to measure the voltage between the shield and the conductor at a certain time interval. The set of data obtained consists of voltage as a function of time.

The oscilloscope displays the simultaneous measurements of time and voltage obtained by the sampler on a liquid crystal display, or the data in a digital form can be directly sent to a PC and displayed there. This generates a curve called the trace.

The evaluation of data obtained by a cable tester is based on the following basic principles (see Nissen and Moldrup, 1995). Any change of impedance in the cable-probe system causes a partial or total reflection of the waves. Therefore, one reflection will be on the cable/probe interface, where the dielectric is suddenly changed, and therefore the impedance must also be changed, while the second reflection is on the open end of the probe, where the impedance tends towards infinity and the wave is reflected in phase.

The reflected waves are superimposed on the waves transmitted from the metallic cable tester. The voltmeter in the sampler detects a change in the voltage between the conductor and the shield, and the timing device in the sampler registers the time interval between the start of the transmission of the waves and the detection of the reflection. Reflected waves can be either in phase with the incoming waves, which happens in the case when the electromagnetic waves meet an increase in impedance, or in counter phase, when a decrease of impedance is met.

Therefore, in our case of a cable-probe system, the reflection on the cable/probe interface can cause either a decrease in the amplitude (if the impedance of the probe is smaller than the impedance of the cable, which is most often the case) or its increase in the opposite case (reflection on the open end always results in an amplitude increase). Thus, three characteristic times can be recognized on the trace: $t = t_1$ is the time when the front of the initial voltage step (always positive), induced by the step pulse generator, reaches the sampler; $t = t_2$, when the voltage step induced by the reflection on the cable/probe interface (mostly negative) reached the sampler; and finally the time $t = t_3$, when the voltage step induced by the reflection on the open end, i.e. on the end of the probe (always positive) reached the sampler. The length of the probe l is known, therefore the velocity of the electromagnetic waves in the probe is equal to

$$v = \frac{2l}{t_3 - t_2}. \tag{5.41}$$

From the basic theory of electromagnetism (see, e.g., Born and Wolf, 1991), the velocity of electromagnetic waves propagation in nonmagnetic materials can be expressed as

$$v = \frac{c}{\sqrt{\hat{\epsilon}_r}}, \tag{5.42}$$

where c is the velocity of electromagnetic waves in vacuum and $\hat{\epsilon}_r$ the complex relative permittivity of the material,

$$\hat{\epsilon}_r = \epsilon_r' - i\epsilon_r''. \tag{5.43}$$

If the imaginary part can be neglected, then combining equations (5.41) and (5.42) we obtain

$$\epsilon_r' = \left[\frac{c(t_3 - t_2)}{2l} \right]^2. \tag{5.44}$$

In the reality, the imaginary part cannot be fully neglected and it affects the measured data to some extent. Also, there is certain frequency dependence of the permittivity of the constituents of a porous material, including water in the working frequency range of the cable tester, although it is not very dramatic. Therefore, the measured value of permittivity we call the apparent permittivity and denote it $\epsilon_{r,a}$,

$$\epsilon_{r,a} = \left[\frac{c(t_3 - t_2)}{2l} \right]^2. \tag{5.45}$$

A variety of TDR probes of different shapes and types have been designed so far. For instance, Topp et al. (1980) used a coaxial cell in a direct (but much larger) extension of the coaxial cable. Nissen and Moldrup (1995) described a three-rod coaxial probe, where the shield of the coaxial cable was replaced by two metal rods of the same length as the conductor. Malicki et al. (1992) designed a two-rod miniprobe suitable for the determination of instantaneous moisture profiles.

Generally, the accuracy of moisture measurement by microwave methods can be affected by the conductivity of the material, polarization on the interfaces of a heterogeneous material, the type of water bond to the porous matrix, wave scattering, standing wave formation, granulation of the material, amount of dissolved salts, temperature, etc.

The microwave methods cannot be applied to good conducting materials or materials containing conducting particles (e.g. reinforced concrete), because the attenuation due to the conductivity of the matrix is higher than that due to the effect of water content. The dielectric loss due to the conductivity depends mainly on the dimension of the conducting part that is parallel to the direction of the wave propagation. If the conducting parts are positioned perpendicular to the direction of wave propagation, the measurement is not affected in a significant way.

The effect of temperature is generally very important, because the change to permittivity due to the change of moisture are of the same order as the changes due to the change of temperature. Therefore, either temperature has to be maintained constant during the measurements or temperature compensation has to be included.

The scattering of microwave energy has substantial effect only for small specimens with dimensions comparable to the cross-section of the waveguide. In such cases, screening has to be provided, to ensure that the energy propagating to the receiver passes through the specimen only. There is a general rule that the specimen should have an area at least four times larger than the cross-section of the horn antenna.

The standing waves are formed by reflection at the air/specimen interface. If these waves are not absorbed by the specimen, they can be added to or subtracted from the propagating wave, which can result in substantial measuring errors. If the material has a sufficiently high absorption, the reflected waves are absorbed and their effect is in fact included in the calibration curve.

In a granular material, a proportion of the energy can be reflected from the surface of the particles, which results in additional losses to those induced by the presence of water. This absorption of energy is related to the dimension of particles. If the diamater of a particle is smaller than one quarter of the wavelength in the material, the absorption is negligible. Larger particles can cause substantial errors in the measured results.

The microwave radiation is usually polarized in one plane. Therefore, anisotropic substances exhibit different energy absorption in different directions, and their orientation during measurements has to be the same as during the calibration.

The presence of dissolved salts in the specimen can affect the conductivity of water in a significant way. However, this effect is substantial only for lower frequencies. With increasing frequency of microwaves the importance of salts for the measured values decreases. The imaginary part of permittivity of water increases with frequency (it has a maximum at the so-called critical frequency, f_{cr}=23.4 GHz), and therefore the relative importance of conductivity decreases.

5.1.1.4 Radiometric Methods

Radiometric methods are based on the absorption of radioactive radiation in a material. The most commonly used are the absorption of fast neutrons or γ radiation.

The neutron method utilizes the deceleration (thermalization) of fast neutrons due to their interaction with nuclei of atoms with low atomic mass. The loss of energy of a neutron due to a collision with an atomic nucleus depends on the mass of the nucleus. It is greatest in a collision of a neutron with a particle of the same mass, which can take over a part of the momentum of the fast neutron. On the other hand, with increasing mass of the nucleus the energy loss due to the collision with a neutron decreases, because a fast neutron is in fact reflected by a large nucleus. The average number of collisions necessary to decrease the energy of fast neutrons (which is typically 9 MeV) to the level of thermal energy ($\sim$ 0.025 eV) is 18 for hydrogen, 114 for carbon, 150 for oxygen, etc.

In common inorganic materials, the most important substance containing hydrogen, which is the most effective moderator among the elements, is water. Therefore, the amount of water in a material substantially affects the absorption of neutrons, which can be utilized in moisture measurements.

The classical experimental set-up for moisture measurements (see, e.g., Van Bavel, 1965) is cylindrically symmetric. A circular hole is bored into the measured medium, and into the hole is placed a probe with the source of fast neutrons. The neutrons are usually emitted in the radial direction, and due to collisions with the surrounding nuclei they gradually lose a part of their kinetic energy. The thermalized neutrons form a spherical cloud around the source. The radius of the cloud depends on the moisture contained in the material. Higher moisture content leads to a faster deceleration of neutrons, and the radius of the thermalized neutrons is smaller but the density of neutrons is higher. Conversely, for dry material the radius of the cloud is larger, and the density of neutrons is lower. An approximation formula for the dependence of the radius r of the cloud on the moisture content w can be written as follows:

$$r = \left(\frac{a}{w}\right)^{1/3}, \tag{5.46}$$

where a is an empirical constant.

The source of neutrons is usually a mixture of an isotope emitting the α particles and beryllium. The neutrons are released due to the reaction

$$\frac{4}{2}\alpha + \frac{9}{4}Be \rightarrow \frac{1}{0}n + \frac{12}{6}C + 5.65 \text{ MeV}.$$

As a source of α particles, there can be used for instance ^{226}Ra, ^{241}Am, ^{239}Pu, ^{210}Po. The most suitable is an Am–Be mixture, which supplies a sufficient number of neutrons, but the secondary γ radiation is negligibly small, which is an important feature from the point of view of health protection while working with these sources.

For the detection of slow neutrons, most frequently are employed boron counters (BF_3), which are cheap, robust and stable. If a thermalized neutron comes in contact with the nucleus of ^{10}B, it is absorbed and an α particle is radiated. An α particle counter then detects the number of slow neutrons that entered the detector chamber.

The evaluation of the moisture content is done using the calibration lines, which express the dependence of moisture on the number of pulses per unit time. The calibration has to be performed for every material, for instance by the gravimetric method.

The experimental set-up described above is not very suitable for a detailed investigation of the moisture profile if a gradient of moisture exists in the specimen. On the other hand, the measured value represents an average moisture content in a sphere of a diameter of about 20 cm. Therefore, the method is suitable for making a moisture balance in larger volumes and for relative comparisons of moisture content.

However, for the neutron method it is also possible to design an experimental set-up that makes it possible to determine moisture profiles, as has been shown by Pel et al. (1993). They used a monochromatic neutron beam of wavelength 0.131 nm (i.e. the energy 47.7 MeV) for the measurements. A boron–cadmium collimator was employed in order to achieve a better spatial resolution, so that the final dimensions of the neutron beam were 1 mm × 30 mm. After transmission through the specimen, the slow neutrons were detected by a proportional ^{3}He detector ($\frac{3}{2}He + \frac{1}{0}n \rightarrow \frac{4}{2}\alpha$), placed behind a further collimator, directing the slow neutrons into a narrow beam. In this way, the detection of scattered neutrons was excluded, and the divergence of the beam after passing through the specimen was reduced.

In the determination of the moisture content in a material, Pel et al. (1993) used a common relation for the radiation absorption in a material,

$$I = I_0 \cdot e^{-\sum_i d_i \mu_i}, \tag{5.47}$$

where μ_i is the macroscopic absorption coefficient of the i-th component and d_i the equivalent thickness of the i-th component, $d_i = d \cdot \frac{\rho_i}{\rho}$. In the case of a nondeformable moist porous material, this relation is simplified to

$$I = I_0 \cdot e^{-d(\mu_{\text{mat}} + w\mu_{\text{w}})},$$
(5.48)

where μ_{mat} is the absorption coefficient of the solid matrix, w the moisture content by volume, μ_{w} the absorption coefficient of water. The coefficients μ_{mat} and μ_{w} have to be determined experimentally for a given experimental set-up. Taking into account that at a wavelength of 0.131 nm $\mu_{\text{w}} = 2.54$ cm^{-1}, while for instance for a dry brick we have $\mu = 0.155$ cm^{-1}, the good sensitivity of the method is apparent.

Generally, the accuracy of the moisture measurements by the neutron methods can be affected by the following factors. Taking into account that thermalization of fast neutrons is due to the hydrogen atoms and not the water molecules, the measurements can effectively be done for materials that contain only negligible amounts of hydrogen in other forms than as water molecules. Alternately, the amount of this "other" hydrogen has to be maintained constant during the experiment. Therefore, the application of the neutron method to organic materials is not appropriate. Also, the method is applicable only to materials that do not contain higher amount of elements which strongly absorb slow neutrons, for instance Cl, K, B, Li, Rh, Ag, Hg, Cd, In, Hf, Re, Ir, Au, Br. On the other hand, temperature and pressure have only negligible effect on the measurements.

The γ-ray attenuation method employs the absorption of γ radiation in materials for the determination of moisture content. The intensity of radiation after passing through a material of thickness d is expressed as

$$I = I_0 \cdot e^{-\mu\rho d} + I_{\text{s}},$$
(5.49)

where I_0 is the intensity of the incident radiation of the source, μ the mass absorption coefficient (in m^2/kg), ρ the density of the material, and I_{s} the intensity of the scattered radiation.

The energy of the basic radiation is in the range of 0.5–3 MeV, while the energy of the scattered radiation 0.1–0.15 MeV. Therefore, how much of scattered radiation is registered it depends on the type of detector. If a less sensitive detector is used, which detects γ radiation of energies higher than 0.15 MeV only, the effect of scattered radiation can be reduced to a minimum. Another possibility to reduce scattered radiation is to use a collimator, which for γ radiation is a lead plate, in most cases. The most frequently used sources of γ radiation are ^{60}Co, ^{137}Cs and ^{241}Am. The last named source can be considered the most safe, because only a few millimetres of lead are quite sufficient for screening. In addition, the ^{241}Am source is relatively stable, with a half life of 458 years.

Measuring the moisture content by the γ-ray attenuation method is based on applying the superposition principle to the effects of dry material and moisture on the γ radiation absorption. The effective absorption coefficient of the mixture μ_{m} is determined according to the relation

$$\mu_{\text{m}} = \mu_{\text{w}}c + \mu_{\text{d}}(1 - c),$$
(5.50)

where μ_w, μ_d are the absorption coefficients of water and dry material, respectively, c the mass concentration of water,

$$c = \frac{m_w}{m_w + m_d} = \frac{\rho_w}{\rho_w + \rho_d} = \frac{\rho_w}{\rho}, \tag{5.51}$$

ρ_w, ρ_d the partial densities of water and dry material, respectively, and ρ the total density of the mixture. Substituting equation (5.50) into (5.49) and neglecting the scattered radiation we obtain

$$I_m = I_0 \cdot e^{-\mu_m \rho d} = I_0 \cdot e^{-[\mu_d + (\mu_w - \mu_d) \cdot c] \rho d}, \tag{5.52}$$

or for the moisture content by mass defined as

$$u = \frac{\rho_w}{\rho_d} \tag{5.53}$$

then

$$I_m = I_0 \cdot e^{-\mu_m \rho d} = I_0 \cdot e^{-(\rho_d \mu_d + \rho_w \mu_w) d} = I_0 \cdot e^{-\rho_d d(\mu_d + \mu_w u)}. \tag{5.54}$$

For a dry material we have

$$I_d = I_0 \cdot e^{-\mu_d \rho_d d}. \tag{5.55}$$

Therefore, comparing equations (5.54) and (5.55) we obtain

$$\ln \frac{I_d}{I_m} = \rho_d \mu_w u d, \tag{5.56}$$

and for the moisture content

$$u = \frac{1}{\rho_d \mu_w d} \cdot \ln \frac{I_d}{I_m}. \tag{5.57}$$

Relation (5.57) is valid only under assumption that the volume does not change with the change of moisture content, otherwise the thicknesses of the dry and the moist materials would differ. For monochromatic γ radiation, the value of μ_w is known or can be determined by a simple experiment. The density of the dry material and the thickness of the specimen are also known. Therefore, the only parameters that have to be measured are the radiation intensities after passing through the dry and the moist material. This is usually performed by a scintillation detector containing NaI activated by thallium. The radiation intensity is determined as the number of photons emitted per unit time, so we register the number of "flashes" – scintillations per second or per minute.

The main advantages of the γ-ray attenuation method are its objective physical character, the unambiguous value of the absorption coefficient μ_w for monochromatic rays, and minimal effect of salts on the measured values of moisture. The main disadvantages are a substantial dependence of results on density, and the need to measure the absorption on both dry and moist material. If the density of the dry material depends on the moisture content and a change in thickness of the specimen occurs during the measurements, it is necessary to measure both density of the material and its moisture content at the same time. This can be done for instance by a simultaneous application of two different sources of γ rays having different values of the absorption coefficients μ_w and μ_d. The absorption coefficient μ is a function of radiation frequency ν, so for the photon energies we have $E = h\nu$, where h is the Planck constant, and therefore an application of sources with different energies leads to different values of absorption coefficients. For example, if we use ^{137}Cs with the energy of 662 keV and ^{241}Am with 60 keV, the radiation from these two sources can be distinguished using two discriminators (filters), where one of them detects only pulses larger than 550 keV, the second detects pulses in the range of 35–85 keV in a differential way. In a practical experimental set-up (see Nofziger and Swartzendruber, 1974), both sources can be placed in series, with a common collimator, and the scintillation counter is connected in parallel to both discriminators.

The need to measure the absorption on both dry and moist material at the same time can be a substantial source of errors, particularly for inhomogeneous materials, where the density ρ_d is generally a function of the position in the specimen. In this case, the installation of a precise positioning device is necessary, which makes it possible to perform first measurements along the whole length of the dry specimen. Only after that it is possible to moisten the specimen and measure the moisture profiles.

The γ-ray attenuation method was used to measure moisture content in concrete, for instance, by Nielsen (1972).

5.1.1.5 The Nuclear Magnetic Resonance Method

Atomic nuclei having an odd number of protons or neutrons possess a magnetic moment. The hydrogen nucleus with a single proton belongs to the most important nuclei with nonzero magnetic moments. While the charge in a hydrogen atom is equally divided between the proton and the electron, practically the whole mass is concentrated in the nucleus. Therefore, the nucleus exhibits gyroscope-like behaviour and also behaves as a rotating rod magnet. If a gyroscope rotates about its axis with one end on a hard base, a precessional movement is induced due to the force of gravity, and the other end of the axis describes a circular motion. A rod magnet exhibits similar behaviour in an external magnetic field. Therefore, if a nucleus possessing nonzero magnetic moment is situated in a magnetic field, its magnetic moment becomes oriented, and the nucleus describes a precessional movement about the axis parallel to the vector of the magnetic field. The angular frequency of the precessional motion ω_0 increases with the value of the amplitude of magnetic induction B_0 of the external magnetic field (see, e.g., Kean and Smith, 1986),

$$\omega_0 = \gamma B_0, \tag{5.58}$$

where the proportionality factor γ_0 is called the gyromagnetic ratio. Among the about 80 elements having nonzero magnetic moments, hydrogen has the highest value of the gyromagnetic ratio, $\gamma = 2.675 \times 10^8$ $T^{-1}s^{-1}$.

The nuclear magnetic moment is quantized, and it can assume $n = (2I+1)$ values, where I is the nuclear spin quantum number. As for a proton $I = 1/2$, we have $n = 2$, and the nuclear magnetic moment of a proton can assume two energy levels. In normal conditions, i.e. without any external magnetic field, these two levels have the same energy; we say they are degenerate. If a uniform direct magnetic field acts on the hydrogen nucleus, the magnetic moments become oriented either in the direction of the field or opposite to the field. The energy of these two orientations is different, and the degeneration disappears.

The thermodynamic equilibrium between the nucleus and the surroundings is not established immediately; it has a relaxation character given by the relation

$$M_z = M_0 \left(1 - e^{-t/\tau_1}\right), \tag{5.59}$$

where M_z is the component of the magnetization along the z-axis, i.e. in the direction of the direct magnetic field, M_0 the equilibrium value of the magnetization in the direction of the field, which is achieved by the nucleus after finishing the transition process, and τ_1 the time constant of the transition process, which is called the spin–lattice (longitudinal) relaxation time. We note that the magnetization is defined by the relation

$$M_0 = \chi_0 B_0, \tag{5.60}$$

where χ_0 is the static nuclear magnetic susceptibility,

$$\chi_0 = \frac{N\mu_0^2 m^2}{3kT} \frac{I+1}{I}, \tag{5.61}$$

N the number of nuclear magnets per unit volume, μ_0 the permeability of vacuum, m the magnetic moment of the nucleus, I the spin quantum number, k the Boltzmann constant, and T the thermodynamic temperature.

If a nucleus making a precessional motion in an external direct magnetic field is exposed to a further, alternating magnetic field of magnetic induction B_1 and angular frequency ω, circularly polarized in the (x, y) plane, perpendicular to the z-axis, where the basic magnetic field B_0 acts, and if the resonance condition $\omega = \omega_0$ is satisfied, the orientation of the magnetic moment of the nucleus is reversed. This is accompanied by energy absorption. This phenomenon is called nuclear magnetic resonance (NMR). A proportion of the nuclei, which before were oriented in the direction of direct magnetic field and had a lower energy, take on the orientation opposite to the direction of magnetic field and have higher energy. This leads to a redistribution of the number of nuclei at the lower and higher energy level. However, this is thermodynamically unstable. Therefore, a proportion of nuclei on the higher energy level transfers the energy gained to the surroundings in the form of heat and returns to the previous lower energy level. The absorption of high-frequency energy is therefore a continuous process.

The energy difference between the levels is (assuming the validity of the resonance condition $\omega = \omega_0$) given by the relation

$$\Delta E = h\nu = h\frac{\omega_0}{2\pi} = h\frac{\gamma B_0}{2\pi}, \tag{5.62}$$

where h is Planck's constant.

The action of the alternating transverse magnetic field B_1 on the nuclear magnetic moments is accompanied by the formation of transverse magnetization, which also has a relaxation character, and is given by the relations

$$\frac{\mathrm{d}M_x}{\mathrm{d}t} = -\frac{M_x}{\tau_2} \tag{5.63}$$

$$\frac{\mathrm{d}M_y}{\mathrm{d}t} = -\frac{M_y}{\tau_2}, \tag{5.64}$$

where M_x, M_y are instantaneous values of the magnetization in the direction of axes x, y, respectively, and τ_2 the time constant of the relaxation process characterizing the rate of the energy change between the nuclei, which is called the transverse spin–spin relaxation time.

The values of the relaxation times have a substantial influence on the form of the NMR signal. For liquids usually $\tau_1 \sim \tau_2$, and the relaxation times are of the order of seconds, for instance for water 2.7 s. In solids $\tau_2 \sim 10^{-4} - 10^{-5}$ s, and τ_1 is of the order of seconds to minutes.

In the NMR theory it is assumed that the magnetic field B_1 is circularly polarized. However, for technical reasons a linearly polarized field is usually applied. Any linearly polarized field can be formally arranged as two rotation fields of the same amplitude and opposite direction of rotations,

$$B_x = B_1 \cos\omega t + B_1 \cos(-\omega t) = 2B_1 \cos\omega t \tag{5.65}$$

$$B_y = B_1 \sin\omega t + B_1 \sin(-\omega t) = 0. \tag{5.66}$$

The rotation in the direction of the precessional motion of nuclei causes the nuclear resonance; the field rotating in the opposite direction has only a negligible influence on the resonance. Therefore, this treatment does not lead to any significant errors.

The action of a magnetic field linearly polarized in the direction of the x-axis, $B_x = 2B_1 \cos\omega t$, can be expressed by Bloch equations using the complex dynamic magnetic susceptibility $\hat{\chi} = \chi' - i\chi''$, which is already a measurable quantity. According to the definition of magnetization, $M = \chi B$, we arrive at two components of magnetization, one of them, $2\chi' B_1 \cos\omega t$, having the same phase as the field B_1, and the other, $2\chi'' B_1 \sin\omega t$, being shifted by 90°.

For the components of susceptibility we have (see Fexa and Široký, 1983, for details)

$$\chi' = \frac{1}{2}\chi_0\omega_0 \frac{\tau_2^2(\omega_0 - \omega)}{1 + \tau_2^2(\omega_0 - \omega) + \gamma^2 B_1^2\tau_1\tau_2} \tag{5.67}$$

$$\chi'' = \frac{1}{2}\chi_0\omega_0 \frac{\tau_2}{1 + \tau_2^2(\omega_0 - \omega) + \gamma^2 B_1^2\tau_1\tau_2}. \tag{5.68}$$

The saturation term $D = \gamma^2 B_1^2 \tau_1 \tau_2$ is usually very small, $D \ll 1$. Therefore, in the state of resonance we have

$$\chi' = 0 \tag{5.69}$$

$$\chi''_{\max} = \frac{1}{2}\chi_0\omega_0\tau_2 = \frac{1}{2}\chi_0\gamma B_0\tau_2. \tag{5.70}$$

Equations (5.67) and (5.68) can be expressed graphically in the form of the dispersion curve $\chi' = \chi'[(\omega_0 - \omega)\tau_2]$ and the absorption curve $\chi'' = \chi''[(\omega_0 - \omega)\tau_2]$. In the absorption part χ'' of the complex susceptibility, χ_0 is related to the number of elementary nuclear magnets N (see equation (5.61)); all other quantities are either known or at least they do not depend on N. Therefore, by measuring χ'' the concentration of nuclear magnets in the measured specimen can be determined.

The absorption curve makes it possible to determine the value of τ_2. From equation (5.68) it follows that for $D \ll 1$ and $\tau_2(\omega_0 - \omega) = 1$ we have $\chi'' = \chi''_{\max}/2$. Given the measured curve $\chi'' = \chi''[(\omega_0 - \omega)\tau_2]$, it is sufficient to find the value of $\Delta\omega = \omega_0 - \omega$ corresponding to $\chi''_{\max}/2$ and calculate $\tau_2 = 1/\Delta\omega$.

Knowing the value of τ_2, we can use the dependence of $\chi''_{\max}$ on the saturation term D in the determination of τ_1. From equation (5.68) it follows, that in the state of resonance, for $D = 1$ we have $\chi''_{\max} = \chi_0\omega_0\tau_2/4$, which is equal to the one half of its value for $D \ll 1$. Therefore, if we increase the value of B_1, so that we obtain the value $\chi''_{\max}(B_1 = B_{\tau_1}) = \chi''_{\max}(D \ll 1)/2$, from the relation for the saturation term $D = \gamma^2 B_1^2 \tau_1 \tau_2$ we can determine $\tau_1 = 1/\gamma^2 B_{\tau_1}^2 \tau_2$.

In the determination of moisture content by the NMR method it is necessary to measure χ'' in the specimen with the known B_0 and τ_2 under the condition $\omega_0 = \omega$. From the magnitude of the absorption signal, χ_0 can be determined, and therefore using equation (5.61) the number of the hydrogen nuclei in the specimen can also be calculated. As with the neutron method, here also the concentration of hydrogen is determined rather than the concentration of water. However, the NMR method makes it possible to distinguish between the hydrogen atoms that are a part of the measured material, and the hydrogen atoms that are present in the form of water. The type of bond of water molecules to the material can also be estimated. The reason is that the relaxation times τ_1, τ_2 depend on the surroundings of the resonating nuclei.

Any NMR moisture meter (see, e.g., Rolwitz, 1965) generally consists of three main components: the magnet, the high-frequency source and the detection system. The magnet is most frequently a permanent one, and generates a uniform magnetic field with B_0 up to 2 T. The high-frequency source generates the magnetic field B_1 necessary to achieve the resonance of nuclei contained in the specimen. As follows from equation (5.58), the resonance frequency increases with increasing B_0; for instance for $B_0 = 2T$ we have $f \sim 8.5$ MHz. With an increase in B_0 and ω_0, the signal/noise ratio is better but the mass of the magnet has to be higher.

The specimen for moisture measurement is put into a coil placed between the poles of the magnet in such a way that its axis is perpendicular to the direction of magnetic induction B_0. The high-frequency voltage with the required frequency is supplied to the coil. The magnetic field B_0 is modulated by additional coils placed on the poles of the magnet in order to measure the signal course in the vicinity of the resonance. The high-frequency voltage supplied to the coil with the specimen can be frequency modulated, but technically this is more difficult. If only the absorption NMR signal (χ'') is to be detected, just one coil is sufficient for both high-frequency energy supply and detection. If also the dispersion signal (χ') is to be detected, one more coil is added, whose axis is perpendicular to both the direction of the magnetic field B_0 and the axis of the first coil.

To measure the NMR signal, a parallel resonance circuit containing the coil with the measured specimen is often used. This circuit is supplied with the high-frequency voltage from an oscillator of the frequency corresponding to the resonance condition. The inductance of the coil is given by the relation

$$L = L_0(1 + 4\pi\xi\hat{\chi}), \tag{5.71}$$

where ξ is a factor expressing the filling of the coil by the specimen. The admittance of the resonance circuit is then

$$Y = \frac{1}{Z} = \frac{1}{R_0} + i\left[\omega C_0 - \frac{1}{\omega L_0(1 + 4\pi\xi\hat{\chi})}\right], \tag{5.72}$$

and its change $\Delta Y = Y - Y(L_0)$ due to the resonance of nuclei can be expressed as

$$\Delta Y = i\left[\frac{1}{\omega L_0} - \frac{1}{\omega L_0(1 + 4\pi\xi\hat{\chi})}\right] = i\frac{4\pi\xi\hat{\chi}}{\omega L_0(1 + 4\pi\xi\hat{\chi})} \sim i\frac{4\pi\xi\hat{\chi}}{\omega L_0}. \tag{5.73}$$

Using $\hat{\chi} = \chi' - i\chi''$, we finally get

$$\Delta Y = -\frac{4\pi\xi}{\omega L_0}\chi'' + i\frac{4\pi\xi}{\omega L_0}\chi'. \tag{5.74}$$

From equation (5.74) it follows that the absorption component χ'' induced by the nuclear resonance changes the conductance, i.e. the real part of the admittance, which can be measured by methods commonly used for resistance measurements. This change of conductance is not constant, because the magnetic field B_0 is modulated in the vicinity of resonance, and the resonance occurs periodically according to this modulation. Therefore, the conductance change has also the frequency of the modulation voltage and causes an amplitude modulation of the high-frequency voltage of the oscillator. The modulation factor is proportional to χ'', and therefore to the concentration of resonating nuclei. The signal is transmitted to an oscilloscope, where it is displayed in either analog or digital form.

The application of NMR methods to measure moisture content in building materials does not have a long history; the first experimental installations appeared within the last decade. Nevertheless, the method is very promising and it can be anticipated that it will be increasingly used in the near future.

Among the experimental installations designed directly for the determination of moisture profiles in building materials, that by Pel et al. (1996) is probably the most advanced to date. The samples used in the Pel et al. (1996) experiments were cylindrical rods with a diameter of 20 mm and length between 2 cm and 20 cm. They could be inserted in a cylindrical coil with an inner diameter of 35 mm, made of 7 turns of 1 mm Cu wire. This coil forms part of a tuned LC circuit, and is placed within a shielded box. To reduce the effect of variations in the dielectric permittivity of the sample, a cylindrical Faraday shield is placed between the coil and the sample. This shield consists of 0.5 mm insulated Cu wires running parallel to the axial direction of the coil. The wires are electrically interconnected and earthed at the lower side of the shield. A small slit in this part of the shield prevents the generation of eddy currents and consequent radio–frequency power losses. The coil is part of a series-tuned circuit, where impedance matching the characteristic impedance of the equipment (50 Ω) is achieved by adjusting a capacitor. The equipment is operating at frequencies near 33 MHz corresponding to an applied field of 0.78 T. The magnet is a conventional water-cooled, iron-cored electromagnet. A magnetic field gradient of 0.25 to 0.5 T/m is generated in the vertical direction by a set of conventional Anderson coils, which makes it possible to achieve a spatial resolution better than 1 mm. To determine the moisture profile over a larger region, the sample can be moved in the vertical direction by a translator driven by a step motor.

Recently, the experimental set-up by Pel et al. (1996) was employed for measuring the moisture content in cement mortar (Brocken, 1998).

5.1.1.6 Infrared Spectroscopy Method

In measuring the moisture content of porous materials by the infrared spectroscopy method, the reflection of infrared radiation from the material surface is employed. The magnitude of the reflected energy depends on the moisture content in the material, and the effect is most pronounced for the wavelengths of 1.4 μm and 1.9 μm. In order to utilize the change of reflectivity in the measurement of absolute moisture content in the materials, all other factors affecting the reflection of infrared radiation, such as the surface roughness, have to be excluded. Therefore, the measurements are performed with two wavelengths, one being 1.4 μm or 1.9 μm, where the effect of moisture is well pronounced, the second 1.3 μm, 1.6 μm or 1.75 μm, where on the contrary the dependence of the reflected energy on the moisture content is very low.

In most experimental installations, the infrared radiation supplied by the source is collimated and transmitted to the measured specimen using a reflecting prism and a lens. The reflected radiation is transmitted using a system of two lenses to a detector (PbS). Two infrared filters are placed either just before the detector or just behind the source in order to detect only the radiation of the two chosen wavelengths.

The main application of the infrared spectroscopy method is in the range of low amounts of moisture up to 10% of the maximum moisture content, where the errors are lowest. Experimental devices based on this method require empirical calibration for every material. The disadvantage of the method is that it determines only the surface moisture content, and not the average moisture content in the specimen.

5.1.1.7 Chemical Methods

Any chemical method for the determination of moisture content is based on the chemical reaction of water in the porous material with a chemical agent. The agent must have the following properties: its reaction with water has to be fast, and it has to make possible an exact quantification of water content from the reaction products; the reaction has to be specific to water only, and its completion has to be easily detectable.

Among the chemical methods, the Fischer's method is the best known. The measured specimen (usually in a fine granular form) is mixed with methanol and titrated with a solution containing iodine, sulphur(IV) oxide, pyridine and methanol. The following reactions take place after mixing:

$$H_2O + I_2 + SO_2 + 3\ C_5H_5N \rightarrow 2\ C_5H_5NHI + C_5H_5NSO_3$$

$$C_5H_5NSO_3 + CH_3OH \rightarrow C_5H_5NHSO_4CH_3.$$

Therefore, one mole of iodine, one mole of sulphur(IV) oxide, three moles of pyridine, and one mole of methanol are consumed per mole of water. The specimen is put in the titration vessel with 50–100 cm^3 of methanol. Then Fischer's agent is slowly added from a burette until it reaches the point of equivalence, i.e. until all the water in the specimen reacted. From the volume of the agent needed for the reaction, the amount of water in the specimen can be calculated from the above reaction equations. Attainment of the point of equivalence is recognized visually. Adding of a further drop of the agent leads to change in the colour of the titrated solution from yellow to red-brown. The reactivity of Fischer's agent decreases with time. Therefore, the titre of the agent, i.e. the amount of the agent in cm^3 per a known amount of water in mg, has to be determined before measurement. The titre is calculated according to the formula

$$T = \frac{mw}{V},$$
(5.75)

where T is the titre in mg/cm^3, m the mass of the standard substance or solution in mg, w the mass fraction of water contained in the standard substance or solution, $w = m_w/(m_w + m_d) = m_w/m$, m_w the mass of water, m_d the mass of the dry material, and V the amount of Fischer's agent in cm^3. The amount of water in the measured specimen can then be calculated as

$$w' = \frac{TV}{m}.$$
(5.76)

Other chemical methods are based on the detection of generated gases by the reaction of water with the added agent. The amount of gas or its pressure in a closed space is then a measure of the amount of water in the material. The reagents most used are CaC_2, CaH_2, $LiAlH_4$, which react with water according to the following equations:

$$CaC_2 + 2\,H_2O \rightarrow Ca(OH)_2 + C_2H_2$$

$$CaH_2 + 2\,H_2O \rightarrow Ca(OH)_2 + 2\,H_2$$

$$LiAlH_4 + 4\,H_2O \rightarrow LiOH + Al(OH)_3 + 4\,H_2.$$

From the point of view of safety, CaC_2 is probably the most appropriate agent of those mentioned above. The CaC_2 method measures the pressure of acetylene in a closed reaction vessel to determine the amount of water in a porous material. The amount of water is determined from a table obtained by empirical calibration. The CaC_2 method was recently employed to determine moisture content in hydrating cement paste (Poděbradská et al., 2000).

For the remaining two reagents, generation of hydrogen violently reacting with oxygen in the surrounding atmosphere presents a serious problem. Therefore, the measurement has to be done in an oxygen-free atmosphere, and the generated hydrogen is collected in a gas burette, where its volume is measured.

5.1.1.8 Ultrasonic Methods

The velocity of ultrasound propagation or its attenuation in a material depends on the composition of the material and its temperature. At constant temperature, the velocity is affected by the amount of the solid phase that can be employed for moisture measurements. The dependence of the sound velocity on moisture is nonlinear for most materials, and its dependence on temperature is so important that either the measurements have to be performed in a conditioning chamber or temperature compensations have to be made. Given these disadvantages, this method is not often used in practice.

5.1.1.9 Azeotropic Distillation Method

Some organic solvents are not miscible with water and form an azeotropic mixture with it. This can be employed to cause a direct separation of water from the analysed material. The material specimen is mixed in a boiling flask with the solvent and boiled under a reflux cooler. A pipe extension is connected between the flask and the cooler. The vapour of the azeotrope condensed in the cooler runs down into a volumetric flask, where water is separated from the solvent, and its volume is directly read on the scale of the volumetric flask. The excess solvent, which is lighter than water, runs back down to the boiling flask. The distillation is finished when the volume of water in the volumetric flask stops changing. As azeotropes, benzene or toluene are usually used. The method can be applied only for materials that do not decompose during boiling with the solvent, or do not chemically react with the solvent.

5.1.1.10 Extraction Method

In some cases, water in a porous material can be extracted using an organic solvent. As such solvents, glycerol, acetone or ethanol are often used. The amount of water in the specimen is determined on the basis of the density change of the solvent, change in its refraction index, permittivity, etc. It should be noted that not all the water contained in the material, but only the amount of water, that is bonded up to a certain limit of bond energy given by the applied solvent, is determined by the method.

5.1.2 Relative Humidity

The relative humidity of the air is measured by the methods that are commonly employed for measuring the moisture content in gases in general. We will first give a short overview of these methods (see, e.g., Fexa and Široký, 1983) and then go into details.

1. Psychrometric method
   The gas flows at a constant velocity around a wetted thermometer. The rate of evaporation of water from the wetted thermometer, and also the temperature of the thermometer (because the thermometer itself provides the heat needed for water evaporation) depend on the relative humidity of the surrounding air.

2. Dew point measurement method
   A small metallic mirror is cooled until dew forms on its surface. The temperature at which this dew appears is called the dew point and it characterizes the relative humidity.

3. Sorption methods
   Many solid materials change their electric resistance, dielectric properties, and length due to the sorption of moisture.

4. Equilibrium electrolytic moisture meter
   A sensor with a thin layer of LiCl on its surface is heated until the water vapour pressure above the LiCl solution is the same as the water vapour pressure in the measured gas.

5. Coulometric method
   A thin P_2O_5 layer absorbs moisture from the measured gas, and this moisture is immediately electrolyzed by an electric current from an external supply. The current required for this is proportional to the moisture content of the gas, on the condition that the rate of flow of the gas is constant.

6. Absorption of infrared radiation
   The absorption of infrared radiation with an appropriate wavelength is proportional to the moisture content of a gas.

7. Absorption of high-frequency energy
   The absorption of microwave energy is proportional to the moisture content of a gas.

8. Chromatographic methods
   The moisture content is determined together with another components of the analysed gas.

We will now describe in more detail each method of measuring the relative humidity of the air. The methods will be treated in order of frequency of their application in practice.

5.1.2.1 Psychrometric Method

The psychrometric method is probably the method most frequently used for the determination of relative humidity. The moisture content in the air is determined using the readings of two thermometers. One of the thermometers, the so-called "wet thermometer", is wetted by water, the other measures the temperature of the air. With decreasing relative humidity, the rate of evaporation of water from the wet thermometer increases, and the difference between the two temperatures, which is called the psychrometric difference or psychrometric depression, is greater. From the value of the psychrometric difference, the water vapour pressure, and therefore also the relative humidity, is determined from tables or from a nomogram.

The tables and nomograms are compiled by using the psychrometric equation, which can be derived as follows. The amount of water vapour $\dot{m}$ that is evaporated per unit time from the surface of the wet thermometer can be expressed as

$$\dot{m} = kSD(p'_{T_w} - p'), \tag{5.77}$$

where S is the area where the evaporation occurs, D the diffusion coefficient of water vapour in air, p'_{T_w} the pressure of saturated water vapour at the temperature of the wet thermometer T_w, p' the water vapour pressure in the air, and k a constant.

For the evaporation of this amount of water, it is necessary to supply heat power $\dot{Q}_1$,

$$\dot{Q}_1 = \dot{m} \cdot L_v = L_v kSD(p'_{T_w} - p'), \tag{5.78}$$

where L_v is the heat of evaporation of water. The heat supplied to the water reservoir of the wet thermometer consists of two parts. One of them is the heat supplied by the measured air, which is given by the heat power $\dot{Q}_2$,

$$\dot{Q}_2 = \alpha S(T - T_w), \tag{5.79}$$

where α is the heat transfer coefficient between the surface of the reservoir of the wet thermometer and the surroundings. The second part is the heat supplied by the thermometer itself, which is given by the heat power $\dot{Q}_3$,

$$\dot{Q}_3 = \lambda S' \frac{T - T_w}{l}, \tag{5.80}$$

where λ is the thermal conductivity of the thermometer, S' the cross-section of the thermometer at the place where it is wetted, T the temperature of the dry thermometer, and l the length of the thermometer.

The temperature of the wet thermometer varies depending on the difference between the heat supplied to the thermometer and the heat consumed in evaporation,

$$\dot{Q}_2 + \dot{Q}_3 - \dot{Q}_1 = C \frac{dT_w}{dt}, \tag{5.81}$$

where C is the heat capacity of the wet thermometer. Substituting for $\dot{Q}_2$, $\dot{Q}_3$, $\dot{Q}_1$ from the above relations we obtain

$$\alpha S(T - T_{\mathrm{w}}) + \lambda S' \frac{T - T_{\mathrm{w}}}{l} - L_{\mathrm{v}} k S D(p'_{T_{\mathrm{w}}} - p') = C \frac{\mathrm{d}T_{\mathrm{w}}}{\mathrm{d}t}, \tag{5.82}$$

and after some simple algebraic modifications

$$T - T_{\mathrm{w}} - \frac{1}{A_1}(p'_{T_{\mathrm{w}}} - p') = \frac{C}{\alpha S + \frac{\lambda}{l}S'} \frac{\mathrm{d}T_{\mathrm{w}}}{\mathrm{d}t}, \tag{5.83}$$

where

$$A_1 = \frac{\alpha + \frac{\lambda}{l}\frac{S'}{S}}{k L_{\mathrm{v}} D} \tag{5.84}$$

is the psychrometric constant.

In steady-state conditions, the temperature of the wet thermometer must be constant,

$$\frac{\mathrm{d}T_{\mathrm{w}}}{\mathrm{d}t} = 0, \tag{5.85}$$

and from equation (5.83) we obtain the psychrometric equation,

$$p' = p'_{T_{\mathrm{w}}} - A_1(T - T_{\mathrm{w}}). \tag{5.86}$$

Using the psychrometric equation (5.86), the pressure of the water vapour in the air can be calculated, provided temperatures T and T_{w} and the psychrometric constant A_1 are known. The psychrometric constant is usually determined empirically.

Error in moisture measurements by the psychrometric method depends on the air temperature, relative humidity, shape and dimension of the wet thermometer, solar radiation, etc. At lower temperatures, the error in the relative humidity increases, because evaporation is lower, and the difference $T - T_{\mathrm{w}}$ is small. Also for higher relative humidities, $T - T_{\mathrm{w}}$ decreases, and the accuracy decreases. On the other hand, for very low relative humidities it is difficult to achieve thorough wetting of the wet thermometer. The measurement error is also negatively affected by increase in the ratio S'/S, and therefore by the heat supplied by conduction along the thermometer. Therefore, resistance thermometers and thermocouples are more suitable than liquid thermometers. The thermal radiation induced by various sources can also have a negative effect on the accuracy of measurement. Consequently, psychrometers have shading shields, and the external parts are polished. Frequent sources of errors are a dirty mantle for wetting the wet thermometer, contaminated water for wetting, inconvenient application of the mantle or using tables for a different atmospheric pressure.

The main advantages of psychrometers are that they are simple, that for temperatures above 0°C they are sufficiently accurate for most applications, that have a relatively small time constant (i.e. measure relatively quickly), and they do not need any special calibration. Their main disadvantages are the decrease in their sensitivity with decreasing temperature, the dependence of the reading on the atmospheric pressure, the need to keep the velocity of air flow constant, that at lower relative humidities the wetting of wet thermometer is difficult, that for lower temperatures the reading depends on the phase of the water on the wet thermometer, and that psychrometers are very sensitive to contamination. These disadvantages have increasingly led to psychrometers being replaced by other types of moisture meters, particularly in recent decades.

5.1.2.2 Dew Point Measurement Method

The method is based on the determination of the temperature of the surface of water, when a dynamic equilibrium between the amount of water evaporated from the surface and the amount of condensed water vapour from the adjacent gas layer is established. In an equilibrium state, the adjacent gas layer will have the same temperature as the water surface and the relative humidity of this layer will be 100%, i.e. the water vapour pressure achieves its maximum value for the given temperature. Therefore, determining the temperature when the equilibrium is established, i.e. the temperature at dew point, makes it possible to determine the water vapour pressure in a gas using tables of maximum pressures.

The basic method for dew point measurement involves cooling the surface of a sensor (usually a metallic mirror), until a condensate appears. Then cooling is stopped and heating started, until the condensate disappears. The mean value between the temperatures at which the condensate appears and disappears is considered to be the equilibrium temperature. This measuring method depends on the type of material, surface roughness of the material, rate of cooling and heating, etc. Therefore, the temperature of the established equilibrium between the condensate and water vapour is sometimes measured. The temperature of the mirror is regulated in such a way that the thickness of the condensate layer does not change. In this way, an equilibrium state between the condensate and the water vapour is established. At that moment, the temperature of the dew point is measured.

The relation between the relative humidity and the dew point can be derived from the Clausius–Clapeyron equation,

$$p' = p_0 \cdot \exp\left[\frac{\Delta H}{R}\left(\frac{1}{T_0} - \frac{1}{T}\right)\right], \tag{5.87}$$

where ΔH is the latent heat of evaporation of water, p_0 the pressure of saturated water vapour at the temperature T_0 (in K), p' the pressure of saturated water vapour at the temperature T (in K), and R the universal gas constant.

If the relative humidity at absolute temperature T is φ, then the water vapour pressure will be

$$p = \varphi p'_T, \tag{5.88}$$

where p'_T is the pressure of saturated water vapour at the temperature T. At the same time, the following relation is valid:

$$p = p'_{T_{dp}}, \tag{5.89}$$

where $p'_{T_{dp}}$ is the pressure of saturated water vapour at the dew point temperature T_{dp}. Therefore,

$$\varphi p'_T = p'_{T_{dp}}. \tag{5.90}$$

Substituting equation (5.87) into (5.90) leads to

$$p_0 \cdot \exp\left[\frac{\Delta H}{R}\left(\frac{1}{T_0} - \frac{1}{T_{dp}}\right)\right] = \varphi p_0 \cdot \exp\left[\frac{\Delta H}{R}\left(\frac{1}{T_0} - \frac{1}{T}\right)\right], \tag{5.91}$$

and therefore

$$\ln \varphi = \frac{\Delta H}{R}\left(\frac{1}{T} - \frac{1}{T_{dp}}\right), \tag{5.92}$$

which results in

$$T_{dp} = \frac{1}{\frac{1}{T} - \frac{R}{\Delta H}\ln \varphi}. \tag{5.93}$$

Differentiation of equation (5.93) with respect to φ then gives

$$\frac{dT_{dp}}{d\varphi} = \frac{\frac{R}{\Delta H}\frac{1}{\varphi}}{\left(\frac{1}{T} - \frac{R}{\Delta H}\ln \varphi\right)^2}. \tag{5.94}$$

Equation (5.94) makes it possible to determine the accuracy of the measurements of dew point temperature needed to achieve the chosen accuracy of the determination of the relative humidity. In the usual temperature range from $-20°C$ to $+40°C$, the precision of the temperature measurements needed to achieve the usual 1% accuracy of relative humidity (in absolute percent) does not differ very much; $0.1°C$ is quite sufficient for both positive and negative temperatures on the Celsius scale. This is a substantial advantage of the method compared to the psychrometric method.

For the temperature measurements, thermocouples, resistance thermometers and thermistors are used most often.

The main advantage of the dew point measurement method is that its precision is sufficient both for low temperatures and low relative humidities, and the measuring devices do not require any special calibration. The main disadvantages are that a device for cooling the sensor is necessary, the precision decreases with increasing relative humidity, and for lower temperatures the phase of the condensed water is important.

5.1.2.3 Equilibrium Electrolytic Moisture Meter

The method is based on the determination of an equilibrium state between a hygroscopic substance and water vapour. It employs the fact that for every ionic salt, the temperature of the equilibrium state between the water vapour pressure above a saturated solution of the ionic salt and the water vapour pressure in the surroundings is exactly defined and known. This can be utilized for the determination of the absolute moisture content of the measured gas. If the saturated salt solution is exposed to surroundings with a lower value of water vapour pressure than the equilibrium pressure above the solution, water evaporation from the solution will occur. On the other hand, at the higher water vapour pressure, the moisture from the surroundings is absorbed by the solution. This process is accompanied by a substantial change in the electric conductivity of the solution, which can be used to provide automatic heating of the solution to the temperature, at which the water vapour pressure in the measured gas would be equal to the equilibrium pressure above the saturated solution of the given salt.

Electrolytic moisture meters working on this principle usually employ LiCl as the ionic salt, because its equilibrium water vapour pressure is sufficiently low at room temperature. The sensor in this type of moisture meter is usually a cylinder with a wound glass fibre cloth saturated with the LiCl solution. The electrodes comprise two wires wound on the cloth and supplied with an alternating voltage. If the LiCl solution is dilute, i.e. the equilibrium water vapour pressure above it is higher than the water vapour pressure in the measured gas, the electric conductivity of the solution is high, and therefore the electric current between the electrodes is also relatively high and warms up the solution. Due to the heating, evaporation of water from the solution starts, the conductivity decreases, and the electric current decreases until the pressure of water vapour above the solution is the same as the measured pressure of water vapour in the air. The temperature of the solution corresponding to this equilibrium state is measured, usually by a resistance thermometer in the cylinder of the sensor. This temperature is a measure of the absolute moisture content of the measured gas. If the relative humidity is also required, an additional resistance thermometer measuring the gas temperature is needed.

The method may be compared to both dew point measurements and psychrometric measurements. Compared to the dew point method, the equilibrium electrolytic moisture meter does not require cooling of the sensor, which is always more difficult than heating. Compared to the psychrometric method, the measurements can also be made for static surroundings without any gas motion.

The range of applicability of the electrolytic moisture meter derives from the fact that it employs the different values of the partial pressure of water vapour on pure water and on the saturated LiCl solution. For instance for a temperature of 27°C it is possible to measure the minimal value of relative humidity 11%. This minimal value increases with decreasing temperature because the pressure of water vapour above the saturated LiCl solution approaches the value of saturated water vapour pressure.

The limited measuring range for lower temperatures is the main disadvantage of the electrolytic moisture meter. Also, the response time increases with decreasing temperature. The main advantages are that the devices are simple and reliable, they can work over a wide range of velocities of the measured gas, and the sensor has a low sensitivity to contamination by electrically nonconducting dust.

5.1.2.4 Coulometric Method

In the coulometric method all moisture contained in the measured gas is absorbed by a film of a hygroscopic substance, and the absorbed moisture is at the same time electrolyzed by a direct current. The hygroscopic substance is usually P_2O_5. In the sensor the following reactions occur:

$$P_2O_5 + H_2O \rightarrow 2\,HPO_3$$

$$2\,HPO_3 \rightarrow H_2 + 0.5\,O_2 + P_2O_5.$$

The film of P_2O_5 is kept almost dry all the time in this way, and so is able to absorb further moisture. According to Faraday's law, the electrolysis of 1 mole of water requires a charge of 1.93×10^5 C. Therefore, on the assumption that the rate of flow of the gas is constant, the current necessary for the electrolysis of the absorbed moisture is a direct measure of the moisture content of the gas.

The moisture meters of this type are suitable for measuring very small amounts of moisture, because all moisture is absorbed by the sensor. They do not require any special calibration, have sufficient accuracy and dynamic behaviour. The sensor can be regenerated several times. The main disadvantage of the method is that its readings depend on the rate of flow of the gas, which has to be known exactly, and that it cannot be used for higher moisture content.

5.1.2.5 Sorption Methods

These methods utilize changes in physical and chemical properties of materials due to change in the amount of absorbed water for moisture measurements. Sensors of sorption type have to satisfy the following requirements:

1. The response both in the sorption and desorption direction has to be fast, with no or very little hysteresis.

2. The change in the measured quantity has to be sufficiently large and if possible should have a linear relation to the relative humidity or to the dew point.

3. Calibration curves have to be stable under common conditions and the sensitive layer should be easy to clean or regenerate.

4. The measurements should be possible over a wide temperature range and the temperature effect on the reading of the device should be negligible.

The moisture absorbed by the material can cause a change in its volume, mass, electric resistance, permittivity, etc. Sorption moisture meters can therefore be of different kinds, such as dilatation, resistance, capacitance, resonance and semiconductor moisture meters.

Dilatation moisture meters are based on the hygric expansion/shrinkage of materials. The most popular representative of this type is the hair moisture meter. A human degreased hair increases its length by 2.5% with change in relative humidity from 0% to 100%. This dependence is nonlinear, approximately logarithmic. The measured range is from 15% to 85% of relative humidity, because outside these limits the calibration curve can be irreversibly shifted. The moisture meter reading depends significantly on temperature. Because of its disadvantages, such a moisture meter is not suitable for industrial applications. Dilatation moisture meters can also use synthetic materials such as nylon, but their properties are not much better than those of the human hair.

Resistance moisture meters determine the amount of absorbed water on the basis of a change of electric resistance. The sensor of a moisture meter utilizing changes in electrolytic conductivity usually consists of a plate with electrodes. The plate can be either ceramic or a suitable polymer, and on its surface, there is a solution of a ionic salt, most often LiCl. The resistance of the sensor changes from 10^8 Ω to 10^4 Ω for changes of relative humidity from 0% to 100%. The resistance also depends on temperature. Therefore, temperature has to be measured as well, and the necessary corrections introduced. The dependence of resistance on the relative humidity is approximately exponential. Besides LiCl, other materials used include SeO_2, BaF_2, ZnO, Cr_2O_3.

Capacitance moisture meters use sensors that are very similar to resistance sensors. With the sorption of moisture by a material, both its resistance and its capacitance are changed. Therefore, in general, its impedance is also changed. In the most cases, one of these prevails, so either resistance or capacitance can be measured. Measuring the capacitance is more difficult than measuring the resistance, but it has some substantial advantages, among them a very high sensitivity and a lower effect of contamination of sensors compared to resistance sensors. These advantages have led to the development of a wide range of capacitance moisture meters. Any capacitance sensor is in principle a capacitor with a dielectric formed by a thin layer of a material, which absorbs moisture from the surroundings in a reversible way. Most often used materials are polymers or metallic oxides, and the usual thickness of the film is about 10 μm. Capacitance moisture meters based on Al_2O_3 are the most frequently used. The impedance of the sensor is generally measured at lower frequencies, usually in the range from 50 Hz to 150 Hz. The advantage of capacitance moisture meters is their fast response, usually better than 1 s.

A gravimetric moisture meter is a moisture meter of sorption type that absorbs moisture in an irreversible way, unlike the types mentioned above. Such a moisture meter is often employed for the calibration of other moisture meters, because it makes it possible to determine the real mass of water vapour in the given volume of a gas.

Generally, the main advantages of sorption moisture meters are that they are simple and relatively cheap, have good sensitivity, and small time constant, particularly in the sorption direction. Also, they do not require gas circulation and can take measurements in stationary conditions, and can operate over a wide moisture range. The main disadvantages are the effect of hysteresis, the temperature dependence of the sorption and desorption, and that the calibration curves are sometimes unstable due to the ageing of the material used in the sensor.

5.1.2.6 Spectral Methods

Water vapour absorbs radiation in many parts of the electromagnetic spectrum. For practical purposes, absorption in infrared range is the most important, particularly for the wavelengths 1.13 μm, 1.38 μm, 1.87 μm and 2.66 μm. The decrease in the intensity of electromagnetic waves propagating in a homogeneous medium depends on the thickness of the absorbing layer and concentration of the absorbing particles, on temperature, and on the radiation wavelength. The intensity decrease can be expressed by the Lambert–Beer law,

$$-\mathrm{d}I = kIc\,\mathrm{d}l, \tag{5.95}$$

where I is the intensity, l the thickness of the absorbing layer, c the concentration of absorbing particles, and k the proportionality constant. By integration of equation (5.95), we obtain

$$\ln\frac{I}{I_0} = -kcl, \tag{5.96}$$

where I_0 is the intensity of the incident radiation. Relation (5.96) can also be expressed as decadic logarithms,

$$\log\frac{I}{I_0} = -\epsilon cl = -A, \tag{5.97}$$

where A is the absorbance, ϵ the molar (linear) absorption coefficient, and c the molar concentration of the absorbing material.

The dependence of absorbance on the water vapour concentration is usually expressed by empirical relations, for instance

$$A = C\sqrt{b} \cdot (p + p')^K, \tag{5.98}$$

where b is the thickness of a layer of liquid water that would be formed by condensation of water vapour in column of the measured gas with the same cross-section, p the total gas pressure, p' the partial pressure of water vapour, K a constant ($K \sim 0.3$), and C a constant depending on the radiation wavelength.

According to equation (5.98), for small concentrations absorption increases as the square root of concentration, while at higher concentrations, when the condition $p >> p'$ is no longer valid, absorption increases more slowly. Therefore, the sensitivity of measurements decreases with increasing water vapour concentration. In order to increase the accuracy, the measurements are usually performed by a comparative treatment, whereby the radiation absorption at some characteristic wavelength is compared with the absorption at another wavelength in its vicinity.

Infrared analysers are either dispersion or dispersionless. Dispersion devices work with monochromatic radiation, which is obtained from dispersion of electromagnetic waves by a prism or a grid. Dispersionless devices delimit the radiation of desired wavelength from an infrared source by means of two narrow-band filters. One filter lets through the radiation of wavelength characteristic for water vapour absorption, usually 1.4 μm or 1.9 μm, the other delimits the wavelength of the reference radiation, where almost no absorption occurs.

Water vapour also absorbs ultraviolet radiation at 121.5 nm, where gases in the air absorb very little. In the microwave range water vapour absorbs at wavelengths 1.33 cm and 1.61 mm.

The advantages of the spectral methods consist in the fact that the measurements are selective for most applications, and can be performed over a wide range of relative humidity. The disadvantages are, that the devices are relatively complicated and the measurements depend on temperature and pressure in a significant way.

5.1.3 Temperature

Temperature change affects most physical parameters of materials. Therefore, a variety of physical principles can be employed for temperature measurements. The methods used most often are based on volume thermal expansion, changes in electric resistance and electric voltage generation.

The thermal expansion of solids, liquids and gases was the first principle applied in practice. The first thermometer was designed by Galileo in 1600 and based on the thermal expansion of air. Historically the second was an alcohol thermometer. The first reproducible scale was introduced by Fahrenheit, who also devised a mercury thermometer. As the zero point on the scale (0°F), the lowest temperature measured at that time in the winter period was chosen, and 100°F was set at the temperature of the human body. Thus, the freezing point of water corresponded to 32°F, water boiling point to 212°F. The second temperature scale was introduced by Reaumur, who denoted the freezing point of water as 0°R, and the boiling point of water as 80°R. He employed an alcohol mixture, with 25% water. As the mixture exhibited an 80% increase in volume between the freezing and boiling points of water, he divided the scale between these two points into 80 divisions.

In 1742, the currently most often used Celsius scale was introduced, which set the freezing point of water at 0°C and its boiling point at 100°C. Celsius used mercury, because mercury exhibits a relatively uniform volume increase with increasing temperature between these two points. The data obtained by the mercury scale are affected by the glass tube, which also changes its volume with temperature. This had to be taken into account in denoting the divisions on the scale.

A gas temperature scale was derived from the change of volume or pressure of a gas with increasing temperature. Physically, this scale was based on the application of Gay-Lussac's law, which in the original statement was expressed in the form

$$\frac{\Delta V}{\Delta \tau} = \text{const. at constant pressure,} \tag{5.99}$$

where V is the volume and τ the temperature. Therefore, we also have

$$\frac{V - V_0}{\tau - \tau_0} = \text{const.,} \tag{5.100}$$

and denoting

$$\text{const.} = \alpha V_0 \tag{5.101}$$

we obtain

$$V - V_0 = \alpha V_0 (\tau - \tau_0), \tag{5.102}$$

and finally

$$V = V_0 [1 + \alpha(\tau - \tau_0)]. \tag{5.103}$$

Gay-Lussac and Dalton discovered that the volume thermal expansion coefficient α is independent of the gas and always has the value

$$\alpha = \frac{1}{273.15} \, (^\circ C)^{-1}. \tag{5.104}$$

Therefore, choosing τ_0 as the freezing point of water under standard conditions,

$$t = \tau - \tau_0 \tag{5.105}$$

is the temperature on Celsius scale. By a modification of equation (5.103), we obtain

$$\frac{V}{V_0} = 1 + \alpha(\tau - \tau_0), \tag{5.106}$$

$$\frac{1}{\alpha}\frac{V}{V_0} = \frac{1}{\alpha} + \tau - \tau_0, \tag{5.107}$$

and therefore

$$T = \frac{1}{\alpha} + \tau - \tau_0 = \frac{1}{\alpha} + t = 273.15 + t = 273.15\frac{V}{V_0}, \tag{5.108}$$

where T is the absolute temperature, with the zero point on its scale 273.15°C lower than the zero point of the Celsius scale. The absolute temperature scale is given in kelvins after Lord Kelvin, who first introduced it. Taking into account the definitional relation (5.108), we have

$$T = t + 273.15 \tag{5.109}$$

and 1 K = 1°C. The Kelvin scale has positive values only. The zero point on its scale cannot be achieved by a finite number of thermodynamic processes, as expressed by the third law of thermodynamics, so negative values are not possible.

Introduction of the absolute temperature consequently made it possible to introduce the absolute thermodynamic scale using the reversible Carnot cycle. The efficiency η of a heat engine in a Carnot cycle is independent of the substances involved. It depends on the upper and lower temperatures T_1, T_2 of the cycle only, and is given as

$$\eta = \frac{Q_1 - Q_2}{Q_1} = \frac{T_1 - T_2}{T_1}, \tag{5.110}$$

and therefore

$$\frac{Q_2}{Q_1} = \frac{T_2}{T_1}, \tag{5.111}$$

where Q_1 is the heat received by the system from the reservoir at temperature T_1, Q_2 the heat transmitted by the system to the reservoir at temperature T_2. Assigning now a certain value, for instance the temperature of the triple point of water, 273.15 K, to the temperature T_1, then the temperature T_2 is given by equation (5.111) and can be determined by measuring the heats Q_1, Q_2.

Thermometers can be classified according to the application method and to the physical principle of measurements.

According to the application method we have:

- Contact thermometers

- Contactless thermometers.

According to the <u>physical principle</u> we have:

- Dilatation thermometers, which are based on the thermal expansion of solids, liquids or gases.

- Electric thermometers, which employ either electric resistance or the thermoelectric effect.

- Special thermometers, which are based on changes in other physical quantities with temperature.

We will now describe the principles of the various types of thermometers in more detail.

5.1.3.1 Contact Dilatation Thermometers

This type of thermometer is historically the oldest and the most widely used until now. They employ mechanical principles of temperature measurement, such as the measurement of volume expansion of liquids, the pressure of gases or linear expansion of solids. The temperature sensors are cheap and undemanding to operate. On the other hand, it is mostly very difficult to get temperature data in a digital form, given the nature and construction of this type of thermometer, which are still made more or less in their historical forms.

<u>Glass liquid thermometers</u> have been the most frequently used in practice over the whole history of thermometers. They measure the elongation of a liquid column in a long capillary tube. The liquid supply is a small flask in direct contact with the measured medium. At the other end of the capillary tube, there is an expansion flask in case of exceeding the measuring range. Temperature data are read directly from the scale on the capillary tube. In the space above the liquid in the capillary tube there is a vacuum to avoid the distortion of data due to the air pressure variations in a closed space.

The basic properties of the glass thermometers depend on the liquid used, on the diameter of the capillary and on the type of glass. The following liquids are the most commonly used: ethanol ($-130°C$ to $+50°C$), mercury ($-30°C$ to $+150°C$), isopentane ($-195°C$ to $+35°C$), pentane ($-130°C$ to $+35°C$), and gallium ($-15°C$ to $+1500°C$).

Glass thermometers are simple, reliable, cheap and precise. Their disadvantages include brittleness, limited possibilities of data digitalization, and sometimes also bad readability. The most frequently used are mercury thermometers.

<u>Liquid pressure thermometers</u> consist of a contact flask, connecting capillary tube and a deformation pressure gauge. The whole system is metallic,

and it is fully filled by the measuring liquid. The deformation part is constructed as a spiral tube to magnify the length changes. The change of volume of the liquid due to the temperature change is converted to a pressure change induced by connecting the end of the tube to a mechanical gear. In this way, the change in position of the deformation part is transmitted by the mechanical gear to an indicator with the temperature scale. The length of the spiral part can be as much as 50 m. The pressure of the liquid in the measuring system is usually about 10 MPa, which increases its boiling point and therefore also the measuring range. The following liquids are used most often: ethanol ($-45°$C to $+150°$C), mercury ($-30°$C to $+550°$C), xylol ($-40°$C to $+440°$C).

The advantages of the pressure thermometers are a wide measuring range, robustness, and applicability in severe conditions. The disadvantages are the corrections necessary with changes of ambient temperature and hydrostatic pressure (if the measuring flask is immersed in a liquid).

Steam pressure thermometers have a similar construction to liquid pressure thermometers. The only differences are in the applied pressure gauge and in the charge. The measuring flask is filled only partially by the measuring liquid. The space above the liquid is filled by its saturated vapour. The thermometers employ the dependence between the temperature and the equilibrium pressure in a system consisting of one component in two phases. This dependence is given by the Clausius–Clapeyron equation (see, e.g., Feynman et al., 1966), which can be written in an integral form as follows:

$$\log p = A - \frac{B}{T} + C \log T. \tag{5.112}$$

The most often used charges for these thermometers are: ethanol ($-10°$C to $+160°$C), methyl chloride ($-50°$C to $+70°$C), ethyl ether ($-20°$C to $+120°$C), and toluene ($+60°$C to $+160°$C). As follows from equation (5.112), the dependence of temperature on the pressure is highly nonlinear, with the scale division increasing towards higher temperatures. As the pressure of saturated vapour on the end of the scale increases rapidly, there is a risk of damage to the pressure gauge through overheating. A possible protection against overheating is to use an exact amount of the charge, so that the whole liquid is vaporized if the maximum temperature range is exceeded. In that case the vapour is superheated and the pressure increases only slowly.

Gas thermometers employ two basic principles: either the change of volume of a gas with temperature at constant pressure, or the change of pressure of a gas with temperature at constant volume. For a gas thermometer working at constant pressure, the temperature is calculated according to the relation

$$t = 100 \cdot \frac{V - V_0}{V_{100} - V_0}, \tag{5.113}$$

where V_0 is the volume of the gas at $0°$C, and V_{100} the volume at $100°$C. Common gases such as hydrogen, helium, nitrogen are used as the charge in such thermometers. The basic thermodynamic rule, that temperature increase of 1 K leads to an increase of $\frac{1}{273.15}$ of the original volume as in the case of ideal gases, is not exactly valid and has to be corrected depending on the gas used.

In the case of gas thermometers working at constant volume, the pressure is measured via its dependence on temperature using an open mercury manometer. The manometer is constructed in such a way that its arms (usually two glass tubes with a flexible connection) can be lifted, and a constant volume of the measuring gas can be maintained. The gas in a flask is closed by the mercury in the glass tubes, and its pressure is equal to the sum of the hydrostatic pressure of mercury, given by the difference of levels in the two arms of the manometer, and the atmospheric pressure. The temperature is calculated according to the relation

$$t = 100 \cdot \frac{p - p_0}{p_{100} - p_0}, \tag{5.114}$$

where p_0 is the pressure of the gas at $0°C$, p_{100} the pressure at $100°C$,

$$p = p_{atm} \pm \rho g h, \tag{5.115}$$

p_{atm} the atmospheric pressure, ρ the density of mercury, g the gravity, h the difference of levels in the arms of the manometers, and the $\pm$ sign expresses the fact that the pressure in the gas can be either higher or lower than the atmospheric pressure.

Bimetallic thermometers employ a bimetal strip consisting of two metals with different thermal expansion coefficients as a sensor. The most often used materials are brass and a chromium–nickel alloy. The materials are bonded by rolling. The strips are usually formed in a spiral shape, because the accuracy of these thermometers depends for practical purposes only on the length of the strip, and a spiral configuration is the most appropriate from this point of view.

5.1.3.2 Electric Thermometers

Electric thermometers employ changes in electric properties of materials that depend on change in temperature.

Thermoelectric thermometers use a thermocouple as a sensor. The physical principle of a thermocouple is an application of the thermoelectric effect. If the ends of two different metals are conductively connected, for instance by welding, so that they form a closed circuit, and these welded ends are put into two media with different temperatures, a thermoelectric voltage is induced in the circuit. This can be measured by a voltmeter.

The thermoelectric voltage U depends on the temperature difference Δt according to the general relation

$$U = \sum_{i=1}^{n} a_i (\Delta t)^i, \tag{5.116}$$

Table 5.1 Marking of thermocouples

Mark	Materials	Colour	Former marking
T	Cu–CuNi	Orange	Copper–constantan
J	Fe–CuNi	Black	Iron–constantan
E	NiCr–CuNi	Brown	Chromel–constantan
K	NiCr–NiAl	Yellow	Chromel–alumel
S	Pt10%Rh–Pt	Green	PtRh–Pt, Pt10Rh–Pt
R	Pt13%Rh–Pt	Green	PtRh13–Pt
B	Pt30%Rh–Pt6%Rh	Violet	PtRh18
N	NiCrSi–NiSi		Nicrosyl–nisil

Table 5.2 Temperature range of thermocouples

Mark	Temperature range		
T	−270°C	to	+400°C
J	−210°C	to	+1200°C
E	−270°C	to	+1000°C
K	−270°C	to	+1370°C
S	−50°C	to	+1760°C
R	−50°C	to	+1760°C
B	0°C	to	+1820°C
N	−270°C	to	+1300°C

where a_i are empirical constants determined experimentally.

The materials for thermocouples are chosen according to several basic criteria: the dependence $U(\Delta t)$ should be as linear as possible; the coefficient a_1, $U = a_1 \Delta t$, should be as high as possible; the value of $U(\Delta t)$ should be as stable as possible during long-term uses; the materials should be resistant to chemical and mechanical effects.

The various types of thermocouple are marked according to the international standard of the IEC (see Brandes and Brook, 1992) as shown in Table 5.1. The temperature range of the different types of thermocouple is given in Table 5.2. The positive side of a thermocouple is always denoted as P, the negative as N.

In laboratory conditions, the simplest temperature measurement by thermocouples is performed by placing one end in a mixture of water and ice, which has the temperature 0°C, and the other end into the measured medium. The temperature difference between the ends is then directly equal to the temperature in °C. The thermoelectric voltage is measured by a voltmeter, and using the inverse relation,

$$t = \sum_{i=1}^{n} b_i U^i,$$ (5.117)

of the relation (5.116). This experimental set-up is currently used only exceptionally, for the demonstration purposes only. Practical measurements are made by measuring control units, which either have a built-in thermostat maintaining a constant temperature at the place where the reference connections are, or electric compensation of the temperature of the reference connections is provided. The measured thermoelectric voltage is converted to a temperature by calculation in a built-in processor, so temperature readings are obtained in a digital form suitable for electronic data transfer.

Resistance thermometers use the increase in resistance of metals with temperature and decrease in resistance of semiconductors, electrolytes and carbon. The dependence of the resistance of a material on temperature is expressed by the temperature coefficient of resistance α,

$$R = R_o(1 + \alpha \Delta t).$$ (5.118)

Over a wider temperature range, α is mostly not constant but varies as $\alpha = \alpha(t)$. Therefore, quadratic or cubic relations are needed.

Metallic resistance thermometers use resistance wires of chemically pure metals coiled on a ceramic, glass or mica base. The diameter of the wire is from 0.02 to 0.06 mm. Platinum is material that is the most suitable for measuring resistors. It has a relatively high coefficient α, $\alpha = (0.385 - 0.391) \times 10^{-2}$ K^{-1}, which does not vary with time in a significant way. The basic measuring range for platinum thermometers is $-200°$C to $+850°$C, and the measuring resistances are $R_0 = 100\ \Omega$ at $0°$C, $R_{100} = 138.5\ \Omega$ at $100°$C. The presence of electric current in the measuring resistor always leads to Joule heat production, which could cause measuring errors due to the resistor heating. Therefore, the currents have to be low enough, typically lower than 10 mA.

Semiconductor resistance thermometers are called thermistors. These are used for temperature measurements in the range of $-200°$C to $+200°$C. Thermistors are about ten times more sensitive than metallic materials but their characteristic (i.e. the dependence of resistance on temperature) is highly nonlinear. There are two types of thermistors, so called "negistors" (NTC) with a negative α coefficient, and "posistors" (PTC) with α positive. For temperature measurements, only the negistors are used, the semiconductors with electron conductivity N. However, for traditional reasons these are mostly called by the more general term thermistors. The main reason for using negistors is better stability and resistance to chemical and physical changes compared to the posistors.

The temperature dependence of the resistance of the most of thermistors is given by the approximate relation

$$R = A \exp\left(\frac{B}{T}\right). \tag{5.119}$$

Taking into account the nonlinearity of equation (5.119), the temperature coefficient of resistance is defined by the differential relation

$$\alpha_R = \frac{1}{R}\frac{dR}{dT}. \tag{5.120}$$

Applying equation (5.119) to the determination of α_R, we obtain first

$$\ln R = \ln A + \frac{B}{T}, \tag{5.121}$$

then differentiating equation (5.121) with respect to temperature

$$\frac{1}{R}\frac{dR}{dT} = -\frac{B}{T^2}, \tag{5.122}$$

and comparing equations (5.120) and (5.122) finally we arrive at

$$\alpha_R = -\frac{B}{T^2}. \tag{5.123}$$

As with all resistance thermometers, the data measured by thermistors is affected by the Joule heat, and the electric currents should be as low as possible. The maximum permitted values of measuring currents are determined by the so-called loading constant D, given in W/K, which is defined as

$$D = \frac{P}{\Delta T} = \frac{RI^2}{\Delta T}. \tag{5.124}$$

As follows from equation (5.124), D is the power input necessary for heating the thermistor by 1 K. If a maximum permittable temperature increase ΔT is specified, the maximum current can be calculated from equation (5.124) as follows:

$$I_{\max} = \sqrt{\frac{D\Delta T}{R}}. \tag{5.125}$$

Basic materials for making thermistors are NiO, Mn_2O_3, Co_2O_3, and other oxides. For series production, powder materials sintering technology is commonly used. This method of production also has the advantage that by mixing semiconducting materials the necessary properties can be obtained. To make the most frequently used bead thermistors, two small PtIr wires are first arranged parallel on a base at a distance of 5 to 10 wire diameters. A drop of water with dispersed fine grains of the basic semiconducting material is then placed on the wires. Finally, after drying the base with the wires and semiconducting oxides, it is fired in a furnace at temperatures of 1000°C to 1400°C. The oxides undergo shrinkage due to the firing, and in this way a rigid joint between the semiconductor and the wires is built. The diameter of the beads is usually 0.15 mm to 2.5 mm, the diameter of the wires 0.025 mm to 0.15 mm. The main disadvantage of all thermistors is that they cannot be produced with exactly the same properties, and therefore they have to be calibrated individually.

5.1.3.3 Pyrometers

The measuring range of contact thermometers is limited by their operation factors, the stability of sensors, mechanical damage, and corrosion. The upper limit of practical applications of thermocouples is about 1800°C. For measuring higher temperatures, pyrometers are mostly used, which determine the amount of energy radiated from the surface of the measured body. Pyrometers have the advantage that they do not disturb the temperature field in the specimen, which cannot be avoided with contact thermometers. Also, they are easily transportable and make pointwise temperature measurements possible.

Radiation pyrometers determine the temperature using the measured total heat flow radiated by the surface of a body in the whole wavelength spectrum. According to the Stefan–Boltzmann law, the total radiation intensity (in W/m^2) emitted by a black body of the temperature T_0 is given by the relation

$$I = \sigma T_0^4, \tag{5.126}$$

where $\sigma = 5.67 \times 10^{-8} \ W/m^2 K^4$ is the coefficient of black body radiation. Real bodies are grey and emit radiation of lower intensity,

$$I = \epsilon \sigma T_0^4, \tag{5.127}$$

where ϵ is the emissivity, $\epsilon \in \langle 0, 1 \rangle$. The emissivity depends on the material of the emitter, its surface quality and its temperature. Therefore, correction of data given by pyrometer for grey bodies is very difficult and radiation pyrometers are used to measure the temperature of black and near-black bodies, i.e. those having emissivity close to 1.

In temperature measurements by radiation pyrometers, the thermal radiation over the whole wavelength range is focussed to a receiver by an optical system. As only the rays from the measured body need to impinge on the receiver to achieve correct measurements, an eyepiece is used. The radiation from the measured source is first amplified about 100 times, and then focussed to a serial thermocouple connected to a millivoltmeter or to a compensator, with a scale divided in °C.

Due to the absorptivity and reflectivity of the optical system, the thermal radiation detected does not correspond exactly to the temperature of the emitter. This effect has to be respected in the calibration of the scale of the particular device. The usual temperature range for application of radiation pyrometers is $-40°C$ to $+300°C$, but they can also be used for significantly higher temperatures, to about $2000°C$ and higher.

Photoelectric (zone) pyrometers detect thermal radiation over a narrow range of wavelengths, which is delimited by filters and by the spectral sensitivity of sensors. Quantum sensors employ the photoelectric effect in semiconductors. Due to the absorption of a photon, an electron jumps into the conduction band, which results in an increase of electric conductivity of the sensor (a photoresistor or a photodiode). This type of sensor is more sensitive than are thermal sensors. For a narrow wavelength range of $\langle \lambda_1, \lambda_2 \rangle$, zone emissivity is introduced, defined by the relation

$$\epsilon_\lambda = \frac{\int_{\lambda_1}^{\lambda_2} I_\lambda \, d\lambda}{\int_{\lambda_1}^{\lambda_2} I_{\lambda_0} \, d\lambda}, \tag{5.128}$$

where I_{λ_0} is the intensity of black body radiation.

The pyrometer shows the temperature T_0 that a black body would have in the same wavelength range of $\langle \lambda_1, \lambda_2 \rangle$ at the same zone radiance. The relation between the real temperature T and the temperature T_0 given by the zone pyrometer can be expressed via the dependence on the zone emissivity and the characteristic wavelength $\lambda_c \in \langle \lambda_1, \lambda_2 \rangle$ as follows:

$$\frac{1}{T} = \frac{1}{T_0} + \frac{\lambda_c}{C_2} \ln \epsilon_\lambda, \tag{5.129}$$

where $C_2 = 0.0143$ Km is the second radiation constant (see Modest, 1993). Therefore, using the characteristic wavelength λ_c, a correction of the measured data can be made. The error due to the uncertainty in the determination of zone emissivity is lower than for radiation pyrometers.

Using a proper sensor, it is possible to choose a working wavelength zone that the absorption of the medium between the measured object and the pyrometer is minimal. The sensitivity of sensors depends on the radiation wavelength. The following sensors are commonly used: selenium photocells (0.3–0.8 μm), silicon photocells (0.5–1.1 μm), germanium photodiodes (0.4–1.8 μm), and PbS photoresistors (0.5–3.6 μm). The output signal of a photoelectric pyrometer is provided in the form of voltage on a resistor, which is connected in a circuit together with the sensor.

Photoelectric pyrometers are suitable for measurements of fast temperature changes. Digitalization of measured data is readily possible, as well as their storage in electronic form.

Spectral (luminance) pyrometers employ a narrow spectral zone in the range of visible radiation to provide the temperature measurements. The basic measuring principle is the comparison of the luminance of the reference source and the luminance of the measured radiator in the part of the spectrum delimited by the applied filter. Two basic methods of comparison are used. The first is the change of luminance of the reference source, which is a lamp in a heating circuit. The luminance of the lamp is adjusted in such a way that the filament of the lamp disappears in the background of the radiator. The magnitude of the heating current, which is measured by a milliammeter, is the measure of the spectral luminance temperature. The milliammeter is usually calibrated in advance, so that it expresses the temperature directly. In the second method of comparison, the luminance of the lamp is constant and the intensity of the detected radiation from the measured source is changed by a grey wedge filter until the filament of the reference source disappears in the background of the radiator. In this case, the position of the wedge filter is the measure of the luminance temperature.

The measured range of temperatures is 700–1500°C. If an absorption grey filter is placed between the lens and the lamp, this range is increased to 1300–3500°C. The most often used luminance pyrometers use the wavelength of 0.65 μm delimited by a red filter.

The measured object is rarely a black radiator. Therefore, correction of data obtained by the pyrometer has to be made taking into account the zone emissivity ϵ_λ in a similar way as with photoelectric pyrometers,

$$\frac{1}{T} = \frac{1}{T_0} + \frac{\lambda}{C_2} \ln \epsilon_\lambda, \tag{5.130}$$

where λ is the effective wavelength of the pyrometer, i.e. the wavelength of the centre of the spectral zone used. To determine the real temperature, a luminance pyrometer usually has a figure which for a given value of ϵ_λ provides a temperature correction factor $\Delta T = T - T_0$ that is added to the pyrometer reading.

Distribution (colour) pyrometers are used in the range of visible radiation. They are suitable for the measurement of luminance of grey radiators in cases when the emissivity is not known. Colour pyrometers give the real temperature using the so-called "colour temperature". The correction to the emissivity of the body is significantly lower than for luminance pyrometers and in the most common cases it is even not necessary at all.

The colour temperature T_c of a grey radiator is equal to the temperature of a black radiator, when the black radiator has over certain part of the spectrum the same relative spectral distribution as the measured grey radiator. The colour temperature can be determined either by a colour sensation induced by mixing two monochromatic radiations or by using the ratio of luminances in two spectral ranges. Therefore, two types of colour pyrometers can be distinguished, namely the comparative pyrometers and the ratio pyrometers.

In comparative colour pyrometers, the thermal rays from the measured object pass through a grey and a red wedge filter, and impinge on a photometer together with the radiation of the pyrometric lamp. A red–green filter is placed between the photometer and the eyepiece. By appropriate positioning of the grey wedge filter, the luminance of the lamp becomes the same as the luminance of the measured radiator. The position of the filter is proportional to the luminance temperature T_0. By appropriate positioning of the red filter, the colour of the measured radiator becomes the same as the colour of the reference light source. Its position is proportional to the colour temperature T_c.

The ratio colour pyrometers work as two independent luminance pyrometers with two wavelengths, for instance $\lambda_1 = 0.65~\mu m$ (red) and $\lambda_2 = 0.55~\mu m$ (green). The colour temperature T_c is determined by the relation

$$\frac{1}{T_c} = \frac{\frac{1}{\lambda_1 T_{01}} + \frac{1}{\lambda_2 T_{02}}}{\frac{1}{\lambda_1} + \frac{1}{\lambda_2}}, \tag{5.131}$$

where T_{01}, T_{02} are the luminance temperatures in K determined for red and green light. The relation between the real temperature T and the colour temperature T_c can be written as

$$\frac{1}{T} = \frac{1}{T_c} - \frac{\lambda_1 \lambda_2}{C_2(\lambda_1 - \lambda_2)} \ln \frac{\epsilon_{\lambda_1}}{\epsilon_{\lambda_2}}. \tag{5.132}$$

The emissivities of the most common radiators are for two close wavelengths λ_1, λ_2 very similar, $\epsilon_{\lambda_1} \sim \epsilon_{\lambda_2}$, and therefore $T \sim T_c$.

5.1.4 Capillary Pressure

Capillary pressure is measured most often by tensiometers. A tensiometer consists of a porous cup, generally of ceramic material, connected through a tube to a manometer, with all parts filled with water (see, e.g., Hillel, 1980). If the cup is placed in a porous material where suction measurements are to be made, the bulk water inside the cup comes into hydraulic contact with the porous material and tends to equilibrate with water in the porous space through the pores in the ceramic walls. When initially placed in the material, the water contained in the tensiometer is generally at atmospheric pressure. Water in the porous material, being generally at subatmospheric pressure (capillary pressure is negative), exercises a suction that draws out a certain amount of water from the rigid and airtight tensiometer, thus causing a drop in its hydrostatic pressure. The pressure in the tensiometer, which after equilibration with the pressure in the porous material is equal to the capillary pressure in the porous material, is measured by a manometer. This may be a simple mercury-filled U-tube, a vacuum gauge or an electrical transducer. The primary output of an electrical transducer is the voltage, and the value of capillary pressure is obtained using the known calibration constant. This recalculation is done by an electronic evaluation circuit, so that a digital value of the pressure is obtained as output.

In practice, many types of tensiometers are used. An example of good construction can be found in Marthaler et al. (1983). The end of the tensiometric probe is formed by a porous ceramic tube making contact with the surrounding environment. The body of the probe is filled with water, and above the water there is an air bubble transmitting the pressure difference. The probe is closed by a neoprene cap. The tensiometer reading is taken after piercing the cap by a hollow needle connected with the pressure sensor and the registration unit. After the needle is removed, the hole in the cap is closed again, and the tensiometer is prepared for further readings.

A mini-tensiometer for laboratory measurements was designed by Plagge et al. (1989) and further developed by Easytest (1999). The installation hole for this tensiometer is 8 mm in diameter, in later versions only 5 mm. The pressure diaphragm is a ceramic cup, the pressure transducer is an integrated, fully active Wheatstone bridge type pressure transducer with four piezoresistive strain gauge resistors diffused into a silicon diaphragm. The air entry pressure is above 90 kPa.

Measurements of capillary pressure by tensiometry are generally limited to values below 100 kPa. This is due to the fact that the vacuum gauge or manometer measures a partial vacuum relative to the external atmospheric pressure. Furthermore, as the ceramic material of the cup is generally made of the most permeable and porous material possible, too high suction may cause air entry into the cup, which would equalize the internal pressure to the atmospheric pressure. In practice, the useful limit of most tensiometers (see Hillel, 1980) is at about 80 kPa. To measure higher capillary pressures (up to 100 kPa), an osmometer with a semipermeable membrane at the wall instead of the ceramic cup may be used (Hillel, 1980).

In measuring capillary pressures higher than 100 kPa, specially designed psychrometers can be employed. A psychrometer is an instrument designed to determine the relative humidity by measuring the difference between the temperatures registered by a wet bulb and a dry bulb thermometer (see Section 5.1.2 for details). Normally, psychrometers are used for measuring relative humidities in usual range up to about 95%. Psychrometers designed to determine the capillary pressure are supposed to measure relative humidity in the range closer to 100%, typically from 95% to 99.9%, and therefore these have to be very accurate. Miniaturized thermocouple psychrometers, which make it possible to perform in situ measurements of capillary pressure, were designed by Dalton and Rawlins (1968) and Brown (1970).

A thermocouple psychrometer (Hillel, 1980) consists of a fine wire thermocouple, one junction of which is equilibrated with the atmosphere in the porous body by placing it inside a hollow porous cup embedded in the porous body, while the other junction is kept in an insulated medium to provide a temperature lag. During operation, the junction exposed to the atmosphere in the porous body is cooled to a temperature below the dew point of that atmosphere, at which point a droplet of water condenses on the junction, allowing it to become, in effect, a wet bulb thermometer. This is a consequence of the Peltier effect. The cooling is then stopped, and as the water from the droplet re-evaporates, the junction attains a wet bulb temperature that remains nearly constant until the junction dries out, after which it returns to the ambient temperature of the porous body. While evaporation takes place, the difference in temperature between the wet bulb and the insulated junction serving as dry bulb generates a voltage signal, which is indicative of the relative humidity and therefore also of capillary pressure. In recalculating relative humidities to give capillary pressure, the Kelvin equation is used in usual way (see Section 1.6).

The measurement of relative humidity is obviously highly sensitive to temperature changes. Hence the need for very accurate temperature control and monitoring. Under field conditions, the accuracy of capillary pressure claimed is of the order of about 50 kPa (Hillel, 1980). The thermocouple psychrometers are thus not practical at low capillary pressure values, but they can be quite useful considerably beyond the capillary pressure range of tensiometers, typically from 200 kPa to 5 MPa.

5.1.5 Salts and Other Chemical Compounds Content

The salts content in concrete may be considered from two viewpoints: salts that are free to move in the porous system of concrete, and salts that have partly reacted with the cement binder.

The content of water-soluble compounds that may freely move in the concrete porous environment without any reaction with the cement binder or aggregates, is important for the study of the transport phenomena, depending on conditions.

For a rough assessment of the presence of some salts and pH values, qualitative "in situ" methods have been devised, which can be used directly on concrete structures. In particular, these are used to determine approximate pH value of the pore solution and the presence of chlorides, which are the most frequently detected constituents.

In the porous system of concrete, salts are either dissolved in water, with their maximum concentration at a given temperature given by their solubility, or they may occur in a solid state. Conversion from the dissolved to the crystallized form can take place in such a way that water evaporates from the solution at elevated temperatures, which may lead to the salt solubility being exceeded, so that salt is deposited in the capillary pores. On the other hand, water-soluble salts may be leached from the cement binder or from the porous aggregates by means of distilled water. Compounds reacting with the cement binder or aggregates are determined by total chemical analysis of the concrete. Interpretation of results is very complex in some cases, and requires high levels of professional knowledge and experience.

The content of chemical compounds in the cement binder pores may also be determined in the pore solution that is extruded from concrete by a very high pressure. The solution obtained in this way does not contain salts deposited in pores in a solid state.

The results of the salt content determination may be expressed per the mass of cement or concrete, or as their concentration in the pore solution. This expression of the concentration may be carried out directly from the determined content of the component in the pore solution or by conversion to the ascertained concrete moisture.

5.1.5.1 Qualitative Methods Used on a Construction

An approximate pH value is often determined on a concrete construction. The measurement proceeds as follows: a 2 mm thick surface layer, which is always carbonated, is removed and a 1% solution of phenolphthalein in 70% ethanol is sprayed on the surface. In some cases, a hole is drilled in the concrete to the depth where the first line of the steel reinforcement is embedded, and the indicator solution is sprayed on the walls. If the surface under the spray turns violet, the pH value exceeds 9.5, and pOH value is less than 4.5. This information is limited however, so in order to determine more accurate pH values for the pore solution on a construction, a set of acid-base indicators is used.

The presence of chlorides is also determined in the field. Here, the principle of argentometric titrimetric determination of chlorides is utilized. A 1% solution of $AgNO_3$ is applied to the concrete fracture surface, and after drying it is sprinkled with a 5% solution of K_2CrO_4. If the concrete contains chlorides above the method sensitivity limit, the colour of the surface will turn from white to light violet by the formation of insoluble AgCl. If free chlorides are not present in the concrete, the insoluble reddish brown Ag_2CrO_4 is formed (Kadleček and Modrý, 1971).

Collepardi (1997) devised another method. The method is based on the reaction of fluorescein with Ag^+ ions. The solution of fluorescein (1 g of fluorescein in 1000 ml of 70% solution of ethanol in water) is sprayed on the concrete fracture surface and after drying, a solution of $AgNO_3$, $c(AgNO_3)$ = 1 mol/l is sprayed on. The insoluble AgCl, which is permanently rose coloured, is formed in the presence of free Cl^- ions. The absence of chlorides or their insoluble form is indicated after exposure to daylight by the formation of dark brown colour of the concrete surface. Evidence of other compounds present in the concrete is not usually detected in the field.

5.1.5.2 Sampling and Sample Preparation for Analysis

The mass of the sample taken for analysis depends on the aggregate grain size. Up to a grain size of 8 mm, the minimum mass of a concrete sample should be 500 g, to 32 mm 1000 g, to 63 mm 2000 g and to 125 mm 3000 g. In most cases, this requirement is not realizable, and therefore a method of sampling by drilling to required depths has been devised (Rovnaníková, 1992).

Table 5.3 Numbers of samples taken and the surface for taking samples

Maximum grain size of aggregate [mm]	Area [mm]	Number of samples
8	200 x 200	8
16	250 x 250	12
32	300 x 300	16
63	400 x 400	20
125	500 x 500	24

Table 5.4 The mass of samples taken for the determination of concrete constituents

Grain size of aggregates [mm]	Sample minimum mass [kg]	Proportional mass [-]
2	2	0.5
4	3	0.5
8	4	0.5
16	5	1
32	10	1
63	15	2
125	25	3

Samples are cut by means of an electric drilling machine fitted with a hard metal drill of 20 mm diameter. The sample is collected in a special collector. The number of drill holes and the surface area for taking samples according to the aggregate grain size are given in Table 5.3.

The mixed sample is properly homogenized and it serves for the preparation of the water leach. The fact that these drill holes may be drilled to different depths, and so used to determine the distribution of the compounds present in the concrete profile from the surface to various depths, is a distinct advantage of this manner of sampling.

In order to specify the content of the constituent that is being determined per unit mass of cement when this mass is unknown, the concrete constituents must be determined. For this purpose, samples of concrete are taken, the masses of which are given in Table 5.4.

The masses of samples for the determination of concrete constituents in real constructions lead to expressions for the salts content by concrete mass. The preparation of a sample is the same as for the determination of the content of chemical compounds in the water extract and for their total content in concrete, if the compounds react with the cement binder. First of all, the concrete sample is crushed to a grain size of 1 mm and 100 g is separated by quartering. The sample is dried at 105°C to its constant mass, and then ground in a vibrating or attrition mill, so that it may pass through a sieve of mesh size 0.090 mm. Then 10 g of the sample prepared in this way is weighed for the preparation of the water extract, or 1 g for the total analysis.

The preparation of the water extract is carried out by dissolving the sample in a beaker containing 130 ml of distilled water and placing on an electromagnetic mixer provided with electric heating. The solution in the beaker covered with a glass lid is brought to a temperature of 90°C, while being mixed for 30 minutes. After reaching the required temperature, the suspension in the beaker is allowed to sedimentate, and the clear solution is decanted to a volumetric flask of 200 ml volume. The sediment is gradually mixed with 20 ml of water three times, solid particles are allowed to sedimentate, and then the solution is decanted to the volumetric flask. The solution in the volumetric flask is allowed to cool down to laboratory temperature. Then it is brought up to the mark with distilled water and thoroughly mixed.

The pore solution is extruded from concrete using a special pressure equipment (Tritthardt and Daňková, 1987). A cone 65 mm in diameter is inserted into the packing in the pressure equipment, and then is inserted in the press. The sample is gradually loaded by a pressure increase rate smaller than 10 bar/s up to the pressure of 1000 bar. About 10 ml of the pore solution is obtained from the sample of the water/cement ratio. The pH value and the content of the ions observed are determined from this sample.

To determine the total content of the constituent both in the free and the bound form, the concrete sample is dissolved in a mixture of hydrochloric and nitric acid. After depositing SiO_2 and the subsequent precipitation of R_2O_3, the total content in the filtrate of the ion being investigated is determined.

5.1.5.3 Determination of pH and Salts Content

To determine the pH value of the water extract from concrete, 30 ml of the solution at a temperature of 20°C is poured from the flask into a small beaker, and pH is determined using combined glass electrodes. For that determination, an electrode suitable for a strongly alkaline environment must be used (to pH 14). Otherwise, there is a danger of etching the glass bulb of the electrode, and the value measured will not be accurate. The device gauging is carried out by calibration solutions of, e.g., pH 9.0 and 12.45. The concentration of OH^- ions is calculated from the determined pH value,

$$pOH = 14 - pH,$$

and the concentration of hydroxide ions is calculated as $c(OH^-) = 10^{-pOH}$.

The <u>concentration of OH^- ions</u> found out in the water extract from concrete also includes the ions occurring in concrete in the form of solid calcium hydroxide. The real concentration of free OH^- ions in the pore solution may be determined only by measuring the pH directly in the extruded pore solution.

Ammonium salts NH_4^+ are soluble in water. Their reaction with the cement binder in a high pH environment leads to the release of ammonia. This partly reacts with water to form ammonium hydroxide, and partly leaves the concrete in the form of a gas. The most frequently used way to determine their content is that ammonia is first distilled to a solution of hydrochloric acid, so that disturbing ions may be separated. Ammonium ions are determined photometrically either with sodium phenoxide and hypochlorite or by Nessler's reagent ($K_2[HgI_4]$). In the first case, the ammonium ions react, adding sodium phenoxide and hypochlorite in an alkaline environment, and the blue indophenol dye is formed. The colour intensity is measured by a photometer at a wavelength of 630 nm. The second method involves the reaction of NH_4^+ in an alkaline environment with $K_2[HgI_4]$, during which a yellow-brown colloid solution or even a precipitate is formed. The colloid solution may be measured photometrically at a wavelength of 400 to 425 nm. In both cases, the calibration curve must be plotted, from which the concentration of ammonium ions in the sample is read.

Magnesium ions Mg^{2+} may be present in the pore solution only in case when concrete has carbonated, and when no free OH^- ions are present in the pore solution. Otherwise, the insoluble magnesium hydroxide $Mg(OH)_2$ is formed.

The total content of magnesium ions penetrating concrete is determined by comparing their content in concrete and in the original cement. Free Mg^{2+} ions, if present, are determined in the water extract. The most frequently used method for their detection is the complexometric determination consisting in the titration of the sum of Ca^{2+} and Mg^{2+} ions by a standard solution of chelatone 3, when indicated by the eriochrome black T and when the colour is changed from red–violet to blue at the equivalence point. The content of Ca^{2+} ions is determined in the other portion of the sample by chelatone 3 titration using fluorexone as an indicator. The equivalence point occurs at the moment when the green fluorescence disappears. The concentration of Mg^{2+} is calculated from the difference in consumption of the standard solution during the titration of the Ca^{2+} and Mg^{2+} sum and Ca^{2+} itself.

In concrete, nitrates NO_3^- are always present in the form of soluble salts. The determination of nitrates is carried out photometrically in the water extract from concrete. The agent used is sodium salicylate, which in an alkaline environment forms a yellow solution in the presence of nitrates. The measurement is made at a wavelength of 410 nm. The concentration of nitrates is determined from the calibration graph. Nitrates in the water extract may also be determined by means of a potentiometer with a nitrate ion selective electrode.

The phosphate ions PO_4^{3-} react in the cement binder with the calcium ions, and the insoluble calcium phosphate $Ca_3(PO_4)_2$ is formed. In leaching water from concrete, phosphates will not be present. Phosphates are determined in the solution formed after the disintegration of the concrete. A crystalline compound of $MgHN_4PO_4.6H_2O$ is precipitated by means of a mixture of $MgCl_2$, NH_4Cl and the ammonium hydroxide from a citrate environment, so that the interfering ions may be masked. The precipitate is annealed at a temperature of 900°C to its constant mass, and after being cooled, the MgP_2O_7 formed is weighed.

At low contents of phosphates, the determination may be carried out photometrically using a molybdenum agent. The absorption is measured at 690 nm, and the concentration of the constituent that is being determined is read from the calibration graph.

Sulphate ions SO_4^{2-} together with free Ca^{2+} ions form slightly soluble calcium sulphate that subsequently may react with the aluminate constituent of the cement binder, with ettringite being formed. The determination of sulphates in the water extract is not valid because SO_4^{2-} ions are bonded to partly soluble $CaSO_4.2H_2O$. Thus the amount of free sulphate ions depends on the amount of water used for the preparation of the leach.

After being disintegrated by an acid, sulphates may be determined gravimetrically or by means of titration. In gravimetric determination, the insoluble barium sulphate $BaSO_4$ is precipitated, and after being annealed, it is weighed.

The volumetric determination of sulphates by a standard solution of $PbSO_4$ using dithizone as an indicator requires the removal of metal cations interfering with the reaction. Before titration, the cations are removed by means of an exchange reaction with the cation exchanger in the H-cycle. The solution colour is changed from green to violet during titration at the equivalence point. If a considerable amount of the $PbSO_4$ precipitate is formed during titration, the solution must be diluted.

The main reason why the determination of chlorides content in concrete is carried out is the fact that chlorides promote the reinforcement corrosion. Chlorides occurring in concrete are in two forms: bonded and free. If added to the mixing water, and setting and hardening takes place in their presence, chlorides are bonded to insoluble compounds, above all to Friedel's salt $3CaO.Al_2O_3.CaCl_2.10H_2O$. This reaction proceeds gradually with the development of the solid concrete structure. It has been found out that at a 0.01% concentration of Cl^- ions by mass of cement, chlorides in concrete are bonded as early as in three days, while at a 0.60% concentration this occurs as late as after 28 days (Collepardi, 1997).

The determination of free chloride ions is carried out in the water leach from concrete. The method of gravimetric and of volumetric analyses may both be used. The chloride ions in the water leach are titrated with a standard solution of mercuric nitrate $Hg(NO_3)_2$ in an environment of nitric acid, in the presence of sodium nitroprusside as an indicator. The equivalence point is reached when turbidity first occurs. The volume of the standard solution, multiplied by an empirical factor, gives the number of Cl^- ions in the titrate.

The chloride ions in the water leach are determined by means of a chloride ion selective electrode connected to a pH meter. The calomel two-case electrode is used as a comparative electrode to prevent the escape of chloride ions from the electrode to the solution being investigated. After the measurement has been evaluated, the calibration graph from the solutions with the rising concentration of Cl^- is measured first, from which the determined concentration is then read. Calibration solutions must contain the constituents that are forming the matrix effect, i.e. the soluble compounds that pass over to the solution.

The content of both free and bound chlorides is determined in the following way: the samples are dissolved in hot diluted HNO_3, in which all the chlorides pass over to the solution. Chlorides are precipitated by a standard solution of silver nitrate. After the precipitate has been cooled down and filtered, the unreacted Ag^+ ions in the filtrate are determined by a standard solution of ammonium thiocyanate, using Fe^{3+} salt as an indicator.

Potentiometric determination by means of an ion selective chloride electrode or by ion chromatography may also be used for the determination of a number of cations and anions.

5.1.5.4 Nondestructive Determination of Salts Content

Free chloride ions may be determined directly in a concrete construction by electrochemical measurement. The active sensor for the determination of chlorides is made of silver wire coated with electrochemically deposited silver chloride AgCl. The silver wire coated with AgCl is inserted in a stainless steel tube so that a part is left projecting. Between the wire and the steel tube there is insulation in the form of a thin polytetrafluoroethylene tube. Epoxy resin is used for packing the steel tube to prevent penetration of the solution. Each sensor is provided with two electric connections, the silver coated wire (chloride sensor) and the steel tube; so the potential and the resistance may be measured, see Elsener et al. (1997).

Salt concentration can also be determined by methods that are primarily used for moisture measurement. In such procedures, either dielectric methods working in the microwave frequency range or nuclear magnetic resonance techniques are the most suitable.

Microwave techniques make it possible in principle to determine the concentration of dissolved salts. In certain ranges of GHz frequencies, the real part of the complex relative permittivity depends on the moisture content only and is quite independent of the salt content in water solutions, while the imaginary part depends on both moisture content and salt concentration. Therefore, using simultaneous measurements of both real and imaginary parts of complex relative permittivity, moisture and concentration of dissolved salts can be determined at the same time.

A microwave technique suitable for simultaneous determination of moisture content and concentration of dissolved salts in building materials was developed by Leschnik (1998). The measuring system requires one bore hole for one aerial, whilst the second aerial is positioned in front of the building component. A plane microwave field with a frequency of 2.45 GHz is radiated from a patch antenna into the building material. The electric field is detected by a small dipole antenna in a bore-hole of 16 mm diameter. The signal detected is analysed by a network analyser, with the electric circuits controlled by a PC. The real and imaginary parts of the complex relative permittivity are determined at the same time via the wave propagation constant and wave attenuation factor, and the moisture content and salt concentration are calculated using measured calibration curves.

Another suitable method based on the pulsed radar technique was designed by Maierhofer and Wöstmann (1998). In their experimental set-up for transmission measurements, the transmitting and receiving transducers are installed on opposite sides of a specimen of a building material. Short microwave pulses with a full width at half maximum of 250 ps at a frequency of 7.24 GHz are radiated by a horn antenna into the building material, and on the receiver side the transmitted electromagnetic wave is analysed by a sampling oscilloscope. Primary outputs from the measurement are the travel time t of the electromagnetic wave through a specimen with a known thickness z, and the emitted and received amplitudes, E_0 and E_z respectively, of the wave. The real and imaginary parts of the relative permittivity, ϵ' and ϵ'' respectively, are calculated using the relations

$$\epsilon' = \left(\frac{c_0 t}{z}\right) \tag{5.133}$$

$$\epsilon'' = -\frac{2c_0}{\omega z}\sqrt{\epsilon'}\ln\frac{E_z}{E_0}, \tag{5.134}$$

where c_0 is the velocity of electromagnetic waves in vacuum, ω the angular frequency. Moisture content and salt concentration are calculated using measured calibration curves.

A method for the determination of salt concentration in building materials using time-domain reflectometry (TDR) operating in the frequency range between 30 MHz and 1.6 GHz was introduced by Plagge et al. (1999). The measurement is based on the determination of the bulk electrical conductivity of the specimen. The bulk electrical conductivity is determined using the potential difference at two resistors. One resistor corresponds to the embedded metal rods of the TDR probe R_p, while the other resistor serves as a reference R_{ref}. The specific resistance of the TDR probe is given by

$$R_p = \frac{U_p \cdot R_{ref}}{U_{ref} - U_p}, \tag{5.135}$$

where U_{ref} and U_p correspond to the amplitude of the voltage at the reference resistor and the metal rods of the TDR probe, respectively. As with TDR for moisture measurements, the electric pulses generated are sent through the cable–sensor system. Instead of interpreting the reflectogram as in the usual TDR methods, the potential drop of the probe is used to measure the specific electrical resistance. TDR probes installed in a building material deliver the bulk electrical conductivity of the system. Since the pore solution contains transportable salts, the liquid part of the electrical conductivity σ_w is of special interest. The part of lesser interest is the surface conductance of the solid phase σ_s, which includes crystallized salts, while the air does not conduct at all. The following empirical relation between the electrical conductivities of the components and the bulk electrical conductivity σ_a was recommended by Plagge et al. (1999):

$$\sqrt{\sigma_a} = \sqrt{\sigma_w w} + s \cdot \sqrt{\sigma_s}, \tag{5.136}$$

where w and s are the volume fractions of the liquid and solid phase in the porous body, respectively.

By measurement of various salt-in-water solutions in particular building materials at different molar concentrations, the electrical conductivity of the bulk material σ_a can be determined. Linear relations between the specific probe resistance and the electrical conductivity hold for about three orders of magnitude of electrical conductivity. If σ_s and its dry solid fraction are measured, the electrical conductivity of the pore solution σ_w can be determined using equation (5.136) by the measurement of water content w and the bulk electrical conductivity σ_a.

A similar method for the determination of salt concentration on the basis of TDR measurements was developed specifically for soils by Malicki and Walczak (1999). In their experimental analysis, the salinity index X_s,

$$X_s = \frac{\partial \sigma_a}{\partial \epsilon_a}, \tag{5.137}$$

where σ_a and ϵ_a are the bulk electrical conductivity and the bulk relative permittivity respectively, was recognized as the basic parameter for the determination of soil salinity, because it was found to be independent of the moisture content and directly proportional to the soil salinity.

A method for the determination of the amount of dissolved salts in building materials using the nuclear magnetic resonance (NMR) technique was proposed recently by Pel et al. (2000). In their experimental set-up, both moisture content and the concentration of a chosen ion are measured quasi-simultaneously. Contrary to classical NMR installations, a specially designed RF circuit is incorporated, which makes it possible to change the capacitors for the resonance frequency and for the impedance matching using two switches that are actuated by a step motor. In this way, the resonance frequency of the tuned LC circuit can be toggled between the hydrogen frequency, necessary for moisture measurements, and the frequency characteristic for the particular ion of the measured salt.

5.2 PORE DISTRIBUTION

The volume of pores and the distribution of their sizes can be determined by the determination of porosity. Pore distribution in concrete is usually determined by the penetration of mercury into the porous system, microscopically, or by using the adsorption isotherms.

The Jeffries method, which involves the direct counting of the pores in a certain limited area of the field of view of the optical microscope (Šašek, 1981), is the simplest. Both the number and the size of pores are measured. The mean pore size is given by the following equation:

$$\bar{d} = \frac{\sum nd^2}{\sum nd},$$
(5.138)

where n is the number of pores per the unit surface and d the pore diameter in meters.

Another possibility for determining the pore volume is optical microscopy. A reflective polished section is prepared from a sample. The sample is examined under microscope, where it is compared with calibrated eyepiece gauzes. By this method, both the mean diameter of pores and the distribution of their sizes are determined.

The method of the integration table involves selecting a particular section on the specimen. Then, the proportion of the pore diameter to the length of the selected section is the same as the proportion of the pore surface to the total surface selected. This method has also been modified for computer processing by means of suitable software.

High-pressure mercury porosimetry is largely used to determine open pores of diameter in the range from 0.007 to 15 μm. Porosimetry is based upon the fact that mercury does not wet the surface of silicate materials, and thus it does not spontaneously penetrate into concrete pores. Penetration of mercury into pores is initiated by the sample evacuation, and consequently uniformly increases with increasing pressure. From the equality of two counteracting forces in the capillary, forces F_1 induced by the mercury surface tension,

$$F_1 = 2\pi r\sigma \cos \Theta,$$
(5.139)

and forces F_2 induced by hydrostatic pressure,

$$F_2 = \pi r^2 \rho g \Delta h = \pi r^2 \Delta p,$$
(5.140)

where r is the radius of capillaries, σ the surface tension of mercury, with a value of 0.4855 N/m at 25°C, ρ the density of mercury, g the gravity acceleration, Δh the difference of the mercury levels before and after its transport to the micropores, Θ the mercury wetting angle 141.3°, and p the hydrostatic pressure in Pa.

Using the condition

$$F_1 = F_2,$$
(5.141)

it is possible to derive a relation for the radius of the capillary r in the form

$$r = \frac{2\sigma \cos \Theta}{\Delta p}. \tag{5.142}$$

In deriving relation (5.142), the circular section of pores and the stability of σ and θ with increasing pressure were considered. By substituting σ and θ of mercury into relation (5.142), the pore radius can be expressed as

$$r = \frac{0.758}{p}. \tag{5.143}$$

The micropore volume is measured in a glass flask, where a sample of the grain size of approximately 3 mm is placed. The flask is closed with a stopper provided with a tube of defined diameter that has been evacuated, and then filled to the mark h_0. Uniformly increasing pressure on the mercury level in the tube causes the penetration of mercury into the micropores, thus reducing its level in the tube. The specific volume of micropores (SVM) is calculated from the difference in the mercury levels in the tube and its diameter,

$$\text{SVM} = \frac{\pi r^2 \Delta h}{m}, \tag{5.144}$$

where m is the sample batch. It is also possible to determine the distribution of micropores according to the size intervals chosen.

Porosity meters are devices that make it possible to measure the volume and the distribution of micropores in porous materials utilizing the above principle. The devices measure the total porosity, i.e. the total volume of pores in the material studied, and also the distribution of micropores over the selected number of size intervals. The measured signals are directed to a PC with software for calculating the total volume and distribution of pores.

The adsorption isotherms can also be used for the determination of concrete porosity. The concrete is classed as the porous material conformable to type IV according to IUPAC classification of adsorption BET isotherms. The adsorption isotherm type IV includes the different adsorption and desorption parts, i.e. hysteresis. The adsorption isotherm is a relation between the adsorption quantity of the adsorbens a and the pressure of the adsorbate p. Using the relative pressure of the adsorbate x, related to the saturation vapour pressure of adsorbate at the temperature of measurement, p_0, the adsorption isotherm can be expressed as the relation

$$a = f(x), \tag{5.145}$$

where a is in mol/g, i.e. it is the molar mass of the adsorbens per unit mass of the adsorbate. The adsorption quantity is very often expressed as the volume of gaseous adsorbate at standard conditions ($0°C$, 101.325 kPa) adsorbed in a unit of the adsorbent mass in cm^3. The BET isotherm is derived for multi-layer adsorption. The molecules of adsorbate are located on the surface of the adsorbens closely side by side. When the first layer is not yet fully occupied, the second and following layers are already beginning to be adsorbed. The term statistical monolayer a_m indicates the mass of adsorbate that is necessary for the creation of just one layer. In fact, a complete monolayer is not formed during the adsorption, and a_m is determined from the adsorption isotherm. The parameter a_m determines the specific surface of the adsorbent. The BET isotherm is expressed by the equation

$$a = \frac{a_m C x}{(1 - x)[1 + (C - 1)x]},$$ (5.146)

where the constant C characterizes the balance in the first layer. The BET isotherm is valid for pressures x such that condensation of the adsorbens does not take place in the pores.

5.3 HYGRIC PROPERTIES

5.3.1 Moisture Storage Characteristics

Moisture storage parameters for porous building materials are usually divided into two groups. In the hygroscopic moisture range, where the transport of water vapour is the dominant mode of moisture transfer, the moisture storage function is called the sorption isotherm. It expresses the dependence of moisture content in the material on relative humidity. In the overhygroscopic moisture range, where liquid water transport is the dominant mode of moisture transfer, the moisture storage function is called the water retention curve. It expresses the dependence of moisture content on capillary pressure. The sorption isotherm and the water retention curve for the same material can be unified into a general moisture storage function using the Kelvin relation (see Section 1.6).

5.3.1.1 Sorption Isotherms

The sorption and desorption isotherms are mostly measured by traditional methods, which have been used in almost the same form for many decades. The basic principle of these methods is very simple. In the case of determination of a sorption isotherm, a dry specimen of a porous material is weighed and put into a chamber with an exactly specified value of relative humidity φ_1. There the specimen is left until equilibrium between the moisture content in the porous space and the relative humidity in the surroundings is achieved, i.e. until the specimen stops absorbing water vapour from the surrounding air. This can take a relatively long time, usually several days or weeks but sometimes even several months. Once it has occurred, the specimen is weighed again and its moisture content w_1 is determined by the gravimetric method. In this way, one point on the sorption isotherm, $[\varphi_1, w_1]$, is determined. Then the same specimen is put into an environment with a higher relative humidity, φ_2, and its corresponding equilibrium moisture content, w_2, is determined. This procedure is repeated until a sufficient number of points on the sorption isotherm is obtained.

The desorption isotherm is determined in a similar way, only the direction of the process is reversed: the initial state is close to 100% relative humidity, and in the subsequent states the value of relative humidity is lowered. It is also possible to begin with the sorption or desorption process at another initial value of relative humidity. Due to sorption hysteresis, a variety of $w = w(\varphi)$ functions can be obtained depending on the direction of the process and the initial value of relative humidity. Nevertheless, these functions always remain in the hysteresis range delimited by the two basic sorption and desorption isotherms beginning from 0% and near 100%. The whole process of the determination of sorption isotherms has to be carried out at constant temperature. If a temperature dependence is required, the whole process has to be repeated for several other temperatures in the required range.

For the determination of sorption isotherms, it is necessary to establish an environment with precisely specified relative humidity for a sufficiently long time in some limited space, for instance a standard desiccator. This is done using specific saturated salt solutions with the known equilibrium values of relative humidity above them. Table 5.5 gives the values of relative humidity above the most frequently used saturated salt solutions, taken from Arai et al. (1976).

5.3.1.2 Water Retention Curves

The water retention curve expresses the dependence of capillary pressure on moisture content. The principle of its determination used in most classical methods is the same. A specimen of a porous material is placed on a semi-permeable plate or membrane and exposed to either underpressure under the plate or overpressure above the plate. Due to the pressure difference between the interior of the porous body and the surroundings, the porous plate either draws in water from the specimen or transfers water to it, until pressure equilibrium is achieved. At that moment, the pressure in the porous space (i.e. the capillary pressure) is equal to the pressure in the measuring device, which can be easily determined. The moisture content is measured by the gravimetric method, i.e. by weighing the specimen before and after the process, or by measuring the volume of the drained or supplied water.

The most important element in laboratory devices for the determination of water retention curves is a porous semi-permeable plate or membrane. The plate should have a K/L ratio, where K is the saturated hydraulic conductiv-

Table 5.5 Relative humidity above the saturated salt solutions at temperature T

Salt/Temperature (°C)	10	20	25	30	40	50
LiBr	7.8	7.2	6.7	6.3	5.2	5.0
LiCl	12.7	11.1	11.1	11.1	11.1	11.0
CH_3COOK	24.5	23.2	22.4	21.6	19.7	19.5
$MgCl_2$	33.8	33.1	32.8	32.4	31.6	30.6
NaI	34.4	40.8	39.2	37.6	33.2	–
K_2CO_3	44.5	44.1	44.0	43.8	43.4	42.8
KNO_2	–	49.1	48.2	47.3	45.5	–
$Mg(NO_3)_2$	57.2	54.3	52.8	51.4	48.5	45.6
$Na_2Cr_2O_7$	57.5	54.7	53.4	52.0	49.2	46.3
NaBr	62.5	59.3	57.8	56.2	53.0	49.6
$NaNO_2$	67.4	65.4	64.3	63.2	61.2	59.2
NaCl	75.8	75.6	75.4	75.2	75.0	74.9
$(NH_4)_2SO_4$	81.7	80.6	80.2	79.9	79.2	78.5
KBr	83.8	81.8	80.7	82.0	80.0	–
KCl	86.8	85.1	84.2	83.5	82.1	80.7
KNO_3	95.5	93.2	92.0	90.7	87.9	85.0
K_2SO_4	98.3	97.9	97.6	97.4	97.0	96.6

ity, and L the plate thickness, as high as possible to allow fast water transport through it. Another characteristic feature of any porous plate is the maximum allowable air pressure difference that the saturated porous plate can bear without allowing air to bubble through its pores (air entry pressure). This pressure value primarily depends on the dimensions of the pores in the membrane. In practical applications, for lower pressures up to 2 MPa, ceramic plates are mostly used. For higher pressures, cellulose acetate membranes are more suitable, because they can achieve air entry pressures over 10 MPa.

The measurement of retention curves is mostly performed on cylindrical specimens with a volume of approximately 100 cm^3. The height of the cylinder should not be greater than 2 cm, otherwise the measurements may take too long. Usually the draining branch of the retention curve is measured, so both specimen and the plate are first saturated with water. Two basic types of devices are employed for such measurements, underpressure devices and overpressure devices.

The underpressure devices are used in the low suction range, for pressures up to 100 kPa, because they are limited by the value of atmospheric pressure. The pressure difference is controlled by either a vacuum pump or a hanging water column.

Tension plate assembly (see, e.g., Hillel, 1980) is one of the most frequently used underpressure devices. The specimen in a fully saturated state is put into a cylindrical vessel on a porous plate, which is in a saturated state as well. The space in the vessel under the plate is filled with water up to the outlet, which is connected by a tube to a water reservoir. The outlet on the lower part of the vessel is adjusted in such a way that the volume of the drained water can be measured. The required underpressure is produced by decreasing the level of free water in the reservoir under the level of the porous plate/measured specimen interface, $p_1 = \rho g h_1$, where h_1 is the height difference between the plate/specimen interface and the level of free water in the reservoir. Sufficient time is allowed until the equilibrium state between the induced underpressure and the capillary pressure in the specimen is established. This is attained when water no longer drains from the sample. If the initial moisture content in the specimen and the amount of drained water are known, the moisture content w_1 in the specimen corresponding to the capillary pressure value p_1 can easily be determined. In other words, we have the first point on the water retention curve. Then the underpressure is increased by setting up greater height differences h_2, h_3, ... , h_i, and the corresponding moisture contents w_2, w_3, ... , w_i are determined. In this way, a pointwise $p(w)$ function is obtained, which is the water retention curve.

A sand tank is another commonly used underpressure device. The semi-permeable membrane is formed by a layer of fine glass sand more than 10 cm thick, with a grain size under 0.1 mm. The maximum allowable pressure of this layer is about 10–15 kPa. The draining system is made of coarse sand, and a perforated wavy tube is connected to the equalizing tank by a polyethylene tube. The measured specimens are fully saturated with water on a wet filter paper. These are then placed on filter papers on the sand surface in the sand tank, which earlier was also saturated with water. The water level in the equalizing tank is decreased by h_1 compared to the level of the specimen, and in this way the underpressure $p_1 = \rho g h_1$ is induced. Finally, time is allowed until equilibrium between the capillary pressure in the specimen and the chosen underpressure is established. Bear in mind that in this method not only the measured specimen but also the sand layer is drained. Therefore, the moisture content w_1 has to be determined by the gravimetric method, i.e. by weighing after the initial water saturation and after the establishment of the equilibrium state at the given underpressure p_1. The whole procedure is repeated with further values of underpressure p_2, p_3, ... , p_i induced by water level differences h_2, h_3, ... , h_i, so that a pointwise $p(w)$ function is finally obtained.

There are several problems with underpressure devices, which can significantly affect the measuring error. The hydraulic contact between the specimen and the plate may be insufficient. The achievement of an equilibrium state may take many hours, for heavier materials even days or weeks. Therefore, the determination of the exact moment when equilibrium was established is very difficult. Also, evaporation of water from the specimens, from the vessel and from the free water reservoir may be important in long-term experiments. Significant volume changes in the specimen can take place during the measurements.

Overpressure devices are used for higher values of capillary pressure, above 100 kPa. Either pressure plate or pressure membrane devices can be used (see, e.g., Hillel, 1980). These devices consist of an overpressure chamber with a semi-permeable plate or membrane. There are two basic types of overpressure device. The first has a ceramic plate, which may have a rubber cloth on its lower side. A metallic net between the membrane and the cloth forms a space with an outlet outside the chamber for retaining the drained water. The second type of overpressure device works with a cellulose acetate membrane. The membrane is placed on a metallic net, which makes space for the drained water, and the outlet is in the bottom of the overpressure chamber under the membrane. To press the specimens to the membrane, a rubber diaphragm may be used with a slightly higher pressure above it than the pressure in the space where the specimens are placed.

The overpressure chambers of all devices are closed by massive covers with large screws. The overpressure is achieved using a compressor unit or by pressure vessels with compressed air, and it is controlled by regulation valves and manometers. The amount of water drained or supplied is measured by burettes connected to the outlets by polyethylene tubes.

In practical measurements, the specimens are first saturated with water, and placed together with a filter paper on the plate or membrane, which was also fully saturated with water. Then, the required overpressure in the chamber is set up and the outflow of drained water into the burette is observed. After the equilibrium state is achieved, the overpressure is released, the specimens are weighed and moisture is determined by the gravimetric method.

5.3.2 Moisture Transport Parameters

In most models describing moisture transport in porous materials, there are two material parameters, one of them for the transport of water vapour, the second for the transport of liquid water.

In modelling the water vapour transport in porous building materials, usually just one generalized thermodynamic force is specified, which is either the gradient of partial pressure or the gradient of concentration of water vapour. We then have two relations for the water vapour flux j_v,

$$j_v = -D \; \mathrm{grad} \; \rho_c, \tag{5.147}$$

or

$$j_v = -\delta \; \mathrm{grad} \; p_v, \tag{5.148}$$

where ρ_c is the mass of water vapour per unit volume of the porous material (the partial density of water), D the diffusion coefficient of water vapour in the porous material (in m^2/s), p_v the partial pressure of water vapour, and δ

the water vapour permeability (in s). Assuming water vapour to be an ideal gas, we can write the equation of state in the form

$$p_v = \frac{\rho_c RT}{M}, \tag{5.149}$$

where T is the temperature in degrees Kelvin, R the universal gas constant, and M the molar mass of water vapour. Under isothermal conditions, we obtain from equations (5.147)–(5.149) the following relation between the diffusion coefficient and the permeability:

$$D = \delta \frac{RT}{M}. \tag{5.150}$$

Besides D and δ, several other coefficients are introduced in building physics for the sake of greater clarity in building practice. Among them, the water vapour diffusion resistance factor μ (e.g. Krischer, 1963), the water vapour resistance Z (e.g. Villadsen et al., 1993) or the equivalent air layer thickness S_D (e.g. Robery, 1988) are among the most frequently used. The definition of these coefficients follows from D or δ, so that we have

$$\mu = \frac{D_a}{D} \tag{5.151}$$

$$Z = \frac{d}{\delta} \tag{5.152}$$

$$S_D = \frac{D_a}{D} d, \tag{5.153}$$

where D_a is the diffusion coefficient of water vapour in air, and d the thickness of the specimen of the porous material.

The simplest way to describe the transport of liquid water in concrete is to use the sorptivity. The sorptivity S (in $m/s^{1/2}$) is defined (see, e.g., Hall and Tse, 1986, Hall, 1989, etc.) as

$$I = S \cdot t^{1/2}, \tag{5.154}$$

where I is the cumulative absorbed volume of water per unit area of supply surface (in m). Equation (5.154) is a simplification of the general expression for the cumulative mass of water in terms of the square-root-of-time rule commonly employed in the theory of transport processes in porous media (see, e.g., Krischer, 1963), which is obtained by dividing the original equation

$$i = A \cdot t^{1/2}, \tag{5.155}$$

by the density of water, ρ_w. In equation (5.155), i is the cumulative mass of water in kg/m^2, and A the water absorption coefficient in $kg/m^2 s^{1/2}$, so that $A = S \cdot \rho_w$.

In modelling the transport of liquid (capillary, overhygroscopic) water by common methods of irreversible thermodynamics, and using just one generalized thermodynamic force, two approaches are usually employed, where either the gradient of partial density of water ρ_m or the gradient of pressure head h are specified. Then, we can write for the water flux j_w either

$$\vec{j}_w = -\kappa(\rho_m)\ \text{grad}\ \rho_m = -\rho_s\kappa(u)\ \text{grad}\ u = -\rho_w\kappa(w)\ \text{grad}\ w \qquad (5.156)$$

or

$$\vec{j}_w = \rho_w\vec{v} = -\rho_w k(h)\ \text{grad}\ h, \qquad (5.157)$$

where κ is the <u>moisture diffusivity</u>, ρ_s the partial density of the porous matrix, u the moisture content by mass, $u = \rho_m/\rho_s$, ρ_w the density of water, w the moisture content by volume, $w = \rho_m/\rho_w$, k the <u>hydraulic conductivity</u> (in m/s), and h the pressure head (in m). In isothermal conditions, simple relations between k and κ can be derived using (5.156) and (5.157),

$$\kappa(u) = k(h) \cdot \frac{\rho_w}{\rho_s} \cdot \frac{\partial h}{\partial u}, \qquad (5.158)$$

or

$$\kappa(w) = k(h) \cdot \frac{\partial h}{\partial w}, \qquad (5.159)$$

where $\frac{\partial h}{\partial u}$, $\frac{\partial h}{\partial w}$ can be obtained from the water retention curves $h = h(u)$ or $h = h(w)$, respectively.

Another way to describe the transport of liquid water is to use the gradient of liquid water pressure, p_w, as the main generalized thermodynamic force. Then, we have for the water flux the following relation:

$$j_w = -\rho_w\frac{K}{\eta_w}k_{rw}(w)\ \text{grad}\ p_w, \qquad (5.160)$$

where K is the <u>water permeability</u> of the porous material (in m^2), which is an intrinsic property of a material independent of water saturation, η_w the dynamic viscosity of water, and k_{rw} the relative permeability (dimensionless) of water, which is a function of water saturation (i.e. of the moisture content).

The relation between hydraulic conductivity and water permeability can immediately be obtained by matching equations (5.157) and (5.160):

$$k(h) = \frac{K}{\eta_w}k_{rw}(w)\rho_w g. \qquad (5.161)$$

5.3.2.1 Water Vapour Permeability

The most frequently used method for the determination of water vapour permeability is the cup method, which is included in many European and American standards (see for instance ASTM E96, CEN/TC 89 N 336 E, DIN 52615, etc.). The theoretical basis of the method is very simple (see for instance Hansen and Lund, 1990, Villadsen et al., 1993, or the standards mentioned). A material specimen is placed on the top of a cup and sealed. The cup should contain sorption material, which is either desiccant, a saturated salt solution, or water. The sealed cup is put in a controlled climate, and weighed periodically. The steady-state mass gain or mass loss makes it possible to determine the steady-state water vapour flux between the internal and external environment through the material specimen.

For the calculation of water vapour permeability, Fick's law (5.148) can be employed, where

$$j_v = \frac{\Delta m}{S \Delta t} = \delta \frac{\Delta p}{d}, \qquad (5.162)$$

Δm is the absolute value of the mass gain or mass loss of the cup during the time Δt, S the cross-sectional area of the specimen in direct contact with the internal environment of the cup, Δp the absolute value of the difference between the water vapour pressure inside the cup and the water vapour pressure in the external climate, and d the thickness of the specimen. From equation (5.162), it directly follows that

$$\delta = \frac{\Delta m}{S \Delta t} \frac{d}{\Delta p}. \qquad (5.163)$$

There is no doubt that the cup method has proved to be a very useful and reliable method in past decades. However, it has a very unpleasant feature, which is the fact that the time necessary for a single measurement is as long as one or two weeks on average. Therefore, some attempts have been made to develop new transient techniques (e.g. Künzel and Kiessl, 1990, Černý et al., 1995), but so far their use has remained limited to the laboratories, where they were developed.

In the method designed by Černý et al. (1995), the measuring device consists of two airtight glass chambers separated by a board-type specimen of the measured material. In the first chamber, a state near to 100% relative humidity is maintained (achieved with the help of a cup of water), while in the second one, there is a state close to 0% relative humidity (established using some desiccant, such as silica gel). Alternately, saturated salt solutions establishing defined relative humidity conditions can be placed in either the wet or the dry chamber or both of them. The change in the mass of water in the cup, Δm_w, and of the desiccant, Δm_a, are recorded by automatic balance in dependence on time. In case steady-state measurements are also required, the validity of the condition $|\Delta m_w| = |\Delta m_a|$ is tested and the experiment continues until this condition is realized. The experiment is carried out under isothermal conditions.

Compared to the cup method, this experimental set-up has the advantage that it is not necessary to maintain constant relative humidity in all the conditioning chamber but only in a relatively small chamber, and also that the flux of incoming water vapour is measured. As a consequence, even the steady-state measurements in this experimental set-up can be significantly faster than in the classical cup method. On the other hand, data evaluation is more complicated because the use of a transient method involves the solution of an inverse problem of transient water vapour diffusion.

The diffusion equation for water vapour in a porous body can be written in the form

$$\frac{\partial \rho_c}{\partial t} = \frac{\partial}{\partial x}\left(D\frac{\partial \rho_c}{\partial x}\right), \tag{5.164}$$

where the symbols are defined as in the introduction to this section.

In formulating initial and boundary conditions for equation (5.164), we take into account that on one surface of the board specimen we have the relative humidity of the air $\varphi_0 \to 100\%$, on the other one $\varphi_d \to 0\%$. The pressure of saturated water vapour as a function of temperature can be expressed by the standard relation

$$p_v = T^C \cdot 10^{-\frac{A}{T}+B}, \tag{5.165}$$

where A, B, C are constants determined experimentally, T is in degrees Kelvin. On the basis of data from Haar et al. (1984), we can specify for pressure in Pa the following values:

A = 2900 K, B = 24.738, C = −4.65.

Using the equation of state of an ideal gas and equation (5.165), we can formulate the boundary conditions on both surfaces:

$$\rho_c(0,t) = \varphi_0 \cdot \frac{p_v M}{RT} = \varphi_0 \cdot \frac{M}{R} \cdot T^{C-1} \cdot 10^{-\frac{A}{T}+B} \tag{5.166}$$

$$\rho_c(d,t) = \varphi_d \cdot \frac{p_v M}{RT} = \varphi_d \cdot \frac{M}{R} \cdot T^{C-1} \cdot 10^{-\frac{A}{T}+B}, \tag{5.167}$$

where d is the thickness of the specimen. As the initial condition we can choose for instance

$$\rho_c(x,0) = \rho_{co}, \tag{5.168}$$

where ρ_{co} is a constant.

In the experimental set-up described above we determine the mass of water evaporated from the cup, in other words the mass of water m_w, which penetrated into the specimen at the point $x = 0$ during the time interval τ. Given that the surface area of the board specimen S is also known, we can write

$$m_w = S \int_0^\tau -D\left(\frac{\partial \rho_c}{\partial x}\right)_{x=0} dt. \tag{5.169}$$

When the experiment is performed in isothermal conditions and the dependence of D on the concentration of water vapour is neglected, we have

$$D = -\frac{m_w}{S \int_0^\tau \left(\frac{\partial \rho_c}{\partial x}\right)_{x=0} dt}. \tag{5.170}$$

In equation (5.169) we do not know $\left(\frac{\partial \rho_c}{\partial x}\right)_{x=0}$, which in the general transient state cannot be measured with sufficient precision. Therefore, the system of equations (5.164), (5.166)–(5.168) has to be solved together with equation (5.169).

Assuming $D = const.$, the system (5.164), (5.166)–(5.168) has an analytical solution of the form (see Carslaw and Jaeger, 1959)

$$\rho_c(x,t) = \varphi_o \frac{M}{R} \cdot T^{C-1} \cdot 10^{-\frac{A}{T}+B} \left(1 - \frac{x}{d}\right)$$

$$+ \varphi_d \frac{M}{R} \cdot T^{C-1} \cdot 10^{-\frac{A}{T}+B} \cdot \frac{x}{d}$$

$$- \frac{2}{\pi} \frac{M}{R} \cdot T^{C-1} \cdot 10^{-\frac{A}{T}+B}$$

$$\times \sum_{n=1}^{\infty} \frac{1}{n} (\varphi_o - \varphi_d \cos(n\pi)) \sin\left(\frac{n\pi x}{d}\right) \exp\left(-\frac{Dn^2\pi^2 t}{d^2}\right)$$

$$+ \frac{2\rho_{co}}{\pi} \sum_{n=1}^{\infty} \frac{1}{n} \sin\left(\frac{n\pi x}{d}\right) \exp\left(-\frac{Dn^2\pi^2 t}{d^2}\right) (1 - \cos(n\pi)). \tag{5.171}$$

From the analytical solution (5.171) it follows that

$$\left(\frac{\partial \rho_c}{\partial x}\right)_{x=0} = -\frac{\varphi_o - \varphi_d}{d} \frac{M}{R} \cdot T^{C-1} \cdot 10^{-\frac{A}{T}+B}$$

$$- \frac{2}{d} \frac{M}{R} \cdot T^{C-1} \cdot 10^{-\frac{A}{T}+B}$$

$$\times \sum_{n=1}^{\infty} (\varphi_o - \varphi_d \cos(n\pi)) \exp\left(-\frac{Dn^2\pi^2 t}{d^2}\right)$$

$$+ \frac{2\rho_{co}}{d} \sum_{n=1}^{\infty} \exp\left(-\frac{Dn^2\pi^2 t}{d^2}\right) (1 - \cos(n\pi)). \tag{5.172}$$

Substituting equation (5.172) into (5.169), we obtain

$$m_{\mathrm{w}}(\tau) + \frac{2Sd}{\pi^2}\left[-\frac{M}{R}\cdot T^{C-1}\cdot 10^{-\frac{A}{T}+B}\right.$$

$$\times \sum_{n=1}^{\infty}\frac{1}{n^2}(\varphi_{\mathrm{o}}-\varphi_{\mathrm{d}}\cos(n\pi))\left(1-\exp\left(-\frac{Dn^2\pi^2\tau}{d^2}\right)\right)$$

$$\left.+\rho_{\mathrm{co}}\sum_{n=1}^{\infty}\frac{1}{n^2}\left(1-\exp\left(-\frac{Dn^2\pi^2\tau}{d^2}\right)\right)(1-\cos(n\pi))\right]$$

$$-\frac{SD}{d}\frac{M}{R}(\varphi_{\mathrm{o}}-\varphi_{\mathrm{d}})\cdot T^{C-1}\cdot 10^{-\frac{A}{T}+B}\cdot\tau = 0. \tag{5.173}$$

Relation (5.173) is a transcendent equation for the diffusion coefficient of water vapour D, which can be solved by iterative methods such as the Newton method. The water vapour permeability δ can be calculated from D using the transformation relation (5.150).

An alternative to the solution described above is to solve the problem at the point $x = d$. Assuming that the mass of water absorbed in the desiccant m_{d} (i.e. the mass of water vapour that left the specimen at point $x = d$ during the time interval τ) is measured, we can write in an analogous way as before

$$m_{\mathrm{d}} = S\int_0^\tau -D\left(\frac{\partial\rho_{\mathrm{c}}}{\partial x}\right)_{x=d}dt, \tag{5.174}$$

and using the analytical solution (5.171)

$$\left(\frac{\partial\rho_{\mathrm{c}}}{\partial x}\right)_{x=d} = -\frac{\varphi_{\mathrm{o}}-\varphi_{\mathrm{d}}}{d}\frac{M}{R}\cdot T^{C-1}\cdot 10^{-\frac{A}{T}+B}$$

$$-\frac{2}{d}\frac{M}{R}\cdot T^{C-1}\cdot 10^{-\frac{A}{T}+B}$$

$$\times \sum_{n=1}^{\infty}(\varphi_{\mathrm{o}}-\varphi_{\mathrm{d}}\cos(n\pi))\cos(n\pi)\exp\left(-\frac{Dn^2\pi^2t}{d^2}\right)$$

$$+\frac{2\rho_{\mathrm{co}}}{d}\sum_{n=1}^{\infty}\exp\left(-\frac{Dn^2\pi^2t}{d^2}\right)\cos(n\pi)(1-\cos(n\pi)). \tag{5.175}$$

After substitution of equation (5.175) into (5.174), we get

$$m_{\mathrm{d}}(\tau) + \frac{2Sd}{\pi^2}\left[-\frac{M}{R}\cdot T^{C-1}\cdot 10^{-\frac{A}{T}+B}\right.$$

$$\times \sum_{n=1}^{\infty} \frac{1}{n^2} (\varphi_o - \varphi_d \cos(n\pi)) \cos(n\pi) \left(1 - \exp\left(-\frac{Dn^2\pi^2\tau}{d^2}\right)\right)$$

$$+ \rho_{co} \sum_{n=1}^{\infty} \frac{1}{n^2} \left(1 - \exp\left(-\frac{Dn^2\pi^2\tau}{d^2}\right)\right) \cos(n\pi) (1 - \cos(n\pi)) \Bigg]$$

$$- (\varphi_o - \varphi_d) \cdot \frac{SD}{d} \frac{M}{R} \cdot T^{C-1} \cdot 10^{-\frac{A}{T}+B} \cdot \tau = 0. \tag{5.176}$$

Equation (5.176) can be again solved by some iterative method.

In the case where steady state is achieved via the experimental set-up by Černý et al. (1995), using equations (5.166) and (5.167) we obtain

$$\left(\frac{\partial \rho_c}{\partial x}\right)_{x=0} = -\frac{(\varphi_o - \varphi_d) p_v M}{RTd}$$

$$= -\frac{(\varphi_o - \varphi_d) M}{Rd} \cdot T^{C-1} \cdot 10^{-\frac{A}{T}+B}, \tag{5.177}$$

and after substitution into (5.170) then

$$D = \frac{m_w Rd}{S\tau M \cdot (\varphi_o - \varphi_d) \cdot T^{C-1} \cdot 10^{-\frac{A}{T}+B}}. \tag{5.178}$$

Given that in the notation corresponding to the calculation formula of the cup method (5.163)

$$\Delta p = (\varphi_o - \varphi_d) \cdot T^C \cdot 10^{-\frac{A}{T}+B}, \tag{5.179}$$

$$\Delta t = \tau \tag{5.180}$$

$$\Delta m = m_w \tag{5.181}$$

and

$$D = \delta \frac{RT}{M}, \tag{5.182}$$

we can immediately see that equation (5.178) is identical with (5.163).

5.3.2.2 Water Sorptivity

A precise procedure for the measurement of sorptivity was described by Hall and Tse (1986). The sample, typically rectangular in section, is dried to constant mass in an air oven at 105°C, and its dry mass noted. After cooling to room temperature the sample is immersed in a tray of water to a depth of 3–5 mm. The sample is removed at intervals (for instance 1, 4, 9, 16, 25 minutes) and weighed. The sorptivity is determined from the straight line (equation (5.154)) obtained by plotting the cumulative volume of water absorbed per unit area I against $t^{1/2}$.

Later, Hall (1989) described three possible test configurations to measure the rate of unidirectional water absorption in a porous medium: (i) horizontal inflow, in which there are no gravitational forces, but absorption is affected by hydrostatic forces; (ii) infiltration, where absorption is partly due to capillary suction and partly to gravitational forces; (iii) the capillary rise case, where the effects of capillarity and gravity forces are opposed. For most building materials including concrete and mortar, the capillary forces are dominant in all these configurations, and due to its simplicity and ease of operation, the capillary rise method is usually employed.

Sabir et al. (1998) presented a simple device for measuring sorptivity of mortar and concrete by the capillary rise method. The device consists of a suspension frame constructed of rigid copper wire, which is attached to the sensor of an electronic balance. The other end of the frame is rigidly attached to a light aluminum tray containing a central hole 45 mm in diameter. The specimen is placed centrally on the aluminum tray with the hole facilitating exposure of the test surface to water. The balance is set on a rigid table, and is controlled by purpose-written software, which is installed on a PC. The test specimen on its suspension mechanism is placed over a reservoir containing water and immersed 2–5 mm into water. The mass gain by the test specimen is automatically recorded at specified intervals.

5.3.2.3 Moisture Diffusivity

One possible way to describe the liquid moisture transport in capillary-porous materials is to use the nonlinear diffusion equation (see Chapter 3)

$$\frac{\partial u}{\partial t} = \text{div}\,(\kappa(u)\,\text{grad}\,u), \tag{5.183}$$

where u is the moisture content by mass (i.e. mass of water per mass of dry specimen) and $\kappa(u)$ the moisture diffusivity. The dependence of κ on the moisture content leads to the nonlinearity of equation (5.183). If $\kappa(u)$ is known, the moisture field in the material and its time evolution can be calculated using equation (5.183) together with appropriate boundary and initial conditions.

The moisture flux $\vec{j}$ can be expressed by Fick's law,

$$\vec{j} = -\rho_s \kappa(u)\,\text{grad}\,u, \tag{5.184}$$

where ρ_s is the partial density of the porous matrix (the mass of the matrix per unit volume of the whole body), and in combination of equation (5.184) with the continuity equation

$$\frac{\partial \rho_m}{\partial t} + \text{div } \vec{j} = 0, \tag{5.185}$$

where ρ_m is the partial density of moisture (the mass of moisture per unit volume of the sample), $\rho_m = \rho_s u$, we immediately get the diffusion equation (5.183).

The moisture diffusivity $\kappa(u)$ is usually measured in the quasi-one-dimensional arrangement (Lykov, 1958). Samples in the form of a rod are used. One end of the sample ($x = 0$) is fed with water, the other end ($x = L$) is left free, and the lateral area of the sample is insulated against water and water vapour. The water flux j at $x = 0$ and the moisture distribution $u(x, t)$ along the sample are measured. Equation (5.183) can then be reduced to the form

$$u_t = (\kappa(u)u_x)_x, \tag{5.186}$$

where subscripts t and x denote partial derivatives with respect to t and x.

Various methods for evaluation of experimental data obtained can be used. The simplest one is the steady-state method introduced by Lykov (1958). In the steady state $u_t = 0$, and therefore $u(x, t) = u(x)$ and the flux j_o of water are time-independent. The moisture diffusivity $\kappa(u)$ is then given according to equations (5.186) and (5.184) by

$$\kappa(u(x)) = -\frac{j_o}{\rho_s u_x(x)}. \tag{5.187}$$

This method has several disadvantages: (i) several months may be required to reach the steady state for some materials; (ii) only one measurement, namely the stationary measurement is used, while all the preceding measurements cannot be; (iii) the steady state for some samples may correspond to the almost homogeneous distribution of the highest possible (saturated) moisture content. The method completely fails in such a case.

Some of these disadvantages (particularly (i) and (iii)) can be eliminated by using transient methods. A simple transient method was proposed by Lykov (1958). Its basic relations can be derived in the following way. By integrating equation (5.186) from 0 to x we obtain

$$\int_0^x u_t(\xi, t) \, d\xi = \kappa(u(x, t))u_x(x, t) - \kappa(u(0, t))u_x(0, t). \tag{5.188}$$

The second term on the right-hand side of equation (5.188) is equal to the diffusion flux $j(0, t)$, as given by Fick's law (5.184) at the wet end of the sample ($x = 0$). Using equation (5.188), the expression for $\kappa(u)$ is

$$\kappa(u(x, t)) = \frac{\int_0^x u_t(\xi, t) \, d\xi - \frac{j(0,t)}{\rho_s}}{u_x(x, t)}, \tag{5.189}$$

which is a generalization of equation (5.187), and it is reduced to it in the steady state ($u_t = 0$).

The two main disadvantages of this method are: (i) the time derivative u_t can usually be determined with lower accuracy and it requires at least two measurements of $u(x)$ at two different times; (ii) it is necessary to measure the water flux $j(0, t)$.

Another transient method was proposed by Matano (1933). It is derived on assumption that

$$\lim_{x \to +\infty} u(x, t) = u_2 \tag{5.190}$$

for all t. Then the Boltzmann transformation (Crank, 1975)

$$\eta = \frac{x}{2\sqrt{t}}, \quad u(x, t) = w(\eta) \tag{5.191}$$

can be used to convert the partial differential equation (5.186) into the ordinary differential equation

$$\kappa(w)w'' + \kappa'(w)w'^2 + 2\eta w' = 0, \tag{5.192}$$

from which

$$\kappa(u) = \frac{2}{w'(\eta)} \int_\eta^\infty \eta' w'(\eta') d\eta' \tag{5.193}$$

or

$$\kappa(u) = -\frac{x'(u)}{2t} \int_{u_2}^u x(u') du', \tag{5.194}$$

where $x(u)$ is the inverse function of $u(x) = u(x, t)$ at time t, and $u(x) = u_2$ for all x at $t = 0$. The results depend sensitively on the accuracy of the measured $u(x, t)$.

The Matano method can easily be modified so as to utilize measurements done at different times t, which makes it possible to reduce the experimental errors. We employ the Boltzmann transformation (5.191) with the value of time variable t_k corresponding to each set of experimental data $u(x_i, t_k)$ separately. As a result we obtain a set of transformed data $(\eta_j, w(\eta_j))$ that can be represented by a smooth function. The moisture diffusivity is then determined from equation (5.193).

Another type of method, which was proposed by Kašpar (1984) and is usually called the gradient method, is based on the measurements of two suction curves $u(x, t_1)$, $u(x, t_2)$, $t_2 > t_1$, calculating the moisture flux from these two curves, and finally using Fick's law (5.184). The following formula for κ is then obtained:

$$\kappa(u_x) = \frac{2}{(t_1 - t_2)(u_x(x_0, t_2) + u_x(x_0, t_1))}$$

$$\times \int_{x_0}^{\infty} (u(x, t_2) - u(x, t_1)) \, dx, \tag{5.195}$$

where $u_x = (u(x_0, t_2) + u(x_0, t_1))/2$.

The accuracy of determining the moisture diffusivity $\kappa(u)$ can be considerably increased, if several consecutive measurements of $u(x, t)$ at different times t are used. This was the idea of the integral method employed by Drchalová and Černý (1998). In this method, the parametrization $x_0(v, t)$ of the curve of the constant moisture content v in the plane of independent variables (x, t) is first introduced. The uniqueness of such parametrization follows from the fact that the function $u(x, t)$ decreases with x for a fixed t, and increases with t for a fixed x. It holds $u(x_0(v, t), t) = v$ for all v and t.

By integrating equation (5.188) with respect to t from 0 to T for $x = x_0(v, t)$, where v is a fixed value of the relative moisture content for which we want to determine the moisture diffusivity κ, we obtain

$$\int_0^T dt \int_0^{x_0(v,t)} d\xi \, u_t(\xi, t)$$

$$= \kappa(v) \int_0^T dt \, u_x(x_0(v, t), t) + \frac{1}{\rho_s} \int_0^T dt \, j(0, t). \tag{5.196}$$

The last term on the right-hand side of equation (5.196) can be expressed using the total amount of water, $M_w(T)$ that flowed into the sample during the time interval $(0, T)$,

$$Q(T) = \frac{1}{\rho_s} \int_0^T dt \, j(0, t) = \frac{M_w(T)}{\rho_s S}, \tag{5.197}$$

where S is the cross-sectional area of the sample.

Changing the order of integrations on the left-hand side of equation (5.196) and after some algebra, we obtain the expression for the moisture diffusivity:

$$\kappa(v) = \frac{1}{\int_0^T dt \, u_x(x_0(v, t), t)} \cdot \left(\int_0^{x_0(v,T)} d\xi \, u(\xi, T) \right.$$

$$\left. - \int_0^{x_0(v,0)} d\xi \, u(\xi, 0) - v[x_0(v, T) - x_0(v, 0)] - Q(T) \right). \tag{5.198}$$

The quantity $Q(T)$ can be determined either directly by measuring $M_w(T)$, or it can be expressed as

$$Q(T) = \int_0^T dt \int_0^L dx \, u(x, t), \tag{5.199}$$

provided no water has evaporated from the free end of the sample $x = L$. This is correct when T is small enough. It should be stressed that it is necessary to determine $Q(T)$ with a high accuracy in order to minimize error in $\kappa(u)$.

An important step in the numerical evaluation of experimental data is the choice of a proper representation of the measured values $u(x_k, t_l)$ of the moisture distribution by a smooth function, which allows us to calculate the derivative u_x and the integrals in equation (5.198) with the least possible error. Drchalová and Černý (1998) employed the continuous analogue of the linear filtration with Gaussian weights (Hamming, 1962)

$$u(x) = \sum_{k=1}^{N} W(x - x_k) u(x_k),$$ (5.200)

where

$$W(x - x_k) = \frac{\exp[-A(x - x_k)^2]}{\sum_{i=1}^{N} \exp[-A(x - x_i)^2]}.$$ (5.201)

The approximation function depends on the parameter A, which can be expressed as

$$A = \alpha \left(\frac{N - 1}{x_{\max}} \right)^2,$$ (5.202)

where $x_{\max}$ is a maximum value of x, which in the case of the method of Drchalová and Černý (1998) was chosen as $x_{\max} = L$. The new dimensionless parameter α determines the character of the interpolation between the experimental points. The approximation function $u(x)$ for large α ($\alpha \gg 1$) turns into a piecewise linear function connecting the points $(x_k, u(x_k))$, while for small α ($0 < \alpha \ll 1$) the function $u(x)$ is reduced to a constant, equal to the arithmetic mean of $u(x_k)$ values. If α is close to 1, equation (5.200) gives a smooth interpolation curve, which fits the experimental data well. The best choice is $0.8 < \alpha < 1.2$.

5.3.2.4 Hydraulic Conductivity

Hydraulic conductivity of saturated porous materials is measured in a similar way to Darcy's original experiment (see Section 3.1.1). Constant head permeameters or falling head permeameters may be used (see Hillel, 1980). The measurements are performed preferably with undisturbed core samples taken from the field or prepared directly for the measurements in the laboratory. In either case, provision must be made to avoid boundary flow along the walls of the container. Reviews of various methods for measuring saturated hydraulic conductivity can be found for example in Klute (1965a) or van Schilfgaarde (1974).

In measuring unsaturated hydraulic conductivity, devices designed for determination of retention curves, such as tension plates or pressure chambers (see Section 5.3.1), can be employed conveniently. Steady flow systems, where flux, gradient and water content are constant in time, or transient flow systems, where they vary, can be utilized (see Hillel, 1980).

In steady flow methods, a constant hydraulic head difference across the sample is applied, and the resulting steady flux of water measured (see Klute, 1965b). Measurements are made at successive levels of suction and wetness, so as to obtain hydraulic conductivity as a function of hydraulic head.

A widely used transient flow method for measuring unsaturated hydraulic conductivity is the outflow method (Gardner, 1956). It is based on measuring the falling rate of outflow from a sample in a pressure cell, when the pressure is increased by a certain increment.

5.3.2.5 Water Permeability

The water permeability (usually understood as the intrinsic permeability) of concrete is measured in a similar way to saturated hydraulic conductivity, as follows from the transformation relation (5.161). The only difference is that the permeability cells are designed directly for concrete measurements, while the devices for determination of hydraulic conductivity discussed in the previous subsection were designed primarily for measuring with soils, which can sometimes lead to problems, e.g. with sealing when they are used with concrete samples.

The most straightforward experimental arrangement for measuring water permeability (see Sosoro et al., 1997, Reinhardt et al., 1997) involves measuring the flow of liquid through a specimen of uniform cross-sectional area with the surfaces parallel to the direction of flow being sealed. In practical laboratory measurements, low-permeability materials are more difficult to test due to the need to maintain higher pressures to produce measurable flows. The higher pressures tend to lead to difficulties with the sealing of the surfaces, through which no flow should occur. An appropriate device for measuring water permeability of concrete is the Hassler cell permeameter (see Sosoro et al., 1997). The Hassler cell is a quick-loading core holder. Cylindrical core specimens are held within a sleeve of nitrile rubber. This sleeve acts as a barrier between the fluid flowing through the core and the pressurized water in the outer chamber of the cell. This pressurized water provides a containing pressure, which ensures axial flow through the core and also prevents the flowing liquid from leaking into the outer chamber. During the operation of the device, the containing pressure must be higher than the flow pressure. A movable plate at the inlet end of the cell enables cores of a range of lengths to be accommodated. A thin disc of plastic mesh is placed between the input plate and the inlet face of the core to ensure liquid access across the full area of the inflow face. Pulse-free flow of the test liquid is provided by a chromatography pump containing a manometric module. Typical flow rates are between 0.01 ml/min and 5 ml/min. The maximum pressure is normally limited by the maximum pressure for the Hassler cell, which is typically 20 MPa. The water permeability of a specimen is measured by maintaining constant liquid flow through the specimen and measuring the pressure of the liquid at the inlet. The pressure data can be logged directly into a dedicated computer terminal. Eluent can be collected from cell to provide an independent measurement of flow rate, and also to provide liquid for further analysis.

Another simple permeameter suitable for measurements on concrete was designed by Hearn and Mills (1991). It consists of a permeability cell, in which concrete disks (100 or 150 mm diameter and 40 mm thick) are sealed with a silicon rubber sleeve. Both inflow and outflow are monitored using a precision piston-cylinder arrangement, and the pressure is applied by the dead loading of the input piston. Electronic sensors monitor the movement of the pistons and the applied pressure, and the data is recorded using a data acquisition system. The intrinsic permeability K (in m^2) is then calculated according to the formula

$$K = \frac{\eta}{\rho g} \frac{QL}{A \Delta h},$$ (5.203)

where η is the dynamic viscosity of the liquid (water in this case), ρ the density of the liquid, g the acceleration due to gravity, Q the flow rate (in m^3/s), A the cross sectional area of the sample (in m^2), Δh the drop in hydraulic head across the sample (in m), and L the thickness of the specimen.

5.3.3 Simultaneous Determination of Moisture Storage and Moisture Transport Parameters

As has been shown in the two previous sections, determination of moisture storage and moisture transport properties may be a very difficult and time consuming task. Sorption isotherms, water retention curves and hydraulic conductivity belong to the parameters that require very significant amounts of time if measured by classical methods. Therefore, it is worthwhile trying to design an experimental set-up that would make it possible to determine moisture storage and moisture transport parameters at the same time from a transient experiment.

A successful attempt in this direction was made recently by Plagge et al. (1999), who developed two sophisticated instantaneous profile methods (IPM) for simultaneous determination of the sorption isotherm, water retention curve and hydraulic conductivity. In their experimental set-up, they use a cylindrical material container, which is standard in measuring hygric properties of soils and can be equipped with different types of sensors to investigate the flow process within the material sample. The development of time-domain reflectometry (TDR) as a new technique to determine water content opened the possibility of an accurate nondestructive laboratory method to determine water content of high temporal and spatial resolution. Therefore, Plagge and various co-authors recently sought to refine the TDR technique using miniaturized probes (see, e.g., Malicki et al., 1992) as a basic tool in the method. Miniaturized pressure transducer tensiometers (Plagge et al., 1990) and relative humidity sensors can be used for determination of capillary pressure in the wet and dry moisture range.

The basic idea for the development of the method is the use of the same material column in a series of experiments, where different flow regimes can be achieved by controlling the initial and boundary conditions. In general, the experimental set-up consists of several components. The vertical measuring sampler is fitted with five pairs of trap holes, into which guide nuts are screwed prior to drill holes to insert specially designed sensors. The material sample is equipped with TDR probes and pressure transducer tensiometers or relative humidity sensors, aligned at equal distances of 20 mm along the 100 mm long container. Movement of water can be achieved by controlling the evaporation, where a ventilator is used to control the evaporation rate and to realize constant relative humidity boundary condition. A lid with sealing material serves for zero-flux boundary conditions at the bottom (IPM I) and the top (IPM II) of the material sample. To monitor, control and register material moisture and water potential, a computer-aided automatic data acquisition system is used. Further information concerning the sensors used, their calibration and additional experimental details are given in Malicki et al. (1992), Plagge et al. (1996), Plagge et al. (1997), and Plagge et al. (1990).

For measurement, the material is sampled, prewetted, fixed into a standard container, and then prepared for sensor installation. For IPM I, the material sample is equipped with mini-tensiometers and TDR probes, and then sealed at the bottom. At the top of the sample, a ceramic plate is fixed, and a controlled suction of −10 hPa is applied. Static equilibrium is achieved typically after several days, which is indicated by time-constant water content and capillary pressure in different positions measured by the TDR and the tensiometer system.

The IPM I experiment is started in the following way: the ceramic plate is removed and the ventilator and/or lids with holes are placed on top of the sample. Thus the material is allowed to evaporate, while water content and capillary pressure are monitored continuously. The end of the experiment is reached when the water content measured by the top TDR sensor is equal to the one measured at 95% relative humidity. Subsequently the tensiometers are removed and the material container is placed in a climatic chamber for equilibration at a relative humidity of 95%.

To start the second experiment (IPM II), the relative humidity sensors are installed in the sample container. The top of the sample is sealed and the sampling container is fixed on the miniaturized climatic chamber box, where drying agents and a ventilator are used to control the relative humidity and dry the material. During the desorption process, water content and relative humidity are continuously monitored and registered by the data acquisition system.

After the experiment is completed, the different sensors are removed and the total water content in the sample is measured by the gravimetric method. From the measurement of the sample mass at the beginning of each IPM run, the TDR water content measurements are corrected using a mixed dielectric component approach (see Plagge et al., 1996). An adjustment of the tensiometers and the relative humidity sensors is made by defining the physical offset at the static equilibrium states before each IPM run.

As a result of the IPM experiments, three different measurement ranges can be distinguished. The tensiometer range is from saturation up to 100 kPa capillary pressure, the second range from 100 kPa up to 6 MPa capillary pressure, and the relative humidity φ range $< 95\%$, which is equivalent to > 6 MPa capillary pressure. Since there is no proper capillary pressure sensor available yet in the range between 100 kPa to 6 MPa (the psychrometric sensors described in Section 5.1.4 are too large to use in the IPM experimental set-up), the capillary pressure is indirectly derived from the static water retention characteristics, and the corresponding water content determined by TDR. For calculation and data interpretation each range is interpreted individually.

During the experimental runs, water content w and capillary pressure p_c are measured (or derived, for the second measurement range) at five different positions along the vertical material sample, resulting in two sets of data for $w(z_i, t_j)$ and $p_c(z_i, t_j)$. Each single measurement corresponds to a triplet, all together define a surface in three-dimensional coordinates. The interpolation of the triplets results in a surface of smoothed data sets. Assuming the measurement error is of purely stochastic nature, part of the measurement error can be removed by the three-dimensional interpolation procedure followed by Bezier (1971).

This smoothing procedure yields the surfaces of the measurement variables for a given position at a defined time. From the corresponding water content and capillary pressure (or relative humidity) time profiles, the storage characteristics $p_c = p_c(w, T)$ for chosen compartments of the material can be determined. As a result of the desorption experiment, the transport coefficients as functions of water content or capillary pressure can be determined directly from their definition relations, namely

$$\vec{j}_w = -\rho_w \kappa(w) \text{ grad } w \tag{5.204}$$

or

$$\vec{j}_w = -\rho_w k(h) \text{ grad } h, \tag{5.205}$$

or

$$j_w = -\rho_w \frac{K}{\eta_w} k_{rw}(w) \text{ grad } p_w, \tag{5.206}$$

where j_w is the water flux, ρ_w the density of water, w the moisture content by volume, κ the moisture diffusivity, k the hydraulic conductivity, h

the pressure head, K the water permeability of the porous material, η_w the dynamic viscosity of water, k_{rw} the relative permeability, and p_w the liquid water pressure. The water flux in the above relations can be calculated using the measured moisture profiles in the following way:

$$j_w(z, t) = \rho_w \int_0^z \frac{\partial w(z, t)}{\partial t} \, dz. \tag{5.207}$$

The instantaneous profile methods (IPM) used by Plagge et al. (1999) are able to measure the water retention characteristic and the unsaturated hydraulic conductivity or moisture diffusivity simultaneously in a reasonably short time. For common building materials, the required measurement time is about four weeks, while only a determination of the water retention characteristics by the static equilibrium reference methods mentioned in Section 5.3.1 usually needs more than one year. Disadvantages of the IPM methods are the need for special sensor installation technologies and (at the moment still not available) capillary pressure sensors in the range between 100 kPa and 6 MPa. However, the latter is more or less a technological problem that could be solved in the near future, for instance by a miniaturization of psychrometric sensors shown in Section 5.3.1.

5.3.4 Hygric Expansion Coefficients

The infinitesimal change in length due to change in moisture u can be expressed in the form (see, e.g., Toman and Černý, 1996)

$$dl_u = l_0 \alpha_u \, du, \tag{5.208}$$

where l_0 is the length at the reference moisture u_0, α_u the linear hygric expansion coefficient. If the moisture content by mass (in kg/kg) is defined by the relation

$$u = \frac{m_m - m_d}{m_d}, \tag{5.209}$$

where m_m is the mass of the moist material, and m_d the mass of the dried material, the hygric expansion coefficient α_u is expressed in $(kg/kg)^{-1}$.

Defining the normal strain due to the change of moisture content ϵ_u as

$$\epsilon_u = \frac{\Delta l_u}{l_0} = \frac{1}{l_0} \int_{l_1}^{l_2} dl_u, \tag{5.210}$$

we obtain

$$\epsilon_u(u) = \int_{u_0}^u \frac{d\epsilon_u}{du} \, du = \int_{u_0}^u \alpha_u \, du. \tag{5.211}$$

From equation (5.211) it directly follows that

$$\alpha_u = \frac{d\epsilon_u}{du}. \tag{5.212}$$

In the case that the material is isotropic, the volume hygric expansion coefficient β_u is defined as

$$\beta_u = 3\alpha_u. \tag{5.213}$$

Measuring the linear hygric expansion coefficient α_u at a constant temperature T_0 with varying moisture content u is, given the assumption that the material is isotropic, relatively easy. The initial length of a rod specimen l_0 is first determined by an extensometer. Then the specimen is conditioned consecutively to the desired moisture contents u_i, $i = 1, \ldots, n$, and the corresponding lengths l_{ui} are measured. In this way, a pointwise function $l_u = l_u(u)$ is obtained. After a regression analysis of the $l_u = l_u(u)$ function, the linear hygric expansion coefficient $\alpha_u = \alpha_u(u)$ can be calculated using equations (5.210) and (5.212).

The $\alpha_u(u)$ function determined by the described procedure is valid only for the temperature T_0. If a temperature dependence of the $\alpha_u(u)$ function is required, which in the case of concrete is almost always necessary because the changes in moisture content are often accompanied by temperature changes, the above treatment is repeated for a set of another ambient temperatures, T_i, $i = 1, \ldots, m$. In this way, a set of functions $\alpha_u = \alpha_u(u, T_i)$ is obtained, which can be transformed into a two-variable function $\alpha_u = \alpha_u(u, T)$ by a regression analysis.

The procedure for measuring the linear hygric expansion coefficient as described above requires only very simple experimental techniques. Practically the only device that is necessary for the measurements is an extensometer. Among the various extensometers used in practice (see, e.g., Kirby, 1992, for a review), the most appropriate for measurements on concrete is a dial gauge. Due to the nonhomogeneity of the material, measurements on concrete require relatively large specimens, and therefore the use of more precise techniques, such as interferometry, is not necessary. Commonly used dial gauges can register a length change of 1 μm. With a 100 mm long specimen, this is the sensitivity for the determination of normal strain of 10^{-5}, which is high enough for hygric expansion of concrete (see, e.g., Černý et al., 2000, for results of practical measurements). Dial gauges have wide measuring ranges, and these are stable. Once they are calibrated with end standards or with accurate screw micrometers, they will stay calibrated unless mistreated. In addition, the currently used dial gauges are no longer the manual devices used before, which was their main disadvantage. Many of these now permit automatic recording of readings and electronic data transfer.

5.4 THERMAL PROPERTIES

5.4.1 Hydration Heat

There are two methods for determining hydration heat: the direct method and the indirect method. Both methods may be carried out under isothermal, adiabatic or isoperibolic regimes.

The direct or physical method continuously follows the development of hydration heat in direct relation to the reaction rate and effects by which it is influenced. In particular, this requires attention to the chemical and mineralogical composition of cement, the grain size, the w/c ratio, the presence of additives and heat.

The isoperibolic method is the most frequently used direct method of measuring the development of cement hydration heat, because it is the simplest and can be carried out in every laboratory. The apparatus consists of a sealable Dewar's flask filled with foam rubber, in which the measuring cell containing the sample is inserted. The flask is placed in a room maintained at constant temperature. Another way of setting up the measuring apparatus is the difference arrangement, where the examined sample of the cement paste or mortar is placed in one flask, and the perfectly hydrated sample is placed in a second flask. The more constant the temperature of the environment, the better the reproducibility of the results. The calorimeters are placed so that the distance between them is approximately equal to their diameter (Gautier, 1973, European Standard, prEN 196-9). In direct methods, the measuring sensor is a thermocouple, a thermistor or a platinum resistance thermometer placed in the middle of the measured sample. The thermocouples may form a battery located on the periphery of the measuring cell. The voltage signal is conducted to a sensitive millivoltmeter connected to a PC via the measuring exchange, or is registered by the millivoltmeter. Measuring is carried out on cement paste, mortar or concrete based on the size of the cell. The calorimeter is placed in a tempered room where a temperature of $20\pm1°C$ is maintained, and the maximum air velocity around the calorimeter is 0.5 m/s.

The isoperibolic method of measuring is used for short-term measurement only, because of heat escape to the surroundings, which must always be measured. The coefficient of the heat loss α is the amount of heat loss from the sample per unit time and per unit temperature difference between the sample and the environment. Its determination rests on the electric heating of a known amount of distilled water by a known output. The cooling of the distilled water brought to the highest temperature attained during the hydration of cement is followed up (specific heat capacity 4 186 J/kgK). In some calorimeters, calibration is performed when the aluminium block is heated by means of the resistance wire wound around it. It is also possible to use the same flask with perfectly hydrated cement, where the resistance wire is placed inside. The resistance wire is connected to the stabilized voltage. The coefficient of heat loss α (in W/K) can be calculated from the relation

$$\alpha = \frac{U^2}{R\Theta_c},\tag{5.214}$$

where U is the stabilized supply voltage in volts, R the resistance of the heating body in ohms and Θ_c the difference in temperature between the heated sample and the surroundings, in the differential calorimeters the comparative sample.

Hydration heat Q for time t is calculated according to the relation

$$Q = \frac{C}{m_c}\Theta_t + \frac{1}{m_c}\int_0^t \alpha\Theta_t \, dt,\tag{5.215}$$

where m_c is the cement mass in the tested mixture in grams, t the time of hydration in hours, C the calorimeter total thermal capacity in J/K, Θ_t the temperature increase of the tested sample, and α the coefficient of heat loss in W/K.

In the <u>adiabatic method</u> of direct measuring the hydration heat of cement, the heat exchange of the system with the surroundings is prevented. The calorimeter is supposed to be adiabatic if the rate of reduction of the sample temperature is less than 0.02 K/h (RILEM, 1997). Adiabatic calorimeters have good thermal insulation, and small heat losses are compensated by electric heating. The calorimeters are constructed with one sensor placed in the sample measured and the other outside the reaction vessel. The automatic control system switches the heating body, and so heats the surroundings of the reaction vessel with the sample until the temperature of the measured sample is reached. The heat released during the cement hydration heats the reaction mixture, which leads to a substantial evaporation of water. The measured heat is smaller by the water evaporation heat, the value of which is 2.66 kJ/g. Therefore, when the adiabatic method of measuring is used, the vessel with the sample must be completely filled and vapour-proof.

The <u>isothermal method</u> compensates for the thermal effect of the reaction studied in a suitable way, and maintains the sample temperature constant throughout the time of measuring. It is not necessary to know the heat capacities of the calorimeter to calculate the reaction heat; the temperature is not measured, because it is constant. The thermal compensation for the reaction heat is carried out in endothermal reactions by Joule heat, heating the sample and the apparatus, so that the temperature may be constant. In exothermal reactions, Peltier's effect is used, which is based on the heat consumption at one of the connections of the thermocouples during the passage of current.

A Calvet type calorimeter consists of a metallic block, in which a metallic vessel is placed and where a glass cell is inserted. Along the sides of the metallic vessel are the thermocouple batteries, both cooling and heating. In the thin-walled glass ampoule is water, which after the ampoule has been broken is mixed with cement. Isothermal calorimeters are mainly used for measuring the hydration heat in the cement paste; the mass of the samples is about 1 g. Mixing cement with water makes it possible to measure the entire heat that is released, including the cement wetting heat (Hansen and Jensen, 1998).

The indirect method called the dissolution or chemical method is based on the determination of the difference in the reaction heat of the nonhydrated and hydrated cement in the mix of acids. This method is based on Hess's law of constant heat summation, which may be represented as follows:

$$-\Delta H_{\text{hydration}} = -\Delta H_{\text{total reaction}} - (-\Delta H_{\text{partial reaction}}), \qquad (5.216)$$

where $-\Delta H_{\text{total reaction}}$ $(-\Delta H_{\text{tr}})$ is the reaction heat of the nonhydrated cement in the mix of acids, $-\Delta H_{\text{partial reaction}}$ $(-\Delta H_{\text{pr}})$ the reaction heat of the residue of nonhydrated cement and $-\Delta H_{\text{hydration}}$ $(-\Delta H_{\text{h}})$ the cement hydration heat for a given time of hydration.

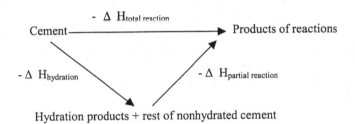

The determination is carried out in the isoperibolic regime in the dissolution calorimeter. The insulated vessel, most frequently a Dewar's flask, is placed in a box of thermal insulated material. The thermal insulation must be such that the temperature of 400 ml of distilled water of temperature 5°C higher than the ambient temperature will not decrease by more than 0.002°C to 1°C of the temperature difference after 30 minutes left standing. The vessel is closed with a cork closure, through which pass a Beckmann's thermometer, a stirrer and a funnel. The internal surface of the vessel and the thermometer are coated with paraffin, and the funnel and stirrer are made of a plastic material to prevent damage by hydrofluoric acid.

The calorimeter calibration is carried out by means of the reaction of annealed zinc oxide ZnO with the mixture of acids. Both the solution and the sample must be tempered to the same temperature. The temperature is read at one-minute intervals on the Beckmann's thermometer five minutes before the reaction starts. Then the sample is added to the mixture of acids and the reaction mixture is uniformly mixed. A reaction proceeds, during which the temperature is recorded every minute. The thermal capacity of the calorimeter C (in J/K) is calculated from the relation

$$C = \frac{m_{ZnO}}{\Delta t_k}[1077.43 + 0.364(30 - t_f) + 0.50(t - t_f)], \tag{5.217}$$

where m_{ZnO} is the sample mass in grams, Δt_k the corrected difference of temperatures in °C, t_f the value on the Beckmann's thermometer at the end of the dissolution period in °C, and t the temperature of ZnO in °C. The coefficient 1077.43 is the reaction heat of ZnO in J/g, -0.364 the temperature coefficient of reaction heat of ZnO, and 0.50 the specific heat capacity of ZnO in J/g.

The preparation of the sample for the determination of hydration heat is as follows: cement is mixed with water, immediately placed in a protective rubber membrane and tightly closed to prevent the escape of water vapour and the penetration of CO_2. The samples are placed in the thermostat tempered to 22 ± 0.5°C and left to hydrate for a specified period of time. Before the hydration heat is determined, the sample is taken out of the membrane, quickly crushed and immediately triturated in the mortar, so that it may pass through a sieve of mesh size 0.500 mm. It is stored in a closed, sealable vessel. In the sample, the loss of ignition is determined.

The sample hydrated for a short period of time, from several hours to two days, contains a considerable amount of unbound water. This water is removed by pure ethanol and consequently diethyl ether being passed through the crushed sample. After being dried, the sample may be triturated and the reaction heat determined.

The determination of the dissolution heat of nonhydrated and partially hydrated cement is carried out in the same way as calibration. The charges of cement should be corrected by the ascertained loss of ignition. The reaction heat of nonhydrated cement, reacting with solutions of acids, is expressed by relation

$$-\Delta H_{tr} = \frac{C\Delta t_k}{m_c} - 0.837(t - t_f) + 0.837(t_f - 20), \tag{5.218}$$

where m_c is the charge of cement in grams corrected for the loss of ignition, -0.837 the temperature coefficient of reaction heat of nonhydrated cement in J/gK, t the temperature of nonhydrated cement when it is put into the calorimeter, and 0.837 the specific heat capacity of nonhydrated cement in J/gK. The reaction heat of the partial hydrated cement is given by the relation

$$-\Delta H_{pr} = \frac{C\Delta t_k}{fm_{hc}} - 1.674(t - t_f) + 1.255(t_f - 20), \tag{5.219}$$

where m_{hc} is the charge of hydrated cement in grams, f the loss of ignition of hydrated cement, t the temperature of hydrated cement, when it is located to the calorimeter, the coefficient 1.674 is the specific heat capacity of hydrated cement in J/gK, and 1.255 the temperature coefficient of heat of dissolving.

The hydration heat expressed in J/g is the difference between the ascertained values,

$$-\Delta H_h = -\Delta H_{tr} - (-\Delta H_{pr}).$$ (5.220)

In the determination of the hydration heat of cement, it is necessary to ensure the dissolution of the whole charge in the mixture of acids. An undissolved portion of the sample causes errors in measuring.

The indirect method of determining the hydration heat of cement was defined more precisely using a thermistor heat-sensing element (Rovnaníková and Brandštetr, 1982). The reaction proceeds under isoperibolic conditions. Measuring is carried out in differential fashion; the same amount of a mixture of acids is placed in both the comparative and the measuring vessel. The plastic reaction cells are put in the Dewar's flasks. Thermistors, stirrers and the plastic powder sample injector are fitted to the sliding plastic cover. Cement is tempered submerged in the injector placed directly in the mixture of acids. The thermistor forms one branch of the Wheatstone bridge, while the voltage signal balancing the bridge is carried to the registration millivoltmeter. The equipment may be connected to an on-line PC. A diagram of the reaction vessel is shown in Fig. 5.1.

Calibration of the equipment is carried out by the reaction of annealed ZnO, whose reaction heat in the mixture of acids is 1077.43 J/g. After finding the voltage that corresponds to the reaction heat of nonhydrated cement and the mixture of acids, the reaction heat is calculated according to the relation

$$-\Delta H_{tr} = \frac{1077.43 \, z'}{S'},$$ (5.221)

where z' in mV is the voltage corresponding to the reaction of 1 g of the sample of or nonhydrated cement, S' in mV the voltage corresponding to the reaction of 1 g of ZnO in the solution of acids. The value of z' is found from the equation

$$z' = \frac{z}{n'},$$ (5.222)

where z in mV is the value of the voltage corresponding to the sample charge n' corrected for the loss of ignition. In the same way the reaction heat of hydrated cement, $-\Delta H_{pr}$ is determined. The hydration heat of cement as a function of time is calculated according to equation (5.220) from the values for the total reaction and the partial reaction.

The values of reaction heat determined after 2, 3, 7, 28 and 90 days give the course of the generation of the hydration heat of cement, as shown in the example of the hydration of ordinary Portland cement (OPC), Fig. 5.2.

Given that the course of the cement hydration manifests itself in the formation of a solid structure in the cement paste in concrete, the development of the strength of concrete may be derived from the course of the generation of the hydration heat of cement.

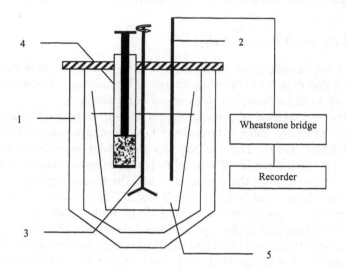

Figure 5.1 A reaction vessel provided with a thermistor
1: Dewar's flask, 2: Thermistor, 3: Stirrer, 4: Batcher of solid samples,
5: Reaction wessel with the mixture of acids

The course of the generation of hydration heat of cement makes it possible to study the effect of the mineralogical composition and the particles size, the temperature during hydration, the water/cement ratio, and the addition of admixtures on the hydration reactions of cement.

5.4.2 Specific Heat Capacity

Specific heat capacity c (in J/kgK) is usually determined according to its definitional relation,

$$c(T) = \frac{\partial h}{\partial T},\tag{5.223}$$

where h is the specific enthalpy (in J/kg) and T the temperature. In some technical applications, the temperature derivative in equation (5.223) is replaced by differential changes of the respective quantities, and the mean specific heat capacity c_0 over a given temperature interval ΔT is obtained,

$$c_0(T) = \frac{\Delta h}{\Delta T}.\tag{5.224}$$

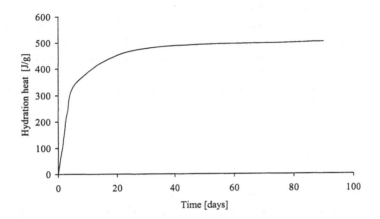

Figure 5.2 Course of the generation of hydration heat of OPC

Strictly taken, equation (5.224) is valid for small temperature intervals only. However, sometimes it may be useful to have some kind of integral value of specific heat capacity, at least for the orientation purposes.

There is a variety of methods for determining specific heat capacity. The main kinds of classical methods are mixing methods, adiabatic heating methods, relative methods and flow methods (see Krempaský, 1969). We will now give a brief overview of these methods.

5.4.2.1 Mixing Methods

These methods have been well known for many years and were very frequently used in many practical applications throughout the last century. The details of the measuring procedure can be found in any textbook on experimental physics, therefore we will not repeat them here. The specific heat capacity (mean specific heat capacity according to the definitional relation (5.224)) is determined using the known heat capacity of a calorimeter K (in J/K), the basic temperatures of the calorimeter T_1 and the measured material T_2, and the equilibrated temperature T_e from the relation

$$c_0 = \frac{K}{m_0} \frac{T_e - T_1}{T_2 - T_e}, \tag{5.225}$$

where m_0 is the mass of the measured material.

The mixing methods are stationary, and therefore the measurement is often time consuming. They are best suited to larger specimens, because for small specimens the increase in the temperature of the calorimeter is too small and the sensitivity of the method is too low. Also, the specific heat capacity is usually determined using a relatively wide temperature range; for instance, temperature difference $\Delta T = T_2 - T_1$ for water calorimeters can be 70–80°C. Use of this type of calorimeter in higher-temperature range is very rare, because it is difficult to find an appropriate liquid. Therefore, the methods are not suitable for measuring the temperature dependence of specific heat capacity.

5.4.2.2 Relative Methods

Relative methods are based on the analysis of either the heating or the cooling of the measured specimen compared to a reference specimen of the same geometrical shape. The specific heat capacity is determined from the temperatures measured at two suitable moments of time or using the known rate of heating or cooling. To distinguish these two ways of data evaluation, we denote the respective methods as the two-temperature method and the microcalorimetric method.

In the two-temperature method, a calorimetric vessel having mass m_0 and specific heat capacity c_0 is heated up or cooled down together with either a reference specimen of mass m_r and specific heat capacity c_r, or with a measured specimen of mass m and specific heat capacity c. In the case of cooling, the heat loss of the calorimeter with the specimen is equal to the heat transferred to the surroundings,

$$-(m_0 c_0 + mc)\frac{\mathrm{d}T}{\mathrm{d}t} = \alpha(T - T_0),\tag{5.226}$$

where α is the heat transfer coefficient, T_0 the temperature of the surroundings. An analogous equation can also be written for the reference specimen,

$$-(m_0 c_0 + m_r c_r)\frac{\mathrm{d}T}{\mathrm{d}t} = \alpha(T - T_0).\tag{5.227}$$

Assuming $T_0 = 0$, we obtain the solutions of equations (5.226) and (5.227) in the form

$$\ln\left(\frac{T_1}{T_2}\right) = \frac{\beta t}{m_0 c_0 + mc}\tag{5.228}$$

$$\ln\left(\frac{T_1}{T_{2r}}\right) = \frac{\beta t}{m_0 c_0 + m_r c_r},\tag{5.229}$$

where T_1 is the initial temperature, T_2, T_{2r} the temperatures of the measured and reference specimens, respectively, at time t. Solving the set of equations (5.228)–(5.229) for the unknowns c, β, we obtain

$$c = \frac{(y-1)m_0 c_0 + y m_r c_r}{m},\tag{5.230}$$

where

$$y = \frac{\ln\left(\frac{T_1}{T_{2r}}\right)}{\ln\left(\frac{T_1}{T_2}\right)}. \tag{5.231}$$

Equation (5.230) is valid on the assumptions that the heat transfer coefficients are the same for both measured and reference specimens, and that the temperatures inside the samples are constant and equal to the temperature of the vessel. Therefore, the rate of change of temperature has to be sufficiently low.

The microcalorimetric method employs the establishment of a so-called "regular state of first type", which is the name introduced by Kondratiev (1957). It refers to cooling or heating procedure of a body in certain environment having temperature T_0, which is independent of both the initial temperature distribution and the position of a point in the body. Mathematically it can be expressed as

$$T(t) = T_0 + A \exp(-Bt), \tag{5.232}$$

where A is a coefficient depending on the initial temperature distribution in the body, and the coefficient B,

$$B = -\frac{\partial}{\partial t} \ln \left| \frac{T - T_0}{T_0} \right| \tag{5.233}$$

is called the rate of cooling or heating.

Both the measured specimen and the reference specimen are cooled or heated in the same conditions until a regular state of the first type is established, and the cooling or heating rates B and B_r are determined. The specific heat capacity of the measured specimen is determined according to the relation

$$c = \frac{K_0}{m} \left[\frac{B_r}{B}(m_0 c_0 + m_r c_r) - m_0 c_0 \right]. \tag{5.234}$$

The coefficient K_0 depends on the shape of the measured specimen and on its thermal diffusivity. For a cylindrical specimen we can write

$$K_0 = 1 - \frac{R^2}{8} \frac{B}{a}, \tag{5.235}$$

where R is the radius of the cylinder, a the thermal diffusivity of the specimen. In most cases, the dependence of K_0 on thermal diffusivity is very slight, and the approximation $K_0 \sim 1$ can be used.

5.4.2.3 Adiabatic Heating Methods

Adiabatic heating methods are based on mostly monotonic heating of the measured specimen under adiabatic conditions, i.e. the temperature of the device surrounding the specimen is maintained equal to the temperature of the specimen, so that no heat loss from the specimen occurs. All the heat supplied by the heat source goes into the temperature increase of the measured specimen. Such a procedure makes it possible to calculate the specific heat capacity in a straightforward way,

$$c = \frac{\Delta Q}{m\Delta T},$$
(5.236)

where ΔQ is the heat supplied by the source, and ΔT the temperature increase of the specimen.

The most important problem with calorimeters of this type is maintaining the adiabatic conditions. This is usually done using electric heating with a classical coil.

5.4.2.4 Nonadiabatic Method for High-Temperature Aplications

Using the classical adiabatic-calorimetry methods for measuring the specific heat capacity of building materials can lead to certain difficulties for the following reasons:

1. The samples of building materials have to be relatively large because of their nonhomogeneity.

2. Due to the low thermal conductivity of most of building materials it can take a relatively long time to achieve temperature equilibration over large dimensions, which results in significant heat loss.

Therefore, nonadiabatic methods with exact calibration of heat loss can be considered as a reasonable alternative. An example of the application of a nonadiabatic method to high-temperature measurements of specific heat capacity of building materials is provided by Toman and Černý (1995).

The nonadiabatic calorimeter used by Toman and Černý (1995) has a mixing vessel 2.5 litres in volume. The volume of the measuring fluid (usually water) is about 1 litre. The maximum volume of the measured samples of building materials is 1 litre.

The amount of heat loss of the nonadiabatic system is determined using calibration. The calorimeter is filled with water at a temperature different from the ambient air. Then, the relation of water temperature to time, $T_c(t)$, is measured. Three thermometers with a relative accuracy of 0.02 K are used to measure the water temperature. The air temperature, T_m, measured by the same thermometers at two points, must be maintained constant (a maximum deviation of about 1 K during the measurement is permitted). Therefore, the experiments are performed in a conditioned room, where sudden changes of temperature can be eliminated. The mixing vessel is located on a laboratory stand ensuring direct contact at the bottom with the ambient air. A low-speed ventilator fan near the calorimeter bucket is used to achieve steady-convection conditions.

Numerous tests have shown that the calibration curve $T_c(t)$ is nearly exponential; small differences in measuring conditions cause only small changes in the calibration curve. The experiments have been found to have a repeatability better than 1%.

The measuring method itself is based on well-known principles. The sample is heated to a predetermined temperature T_s in a furnace and put into the calorimeter with water. Then, the relation of water temperature to time $T_w(t)$ is measured, water being slowly stirred all the time, until the temperatures of the measured sample and the calorimeter are equal. The duration of temperature equilibration ranges from 20 minutes to one hour, depending on the thermal conductivity and size of the material sample being measured.

As mentioned above, the heat loss calibration curve is used to calculate the theoretical equilibrated temperature (in absence of heat loss), T_e, of the sample–calorimeter system. The principle of this approach is as follows. The corrected water temperature $T_r(t_i)$ taking heat loss into account can be calculated as:

$$T_r(t_i) = T_w(t_i) + \Delta T(t_i) \tag{5.237}$$

$$\Delta T(t_i) = \sum_{j=1}^{i} \Delta T(\Delta t_j) \tag{5.238}$$

$$t_i = \sum_{j=1}^{i} \Delta t_j, \tag{5.239}$$

where T_r is the reduced temperature, $\Delta T(t_i)$ the cumulative temperature reduction in time t_i from the beginning of the measurement, and $\Delta T(\Delta t_j)$ are the particular temperature reductions for the various time intervals t_j, which are determined using the heat loss calibration curve $T_c(t)$. The value of Δt_j usually ranges from 10 s to 60 s. For the equilibrated temperature T_e we can write

$$T_e = \lim_{t \to \infty} T_r(t). \tag{5.240}$$

The heat balance of the sample–calorimeter system has the following form:

$$mc(T_s - T_e) = (K + m_w c_w)(T_e - T_{wo}) + \Delta m \cdot L - Q_r, \tag{5.241}$$

where m is the mass of the sample, T_s the temperature of the sample prior to being put into the calorimeter, c the specific heat capacity of the sample in the temperature interval $\langle T_e, T_s \rangle$, K the heat capacity of the calorimeter, m_w the mass of water, c_w the specific heat capacity of water, T_{wo} the initial water temperature, L the latent heat of vaporization of water, Δm the mass of evaporated water,

$$\Delta m = m + m_{cw} - m_s - \Delta m_N - \Delta m_{sc}, \tag{5.242}$$

m_{cw} the mass of the calorimeter with water before measurement, m_s the mass of the calorimeter–water–sample system after the measurement, Δm_N the mass of water naturally evaporated during measurement (this heat loss is already included in the heat loss calibration curve), Δm_{sc} the change of mass due to the chemical reaction of the sample with water (e.g. hydrolysis; this value can be obtained as $\Delta m_{sc} = m - m_D$, where m_D is the mass of the dried sample after the measurement), and Q_r the reaction heat.

Determining the specific heat capacity c directly from equation (5.241), we obtain a mean value of specific heat capacity, c_0, in the interval $\langle T_e, T_s \rangle$ by

$$c_0 = \frac{(K + m_w c_w)(T_e - T_{wo}) + \Delta m \cdot L - Q_r}{m(T_s - T_e)}. \tag{5.243}$$

However, as has been mentioned in the introduction to this subsection, from the physical point of view it is more correct to determine the value of the specific heat capacity pointwise, in accordance with the definition of specific heat capacity,

$$c(T_i) = \left. \frac{\partial h}{\partial T} \right|_{T=T_i}, \tag{5.244}$$

where h is the specific enthalpy.

Using relation (5.241) to determine the specific heat capacity according to the definition (5.244), we have to specify the zero point of the enthalpy scale, i.e. we have to ensure that all the enthalpy calculations are related to a certain constant temperature. This reference temperature can be, for example, $T_k = 0°C$. Upon adding

$$Q = m \cdot c_x \cdot (T_e - T_k), \tag{5.245}$$

where c_x is the mean specific heat capacity of the sample in the temperature interval $\langle 0, T_e \rangle$, to both sides of equation (5.241) and dividing by m, we obtain the following:

$$h(T_s) = \frac{(K + m_w c_w)(T_e - T_{wo}) + \Delta m \cdot L - Q_r}{m} + c_x(T_e - T_k). \tag{5.246}$$

The value of c_x is considered to be constant, taking into account the condition

$$T_s - T_e >> T_e - T_k. \tag{5.247}$$

It can be measured, for example, by using some of the classical adiabatic methods.

Performing a set of measurements for various sample temperatures T_i, we obtain a set of points $[T_i, h(T_i)]$. A regression analysis of this pointwise function results in a functional relationship for $h = h(T)$, and using relation (5.244) also in the function $c = c(T)$ as the first derivative of h with respect to T.

5.4.3 Thermal Diffusivity

Mathematically, all methods for determining the thermal diffusivity are based on the solution of the heat conduction equation

$$\rho c \frac{\partial T}{\partial t} = \text{div} \left(\lambda \, \text{grad} \, T \right) + Q, \tag{5.248}$$

where λ is the thermal conductivity, c the specific heat capacity, ρ the density, and Q the heat source (heat power density, in W/m^3). For $\lambda \sim const.$, i.e. for small temperature changes, we can write

$$\frac{\partial T}{\partial t} = a \, \text{div} \left(\text{grad} \, T \right) + \frac{Q}{\rho c}, \tag{5.249}$$

where a is the thermal diffusivity,

$$a = \frac{\lambda}{\rho c}. \tag{5.250}$$

Equation (5.249) is a diffusion-type equation with a constant transport parameter, which can be solved analytically for many combinations of initial and boundary conditions in 1-D, 2-D and 3-D cases. A variety of such solutions can be found in Carslaw and Jaeger (1959) and Krempaský (1969). Given an analytical expression for the temperature field and knowing the initial and boundary conditions, the thermal diffusivity a can be determined on the basis of temperature measurements in the particular experimental situation.

The limitations of the methods for direct determination of thermal diffusivity are given by equation (5.249). Only one single value of a can be determined in a single measurement. If a temperature dependence of thermal diffusivity, $a = a(T)$, is required, a set of different boundary and initial conditions has to be chosen, and a set of particular measurements has to be made. In this way, on the condition that the temperature changes during a single experiment are small, it is possible to approximate the $a(T)$ function by a piecewise constant function, i.e. by a function that is supposed to be constant over a chosen temperature interval, $\Delta T_i = T_{2i} - T_{1i}$, where T_{1i}, T_{2i} are the temperature minimum and maximum, respectively, in the i-th experiment.

As follows from equation (5.249), thermal diffusivity can be directly determined on the basis of transient methods only when the condition

$$\frac{\partial T}{\partial t} \neq 0 \tag{5.251}$$

is valid. The steady-state equivalent of the heat conduction equation (5.249),

$$a \operatorname{div} (\operatorname{grad} T)) + \frac{Q}{\rho c} = 0, \tag{5.252}$$

leads to a direct application of Fourier's law,

$$q = -\lambda \operatorname{grad} T, \tag{5.253}$$

where q is the heat flux (in W/m^2), and therefore to a direct determination of thermal conductivity.

Thermal diffusivity can also be determined indirectly on the basis of the definitional relation (5.250) and the known values of λ, ρ and c. In this section, we will deal only with direct methods for the determination of thermal diffusivity. The methods for direct determination of thermal conductivity and specific heat capacity are given in other sections of this chapter.

The form of the analytical solution to equation (5.249), which is fundamental to any experimental method for the determination of thermal diffusivity, depends primarily on the heat source term. Therefore, it seems logical to classify these methods according to the existence and shape of the heat source (see Krempaský, 1969, for details).

If the heat source is absent, i.e. $Q = 0$, the respective methods are called sourceless methods. In this case, the temperature of the specimen is modulated by its contact with another substance or environment, which plays the role of an infinite heat reservoir.

For $Q \neq 0$ we call the respective methods as source methods. They are characterized by the existence of a heat source either inside the specimen or on its surface. Source methods can be further classified according to the geometrical shape of the source, the time shape of the heat power of the source, and the geometrical shape of the specimen. By their geometrical shape we distinguish point sources, line sources, planar sources, volume sources, and combined sources. According to the time shape of the heat power function, we have pulse sources, constant sources, and periodic sources. By the geometric shape of the specimens we distinguish infinite environments, semi-infinite environments, specimens of exactly defined geometrical shape (plates, cylinders, spheres, etc.), and thin plates and layers.

For any experimental method for the determination of thermal diffusivity, the initial and boundary conditions chosen are of fundamental importance. From a practical point of view the initial conditions can be divided into just two groups, constant conditions,

$$T(x, y, z, 0) = T_0, \tag{5.254}$$

or spatial point-dependent conditions,

$$T(x, y, z, 0) = T_0(x, y, z). \tag{5.255}$$

The constant conditions are the more often used case.

The boundary conditions at the particular part of the boundary of the specimen Γ are usually divided into three basic groups: Dirichlet conditions,

$$T(\Gamma) = T_1(t), \tag{5.256}$$

where the temperature on the boundary is prescribed, the Neumann conditions,

$$-\lambda \frac{\partial T}{\partial \vec{n}} = q_0(t) \cdot \vec{n}, \tag{5.257}$$

where the prescribed heat flux on the boundary $q_0(t)$ is the main external parameter, and $\vec{n}$ is the normal vector to the boundary, and finally the Newton conditions,

$$-\lambda \frac{\partial T}{\partial \vec{n}} = \alpha(T_s - T_e) \cdot \vec{n}, \tag{5.258}$$

where α is the heat transfer coefficient, T_s the surface temperature, and T_e the temperature of the surroundings.

If heat conduction in the form of equation (5.249) is used, which is the case of thermal diffusivity measurements, the Neumann and Newton conditions (5.257) and (5.258) can be formally rewritten in the form

$$-a \frac{\partial T}{\partial \vec{n}} = \frac{q_0(t)}{\rho c} \cdot \vec{n}, \tag{5.259}$$

$$-a \frac{\partial T}{\partial \vec{n}} = \frac{\alpha(T)}{\rho c}(T_s - T_e) \cdot \vec{n}. \tag{5.260}$$

Therefore, if the Neumann or Newton conditions are operative, we can theoretically determine two thermophysical quantities from a single experiment, namely the thermal diffusivity a and the specific heat capacity c, provided that the functions $q_0(t)$ or $\alpha(T)$ are known. This is a fundamental feature of transient methods, which are therefore being used with increasing frequency for the practical measurement of thermophysical properties of building materials.

A variety of experimental set-ups for the determination of thermal diffusivity have been devised over the last hundred years or so. We will give now a short overview of the most frequently used methods with a brief explanation of their physical and mathematical principles, in order to provide some general guidance on the currently available devices for the determination of thermophysical properties. A more detailed survey can be found for instance in Krempaský (1969).

5.4.3.1 Transient Sourceless Methods

The following methods are most commonly employed, classified by their basic measuring principles (Krempaský, 1969): methods of regular state of the first and second type, constant temperature methods, periodic temperature change methods, radiation waves methods, and methods of general temperature change.

The methods of regular state of the first type are based on the establishment of the so-called regular state of first type, which is the term introduced by Kondratiev (1957). It describes a cooling or heating procedure of a body in certain environment having temperature T_0, which is independent of both the initial temperature distribution and the position of a point in the body. The mathematical description of a regular state of first type is given by the equation

$$T(t) = T_0 + A \exp(-Bt), \tag{5.261}$$

where A is a coefficient depending on the initial temperature distribution in the body, and the coefficient B,

$$B = -\frac{\partial}{\partial t} \ln \left| \frac{T - T_0}{T_0} \right|, \tag{5.262}$$

is called the rate of cooling or heating. For easier manipulation, equation (5.261) can be rewritten in the form

$$\ln \left| \frac{T - T_0}{T_0} \right| = -Bt + G(x, y, z). \tag{5.263}$$

In practical applications, the $\ln |T - T_0|$ function is plotted as a function of time, and its linear part, which is established after some time and corresponds to the regular state of first type, is identified. The cooling rate B can then be calculated using the known line slope $\tan \phi$,

$$B = \frac{\tan \phi}{T_0}. \tag{5.264}$$

There are several methods of calculating the thermal diffusivity using the known cooling rate B and the known analytic solution of the heat conduction equation corresponding to the given experimental set-up. The details can be found in Kondratiev (1954) or Kondratiev (1957).

The methods of regular state of the second type are based on the establishment of the so-called regular state of the second type (see Kondratiev, 1957), which can be expressed by the condition

$$\frac{\mathrm{d}T}{\mathrm{d}t} = b_0 = const. \tag{5.265}$$

This state is established in a specimen some time after putting it in an environment where the temperature T_0 is changing at a constant rate,

$$\frac{dT_0}{dt} = b_0 = const. \tag{5.266}$$

The characteristic feature of this state is that the temperature difference between any point in the specimen and the surroundings is constant. Again, on the basis of the known cooling or heating rate b_0, the measured temperatures at some characteristic points in the specimen and the known analytic solution of the heat conduction equation corresponding to the given experimental set-up, the thermal diffusivity a can be calculated.

The constant temperature methods employ an experimental set-up where the specimen is in contact with an infinite heat supply having a constant temperature. The temperature response of the specimen is analysed over a short time period after beginning the experiment. As the analytic solutions for Dirichlet conditions are mostly well known, and the temperatures in characteristic points of the specimen can be determined by measurement, the thermal diffusivity can be determined as the only unknown parameter.

The periodic temperature change methods employ periodic boundary conditions (mostly harmonic) on a part of the boundary of the specimen. The analytic solutions are again well known in most cases. Therefore, measurements of temperature field in the specimen (recording temperature as a function of time at just two spatial points is sufficient) lead to the identification of thermal diffusivity as the only unknown parameter (the Dirichlet conditions are assumed) by matching the experimental data with the analytic solutions.

Radiation waves methods work with an experimental set-up in which specimens of given geometrical shape are put into an environment where the temperature can be changed in a periodic way. Contrary to the constant temperature methods and the periodic temperature change methods described above, the boundary conditions apply to the whole boundary of the specimen. Newton boundary conditions with a finite value of the heat transfer coefficient α may be considered. If specimens with sufficiently regular shape are used, for instance plates, cylinders or spheres, the analytic solutions of the heat conduction equation are known. Measuring the amplitude of the waves at two spatial points in the specimen, or the phase shift of a given temperature at two points, the thermal diffusivity can be determined.

The methods of general temperature change consist in a generalization of the methods of regular state of the first and second type. They were devised in view of the experimental difficulties in maintaining the linear temperature change in the surroundings. The method consists in putting a specimen in contact with a metal block heated at an arbitrary rate. The temperature at two spatial points in the specimen is recorded as function of time, as well as the rate of temperature change of the surrounding block. A combination of measured data with the known analytical solution then gives the value of thermal diffusivity.

5.4.3.2 Point Pulse Source Methods

The main advantages of point sources are that they do not require any special specimen preparation and that it is possible to perform the measurements on almost every point in the specimen. An ideal point heat source should transfer heat to the specimen by a contact area as small as possible, and in an ideal experimental set-up, the heat transferred from the source to the specimen should be absorbed in a very thin surface layer. Therefore, a concentrated electron beam or a concentrated beam of electromagnetic radiation may be considered the most suitable heat sources.

Among the point pulse source methods currently used in practice, the laser flash method is probably used most often (see, e.g., Parker et al., 1961). The method is based on measuring and analysing the temperature response on the rear face after applying an instantaneous heat pulse to its front face. Using the known dependence of the heat flux on time, given by the properties of the laser pulse and the sample surface, and the known temperature response on the rear side, the thermal diffusivity can be calculated using an appropriate analytic solution to the heat conduction equation.

Among other methods (see Krempaský, 1969), two-point methods employ two point contacts on the specimen. The first is the heat source, the second the temperature sensor. The experiments can be set up for a semi-infinite case or for plates with finite thickness. If the heat power of the source and the pulse duration are known, analytical solutions can be derived, which makes it possible to calculate the thermal diffusivity. In some experimental set-ups, it is also possible to use the measured time to achieve a temperature maximum at certain point in the specimen at a known distance from the source to determine the thermal diffusivity.

Four-sensor methods involve the use of two pairs of sensors. The external sensors transfer electric impulse from an electric circuit in such a way that different amounts of heat are produced at the contacts of these sensors with the specimen, which can be done for instance by using contacts with different resistances in both sensors. The internal sensors indicate the temperature. Again, the time to achieve the temperature extreme can be identified, which makes it possible to calculate the thermal diffusivity.

5.4.3.3 Continuous Point Source Methods

The continuous point sources are less demanding on the measuring equipment for the determination of voltage than the pulse sources. Their power is either constant or periodic. The measurements are usually performed a in transient state or quasi-steady state.

Among the continuous point source methods, commonly used in practice, the step-heating method (see, e.g., Bittle and Taylor, 1984) is among the most often used. It is an experimentally simple transient photothermal technique, in which the front face of a small thermally insulated disk-shaped sample is exposed to a constant heat flux. The resulting temperature rise on the rear face of the sample is recorded, and the value of thermal diffusivity is calculated from this temperature rise versus time data. The ideal model for the step-heating method is based on the behaviour of a homogeneous, thermally insulated, infinite slab, with uniform and constant thermal properties and density, initially at constant (zero) temperature, exposed to a constant heat flux, uniformly applied over its front face from time zero. The nonideal model uses a disk-shaped sample and takes into account linearized heat losses from the sample governed by Biot numbers related to the front, rear and lateral faces.

A step-heating device suitable for measuring thermal diffusivity of building materials including concrete was built recently by Vozár and Šrámková (1997). It consists of the heat source, a halogen lamp (12V/100W) with a parabolic reflector, and an electrically controlled mechanical light shutter. The lamp is powered by a direct current produced by stabilized current/voltage supply Z-YE-2T-X (Mesit) with the unit of remote control JDR-1 (Mesit). The optimal current setting depends on the thermal properties and dimensions of the sample, and is chosen so that the rear face temperature rise reaches about 1–3°C in the measurement. The temperature rise at the rear face is measured by a spot-welded K type thermocouple (NiCr–Ni wires 0.1 mm in diameter). The reference cold junction is immersed in a Dewar's cup at 0°C. The transient voltage signal of the thermocouple is amplified and digitized by a standard 12-bit A/D converter. The apparatus is fully controlled by a PC.

Point sources with constant power can employ the same experimental set-ups as the pulse point sources. Data evaluation is similar, only the time of the impulse is set equal to the time of measurement, which for transient methods is several tens of seconds, for the quasi-steady methods several tens of minutes. As heat sources, focused electron beam or focused electromagnetic radiation are usually chosen. For data evaluation, in the case of thin plate specimens or the infinite or semi-infinite specimens the temperature at two points in the specimen are usually measured at two separate times, and using the analytical solution of the heat conduction equation, the thermal diffusivity can be calculated.

Periodic point sources can be used for bigger specimens. The experimental set-ups usually take the form of infinite or semi-infinite specimens, though thin plate specimens are also used. In data evaluation, either the phase shift or the ratio of the amplitudes in two different distances from the source are used for calculation of thermal diffusivity from the analytical solutions.

5.4.3.4 Line Source Methods

The most frequently used line sources are based on Joule heat production, i.e. a thin wire is heated by an electric current. The Peltier heat due to the transfer of direct electric current through contact between the source and the specimen can also be used for the specimens with a high Seebeck coefficient. The sources usually have the shape of a circle or a straight line. As with the point sources, also here either pulse sources or continuous power sources can be used. For line sources with circular shape, similar evaluation formulas to those for the point sources apply, if the temperature is detected in the centre of the circle and the power of the line source is the same as for the point source. Straight line pulse sources are the most popular. Line sources can also have a shape of a coil. Data evaluation methods are similar to those for point source methods. For pulse sources the time to achieve temperature maximum at a certain point together with the power of the source and its time duration can be used for the determination of thermal diffusivity. For continuous constant heat power sources the temperatures at two points in the specimen are measured at two separate times, while for periodic sources it is the phase shift or the ratio of the amplitudes at two different distances from the source.

5.4.3.5 Planar Source Methods

Planar sources can be devised using similar principles to those used in line sources, i.e. they can use Joule heat generated by the electric current in a thin plate or Peltier heat generated due to the transfer of electric current through planar contact with the measured specimen. The planar sources usually have the shape of a thin-walled hollow sphere, a thin-walled hollow cylinder or a thin plate. Both pulse and continuous sources are used, as with point sources or line sources. The mathematical principles for the determination of thermal diffusivity for the particular experimental set-ups are also quite similar, though only with regard to the respective analytical solutions to be used.

An example of practical application of a pulse planar source method can be found in Kubičár (1990). The Thermophysical Transient Tester RT 1.02 constructed to the theoretical design in Kubičár (1990) is based on the following principle. A heat pulse inside the specimen generates a dynamic temperature field. From the parameters of the temperature response, namely the time t_m and the magnitude T_m of the temperature response to the heat pulse, the thermal diffusivity and the specific heat capacity can be calculated. Technically the specimen is cut into three pieces and the measuring probes are inserted between the cut surfaces. A planar heat source made of thin metallic foil is placed between the first and the second pieces of the specimen, while a thermocouple is placed between the second and the third pieces. The heat pulse is generated by the passage of an electrical current through the metallic foil for a short time. The thermal diffusivity a is calculated as

$$a = \frac{h^2}{2t_{\mathrm{m}}}, \tag{5.267}$$

where h is the distance between the planar source and the temperature sensor, i.e. the thickness of the second piece of the specimen, the specific heat capacity c is given by the relation

$$c = \frac{Q}{\sqrt{2\pi e}Sh\rho T_{\mathrm{m}}}, \tag{5.268}$$

where Q is the energy of the heat pulse,

$$Q = UIt_0, \tag{5.269}$$

U is the voltage, I the magnitude of the electric current, t_0 the duration of the rectangular pulse, and S the cross-sectional area of the specimen.

The experimental arrangement of the Thermophysical Transient Tester RT 1.02 by Kubičár (1990) is as follows. The instrument takes measurements in specified surroundings (air, vacuum, inert atmosphere). The device consists of a chamber, thermostat, vacuum pump, electronic unit, and PC. A specimen holder is placed in the chamber, in which the specimen set including measuring probes is mounted. An isothermal shield is used to suppress any temperature gradient along the specimen set. A thermostat controls the temperature of the specimen holder. Low heating and cooling rates are used to maintain quasi-isothermal measurement conditions. A vacuum cover makes it possible to perform measurements in vacuum up to 0.1 Pa or in any atmosphere. The temperature range is from $-40°C$ to $200°C$.

The applicability of the RT 1.02 tester to building materials is limited by the relatively small allowable dimensions of the specimens, which is either a cylinder of diameter up to 30 mm or a rectangle with cross-section up to 30×30 mm and length up to 45 mm. Therefore, another tester based on the same theoretical principles was designed directly for use with building materials, model RTB 1.01 (SAS, 2000), which allows rectangular specimens with a cross section of either 100×100 mm or 150×150 mm. However, the need to use larger measuring chamber has led to limitations on the temperature range, from $-40°C$ to $70°C$.

5.4.3.6 Volume and Combined Source Methods

Volume sources offer much fewer variations in measuring methods than the point, line or planar sources. There can be either volume sources directly in the measured specimen, or sources that are in fact sensors for the determination of thermal parameters. Volume sources in the specimen are most often based on Joule heat production. These can take the form of pulse sources, continuous constant sources or periodic sources. The heated sensors can have plate, cylindrical or spherical shapes. The heat capacity of the sensors should be as low as possible, so that they do not affect the measurements. Combined sources are based on forming both planar and volume effects at the same time, for instance by a combination of Joule and Peltier effects.

5.4.4 Thermal Conductivity

In the most straightforward way, the thermal conductivity λ can be determined directly from Fourier's law,

$$q = -\lambda \operatorname{grad} T, \tag{5.270}$$

where q is the heat flux (in W/m^2). However, indirect methods based on the determination of thermal diffusivity and specific heat capacity are also frequently used. We will give a short overview of both direct and indirect methods for λ determination in what follows.

5.4.4.1 Steady-State Methods

For the application of relation (5.270) to the measurement of thermal conductivity, steady-state methods are the most convenient. These methods are considered classical, because they have been used for a long time, and in principal are very simple. The main advantage of steady-state methods is they make it possible to determine temperature fields and heat fluxes easily and precisely, because these quantities have steady values. On the other hand, the long measuring times induce problems with heat losses, which may cause significant systematic errors. The measured specimens usually have a plate, spherical or cylindrical shape.

The basic experimental set-up is very simple. The specimen is placed between two environments, one functioning as a heater and the other as a cooler. The heater is usually a block of a good conducting substance with electric heating built in. The cooler can for instance be a thermostatic liquid or water-ice mixture. If the temperatures of the heater T_1 and the cooler T_2, and the heat flux q are measured, the thermal conductivity can be easily calculated. For a plate specimen the calculation formula is as follows:

$$\lambda = \frac{Q'h}{S(T_1 - T_2)}, \tag{5.271}$$

where Q' is the heat power (in W), S the surface area of the plate specimen, and h the specimen thickness. This relation is valid only on the assumption that there is no heat loss from the sides of the specimen. This is the most serious problem with steady-state methods, though it has been solved using more or less sophisticated equipment. As a result, steady-state methods based on the simplest physical measuring principles tend to use the most complicated measuring devices.

Among the practical experimental arrangements for applying steady-state methods to determine thermal conductivity, the guarded hot plate method is the most frequently used. A design for a reference guarded hot plate apparatus in accord with the ISO 8302 standard was developed by de Ponte et al. (1992). Their device is a conventional "two-specimen apparatus". The assembly is placed in a temperature and relative humidity controlled environment, with surrounding air flowing along a "cold" and "hot" source in a closed test chamber. The hot plate is made of photo-etched heaters sandwiched between 10 mm aluminum plates. It is positioned horizontally using a system of crossbars fixed at the bottom of the test chamber and lies at four points on teflon supports of small cross-section (10^{-4} m^2). The two cold plates are mobile with a symmetrical displacement with respect to the hot plate, and are moved by means of a mechanical device operated from the outside of the test chamber with a hand wheel. A rectilinear potentiometer allows for the electrical measurement of the displacement. The metering area is surrounded by two guards with just eight epoxy mechanical connections embedded in the thickness of the two repartition plates. Temperature measurements are taken with type K thermocouples. Platinum resistance thermometers are used as temperature control sensors. Imbalance between the metering area and the first guard, as well as between the first and the second guard, is controlled by thermopiles with 100 and 200 thermocouple junctions, respectively. The minimum cold plate temperature in the device described by de Ponte et al. (1992) is limited only by the power of the chilling units and can be lowered to 210 K. The maximum hot plate temperature is limited to 340 K by the maximum service temperature for the epoxy used in the mechanical junctions between the guard and the metering area. Other mechanical junctions withstanding higher temperatures could have been used, but a substantial limit only a few tens of kelvin higher comes from the use of photo-etched heaters. If this type of heater is not used, achieving the required hot plate flatness becomes much more critical.

Steady-state methods for determination of thermal conductivity are considered reference standard methods. However, their application is limited to a very narrow range of temperatures. Therefore, if high-temperature measurements or measurements at cryogenic temperatures are necessary, another type of methods must be used.

5.4.4.2 Transient Methods

The most frequently used transient method for direct thermal conductivity measurements is the <u>hot wire method</u>. It is a dynamic technique based on the measurement of the temperature rise at a defined distance from a linear heat source embedded in the test material. If the heat source is assumed to have a constant and uniform output along the length of the test piece, the thermal conductivity can be derived directly from the resulting change in the temperature over a known time interval (see, e.g., Davis, 1984, for details):

$$\lambda = \frac{q}{4\pi K}, \tag{5.272}$$

where q is the constant heat production per unit time and per unit length of the wire (in W/m), and K a constant (in K) calculated as the slope of the linear portion of the temperature rise versus natural logarithm of time function.

There is a variety of practical applications of the hot wire method suitable for measuring concrete. One of the most convenient is the computer controlled hot wire device developed by Vozár (1996), which is designed for the determination of thermal conductivity of solids, powders and granular materials. The device allows for three measuring techniques: the standard cross wire technique, the resistance potential lead method, and the probe modification of the hot wire method.

In the cross wire technique, a platinum wire 0.35 mm in diameter or a kanthal wire 0.4 mm in diameter, depending on the temperature range of measurement, act as the linear heat source. The temperature rise versus time evolution is measured by spot welded K type thermocouple, made from Ni–NiCr wires 0.2 mm in diameter, or S type (PtRh 90/10%) 0.1 mm. The hot spot of the thermocouple is in direct contact with the heating wire, and is placed in the centre of sample. The reference – cold junction is immersed in a Dewar's cup at 0°C. In the resistance technique, a platinum wire 0.35 mm in diameter is used as the heating wire, which at the same time functions as the temperature sensor. Potential leads consist of platinum wires 0.1 mm in diameter, fixed to the heating wire at about 1.5 cm from the end of the sample. The probe method is based on the original very simple cylindrical probe construction, which consists of a heating wire and a temperature sensor, both placed in a ceramic microcapillary 12 cm long and 2 mm in diameter. Heating wire is located in the first hole, and the temperature sensor, a spot welded thermocouple, is in the second one, in order for the measurement point to be near to the centre of the probe. The wires used are the same as described in the cross wire technique.

The method of Vozár (1996) is quite general concerning its applicability to different kinds of materials. If a building material is to be measured, the sample usually has the shape of a cylinder 10 cm in length and diameter depending on thermal properties (4–10 cm), which is cut into two parts along the axis. The sample is placed in a furnace in a horizontal position, so that the bottom half carrying the grooves for wires (or probe) is put in first. Then the wires are put in position. In order to ensure good embedding, loose finely ground test material is put on the upper surface of the lower cylinder with the wires before inserting the upper half cylinder. Here, the use of cylindrical specimens rather than two bricks is preferred, because of the decreased time required to reach thermal equilibrium. The sample is heated by an electro-resistance furnace, which uses four silit rods as heating elements and alporite for the thermal insulating layer. This arrangement allows measurement in the air or in a reducing environment under atmospheric pressure, in the temperature range from room temperature up to 1200°C.

5.4.4.3 Indirect Methods

Thermal conductivity can also be determined indirectly, using the measured values of thermal diffusivity a and specific heat capacity c. This can be done using most of methods described in the previous subsection, which enable not only the determination of thermal diffusivity but also the determination of the specific heat capacity at the same time. The thermal conductivity can then be calculated from the relation

$$\lambda = a\rho c, \tag{5.273}$$

where ρ is the density.

5.4.4.4 Solution of Inverse Problems with Temperature Dependent Thermal Conductivity

All methods for the determination of thermal diffusivity and thermal conductivity so far discussed suffer from the basic limitations mentioned before. Only a single value of a, c or λ can be determined in a single measurement. If a temperature dependence of thermal diffusivity or thermal conductivity is required, a series of measurements under various temperature conditions has to be made, which may often be a very time consuming task, particularly if high temperature data is required.

A reasonable alternative to the classical methods described above is a direct solution of an inverse problem of heat conduction, supposing that the thermal conductivity λ is a function of temperature, $\lambda = \lambda(T)$. In this way, the temperature dependence of thermal conductivity can be determined in a single experiment using the measured temperature field. However, application of these methods requires knowledge of the specific heat capacity as a function of temperature, $c = c(T)$, which might be a problem. Nevertheless, the specific heat capacity can be determined using methods given in Section 5.4.2, so the problem is not a major one. We will give an overview of the most often used methods of this type in what follows.

Consider the heat conduction equation in one dimension in the form

$$\rho c \frac{\partial T}{\partial t} = \frac{\partial}{\partial x}\left(\lambda \frac{\partial T}{\partial x}\right). \tag{5.274}$$

Suppose that the temperature field $T(x,t)$ is known from the experimental measurements, and the functions $\rho(T)$, $c(T)$ were measured in previous experiments. Also, the initial and boundary conditions of the experiment are known. Solving an inverse problem means in this particular case the determination of the $\lambda = \lambda(T)$ function using the above data.

Solution of inverse problems of heat conduction is generally more complicated than the solution of partial differential equations. Even for relatively simple problems, the solution either does not exist at all, or it is not unique. Therefore, a "physical solution" is often used, when a sufficiently simple problem is chosen, which is then solved together with certain complementary conditions. We will describe several such methods for solving inverse problems of heat conduction in what follows.

One-curve methods employ just one $T(x, t_0)$ or $T(x_0, t)$ curve, where t_0 is a constant time, and x_0 a constant position, for determination of the $\lambda = \lambda(T)$ function. The inverse problem of heat conduction is mostly defined by equation (5.274), with Dirichlet boundary conditions

$$T(0, t) = T_1 = const. \tag{5.275}$$

$$T(\infty, t) = T_2 = const., \tag{5.276}$$

and the constant initial condition

$$T(x, 0) = T_2 = const. \tag{5.277}$$

A practical experimental set-up obeying these conditions can be realized in a relatively simple way. The constant temperature T_1 required by the condition (5.275) is maintained in a furnace. The surrounding air has a constant temperature T_2, so that the boundary condition (5.276) is satisfied. The specimen has an initial temperature equal to the temperature of the surroundings, which satisfies condition (5.277). The specimen is thermally insulated on all sides, fitted with temperature sensors along its longitudinal axis, and then installed in the door of the furnace. The temperature field in the specimen is recorded at given time intervals, so that a pointwise temperature field $T(x_i, t_i)$ is obtained.

The classical Boltzmann–Matano treatment (Matano, 1933), transforming the heat conduction equation into an ordinary second-order differential equation for T, is one of the most popular one-curve methods. In solving the inverse problem (5.274)–(5.277), the Boltzmann transformation

$$\eta = \frac{x}{2\sqrt{t}} \tag{5.278}$$

$$w(\eta) = T(x, t) \tag{5.279}$$

is used first. Substituting equations (5.278)–(5.279) into equations (5.274)–(5.277) using the relations

$$\frac{\partial T}{\partial x} = \frac{dT}{d\eta} \frac{\partial \eta}{\partial x} = \frac{dT}{d\eta} \frac{1}{2\sqrt{t}} \tag{5.280}$$

$$\frac{\partial T}{\partial t} = \frac{dT}{d\eta} \frac{\partial \eta}{\partial t} = \frac{dT}{d\eta} \cdot \left(-\frac{x}{4t\sqrt{t}}\right) = \frac{dT}{d\eta} \cdot \left(-\frac{\eta}{2t}\right), \tag{5.281}$$

we obtain

$$\frac{d}{d\eta}\left(\lambda \frac{dw}{d\eta}\right) + 2\rho c \, \eta \, \frac{dw}{d\eta} = 0 \tag{5.282}$$

$$w(0) = T_1 \tag{5.283}$$

$$w(+\infty) = T_2. \tag{5.284}$$

In this way, the partial differential equation (5.274) is transformed into the ordinary differential equation (5.282). Assuming now that for a given time $t_0 = const.$, the temperature field $T(x, t_0)$ is known, the following second transformation can be used:

$$z = \eta \cdot 2\sqrt{t_0} \tag{5.285}$$

$$T(z) = w(\eta). \tag{5.286}$$

Substituting equations (5.285)–(5.286) into equations (5.282)–(5.284), we get

$$\frac{d}{dz}\left(\lambda \frac{dT}{dz}\right) + \rho c \, \frac{z}{2t_0} \frac{dT}{dz} = 0 \tag{5.287}$$

$$T(0) = T_1 \tag{5.288}$$

$$T(+\infty) = T_2. \tag{5.289}$$

Further solutions to the system (5.287)–(5.289) can be found in two main ways. The first is an integral treatment, which was used in the original Matano (1933) work, the second is to express the differential equation (5.287) for λ as has been shown by Toman (1986). We will describe both methods here.

In the differential treatment, we first perform some simple algebraic modifications of equation (5.287), which will make the solution more straightforward, namely

$$\frac{d\lambda}{dz} + \lambda \frac{\frac{d^2T}{dz^2}}{\frac{dT}{dz}} = -\frac{\rho c z}{2t_0}, \tag{5.290}$$

or schematically

$$\frac{d\lambda}{dz} + A(z)\,\lambda = B(z). \tag{5.291}$$

Equation (5.291) is an ordinary differential equation of first order for λ, which in principle can be solved analytically. However, the particular solution depends on the type of $A(z)$, $B(z)$ functions, which are not known in general. The $T(z)$ function is usually given in a pointwise form $[z_i, T_i]$, obtained by experiment. Therefore, a regression analysis is always necessary and the regression functions can differ from case to case. The problem of the generality of the solution to equation (5.291) can be overcome by application of a suitable numerical method, such as the finite element method. We will show the basic features of this solution.

Choosing the approximation of the $\lambda(z)$ function on one finite element in the form

$$\lambda = \langle N \rangle\,\{\lambda\}_e = \left\langle 1 - \frac{z}{l}, \frac{z}{l} \right\rangle \left\{ \begin{array}{c} \lambda_1 \\ \lambda_2 \end{array} \right\}, \tag{5.292}$$

then the system of residual equations in the Galerkin method of solution (see, e.g., Zienkiewicz, 1971, for details) can be written in the form

$$\int_0^l \{N\} \left(\frac{\partial \langle N \rangle}{\partial z}\,\{\lambda\}_e + A(z)\,\langle N \rangle\,\{\lambda\}_e - B(z) \right)\,dz = \{0\} \tag{5.293}$$

or

$$([H_1]_e + [H_2]_e)\,\{\lambda\}_e = \{R\}_e, \tag{5.294}$$

where

$$[H_1]_e = \int_0^l \{N\}\,\frac{\partial \langle N \rangle}{\partial z}\,dz \tag{5.295}$$

$$[H_2]_e = \int_0^l \{N\}\,\langle N \rangle\,A(z)\,dz \tag{5.296}$$

$$\{R\}_e = \int_0^l \{N\}\,B(z)\,dz. \tag{5.297}$$

According to the general principles of the finite element method (Zienkiewicz, 1971), the element matrices and vectors are then localized into domain ones (for $z = +\infty$, we have to choose a point $z = z_1$, which is sufficiently distant from the heat source, so that for $z \geq z_1$ we have $T(z) = T_2$). Then, the boundary conditions are introduced into the system of algebraic equations obtained by the localization, which in the given case means an application of the Dirichlet condition

$$\lambda(z_0) = \lambda_0, \tag{5.298}$$

where z_0 is one of the boundaries of the spatial solution domain. The determination of λ_0 is usually not a very difficult problem, because in the common experimental set-up described above, we usually have temperature T_2 equal to the room temperature, say $T_2 = 25°C$, at one of the boundaries, and for this temperature, λ can easily be determined by some other method, for instance the guarded hot plate. Solving the resulting system of algebraic equations, a pointwise function $\lambda = \lambda(z)$ is obtained, which can easily be converted into the required $\lambda = \lambda(T)$ relationship because $T(z)$ is a known function given by the measurements.

In solving the system (5.287)–(5.289) by an integral treatment (Matano, 1933), we first integrate equation (5.287) from the chosen point $z = z_0$ to $z \to +\infty$, so that we obtain

$$\left[\lambda \frac{dT}{dz} \right]_{z_0}^{\infty} + \frac{1}{2t_0} \int_{z_0}^{\infty} \rho c \, z \, \frac{dT}{dz} \, dz = 0. \tag{5.299}$$

For $z \to \infty$ we assume a zero value of heat flux (the point is far enough from the heat source and at the same temperature as the surroundings),

$$j_Q \to 0, \tag{5.300}$$

and therefore

$$\lambda \frac{dT}{dz} \to 0. \tag{5.301}$$

Equation (5.299) is then modified into the form

$$\lambda(T_0) = \frac{1}{2 \, t_0 \, \left(\frac{dT}{dz}\right)_{z=z_0}} \int_{z_0}^{\infty} \rho c \, z \, \frac{dT}{dz} \, dz, \tag{5.302}$$

where $T_0 = T(z_0, t_0)$. The integral in equation (5.302) is solved by common numerical methods, such as Simpson's rule.

In some experimental cases we might not know the $T(x)$ curves at a time t_0, but do know $T(t)$ at a fixed spatial point x_0, i.e. we will have just one temperature sensor for measurements. The previously derived methods with the following modifications can then be used.

Our starting point will be the formulation (5.282)–(5.284), but we will assume that for a fixed point $x_0 = const.$, we know $T(x_0, t)$. Then the following transformation can be used:

$$t = \left(\frac{x_0}{2\eta} \right)^2 \tag{5.303}$$

$$T(t) = w(\eta), \tag{5.304}$$

so that we obtain

$$\frac{d}{dt} \left(\lambda \, t \, \frac{dT}{dt} \right) = \frac{x_0^2}{4t} \, \rho c \, \frac{dT}{dt} \tag{5.305}$$

$$T(0) = T_2 \tag{5.306}$$

$$T(+\infty) = T_1. \tag{5.307}$$

As in the previous case, we can choose either a differential or an integral treatment for the solution of the system (5.305)–(5.307).

The differential treatment leads to the equation

$$\frac{\mathrm{d}\lambda}{\mathrm{d}t} + A(t)\,\lambda = B(t), \tag{5.308}$$

where

$$A(t) = \frac{1}{t} + \frac{\frac{\mathrm{d}^2 T}{\mathrm{d}t^2}}{\frac{\mathrm{d}T}{\mathrm{d}t}} \tag{5.309}$$

$$B(t) = \frac{x_0^2}{4t^2}\,\rho c. \tag{5.310}$$

Equation (5.308) is solved in the same way as equation (5.291).

In the integral treatment we have

$$\left[\lambda(t)\,t\frac{\mathrm{d}T}{\mathrm{d}t}\right]_{t_x}^{t_0} = \frac{x_0^2}{4}\int_{t_x}^{t_0}\frac{\rho c}{t}\frac{\mathrm{d}T}{\mathrm{d}t}\,\mathrm{d}t, \tag{5.311}$$

and therefore

$$\lambda(T_0) = \frac{x_0^2}{4\,t_0\left(\frac{\mathrm{d}T}{\mathrm{d}t}\right)_{t=t_0}}\int_{t_x}^{t_0}\frac{\rho c}{t}\frac{\mathrm{d}T}{\mathrm{d}t}\,\mathrm{d}t + \frac{\lambda(T_x)\,t_x\left(\frac{\mathrm{d}T}{\mathrm{d}t}\right)_{t=t_x}}{t_0\left(\frac{\mathrm{d}T}{\mathrm{d}t}\right)_{t=t_0}}, \tag{5.312}$$

where $\lambda(T_x)$ is the known value of the thermal conductivity at temperature $T_x = T(x_0, t_x)$ (for instance $T_x = 25^\circ C$, $T_x \to T_2$), $T_0 = T(x_0, t_0)$.

One-curve methods are very suitable for practical applications because of their simplicity. However, these have one basic disadvantage: they strictly require the validity of the $\sqrt{t}$ rule, i.e. after performing the Boltzmann transformation $\eta = x/2\sqrt{t}$ we should obtain just one curve $T(\eta)$. This condition is not satisfied in all practical cases. Generally, the $\sqrt{t}$ rule can be applied for relatively short time intervals when the process is still far enough from the steady-state. On the other hand, when the time intervals are too short, the $\sqrt{t}$ rule should not be used because the sudden application of the Dirichlet condition (5.275) is not very realistic for real experimental cases, and some deviations may appear.

Two-curves methods are based on an analysis of differences between the temperature versus position curves in two close time intervals, $T(x, t_2)$ and $T(x, t_1)$. The gradient method is based on the determination of the $\lambda(T)$ function by a direct application of Fourier's law,

$$j_Q = -\lambda\frac{\partial T}{\partial x}. \tag{5.313}$$

We choose a point x_0 in the solution domain and assign it to the average temperature between the two $T(x, t_2)$, $T(x, t_1)$ curves,

$$T_0 = \frac{1}{2}(T(x_0, t_1) + T(x_0, t_2)). \tag{5.314}$$

The heat flux at point x_0 can be approximated by the relation

$$j_Q(x_0) = \frac{1}{t_2 - t_1} \int_{x_0}^{\infty} [\rho(T)c(T)T(x, t_2) - \rho(T)c(T)T(x, t_1)] \, dx. \tag{5.315}$$

The last term in equation (5.313), namely the derivative $\frac{\partial T}{\partial x}$, is again determined as an average value between the two $T(x, t_2)$, $T(x, t_1)$ curves,

$$\left(\frac{\partial T}{\partial x}\right)_{x=x_0} = \frac{1}{2}\left(\frac{\partial T(x_0, t_2)}{\partial x} + \frac{\partial T(x_0, t_1)}{\partial x}\right). \tag{5.316}$$

Substituting equations (5.315) and (5.316) in Fourier's law (5.313), we obtain

$$\lambda(T_0) = -\frac{2}{(t_2 - t_1)\left[\frac{\partial T(x_0, t_2)}{\partial x} + \frac{\partial T(x_0, t_1)}{\partial x}\right]}$$

$$\times \int_{x_0}^{\infty} [\rho(T)c(T)T(x, t_2) - \rho(T)c(T)T(x, t_1)] \, dx. \tag{5.317}$$

For a solution to a particular experimental case, it is important to choose suitable time intervals t_1 and t_2. If the difference $t_2 - t_1$ is too large, the accuracy of the calculation of the derivative $\frac{\partial T}{\partial x}$ and also of the temperature T_0 is too low. On the other hand, for $t_2 - t_1$ too small the accuracy of the calculation of the heat flux j_Q might not be sufficient. Generally it can be stated that optimal results are obtained if the temperatures $T(x, t_2)$ and $T(x, t_1)$ differ by approximately 10%.

Another two-curve method is the difference method of Černý and Toman (1997), which for sufficiently accurate data can work directly with the experimentally determined temperature data, without any further approximations. We again solve the system (5.274)–(5.277), but in (5.276) the length pertains to the one-dimensional domain D instead of the semi-infinite domain $\langle 0, +\infty\rangle$.

We assume that from experimental measurements, two pointwise temperature profiles $T(x, t_1)$, $T(x, t_2)$ are known, which obey the following conditions:

1. $\Delta t = t_2 - t_1 > 0$, $\Delta t << t_1$

2. The distance $\Delta x = x_{i+1} - x_i, i = 1, \ldots, n - 1$ of consecutive points in the x-direction is constant.

From equations (5.275)–(5.277) it also follows that both $T(x, t_1)$ and $T(x, t_2)$ are decreasing functions of the space coordinate, and that $T(x, t_1) < T(x, t_2)$, $x \in \langle 0, D\rangle$. Our task is to determine the unknown $\lambda(T)$ function.

In the solution of this formulation of the inverse problem of heat conduction, we employ a finite-difference treatment. In order to increase the precision of the finite-difference scheme, we express the time derivative in half of the time step Δt,

$$\frac{\partial T}{\partial t}(x, t + \Delta t/2) = \frac{T(x, t + \Delta t) - T(x, t)}{\Delta t} + O(\Delta t^2). \tag{5.318}$$

In an analogous way, we can write for the first spatial derivatives in the times t and $t + \Delta t$

$$\frac{\partial T}{\partial x}(x + \Delta x/2, t) = \frac{T(x + \Delta x, t) - T(x, t)}{\Delta x} + O(\Delta x^2), \tag{5.319}$$

$$\frac{\partial T}{\partial x}(x + \Delta x/2, t + \Delta t)$$

$$= \frac{T(x + \Delta x, t + \Delta t) - T(x, t + \Delta t)}{\Delta x} + O(\Delta x^2), \tag{5.320}$$

and after averaging over time, we obtain the first spatial derivative in half of the time interval,

$$\frac{\partial T}{\partial x}(x + \Delta x/2, t + \Delta t/2) = \frac{1}{2\Delta x}(T(x + \Delta x, t) - T(x, t)$$

$$+ T(x + \Delta x, t + \Delta t) - T(x, t + \Delta t)) + O(\Delta x^2). \tag{5.321}$$

Using equations (5.318)–(5.321), the finite-difference scheme of (5.274) can be written as follows:

$$(\rho(x, t) + \rho(x, t + \Delta t)) \cdot (c(x, t) + c(x, t + \Delta t))$$

$$\times \frac{T(x, t + \Delta t) - T(x, t)}{\Delta t}$$

$$= \frac{2}{(\Delta x)^2}[\lambda(x + \Delta x/2, t + \Delta t/2) \cdot (T(x + \Delta x, t) - T(x, t)$$

$$+ T(x + \Delta x, t + \Delta t) - T(x, t + \Delta t))$$

$$- \lambda(x - \Delta x/2, t + \Delta t/2) \cdot (T(x, t) - T(x - \Delta x, t)$$

$$+ T(x, t + \Delta t) - T(x - \Delta x, t + \Delta t))]. \tag{5.322}$$

Equation (5.322) can be solved explicitly for $\lambda(x - \Delta x/2, t + \Delta t/2)$ with the result

$$\lambda(x - \Delta x/2, t + \Delta t/2) = \lambda(x + \Delta x/2, t + \Delta t/2) \cdot \frac{A}{C} - \frac{B}{C}, \tag{5.323}$$

where

$$A = T(x + \Delta x, t) - T(x, t) + T(x + \Delta x, t + \Delta t) - T(x, t + \Delta t) \quad (5.324)$$

$$B = \frac{(\Delta x)^2}{2\Delta t}(\rho(x,t) + \rho(x, t + \Delta t))$$

$$\times (c(x,t) + c(x, t + \Delta t)) \cdot (T(x, t + \Delta t) - T(x,t)) \quad (5.325)$$

$$C = T(x,t) - T(x - \Delta x, t) + T(x, t + \Delta t) - T(x - \Delta x, t + \Delta t). \quad (5.326)$$

Equation (5.323) is a recursive relation for determining the thermal conductivity as dependent on temperature, which can be used for practical calculations providing $\rho(T)$, $c(T)$ and $\lambda(T_2)$ are known. These conditions are, however, not very restrictive. The measurements of $\rho(T)$, $c(T)$ are commonly performed in the high-temperature region also (Toman and Černý, 1995), the value of thermal conductivity λ at the lower temperature T_2 can be determined by some independent method suitable for the lower-temperature region, such as the hot-wire method.

There are also methods that use more than one or two curves for the determination of the $\lambda(T)$ function. An example of multiple-curves methods is the double integration method by Černý and Toman (1995). We assume, that n curves $T(x, t_i), i = 1, 2, ..., n$ are known from temperature measurements, and again solve the system (5.274)–(5.277), where in (5.276) we have the length of the one-dimensional domain D instead of the semi-infinite domain with $+\infty$. In the solution to the defined inverse problem, we suppose $T(t)$ and $T(x)$ to be monotonic functions, and first choose a constant value of temperature, $\tau = T(x,t)$. There must then exist one-to-one parametrizations $x = x_0(\tau, t)$, $t = t_0(\tau, x)$, where both x_0 and t_0 are monotonic functions. In other words, on the above assumptions it is possible to express x as a function of the chosen temperature τ and time t, and t as a function of τ and position x. The construction of the $\tau = T(x,t)$ curve in the (x,t) plane, i.e. the derivation of the shape of the integration region, is as follows. We choose a time t_A, $t_1 \leq t_A \leq t_n$, and using an interpolation procedure on the system of measured temperature curves $T(x, t_i)$, $i = 1, 2, ..., n$, we find the corresponding position x_A for temperature τ. The point $A = [x_A, t_A]$ lies on the boundary of the integration region. Then we choose a sufficient number of other points t_j from the interval $\langle t_1, t_n \rangle$, and in this way obtain a pointwise function $x = x_0(\tau, t)$, which forms the boundary of the integration region. We now integrate the heat conduction equation (5.274) over x and t on the integration region we just have found,

$$\int_{t_1}^{t_n} \int_0^{x_0(\tau,t)} \rho c \frac{\partial T}{\partial t} \, dx dt = \lambda(\tau) \int_{t_1}^{t_n} \frac{\partial T}{\partial x}(x_0(\tau, t), t) \, dt$$

$$- \int_{t_1}^{t_n} \lambda[T(0,t)] \frac{\partial T}{\partial x}(0,t) \, \mathrm{d}t. \tag{5.327}$$

The left-hand side of equation (5.327) would be difficult to solve directly. Therefore, we change the sequence of integration, accounting for the shape of the integration area,

$$LS = \int_{t_1}^{t_n} \int_0^{x_0(\tau,t)} \rho c \frac{\partial T}{\partial t} \, \mathrm{d}x \mathrm{d}t$$

$$= \int_0^{x_0(\tau,t)} \mathrm{d}x \int_{t_1}^{t_n} \rho c \frac{\partial T}{\partial t} \, \mathrm{d}t$$

$$+ \int_{x_0(\tau,t_1)}^{x_0(\tau,t_n)} \mathrm{d}x \int_{t_0(\tau,x)}^{t_n} \rho c \frac{\partial T}{\partial t} \, \mathrm{d}t. \tag{5.328}$$

Denoting

$$\int \rho c \frac{\partial T}{\partial t} \, \mathrm{d}t = \int \rho(T) \, c(T) \, \mathrm{d}T = I_T(T), \tag{5.329}$$

we obtain

$$LS = \int_0^{x_0(\tau,t_1)} [I_T(T(x,t_n)) - I_T(T(x,t_1))] \, \mathrm{d}x$$

$$+ \int_{x_0(\tau,t_1)}^{x_0(\tau,t_n)} [I_T(T(x,t_n)) - I_T(\tau)] \, \mathrm{d}x$$

$$= \int_0^{x_0(\tau,t_n)} I_T(T(x,t_n)) \, \mathrm{d}x$$

$$- \int_0^{x_0(\tau,t_1)} I_T(T(x,t_1)) \, \mathrm{d}x - I_T(\tau)[x_0(\tau,t_n) - x_0(\tau,t_1)]. \tag{5.330}$$

Taking into account that

$$-\lambda[T(0,t)] \frac{\partial T}{\partial x}(0,t) = j_Q(0,t), \tag{5.331}$$

and substituting equations (5.330) and (5.331) into equation (5.327), we get

$$\lambda(\tau) = \frac{1}{\int_{t_1}^{t_n} \frac{\partial T}{\partial x}(x_0(\tau,t),t) \, \mathrm{d}t} \left(\int_0^{x_0(\tau,t_n)} I_T(T(x,t_n)) \, \mathrm{d}x \right.$$

$$- \int_0^{x_0(\tau,t_1)} I_T(T(x,t_1)) \, \mathrm{d}x - I_T(\tau)[x_0(\tau,t_n) - x_0(\tau,t_1)]$$

$$- \int_{t_1}^{t_n} j_Q(0, t) \, dt \bigg), \tag{5.332}$$

where $j_Q(0, t)$ can be determined in a similar way to equation (5.315),

$$j_Q\left(0, \frac{t_j + t_i}{2}\right) = \frac{1}{t_j - t_i} \int_0^D [\rho(T) \, c(T) \, T(x, t_j)$$

$$- \rho(T) \, c(T) \, T(x, t_i)] \, dx. \tag{5.333}$$

The double integration method is the most universal of the methods for solving the inverse problem of heat conduction given above. Practically the only limitation of this method is the requirement the $T(x)$ and $T(t)$ functions to be monotonic.

5.4.5 Thermal Expansion Coefficients

The infinitesimal change of length due to a change of temperature T can be expressed in the form (see, e.g., Toman and Černý, 1996)

$$dl_T = l_0 \alpha_T \, dT, \tag{5.334}$$

where l_0 is the length at the reference temperature T_0, and α_T the linear thermal expansion coefficient (in K^{-1}).

Defining the normal strain due to a change of temperature ϵ_T as

$$\epsilon_T = \frac{\Delta l_T}{l_0} = \frac{1}{l_0} \int_{l_1}^{l_2} dl_T, \tag{5.335}$$

we obtain

$$\epsilon_T(T) = \int_{T_0}^{T} \frac{d\epsilon_T}{dT} \, dT = \int_{T_0}^{T} \alpha_T \, dT. \tag{5.336}$$

The definitional relation for α_T using the normal strain ϵ_T then follows directly from equation (5.336),

$$\alpha_T = \frac{d\epsilon_T}{dT}. \tag{5.337}$$

In the case where the material is isotropic, the volume thermal expansion coefficient β_T is defined as

$$\beta_T = 3\alpha_T. \tag{5.338}$$

The easiest way to determine the linear thermal expansion coefficient α_T of an isotropic material is first to dry it to zero moisture content (the moisture-induced length changes are quite comparable with those induced by temperature, see Toman and Černý, 1996) and then to measure the length changes due to the temperature changes using a proper extensometer. At first, the initial length of a rod specimen l_0 is determined. Then, the specimen is conditioned to the desired set of temperatures T_i, $i = 1, \ldots, n$, the material being maintained in the dry state, and the corresponding lengths l_{Ti} are measured. The pointwise function $l_T = l_T(T)$ is obtained in this way. After a regression analysis of the $l_T = l_T(T)$ function, the linear thermal expansion coefficient $\alpha_T = \alpha_T(T)$ can be calculated using equations (5.335) and (5.337).

A determination of α_T for another constant moisture content than the dry state is technically difficult, because keeping constant the absolute moisture content in a porous body while the temperature changes is a complicated task. The sorption isotherm $w = w(\varphi, T)$ is a function of temperature, and before measuring the length changes it would also be necessary to determine the $w = w(\varphi, T)$ function over the whole temperature range in order to set the relative humidity in the conditioning chamber. Therefore, it seems reasonable to use a measuring procedure that would consider the length changes due to both temperature and moisture changes at the same time.

The simplest way to couple the temperature and moisture induced length changes is the superposition principle, as has been demonstrated by Toman and Černý (1996). The total change of length can be expressed in differential form as

$$dl = dl_T + dl_u = l_0(\alpha_T \, dT + \alpha_u \, du). \tag{5.339}$$

The total normal strain due to changes of both temperature and moisture ϵ can be then written as

$$\epsilon(u, T) = \epsilon_T(u_0, T) + \epsilon_u(u, T_0)$$

$$= \int_{T_0}^{T} \left(\frac{\partial \epsilon}{\partial T}\right)_u \, dT + \int_{u_0}^{u} \left(\frac{\partial \epsilon}{\partial u}\right)_T \, du$$

$$= \int_{T_0}^{T} \alpha_T \, dT + \int_{u_0}^{u} \alpha_u \, du, \tag{5.340}$$

which already includes the definitional relations for α_T and α_u,

$$\alpha_T(u, T) = \left(\frac{\partial \epsilon(u, T)}{\partial T}\right)_u \tag{5.341}$$

$$\alpha_u(u, T) = \left(\frac{\partial \epsilon(u, T)}{\partial u}\right)_T. \tag{5.342}$$

For the simultaneous determination of the coefficients α_T and α_u, Toman and Černý (1996) proposed the following simple procedure. First, they determined the dependence of the normal strain on the moisture content at room temperature T_0. The material was moistened either to the saturated moisture content, maximum hygroscopic moisture content or to the equilibrium moisture content in the given environment, and then it was slowly dried, while the length changes were simultaneously measured. Thus a relation $\epsilon_u = \epsilon_u(u)$ was obtained over a wide range of moistures at constant temperature T_0. The function $\epsilon_u(u)$ was represented by point values. Therefore, a regression analysis was necessary to obtain a continuous function. In accordance with equation (5.342), α_u was obtained as the first derivative of $\epsilon_u(u)$ with respect to moisture.

Knowing the function $\alpha_u(u)$, the measuring process can continue with the temperature changes. In accord with the superposition principle, the length changes caused by the changing temperature are measured first, while moisture changes are monitored simultaneously. Then, these length changes are recalculated for zero moisture content using the $\epsilon_u(u)$ functions. In this way, a pointwise function $\epsilon_T = \epsilon_T(T)$ for a constant moisture ($u = 0$) is obtained, as required by equation (5.341). Regression analysis and calculation of the first derivative of $\epsilon_T(T)$ with respect to T leads directly to the function $\alpha_T = \alpha_T(T)$.

In the common temperature range up to $80°C$, the methods for measuring the temperature induced length changes of concrete specimens are the same as in the case of moisture induced changes. Again, dial gauge may be considered the most appropriate extensometer for that purpose.

In measuring the high-temperature linear thermal expansion of concrete, the method developed recently by Toman et al. (1999) can be used. The measuring device (the extensometer) consists of a cylindrical, vertically oriented electric furnace with two bar samples placed in the furnace. The first sample is the measured material, the second sample is a reference material for which the dependence of thermal expansion coefficient on temperature is known. The length changes of the samples are measured mechanically by dial gauges outside the furnace, using thin ceramic rods that pass through the furnace cover and are fixed on the top of the measured sample. These ceramic rods pass through an undefined temperature field, so their normal strain cannot be determined mathematically, and a comparative method of determining the normal strain of the rod is used instead.

A practical measurement of the linear thermal expansion coefficient of a building material using the extensometer carried out by Toman et al. (1999) can be described as follows. The measured sample and the standard are put into the furnace, fitted with the contact ceramic rods, and the initial reading on the dial gauges is taken. The furnace control system is then adjusted to the desired temperature T_i, the length changes are continuously monitored on the dial gauges and recorded by a computer. After the steady state is achieved, i.e. no temperature changes in the furnace and no length changes of both measured sample and the standard are observed, the final readings of length changes are taken. The length change of the measured sample is calculated according to the formula

$$\Delta l(T_i) = \Delta l_{\mathrm{m}}(T_i) - \Delta l_{\mathrm{s}}(T_i) + l_{0,\mathrm{s}} \int_{T_0}^{T_i} \alpha_{\mathrm{s}}(\tau)\, d\tau, \qquad (5.343)$$

where Δl_{m}, Δl_{s} are the final readings of total length changes of the measured sample and of the standard, respectively, including the length changes of the ceramic rods, $l_{0,\mathrm{s}}$ is the initial length of the standard, and $\alpha_{\mathrm{s}}(T)$ is the known linear thermal expansion coefficient of the standard.

The corresponding value of normal strain can be expressed in the form

$$\epsilon(T_i) = \frac{\Delta l(T_i)}{l_{0,\mathrm{m}}}, \qquad (5.344)$$

where $l_{0,\mathrm{m}}$ is the initial length of the measured sample.

The measurements are then repeated with other chosen values of furnace temperatures T_j, $j = 1, \ldots, n$, the pointwise function $\epsilon_{\mathrm{T}}(T)$ is obtained, and after a regression analysis of the $\epsilon_{\mathrm{T}}(T)$ function, the determination of the $\alpha(T)$ function of the measured material is performed according to equation (5.337).

The described method by Toman et al. (1999) was recently used by Černý et al. (2000) for measuring the high-temperature thermal expansion of cement mortar.

5.5 OTHER PARAMETERS

5.5.1 Gas Permeability

Gas permeability is defined in the same way as water permeability, namely

$$\vec{j}_{\mathrm{g}} = -\rho_{\mathrm{g}} \frac{K}{\eta_{\mathrm{g}}} k_{\mathrm{rg}}(c)\, \mathrm{grad}\, p_{\mathrm{g}}, \qquad (5.345)$$

where $\vec{j}_{\mathrm{g}}$ is the gas flux (in kg/m^{2}s), K the gas permeability of the porous material (in m^2), which is again an intrinsic property of the material independent of the gas concentration, η_{g} the dynamic viscosity of the gas, k_{rg} the relative permeability (dimensionless) of gas, which is a function of gas concentration c, and p_{g} the gas pressure.

The measuring principles are therefore similar to those used for the determination of water permeability in Section 5.3.2.5.

Figg (1973) developed a method for the determination of gas permeability, which was later improved by Dhir et al. (1987). The method is based on applying underpressure to hole drilled in the concrete specimen through a hypodermic needle using a hand generated vacuum. A test hole, 50 mm in depth and 13 mm in diameter, is drilled into the concrete. After thorough cleaning, the hole is plugged, to a depth of 20 mm from the outside surface, by polyether foam, and then sealed with catalyzed silicon rubber. When the rubber is hardened, a hypodermic needle is pushed through the silicon rubber plug. Connections are made to the hypodermic needle to introduce air under vacuum using a handheld digital electronic manometer. The vacuum applied is 45 kPa and the permeation index is taken as the time taken for the decrease of the applied pressure from 45 kPa to 55 kPa.

Claisse et al. (1999) performed a theoretical analysis, making it possible to calculate the gas permeability K on the basis of the Figg test. The resulting formula reads as follows:

$$\frac{(P + P_a)(P_i - P_a)}{(P - P_a)(P_i + P_a)} = \exp\left[\frac{2KP_a t}{\mu x_0^2 \ln(X/x_0)}\right], \tag{5.346}$$

where P_a is the atmospheric pressure (100 kPa), P_i the applied test pressure (45 kPa), P the pressure after the time t (55 kPa), μ the dynamic viscosity of the gas (in Pa.s), x_0 the radius of the evacuated volume, and X the distance from the reservoir of gas at atmospheric pressure.

RILEM (1999) recommends the Cembureau method (see Kollek, 1989, for details) for measuring of gas permeability. In the Cembureau method, the measuring device consists of a tool to measure the specimen dimensions with a precision of 0.1 mm, a Cembureau permeameter with a pressure cell to accommodate concrete disks with a diameter of 150 mm and a thickness of 50 mm, a gas supply with pressure regulation from 0.1 MPa to 0.6 MPa and a precision of 0.01 MPa, calibrated soap bubble volumetric gas flow meters, e.g., 150 ml, 15 ml, 5 ml, 1 ml, a chronometer sensitive to 0.1 s, and oxygen or nitrogen bottles.

The Cembureau permeameter cell (see Kollek, 1989) consists of a rubber tube designed to 0.5–1.5 MPa, polyurethane rubber ring surrounding the tube, gas inlet and outlet holes, and a valve. The specimen is put into the rubber tube, sealed and exposed to a defined pressure generated by a gas supply with a precision pressure regulator. The gas flow through the specimen is determined by the soap bubble volumetric gas flow meters.

Practical measurements are performed in the following way. First, the diameter of the test specimen is determined with a precision of 0.1 mm. Then the test specimen is placed in the cell, and the measuring device is assembled. A minimum lateral pressure of 0.7 MPa is exerted on the rubber tube in order to avoid gas leakage through its sides, and three pressure stages are selected for measurements of gas permeability. Typically these start with 0.15 MPa, then the pressure is increased first to 0.20 MPa, and finally to 0.30 MPa absolute gas pressure. The first flow measurements are performed after 30 minutes. The flows at each pressure stage are measured, until constant readings are achieved. The flow measurements are performed as follows. First the capillary of the soap bubble flow meter is moistened 1 minute before creating the bubble for measurement. The time measurements are started when the bubble is at the lowest marking on the calibrated tube. The measuring volume is selected by choosing the appropriate soap bubble flow meter, such that the time reading is more than 20 seconds. Then provisional readings of the flow rate are taken. If the difference between successive readings taken within 5 to 15 minutes is less than 3%, at least two readings in quick succession are taken, and the flow rate Q_i is determined for the given pressure stage,

$$Q_i = \frac{V_i}{t_i}, \tag{5.347}$$

where V_i is the volume of the gas, t_i the time reading. Then the pressure is increased to the next pressure level, and the flow rate determination procedure is repeated.

The mean gas permeability coefficient K (in m^2) is determined as the average of three values of K_i, calculated for the three different pressures (see Kollek, 1989, RILEM, 1999) using the following formula:

$$K_i = \frac{2P_a Q_i L \mu}{A(P_i^2 - P_a^2)}, \tag{5.348}$$

where P_a is the atmospheric pressure (100 kPa), P_i the applied test pressure, Q_i the gas flow rate (in m^3/s), L the thickness of the specimen (in m), μ the dynamic viscosity of the gas (in Pa.s), and A the cross-sectional area of the specimen (in m^2).

5.5.2 Ion Binding Isotherms

In concrete, cement mortar and cement paste, the determination of ion binding isotherms is chiefly concerned with chlorides. The common method of determining the chloride binding capacity, and therefore also the chloride binding isotherms, involves dissociating the free chloride fraction from the total chloride content by analysing the pore solution squeezed out from the porous material under high pressure (Barneyback and Diamond, 1981). The total chloride content may be determined using acid-soluble extraction (see, e.g., Dhir et al. 1990). However, it has been reported (Glass and Buenfeld, 1997) that pore solution expression under pressure results in the release of some loosely bound chlorides, which in turn may result in an overestimation of the level of free chlorides. Acid-soluble extraction may underestimate the total chloride content.

Tang and Nilsson (1993) proposed a method for the determination of chloride adsorption isotherms based on the adsorption from solution. A sample of cement mortar or cement paste dried at 11% relative humidity was put into a cup, then the cup was vacuumed in a desiccator for 2 hours, before being filled with a specific concentration NaCl solution saturated with $Ca(OH)_2$. The volume of the solution inside the cup was calculated from the increment of the mass of the cup and the density of the solution. The cup was covered and stored at 20°C to reach equilibrium. Adsorption equilibrium was typically achieved after 7 days for 25 g samples. Then the inside solution was pippetted to determine the chloride concentration by potentiometric titration using 0.01 N $AgNO_3$ and a chloride selective electrode. The bound chloride content C_b (in mg/g) was calculated from the equation

$$C_b = \frac{M_{Cl}V(C_0 - C_1)}{W}, \qquad (5.349)$$

where M_{Cl} is the molar mass of chlorine, V the volume of the solution (in ml), C_0, C_1 the initial and equilibrium concentrations, respectively, of chloride solution (in mol/l), and W the mass of the dry sample, which can be calculated from the difference in mass of the sample dried in a desiccator at 11% relative humidity and in an oven at 105°C. The free chloride content C_f (in mol/l), corresponding to the value of C_b calculated from equation (5.349) was given by

$$C_f = C_1. \qquad (5.350)$$

By performing the experiment with different values for the initial salt concentration C_0, a pointwise function $C_b = C_b(C_f)$ can be obtained, which is the ion binding isotherm.

Tang and Nilsson (1993) also proposed a similar method for the determination of chloride desorption isotherms. After the adsorption test, the surplus solution in the cup was removed as far as possible, and about 200 ml de-ionized water saturated with $Ca(OH)_2$ was added to the cup. Then the cup was stored at 20°C to reach a new equilibrium before the inside solution was pippetted and analysed in the same way as in the case of adsorption isotherms. The bound chloride content in this case was calculated from the formula

$$C_b = \frac{M_{Cl}[C_0 V - C_1 V' - C_2(V + V'' - V')]}{W}, \qquad (5.351)$$

where C_2 is the equilibrium concentration of chloride solution (in mol/l) after equilibration at desorption, V' the volume of solution removed, including the first pipetted, and V'' the volume of de-ionized water added. The free chloride content C_f (in mol/l), corresponding to the value of C_b, is now

$$C_f = C_2, \qquad (5.352)$$

and the desorption isotherm $C_b = C_b(C_f)$ is drawn again from the data of C_b and C_f, determined in a series of tests with different initial concentrations.

5.5.3 Salt Diffusion Coefficient

The most frequently used methods for the determination of salt diffusion coefficients in concrete use the experimental set-up designed for the so-called

"Rapid Chloride Permeability Test" (sometimes called as "Migration Test"), which was developed at the beginning of the 1980s in USA (Whiting, 1981, AASHTO, 1983). The experiment uses two cells separated by a plate specimen of the measured material. In measuring chloride diffusion, cell 1 normally contains a NaCl solution, and cell 2 is filled with a NaOH solution of the same molar concentration as that of the chloride source solution. Two mesh electrodes are fitted, one on each side of the concrete specimen in such a way that the electric field is applied primarily across the test specimen.

After establishing steady-state conditions, the diffusion coefficient can be determined using the Nernst–Planck equation

$$\vec{j}_i = \frac{z_i F}{RT} D_i \rho_i \vec{E},\tag{5.353}$$

where j_i is the mass flux (in kg/m^2s) of the particular ion, z_i its valence, F the Faraday constant (96487 C/mol), D_i the diffusion coefficient, ρ_i the partial density of the ion i (in kg/m^3), and $\vec{E}$ the electric field intensity.

Taking into account the one-dimensional geometry of the experiment, the mass flux j_i can be expressed as

$$j_i = \frac{V_0}{A_0} \frac{d\rho_i}{dt},\tag{5.354}$$

where V_0 is the volume of the cell, $\frac{d\rho_i}{dt}$ the rate of change of partial density of the particular ion in the cell determined using the measured change of concentration, and A_0 the surface area of the specimen that is in contact with the solution in the cell (i.e. the cross-section of the specimen). The electric field intensity can be expressed as

$$E = \frac{\Delta\varphi}{L},\tag{5.355}$$

where $\Delta\varphi$ is the potential difference between the two electrodes, L the thickness of the specimen.

Substituting equations (5.354) and (5.355) into equation (5.353), we obtain

$$D_i = \frac{j_i RTL}{z_i F \rho_i \Delta\varphi},\tag{5.356}$$

where ρ_i is understood as the partial density of the particular ion in the source cell.

The above treatment for the determination of the salt diffusion coefficient was described by Andrade et al. (1997) or Zhang and Gjørv (1997). It can be applied by assuming the following simplifications (Andrade, 1993). First, the ionic mobility is much higher in the solution than in the concrete specimen, and therefore the "distance" of the experiment is the specimen thickness. Second, mass transport by convection is negligible. Third, classical Fick's diffusion is not considered because its influence is negligible compared to the migration term; this is due to the high value of the expression $\frac{z_i F}{RT}$ when multiplied by the applied potential difference (usually higher than 10 V). Finally, the electric field along the concrete specimen follows a linear course.

These assumptions are relatively strong. Hence, the reliability of "Migration Tests" is often discussed in the scientific literature. A direct comparison with the classical Ficks's diffusion test, which was carried out under identical experimental arrangements as the "Accelerated Diffusion Test" described before (apart from the application of electric field), has shown (Arsenault et al., 1997), that the results of the "Conventional Diffusion Test" were systematically lower than those of "Accelerated Diffusion Test", typically by a factor difference of about 1.6 to 2.2.

It should be noted (see Andrade et al., 1997) that the duration of the "Accelerated Diffusion Test" is significantly longer than the classical "Rapid Chloride Permeability Test". Typically one week may be enough for ordinary concretes with specimens 1 cm thick. Nevertheless, the test is significantly shorter compared to the "Conventional Diffusion Test", which may take several months.

Another possibility for determining the salt diffusion coefficient is to use a mathematical analysis of measured salt profiles. A typical experiment for the determination of the salt profile in a concrete specimen was described by Baroghel-Bouny et al. (1997) for chloride penetration. Concrete samples cured in wet conditions were exposed to a chloride solution from one side in a one-dimensional transport arrangement. After a specified time, slices 3–4 mm thick were cut in the near-surface layer for chloride analysis. The total chloride content was extracted from the specimens using a boiling nitric acid solution for 30 minutes. The free chloride content was determined after a contact time of 3 minutes between water and concrete. Chloride concentrations were measured by potentiometry using a titrated silver nitrate solution.

The mathematical analysis of experimentally determined salt profiles depends on the assumed mode of salt transport in the material. If purely diffusion transport is assumed, common methods for solving the inverse problems for parabolic equations can be used.

The simplest method makes the assumption that the diffusion coefficient is constant, the domain under solution is semi-infinite, and the boundary condition on the remaining side of a one-dimensional arrangement is Dirichlet-type. Then diffusion coefficient can be identified using a simple analytical solution of the parabolic problem with an error function (see Section 4.1). This method was employed for chloride diffusion by Baroghel-Bouny et al. (1997).

The dependence of the diffusion coefficient on salt concentration can be found if some more sophisticated methods for the analysis of measured salt profiles are used. However, such methods are not yet commonly used in analysing salt profiles. One of the few exceptions is the work by Tumidajski et al. (1995), where a classical Boltzmann–Matano analysis (see Section 5.3.2 for details) was employed to determine chloride diffusion coefficients. It should be noted in this respect, that if only the diffusion mode of salt transport is assumed, the calculation of concentration-dependent diffusion coefficients from the measured salt profiles can be done using the same methods as those for the determination of moisture-dependent moisture diffusivity and temperature-dependent thermal conductivity, which were presented in Sections 5.3.2 and 5.4.4 (see, e.g., Černý and Toman, 1995, Černý and Toman, 1997).

Pel et al. (2000) employed for the description of salt transport in a porous material the diffusion–advection equation of the form

$$\frac{\partial(wC_f)}{\partial t} = \text{div}\,(wD\,\text{grad}\,C_f) - \text{div}\,(wC_f\vec{v}) - \frac{\partial C_b}{\partial t}, \tag{5.357}$$

where C_b is the bound ion concentration, C_f the free ion concentration, w the moisture content by volume, D the diffusion coefficient of the particular ion in water (in some references it is called hydrodynamic dispersion coefficient), and $\vec{v}$ the velocity of the liquid phase.

They measured moisture profiles and concentration profiles of both free and bound Na ions at the absorption of a 4M NaCl solution using the NMR method at the same time. In the analysis of the measured data, Pel et al. (2000) first modified equation (5.357) using a relation for moisture flux j of the form

$$j = -\rho_w \kappa\,\text{grad}\,w = \rho_w\vec{v}. \tag{5.358}$$

They obtained

$$\frac{\partial(wC_f)}{\partial t} = \text{div}\,(wD\,\text{grad}\,C_f) + \text{div}\,(wC_f\kappa\,\text{grad}\,w) - \frac{\partial C_b}{\partial t}. \tag{5.359}$$

Applying the Boltzmann transformation

$$\eta = \frac{x}{\sqrt{t}} \tag{5.360}$$

to the one-dimensional analogue of equation (5.359), they arrived at

$$2\frac{d}{d\eta}\left(wD\frac{dC_f}{d\eta}\right) + 2\frac{d}{d\eta}\left(wC_f\kappa\frac{dw}{d\eta}\right) + \eta\frac{d(wC_f)}{d\eta} + \eta\frac{dC_b}{d\eta} = 0. \tag{5.361}$$

Equation (5.361) contains two unknown parameters, namely the diffusion coefficient D and the moisture diffusivity κ. Therefore, it cannot lead to a unique solution without applying further restricting relations. Pel et al. (2000) assumed that the hydrodynamic dispersion term can be neglected, i.e. $D \to 0$, and used equation (5.361) to $\sqrt{t}$ scaling of the ion profiles only. However, equation (5.361) together with the liquid moisture transport equation

$$\frac{\partial w}{\partial t} = \text{div}\,(\kappa\,\text{grad}\,w), \tag{5.362}$$

which in the one-dimensional case after Boltzmann transformation leads to

$$2\frac{d}{d\eta}\left(\kappa\frac{dw}{d\eta}\right) + \eta\frac{dw}{d\eta} = 0, \tag{5.363}$$

also offers some other possibilities for data analysis. The solution of equation (5.363) leads to a $\kappa(w, C_f)$ function, which is not calculated in the full (w, C_f) plane but only for discrete points $[w_i, C_{f,i}]$. Nevertheless, for an application with equation (5.361) it is sufficient, because it works with the same moisture and concentration profiles as equation (5.363). Thus, using equation (5.361) with the known $\kappa(w, C_f)$ function, we can calculate the $D(C_f)$ function.

References

AASHTO Designation T277-83, 1983, *Standard Method of Test for Resistance of Concrete to Chloride Ion Penetration.* (Washington, D.C.: American Association of State Highway and Transportation Officials).

Ambros, F., 1993, *Experimental Methods in Engineering.* (Praha: CTU Press), (in Czech).

Andrade, C., 1993, Calculation of chloride diffusion coefficients in concrete from ionic migration experiments. *Cement and Concrete Research*, **23**, pp. 724–742.

Andrade, C., Castellote, M., Cervigon, D., Alonso, C., 1997, Fundamentals of migration experiments. In: *Proceedings of the RILEM International Workshop Chloride Penetration into Concrete*, St-Remy-les-Chevreuse, edited by Nilsson, L.O., Ollivier, J.P. (St-Remy-les-Chevreuse: RILEM), pp. 95–114.

Arai, C., Hosaka, S., Murase, K., Sano, Y., 1976, Measurements of the relative humidity of saturated aqueous salt solutions. *J. Chem. Eng. Jap.*, **9**, pp. 328–342.

Arsenault, J., Bigas, J.P., Ollivier, J.P., 1997, Determination of chloride diffusion coefficient using two different steady-state methods: influence of concentration gradient. In: *Proceedings of the RILEM International Workshop Chloride Penetration into Concrete*, St-Remy-les-Chevreuse, edited by Nilsson, L.O., Ollivier, J.P. (St-Remy-les-Chevreuse: RILEM), pp. 150–160.

Barneyback, R.S., Diamond, S., 1981, Expression and analysis of pore fluids from hardened cement pastes and mortars. *Cement and Concrete Research*, **11**, pp. 279–285.

Baroghel-Bouny, V., Chaussadent, T., Raharinaivo, A., 1997, Experimental investigations on binding of chloride and combined effects of moisture and chloride in cementitious materials. In: *Proceedings of the RILEM International Workshop Chloride Penetration into Concrete*, St-Remy-les-Chevreuse, edited by Nilsson, L.O., Ollivier, J.P. (St-Remy-les-Chevreuse: RILEM), pp. 290–301.

Bezier, P.E., 1971, Example of an existing system in the motor industry: the unisurf system. *Proceedings Roy. Soc., London*, **A 321**, pp. 207–218.

Bittle, R.R., Taylor, R.E., 1984, Step-heating technique for thermal diffusivity measurements of large-grained heterogeneous materials. *J. Amer. Ceram. Soc.*, **67**, pp. 186–190.

Born, M., Wolf, E., 1991, *Principles of Optics*, 6th Edition. (Oxford: Pergamon Press).

Brandes, E.A., Brook, G.B. (eds.), 1992, *Smithells Metals Reference Book*, 7th Edition. (Oxford: Butterworth-Heinemann).

Brocken, H.J.P., 1998, *Moisture Transport in Brick Masonry: The Grey Area between Bricks*, PhD Thesis. (Eindhoven: TU Eindhoven).

Brown, P.A., 1970, Measurement of water potential with thermocouple psychrometers: construction and application. *USDA Forest Service Research Rep., INT-80*.

Carslaw, H.S., Jaeger, J.C., 1959. *Conduction of Heat in Solids*. (Oxford: Clarendon Press).

Černý, R., Hošková, Š., Toman, J., 1995. A transient method for measuring the water vapor diffusion in porous building materials. *Proceedings of International Symposium on Moisture Problems in Building Walls*, V.P. de Freitas, V. Abrantes (eds.). (Porto: University of Porto), pp. 137–147.

Černý, R., Maděra, J., Poděbradská, J., Toman, J., Drchalová, J., Klečka, T., Jurek, K., Rovnaníková, P., 2000, The effect of compressive stress on thermal and hygric properties of Portland cement mortar in wide temperature and moisture ranges. *Cement and Concrete Research*, **30**, pp. 1267–1276.

Černý, R., Toman, J., 1995, Determination of temperature- and moisture-dependent thermal conductivity by solving the inverse problem of heat conduction. *Proceedings of International Symposium on Moisture Problems in Building Walls*, V.P. de Freitas, V. Abrantes (eds.). (Porto: University of Porto), pp. 299–308.

Černý, R., Toman, J., 1997, A Difference Method for Determining the Thermal Conductivity of Porous Materials in a Wide Temperature Range. *High Temp.-High Press.*, **29**, pp. 51–57.

Černý, R., Toman, J., Šesták, J., 1996, Measuring the effective specific heat of building materials. *Thermochimica Acta*, **282/283**, pp. 239–250.

Černý, R., Tydlitát, V., Klečka, T., Bouška, P., Rovnaníková, P., 2001a, Determination of moisture content in hydrating cement paste. *Proceedings of CIB W40 Meeting.* (Wellington, CIB), pp. 29–36.

Černý, R., Tydlitát, V., Pavlík, J., Klečka, T., Bouška, P., Rovnaníková, P., 2001b, Monitoring free water content in hydrating cement paste by a microwave impulse technique. *Proceedings of Fourth International Conference on Electromagnetic Wave Interaction with Water and Moist Substances*, K. Kupfer, Ch. Huebner (eds.). (Weimar, MFPA Weimar), pp. 251–258.

Claisse, P.A., Elsayad, H.I., Shaaban, I.G., 1999, Test methods for measuring fluid transport in cover concrete. *Journal of Materials in Civil Engineering*, **11**, pp. 138–143.

Collepardi, M., 1997, Quick method to determination of free and bound chlorides in concrete. In: *Proceedings of Chloride Penetration into Concrete*, St-Remy-les-Chevreuse, edited by Nilsson, L.O., Ollivier, J.P. (St-Remy-les-Chevreuse: RILEM), pp. 10–16.

Crank, J., 1975, *The Mathematics of Diffusion.* (Oxford: Clarendon Press).

Dalton, F.N., Rawlins, S.L., 1968, Design criteria for Peltier effect thermocouple psychrometers. *Soil Sci.*, **105**, pp. 12–17.

Davis, W.R., 1984, Hot-wire method for the measurement of the thermal conductivity of refractory materials. In: *Compendium of Thermophysical Properties Measurement Methods*, Vol. 1, edited by K.D. Maglic, A. Cezairliyan, V.E. Peletsky. (New York: Plenum Press), pp. 231–253.

Dhir, R.K., Hewlett, P.C., Chan, Y.N., 1987, Near surface characteristics of concrete: Assessment and development of in-situ test methods. *Mag. of Concrete Res.*, **39**, pp. 183–195.

Dhir, R.K., Jones, M.R., Ahmed, H.E.H., 1990, Determination of total and soluble chlorides in concrete. *Cement and Concrete Research*, **20**, pp. 579–590.

Drchalová, J., Černý, R., 1998, Non-steady-state methods for determining the moisture diffusivity of porous materials. *Int. Comm. in Heat and Mass Transfer*, **25**, pp. 109–116.

Easytest, 1999, *Soil water, salinity and oxygenation status monitoring devices.* (Lublin: Easytest, Ltd.).

Elsener, B., Zimmermann, L., Flückiger, D., Bürchler, D., Böhni, H., 1997, Chloride penetration - non destructive determination of the free chloride content in mortar and concrete. In: *Proceedings of Chloride Penetration into Concrete*, St-Remy-les-Chevreuse, edited by Nilsson, L.O., Ollivier, J.P. (St-Remy-les-Chevreuse: RILEM), pp. 17–26.

European Standard prEn 196-9, 2000, *Methods of testing cement – Part 9: Determination of heat hydration by semi-adiabatic method.*

Fexa, J., Široký, K., 1983, *Moisture measurement.* (Praha: SNTL), (in Czech).

Feynman, R.P., Leighton, R.B., Sands, M., 1966, *The Feynman Lectures on Physics*, 4th Edition. (New York: Addison-Wesley).

Figg, J.W., 1973, Methods of measuring air and water permeability of concrete. *Mag. of Concrete Res.*, **25**, pp. 213–219.

Gardner, W.R., 1956, Calculation of capillary conductivity from pressure plate outflow data. *Soil Sci. Soc. Am. Proc.*, **20**, pp. 317–320.

Gautier, G., 1973, Deux méthodes de mesure de la chaleur d'hydratation des ciments. *Revue des materiaux de construction et travaux publics*, No. 677, pp. 17–27.

Glass, G.K., Buenfeld, N.R., 1997, The determination of chloride binding relationships. In: *Proceedings of the RILEM International Workshop Chloride Penetration into Concrete*, St-Remy-les-Chevreuse, edited by Nilsson, L.O., Ollivier, J.P. (St-Remy-les-Chevreuse: RILEM), pp. 3–9.

Gorur, K., Smit, M.K., Wittman, F.H., 1982, Microwave study of hydrating cement paste at early age. *Cement and Concrete Research*, **12**, pp. 447–454.

Haar, L., Gallagher, J.S., Kell, G.S., 1984. *NBS/NRC Steam Tables.* (New York: Hemisphere).

Hall, C., 1989, Water sorptivity of mortars and concretes: a review. *Magazine of Concrete Research*, **41**, pp. 51–61.

Hall, C., Kam Ming Tse, T., 1986, Water movement in porous building materials – VII. The sorptivity of mortars. *Building and Environment*, **21**, pp. 101–108.

Hamming, R.W., 1962, *The Numerical Methods for Scientists and Engineers.* (New York: McGraw-Hill).

Hansen, K.K., Lund, H.B., 1990, Cup method for determination of water vapour transmission properties of building materials. Sources of uncertainty in the method. *Proceedings of the 2nd Symposium Building Physics in the Nordic Countries*, J.V. Thue (ed.). (Trondheim: Tapir), pp. 291–298.

Hansen, P.F. and Jensen, O. M., 1998, A sample holder for the study of isothermal heat of hydration of cement. *Materials and Structures*, **31**, pp. 133–136.

Hearn, N., Mills, R.H., 1991, A simple permeameter for water or gas flow. *Cement and Concrete Research*, **21**, pp. 257–261.

Hillel, D., 1980, *Fundamentals of Soil Physics*. (New York: Academic Press).

Kadleček, V. and Modrý, S., 1971, Determination method of chloride penetration to the surface of concrete. CZ Patent, No. 143204 (In Czech).

Kašpar, I., 1984, *Moisture Transport in Building Materials* (in Czech), DSc. Thesis. (Praha: CTU), (in Czech).

Kean, D.M., Smith, M.A., 1986, *Magnetic Resonance Imaging, Principles and Applications*. (London: Heinemann).

Kirby, R.K., 1992, Methods of measuring thermal expansion. In: *Compendium of Thermophysical Property Measurement Methods*, Vol. 2, edited by K.D. Maglic, A. Cezairliyan, V.E. Peletsky. (New York: Plenum Press), pp. 549–567.

Klute, A., 1965a, Laboratory measurement of hydraulic conductivity of saturated soil. In: *Methods of Soil Analysis*. (Madison: Am. Soc. Agron.), pp. 210–221.

Klute, A., 1965b, Laboratory measurement of hydraulic conductivity of unsaturated soil. In: *Methods of Soil Analysis*. (Madison: Am. Soc. Agron.), pp. 253–261.

Kollek, J.J., 1989, The determination of permeability of concrete by CEMBUREAU method. *Materials and Structures*, **22**, pp. 225–230.

Kondratiev, G.M., 1954, *Regularnyj teplovoj rezhim*. (Moskva: Gostechizdat), (in Russian).

Kondratiev, G.M., 1957, *Teplovyje izmerenija*. (Moskva: Mashgiz), (in Russian).

Krempaský, J., 1969, *Measurements of Thermophysical Quantities*. (Bratislava: Slovak Academy of Sciences Press), (in Slovak).

Krischer, O., 1963, *Die wissentschaftlichen Grundlagen der Trocknungstechnik*. (Berlin: Springer Verlag).

Kubičár, L., 1990, *Pulse Method of Measuring Basic Thermophysical Parameters*. (Amsterdam: Elsevier).

Künzel, H.M., Kiessl, K., 1990. Bestimmung des Wasserdampfdiffusionwiderstandes von mineralischen Baustoffen aus Sorptionsversuchen. *Bauphysik*, **12**, pp. 140–143.

Leroy, R., 1954, Une methode correcte de dosage de l'eau. *Chemical Analysis* **36**, pp. 294–302.

Leschnik, W., 1998, Measurement of the moisture content of building materials. In: *Proceedings of New Requirements for Materials and Structures.* (Prague: CTU Press), pp. 13–22.

Leschnik, W., Hauenschild, C., Knöchel, R., Menke, F., Boltze, T., 1995, A microwave sensor for building components. *Proceedings of Non-Destructive Testing in Civil Engineering,* G. Schickert, H. Wiggenhauser (eds.). (Berlin: DGZfP), pp. 151–155.

Lykov, A.W., 1958, *Transporterscheinungen in Kapillarporösen Körpern.* (Berlin: Akademie Verlag).

Maierhofer, Ch., Wöstmann, J., 1997, Investigation of dielectric properties of brick materials as a function of moisture and salt content using a microwave impulse technique at very high frequencies. *Proceedings of Non-Destructive Testing in Civil Engineering,* J.H. Bungey (ed.). (Northampton: The British Institute of NDT), pp. 743–754.

Maierhofer, Ch., Wöstmann, J., 1998, Investigation of dielectric properties of brick materials as a function of moisture and salt content using a microwave impulse technique at very high frequencies. *NDT & E International,* **31**, pp. 259–263.

Marthaler, W., Vogelsanger, W., Richard, F., Wierenga, P., 1983, A pressure transducer for field tensiometers. *Soil Science Society of America Journal,* **47**, pp. 624–627.

Malicki, M.A., Plagge, R., Renger, M., Walczak, R.T., 1992, Application of time-domain reflectometry (TDR) soil moisture miniprobe for the determination of unsaturated soil water characteristics from undisturbed soil cores. *Irrigation Science,* **13**, pp. 65–72.

Malicki, M.A., Walczak, R.T., 1999, Evaluating soil salinity status from bulk electrical conductivity and permittivity. *European Journal of Soil Science,* **50**, pp. 505–514.

Matano, C., 1933, On the relation between the diffusion coefficient and concentration of solid metals. *Jap. J. Phys.,* **8**, pp. 109–115.

Mc Intire, P. (ed.), 1986, *Nondestructive Testing Handbook,* 2nd Edition, Vol. 4. (New York: American Society for Nondestructive Testing).

Modest, M.F., 1993, *Radiative Heat Transfer.* (New York: McGraw- Hill).

Moore, W.J., 1972, *Physical Chemistry,* 4th Edition. (Englewood Cliffs: Prentice-Hall).

Mouhasseb, H., Suhm, J., Garrecht, H., Hilsdorf, H., 1995, A new dielectric method to measure moisture in masonry. *Proceedings of Non-Destructive Testing in Civil Engineering,* G. Schickert, H. Wiggenhauser (eds.). (Berlin: DGZfP), pp. 159–166.

Moukwa, M., Brodwin, M., Christo, S., Chang, J., Shah, S.P., 1991, The influence of the hydration process upon microwave properties of cements. *Cement and Concrete Research,* **21**, pp. 863–872.

Nielsen, A., 1972, Gamma-ray attenuation used for measuring the moisture content and homogeneity of porous concrete. *Building Science,* **7**, pp. 257–265.

Nissen, H.H., Moldrup, P., 1995, Theoretical background for the TDR methodology. *SP Report No. 11.* (Lyngby: Danish Institute of Plant and Soil Science), pp. 9–23.

Nofziger, D.L., Swartzendruber, D., 1974, Material content of binary physical mixtures as measured with a dual energy beam of γ-rays. *Journal of Applied Physics,* **45**, pp. 5443–5449.

Pande, A., 1974, *Handbook of Moisture Determination and Control.* (New York: Marcel Dekker).

Parker, W.J., Jenkins, R.J., Butler, C.P., Abbott, G.L., 1961, Flash method for determining thermal diffusivity, heat capacity and thermal conductivity. *J. Appl. Phys.,* **32**, pp. 1679–1684.

Pel, L., Ketelaars, A.A.J., Adan, O.C.G., van Well, A.A., 1993, Determination of moisture diffusivity in porous media using scanning neutron radiography, *International Journal of Heat and Mass Transfer,* **36**, pp. 1261–1267.

Pel, L., Kopinga, K., Brocken, H., 1996, Determination of Moisture Profiles in Porous Building Materials by NMR. *Magnetic Resonance Imaging,* **14**, pp. 931–934.

Pel, L., Kopinga, K., Kaasschieter, E.F., 2000, Saline absorption in calcium-silicate brick observed by NMR scanning. *J. Phys. D: Appl. Phys.* **33**, pp. 1380–1385.

Plagge, R., Renger, M., Häupl, P., 1997, Effect of transient conditions on hydraulic properties of porous media. In: *Proceedings of the International Workshop on Characterization and Measurement of the Hydraulic Properties of Unsaturated Porous Media,* edited by van Genuchten et al. (Riverside, California).

Plagge, R., Renger, M., Roth, C., 1990, A new laboratory method to quickly determine the unsaturated hydraulic conductivity of undisturbed soil cores within a wide range of textures. *J. Plant Nutrit. Soil Sci.*, **153**, pp. 39–45.

Plagge, R., Roth, C.H., Renger, M., 1996, Dielectric soil water content determination using time-domain reflectometry (TDR). In: *Proceedings of Second Workshop on Electromagnetic Wave Interaction with Water and Moist Substances at the 1996. IEEE Microwave Theory and Techniques Society International Microwave Symposium*, A. Kraszewski (ed.). (San Francisco, California), pp. 59–62.

Plagge, R., Grunewald, J., Häupl, P., 1999, Application of time domain reflectometry to determine water content and electrical conductivity of capillary porous media. *Proceedings of the 5th Symposium Building Physics in the Nordic Countries*, C.E. Hagentoft, P.I. Sandberg (eds.).(Goteborg: Vasastaden AB), pp. 337–344.

Poděbradská, J., Maděra, J., Tydlitát, V., Rovnaníková, P., Černý, R., 2000, Determination of moisture content in hydrating cement paste using the calcium carbide method. *Ceramics*, **44**, pp. 35–38.

de Ponte, F., Langlais, C., Klarsfeld, S., 1992, Reference guarded hot plate apparatus for the determination of steady-state thermal transmission properites. In: *Compendium of Thermophysical Property Measurement Methods*, Vol. 2, edited by K.D. Maglic, A. Cezairliyan, V.E. Peletsky. (New York: Plenum Press), pp. 99–131.

Reboul, J.P., 1978, The hydraulic reaction of tricalcium silicate observed by microwave dielectric measurements. *Revue de Physique Apliquee*, **13**, pp. 383–386.

Reinhardt, H.W., Hearn, N., Sosoro, M., 1997, Transport properties of concrete. In: *Penetration and Permeability of Concrete: Barriers to organic and contaminating liquids*, edited by H.W. Reinhardt. (London: E & FN Spon), pp. 213–264.

Rhim, H.C., Büyüköztürk, O., 1998, Electromagnetic properties of concrete at microwave frequency range. *ACI Materials Journal*, **95**, pp. 262–271.

RILEM TC 116-PCD, 1999, Permeability of concrete as a criterion of its durability. *Materials and Structures*, **32**, pp. 174–179.

RILEM TC 119-TCE 1 Recommendations, 1997, Adiabatic and semi-adiabatic calorimetry to determine the temperature increase in concrete due to hydration heat of cement. *Materials and Structures*, **30**, pp. 451–464.

Robery, P.C., 1988, Requirements of Coatings. *JOCCA*, **71**, pp. 403–406.

Rolwitz W.L., 1965, Nuclear magnetic resonance as a technique for measuring moisture in liquids and solids. In: *Humidity and Moisture*, A. Wexler (ed.). (New York: Reinhold).

Rovnaníková, P., 1992, *Chemical Reconnaissance of Steel Reinforced Construction*. Habilitation Thesis. (Brno: FCE BTU), (in Czech).

Rovnaníková, P., Brandštetr, J., 1982, Utilization of hydration heat of cement during acceleration of its hardening. *Stavivo*, **12**, pp. 484–486 (in Czech).

Sabir, B.B., Wild, S., O'Farrell, M., 1998, A water sorptivity test for mortar and concrete. *Materials and Structures*, **31**, pp. 568–574.

Šašek, L., et al., 1981, *Laboratory Methods in the Field of Silicates*. (Prague: SNTL), (in Czech).

SAS, 2000, *Thermophysical Transient Tester Model RTB 1.01*. (Bratislava: Institute of Physics SAS).

van Schilfgaarde, J. (ed.), 1974, *Drainage for Agriculture*. (Madison: Am. Soc. Agron.).

Smythe, W.R., 1968, *Static and Dynamic Electricity*, 3rd Edition. (New York: McGraw-Hill).

Sosoro, M., Hoff, W.D., Wilson, M.A., 1997, Testing methods. In: *Penetration and Permeability of Concrete: Barriers to organic and contaminating liquids*, edited by H.W. Reinhardt. (London: E & FN Spon), pp. 187–211.

Thompson, F., 1997, Measurement of moisture and permittivity in concrete samples during curing using microwave open transmission line techniques. *Proceedings of Non-Destructive Testing in Civil Engineering*, J.H. Bungey (ed.). (Northampton: The British Institute of NDT), pp. 287–301.

Tang, L., Nilsson, L.O., 1993, Chloride binding capacity and binding iso therms of OPC pasted and mortars. *Cement and Concrete Research*, **23**, pp. 247–253.

Toman, J., 1986, *Influence of External Conditions on Building Materials and Constructions*. DSc. Thesis. (Prague: CTU), (in Czech).

Toman, J., Černý, R., 1995, Calorimetry of building materials. *J. Thermal Anal.*, **43**, pp. 489–496.

Toman, J., Černý, R., 1996, Coupled thermal and moisture expansion of porous materials. *Int. J. Thermophysics*, **17**, pp. 271–277.

Toman, J., Koudelová, P., Černý, R., 1999, A measuring method for the determination of linear thermal expansion of porous materials at high temperatures. *High Temp.-High Press.*, **31**, pp. 595–600.

Topp, G.C., Davis, J.L., Annan, A.P., 1980, Electromagnetic determination of soil water content: Measurements in coaxial transmission lines. *Water Resources Research*, **16**, pp. 574–582.

Tritthardt, J. and Daňková, M., 1987, Zum Einfluss des Zuschlags auf die Chlorid- und Hydroxidkonzentration des Porenwasser von Beton. *Zement und Beton*, **32**, pp. 73–77.

Tydlitát, V., Semerák, P., Maděra, J., Poděbradská, J., Černý, R., 2000, Application of capacitance method for the monitoring of moisture content in early hydration stages of cement mortar. *Stavební obzor*, **9**, pp. 151–153 (in Czech).

Van Bavel, C.H.M., 1965, Neutron scattering measurement of soil moisture: development and current status, in: *Humidity and Moisture*, A. Wexler (ed.). (New York: Reinhold).

Villadsen, J., Hansen, K.K., Wadsö, L., 1993, Water vapour transmission properties of wood determined by the cup method. *Proceedings of the 3rd Symposium Building Physics in the Nordic Countries*, Vol. 2, B. Saxhof (ed.). (Copenhagen: TU Lyngby), pp. 685–694.

Vozár, L., 1996, A computer-controlled apparatus for thermal conductivity measurement by the transient hot wire method. *Journal of Thermal Analysis*, **46**, pp. 495–505.

Vozár, L., Šrámková, T., 1997, Step-heating method for thermal diffusivity measurement. *Proceedings of the XIV IMEKO World Congress*. (Tampere, Finland), pp. 179–184.

Wittmann, F. H., Schlude, F., 1995, Microwave absorption of hardened cement paste. *Cement and Concrete Research*, **5**, pp. 63–71.

Whiting, D., 1981, Rapid Determination of the Chloride Permeability of Concrete. *Report No. FHWA/RD-81/119*. (Portland Cement Association, NTIS DB No. 82140724).

Zhang, T., Gjørv, O.E., 1997, An electrochemical method for accelerated testing of chloride diffusivity in concrete. In: *Proceedings of the RILEM International Workshop Chloride Penetration into Concrete*, St-Remy-les-Chevreuse, edited by Nilsson, L.O., Ollivier, J.P. (St-Remy-les-Chevreuse: RILEM), pp. 105–114.

Zienkiewicz, O.C., 1971, *The Finite Element Method in Engineering Science*. (London: McGraw-Hill).

## CHAPTER SIX

Examples of Practical Applications of Computational Models of Heat, Moisture and Salt Transport in the Design of Concrete Structures

6.1 BUILDING PHYSICS RELATED ASSESSMENT OF THE ENVELOPE

Concrete is one of the most commonly used materials for load-bearing structures. Therefore, it often features as a part of building envelope. However, any envelope has to be assessed not only from the point of view of its load bearing capabilities but also from the point of view of building physics, which generally means its hygrothermal performance. In this section, we have chosen an example that is relatively difficult to solve from the building physics point of view, namely a concrete structure with an interior thermal insulation system. All calculations have been done by the computer code DELPHIN 4.1, which was introduced in Chapter 4.

Thermal insulation systems are mostly used on the exterior side of buildings. This is a logical solution from the point of view of building physics. The insulation materials usually have higher water vapour permeability and lower thermal conductivity than the materials of the load bearing structures, so that the risk of water condensation in the winter period is low. In addition, the highest temperature gradients appear in the insulation material, so that the bearing structure is exposed to lower thermal stress.

The use of interior thermal insulation systems on building envelopes is not a natural solution but sometimes there is no other option available. A typical example is a group house, where the facade has to maintain its original appearance as far as possible, and exterior insulation systems are excluded for that reason. In that case the development of such an interior insulation system would make it possible to prevent moisture damages and to upgrade the thermal properties of the envelope as the only reasonable option. In this section, we introduce several interior thermal insulation systems based on mineral wool.

Fig. 6.1 shows the composition of a designed thermal insulation system used in carrying out hygrothermal calculations. On the internal side, there is a light plaster. Mineral wool boards are used as the thermal insulation

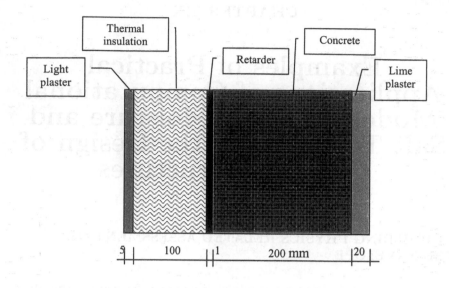

Figure 6.1 Composition of the interior thermal insulation system used to calculations

material. Between mineral wool and concrete wall, there is a water vapour retarder. We assume no air gap between the retarder and the old structure. The reason why we could afford this is the chosen technology. We consider a glue or a mastic as the retarder, which is applied directly to the old structure. As its workability is supposed to be several hours, it can glue together the old structure with the thermal insulations boards. On the external side of the brick wall is a lime plaster. The thickness of the concrete wall is intended to be 200 mm. The physical properties of the basic materials are given in Table 6.1. These properties were obtained from the material database of the DELPHIN computer code. The values for light internal plaster were estimated to be close to those of calcium silicate. The physical properties of mineral wools and retarder were measured in the laboratories of CTU Prague, and the results are given in Table 6.2. The particular mineral wool materials were products of Rockwool, SA, which are currently under development. A part of them was hydrophobized (CL, TR90), and the material FR was provided with hydrophilic chemicals enhancing its capillary activity. The water vapour retarder was based on acetate and was a product of Karlomix, Ltd. that is currently under development.

An unavoidable preliminary condition for any reasonable building physics related computational design is the application of an appropriate mathematical model and the respective computer code. We employed the advanced

Table 6.1 Physical characteristics of common materials

Material	$\rho[\text{kg/m}^3]$	Θ_{hyg} [m^3/m^3]	λ [W/mK]	κ [m^2/s]	μ
Light plaster	230	0.01	0.06	5.00e-6	5
Lime plaster	1800	0.03	1.05	1.00e-9	21
Concrete	2100	0.06	2.10	2.00e-9	65

Table 6.2 Physical characteristics of mineral wools and retarder

Material	$\rho[\text{kg/m}^3]$	Θ_{hyg} [% m^3/m^3]	μ	λ [W/mK]
Retarder	1565	2.00	5.2	0.224
CL	273	1.13	2.9	0.055
PR	1054	9.56	7.8	0.134
TR 90	91	4.49	3.0	0.038
FR	78	0.90	10.8	0.041

Table 6.2 Physical characteristics of mineral wools and retarder (continued)

Material	c [J/kgK]	κ [m^2/s]	Θ_{por} [% m^3/m^3]
Retarder	330	1.00e-10	32
CL	560	2.04e-10	40
PR	910	2.58e-10	42
TR 90	736	1.15e-11	99
FR	938	1.35e-6	98

software tool DELPHIN4.1, as mentioned before. Unlike some other software packages for modelling heat and moisture transport working with temperature and relative humidity only, DELPHIN4.1 also includes capillary pressure as one of the basic state variables, which makes it possible to determine the transport of liquid moisture with higher precision and reliability. For the interior insulation systems this is a crucial factor, because liquid moisture is almost always the basic cause of defects and failures.

The hygrothermal calculations have to be performed over a sufficiently long time period. Under certain circumstances, the hygrothermal properties of the system can worsen slowly and gradually, so that the damage may appear only after a long time period. Therefore, not only the actual values of the main hygrothermal properties are important but also their development in subsequent years. A time period of 10 or 20 years seems reasonable for deciding whether the designed system will perform without substantial damage.

The proper initial and boundary conditions of the model are another cru-

Inside   **Outside**

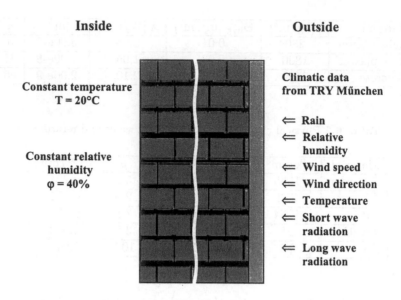

Constant temperature
T = 20°C

Constant relative
humidity
φ = 40%

Climatic data
from TRY München

⇐ Rain
⇐ Relative
humidity
⇐ Wind speed
⇐ Wind direction
⇐ Temperature
⇐ Short wave
radiation
⇐ Long wave
radiation

Figure 6.2 Scheme for the analysis of initial conditions

cial factor affecting the reliability of the calculations. Therefore, the calculations should be done for exactly the same conditions as will be encountered in practical reconstruction on the building site. First, the boundary conditions for the external side should be as accurate as possible. This can be achieved by using the meteorological data for the locality as close as possible to the real object. From the point of view of long term reliability, the use of so-called "reference year" data should be preferred. Second, the initial conditions should be realistic. To ensure this, the calculations should first be done for a construction without the interior insulation system in order to find the long-term conditions in the wall before the reconstruction (see Fig. 6.2). Third, the calculations with the interior insulation system should be started at the same time of year when the real reconstruction will begin. In our case, we chose the 1 June as the initial point, when we assumed the installation of the insulation system on the load-bearing structure was done.

Fig. 6.3 shows the initial conditions calculated for the scheme in Fig. 6.2. These initial conditions were then used in common calculations of the hygrothermal performance of the envelopes with interior thermal insulation systems. The scheme for these calculations is given in Fig. 6.4.

The long-term performance of a system should be characterized not only by calculated temperature and relative humidity. Overhygroscopic moisture fields also have to be determined, which decide about the real hygrothermal

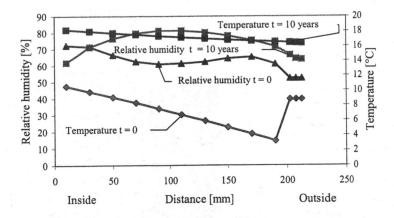

Figure 6.3 Calculated initial conditions

performance of the system. The critical places where long-term water accumulation can occur should be detected. We have chosen two critical profiles for the evaluation of the performance of the envelope, A–A', B–B', shown in Fig. 6.4. For these profiles, we calculated the dependence of the overhygroscopic water mass content (OWMC), relative humidity and temperature on time. The calculations were done for four different thermal insulation materials, developed by Rockwool (see Table 6.2).

In Figs. 6.5–6.7, we present the calculated relative humidity, temperature and water content fields for the thermal insulation material FR. It is apparent that no overhygroscopic moisture appeared in the system, and the relative humidity values were well below the danger level for condensation.

Figs. 6.8–6.10 show the relative humidity, temperature and water content fields in the system using the thermal insulation material PR. We can see that again no overhygroscopic moisture appeared, but the values of relative humidity were higher than in the case of FR. However, the system was found to be still relatively safe against condensation.

Figs. 6.11–6.15 show the relative humidity, temperature, water content and overhygroscopic water mass content fields in the system using the thermal insulation material CL. We can see that overhygroscopic moisture was present for substantial time periods in ten consecutive years, and these periods lengthened with time, so that in the tenth year there was overhygroscopic moisture in the concrete wall for more than 300 days per year. We may conclude that the system performed badly from the hygrothermal point of view. The primary reasons were that the retarder had too low a value of μ, and the insulation board had too low a value of κ.

Figs. 6.16–6.20 show the relative humidity, temperature, water content

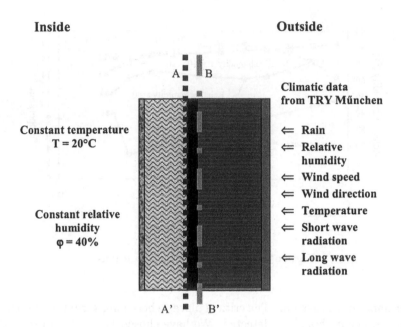

Figure 6.4 Scheme of typical envelope for hygrothermal calculations

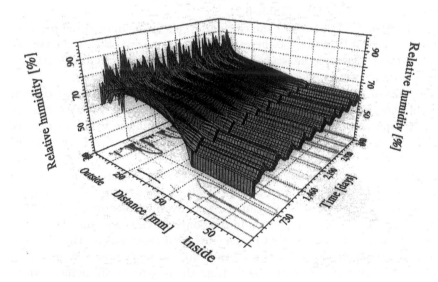

Figure 6.5 Relative humidity in the concrete wall-FR system

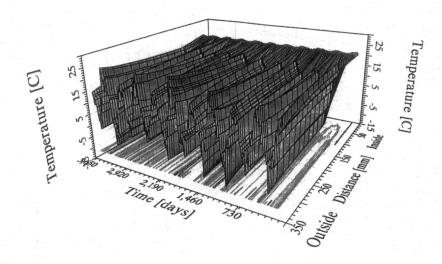

Figure 6.6 Temperature in the concrete wall-FR system

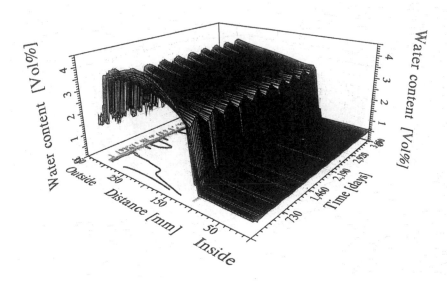

Figure 6.7 Water content in the concrete wall-FR system

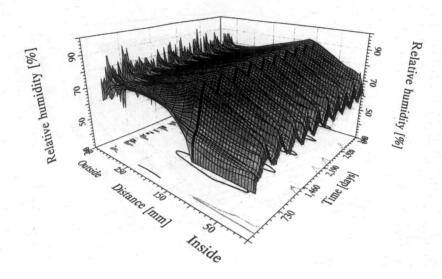

Figure 6.8 Relative humidity in the concrete wall-PR system

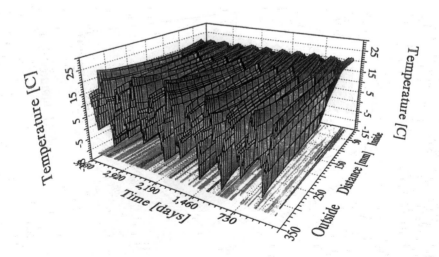

Figure 6.9 Temperature in the concrete wall-PR system

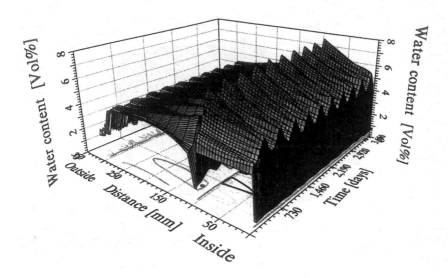

Figure 6.10 Water content in the concrete wall-PR system

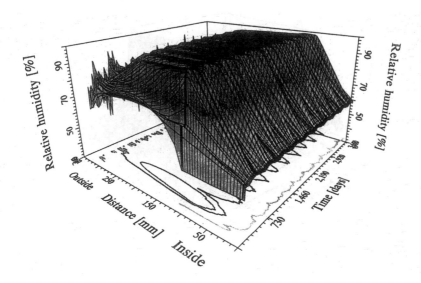

Figure 6.11 Relative humidity in the concrete wall-CL system

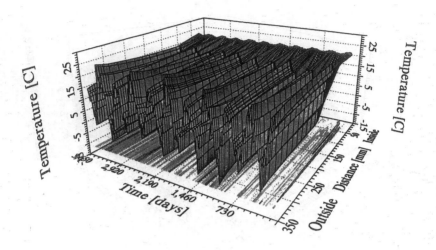

Figure 6.12 Temperature in the concrete wall-CL system

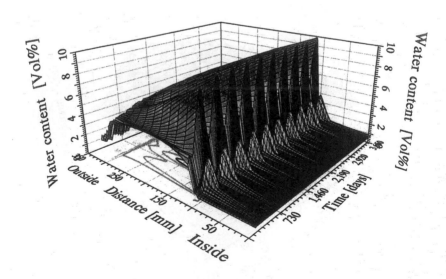

Figure 6.13 Water content in the concrete wall-CL system

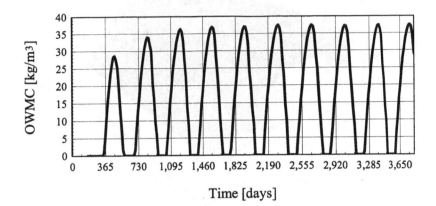

Figure 6.14 Overhygroscopic water mass content in the concrete wall-CL system, profile B–B'

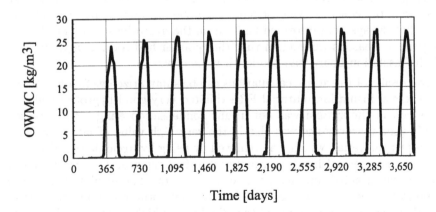

Figure 6.15 Overhygroscopic water mass content in the concrete wall-CL system, profile A–A'

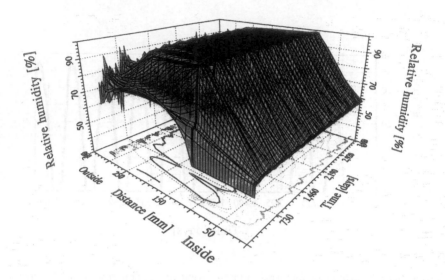

Figure 6.16 Relative humidity in the concrete wall-TR90 system

and overhygroscopic water mass content fields in the system using the thermal insulation material TR90. We can see that overhygroscopic moisture was present for substantial time periods in ten consecutive years, and these periods lengthened with time, so that in the tenth year there was overhygroscopic moisture in the concrete wall for more than 300 days per year. We can conclude that the system performed badly from the hygrothermal point of view. The primary reasons were again that the retarder had too low a value of μ, and the insulation board had too low a value of κ.

The results of the above calculations can be summarized as follows. Rockwool thermal insulation material FR exhibited very good hygrothermal performance and the system was not in any condensation danger. The reason for this good performance was clearly the very high value of moisture diffusivity of the material, which was achieved by using hydrophilic chemicals. In this case, the thermal insulation material in fact replaced the water vapour retarder, which acted only as a glue in the system. The insulation material PR performed well in concrete wall from the hygrothermal point of view as well. However, it should be noted that PR has too high a density, which may lead to technological problems on the building site. The Rockwool insulation materials CL and TR90 behaved in quite similar ways from the hygrothermal point of view, and the difference in densities between these two groups had no effect on their performance. The materials performed very badly, and the condensed water remained in the system for almost the whole year. The primary reason for the appearance of liquid moisture in the load bearing structure was

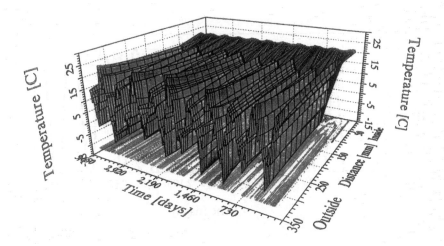

Figure 6.17 Temperature in the concrete wall-TR90 system

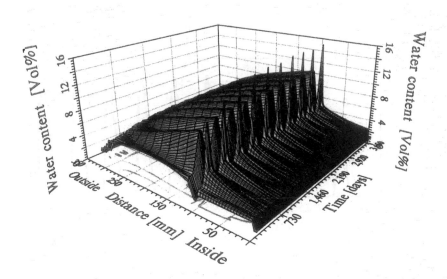

Figure 6.18 Water content in the concrete wall-TR90 system

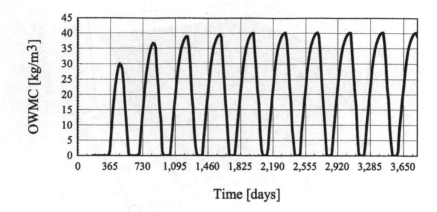

Figure 6.19 Overhygroscopic water mass content in the concrete wall-TR90 system, profile B–B'

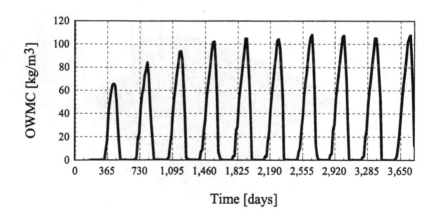

Figure 6.20 Overhygroscopic water mass content in the concrete wall-TR90 system, profile A–A'

the low value of the water vapour diffusion resistance factor μ of the retarder. It can be anticipated that for retarders with higher μ, liquid moisture in the load bearing structures would disappear. However, in this case the problem of water condensation would only be transferred from the load bearing structure to the thermal insulation boards, which have a very low value of moisture diffusivity, so that they would not be able to transport the condensed water back to the interior. Hydrophobization of the thermal insulation materials would make the situation even worse, because due to the gravity forces water would accumulate in the lower parts of the insulation boards, which is very dangerous.

6.2 HYDRATION HEAT RELATED PROBLEMS IN LARGE-SCALE CONCRETE STRUCTURES

Monolithic large-scale concrete structures (i.e. large concrete blocks in most cases) always suffer problems caused by hydration heat production because of their limited ability to remove heat into the surroundings. Temperature increase induces thermal stress, which can lead to the development of cracks, lower final quality of concrete, and consequently also to worse material properties and worse structure performance. If sprayed concrete technology is used, which usually includes application of rapid cementing agents, these effects may be even more serious, because the concrete attains material properties, which are close to their "final" values faster. In normal practice, the only way to deal with the negative consequences of hydration heat production, is slower rate of building these structures. The question, what is slow enough, has to be answered case by case in terms of the properties of the concrete mixture. It could be said that the most effective way to learn how long the breaks between concreting cycles should be is computer simulation. In this way, different combinations of geometric parameters of building structure and applied technology can be easily and cheaply modelled, and temperature and hygrothermal stress fields can be obtained. Obviously, each computer model needs input values that are sufficiently accurate, and the accuracy of the computer simulation results cannot be greater than the accuracy of the input values. In this particular case, the material properties of concrete and its surroundings need to be considered as functions of time, i.e. over the full extent of the hydration process. The most important parameter in this respect is the hydration heat.

In this section, computational modelling of temperature, moisture and stress fields is performed for a case study of a gas-tight concrete seal for an underground natural gas reservoir built using sprayed concrete technology (geometrically a cylinder 3 m long, with a radius of 1–3.15 m, in the calculated variants). The model is formulated using the methods of continuum physics. It accounts for the heat, moisture and momentum transport in the hydrating concrete. The numerical solution of the model is found by the finite element method. The computer implementation is performed using computer code in C++, which is employed for the practical calculations. As for the temperature fields, we had some feedback in the form of measured data on the concrete seal structure that was built in Czech Republic at the same time. Therefore, these results can be considered as reasonably precise and reliable. On the other hand, our calculations of hygrothermal stress did not achieve similar accuracy because we did not have sufficiently precise values for the material properties, so its modelling should be regarded only as an indicative demonstration.

6.2.1 Mathematical Model

6.2.1.1 Temperature Fields

The heat conduction problem under consideration possesses the following main features:

1. Internal heat source

2. Heat removal to the surrounding environment

3. Special geometry of the object.

The heat conduction equation in a three-dimensional case has the following form:

$$\rho c \frac{\partial T}{\partial t} = \frac{\partial}{\partial x}\left(\lambda \frac{\partial T}{\partial x}\right) + \frac{\partial}{\partial y}\left(\lambda \frac{\partial T}{\partial y}\right) + \frac{\partial}{\partial z}\left(\lambda \frac{\partial T}{\partial z}\right) + I(x, y, z, t), \quad (6.1)$$

where ρ is the density, c the specific heat capacity, λ the thermal conductivity, T the temperature, t the time, and x, y, z the space coordinates. The term $I(x, y, z, t)$ represents the internal heat source, which in general can depend on both space coordinates and time. Boundary conditions can be characterized as follows:

- Newton conditions for contact of the concrete structure with the air

- Ideal thermal contact on the concrete-rock interface

- Thermal insulation conditions (zero temperature gradient) on the external part of the boundary (solid rock).

In the practical computations, equation (6.1) in the following matrix form was employed (see, e.g., Zienkiewicz, 1971, for details):

$$[HP]\{T\}_1 = [PS]\{T\}_0 + \{A\}.$$

6.2.1.2 Moisture Fields

Due to the special conditions applying in our case study, we could assume the following:

- The only factor affecting the moisture loss in the concrete block is the hydration process.

- The local change in moisture content due to loss induced by the hydration process is much faster than the moisture conduction.

- There is no moisture transfer from concrete to the surrounding rock because cracks in the rock are already full of water under relatively high pressure of about 1–2 MPa.

- There is no moisture transport from the concrete block to the air in the tunnel (the relative humidity of the air is about 90%).

Therefore, the moisture transfer equation can be simplified to the following form:

$$\frac{\partial w}{\partial t} = J(x, y, z, t), \tag{6.2}$$

where w is the moisture content, and $J(x, y, z, t)$ a moisture sink term fitted using the differential function of hydration heat production $I(x, y, z, t)$. The initial value of the moisture content is determined from the known water/cement ratio of the concrete mixture, and the final value is equal to the maximum hygroscopic moisture content.

6.2.1.3 Hygrothermal Stress Fields

As a consequence of hydration heat production, hygrothermal stress is induced in the concrete structure, as well as in the rock that surrounds it. In the mathematical modelling of this process, we employed the simplified quasi-steady version of the momentum equation of the form (see, e.g., Wrobel, 1997)

$$(\Lambda + \mu) \, \text{grad div} \, \vec{u} + \mu \nabla^2 \vec{u} - (3\Lambda + 2\mu) \, \alpha_T \, \text{grad} \, T$$

$$- (3\Lambda + 2\mu) \, \alpha_w \, \text{grad} \, w = 0, \tag{6.3}$$

where $\vec{u}$ stands for the displacement vector, α_T is the linear thermal expansion coefficient, α_w the linear hygric expansion coefficient, and material properties Λ and μ are Lamé constants, which are related as follows to E a ν (Young's modulus and Poisson number):

$$\Lambda = \frac{E\nu}{(1 + \nu)(1 - 2\nu)}, \tag{6.4}$$

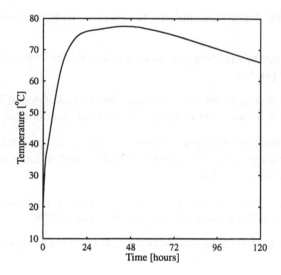

Figure 6.21 The time development of the maximum temperature value
for $r = 2$ m

$$\mu = \frac{E}{2(1 + \nu)}. \tag{6.5}$$

The matrix form of equation (6.3) applied in the computer modelling is the following:

$$[M_{rr}]\{u_r\}^e + [M_{rz}]\{u_z\}^e + [M_{rT}]\{T\}^e = 0 \tag{6.6}$$

$$[M_{zr}]\{u_r\}^e + [M_{zz}]\{u_z\}^e + [M_{zT}]\{T\}^e = 0. \tag{6.7}$$

There are no time derivatives in these two equations, for translation relates only to momentary values of temperature, not to its time development. However, the temperature does vary with time, and therefore this problem can be considered not as simply a steady state, but rather quasi-steady. Consequently, the practical computations need not to be done at every time step at which we calculate the temperature and moisture fields, but only in selected times that are of interest to the designer.

In calculating stress fields from the displacement values determined on the basis of equation (6.3), we employed Hooke's general theorem, which defines the relation between the stress tensor σ and the deformation tensor ϵ:

$$\sigma_{ik} = \left[\Lambda \sum_m \epsilon_{mm} - \alpha_T (3\Lambda + 2\mu)(T - T_0) \right.$$

$$\left. - \alpha_w (3\Lambda + 2\mu)(w - w_0) \right] \delta_{ik} + 2\mu\epsilon_{ik}. \tag{6.8}$$

6.2.2 Computational Results

The results of our computations can be summarized as follows. The maximum temperature values in the adiabatic zone (almost 80°C) occur on the longitudinal axis of the structure near the middle of its length, but not exactly in the middle.

If the radius of the cylindrical structure is 3.15 m, the temperature is constant until a distance of about 2.0 m from the longitudinal axis. This leads us to conclude that if radius increases, the maximum temperature will not continue to increase with it.

In the longitudinal direction the results are very similar, only the temperature is not constant over such a large region of space. This is a consequence of building the structure layer by layer (a 12 cm thick single layer was assumed) with relatively long construction breaks of 150 minutes.

The hygrothermal stresses have similar values and similar shapes in all directions. In this case, all directional dilatation of the material competes with the material resistance both in radial and tangential directions.

If we take into account only thermal stress (i.e. the length changes due to thermal expansion only), the maximum stress near the centre of the structure increases more than 20–30 MPa, which is much higher than the plasticity limit at that stage of development of concrete material properties. Therefore, it seems that in this region, the material begins to plastify much sooner than stress can grow to these values, and higher stress values must be shifted further from the centre of the structure. The stress may even be almost constant over the whole structure.

If the influence of moisture induced shrinkage during the hydration process is also accounted for, the stress fields become completely different. The tensile stress appears near to the boundary of the structure, where the influence of moisture loss is more significant than that of temperature increase.

This summary can be illustrated by some figures chosen for the case where the radius of the seal structure $r = 2$ m. Fig. 6.21 shows the time development of the maximum temperature value, Figs. 6.22 and 6.23 the characteristic profiles of σ_r without (Fig. 6.22) and with (Fig. 6.23) the influence of the moisture induced shrinkage in the $z = 1.5$ m cross-section, i.e. in the middle of the cylinder, perpendicular to its longitudinal axis.

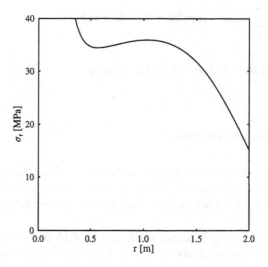

Figure 6.22 σ_r in the cross-section $z = 1.5$ m, induced by thermal expansion only, for $r = 2$ m

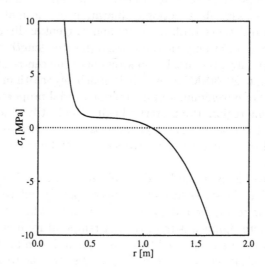

Figure 6.23 σ_r in the cross-section $z = 1.5$ m, induced by both thermal expansion and moisture shrinkage, for $r = 2$ m

The most important result of our computations is that the maximum temperature and hygrothermal stress values and also the time development of these parameters are determined by two main factors: the time development of hydration heat and the dimensions of the concrete block. If the dimensions of the concrete block exceed certain limit (in our case it was about 1 m in length or radius), an adiabatic region appears in the central part of the structure. This means that over a certain time interval, which is characteristic of the process, there is no heat exchange between this region and the surroundings of the structure. In the adiabatic region, the temperature and hygrothermal stress fields are not affected by the boundary conditions, and therefore cannot be influenced from outside the structure, for instance by such well-known methods as water spraying. In this region, processes are practically the same as in an adiabatic calorimeter during laboratory experiments.

Comparison of the stress fields in the two basic cases calculated, namely if the influence of moisture variations during the hydration process is or is not taken into account, shows very significant differences. Contrary to the results obtained when accounting for thermal stress only, hygrothermal stress calculations show evidence of tensile stress near the boundary of the structure, which may be considered very dangerous. Therefore, it leads us to the con clusion that taking into account moisture induced shrinkage is very significant in practical calculations.

6.3 CONCRETE IN HIGH TEMPERATURE CONDITIONS

Concrete is a material that can survive severe thermal conditions. There are examples of concrete structures that were exposed to a major fire but after reconstruction were quite serviceable (for instance the Great Exhibition Palace in Prague). In the determination of fire resistance of building structures, the time period for which the construction is capable of performing its heat-insulating function and protecting the other parts of the building from rapid temperature increase is one of the most important parameters. The duration of this period depends primarily on the external conditions, such as the temperature of the fire. However, variation in thermal material parameters, such as thermal conductivity or specific heat capacity, with temperature can play also an important role, because these can be so significant, that calculations performed using room temperature data are not of any use.

In this section, we have chosen an example of a high performance concrete I beam exposed to a fire (see Gawin et al, 2001, for details). The model of transport processes in concrete exposed to high temperatures developed by Gawin et al. (1999), which was described in Section 4.1, was used for the practical calculations. The geometry of the concrete beam, including the finite element discretization by 173 eight node serendipity elements, is shown in Fig. 6.24.

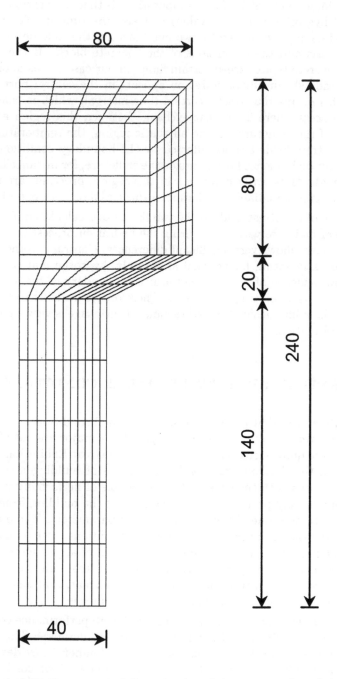

Figure 6.24 Geometry and discretization of the cross-section of the I beam

Constant initial conditions were chosen in the modelled example, the relative humidity being 50%, and temperature 298.15 K. On the external surface mixed convective and radiative boundary conditions for energy exchange, with heat transfer coefficient $\alpha_c = 18$ W/m²K and surface emissivity $e\sigma_0 = 5.1 \cdot 10^{-8}$ W/m²K⁴ were considered, and the ambient temperature was assumed to increase following the standard ISO fire curve. For mass exchange, convective boundary conditions with the partial vapour pressure $p_v = 1000$ Pa and moisture transfer coefficient $\beta_c = 0.018$ m/s were assumed.

The analysed material was high-performance concrete of the class 80 MPa compressive strength. In the initial state it was characterized by the following parameters: dry state apparent density $\rho_0 = 2590$ kg/m³, initial porosity $\Phi = 0.12$, initial thermal conductivity of dry material $\lambda_{\text{dry}} = 1.5$ W/mK, initial intrinsic permeability $K_0 = 10^{-21}$ m², compressive strength $f_c = 80$ MPa, Young's modulus $E = 44$ GPa and Poisson's ratio $\nu = 0.20$. All parameters were assumed to be temperature dependent.

Characteristic results of computational simulations are presented in Figs. 6.25–6.29. In Fig. 6.25 and Fig. 6.26 we can see that at the beginning of the simulation, after 2 minutes, only the surface layer and the zone close to the corner are subjected to an increase of temperature and a desaturation process due to external heating. In the relative humidity distribution map (Fig. 6.26), the thermodiffusion phenomenon is already visible at this time causing a rise of the values of relative humidity inside the beam. The vapour pressure had not yet reached significant values.

Figs. 6.27 and 6.28 show that after 10 minutes a larger part of the beam is subjected to the desaturation process, the thermal front has advanced into the core of the beam, and the temperature in the corner has already passed the critical point for water. Fig. 6.29 shows that vapour pressure has already increased in a significant way, with peak values over 2 MPa.

The results of example calculations using the finite element model of transport processes in concrete at high temperature based on a mechanistic approach developed by Gawin et al. (1999) show its good practical applicability. It should be noted that in the Gawin et al. (1999) model, concrete is considered as a multiphase porous material, in which adsorption–desorption, hydration–dehydration, evaporation–condensation, different fluid flows and nonlinearities with temperature are all taken into account (see Section 4.1 and the original papers for details). This approach makes it possible to consider the different phenomena that occur in concrete during heating and their interrelations. The resulting model is attractive for predicting the thermo-hygro-mechanical behaviour of concrete structures under such severe conditions, as shown in the two-dimensional example above. Phenomena concerning real behaviour of gases close to critical point of water, behaviour of concrete in a range of temperatures mostly above 374.15°C, and the saturation plug process, are also introduced in the recent version of the model.

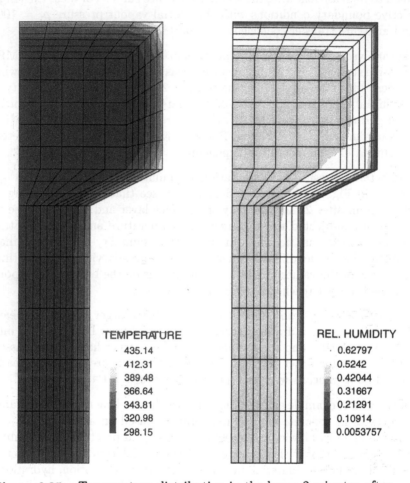

Figure 6.25   Temperature distribution in the beam 2 minutes after
the beginning of the heating process
Figure 6.26   Relative humidity distribution in the beam 2 minutes
after the beginning of the heating process

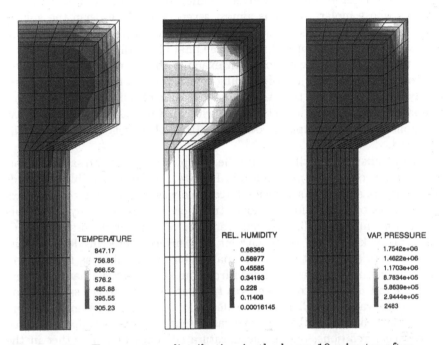

Figure 6.27 Temperature distribution in the beam 10 minutes after the beginning of the heating process

Figure 6.28 Relative humidity distribution in the beam 10 minutes after the beginning of the heating process

Figure 6.29 Vapour pressure distribution in the beam 10 minutes after the beginning of the heating process

6.4 SALT TRANSPORT INDUCED EMBEDDED STEEL CORROSION

As it has been already mentioned in Section 1.7, chlorides are the most dangerous salts from the point of view of embedded steel corrosion. The presence of chloride ions disrupts the passivating film that is normally formed on the surface of steel embedded in concrete and maintained in the highly alkaline environment produced by cement hydration. Consequently, the process of corrosion of the reinforcement is started, leading to the formation of corrosion products whose molar volume is almost seven times larger than the volume of the original iron. This change induces pressures on the concrete covering layer, where cracks are formed, and in the final phase this layer peels off.

In this section we have chosen an example that is common in calculating the predicted service life of building structures in a marine environment, namely modelling of the time to initiate corrosion of reinforcement in a concrete structure that is in direct contact with seawater. All calculations have been done using the computer code DELPHIN 4.1, which was introduced in Chapter 4.

For the sake of simplicity, we modelled coupled moisture and salt transport in isothermal conditions, although DELPHIN 4.1 makes it possible to include heat transport as well. As the chosen example was calculated for demonstration purposes only, we did not model transport processes in any real structure, and assumed one-dimensional transport of water with dissolved NaCl into 3 cm thick concrete cover without any surface protecting layer. Moisture transport and moisture storage parameters were taken as functions of moisture content, and measured data from the DELPHIN 4.1 database were used. The salt diffusion coefficient was taken to be equal to the salt diffusion coefficient in water (see Chapter 4 for details of modelling heat, moisture and salt transport in DELPHIN 4.1), the salt retention curve was approximated by two linear relations, one in the unsaturated range and the other in the saturated range.

The initial conditions were chosen as follows: the relative humidity in the concrete structure was constant and equal to 50%, which is equivalent to the moisture content of about 2% by volume, and the salt content was zero. The boundary conditions represented direct contact of concrete with seawater. We assumed an infinite source of NaCl solution with the relative concentration of 13% (100% means a saturated solution), and a Newton-type condition with a very high value of surface moisture transfer coefficient, which is in fact almost equivalent to an application of the Dirichlet condition. The calculations were run until the NaCl concentration at the steel surface achieved the corrosion threshold limit, and then were continued for another few days to show the further development of transport phenomena. For the corrosion threshold limit, we chose the most often used value for chlorides, 0.71 kg/m^3 (see Weyers, 1998).

The calculated results are presented in Figs. 6.30–6.35. Fig. 6.30 shows

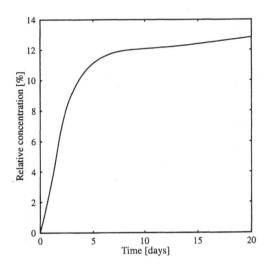

Figure 6.30 Relative concentration of dissolved NaCl on the reinforcement surface

that the corrosion threshold limit on the steel surface was achieved very fast, during the first day the concrete was exposed to the NaCl solution. Water transport was apparently slower than salt transport, as shown in Fig. 6.31. Therefore, the dominant mode of salt transport in the initial phase was diffusion in water already present in concrete, which initially did not contain any salts. The amount of precipitated salts close to the steel surface over the studied period of time was not very high, remaining below 0.3% of the volume of concrete (see Fig. 6.32). Figs. 6.33–6.35 show the distribution of the concentration of dissolved salts, water content and volumetric content of precipitated salts in the concrete cover at the time when the corrosion threshold limit was achieved on the steel surface. It is apparent that convection-dominated salt transport processes, characterized by liquid water transport, prevailed only within the first 10 mm under the concrete surface. In deeper regions, salt diffusion in the water already present in the concrete dominated. The amount of precipitated salts in fact reflected the profile of dissolved salts, which is a consequence of the fact that the salt retention curve in the unsaturated range was chosen as linear.

The calculations of coupled moisture and salt transport in this section were done with the primary aim of demonstrating the capabilities of the mathematical model, which we recommended for practical applications in Chapter 4. Therefore, the particular results obtained above should not be considered as

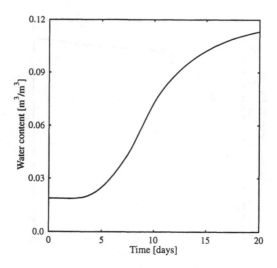

Figure 6.31 Water content on the reinforcement surface

a guide for the design of real concrete structures. The concrete structures in marine environments are usually more complicated. For instance they are usually given a protective surface layer, the reinforcement is mostly laid deeper into the concrete bulk, the type of concrete can be different from the concrete used in our calculations, etc. Also, some simplifications were assumed in our model, which should not be made in practical calculations with real concrete structures; for instance, the isothermal transport or ideal surface water and salt transfer without surface resistance. The main reason for these simplifications was the fact that measured data for salt transport in concrete is still relatively sparse in the literature, and it is difficult to put together hygric, thermal and salt storage and transport parameters for the same type of concrete. However, in recent years, the salt-related parameters have become the subject of increasing interest in the concrete community, and it can be anticipated that the necessary database may be available within the next few years. Then, the capabilities of the model will be used to much higher direct practical impact on the design process for concrete structures exposed to salty environments.

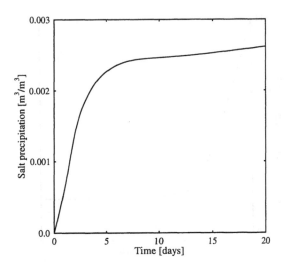

Figure 6.32 Volumetric content of precipitated NaCl on the reinforcement surface

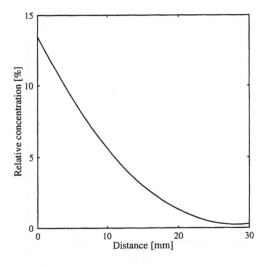

Figure 6.33 Relative concentration of dissolved NaCl in the concrete cover at the time when the corrosion threshold limit was achieved on the steel surface

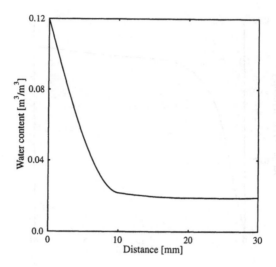

Figure 6.34 Water content in the concrete cover at the time when the corrosion threshold limit was achieved on the steel surface

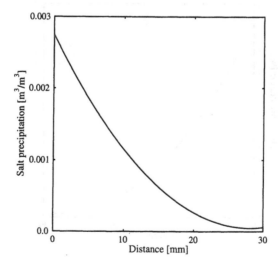

Figure 6.35 Volumetric content of precipitated NaCl in the concrete cover at the time when the corrosion threshold limit was achieved on the steel surface

6.5 SERVICE LIFE PREDICTION MODELS

The service life of a building component or material is defined (see ASTM, 1991) as the period of time after installation during which all properties exceed the minimum acceptable values, when routinely maintained. The data on the service life of building materials and components are essential to achieving longevity through the effective selection, use and maintenance of materials. Service life data are also needed to assess performance as a function of cost, and thereby permit selection of the most economically attractive option.

Modern durability technology for concrete structures profits from consistent engineering models describing the deterioration mechanisms threatening safety, serviceability and durability to ensure a rational and coherent service life design. Consistent engineering models incorporate knowledge from a wide range of technical disciplines, such as statics (structural behaviour), materials technology (materials composition), design (structural form, design tradition, codes), construction (workmanship, local traditions), statistics and economics. Experience from inspection, maintenance and repair of existing structures can be used to identify and calibrate the critical parameters governing these engineering models. Based on these models, durability performance can be expanded to include the whole range of multidisciplinary structural engineering problems from operation, maintenance, inspection, assessment, repair and re-design of existing structures to design and construction of new structures.

6.5.1 Approaches for Predicting the Service Life of Concrete

There is a variety of approaches that have been used for predicting the service life of concrete. These can be formally classified into several groups (see Clifton, 1993):

1. Estimates based on the experience

2. Deductions from performance of similar materials

3. Accelerated testing

4. Mathematical modelling based on the chemistry and physics of degradation processes

5. Application of reliability and stochastic concepts.

The estimates based on the experience employ an expert judgment using accumulated knowledge from laboratory, field testing and experience. This treatment includes both empirical knowledge and heuristics, but its reliability is not very high (see, e.g., Fagerlund, 1985), particularly if new environments are encountered, new concrete materials are used, and if the concrete is required to be durable for a time that exceeds current experience with concrete.

Deductions from performance of similar materials are based on a comparative approach, which employs an assumption that if a kind of concrete has been durable for a certain time, a similar concrete exposed to similar environment will have the same life. A problem with this approach is (Clifton, 1993) that each concrete structure has a certain uniqueness due to variability in materials, geometry and construction practices. Also, over the years, properties of concrete materials have changed. The cement particles are finer than before, the new chemical and mineral admixtures change the properties of concrete, etc. Therefore, comparisons between the durabilities of old and new concretes are not straightforward.

Accelerated testing methods for service life prediction consist in an application of elevated loads, such as concentration of reactants, temperature, relative humidity, and overhygroscopic moisture, to accelerate degradation. An important requirement for using accelerated testing is that the degradation mechanism in the accelerated test must be the same as that responsible for the in-service deterioration. If this condition is satisfied, the acceleration factor, i.e. the ratio between the rate of degradation in accelerated tests and the rate of degradation in the long-term in-service testing, can be determined. However, the relationship between the rates is mostly nonlinear, and the lack of long-term data on the in-service performance of concrete leads to major difficulties in a practical determination of the acceleration factor.

Methods based on mathematical modelling of degradation processes can provide a sound basis for predicting the performance of concrete and assist in predicting service life, if they are properly designed. There are only a few truly significant deterioration mechanisms and only the following four are really important (Rostam, 1993): reinforcement corrosion, alkali-aggregate reactions, chemical attacks (e.g. sulphates), and freeze–thaw bursting. Corrosion primarily destroys the reinforcement, and subsequently cracks and spalls the concrete. The other three primarily destroy the concrete.

Water is the most important substance from the point of view of concrete degradation. Most of major deterioration mechanisms require a sufficient amount of water. Only temperature-conditioned cracking, shrinkage cracking and mechanical wear can take place without water. Nevertheless, such crack formation opens the way for water to penetrate into the bulk material more easily, which leads to an increased risk of ingress of harmful materials dissolved in water. Therefore, modelling the water transport in concrete is a principal task.

Chloride-based salts belong to the most dangerous materials for concrete. They present a serious risk of local pit corrosion wherever they contact reinforcement. Also, if the salt contains alkali metal ions, these increase the risk of alkali-aggregate reactions in concrete containing reactive particles (Rostam, 1993). There are also indirect negative effects of chlorides. If a salt is used as a de-icing agent, the concrete surface is subject to freeze shock, which leads to crack formation. The hygroscopicity of salts can also be considered as a dangerous factor, because the presence of salts makes it more difficult to dry out the concrete to prevent possible further deterioration. Modelling of salt transport is an effective tool in service life assessment because it makes it possible to predict the salt profiles in a concrete element. However, modelling just the salt transport may not be sufficient, because the transportable salts are always dissolved in water. Therefore combined water and salt transport should be used in mathematical models of degradation processes.

Among the chemical attacks, the sulphates present the most serious danger for concrete structures. Sulphate ions penetrate from the environment into the concrete and react expansively with aluminates, which results in cracking and delamination of concrete surfaces.

Freeze–thaw cycles can affect negatively all concrete surfaces via cracking. Thus, if some other deterioration mechanism described above leads to crack formation, the freeze–thaw cycles will enhance the deterioration process in a mechanical way.

Among the other chemical and physical mechanisms negatively affecting the durability of concrete are carbonation and cyclic wetting and drying. Carbonation can depassivate reinforcing steel in the concrete by reducing the pH in concrete below the protective 11.5 level needed to form and maintain the protective γ ferric oxide coating. However, under most circumstances, the time required for carbonation to proceed to the depth of concrete where the reinforcing steel is located can be expected to be much longer than the probable useful life of the structure (Somerville, 1986, Cady and Weyers, 1992). Cyclic wetting and drying can greatly accelerate the rate at which a dissolved aggressive substance penetrates into the bulk concrete from the surface region where it originally appeared (Rostam, 1993). This is a further sound argument for modelling not just chemical compounds transport in concrete but combined water and chemical compounds transport in order to get more precise information on the profiles of the chemical compounds concentration in concrete elements.

Service life models using <u>stochastic methods</u> are based on the premise that service lives cannot be precisely predicted (Siemes et al., 1985). A large number of factors affect the service life of concrete, and their interactions are not well known. These factors include the extent of adherence to design specifications, randomness of the in-service environment, and material response to microclimates. Basically, there are two stochastic approaches (Clifton, 1993): the reliability method and the combination of statistical and deterministic models.

The reliability method takes into account the expectation that supposedly identical concrete specimens exposed to the same conditions will show a distribution of times to failure. This distribution can be obtained by testing multiple specimens via an accelerated testing method and using a time transformation function (Martin, 1985). Combination of statistical and deterministic models can consist for example in predicting the mean service life by using mean values of the parameters in deterministic models (Siemes et al., 1985) and calculating the standard deviation of the service life.

The approaches for prediction of service life of concrete given above are often used in a combination. Clifton (1993) considers a combination of models and accelerated testing as the most promising approach. In this case, the relationship between the rate of degradation in accelerated tests and the rate of degradation in long-term in-service testing can be determined by the mathematical modelling of the degradation mechanisms, and its nonlinearity no longer causes any problems in the service life assessment. This approach has several advantages: it leads to more reliable predictions, far less data is needed, and wider applications of the theoretical models are possible, for example applicability to a broad range of environmental conditions. Also, deterministic models can be combined with statistical approaches to give realistic predictions of service life, including their uncertainty.

As follows from the above analysis of service life prediction models, the exact identification and proper physical-chemical and mathematical description of all important deterioration mechanisms are crucial factors for obtaining realistic service life estimates for concrete structures. Nearly all deterioration mechanisms progress through two phases (Rostam, 1993). In the initiation phase, no noticeable weakening of the material or of the function of the structure occurs, but some protective barrier is broken down or overcome by the aggressive media. In the propagation phase, active deterioration becomes apparent. The initiation period is usually much longer than the propagation period (Tuutti, 1982). Therefore, a conservative estimate of the service life is usually made by considering only the initiation period.

Practically all deterioration mechanisms in concrete involve transport processes of heat, momentum, various phases of water, and chemical compounds. For example, a chemical substance from the surroundings first contacts the surface of concrete, and concentrates in the surface region. It may then be dissolved in water and transported in the form of a solution into the concrete bulk. The rate of penetration of this substance into the bulk depends on a variety of physical and chemical quantities, for instance on temperature, moisture content in concrete, the level of stress in the concrete element, the structure of the concrete porous matrix and the level of its deterioration, etc. Also, the external conditions are very important, for instance the supply of the particular substance in the surroundings, and temperature and moisture variations in those surroundings.

An accurate description of all these effects is not an easy task. First, it requires an application of a proper model of transport processes, because the mechanisms governing the heat, momentum, moisture and chemical compounds transport are coupled. In many cases, these mechanisms cannot be completely separated from each other and treated individually without a significant loss of accuracy. The possible departures from physical-chemical reality might be too significant. Also, the mathematical models of heat, momentum, water and chemical compounds transport contain a variety of storage and transport material parameters, which have to be determined in a wide range of field quantities, such as temperature, moisture content and concentration of the particular chemical compound. This is not a major problem, as it has been demonstrated in Chapter 5, but it can be a very time consuming task, particularly if the material parameters functions also include the time factor, i.e. these are dependent on the level of concrete deterioration.

Therefore, the choice of a mathematical model for describing a specific problem of concrete degradation is always a compromise between the required physical-chemical suitability and the accuracy on one side, and the availability of material data for performing the calculations on the other. The models should be simple enough to allow the required data to be obtained with a minimum of necessary measured transport and storage material parameters. Otherwise, the measuring procedure for the input data determination could take too long to have a practical importance. On the other hand, application of oversimplified models could lead to serious errors, which might harm the service life prediction process and result in unrealistic service life estimates.

One of the characteristic examples of an oversimplification is modelling the chloride diffusion into concrete to predict the service life of concrete subject to reinforcement corrosion using Fick's diffusion equation for the chloride concentration with a constant diffusion coefficient and constant boundary conditions (see Section 4.1, for details). This oversimplified model has been used for many years in service life prediction models (see, e.g., Tuutti, 1982, Funahashi, 1990, Cady and Weyers, 1992, Sandberg, 1996, Sarja and Vesikari, 1996, Weyers, 1998, Siemes et al., 1998, and many others). In fact, it is the simplest model that can be used. The common assumption of a semi-infinite space domain together with the assumptions of a constant diffusion coefficient and constant Dirichlet boundary conditions leads to a very simple analytical solution using the error function. The simplicity of the solution function can be then utilized in solving the inverse diffusion problem for a constant diffusion coefficient, i.e. for the identification of the diffusion coefficient from the measured concentration profiles. For orientation purposes, i.e. to get just an order of magnitude estimate of service live that can distinguish whether the service life is 10 or 100 years, this treatment is certainly sufficient. However, for more precise estimates of service life of concrete structures it is not suitable at all.

There are several reasons for this:

1. Constant boundary conditions only seldom occur in practice. It would require an environment with constant properties, which is excluded taking into account changes between summer and winter periods. Some authors are aware of this problem (e.g. Weyers, 1998), and call the diffusion coefficients determined from experiments as "effective diffusion constants", but they do not pursue the problem any further, choosing to use the oversimplified formulations in any event.

2. Diffusion coefficients of aqueous electrolytes are often functions of con-

centration (see, e.g., Robinson and Stokes, 1959), and this dependence can be quite remarkable even for ideal mixtures of simple substances.

3. Diffusion coefficients are functions of temperature, which can lead to significant differences between summer and winter periods.

4. Salt penetration into concrete is always governed by moisture transport because salts are mostly transported into the concrete bulk in the form of a solution. Cyclic wetting and drying processes can lead to a significant departure from the simple diffusion model because the moisture front does not move uniformly. The alternation of summer and winter periods can again serve as a characteristic example in this sense, as well as the alternation of rainy periods and periods without raining.

5. The concrete structure only rarely has a uniform constitution. For example, surface sealers and coatings for concrete protection are often employed. These surface layers have quite different properties from the point of view of water and gas permeability compared to concrete, and the diffusion coefficients are therefore quite different. Also, cracked surface layers have different properties to the still unaffected concrete bulk. Some concrete structures are completed by thermally insulating layers and renders. For these reasons, most concrete structures should be treated as multilayered systems.

6.5.2 Recommendations for the Improved Service Life Analyses from the Point of View of Transport Phenomena

As already mentioned, models of transport processes serving as a basis for service life analyses should not be too simple, but too much complexity can often be impractical. A natural question is: where is the limit? We will try to give some recommendations to improve current practice.

We begin with only slight improvements to the oversimplified model with a constant diffusion coefficient and constant Dirichlet boundary conditions. A minimum requirement for the diffusion coefficient of chemical substances in concrete should be its determination as a function of concentration. This may not be a serious problem. Taking specimens from a concrete structure and performing laboratory measurements on the diffusion coefficient is a common procedure (see Chapter 5). If the diffusion coefficient is known as a function of concentration, and the level of pollution in the surroundings and contact conditions of concrete with harmful substances are known (this may require special monitoring), the concentration profiles can be determined using numerical modelling. There are plenty of computational models for solving diffusion problems, therefore here too the solution is straightforward.

Although the described procedure presents already a significant improvement compared to the oversimplified model with constant diffusion coefficient and constant Dirichlet boundary conditions, it should be noted that it is still based on pure diffusion. Therefore, it seeks to express the combined water, water vapour, energy and chemical compounds transport using only the balance equation of mass for the particular chemical substance. The effect of temperature is not included at all. The effect of moisture is included only indirectly, and under very strong assumptions. In laboratory measurements of the diffusion coefficient, we use either water solutions of the particular chemical substance penetrating into the dry specimen, or the specimen is first saturated with water, and then put in contact with the salt solution. For calculation on a real structure this would mean that in the first case, combined water and chemical compound transport would take place continuously on the condition that the surface of the concrete is in full contact with water, i.e. we have a problem of salt solution penetration into the bulk concrete. In the second case it would mean that liquid (overhygroscopic) water would be continuously present in all the concrete bulk. These are very special cases of moisture conditions in concrete. Overhygroscopic water is not always present in concrete structures. During a characteristic year, there are wetting and drying periods according to the weather conditions. Therefore, if the diffusion coefficients are measured at room temperature conditions, which is the usual practice, for most concrete structures (the structures in contact with seawater belong to the few exceptions to this statement), the procedure just described presents the worst scenario, and the service life estimates are on the safe side.

A further improvement in modelling transport phenomena during the concrete deterioration process consists in inclusion of moisture transport. Then we have the balance equation of mass for water, which includes water and water vapour transport and requires a knowledge of the moisture storage function in order to express the phase change processes between water and water vapour, and the balance equation of mass for the chemical compound dissolved in water. The necessary transport and storage coefficients can be determined at room temperature conditions as described in Chapter 5, and a numerical solution can be derived to calculate profiles of moisture and concentration of the chemical compound. The requirements on boundary conditions are higher than in the previous case. We need to know not only the concentration of the chemical compound on the surface or in the surroundings but also some weather conditions, namely the relative humidity, wind and rain data. This however is not a serious problem; the data is usually available for the particular location, because it is commonly measured by meteorologists.

The model of coupled water, water vapour and chemical compounds transport already involves a significant improvement compared to the models based just on the chemical compounds transport, because it directly links the transport of chemical compounds to the water transport. However, neglecting heat transport may lead to errors in calculating water vapour fluxes, because the direction of the water vapour flux is given by the direction of the gradient of partial pressure of water vapour, not the gradient of relative humidity. If a temperature gradient is present, in some situations the gradients of partial pressure of water vapour and of relative humidity may have opposite directions. Consequently, errors in calculated overhygroscopic moisture content may lead to errors in calculated concentrations of chemical compounds. Nevertheless, if proper meteorological data and room temperature data for the transport and storage coefficients are used, for most concrete structures the service life estimates should remain on the safe side. For instance, the calculations of chloride concentrations will lead to higher service life estimates than in the case of the previous diffusion model but the service life prediction should still be lower than in reality for most structures (with the few exceptions mentioned above).

Inclusion of heat transport in the models of water, water vapour and chemical compounds transport puts the service life models on a higher level. The data required in addition to that needed for models of water, water vapour and chemical compounds transport are the thermal transport and storage coefficients, which can be measured in relatively easy way (see Chapter 5), and meteorological data for air temperature and solar radiation, which are commonly available. However, it has to be taken into account that if this model is to present a real improvement, all transport and storage coefficients should be determined as functions of temperature, moisture and concentration of chemical compounds, which might be a very time consuming task. It is certainly possible to use room temperature data for the necessary coefficients, and not to take into account the above effects, but in that case the improvement compared to the previous model neglecting heat transport will not be so remarkable. In addition, the use of such a complex model with such a high uncertainty of input data is always questionable. Therefore, before a practical application of a complex model of this type it is necessary to decide, if for the particular structure it is really worth performing the lengthy measurements. Generally, in the case of common concrete structures such as private houses or administrative buildings, it will probably not be necessary. Applications of such complex models should be directed to concrete structures exposed to severe conditions, for instance nuclear safety related structures or structures exposed to aggressive environments, where the use of simple models could result in a significant departure from reality.

For the most complex concrete structures in harsh environments that are supposed to have a very long service life, such as nuclear waste storage sites, it might be wise to employ even more complex models of transport phenomena in concrete for service life predictions. These models would include not only water, water vapour, energy and chemical compounds transport but also momentum transport (see Chapters 2 and 3). Inclusion of mechanistic effects in the form of the momentum balance equation makes it possible to calculate stress and strain due to external loads, thermal and hygric movements, chemical reaction induced movements (such as in the case of reinforcement corrosion), and therefore crack appearance can be exactly predicted. However, the number of new coefficients required is quite high, for instance thermal and hygric expansion coefficients, Biot's coefficient, and chemical expansion coefficients, and all of them, including the previous coefficients, should be determined as functions of temperature, moisture, concentration of chemical compounds, and stress level. This might be an extremely time consuming task.

We were talking mostly about service life predictions for new structures, until now. Service life predictions of existing structures in conjunction with the application of repair technologies for service life increase is equally important. Modelling the transport processes in old concrete is in principle the same as in the new structures. The specimens have to be taken from the old structure by a common boring technology but before measuring the transport and storage characteristics, moisture and chemical compounds concentration profiles should be determined. This makes it possible to estimate the penetration characteristics in the previous life of the structure.

At present, there is still a common method of determining just the "effective diffusion constant" from the measured concentration profiles (see, e.g., Weyers, 1998). This is a very crude treatment and a classical example of oversimplification. If the measured concentration data (for instance the chloride concentrations in the data collected by Weyers, 1998, for bridge decks, but many other examples may be found) exhibit at fixed spatial points oscillations of a remarkable magnitude, so that for instance the concentration can decrease for periods of several years, this is definitely not a simple diffusion problem. The more complex models proposed above, which include water and water vapour transport and possibly also heat transport, offer much wider possibilities for inverse modelling. For example, with measured water and water vapour transport and storage coefficients, and given meteorological data for the modelled time period, the diffusion coefficient of chemical compounds can be determined using a computational model of heat, water, water vapour and chemical compounds transport by inverse modelling with a simple fitting procedure. The results will be much closer to reality than in the case of using the rough fitting procedure based on the analytical solution of Fick's diffusion equation with a constant diffusion coefficient and constant boundary conditions.

If the history of a particular structure is known with reasonable accuracy in the form of precisely determined storage and transport coefficients, there is a sound basis for both repair decisions and for further service life estimates. Therefore, it is worth taking into account application of more complex models of transport processes in the analysis of the previous service life of the concrete structure before a particular repair technique is chosen. Nevertheless, the level of complexity has again to be compromised as mentioned before.

References

ASTM, 1991, ASTM E632-82 *Standard practice for developing accelerated tests to aid prediction of the service life of building components and materials.* (Philadelphia: American Society for Testing and Materials).

Cady, P.D., Weyers, R.E., 1992, Predicting service life of concrete bridge decks subject to reinforcement corrosion. In: *Corrosion Forms and Control for Infrastructure*, ASTM STP 1137. (Philadelphia: American Society for Testing and Materials), pp. 328–338.

Clifton, J.R., 1993, Predicting the service life of concrete. *ACI Materials Journal*, **90**, pp. 611–617.

Fagerlund, G., 1985, Essential data for service life prediction. In: *Problems in Service Life Prediction of Building and Construction Materials*, L.W. Masters (ed.). (Dordrecht: Martinus Nijhoff Publishers), pp. 113–138.

Funahashi, M., 1990, Predicting corrosion-free service life of a concrete structure in a chloride environment. *ACI Materials Journal*, **87**, pp. 581–587.

Gawin, D., Majorana, C.F., Schrefler, B.A., 1999, Numerical analysis of hygro-thermal behaviour and damage of concrete at high temperature. *Mechanics of Cohesive Frictional Materials*, **4**, pp. 37–74.

Gawin, D., Majorana, C.E., Pesavento, F., Schrefler, B.A., 2001, Modelling thermo-hygro-mechanical behaviour of high performance concrete in high temperature environments. In: *Proceedings of Fourth International Conference on Fracture Mechanics of Concrete and Concrete Structures*, Cachan, 2001.

Hrstka, O., Černý, R., Rovnaníková, P., 1999, Hygrothermal stress induced problems in large scale sprayed concrete structures. In: *Specialist Techniques and Materials for Concrete Construction*, R.K. Dhir, N.A. Henderson (eds.), (London: Thomas Telford), pp. 103–109.

Martin, J., 1985, Service life predictions from accelerated aging tests using reliability theory and life testing analysis. In: *Problems in Service Life Prediction of Building and Construction Materials*, L.W. Masters (ed.). (Dordrecht: Martinus Nijhoff Publishers), pp. 191–212.

Robinson, R.A., Stokes, R.H., 1959, *Electrolyte Solutions.* (London: Butterworths).

Rostam, S., 1993, Service life design – the European approach. *Concrete International*, **15**, pp. 24–32.

Sandberg, P., 1996, Cost-effective service life design of concrete structures in saline environment. In: *Durability of Building Materials and Components 7*, Volume 2, edited by C. Sjöström. (London: E & FN Spon), pp. 1195-1204.

Sarja, A., Vesikari, E. (eds.), 1996, *Durability Design of Concrete Structures.* Report of RILEM Technical Committee 130-CSL. (London: E& FN Spon).

Siemes, A., Vrouwenvelder, A., Beukel, A., 1985, Durability of buildings: a reliability analysis. *Heron,* **30**, pp. 3–48.

Siemes, T., Polder, R., de Vries, H., 1998, Design of concrete structures for durability. *Heron,* **43**, pp. 227–244.

Somerville, G., 1986, The design life of concrete structures. *Structural Engineer,* **64A**, pp. 60–71.

Tuuti, K., 1982, *Corrosion of Steel in Concrete.* (Stockholm, Swedish Cement and Concrete Research Institute).

Weyers, R.E., 1998, Service life model for concrete structures in chloride laden environments. *ACI Materials Journal,* **95**, pp. 445–453.

Wrobel, M.A., 1997, Heat and mass flows coupled with stress in a continuous medium. *Int. J. Heat Mass Transfer,* **40**, pp. 191–207.

Zienkiewicz, O.C., 1971, *The Finite Element Method in Engineering Science.* (London: McGraw-Hill).

APPENDIX 1

Basic Mathematical Relations

A1.1 COORDINATE TRANSFORMATIONS

Suppose we have two coordinate systems, $\vec{x} = (x_1, x_2, x_3)$ and $\vec{X} = (X_1, X_2, X_3)$, in three-dimensional Euclidean space. The transformation relations

$$X_i = X_i(x_1, x_2, x_3) \tag{1}$$

define for an arbitrary point (x_1, x_2, x_3) of the system $\vec{x}$ its coordinates in the system $\vec{X}$, (X_1, X_2, X_3). The representation (1) is one-to-one, i.e. to every point (x_1, x_2, x_3) of the system $\vec{x}$, there corresponds just one point (X_1, X_2, X_3) in the system $\vec{X}$, if the Jacobian J of the transformation is not equal to zero,

$$J = \begin{vmatrix} \frac{\partial X_1}{\partial x_1} & \frac{\partial X_1}{\partial x_2} & \frac{\partial X_1}{\partial x_3} \\ \frac{\partial X_2}{\partial x_1} & \frac{\partial X_2}{\partial x_2} & \frac{\partial X_2}{\partial x_3} \\ \frac{\partial X_3}{\partial x_1} & \frac{\partial X_3}{\partial x_2} & \frac{\partial X_3}{\partial x_3} \end{vmatrix}, \tag{2}$$

or in short notation

$$J = \left| \frac{\partial X_i}{\partial x_j} \right|. \tag{3}$$

If both $\vec{x}$, $\vec{X}$ are orthogonal Cartesian coordinate systems with a common origin, we can write the following transformation relation:

$$X_i = \sum_{j=1}^{3} \alpha_{ij} x_j, \tag{4}$$

where α_{ij} is the cosine of the angle between the i-th axis of the $\vec{X}$ system and the j-th axis of the $\vec{x}$ system,

$$\alpha_{ij} = \cos(X_i, x_j). \tag{5}$$

For expressions with summations, such as (4), we often use a short-cut notation, which is usually called the Einstein summation convention. In this short notation we omit the summation sign and consider the index, which is duplicated on one side of the expression, as a summing index. Therefore, the expression in (4) could be rewritten in the form

$$X_i = \alpha_{i1} x_1 + \alpha_{i2} x_2 + \alpha_{i3} x_3 = \alpha_{ij} x_j. \tag{6}$$

Another possibility for how the transformation relation (4) could be written, is in a matrix form,

$$\left\{\begin{array}{c} X_1 \\ X_2 \\ X_3 \end{array}\right\} = \left[\begin{array}{ccc} \alpha_{11} & \alpha_{12} & \alpha_{13} \\ \alpha_{21} & \alpha_{22} & \alpha_{23} \\ \alpha_{31} & \alpha_{32} & \alpha_{33} \end{array}\right] \left\{\begin{array}{c} x_1 \\ x_2 \\ x_3 \end{array}\right\} \tag{7}$$

or

$$\vec{X} = \underline{\alpha} \cdot \vec{x}, \tag{8}$$

where $\underline{\alpha}$ is the transformation matrix.

Every vector $\vec{v}$ in a Cartesian coordinate system can be written in component form,

$$\vec{v} = v_1 \vec{e}_1 + v_2 \vec{e}_2 + v_3 \vec{e}_3 = v_i \vec{e}_i, \tag{9}$$

where v_i is the component of the vector $\vec{v}$ in the direction of the x_i axis, $\vec{e}_i$ is the unit vector in the direction of the x_i axis. The unit vectors $\vec{e}_1, \vec{e}_2, \vec{e}_3$ together form a base for the coordinate system, which is commonly denoted as $\vec{e}$. Denoting $\vec{e}$ the base of the Cartesian coordinate system $\vec{x}$ and $\vec{E}$ the base of the Cartesian coordinate system $\vec{X}$, we can write for the bases the same transformation relation as for the coordinates, i.e.

$$\vec{e}_i = \alpha_{ij} \vec{E}_j. \tag{10}$$

A1.2 TENSORS AND THEIR BASIC PROPERTIES

We consider a quantity T, which in the $\vec{e}$ base is expressed by the set of numbers $t_{i_1 i_2 i_3 \ldots i_n}$ $(i_k = 1, 2, 3; k = 1, 2, \ldots, n)$ and in the $\vec{E}$ base by the set of numbers $T_{I_1 I_2 \ldots I_n}$ $(I_k = 1, 2, 3; k = 1, 2, \ldots, n)$. If these numbers transform from the coordinate system with the $\vec{e}$ base into the coordinate system with the $\vec{E}$ base according to the relation

$$T_{I_1 I_2 \ldots I_n} = \alpha_{I_1 i_1} \alpha_{I_2 i_2} \ldots \alpha_{I_n i_n} t_{i_1 i_2 i_3 \ldots i_n}, \tag{11}$$

where $\alpha_{I_k i_k}$ are the transformation matrices, then the quantity T is called n-th order tensor.

In practice, the most often used are zero-, first- and second-order tensors. The zero-order tensor is called a scalar. It is a quantity, which in both the $\vec{e}$ base and the $\vec{E}$ base is expressed by a single same number, and is transformed by the relation

$$S = s. \tag{12}$$

The first-order tensor is called a vector, and its transformation can be expressed according to (11) as

$$V_I = \alpha_{Ii} v_i. \tag{13}$$

The second-order tensor is transformed by the relation

$$T_{IJ} = \alpha_{Ii}\alpha_{Jj}t_{ij}, \tag{14}$$

and usually it is called a tensor for short.

Among the second-order tensors, the unit tensor δ plays a specific role. Its components are expressed as

$$\delta_{ij} = \vec{e}_i \cdot \vec{e}_j, \tag{15}$$

where $\vec{e}_i$, $\vec{e}_j$ are base vectors, and therefore

$$\delta_{ij} = \begin{cases} 1 & \text{for } i = j \\ 0 & \text{for } i \neq j \end{cases}. \tag{16}$$

A single component of the unit tensor of second order, δ_{ij}, is usually referred to as Kronecker delta.

In component form, a second-order tensor $\mathbf{T}$ can be expressed as

$$\mathbf{T} = \vec{e}_i \otimes \vec{e}_j \, T_{ij}, \tag{17}$$

where in accordance with the Einstein summation rule, both i and j are summing indices, and the symbol $\otimes$ means a tensor product (see (38)).

The tensor $\mathbf{T}^T$, transposed to the tensor $\mathbf{T}$, is defined as

$$\mathbf{T}^T = \vec{e}_i \otimes \vec{e}_j \, T_{ji}. \tag{18}$$

A tensor $\mathbf{T}$, defined by the relation (17), is called symmetric if its components obey the condition

$$T_{ij} = T_{ji}, \tag{19}$$

and antisymmetric if the following relation is valid:

$$T_{ij} = -T_{ji}. \tag{20}$$

Every tensor can be expressed by its symmetric and antisymmetric parts:

$$\mathbf{T} = \mathbf{T}^S + \mathbf{T}^A, \tag{21}$$

where

$$\mathbf{T}^S = \frac{1}{2}(\mathbf{T} + \mathbf{T}^T) \tag{22}$$

$$\mathbf{T}^A = \frac{1}{2}(\mathbf{T} - \mathbf{T}^T), \tag{23}$$

or in a component form

$$T_{ij}^S = \frac{1}{2}(T_{ij} + T_{ji}) \tag{24}$$

$$T_{ij}^A = \frac{1}{2}(T_{ij} - T_{ji}). \tag{25}$$

The symmetric part $\mathbf{T}^S$ of a second-order tensor can be further expressed by a scalar and a deviator,

$$\mathbf{T}^S = T + {}^d\mathbf{T}^S, \tag{26}$$

where the scalar part can be written as

$$T = \frac{1}{3}T_{ii}, \tag{27}$$

and the deviatoric part

$${}^d\mathbf{T}^S = \frac{1}{2}(\mathbf{T} + \mathbf{T}^T) - T \cdot \delta. \tag{28}$$

In component form, relation (26) can be expressed as

$${}^dT_{ij}^S = \frac{1}{2}(T_{ij} + T_{ji}) - \frac{1}{3}\delta_{ij}T_{kk}. \tag{29}$$

The sum of the diagonal components of a tensor $\mathbf{T}$ is called the trace of the tensor,

$$\mathrm{tr}\,\mathbf{T} = T_{ii}. \tag{30}$$

Therefore,

$$T = \frac{1}{3}\mathrm{tr}\,\mathbf{T}. \tag{31}$$

The antisymmetric part of a second-order tensor can be expressed using the pseudovector $\vec{T}^A$, which is defined as

$$T_k^A = \frac{1}{2}\epsilon_{ij}T_{ij}^A, \tag{32}$$

where ϵ_{ijk} are components of the Levi-Civita tensor, which is the antisymmetric unit tensor of third order, having the following properties:

$$\epsilon_{ijk} = \begin{cases} 0 & \text{if at least two indices are the same} \\ 1 & \text{if the indices have the values} \quad ijk = 123, 231, 312 \\ -1 & \text{if the indices have the values} \quad ijk = 321, 132, 213. \end{cases} \tag{33}$$

A1.3 VECTOR AND TENSOR OPERATIONS

The scalar product of two vectors is defined as

$$c = \vec{a} \cdot \vec{b} = a_i b_i. \tag{34}$$

For a scalar product of two tensors we have

$$C = \mathbf{T} : \mathbf{S} = T_{ij} S_{ji}. \tag{35}$$

If at least one of the two tensors $\mathbf{T}$, $\mathbf{S}$ is symmetric, we can write

$$C' = \mathbf{T} : \mathbf{S} = T_{ij} S_{ij}. \tag{36}$$

The vector product is defined by the relation

$$\vec{c} = \vec{a} \times \vec{b} = \epsilon_{ijk} \, \vec{e}_k \, a_i b_j. \tag{37}$$

The tensor product of two vectors is a second-order tensor, and we have

$$\mathbf{c} = \vec{a} \otimes \vec{b} = \vec{e}_i \otimes \vec{e}_j a_i b_j, \tag{38}$$

or in a clearer arrangement

$$\mathbf{c} = \begin{pmatrix} a_1 b_1 & a_1 b_2 & a_1 b_3 \\ a_2 b_1 & a_2 b_2 & a_2 b_3 \\ a_3 b_1 & a_3 b_2 & a_3 b_3 \end{pmatrix}, \quad c_{ij} = a_i b_j. \tag{39}$$

For the multiplication of a tensor and a vector we can write

$$\vec{a} = \mathbf{T} \cdot \vec{b} = \vec{e}_i \, T_{ij} b_j. \tag{40}$$

In an analogous way (note that this multiplication procedure is not commutative) we have

$$\vec{d} = \vec{b} \cdot \mathbf{T} = \vec{e}_i \, b_j T_{ji}, \tag{41}$$

so that $\vec{a} = \vec{d}$ is valid for symmetric tensors $\mathbf{T}$ only.

For a representation of vector differential operations, we define the Hamilton operator $\vec{\nabla}$,

$$\vec{\nabla} = \vec{e}_i \frac{\partial}{\partial x_i}. \tag{42}$$

Using $\vec{\nabla}$, the basic vector operations gradient, divergence and curl can be specified.

For the gradient of a scalar field we can write

$$\vec{a} = \operatorname{grad} s = \vec{\nabla} \cdot s = \vec{e}_i \frac{\partial s}{\partial x_i}, \tag{43}$$

for the gradient of a vector field we have

$$\mathbf{T} = \operatorname{grad} \vec{v} = \vec{\nabla} \otimes \vec{v} = \vec{e}_i \otimes \vec{e}_j \frac{\partial v_j}{\partial x_i}. \tag{44}$$

The divergence of a vector field is defined as

$$c = \operatorname{div} \vec{v} = \vec{\nabla} \cdot \vec{v} = \frac{\partial v_i}{\partial x_i}, \tag{45}$$

and the divergence of a tensor field

$$\vec{b} = \text{div } \mathbf{T} = \vec{\nabla} \cdot \mathbf{T} = \vec{e}_i \frac{\partial T_{ij}}{\partial x_j}. \tag{46}$$

For the curl of a vector field we have

$$\vec{d} = \text{curl } \vec{v} = \vec{\nabla} \times \vec{v} = \epsilon_{ijk} \, \vec{e}_k \, \frac{\partial v_j}{\partial x_i}. \tag{47}$$

On the basis of the above-defined multiplication formulas and vector differential operations, we can formulate the following differential relations:

$$\text{grad} (a \cdot b) = a \cdot \text{grad } b + b \cdot \text{grad } a \tag{48}$$

$$\text{grad} (\vec{a} \cdot \vec{b}) = \vec{a} \times \text{ curl } \vec{b} + \vec{b} \times \text{ curl } \vec{a} + \vec{b} \cdot \text{ grad } \vec{a} + \vec{a} \cdot \text{ grad } \vec{b} \tag{49}$$

$$\text{div} (a \cdot \vec{b}) = a \cdot \text{ div } \vec{b} + \text{ grad } a \cdot \vec{b} \tag{50}$$

$$\text{div} (\vec{a} \times \vec{b}) = \vec{b} \cdot \text{ curl } \vec{a} - \vec{a} \cdot \text{ curl } \vec{b} \tag{51}$$

$$\text{curl} (a \cdot \vec{b}) = \text{ grad } a \times \vec{b} + a \cdot \text{ curl } \vec{b} \tag{52}$$

$$\text{curl} (\vec{a} \times \vec{b}) = \vec{a} \cdot \text{ div } \vec{b} - \vec{b} \cdot \text{ div } \vec{a} + \vec{b} \cdot \text{ grad } \vec{a} - \vec{a} \cdot \text{ grad } \vec{b} \tag{53}$$

$$\text{curl grad } a = 0 \tag{54}$$

$$\text{div curl } \vec{a} = 0 \tag{55}$$

$$\text{curl curl } \vec{a} = \text{ grad div } \vec{a} - \nabla^2 \vec{a} \tag{56}$$

$$\text{div grad } a = \Delta a = \vec{\nabla}^2 a = \sum_i \frac{\partial^2 a}{\partial x_i^2} \tag{57}$$

$$\text{div} (a \cdot \delta) = \text{ grad } a \tag{58}$$

$$\text{div} (\vec{a} \otimes \vec{b}) = \vec{a} \cdot \text{ grad } \vec{b} + \vec{b} \cdot \vec{a} \tag{59}$$

$$(a \cdot \delta) : \text{ grad } \vec{b} = a \cdot \text{ div } \vec{b} \tag{60}$$

$$\text{div} (\mathbf{T} \cdot \vec{a}) = \vec{a} \cdot \text{ div } \mathbf{T} + \mathbf{T}^T : \text{ grad } \vec{a} \tag{61}$$

$$(\text{curl } \vec{a}) \times \vec{b} = \vec{e}_i \left(\frac{\partial a_i}{\partial x_j} b_j - \frac{\partial a_k}{\partial x_i} b_k \right) \tag{62}$$

$$\text{div grad } \vec{a} = \Delta \vec{a} = \vec{\nabla}^2 \vec{a} = \sum_i \frac{\partial^2 a_j}{\partial x_i^2} \vec{e}_j. \tag{63}$$

A1.4 INTEGRAL RELATIONSHIPS

We assume a space domain with volume V, bounded by a closed surface S. The unit vector in the direction of the outward normal to the surface S will be denoted by $\vec{n}$. If in this domain the scalar field s, the vector field $\vec{v}$ and the tensor field $\mathbf{T}$ are defined, then for these fields the Gauss theorems can be formulated:

$$\int_V \text{grad } s \, dV = \int_S (\vec{n} \cdot s) \, dS \tag{64}$$

$$\int_V \text{div } \vec{v} \, dV = \int_S (\vec{n} \cdot \vec{v}) \, dS \tag{65}$$

$$\int_V \text{curl } \vec{v} \, dV = \int_S (\vec{n} \times \vec{v}) \, dS \tag{66}$$

$$\int_V \text{div } \mathbf{T} \, dV = \int_S (\vec{n} \cdot \mathbf{T}) \, dS. \tag{67}$$

If we have a surface S bounded by a closed curve c, which can be characterized at every point by its tangential unit vector $\vec{t}$, for a scalar field s and a vector field $\vec{v}$, the Stokes theorems can be formulated:

$$\int_S (\vec{n} \times \text{grad } s) \, dS = \int_c (\vec{t} \cdot s) \, dc \tag{68}$$

$$\int_S (\vec{n} \cdot \text{curl } \vec{v}) \, dS = \int_c (\vec{t} \cdot \vec{v}) \, dc \tag{69}$$

$$\int_S (\vec{n} \times \text{grad}) \times \vec{v} \, dS = \int_c (\vec{t} \times \vec{v}) \, dc. \tag{70}$$

Recommended Data for Material Parameters of Concrete

From the point of view of hygrothermal properties, what is called generally "concrete", comprises a variety of different materials. Therefore a simple survey of material properties of concrete cannot be given without any further specification. Most hygrothermal parameters of concrete depend in a very significant way on its composition (e.g. type of aggregates, water/cement ratio, additives, etc.). Further, the properties of mature concrete are very different from those of fresh concrete. Finally, all hygrothermal properties of concrete are affected by temperature, moisture content and salt content.

There is much data on thermal and hygric properties of various types of concretes in the scientific literature. However, this data is mostly disaggregated and dispersed, and finding a complete set of data for one particular concrete type is very difficult.

Therefore, the data survey, which follows, cannot pretend to completeness in any case. We have tried to collect a set of typical data, measured by various investigators, and then on the basis of its analysis to formulate a recommended set of data that could be used in hygrothermal design for orientation purposes, i.e. for preliminary calculations. The data is organized in the following way. First, a basic set of data for room temperature (RT) conditions is given. Then selected specific data sets, such as in high temperature (HT) conditions or data for cracked concrete, are presented. Finally, a summary of recommended data is given. Unless specified otherwise, the data is for mature concrete after 28 days or more of curing.

Among another material parameters data surveys published in the past, the authors can particularly recommend the following: for concrete in normal conditions Neville (1973), for high temperature data Schneider (1982), Bažant and Kaplan (1996).

Taking into account the large differences among different authors in measuring the same material parameters of concretes, which at first glance look very similar, we strongly recommend regarding each particular concrete as a unique material. In the precise scientific calculations of heat, moisture and chemical compounds transport in concrete elements and structures, we recommend a complex approach to any modelling. This approach consists first in measuring all necessary hygric and thermal properties, at least in their dependence on temperature, moisture and salt content. For specific situations, such as modelling the processes in fresh concrete or doing long-term calculations, measuring the dependence of material parameters on the degree of hydration is also necessary. The basic calculations can be then done. Finally, a verification experiment, consisting of measuring the temperature, moisture and salt concentration fields and comparing them with the computational results, should be performed and necessary adjustments of the free parameters of the model should be made.

This approach is certainly very time consuming. Therefore, we suggest it to be adopted in complicated cases only, where there is not enough experience of the concrete behaviour and where great damage can be anticipated in the event of malfunction of the concrete structure. Examples of such structures are nuclear safety related concrete structures in nuclear power plants, and more generally concrete exposed to severe conditions.

A2.1 DENSITY AND BULK DENSITY

Table A2.1.1 Basic data for density of clinker minerals (in kg/m^3)

Specification	Value	Source
C_3S	3120–3150	Lea (1971)
C_2S	3280	Lea (1971)
C_3A	3040	Lea (1971)
C_4AF	3970	Lea (1971)
CaO	3320	Lea (1971)
MgO	3580	Lea (1971)
$Ca(OH)_2$	2230	Lea (1971)
$3CaO.2SiO_2.3H_2O$	2630	Lea (1971)
$Mg(OH)_2$	2400	Lea (1971)
$3CaO.Al_2O_3.6H_2O$	2520	Lea (1971)

Table A2.1.2 Basic data for bulk density of cement pastes, cement mortars and concretes (in kg/m^3)

Specification	Value	Source
OPC paste, w/c = 0.5	1700–1800	Dědek (1989)
OPC mortar, w/c = 0.5	2100–2200	Dědek (1989)
OPC concrete, w/c = 0.5	2250–2350	Dědek (1989)
Light concrete depending on quality of light aggregate	500–2000	Dědek (1989)
Baryte concrete	3500	Dědek (1989)

A2.2 HYDRATION HEAT OF CLINKER MINERALS AND CEMENTS

Table A2.2.1 Basic data for clinker minerals and cements (in J/g)

Specification	Value	Source
C_3S	517	Jolicoeur and Simard (1998)
	500	Lea (1971)
	567	Woods et al. (1932)
	501	Lerch and Bogue (1935)
C_2S	262	Jolicoeur and Simard (1998)
	258	Lea (1971)
	259	Woods et al. (1932)
	259	Lerch and Bogue (1935)
C_3A	1144	Jolicoeur and Simard (1998)
	863	Lea (1971)
	836	Woods et al. (1932)
	865	Lerch and Bogue (1935)
C_4AF	418	Jolicoeur and Simard (1998)
	417	Lea (1971)
	125	Woods et al. (1932)
	418	Lerch and Bogue (1935)
OPC/28 d	385	Livesey et al. (1991)
	400	Lea (1971)
Sulphate resistant PC	345	Livesey et al. (1991)
Fly ash PC (50:50)	305	Livesey et al. (1991)
High slag PC (30:70)	285	Livesey et al. (1991)

Table **A2.2.2** Recommended data (in J/g)

Specification	Value
C_3S	510
C_2S	259
C_3A	860
C_4AF	418
OPC/28 d according to mineralogical comp.	380–500
Sulphate resistant PC	350
Fly ash PC according to PC/FA ratio	250–350
High slag PC according to PC/slag ratio	280–380

A2.3 THERMAL CONDUCTIVITY

Table A2.3.1 Basic room temperature (RT) data (in W/mK)

Specification	Value	Source
Baryte aggregates typical value	1.38	Mitchell (1956)
Igneous aggregates typical value	1.44	Mitchell (1956)
Dolomite aggregates typical value	3.68	Mitchell (1956)
Siliceous aggregates saturated	2.4–3.6	Blundell et al. (1976)
Igneous cryst. aggr. (limestone, dolomite) saturated	1.9–2.8	Blundell et al. (1976)
Igneous amorph. aggr. (basalt) saturated	1.0–1.6	Blundell et al. (1976)
Quartz aggregates dry	2.6	Harmathy (1970)
Anorthosite aggregates dry	1.2	Harmathy (1970)
Limestone aggregates natural cond.	2.1	Crispino (1972)
Baryte aggregates natural cond.	2.1	Crispino (1972)
Gravel aggregates natural cond.	2.3	Abe et al. (1972)
Quartzit aggregates natural cond.	2.7	Marechal (1972)
Density 1600–2400 kg/m^3 protected from weather typ. moisture cont.	0.71–2.27	Loudon and Stacy (1966)
Density 1600–2400 kg/m^3 exposed to weather typ. moisture cont.	0.81–2.56	Loudon and Stacy (1966)
Ordinary concrete dry	1.2–2.0	Kumaran (1996)
Ordinary concrete w – moisture in kg/m^3	2.74+0.0032 w	IEA (1991)
Hydrating concrete 6 hours–7 days	2.28–1.51	Brown and Javaid (1970)

Table A2.3.1 Basic RT data (in W/mK) (continued)

Specification	Value	Source
30 MPa	1.17	Khan et al. (1998)
70 MPa	1.59	Khan et al. (1998)
100 MPa	1.77	Khan et al. (1998)
Quartzite aggregates typical value	3.5	Mehta and Monteiro (1993)
Dolomite aggregates typical value	3.2	Mehta and Monteiro (1993)
Limestone aggregates typical value	2.6–3.3	Mehta and Monteiro (1993)
Granite aggregates typical value	2.6–2.7	Mehta and Monteiro (1993)
Rhyolite aggregates typical value	2.2	Mehta and Monteiro (1993)
Basalt aggregates typical value	1.9–2.2	Mehta and Monteiro (1993)
Cement paste	0.53	Xu and Chung (2000)
Cement mortar	0.58	Xu and Chung (2000)
Cement paste dry	1.6	Harmathy (1970)
Cement paste	1.1–1.6	Zoldners (1971)
Cement mortar u in %kg/kg hygrosc. range	2.30+0.12 u	Černý et al. (2000)
HPC, siliceous aggr. u in %kg/kg	1.5+0.1 u	Toman and Černý (2001b)
HPC, calcareous aggr. u in %kg/kg	2.3+0.16 u	Toman and Černý (2001b)
Cement mortar	1.16	Černý et al. (2001b)
Cement mortar w in m^3/m^3	1.34+0.56 w	Grunewald (2000)
Ordinary concrete w in m^3/m^3	2.10+0.56 w	Grunewald (2000)

Table A2.3.2 High temperature (HT) data (in W/mK)

Specification	Value	Source
Quartz aggregates dry, 0–1000°C	2.6–1.3	Harmathy (1970)
Anorthosite aggregates dry, 0–1000°C	1.3–1.1	Harmathy (1970)
Limestone aggregates natural cond. 20–200°C	2.1–1.6	Crispino (1972)
Baryte aggregates natural cond. 20–180°C	2.1–1.2	Crispino (1972)
Gravel aggregates natural cond. 20–200°C	2.3–1.8	Abe et al. (1972)
Quartzit aggregates natural cond. 20–300°C	2.7–1.5	Marechal (1972)
Limestone aggregates 600–1000°C	1.0–1.7	Hildenbrand et al. (1978)
Siliceous aggregates 600–1000°C	1.5–1.8	Hildenbrand et al. (1978)
Cement mortar 20–400°C	2.3–0.9	Černý et al. (2000)
Cement mortar 400–800°C	0.9–2.2	Černý et al. (2000)
HPC, siliceous aggr. 0–400°C	1.5–1.0	Toman and Černý (2001b)
HPC, siliceous aggr. 400–800°C	1.0–2.8	Toman and Černý (2001b)
HPC, calcareous aggr. 0–400°C	2.2–1.1	Toman and Černý (2001b)
HPC, calcareous aggr. 400–800°C	1.1–2.7	Toman and Černý (2001b)

Table A2.3.3 Cracked concrete (in W/mK)

Specification	Value	Source
Cement mortar exp. to 800°C	0.27	Černý et al. (2001b)
Cement mortar comp. stress 90% ult. load	1.05	Černý et al. (2001b)

Table A2.3.4 Recommended data (in W/mK)

Specification	Value
Siliceous aggr. RT, u in %kg/kg	1.5+0.1 u
Calcareous aggr. RT, u in %kg/kg	2.2+0.16 u
Cement mortar RT, u in %kg/kg	1.2 + 0.12 u
Cement paste RT	1.6
Siliceous aggr. 0–400°C	1.5–1.0
Siliceous aggr. 400–1000°C	1.0–1.8
Calcareous aggr. 0–400°C	2.2–1.1
Calcareous aggr. 400–1000°C	1.1–1.7
Cement mortar 0–400°C	1.2–0.9
Cement mortar 400–1000°C	0.9–1.5
Cement mortar RT, cracked exp. to 800°C	0.27
Cement mortar RT, cracked comp. stress 90% ult. load	1.05

A2.4 SPECIFIC HEAT CAPACITY

Table A2.4.1 Basic RT data (in J/kgK)

Specification	Value	Source
Ordinary concrete typical values	840–1170	Neville (1973)
Ordinary concrete dry, typical values	500–1130	Browne (1972)
Cement paste dry, typical values	630–1720	Browne (1972)
Ordinary concrete saturated, typical values	700–1500	Browne (1972)
Cement paste	1000	Harmathy (1970)
Gravel aggregates	960	Zoldners (1960)
Limestone aggregates	990	Zoldners (1960)
Sandstone aggregates	980	Zoldners (1960)
Limestone aggregates	780	Harmathy and Allen (1973)
Siliceous aggregates	800	Harmathy and Allen (1973)
Limestone aggregates	900	Hildenbrand et al. (1978)
Siliceous aggregates	980	Hildenbrand et al. (1978)
Granite aggregates	770	Ödeen (1968)
Ordinary concrete	840–940	Kumaran (1996)
Hydrating concrete 6 hours–7 days	1150–890	Brown and Javaid (1970)
Ordinary concrete typical values	800–1200	Mindess and Young (1981)
Ordinary concrete saturated, typ. values	650–2700	Khan et al. (1998)
Ordinary concrete dry, typ. values	650–1100	Khan et al. (1998)
Cement paste	736	Xu and Chung (2000)
Cement mortar	642	Xu and Chung (2000)

Table A2.4.1 Basic RT data (in J/kgK) (continued)

Specification	Value	Source
HPC, siliceous aggr. u in %kg/kg	800 + 32 u	Toman and Černý (2001a)
HPC, calcareous aggr. u in %kg/kg	780 + 35 u	Toman and Černý (2001a)
Cement mortar	850	Černý et al. (2000)
Cement mortar	1000	Grunewald (2000)
Ordinary concrete	1050	Grunewald (2000)

Table A2.4.2 HT data (in J/kgK)

Specification	Value	Source
Limestone aggregates 20–700°C	780–1300	Harmathy and Allen (1973)
Siliceous aggregates 20–700°C	800–1450	Harmathy and Allen (1973)
Limestone aggregates 20–700°C	750–1600	Hildenbrand et al. (1978)
Siliceous aggregates 20–700°C	980–1800	Hildenbrand et al. (1978)
Granite aggregates 20–700°C	770–900	Ödeen (1968)
Siliceous aggregates 0–1000°C	980–1750	Černý et al. (1996)
Calcareous aggregates 0–1000°C	900–1800	Černý et al. (1996)
Cement mortar 0–1000°C	850–1600	Černý et al. (2000)
HPC, siliceous aggr. 0–1000°C	800–1500	Toman and Černý (2001a)
HPC, calcareous aggr. 0–1000°C	780–1750	Toman and Černý (2001a)

Table A2.4.3 Recommended data (in J/kgK)

Specification	Value
Cement paste RT, u in %kg/kg	900 + 30 u
Cement mortar RT, u in %kg/kg	850 + 30 u
Siliceous aggr. RT, u in %kg/kg	800 + 32 u
Calcareous aggr. RT, u in %kg/kg	780 + 35 u
Cement mortar 0–1000°C	850–1600
Siliceous aggr. 0–1000°C	800–1500
Calcareous aggr. 0–1000°C	800–1750

A2.5 WATER PERMEABILITY K, HYDRAULIC CONDUCTIVITY k AND MOISTURE DIFFUSIVITY κ

Table A2.5.1 Basic data (K in m^2 intrinsic, k in m/s saturated, κ in m^2/s)

	Specification	Value	Source
k	Cement paste	$\sim 10^{-12}$	Powers, Brownyard (1948)
κ	Ordinary concrete typical values	$(1.2 - 5) \cdot 10^{-10}$	Bažant and Najjar (1972)
k	Cement paste dep. on porosity	$10^{-10} - 10^{-14}$	Powers (1958)
k	Fresh cement paste	$2 \cdot 10^{-6}$	Powers et al. (1954)
κ	Ordinary concrete w in kg/m^3	$1.8 \cdot 10^{-11}$ $\times \exp(0.0582\,w)$	Kumaran (1996)
K	Cement paste	$(0.85 - 15) \cdot 10^{-19}$	Powers et al. (1954)
k	HPC 74.6 MPa	$(1.9 - 2.0) \cdot 10^{-16}$	El-Dieb, Hooton (1995)
k	Ordinary concrete 42 MPa	$(3.7 - 28) \cdot 10^{-14}$	El-Dieb, Hooton (1995)
k	Ordinary concrete 29.6 MPa	$(2.0 - 3.7) \cdot 10^{-12}$	El-Dieb, Hooton (1995)
k	Cement paste w/c 0.47-1.0	$10^{-13} - 10^{-9}$	Christensen et al. (1996)
k	Cement paste Φ – porosity	$\ln k = 9.50+$ $+5.95 \ln \Phi$ k in 10^{-13} m/s	Breysse and Gerard (1997)
k	Ordinary concrete $\Phi' = \Phi/(1 - v_{\mathrm{a}})$ v_{a} – vol. ratio of aggr.	$\ln k = 10.80+$ $+6.41 \ln \Phi'$ k in 10^{-13} m/s	Breysse and Gerard (1997)
K	Cement mortar w/c=0.4 dep. on hydr. deg.	$10^{-18} - 10^{-20}$	Halamickova et al. (1995)
K	Cement mortar w/c=0.5 dep. on hydr. deg.	$10^{-17} - 10^{-19}$	Halamickova et al. (1995)
K	Cement paste fit. proc.	10^{-21}	Baroghel-Bouny (1999)
K	HP cement paste fit. proc.	$0.3 \cdot 10^{-22}$	Baroghel-Bouny (1999)
K	Ordinary concrete fit. proc.	$3 \cdot 10^{-21}$	Baroghel-Bouny (1999)
K	HP concrete fit. proc.	$5 \cdot 10^{-22}$	Baroghel-Bouny (1999)

Table A2.5.1 Basic data (continued) (K in m^2 intrinsic, k in m/s saturated, κ in m^2/s)

	Specification	Value	Source
κ	Cement mortar	$3 \cdot 10^{-9}$	Černý et al. (2000)
κ	HPC, siliceous aggr.	$1.0 \cdot 10^{-8}$	Černý et al. (2001a)
κ	HPC, calcareous aggr.	$7 \cdot 10^{-9}$	Černý et al. (2001a)
k	Cement paste	$5 \cdot 10^{-11}$	Aldea et al. (1999b)
k	Cement mortar	$3 \cdot 10^{-11}$	Aldea et al. (1999b)
k	Ordinary concrete	$1.0 \cdot 10^{-11}$	Aldea et al. (1999b)
k	HPC	$4 \cdot 10^{-12}$	Aldea et al. (1999b)
k	45 MPa concrete	$1.0 \cdot 10^{-11}$	Wang et al. (1997)
κ	Cement mortar	$5.0 \cdot 10^{-10}$	Grunewald (2000)
κ	Ordinary concrete	$2.0 \cdot 10^{-9}$	Grunewald (2000)

Table A2.5.2 Cracked concrete (multiplication factor – increase of K, k, κ compared to uncracked material)

	Specification	Value	Source
k	30MPa concrete comp. stress 70% ult. load	$10^2 - 10^4$	Kermani (1991)
k	Ordinary concrete 100°C	10^2	Bažant and Thonguthai (1978)
κ	Ordinary concrete bending stress 0.1 mm	2.25	Bažant et al. (1987)
k	Cement paste tensile stress 110 μm	14	Aldea et al. (1999b)
k	Cement mortar 130 μm tensile stress	10	Aldea et al. (1999b)
k	Ordinary concrete 130 μm tensile stress	$2 \cdot 10^3$	Aldea et al. (1999b)
k	HPC 110 μm tensile stress	10^2	Aldea et al. (1999b)
k	45 MPa concrete 350 μm tensile stress	10^7	Wang et al. (1997)
k	45 MPa concrete 550 μm tensile stress under load	10^7	Wang et al. (1997)
k	Ordinary concrete 350 μm tensile stress under load	$2.5 \cdot 10^3$	Aldea et al. (1999a)
k	HPC 300 μm tensile stress under load	35	Aldea et al. (1999a)
κ	Cement mortar comp. stress 90% ult. load	16	Černý et al. (2000)

Table A2.5.2 Cracked concrete (continued) (multiplication factor – increase of K, k, κ compared to uncracked material)

	Specification	Value	Source
κ	HPC, siliceous aggr. exp. to 800°C	$3 \cdot 10^2$	Černý et al. (2001a)
κ	HPC, calcareous aggr. exp. to 800°C	10^4	Černý et al. (2001a)
κ	Cement mortar exp. to 800°C	10^3	Černý et al. (2001b)
κ	HPC, siliceous aggr. exp. to 400 frost cyc.	4	Černý et al. (2001a)
κ	HPC, calcareous aggr. exp. to 300 frost cyc.	20	Černý et al. (2001a)

Table A2.5.3 Recommended data (K in m^2 intrinsic, k in m/s saturated, κ in m^2/s), for cracked concrete multiplication factor – increase of K, k, κ compared to uncracked material)

	Specification	Value
κ	Cement mortar	$3 \cdot 10^{-9}$
κ	Siliceous aggr.	$1 \cdot 10^{-8}$
κ	Calcareous aggr.	$7 \cdot 10^{-9}$
k	Cement paste	$5 \cdot 10^{-11}$
k	Cement mortar	$3 \cdot 10^{-11}$
k	Ordinary concrete	$1 \cdot 10^{-11}$
K	HPC	$4 \cdot 10^{-12}$
K	Cement paste	$9 \cdot 10^{-21}$
K	Cement mortar	$6 \cdot 10^{-21}$
K	Ordinary concrete	$2 \cdot 10^{-21}$
K	HPC	$7 \cdot 10^{-22}$

Table A2.5.3 Recommended data (continued) (K in m^2 intrinsic, k in m/s saturated, κ in m^2/s), for cracked concrete multiplication factor – increase of K, k, κ compared to uncracked material)

	Specification	Value
k	Cement paste cracked tensile stress 110 μm	14
k	Cement mortar cracked 130 μm tensile stress	10
k	Ordinary concrete cracked 130 μm tensile stress	$2 \cdot 10^3$
k	HPC cracked 110 μm tensile stress	10^2
κ	Cement mortar cracked comp. stress 90% ult. load	16
κ	Siliceous aggr. cracked exp. to 800°C	$3 \cdot 10^2$
κ	Calcareous aggr. cracked exp. to 800°C	10^4
κ	Cement mortar cracked exp. to 800°C	10^3

A2.6 GAS PERMEABILITY K, GAS DIFFUSION COEFFICIENT D, WATER VAPOR PERMEABILITY δ, WATER VAPOR DIFFUSION RESISTANCE FACTOR μ

Table A2.6.1 Basic data (K in m^2 intrinsic, D in m^2/s, δ in s, μ dimensionless)

	Specification	Value	Source
K	Ordinary concrete nitrogen porosity 6–20%	$(1-15)\cdot 10^{-17}$	Sugiyama et al. (1996)
K	33 MPa concrete oxygen dep. on sat. deg.	$(2.5-230)\cdot 10^{-19}$	Abbas et al. (1999)
K	Ordinary concrete air 120–20 MPa	$10^{-18}-10^{-15}$	Abbas et al. (2000)
K	30 MPa concrete air	$(2-14)\cdot 10^{-16}$	Austin, Al-Kindy (2000)
K	50 MPa concrete air	$(1-2)\cdot 10^{-16}$	Austin, Al-Kindy (2000)
D	Cement paste oxygen dep. on porosity	$(10-22)\cdot 10^{-12}$	Ngala and Page (1997)
μ	Ordinary concrete φ as frac.	$1/[6.8\cdot 10^{-3}+8.21\times \times 10^{-5}\exp(5.66\,\varphi)]$	Kumaran (1996)
δ	Cement mortar	$3.4\cdot 10^{-12}$	Černý et al. (2000)
μ	Cement mortar	45	Grunewald (2000)
μ	Ordinary concrete	65	Grunewald (2000)
μ	HPC	100	Grunewald (2000)

Table A2.6.2 Cracked concrete (K in m^2 intrinsic, D in m^2/s, δ in s, μ dimensionless)

	Specification	Value	Source
δ	Cement mortar comp. stress 90% ult. load	$3.9\cdot 10^{-12}$	Černý et al. (2000)
δ	Cement mortar exp. to 800°C	$4.0\cdot 10^{-12}$	Černý et al. (2001b)

Table A2.6.3 Recommended data (K in m^2 intrinsic, D in m^2/s, δ in s, μ dimensionless)

	Specification	Value
K	Ordinary concrete air 120–20 MPa	$10^{-18} - 10^{-15}$
μ	Ordinary concrete φ as frac.	$1/[6.8 \cdot 10^{-3} + 8.21 \times \times 10^{-5} \exp(5.66\,\varphi)]$
δ	Cement mortar	$3.4 \cdot 10^{-12}$
δ	Cement mortar cracked comp. stress 90% ult. load	$3.9 \cdot 10^{-12}$
δ	Cement mortar cracked exp. to 800°C	$4.0 \cdot 10^{-12}$

A2.7 SORPTION ISOTHERMS AND WATER RETENTION CURVES

Table A2.7.1 Sorption isotherms – Hansen (1986)

w/c	A	B	C
0.40	4.82	1.51	0.117
0.44	4.65	2.67	0.0489
0.48	4.76	0.18	4.85
0.52	4.05	1.63	0.130
0.55	6.03	3.11	0.0265
0.58	4.93	1.40	0.114
0.61	4.60	1.62	0.0913
0.65	5.16	2.76	0.0314
0.66	4.40	2.04	0.0537
0.72	6.33	2.40	0.0116

Ordinary concrete 2300 kg/m^3

$$u = A \exp\left[-\tfrac{1}{B} \ln\left(1 - \tfrac{\ln \varphi}{C}\right)\right]$$

u in %kg/kg, φ in %

Table A2.7.2 Sorption isotherms – Kumaran (1996)

	A	B	C
Adsorption	147.5	0.0453	1.67
Desorption	147.5	0.570	0.64

Ordinary concrete

$$w = A \left[1 - \tfrac{\ln \varphi}{B}\right]^{-\tfrac{1}{C}}$$

w in kg/m^3, φ as a fraction

Table A2.7.3 Water retention curves – Baroghel-Bouny et al. (1999)

Specification	a	b
Cement paste	37.5	2.17
HP cement paste	96.3	1.95
Ordinary concrete	18.6	2.27
HPC	46.9	2.06

$$p_c = a(S_l^{-b} - 1)^{1-\frac{1}{b}}$$

S_l liquid water saturation

Table A2.7.4 Recommended data

Specification	Value
Sorption isotherms	Hansen (1986)
Water retention curves	Baroghel-Bouny et al. (1999)

A2.8 SALT DIFFUSION COEFFICIENTS

Table A2.8.1 Basic data for chlorides (in m^2/s)

Specification	Value	Source
Ordinary concrete accelerated test	$(10.6 - 23.1) \cdot 10^{-13}$	Sugiyama et al. (1996)
Ordinary concrete profiles in real str.	$(3.9 - 94.5) \cdot 10^{-13}$	Costa and Appleton (1999)
NaCl in water	$1.67 \cdot 10^{-9}$	Johannesson (1999)
KCl in water	$2.06 \cdot 10^{-9}$	Johannesson (1999)
Ordinary concrete water suction	$(1 - 3) \cdot 10^{-11}$	Baroghel-Bouny et al. (1995)
Ordinary concrete w/c 0.35–0.75	$(9.4 - 91.8) \cdot 10^{-12}$	Tang (1995)
Ordinary concrete convent. diffusion	$(2.4 - 2.9) \cdot 10^{-12}$	Arsenault et al. (1995)
Ordinary concrete migration	$(3.8 - 4.7) \cdot 10^{-12}$	Arsenault et al. (1995)
Cement paste porosity 5–25%	$(4 - 20) \cdot 10^{-12}$	Ngala and Page (1997)
Cement mortar w/c=0.4 dep. on hydr. deg.	$(0.82 - 2.33) \cdot 10^{-11}$	Halamickova et al. (1995)
Cement mortar w/c=0.5 dep. on hydr. deg.	$(0.84 - 7.54) \cdot 10^{-11}$	Halamickova et al. (1995)

Table A2.8.2 Cracked concrete – data for chlorides (multiplication factor – increase of D compared to uncracked material)

Specification	Value	Source
Ordinary concrete	8	Gerard and Marchand (2000)

Table A2.8.3 Recommended data for chlorides (in m^2/s)

Specification	Value
Ordinary concrete w/c 0.35–0.75	$(9 - 90) \cdot 10^{-12}$

A2.9 ION BINDING ISOTHERMS

Table A2.9.1 Basic data for chlorides

Specification	Value	Source
Cement paste $c > 0.01$ mol/l desorption	$\log C_b = 0.3788 \log c + 1.140$ C_b in mg/g-gel, c in mol/l	Tang and Nilsson (1993)
Cement paste $c > 0.01$ mol/l adsorption	$\log C_b = 0.3864 \log c + 1.412$ C_b in mg/g-gel, c in mol/l	Tang and Nilsson (1993)
Cement paste $c < 0.05$ mol/l	$\frac{1}{C_b} = \frac{0.002438}{c} + 0.1849$ C_b in mg/g-gel, c in mol/l	Tang and Nilsson (1993)

Table A2.9.2 Recommended data for chlorides

Specification	Value
Cement paste	Tang and Nilsson (1993)

References

Abbas, A., Carcasses, M., Ollivier, J.P., 1999, Gas permeability of concrete in relation to its degree of saturation. *Materials and Structures*, **32**, pp. 3–8.

Abbas, A., Carcasses, M., Ollivier, J.P., 2000, The importance of gas permeability in addition to the compressive strength of concrete. *Magazine of Concrete Research*, **52**, pp. 1–6.

Abe, H., Kawahara, T., Ito, T., Haraguchi, A., 1972, Influence factors of elevated temperatures on thermal properties and inelastic behaviour of concrete. In: *International Seminar on Concrete for Nuclear Reactors. ACI Special Publication*, No. 34, Vol. 2. (Detroit: American Concrete Institute), pp. 847–870.

Aldea, C.M., Shah, S.P., Karr, A., 1999a, Effect of cracking on water and chloride permeability of concrete. *Journal of Materials in Civil Engineering*, **11**, pp. 181–187.

Aldea, C.M., Shah, S.P., Karr, A., 1999b, Permeability of cracked concrete. *Materials and Structures*, **32**, pp. 370–376.

Arsenault, J., Bigas, J.P., Ollivier, J.P., 1995, Determination of chloride diffusion coefficient using two different steady-state methods: influence of concentration gradient. In: *Proceedings of the RILEM International Workshop Chloride Penetration into Concrete*, St-Remy-les-Chevreuse, edited by Nilsson, L.O., Ollivier, J.P. (St-Remy-les-Chevreuse: RILEM), pp. 150–160.

Austin, S.A., Al-Kindy, A.A., 2000, Air permeability versus sorptivity: effects of field curing on cover concrete after one year of field exposure. *Magazine of Concrete Research*, **52**, pp. 17–24.

Baroghel-Bouny, V., Chaussadent, T., Raharinaivo, A., 1995, Experimental investigations on binding of chloride and combined effects of moisture and chloride in cementitious materials. In: *Proceedings of the RILEM International Workshop Chloride Penetration into Concrete*, St-Remy-les-Chevreuse, edited by Nilsson, L.O., Ollivier, J.P. (St-Remy-les-Chevreuse: RILEM), pp. 290–301.

Baroghel-Bouny, V., Mainguy, M., Lassabatere, T., Coussy, O., 1999, Characterization and identification of equilibrium and transfer moisture properties for ordinary and high-performance cementitious materials. *Cement and Concrete Research*, **29**, pp. 1225–1238.

Bažant, Z.P., Kaplan, M.F., 1996, *Concrete at High Temperatures: Material Properties and Mathematical Models*. (Harlow: Longman).

Bažant, Z.P., Najjar, L.J., 1972, Nonlinear water diffusion in nonsaturated concrete. *Materials and Structures*, **5**, pp. 3–20.

Bažant, Z.P., Sener, S., Kim, J.K., 1987, Effect of cracking on drying permeability and diffusivity of concrete. *ACI Materials Journal*, **84**, pp. 351–357.

Bažant, Z.P., Thonguthai, W., 1978, Pore pressure and drying of concrete at high temperature. *Journal of Engineering Mechanics Division ASCE*, **104**, pp. 1059–1079.

Blundell, R., Diamond, S., Browne, R.G., 1976, The properties of concrete subjected to elevated temperatures, *Technical Note No. 9*. (London: CIRIA Underwater Engineering Group).

Breysse, D., Gerard, B., 1997, Modelling of permeability in cement-based materials. Part 1: uncracked medium. *Cement and Concrete Research*, **27**, pp. 761–775.

Brown, T.D., Javaid, M.Y., 1970, The thermal conductivity of fresh concrete. *Materials and Structures*, **3**, pp. 411–416.

Browne, R.D., 1972, Properties of concrete in reactor vessels. *Concrete*, **6**, pp. 51–53.

Costa, A., Appleton, J., 1999, Chloride penetration into concrete in marine environment. Part I: Main parameters affecting chloride penetration. *Materials and Structures*, **32**, pp. 252–259.

Černý, R., Drchalová, J., Rovnaníková, P., 2001a, The effects of thermal load and frost cycles on the water transport in two high performance concretes. *Cement and Concrete Research*, **31**, 1129–1140.

Černý, R., Maděra, J., Poděbradská, J., Toman, J., Drchalová, J., Klečka, T., Jurek, K., Rovnaníková, P., 2000, The effect of compressive stress on thermal and hygric properties of Portland cement mortar in wide temperature and moisture ranges. *Cement and Concrete Research*, **30**, pp. 1267–1276.

Černý, R., Totová, M., Poděbradská, J., Toman, J., Drchalová, J., Rovnaníková, P., 2001b, Thermal and hygric properties of Portland cement mortar after high temperature exposure combined with compressive stress. *Cement and Concrete Research*, submitted for publication

Černý, R., Toman, J., Šesták, J., 1996, Measuring the effective specific heat of building materials. *Thermochimica Acta*, **282/283**, pp. 239–250.

Crispino, E., 1972, Studies on the technology of concretes under thermal conditions. In: *International Seminar on Concrete for Nuclear Reactors. ACI Special Publication*, No. 34, Vol. 1. (Detroit: American Concrete Institute), pp. 443–479.

Christensen, B.J., Mason, T.O., Jennings, H.M., 1996, Comparison of measured and calculated permeabilities for hardened cement pastes. *Cement and Concrete Research*, **26**, pp. 1325–1334.

Dědek, M., 1989, *Building materials*. (Praha: MVS), pp. 172–179 (in Czech).

El-Dieb, A.S., Hooton, R.D., 1995, Water permeability measurement of high performance concrete using a high pressure triaxial cell. *Cement and Concrete Research*, **25**, pp. 1199–1208.

Gerard, B., Marchand, J., 2000, Influence of cracking on the diffusion properties of cement-based materials. Part I: Influence of continuous cracks on the steady-state regime. *Cement and Concrete Research*, **30**, pp. 37–43.

Grunewald, J., 2000, *DELPHIN 4.1 – Documentation, Theoretical Fundamentals*. (Dresden: TU Dresden).

Halamickova, P., Detwiler, R.J., Bentz, D.P., Garboczi, E.J., 1995, Water permeability and chloride diffusion in Portland cement mortars: relationship to sand content and critical pore diameter. *Cement and Concrete Research*, **25**, pp. 790–802.

Hansen, K.K., 1986, *Sorption Isotherms*, Technical Report 163/86. (Lyngby: The Technical University of Denmark).

Harmathy, T.Z., 1970, Thermal properties of concrete at elevated temperatures. *ASTM Journal of Materials*, **5**, pp. 47–74.

Harmathy, T.Z., Allen, L.W., 1973, Thermal properties of selected masonry unit concretes. *J. Amer. Concr. Inst.*, **70**, pp. 132–142.

Hildenbrand, G., Peeks, M., Skokan, A., Reimann, M., 1978, Investigations in Germany of the barrier effect of reactor concrete against propagating molten Corium in the case of a hypothetical core meltdown accident of an LWR. In: *ENS/ANS Int. Meeting on Nuclear Power Reactor Safety*, Vol. 1. (Brussels: ENS/ANS), pp. 16–19.

IEA-Annex XIV, 1991, *Condensation and Energy*, Volume 3, Material Properties. (Leuven: International Energy Agency).

Johanesson, B.F., 1999, Diffusion of a mixture of cations and anions dissolved in water. *Cement and Concrete research*, **29**, pp. 1261–1270.

Jouliceour, C. and Simard, M.A., 1998, Chemical admixture–cement interaction: phenomenology and physico-chemical concepts. *Cement and Concrete Composites*, **20**, pp. 87–101.

Kermani, A., 1991, Permeability of stressed concrete. *Building Research and Information*, **19**, pp. 360–366.

Khan, A.A., Cook, W.D., Mitchell, D., 1998, Thermal properties and transient thermal analysis of structural members during hydration. *ACI Materials Journal*, **95**, pp. 293–303.

Kumaran, M.K., 1996, IEA-Annex XXIV: *Heat and Moisture Transfer in Insulated Envelope Parts*, Volume 3, Material Properties. (Leuven: International Energy Agency).

Lea, F.M., 1971, *The Chemistry of Cement and Concrete*. (New York: Chemical Publishing Co., Inc.), p. 270.

Lerch, W. and Bogue, R.H., 1935, Zement, **24**, p. 155.

Livesey, P., Donnelly, A. and Tomlinson, C., 1991, Measurement of the heat of hydration of cement. *Cement and Concrete Composites*, **13**, pp. 177–185.

Loudon, A.G., Stacy, E.F., 1966, The thermal and acoustic properties of lightweight concretes. *Structural Concrete*, **3**, pp. 58–95.

Marechal, J.C., 1972, Thermal conductivity and thermal expansion coefficients of concrete as a function of temperature and humidity. In: *International Seminar on Concrete for Nuclear Reactors. ACI Special Publication*, No. 34, Vol. 2. (Detroit: American Concrete Institute), pp. 1047–1057.

Mehta, P.K., Monteiro, P.J.M., 1993, *Concrete. Structure, Properties and Materials*. (Englewood Cliffs: Prentice-Hall).

Mindess, S., Young, J.F., 1981, *Concrete*. (Englewood Cliffs: Prentice-Hall).

Mitchell, L.J., 1956, Thermal properties. *ASTM Sp. Tech. Publicn.*, **169**, pp. 129–135.

Neville, A.M., 1973, *Properties of Concrete*. (London: Pitman).

Ngala, V.T., Page, C.L., 1997, Effects of carbonation on pore structure and diffusional properties of hydrated cement pastes. *Cement and Concrete Research*, **27**, pp. 995–1007.

Ödeen, K., 1968, *Fire Resistance of Prestressed Concrete Double T Units*. (Stockholm: National Swedish Institute for Materials Testing).

Page, C.L., Ngala, V.T., 1995, Steady state diffusion characteristics of cementitious materials. In: *Proceedings of the RILEM International Workshop Chloride Penetration into Concrete*, St-Remy-les-Chevreuse, edited by Nilsson, L.O., Ollivier, J.P. (St-Remy-les-Chevreuse: RILEM), pp. 77–84.

Perrin, B., Bonnet, S., 1995, Experimental results concerning combined transport of humidity and chloride in non steady state. In: *Proceedings of the RILEM International Workshop Chloride Penetration into Concrete*, St-Remy-les-Chevreuse, edited by Nilsson, L.O., Ollivier, J.P. (St-Remy-les-Chevreuse: RILEM), pp. 302–314.

Powers, T.C., 1958, Structure and physical properties of hardened Portland cement paste. *J. Amer. Ceramic Soc.*, **41**, pp. 1–6.

Powers, T.C., Brownyard, T.L., 1948, Studies of the physical properties of hardened cement paste. *Research Department Bulletin No. 22.* (Chicago: Portland Cement Association).

Powers, T.C., Copeland, L.E., Hayes, J.C., Mann, H.M., 1954, Permeability of Portland cement paste. *J. Amer. Concr. Inst.*, **51**, pp. 285–298.

Reinhardt, H.W., Hearn, N., Sosoro, M., 1997, Transport properties of concrete. In: *Penetration and Permeability of Concrete: Barriers to organic and contaminating liquids*, edited by H.W. Reinhardt. (London: E & FN Spon), pp. 213–264.

Reinhardt, H.W., Sosoro, M., Zhu, X.F., 1998, Cracked and repaired concrete subject to fluid penetration. *Materials and Structures*, **31**, pp. 74–83.

Schiessl, P., Wiens, U., 1995, Rapid determination of chloride diffusivity in concrete with blending agents. In: *Proceedings of the RILEM International Workshop Chloride Penetration into Concrete*, St-Remy-les-Chevreuse, edited by Nilsson, L.O., Ollivier, J.P. (St-Remy-les-Chevreuse: RILEM), pp. 115–125.

Schneider, U., 1982, Behaviour of concrete at high temperatures. *Deutscher Ausschuss für Stahlbeton*, Heft 337. (Berlin: W. Ernst and Sohn), pp. 1–122.

Sugiyama, T., Bremner, T.W., Tsuji, Y., 1996, Determination of chloride diffusion coefficient and gas permeability of concrete and their relationship. *Cement and Concrete Research*, **26**, pp. 781–790.

Tang, L., 1995, On chloride diffusion coefficients obtained using the electrically accelerated methods. In: *Proceedings of the RILEM International Workshop Chloride Penetration into Concrete*, St-Remy-les-Chevreuse, edited by Nilsson, L.O., Ollivier, J.P. (St-Remy-les-Chevreuse: RILEM), pp. 126–134.

Tang, L., Nilsson, L.O., 1993, Chloride binding capacity and binding isotherms of OPC pastes and mortars. *Cement and Concrete Research*, **23**, pp. 247–253.

Tang, L., Nilsson, L.O., 1995, Chloride binding isotherms – an approach by applying the modified BET equation. In: *Proceedings of the RILEM International Workshop Chloride Penetration into Concrete*, St-Remy-les-Chevreuse, edited by Nilsson, L.O., Ollivier, J.P. (St-Remy-les-Chevreuse: RILEM), pp. 36–42.

Toman, J., Černý, R., 2001a, Temperature and moisture dependence of the specific heat of high performance concrete. *Acta Polytechnica*, **41**, pp. 5–7.

Toman, J., Černý, R, 2001b, Thermal conductivity of high performance concrete in wide temperature and moisture ranges. *Acta Polytechnica*, **41**, pp. 8–10.

Vodák, F., Černý, R., Drchalová, J., Hošková, Š., Kapičková, O., Michalko, O., Semerák, P., Toman, J., 1997, Thermophysical properties of concrete for nuclear-safety related structures. *Cement and Concrete Research*, **27**, pp. 415–426.

Wang, K., Jansen, D.C., Shah, S.P., 1997, Permeability study of cracked concrete. *Cement and Concrete Research*, **27**, pp. 381–393.

Woods, H., Steinour, H.H. and Starke, H.R., 1932, *Journal Ind. and Eng. Chemistry*, **24**, p. 1207.

Xu, Y., Chung, D.D.L., 2000, Effect of sand addition on the specific heat and thermal conductivity of cement. *Cement and Concrete Research*, **30**, pp. 59–61.

Xu, Y., Chung, D.D.L., 1999, Increasing the specific heat of cement paste by admixture surface treatments. *Cement and Concrete Research*, **29**, pp. 1117–1121.

Zoldners, N.F., 1960, Effect of high temperatures on concretes incorporating different aggregates. *Proceedings of the ASTM*, **60**, pp. 1087–1108.

Zoldners, N.F., 1971, Thermal properties of concrete under sustained elevated temperatures. In: *Temperature and Concrete*, ACI Special Publication **25**. (Detroit: American Concrete Institute), pp. 1–31.

Subject Index

Milton Keynes UK
Ingram Content Group UK Ltd.
UKHW021928071024
449327UK00022B/1724

> 2B30? 44FDA2